Mechanisms and Machine Science

Volume 23

Series editor

Marco Ceccarelli, Cassino, Italy

More information about this series at http://www.springer.com/series/8779

Jyoti K. Sinha
Editor

Vibration Engineering and Technology of Machinery

Proceedings of VETOMAC X 2014,
held at the University of Manchester, UK,
September 9–11, 2014

Springer

Editor
Jyoti K. Sinha
School of Mechanical, Aerospace and Civil Engineering (MACE)
The University of Manchester
Manchester
UK

ISSN 2211-0984 ISSN 2211-0992 (electronic)
ISBN 978-3-319-09917-0 ISBN 978-3-319-09918-7 (eBook)
DOI 10.1007/978-3-319-09918-7

Library of Congress Control Number: 2014946957

Springer Cham Heidelberg New York Dordrecht London

Printed on acid-free paper

Springer is part of Springer Science+Business Media (www.springer.com)

Preface

The 10th International Conference on Vibration Engineering and Technology of Machinery (VETOMAC X) 2014 in Manchester is the 10th instalment in a series of conferences being held in an effort to bring together researchers and engineers working in the field of vibration. This conference is organised by the University of Manchester (in collaboration with the Vibration Institute of India) and held in Manchester, UK, during 9–11, September 2014. The conference aims to provide engineers and researchers a platform for sharing their innovative and creative research findings aimed to benefit the engineering and technology of machinery.

VETOMAC X covered a holistic plethora of relevant topics in vibration and engineering technology including (but surely not limited to); Condition Monitoring, Machinery and Structural Dynamics, Rotor Dynamics, Experimental Techniques, Finite Element Model updating, Industrial case studies, Vibration control and Energy harvesting, Signal processing, etc. The conference concentrated on nearly 100 papers from key researchers and engineers from universities, private and public research centres and industries from more than 20 countries worldwide. The papers were subjected to a unanimous peer review by experts from different fields covered by the conference before being accepted for oral presentation and publication. The full proceeding also includes six invited lectures from renowned experts: Prof. K. Gupta, Mr. W. Hahn, Prof. A.W. Lees, Prof. John Mottershead, Prof. J.S. Rao and Dr. P. Russhard.

The organisers would like to thank the keynote speakers, authors, presenters and session chairs for their participation. Special gratitude must be extended to a number of individuals whose invaluable help enabled the organisation of the VETOMAC X 2014 Conference in Manchester, UK. Sincere gratitude is also expressed to the scientific committee and all the reviewers. The consistent support of the School of Mechanical, Aerospace and Civil Engineering (MACE) of The University of Manchester in every respect in organising VETOMAC X 2014 is gratefully acknowledged.

I particularly wish to thank Prof. Andrew Gibson (Head of School), Prof. Paul Mummery (Line Manager), Ms. Anne McGourlay (Head of School Administration), Mrs. Vera Sokolovski (Web Manager), Ms. Alison Lawton (School Finance

Manager), Miss. Michelle Mallon (Conference Secretary) from School of MACE, my Research Associate and Ph.D. students; Mr. Erfan Asnaashari, Dr. Keri Elbhbah, Mr. Ahmed Gubran, Mr. Akilu Kaltungo and Mr. Adrian Nembhard for their day-to-day support. Lastly, I acknowledge my wife Sarita and my son Aarambh as my real strengths throughout.

Manchester, UK Jyoti K. Sinha

Committee

Conference Chairs

Dr. Jyoti K. Sinha
The University of Manchester, UK

Prof. J.S. Rao (Retired from IIT Delhi, India)
President-Academics, Kumaraguru College of Technology, India
President, The Vibration Institute of India
Chief Editor, Advances in Vibration Engineering
Chief Science Officer (consulting)
Altair Engineering (India)

Scientific Committee

Prof. Jamil Abdo, Sultan Qaboos University, Oman
Prof. Douglas E. Adams, Vanderbilt University, USA
Prof. S. Adhikari, Swansea University, UK
Prof. A. Ball, University of Huddersfield, UK
Prof. W. Bartelmus, Wroclaw University of Technology, Poland
Prof. Rama Bhat, Concordia University, Canada
Dr. Philip Bonello, The University of Manchester, UK
Dr. P. Boskoski, Jozef Stefan Institute, Slovenia
Prof. Sanjin Braut, University of Rijeka, Croatia
Prof. K.L. Cavalca, University of Campinas, Brazil
Prof. S. Chatterton, Politecnico di Milano, Italy
Prof. A.K. Darpe, Indian Institute of Technology Delhi, India
Prof. R. Dufour, INSA-Lyon, France
Prof. M. El Badaoui, Jean Monnet University, France

History of VETOMAC Series of Conferences
(in collaboration with the Vibration Institute of India)

- **VETOMAC I, 2000**: Indian Institute of Science, Bangalore, India.
- **VETOMAC II, 2002**: Bhabha Atomic Research Centre, Mumbai, India.
- **VETOMAC III, 2004**: Indian Institute of Technology, Kanpur, India.
- **VETOMAC IV, 2007**: University College of Engineering, Osmania University, Hyderabad, India, Bharat Heavy Electricals Limited, Hyderabad, India.
- **VETOMAC V, 2009**: Huazhong University of Science and Technology, Wuhan, China and City University of Hong Kong, China.
- **VETOMAC VI, 2010**: Indian Institute of Technology, New Delhi, India.
- **VETOMAC VII, 2011**: Shanghai Jiao Tong University and City University of Hong Kong, China.
- **VETOMAC VIII, 2012**: Institute of Fluid Flow Machinery, Polish Academy of Sciences, Gdansk, Poland.
- **VETOMAC IX, 2013**: Nanjing University of Aeronautics and Astronautics, Nanjing, China.

Contents

Part I Keynote Lectures

Ancient Temple Carts—Modifications for Structure and Steering 3
J.S. Rao and Bigil Kumar

The Rise and Fall of the Rotor Blade Strain Gauge 27
Peter Russhard

Managing Life Extension for Large Rotating Plant in Power Stations 39
Wolfgang Hahn

Recent Advances and Prospects in Condition Monitoring 51
Arthur W. Lees

The Sensitivity Method in Stochastic Model Updating 65
John E. Mottershead, Michael Link, Tiago A.N. Silva, Yves Govers and Hamed Haddad Khodaparast

Composite Shaft Rotor Dynamics: An Overview 79
K. Gupta

Part II Contributed Papers/Chapters

Investigation of the Synchronous Response in a Back-Back Centrifugal Compressor Due to Imbalance and Seal Distortion 97
Mohamed Al Yahyai and David Mba

Comparison of On-Bearing and On-Casing Vibration for Blade Health Monitoring in Rotating Machine 107
Ahmed A. Gubran and Jyoti K. Sinha

Vibration-Based Condition Monitoring for Rotating Machinery with Different Flexible Supports 119
Adrian D. Nembhard and Jyoti K. Sinha

Off-Line Crack Detection in Rotors 129
Erfan Asnaashari and Jyoti K. Sinha

Application of Composite Spectrum in Steam Turbo-Generator Set . . . 139
Keri Elbhbah, Jyoti K. Sinha, W. Hahn and G. Tasker

Coherent Composite HOS Analysis of Rotating Machines with Different Support Flexibilities 145
Akilu Yunusa-Kaltungo and Jyoti K. Sinha

Vibration Behaviour of a Turbo-Generator Set 155
Wolfgang Hahn and Jyoti K. Sinha

A Comparison Between Hilbert Transform and a New Method for Signal Enveloping 163
Abdullah Al-Ahmari, Jyoti K. Sinha and Erfan Asnaashari

The Detection of Shaft Misalignments Using Motor Current Signals from a Sensorless Variable Speed Drive 173
Samieh Abusaad, Ahmed Benghozzi, Ann Smith, Fengshou Gu and Andrew Ball

Analysis of SGTF Vertical Sodium Pump Using ARMD 183
Hemant Prakash Agnihotri, V. Prakash and K.K. Rajan

Qualification of Large Structures for Shock Loads 195
K.K. Meher and A. Rama Rao

Hybrid Viscous-Structural Damping Identification Method 209
Vikas Arora

Effectiveness of Adaptive Filter Algorithms and Spectral Kurtosis in Bearing Faults Detection in a Gearbox 219
Faris Elasha, David Mba and Cristobal Ruiz-Carcel

Pitting Detection in Worm Gearboxes with Vibration Analysis 231
Faris Elasha, David Mba and Cristobal Ruiz-Carcel

Impact Vibrations of Ultrasonic Multi-striker Hand Tool 243
Makhmut Ganiev, Ilshat Gafurov and Ildar Vagapov

Use of the Roving Mass Technique and SWT to Identify and Locate Multiple Cracks in Simply-Supported Beams 253
Amirabbas Bahador and S. Olutunde Oyadiji

Single and Multiple Crack Detection in Simply-Supported Beams Using SWT 265
Amirabbas Bahador and S. Olutunde Oyadiji

Modal Approach for Optimal Design of Two-Layer Piezoelectric Vibration Energy Harvesters 277
S. Olutunde Oyadiji and Xingyu Xiong

Synthesis of Vibration Gear Continuously Variable Transfer 289
Konstantin S. Ivanov, Roza K. Koilibaeva, Gachip Ualiev and Baurjan Tultaev

Crack Propagation in a 500 MW LP Turbine in Service 297
Wolfgang Hahn and David Futter

Effect of Ultrasonic Oscillations on the Interaction of al with Ga-In Eutectic Alloy 309
Alexey A. Novikov, Alexandr I. Nizovskii, Mikhail V. Trenikhin and Valerii I. Bukhtiyarov

Nonlinear Control of an Aeroelastic System with a Non-smooth Structural Nonlinearity 317
Shakir Jiffri and John E. Mottershead

A Block Decoupling Control Algorithm for Vibration Suppression of Linear Structures 329
Xiaojun Wei and John E. Mottershead

Measurement and Simulation of the Vibroacoustic Performance of an Electric Motor 339
Alex McCloskey, Xabier Arrasate, Xabier Hernandez and Oscar Salgado

Forced Vibration of Eight Mistuned Bladed Disks on a Solid Shaft—Excitation of the First Compressor Bladed Disc 349
Romuald Rzadkowski and Artur Maurin

Dynamic Analysis of Parametrically Excited Piezoelectric Bimorph Beam for Energy Harvesting 363
S.K. Dwivedy, Anvesh K. Reddy and Anshul Garg

About Multi-parametric Analysis of Drill String Vibrations 373
Almaz Sergaliyev, Julius Kaplunov and Lelya Khajiyeva

Tuned Mass Damper Inverted Pendulum to Reduce Offshore Wind Turbine Vibrations 379
P.V.B. Guimarães, M.V.G. de Morais and S.M. Avila

Vibration Control of Active Vehicle Suspension System Using Fuzzy Logic Controller 389
A. Shehata, H. Metered and Walid A.H. Oraby

Two-Stage Data Driven Filtering for Local Damage Detection in Presence of Time Varying Signal to Noise Ratio 401
Jakub Obuchowski, Agnieszka Wylomanska and Radoslaw Zimroz

Energy Harvesting from Near Periodic Structures 411
P.V. Malaji and S.F. Ali

Dynamic Analysis of Multilayered Sandwich Beam with MRE Core 421
S.K. Dwivedy and S.L.V.N. Avinash

Analysis of Coupled Transverse and Axial Vibrations of Euler Bernoulli and Timoshenko Beams with Longitudinal Crack for Its Detection 433
Jeslin Thalapil and S.K. Maiti

Detection of Arbitrary Number of New Cracks Appearing in a Beam with Pre-existing Cracks 449
Dhanjee K. Mahto and Surjya K. Maiti

Acoustic Technique for Diverse Safety Rod Drop Time Measurement in PFBR 463
V. Prakash, Hemant Prakash Agnihotri, P. Anup Kumar, R. Ramakrishna and K..K. Rajan

Large Amplitude Free Vibration Analysis of Axially Functionally Graded Tapered Rotating Beam by Energy Method 473
Saurabh Kumar and Anirban Mitra

A Frequency-Based Criterion for Automatic Selection of the Optimal Intrinsic Mode Function in Diagnostics of Localized Gear Tooth Faults 485
Pawel Rzeszucinski, Michal Juraszek and James R. Ottewill

Vehicle Suspension Performance Analysis Based on Full Vehicle Model for Condition Monitoring Development 495
Moamar Hamed, Belachew Tesfa, Fengshou Gu and Andrew D. Ball

Effects of Monograde and Multigrade Oils on the Friction Force in Four-Stroke Motor Engine: An Experimental and Analytical Approach 507
Anastasios Zavos and Pantelis G. Nikolakopoulos

Delamination Detection in a Laminated Composite Beam Based on Changes in Natural Frequencies 519
Kajal Khan and Surjya K. Maiti

Investigation of the Effect of Biodiesel Blends on Fuel Injection Pumps Based on Vibration and Pressure Measurements 535
Ali Abruss, Yangchun Guo, Tie Wang, Fengshou Gu, Andrew D. Ball and D. Brown

Robustness Evaluation of Adhesively Bonded Ceramic Quad Flat Chips for Space Applications 547
Lassaad Ben Fekih, Georges Kouroussis, Christophe De Fruytier and Olivier Verlinden

Coupling Analysis for the Thermo-Acoustic-Vibration Response of a Thin-Walled Box 559
Yan Yang and Chen-Wu Wu

Chatter Detection in CNC Milling Processes Based on Wiener-SVM Approach and Using Only Motor Current Signals 567
M. Lamraoui, M. El Badaoui and F. Guillet

Dynamics of the Unbalanced Support Centrifuge with Cavity Partially Filled with a Multiphase Liquid Medium 579
Almatbek Kydyrbekuly and Lelya Khajiyeva

Design and Test of a Piezoelectric Inertial Mass Actuator for Active Vibration Control 587
Sven Herold, Dirk Mayer, Tobias Melz and Tobias Röglin

Heat Transfer and Thermoelastic Dynamics of a Rotating Flexible Disc in a Hard Disc Drive 599
Yong-Chen Pei and Huajiang Ouyang

A New Direct Method for Updating Mass and Stiffness Matrices with No Spillover 609
Jiafan Zhang, Huajiang Ouyang and Jun Yang

Wave Propagation in a Honeycomb Composite Sandwich Structure in the Presence of High-Density Core Using Bonded PZT-Sensors 619
Shirsendu Sikdar and Sauvik Banerjee

Health Monitoring of Stiffened Metallic Plates Using Nonlinear Wave Interaction and Embedded PZT Transducers 629
Deba Datta Mandal, Debashis Wadadar and Sauvik Banerjee

Using Roving Disc and Natural Frequency Curves for Crack Detection in Rotors 639
Zyad N. Haji and S.O. Oyadiji

Vibration Based Diagnosis of Distributed Bearing Faults 651
Boštjan Dolenc, Pavle Boškoski, Jurij Pfajfar and Đani Juričić

Freight Car Roller Bearing Fault Detection Using Artificial Neural Networks and Support Vector Machines 663
Daniel Maraini and C. Nataraj

Light Rotor-Stator Partial Rub Characterization Using Instantaneous Angular Speed Measurement 673
Sanjin Braut, Roberto Zigulic, Goranka Stimac and Ante Skoblar

Sequential Recurrence Analysis of Experimental Time Series of a Rotor Response with Bearing Outer Race Faults 683
C.A. Kitio Kwuimy, M. Samadani, K. Kappaganthu and C. Nataraj

Adhesive Layer Influence on Compressive and Tension Stiffness of Thin-Layer Rubber-Metal Elements 697
Vladimirs Gonca, Svetlana Polukoshko and Galina Hilkevica

Multi Speed Model Updating of Rotor Systems 709
Manoj Chouksey, Jayanta K. Dutt and Subodh V. Modak

Vibration Control of Semi-active MR Seat Suspension for Commercial Vehicles Using Genetic PID Controller 721
S. Gad, H. Metered, A. Bassuiny and A.M. Abdel Ghany

Identification of Hysteretic Behavior of Magnetorheological Dampers Using NLARX Model 733
A. Shehata, H. Metered and W. Oraby

Online Roundness Error Identification and Model-Based Monitoring for Rotors 743
Markus Rossner, Thomas Thuemmel and Heinz Ulbrich

Study on Dynamic Load Sharing Characteristics of Double-Row Planetary Gear Train 753
Rupeng Zhu, Dongping Sheng, Guanghu Jin, Fengxia Lu, Heyun Bao and Miaomiao Li

A Novel Method to Improve the Resolution of Envelope Spectrum for Bearing Fault Diagnosis Based on a Wireless Sensor Node 765
Guojin Feng, Dong Zhen, Xiange Tian, Fengshou Gu and Andrew D. Ball

Vibrational Characteristics of a Cracked Rotor Subjected to Periodic Auxiliary Axial Excitation 777
Heba H. El-Mongy, Younes K. Younes and Mohamed S. El-Morsy

Parametric Modeling of Main Excitation Sources on Board Vessels . . . 789
Christoph Tamm, Georg Stoll and Sven Herold

Investigation of Motor Current Signature Analysis in Detecting Unbalanced Motor Windings of an Induction Motor with Sensorless Vector Control Drive 801
M. Lane, D. Ashari, F. Gu and A.D. Ball

Analysis of Locally Nonlinear Two Dimensional Periodic Structures Using NOFRFs 811
C.M. Cheng, Z.K. Peng, W.M. Zhang and G. Meng

Research on Multi-correlative Subsystems Analysis for Vibration Source Identification 825
S.M. Li, X.X. Jiang and C.W. Lim

A Nut-Type Ultrasonic Motor Driven with Single Phase Signal 835
Zhuzi Chen, Yu Chen, Tieying Zhou and Deyong Fu

Theoretical and Experimental Investigation into 'Efficient Modal Control Strategies' as Applied on a Plate Structure 845
Manu Sharma and S.P. Singh

Smart Machines with Flexible Rotors 855
Arthur W. Lees and Michael I. Friswell

Vibration Localization of Rotationally Periodic Structures 865
Arun Chandrashaker and Sondipon Adhikari

Forward and Backward Whirling of a Rotor with Gyroscopic Effect 879
Ali Fellah Jahromi, Rama B. Bhat and Wen-Fang Xie

Analysis of Nonlinear Dynamic Second-Order Effect of a Large-Scale Container Crane Under Seismic Excitations 889
Dong Wang, Gongxian Wang, Yeping Xiong and Jiquan Hu

Predictive Coefficient Method for Establishing Similarity Model of Jumbo Container Cranes 901
Zhe Li, Gongxian Wang and Jiquan Hu

Stability and Bifurcation Analysis of an Axially Accelerating Beam . . . 915
Bamadev Sahoo, L.N. Panda and Goutam Pohit

Analysis of Indirect Measurement of Cutting Forces Turning Metal Cylindrical Shells 929
Kirill Kondratenko, Alexandre Gouskov, Mikhail Guskov, Philippe Lorong and Grigory Panovko

Hybrid Permanent Magnet and Foil Bearing System for Complete Passive Levitation of Rotor 939
Siddappa Iranna Bekinal, Tumkur Ramakrishna rao Anil, Sadanand Subhas Kulkarni and Soumendu Jana

Amplitude Demodulation of Instantaneous Angular Speed for Fault Detection in Multistage Gearbox 951
Sankar K. Roy, A.R. Mohanty and C.S. Kumar

Experimental Grey Box Model Identification and Control of an Active Gas Bearing 963
Lukas Roy Svane Theisen, Fabián G. Pierart, Henrik Niemann, Ilmar F. Santos and Mogens Blanke

Monitoring of the Damage in Rolling Element Bearings 977
Steven Chatterton, Pietro Borghesani, Paolo Pennacchi and Andrea Vania

Effect of Structural Dynamics on the Shaft Line Rotor Response in Turbomachines 987
E. Meli, G. Pallini, A. Rindi, F. Capanni and S. Rossin

Nonstationary Processes Studying Based on "Caterpillar"—SSA Method 999
Svetlana Polukoshko, Galina Hilkevica and Vladimirs Gonca

Vibration Measurements on PFBR Fuel Subassemblies 1009
P. Anup Kumar, R. Vidhyalakshmi, H.P. Agnihotri, Sudheer Patri, S. Chandramouli, V. Prakash and K.K. Rajan

Multiple Fault Classification Using Support Vector Machine in a Machinery Fault Simulator 1021
S. Fatima, A.R. Mohanty and V.N.A. Naikan

A Method for Vibro-Acoustic FE Model Updating of Cavities Using Frequency Response 1033
D.V. Nehete, S.V. Modak and K. Gupta

Experimental and Numerical Study of Friction-Induced Noise of Brake Pad Materials Having Grooved Surface 1043
X.C. Wang, J.L. Mo, D.W. Wang, H. Ouyang, J. Zhao, G.X. Chen and M.H. Zhu

Neural Network Modeling of Submarine Shell 1055
Fei Wang, Yeping Xiong and Zhenping Weng

Novel Approach for Structural Dynamic Topology Optimizations Based on Power Flow Mode Theory 1065
Ye-Ping Xiong

Accurate Calculation of the Phase Shift Angle of Sinusoidal Signals Using Trapezoidal Rule 1077
Luis E. Monterrubio

Effect of the Rail Support Stiffness on the Occurrence Propensity of Rail Corrugation Based on the Self-excited Vibration of a Wheel-Rail System 1083
G.X. Chen, W.J. Qian, J.L. Mo and M.H. Zhu

Modeling and Dynamic Force Simulation for Detection of Profile Error in Spur Gear Pair 1091
Atul B. Andhare and Manish Kumar Verma

Availability Improvement by Early Detection of Motor Bearing Failure Using Comprehensive Condition Monitoring Techniques at DTPS 1101
Hemant M. Bari, Atul A. Deshpande and Suhas S. Patil

Base Isolator of Vertical Seismic Vibration Using a Negative Stiffness Mechanism. 1113
Yusuke Mochida, Naohiro Kida and Sinniah Ilanko

A Novel Piezoelectric Rotational Speed Meter. 1121
Wei-qing Huang, Meng-xin Sun, Jian-hui Zhang, Qing-jun Ding and Yin Wang

A New Linear Motor with Two Piezoelectric Stacks Based on Differential Friction Effects 1133
Weiqing Huang, Xifu Chen and Jianhui Zhang

Part I
Keynote Lectures

Ancient Temple Carts—Modifications for Structure and Steering

J.S. Rao and Bigil Kumar

Abstract Temple carts from ancient times are used to bring out Deities on festival days in a procession. The origin of these carts and their construction is discussed in this paper together with current design practices befitting with changing times.

Keywords Rig-Veda · Hindu gods or extra-terrestrial visitors · Temple carts · Steering · Globally elastic but locally plastic structures

1 Introduction

Indian temples have their origins from Rig-Veda times, a period exactly not identifiable from modern historical dating methods, but derived from inference to be at least 4,000 years old. The epic Ramayana contains description of battles with incredible weapons cited as Brahma weapon.

The Gāyatrī Mantra a highly revered mantra, based on a Vedic Sanskrit verse from a hymn of the Rig-Veda (3.62.10), attributed to the rishi (sage) Viśvāmitra, see Griffith [11].

This is 16th stanza of Upanishads in Devanagari script, a translation into English is provided by several, among them that of Easwaran [10] is "O nourishing Sun, solitary traveler, controller, source of life for all creatures, spread your light and subdue your dazzling splendor so that I may see your blessed Self. Even that very Self am I!".

Jones [12] on February 2nd 1786 addressed the Asiatic Society of Bengal about the origin of Sanskrit, having similarities with Latin and Greek; commonly used words are: Father, Mother and Horse.

J.S. Rao (✉) · B. Kumar
Kumaraguru College of Technology, Coimbatore 641049, India
e-mail: jsrao@kct.ac.in

B. Kumar
e-mail: bigil.kumar@pricoltech.com

J.K. Sinha (ed.), *Vibration Engineering and Technology of Machinery*,
Mechanisms and Machine Science 23, DOI 10.1007/978-3-319-09918-7_1

Ramayana refers to some Vedic Gods like Indra, Agni and Varuna among others, see Didhiti Biswas article in Dodiya [9]. The Rig-Veda and the life of ancient Indians during that period are described by Srinivasa Iyengar [25].

Mohenjo-Daro Harappa (Mound of the Dead) was discovered in 1922. Kenoyer [14] places the settlements of the Indus around 3750 BC. At this site, 37 skeletons were found in contorted positions, in a residential district of the town, see Fig. 1. In all there were just 44 skeletons found at this site.

At the time of Mohenjo-Daro, mankind has not known an atomic weapon. No conventional weapons such as swords or spears were found. No scavenging animals and their skeletons were discovered at this site. This is despite the well recorded and understood destruction of the city of Pompeii in the eruption of Mount Vesuvius in 79 AD, see De Carolis et al. [8]. Just as we ignored completely the battles described in Hindu epics Ramayana and Mahabharata, here too in Mohenjo-Daro the mysterious circumstances were ignored.

The first atomic bomb was detonated on July 16, 1945, in the Trinity test in New Mexico; its creator, Oppenheimer [20] remarked later "it brought to mind words from the Bhagavad Gita: Now I am become Death, the destroyer of worlds."

According to Dales [5], the evidence given earlier suggesting that Mohenjo-Daro was destroyed by armed invaders was a quick conclusion drawn.

Suggestions were made that this might be the site where Ramayana epic battles took place. Ramayana describes in the battle that Brahma Astra was used which generated brightness more than 1,000 Suns, trees were up in flames; more significantly the epic describes people who survived the battle were losing hair and nails started to fall out—typical effects of radiation observed after the present day atomic explosions, see Tsoukalos [26]. Childress [4] discusses Vimanika Sasthras (flying

Fig. 1 Skeletons at Mohenjo-Daro attributed to Massacre theory

machine technologies) described 6,000 years ago in Sanskrit texts indicating that this technology existed in the olden days. The texts describe three giant cities orbiting the earth with gleaming metal and iron with them going to war with each other. Tsoukalos and others are proponents of the idea that ancient astronauts interacted with ancient humans. What is it our ancestors are describing in the ancient Sanskrit scripts, some type of technology that was witnessed; they didn't understand nuts and bolts aspects of this technology, yet they created divine representation; the practices that they have developed are so effective, they remain in force even today in India.

Davenport [7] spent 12 years in studying at the site; according to him Mohenjo-Daro corresponds exactly to Nagasaki. An epicenter about 50 yards wide where everything was crystallized, fused or melted was found. Sixty yards from the center the bricks are melted on one side indicating a blast. At this site even today the radiation levels are high suggesting that there was an atomic attack; see Childress, History Channel Video [4].

According to recent findings based on radio-metric dates from Bhirrana (Haryana) the origin of the Indus Valley Civilization emerged in the 8th millennium BC in the Ghaggar-Hakra and Baluchistan area, see Khandekar [15].

Rig Veda cites chariots in many places: The chariot carrying the God is moving through the air with very swift motion; (Rig Veda V. 77.3). It is hard to overtake (Rig Veda. V.35.7), see Griffith [11]. Parts of *Ratha* are mentioned in Vedas.

Pallavas (600–900 AD) are amongst the earlier patrons who built rock-cut chariot-shaped temples of Mahabalipuram: They are called locally as *Ratha* (Chariot) and the Gopuram (superstructure) called Vimanas, see Fig. 2.

In Tanjore Brihadeshvara Temple (1009 AD), Fig. 3, there is an elaborate mural as given in Fig. 4; see Mohamed and Iqbal [19].

Lord Siva on his chariot with his eight arms wielding different weapons is driven by Lord Brahma. The *Ther* is Earth with the Sun and Moon as wheels. Nandi, Siva's *vahana* is pulling the chariot. Brahma is smiling at Siva looking at his

Fig. 2 Five *Rathas* of Mamallapuram (Mahabalipuram) (630–638 AD)

Fig. 3 Brihadeshvara Temple

Fig. 4 Destruction of Demons by Lord Siva as Tirupurantaka on his chariot with eight arms

inability to destroy the three demons with his bow and arrow, being closely watched by other Gods. Siva observing Brahma's smile turns the bow backwards and opens his third eye to destroy the asuras. This mural is recalling what our ancestors saw during Rig-Veda and Ramayana periods.

Fig. 5 First temple with a Wheel (1117–1135 AD)

This mural clue that Gods were flying but not resident on earth is taken by Vikrama Chola (1117–1135 AD) who built the first temple in Thukkachi with a wheel, see Fig. 5. This was followed by Rajaraja Chola II who built the Airavatesvara temple (1146 AD), see Fig. 6 which has a horse-drawn chariot carved on the front of the mandapam. This trend followed in Konark in Odisha and Hampi.

Konark Sun Temple (1278 AD) was conceived as a gigantic chariot of the Sun God, Fig. 7 shows one of these wheels. The next step seems to be taking the Gods as they are known in the scriptures to flying chariots, see Fig. 8 as depicted by History Channel. Rig-Veda also describes battles using Vimanas, as depicted in History Channel and described best by our ancestors in Fig. 9; that's the reason why the superstructures of temples are called Vimanas.

Stone chariots were built in Konark (1278 AD), see Fig. 10. Vijayanagara Kings built several temples in Hampi and one stone chariot in Vittala temple in 16th century, see Fig. 11.

It is difficult to date when the practice of chariots taking Gods on festival days began; it is probably during the Vijayanagara period in 16th century. *Ther* itself is considered as the manifestation of the temple housing a God. Those pulling the *Ther* consider to be rewarded by divine blessing. In to days practice, Rathotsava is

Fig. 6 Airavatesvara Temple and spoked chariot wheel

Fig. 7 Konark Sun Temple

Fig. 8 Battles from flying chariots

Fig. 9 Vedic battles in Sky from vimanas

Fig. 10 Konark Temple chariot (1278 AD)

Fig. 11 Stone chariot in hampi (16th century)

an important social gathering enhancing living in harmony. An ancient Rathotsava is depicted in Fig. 12; see Dallapiccola [6].

A special section of Vedas deal with the construction of temples and *Rathas*. Silpa Shastra deals with *Ther* construction (Maha Viswakarmiyam) script of which is put on palm leaves between 8 and 10th AD is shown in Fig. 13 see Umapathy [27].

Shastras (Rules) to design the *Thers* are probably developed as early as 8th century AD; 1. Viswakarma *Ratha*laksanam, 2. Kumara Tantra, 3. Maha Viswakarmiyam and 4. Karanagama [13] see, Acharya [1, 2] and Mankad [18]. The *Ther* used in Brihadeshvara temple in Tanjore is shown in Fig. 14. In general a *Ther* should have a *Simmasanam* (Crown Chair) for the Deity, a *Devasanam* (Seat of Devas) a *Narasanam* (Seat for priests), different *Boothabars* from which the Crown Chair is reached, see Fig. 15.

The *Ther* has a superstructure like a Vamana; see Fig. 16; but this is not essentially a load bearing structure but has all decorations befitting a festive mood.

Fig. 12 Ancient Rathotsava and procession

Fig. 13 Silpa Shastras on palm leaves

Fig. 14 Tanjore Temple car

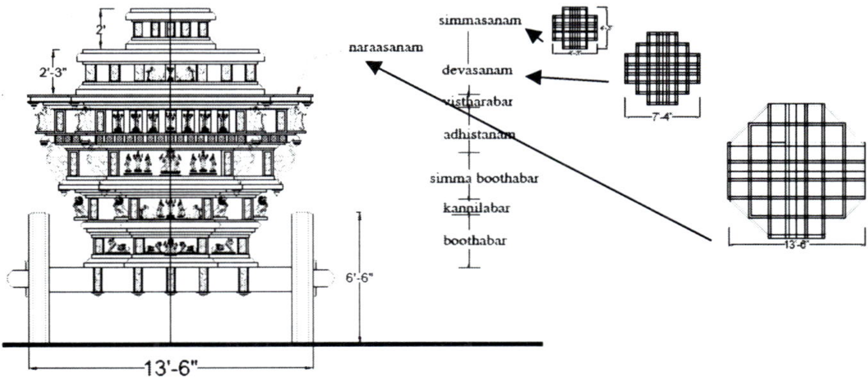

Fig. 15 A typical temple *Ratha* (car)

Fig. 16 Tanjore Temple car in procession

These *Thers* do not have any steering system or a braking system. The *Thers* are pulled by devotees and being very heavy about 300 tons in the case of Tiruvarur temple; move inch by inch. Steering, sometimes breaking is achieved by using wedges placed by trained volunteers. The state of Tamilnadu alone has nearly 500 such *Thers* in different temples; each of these *Thers* cost around 600,000 US$.

A review of the ancient *Thers* is given by Rao, Bhonsle and Kumar [23].

Here the modern approach to determine the stresses, a simple steering linkage and weight reduction through topology optimization is presented.

2 *Ther* Structure and FE Analysis

Traditionally *Thers* are designed without the application of any structural mechanics principles; they tend to be very heavy. The *Ther* shown in Figs. 14 and 15 is redesigned keeping the Vedic principles intact.

The CAD model of this *Ther* is given in Fig. 17. It is made of Madhuca Latifolia wood with $G_{11} = 23{,}800$ MPa, $G_{22} = G_{23} = 238$ MPa and Density $= 9.2 \times 10^{-10}$ ton/mm^3. Bending strength is 171 MPa. The weight of the *Ther* is found from CAD model to be 26 tons.

Overlapping of each section is merged together in the model, i.e. no bolts are represented. The superstructure or Vimana or Gopuram mass is taken as 50 tons. The top of the Gopuram is at 30.398 ft. The gravity load is taken to be 9,810 mm/s^2. The FE model with Hex mesh is shown in Fig. 18. Total number of elements is 256,288 with 74,117 nodes. Including 500 kg mass for the Deity and 70 kg mass per each of 22 persons performing offerings etc., the total mass is taken as 26.665 tons. As the cart moves at a very slow speed, 1 g load is applied with boundary conditions as shown.

The maximum von Mises stress is on axle bars as shown in Figs. 19 and 20. The maximum stress is very low and is located on the axle bars as shown in Fig. 20.

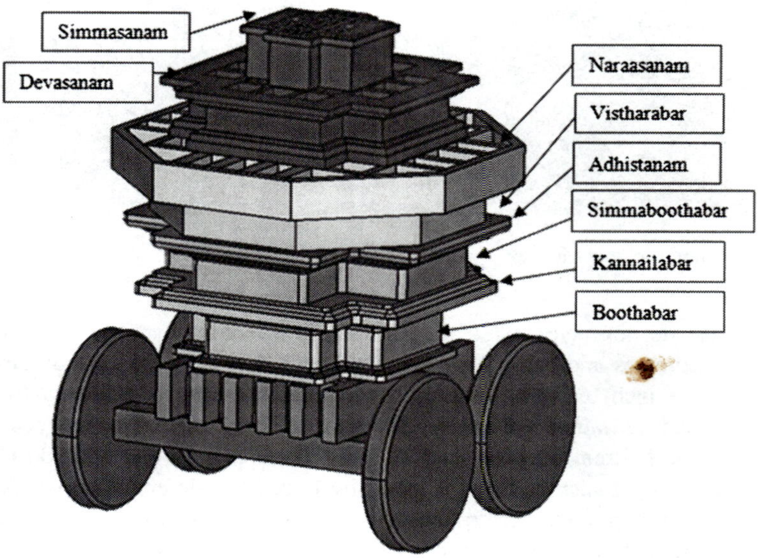

Fig. 17 CAD model of a Tanjore *Ther*

Fig. 18 Hex mesh FE model of the *Ther*

3 Wedging for Cart Turning

In absence of mechanized means, wedges, see Fig. 21, are used by skilled personnel for negotiating turning as well as breaking. This practice has led to accidents.

A wedge is simulated under elastic conditions with the cart moving at 2 km/h, as shown in Fig. 22. The transient stress distribution under elastic conditions is obtained by using steel material for the axle bars. The total load is directly applied at the supports on the axle bars. Total number of elements is 54,357 with 62,570 nodes. Young's modulus is 210 GPa, Density = 0.0078 ton/m^3 and μ = 0.3, Yield strength = 225 MPa and Ultimate Strength = 411 MPa. The von Mises stress obtained by LSdyna is given in Fig. 23.

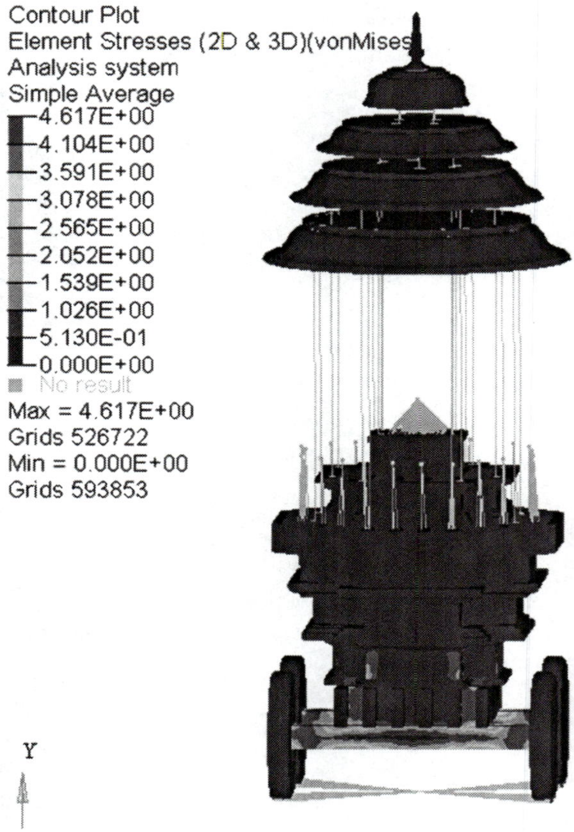

Fig. 19 von Mises stress distribution

The peak stress observed is 277.5 MPa at 0.055 s and decreases subsequently. This value is just above the yield and if an elasto-plastic analysis this value will be falling on the elasto-plastic line just a shade higher than 225 MPa. Actually this stress is edge of contact stress, see Chan and Tuba [3] and it is somewhat spurious and it is arising because of the modeling that is adopted to decrease the number of elements and computational time. This can be safely ignored.

4 Steering Linkage

A steering linkage based on Ackerman principle; see Rao [21] is schematically shown in Fig. 24. The CAD model of this linkage is given in Fig. 25. Figure 26 shows the modifications for the *Ther* front wheels. The material is taken as Material —IS:961 1975 steel.

Fig. 20 Maximum von Mises stress 4.6 MPa

It may be noted that a right movement of the handle of the coupler rod will produce a left turn and vice versa. This understanding keeps the linkage with a minimum of only four pivots. Otherwise the linkage becomes more complex and is considered unsafe. Also no sliding joints are allowed.

The model for stress evaluation of the steering linkage is so chosen to minimize the effort of computation and given in Fig. 27. Total number of elements is 111,352 with 131,365 nodes. Young's modulus is 210 GPa, Density = 0.0078 ton/m^3 and $\mu = 0.3$, Yield strength = 225 MPa and Ultimate Strength = 411 MPa.

The maximum value of stress obtained is 31.471 MPa; see Fig. 28 which is well below the Yield value of 225 MPa.

5 Steering Linkage with Wedging

Figure 29 gives the *Ther* model of Fig. 27 with a wedge subjected to 2 km/h speed. The transient stress distribution under elastic conditions is obtained using LSdyna and the result obtained for stress is shown. The material at discontinuities at element 1,741,318 goes into plastic range given by 1,048 MPa.

Under elastic analysis, this will be the kind of result as found by Rao [22]. Globally elastic but locally plastic structures, a rotating turbine blade has shown 1,825 MPa with average stress of 256 MPa; with an elasto-plastic analysis the peak stress was reduced to 768 MPa following the elasto-plastic law, though the average stress was reduced to a value 216.86 MPa. Here the peak stress observed at 0.22 s is

Fig. 21 Use of wedges for cart turning and braking

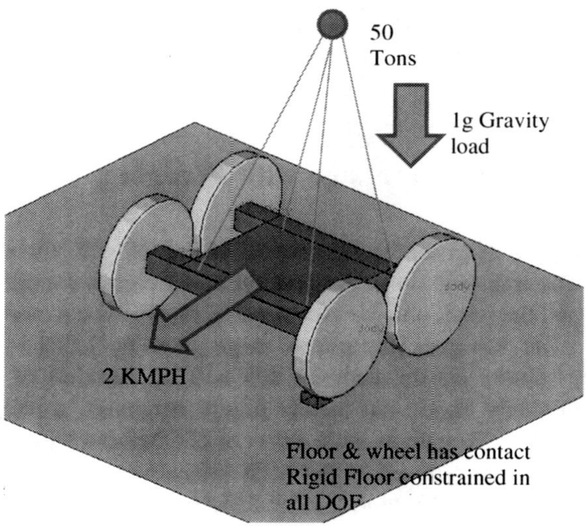

Fig. 22 Simulation for wedging

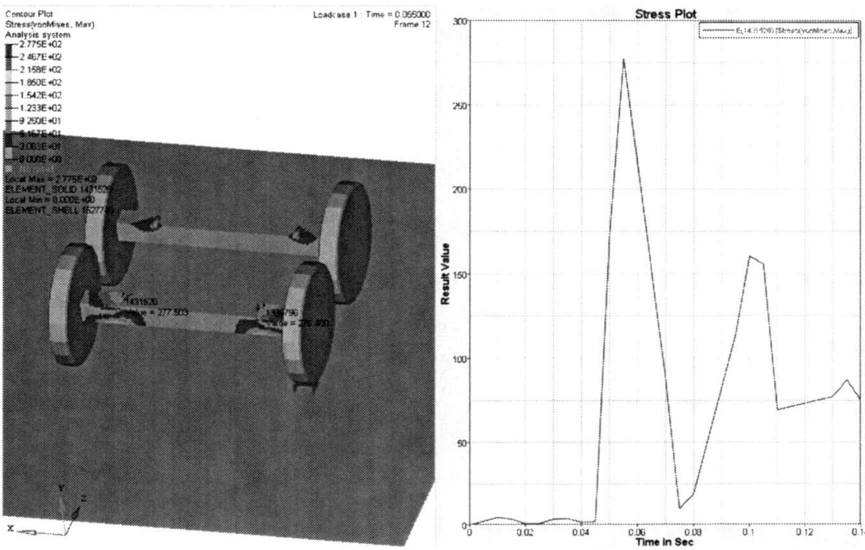

Fig. 23 Stress field under wedging

Fig. 24 Schematic of Ackerman linkage for the *Ther*

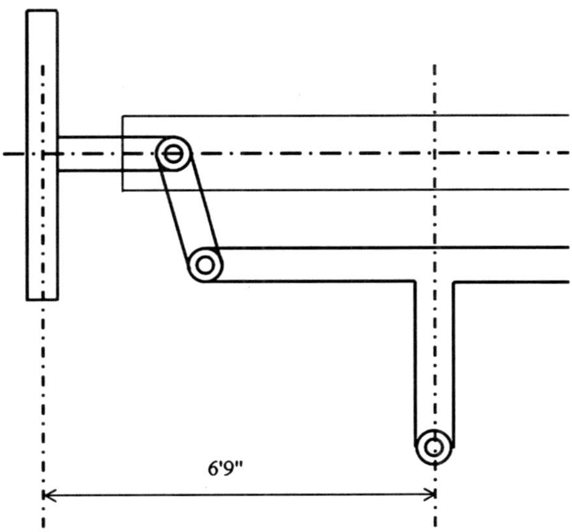

1,048 MPa as shown in Fig. 30. Note that no damping was assumed in this run. Actually at this location the stress will follow the elasto-plastic law in the local stress riser region. It will be necessary to determine true stress and strain using

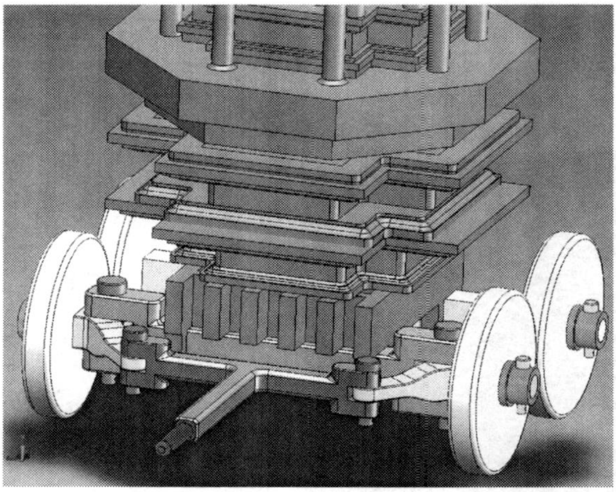

Fig. 25 Proposed Ackerman steering linkage

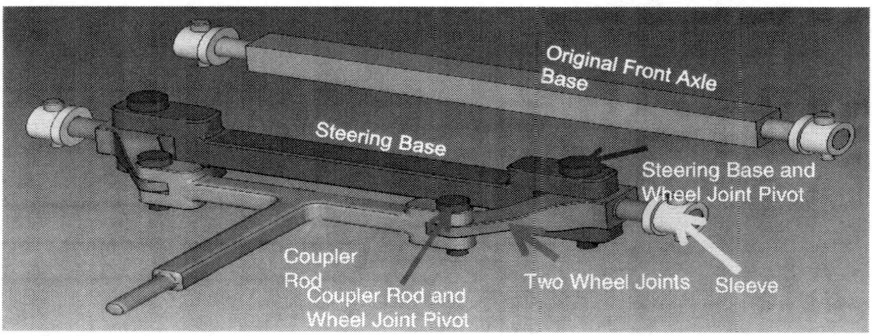

Fig. 26 Modifications for steering linkage

Neuberization process and the life for each wedging can be calculated using Strain Based Lifing process. That will determine the number of actual wedging operations that can be performed before a crack can initiate at this location. This process is described by Rao [22] for similar globally elastic and locally plastic structures. With the proposed steering mechanism, one can avoid the wedging operations for steering. The lifing process is left for a future exercise.

We next look at weight reduction of this heavy structure following topology optimization procedures, see Rao [22]. The procedure is explained by Rao et al. [24], while optimizing a gas turbine blade for the weight.

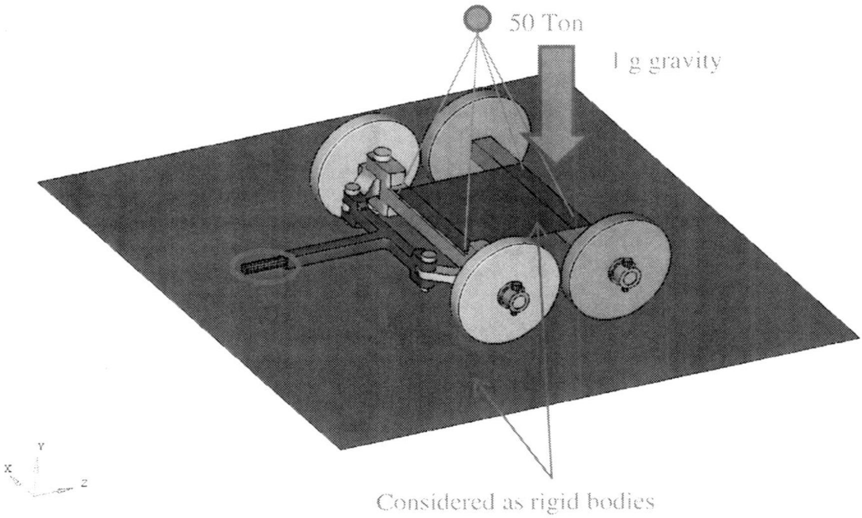

Fig. 27 Simple model for stress evaluation in the steering linkage

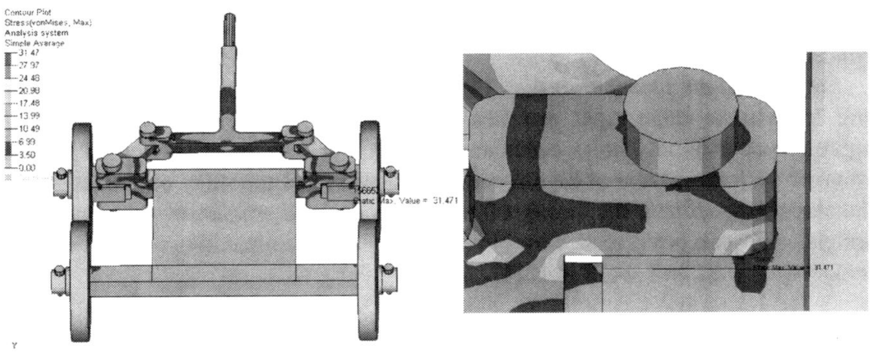

Fig. 28 Maximum von Mises stress 31.471 MPa

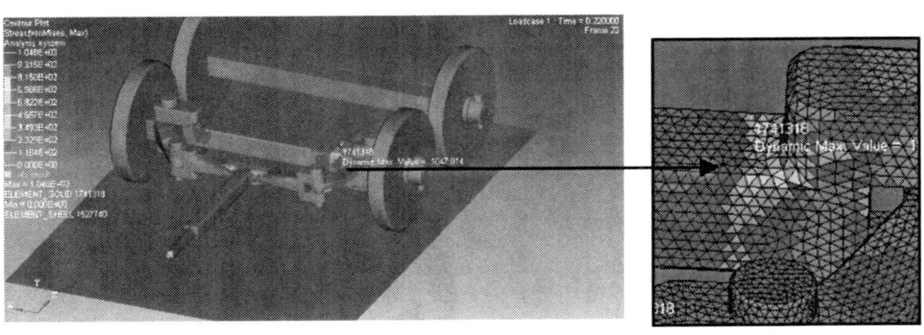

Fig. 29 Steering linkage with a wedge and peak stress location

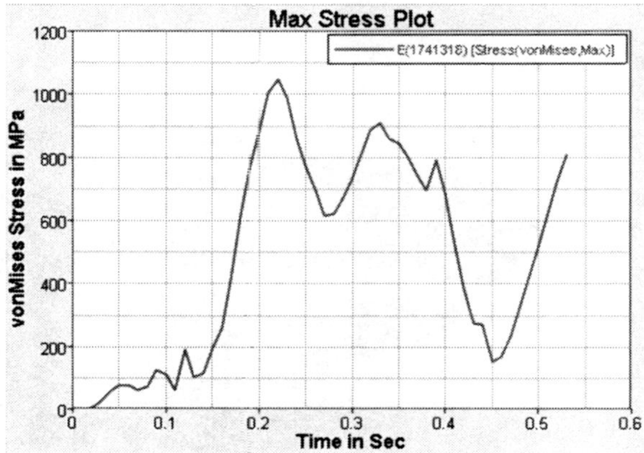

Fig. 30 Peak stress time history

6 Wight Reduction

Figures 19 and 20 give the distribution of von Mises stress in the *Ther* structure. The maximum value is found to be only 4.6 MPa. There is obviously a considerable material sitting idle in the structure which can be removed to the extent allowed according to Vedic rules and keep the structure still safe under the prevailing conditions.

Earlier designs like the present *Ther* were made based on durability condition and tend to be more rigid and heavy in weight. The designs were based on approximate methods, stress concentration charts and factor of safety. All this has changed with the advent of high speed computation and topology optimization was developed to address the needs of advanced designs consistent with minimum weight, optimum performance for different physics state quantities such as pressure, temperature etc., and see Rao [22].

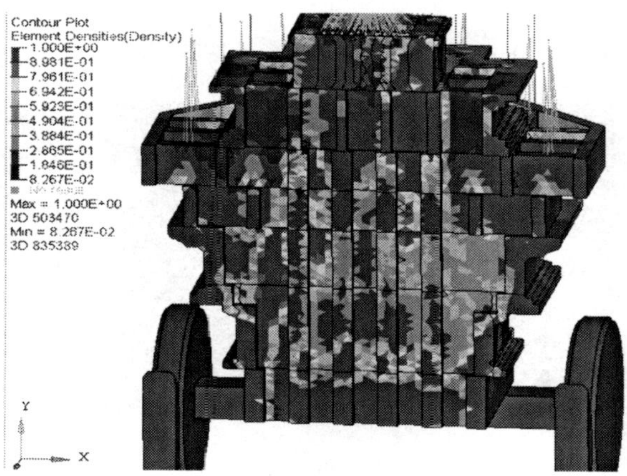

Fig. 31 von Mises stress in the middle cross-section across the *Ther*

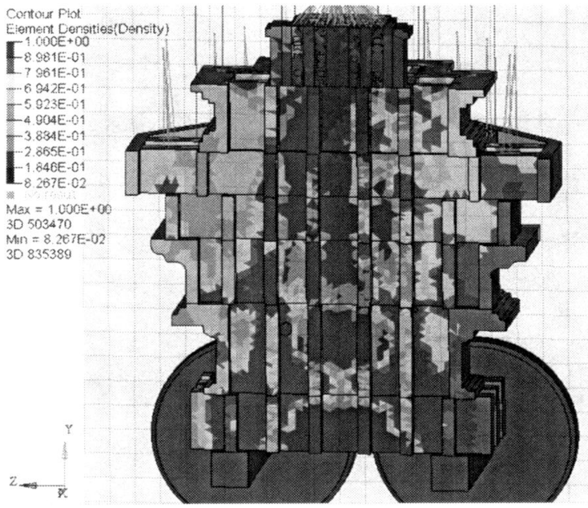

Fig. 32 von Mises stress in the middle cross-section along the *Ther*

Following this procedure Figs. 31 and 32 show the relative densities of the main structure of the *Ther* at the middle sections along and across the *Ther*. The material in blue color region can be removed and restructured to reduce the weight.

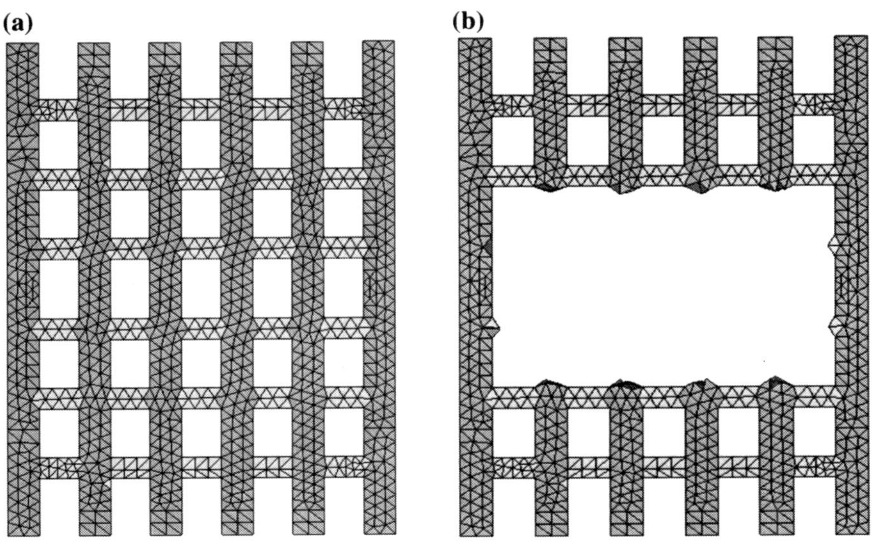

Fig. 33 Rearrangement of the cross bars on wheel chassis before and after optimization; thickness (*height*) maintained the same so as not to change the overall dimensions. **a** Mass of the component = 2.164 tons. **b** Mass of the component = 1.562 tons

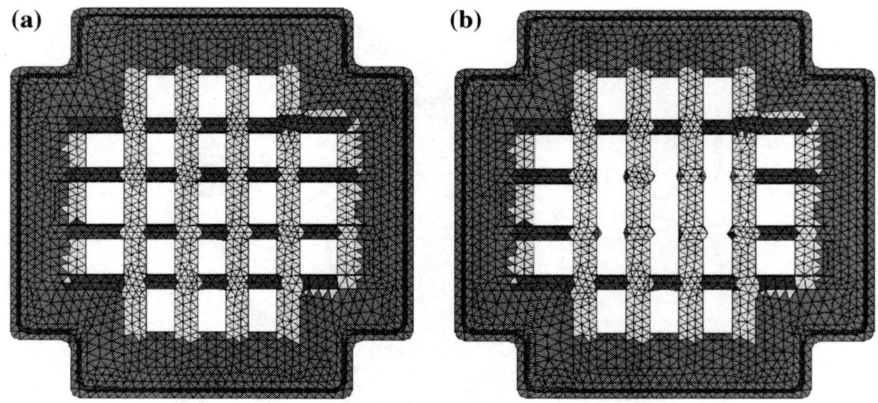

Fig. 34 Rearrangement of the boothabar. **a** Mass of the component = 3.577 tons. **b** Mass of the component = 3.434 tons

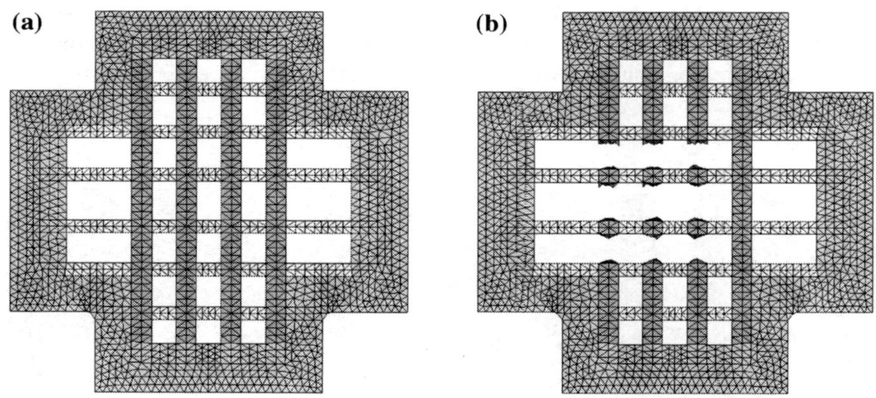

Fig. 35 Rearrangement of the simma boothabar. **a** Mass of the component = 3.139 tons. **a** Mass of the component = 2.839 tons

The rearrangement of the structure by removing the material in low relative density blue regions thus obtained is shown in Figs. 33, 34, 35, 36, 37, 38 and 39.

The original structure weight 17.383 tons is reduced to 15.272 tons. The reduction achieved here is 11.76 %. A reduction in the weight helps in reducing the peak stresses, particularly while steering.

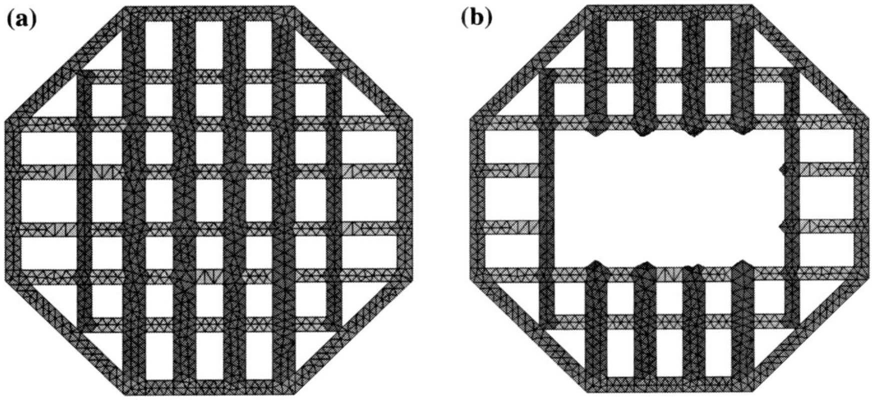

Fig. 36 Rearrangement of the Adhistanam. **a** Mass of the component = 2.231 tons. **b** Mass of the component = 1.756 tons

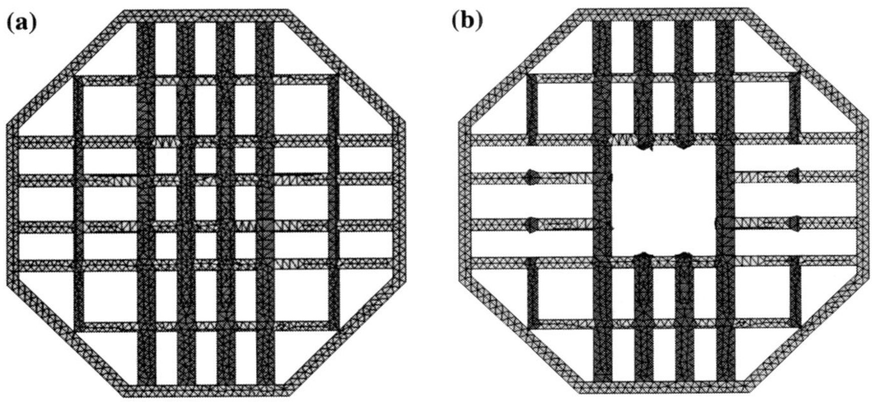

Fig. 37 Rearrangement of the Narasanam. **a** Mass of the component = 2.792 tons. **b** Mass of the component = 2.467 tons

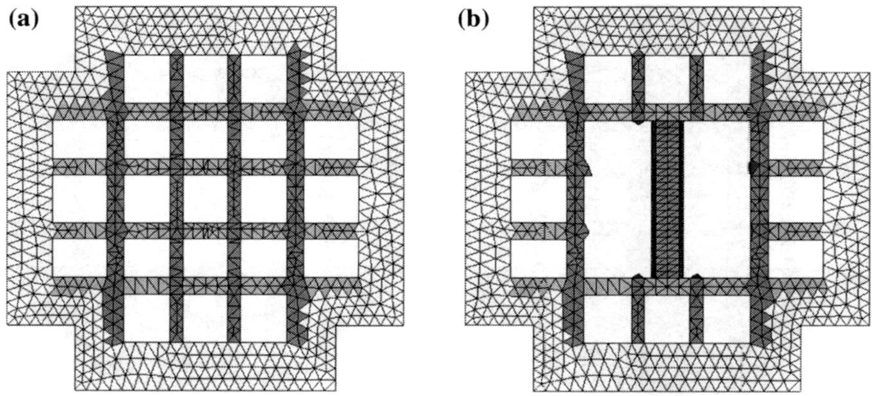

Fig. 38 Rearrangement of the Devasanam. **a** Mass of the component = 2.035 tons. **b** Mass of the component = 1.887 tons

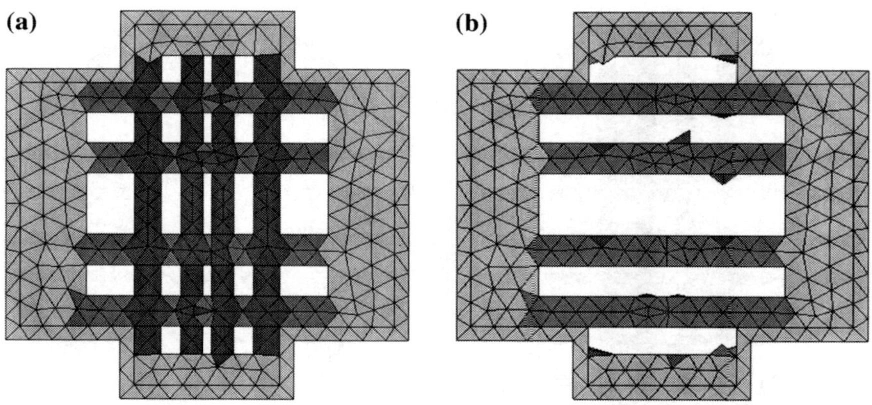

Fig. 39 Rearrangement of the Simmasanam. **a** Mass of the component = 0.633 tons. **b** Mass of the component = 0.5121 tons

7 Conclusion

The origin of temples and *Thers* is traced back from Rig-Veda times. The *Thers* designed by Vedic rules appear to be good structures according to modern finite element analysis. The wedging practiced to steer the *Ther* can be unsafe that can lead to failures. A steering linkage is designed and the details presented. The weight of the structure is reduced by following topology optimization.

Acknowledgments The authors are grateful to Chairman Padmabhushan Dr. Mahalingam of Kumaraguru College of Technology and Chief Executive Officer Mr. Bobby John of Pricol Technologies, Coimbatore for their support in this work.

References

1. Acharya PK (1934a) Manasara vol. III, P-286-292. DK Publishers, New Delhi. ISBN:81-86142-71-1
2. Acharya PK (1934b) Manasara vol. V, p. 20. DK Publishers, New Delhi. ISBN7536-136-0
3. Chan SK, Tuba IS (1971) A finite element method for contact problems of solid bodies—Part II: applications to turbine blade fastenings. Int J Mech Sci 13:627
4. Childress D (2012) Ancient Hindu technology by history channel 2. http://www.youtube.com/watch?v=yT7FRbstJCI
5. Dales GF (1964) The mythical massacre at mohenjo-daro. Expedition 6:36
6. Dallapiccola AL (2010) South Indian paintings. A Catalogue of British Museum Collection, British Museum Press. http://collections.vam.ac.uk/item/O68238/procession-with-festival-car-painting-unknown/
7. Davenport DW, Vincenti E (1979) 2000 a.C., distruzione atomica, SugarCo, Milan, Italy
8. De Carolis E, Patricelli G (2003) Vesuvius, A.D. 79: the destruction of Pompeii and Herculaneum. L'erma Di Bretschneider
9. Dodiya JK (ed) (2001) Critical perspectives of ramayana. Sarup and Sons, New Delhi
10. Easwaran E (2007) The upanishads. The Blue Mountain Center of Meditation, Tomales
11. Griffith RTH (1890) The hymns of the Rig-Veda. E.J. Lazarus, Benares
12. Jones SW (1824) Discourses delivered before the asiatic society: and miscellaneous papers, on the religion, poetry, literature, etc., of the nations of India. Printed for C. S. Arnold, p. 28
13. Karanagama (MS.No:T1035Aa), Palm leaf documents in French Institute of Pondicherry, Pondicherry
14. Kenoyer JM (1996) Craft traditions of the indus civilization and their legacy in modern Pakistan. Lahore Museum Bull IX:2
15. Khandekar N (2012) Indus valley 2,000 years older than thought, Hindustan Times New Delhi, November 04, 2012
16. Kumara Tantra: (MS.No:T0472), Palm leaf documents in French Institute of Pondicherry, Pondicherry
17. Maha Viswakarmiyam: (MS.NoT167), Palm leaf documents in French Institute of Pondicherry, Pondicherry
18. Mankad PA (Ed) (1950) Aparajitaprccha, Gaekwad's oriental series. Oriental Institute, Baroda, p. 115.
19. Mohamed IK, Iqbal A (2010) Vibrant at 1000 big temple, Tanjore, India. Productivity and Quality Publishing, Chennai
20. Oppenheimer JR (1965). http://www.atomicarchive.com/Movies/Movie8.shtml
21. Rao JS (2011) Kinematics of machinery through hyperworks. Springer, New York
22. Rao JS (2011') History of rotating machinery dynamics. Springer, New York
23. Rao JS (2013) Hindu temple carts—*Ratha*ms, IFToMM Workshop on History of MMS, Palermo
24. Rao JS, Bhaskar Kishore Ch, Vasanthakumar M (2008) Weight optimization of turbine blades. In: Proceedings of 12th international symposium on transport phenomena and dynamics of rotating machinery, Honolulu, Hawaii, ISROMAC 12–2008, 20020
25. Srinivasa Iyengar PT (1912) Life in ancient India in the age of mantras, Srinivasa Varadachari & co., Madras. Original from University of California
26. Tsoukalos G (2012) In Ancient Hindu technology by history channel. www.youtube.com/watch?v=BCb40zQb5rc
27. Umapathy Acharya (2013) Family collection. Kumbakonam, India
28. V&A Collection (1800) Museum number: AL.8806
29. Viswakarma *Ratha*laksanam: MS.No:11066, Saraswati Mahal Library, Tanjore

The Rise and Fall of the Rotor Blade Strain Gauge

Peter Russhard

Abstract In any gas turbine the understanding of the way in which the rotor blades vibrate is essential for both the development and production phases. For many years the standard method of obtaining such data has been with the use of strain gauges in conjunction with radio telemetry units or slip rings. With the complexity of the machine increasing and as the operating environment becomes more and more hostile for such systems, new methods are being investigated that will ultimately lead to a non-intrusive technology being developed. Here we review the progress and pioneers of strain gauge measurement dating back to the 1930s and look at a number of technologies that have tried to displace them from their positions as the mainstay of rotor blade vibration measurement.

Keywords Strain gauge · Radio telemetry · Rotor blade vibration

1 Introduction

The measurement of rotor blade vibration in turbine engine development is made to ensure a physical understanding of how blades behave. Advancement in acquisition and analysis tools has the potential to reduce the amount of development time required to validate new designs. Blades fail due to the stresses in them being too high and there are a number of ways of generating stress. All of the many mechanisms for blade failure during engine operation are the result of excessive blade vibration. By quantifying blade vibration through measurement it is possible to define a set of criteria to protect components from these problems. There are a number of reasons for requiring a new technology and most hinges on both the lead-time and cost of application. Today's gas turbine development programs are up to 50 % shorter than those of a decade ago and the machines operate at increasingly higher speeds and temperatures. Instrumentation failure is high, as is the mortality

P. Russhard (✉)
Rolls-Royce Plc Derby, Derby DE24 8BJ, UK
e-mail: Peter.russhard@rolls-royce.com

J.K. Sinha (ed.), *Vibration Engineering and Technology of Machinery*,
Mechanisms and Machine Science 23, DOI 10.1007/978-3-319-09918-7_2

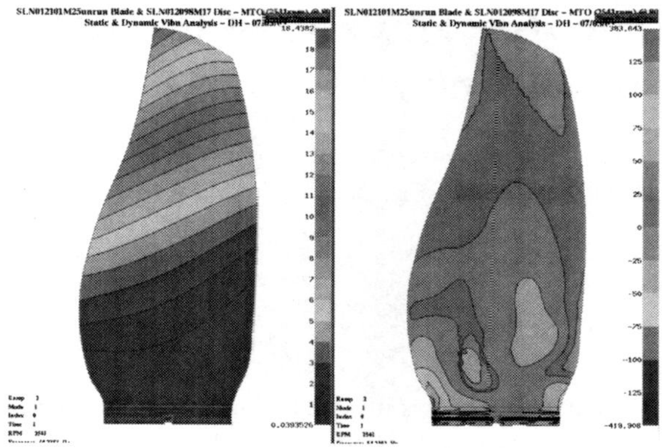

Fig. 1 Rotor blade modelling

of a strain gauge in such an environment. Similarly, in a production environment, it is desirable to have an indication of blade health for the most vulnerable rotors of a gas turbine. This leads to lower maintenance costs and fewer removals and routine inspections. With significant profit now coming from the Total Care™ packages offered by OEM's (Original Equipment Manufacturers) the desire to have a reliable Blade Health Monitoring system (BHM) is high. Unfortunately strain gauges are not suitable for long-term health monitoring of rotor blades [1].

During the design stage of a compressor/turbine rotor blade, a number of blade simulations provide predictions of blade deflection and stress distribution for all of the blade natural frequencies. These levels are often the maximum allowable, but not necessarily that which will be encountered in operation. Factors affecting the prediction of rotor blade stresses include the uncertainty in calculating both the structural and the aerodynamic damping associated with each mode of vibration. Current design simulation software is unable to accurately calculate these values and direct measurement is often required to obtain them. A measurement system requires the ability to confirm expected blade behaviour and to provide additional data to allow the system to be accurately modelled (Fig. 1).

2 Measurement of Rotor Blade Strain

Between the years 1931 and 1938, aircraft propeller failures in flight ran anywhere from 8 to 41 times a year. By 1938 OEM's had begun to study all of the installations in the field and restricting them so that they did not operate under conditions of severe vibrations. In 1939 there were 6 failures and in 1940 there were none. The result of this was that aircraft and pilot losses went from 40 to zero. The necessity for measuring and understanding rotor blade strains had become imperative and was the main driver for the development of rotor blade strain measurement technology [2].

A number of alternate technologies have challenged the dominant position of strain gauges with one in particular starting to displace it.

3 Strain Gauges

Strain gauges have been described as the preferred transducer for blade vibration measurement [3], strain gauges are a field proven technology for extracting accurate real-time information [4] relating to the alternating strains arising in critical areas of a blade. In addition, strain gauges have proven to be well established and reliable when applied to non-rotating components for most applications.

One of a number of strain gauges pioneers was Charles M. Kearns, Jr. who invented the bonded carbon strain gauge whilst working at Hamilton Standard Propeller Co [5]. Within 4 years, in-flight propeller failures went from dozens per year to zero. The carbon gauge is sold, copied, and spreads like wild fire worldwide. Prior to this Kearns experimented with a mixture of sulphur and graphite (Aquadag) that changed its resistance with stress but also with just about everything else. The first carbon based gauge was simply a ground down carbon resistor cemented to a beam which appeared to have a linear relationship between strain and resistance. Kearns achievements included;

- Developing and characterising carbon resistor strain gauges
- Attached carbon resistor strain gauges to HS propellers
- Designed and fabricated a turntable silver graphite slip ring
- Determined necessary signal conditioning
- Collected data for the strain gauges
- Solved a propeller vibration problem
- Implemented changes in propeller design and manufacture

This is a process that is still repeated today in the development of measurement technology.

The carbon gauge was replaced around 1945 with the application of SR-4 resistance wire strain gauge and finally Peter Scott Jackson [6] (UK RAF) invented the foil strain gauge in 1952 (Fig. 2).

Developments in the size, sensitivity and adhesive technologies have kept the application of rotor blade strain gauges current since then. This combined with their ease of conditioning and processing makes it a simple but reliable method of measuring rotor blade strain.

Fig. 2 Foil strain gauge

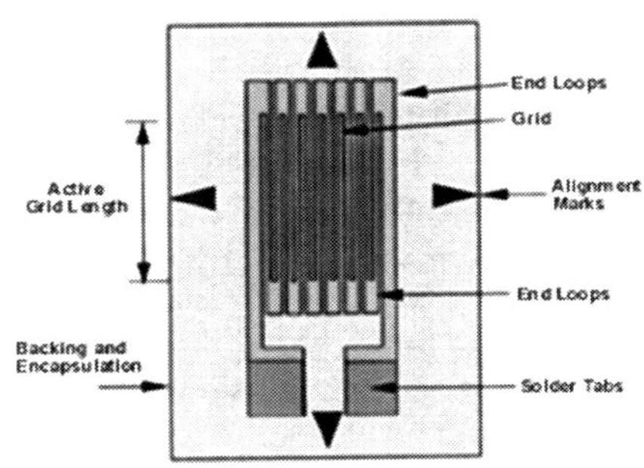

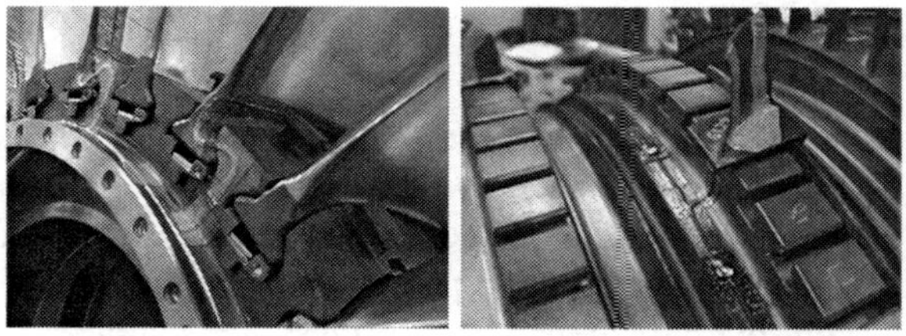

Fig. 3 Typical rotor strain gauge installation

However for rotating components, strain gauge measurement systems suffer from a low operating life. This limited life expectancy is often due to the use of the slip ring or telemetry systems to transmit the strain gauge signals from the rotating structure to the stationary analysis hardware and the ever increasing temperatures and flows that the gauges have to operate in.

In addition, the strain gauges attached to the component under observation, Fig. 3, and the wire paths can alter the blade structural characteristics as well as interfering with the local fluid flow field, hence restricting the number of strain gauges to a few blades per rotor stage. Associated with this test method, the strain gauge measurements can provide valuable information on certain parts of the blades but are statistically inaccurate due to blade-to-blade differences in response caused by mistuning. The introduction of blisks in compressor designs and the mistuning characteristics associated with them compounds the problem of relying on a small number of strain gauges for design validation.

4 FM Grids

In April 1962 Bristol Siddeley Engines Ltd published a paper [7] describing a new method of quantifying the stress in a rotor blade by an indirect method that measured blade displacement. It was shown that the maximum bending stress in a blade is proportional to the amplitude (a) of vibration and its frequency (f). This was an approximation that held for blades regardless of their form and applies to torsional and flexural modes of vibration. The principle, which is still in use today in some industries, is that the product af to cause a blade failure is constant for any given material.

The measurement of blade displacement is achieved by attaching a magnet into a blade tip and having a conductive wire grid arrangement embedded in the rotor liner (Fig. 4). A combination of the grid shape, magnet and rotor speed causes a signal to be induced into the grid. The frequency of this signal is affected by blade vibrations which cause frequency modulation that is directly proportional to the blade af. This arrangement is known as the Frequency Modulated (FM) Grid System.

Today FM demodulation is an easy technique to apply and the resulting demodulated signal can be used directly in analysis. A comparison made in 1962

Fig. 4 FM Grid arrangement

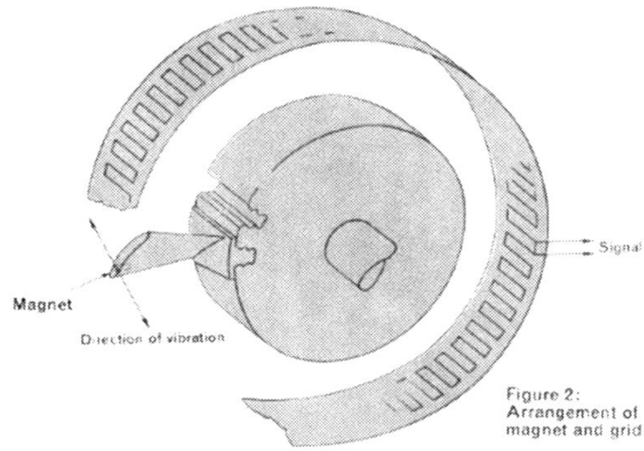

shows that both strain gauge and FM grid measurement techniques give similar results. The calibration of the FM grid system was often completed by comparison with a strain gauge allowing further and future engine testing to be completed without referring back to the strain gauge. Blade modelling methods were not mature enough to do this otherwise (Fig. 5).

Today we can repeat the comparison for the FM grid system with a more recent technique, Blade Tip Timing (BTT), and demonstrate that again we have similar results. In this case both techniques measure *af* directly. The use of strain gauge data in this type comparison requires the use of a validated model to convert stress at the measurement point into displacement at the tip (Fig. 6).

By fitting more than one grid in the rotor path it is possible to obtain data from multiple points on the blade. This data can be used to confirm the mode shapes at operating conditions. Figure 7 shows the attachment of a magnet in the blade tip.

5 Blade Tip Timing

In July 1984, Ray Chi from the United Technologies Research Centre, Connecticut USA published one of the first papers [8] that dealt with the practical application of a non-contact strain measurement system (NSMS). It provided the mathematical background to explain how data collected from casing mounted sensors could be used to infer rotor blade stresses. This was followed in 1988 by a paper

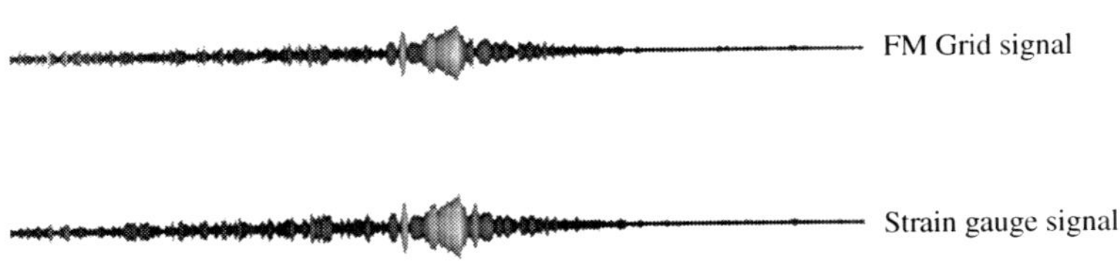

Fig. 5 Visual comparisons of strain gauge and FM grid output

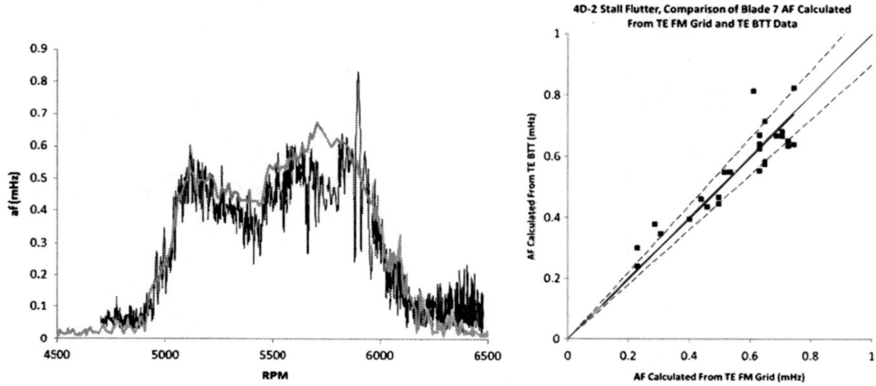

Fig. 6 Quantitative comparison of tip deflection measurements

Fig. 7 Blade magnet arrangement

demonstrating and reporting results from a practical application during a rig test of a first stage turbine [9]. At this time the other Gas Turbine OEM's were also developing the same technology and it was not until 2002 when the US government funded an engine demonstration test at Rolls-Royce Indianapolis that three separately developed systems were finally brought together for comparison.

5.1 Overview of the XTL17 Technology Demonstrator

An experiment, designated XTL17, was designed to quantify the amplitude of the mode 8 response, distinguish it from the mode 7 response and detect a possible flutter event (Fig. 8). This was part of a High Cycle Fatigue (HCF) experiment using blade coatings to dampen responses and to understand the processing capabilities of several tip timing systems fitted to the vehicle.

The data collected from a variety of instrumentation was compared against strain gauge data obtained via a radio telemetry system. The result being that BTT and strain gauge correlation had a ±25 % uncertainty value associated with it. This is not consistent with BTT being a potential replacement for strain gauges (Fig. 9).

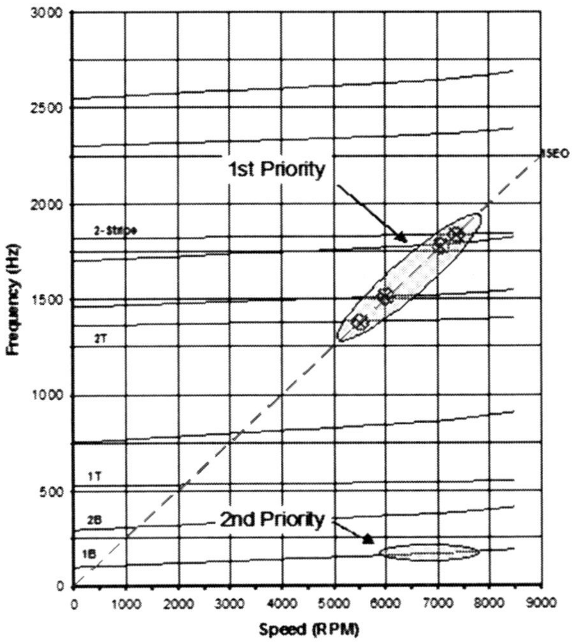

Fig. 8 XTL17 analysis requirements

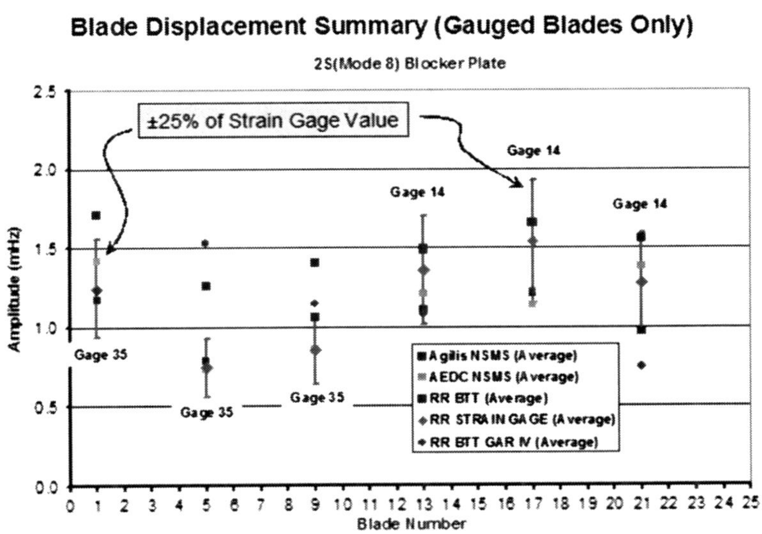

Fig. 9 BTT comparisons circa 2002

The secrecy under which BTT technology had been developed contributed to the significant uncertainty between systems. Over the next few years gas turbine OEM's formed partnerships in a number of military programs and this led to a desire to standardize the techniques which has been driven by the US and the UK defence departments.

Improvements in computing power and investment is academic research has moved the technology to a point where much of the validation is complete, the process of application to analysis is understood, and some of the OEM's have achieved the long promised cost savings associated with BTT. A standards committee has been formed and BTT is now well on the way to becoming a technique that all turbine manufacturers can begin to adopt. Currently the detailed knowledge is held by a small group of individuals working across the gas turbine industry but the release of detailed algorithms and best practice aim to spread the knowledge and drive down the cost of application.

5.2 How Does It Work: In Brief

Typically, tip deflections are obtained by measuring the blade arrival time at a number of probes situated around the casing with reference to a stable non-vibrating Once per Revolution (OPR) signal obtained from a shaft or a disc. A typical system for doing this was described by Heath [5]. The portion of blade illuminated by the laser light is the position where the deflection is measured and is the positions where calibration data must be relevant in any later analysis. The laser spot size may be different for different size of blade but is typically between 100 μm and 1 mm in diameter. A number of probe technologies can be employed but each has an effect on the measurement capability of BTT. Figure 10 shows typical outputs for a blade passing under probes using different sense technologies.

At constant speed, non-vibrating blade tips pass the probes at equal intervals of time on each revolution. However, vibrating blades will reach the probes slightly

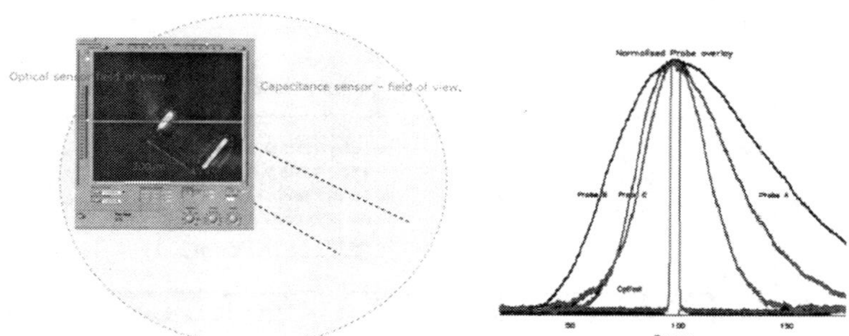

Fig. 10 Probe waveforms

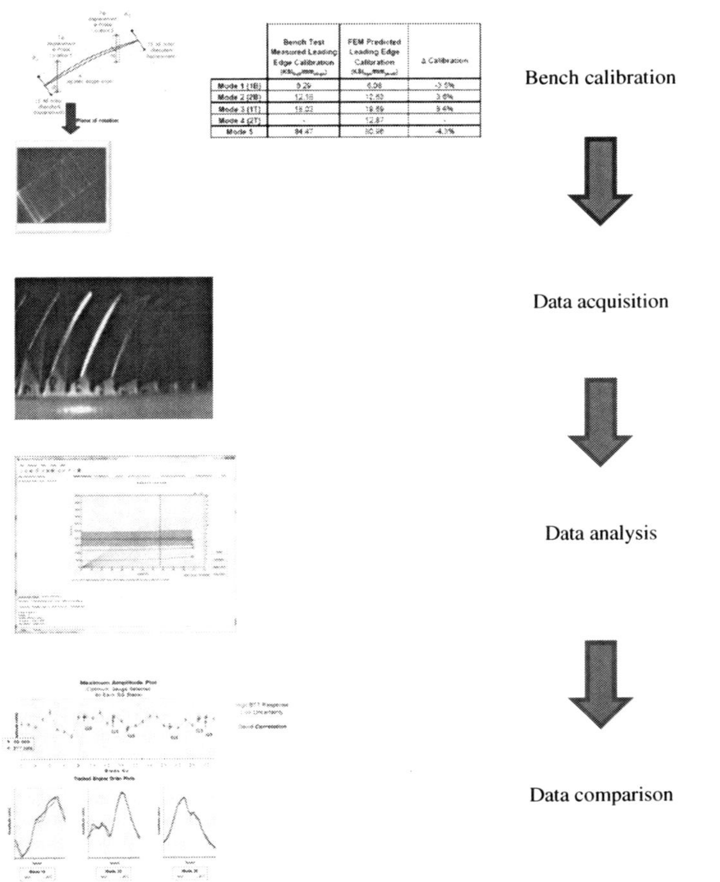

Fig. 11 Simplified BTT—strain gauge correlation process

later or earlier. Taking the difference between the expected and actual arrival times allows the calculation of individual blade tip displacements at the probe. Combine the test data with calibration information and proprietary data analysis methods to resolve blade stress. Figure 11 shows a simplified BTT process for performing strain gauge correlation.

5.3 *Current BTT Capabilities*

The large Gas Turbine OEM's have the resources to complete hundreds of strain gauge and FEM correlations with BTT data, the gradual release of this data allows the technique to be adopted outside of the aerospace sector. The ability to

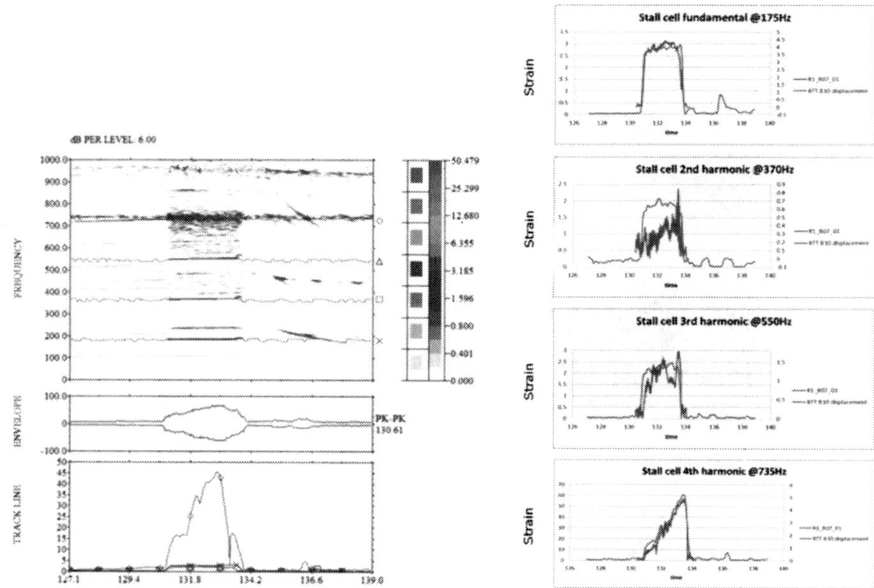

Fig. 12 Stall analysis and comparison

characterise both integral and non-integral vibrations is common, as is the analysis of stall events. The differences now seen in the correlation between strain gauge and BTT are mostly associated with the proprietary analysis and comparison techniques used by different companies. These differences are usually built into any safety margins when the data is used for component certification. Standardisation of the technique will help to narrow these differences.

A complex stall event (Fig. 12) shows the correlation between a strain gauge and BTT analysis [10]. Detection capability is limited by the acquisition process and typical end to end figures of 20 μm are common.

6 Conclusions

The use of strain gauges for rotor blade vibration measurement has declined due to the restrictions imposed by the latest standard of machinery. The requirement for more statistical data to support both the design and in-service phase's favours a technology that can be applied from the casing captures data from all blades, has line replaceable units and can build on data collected during the development cycle.

Blade tip timing alone will not replace rotor strain-gauges but combined with FE models and fatigue test data it has displaced numerous expensive telemetry tests in the past few years and has provided superior statistical data on blade behaviour.

Data gathered during the development phase is directly applicable to long term health monitoring and sub sets of data can be used to develop detection and trending techniques prior to entry into service.

More robust probe technologies are available and the standardisation of techniques is beginning to make this available to all.

In the development of rotor machinery BTT has met the goal of replacing telemetry tests and shortening development cycles. In particular modifications to existing equipment have benefitted most. The next phase is to understand the potential of in-service data and to exploit it.

References

1. Russhard P (2010) Development of a blade tip timing based engine health monitoring system—thesis, Faculty of Engineering and Physical Sciences, School of Mechanical Aerospace and Civil Engineering, Manchester University, Sept 2010
2. Niska H (2012) Strain Gage Pioneers and Propulsion Engine History Joint EVI-GTI and PIWG conference, Oct 2012
3. Kadambi JR, Quinn RD, Adams ML (1989) Turbomachinery blade vibration and dynamic stress measurements utilizing nonintrusive techniques. J Turbomach 111(4):468–474
4. Heath S (1996) A study of tip-timing techniques for the determination of bladed disk vibration characteristics, Imperial College of Science and Technology, University of London, PhD Thesis, 1996
5. Kearns CM Jr (1936) Stress measuring means—US patent application
6. Jackson PS (1953) Foil strain gauge England—patent application
7. ThomasJC (1962) A new technique for investigating blade vibrations. Bristol Siddeley J
8. Chi RM et al (1984) Conceptual design of a non-interference stress measurement system for gas turbine engines. Arnold Engineering Development Centre 1984
9. Chi RM et al (1988) Demonstration of testing of a non-interference technique for measuring rotor blade stress, AIAA/ASME/SAE/ASEE 24th joint propulsion conference 1988
10. Russhard P (2012) Analysis of rotating stall in a contra-rotating system using blade tip timing, IIS conference June 2012

Managing Life Extension for Large Rotating Plant in Power Stations

Wolfgang Hahn

Abstract Risk and reliability management of rotating plant in power stations under life extension have become a major challenge for industry. The increasing numbers of catastrophic failures are calling for a more rigorous regime to address integrity through qualitative as well as quantitative methods. Current condition based assessments as well as condition monitoring techniques need further development into the future to assist power generators manage into unknown territory during life extension.

Keywords Life extension · Rotating plant · Risk · Reliability · Turbines · Cracking

1 Introduction

A Large number of the Power Stations in the United Kingdom and Europe are facing challenges with underwriting large rotating plant integrity on plants which have already doubled the expected design life in running hours and number of starts. The importance of safe reliable plant can not be over-stressed, specifically on the large rotating machinery such as steam turbines at risk of catastrophic failure. Whilst this risk is real and with an increasing number of turbine rotors failing in service, industry continue to extend the life of rotors into the future. Management of the risk therefore requires careful consideration in which best available techniques to use. A combination of qualitative and quantitative risk management techniques are needed to demonstrate the risk mitigation during the life extension process.

Risks related to turbine bore cracking, shaft cracking and blade cracking are case studies covered in this paper during life extension.

W. Hahn (✉)
EDF Energy, West Burton Power Station, Retford, Nottinghamshire DN22 9BL, UK
e-mail: wolfgang.hahn@edfenergy.com

J.K. Sinha (ed.), *Vibration Engineering and Technology of Machinery*,
Mechanisms and Machine Science 23, DOI 10.1007/978-3-319-09918-7_3

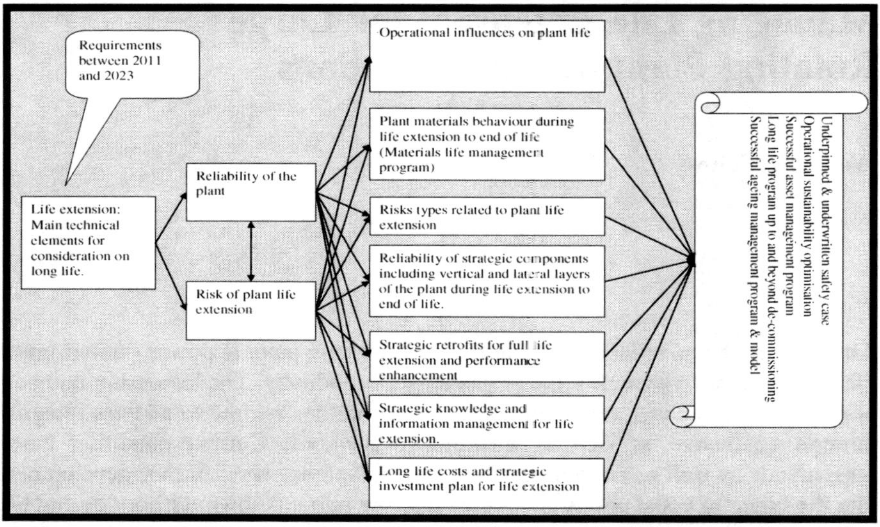

Fig. 1 Plant life extension considerations for long life during plant lifecycles

2 Theory Related to Life Extension

A holistic approach to life extension is required considering all components and influencing variables that may contribute to the overall life expectancy of critical component behavior as indicated in Fig. 1. Plant long life is a term introduced by Riddell referring to the investment costs for plant wanting to extend the operating life further. This is a very important consideration for power plants as investment in life extension may only cost around one third of a new built power station of the same capacity.

2.1 *Reliability*

Risk and reliability related to failure was demonstrated in the asset management strategy as published by Hahn [1], the Gaussian distribution of the failures relate directly back to the number of like for like components exposed to similar variable conditions within a margin of tolerance. The six basic accumulative influences of plant degradation mechanisms are creep, fatigue, corrosion, erosion and vibration with a combination of any of these.

Risk and reliability form a unique combination in the top event probability of failure with consequential outcome. This assessment can normally be seen in the qualitative approach used in the form of a five by five matrix; however, for components life predictions coupled to changing conditions associated with one or more

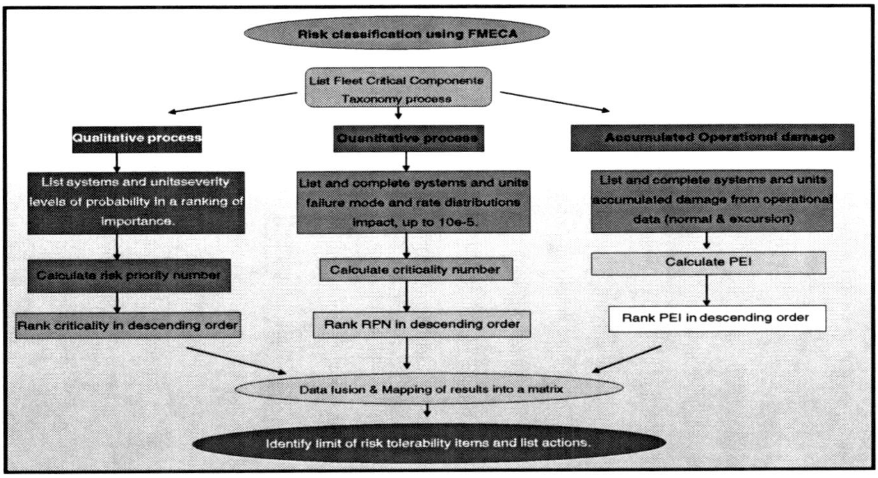

Fig. 2 Failure mode and effects criticality analysis for major components

variables such as temperature, pressure and/or fluid behaviour, this approach alone is unacceptably coarse.

Figure 2 reflects an approach more closely coupled with the quantitative, qualitative and related sensitivity factor for the associated operational impacts of the plant. The original two stream approach from Barringer [2], has been improved by the author to a three stream approach fusing data using probability statistics, in this case using performance effectiveness index, PEI as an independently defined outcome from system expediency [3].

Accurate taxonomy of the plant and systems is of utmost importance for realistically addressing the components at risk as well as classification of components for laboratory testing and validation of deterioration and remaining life left. Levels 1–5 and levels 6–9 as defined by British Standard 14224, defines the high level categorization related to the industries and plant applications as well as the low level related to the equipment, unit and lower indenture levels respectively. These lower levels become more applicable in the parent to child relationship for reliability in the sub units and components such as steam turbines [4].

An example of criticality selection in the turbine island taxonomy process is reflected in Fig. 3. This reflects back on large rotating plant in terms of a qualitative and quantitative approach for equipment selection. Again, in each case the component history, laboratory testing results and modeling results must support confidence levels for life assessments. There is nothing linking true differentiation between the layers from a quantitative perspective in Fig. 2. The true differentiation in a taxonomy process is demonstrated in the criticality ranking of the component and its failure mode impact on the plant.

The process of using risk and reliability as referred to in Fig. 1 is now complimented using Figs. 2 and 3 respectively, building the risk register by area and populating the fused data according to the risk ranking for taxonomy in a risk log.

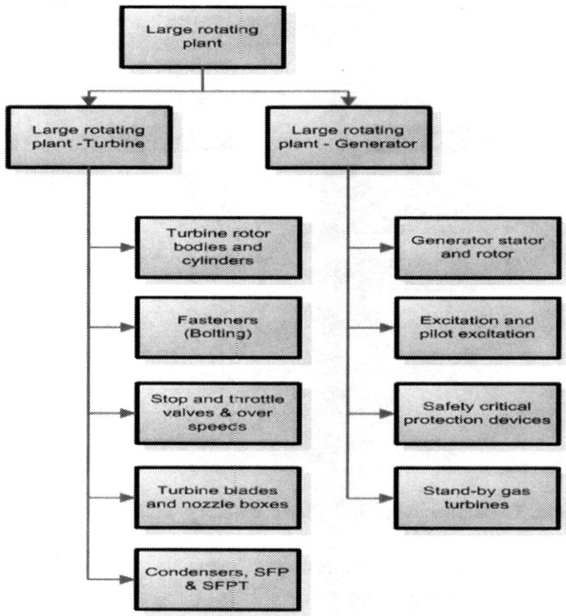

Fig. 3 Large rotating plant taxonomy example for sub-components

2.2 Differentiation Between Life Extension and Wear Out

Plants running in the wear out phase have been well defined and investigated as per previous paper produced by Hahn [1], however a plant running the life extension period, phase 4, may have different variables defined for changing the wear out rate and thus influencing the hazard phase curve. An example of the bathtub curve represents the hazard phase plot of the plants through the whole lifecycle phase as per Fig. 4. The hazard plot, Z (t), can thus be influenced at point A, when life extension is defined and announced on a unit, sub-unit or component level and the necessary operational, maintenance and engineering integrity policies incorporated into the life management programme.

The Gaussian distribution for the items under consideration for life extension needs to meet the beta value in order to qualify for the life extension management programme.

This means that the following rules would apply:

1. Beta < 1: Running in failure (no life extension possible)—re-design component
2. Beta = 1: Random failure mode (useful life)—No life extension possible—complete FMEA.
3. Beta > 2: Wear out failure—Life extension(LE) applicable—exercise optionality (LE or retrofit)

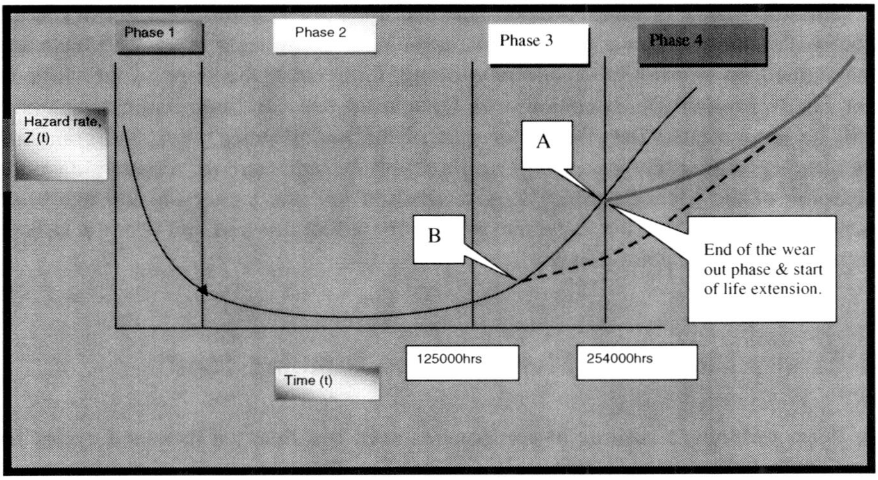

Fig. 4 Example of the hazard phase plot during the plant lifecycle

Fig. 5 Probability density function with beta values representation

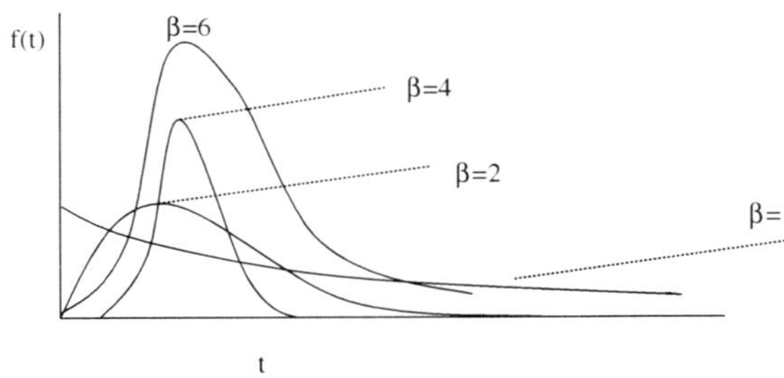

It can be seen from Fig. 5 that the approach has some rules that need to be applied as part of the analysis during the failure of a component in association to the beta component. The rules that need to be applied for assessment of components failure includes: Answers to the following to be incorporated into the beta value evaluation.

- What is the components original design life? (operating hours)
- What was the designed mode of operation? (start -stop or continuous running)
- What was the operational capacity design intent? (components utilised beyond design)
- What degradation mechanisms have been allowed for in the original design?
- Is the beta value for this component typical for other operators with similar conditions?

The hazard rate Z (t), plays a significant role in the wear out phase as the slope of the curve needs to be managed by the variables that could influence the curve.

The importance of differentiation between life extension and wear out lies in the decision making of how a system, unit, sub-unit or component needs to behave and maintained up to point of de-commissioning, influencing the slope of the curve as per Fig. 4. During life extension, it is recognised that condition based assessment will be incorporated into the major part of the maintenance work and condition monitoring during the operating regimes will be adjusted to accommodate the outcome of the life expectancy It is recognised that the earlier the life extension period start the better the outcome will be for reliability and risk from a holistic asset management point of view.

3 Failure Modes and Monitoring on Rotating Plant

In steam turbines, a number of components need less than ten thousand cycles for low cycle fatigue related failures to occur. The thermal origin of the low cycle fatigue is a result of the thermal stresses manifesting as a result of thermal expansion in the turbine. In this case, fatigue results from the cyclic strain rather than the stress cycle [3]. The structural changes in the material that occur during cyclic stress related failures can be divided into three distinct phases, they are:

1. Crack initiation—this includes early development of fatigue damage and could in certain instances be removed by annealing the materials. Crack propagation takes place along slip bands, these are normally a few nm per cycle.
2. Slip band crack growth—this involves the deepening of the initial micro crack on planes of high shear stress also known as first stage crack growth. The crack propagation is normally a few microns per cycle with visible striations per tensile stress cycle.
3. Ultimate ductile failure—this occurs when the crack has propagated to a sufficient length so that the remaining cross sectional area can no longer support the applied load.

Cold end materials in the risk register are examined from a metallurgical and mechanical properties perspective as well as tested in the laboratory to establish endurance fatigue life where applicable.

4 IP Turbine Bore Cracking: Life and Condition Management

In 2011, the IP rotor depicted in Fig. 6 was removed after 12 years of operation between inspections with the intent of completing a health check, executing remedial work and returning the rotor to service in another unit as part of an exchange program for another 12 years of operation.

Fig. 6 IP rotor on supports in off-site works

Table 1 Position of major through wall extent (TWE) flaw indications

Flaw identified	Length (mm)	Depth from Bore (mm)	Circumferential Position	Through Wall extent (mm)
BF	75	18–44	212–232	11
AA	43	5	18	18
AJ	37	9	352–364	30

The 2011, non destructive testing on the West Burton Unit 4 IP Rotor reported 68 indications, most of these were relatively small with radial through-thickness measurements reported to be <3 mm [5]. A number of indications were reported to be more significant, three of which were highlighted below in Table 1.

Three major defects were found at distances away from the bore centre line as indicated in Fig. 7. These defects had grown from information gathered between the 12 year inspection intervals. The forging quality in the 1950s and 1960s was very variable, which led to a large number of embedded defects caused by impurities and other constituents in the rotor pencil shafts as can be seen in Fig. 8.

The indications all have different circumferential positions but have been plotted in one circumferential plane to simplify visualization. Figure 8 includes the location

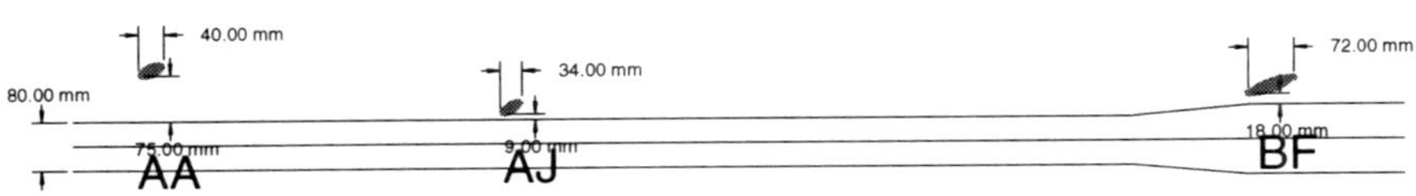

Fig. 7 Three major defects locations depicted with relation to the bore position

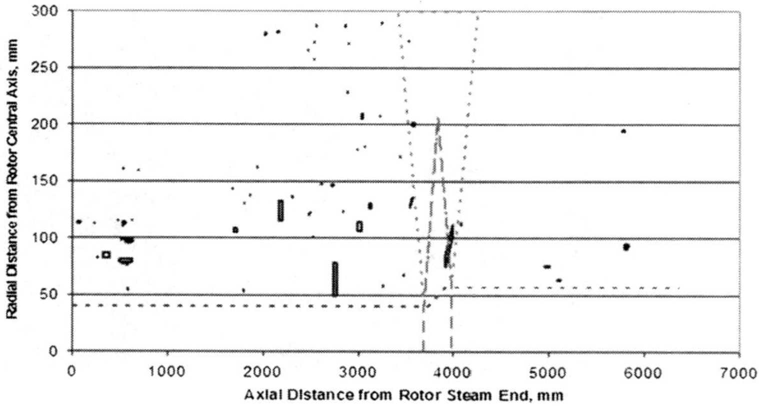

Fig. 8 Embedded defects from the original forging process in the rotor body

of the bore (dashed blue line) with respect to distance from the centre and axial distance from the steam end [6].

The rotor was bottle bored and all three major defects removed, limiting future life to 1,000 starts before the next inspection for underwriting the integrity [5].

5 LP Rotor Pencil Shaft Cracking: Root Cause Analysis and Life Extension Options

The LP rotor depicted in Fig. 9 was removed from service after detecting a crack in the pencil shaft from vibration monitoring during transient conditions (Start up and shut down vibration signatures). Further confirmation of the cracked rotor was completed through a series of truth checks and non destructive testing.

The rotor is of dual flow type and construction is made up of a pencil shaft with shrunk on disc and button drive for each stage with centre collar as per Fig. 10. The

Fig. 9 Cracked low pressure English Electric rotor on supports

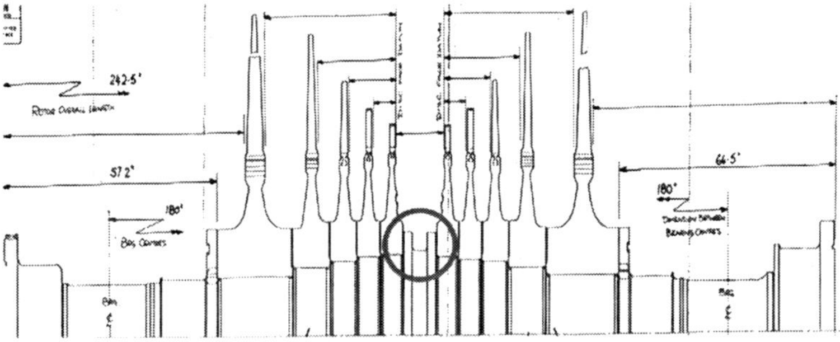

Fig. 10 English Electric rotor body with shrunk on discs and blade fixings

location of the collars and discs made the non destructive testing difficult during crack detection by virtue of the angles and distances from testing surface with the crack location in the circled area.

The cracked rotor covered more than 200,000 operating hours and 3,000 start ups to point of the large crack detection. The extent of the crack covered approximately 270 degrees of the circumference and the measured value from the non destructive testing indicated a crack depth of approximately 200 mm deep [7].

Further metallurgical and destructive examination was planned at the time on the rotor to determine exact root cause and origin of the crack to ensure future life management mitigation measures are incorporated into the life plans.

Fig. 11 Low pressure turbine with cracked blade removed and rotor in situ

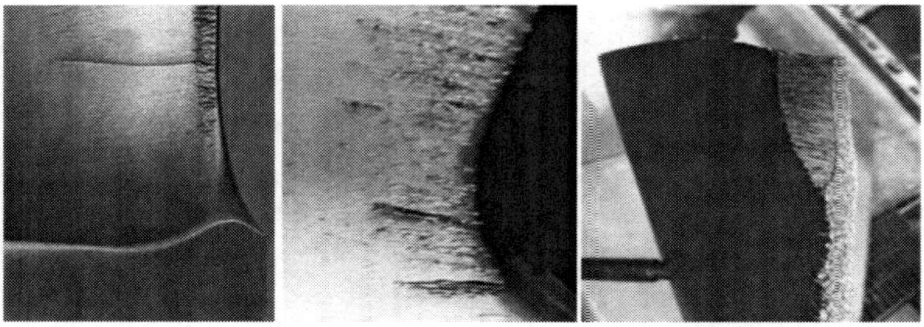

Fig. 12 Erosion damage on low pressure turbine blade with cracking in the trailing edge

6 LP Turbine Erosion and Blade Cracking: Life Extension Through Monitoring

As part of life extension and efficiency enhancement on coal fired power stations, retrofitting was completed on Low Pressure turbines and then High Pressure turbines. As a result of the aggressive start up regime, blade erosion on the leading edges and trailing edges of retrofitted LP Turbines cause erosion damage on the blade surfaces [8]. The low pressure turbine last stage blade damage and crack propagation are mainly starts related with the cold starts delivering most damage due to the exposure time (Fig. 11).

The trailing edges in this case, develop cracking above the root platform from the notched effect of the erosion and propagate through high cycle fatigue as a result of blade vibration [9] (Fig. 12).

An accumulative pitting damage model was created from research to monitor turbine blade health through on-line rotation counting and in situ blade inspections [8].

7 Monitoring Requirements During Life Extension

For all three case studies covered under large rotating plant in this document, it is unavoidable to ignore monitoring methods in order to prevent catastrophic failure. Cracking in these cases are associated with early warning signs which can be picked up through vibration monitoring techniques. Methods such as on-line shaft vibration monitoring and blade tip timing are examples of early warning methods; however, there are limitations in each of the methods and the practicality of the applications. There remains a need in industry, for a health monitoring system that covers the complete rotating unit in its entirety, covering bores, shafts, discs and blades.

8 Conclusion

The challenges of plant life extension on large rotating plant remain to be a concern as a result of Power plant generators pushing the boundaries of design life envelope out into the unknown territory of life extension. Operators require a rigorous risk and reliability management plan demonstrating control over potential catastrophic failures against an aggressive commercial market. Qualitative and quantitative methods are required to benchmark best practice in health assessments and condition monitoring of large rotating plant. Further development is needed in system health monitoring such as vibration and dynamics, to assist power plant generators with effective early warning support on large rotating plant in order to prevent catastrophic failure. All examples in this document were pro-actively managed through qualitative and quantitative methods, however, further work is needed to simplify methods and reduce costs effectively.

References

1. Hahn W, Kidd M (2012) Asset management for a coal fired power station in the wear out phase. School of Mechanical, Aerospace and Civil Engineering, University of Manchester, Manchester United Kingdom
2. BS 5760-4:2003, reliability of systems, equipment and components, part 4, guide to the specification of dependability requirements
3. Gnedenko BV, Ushakov IA (1995) Probabilistic reliability engineering. Wiley, New York
4. BS EN ISO 14224:2006, petroleum, petrochemical and gas industries—collection and exchange of reliability and maintenance data for equipment
5. Patterson IR, Zhang X (2012) Remnant life IP turbine rotors, E. on new build and technology, Ratcliffe on Soar, Nottinghamshire, NG11 0EE
6. Skinner D (2011) West Burton Power Station, Ex. unit 2 IP Rotor, IPWn19006 Interim report, Eon—PES, Hamms Hall, United Kingdom
7. Mikkelsen T Siemens, NDT03-2013, automated UT examination of the West Burton Unit 4 LP3 Turbine and the retired LP1 spindle, 70PO03636
8. Hahn W, Tasker G, Naylor E, Kidd M (2014) Crack initiation in 14 % Cr Low pressure Turbine Blade Steel, ASME, Journal of Engineering for Gas Turbines and Power, GTP -13-1425
9. Sinha J, Hahn W, Elbhbah K et al (2012) Vibration investigation for low pressure turbine last stage blade failure in steam turbines of a power plant, proceedings of the ASME Turbo Expo 2012, Copenhagen, Denmark

Recent Advances and Prospects in Condition Monitoring

Arthur W. Lees

Abstract The underlying reason for pursuing Condition Monitoring activities is to enhance the overall performance of plant either by accurately predicting the end of effective life or by diagnosis and location of incipient faults. The achievement of both these objectives requires the use of a combination of measurement, modelling and statistical techniques, but the balance of these three strands varies with the plant under study. Practitioners of these different disciplines sometimes develop rivalries, but in reality all three are essential. The fundamental problem involved is the determination of the physical process within a machine which cannot be measured directly and so must be inferred from external measurements, the most common of which is vibration. This inference from measurement to diagnosis requires some form of model, whether physics or statistics based. The paper describes some work on the monitoring of large turbines and in particular, the ways in which measurements and models can be combined to enhance insight into a machine's operation. Similarly, our understanding of complex data can be enhanced by the use of Artificial Neural Networks and these can be used to enhance understanding. Combining various approaches leads to some new possibilities which are briefly outlined.

Keywords Rotating machine · Condition monitoring · System identification · Modelling · Diagnostics

1 Introduction

Over recent decades there have been some considerable advances in Condition Monitoring and it is appropriate to form an assessment of some of the prospects for the foreseeable future. In forming a view of this two factors are important firstly what is possible/feasible and secondly, what is advantageous in an industrial

---

A.W. Lees (✉)
College of Engineering, Swansea University, Singleton Park, Swansea SA2 8PP, UK
e-mail: a.w.lees@swansea.ac.uk

J.K. Sinha (ed.), *Vibration Engineering and Technology of Machinery*,
Mechanisms and Machine Science 23, DOI 10.1007/978-3-319-09918-7_4

setting. It is important to consider the full context of CM: it ranges from the collection of data to some form of assessment of the condition of the machine. Fundamentally, the critical aspect of CM is the extraction of information from a set of data. As discussed later, the distinction between information and data is important: a relevant example can be quoted for rotating machinery—whilst there is an abundance of on load data, it carries limited information (at least, given present knowledge and analysis methods). Machine transients, by contrast, give far more limited amount data but much more information about a machine's condition.

The mere collection of data is, of course, of little consequence unless it is useful. There are several ways in which data can be a useful resource but all depend crucially on some connection between the data and the physics of the operation of the machine. This may take the form of a trend line, a statistical model or a model based on the physics either analytical or a more detailed numerical model. In fact all these types of approach have a role to play.

Operators of machinery have used trending information for many years and this continues to be an important contribution to CM. But for the diagnosis and localisation of many types of fault in rotation machinery some physically based model is needed and increasingly over the last 50 years, an increasing proportion of these have been numerical models. In the early years progress was made with the transfer matrix technique which made very efficient use of limited computer storage. More recently however, the Finite Element Method (FEM) has become the dominant approach to both linear and non-linear analyses.

From about 1970 onwards, FEM became well established and significant advances were made in the modelling of complex machinery such as power turbo-alternators. The main features of machine behaviour were reflected in model predictions, but closer inspection revealed some discrepancies. This led some engineers to be a little dismissive of the potential of modelling whilst others sought refinements of the modelling techniques employed. It was gradually appreciated that a very significant part of model imprecision stems not from any deficiencies of the mathematical approach, but rather the inaccuracies of the data used be the models, as discussed by [8, 14].

An example of this may be found on a power station operating four turbo-alternators, all built to the same engineering drawings. The vibrational behaviour of these machines will have general features in common, but there will be subtle (and not so subtle) differences in their properties. Why should this be so? There are several areas of uncertainty, the main ones being,

(a) Accumulation of minor variations in build (albeit all within specified tolerance)
(b) Variation of joint tightness (again, all within tolerance)
(c) Differences in subsoil conditions
(d) Variable attachment of auxiliary plant items
(e) Variation in thermal conditions.

The list could go on: the essential point is that industrial machines are complex and no two examples will be identical and perhaps surprisingly, it was some time before this was fully appreciated. This does mean, however, plant data is needed to refine machine models; design data alone is not adequate for the development of detailed models for the diagnosis of plant problems and this paper will discuss a number of ways in which improvement can be achieved. Of course, there is a degree of judgement here—how refined a model is needed?

There are a number of consequences though to the relatively limited scope of currently used machine models. One important issue is that operators have a vast amount of on-load data, but can make very little use of it apart from defining some statistical limits on safe operation. This is partly because the on-load variation of vibration are comparable in scale with the observed differences between nominally identical units and consequently the 'standard' models of these machines are not useful in investigating the on-load variations. In Sect. 2 an approach is presented to derive accurate models of individual machines, then sections three and four discuss regularisation with some examples from simulation, rigs and real plant. In sections five and a discussion is given on the use of on-load data whilst section six outlines some general conclusions.

2 System Identification

Soon after the development of the Finite Element Method, the techniques were applied to the study of the dynamics of turbo-generators and their supporting structures. Lees and Simpson [8] described some of these early attempts and their contributions to understanding, albeit with substantial inaccuracies. It was soon appreciated that the lack of precision was not the result of any shortcoming in the numerical method, but rather a lack of detailed knowledge of various parameters. In particular, it was realised that small differences between units can have a substantial influence on vibrational behaviour and a method to apply System Identification to identify the parameters of the supporting structure was proposed by [7] whilst at about the same time another approach to the same problem was put forward by [2].

The method proposed by [7] assumes that a good numerical model for the rotor exists. The objective now is to determine the elements of the three matrices describing the support structure. In a series of papers [9, 16, 17, 19, 20] have developed an approach to estimate faults on a machine when the stator or foundation is flexible and unknown. The equations of motion of the machine may be written as

$$\begin{bmatrix} \mathbf{Z}_{R,ii} & \mathbf{Z}_{R,ib} & \mathbf{0} \\ \mathbf{Z}_{R,bi} & \mathbf{Z}_{R,bb} + \mathbf{Z}_B & -\mathbf{Z}_B \\ \mathbf{0} & -\mathbf{Z}_B & \mathbf{Z}_B + \bar{\mathbf{Z}}_F \end{bmatrix} \begin{Bmatrix} \mathbf{r}_{R,i} \\ \mathbf{r}_{R,b} \\ \mathbf{r}_{F,b} \end{Bmatrix} = \begin{Bmatrix} \mathbf{f}_u \\ \mathbf{0} \\ \mathbf{0} \end{Bmatrix} \tag{1}$$

where $\mathbf{Z}$ is the dynamic stiffness matrix, the subscripts b and i refer to internal and bearing (connection) degrees of freedom respectively, and the subscripts F, R, and B refer to the foundation, the rotor and the bearings. $\mathbf{r}$ are the responses and $\mathbf{f}_u$ are the force vectors, which are assumed to be applied only at the rotor internal degrees of freedom. The dynamic stiffness matrix of the foundation, $\bar{\mathbf{Z}}_F$, is defined only at the degrees of freedom connecting the bearings and the foundation. In practice this will be a reduced order model, where the internal foundation degrees of freedom have been eliminated. The dynamic stiffness matrix of the bearings is given by $\mathbf{Z}_B$. It has been assumed that the inertia effects within the bearings are negligible, although these could be included if required.

Solving Eq. (1) to eliminate the unknown response of the rotor gives,

$$\bar{\mathbf{Z}}_F \mathbf{r}_{F,b} + \mathbf{Z}_B \mathbf{P}^{-1} \mathbf{Z}_{R,bi} \mathbf{Z}_{R,ii}^{-1} \mathbf{f}_u = \mathbf{Z}_B \left[\mathbf{P}^{-1} \mathbf{Z}_B - \mathbf{I} \right] r_{F,b} \tag{2}$$

where $\mathbf{P} = \mathbf{Z}_{R,bb} + \mathbf{Z}_B - \mathbf{Z}_{R,bi} \mathbf{Z}_{R,ii}^{-1} \mathbf{Z}_{R,ib}$. For the identification it remains to determine which quantities are known and which are to be estimated. Of course all component models in Eq. (2) are functions of the shaft speed, and the measurements are the synchronous responses during a run-down. It is assumed that good models for the rotor and the bearings, $\mathbf{Z}_R$ and $\mathbf{Z}_B$, are known a priori and that the response $\mathbf{r}_{F,b}$ is measured. Here we assume that the synchronous force vector (unbalance or bend), $\mathbf{f}_u$, foundation model parameters, $\bar{\mathbf{Z}}_F$, are unknown. The foundation model will be modelled using mass, damping and stiffness matrices, and the frequency range will be split if required. There are several ways of formulating this problem and Eq. (2) shows the case in which only the pedestal motion is measured. The penalty paid for this is that the method requires a model for the bearing behaviour, although it has been shown that the results of the calculation are not very sensitive to errors in this model.

However, for the case in which shaft motion is also monitored, a rather simpler formulation may be derived from Eq. (1) to give

$$\begin{bmatrix} \mathbf{Z}_{R,ii} & \mathbf{Z}_{R,ib} & 0 \\ \mathbf{Z}_{R,bi} & \mathbf{Z}_{R,bb} & 0 \\ 0 & 0 & \mathbf{Z}_F \end{bmatrix} \begin{Bmatrix} r_{R,i} \\ r_{R,b} \\ r_F \end{Bmatrix} = \begin{Bmatrix} f_u \\ f_b \\ -f_b \end{Bmatrix} \tag{3}$$

At the stage, of course, the force exerted on the shaft at the bearings, f_b is unknown, but it is easily eliminated and recovered from the calculation later. By adding the second and third rows

$$\begin{bmatrix} \mathbf{Z}_{R,ii} & \mathbf{Z}_{R,ib} \\ \mathbf{Z}_{R,bi} & \mathbf{Z}_{R,bb} \end{bmatrix} \begin{Bmatrix} r_{R,i} \\ r_{R,b} \end{Bmatrix} + \begin{bmatrix} 0 & 0 \\ 0 & Z_F \end{bmatrix} \begin{Bmatrix} 0 \\ r_F \end{Bmatrix} = \begin{Bmatrix} f_u \\ 0 \end{Bmatrix} \tag{4}$$

It is worth considering each of the terms in some detail: all the values of r are known since these are the measured quantities. In the first matrix, all four terms $\mathbf{Z}_R$ are known since this is just the model of the rotor which (it is assumed) has been validated separately. $\mathbf{Z}_F$ is just the model of the foundation which is to be determined. The first row of Eq. (4) is now used to eliminate the internal rotor degrees of freedom using dynamic condensation. The imbalance force may be either known (by taking the difference of two balancing runs) or treated as an extra unknown vector to be determined. The unknowns in Eq. (4) can now be expressed as the elements of the stiffness, mass and damping are N bearings, if all matrices are full the total number of degrees of freedom in the problem will be

$$Ndf = 3N(2N + 1) + m$$

Whilst this can be a large number for a power turbine, during a machine run-down a large volume of data is available: for each bearing there will be measurements in two directions at each rotor speed during the rundown. Consequently there will typically be several thousand equations from which the unknown parameters are derived and hence the problem is over specified. Allowing for the fact that the measured data will be contaminated with noise a least squares approach is appropriate to this problem. The required solution can be obtained by first re-expressing Eq. (4) in the slightly unusual form

$$[\mathbf{R}(\omega)]\{\vartheta\} = \omega^2\{e\} \tag{5}$$

Or alternatively, if the imbalance is itself to be estimated [3]

$$[\mathbf{R}'(\omega)]\{\,\vartheta \quad e\,\} = 0 \tag{5a}$$

In this equation the matrix **R** comprises all the measured vibration readings at all the frequencies measured whilst the vector $\{\vartheta\}$ has all the unknown elements of the mass, stiffness and damping matrices. Because Eq. (5, 5a) are homogeneous, a scaling factor is required for the solution. Any readily available parameter, such as the total rotor mass can be used for this purpose.

It must emphasized that these equations define the physics of what is expected to occur, but given errors in modelling and noise in readings, the set of equations will not be obeyed precisely. To express this fact the zero equation is on the right hand side of the equation is replaced by a vector of small quantities so that the equation becomes

$$[\mathbf{W}][\mathbf{R}'(\omega)]\{\,\vartheta \quad e\,\} = [\mathbf{W}]\{\varepsilon\} \tag{6}$$

Note that a diagonal matrix **W** has been includes multiplying each side of the equation. This is to express the fact that not all the individual measurements are equally reliable—some, such as those involving low vibration levels may well be subject to substantial noise. This is the first step in seeking to optimise the accuracy of the solution, a process known as regularisation, which is discussed in the next section.

The aim now is to determine the vector $\{\,\vartheta \quad e\,\}^T$ which gives the 'best fit' which is interpreted as minimising the scalar quantity $\{\varepsilon\}^T[W]\{\varepsilon\}$. This may be achieved by writing

$$\{\vartheta \quad e\}[\mathbf{R}'(\omega)]^T[\mathbf{W}][\mathbf{R}'(\omega)]\{\vartheta \quad e\}^T = \{\varepsilon\}^T[\mathbf{W}]\{\varepsilon\} \tag{7}$$

This is readily solved with the Moore-Penrose inverse as

$$\begin{Bmatrix} \vartheta \\ e \end{Bmatrix} = \left[\mathbf{A}^{\mathrm{T}}\mathbf{A}\right]^{-1}[\mathbf{A}]^T[\mathbf{W}]\{\varepsilon\} \tag{8}$$

where $\mathbf{A} = [\mathbf{W}][\mathbf{R}'(\omega)]$.
A somewhat different approach has been developed by Pennacchi et al. [11, 12, 13, 15, 21]. Whilst the mathematical approach is slightly different the physical concepts are similar. A distinguishing feature of the Swansea group's approach is their reliance on a validated rotor model. Other researchers, [1, 4, 5] have also followed this approach.

3 Regularisation

A common difficulty with all inverse problems is one ill of conditioning resulting in inaccurate solutions: a closely related issue is non-uniqueness of the solution [6, 18]. Techniques to overcome these difficulties are given the name regularisation and the methods may be divided into three categories.

(a) Appropriate weighting of the equations
(b) Use of known physics to minimise the number of unknown variables
(c) Selection of stable solution routines.

In setting up Eq. (8) it is relatively straight forward to impose symmetry on the mass, stiffness and damping matrices. For the case of a turbo-generator with 14 bearings, at each of which two directions of vibration are monitored, then each of the three matrices will have 29*28/2 unknown parameters, in addition to a complex unknown for each balance plane. This would present a formidable problem, but the number of variables can be reduced substantially by some reliance on the known physical situation. For instance, for many structures there is little interaction between the two orthogonal directions: this would immediately reduce the numbers of unknowns in each matrix to 15*14, but further reductions are desirable. To proceed further, observe that most of the damping in a large machine arises in the oil film bearings and hence detailed modelling of the damping in the support structure is largely irrelevant. Recognising this, the damping matrix may be restricted to a diagonal, or even eliminated altogether. A further reduction can be brought about by restricting the mass and stiffness matrices to a banded structure. This is justified by the observation that in a large machine, bearings which are remote from each other have negligible cross influence. In all these cases, constraints are applied simply be eliminating rows and columns from the matrix equations.

Table 1 Comparison of estimated and actual imbalance

Test no.	Bal. plane	Magnitude (kgm) actual	Phase (Deg) actual	Magnitude (kgm) estimated	Phase (Deg) estimated
1	3	0.160	45	0.1678	44.4
2	1	0.160	195	0.1797	188.2
	3	0.160	45	0.1606	37.5
3	3	0.160	45	0.1715	48.4
	4	0.160	150	0.2108	149.1
4	1	0.160	195	0.1737	195.9
	5	0.160	0	0.1811	−0.02
5	1	0.160	195	0.1624	182.3
	3	0.160	45	0.1318	42.9
	4	0.160	150	0.1941	154.2
6	1	0.160	195	0.1852	195.7
	3	0.160	45	0.1829	42.2
	5	0.160	0	0.1830	2.77

Table 1 shows some actual estimates of imbalance on a large laboratory test rig and it is clear that accurate estimates of imbalance have been obtained. On some machine the estimates can be further improved by dividing the frequency range into a number of sections and performing a constrained estimate in the separate sections.

The final means of regularisation is by appropriate inversion of the matrix in Eq. (8). Rather than a direct inversion, it is preferable to consider the Singular Value Decomposition of the matrix A, or equivalently, the eigenvalues of the matrix $\mathbf{A}^T\mathbf{A}$. Low singular values (or eigenvalues) indicate that the combination of parameters in the corresponding eigenvectors have very little influence on the total residue and consequently, inclusion of these terms can lead to significant errors. The way round this is to substitute a set value for the lower eigenvalues at some cut-off point.

4 Some Examples

Figure 1 shows some good agreement between specified and estimated values in a simulation exercise. Note that the criterion of success is not the degree of fit between the observed and model data. A good fit can always be achieved merely by increasing the number of variable parameters. The real test is whether the derived model has predictive capability and to achieve this the requirement is that the parameters are tracking the underlying physics rather than noise. It is important that the fit is derived with enough, but not too many, parameters. Hence the real test of the approach is to derive a model using one set of data, then compare the predictions of this model for a different scenario, say a different imbalance. The results of

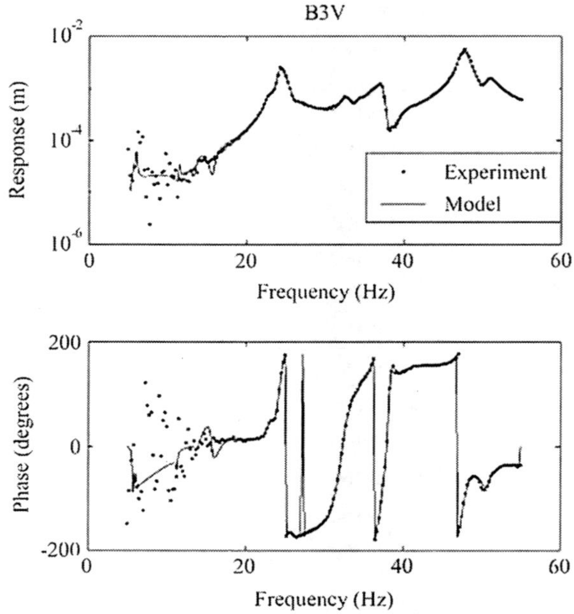

Fig. 1 Comparison of measurements with derived model predictions (simulated)

such a test are shown in Fig. 1. The agreement is good except at low frequencies (which are of little operational significance in any case).

On a real turbo-alternator, as a test of the approach we were asked to assess the location and magnitude of imbalance following a blade failure. Part of the assessment is shown in Table 2. Subsequent inspection revealed that our estimate of magnitude was 11 % in error and the location was about 50 cm from true. This is considered a very promising result. Of course, the analysis also generated extensive other information on foundation stiffness and mass coefficients and bearing loading, but these in themselves do not offer the prospect of direct comparison for verification. Nevertheless, they would be an important part of an on-going monitoring strategy.

The real potential of this approach to monitoring is that whereas current methods record amplitude and phase of vibration as a function of speed, these in themselves

Table 2 Imbalance on a 350 MW Turbo-generator

Location	Mag (kg.m) plane 1	Phase (degrees) plane 1	Mag (kg.m) plane 2	Phase (degrees) plane 2
HP Frt/Rr	4.004	327	2.697	110
LP1 Frt/Rr	0.511	300	0.436	243
LP2 Frt/Rr	1.328	266	1.139	88
LP3 Frt/Rr	0.486	70	0.77	70/252
GEN Frt/Rr	0.539	204	0.656	29

are rather obscure items of data, require the intervention of expert staff to ascribe some physical meaning. With appropriately developed software, the new methods would provide following each transient record:

(1) Stiffness and Mass parameters for the foundation
(2) The state of imbalance
(3) Presence/absence of shaft bend and appropriate parameters
(4) Bearing loadings—which is in effect the alignment.

The point to emphasise here is that the raw data has been interpreted into meaningful physical parameters and any changes or trends in these parameters will be of direct value to the operator of the plant.

A discussion of these methods was given in a recent review paper [9].

5 On Load Variations

With enhanced modelling capabilities, a laudable objective is to gain some understanding of the variations of vibration when turbo-generators are operating under pseudo steady state conditions. It is an irony that operator have abundant data on such conditions yet this data is relatively low in information content, but nevertheless there is some valuable insight to be gained. Figure 2 shows a carpet plot of

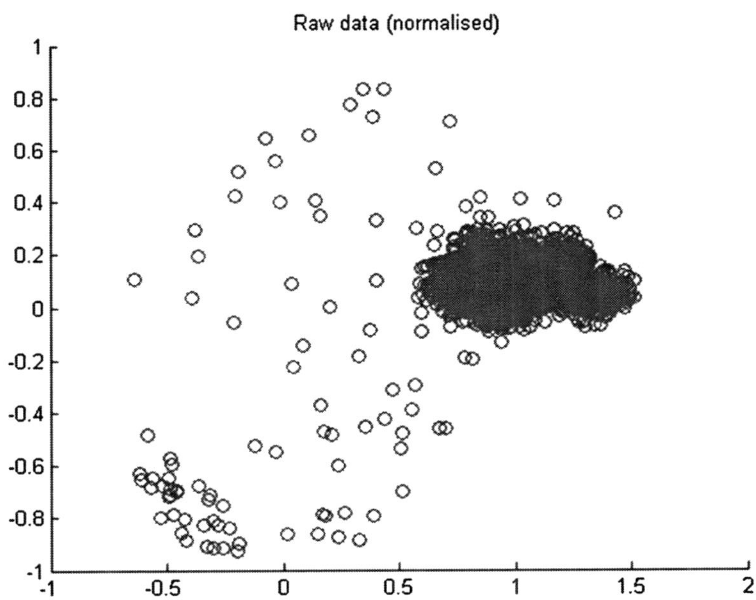

Fig. 2 Normalised carpet plot

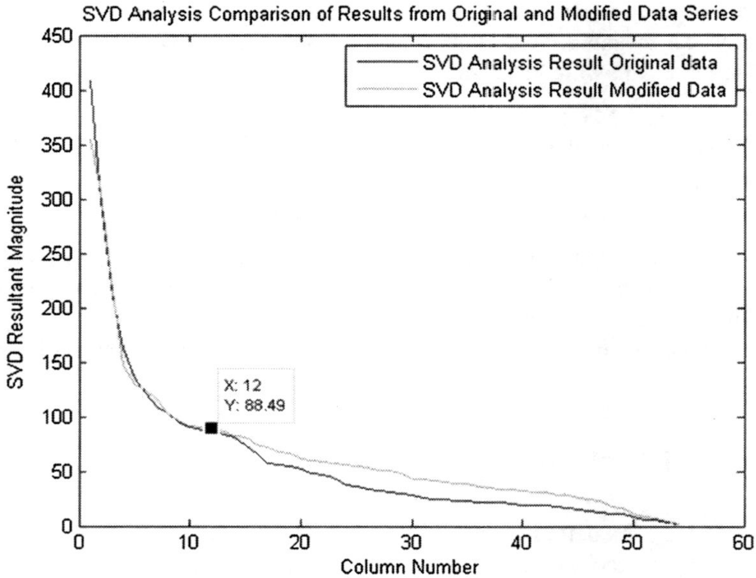

Fig. 3 Singular value study

some normalised vibration values taken from a turbo-generator every 2 min over a period of 12 days. The log provides some 54 parameters for each time step relating to turbine and generator operation, but clearly not all of these have a significant influence on the machine's vibrational behaviour. A good test of the relative importance of data is to perform a Singular Value Analysis of the array of data. In this case the matrix analysed had dimension 8,708 × 54 and the singular values are plotted in Fig. 3.

The Singular Value Decomposition gives some valuable insight into what is going on, but it does require some interpretation. It easily shown that high eigenvectors corresponding to high singular values have a high influence on the residues, but those with a low singular value are of little importance, but they may well contaminate the resulting matrix inversion with noise. The way around this issue is to limit the value of the lowest singular value. In the case shown above a suitable threshold appears to be around 80–90 corresponding to 7–11 sets of variables. This is saying that most of the 54 variables monitored have little influence on the vibration levels. It is evident that only a few of the singular values and their corresponding functions are significantly contributing to the solution, but translating this information to specific parameters is a little more involved.

One way of making some sense of this data is to again use system identification techniques to evaluate the derivatives with respect to each of the parameters. Again it was found that only a few of these derivatives were significant. The procedure is based on first order expansion and, of course, once the derivatives have be

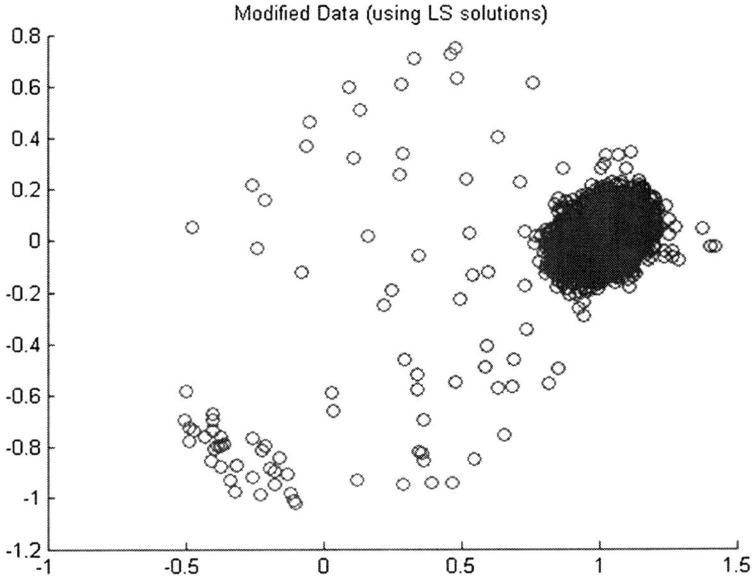

Fig. 4 Revised carpet plot

estimated, modified values of the pseudo-vibration values are readily calculated. The results of this process are shown in Fig. 4. It is interesting to note that the grouping of the majority of the data points is considerably tighter, enabling a more precise monitoring to be devised.

Nevertheless, there are still a significant number of outliers which need some study. The reasons for apparent changes are:

(a) Dependence on some unmonitored parameter, in this case possible external temperature.
(b) Dynamic or transient effects at the measurement time
(c) A real change in the plant
(d) Dependence on higher order effects, but in the case studies the SVD suggests that this is not the case.

A variety of approaches are available for a more extensive use of on load data. The extent to which this area is developed will be determined by industrial needs.

Artificial Neural Networks (ANN) offer one route into this area. Figure 5 shows the variation of vibration compared with its prediction. It is clear that the main variations are well modelled. Used as a direct monitoring scheme (see [10] such a system can be used to recognise pre-determined faults). The limitation is the paucity of data under fault conditions. Perhaps more importantly, ANNs can be used to investigate complex data relationships to help formulate functional dependencies thus assisting in extending understanding.

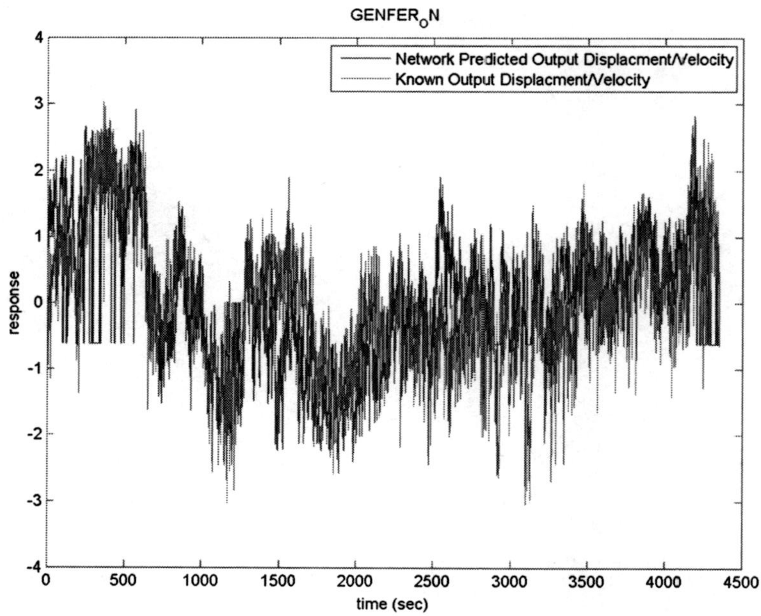

Fig. 5 Monitoring with ANN

6 The Logical Conclusion

There is sometimes heated discussion among engineers about the relative importance of modelling, statistics, measurements, ANNs and other approaches, but in the opinion of the author all these approaches are needed in attempts to understand the behaviour of complex machinery. The fundamental problem of all monitoring is to gain insight into internal processes from the limited external measurements. In this paper only a brief overview of available techniques and question have been outlined but clearly there is now considerable potential for progress.

Acknowledgment The author thanks EDF, BNFL(Magnox) and RWE for fruitful collaboration over a number of years. He also thanks Mr Andrew Little, a former student, for his study on Artificial Neural Networks.

References

1. Cavalca KL, Cavalcante F, Okabe EP (2005) An investigation on the influence of the supporting structure on the dynamics of the rotor system. Mech Syst Signal Proc 19 (1):157–174
2. Diana G, Chelli F, Vania A (1988) Foundations from run-down records—a preliminary experimental study, I.Mech.E. conference on vibrations in rotating machinery, Edinburgh

3. Edwards S, Lees AW, Friswell MI (2000) Experimental identification of excitation and support parameters of a flexible rotor-bearing-foundation system from a single run-down. J Sound Vib 232(5):963–992
4. Feng NS, Hahn EJ (1995) Including foundation effects on the vibration behaviour of rotating machinery. Mech Syst Signal Process 9(3):243–256
5. Feng NS, Hahn EJ (2003) Identification of the modal parameters of a flexibly supported rigid casing. J Eng Gas Turbines Power-Transactions ASME 125(1):184–192
6. Friswell MI, Mottershead JE (1995) Finite element model updating in structural dynamics. Kluwer Academic Publishers, Dordrecht
7. Lees AW (1988) The least squares method applied to identify rotor foundation parameters, I. Mech.E. conference on vibrations in rotating machinery, Edinburgh
8. Lees AW, Simpson IC (1983) Dynamics of turbo-alternator foundations, I.Mech.E. conference on vibrations in rotating machinery, London, paper C6/83
9. Lees AW, Sinha JK, Friswell MI (2009) Model based identification of rotating machines. Mech Syst Signal Process 23:1884–1893
10. Mayes IW (1994) Use of neural networks for on-line vibration monitoring. Proc Inst Mech Eng: Part A 208:267–274
11. Pennacchi P, Bachschmid N, Vania A et al (2006) Use of modal representation for the supporting structure in model-based fault identification of large rotating machinery: part 1—theoretical remarks. Mech Syst Signal Process 20(3):662–681
12. Pennacchi P, Vania A (2004) Accuracy in the identification of a generator thermal bow. J Sound Vib 274(1–2):273–295
13. Pennacchi P, Bachschmid N, Vania A (2006) A model-based identification method of transverse cracks in rotating shafts suitable for industrial machines. Mech Syst Signal Process 20(8):2112–2147
14. Pons A (1986) Experimental and numerical analysis on a large nuclear steam turbo-generator. IFToMM, Int Conf Rotor Dyn 1986:269
15. Provasi R, Zanetta GA, Vania A (2000) The extended Kalman filter in the frequency domain for the identification of mechanical structures excited by sinusoidal multiple inputs. Mech Syst Signal Process 14(3):327–341
16. Sinha JK (2002) Health monitoring techniques for rotating machinery, Ph.D. Thesis, University of Wales Swansea, Swansea
17. Sinha JK, Lees AW, Friswell MI (2004) Estimating the static load on the fluid bearings of a flexible machine from run-down data. Mech Syst Signal Process 18:1349–1368
18. Smart MG (1998) Identification of flexible turbogenerator foundations. PhD Thesis, Department of Mechanical Engineering, University of Wales Swansea
19. Smart MG, Friswell MI, Lees AW (2000) Estimating turbogenerator foundation parameters—model selection and regularisation. Proc Roy Soc A 456:1583–1607
20. Smart MG, Friswell MI, Lees AW, Prells U (2000) Estimating turbogenerator foundation parameters from run down data, I.Mech.E. conference on vibrations in rotating machinery, C/212, pp 653–665
21. Zanetta GA (1992) Identification methods in the dynamics of turbogenerator rotors, I.Mech.E international conference on vibrations in rotating machinery, Bath, C432/092, pp 173–181

The Sensitivity Method in Stochastic Model Updating

John E. Mottershead, Michael Link, Tiago A.N. Silva, Yves Govers and Hamed Haddad Khodaparast

Abstract Probabilistic and interval model updating methods are described, with particular attention paid to variability in nominally identical test structures due, for example, to the effect of accumulated manufacturing tolerances, or degradation of performance caused by wear of engineering components. In such cases the updating parameter distributions are meaningful physically either as PDFs or as intervals. Stochastic model updating is an inverse problem, generally requiring multiple forward solutions, which may be carried out very efficiently by the use of surrogates, in place of full FE models. The procedure is illustrated by experimental examples, including model updating of (i) a frame structure with uncertain locations of two internal beams and (ii) the DLR AIRMOD structure, which displays vibration characteristics very similar to those of a real aircraft.

Keywords Model updating · Sensitivity method · Probabilistic · Interval

1 Introduction

Deterministic model updating of finite element models [1–4] has become a mature technology. It is a classical inverse problem in the sense that a measurable output is used to correct a set of analytical parameters that are themselves inaccessible to

J.E. Mottershead (✉)
Centre for Engineering Dynamics, University of Liverpool, Liverpool L69 3GH, UK
e-mail: j.e.mottershead@liverpool.ac.uk

M. Link
Institute for Statics and Dynamics, University of Kassel, 34109 Kassel, Germany

T.A.N. Silva
LAETA, IDMEC, Instituto Superior Técnico, Universidade de Lisboa, Lisboa, Portugal

Y. Govers
German Aerospace Centre (DLR), Institute of Aeroelasticity, 37073 Göttingen, Germany

H.H. Khodaparast
College of Engineering, University of Swansea, Swansea SA2 8PP, UK

J.K. Sinha (ed.), *Vibration Engineering and Technology of Machinery*,
Mechanisms and Machine Science 23, DOI 10.1007/978-3-319-09918-7_5

measurement. As with most inverse problems the resulting system of equations are ill-posed and require regularisation, discussed in full by Ahmadian et al. [5].

Variability in performance has become one of the major challenges facing researchers and engineering scientists concerned with the robust reliability of engineering systems. This has resulted in the emergence of powerful probabilistic and non-probabilistic (interval, fuzzy etc.) methods combined with modern computer systems and codes. An aspect of uncertainty of particular interest is the inherent variability that arises from the effect of accumulated manufacturing tolerances or from wear so that nominally identical engineering systems behave differently.

The problem of variability in the dynamics of nominally identical test structures was addressed by Mares et al. [6, 7] using a multivariate gradient-regression approach combined with a minimum variance estimator. Haddad Khodaparast et al. [8] developed a first-order perturbation approach whereas the method of Hua et al. [9] required the computation of second-order sensitivities. Govers and Link [10] defined an objective function for the identification of updating-parameter covariances with forward propagation of parameters at each iteration. This approach was modified by Rui et al. [11] who used a polynomial chaos expansion (PCE) as a surrogate for the full finite element model. Fang et al. [12] used the analysis of variance (ANOVA) for the selection of updating parameters together with Monte-Carlo simulation (MSC). Adhikari and Friswell [13] used a sensitivity approach and the Karhunen-Loève expansion to represent distributed parameters.

Non-probabilistic model updating methods were developed by Haddad Khodaparast et al. [14] and Erdogan and Bakir [15] using interval and fuzzy-arithmetic-based procedures. In the former, a Kriging predictor is used as a surrogate and, in the latter, the membership functions of each updating parameter is determined by minimising an objective function using a genetic algorithm.

This paper gives a brief overview of probabilistic and interval updating methods based on first-order output sensitivities. This is followed by experimental example problems designed to illustrate the application of stochastic model updating methods, including the updating of geometrical configuration, made possible by the use of a trained surrogate model. Finally the example of probabilistic and interval updating of the DLR AIRMOD structure is presented using modal data obtained by multiple disassembly and reassembly of bolted joints—resulting in variability due principally to slight differences in joint tightness in each of the separate tests.

2 Deterministic Model Updating

The deterministic model updating problem is usually cast as,

$$\left(\mathbf{z}^m - \mathbf{z}_j\right) = \mathbf{S}_j\left(\boldsymbol{\theta}_{j+1} - \boldsymbol{\theta}_j\right) \tag{1}$$

where typically $\mathbf{z}^m$ is a vector of measured natural frequencies and mode-shape terms, $\mathbf{z}_j$ is the corresponding vector of predictions determined from an analytical model with parameters $\boldsymbol{\theta}_j$ and j denotes the iteration index. The updated vector of parameters is then given by,

$$\boldsymbol{\theta}_{j+1} = \boldsymbol{\theta}_j + \mathbf{T}_j\left(\mathbf{z}^m - \mathbf{z}_j\right) \tag{2}$$

where the transformation matrix $\mathbf{T}_j$ is generally the weighted pseudo inverse of the sensitivity matrix $\mathbf{S}_j$.

In the following analysis we consider a 'cloud' of data points in multi-dimensional space so that each point represents a separate test. There is a corresponding cloud of predicted points and the objective is to adjust the statistics of the parameters so that the position and orientation of the cloud, determined according to the mean values and covariances, or alternatively the upper and lower bounds, of the predictions, is made to agree with the data cloud.

3 Probabilistic Model Updating

The probabilistic model updating approach proceeds by iterative correction to parameter means and covariances. An assumption of statistical normality is therefore implied. However, this is a mathematical constraint rather than a practical drawback, because structural measurements are likely to be limited to mean values and standard deviations (or upper and lower bounds as addressed by interval updating, described in Sect. 4). Equivalently to Eq. (2) the stochastic model updating equation may be written as,

$$\bar{\boldsymbol{\theta}}_{j+1} + \Delta\boldsymbol{\theta}_{j+1} = \bar{\boldsymbol{\theta}}_j + \Delta\boldsymbol{\theta}_j + \left(\bar{\mathbf{T}}_j + \Delta\mathbf{T}_j\right)\left(\bar{\mathbf{z}}^m + \Delta\mathbf{z}^m - \bar{\mathbf{z}}_j - \Delta\mathbf{z}_j\right) \tag{3}$$

where the overbar and Δ denote the mean and variability on the mean respectively. Then by separating the zeroth-order and first-order terms, Khodaparast et al. [8] obtained the following two expressions:

$$O\left(\Delta^0\right): \quad \bar{\boldsymbol{\theta}}_{j+1} = \bar{\boldsymbol{\theta}}_j + \bar{\mathbf{T}}_j\left(\bar{\mathbf{z}}^m - \bar{\mathbf{z}}_j\right) \tag{4}$$

$$O\left(\Delta^1\right): \quad \Delta\boldsymbol{\theta}_{j+1} = \Delta\boldsymbol{\theta}_j + \bar{\mathbf{T}}_j\left(\Delta\mathbf{z}^m - \Delta\mathbf{z}_j\right) + \Delta\mathbf{T}_j\left(\bar{\mathbf{z}}^m - \bar{\mathbf{z}}_j\right) \tag{5}$$

Equation (4) is the updating equation for the parameter mean values, exactly the same as Eq. (12) given by Govers and Link [10].

Using assumptions of (i) small parameter variability, (ii) neglecting the third right-hand-side term of Eq. (5) and (iii) statistical independence of the updated parameters $\Delta\boldsymbol{\theta}_j$ (and hence the predictions $\Delta\mathbf{z}_j$) from the measurements $\Delta\mathbf{z}^m$, Silva et al. [16] showed three previously developed expressions for the updated covariance $Cov\left(\Delta\boldsymbol{\theta}_{j+1}, \Delta\boldsymbol{\theta}_{j+1}\right)$ to be equivalent. The first of the three expressions, due to Khodaparast et al. [8] is expressed as,

$$Cov(\Delta\boldsymbol{\theta}_{j+1}, \Delta\boldsymbol{\theta}_{j+1}) = Cov(\Delta\boldsymbol{\theta}_j, \Delta\boldsymbol{\theta}_j) - Cov(\Delta\boldsymbol{\theta}_j, \Delta\mathbf{z}_j)\bar{\mathbf{T}}_j^T - \bar{\mathbf{T}}_j Cov(\Delta\mathbf{z}_j, \Delta\boldsymbol{\theta}_j) + \bar{\mathbf{T}}_j Cov(\Delta\mathbf{z}_j, \Delta\mathbf{z}_j)\bar{\mathbf{T}}_j^T + \bar{\mathbf{T}}_j Cov(\Delta\mathbf{z}^m, \Delta\mathbf{z}^m)\bar{\mathbf{T}}_j^T \tag{6}$$

The second expression, developed by Govers and Link [10] using the Frobenius norm for the minimisation of the difference between measured and analytical output covariances, is given by,

$$Cov(\Delta\boldsymbol{\theta}_{j+1}, \Delta\boldsymbol{\theta}_{j+1}) = Cov(\Delta\boldsymbol{\theta}_j, \Delta\boldsymbol{\theta}_j) + \bar{\mathbf{T}}_j Cov(\Delta\mathbf{z}_m, \Delta\mathbf{z}_m)\bar{\mathbf{T}}_j^T - \bar{\mathbf{T}}_j Cov(\Delta\mathbf{z}_j, \Delta\mathbf{z}_j)\bar{\mathbf{T}}_j^T \tag{7}$$

Forward propagation is required to determine $Cov(\Delta\mathbf{z}_j, \Delta\mathbf{z}_j)$ in both Eqs. (6) and (7), and $Cov(\Delta\boldsymbol{\theta}_j, \Delta\mathbf{z}_j)$and $Cov(\Delta\mathbf{z}_j, \Delta\boldsymbol{\theta}_j)$in Eq. (6). This is achieved by application of the relationships,

$$Cov(\Delta\mathbf{z}_j, \Delta\mathbf{z}_j) = \bar{\mathbf{S}}_j Cov(\Delta\boldsymbol{\theta}_j, \Delta\boldsymbol{\theta}_j)\bar{\mathbf{S}}_j^T \tag{8}$$

$$Cov(\Delta\boldsymbol{\theta}_j, \Delta\mathbf{z}_j) = Cov(\Delta\boldsymbol{\theta}_j, \Delta\boldsymbol{\theta}_j)\bar{\mathbf{S}}_j^T \tag{9}$$

$$Cov(\Delta\mathbf{z}_j, \Delta\boldsymbol{\theta}_j) = \bar{\mathbf{S}}_j Cov(\Delta\boldsymbol{\theta}_j, \Delta\boldsymbol{\theta}_j) \tag{10}$$

However, direct substitution of Eqs. (8), (9) and (10) into either Eqs. (6) or (7) leads immediately to the third expression, given for example by Hart [17], page 47 (for the general case of linearly related sets of random variables),

$$Cov(\Delta\boldsymbol{\theta}_{j+1}, \Delta\boldsymbol{\theta}_{j+1}) = \bar{\mathbf{T}}_j Cov(\Delta\mathbf{z}_m, \Delta\mathbf{z}_m)\bar{\mathbf{T}}_j^T \tag{11}$$

where it is seen that the updated parameter covariance matrix is computed once using the final transformation matrix (assuming $\bar{\mathbf{T}}_j = \mathbf{T}(\bar{\theta}_j)$) from the converged solution of the means. There is no need for forward propagation.

4 Interval Model Updating

Interval model updating is carried out by updating a large number of deterministic models. This would be prohibitively expensive were the full FE model to be used and it is for this reason that a meta-model is used to provide a highly efficient mapping between the input parameters and the output measurements. No assumption is placed on the probability distribution which is described only in terms of upper and lower bounds on the updating parameters. The interval model updating problem is formally described as,

$$\tilde{\mathbf{z}}_m - \tilde{\mathbf{z}}_j = \tilde{\mathbf{S}}_j\left(\tilde{\boldsymbol{\theta}}_{j+1} - \tilde{\boldsymbol{\theta}}_j\right) \tag{12}$$

where $(\tilde{\bullet})$ represents an interval vector or matrix. An advantage of the meta-model approach, developed by Haddad Khodaparast et al. [14] is that it avoids the use of interval arithmetic, which is known to be unduly conservative. The accuracy of the technique depends on the type of meta-model, the sampling used and the behaviour of the outputs within the range of parameter variation. The Kriging meta-model was used due to its excellent performance in the prediction of FE model behaviour (particularly with non-smooth behaviour) and also the provision of a probabilistic interpretation of the residual between the outputs of the FE model and the surrogate itself. The interval model updating procedure is illustrated in Fig. 1. Firstly, the mean values of updating parameters are found by deterministic model updating using the mean values of measured data. If the solution is unique, the vector of updated parameter mean values can be represented by a point in the parameter space. In the next step, an initial hypercube around this point is constructed and the meta-model (the Kriging predictor in the present case) is then used to map the space of the initial hypercube of updating parameters to the space of outputs. Care should be taken to make sure that the mapping is good enough to represent the relationship between inputs and outputs with sufficient accuracy. This can be achieved by increasing the number of training samples until the Mean Squared Error (MSE) at an unsampled point falls below a threshold, when the Kriging model is deemed to be accurate enough. Equation 6 may then be solved by taking all the measured data, represented by circles in Fig. 1, and mapping them in the reverse direction (using the Kriging predictor) to find the vertices of the updated parameter hypercube. The

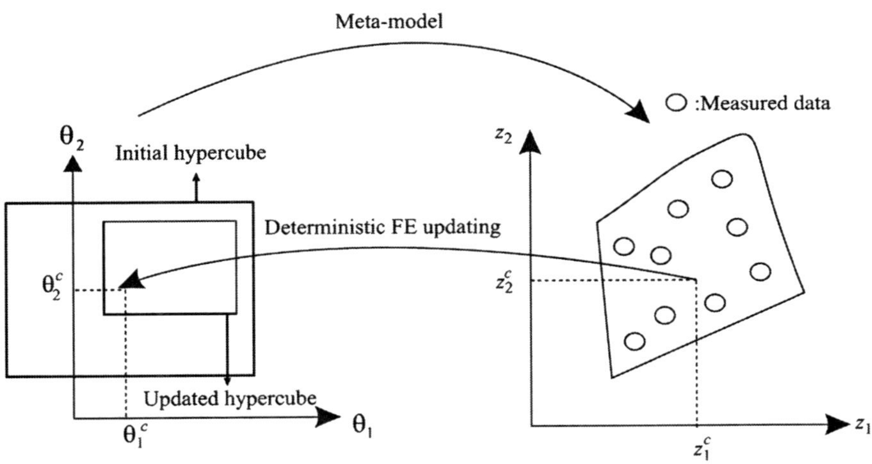

Fig. 1 Interval model updating using the Kriging predictor (Reprinted from [14] with permission from Elsevier)

need for a highly computationally efficient meta-model is illustrated by the fact that a large number of inversions are required for interval model updating. Further details on the mathematical development of the method are provided by Haddad Khodaparast et al. [14].

5 Experimental Examples

Stochastic model updating procedures and their effectiveness are illustrated by two experimental examples, both of which are designed to reproduce the conditions that would be experienced in the case of multiple nominally-identical structures each with variability arising from, for example, manufacturing tolerances or degradation due to wear. The objective is to estimate the spread of uncertainty in parameters deemed to be responsible for observed variability in dynamic behaviour from modal tests.

In the first example, a frame structure is tested nine times to determine the natural frequencies. In each of the nine tests, two internal beams were placed in different configurations and intervals were determined on the extreme positions of the two beams. The use of a meta-model enabled the parameterization of beam locations, which could not have been done by conventional model updating without re-meshing of the FE model at each iteration.

The second example is the DLR AIRMOD structure, a replica of the GARTEUR SM-AG19 benchmark described by Balmes [18]. Probabilistic and interval model updating was carried out using data selected from 130 disassemblies and reassemblies of the structure with a modal test carried out at each stage. Details of the test procedure, carried out at DLR Göttingen, can be found in [19, 20].

5.1 Frame Structure with Internal Beams

The two internal beams of the test structure shown in Fig. 2 can each be placed independently in 3 locations (1, 2, 3 in Fig. 2b), making a total of nine different configurations. In each case natural frequencies were determined in modal tests described by Haddad Khodaparast [14]. The FE model, shown in Fig. 2c, was constructed using 8-noded CHEXA elements and a preliminary deterministic model updating exercise was completed in order correct the FE boundary conditions. A Kriging model was then constructed based on a CCD design using the 9 samples corresponding to the 9 different beam positions. This model was deemed sufficiently accurate according to a tolerance placed on the mean square error (MSE) at beam locations (θ_1 and θ_2 shown in Fig. 2b) away from the 9 training points. Interval model updating was then carried out using the continuous mapping from updating parameters, θ_1 and θ_2, to natural frequencies provided by the mean prediction of the trained Kriging model.

Six natural frequencies were used for interval updating, these being the first and second in-plane bending, first and second out-of-plane bending and the first and

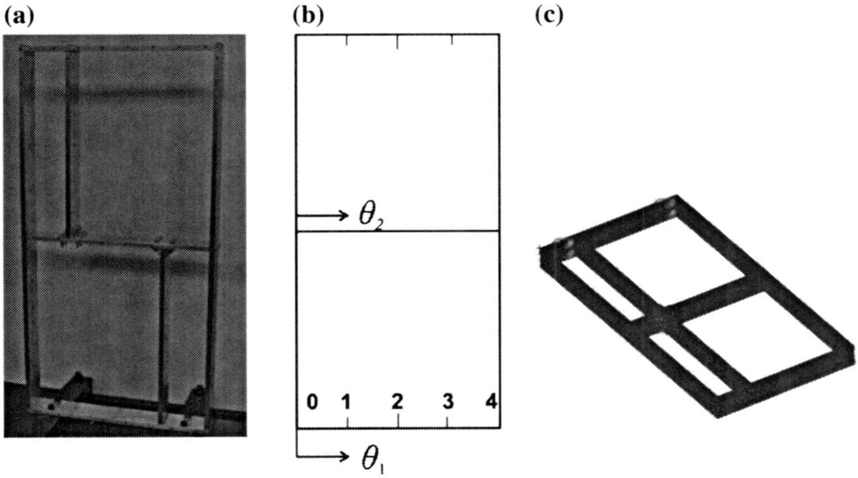

Fig. 2 Test structure and FE model (Reprinted from [14] with permission from Elsevier). **a** Test structure, **b** Beam locations, **c** FE Model

Table 1 True, initial and updated beam locations

True parameters		Initial parameter error %		Updated parameter error %	
θ_1	θ_2	θ_1	θ_2	θ_1	θ_2
1.0	1.0	60	60	3.7	2.0
1.0	2.0	60	20	−0.2	7.6
1.0	3.0	60	−20	0.2	2.8
2.0	1.0	−20	60	1.8	−9.8
2.0	2.0	20	20	6.5	−0.1
2.0	3.0	20	−20	−2.4	3.0
3.0	1.0	−20	60	−0.6	−11.0
3.0	2.0	−20	−20	−0.3	−8.4
3.0	3.0	−20	−20	−2.2	−0.6

Table 2 Measured, initial and updated frequency intervals (Hz)

	Measured	Initial FE	Updated FE	Initial FE error %	Updated FE error %
1	22.5–24.3	21.6–24.6	22.6–24.6	[−4.1, 1.1]	[0.1, 1.1]
2	24.4–27.8	23.7–35.5	23.9–27.5	[−3.0, 27.6]	[−2.1, −1.3]
3	47.1–49.9	43.7–67.6	45.1–50.6	[−7.2, 35.6]	[−4.2, 1.4]
4	74.4–81.1	71.1–82.5	74.0–81.4	[−4.4, 1.6]	[−0.5, 0.2]
5	219.5–256.4	224.1–267.3	224.1–259.5	[2.1, 4.3]	[2.1, 1.2]
6	299.7–312.4	300.3–339.7	303.6–317.2	[0.2, 8.7]	[1.3, 1.5]

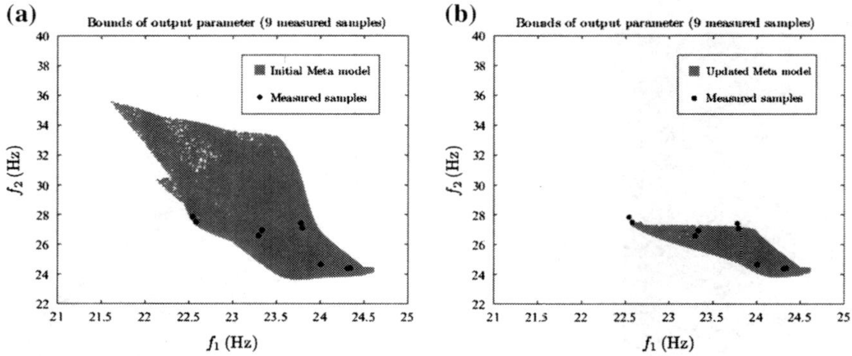

Fig. 3 Space of first and second natural frequencies (Reprinted from [14] with permission from Elsevier). **a** Initial space, **b** Updated space

second twisting modes. Interval updating is carried out in practice by a series of deterministic updates using the measured natural frequencies at each 9 different beam configurations. The true, initial and updated parameters are listed in Table 1 whereas Table 2 shows the measured, initial and updated natural frequencies. It is seen that the maximum error of 35.6 % in the natural frequencies is reduced to 4.2 %. A sample showing the initial and updated spaces in the plane of the first and second natural frequency is shown in Fig. 3 (black dots denote measurements).

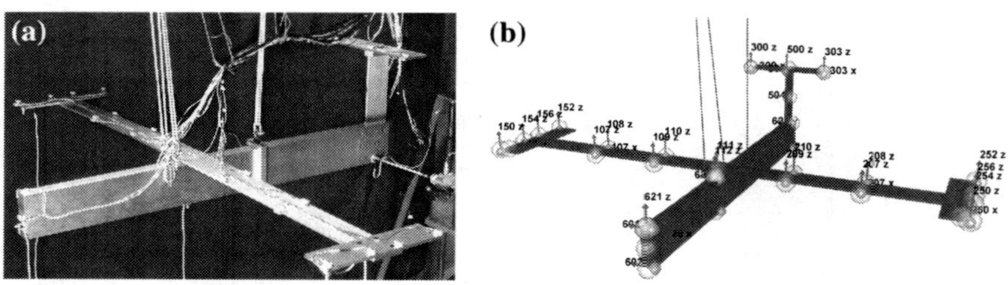

Fig. 4 AIRMOD structure and FE model. **a** Structure, **b** FE Model (Reprinted from [21] with permission from Elsevier)

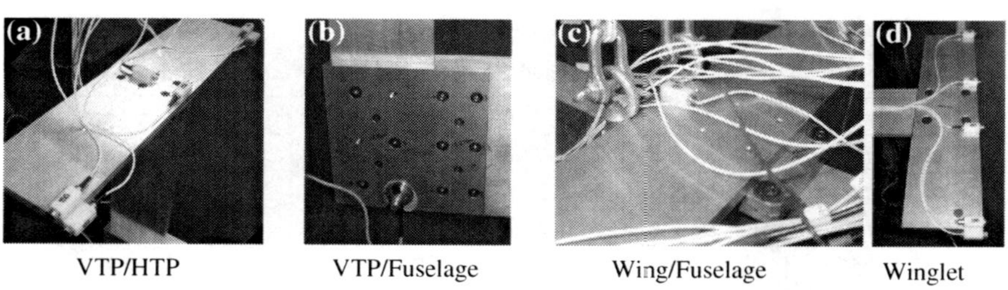

Fig. 5 Bolted joints. **a** VTP/HTP, **b** VTP/Fuselage, **c** Wing/Fuselage, **d** Winglet (Reprinted from [21] with permission from Elsevier)

5.2 DLR AIRMOD Structure

The AIRMOD structure, shown in Fig. 4 is made of aluminium and consists of six beam-like components connected by five bolted joints, illustrated in Fig. 5. The wing span of AIRMOD is 2.0 m, the length of the fuselage is 1.5 m and the height is

Table 3 Mean value updating results (*Italics rows* active modes, *Non-italics rows* passive modes)

	f_{test}/Hz	f_{FE_ini}/Hz	Δf_{FE_ini}/%	Δf_{FE_upd}/%
RBM Yaw	*0.23*	*0.17*	–	*0.01*
RBM Roll	*0.65*	*0.56*	–	*–0.01*
RBM pitch	*0.83*	*0.82*	*–1.00*	*–0.00*
RBM Heave	*2.17*	*2.14*	*–1.62*	*–0.02*
2nWingBending	*5.50*	*5.65*	*2.78*	*0.40*
3nWingBending	*14.91*	*15.11*	*1.35*	*–0.01*
WingTorsionAnti	*31.96*	*33.31*	*4.25*	*0.25*
WingTorsionSym	*32.33*	*33.62*	*3.98*	*–0.42*
VtpBending	34.38	35.39	2.94	1.14
4nWingBending	*43.89*	*44.66*	*1.77*	*–0.08*
1nWingForeAft	*46.71*	*47.21*	*1.08*	*0.02*
2nWingForeAft	*51.88*	*52.91*	*1.99*	*0.05*
5nWingBending	58.59	60.59	3.43	1.29
VtpTorsion	*65.93*	*67.69*	*2.67*	*–0.05*
2nFuseLat	100.05	102.59	2.55	2.07
2nVtpBending	124.56	128.62	3.26	1.48
6nWingBending	129.38	132.08	2.08	–0.14
7nWingBending	141.47	145.91	3.14	0.76
2nHtpBending	*205.59*	*206.73*	*0.56*	*–0.04*
HtpForeAft	*219.07*	*225.73*	*3.04*	*0.11*
WingBendingRight	254.73	261.53	2.67	–0.02
WingBendingLeft	255.02	262.64	2.99	0.32
3nWingForeAft	272.08	278.71	2.44	1.44
WingletBendingLeft	303.96	320.15	5.33	2.24
WingletBendingRight	304.32	321.64	5.69	2.35
3nFuseLat	313.68	324.12	3.33	2.39
WingTorsionSym2	328.55	336.31	2.36	0.60
WingTorsionAnti2	331.18	341.15	3.01	0.71
4nWingForeAft	336.21	343.55	2.18	–0.15
2nFuseVert	348.68	359.54	3.12	1.77

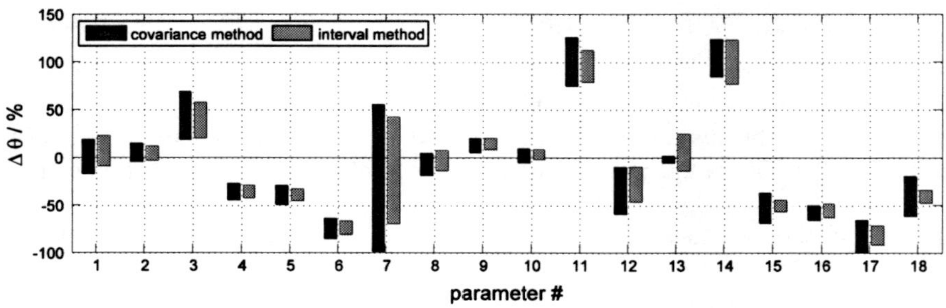

Fig. 6 Updated parameter intervals: Changes with respect to the initial finite element model (Reprinted from [21] with permission from Elsevier)

0.46 m. The total weight of the structure is 44 kg. Two additional masses of 167 grams each are installed at the forward tips of the winglets to ensure better excitation of the wing torsion modes. The complete FE model shown in Fig. 4b, including the main structure, cables, sensors and bungee cords consists of 1440 CHEXA, 6 CPENTA and 561 CELAS1, 55 CMASS1, 18 CONM2 and 3 CROD elements. The structural behaviour of AIRMOD shows similarities to the dynamical characteristics of a real aircraft, including mode shapes that show either symmetric or antisymmetric deformations and closely space modes are found at 32, 255 and 330 Hz.

Modal testing of the AIRMOD structure and assessment of sources of uncertainty is discussed in detail in a series of papers by Govers and Link [10, 19, 20]. Model updating using both probabilistic and interval methods is described in full by Govers et al. [21]. The means of the parameters were estimated by deterministic model updating of 18 parameters using both the eigenvalue and mode-shape residuals of the 14 active modes shown in Table 3. It is seen from in the table that the mean frequency values of the updated model match perfectly with the test data for the active mode set. Also, the passive modes (not included in the residual) are significantly improved, thereby providing validation of the updated FE model.

The parameter covariances and intervals were estimated using the natural frequencies of the same 14 modes (the mode shapes were not included) and results are shown in Fig. 6. Three standard deviations are shown in the case of covariance updating.

Updated covariance and interval results for the 14 natural frequencies are presented in Fig. 7. The size and orientation of the 105 frequency clouds show a very good agreement between test (red) and updated analysis clouds (green) for the active range. The interval-updated regions (black) are generally greater than those calculated by the covariance method. This is to be expected because the number of terms that define the multi-variate probability distributions on the updating parameters (the means and fully populated covariance matrix) is greater than the set of upper- and lower-bounds that define the interval model. The probabilistic method starts with the diagonal covariance matrix at the first iteration (assuming uncorrelated updated parameters), but becomes a fully populated matrix after convergence. The interval

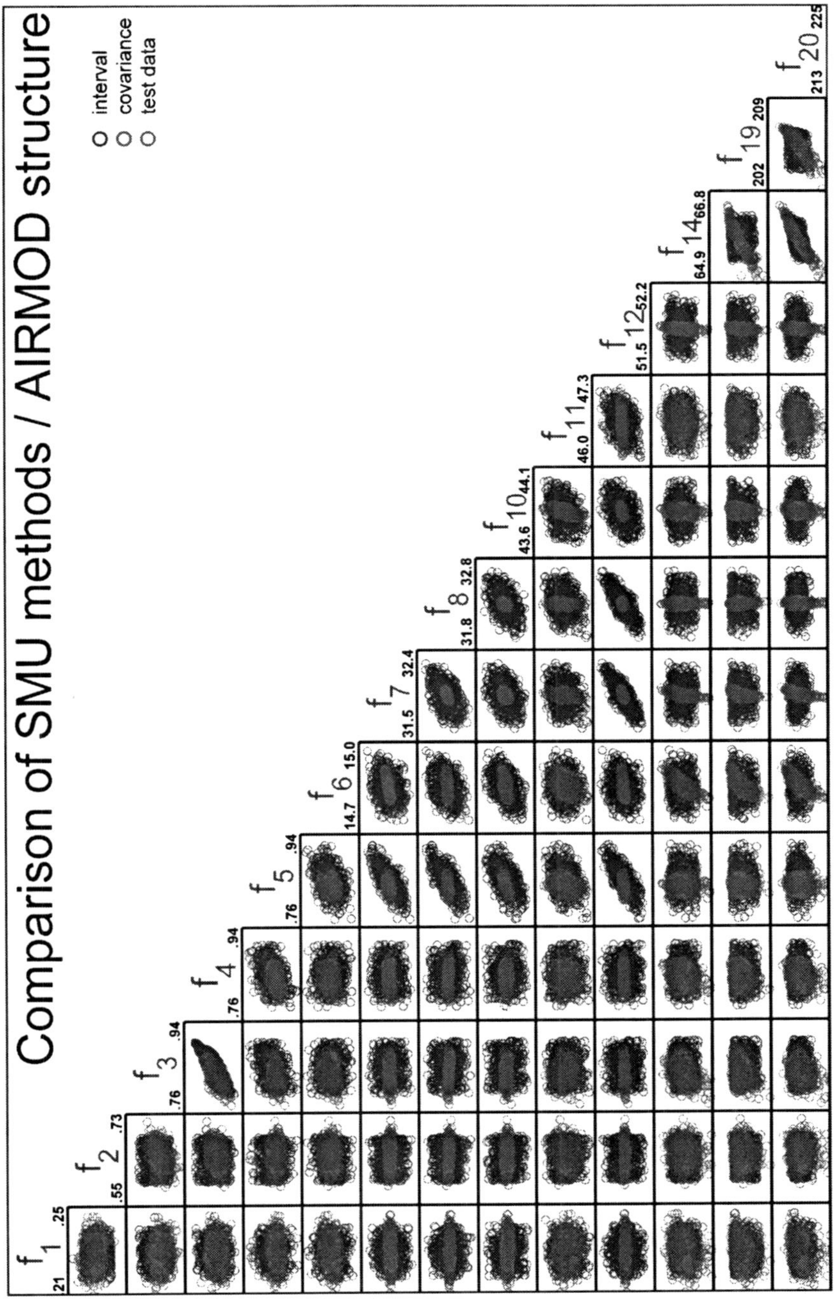

Fig. 7 Frequency distributions from updated models (Reprinted from [21] with permission from Elsevier)

method makes no assumption about the correlation of the updating parameters, which are therefore confined to a hypercube, whereas the updated fully-populated covariance matrix defines a space of hyper-elliptic parameters. The hypercube creates a greater sample space than that of the hyper-ellipse. Another observation from Fig. 6 is that the degree of overestimation of the interval results is not of the same order for all the frequencies, as can be seen by comparing plane f_8-f_{10} to plane f_1-f_2. One may relate this to the sensitivity of the natural frequencies with respect to the parameters which are located outside the hyper-ellipse but inside the hypercube. In other words, the greater the sensitivities of the frequencies to parameter variation, the higher will be the degree of over-estimation seen in the results.

6 Conclusions

An overview of probabilistic and interval methods in finite element model updating is presented. The study is focused on the problem of multiple nominally-identical test structures where the problem is to quantify the certainty in parameters deemed responsible for observed variability in dynamic behavior. The overview is accompanied by two experimental example problems designed to demonstrate the working of the techniques involved.

References

1. Natke HG (1992) Einführung in Theorie und Praxis der Zeitreihen und Modalanalyse. Vieweg Verlag, Braunschweig/Wiesbaden, Germany
2. Mottershead JE, Friswell MI (1993) Model updating in structural dynamics: a survey. J Sound Vib 167(2):347–375
3. Friswell MI, Mottershead JE (1995) Finite element model updating in structural dynamics. Kluwer Academic Publishers, Dordrecht
4. Mottershead JE, Link M, Friswell MI (2011) The sensitivity method in finite element model updating: a tutorial. Mech Syst Signal Process 25(7):2275–2296
5. Ahmadian H, Mottershead JE, Friswell MI (1998) Regularisation methods for finite element model updating. Mech Syst Signal Process 12(1):47–64
6. Mares C, Mottershead JE, Friswell MI (2006) Stochastic model updating: part 1- theory and simulated examples. Mech Syst Signal Process 20(7):1674–1695
7. Mottershead JE, Mares C, James S, Friswell MI (2006) Stochastic model updating: part 2-application to a set of physical structures. Mech Syst Signal Process 20(8):2171–2185
8. Haddad Khodaparast H, Mottershead JE, Friswell MI (2008) Perturbation methods for the estimation of parameter variability in stochastic model updating. Mech Syst Signal Proc 22 (8):1751–1773
9. Hua XG, Ni YQ, Chen ZQ, Ko JM (2008) An improved perturbation method for stochastic finite element model updating. Int J Numer Meth Eng 73:1845–1864
10. Govers Y, Link M (2010) Stochastic model updating—covariance matrix adjustment from uncertain experimental modal data. Mech Syst Signal Process 24(3):696–706

11. Rui Q, Ouyang H, Wang HY (2013) An efficient statistically equivalent reduced method on stochastic model updating. Appl Math Model 37:6079–6096
12. Fang S-E, Ren W-X, Perera R (2012) A stochastic model updating method for parameter variability quantification based on response surface models and Monte-Carlo simulation. Mech Syst Signal Proc 33:83–96
13. Adhikari S, Friswell MI (2010) Distributed parameter model updating using the Karhunen-Loève expansion. Mech Syst Signal Process 24:326–339
14. Haddad Khodaparast H, Mottershead JE, Badcock KJ (2011) Interval model updating with irreducible uncertainty using the Kriging predictor. Mech Syst Signal Proc 25(4):1204–1226
15. Erdogan YS, Bakir PG (2013) Inverse propagation of uncertainties in finite element model updating through use of fuzzy arithmetic. Eng Appl Artif Intell 26:357–367
16. Silva TAN, Mottershead JE, Link M, Maia NMM Updating the parameter covariance matrix of a finite element model, in preparation
17. Hart GC (1982) Uncertainty analysis, loads and safety in structural engineering. Prentice-Hall, Englewood Cliffs
18. Balmes E (1997) Garteur group on ground vibration testing: results from the test of a single structure by 12 laboratories in Europe, proceedings of international modal analysis conference, 1997, pp 1346–1352
19. Govers Y, Link M (2010) Stochastic model updating of an aircraft like structure by parameter covariance matrix adjustment, proceedings of international conference on noise and vibration engineering, Leuven, Belgium, 2010, pp 2639–2656
20. Govers Y, Link M (2012) Using stochastic experimental modal data for identifying stochastic finite element parameters of the AIRMOD benchmark structure, proceedings of the international conference on noise and vibration engineering, USD2012, Leuven, Belgium, 2012, pp 4697–4715
21. Govers Y, Haddad Khodaparast H, Link M, Mottershead JE (2014) A comparison of two stochastic model updating methods using the DLR AIRMOD test structure, mechanical systems and signal processing, http://dx.doi.org/10.1016/j.ymssp.2014.06.003

Composite Shaft Rotor Dynamics: An Overview

K. Gupta

Abstract This paper summarizes research on dynamics of composite shafts, with an objective to develop light weight rotor systems. A review of available literature post-1996 is presented here, since a comprehensive review on similar topic was published in 1997 by Singh et al. Specifically the various aspects covered are, theories for dynamic analysis of fiber reinforced composite shaft, their modeling and analysis, experimental work, and design optimization procedures. All these aspects are important for full development of 'Composite Rotors' for light weight high performance transmission systems. Directions for future research are outlined.

Keywords Composite rotors · Composite shafts · Fibre reinforced shaft · Drive shafts · Optimization of shafts · Composite rotor instability · Laminate theory · EMBT · LBT · Smart rotors

1 Introduction

Fibre reinforced composites are now a days used extensively in large variety of components, primarily because of their higher specific strength and stiffness, and because of the possibility of developing light weight tailor made designs. Their use in machines and structures for the non-load-bearing parts such as the covers, doors, lids etc., is very common. However there use in critical load bearing components, such as a super-critical rotor shaft of a propulsion system, is very limited. Apart from difficulties in manufacturing, the primary reason is lack of confidence in designing flexible composite rotors which operate at high speeds, and have material heterogeneity and stringent torque transmission requirements. Reliability of these

K. Gupta (✉)
Mechanical Engineering Department, Indian Institute of Technology (IIT) Delhi, New Delhi 110016, India
e-mail: kgupta@mech.iitd.ac.in

J.K. Sinha (ed.), *Vibration Engineering and Technology of Machinery*, Mechanisms and Machine Science 23, DOI 10.1007/978-3-319-09918-7_6

systems during operation is also of concern. The developmental efforts in 'composite rotors' have been in progress for several decades now. The primary design objective is to minimize the shaft or system weight while satisfying other requirements. Application of composite to automotive drive shaft, and to helicopter tail rotor has been extensively studied and reported in literature. Some of the other applications reported [45] are to, propeller shafts in small ships, an aircraft power take of (PTO) shaft, high speed machine tool spindles, transmission/quill shafts, and long boring bars. Another interesting example is discussed by Vance [53] in which the possibility of using a composite shaft with much higher specific stiffness (E/ρ) is explored, as an alternative to a super-critical power turbine shaft of a small front drive aero gas turbine engine. Various applications can be categorized in two sub-groups i.e., the sub-critical operation (rigid rotor) or the super-critical operation (flexible rotor). For sub-critical systems the primary focus is on maximum weight saving and torque transmission. In super-critical operation additional factors like shaft bending vibrations and dynamic stresses, stability and fatigue life acquire importance. For some systems like the automotive drive shaft, minimization of cost is also an important consideration. Singh et al. [45] have presented a comprehensive review of literature on various aspects of 'composite shafts'.

The examples enumerated above underline the fact that in many critical applications with stringent requirements, composite shaft may be the only feasible solution. Undoubtedly there can be very many applications of composite shafts, for example in aero-propulsion and other transmission systems, provided the designers and the users of these systems are ready to accept the challenges and risks involved.

1.1 A 'Composite Rotor'

A 'composite rotor' would necessarily have a tubular composite shaft. The wall thickness will have several (n) layers each of certain thickness (t_i) and fibre orientation (α_i) stacked in a particular order (stacking sequence). Here continuous filament wound composite shaft is considered, in which fibers are placed in helical direction at particular winding angle (α_i). Epoxy is used as resin, while the carbon and glass fibers amongst several others (boron, kevlar etc.) are commonly used.

The shaft is assumed to be supported on two or more bearings. Each bearing is modeled by springs and dampers (typically an eight coefficient bearing). The shaft would carry several discs, each having some unbalance. In rotor dynamic analysis, the designer would be interested in knowing system natural frequencies and mode shapes, critical speeds, vibration response (Bode, cascade, orbital plots etc.), and instability threshold speeds. Rotor critical speeds must be sufficiently away from the range of operating speeds. From vibration response, dynamic stresses and shaft fatigue life can be estimated.

To model the composite shaft, several theories have been proposed in literature, salient features of these are presented in next section.

2 Theories for Composite Shaft Dynamic Analysis

The composite shaft wall thickness comprises of n number of layers (laminae, plies). Each ply has a thickness t_i and the fibers wound in the helical direction at respective angles α_i. The total wall thickness is obtained by stacking together the number of plies. This is called a laminate. In composites literature the analysis of a laminate to determine stresses, strains, and deformations i.e., stiffness and strength, is performed by 'laminate theory (LT)' [27, 52]. The essence of LT is that the macro-mechanical behavior of the laminate is evaluated by first analyzing each lamina treating it as a thin 2-dimensional orthotropic plate, and then appropriately combining the characteristics of all the individual laminae. Important assumptions are of perfect adhesion between the plies (laminae), continuity of displacement at the interface of two adjacent laminae, and linear variation of displacement across the thickness of each lamina. In the simpler version of LT known as 'classical laminate theory (CLT)', the Kirchoff-Love hypothesis is applicable, whereas in general the Mindlin theory will be applicable accounting for first order shear deformation effect. Various approaches adopted for modeling of composite shaft, for the purpose of dynamic analysis of composite rotors, are summarized below.

2.1 The Equivalent Modulus Beam Theory (EMBT)

In this approach the layered composite shaft is replaced by a homogeneous shaft. The equivalent longitudinal modulus E_x and the shear modulus $G_{x\theta}$ are obtained [52] from an analysis of macro-mechanical behavior of the composite shaft and are given in terms of the laminate invariants U_i ($i = 1, 5$), ply thickness t_i and fiber angle α_i ($i = 1, n$), n being the number of plies/laminae. Here x represents the coordinate along the shaft axis (length), and θ is along the circumferential direction. Thus the modulii E_x and $G_{x\theta}$ control the shaft bending and torsional behavior respectively. The expressions for equivalent modulii for composite shaft are concisely given by [46].

It is evident that the equivalent modulii E_x, $G_{x\theta}$ will have numerical values in between that of matrix and fiber ($E_m < E_x < E_f$) and ($G_m < G_{x\theta} < G_f$), which is a large range since the modulii of the matrix are usually much lower than that of the fiber. Also the equivalent modulii will strongly depend on fiber angles in the various laminae. For example a fiber angle of 0° will lead to higher value of longitudinal modulus E_x. Similarly E_x and $G_{x\theta}$ can be increased by increasing the fiber volume fraction V_f.

The EMBT though simple and quite adequate for approximate estimation of natural frequencies has severe limitations. It can not take into account effect of stacking sequence and various coupling mechanisms commonly present in composite laminates. These limitations are overcome by the 'modified EMBT' [13, 18], simplified homogenized beam theory SHBT [50], and the 'layerwise beam theory (LBT)' [44, 48].

2.2 The Modified EMBT

Gubran [13] modified EMBT in order to overcome the limitations of the EMBT, yet maintaining its simplicity. Specifically limitations of EMBT can be summarized as its inability to take into account (i) the effect of radial position of a lamina and (ii) the shear-normal, bending-twisting couplings in unbalanced configurations, and bending-stretching coupling effect in unsymmetric configurations. Gubran and Gupta [18] have given derivations to incorporate modifications in classical EMBT in order to overcome the above limitations.

The strain energy of the shaft including bending in two planes, shear deformation, shaft torsion, and bending-stretching effects can be obtained in terms of the stiffness coefficients associated with bending ($C_B = E_x I$), shear deformation ($C_S = k'AG_{x\theta}$), torsion ($C_T = G_{x\theta}I_p$), and bending-stretching coupling (C_{BT}), in addition to slopes of deflections, curvatures in both the transverse planes, and the gradient of twist angle [18, Eq. 21]. I and I_p are 2nd area moment of cross-section about diametral (neutral) and shaft axes passing through the shaft centre. A is the cross-sectional area and the shear correction factor $k' = ½$ for tubular shaft. C_{BT} is a function of stiffness Q_{16}, and the inner and outer radii of each lamina, summed up over all the n laminae.

Shear-normal coupling effect ($Q_{16} \neq 0$) implies coupling between normal stress σ_{xx} with shear strain $\varepsilon_{x\theta}$, and shear stress $\tau_{x\theta}$ with normal strain ε_{xx}. E_x and $G_{x\theta}$ to account for this effect are given by expressions, which are functions of elements A_{ij} and the total shaft wall thickness t [18, Eqs. 16(a) and 16(b)]. Here A_{ij} (i, j = 1 and 2 correspond to x and θ directions, and $i, j = 6$ correspond to $x\theta$ plane) are elements of the stiffness matrix [27] which relates the in-plane laminate forces $\{N_x, N_\theta, N_{x\theta}\}^T$ with in-plane strains $\{\varepsilon\} = \{\varepsilon_{xx}, \varepsilon_{\theta\theta}, \varepsilon_{x\theta}\}^T$ i.e., $\{N\} = [A]\{\varepsilon\}$.

Bending–twisting coupling is represented by the nonzero terms D_{16} and D_{26} in the bending-stiffness matrix $[D]$ [52] for the laminate. The associated stiffness C_{BT} to account for this effect is given by [5] (also see Eq. 20 in [18]). In EMBT another important modification is to Q_{11} to account for the Poisson's coupling effect according to which the realistic condition that $\sigma_{\theta\theta} = 0$ and $\varepsilon_{\theta\theta} \neq 0$ is enforced. The modified value of Q_{11} is given by [18, Eq. 13(b)].

2.3 Simplified Homogenized Beam Theory (SHBT), and ECMBT

This approach by Sino et al. [50] closely follows from the modified EMBT discussed in Sect. 2.2. The homogenized beam parameters such as flexural (*EI*) and shear stiffness (*GS*) are evaluated using an energy formulation that simultaneously considers Young's modulus, shear modulus, specific damping capacity, the distance to the shaft axis orientation and the thickness of each layer of the shaft. Flexural warping is determined by solving a boundary value problem defined on the cross-section, then the shear corrector factor is evaluated. However, the formulation is simplified by assuming that the mechanical coupling effects, induced by the non-symmetrical stacking of layers, are negligible. The SHBT is used to analyze the sensitivity of the frequencies and instability thresholds to such parameters as shear effect, stacking order and fiber orientation. In [25, 50] an appreciable difference is shown between the improved values of the fundamental natural frequency and the instability threshold speed from SHBT as compared to values obtained from EMBT.

In another variation [39] the classical EMBT which is based on storage modulii (ignores damping), is modified to include the effect of loss modulii. The equivalent modulii obtained are complex, hence the modified theory is termed as 'Equivalent Complex Modulus Beam Theory (ECMBT)'.

2.4 The Layerwise Beam Theory (LBT)

In LBT [44, 48], the layerwise character of the shaft is preserved. The shaft wall thickness is made up of several layers with different individual thicknesses and fibre orientations. Mindlin type conditions are applied for each lamina i.e., the cross-section of each layer remains straight (constant slope) which implies linear variation of displacement across the ply thickness. There is continuity of displacement (due to perfect adhesion assumed between plies) at the interface of two layers. However there can be a discontinuity in slope i.e., an abrupt change in the slope from one ply to other.

In LBT during deflection, there is no distortion of the shaft cross-section (a circular section would remain so) in order to retain a beam like character of the shaft. Thus a transverse (vertical and/or horizontal) deflection of shaft cross-section (due to shaft bending) can also be represented by two components in cylindrical coordinates i.e., the in-plane (circumferential θ direction) and the normal (through the thickness direction) displacements. Displacement in axial direction x will be due to rotation of cross-section about a diametral (neutral axis in bending) axis.

Using this displacement field in strain displacement relations for a cylindrical shell, strain field is obtained. Then the expression for shaft strain energy is setup. This involves integration over the entire volume ($\theta = 0$ to 2π, $x = 0$ to ℓ and wall thickness). Note that integration will be carried out across thickness of a ply, and then summation over all plies ($i = 1, n$). Next the kinetic energy is setup, and equations of motion are obtained which are solved for estimation of natural frequencies as well as the rotor unbalance response. Singh and Gupta [48] have shown the improvement obtained in the values of shaft natural frequency by use of LBT, through comparisons with results available from other theories, as well as from experiments.

A review of available literature on various aspects of composite shaft dynamic design is presented in the following sections. Singh et al. [45] presented a comprehensive review on similar topic. Hence the focus here is for period post-1996. The various aspects covered are modeling and analysis including Finite Element

(FE) applications in Sect. 3, details of experimental work in Sect. 4, and design optimization issues in Sect. 5. Concluding remarks and directions for future research are given in Sect. 6.

3 Modeling and Analysis

Salzar [43] analyzed a metal matrix composite (MMC) shaft for a potential aerogas turbine application. An inelastic axisymmetric generalized plane-strain model of rotating shaft is developed, in which stresses and strains due to centrifugal loading are estimated. The inelastic analysis is performed through the method of successive elastic solutions. Stresses in uniform as well as variable wall thickness rotating simply supported shafts due to transmitted torque and unbalance excitation are evaluated by Gubran et al. [15, 20]. Shaft strain and kinetic energy including unbalance effect, are obtained. Lagrangian is setup and Rayleigh-Ritz approach is adopted to setup the solution equations. These are solved for natural frequencies and displacement response from which the stresses are evaluated. Effect of shear deformation, rotary inertia and gyroscopic moments is accounted for. It is shown that by using variable wall thickness configuration and without altering the shaft mass, (i) 12–17 % increase in fundamental natural frequency can be achieved (ii) more uniform stress distribution along the shaft length is obtained.

Chang et al. [7] have performed Vibration analysis of rotating composite shafts containing randomly oriented reinforcements. Four cases are considered in which short fibers are aligned (i) axially (ii) in plane of composite tube (iii) in plane formed by axial and through the thickness directions and (iv) in all the three directions. Effect of volume fraction on shaft natural frequencies and the whirl frequency for all the four cases is studied. In this analysis Mori–Tanaka theory is used for modeling the discrete random non-dilute concentration of fiber reinforcement. Na et al. [40], have analyzed a non-uniform thin-walled, tapered, spinning composite beam subjected to a compressive force. Expressions for strain and kinetic energies are setup, and then applying Hamilton's principle equations of motion are obtained, which are solved for eigenvalues by extended Galerkin method. It is shown that the shaft can experience flutter and divergence type instabilities. A detailed parametric study is carried out with fiber angle varying from 0° to 90°, and the taper and compressive force parameters varying in a sufficiently large range. It is shown that by suitably adjusting the various parameters, the critical speed and the region of stability can be increased considerably.

As against the analytical approaches mentioned above, the finite element (FE) modeling due to its versatility and ease of application is quite prevalent. Wettergren [54] has performed FE modeling on composite shaft to study the effect of imperfections (variations in winding angles, volume fraction etc.). A 10 degree of freedom thin walled element developed by [55] which was further developed by the author (in an earlier publication) to allow for transverse shear and warping, has been used. Gyroscopic moments are accounted for. Also to determine deformation due to dynamic forces ABAQUS is used with double curved Shell element S8R5. The element has 8 nodes with 5 degrees of freedom per node. Weak elements have been used at the ends to avoid rigid body motion. The change in the geometry (radius of different elements on shaft cross section) due to shaft imperfections at high rotational speeds, and its effect on the shaft natural frequency, has been assessed. The changes are small but not negligible. The changes in shaft geometry under rotating condition due to imperfections may cause localized bow in the shaft, and has been explained to be the cause of increased sensitivity of composite shafts to unbalance as observed experimentally [47, 49, 57]. Chen and Peng [10, 11] have performed stability analysis under compressive steady and periodic axial force. FE model is based on Timoshenko beam theory with 5 degrees of freedom (excluding displacement in axial direction) at each node.

Chatelet et al. [9] have performed 3 dimensional FE modeling of a shaft-disc-blade system, and determined its natural frequencies and mode shapes. In Ist part of the paper FE modeling of only a composite shaft is performed. An 8 noded Ahmed type (see [1]) element with 5 dof at each node is taken. The element is based on Reissner-Mindlin hypothesis. Shear deformation is taken into account. Computational effort is reduced as follows. Firstly undamped natural frequencies and mode shapes under non-rotating condition are determined only once. These are then used in modal analysis (mode superposition). Also the structure is assumed to have cyclic symmetry, so only a sector i.e., 1/6th (60°) is discretized. The composite shaft natural frequencies are compared with three test cases. In 2nd part of the paper modeling is extended to a shaft-disc-blade system. A steel disc with integrated blades is mounted overhanging on a fiber reinforced composite shaft. Here a sector 1/8th (45°) only is meshed assuming cyclic symmetry, resulting in 711 elements and 1848 dof. Analysis showed that 1 nodal diameter disc mode exhibited strong coupling with the shaft bending modes.

Chang et al. [8] have analyzed a composite shaft rotor. The shaft strain energy is determined using the three-dimensional constitutive relations of material and the first-order shear deformable beam theory. Coupling effects due to lamination scheme, along with rotary inertia and gyroscopic effects are taken into account. Kinetic energy of discs, and strain energy from stiffness of bearings are included. Solution is obtained by FEM in which one-dimensional 3 node element with 6 dof at each node is used. Galerkin procedure is applied to obtain equations of motion which are then solved for eigenvalues and eignvectors. The rotor natural frequencies are compared with available results. Gubran [14] and Gubran and Gupta [19] have also used an Ahmed type 9 noded degenerated shell element with 5 dof per node, for analysis of thin tubular composite shafts. The element takes into account shear deformation and geometric non-linearity. The FE formulation is used for analyzing hybrid shafts with different material combinations [14] and is integrated into a simulated annealing (SA) based optimization algorithm [19]. A high speed ($\sim$ 1,20,000 rpm) carbon fiber reinforced air spindle is analyzed and designed in [2]. Dynamic analysis for estimation of bending critical speeds is performed in ANSYS using one dimensional beam elements. Static characteristics of the shaft under bending and centrifugal loading are evaluated using ABAQUS to optimize the stacking pattern and sequence. Effect of temperature rise is also taken into account in modeling.

Sino et al. [50, 51], Jacquet-Richardet et al. [25], and Montagnier and Hochard [39] have examined the effect of internal damping on dynamic behavior and stability of composite rotors. Consideration of internal damping which is associated with the rotating parts is important in rotor design, because it can be a source of instability in super-critical rotors. Composite shafts are even more susceptible because of their relatively higher internal damping coming from the matrix, and preference for super-critical operation to derive maximum benefit. FE modeling of the shaft has been performed and instability threshold speeds for the various cases are determined. These are found to be significantly sensitive to the stacking sequence, because of strong dependence of internal damping on fiber layup/stacking sequence. Provision of external passive damping is necessary to raise the instability threshold. It is shown from analysis that by proper selection of parameters, instability threshold can be raised beyond the 3rd rotor critical speed. Jacquet-Richardet et al. [25] have reported good correlation with experimental values.

Montagnier and Hochard [39] have applied FE method in ABAQUS using the thin shell type element (S4R) for the tubular shaft. The shaft of Zinberg and Symmonds is modeled. In order to refine the obtained natural frequencies, the end fitting in the shaft is modeled using linear hexahedral C3D8R type elements. Appropriate (pinned and slider) boundary conditions were applied at the center of these end fittings. The fundamental natural frequency obtained in both cases was slightly different, but matched reasonably well with other results on Zinberg and Symmond shaft reported in literature. They have also studied in detail the rotor dynamic aspects including rotor on visco-elastic supports. A comprehensive comparison of critical speed from various theories is presented isolating the effects of various factors such as rotary inertia, gyroscopic moments, transverse shear, coupling mechanisms, and rotor support conditions. Instability due to rotor internal damping is also studied. Regimes of unstable rotor behavior are obtained and the effect of support stiffness, the shaft length and stacking sequence is analyzed systematically. It is concluded that ±30° to ±45° fiber angle plies would not be suitable for super-critical operation from instability considerations. Another interesting comparison presented by authors is between the results of EMBT and ECMBT for different stacking sequences. The fact that EMBT can not predict instability (due to internal damping) threshold speed is brought out.

4 Experiments on Rotating Composite Shafts

Significant contributions have been by [56] on determination of critical speeds of a helicopter tail rotor, in [3, 4] where bending strains were measured under unbalance excitation, [47, 49, 57] showing presence of super-synchronous components at ½ and 1/3 of first critical speed, and increased sensitivity to unbalance near the critical speeds, [23, 33] applying a unified balancing approach to balance the shaft up to

2nd critical speed, and demonstrating the use of external damper to cross critical speeds, [12] achieving super-critical operation under torsional loading, [6] evaluating material constants of Paris law from delamination propagation measurements, [47] on interaction between shaft bow and unbalance which is common in composite shafts, and estimation of damping by Gupta and Singh [22].

Lee at al. [31] designed and fabricated long composite boring bars and tested them on a machining centre for condition of chatter. The rotational speed was 2,500 rpm, which was the same as for the conventional (metallic) boring bar. The overhang length of the composite bars varied in a large range from 160 to 420 mm. The onset of condition of chatter was detected by sharp audible noise, and the same was confirmed by the surface roughness measurements. Chatter was observed under identical cutting conditions, at *L/d* ratios of 8 and 10.7 for the conventional and the composite boring bars respectively. Khalid et al. [29] examined bending fatigue behavior of hybrid aluminum/composite drive shafts with various stacking sequences. The specimen was mounted in a rotating fatigue bending machine. The speed was fixed at 200 rpm with bending load in the range of 50–250 N. The important findings from the experiments are, (i) improvement in fatigue strength with increasing number of composite layers (ii) no fiber breakage observed (iii) stacking sequence $[+45/-45]_{3s}$ provided highest fatigue strength, up to 40 % more compared to other cases.

Montagnier and Hochard [34, 35] have performed extensive experiments on super-critical composite shafts with a primary objective to validate the theoretical results. A reasonable correlation between theory and experiment is reported. Ingle and Ahuja [24] fabricated a 16 layer carbon fiber composite shaft with a stacking sequence of $[0/90/45/-45]_{2s}$, and placed it on an appropriately designed aerostatic bearing. They performed vibration experiments in the range of 10,000–60,000 rpm to study the dynamic behavior. Jacquet-Richardet et al. [25], Montagnier and Hochard [39] have described a test setup for estimation of instability threshold speeds. Shafts with length ranging from 0.5 to 3 m and made of different materials, can be tested for instability.

5 Composite Shaft Optimization

In one of the earliest studies carried out by Zinberg and Symmonds [56], a boron/epoxy helicopter tail rotor drive shaft was optimized for minimum weight with critical speed, torsional buckling and strength as design criteria, which resulted in weight savings of 28 %. In early work, the focus was on rigid rotors i.e., operation in sub-critical region, therefore shaft bending stresses, vibrations and stability, and fatigue were not of significance. These factors began to acquire importance as designers explored super-critical operation (flexible rotor) in order to achieve higher weight savings. Most of the optimization work in early period was based on heuristics. A comprehensive review of literature up to 1996 is presented by Singh et al. [45].

Kim et al. [28] developed a model of a tapered rotating composite shaft using Timoshenko beam theory. Equations of motion are solved by Galerkin method for free vibration response. The model represents a high speed extended length cutting tool. It is shown from analysis that the natural frequency as well as the static stiffness can be increased considerably by providing a taper. The idea of a flexible exponential taper to meet the specific requirements of a cutting tool is explored. Salzer [43] examined design of metal matrix composite (MMC) shafts for high temperature applications, such as an aerogas turbine engine shaft. High strength titanium alloy is considered as MMC in two forms, the monolithic and with 35 % SiC fibers in which case a weight saving of 11.6 % (compared to the monolithic case) is achieved. The approach for developing an MMC shaft for high temperature applications appears promising.

Perhaps the first systematic attempt towards rigorous mathematical procedure was by Gubran [13] where he formulated the composite shaft optimization problem in terms of multi-objective functions and constraints. Simulated annealing (SA) was used to arrive at an optimum solution. Gubran and Gupta [16, 17] have given details of the optimization procedures described above with case studies. In the 1st paper, a two stage optimization procedure is illustrated i.e., maximization of natural frequencies in 1st stage followed by minimization of weight in 2nd stage. In 1st stage the constraints are imposed on shaft buckling torque and torsional strength. The maximized natural frequency obtained in the 1st stage forms the constraint in the 2nd stage. Four cases from the various combinations of uniform/variable (U/V) wall thickness, and uniform/variable (U/V) fibre lay up are explored. Results show maximum advantage obtained from the variable wall thickness and variable fiber lay up (VV) design. In 2nd paper the strength criteria is added as an additional constraint to weight minimization problem. Tsai-Wu failure criterion is applied to stresses due to torque transmission and unbalance (bending) response. Above approach is further extended to setup a multi-objective optimization problem, where in addition to minimization of shaft weight, an objective of minimization of maximum stress along shaft length is also added. The constraints are on buckling torque and strength.

High speed aerostatic bearing systems are used because of their low friction and low heat generation. However air being the medium, their radial and axial stiffness is low and therefore heavy metallic spindles are not suitable. Lee and Choi [30] developed a light weight spindle with carbon fiber expoxy. Its mass and mass moment of inertia were only 36 and 29 % of that of a steel spindle. With proper stacking sequence the fundamental mode frequency could be raised beyond the operating speed of 40,000 rpm. The damping could be increased 5 times. Dynamic characteristics of the composite spindle were shown to be much superior to a metallic (steel) spindle. The specific requirements of a high speed spindle designed by Bang and Lee [2] to operate at 1,20,000 rpm are its low inertia, high damping and high stiffness; composite material being most suitable for all the three required characteristics. The thickness and the stacking sequence were worked out on the basis of bending stresses and maximization of critical speeds. Shaft was reinforced in the circumferential direction to increase its radial stiffness. Following an FE analysis in ABAQUS, stacking sequence of $[(90)_2/(\pm 5)_4/(90)_2/(\pm 5)_4]_T$ was arrived at. Safety of the high speed composite shaft was evaluated on the basis of residual thermal stresses, bending loads and centrifugal force.

In order to be able to bore deeper holes Lee et al. [31] have designed a composite shaft aimed to have a *L/d* ratio more than 10 with no chatter. According to the authors, conventional boring bars usually begin to chatter when *L/d* ratio exceeds a value of about 5. Typically the boring bar is designed to have a damping core, covered by carbon fiber composite layers in sufficient thickness, and an outer steel cover with a layer of adhesive in between. Thus it is attempted to maximize its damping and stiffness. Thirteen core materials were explored. Finally a lead core inside a thin stainless steel pipe of 0.5 mm thickness was found to provide the largest damping ratios. Steel pipe was necessary because a bare lead surface was not suitable for directly wrapping of prepregs. An increase of 72, 168 and 28 % (compared to a conventional boring bar) in natural frequency, damping ratio and dynamic stiffness respectively was achieved. The composite boring bar could be used up to *L/d* ratio of 10.7 without any indication of chatter. A one piece hybrid aluminum/composite drive shaft for automotive applications is developed [32]. The unique feature of the design is that the carbon fiber epoxy composite layer was stacked on the inner surface of the aluminum tube, with an epoxy glass fiber prepreg layer in between as an insulator for possible galvanic corrosion between aluminum and the carbon composite layer. A 75 % mass reduction, and 160 % increase in torque capability compared with a conventional two-piece steel drive shaft, was achieved.

The question of sub-critical versus the super-critical design solutions for drive shafts is addressed by Montagnier and Hochard [37]. It is brought out clearly that the super-critical designs provide a superior optimum solution as compared to sub-critical designs. However the super-critical shaft has to negotiate a resonance to reach its operating speed, and is also susceptible to instability. Genetic algorithm based approach is adopted for optimum design of super-critical drive shafts [36]. Several case studies are performed. The well established rules of fiber lay up for drive shafts are reconfirmed. According to these general rules provision should be made for high strength 45° plies for maximum torsional strength, 90° plies for maximum buckling torque, and high modulus 0° plies for maximum bending stiffness and minimum damping.

Roos and Bakis [42] used genetic algorithm approach for a helicopter driveshaft optimization which involves determination of the optimal stacking sequence, number of plies and number of in-span bearings for minimum overall weight. Various practical conditions such as variable torque and speed, shaft misalignment and different load scenarios are taken into consideration in analysis. Authors claim this analysis to be the first one, where self-heating of a driveshaft is evaluated, and dependence of material properties on frequency and temperature is accounted for. Optimization is performed on two systems, the Blackhawk and Chinook drivelines. Weight savings of the order of 20 % are shown. Some important findings from the paper are, (i) whirling instability being dependent on maximum bearing span, which decides upon the minimum number of in-span bearings required (ii) that the minimum number of plies required is governed by maximum stresses and buckling (iii) shaft overheating strongly effects stacking sequence, and typically precludes fiber angle orientations of about 20°. It is also shown that for the case of constant power transmission in Blackhawk driveline, a 30 % weight saving is possible by suitably adjusting ratio of rotational speed and driving torque.

Montaignier and Hochard [38] performed optimization of a hybrid helicopter tail rotor composite shaft operating at sub- and super- critical speeds. Genetic algorithm is used. Results show the feasibility of reduction in number of supports, thereby considerably reducing the overall weight of the driveline. Following general guidelines for fiber lay up are suggested to facilitate the optimization procedure i.e., provision of (i) ±45° high strength (HS) plies for torque transmission (ii) 0° high modulus (HM) fibers for maximum axial stiffness and minimum damping (iii) 90° high modulus (HM) plies farthest from the shaft middle surface for buckling torque. Simulated annealing (SA) approach has been applied in [19] for weight optimization of automotive composite propeller shaft. The constraints are on shaft bending natural frequency, buckling and transmission torque capacity and the shaft outer diameter (limited to 90 mm). Design variables are ply thickness, angles and the stacking sequence. A comparison is presented between six shafts of different materials (steel, aluminum, graphite/epoxy and glass/epoxy) and combinations. The primary requirements are a critical speed more than 6,500 rpm and torque capacity of at least 3,500 Nm, shaft length of 1.25 m. These are same as used by Rangaswamy et al. (2002). Results show that (i) a weight saving of 88 % for carbon-epoxy, and 73 % for glass epoxy shaft is possible (ii) a single piece hybrid (glass-epoxy/aluminum) is feasible since its natural frequency is 40 % higher than that of the metallic shaft.

6 Future Directions

Research on composite rotors has come a long way over the past several decades. As outlined above efforts are continuing on various aspects which are important for full development of composite rotor systems. Considerable progress has been achieved in understanding of several phenomena. In spite of these, the applications of composite rotors to engineering systems even today remain as limited as ever before. There is a need to examine if gaps between theory and practice persist. Improvements are required to make modeling of composite shafts more realistic. Efforts are required in the following directions.

(i) Presently FE modeling of composite shafts is mostly through conventional beam, plate and shell elements. There is a need to develop larger variety of finite elements specifically for laminated composite shafts.
(ii) In present studies the focus is on predicting dynamic characteristics of the shaft. While the dynamic phenomenon is predicted well, not much is known about any happenings in the interior of wall thickness. A much finer mesh, compared to ones being used in present dynamic studies, can be generated which could model special features of a composite in the interior and predict simultaneously the state therein. These could be various defects like the

delamination, fiber breakage, matrix degradation, imperfections in the form of variations in fiber angles and volume fraction etc. Whether this approach of moving from 'macro-mechanical' to 'micro-mechanical' modeling for composite shaft dynamics will be worth the additional complexity and computational effort, needs to be assessed.

(iii) To assess the reliability of shaft design, the modeling must include its interaction with the surrounding medium. Such factors as heat generated with in the composite shaft, its dissipation in the surrounding medium, resulting temperature profiles on the surface and in the interior, dependence of material properties on temperature, and material degradation are important factors which are required to be in-built in the rotor dynamic system model.

(iv) More research is required on interaction of shell modes and flexural (bending) modes in thin walled rotating composite shafts. There are several phenomena which have not received attention. For example mounted parts such as discs will locally suppress the shell modes at several axial locations along the shaft; what are its implications. The precise dynamic behavior, and any happenings in the interior are not fully understood when a thin walled shaft is made to rotate at extremely high speed.

(v) There should be more research in the direction of integral rotors (composite shaft-disc-blade system). Apart from the fact that such systems being large because of their size and complex geometry, pose the well known problems associated with the very large dof, there are important issues specific to composite rotors. The nature of fiber reinforcement in the shaft, disc and blades is very different, requiring special consideration. Choice of finite elements for different parts will be crucial for an accurate and realistic modeling. Similarly metal matrix composites (MMC) are inherently different from resin (epoxy) based composites, and will require special considerations in modeling.

(vi) The focus of research on composite rotors till now has been on epoxy based composite shafts, and not so much towards MMC shafts. In foreseeable future the applications of MMC shafts in various systems involving high temperatures (for example gas turbines), are expected to increase. That will demand greater attention of researchers towards modeling of MMC. Is a 'nonmetallic aeroengine' [26], or an 'all composite aero gas turbine' [21] realizable?

For high temperature applications, in addition to metal matrix composites, other types like ceramic or glass matrix based, or carbon-carbon composites will require careful modeling. Considerable efforts will be required from material scientists in developing these materials, and in evaluating their mechanical properties.

(vii) Greater application of smart materials such as shape memory alloy (SMA), magnetorheological fluids (MRF), piezoelectric fibers/patches etc. to composite rotors, will receive the attention of researchers. The powerful combination of 'composite' and 'smart' materials will pave the way for development of what may be called as 'intelligent/smart rotors'.

References

1. Ahmad S, Zienkiewicz OC, Irons B (1970) Analysis of thin and thick shell structures by curved finite elements. Int J Numer Meth Eng 2:419–451
2. Bang KG, Lee DG (2002) Design of carbon fiber composite shafts for high speed air spindles. Compos Struct 55:247–259
3. Bauchau OA (1981) Design, manufacturing and testing of high speed rotating graphite/epoxy shafts, Doctorate thesis, Department of Aeronautics and Astronautics, MIT
4. Bauchau O (1983) Optimal design of high speed rotating graphite/epoxy shafts. J Compos Mater 17(3):170–181
5. Bert CW (1993) The effect of bending-twisting coupling on the critical speed of a driveshaft. In: Proceedings 6th Japan-US conference on composite materials, Orlando, FL, Techonomic, Lancaster, PA, pp 29–36
6. Chandramouli G, Gupta K, Pandey RK (1994) Delamination propagation in rotating carbon-epoxy composite shafts. Eng Fract Mech 49(1):121–132
7. Chang C-Y, Chang M-Y, Huang JH (2004) Vibration analysis of rotating composite shafts containing randomly oriented reinforcements. Compos Struct 63:21–32
8. Chang M-Y, Chen J-K, Chang C-Y (2004) A simple spinning laminated composite shaft model. Int J Solids Struct 41:637–662
9. Chatelet E, Lornage D, Jacquet-Richardet G (2002) A three dimensional modeling of the dynamic behavior of composite rotors. Int J Rotating Mach 8(3):185–192
10. Chen L-W, Peng W-K (1998) Dynamic stability of rotating composite shaft under periodical axial compressive loads. J Sound Vib 212(2):215–230
11. Chen L-W, Peng W-K (1998) The stability behavior of rotating composite shaft under axial compressive loads. Compos Struct 41:253–263
12. Frederick JR, Darlow MS (1991) Operation of torsionally loaded composite shaft above two flexural critical speeds. In: 13th biennial conference on mechanical vibration and noise, presented at 1991 ASME design technical conference, Miami, FL, USA, vol 36, pp 247–255
13. Gubran HBH (2000) Dynamic stress analysis and optimization studies on fibre- reinforced composite shafts. In: Ph.D. thesis, mechanical engineering department, Indian Institute of Technology (IIT) Delhi, India
14. Gubran HBH (2005) Dynamics of hybrid shafts. Mech Res Commun 32(4):368–374
15. Gubran HBH, Bahashwan AA, Gupta K (2002) Dynamic stress analysis in fibre- reinforced composite shafts with axially varying fibre layup and wall thickness. Univ Aden J Nat Appl Sci 6(1):131–146
16. Gubran HBH, Gupta K (2002) Composite shaft optimization using simulated annealing part i: natural frequency. Int J Rotating Mach IJRM 8(4):275–284
17. Gubran HBH, Gupta K (2002c) Composite shaft optimization using simulated annealing part ii: strength and stresses. Int J Rotating Mach IJRM 8(4):285–293
18. Gubran HBH, Gupta K (2005) The effect of stacking sequence and coupling mechanisms on the natural frequencies of composite shafts. J Sound Vib 282:231–248
19. Gubran HBH, Gupta K (2014) Design optimization of automotive propeller shafts. J Vib Eng Technol 2(1):35–45
20. Gubran HBH, Singh SP, Gupta K (2000) Stresses in composite shafts subjected to unbalance excitation and transmitted torque. Int J Rotating Mach 6(4):235–244
21. Gupta K (1998) Some issues relating to design and development of an all composite aero gas turbine engine rotor. Def Sci J 48(1):93–98
22. Gupta K, Singh SP (1998) Damping measurements in fiber reinforced composite rotors. J Sound Vib 211(3):513–520
23. Hetherington EL, Kraus RE, Darlow MS (1990) Demonstration of a super critical composite helicopter power transmission shaft. J Am Helicopter Soc 35(1):23–28
24. Ingle RB, Ahuja BB (2006) An experimental investigation on dynamic analysis of high speed carbon–epoxy shaft in aerostatic conical journal bearings. Compos Sci Technol 66:604–612

25. Jacquet-Richardet G, Chatelet E, Nouri-Baranger T (2011) Rotating internal damping in the case of composite shafts. In: Proceedings of IUTAM symposium on emerging trends in rotor dynamics, IIT Delhi, India
26. Jeal RH (1988) Meeting the high temperature challenge-the nonmetallic aeroengine. Metals Mater 4:539–542
27. Jones RM (1999) Mechanics of composite materials, 2nd edn. Taylor & Francis Inc, Philadelphia
28. Kim W, Argento A, Scott RA (1999) Free vibration of a rotating tapered composite timoshenko shaft. J Sound Vib 226(1):125–147
29. Khalid YA, Mutasher SA, Sahari BB, Hamouda AMS (2007) Bending fatigue behavior of hybrid aluminum/composite drive shafts. Mater Des 28:329–334
30. Lee DG, Choi JK (2000) Design and manufacture of an aerostatic spindle bearing system with fiber-epoxy composites. J Comput Mater 34:1150–1175
31. Lee DG, Hwang HY, Kim JK (2003) Design and manufacture of a carbon fiber epoxy rotating boring bar. Compos Struct 60:115–124
32. Lee DG, Kim HS, Kim JW, Kim JK (2004) Design and manufacture of an automotive hybrid aluminum/composite drive shaft. Compos Struct 63:87–99
33. Lim JW, Darlow MS (1986) Optimal sizing of composite power transmission shafting. J Am Helicopter Soc 31(1):75–83
34. Montagnier O, Hochard C (2005a) Design of supercritical composite helicopter driveshafts: theoretical and experimental study. In: European conference for aerospace sciences (EUCASS)
35. Montagnier O, et Hochard C (2005b) Etude théorique et expérimentale de la Dynamique des arbres de transmission supercritiques, 17èmecongrès Français de Mécanique, Troyes, 29 Août-2
36. Montagnier O, Hochard Ch (2008) Optimization of supercritical carbon/epoxy drive shafts using a genetic algorithm. In: 13th European conference on composite materials, Stockholm, Sweden
37. Montagnier O, Hochard Ch (2006) Design of hybrid high modulus/high resistance carbon fibers driveshafts—subcritical and supercritical solutions. In: 12th European conference on composite materials, Biarritz, France
38. Montagnier O, Hochard C (2013) Optimisation of hybrid high-modulus/high-strength carbon fibre reinforced plastic composite drive shafts. Mater Des 46:88–100
39. Montagnier O, Hochard C (2014) Dynamics of a supercritical composite shaft mounted on viscoelastic supports. J Sound Vib 333:470–484
40. Na S, Yoon H, Liviu L (2006) Effect of taper ratio on vibration and stability of a composite thin-walled spinning shaft. Thin Walled Struct 44:362–371
41. Rangaswamy T, Vijayarangan S, Chandrashekar RA, Anantharaman K (2004) Optimal design and analysis of automotive composite drive shaft. In: Proceeding of international symposium of research students on materials science and engineering, Madras, India
42. Roos C, Bakis CE (2011) Multi-physics design and optimization of flexible matrix composite driveshafts. Compos Struct 93:2231–2240
43. Salzar RS (1999) Design considerations for rotating laminated metal-matrix-composite shafts. Compos Sci Technol 59:883–896
44. Singh SP (1992) Some studies on dynamics of composite shafts. In: Ph.D. thesis, mechanical engineering department, Indian Institute of Technology (IIT) Delhi, India
45. Singh SP, Gubran HBH, Gupta K (1997) Developments in dynamics of composite material shafts. Int J Rotating Mach 3(3):189–198
46. Singh SP, Gupta K (1994) Free damped flexural vibration analysis of composite cylindrical tubes using beam and shell theories. J Sound Vib 172(2):171–190
47. Singh SP, Gupta K (1995) Experimental studies on composite shafts. In: Proceedings of the international conference on advances in mechanical engineering, Bangalore, India, pp 1205–1221
48. Singh SP, Gupta K (1996) Composite shaft rotordynamic analysis using a layerwise theory. J Sound Vib 191(5):739–756

49. Singh SP, Gupta K (1996) Rotordynamic experiments on composite shafts. ASTM J Compos Technol Res 18(4):256–264
50. Sino R, Baranger TN, Chatelet E, Jacquet G (2008) Dynamic analysis of a rotating composite shaft. Compos Sci Technol 68:337–345
51. Sino R, Chatelet E, Baranger TN, Jaquet-Richardet G (2006) Stability analysis of internally damped rotating composite shafts considering transversal shear. In: Proceedings of ISROMAC—11, Honolulu, Hawaii USA
52. Tsai SW (1988) Composites design, 4th edn. Ohio, USA
53. Vance John M (1988) Dynamics of turbomachinery. Wiley, New York
54. Wettergren HL (1997) Influence of imperfections on the eigenfrequencies of a rotating composite shaft. J Sound Vib 204(1):99–116
55. Wu XX, Sun CT (1989) Vibration of thin walled composite structures. In: Vibration of Composite Structures, 14. ASME, New York, pp 47–52
56. Zinberg H, Symmonds MF (1970) The development of an advanced composite tail rotor driveshaft. In: Presented at 26th annual national forum of American helicopter society, Washington, DC
57. Zorzi ES, Giordano JC (1985) Composite shaft rotor dynamic evaluation. In: Presented at ASME design engineering conference on mechanical vibrations and noise, Sept. 1985, ASME paper no. 85-DET-114, pp 1–8

Part II
Contributed Papers/Chapters

Investigation of the Synchronous Response in a Back-Back Centrifugal Compressor Due to Imbalance and Seal Distortion

Mohamed Al Yahyai and David Mba

Abstract This paper presents a case study on the effect of fouling and damaged long-span labyrinth seals on the rotor dynamics of a high-speed gas centrifugal compressor. This back to back, two stage, nine-impeller compressor is used to compressor natural gas associated with oil production in a particular field within the Sultanate of Oman. The case study investigated involved replicating vibration problems experienced in the field with a rotor-dynamic model so as to understand and demonstrate the influences of seal wear and hydrocarbon fouling on the compressor synchronous vibration.

Keywords Gas compressor · Rotordynamics · Fouling · Seal wear · Vibration

1 Introduction

Mature and depleted oil fields require unconventional technical solutions for Enhanced Oil Recovery (EOR). One such solution requires the use of turbomachinery particularly centrifugal compressors. The increasingly difficult production environments due to deep oil wells, reservoir formation, gas properties, viscous crude oil and high Hydrogen Sulfide (H_2S) contents all influence the need for more powerful and reliable turbomachinery that can handle extreme head, flow rates and unpredictable fluid properties. The cost associated with unplanned downtime of gas lift compressors is extremely high [5] and causes of such downtime include high vibration levels.

M.A. Yahyai (✉) · D. Mba
Department of Power and Propulsion, School of Engineering, Cranfield University, Cranfield, Beds MK43 0AL, UK
e-mail: m.k.alyahyai@cranfield.ac.uk

D. Mba
e-mail: D.Mba@cranfield.ac.uk

J.K. Sinha (ed.), *Vibration Engineering and Technology of Machinery*, Mechanisms and Machine Science 23, DOI 10.1007/978-3-319-09918-7_7

Generally centrifugal compressor rotor dynamic problems are caused by two main sources, forced response and self-induced vibration. The former source is driven by the residual unbalance and process fouling while the latter is driven by the cross coupling forces generated at the impeller stages and labyrinth seals. A detailed explanation of how to undertake forced response and self-induced vibration analysis in centrifugal compressor is provided by the American Petroleum Institute [2]. The open literature reveals many case studies where forced response and self-induced vibration (rotor instability) are reported due to a variety of causes such as poor balancing, fouling, seal design degradation and ineffective or worn bearings [1, 4, 7].

2 Case Study: Compressor Description

The gas lift centrifugal compressor train consists of two back-to-back Low Pressure (LP) and High Pressure (HP) compressors driven by a 9.24 MW brushless synchronous electric motor via a speed increasing fluid coupling gearbox (Fig. 1). The compressor train has a rated power of 9 MW with a Rated Maximum Continuous Operating Speed (RCOS) of approximately 8,800 rpm while the Maximum Continuous Operating Speed (MCOS) is approximately 9,500 rpm with an overall pressure ratio of 45 (LP ratio of 5 and HP ratio of 9). The LP compressor is divided into two stages (1st and 2nd) with four closed impellers at each stage acting in opposite directions (back-to-back configuration). The HP Compressor is also divided into two opposed stages (3rd and 4th) having four impellers on the 3rd stage and five impellers on the 4th stage. As both compressors utilise the back-to-back rotor configuration they employ an interlocked mid-centre labyrinth seal between the two stages and two long span abradable Tooth-on-Rotor (TOR) labyrinth seals located at the rotor's Drive End (DE) and Non Drive End (NDE). Dry Gas Seals (DGS) are used to seal the compressed gas from leaking into the atmosphere,and the bearings. The DGS is located in-between the TOR seal and the journal bearing. For both compressors the impeller shroud uses Tooth on Stator (TOS) labyrinth seals with no swirl brakes. The LP and HP compressors both use 4-pad tilting

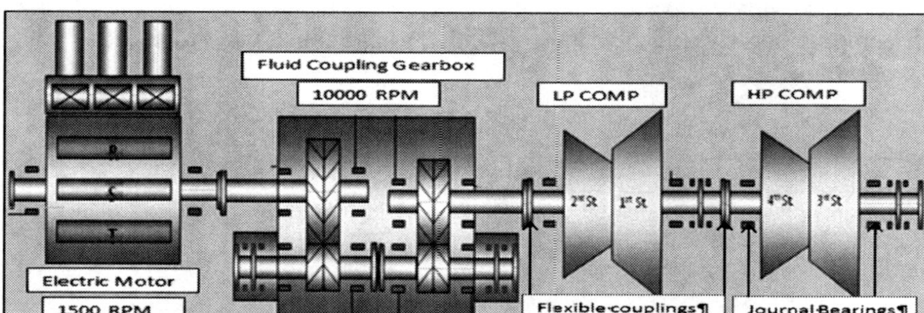

Fig. 1 Gas lift compressor layout

journal bearings with a Load-Between-Pivot (LBP) configuration. The thrust bearings employed were tilting pad (8-pads) journal bearing with the Load-Between-Pivot (LBP). Finally, three flexible couplings were used between the prime mover (Motor) and Gearbox, Gearbox and LP compressor, and, LP and HP compressor see Fig. 1.

3 Vibration Problems and Field Observation

After commissioning, the unit operated for a year without any major performance or vibration problems. At 12-months of operation, the compressor performance degraded to a point where it entered intermittent surge with high vibration levels noted at HP NDE (Table 1 and Fig. 3). Table 1 shows the compressor thermodynamic performance at optimum and degraded conditions, the optimum data was taken directly after a major overhaul (December 2011) where most critical parts such as compressor rotor, dry gas seals and bearings were replaced while the degraded data was taken prior to the unscheduled major overhaul (November 2012). Pressure, temperature and overall efficiency of each stage were used to monitor the compressor performance on a daily basis. The data in Table 1 shows the efficiencies of the various stages under optimum and degraded conditions.

The compression train had several vibration problems with the HP compressor throughout the degradation period (Sept 11 to Dec 12) whilst the LP compressor maintained relative stable rotordynamic behaviour. For the HP compressor, vibration problems were associated with start-ups and shutdowns such that whilst traversing the HP compressor critical speed a temporary trip multiplier was used to ensure continuous operation irrespective of the high vibration amplitude. Figures 2 and 3 show corresponding bode plot of the HP Drive End (DE) and Non Drive End (NDE) respectively taken from proximity probes in January 2012, at the start of the gradual reduction in efficiency and prior to the overhaul.

From Fig. 2, it can be seen that the amplitude at critical speed at the DE side was approximately 38 μm, below the specified alert setting of 40 μm whilst at the NDE probes the vibration amplitude reached a value of 52 μm close to the trip setting of 55 μm. Moreover, Fig. 2 shows that the vibration amplitude at Maximum Continuous Operating Speed (MCOS) at DE side was 13 μm whilst the amplitude at the NDE side was relatively high at 30 μm. Figure 3 shows a similar bode plot in Nov

Table 1 Compressor thermodynamic performance degradation

Stg	Efficiency at (optimum condition (December 2011)) (%)	Efficiency at degraded condition (November 2012) (%)
LP	93	86
LP2	91	84
LP3	70	51
HP	66	46

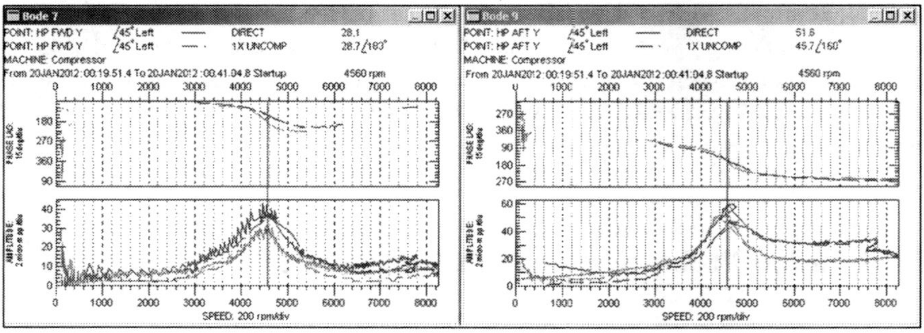

Fig. 2 Bode plot for HP compressor DE (*right*) and NDE (*left*)

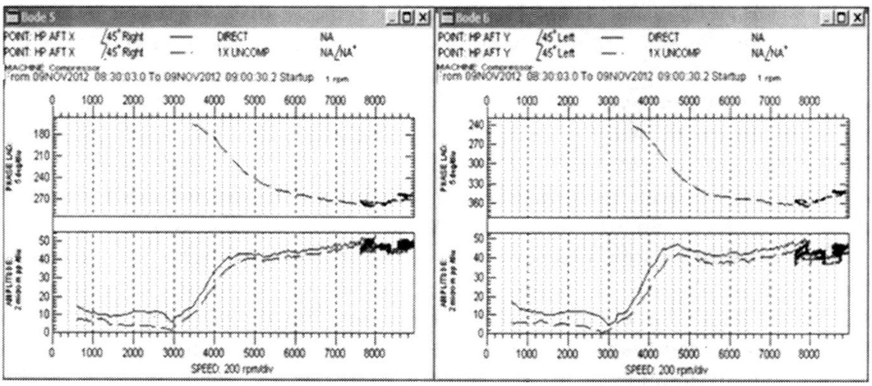

Fig. 3 Bode plot for HP compressor NDE

2012, one month before shutdown for overhaul (Nov 2012). By this period at the MCOS, the vibration amplitude at the NDE had reached 48 μm. Essentially the run-up vibration data in Nov 2012 showed vibration levels at MCOS exceeded that measured at critical speed (NDE) see Fig. 3. The vibration amplitude under this condition of operation is intolerable for long-term operation with high probability of rotor damage (impellers, seals and bearings) as well as other sealing components such as dry gas seal. Also noted in November 2012 was that the vibration level at the critical speed was lower than that measured in January 2012.

4 Rotordynamic Model and Results

A rotor–bearing model was generated to understand the root cause of the high vibration levels from the HP compressor. The rotor model was built and analysed using 'XLRotor' which employs a Polynomial transfer matrix method to solve for

Undamped Critical Speeds (UCS), Eigenvalue and Eigenvector, Imbalance Response... etc. [12]. The Transfer Matrix method is widely used by turbomachinery designers and manufacturers for rotordynamics analyses. Hozler [6], initially developed the Linear Transfer Method then Myklestad improved it in 1944 for airplane wings [10] Prohl [11] in 1945 adapted it for steam turbines. In 1974 Lund applied it with adaptation for rotor bearing system analyses. Finally, Murphy improved the transfer matrix method using the Polynomial method overcoming some limitation in application [9]. The aerodynamic excitation forces from the impellers, which influence rotor dynamic characteristics, were calculated using the Dimensional form of Wachel's equation [13]. The HP rotor was modelled using 69 stations (Fig. 4) comprising the two stages, nine impellers, seals and mass and inertias of each impeller. Bearings Characteristics were analysed using 'XLHydroDyn' which provides a full film solution of Reynolds equation [8]. The bearing stiffness and damping terms were calculated for different operating conditions in the rotor dynamic analysis. Labyrinth seals forces were determined by employing the 'one control volume' model presented by Childs [3]. Seal clearances were also varied from a minimum of 0.2 mm to a maximum of 0.9 mm.

An unbalance response analysis was conducted in accordance with the American Petroleum Institute API 617, 7th Edition [1]. The level of unbalance used for the rotor weight (W) of 683 kg and speed (N) of 8,600 RPM is 504 g-mm based on the API specification. This unbalance mass was placed at an anti-node location so that the first mode was excited. The anti-node location is the mid span of the rotor, which corresponds to station 38. The unbalance response analysis was carried without considering the effect of impellers and labyrinth seals forces in order to compare the results against the Factory Acceptance Test (FAT) (referred to as 'no-load' test). The results showed peak displacement amplitude of approximately 2.7 μm at DE, and at the NDE was 2.2 μm whilst 22 μm was noted at rotor mid-span. Table 2 shows that both critical speed and Amplification Factor (AF) compares well with the Factory test results and within the API recommended numerical

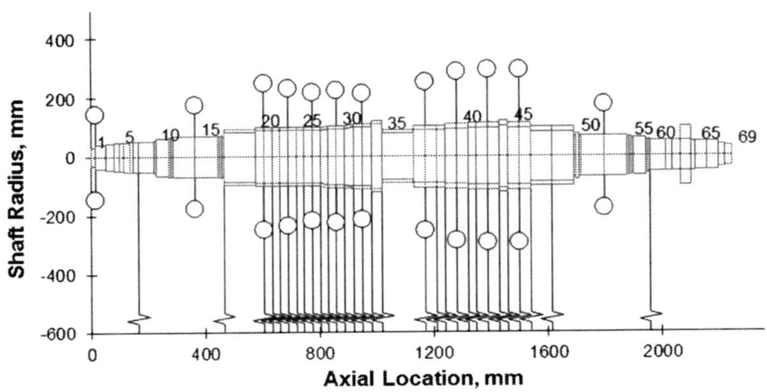

Fig. 4 HP rotor model

Table 2 FAT test against numerical model comparison

	Factory acceptance test (FAT)	Numerical model
1st critical speed	4,802 (RPM)	4,833 (RPM)
Amplification factor	3.75	3.92

model variance of 5 %. The calculated first critical speed was estimated at 4,833 rpm with an Amplification Factor (AF), based on half power point of 3.92.

To simulate the condition of the HP compressor that represents the inoperable state (trip) as noted in November 2012, different scenarios were considered which included simulations for varying bearing and seal clearance conditions (indicative of wear states). The bearing clearance ranged from 0.15 mm to 0.18 mm. A sensitivity analyses was conducted to explore the effect of impellers and seals cross coupling forces on the rotor synchronous response while maintaining the same initial unbalance amount (504 g-mm). Figure 5 shows the synchronous response of the DE for a variety of conditions on the HP that included the response without inclusion of the impellers and seal forces at minimum bearing clearances and an unbalance of 504 gm-mm. A maximum displacement of 5.6 μm was noted for the DE under rotor model conditions that included the influence of seal and impeller forces The inclusion of the long span labyrinth seals and impeller forces reduced the critical speed to 4416 rpm which compares well with the field data critical speed of 4570 rpm and with a corresponding measured amplitude of 6.5 μm at the DE. Figure 5 shows that the labyrinth seals and impeller forces increased the response amplitude prior to and whilst traversing critical speed but had no effect at Maximum Continuous Operating Speed (MCOS).

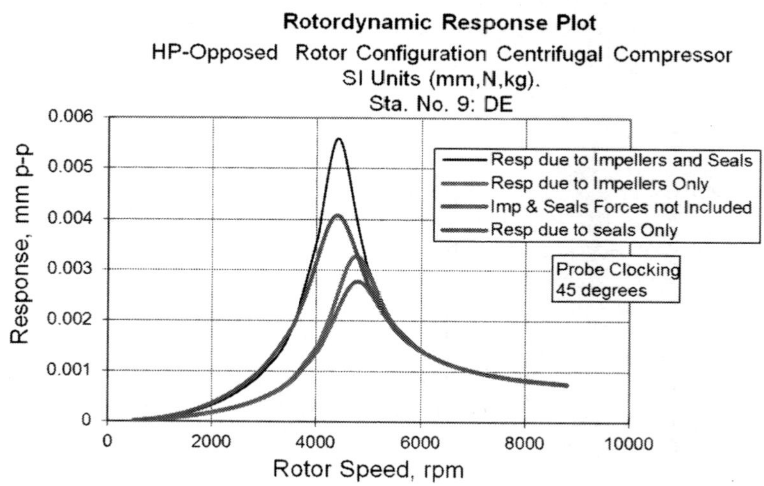

Fig. 5 Synchronous response analysis at varying conditions

The model allowed for an investigation on the effect of minimum and maximum labyrinth seals clearance for the same amount of unbalance and bearing clearance used in the previous analysis. From analysis it was observed that higher vibration amplitude levels were noted at the critical speed for minimum seal clearances as compared to that of worn seals (maximum) while the same amplitude was observed after traversing the critical speed for both cases. An increase in amplitude of up to 400 % at the critical speed was noted for the case with minimum seal clearance as opposed to maximum seal clearance.

To simulate the rotor dynamic conditions of the degraded compressor the DE bearing was kept at its minimum clearance while the NDE bearing was kept at its maximum clearance. Simultaneously all seals clearances were set to an out-of-tolerance value (0.9 mm) while varying the amount of unbalance along the rotor. Different unbalance values and associated locations were considered starting with an equal unbalance distribution along the rotor, concentrated unbalance at mid-span position, single sided unbalance locations, etc. The simulation which provided the closest match of the compressor rotordynamic behaviour prior to the inoperable condition was achieved by placing a total imbalance amount of 6,900 g-mm distributed between the 3rd stage first impeller and 4th stage first impeller, with 90 % of the unbalance placed on the 3rd stage (1st impeller) and 10 % on the 4th stage 1st impeller. Comparative results of the distributed and targeted unbalance analysis are presented in Figs. 6 and 7. This shows that the equally distributed unbalance produces higher amplitude at critical speed than the targeted condition. Also the equally distributed unbalance condition showed decreasing vibration amplitude after critical speed whilst the targeted unbalance showed an increasing vibration level away from the critical speed as was noted on the operational data (see Fig. 3) particularly at the NDE.

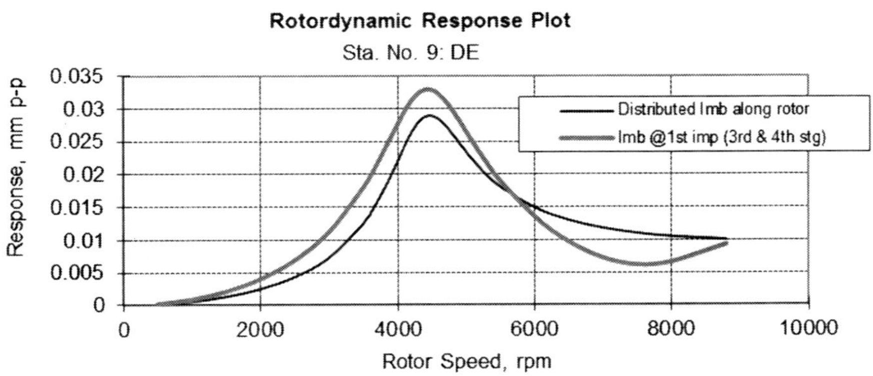

Fig. 6 Equal and sided distributed imbalance response plot at DE

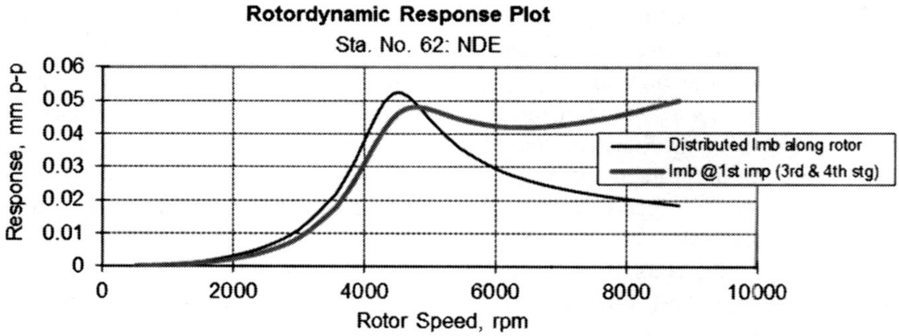

Fig. 7 Equal and sided distributed imbalance response plot at NDE

5 Discussion

A major overhaul was undertaken in January 2013 and different observations were noted such as carbon deposit on the impellers and dry gas seal oil contamination. In addition, seal clearances across the rotor were measured at 1.2 mm, above the 0.9 mm maximum limit, which is indicative of wear. Figure 8 (left) shows the condition of the NDE Stationary seal after 1 year of operation and its corresponding new seal (right); comparing the two conditions it can be seen that the de-swirl chambers in the old seal are fully blocked by the carbon deposit.

Figure 9 shows both DE and NDE long span Labyrinth seals with carbon deposits evident at both ends especially at the impeller eye. This observation explains the HP compressor performance degradation as well as the high vibration amplitude. The formation of the carbon deposit within the impeller eye restricted the inlet flow and this in turn affected the efficiency of the unit. In addition, the carbon deposit affected the rotor balance state, which resulted in high vibration amplitude. Typically, fouling occurs mainly in the inlet of the 3rd and 4th stage impellers and both are located close to the long span end labyrinth seals.

Fig. 8 Fouled stationary NDE seal (*Left*), New NDE seal (*Right*)

Fig. 9 Worn DE (*left*) and NDE (*right*) Labyrinth seal and carbon deposit formation on the impeller eye

Using the observations discussed above, the sequence of the compressor degradation and failure becomes clearer. At the earliest stages of degradation, the initial high vibration amplitude was caused by unbalance and cross coupling forces generated by the labyrinth seals. At this initial stage, the compressor operators had to engage the 'trip multiplier' to pass the critical speed. The trip multiplier essentially allows the operators to increase the vibration trip limit. However, with time, the degradation of performance continued to such an extent that the compressor went in and out of surge and to overcome this process instability the compressor speed automatically varied from maximum continuous operating speed to 3,000 rpm, i.e., transversing the critical speed numerous times (about twice a week) in an attempt to overcome the surge problem. However, by continuously traversing the critical speed the compressor experienced the high vibration levels described and this in turn contributed to the gradual wear of the labyrinth seals along the rotor span. As the seals gradually lost their optimum clearances, the cross coupling forces were altered and this in turn reduced the rotor vibration at critical speed, allowing the operators to transverse the critical speed without using the 'trip-multiplier'. However, the continuous accumulation of deposits on the impellers caused an increase in unbalance that eventually reached a point where at MCOS the vibration amplitude exceeded the allowable limit.

6 Conclusion

A variety of rotor dynamic models were used to accurately predict the high pressure centrifugal compressor synchronous response where the effect of rotor unbalance due to natural gas fouling and seal clearances were considered. The numerical model developed compared well with the compressor field data for the variety of conditions analysed. The results show that rotor unbalance response behaviour depends on the unbalance amount, unbalance locations as well as the labyrinth seals clearance. The seal wear rate plays a vital role in increasing or decreasing the vibration amplitude particularly when traversing the compressor critical speed. The high level of vibration at MCOS was attributed to the unbalance mass due to fouling at the inlet stages.

References

1. API 617, 7. E. (2002) Axial and centrifugal compressors and expander-compressors for petroleum chemical and gas industry services, 7th edn. American Petroleum Institute, USA
2. API684, 5. E. (2005) Rotordynamic tutorial: lateral critical speeds, unbalance response, stability, train torsionals, and rotor balancing, 5th ed, American Petroleum Institute, USA
3. Childs D (1993) Turbomachinery rotordyanmics phenomena modeling, & analysis. Wiley, USA
4. Collins D, Teixeira J, Crudgington P (2008) The degradation of abradable honeycomb labyrinth seal performance due to wear. Sealing Technol 8:7–10
5. Ertas BH, Delgado A, Vannini G (2012). Rotordynamic force coefficients for three types of annular gas seals with inlet preswirl and high differential pressure ratio. J Eng Gas Turbines Power 134:4
6. Heinrich H (1921) Die Berechnung der Drehschwingungen und ihre Anwendung im Maschinenbau, J, Springer
7. Lees AW (2006) Agnes Muszynska, Rotordynamics (2005) CRC Press, Boca Raton 0-8247-2399-6 (1075 pp £97). J Sound Vib 298(1–2):477
8. Murphy BT (2013) XLHydroDyn. http://www.xlrotor.com/xlrotor/xlhydrodyn.shtml. Accessed 13 June 2013
9. Murphy BT, Vance JM (1982) An improved method for calculating critical speeds and rotordynamic stability of turbomachinery
10. Myklestad N (1945) New method of calculating natural modes of coupled bending-torsion vibration of beams. Trans ASME 67(1):61–67
11. Prohl M (1945) A general method for calculating critical speeds of flexible rotors. J Appl Mech 12(3):142–148
12. Vance J, Zeidan F, Murphy B (2010) Machinery vibration and rotordynamics. Wiley, Hoboken, New Jersey
13. Wachel J, von Nimitz WW (1981) Ensuring the reliability of offshore gas compression systems. J Pet Technol 33(11):2252–2260

Comparison of On-Bearing and On-Casing Vibration for Blade Health Monitoring in Rotating Machine

Ahmed A. Gubran and Jyoti K. Sinha

Abstract Rotating blades are considered as the most common cause of failures in rotating machinery. In the present research study, the dynamics of the blades both in the healthy and crack conditions are studies on a small experimental rig using the on-bearing and on-casing vibration, which are measured using the accelerometer on bearing pedestals and on casing cover towards on blades. The measured vibration data are analyzed by computing the responses at different engine orders (EOs) related to the blade resonance frequencies and their higher harmonics to understand the blade(s) dynamics behavior. The observations suggest that the on-bearing and on-casing vibration measurements can be extended to the non-intrusive method for the blade health monitoring (BHM). Comparative study between the on-bearing and on-casing vibrations on the simple experimental rig further allow to understand which method is more effective for blade damage detection.

Keywords Blade vibration · Blade health monitoring (BHM) · Blade faults · Engine orders (EO's) · Order tracking

1 Introduction

Rotating blades are considered as the most common cause of failures in rotating machinery which is nearly 42 % of the total failures for gas turbines [1]. The blade failures normally occurred as a result of crack due to unexpected operating conditions and also due to variable of external loads. For this reason, the early detection of damage in the blades is important to reduce the machine downtime, maintenance

A.A. Gubran (✉) · J.K. Sinha
School of Mechanical, Aerospace and Civil Engineering, The University of Manchester, Oxford Road, Manchester M13 9PL, UK
e-mail: ahmed.gubran@manchester.ac.uk

J.K. Sinha
e-mail: jyoti.sinha@manchester.ac.uk

J.K. Sinha (ed.), *Vibration Engineering and Technology of Machinery*, Mechanisms and Machine Science 23, DOI 10.1007/978-3-319-09918-7_8

overhead and from the safety consideration. The vibration based monitoring is the most accepted technique used in condition monitoring of the rotating machines. Different vibration-based methods are attempted for the blade health monitoring (BHM) can be found in the literature which includes the vibration measurement on turbine casing, torsional shaft vibration, etc. [2–6]. The recent study [2] also highlights the potential of the use of shaft torsional vibration for monitoring the blade vibration behavior. A number of studies are found in the literature to understand blade excitation and estimation of blade natural frequencies in order to detect blade health condition [7–10]. In this study, the dynamics of the blades both in the healthy and crack conditions are studies on a small experimental rig using the on-bearing and on-casing vibration measurement techniques, which are measured signals using the accelerometers. The instrumentations needed for the on-bearing and on-casing vibration measurement are simples and non-intrusive. The main purpose of this study is to compare between these 2 methods in order to bring out which of these methods is better for the BHM. The experiments are conducted on a test rig having 8-bladed disc for blade conditions; healthy, crack in one blade and crack in two blades and the acceleration data are recorded into the computer during the machine run-up and stored for a further signal processing analysis. The measured vibration data which is obtained by accelerometers from both positions, front bearing pedestal in vertical direction and on casing towards the blades in the mid way between two bearings are analyzed using Fast Fourier Transform (FFT) and by computing the responses at different engine orders (EO's) related to the blade resonance frequencies and their higher harmonics to understand the dynamic behavior of the blades under different blade fault cases.

2 Test Rig

A small test rig was designed and manufactured for the proposal study is as it shown in Fig. 1. The rig details, dimensions and material properties are discussed in [2, 3]. Initially shaft-disc-blades system is designed and analysed using the finite element (FE) model constructed in ABAQUS code to understand the natural frequencies and mode shapes of the shaft-disc-blades system to determine the suitable dimensions for the rig. Figure 2 shows the pre-design mode shapes and natural frequencies of shaft-disc-blades system. Small mistuning among the 8 blades is also introduced into the FE model. Blades frequencies were estimated in the range of 184.5–201.65 and 289.49 Hz for the rotor. The rig was then manufactured using the dimensions in accordance to this model. Modal tests were also conducted on the blades and shaft to know the natural frequencies and compare with the FE prediction during design stage.

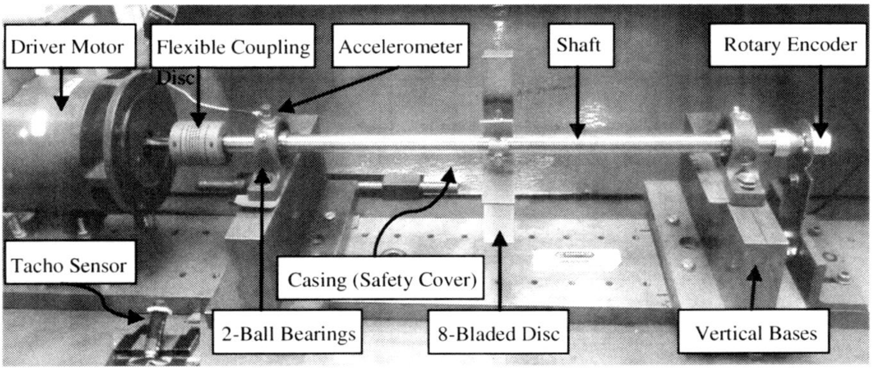

Fig. 1 Photograph of the test rig

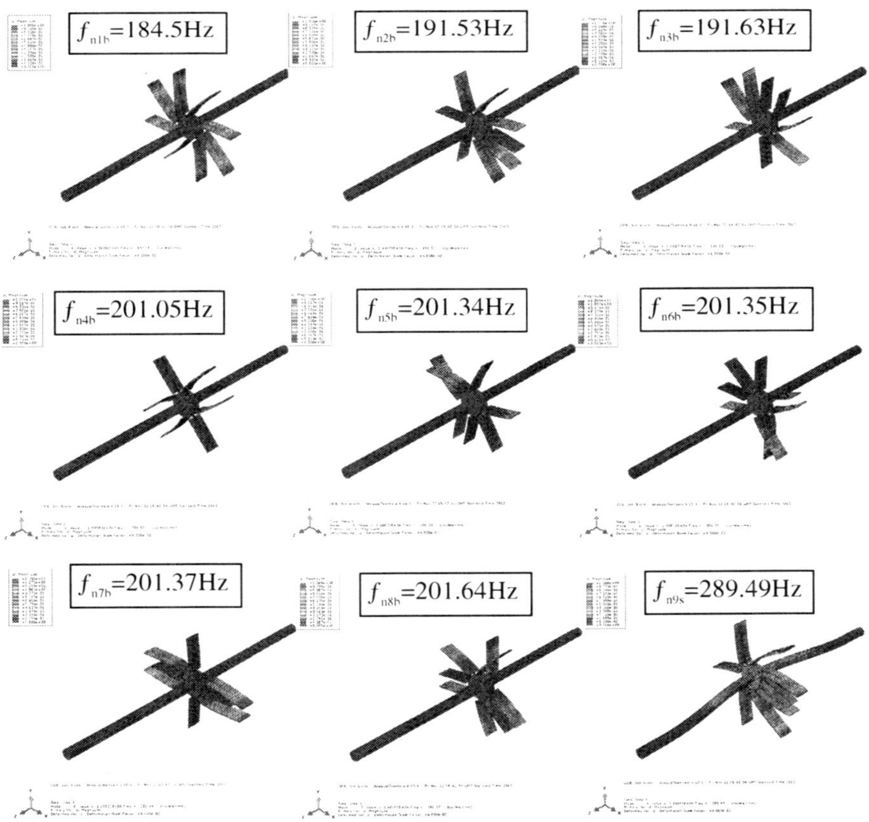

Fig. 2 Mode shapes and natural frequencies of shaft-disc-blades system using Finite Element (FE) method

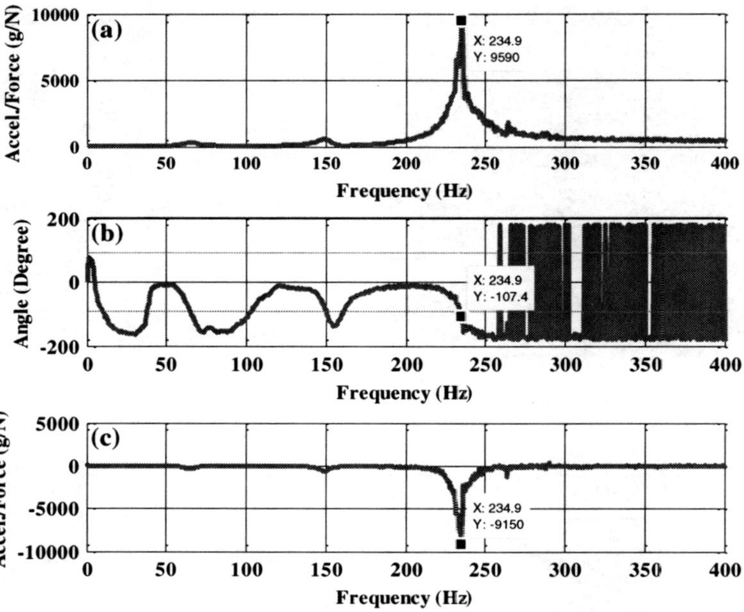

Fig. 3 A typical FRF plots of blade first natural frequency at 234.90 Hz: **a** Amplitude, **b** Phase and **c** Imaginary part

3 Modal Testing

The modal testing for blades has been conducted on the test rig by the Impulse-Response modal test method [11]. The experimentally identified first mean natural frequencies of the eight blades are 235.47 Hz with small deviation ±5.53 Hz due to blade mistuned effect due to the blade manufacturing and/or fitting. A typical frequency response function (FRF) plot of magnitude, phase and imaginary, it shows the blade frequency peak at 234.90 Hz as shows in Fig. 3.

4 Experiments and Fault Simulation

The experiments are conducted during the machine run-up from speed of 600 RPM (10 Hz) to 1800 RPM (30 Hz). The run-up rate was for kept equal to 40 RPM's. The data were then collected at 30,000 Samples/s and the data stored on PC for further signal processing analysis in the MatLab.

Three different conditions of blade faults were simulated for the experimental tests. These conditions of blade health conditions are healthy with mistuned effects, crack in one blade and crack in two blades. Experiments conducted are summarised in Table 1.

Table 1 Blade faults conditions

Case	Description	Blade no.
1	Healthy blade with mistuned effect	
2	Crack in one blade	5
3	Crack in two blades	5 and 7

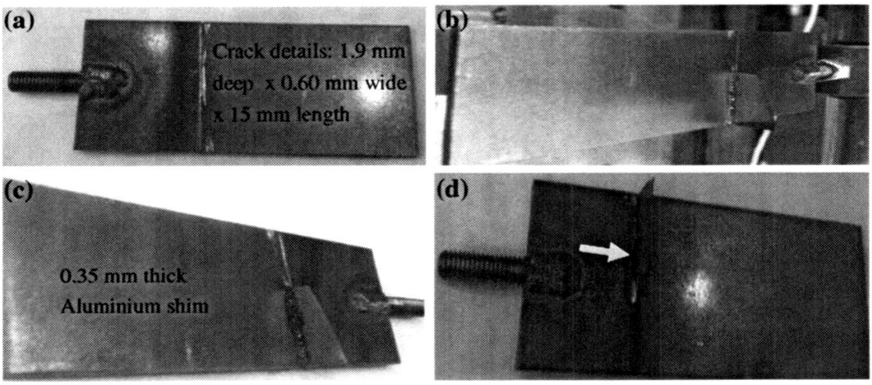

Fig. 4 Blade faults simulation: **a** Crack dimensions, **b** Cracked blade with shim location on bladed disc on rig, **c** shim in cracked adhesive side, **d** shim in crack non adhesive side

The three conditions of blade conditions were discussed in details in the references [2, 3]. A photograph of blade fault simulation is shown in Fig. 4. The vibration experiments were carried out for all fault cases listed in Table 1.

5 Data Analysis, Observation and Discussion

The measured vibration signals from on-bearing and on-casing acceleration signals are processed with the different engine orders (EO's) accelerations to understand the appearance of blade resonances and its dynamics behavior. The accelerometer locations on the bearing and casing are shown in Fig. 5. Typical measured acceleration signals for the on-bearing and on-casing vibration are shown in Fig. 6. The processed measured vibration acceleration signals for the EO10 and EO20 for both on-bearing and on-casing vibrations are shown in Figs. 7, 8, 9 and 10 for all cases. The EO10 is likely to contain the first resonance frequencies of blades and its 2nd harmonics in the EO20. Figures 7a–c and 8a–c) related to the EO10 are shown in the frequency range 220 Hz to 260 Hz related to the first blade resonance (1 × BR) frequency band and it's higher harmonics (2 × BR) in Fig. 9a–c and 10a–c for both on-bearing and on-casing measurements for the easy comparison.

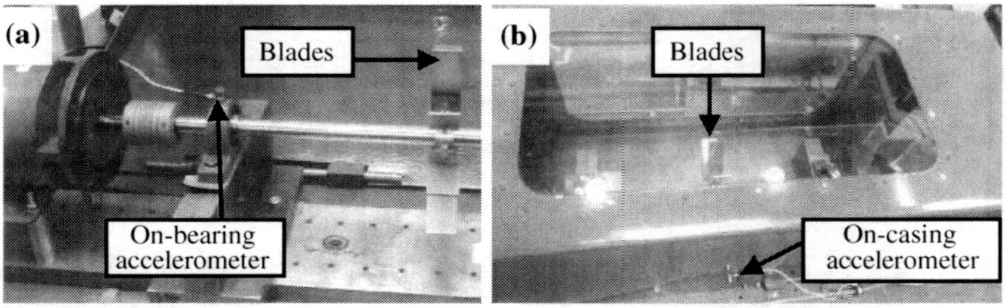

Fig. 5 Measurement locations on test rig: **a** On-Bearing, **b** On-Casing

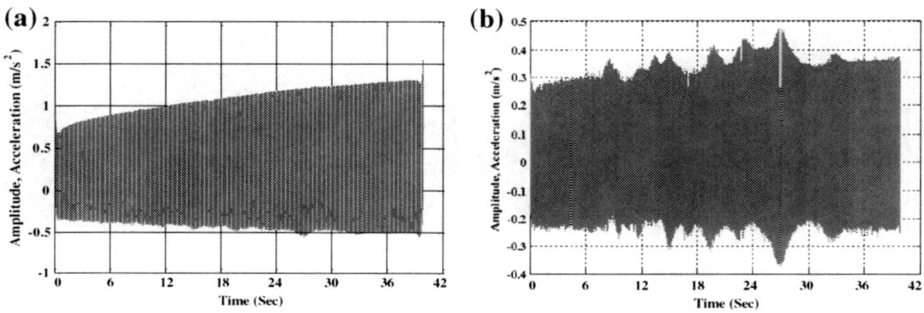

Fig. 6 Typical measured acceleration signals at machine run up from 600 to 1,800 RPM: **a** On-Bearing, **b** On-Casing

The results was obtained from on-bearing and on-casing vibration for blade conditions of Healthy blades with mistuned effect, crack in 1 blade and crack in 2 blades are discussed and compared on the Table 2.

From the comparison in the Table 2, it is observed that the on-bearing measurements gave a good indication for the cases of more than one crack in blades on bladed disc. Hence the presence of the distinct multiple peaks in the BR region and its higher harmonics for cases of 2 cracked blades or more can be considered as the feature of the BHM. On the other hand, the on-casing signal seems to be given much better blade resonance responses but no indication for BHM when comparing 3 cases.

Fig. 7 On-Bearing acceleration measured data for engine order spectra of EO10: **a** healthy, **b** crack in one blade and **c** crack in 2 blades

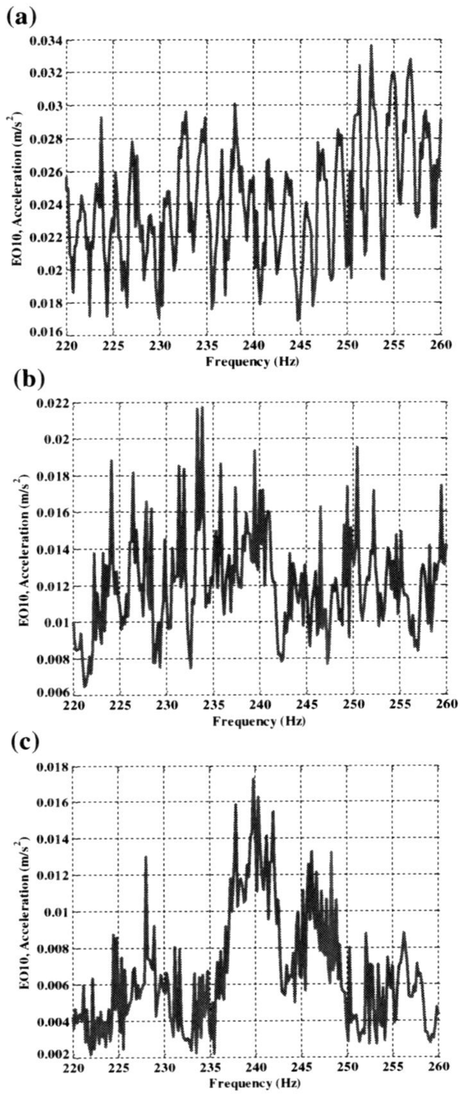

Fig. 8 On-Casing acceleration measured data for engine order spectra of EO10: **a** healthy, **b** crack in one blade and **c** crack in 2 blades

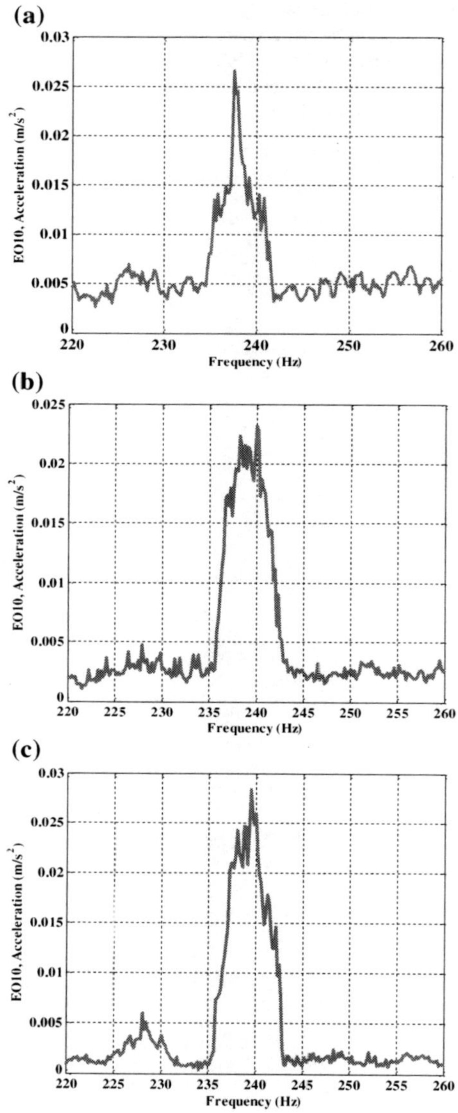

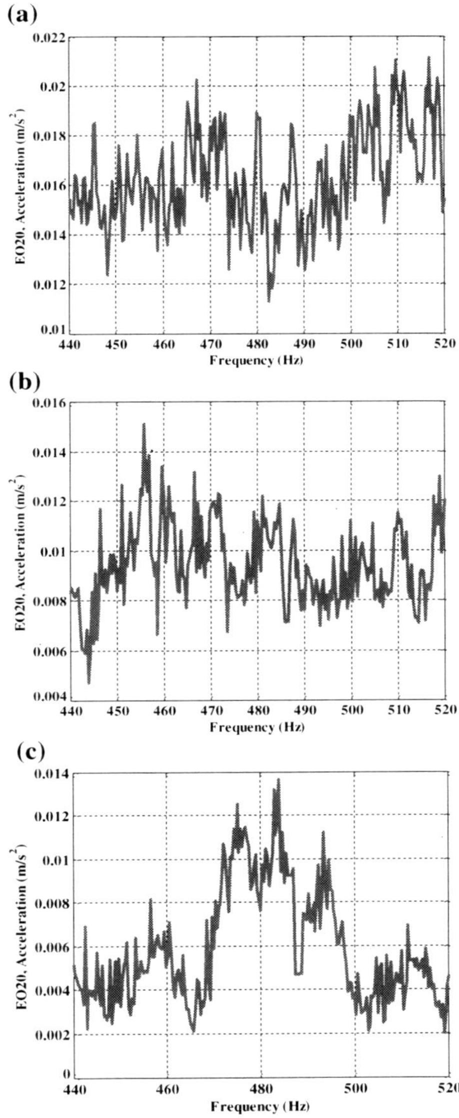

Fig. 9 On-Bearing acceleration measured data for engine order spectra of EO20: **a** healthy, **b** crack in one blade and **c** crack in 2 blades

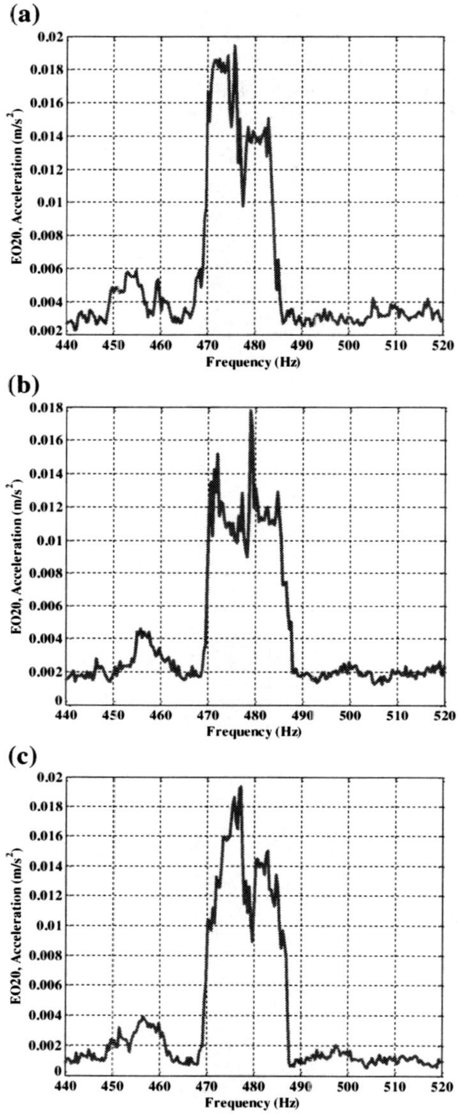

Fig. 10 On-Casing acceleration measured data for engine order spectra of EO20: **a** healthy, **b** crack in one blade and **c** crack in 2 blades

Table 2 Comparison of the results between on-bearing and on-casing vibrations

Blade condition	On-bearing vibration	On-casing vibration
Case 1: healthy with mistuned effect	There are no responses in the blade resonance regions of (1 × BR) and their higher harmonics (2 × BR). As shown in Figs. 7a and 9a respectively	There are a one distinct peak around the frequency 238 Hz confirms the excitation of the blade resonance of (1 × BR) and also at (2 × BR). As in Figs. 8a and 10a
Case 2: Crack in one blade	Also, here no BR and no changes in the harmonics regions of (1 × BR and 2 × BR) compared to the healthy case condition in both Engine orders EO10 and EO20. As shown in Figs. 7b and 9b respectively	There are no changes in the blade behavior related to EO10 & EO20 in (1 × BR and 2 × BR) regions compared to healthy condition. Which are appears only single distinct peak as shown in Figs. 8b and 10b respectively
Case 3: Crack in two blades	There is a distinct multiple peaks in the (1 × BR and 2 × BR) regions related to the EO10 and EO20 responses respectively. As shown in Figs. 7c and 9c	It's the same to the case 2, which are no changes in the blade harmonics regions of (1 × BR and 2 × BR) for both EO10 and EO20 compared to the healthy case. As shows in Figs. 8c and 10c

6 Conclusion

In this paper the measured vibration data was obtained from on-bearing pedestals and on-casing towards to blades during the machine run-up was used to understand the dynamics behavior of rotating blades with and without faults. Experiments were conducted for three different types of blade conditions, (a) Healthy with mistuned effects, (b) crack in 1 blade and (c) crack in 2 blades. From the results observed that on-casing vibration when order tracked with EO10 and their higher harmonic EO20 shows the existence of BR in all 3 cases. The appearance of the multiple peaks only observed in Case 3 (crack in 2 blades) compared to single peak in Case 1 (healthy) and it could be useful for the BHM, only when crack in 2 blades or more in bladed disc. Note that the on-casing signal seems to be given much better blade resonance responses but unfortunately no indication for BHM when comparing 3 cases in the present experimental rig.

References

1. Meher-Homji CB (1995b) Blading vibration and failures in gas turbines: Part B, compressor and turbine airfoil distress. ASME paper no. 95- (1995) GT-419
2. Gubran AA, Sinha JK (2014) Shaft instantaneous angular speed for blade vibration in rotating machine. Mech Syst Sig Process 44(1–2):47–59 (Special issue on Instantaneous Angular Speed (IAS) processing and angular applications)

3. Gubran AA, Sinha JK (2014) Comparison between long and short blade vibration using shaft instantaneous angular speed in rotating machine. In: Proceedings for the ASME 2014: turbine technical conference and exposition, Dusseldorf, Germany, 16–20 June (in Press)
4. Abdelrhman AM et al (2012) A Review of vibration monitoring as a diagnostic tool for turbine blade faults. Appl Mech Mater 229:1459–1463
5. Rao AR, Sinha RK (2000) Insitu measurement on turbo-generators for detection of blade vibrations. In: Proceedings of VETOMAC-1, Bangalore, India, Oct
6. Maynard KP, Trethewey MW (2000) Blade and shaft crack detection using torsional vibration measurement Part 1: feasibility studies. Noise Vib 31(11):9–15
7. Forbes GL, Randall RB (2009) Gas turbine casing vibrations under blade pressure excitation. In: MFPT, Dayton, Ohio
8. Forbes GL, Randall RB (2007) Simulated gas turbine casing response to rotor blade pressure excitation. In: Proceedings of the 5th Australasian Congress on applied mechanics, Australia
9. Forbes GL, Randall RB (2013) Estimation of turbine blade natural frequencies from casing pressure and vibration measurements. MSSP 36:549–561
10. Forbes GL, Randall RB (2008) Detection of a Blade fault from simulated gas turbine casing response measurements. In: 4th European workshop on SHM, Krakow, Poland
11. Ewins DJ (2000) Modal testing-theory, practice and applications, 2nd edn. Research Studies Press Ltd, England

Vibration-Based Condition Monitoring for Rotating Machinery with Different Flexible Supports

Adrian D. Nembhard and Jyoti K. Sinha

Abstract In a previous study, a combined vibration and temperature analysis technique for the diagnosis of commonly encountered rotor related faults produced good practicable results even without the use of temperature. This was, however, developed on an experimental rig with relatively rigid supports. The objective of the current study is to investigate the transferability of the said approach on rotating machines with relatively flexible supports. A fault diagnosis (FD) method which compares vibration data acquired from similar machines with different flexible supports is also considered here. The experimental rig used in the previous study was modified to accommodate two different relatively flexible supports. The results of the previously proposed technique showed that diagnosis is insensitive to the support types used. The result of the newly introduced combined FD approach gave improved diagnosis over the previously proposed technique. The observations made are presented in this paper.

Keywords Vibration monitoring · Condition monitoring · Principal component analysis · Rotating machinery · Fault diagnosis

1 Introduction

In a previous study, Nembhard et al. [8, 9] applied a combined vibration and temperature analysis technique for the diagnosis of commonly encountered rotor (also referred to as shaft) related faults on a small experimental rig with relatively rigid supports. The study had several objectives, key among which were; to develop a simple but robust diagnostic method that can use data from a single vibration sensor and a single temperature sensor at each bearing while preserving moderate computational load for the diagnosis of crack misalignment and rub faults.

A.D. Nembhard (✉) · J.K. Sinha
School of Mechanical, Aerospace and Civil Engineering (MACE),
The University of Manchester, Manchester M13 9PL, UK
e-mail: adrian.nembhard@manchester.ac.uk

J.K. Sinha (ed.), *Vibration Engineering and Technology of Machinery*,
Mechanisms and Machine Science 23, DOI 10.1007/978-3-319-09918-7_9

Condition indicators computed from the measured vibration data were processed with and without bearing temperature measurements. It was concluded that the proposed technique produced an easily interpreted graphical representation which had good separation of the healthy condition from the faults tested both with and without the use of temperature.

The objective of the current study is to investigate the transferability of the approach proposed by Nembhard et al. [8, 9] on two rotating machines with relatively flexible supports (FS1 and FS2). It is possible that the same rotating machine when installed at different plant sites exhibit different flexibility with different natural frequencies. However, the vibration data history requisite for condition monitoring and subsequent fault diagnosis may exist at one site and be lacking in the other. A combined analysis for directly comparing the two different flexible supports tested (FS1 and FS2) was carried out to explore whether the method can be used for diagnosis of machines where such a scenario exists.

Thus, the experimental rig used by Nembhard et al. [8, 9] was modified to accommodate two different relatively flexible bearing supports. Several samples of on-bearing vibration data were collected at a steady state speed for a residual misalignment with residual unbalance (RMRU) baseline condition as well as for shaft crack, coupling misalignment and shaft rub fault conditions. Testing was done with one condition existing at a time, first with the rig set up for FS1 and afterwards for FS2. Widely applied health indicators (referred to as features) that best describe the state of the machine were computed from the measured vibration data. Three sets of data processing were done: previous technique on FS1 features, previous technique on FS2 features and then the newly introduced technique on FS1 and FS2 features combined in a single analysis step. The results of the previous technique for both supports gave good separation between the different conditions tested, thus suggesting the method is transferable to machines with relatively flexible of supports and hence independent of machine supports. The results from newly proposed combined technique showed improved clustering and separation between the conditions tested. The results obtained for the newly proposed technique suggested the method can be useful in practice for the comparison of vibration data acquired from similarly configured rotating machines with different natural frequencies. The observations made are presented in this paper.

2 Experimental Set Up and Experiments Conducted

Figure 1 shows a schematic of the experimental rig and instrumentation used in the current study. The rig is a fundamental model of a multi-stage rotating machine that was modified from the previous study done by Nembhard et al. [8, 9] to accommodate relatively flexible supports. The rig's rotating assembly consists of two 20 mm diameter rigidly coupled (C2) dissimilar length bright mild steel shafts of 1,000 and 500 mm length. The long shaft is connected to a three phase 0.75 kW 2800 rpm electric motor by a semi-flexible coupling (C1). The rotating assembly is supported by four grease lubricated 20 mm internal diameter flanged ball bearings (B1–B4). As shown in Fig. 2a, b, each bearing is mounted by four flexible rods that are secured to a flexible bearing pedestal. Each bearing pedestal is bolted to a large lathe bed that serves as the rig's foundation. Machine sections which act as balancing disks (125 mm in diameter and 14 mm thick) are mounted to each rotor; two disks (D1–D2) for the long shaft and one disk (D1) for the short shaft.

The dynamic response of the rig was measured by four ICP type 100 mV/g accelerometers; one each mounted at a 45^0 position on the flange of each bearing (see Fig. 2b). This mounting position was adopted to simultaneously account for the measurement of both the vertical and horizontal responses from the sole accelerometer used at each bearing. The accelerometers were mounted on studs that were attached to the bearings with super glue. The vibration data are transmitted from the accelerometers to a four channel signal conditioner with a 1:1 gain and afterwards to a National Instrument 16 bit analogue to digital (A/D) data acquisition (DAQ) system which is connected to a personal computer via a USB cable. The vibration data are then recorded and stored on a personal computer for later processing and analysis.

Two different flexible supports were investigated. The support types were modified by changing the diameter of the bearing mounting rods. Therefore, the first support tested, Flexible Support 1 (FS1) and second support tested, Flexible Support (FS2) consists of 10 mm and 6 mm diameter bearing mounting rods respectively. The impact-response method of modal testing [4] was conducted to confirm that FS1 and FS2 were dynamically different from each other and also from the supports used in Nembhard et al. [8, 9]. The natural frequencies of the previous rig used by Nembhard et al. [8, 9] were identified by Elbhbah and Sinha [3] at 68, 144 and 352.5 Hz. By appearance, the first four natural frequencies for FS1 are 50.66, 56.76, 59.2 and 127 Hz. Similarly, the first four natural frequencies for FS2 are 47, 55.54, 57.98 and 127 Hz.

In order to establish baseline operating conditions for the experimental rig, data was first collected for a baseline condition. In practice, it is best for the baseline condition to be as healthy as possible to be used as a reference [1] however, there were some difficulty in getting the rig sufficiently aligned to be deemed fully healthy. As demonstrated in Fig. 3a, b, the baseline condition was characterised by some residual misalignment with residual unbalance, and thus referred in the present study as the Residual Misalignment with Residual Unbalance (RMRU) case. With one fault condition existing at a time, testing was done for three rotor fault conditions; a transverse shaft crack, coupling misalignment and full shaft rub. Shaft crack was simulated by introducing a 0.33 mm wide by 4 mm (20 % of shaft diameter) deep transverse notch 160 mm from bearing 1 along the length of the longer shaft. The Electro Discharge Machining (EDM) process was used to produce the notch. Misalignment fault was introduced at C1 by positioning 0.8 mm shims under either side of B1 pedestal (that is between the base of the pedestal and the lathe bed). Rub condition was simulated using a rub apparatus. Full annular rub was produced. Experiments were performed at a steady state rotational speed of 2,400 rpm (40 Hz). Vibration data were acquired for 20 samples in a total combined

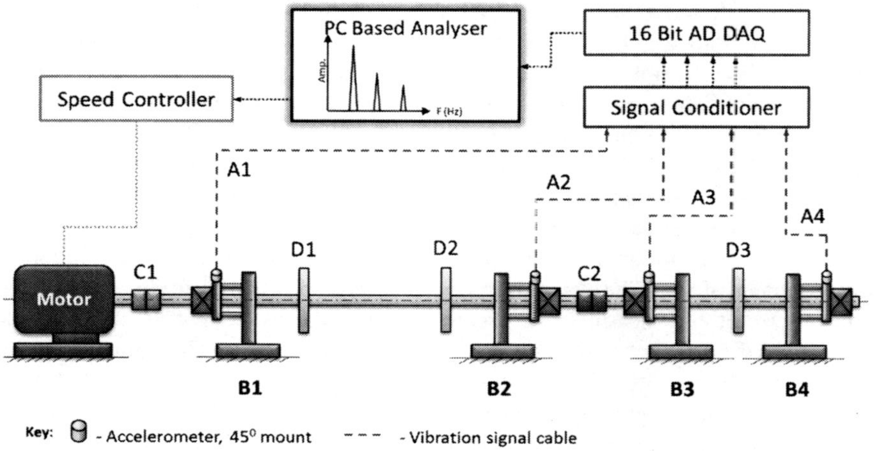

Fig. 1 Schematic of experimental rig with instrumentation

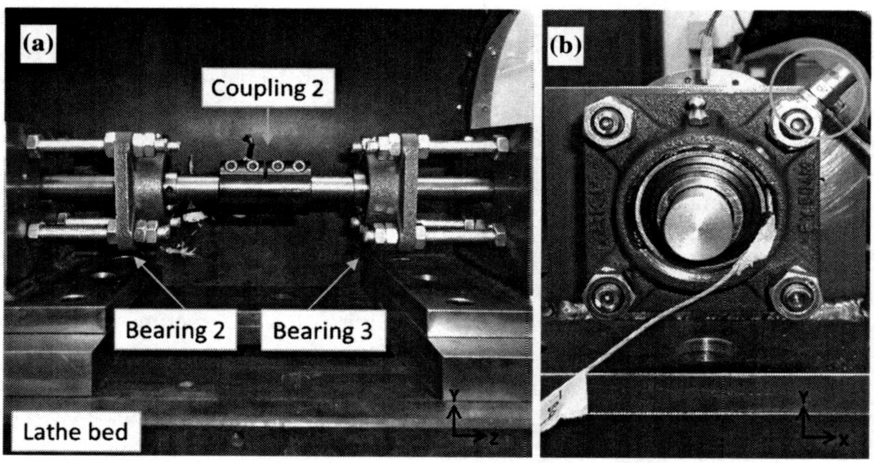

Fig. 2 Close up of: **a** front elevation of rigid coupling area showing bearing supports with flexible rods and **b** end elevation of bearing assembling showing 45° accelerometer mount

sample time of 2 min using a sampling frequency of 10,000 Hz. After data collection was complete for the FS1 set up, all four bearings on the rig were reconfigured to the FS2 set up and the experiments were repeated.

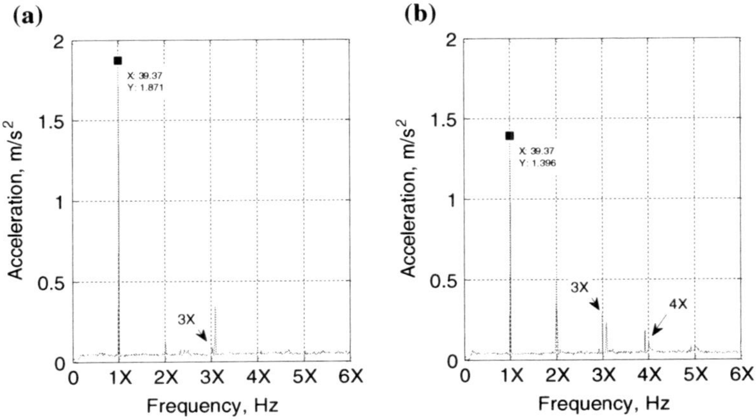

Fig. 3 Amplitude spectra for RMRU at B2 for **a** FS1 and **b** FS2

3 Comparison of Dynamic Behaviour

Figure 3a, b present samples of the amplitude spectra that were computed for the baseline RMRU condition at Bearing 2 for both FS1 and FS2 when operated at 2,400 rpm (40 Hz) respectively. For FS1 a fundamental frequency at 39.37 Hz is seen in addition to 2x and 3x harmonic components. The presence of the 2x and 3x harmonic components suggest that this condition is characterised by some residual misalignment. The conspicuous peak observed to the right of the 3x harmonic component was generated due to the excitation of the 127 Hz natural frequency of FS1. Similar observations were made from the spectral plot of FS2, where a distinct fundamental frequency of 39.37 Hz is observed in addition to multiple harmonic components of the fundamental frequency up to the 6x harmonic. The peaks seen in close proximity to the 3x and 4x harmonic components were due to excitation of the rig's natural frequency when configured to FS2. Again, there is some in built misalignment with rig for FS2.

The observations made here is typical of all the spectra generated for the different cases tested for FS1 and FS2. By extension, the observations made in this section demonstrate what is typical of machines with different natural frequencies, in that they may possess different features due to the excitation of the natural frequencies by the harmonics present in the spectrum even when the machine states (health) are the same. Then again, it is widely accepted that different faults may generate similar features [10]. Such is the nature of diagnosis with the simple amplitude spectra, which is why a skilled and experienced analyst and more advanced processing techniques are usually needed to extract some useful information from the measured vibration data that can lead to diagnosis [5].

4 Data Analysis and Processing

Pattern recognition by Principal Component Analysis (PCA) adopted earlier [8, 9] was again used for diagnosis in the present study. PCA is an orthogonal linear transformation that is used to reduce the dimensionality of a data set by transforming to a new set of ordered and uncorrelated variables called Principal Components (PCs), which retain as much variation present in all the original data set [6]. Therefore, for each support tested PCA is used to transform a matrix of the health indicators computed from the samples of each bearing of each of the four scenarios tested (RMRU, crack, misalignment and rub) to a set of PCs [7] for interpretation. To produce an accurate representation of the state of the machinery, the features used in the present study are root mean square (r.m.s.) amplitude, amplitudes of the 1x, 2x, 3x, 4x and 5x harmonic components and the spectrum energy (S.E.). R.M.S. and S.E. give overall indications of the machine in the time and frequency domains respectively, while the amplitudes of the harmonic components can provide further insight to the specific machine state present. Thus one time domain and six frequency domain features computed for each bearing location were applied. The features derived from the frequency domain were computed using; 16,384 spectral lines, high pass filtering at 5 Hz, a Hanning window and 35 averages.

Transformation of the features to the Principal Component space is done using a Matlab algorithm in which the matrix of features (health indicators) is input for processing. The input feature matrix for the independent analysis consists of 28 rows and 80 columns while that for the combined analysis consist of data from FS1 and FS2 combined in a single matrix of 56 rows and 80 columns.

5 Fault Diagnosis

5.1 Independent Analysis of Flexible Supports, FS1 and FS2

The results of the independent analysis on FS1 and FS2 are shown in Figs. 4 and 5 respectively. Figure 4a, b show two dimensional representations of the second principal component (PC2) versus the first principal Component (PC1) [8, 9] in full view and in a zoomed view respectively. The full view shows clear separation of the rub and crack cases from the RMRU condition. It did not seem like the technique was able to discriminate the misalignment case from the RMRU condition. However, on closer inspection of the apparent overlap between RMRU and the misalignment cases in Fig. 4b, good separation is observed between the two cases. Similar observations were made for FS2 as shown in Fig. 5a, b. However, it was noted that less separation was seen between all the faulty cases and the RMRU case for FS2 when compared to FS1. Decreased separation between the different cases suggests there are stronger correlations between the sets of data used in the analysis [6].

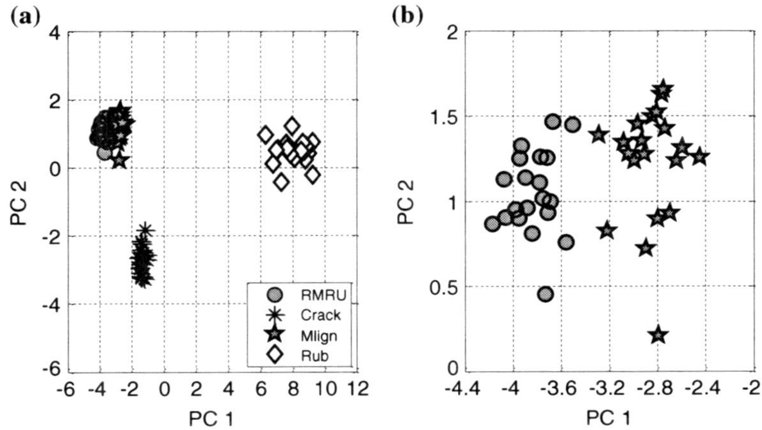

Fig. 4 Fault diagnosis of FS1 at **a** full view and **b** zoomed view

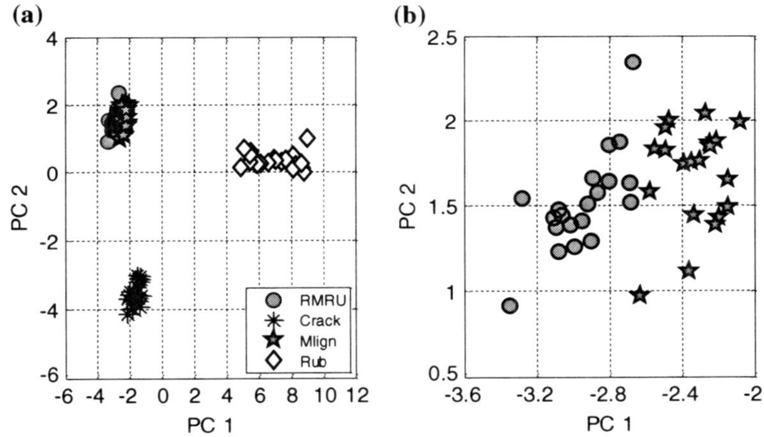

Fig. 5 Fault diagnosis of FS2 at **a** full view and **b** zoomed view

5.2 Combined Analysis of FS1 and FS2

The results of the combined analysis are shown in full view and zoomed view in Fig. 6a, b respectively. From a cursory glance of the full view of the results, the different fault conditions tested were conspicuously separated from the RMRU baseline condition and also from each other. There was also better clustering of the different conditions which is desirable for the diagnosis process [8, 9]. It was apparent that the combined approach was able to utilise the information available from the two separate support types to improve its efficacy. This analysis is based

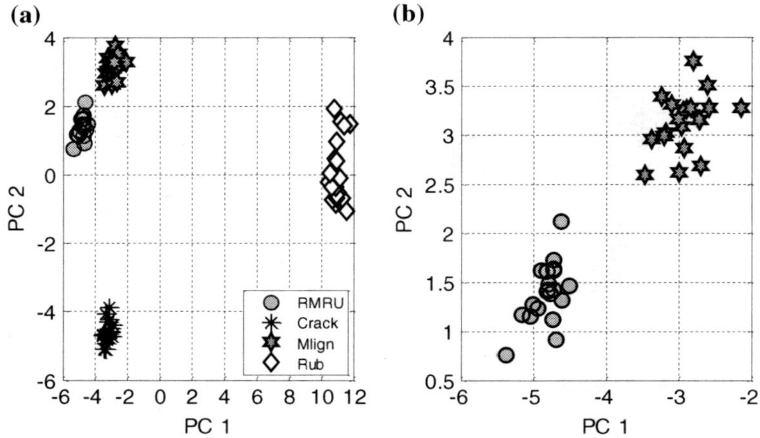

Fig. 6 Fault diagnosis with combined classification of FS1 and FS2 in **a** full view and **b** zoomed view

on the recognition of patterns (similarities and differences) between the sets of data from the FS1 and FS2. The current analysis differs from vibration analysis in the traditional sense where the amplitude spectrum for a given fault condition on a given machine is interrogated for changes relative to a previously acquired "healthy" baseline condition from that said machine [2]. Since "traditional" vibration-based condition monitoring requires comparison with a baseline condition it becomes apparent that the baseline amplitude spectra have to remain constant. Consequently, if the amplitude spectra of the baseline conditions differ, then comparison of vibration data between machines with different dynamic characteristics would not be useful for condition monitoring.

6 Conclusion and Future Work

The transferability of a previously developed FD technique to rotating machines with relatively flexible supports was investigated in the current study. A FD method for comparing vibration data from similar machines with different flexible supports was also introduced. The experimental rig used to develop the previous technique had relatively rigid supports and was modified to accommodate two different relatively flexible supports. Samples of on-bearing vibration data were acquired for a baseline residual misalignment with residual unbalance (RMRU), crack, misalignment and rub conditions and used to compute machine health indicators for further processing. Three sets of processing were done: previous technique on FS1, previous technique on FS2 and the newly introduced combined technique on FS1 and FS2. The previously proposed technique gave good separation between the conditions tested and is therefore confirmed insensitive to the support types used. The result of the newly introduced combined analysis gave improved separation between the different conditions tested and thus may be useful in a practical case where similar machines are installed at different sites. There is need for further validation of both techniques with additional faults.

Acknowledgments A.D. Nembhard sincerely thanks the Commonwealth Scholarship and Fellowship Plan (CSFP) for affording him the Ph.D. Commonwealth Scholarship to the UK from Jamaica to pursue the current research.

References

1. British Standards Institution (2007) BS ISO 13374-2:2007 Condition monitoring and diagnostics of machines—data processing, communication and presentation—Part 2: data processing. BSI, London
2. British Standards Institution (2009) BS ISO 10816-1:1995 + A1:2009 Mechanical vibration—evaluation of machine vibration by measurement of on non-rotating parts—Part 1: general guidelines. BSI, London
3. Elbhbah K, Sinha JK (2013) Vibration-based condition monitoring of rotating machines using a machine composite spectrum. J Sound Vib 332(11):2831–2845
4. Ewins DJ (2000) Modal testing—theory, practice and application, 2nd edn. Research Studies Press, Baldock
5. Jardine AKS, Lin D, Banjevic D (2006) A review of machinery diagnostics and prognostics implementing condition-based maintenance. Mech Syst Signal Process 20(7):1483–1510
6. Jolliffe IT (2002) Principal component analysis, 2nd edn. Springer, New York
7. Li W, Shi T, Liao G, Yang S (2003) Feature extraction and classification of gear faults using principal component analysis. J Qual Maint Eng 9(2003):132–132
8. Nembhard AD, Sinha JK, Pinkerton A, Elbhbah K (2013) Fault diagnosis of rotating machines using vibration and bearing temperature measurements. Diagn Appl Struct Usage Cond Monit 14(3):45–51
9. Nembhard AD, Sinha JK, Pinkerton AJ, Elbhbah K (2013b) Combined vibration and thermal analysis for the condition monitoring of rotating machinery. Struct Health Monit 13 (3):281–295
10. Patel TH, Darpe AK (2009) Experimental investigations on vibration response of misaligned rotors. Mech Syst Signal Process 23(7):2236–2252

Off-Line Crack Detection in Rotors

Erfan Asnaashari and Jyoti K. Sinha

Abstract This paper intends to extend the application of the residual operational deflection shape (R-ODS) to a simplified model of the LP turbine of a steam turbo generator. The numerical model consists of a stepped shaft which accommodates some balance disks to represent the blades at different stages of the LP turbine. The ability of the R-ODS method in detecting cracks in beam-like structures has already been presented. The method removes the effect of the first harmonic component (1x) from the ODSs at higher harmonic components to highlight the non-linear behaviour of cracks. In this paper, single and multiple cracks are introduced at different locations to the finite element model of the rotor. It is assumed the rotor can be taken out from the rig and mounted on a simply supported boundary condition. After carrying out the modal analysis, the rotor is excited at its first resonant frequency and acceleration responses along the rotor are computed to construct the ODSs. The R-ODS method uses the normalised ODSs for the purpose of crack detection. The curvature of R-ODS method is also presented for further improvement in indication of crack locations especially for multiple-crack rotors.

Keywords Cracked rotor · Crack detection · Crack breathing · Residual operational deflection shape

1 Introduction

Shafts play an important role in power transmission in rotating machines. Initiation and propagation of fatigue cracks are very common in rotating shafts since they are constantly subjected to fluctuating stresses and broad variations in temperature and

E. Asnaashari (✉) · J.K. Sinha
School of Mechanical, Aerospace and Civil Engineering, University of Manchester, Sackville Street, Manchester M13 9PL, UK
e-mail: erfan.asnaashari@postgrad.manchester.ac.uk

J.K. Sinha
e-mail: jyoti.sinha@manchester.ac.uk

J.K. Sinha (ed.), *Vibration Engineering and Technology of Machinery*, Mechanisms and Machine Science 23, DOI 10.1007/978-3-319-09918-7_10

environmental conditions. Early detection of fatigue cracks in the shaft of a rotor system is vital to prevent fatigue failure and consequent potential damage to other equipment and huge costs in down time.

Crack development changes the dynamic characteristics of the shaft. These changes are the main idea of many vibration-based methods for crack detection. A comprehensive overview of the vibration-based crack detection in shafts is presented in [1]. Breathing is the key feature of fatigue cracks which produces non-linear behaviour. Pennacchi et al. [2] presented a model-based identification method for transverse breathing cracks in industrial machines. They simulated the crack effect on static and dynamic behaviour of the rotor in the frequency domain by applying equivalent forces of the cracked beam element to the rotor. The crack was then identified by an external force identification procedure. Chasalevris and Papadopoulos [3] investigated the cross-coupled bending vibrations of a rotating shaft with a breathing crack and resilient bearings. A continuous model for the cracked rotor-bearing system was offered to detect the existence of the crack by monitoring the change in the stiffness due to the change of crack local compliance. Darpe [4] proposed a detection method based on the breathing of the crack. The method uses the presence or absence of bending and torsional vibrations coupling at various orientations of the crack. Darpe applied a transient torsional excitation for a very short period of time and investigated the lateral vibrations. The transient features of the resonant bending vibrations were revealed by wavelet transforms in order to detect the crack. Sinou [5] used higher harmonic components of exciting frequency and the crack-unbalance interactions to identify the presence of a single breathing crack in a rotor system. By means of numerical simulations, it was demonstrated that for a given crack depth, the unbalance affects the amplitude of the $\frac{1}{2}$ and $\frac{1}{3}$ sub-critical resonant peaks as well as the 1x component which could be used to identify the crack. Sawicki et al. [6] presented the modelling and analysis of rotors with breathing cracks. They utilised sinusoidal excitations to the shaft through active magnetic bearings (AMB). These excitations generated response frequencies combining the rotor spin speed and excitation frequency. The system was finally analysed using an approach based on the harmonic balance method and could be used with inverse methods for the purpose of damage detection.

Recently, the residual operational deflection shape (R-ODS) method has been proposed for crack detection in beam-like structures [7]. The R-ODS method removes the effect of first harmonic component (1x) from the ODSs at higher harmonics (2x, 3x, etc.) to highlight the non-linear behaviour of the crack. This study intends to extend the application of the method by using a very simplified model of the LP turbine of a steam turbo generator. The model consists of a stepped shaft which accommodates some balance disks to represent the blades at different stages of the LP turbine. The reliability of the R-ODS method in detecting the location of single and multiple cracks in the shaft is investigated.

2 Finite Element Simulation

A rotor consisting of a stepped shaft with mounting balance disks is considered. The density and Young's modulus of the rotor are 7,850 kg/m^3 and 210 GPa respectively. Figure 1 shows the dimensions of the rotor with a single crack. Here, it is assumed the rotor can be taken out from the rig and mounted on a simply supported boundary condition. Single and multiple cracks are introduced at different locations with depth ratio of 20 % ($\frac{d_c}{D} = 0.2$). Table 1 shows the details of all the investigated cases.

A finite element model was created for the rotor using Euler-Bernoulli beam elements. There are two nodes defined per element and each node has two degrees of freedom; translational displacement and bending rotation. Stiffness proportional damping is obtained using the constructed stiffness matrix. The approach proposed in [8] is used here to introduce cracks to the finite element model of the rotor. This approach modifies the local flexibility in the vicinity of the crack within Euler-Bernoulli beam elements based on the concept explained in [9]. It assumes that the flexibility from uncracked to cracked section of the beam varies linearly and considers triangular reduction in the stiffness of cracked element. If this reduction happens over more than one element, the nodes of the model will be moved so that the crack effect remains within a single element.

The breathing of the crack(s) is modelled through sign recognition of nodal displacements of the cracked element(s). This means that when the displacement of nodes of the cracked element is greater than zero, the crack is open. Otherwise, the crack is assumed to be closed. The equation of motion for the cracked rotor under external excitation is:

$$M\ddot{x}(t) + C\dot{x}(t) + K(t)x(t) = F(t) \tag{1}$$

where M, C and K are the mass, damping and stiffness matrices for the rotor respectively and F is the external force.

In Eq. 1, the stiffness matrix is time dependent because it changes with opening and closing of the crack(s) during the external excitation. When the crack is closed, a healthy (global) stiffness matrix is used in Eq. 1. However, the existence of cracks reduces the stiffness of the cracked elements and consequently the stiffness of the whole rotor. In order to consider the stiffness reduction corresponding to each open crack in the global stiffness matrix, the number and location of all open cracks are determined at every time step. The stiffness matrix in the presence of open cracks can be written as:

$$K_c = K - \sum_{p=1}^{n} \Delta K_{c,p} \tag{2}$$

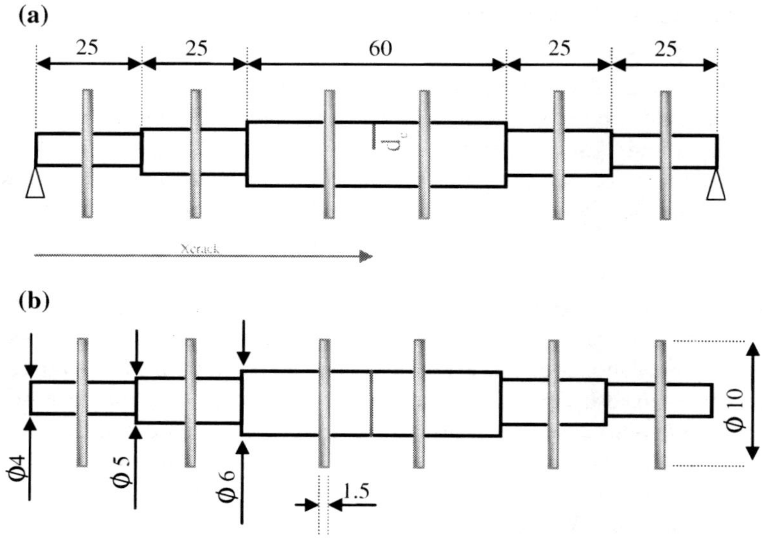

Fig. 1 Schematic view of the simply supported cracked rotor, **a** side view, **b** top view

Table 1 Number and locations of cracks used in the rotor for numerical simulations

	No. of cracks	Crack location (m)	Crack depth (cm)
Case 1	1	0.35	1.0
Case 2	1	0.85	1.2
Case 3	2	0.4, 0.9	1.0, 1.2
Case 4	2	0.4, 1.2	1.0, 1.0

where K_c is the stiffness matrix for the cracked rotor, K is the stiffness matrix for the healthy rotor (global), n is the number of open cracks and $\Delta K_{c,p}$ is the maximum reduction in the stiffness due to p-th open crack.

Equation 1 is solved numerically using the Newmark-β method with time steps of 200 μs assuming that the initial displacement and velocity is zero. Modal analysis is carried out to compute the natural frequencies of the rotor assuming all the existing cracks are open. The first three calculated natural frequencies of the rotor in the vertical direction for all the cases are shown in Table 2. The rotor is then excited by a sinusoidal input at its first natural frequency. The calculated acceleration responses are polluted with normalised random noise of 20 dB in order to simulate experimental condition. By using the Fast Fourier transform (FFT) the time-domain acceleration responses are converted into the frequency-domain with sampling frequency of 5 kHz. Figure 2 shows a typical noisy acceleration response in time domain and its frequency spectrum from a specific node at mode 1 where the higher

Table 2 Computed natural frequencies

	Mode 1 (Hz)	Mode 2 (Hz)	Mode 3 (Hz)
Case 1	38.88	137.22	321.76
Case 2	38.77	138.81	315.49
Case 3	38.35	136.95	311.59
Case 4	38.50	134.67	314.66

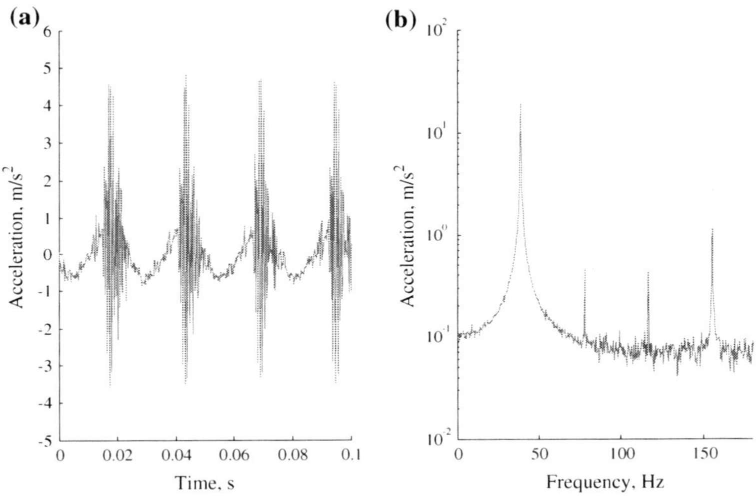

Fig. 2 Calculated acceleration response with noise in **a** time domain and **b** frequency domain for the cracked rotor at mode 1

harmonics of the exciting frequency can be observed clearly. These higher harmonics confirm the breathing of the crack during external excitation.

3 The Curvature of R-ODS Method

The curvature of R-ODS method is proposed in order to magnify the effect of crack in the R-ODS signals which is expressed in Eq. 3:

$$d^2(R - ODS_j) = \frac{(R - ODS)_{j+1} - 2(R - ODS)_j + (R - ODS)_{j-1}}{(\Delta x)^2}, \quad j = 2, 3, \ldots \tag{3}$$

where j refers to each R-ODS data point and ΔX is the difference between adjacent data points.

In fact Eq. 3 is the second derivative of the R-ODS data which removes the effect of curvature for better indication of discontinuities related to the crack(s) in the R-ODS signal. The R-ODS data can be obtained as follows:

$$(R-ODS)_{p,m} = (NODS)_{p,m} - (NODS)_{1,m}. \quad p = 2, 3, 4, \ldots \tag{4}$$

where $(NODS)_{p,m}$ is the normalised ODS at p-th harmonic for mode m.

Generally, as stated in Eq. 4, the R-ODS method eliminates the effect of the exciting frequency (1x) from the normalised ODSs at the higher harmonic components. Therefore, The R-ODS only contains the non-linearity due to the crack(s) breathing. Indices p and m in Eq. 4 are equal to 2 and 1 respectively in this paper because firstly the R-ODSs are calculated for 2x data and secondly the cracked rotors are excited at their first resonant frequency.

4 Results and Discussion

Figures 3 and 4 demonstrate the obtained R-ODS and their second derivatives for shafts with single and multiple cracks respectively. The method explained in [10] was used to obtain normal ODSs at the exciting frequency and its higher harmonic components. In the case of the rotor with a single crack (Fig. 3a, c) the R-ODS plot itself indicates the crack location clearly and the curvature magnifies the

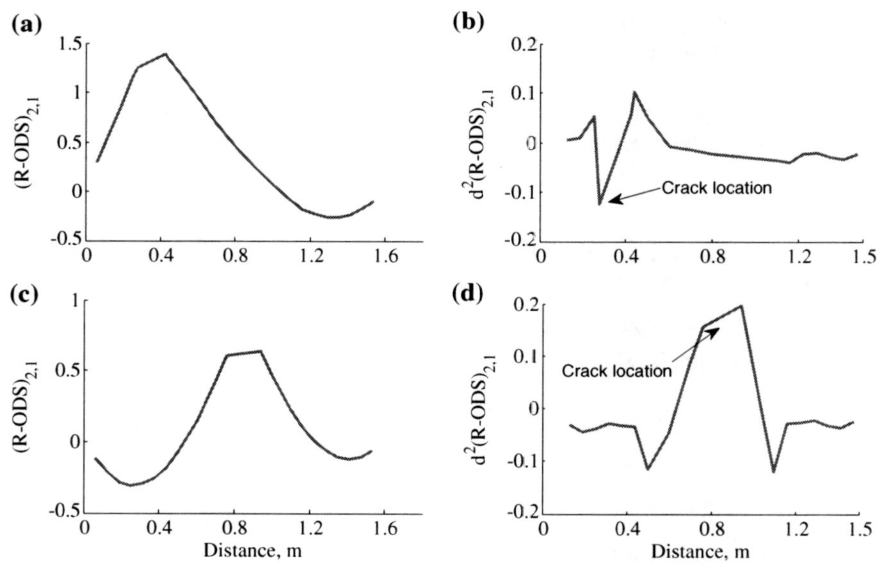

Fig. 3 The $(R\text{-}ODS)_{2,1}$ for Cases 1 and 2 (**a**, **c**) and their $d^2(R\text{-}ODS)_{2,1}$ (**b**, **d**) respectively

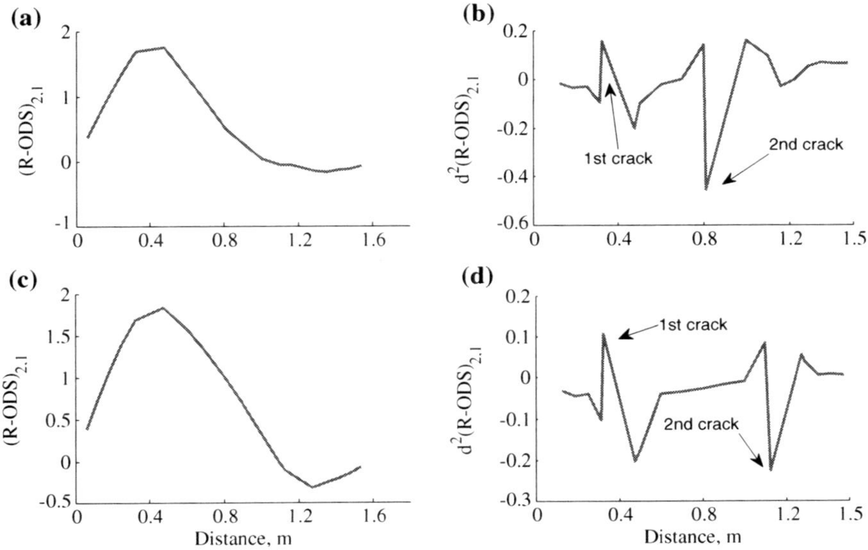

Fig. 4 The $(\text{R-ODS})_{2,1}$ for Cases 3 and 4 (**a**, **c**) and their $d^2(\text{R-ODS})_{2,1}$ (**b**, **d**) respectively

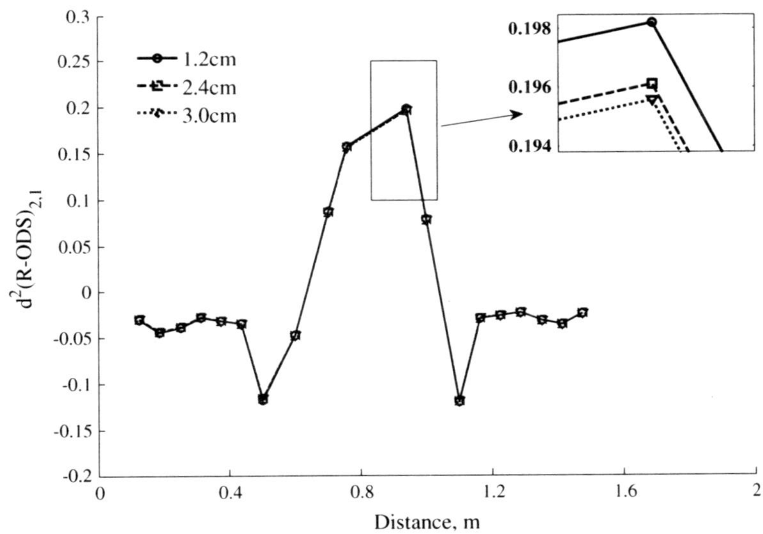

Fig. 5 Effect of crack size on the curvature of RODS

discontinuity even further. However, when there is more than one crack in the rotor the R-ODS plot may not provide a clear indication of all the crack locations, since the amplitude of discontinuity depends on the breathing intensity of each crack.

This fact can be seen in Fig. 4 for the rotor with multiple cracks. In such cases, the curvature of the R-ODS (Fig. 4b, d) plots presents encouraging results for detecting the crack locations.

The rotor in Case 2 is used here to assess the ability of the curvature of R-ODS method to detect the severity of cracks. For this purpose crack sizes of 1.2, 2.4 and 3.0 cm were used in the rotor. As Fig. 5 demonstrates, increase in the crack depth reduces the magnitude of the R-ODS curvature.

5 Conclusion

The application of the curvature of R-ODS method to the simplified LP turbine of a steam generator is presented in this paper. The numerical model consists of a stationary stepped shaft with multiple balance disks representing the blades of the turbine. The details of modelling the breathing fatigue crack(s) are discussed. Shafts with different numbers of cracks at various locations were excited at their resonant frequency. Afterwards, the acceleration response at different locations along the shafts was computed to construct the ODS of the cracked shafts. Normalised ODSs were then utilised to obtain the R-ODSs. The numerical results illustrated that the R-ODS method indicates the location of single-crack shaft by showing a sharp discontinuity. However, when multiple cracks exist in the shaft, the amplitudes of discontinuity relating to each crack may not have the same magnitude in the R-ODS plot. In other words, depending upon the intensity of breathing, the discontinuity generated by one crack might be significantly smaller than the other crack in the R-ODS plot. Therefore, the second derivative is used to remove the curvature of R-ODS plots so that all the discontinuities relating to the crack locations can be observed clearly.

References

1. Sabnavis G et al (2004) Cracked shaft detection and diagnostics: a literature review. Shock Vib Digest 36:287–296
2. Pennacchi P et al (2006) A model-based identification method of transverse cracks in rotating shafts suitable for industrial machines. Mech Syst Signal Process 20:2112–2147
3. Chasalevris AC, Papadopoulos CA (2009) A continuous model approach for cross-coupled bending vibrations of a rotor-bearing system with a transverse breathing crack. Mech Mach Theory 44:1176–1191
4. Darpe AK (2007) A novel way to detect transverse surface crack in a rotating shaft. J Sound Vib 305:151–171
5. Sinou J-J (2008) Detection of cracks in rotor based on the 2 × and 3 × super-harmonic frequency components and the crack–unbalance interactions. Commun Nonlinear Sci Numer Simul 13:2024–2040
6. Sawicki JT et al (2011) Detecting cracked rotors using auxiliary harmonic excitation. J Sound Vib 330:1365–1381

7. Asnaashari E, Sinha JK (2013) Development of residual operational deflection shape for crack detection in structures. Mech Syst Sig Process 43:113–123
8. Sinha JK et al (2002) Simplified models for the location of cracks in beam structures using measured vibration data. J Sound Vib 251:13–38
9. Christides S, Barr ADS (1984) One-dimensional theory of cracked Bernoulli-Euler beams. Int J Mech Sci 26:639–648
10. Friswell MI, Mottershead JE (1995) Finite element model updating in structural dynamics. Kluwer Academic Publishers, Dordrecht

Application of Composite Spectrum in Steam Turbo-Generator Set

Keri Elbhbah, Jyoti K. Sinha, W. Hahn and G. Tasker

Abstract Composite spectrum and composite bispectrum techniques have been recently developed for identifying fault(s) in rotating machines. The fusion of the measured vibration data at all the machine bearings in the frequency domain is used to construct such a composite spectrum and composite bispectrum for the representation of the entire machine. These techniques are now applied to a Turbo-generator (TG) set which consists of a high pressure (HP) turbine, an intermediate pressure (IP) turbine and 3 low pressure (LP) turbines in addition to an exciter and a generator. Earlier study using the spectrum analysis of the measured vibration acceleration data on the bearings of the TG set during machine normal operation indicates the possibility of the rotating stall in LP turbines and misalignment between the IP and LP rotors. The observations and results of the current study also indicate the presence of localised misalignment and suspected rotating stall in the composite spectrum.

Keywords Rotating machines · Condition monitoring · Spectrum · Data fusion · Composite spectrum

K. Elbhbah (✉) · J.K. Sinha
School of Mechanical, Aerospace and Civil Engineering, The University of Manchester, Oxford Road, M13 9PL Manchester, UK
e-mail: keri.elbhbah@postgrad.manchester.ac.uk

J.K. Sinha
e-mail: jyoti.sinha@manchester.ac.uk

W. Hahn · G. Tasker
West Burton Power Station, EDF Energy, DN22 9BL Retford, Nottinghamshire, UK
e-mail: Wolfgang.Hahn@edfenergy.com

G. Tasker
e-mail: Geoff.Tasker@edfenergy.com

J.K. Sinha (ed.), *Vibration Engineering and Technology of Machinery*, Mechanisms and Machine Science 23, DOI 10.1007/978-3-319-09918-7_11

1 Introduction

West Burton power plant is a plant owned by EDF Company and consists of 4 steam (TG) units for power generation. Each TG unit consists of a High Pressure (HP) turbine, an Intermediate Pressure (IP) turbine, and three Low Pressure (LP) turbines, together with a generator and an exciter. The schematic of the TG set consisting of 14 journal bearings are shown in Fig. 1.

Vibration based diagnosis often requires vibration measurements on all bearing pedestals in both horizontal and vertical directions for the fault diagnosis. Such measurement approach is often necessary for the bigger machines/structures so that local defect/fault can be detected reliably. However, this process generally leads to the analysis of a number of data and make the diagnosis a complex process. Similarly, the in situ vibration measurement at all the 14 bearings on the TG set during machine normal operation at 3,000 RPM were carried out and data were analysed earlier., so as to understand the dynamic behaviour of the TG set [1]. The misalignment was identified at the coupling between the IP and the LP1 rotor, based on the appearance of 2nd harmonic peak (2×) of 3,000 RPM in the vibration spectra at bearings 4 and 5. The appearance of a band peak around 7–12 Hz and its modulation with 50, 100 Hz and higher multiple on the bearings related to the LP turbines only. Such a phenomenon generally indicates the possibility of rotor stall, for which the LP turbines seem to be the source [1]. This effect is not seen on the HP, IP and the generator bearings.

Recently, data fusion technique is proposed to construct a single spectrum representing the dynamic behaviour of the complete rotating machine condition from the measured vibration data from all bearings [2]. The composite spectrum is observed to be useful for fault diagnosis through initial experiments conducted on a small experimental rotating rig [3]. This is expected to simplify the condition monitoring procedure and fault diagnosis process. This technique is now applied to the TG set, so as to observe weather the single composite spectrum of the TG set can detect the localized misalignment and possible stall phenomenon. The observations and the results are discussed in the paper.

2 Spectrum

The conventional power spectrum density (PSD) provides information on the second-order properties (i.e., energy) of a signal. The PSD of a time series $x(t)$is computed by the discrete Fourier transform (DFT) of the signal as;

$$\text{PSD, } S_{xx}(f_k) = E[X(f_k)X^{*}f_k)], \quad k = 1, 2, 3, \ldots, N \tag{1}$$

where, $S_{xx}(f_k)$ is the power density, $X(f_k)$ is the DFT at frequency f_k for the time series$x(t)$ and $X^{*}(f_k)$ is its complex conjugate. N is the number of the frequency points. E[] denotes the mean operator, here it means that the PSD is the averaged spectrum over a time length of the signal, say, t.

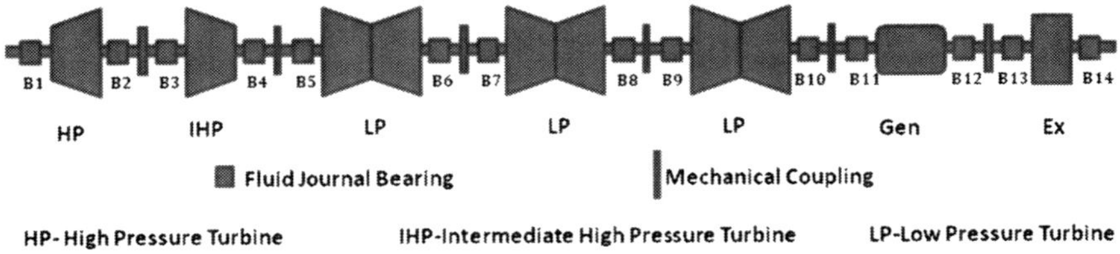

Fig. 1 Schematic diagram of Turbo generator set

Fig. 2 Typical amplitude spectrum at bearing 4

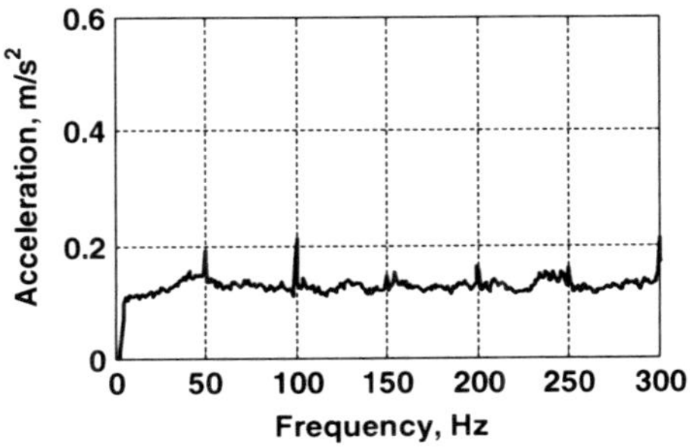

Fig. 3 Typical amplitude spectra at bearing 11

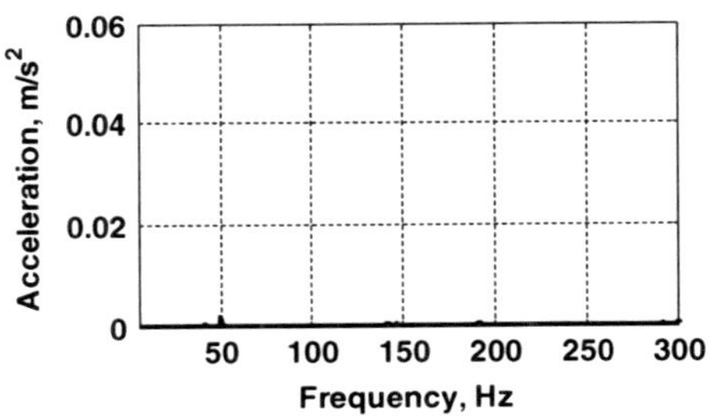

2.1 Spectrum Analysis [1]

Typical measured amplitude spectra at bearing 4 (shown in Fig. 2) shows peaks at the first two harmonics 1× (rotation speed- 50 Hz) and 2 × (100 Hz). This generally indicates the presence of misalignment. It is possible that a small degree of misalignment between the HP and IP rotors is deliberately kept, as it is expected that the rotor well self-align during operation.

Similarly, the measured spectra at the bearings related to the LP turbines shown in Fig. 4 show the appearance of a banded peak around 7–12 Hz and its modulation at 50, 100, 150 Hz.... Which is likely to be related to the stall phenomenon [1]. Figure 3 show that the measured vibration spectrum at the generator (i.e. bearing 11) does not show any indication of this phenomenon.

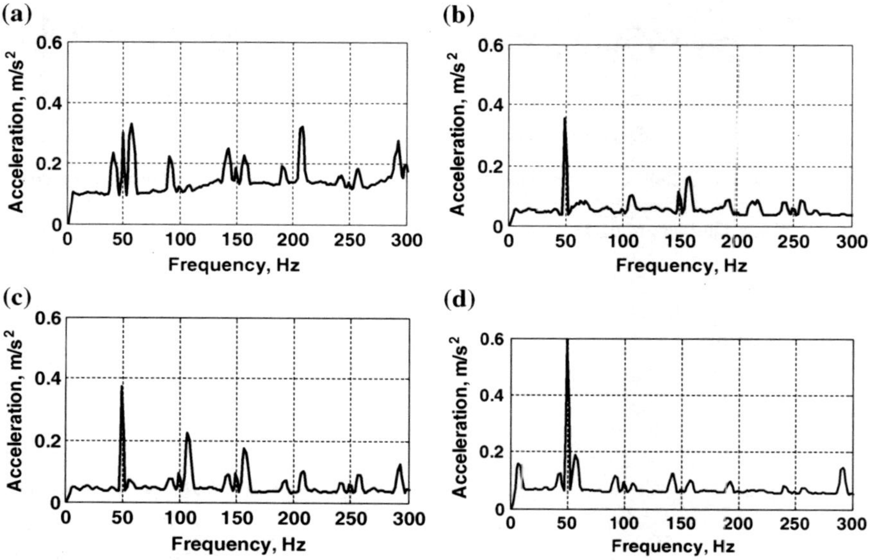

Fig. 4 Typical amplitude spectra at 3,000 RPM. **a** Bearing 5. **b** Bearing 6. **c** Bearing 7. **d** Bearing 8

3 Composite Spectrum

The composite spectrum is a Data Fusion of data gathered from a number of sensors in the frequency domain. The measured vibration data are divided into "n" numbers of equal segments for all the bearings and the composite spectrum is calculated as [3];

$$S_{CS}(f_k) = \frac{\sum_{r=1}^{n} X_{CS}^{r}(f_k) X_{CS}^{*}(f_k)}{n} \tag{2}$$

where, $X_{CS}^{r}(f_k)$ is called the Composite Fourier Transformation (FT) for the rth segment of data from all bearings at the frequency, f_k, and $X_{CS}^{*}(f_k)$ is its complex conjugate. $X_{CS}^{r}(f_k)$ is computed as;

$$X_{CS}^{r}(f_k) = \sqrt{\left(S_{x_1\gamma_{12}x_2}(f_k) S_{x_2\gamma_{23}x_3}(f_k) \ldots S_{x_{(b-1)}\gamma_{(b-1)b}x_b}(f_k)\right)^{1/(b-1)}} \tag{3}$$

where, $S_{x_1\gamma_{12}x_2}$ is the Cross-power Spectral Density (CSD) between the two signals $x_1(t)$ and $x_2(t)$ collected at two bearings. It is also computed by using the

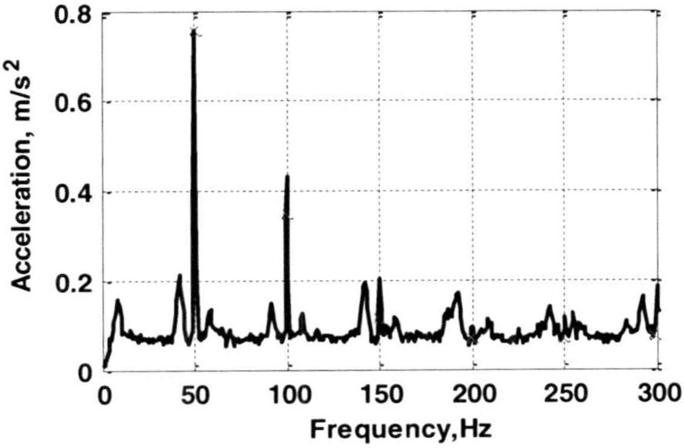

Fig. 5 Typical amplitude composite spectra at 3,000 RPM

Discrete Fourier transform (DFT) of these signals as;

$$\text{CSD, } S_{x_p \gamma_{p(p+1)} x_{p+1}}(f_k) = [X_p(f_k)\gamma^2_{p(p+1)}(f_k)X^*_{p+1}(f_k)] \tag{4}$$

where, $\gamma^2_{p(p+1)}$ is the ordinary coherence [4] between two signals averaged over the time length.

3.1 Composite Spectrum Analysis

Composite spectrum (CS) has been computed for the measured vibration data from all bearings. It will be good to determine whether the CS provides a single representative spectrum, comprising of all the diagnosis features of both faults. Figure 5 represents a typical CS, which shows the peaks at 1×, 2×, and 3×, in addition to the appearance of the banded peak around 7–12 Hz and its modulation with 1×, 2× and 3×,.... components. The presence of high harmonics and banded peak and its modulation clearly confirm that the composite spectrum is capable of identifying the local defects.

4 Conclusion

Data fusion (composite spectrum) proposed in an earlier study is applied on a vibration data collected at a TG set at West Burton power plant. The study confirms that the single composite spectrum for a complete machine is capable of reliable fault(s) diagnosis. The study also highlights the fact that the use of the composite spectrum makes the diagnosis easier compared to the adopted practice in the vibration-based diagnosis in rotating machines.

References

1. Sinha JK, Hahn W, Elbhbah K, Tasker G, Ullah I (2012) Vibration investigation for low pressure turbine last stage BLADE failure in steam turbines of a power plant. ASME TurboExpo 2012, June 11–15. Copenhagen, Denmark
2. Sinha JK, Elbhbah K (2013) A future possibility of vibration based condition monitoring of rotating machines. Mech Syst Sig Process (MSSP) 34(2013):231–240
3. Elbhbah K, Sinha JK (2013) Vibration-based condition monitoring of rotating machines using a machine composite spectrum. J Sound Vib (JSV) 332(2013):2831–2845
4. Ewins DJ (2000) Modal testing-theory practice and application, 2nd edn. Research Studies Press, UK

Coherent Composite HOS Analysis of Rotating Machines with Different Support Flexibilities

Akilu Yunusa-Kaltungo and Jyoti K. Sinha

Abstract Earlier studies have shown the possibilities of detecting different faults on a rigidly supported rotating machine through coherent composite spectrum (CCS) and coherent composite bispectrum (CCB) data fusion techniques. The current paper is also related to the use of CS, CB and a newly introduced coherent composite trispectrum (CCT) for faults identification in flexibly supported rotating machines. Six experimentally simulated cases have been considered on two sets of flexibly supported (FS1 and FS2) rotating machines, at 1,200 RPM (20 Hz) speed. Vibration data were then collected from both FS1 and FS2, using only four accelerometers (one per bearing pedestal in the diagonal direction), for the computation of CCS, CCB and CCT. Some combinations of CCB and CCT components were then used to further test the robustness of the techniques in rotating machines' faults differentiation. The observations and results of the experiments are hereby discussed here.

Keywords Rotating machines · Flexible supports · Condition monitoring · Fault diagnosis · Data fusion · Coherent composite spectrum · Coherent composite bispectra · Coherent composite trispectra

1 Introduction

Rotating machines are often subjected to a wide variety of operational conditions, which in turn leads to the emergence of different faults (e.g. shaft misalignment, bent shaft, looseness, cracked shaft, shaft rub, etc.). The presence of such faults

A. Yunusa-Kaltungo (✉) · J.K. Sinha
School of Mechanical, Aerospace and Civil Engineering, The University of Manchester, Oxford Road, Manchester M13 9PL, UK
e-mail: Akilu.kaltungo@postgrad.manchester.ac.uk

J.K. Sinha
e-mail: jyoti.sinha@manchester.ac.uk

J.K. Sinha (ed.), *Vibration Engineering and Technology of Machinery*, Mechanisms and Machine Science 23, DOI 10.1007/978-3-319-09918-7_12

(irrespective of their severities) usually alters the dynamic characteristics of this class of valuable and versatile machines, which emphasizes the significance of detection at incipient stages. Over the years, the application of several sensors (accelerometers, proximity probes, tachometers, etc.) at individual bearing pedestals for the collection and diagnosis of vibration data related to rotating machines has been the norm [1]. However, when dealing with large and complex rotating machines, such as multi-staged drive assemblies and turbo-generator sets (which are often characterized by the existence of numerous bearings), there is bound to be a correspondingly large number of vibration measurement sensors (usually 2 or 3 accelerometers per bearing pedestal, 2 eddy current probes per bearing pedestal and a tachometer for shaft speed reference measurement) for collecting data that will be used to accurately detect and differentiate changes in machine characteristics due to emerging fault(s).

Sinha [2] comprehensively studied various health monitoring techniques for rotating machines. Studies by Jardine et al. [3] and Mehrjou et al. [4] respectively provided extensive reviews of the widely applied condition monitoring techniques for rotating machines and squirrel-cage induction machines. Similarly, a host of other studies [5–13] have also given details of different vibration-based condition monitoring techniques for faults detection in rotating machines, ranging from full spectrum analysis [5]; wavelet analysis [6]; intelligent order tracking [7]; vibration symptoms trend analysis [8]; higher order spectra (HOS) analysis [9–11]; etc., through the application of measured vibration data.

Relatively recent efforts aimed at simplifying rotating machines condition monitoring as well as reducing the number of sensors needed were explored through the development of the coherent composite spectrum (CCS) and coherent composite bispectrum (CCB) data fusion (in the frequency domain) techniques [12, 13]. Both CCS and CCB techniques showed meaningful diagnosis and fault differentiation potentials on a relatively rigid rotating machine with several simulated cases (healthy, misalignment, crack and rub). The current study is also related to the diagnosis of faults in flexibly mounted rotating machines, through the inclusion of the new coherent composite trispectrum (CCT) technique (in addition to the earlier CCS and CCB techniques), with the sole aim of minimizing the number of vibration sensors needed for diagnosis.

In this study, different operational conditions (healthy with residual misalignment, shaft misalignment, cracked shaft, bearing looseness, bent shaft and shaft rub) were simulated on two experimental rigs with varying support mounting flexibilities (i.e. flexible support 1 and flexible support 2) at 1,200 RPM (20 Hz) rotational speed, so as to observe the consistency and versatility of the proposed technique. Furthermore, vibration data from both flexible supports (FS1 and FS2) were collected using a single sensor (accelerometer) at each bearing pedestal in the diagonal direction. The measured vibration data were then used for computing the CCS, CCB and CCT for the individual machines through data fusion in the frequency domain, which aims to eliminate the need for computations at individual machine bearings. Hence, the results and observations of the CCS, CCB and CCT for FS1 and FS2 are discussed in the paper.

2 CCS, CCB and CCT Computations

The potentials of CCS and CCB data fusion techniques for the diagnosis of faults related to rotating machines were initially presented in earlier studies by Elbhbah et al. [12] and Sinha et al. [13]. In the current study however, the use of CCT is proposed for the diagnosis of different simulated cases on two flexibly supported experimental rigs (FS1 and FS2), with the aim of further simplifying faults diagnosis. Hence, the computational approaches of CCS, CCB and CCT are discussed here.

Assuming that vibration measurements were collected at "b" number of bearing locations on a rotating machine, and the measured data were divided into "n_s" number of equal segments, then the CCS for the entire machine was computed as [12, 13];

$$S_{CCS}(f_k) = \frac{\sum_{r=1}^{n_s} X_{CCS}^{r}(f_k) X_{CCS}^{r^*}(f_k)}{n_s} \quad (1)$$

where $X_{CCS}^{r}(f_k)$ and $X_{CCS}^{r^*}(f_k)$ are respectively the coherent composite Fourier Transformation (FT) and its complex conjugate for the rth segment of the measured vibration data collected from "b" number of bearings at frequency, f_k. $X_{CCS}^{r}(f_k)$ is thus computed as [12, 13];

$$X_{CCS}^{r}(f_k) = \sqrt{\left(S_{x_1\gamma_{12}^2 x_2}^{r}(f_k) S_{x_2\gamma_{23}^2 x_3}^{r}(f_k) S_{x_{(b-1)}\gamma_{(b-1)b}^2 x_b}^{r}(f_k)\right)^{\frac{1}{(b-1)}}} \quad (2)$$

In Eq. 2, $\gamma_{12}^2, \gamma_{23}^2$ and $\gamma_{(b-1)b}^2$ respectively represent the coherence between bearings 1–2, 2–3, ..., (b−1)−b. Also, $S_{x_1\gamma_{12}^2 x_2}^{r}(f_k)\, S_{x_2\gamma_{23}^2 x_3}^{r}(f_k)\, S_{x_{(b-1)}\gamma_{(b-1)b}^2 x_b}^{r}(f_k)$ respectively denote the coherent cross-power spectrum between bearings 1–2, 2–3, ..., (b−1)−b, which was computed as [12, 13];

$$S_{x_1\gamma_{12}^2 x_2}^{r}(f_k) = \left[X_1^{r}(f_k)\gamma_{12}^2 X_2^{r^*}(f_k)\right] \quad (3)$$

HOS [9, 10], mainly bispectrum and trispectrum provide information about the relationship between frequency components in a time domain signal $x(t)$. The bispectrum represents a combination of two frequencies (each having amplitude and phase), f_l and f_m with a third frequency $f_l + f_m$ which is equal to the sum of the prior two. Similarly, the trispectrum represents a combination of three frequencies (each having amplitude and phase), f_l, f_m and f_n with a fourth frequency $f_l + f_m + f_n$ which is equal to the sum of the prior three. Hence, the CCB and CCT for the measured vibration data were computed as [11, 13];

$$\mathrm{B}(f_l, f_m) = \frac{\sum_{r=1}^{n_s}\left(X_{CCS}^{r}(f_l) X_{CCS}^{r}(f_m) X_{CCS}^{r\,*}(f_l + f_m)\right)}{n_s} \quad (4)$$

$$T(f_l,f_m,f_n) = \frac{\sum_{r=1}^{n_s} \left(X_{CCS}^r(f_l) X_{CCS}^r(f_m) X_{CCS}^r(f_n) X_{CCS}^{r\ *} (f_l + f_m + f_n)\right)}{n_s} \tag{5}$$

3 Experimental Rigs

Two sets of flexibly supported experimental rigs (Figs. 1, 2) have been used to simulate different cases of typical rotating machines. Although both experimental rigs are made up of similarly dimensioned components, however, their dynamic properties differ significantly, due to the varying stiffness of their supports. In both set-ups, two mild steel shafts (1,000 and 500 mm lengths respectively) of 20 mm diameters are rigidly coupled together. The 1,000 mm shaft is then coupled to a 0.75 kW electric motor, while the entire rotor assemblies are supported by 4 anti-friction bearings, using 10 and 6 mm diameter threaded bars respectively. Also, there are three mild steel balance discs of dimensions 125 mm (outside diameter) × 15 mm (thickness), with two of the discs fitted on the long shaft (first disc is 300 mm from the drive motor and the second is 190 mm from the second bearing) and the third on the shorter shaft (210 mm from both bearings 3 and 4). In Fig. 2, a–m respectively represent accelerometer, rigid coupling, shaft, balance disc, bearing flange, threaded bar, anti-friction ball bearing, flexible coupling, tachometer, electric motor, motor base, lathe bed and neoprene rubber pad.

4 Experimental Simulation of Faults

The experiments were conducted on two different set-ups (FS1 and FS2) at a rotational speed of 1,200 RPM (20 Hz) for a total of six cases, namely; healthy with residual misalignment (HRM), shaft misalignment (SM), cracked shaft (CS), bearing looseness (BL), bent shaft (BS) and shaft rub (SR). According to the authors' experience with long slender shafts, it is usually very difficult to achieve a perfect alignment; owing to the fact that such shafts are sometimes slightly bowed, which could cause alignment problems. Therefore, the HRM case represents the baseline case. A 0.4 mm shim was used to introduce misalignment in the vertical direction near bearing 1 (SM case). A 4 mm (depth) × 0.25 mm (width) breathing crack was induced on the 1,000 mm shaft at a distance of 160 mm from bearing 1 (CS case). The BL case was studied by loosening the threaded bar nuts on bearing 3. A centre line run-out of 3.4 mm was created on the 1,000 mm shaft for the BS case. The SR case was simulated using 2 Perspex blades placed at 275 mm from bearing 1. Vibration responses were then collected from both set-ups, using 4 accelerometers per set-up (i.e. one accelerometer per bearing in the diagonal direction, so as to acquire vibration data representing both vertical and horizontal directions), and then recorded on to a PC for eventual analysis using a MATLAB code.

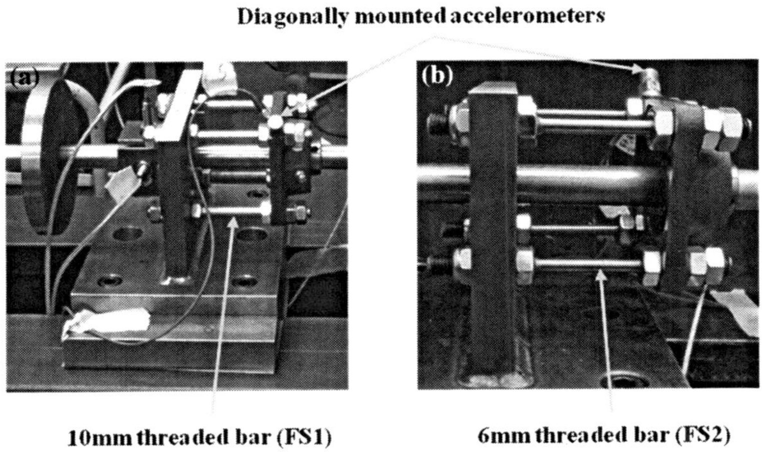

Fig. 1 Typical flexible bearing pedestals with anti-friction ball bearings: **a** FS1 (10 mm threaded bars) and **b** FS2 (6 mm threaded bars)

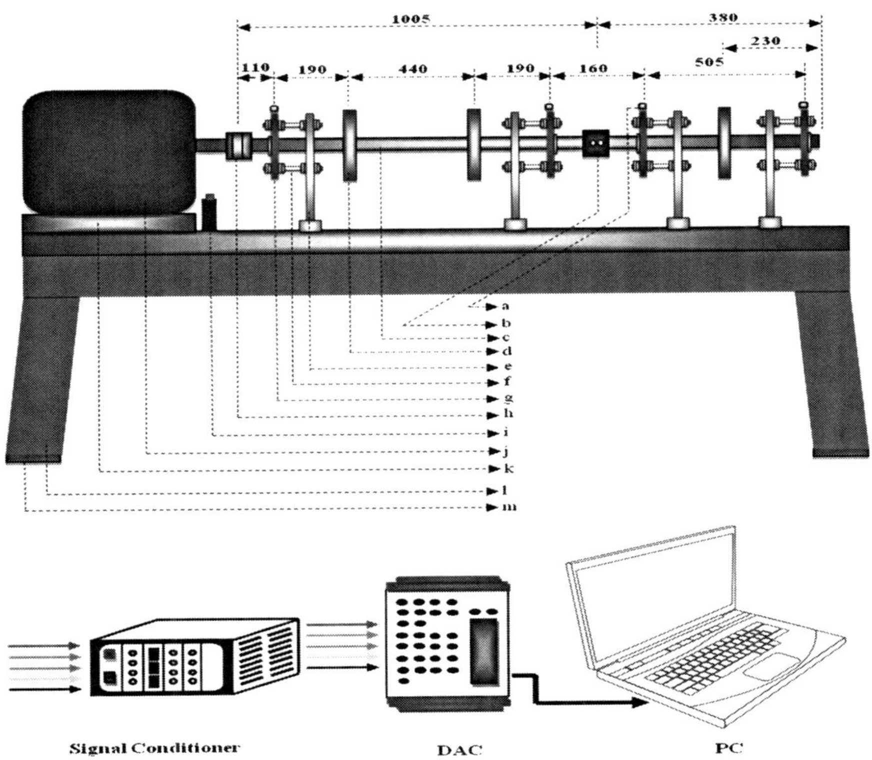

Fig. 2 Schematic representation of experimental rig (all dimensions in mm)

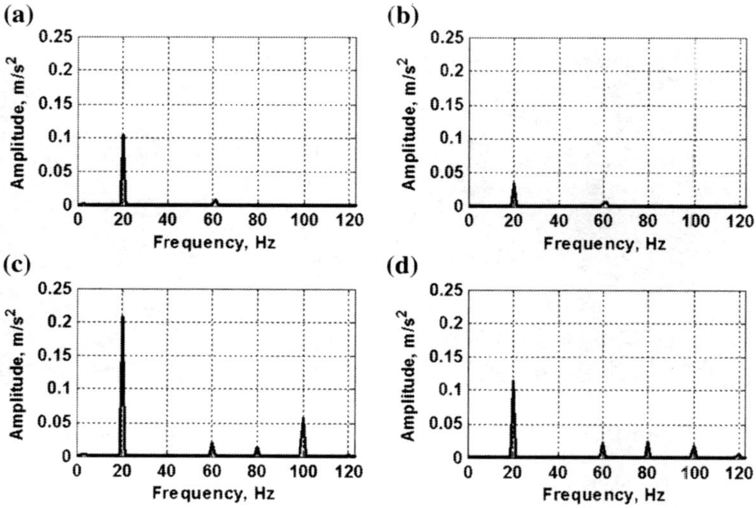

Fig. 3 Typical coherent composite spectra (CCS): **a** HRM (FS1), **b** HRM (FS2), **c** BL (FS1), and **d** BL (FS2)

5 Data Analysis and Results

Figures 3, 4 and 5 show typical CCS, CCB and CCT plots of the measured vibration data for two (HRM and BL) of the six experimentally simulated cases, on both set-ups at 20 Hz speed, which were computed as per Eqs. (1)–(5). The measured vibration data were analysed using a 95 % overlap, frequency resolution (df) = 1.2207 Hz, sampling frequency (f_s) = 10,000 Hz and number of data points (N) = 8,192.

The CCS plots (Fig. 3), clearly show a distinction between the two cases (HRM and BL) on both FS1 (Fig. 3a, c) and FS2 (Fig. 3b, d). On both set-ups, BL case was characterized by several higher harmonics of the machine speed with correspondingly higher amplitudes when compared to the HRM case. The composite HOS (CCB and CCT) plots in Figs. 4 and 5 show the relation between different frequency components (each having amplitude and phase) in the measured vibration data. For instance, B_{11} CCB component (Fig. 4) denotes the relation between 1× (twice) and 2× frequency components, while B_{12} denotes the relation between 1×, 2× and 3× frequency component, and so on. Similarly, T_{111} CCT component (Fig. 5) denotes the relation between 1× (thrice) and 3× frequency components in the measured vibration data and so on. It can be seen that CCS, CCB and CCT plots provides a distinction between the different experimentally simulated cases. Additionally, a set of 20 measurements have been collected for each of the experimentally simulated cases at 20 Hz speed, for which the CCB and CCT for each set of data was computed as per Eqs. (1)–(5). Figure 6 shows plots of the

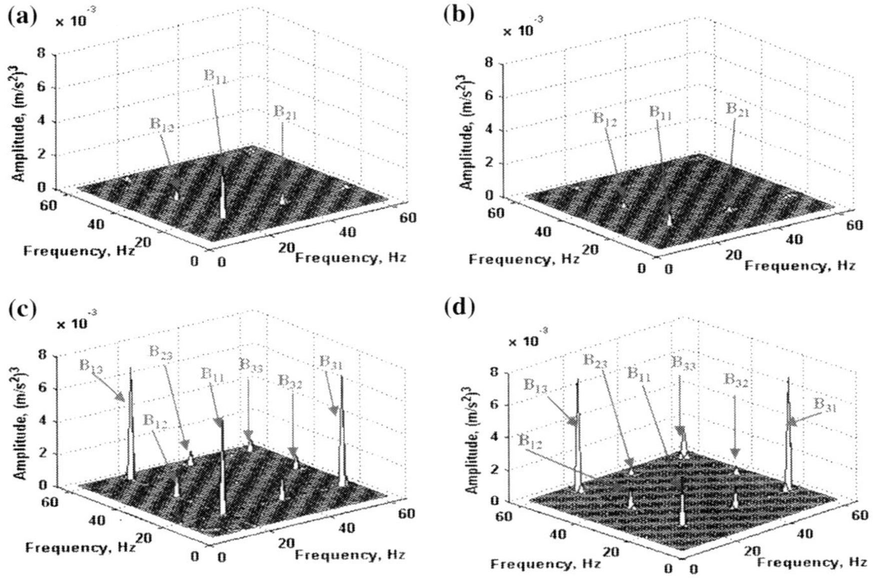

Fig. 4 Typical coherent composite bispectra (CCB): **a** HRM (FS1), **b** HRM (FS2), **c** BL (FS1), and **d** BL (FS2)

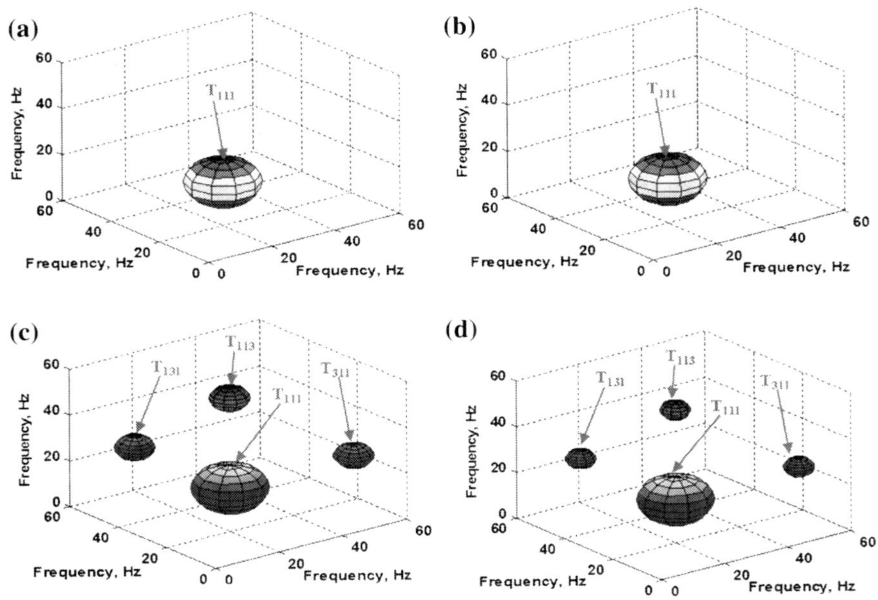

Fig. 5 Typical coherent composite trispectra (CCT): **a** HRM (FS1), **b** HRM (FS2), **c** BL (FS1), and **d** BL (FS2)

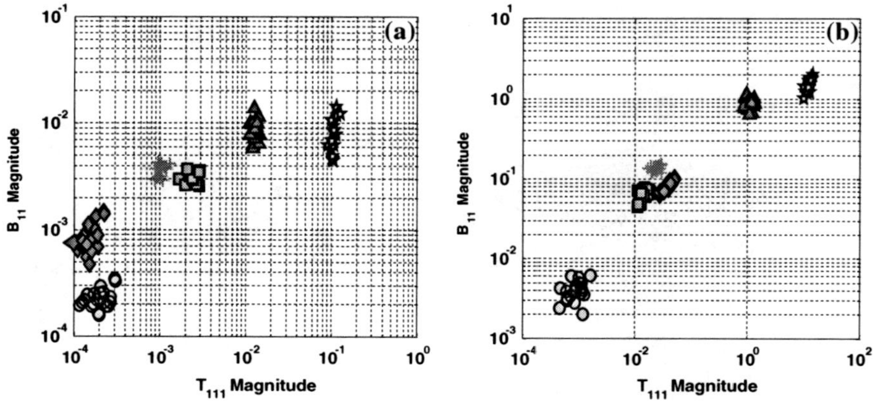

Fig. 6 Typical plots of T_{111} CCT versus B_{11} CCB components amplitudes: **a** FS1 and **b** FS2

magnitudes of T_{111} CCT component against the magnitudes of B_{11} CCB component for all cases, so as to ascertain the capabilities of CCB and CCT in faults diagnosis, which has been observed to show encouraging classification potentials for the different cases.

6 Conclusion

The current research has explored the possibilities of using CCS, CCB and the newly introduced CCT data fusion techniques for the diagnosis of different experimentally simulated cases on flexibly supported rotating machines, using a reduced number of vibration sensors. The experimental results have also shown encouraging possibilities of rotating machines' faults identification and differentiation, through the application of a combination of very few CCB and CCT components amplitudes. However, further experiments are planned at different machines' speeds, so as to further boost the certainty of the techniques.

References

1. Yunusa-Kaltungo A, Sinha JK (2012) Condition monitoring: a simple and practical approach (dissertation converted into a book). Lambert Academic Publishing (LAP) GmbH & Co. KG. ISBN: 978-3-8473-7807-3109
2. Sinha JK (2002) Health monitoring techniques for rotating machinery. Ph.D. thesis, University of Wales (Swansea University), Swansea, UK
3. Jardine AKS, Lin D, Banjevic D (2006) A review on machinery diagnostics and prognostics implementing condition-based maintenance. Mech Syst Signal Process 20(2006):1483–1510

4. Mehrjou MR, Mariun N, Marhaban MH, Misron N (2011) Rotor fault condition monitoring techniques for squirrel-cage induction machine—a review. Mech Syst Signal Process 25 (2011):2827–2848
5. Goldman P, Muszynska A (1999) Application of full spectrum to rotating machinery diagnostics. Orbit, 17–21, Bently Nevada Corporation, USA, 1st Quarter
6. Peng ZK, Chu FL (2004) Application of the wavelet transform in machine condition monitoring and fault diagnostics: a review with bibliography. Mech Syst Signal Process 18 (2004):199–221
7. Bai M, Huang J, Hong M, Su F (2005) Fault diagnosis of rotating machinery using an intelligent order tracking system. J Sound Vib 280(3–5):699–718
8. Galka T, Tabaszewski M (2011) An application of statistical symptoms in machine condition diagnostics. Mech Syst Signal Process 25(2011):253–265
9. Howard IM (1997) Higher-order spectral techniques for machine vibration condition monitoring. Proc Inst Mech Eng Part G: J Aerospace Eng 211(4):211–219
10. Collis WB, White PR, Hammond JK (1998) Higher-order spectra: the bispectrum and trispectrum. Mech Syst Signal Process 12(3):375–394
11. Yunusa-Kaltungo A, Sinha JK, Elbhbah K (2013) HOS analysis of measured vibration data on rotating machines with different simulated faults. Advances in condition monitoring of machinery in non-stationary operations, Chap. 6 (Proceedings of the International Conference on Condition Monitoring of Machinery in Non-Stationary Operations 2013). Springer Publishing, Series: 11236. ISBN: 978-3-642-39347-1
12. Elbhbah K, Sinha JK (2013) Vibration-based condition monitoring of rotating machines using a composite spectrum. J Sound Vib 332(11):2831–2845
13. Sinha JK, Elbhbah K (2013) A future possibility of vibration-based condition monitoring of rotating machines. Mech Syst Signal Process 34(1–2):231–240

Vibration Behaviour of a Turbo-Generator Set

Wolfgang Hahn and Jyoti K. Sinha

Abstract In situ vibration measurements are carried out on a a typical steam turbo-generator (TG) unit at the West Burton Power Plant UK during the steady state operation. A typical phenomenon of appearance of low frequencies in band of 7–12 Hz is observed mainly related to the vibration measurements on low pressure (LP) turbines bearing pedestals. This band of frequency observed to be modulating with vibration at the machine RPM (50 Hz) and its higher harmonics. The paper summaries the possible causes for such dynamic behaviour.

Keywords Steam turbine · LP blade cracking · Vibration measurement · Vibration spectrum · Rotor stall

1 Introduction

West Burton Power Plant, UK owned by EDF Energy has 4 steam turbo-generator (TG) units for the power generation. Each TG unit consists of a high pressure (HP) turbine, an intermediate pressure (IP) turbine and three low pressure (LP) turbines together with a generator and an exciter [1, 2]. The 2 TG units have been suffering from blade vibration problems in stages 3, 4 and 5 on the LP turbines and blade cracking in the last stage blades [1]. The cracking in the turbine blades have been found just above the blade root which is expected to be high stress location for the first bending mode for the blade. The last stage bladed disc is shown in Fig. 1. Typical blades with erosion and with crack are shown in Fig. 2. The blade erosion and then crack propagation is suspected as a result of blade excitation during the

W. Hahn
West Burton Power Station, EDF Energy, Nottinghamshire DN22 9BL, UK
e-mail: Wolfgang.hahn@edfenergy.com

J.K. Sinha (✉)
School of MACE, The University of Manchester, Manchester M13 9PL, UK
e-mail: jyoti.sinha@manchester.ac.uk

J.K. Sinha (ed.), *Vibration Engineering and Technology of Machinery*,
Mechanisms and Machine Science 23, DOI 10.1007/978-3-319-09918-7_13

Fig. 1 Typical LP1 last stage front bladed disc (*new design blades*) near IP turbine (*steam end*)

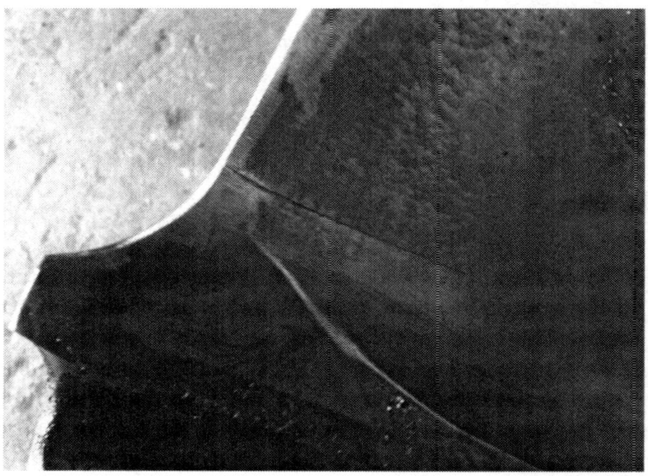

Fig. 2 Typical LP last stage blade with a crack

normal and transient (run-up or run-down) operations [1]. The in situ blade tip timing (BTT) measured data on the last stage blades (steam side) of the Unit 3 LP1 turbine also confirm the high vibration amplitude of a few blades which could cause the blade cracking [2].

In situ vibration measurements are carried out on the Unit 3 during the steady state operation at different power generation outputs to understand the machine vibration behaviour. A typical phenomenon of appearance of low frequencies in

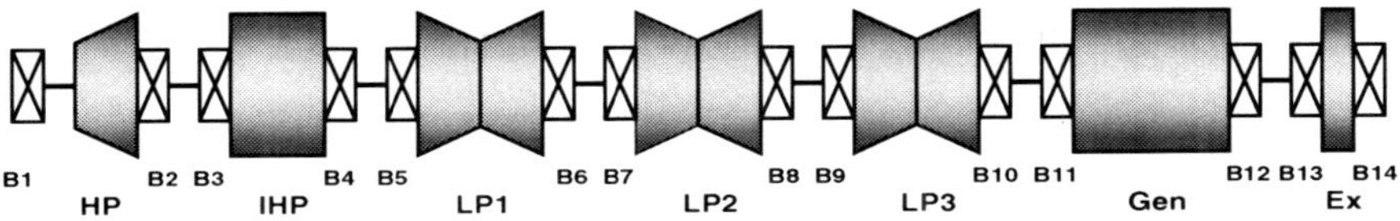

Fig. 3 A simple schematic of TG unit

band of 7–12 Hz is observed mainly related to the vibration measurements on LP turbines. This band of frequency observed to be modulating with vibration at the machine RPM (50 Hz) and its higher harmonics. Such observations are also found in literature related to rotating machines mainly compressors often related to the stall phenomena [3–9] which generally has high potential of damage. The paper summaries the possible causes for such dynamic behaviour of the TG set.

2 Vibration Measurements

A simple schematic of the TG set is shown in Fig. 3. A total of 14 numbers of fluid bearings in each TG unit are used in a TG unit to support the rotors of the steam turbines, generator and exciter. In situ vibration measurements were done on the bearing housings mainly in the vertical direction and also in the horizontal direction at the bearing B5 and B6 for Unit 3 during machine operation but at different power outputs.

3 Spectrum Analysis

The spectrum analysis of all the measured vibration data are carried out to understand the vibration behaviour of the TG set (Unit 3). It is observed that TG unit has in general strong peaks at 1× (1 times RPM) and 2× (2 times RPM) at all the bearings. This indicates there could be some misalignment in the shafts of HP, IP, LP turbines and the generator rotors. Typical 3D waterfall diagram of the measured vibration spectra at bearing B5 in the vertical direction are shown in Fig. 4. Further inspection of spectra also indicates a presence of frequency peaks of 7–12 Hz and this frequency band gets modulated with 50 Hz (1×) and higher harmonics. The averaged spectrum is also computed for all data. A few typical averaged spectra are also shown in Figs. 5 and 6. It is clear from the spectra that the existence and the modulation of 7–12 Hz in several occasions during machine operation. This effect is generally prominent in the LP turbines and often insignificant in the HP-IP turbines (see Fig. 7) and the generator.

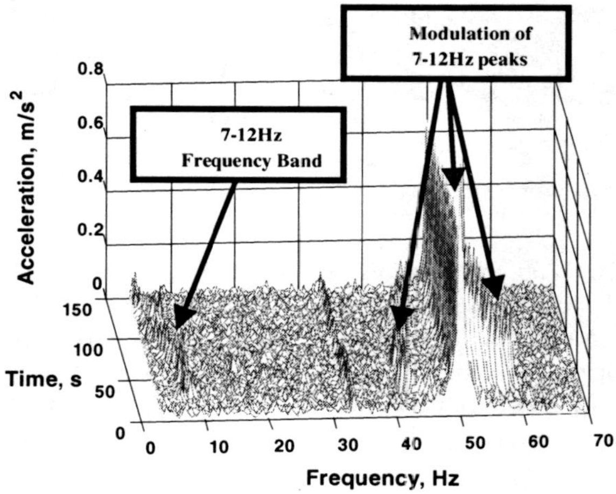

Fig. 4 Typical waterfall diagram of the measured acceleration vibration spectra at the bearing B5 in vertical direction

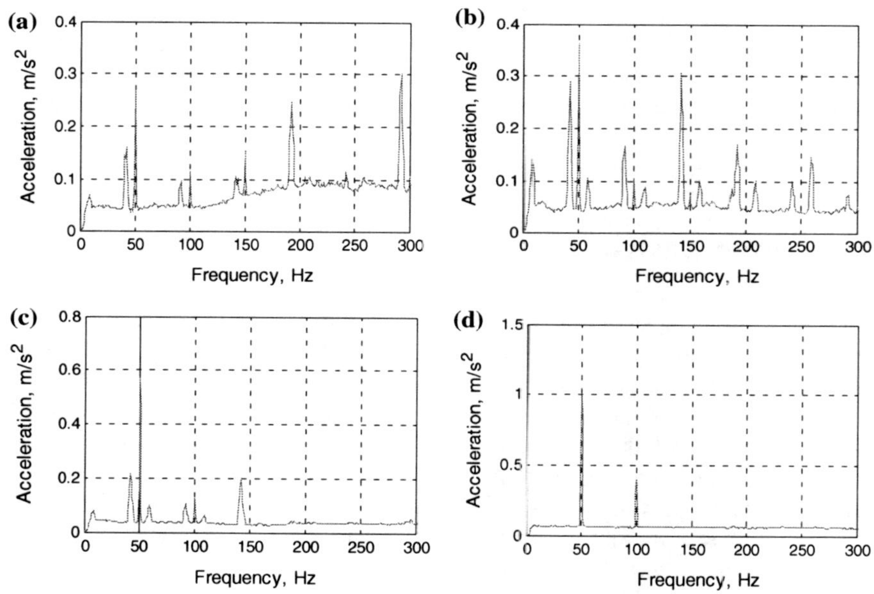

Fig. 5 Measured vibration spectra at the bearing pedestals in the vertical direction, **a** B5, **b** B7, **c** B9, **d** B11

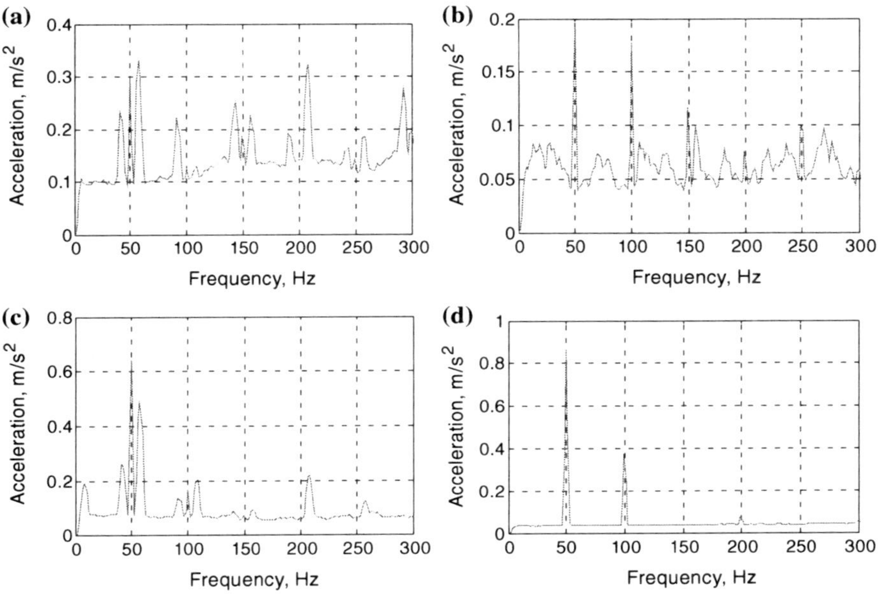

Fig. 6 Measured vibration spectra at the bearing pedestals in the vertical, **a** B5, **b** B7, **c** B9, **d** B11

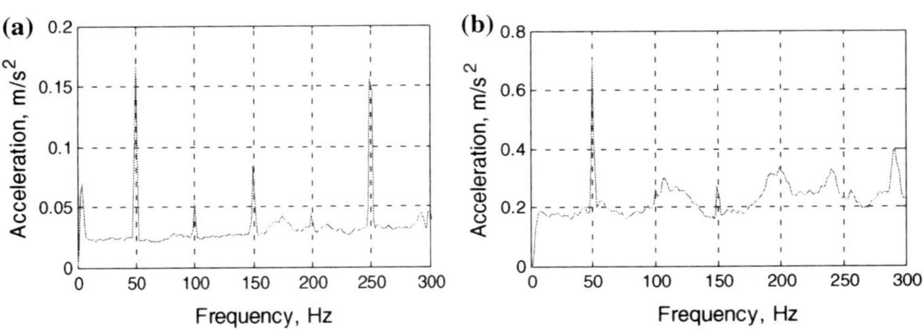

Fig. 7 Typical Measured vibration spectra at the bearing pedestals in the vertical direction **a** B2, **b** B4

4 Further Investigation

The phenomenon of low frequency 7–12 Hz peak and its modulation with the 1× and higher harmonic components is not a usual feature for the steam turbine. It is also important to note that it is not exiting continuously and not clearly observed in the HP-IP turbines and the generator. The phenomenon is often intermittent and may be depending upon several operating parameters and possibly related with the LP turbines.

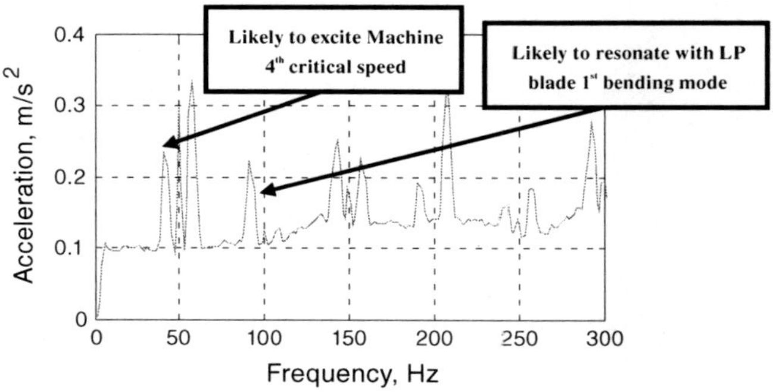

Fig. 8 Zoom view of Fig. 6a with comments to highlight the potential resonance at 4th machine critical speed and at blade 1st bending mode during stall condition

Often the fluid induced instability in the fluid bearing generates a frequency peak below sub-harmonic frequency (0.5×) [3], however the present observations are much different. The experimental study on a compressor by Bently et al. [4] suggests that this phenomenon could be related to the rotating stall which may often have potential for any damage. This suggests the suspicion on the rotating stall in the LP turbines. The rotating stall in the compressor is generally related to the insufficient vacuum for the smooth air flow or fluid bearing related problem [4].

Another study by Wu and Qu [5] on the compressor also observed the modulation around the rotating speed due to inappropriate angle of attack of the blade inlet causing the rotating stall. If the inlet steam relative velocity is not tangential to the blade then it may result in the rotating stall. The inappropriate angle of attack for the blade inlet seems to be the common problem for the rotating stall, back pressure pulse or flutter in the last stage LP turbine blades [6, 7]. Owczarek [8] gave an excellent overview of the pressure pulse reflecting phenomenon between the blades and rotating stall that can cause the blade failure.

Hence, considering the earlier studies on the rotordynamics and the present observations in the spectra of the TG unit, the possibility of rotating stall cannot be ruled out. The presence of the low frequency 7–12 peaks and their modulation with the machine 2× and higher harmonic components in vibration is definitely results into the resonance of the machine at 4th critical speed (46.88 Hz [1]) and at the blade 1st bending natural frequency around 90 Hz [1, 2]. It is typically marked in a measured vibration spectrum at the bearing B5 in the vertical direction shown in Fig. 8 for the clarity. This phenomenon can have high potential for the failure.

5 Concluding Remarks

In situ vibration measurements were carried on the bearing housing on steam turbines and the generator of a typical TG set with a problem of the LP turbines stage blades cracking. The spectrum analysis during different operating times and powers indicate the presence of the frequency peaks in the band of 7–12 Hz and their modulation with the machine vibration at 1× and the higher harmonic components in a few occasion during the measurement process. The modulated frequencies with the 1× and 2× vibration components likely to trigger the resonance with the machine 4th critical speed and the blade 1st bending mode during machine normal operating speed which may have high potential of damage. The case studies in the literature pointing the suspicion towards the possible stall in the TG set. The stall is generally having sufficient potential for the damage. Plant is planning to carry out in-depth study on the TG set by the online strain, blade tip timing measurements and possibly installation of camera to confirm the suspicion of the stall existence.

Acknowledgements Authors acknowledge Dr. Keri Elbhbah, the University of Manchester and Mr. Joeff Tasker, West Burton EDF plant for their excellent help during the vibration measurements.

References

1. Sinha JK, Hahn W, Elbhbah K, Tasker J, Ullah I (2012) Vibration investigation for low pressure turbine last stage blade failure in steam turbines of a power plant. In: Proceedings of the ASME TURBO EXPO conference, Copenhagen, Denmark, 11–15 June 2012
2. Hahn W, Sinha JK (2013) Comparative study between in situ measured vibration data at bearing and BTT on a LP turbine last stage blades in a steam turbo-generator set. Paper number GTINDIA2013-3614. In: Proceedings of the of the ASME gas turbine India conference, Bangalore, India, 5–6 Dec 2013
3. Muszyska A (1988) Stability of whirl and whip in rotor/bearing systems. J Sound Vib 127 (1):49–64
4. Bently DE, Goldman P, Yuan J (2001) Rotor dynamics of centrifugal compressors in rotating stall. Orbit, 2nd Quarter
5. Wu F, Qu L (2009) Diagnosis of subharmonic faults of large rotating machinery based on EMD. Mech Syst Signal Process 23:467–475
6. Rieger NF (2012) Progress with the solution of vibration problems in steam turbine blades. www.sti-tech.com/dl/vibnfr.pdf. Accessed 10 Jan 2012
7. Jonas O (2008) Steam turbine corrosion and deposits problems and solutions. In: Proceedings of 37th turbomachinery symposium, Houston, Texas, 8–11 Sept 2008, pp 211–228
8. Owczarek JA (2011) On the phenomenon of pressure pulses reflecting between blades of adjacent blade rows of turbomachines. ASME J Turbomach 133:021016-1-021016-11
9. Kita M, Iwamoto S, Kiuchi D, Kawashita R (2008) Prediction of subsynchronous rotor vibration amplitude caused by rotating stall. In: Proceedings of 37th turbomachinery symposium, Houston, Texas, 8–11 Sept 2008, pp 97–102

A Comparison Between Hilbert Transform and a New Method for Signal Enveloping

Abdullah Al-Ahmari, Jyoti K. Sinha and Erfan Asnaashari

Abstract Envelope analysis of vibration signals is a well known tool for amplitude demodulation and diagnosis of a number of vibration problems in machines and structures. The typical application is the fault diagnosis in the anti-friction bearings and gearboxes. Hilbert transformation (HT) is often used to extract the envelope signals (upper and lower) from a time domain signal. However it is observed that the envelope signals obtained by the HT are not always without any error. In this paper, 4 different signals, 3 simulated; sine, amplitude modulated and random, and measured vibration data on anti-friction bearings are analyzed using the HT. The paper compares the accuracy of the envelope data obtained for these different signals using the HT. A new method called three-point moving window (TPMW) method is also developed to generate envelope signals and applied to the 4 different signals that gives better results.

Keywords Envelop analysis · Bearings · Condition monitoring

1 Introduction

Envelope analysis is a well-known technique used for detecting incipient failure of rolling element bearing. This method is also known by other names such as, amplitude demodulation [1], demodulated resonance analysis, narrow band envelope analysis [2, 3] or high frequency resonance technique [4–7]. Envelope analysis

A. Al-Ahmari (✉) · J.K. Sinha · E. Asnaashari
School of Mechanical, Aerospace and Civil Engineering, The University of Manchester, Oxford Road, Manchester M13 9PL, UK
e-mail: Abdullah.al-ahmari@postgrad.manchester.ac.uk

J.K. Sinha
e-mail: jyoti.sinha@manchester.ac.uk

E. Asnaashari
e-mail: Erfan.asnaashari@postgrad.ac.uk; Erfan.asnaashari@postgrad.manchester.ac.uk

J.K. Sinha (ed.), *Vibration Engineering and Technology of Machinery*, Mechanisms and Machine Science 23, DOI 10.1007/978-3-319-09918-7_14

or amplitude demodulation is the technique that can be used to extract the periodic impacts from the modulated random noise generated within a faulty or deteriorating rolling element bearing. This is a possible process even when the signal from the rolling element bearing is relatively low in energy and buried within the other vibrations produced from the machine.. The effectiveness of the envelope analysis has been assessed by a number of researchers [2–9]. Hilbert transform is an important tool in envelope analysis which can be used for machine fault detection and diagnostics [10–12]. There are also some literatures which used the HT combined with other techniques to make the fault diagnosis easier [7, 13, 14].

It has been observed that the envelope signals obtained by the HT are not always without any error. In this paper, 4 different signals, 3 simulated; sine, amplitude modulated, random, and the measured vibration data for anti-friction bearings are examined by the HT. The paper compares the accuracy of the envelope data obtained for these signals using the HT. A new method called three-point moving window (TPMW) method is then developed to generate the envelope data for the same signals. The new method is shown to be much more accurate compared to the HT.

2 Hilbert Transformation

The HT is used for solving integral equations. The mathematical definition of HT of the signal $x(t)$ is defined as:

$$H[x(t)] = \tilde{x}(t) = \frac{1}{\pi} \int_{-\infty}^{+\infty} \frac{x(\tau)}{t - \tau} d\tau \tag{1}$$

The integral is considered as a Cauchy principal value due to possible singularity at $t = \tau$. The HT can be expressed physically as a special type of a linear filter which keeps all the amplitudes of spectral components but shifts their phases by $\frac{-\pi}{2}$. Therefore, the HT of the original signal is equal to the convolution integral of $x(t)$ with function $(\pi t)^{-1}$ which can be written as $\tilde{x}(t) = x(t) * (\pi t)^{-1}$.

The HT has attracted great attention in vibration analysis over the past two decades [15]. It can be applied to vibration signals to present useful information regarding the amplitude, instantaneous phase and frequency of vibrations. Simon et al. [16] and Tomlinson [17] utilised HT for identification of structural non-linearities using sinusoidal input excitation in frequency domain. They identified common types of nonlinearity from the frequency response function (FRF) distortions. Feldman [18] employed HT of input and output time domain measurements for identification of non-linearities in stiffness and damping characteristics of a vibration system.

Signal demodulation is another area which the HT can be used. An amplitude modulated wave can be represented as:

$$x(t) = u(t)\cos\psi(t) \tag{2}$$

where $u(t)$ is the modulating envelope function or message waveform and $\cos\psi(t)$ is the oscillations or carrier waveform. To demodulate waveform $x(t)$, the carrier waveform should be removed while the message waveform is preserved. The HT shifts the phase of every frequency component of the modulated wave by 90° to obtain a quadrature signal:

$$H[x(t)] = \tilde{x}(t) = u(t)\sin\psi(t). \tag{3}$$

The message waveform can then be acquired by adding the squares of waveforms 2 and 3:

$$u^2(t) = u^2(t)\cos^2\psi(t) + u^2(t)\sin^2\psi(t) \tag{5}$$

3 Application of HT

In order to visualize the envelope using the HT, this section demonstrates the upper and lower envelopes for the simulated and measured vibration data extracted experimentally on an anti-friction bearing.

3.1 Sine and Multi-sine Signals

A sine wave;

$$x(t) = A\sin(2\pi ft) \tag{5}$$

was generated with a sampling frequency of 20 kHz. Amplitude A and frequency f were chosen to be 1.5 and 60 Hz respectively. Figure 1 illustrates $x(t)$ and its upper and lower envelopes using the Hilbert transform.

To demonstrate more elaborations, a combination of two sine waves was investigated:

$$y(t) = A\sin(2\pi f_1 t) + B\sin(2\pi f_2 t) \tag{6}$$

where amplitude B and frequency f_2 were chosen to be 1.5 and 80 Hz respectively. The upper and lower envelopes of this combined signal are shown in Fig. 2. As it can be seen, at some locations, the upper envelope is tangent to one side of the peaks instead of the peak tips.

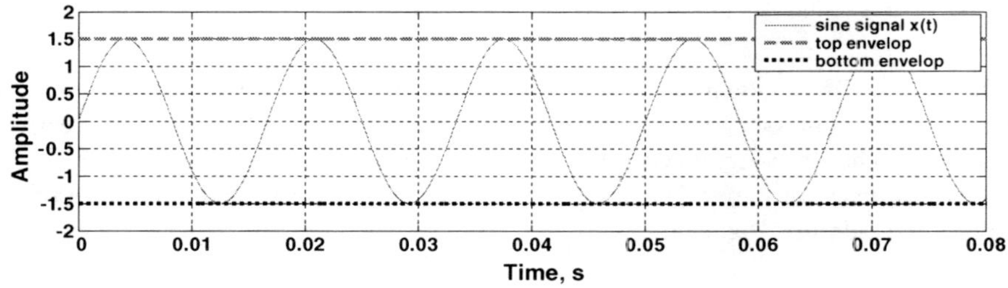

Fig. 1 Envelope using HT for the sine wave $x(t)$

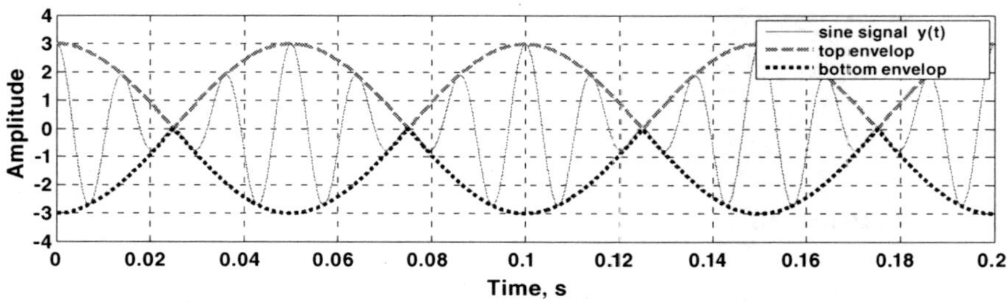

Fig. 2 Envelope using HT for $y(t)$

3.2 Amplitude Modulated Signals

Three different amplitude modulated signals were tested separately. The first signal is a sine wave considered as the original signal (Fig. 3). The second and third signals are superposition of two and three sinusoidal wave components respectively with different frequencies (Figs. 4 and 5). In all the cases, the modulation index is 50 % while the sampling frequency and the frequency of the carrier signal are 10 and 2 kHz respectively.

3.3 Random Signal

The HT was also performed on a random signal to generate upper and lower envelopes as can be seen in Fig. 6. Samples were taken at the frequency of 20 kHz. Figure 6 shows that the envelope does not precisely cross the local maxima and local minima for upper and lower envelopes respectively.

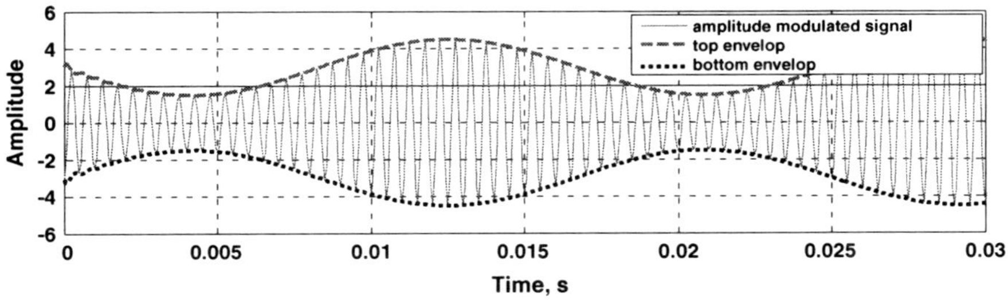

Fig. 3 Envelope using HT for an amplitude modulated signal with 1 sine wave

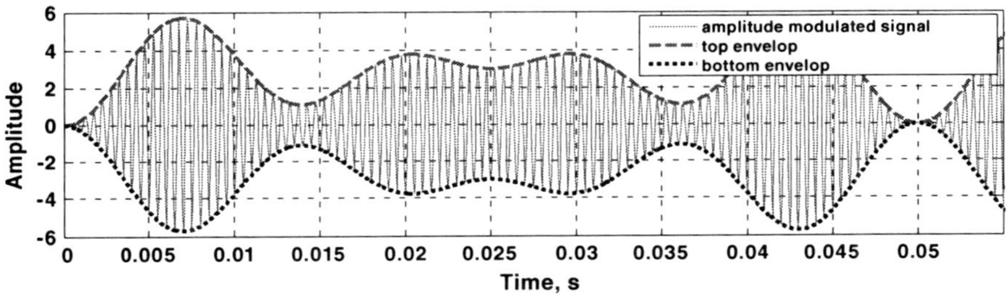

Fig. 4 Envelope using HT for an amplitude modulated signal with 2 sine waves

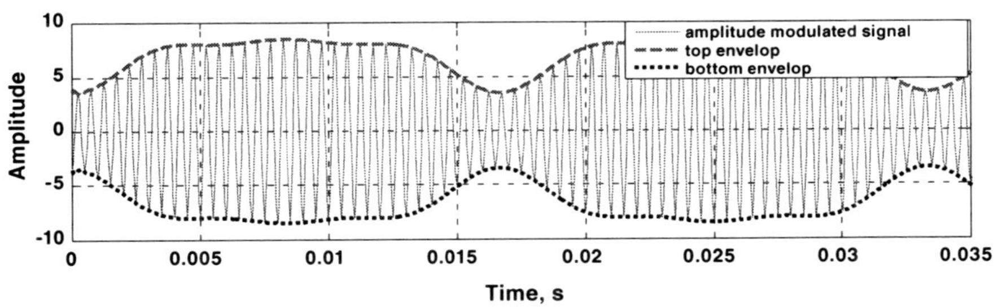

Fig. 5 Envelope using HT for an amplitude modulated signal with 3 sine waves

3.4 Measured Signal

The HT was applied to measured vibration data of an anti-friction bearing to produce upper and lower envelopes as seen in Fig. 7. Similarly, the same error as in the random signal appears in the measured vibration signal.

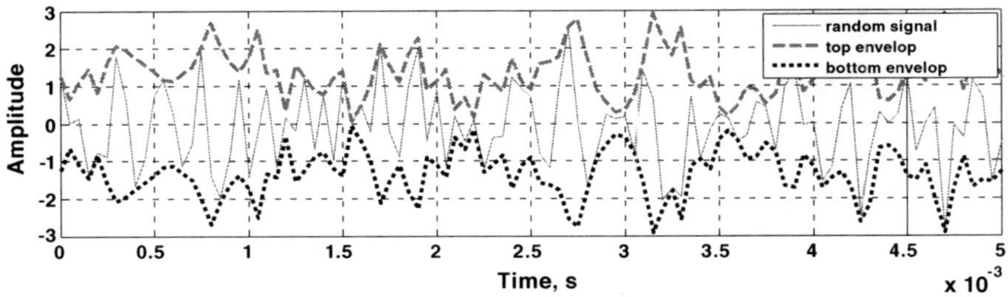

Fig. 6 Envelope using HT for a random signal

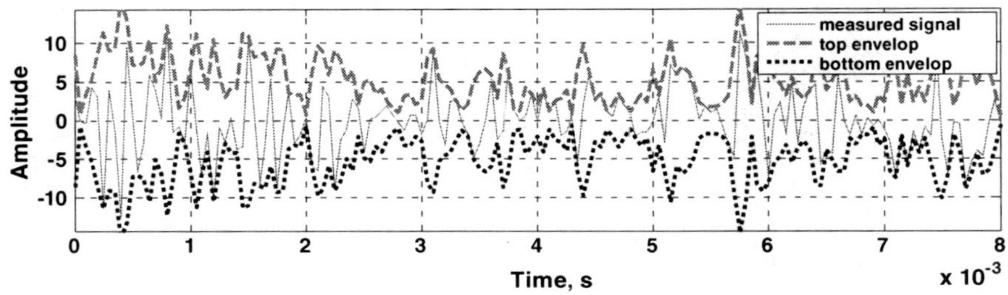

Fig. 7 Envelope using HT for a measured vibration signal on anti-friction bearing

4 Proposed Method

In order to avoid the error associated with the HT in random and measured vibration data, a three-point moving window method (TPMW) was developed. To find the upper envelope, the TPMW method scans the whole signal and search for the local maximum points. When the value of the middle point is greater than its two adjacent points, both the index and value of the middle point are stored. By moving the scanning window over the whole signal, all maximum points and their indices can be stored. After finding the local maximum points, they are connected linearly by interpolating the points in between. To produce the lower envelope, the TPMW searches for local minimum points and then connect those points linearly as it is done with the upper envelope.

In other words, Let's assume x_{i-1}, x_i and x_{i+1}, are three consecutive data points that form a three-point moving window in the time domain, where $i = 2, 3, \ldots, n$ and n is the number of samples collected. To consider a point as the local minimum or local maximum the following conditions should be satisfied simultaneously:

$$\begin{cases} x_i = local\, minimum, & x_i <; 0\, and\, x_i < x_{i-1}\, and\, x_i < x_{i+1} \\ x_i = local\, maximum, & x_i > 0\, and\, x_i > x_{i-1}\, and\, x_i > x_{i+1} \end{cases}$$

If the conditions for either local minimum or maximum are satisfied, the value and index of the middle point (x_i) are stored and the window is shifted by two indices. Otherwise, the window will move one index forward and test the new points.

5 Application of the TPMW Method

The TPMW was applied to the same signals which were previously tested by the HT in Sect. 3. Figure 8 shows the upper and lower envelopes for the signal $x(t)$ and Fig. 9 displays it for $y(t)$ where the envelopes cross the local maxima and local minima of the signal.

5.1 Amplitude Modulated Signal

The amplitude modulated signals generated in 3.2 is again utilized here, but the envelope was made using the TPMW method (Figs. 10, 11, 12, 13 and 14). In both the cases of random and measured vibration signals (Figs. 12–14), the TPMW

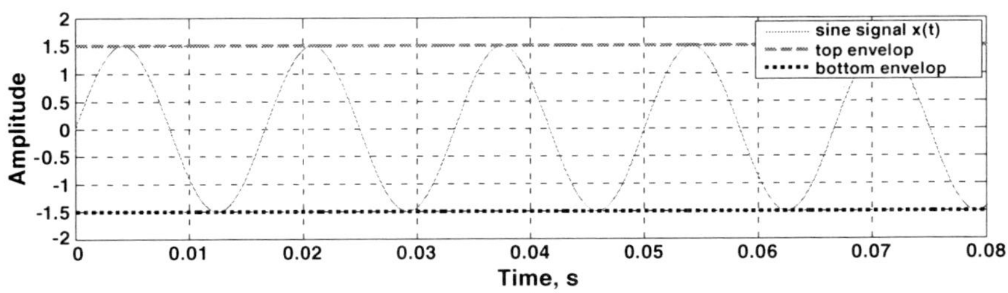

Fig. 8 Envelope using HT for the sine wave $x(t)$

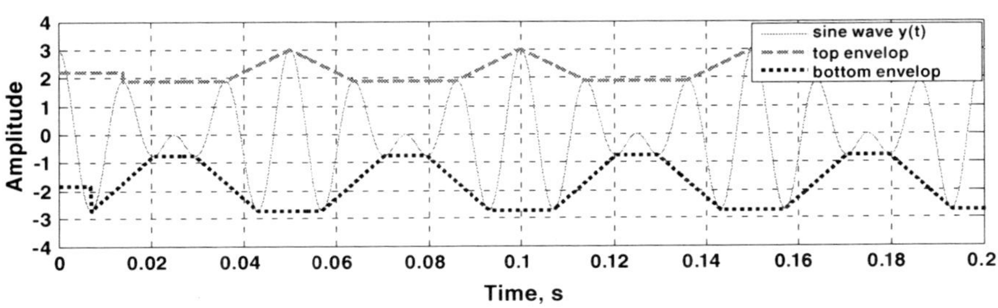

Fig. 9 Envelope using HT for the sine wave $y(t)$

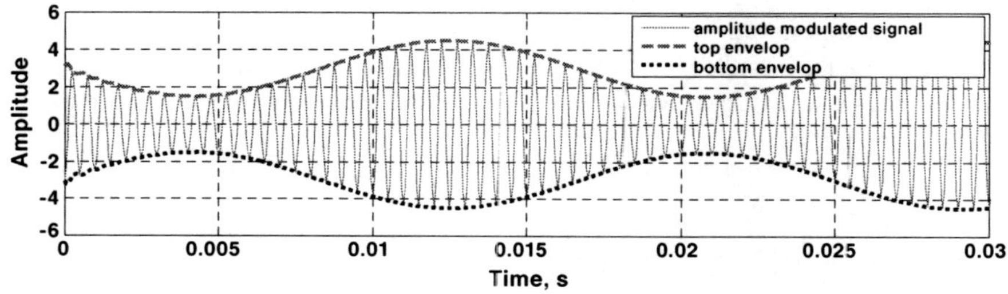

Fig. 10 Envelope using TPMW for an amplitude modulated signal with 1 sine wave

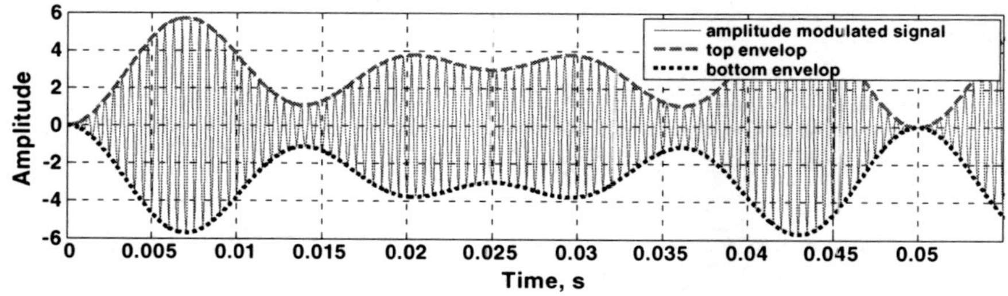

Fig. 11 Envelope using TPMW for an amplitude modulated signal with 2 sine waves

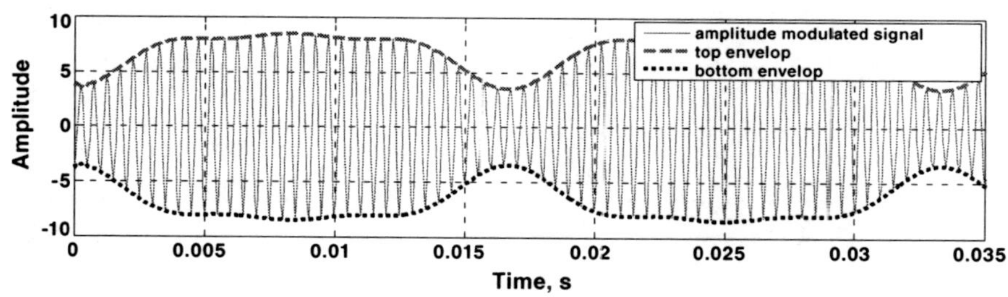

Fig. 12 Envelope using TPMW for an amplitude modulated signal with 3 sine waves

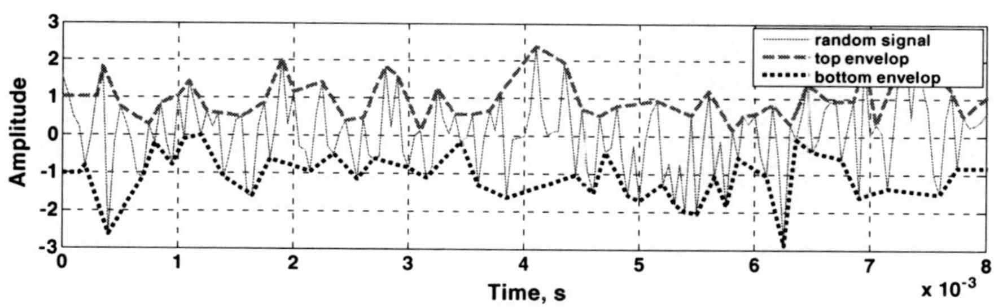

Fig. 13 Envelope using TPMW for a random signal

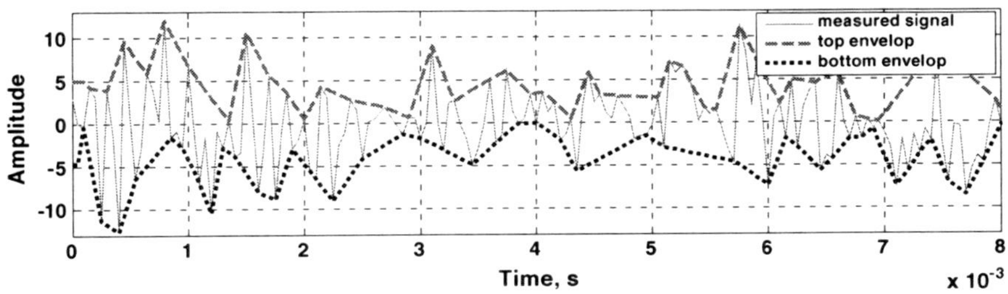

Fig. 14 Envelope using TPMW for a measured vibration signal on anti-friction bearing

clearly connects the local maxima and local minima to generate upper and lower envelopes respectively.

6 Conclusion

In conclusion, the HT was found to produce good envelope for sine and modulated signals; however, it was unable to accurately connect the local maxima and local minima of the random and measured vibration signals. The developed method, TPMW, proposed in this paper, generated better envelopes that can be used in envelope spectrum analysis.

References

1. White G (1991) Amplitude demodulation—a new tool for predictive maintenance. Sound Vib 25:14–19
2. McMahon SW (1991) Condition monitoring of bearing using ESP. Cond Monit Diagn Technol 2:21–25
3. Azovtsev YA, Barkov AV, Yudin IA (1994) Automatic diagnostics and condition prediction of rolling element bearing using enveloping methods. In: Vibration Institute 18th annual meeting
4. McFadden PD, Smith JD (1984) Vibration monitoring of rolling element bearings by the high-frequency resonancetechnique—a review. Tribol Int 17:3–10
5. Prashad H, Ghosh M, Biswas S (1985) Diagnostic monitoring of rolling-element bearings by high-frequency resonance technique. ASLE Trans 28:439–448
6. Su YT, Lin SJ (1992) On initial fault detection of a tapered roller bearing: Frequency domain analysis. J Sound Vib 155:75–84
7. Segla M, Shaoping W, Fang W (2012) Bearing fault diagnosis with an improved high frequency resonance technique. In: 2012 10th IEEE International Conference on Industrial Informatics (INDIN), pp 580–585
8. Martin KF, Thorpe P (1992) Normalised spectra in monitoring of rolling bearing elements. Wear 159:153–160

9. Randall RB, Antoni J, Chobsaard S (2001) The relationship between spectral correlation and envelope analysis in the diagnostics of bearing faults and other cyclostationary machine signals. Mech Syst Signal Process 15:945–962
10. Tandon N, Nakra BC (1992) Vibration and acoustic monitoring techniques for the detection of defects in rolling element bearings—a review. Shock Vib Dig 24:3–11
11. Yu D, Cheng J, Yang Y (2005) Application of EMD method and Hilbert spectrum to the fault diagnosis of roller bearings. Mech Syst Sig Process 19:259–270
12. Yu B, Liu J, Wang C (2007) Rolling bearing fault diagnosis using refinement envelope analysis based on the EMD method. In: Huang D-S, Heutte L, Loog M (eds) Advanced intelligent computing theories and applications: with aspects of contemporary intelligent computing techniques, vol 2. Springer, Berlin, pp 562–570
13. Wan Z, Liao XZ, Xiong X, Han JC (2013) The rolling bearing fault diagnosis research based on improved Hilbert-Huang transformation. Appl Mech Mater 300–301:344–350
14. Wang D, Miao Q, Fan X, Huang H-Z (2009) Rolling element bearing fault detection using an improved combination of Hilbert and wavelet transforms. J Mech Sci Technol 23:3292–3301
15. Feldman M (2011) Hilbert transform in vibration analysis. Mech Syst Sig Process 25:735–802
16. Simon M, Tomlinson GR (1984) Use of the Hilbert transform in modal analysis of linear and non-linear structures. J Sound Vib 96:421–436
17. Tomlinson GR (1987) Developments in the use of the Hilbert transform for detecting and quantifying non-linearity associated with frequency response functions. Mech Syst Sig Process 1:151–171
18. Feldman M (1994) Non-linear system vibration analysis using Hilbert transform-II. Forced vibration analysis method Forcevib. Mech Syst Sig Process 8:309–318

The Detection of Shaft Misalignments Using Motor Current Signals from a Sensorless Variable Speed Drive

Samieh Abusaad, Ahmed Benghozzi, Ann Smith, Fengshou Gu and Andrew Ball

Abstract Shaft misalignments are common problems in rotating machines which cause additional dynamic and static loads, and vibrations in the system, leading to early damages and energy loss. It has been shown previously that it is possible to use motor current signature analysis to detect and diagnose this fault in motor drives. However, with a variable speed drive (VSD) system, it becomes difficult to detect faults as the drive compensates for the small changes from fault effects and increased noise in the measured data. In this paper, motor current signatures including dynamic and static data have been investigated for misalignment diagnosis in a VSD system. The study has made a systemic comparison of different control parameters between two common operation modes: open loop and sensorless control. Results show that fault detection features on the motor current from the sensorless mode can be the same as those of the open loop mode, however, the detection and diagnosis is significantly more difficult. In contrast, because of the additional frictional load, features from static data show results of early detection and diagnosis of different degrees of misalignment is as good as that from conventional vibration methods.

Keywords Sensorless VSD · Misalignment · Current signature · Condition monitoring · Vibration

1 Introduction

Shaft misalignments give rise to major problems in rotating machines which often cause additional dynamic loads and vibrations in the system, leading to early damages and energy loss. Studies show that badly aligned machines can cost a

S. Abusaad · A. Benghozzi · A. Smith · F. Gu (✉) · A. Ball
Centre for Efficiency and Performance Engineering, University of Huddersfield, Huddersfield, UK
e-mail: f.gu@hud.ac.uk

J.K. Sinha (ed.), *Vibration Engineering and Technology of Machinery*, Mechanisms and Machine Science 23, DOI 10.1007/978-3-319-09918-7_15

factory 20–30 % of machines downtime [1]. Other studies show that misalignment is the source of 70 % of vibration problems [2]. Moreover, it also consumes additional energy [3]. Many studies have been made using different diagnosis and detection techniques for early and effective monitoring of shaft misalignments. For instance vibration [4, 5], wireless sensors [6, 7], model based fault diagnosis [8, 9], and motor current signature analysis [10, 11]. However, most of these studies were performed on motors fed through either direct mains or V/Hz drives.

Previous studies on the detection of faults in a mechanical system driven by a sensorless VSD have been limited. One of the main reasons for this is that the drive's control system compensates for any dynamic effects caused by the fault and introduces more noise to the measured data. Therefore, frequency components commonly used could be masked by the drive actions. In addition, the harmonic contents at the terminal output parameters from the drive, and in particular the stator current and voltage, is high, which makes it more difficult to identify the small changes in current signatures [12, 13].

In this study, motor current signature analyses have been conducted to study the behaviour of current signals under two control modes: V/Hz open loop and sensorless mode are subject to parallel misalignments occurring with different degrees of severity in a gearbox test system. Both instantaneous or dynamic current signals from an external sensor and average or static current data from the drive are examined to benchmark the performance of these two data sources.

2 Sensorless VSD

Sensorless flux vector control drives are very common nowadays in industry due to their improved dynamic performance, accurate speed control, higher efficiency, lower operating cost and higher reliability. Sensorless VSDs represent three phase AC motor currents in a single phase d-q reference frame. Both d and q current components are regulated separately copying the way the DC machines work. The flux is regulated by controlling the d current component while the torque is regulated by adjusting the q current component. The motor speed is estimated and compared with the speed set point in order to generate the reference voltage to the pulse width modulator which sets the duty cycles for the inverter. Figure 1 represents the general structure of sensorless VSDs. However, when the system operates at or higher than the rated speed, the output voltage must be limited to the DC-Bus voltage of the drive and the output current is saturated to the rated motor current.

In general, mechanical faults such as misaligned shafts will cause the load torque of the motor to oscillate at a frequency relating to the mechanical faults. This torque oscillation will lead to speed oscillation. To minimize this speed oscillation the VSD drive will adjust i_q to react against any speed changes. The drive therefore will continually adjust the terminal supply to maintain the system in stable conditions. As shown in [14], this type of fault can be identified by inspecting the spectral peaks

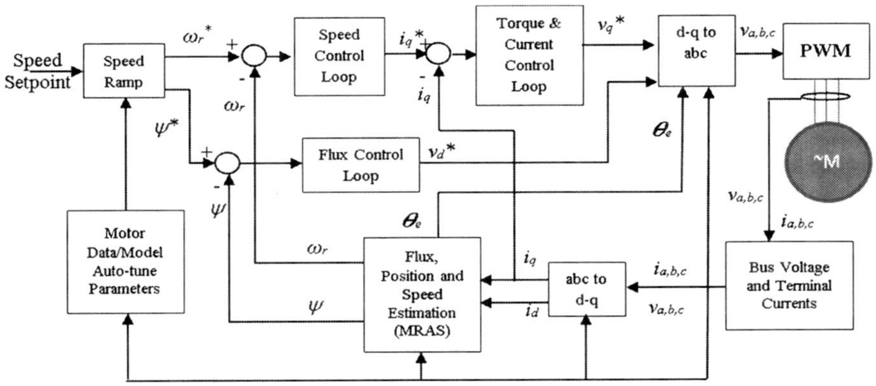

Fig. 1 General structure of sensorless VSD

at frequencies that are related to the rotor mechanical speed in the motor current spectrum at the following frequencies:

$$f_x = f_s \pm kf_r \quad k = 1, 2, 3, \ldots \tag{1}$$

where f_x the feature fault frequency component, f_s supply frequency, f_r rotor mechanical frequency. In addition to the dynamic oscillations, the misalignment will also cause additional static friction. As reported in [15] there will be up to 1.2 % energy loss due to the frictional loss caused by misalignment, which will require an increase in electrical torque and hence an increase in current and voltage to overcome the additional load and maintain the operation at setpoints. Therefore it is also possible to detect misalignment by the static current values.

Where ω_r rotor angular speed, d-q refers to the direct and quadrature axis respectively in the d-q plane, i_d and v_d current and voltage field components respectively, i_q and v_q current and voltage torque components respectively, ψ flux linkage, $i_{a,b,c}$ and $v_{a,b,c}$ the three phase line currents and voltages respectively, θ_e supply phase angle, MRAS model reference adaptive system estimator, the subscript * refers to the reference values, and PWM the pulse width modulator in the VSD.

3 Test Facility

The system employed for this study consists of two main parts: the mechanical part and the control part, as shown in Fig. 2. The mechanical part consists of one 15 kW AC induction motor as the prime driver, two back-to-back 2 stage helical gearboxes for coupling the DC motor with the AC motor, flexible spider couplings link the components. The first gearbox operates as a speed reducer while the second is a speed increaser so that the system maintains sufficient speed for the DC motor to generate sufficient load to the AC motor. The control system consists of a programmable logic controller (PLC) for setting up test profiles requested by the operator, an AC VSD that can be set either to sensorless flux vector control or V/Hz modes for adjusting the speed of the system, a DC variable speed drive providing a controlled load to the AC motor by regulating the torque of the DC motor/generator.

In addition, a low speed data log system is employed to log the static data from the VSD at the rate of 1 Hz. In the meantime a high speed data acquisition system operates at 96 kHz to log dynamic data from the system, which includes three accelerometers attached to the two gearboxes and the AC IM for monitoring the vibration levels; one shaft encoder attached to the AC motor. Three terminals measure voltage and current using Hall Effect sensors. The accelerometers are placed horizontally in parallel with shafts centerlines, as shown in Fig. 2. All data sets from the data acquisition systems are stored in a PC for post processing and analysis in the Matlab environment. This dynamic data is for both the evaluation of the performance of conventional analysis methods and the benchmark of the performance developed from static data.

Four different degrees of severities of shaft misalignment are simulated at 0.5, 0.7, 1.00 and 1.30 mm successively which are induced by raising up the gearbox next to the AC motor using precut stainless steel standard shims placed under the gearbox as indicated in Fig. 2. The test rig is firstly aligned at the lowest possible vibration level. A cycle of 5 different load settings was applied, 0, 25, 50, 75 and 100 % of the full load. Each load applied for a period of 2 min. The AC motor speed is set at its full speed during the tests. A set of 10 s of dynamic data is collected for each load, while the data from the VSD is collected for the entire cycle.

Figure 3 shows the vibration signals from gearbox 1 under different misalignment severities and loads. It can be seen that the vibration amplitudes at 3X of f_{r1} and of f_{r3} increasing consistently with the degrees of the misalignment, showing that the fault cases are induced appropriately. Therefore, these results can be a benchmark of the current signals analysis and diagnosis method developed in this paper.

4 Results and Analysis

To investigate the performance of the current signals in differentiating the misalignment, the dynamic current signals are converted into the frequency domain and the amplitudes extracted and compared at corresponding feature frequencies: f_{r1} and f_{r3}. Meanwhile a direct comparison of the static current signals is made between the baseline and different misalignment cases to examine their detection performance in line with the results from the dynamic data and vibration signals.

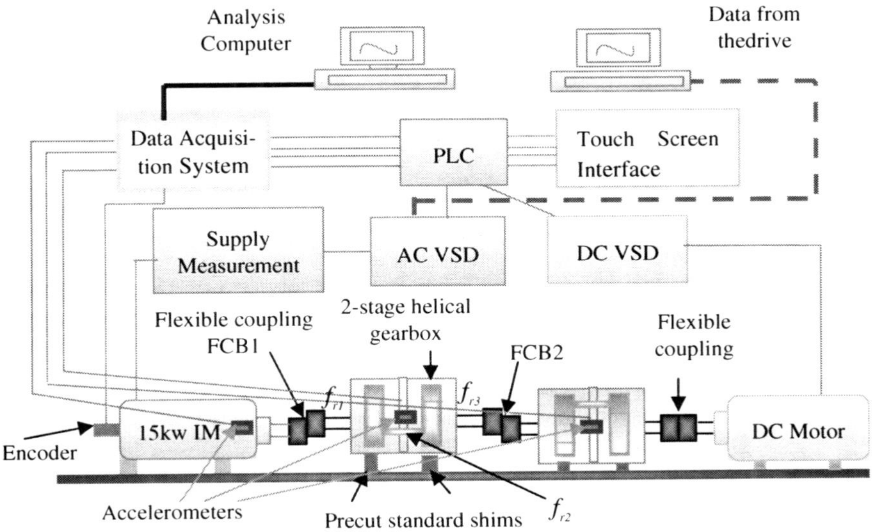

Fig. 2 Schematic represents the test rig layout and fault induced

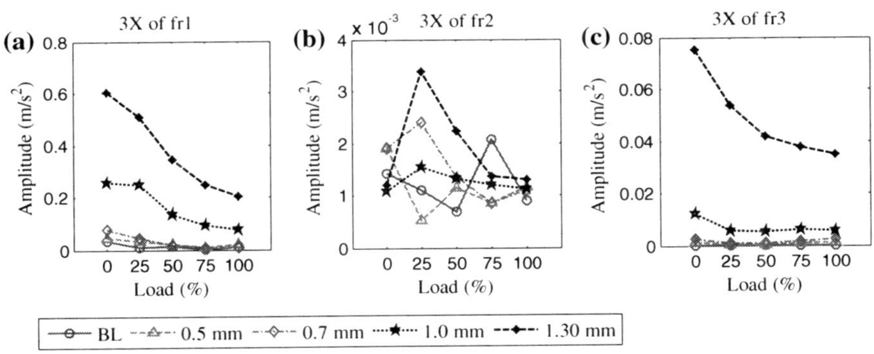

Fig. 3 Vibration signals comparison at 3X of f_{r1} and of f_{r3} under different severities

4.1 Detection and Diagnosis Under Open Loop

To show the general characteristic, Fig. 4 shows a comparison of current spectrum between 1.0 mm misalignment and the baseline under 50 and 100 % load. It can be seen that the components at f_{r3} are distinctive and the 3X components are higher than the first and the second, showing the basic feature of the three jaw coupling. Moreover, the amplitude at f_{r3} sidebands increases with the degree of misalignment, which is consistent with fault severities. However, the amplitude decreases with load due to the effect of the nonlinear behavior of the coupling.

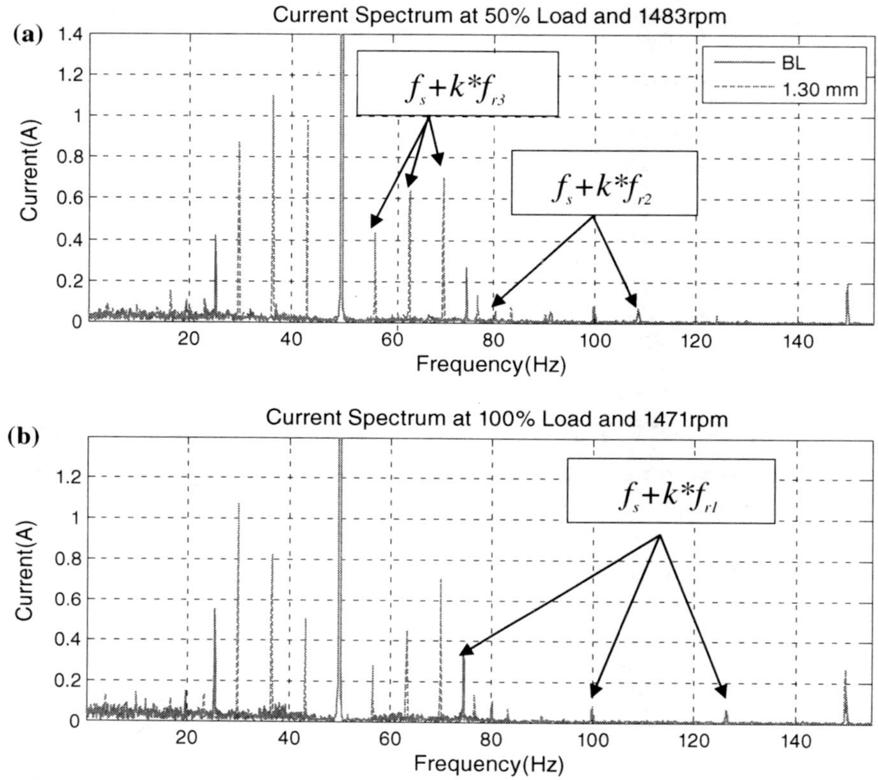

Fig. 4 Current spectra under different load and cases

Components at f_{r1} increase with the misalignment but the amplitudes are not clearly identifiable due to high background noise from the VSD. The component at 1X of f_{r2} shows a little change with fault severity as the shaft is the internal one of the gearbox.

To diagnose the fault severity, sideband components at the 3X f_{r1} and f_{r3} are extracted and presented against loads in Fig. 5. Figure 5a shows the misalignment on the first shaft is just detectable by the current spectrum at the high severity of 1.3 mm. Meanwhile the misalignment on the third shaft can be detected and quantified starting from 0.7 mm, as shown in Fig. 5b. The reason for higher sensitivity to the third shaft is that it undertakes a torsional load 3.6 times of the first shaft.

Under open-loop mode, there is no speed feedback to the drive. Instead the drive provides a constant *V/Hz* ratio. In the case of the misalignment, there is early indication that additional fractional torque is induced. This friction adds additional static load on the motor which in turn causes an additional electromagnetic torque and more current in the motor. Consequently, static torque and current signals are

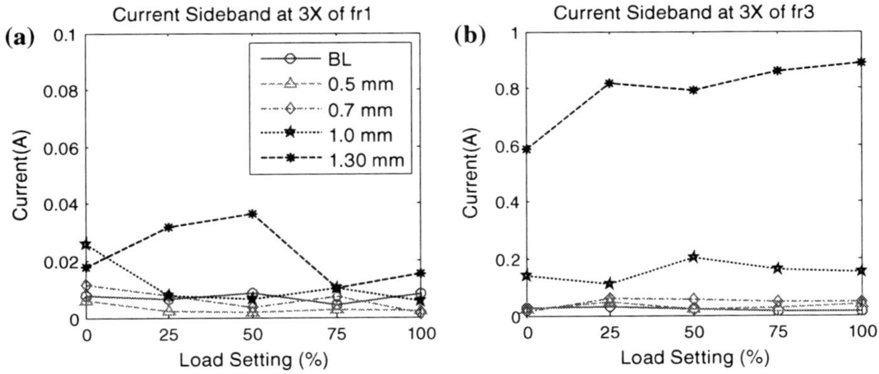

Fig. 5 Current 3X of f_{r1} and f_{r3} comparison under different misalignments and loads

most likely to be affected by such faults. Figure 6 represents the average of current, torque, i_d and i_q signals under different load and misalignment conditions. Both the torque and torque current component i_q show a gradual increase with an increase in misalignment levels. So they can be used for fault classification. Nevertheless, the field current component i_d has no clear change related to the fault severity. Authors reported in [16] a new cos effective technique utilising data from sensorless VSDs for mechanical misalignment detection. Noticeably even when the misalignment is as small as 0.5 mm it can be clearly detected from both torque and i_q signals. This shows that the static data has a better sensitivity than that of the dynamic data under the open loop operation.

4.2 Detection and Diagnosis Under Sensorless Mode

In sensorless control mode, the drive adjusts the motor supply parameters keeping the speed of the machine stable and follows the reference without feedback encoder. When a fault occurs due to an abnormality such as misalignment, it changes the stability conditions. The drive's regulators will adjust the machine's supply to compensate for the effects of defects on the load and speed. As a consequence of this control processes and that of the level of noise from the IGBs on the PWM which is very high, the sideband frequency components of current signal cannot be easily observed. As depicted in Fig. 7a, rotor frequency f_{r1} of the current spectrum and its sidebands do not show any clear changes correlated with the fault severities.

On the other hand, the components at 3X of f_{r3} show identifiable changes at 1.0 and 1.30 mm, as detailed in Fig. 7b. However, severities at and below the 0.7 mm are difficult to discriminate. This shows that it is less possible to detect and diagnose the small faults under the sensorless mode, compared with that of open loop mode.

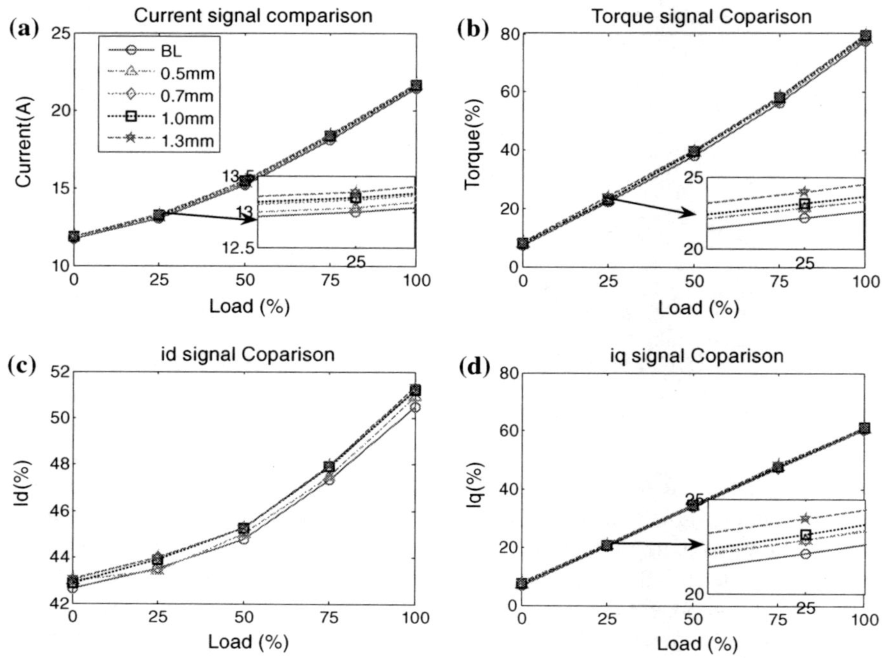

Fig. 6 Current, torque, i_q and i_d comparison at different fault severities and loads

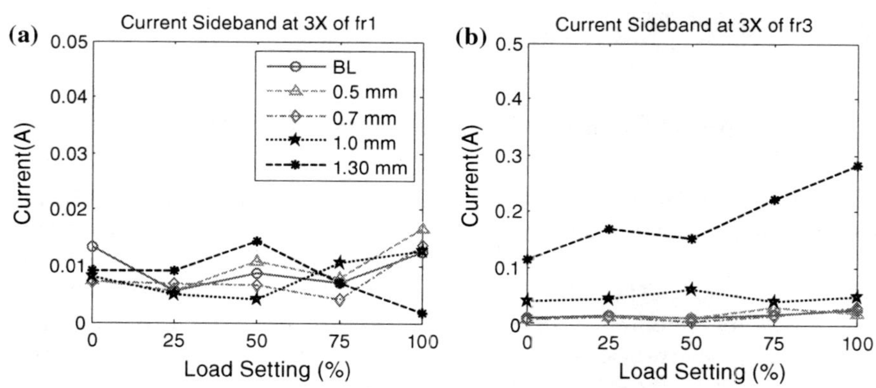

Fig. 7 Current sidebands 3X of f_{r1} and f_{r3} with different fault and load conditions

Data from the drive was acquired at the same time as the dynamic data was collected under the sensorless operation. Figure 8 shows the system variables under different degrees of severity and loads. It shows that current, i_q components and torque signals vary with the fault severity. This is due to the fact that the motor needs more torque to recover from the disturbances from the misalignment.

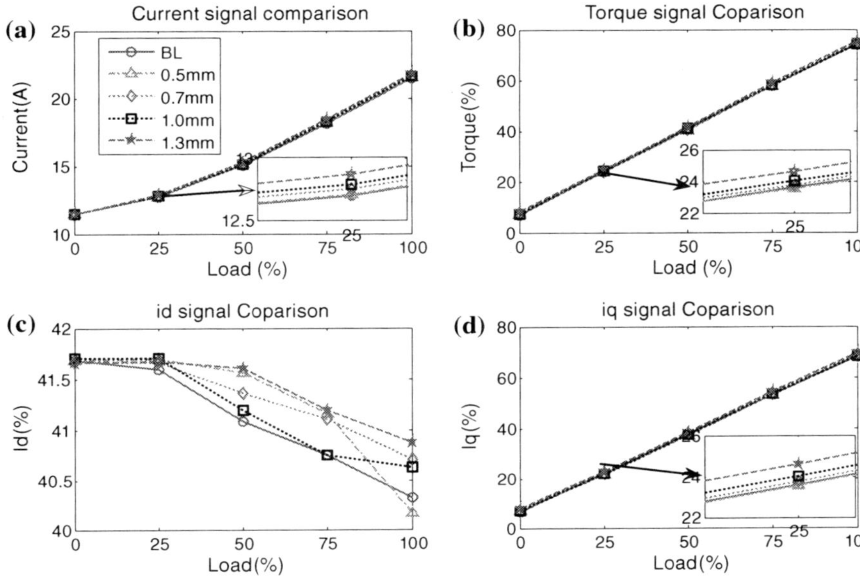

Fig. 8 Data from sensorless VSD comparison under different misalignment severities and loads

This shows that misalignment has induced a static load in addition to the dynamic oscillations. Conversely, i_d is changing slightly, but not correlating with the fault severity. In order to maintain the speed at the desired value the drive feeds more current compensating for any disturbances due to misalignment. Therefore most torque related signals have been changing with the fault severity. Authors reported in [16] a new cost effective technique utilising static data for detecting mechanical misalignment in systems driven by sensorless VSDs. Polynomial models have been developed and trained for baseline conditions. In cases of abnormalities, it generates residuals related to the fault severities.

5 Conclusion

The study shows that motor current data can be used for shaft misalignment detection under sensorless operation. The detection features from dynamic data are consistent with those of open-loop, i.e. the sidebands around the supply frequency due to the effect of load and speed oscillation. However, the detection performance is lower than that of open loop operation because of the effect of oscillation compensation and noise contamination from the drive.

On the other hand the misalignment creates additional frictional loads to the systems which need more electrical power to overcome it. Based on this understanding, the static data from the drive under both the open and sensorless operations can be used to detect and diagnose all different levels of misalignments. In particular, the torque related variables such as i_q current and torque signals show consistent changes correlated with the degree of severities. This has demonstrated that this static based misalignment detection outperforms that of the dynamic data and shows the same performances as that of conventional vibration analysis.

References

1. Ahmed M, Ahmed K, Imran M, Khan S, Akram T, Jawad M (2010) Spectral analysis of misalignment in machines using sideband components of broken rotor bar, shorted turns and eccentricity. Int J Electr Comput Sci IJECS-IJENS 10(6):85–93
2. Bognatz SR (1995) Alignment of critical and non-critical machines. Orbit 4:23–25
3. Howard AG, Ray C (1999) Energy losses caused by Machinery misalignment and unbalance. In: 17th international model analysis conference-IMAC XVII
4. Tejas H, Ashish KD (2009) Experimental investigations on vibration response of misaligned rotors. Mech Syst Sig Process 23(7):2236–2252
5. Nagrani S, Pathan S, Bhoraniya I (2012) Misalignment fault diagnosis in rotating machinery through the signal processing technique- signature analysis. Int J Adv Eng Res Stud E-ISSN 2249–8974 (2012)
6. Arebi L, Gu F, Ball A (2012) A comparative study of misalignment detection using a novel Wireless Sensor with conventional Wired Sensors. In: 25th international congress on condition monitoring and diagnostic engineering
7. Arebi L, Gu F, Hu N, Ball A (2011) Misalignment detection using a wireless sensor mounted on a rotating shaft. In: 24th international congress on condition monitoring and diagnostics engineering management. COMADEM, Stavanger, Norway, pp 1289–1299
8. Jalan Kr, Mohanty AR (2009) Model based fault diagnosis of a rotor–bearing system for misalignment and unbalance under steady-state condition. J Sound Vib 327:604–622
9. Tichate W, Poj T, Cherdchai P, Mongkol P (2006) Motor misalignment detection based on hidden markov model. Struct Health Monit 6:325–334
10. Blodt M, Chabert M, Regnier J, Faucher J, Dagues B (2005) Detection of mechanical load faults in induction motors at variable speed using stator current time-frequency analysis. In: 5th IEEE international symposium on diagnostics for electric machines, power electronics and drives, 2005, SDEMPED, pp.1–6
11. Obaid RR, Habetler TG, Tallam RM (2003) Detecting load unbalance and shaft misalignment using stator current in inverter-driven induction motors. In: IEEE international electric machines and drives conference (IEMDC'03), USA, vol 32003, pp 1454–1458
12. Bellini A, Filippetti F, Franceschini G, Tassoni C (2000) Closed-loop control impact on the diagnosis of induction motors faults. IEEE Trans Ind Appl 36:1318–1329
13. Kolodziejek P, Elzbieta B (2009) Broken rotor bar impact on sensorless control of induction machine. Int J Comput Math Electr Electron Eng 28:540–555
14. Schoen RR, Habetler TG (1995) Effects of time-varying loads on rotor fault detection in induction machines. IEEE Trans Ind Appl 31:900–906
15. Howard AG, Ray C (1999) Energy losses caused by machinery misalignment and unbalance. In: 17th international model analysis conference-IMAC XVII
16. Abusaad S, Benghozzi A, Shao Y, Gu F, Ball A (2013) Utilizing data from a sensorless AC variable speed drive for detecting mechanical misalignments. Key Eng Mater 569:465–472. ISSN 1013-9826

Analysis of SGTF Vertical Sodium Pump Using ARMD

Hemant Prakash Agnihotri, V. Prakash and K.K. Rajan

Abstract In sodium cooled fast breeder reactors (FBR), centrifugal sodium pumps are widely used to circulate the hot sodium in primary as well as in secondary loops. Performance of this critical component is very important in reactor circuits as it affects the load factor and availability of the plant. Operation of the centrifugal sodium pumps, limiting its vibration is a challenge in view of both safety and economy. Various experimental facilities have been set up at IGCAR for testing of FBR components. Vertical centrifugal sodium pumps are used to circulate hot sodium in these facilities. The pump shaft is supported by hydrostatic bearing (HSB) at the bottom and by tapper roller bearing at the top. As the HSB runs with the close tolerance, rotor dynamic analysis was carried out for ensuring its effectiveness and proper functioning of the pump. The results obtained from the analysis are also verified by conducting experiments. This paper discusses the rotor dynamic analysis carried out on a vertical sodium pump used to circulate sodium in steam generator test facility (SGTF) using the rotor dynamic analysis software. In this pump, shaft is supported at the top by tapper roller bearing and hydrostatic bearing at the bottom. Critical speeds and corresponding mode shapes were obtained in the operating range of the rotor speed. Stability analysis was performed for the rotor-bearing system to get the threshold speed of the rotor. Unbalance response of the pump was also carried out to estimate the amplitude of vibration at the bearing locations and compared with experimental results.

Keywords Rotor dynamic analysis · Sodium pump · FEM · Fluid film bearing · Critical speed

H.P. Agnihotri (✉) · V. Prakash · K.K. Rajan
Fast Reactor Technology Group, Indira Gandhi Centre for Atomic Research (IGCAR), Kalpakkam, TN, India
e-mail: agnihotri@igcar.gov.in

V. Prakash
e-mail: prakash@igcar.gov.in

K.K. Rajan
e-mail: kkrajan@igcar.gov.in

J.K. Sinha (ed.), *Vibration Engineering and Technology of Machinery*, Mechanisms and Machine Science 23, DOI 10.1007/978-3-319-09918-7_16

1 Introduction

In sodium cooled fast reactors, centrifugal sodium pumps are widely used to circulate the hot sodium in primary as well as in secondary loops. Performance of this critical component is very important in reactor circuits as it affects the load factor and availability of the plant. Operation of the centrifugal sodium pumps, limiting its vibration is a challenge in view of both safety and economy. For carrying out various experiments related to development of fast breeder reactor technology, testing of sodium components and to have experience in safe operation and maintenance of sodium systems, various experimental facilities are in operation at IGCAR. Vertical centrifugal sodium pump is used to circulate hot sodium in these facilities. Rotor dynamic analysis was carried out for ensuring its effectiveness and proper functioning of the pump [1]. The results obtained from the analysis are also verified by conducting experiments.

This paper discusses the rotor dynamic analysis carried out on a vertical sodium pump used to circulate sodium in steam generator test facility (SGTF) using the rotor dynamic analysis software ARMD (Advanced Rotating Machinery Dynamics) [1]. HSB was modeled and stiffness and damping values were estimated. Using these values as boundary conditions, rotor dynamic analysis was carried out and critical speeds and corresponding mode shapes were obtained. Rigid boundary condition was applied at roller bearing location as its stiffness is very high as compared to HSB. Instability phenomenon can occur in hydrostatic bearing at higher speed ranges and hence stability analysis was performed for the rotor-bearing system to get the threshold speed of the pump. Unbalance response of the pump was also carried out to estimate the amplitude of vibration at the bearing locations.

2 SGTF Centrifugal Sodium Pump

Steam Generator Test Facility (SGTF) was setup at IGCAR, to test and qualify the steam generator design of Prototype Fast Breeder Reactor (PFBR), which is under construction at Kalpakkam, India. Vertical centrifugal sodium pump is one of the key components of the sodium system in SGTF and is shown in Fig. 1. This pump consists of side suction, bottom discharge, mechanical seal, cover gas, large rotor assembly, a vessel to house the shaft, impeller and diffuser casing and other necessary parts. The span length of the shaft is 1.7 m and diameter is 62.8 mm. This shaft is supported by roller bearing at the top and hydro-static bearing (HSB) lubricated by liquid sodium at the bottom. Specifications of this pump are given in Table 1.

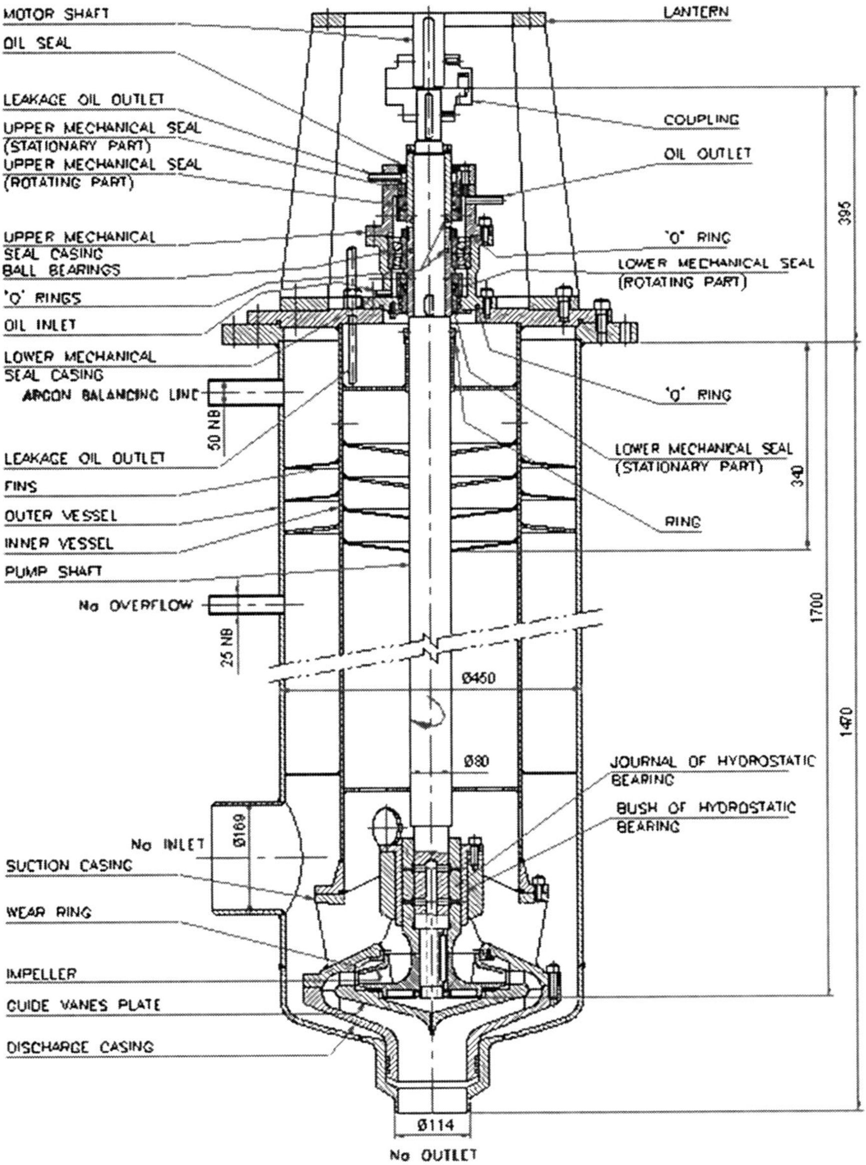

Fig. 1 Sodium pump in SGTF

Table 1 Specifications of the pump

Type	Centrifugal pump with vertical shaft
Operating flow	105 m^3/h
Power	37 kW
Head	50 mlc
Critical speed	3,848 RPM
Bottom bearing	Hydrostatic bearing
Top bearing	Tapper roller bearing
Operating fluid	Sodium
Operating speed	Variable (20–100 % of 2,100 RPM)

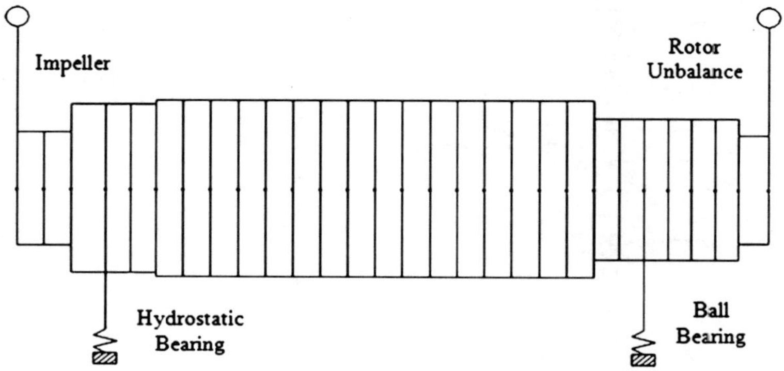

Fig. 2 Model of the sodium pump shaft in ARMD

3 Rotor Dynamic Analysis and Modeling

Centrifugal Sodium Pump of SGTF was modeled in ARMD software. Various inputs data to the model are as follows. The rotor shaft model including discs and bearing locations is as shown in Fig. 2.

3.1 Discs

Two discs have been modeled as shown in Fig. 2. One corresponds to the impeller and the other corresponds to unbalance in the rotor. As per ISO 1940/1, balance quality grade of this shaft is G 6.3 and the residual unbalance of the shaft was calculated as 889 gm·mm (For 31 kg shaft at 2,100 RPM). This residual unbalance comes from imperfections in machining the shaft and leads to unbalance vibration.

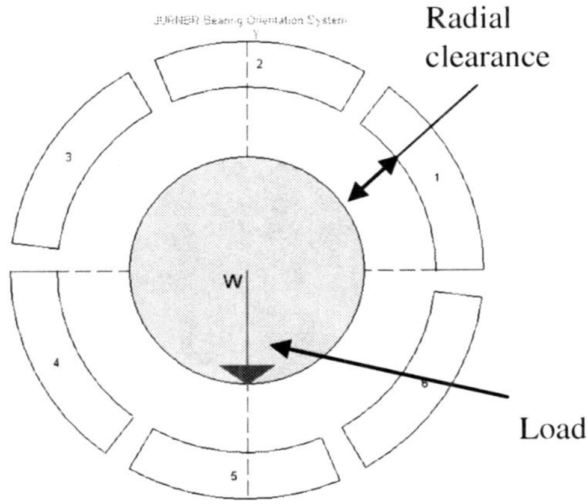

Fig. 3 Model of hydrostatic bearing

3.2 *Bearing*

Hydrostatic bearing has been modeled using journal bearing module (JURNBR) of ARMD and is shown in Fig. 3. The modeling technique employed in this program is based on finite difference cell approach. Following are the inputs given:

Diameter: 99.0 mm
Length: 120 mm
Radial Clearance: 100 μm
No of grooves: 6.

Pressure distribution (Pascal) obtained on the bearing pads is shown in Fig. 4. This pressure distribution is used in determining the coefficients of the bearing by suitably integrating the pressure in the bearing pads with the area of journal of the bearing. It can be seen that lubricant pressure is equally distributed over the pads of the journal. This is due to vertical orientation of the pump and equal amount of load is acting on any circumferential direction of the bearing. From the journal bearing analysis, stiffness coefficient matrix (Nm) and damping coefficient matrix (Ns/m) have been obtained at operating speed and are given in Tables 2 and 3.

4 Results

4.1 *Mode Shapes*

Mode shapes and corresponding critical speeds obtained from the analysis are shown in Fig. 5a, b and c. These mode shapes indicate the pattern of vibration of shaft at operating speed equal to any of these critical speeds. Shaft operating at a speed near to critical speed can produce very large amplitude of vibration and thus it is desirable that operating speed should be 25 % above or below the critical speed.

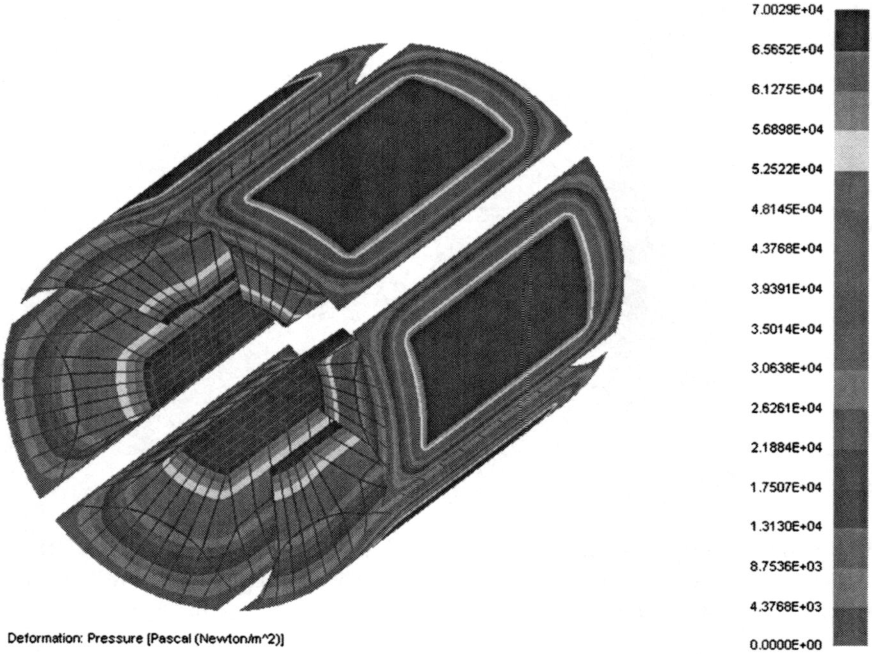

Fig. 4 Pressure distribution (Pa) in the bearing pads

Table 2 Stiffness coefficients (N/m)

2.540E+06	5.442E+05
5.448E+05	2.540E+06

Table 3 Damping coefficients (Ns/m)

0.548E+03	−1.827E+01
−9.133E+00	2.556E+03

4.2 Unbalance Response Analysis

Unbalance response analysis was performed and the amplitude of displacement at hydrostatic bearing location is plotted in Fig. 6 corresponding to operating speed. It is observed that the unbalance at operating speed (2,100 RPM) is 6.5 μm which corresponds to vibration velocity of 0.35 mm/s (rms). It may also be inferred from the Fig. 6 that the first critical speed of the pump is 4,200 RPM as peak amplitude of vibration occurs at this speed. This obtained value of the critical speed i.e. 4,200 RPM is very near to rated critical speed 3,848 RPM (theoretical estimation) as given in Table 1.

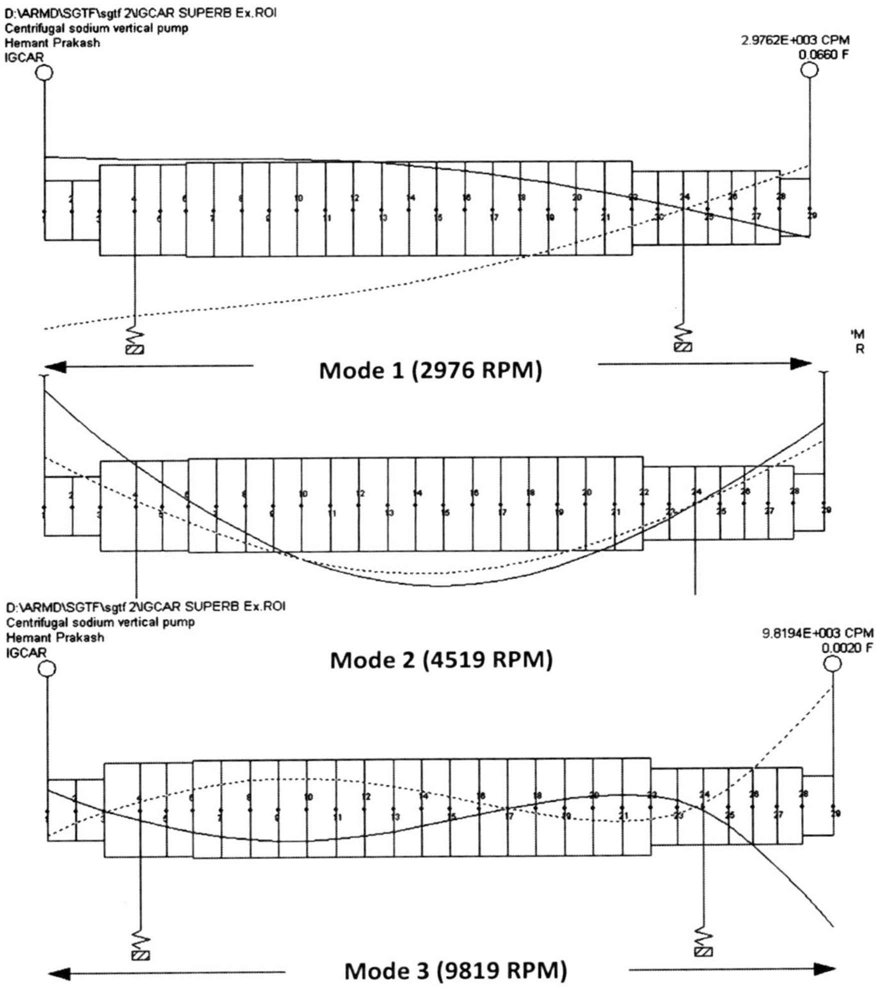

Fig. 5 **a** Mode 1 (2,976 RPM). **b** Mode 2 (4,519 RPM). **c** Mode 3 (9,819 RPM)

4.3 Stability Analysis

Stability analysis of the rotor shaft was carried out to get threshold speed of the instability. The rotor system becomes unstable and large vibration may occur if the operating speed of the shaft crosses the threshold speed of instability. This is due to damping from the hydrostatic bearing reduces to zero after certain speed. So it is desirable to keep the threshold speed of instability as high as possible. A stability chart obtained from analysis is shown in Fig. 7. This chart shows the variation of the damping value in logarithmic decrement Vs rotor speed. It is observed from the

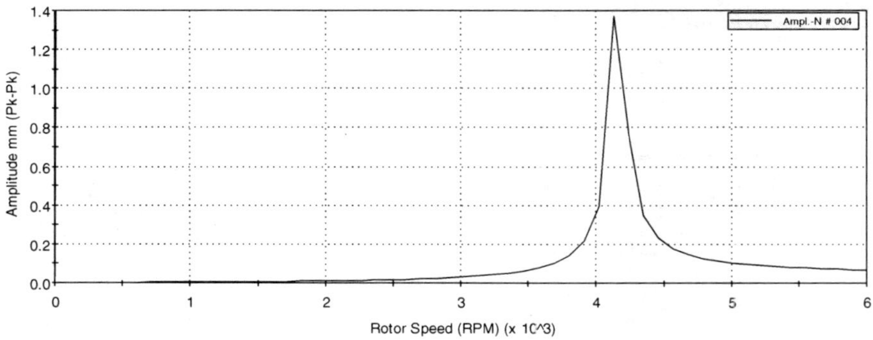

Fig. 6 Unbalance response at hydrostatic bearing support

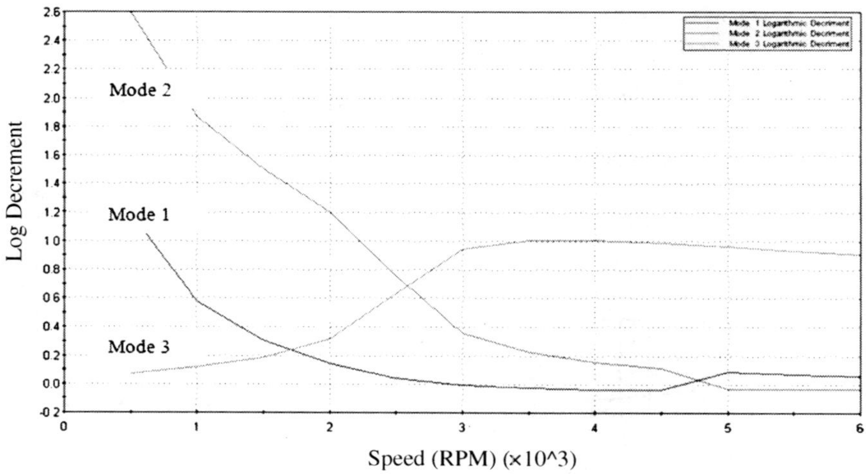

Fig. 7 Stability chart

stability chart that the logarithmic decrement becomes zero at 3,000 RPM for the mode 1 and at 5,000 RPM for mode 2. Thus the threshold speed of the rotor is 3,000 RPM which is much higher than operating speed. Also damping value for mode 3 increases with speed of the shaft and so this is stable mode of vibration.

5 Experimental Procedure

Experiments were conducted on the pump to measure the vibration readings. For this, accelerometers were mounted on the measurement locations on the pump vessels near hydrostatic bearing, pedestal, pump inlet and discharge pipes as shown

Fig. 8 Measurement points in pump assembly

in Fig. 8. Accelerometer signals were fed to charge amplifiers and filters, where it was amplified and filtered. Filtered signals were fed to a two channel FFT analyzer for frequency domain analysis and to a digital oscilloscope for real time monitoring. Vibration velocities were recorded in three orientations (axial and radial). Response spectrum at the inner vessel flange near hydrostatic bearing was obtained from the experiment and is shown in Fig. 9. It can be inferred from the spectrum that vibration of the pump is maximum at a frequency of 35 Hz which corresponds to operating speed (2,100 RPM) of pump. The vibration value at this frequency is 0.44 mm/s. The figure shows the frequency values which corresponds to peak vibration in the spectrum.

The measured value of vibration was compared with standard vibration severity chart as per ISO 10816 [2]. The power rating of the pump is approximately 37 kW

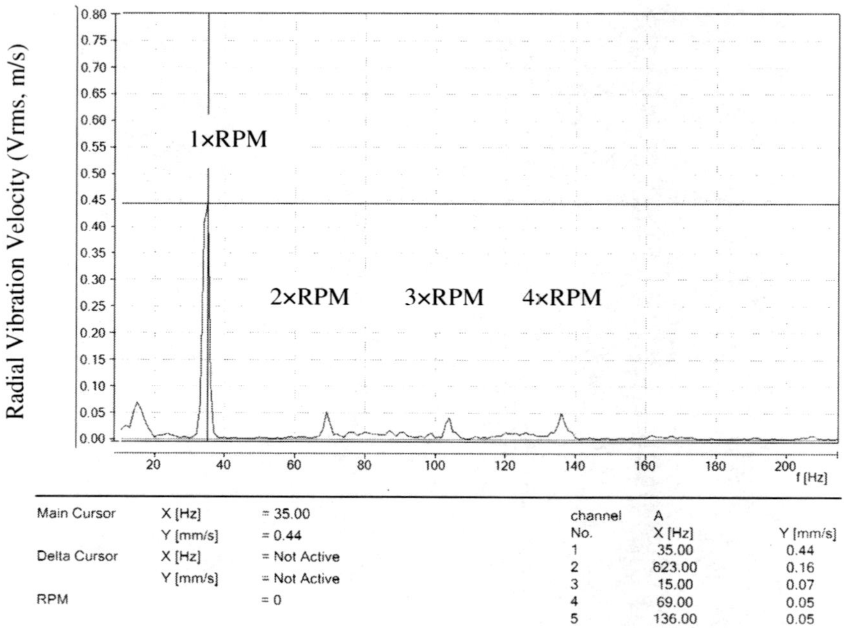

Fig. 9 Response spectrum at hydrostatic bearing location

and belongs to Class II category as per ISO 10816. The calculated and measured vibration values are well within the "Good" limit specified for Class II. This will ensure long term safe operation of the pump.

6 Conclusion

Rotor dynamic analysis was carried out for a vertical centrifugal sodium pump, which is used to circulate sodium in steam generator test facility, to estimate its natural frequencies, mode shapes, and unbalance response. Calculated value of the critical speed is 4,200. It is observed that the unbalance response at hydrostatic bearing location at normal operating speed (2,100 RPM) is 6.5 μm, which corresponds to vibration velocity of 0.35 mm/s (rms). Vibration amplitude was also measured on the pump and was found to be 0.44 mm/s, which is in close agreement with the estimated value. The measured values of vibration was compared with vibration standard ISO 10816 and was found to be well within the "Good" limit specified for similar rated machineries.

References

1. ARMD Software manual (Version-5.1.2600.2180), RBTS Inc, USA
2. ISO 10816–1 (1998) Mechanical vibration—evaluation of machine vibration by measurements on non-rotating parts—part 1: general guidelines. International Organization for Standardization, Switzerland

Qualification of Large Structures for Shock Loads

K.K. Meher and A. Rama Rao

Abstract Many critical equipments and components have to withstand shock and vibration during its life time. In defence, space and nuclear industries in particular, it is mandatory to test and analyze such components before acceptance. Analytical and experimental qualification of these components is an important exercise to be fulfilled as per the applicable codes and practices. Test for shock qualification is particularly important for components designed for defence and space applications. When the structure is large, generally an analytical approach is followed and a small scale model is experimentally validated for such qualification and the result is extended to the prototype. This approach may or may not be true always. The paper discusses about the phases of qualifying a large component for a certain level of shock. Achieving the required shock level and the pulse duration experimentally with repeatability is a challenging task. Besides, when the size and shape of the structure is odd, the qualification procedure becomes more complicated and at times difficult to achieve the desired objective. The objective was to achieve the required shock pulse by dropping the test structure from a predetermined height on a target made of commercially available rubber material. The test object is a long stainless steel vessel weighing 805 kg and 2.8 m long. The vessel is housed in a rigid cage with arrangement to lift with a magnetic clutch for instantaneous release for dropping. It was required to subject the vessel to both vertical and horizontal shock of 25 g of 35 ms duration. For dropping the vessel horizontally, the housing cage was dropped on more than one target from a predetermined height. The procedure followed in arriving at the drop height, actual drop height and the achieved shock levels are discussed in the paper.

Keywords Shock test · Shock pulse · Shock signal · Shock qualification · SRS

K.K. Meher (✉) · A. Rama Rao
Vibration Laboratory Section, Reactor Engineering Division, Bhabha Atomic Research Centre, Mumbai 400 085, India
e-mail: kkmeher@barc.gov.in

A. Rama Rao
e-mail: arr@barc.gov.in

J.K. Sinha (ed.), *Vibration Engineering and Technology of Machinery*, Mechanisms and Machine Science 23, DOI 10.1007/978-3-319-09918-7_17

1 Introduction

Many critical components and equipments designed for space, defence and nuclear applications have to undergo mandatory qualifications [1, 2] test for shock and vibration prior to their actual acceptance. Qualification for shock is one of the most important tests done both by analytical and/or experimental tools [2–4]. When the structure is large, generally an analytical approach is followed and a small scale model is experimentally validated and the result is extended to the prototype. This approach may or may not be true in all the cases. There are many guidelines and standards [1–5] available for performing such tests for design qualifications. Though there are many standard procedure and facilities available to cater for small size component testing [5], one has to tailor the procedure to fit the design requirement especially when the components are very big and cannot be tested by the available conventional facilities. There are many references in this area [6–9] which address specific problems related to structural dynamics. A Bertram [9] suggests a method for dynamic qualification of large structure.

The test discussed in this paper is experimental shock testing [10, 11] of a vessel weighing 805 kg and 2.8 m long for a 25 g shock level and 35 ms pulse duration as shown in Fig. 1. Achieving the required shock level and the pulse duration experimentally with repeatability was a challenging task.

Over and above, the shape and size of the structure to be tested made the procedure more complicated and at times difficult to achieve the required shock throughout [10, 11]. The vessel housed in a rigid cage is lifted to a predetermined height [12] and dropped on a specially designed target.

The vessel was tested in all three directions (one vertical and two horizontal) for the required level of shock [10]. The signals acquired from different parts of the test vessel show different levels for the same drop height due to secondary interaction and complex response of different parts. The complexity of the tests was not only

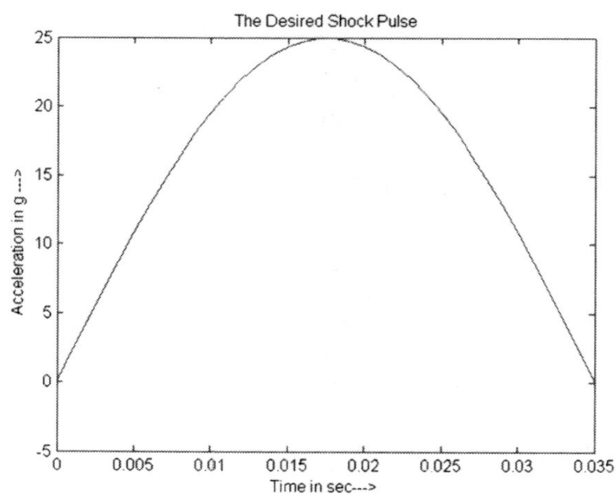

Fig. 1 The required ideal shock pulse

with the levels and repeatability, but also with the high frequency content of the signal emanated by metallic interaction during and at the time of touching the target. The signal processing [3] needs special filtering techniques to get useful signal from the acquired data. Also, the tests appears more inconsistent in certain directions as regard to the level, duration and repeatability and need an engineering decision to validate the qualification for the required goal.

This paper addresses the issues of consistency in shock level and duration, repeatability and noisy signal during such a shock test as experienced in the case [10].

2 The Test Vessel and Dropping Arrangement

The test vessel [10] is a cylindrical housing for testing individual components. The components are introduced in the vessel and assembled such that it simulates the conditions surrounding the component in actual use. Figure 2 shows the schematic of the test vessel with its supporting cage structure for tests in vertical direction and Figs. 3 and 4 for horizontal direction. The test vessel along with the supporting cage weighs 805 kg and 2.8 m long. The test vessel has two flanges one on top and one

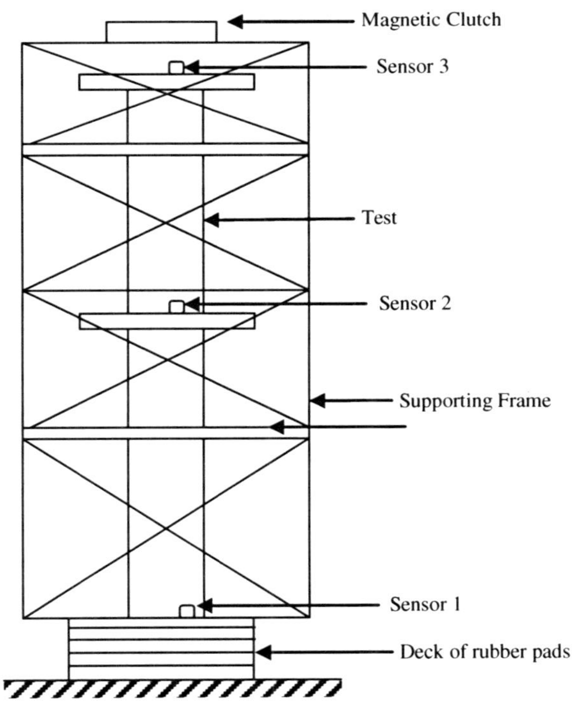

Fig. 2 Schematic of the test vessel with supporting structure for test in vertical direction

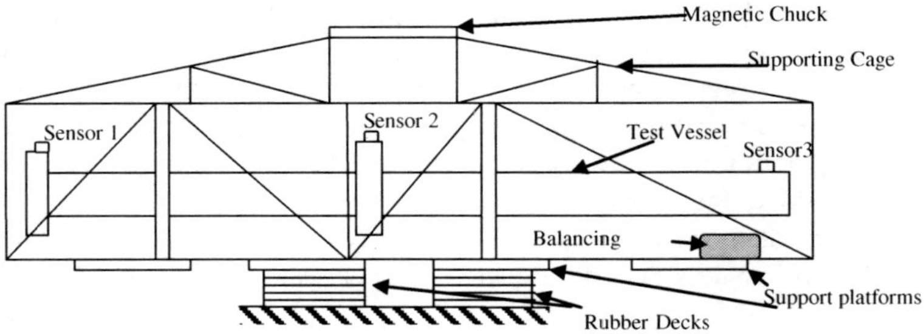

Fig. 3 Schematic of the test vessel with supporting structure for horizontal tests

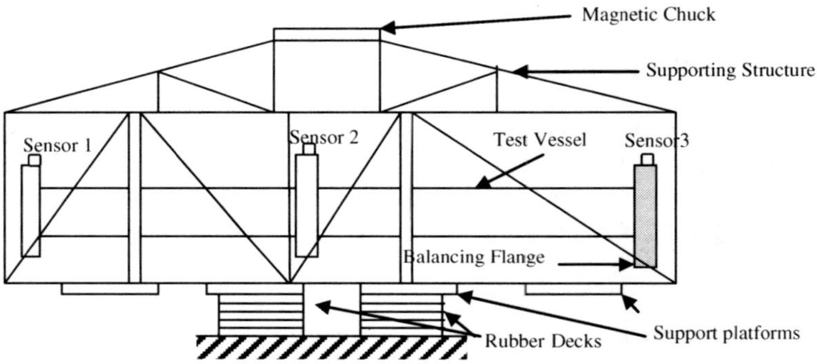

Fig. 4 Schematic of the test vessel with supporting structure with balance flange

in the middle. Provision for rigid mounting of shock measuring accelerometers for sensing vertical and horizontal shock was made on the two flanges of the vessel and at the bottom. Mounting positions are accordingly chosen. The vessel is locked inside the cage allowing minimum relative motion between the cage and the vessel. A magnetic clutch lifts the cage and releases it as and when required. In vertical direction the cage gets guided to fall vertically, however for horizontal test there was no guiding arrangement.

3 Analytical Estimation of Drop Height

Theoretical calculations assuming rigid floor, rigid cage, and ideal fall shows a drop height of around 0.35 m required to achieve the shock of 25 g and 35 ms duration. An analytical estimation of the drop height has been given below based on theory and practical experiences.

The cage has been idealized as a free falling mass from a height (h) on a grounded spring (simulated ground condition with stiffness k) for analyzing its

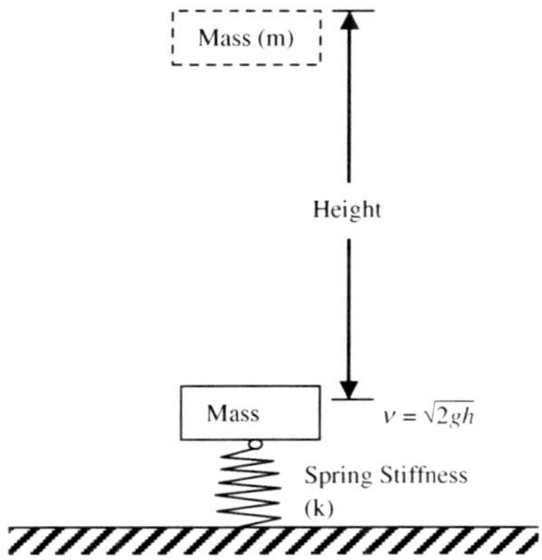

Fig. 5 Drop height estimation for an idealized spring-mass system

dynamic behavior. The idealized system is shown in Fig. 5. The response [13] from such a system with approach velocity of $\sqrt{2gh}$ is given by

$$\boldsymbol{x}(\boldsymbol{t}) = \sqrt{\frac{2\boldsymbol{gh}}{\omega^2} + \left(\frac{\boldsymbol{g}}{\omega^2}\right)^2} \boldsymbol{sin}(\omega \boldsymbol{t} - \phi) + \frac{\boldsymbol{g}}{\omega^2} \tag{1}$$

and

$$\ddot{x}(\boldsymbol{t}) = -\omega^2 \sqrt{\frac{2\boldsymbol{gh}}{\omega^2} + \left(\frac{\boldsymbol{g}}{\omega^2}\right)^2} \boldsymbol{sin}(\omega \boldsymbol{t} - \phi) \tag{2}$$

where,

$\boldsymbol{x}(\boldsymbol{t})$	Displacement response with time (t)
$\ddot{x}(\boldsymbol{t})$	Acceleration response with time (t)
$\boldsymbol{g}$	Acceleration due to gravity
ω	Natural frequency of the idealized spring mass system
ϕ	Phase angle

Recognizing the fact that, $\frac{g}{\omega^2} = \delta_{st}$ (the statical deflection) and the maximum acceleration occurs when the sine function is 1, which leads to a direct relationship [13] between the drop height and the acceleration as

$$\frac{\ddot{x}}{g} = -\sqrt{\frac{2h}{\delta_{st}} + 1} \tag{3}$$

For the shock pulse duration of 35 ms (half sine pulse), the time period is 70 ms. This leads to natural frequency (ω) of 89.76 rad/s. With a total drop weight of 805 kg (vessel along with the supporting cage), the calculated stiffness (k) is 6485.77 KN/m. From relationship 3, the drop height (h) for 25 g shock level is 0.35 m.

However, due to variations in as installed conditions, the drop height is expected to vary [12]. In the above idealized calculation, damping has not been considered which will appear in the form of stain energy, heat, sound and structural damping of the falling mass. These considerations will ask for a higher drop height than the ideal calculation [12]. The calculated stiffness from the above relationship is the dynamic stiffness which is 1.4 times higher than the static stiffness [12].

$$K_{shock} = 1.4K_{static} \tag{4}$$

From relationship 4, the static stiffness K_{static} comes out to be 4632.7 KN/m. Using the static stiffness for calculating static deflection δ_{st}, leads to the new drop height of 0.49 m. The actual and theoretical drop heights also have a similar relationship as shown below.

$$H_{actual} = 1.4H_{ideal} \tag{5}$$

The target is a stack of commercial grade rubber pads that together gives dynamic stiffness close to calculated dynamic stiffness. The calibration trials were carried out to account for all the ground variables to arrive at a conservative drop height to achieve the required shock pulse.

4 The Shock Test Procedure

As shown in Figs. 2, 3, and 4, all the three accelerometers measure the shock in the direction of test and so responses from all the three sensors are analyzed to assess uniformity of the shock loading. Variations in the amplitude of shock amplitude at the three locations were expected due to unsymmetrical participation of masses.

The recorded signals were analyzed online, with a standard FFT analyzer and MATLAB. The undulations in the signals are due to the high frequency components arising during impact. With proper filtering of the signal, the actual half sine signal is obtained. Tables 1 and 2 show the trial data of the tests and Figs. 6, 7, 8, and 9 show both unfiltered and filtered signals from the tests.

5 Observation from the Vertical Test

For the vertical tests, the assembly was made to fall on a deck of rubber pads (each layer having a thickness of 12 mm). The acceleration levels and the pulse durations for different height are tabulated in Table 1. The final height of drop required for achieving the required amplitude of shock on 13 layers of rubber pads was 0.5 m. This is close to the calculated height in Para 3. This height and the rubber pad

Table 1 Vertical test trials

Trial no.	Drop height (mt)	Sensor location	Unfiltered		Filtered	
			Pulse amp. level (g)	Pulse time (ms)	Pulse amplitude level (g)	Pulse time (ms)
1	0.35	1	22	35	19	35
		2	27	35	20	35
		3	22	35	21	35
2	0.38	1	23	35	20	38
		2	27	35	22	38
		3	22	35	22	38
3	0.45	1	25	30	23	38
		2	31	30	23	38
		3	25	30	24	35
4	0.5	1	27	35	24	40
		2	31	35	24	40
		3	28	35	24	40
5	0.5	1	28	35	24	35
		2	32	35	26	35
		3	28	35	26.5	35
6	0.5	1	28	35	23.5	36
		2	31	35	25	35
		3	26	35	26	35

For sensor location see Fig. 2

combination give an acceleration level of 25 g and time duration of 35 ms at all the three response locations. A typical plot is shown in Fig. 6 for all three locations of measurement.

6 Observation from the Horizontal Test

The horizontal tests were performed following similar procedure. However, the response from the three locations showed wide variation in amplitude of the pulse and its duration. As the test vessel is unsymmetrical in horizontal position, a counter weight at the bottom end (non-flanged end) was provided to balance its horizontal fall. A balancing weight was placed on the supporting frame as shown in Fig. 3.

Under this condition, the shock experienced by the three locations had wide variations [10] as shown in first few rows of Table 2 (trials 1–3). Figure 7 shows typical plots from horizontal tests of the vessel with the balance weight for the three

Table 2 Horizontal test trials

Trial no.	Drop height (mt)	Sensor location	Unfiltered		Filtered	
			Pulse Amplitude Level (g)	Pulse Time (ms)	Pulse Amplitude Level (g)	Pulse Time (ms)
1	0.25	1	32	30	29	32
		2	25	33	19	33
		3	25	25	24	25
2	0.28	1	31	33	28	35
		2	26	35	19.5	35
		3	25	30	23	35
3	0.3	1	36	25	35	30
		2	22	25	19	30
		3	38	20	35	32
4	0.35	1	38	28	36	30
		2	25	30	23	35
		3	40	30	38	30
5	0.35	1	37	30	35	32
		2	25	25	22	30
		3	40	30	37	30

For sensor locations, refer Figs. 3 and 4

locations for 0.28 m drop height and 17 layers of rubber pads in two stacks close to the center. The shock amplitudes at the ends were high as compared to shock in the centre. The pulse showed a double peak feature as shown in Fig. 8, which is a distorted shape of half sine shape. It also had longer pulse duration.

The assembly was then made to fall on four stacks of rubber pads. Due to different contact time of different portions of the assembly on the rubber pads and the shock traversal from ends towards the centre, variation in the shock amplitude showed more distortions. So, this scheme was not pursued further.

The assembly was then made to fall on two decks of rubber pads positioned further close to the centre for a uniform fall and response. Though the double peak feature at the centre improved to a single peak like pattern, the pulse duration was very long with reduced peak acceleration levels. Shock amplitudes at the ends were very high as compared to in the centre as shown in Table 2 (trials 7–8). The reason was recoil motion of the overhung end portion. Some typical plots are shown in Figs. 7 and 8 for the horizontal test with balance weight and two decks of rubber pads with various arrangements.

For better symmetry in mass distribution, a counter flange was fitted to the bottom end of the vessel instead of the counter weight tried earlier on the supporting structure. To minimize the distortion at the middle and at the end, the assembly was

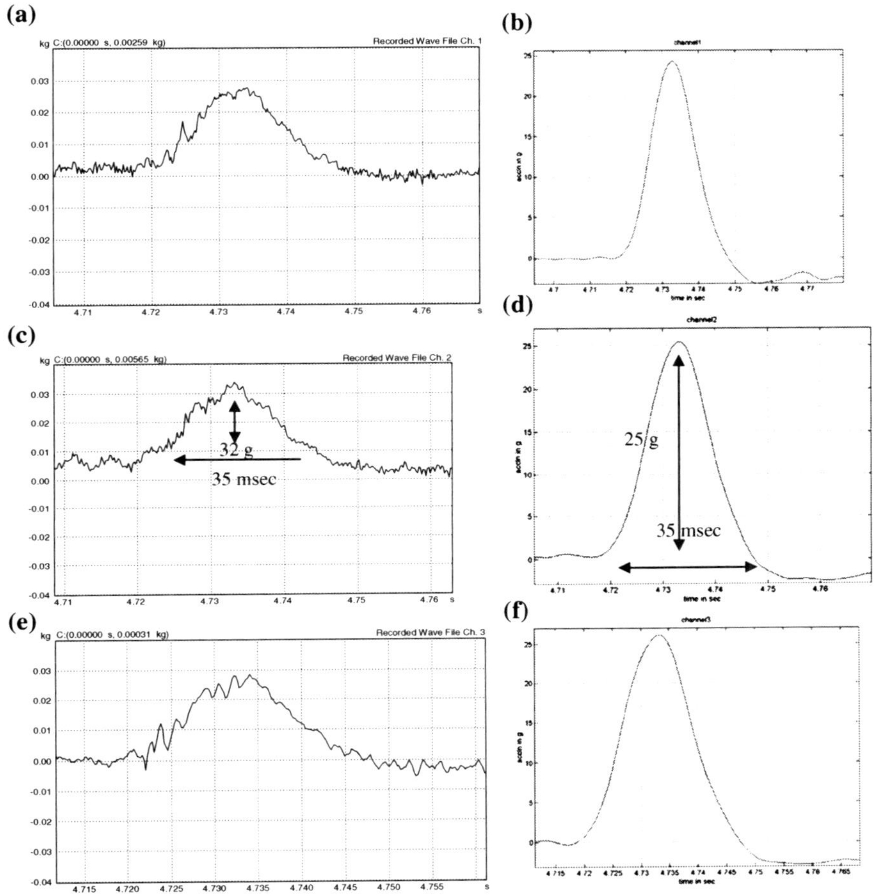

Fig. 6 Vertical drop test with 13 layers of rubber pads and 0.5 m drop height (trial 5, Table 1; **a**, **c**, **e** unfiltered signal, **b**, **d**, **f** filtered signal), x-axis → time and y-axis → acceleration in g. **a** Response from *bottom* (unfiltered). **b** Response from *bottom* (filtered). **c** Response from *middle lange* (unfiltered). **c** Response from *middle lange* (unfiltered). **e** Response from *top flange* (unfiltered). **f** Response from *top flange* (filtered)

made to fall on two stacks of rubber pads instead of four (which showed more distortion in the response) for the horizontal tests. The two ends of vessel recorded higher 'g' level of shock pulse duration, where as the middle showed lower 'g' and longer pulse. To maintain the minimum required 'g' level throughout, the height and drop targets were adjusted. However, the level shown by the end responses were always more than the required level.

By repeated adjustment of drop height and the position of the target, the shock levels were brought to acceptable value as shown in Fig. 9 and in Table 2 (trials 9–10).

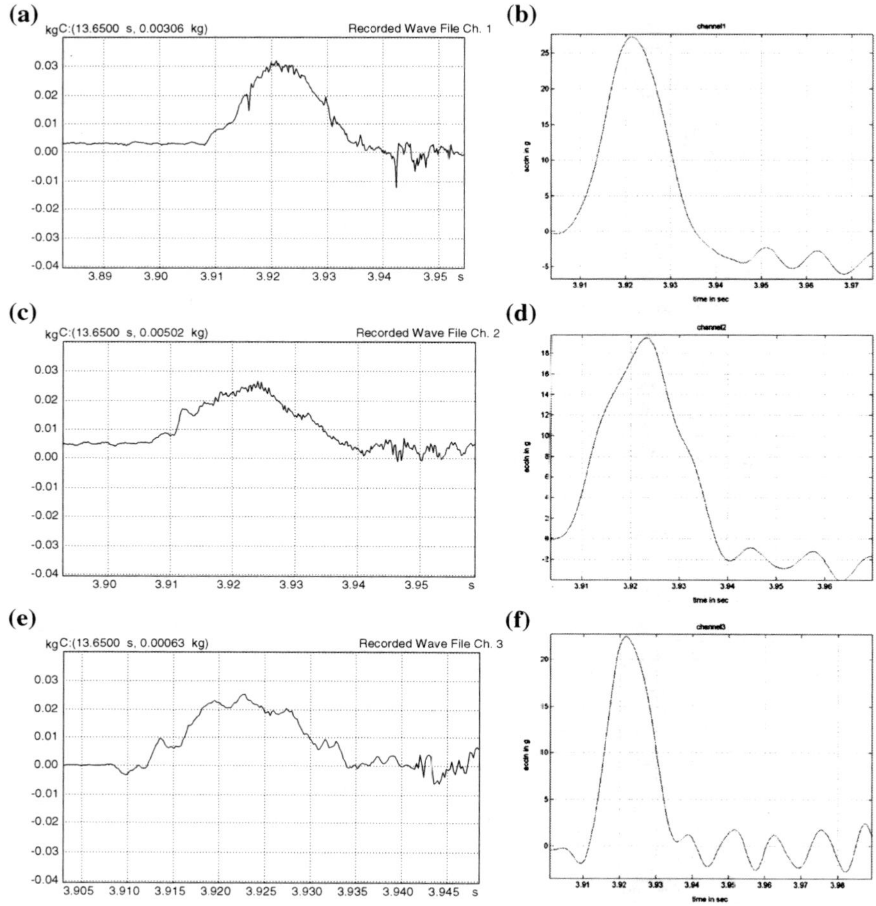

Fig. 7 Horizontal drop test of the vessel with balance weight with 17 layers of rubber pads in two decks located near the centre and 0.28 mt drop height (Trial 2, Table 2). (**a**, **c**, **e** unfiltered signal and **b**, **d**, **f** filtered signal). **a** Response from *top* (unfiltered). **b** Response from *top* (filtered). **c** Response from *middle* (unfiltered). **c** Response from *middle* (unfiltered). **e** Response from *bottom* (unfiltered). **f** Response from *bottom* (filtered)

7 Analysis

The signals were analyzed with a standard analyzer and the data were recorded for further conditioning. The raw signals normally consist of high frequency contents in the form of spikes [10]. The high frequency noises were filtered out using MATLAB to arrive at the decisive pulse.

The drop height and number of targets to be used in horizontal test was arrived at by several adjustments. The repeatability of the tests was performed to confirm their

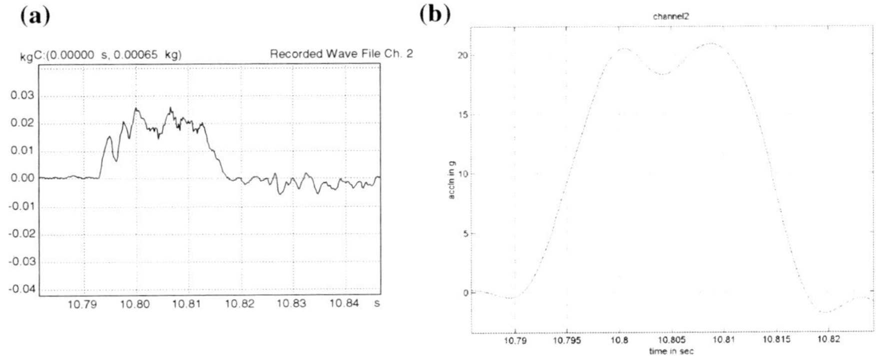

Fig. 8 Typical time spectra showing double peak feature of the response at the middle flange (with balance weight). **a** Horizontal test (unfiltered) shock at *middle flange* with drop height 0.35 mt and 17 layers of target pads. **b** Horizontal test (filtered) shock at *middle flange* with drop height 0.35 m and 17 layers of target pads

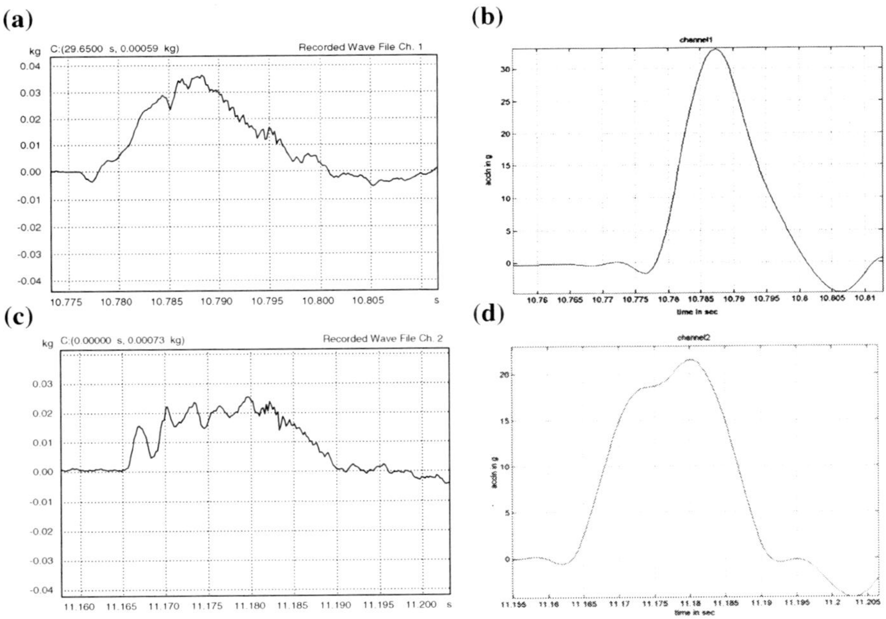

Fig. 9 Typical shock pulses for the horizontal test of the assembly from sensor 1 and 2 of the test vessel with the balancing flange at the *bottom* (other end). (**a**, **c**, **e** unfiltered signal and **b**, **d**, **f** filtered signal). **a** Horizontal test (unfiltered) (height 0.3 mt, 17 layers sensor 1). **b** Horizontal test (filtered) (height 0.3 mt, 17 layers pad, sensor 1). **c** Horizontal test (unfiltered) (height 0.35 mt, 17 layers pad, sensor 2). **d** Horizontal test (filtered) (height 0.35 mt, 17 layers pad, sensor 2)

reproducibility. Though the results of vertical tests were acceptable, the horizontal tests showed wide variations as regard to the pulse amplitude and duration for the three locations. Owing to the huge horizontal span, the levels were difficult to adjust to a uniform level at all the locations. However, efforts were made to make the level such that, 25 g acceleration is measured on the vessel with reasonable time duration in all the portions.

8 Analysis for Shock Response Spectrum (SRS)

Shock response spectrum [13] is a plot of the maximum response to a specified shock load for all possible single degree of freedom systems in the frequency range of consideration. This is a useful tool in analyzing systems for design qualification, especially for seismic loading.

Analytical approach was followed to ensure that the achieved shock level envelopes the response spectrum of the desired shock at least up to the desired frequency range. A typical shock response spectrum (SRS) for the ideal shock level of 25 g and 35 ms duration is shown in Fig. 10. The analyzed SRS is up to 250 Hz for single degree of freedom systems having a viscous damping ratio of 10 %. The behavior will be similar for any other damping value. Figure 11 shows the enveloping trend of the SRS of a 26 g–34 ms shock pulse. Though both the SRS trends look similar up to 15 Hz, the 26 g–34 ms SRS clearly overshoots the ideal SRS beyond 15 Hz. To prevent under qualification, the achieved SRS should envelope the ideal SRS in the desired frequency range. Similar analysis was done for all the achieved shock levels.

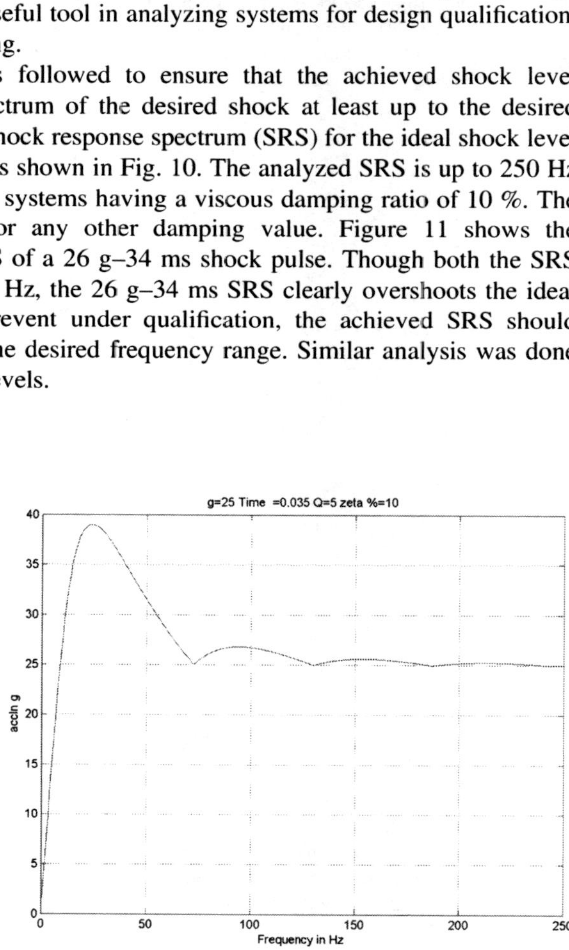

Fig. 10 Shock response spectrum for 25 g and 35 ms shock pulse with a system damping ratio of 10 %

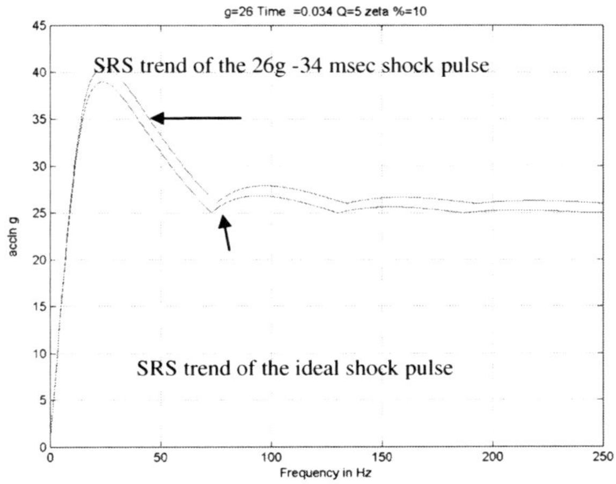

Fig. 11 Shock response spectrum for a shock load of 26 g and 34 ms time showing enveloping of the ideal SRS

9 Conclusion

(a) Shock testing using the drop test method may show inconsistencies as regard to the level and duration of the shock pulse, when seen from different response locations on the test specimen especially when the test structure is large.

(b) Several response locations on the test piece are advised, as the reflection from one location may not be the actual representation of the whole structure.

(c) Test specimen undergoing restricted number of shock cycles during test, need special care in standardizing the test procedure for shock level and duration prior to subjecting the specimen for actual test.

(d) The drop height, ground condition and the drop target are sensitive parameters to control and standardize the procedure. They need validation from time to time to account for the changing conditions during the tests.

(e) Normally, the increase in drop target thickness (rubber pads) was found to increase the time duration of pulse and an increase in drop height increases the shock level.

(f) Any change in shape and size of the specimen need validation of the drop height and target.

(g) Consistent reproducibility is important and need constant vigilance to avoid under or over qualification of part or whole of the structure.

(h) If the noisy signal is not properly handled by appropriate filtering technique, getting to acceptable level and duration of shock may be difficult even for straight forward tests like in vertical direction.

References

1. ASME (2010) Boiler and pressure vessel code, Section I, IV, III, XI, VIII, X, XII
2. ISO 4866:2010—Mechanical vibration and shock—vibration of fixed structures—guidelines for the measurement of vibrations and evaluation of their effects on structures
3. ISO 18431:2007—Mechanical vibration and shock—signal processing—Part 4: Shock-response spectrum analysis
4. Harris CM, Peirsol AG (2001) Shock and vibration handbook, 5th edn. McGraw Hill, New York
5. ICC-ES AC156, Seismic qualification by shake-table testing of nonstructural components and systems. www.icc-es.org/criteria/pdf_files/ac156.pdf
6. Rittweger A, Beuchel W, Andersen MG, Albus J (2005) Lessons learned from the dynamic identification/qualification tests on the ESC-A upper stage model. Acta Astronaut 57 (12):877–886
7. Meher KK, Rao AR (2006) Optimization of the number of spacers in a nuclear fuel bundle with respect to flow-induced vibration. Nucl Eng Des 236(22):2348–2355
8. Moorthy RIK, Sinha JK (1998) Dynamic qualification of complex structural components of nuclear power plants. Nucl Eng Des 180(2):147–154
9. Bertram A (1980) Dynamic qualification of large space structures by means of modal coupling techniques. Acta Astronaut 7(10):1179–1190
10. Meher KK, Narayan AK, RamaRao A (2010) Drop test of assembly with CCG and PCG, RED/VLS/2010, dated Jan 27, 2010. BARC internal report
11. Meher KK, Narayan AK, RamaRao A (2010) Re-calibration of K-200 shock testing machine. RED/DS/VLS/2010/567, dated Feb 25, 2010, BARC internal report
12. Theory of vibration/shock isolators, www.rpmmech.com/pdf/selecting-a-vibration-shock-isolator.pdf
13. Thomson WT, Dillon Dahleh M (2005) Theory of vibration with applications, 5th edn. Pearson Education, Singapore

Hybrid Viscous-Structural Damping Identification Method

Vikas Arora

Abstract Damping still remains one of the least well-understood aspects of general vibration analysis. The effects of damping are clear, but the characterization of damping is a puzzle waiting to be solved. A major reason for this is that, in contrast with inertia and stiffness forces, it is not clear which state variables are relevant to determine the damping forces. In this paper, a new hybrid viscous-structural damping identification method is proposed. The proposed method is a direct method and gives explicit structural and viscous damping matrices. The effectiveness of the proposed structural damping identification method is demonstrated by two numerical examples. First, numerical study of lumped mass system is presented which is followed by a numerical study of fixed-fixed beam. The effects of coordinate incompleteness and different level of damping are investigated. The results have shown that the proposed method is able to identify accurately the damping of the system.

Keywords Damping identification · Hybrid viscous-structural dynamic system · Direct method

1 Introduction

The modeling of damping is a very complex and still considered somewhat an unknown or grey area. A major reason for this is that, in contrast with inertia and stiffness forces, it is not in general clear which state variables are relevant to determine the damping forces. A commonly used model originated by Rayleigh [1] assumes that instantaneous generalized velocities are the only variables. However, it is important to avoid the misconception that, this is the only model of vibration

V. Arora (✉)
Institute of Technology and Innovation, University of Southern Denmark, Campusvej 55, 5230 Odense, Denmark
e-mail: viar@iti.sdu.dk

J.K. Sinha (ed.), *Vibration Engineering and Technology of Machinery*, Mechanisms and Machine Science 23, DOI 10.1007/978-3-319-09918-7_18

damping. Any model, which guarantees that the energy dissipation rate is non-negative, can be a potential candidate to represent the damping of a given structure. In broad terms, damping mechanisms can be divided into three categories namely material damping, boundary damping and viscous damping. Material damping can arise from variety of micro structural mechanisms [2] but for small strains it is often adequate to represent it through an equivalent linear, visco-elastic continuum model of the material. Heckl [3] found that linear theory produce acceptable response predictions for panels whose damping mechanism arose from a bolted joint on beam. Oliveto and Greco [4] conducted a study on how the modal damping ratios change with different boundary conditions and they found that Rayleigh-type damping is actually independent of the boundary conditions and modal damping ratios can be easily converted from one boundary condition to another. Maia et al. [5] discussed the use of fractional damping concept for the modeling the dynamic behavior of the linear systems and showed how this concept allows for clearer interpretation and explanation of the behavior displayed by common viscous and hysteric damping models. Agrawal and Yuan [6] modeled the damping forces proportional to the fractional derivative of displacements and the fractional differential equations governing the dynamics of a system. Adhikari and Woodhouse [7] developed four indices to quantify non-viscous damping in discrete linear system. Chen et al. [8] presented a method for getting the spatial model from complex frequency response function. Unfortunately, it is unrealistic to assume that all pertinent information is given to solve for damping matrix. Actually, data from testing is neither complete nor error free. Minas and Inman [9] proposed a method which assumes that analytical mass and stiffness matrices are determined a priori from a finite element model. The identified damping matrix is assumed to be real, symmetric and positive definite. The structure must exhibit complex modes for this procedure and the solution is limited to real symmetric positive definite damping matrices. Beliveau [10] uses natural frequencies, damping ratios, mode shapes and phase angles to identify parameters of viscous damping matrix. The identification is performed iteratively. Lancaster [11] proposed a method of identifying the mass, stiffness and damping matrices of a system directly given only the eigenvalues and eigenvectors. This method is only for calculating the viscous damping. The shortfall of this method comes in normalizing the eigenvectors, which requires knowledge of the very same damping matrix which we wish to, find in the end. Pilkey [12, 13] proposed two methods for computing the damping matrix using modal data. The first method is an iterative method which requires prior knowledge of the mass matrix and eigenvalues and eigenvectors. The second method requires more information but less computationally intensive. Both the methods developed from the Lancaster [11] concept. Oho et al. [14] proposed a method of identifying experimental set of spatial matrices valid only for the hysterical damping for the entire frequency range of interest using FRFs. The limitation of this method is that it is unable to predict correctly in the modal domain. Lee and Kim [15] proposed an algorithm for the identification of the damping matrices which identifies the viscous and structural damping matrices using frequency response matrix. The accuracy of the identified damping matrices depends almost entirely on the accuracy of the measured FRFs, especially their phase angles. Adhikari and Woodhouse [16] identified the damping of the system as viscous damping. However, viscous damping is by no means the only damping model. Adhikari and Woodhouse [17] identified non-viscous damping model using an exponentially decaying relaxation function. Phani and Woodhouse [18] proposed that complex modes arising out of non-proportional dissipative matrix hold the key to successful modeling and identification of correct physical damping mechanisms in the vibrating systems but these identified complex modes are very sensitive to experimental errors and errors arising out from fitting algorithms. Some research efforts have also been made to update the damping matrices. Imregun et al. [19] extended the response function method (RFM) given by Lin and Ewins [20] to update proportional viscous and structural damping matrices by updating the coefficients of viscous and structural damping matrices. Arora et al. [21] identified the structural damping matrix using complex frequency response functions (FRFs) of the structure. Arora et al. [22] proposed a viscous damping identification method in which viscous damping is identified explicitly. This procedure is a two steps procedure. In the first step, mass and stiffness matrices are updated and in the second step, viscous damping is identified using updated mass and stiffness matrices obtained in the previous step. Pradhan and Modak [23] used the normalized FRFs for updating damping matrices. The normalized FRFs are estimated from the complex FRFs. Arora [24] proposed a structural damping identification method in which normalized FRFs are used. The proposed method is a direct method and gives explicit structural damping matrix.

The proposed hybrid viscous-structural damping identification method requires prior knowledge of accurate mass and stiffness matrices. The identified viscous-structural damping matrices are both symmetric and positive definite. The effectiveness of the proposed structure damping identification method is demonstrated by two numerical simulated examples. Firstly, a study is performed using a lumped mass system. The lumped mass system study is followed by case involving numerical simulation of fixed-fixed beam. The effect of coordinate incompleteness and different levels of damping in a structure are investigated. The results have shown that the proposed method able to identify the damping matrices accurately with all levels of damping.

2 Hybrid Viscous-Structural Damping Identification Method

In the following section, the theory and procedure of proposed direct structural damping identification method is developed. The governing equations of motion of a multi degree of freedom (DOF) structure with structural damping can be written in matrix form as:

$$[M]\ddot{x}(t)^C + [K]x(t)^C + [C]\dot{x}(t)^C + [D]x(t)^C = f(t) \tag{1}$$

where [M],[K] and [D] are n × n DOF mass, stiffness and structural damping matrices respectively. All these are real matrices. $j = \sqrt{-1}$. x(t) and f(t) are n × 1 vectors of time-varying displacements and forces and 'N' is the total number of degrees of freedom in the finite element model. The superscript C indicates complex value corresponding to damped system. For harmonic excitation, $f(t) = F(\omega)e^{j\omega t}$ and $x(t) = X(\omega)e^{j\omega t}$. The Eq. (1) becomes:

$$([K] - \omega^2[M])X(\omega)^C + j([C]\omega + [D])X(\omega)^C = F(\omega) \tag{2}$$

Real or normal dynamic stiffness matrix (DSM) is given by:

$$\left[Z(\omega)^N\right] = [K] - \omega^2[M] = \left[\alpha(\omega)^N\right]^{-1} \tag{3}$$

where $\left[\alpha(\omega)^N\right]$ represents real or normal FRF matrix. Using above equation, Eq. (1) becomes:

$$\left[Z(\omega)^N\right]X(\omega)^C + j([C]\omega + [D])X(\omega)^C = F(\omega) \tag{4}$$

Multiply both sides by real or normal FRF matrix $\left[\alpha(\omega)^N\right]$. The Eq. (4) becomes:

$$X(\omega)^C + j\left[\alpha(\omega)^N\right]([C]\omega + [D])X(\omega)^C = \left[\alpha(\omega)^N\right]F(\omega) \tag{5}$$

G(ω) is a transformation matrix:

$$G(\omega) = \left[\alpha(\omega)^N\right]([C]\omega + [D]) \tag{6}$$

The Eq. (5) becomes:

$$X(\omega)^C + jG(\omega)X(\omega)^C = \left[\alpha(\omega)^N\right]F(\omega) \tag{7}$$

Equation (7) can be rewritten as:

$$(ID + jG(\omega))X(\omega)^C = \left[\alpha(\omega)^N\right]F(\omega) \tag{8}$$

where ID represents identity matrix. The displacement vector $X(\omega)^C$ is related to input force vector and also to complex FRF matrix $\left[\alpha(\omega)^C\right]$ as:

$$x(\omega) = \left[\alpha(\omega)^C\right]f(\omega) \tag{9}$$

complex FRF matrix $\left[\alpha(\omega)^C\right]$ consist of real and imaginary parts as:

$$\left[\alpha(\omega)^C\right] = \left[\alpha(\omega)_R^C\right] + j\left[\alpha(\omega)_I^C\right] \tag{10}$$

where $\left[\alpha(\omega)_R^C\right]$ and $\left[\alpha(\omega)_I^C\right]$ represents real and imaginary part of complex FRF matrix $\left[\alpha(\omega)^C\right]$. The subscripts R and I represents real and imaginary values respectively. The complex FRF $\left[\alpha(\omega)^C\right]$ is available from experimentation. Substituting Eq. (10) in Eq. (8), we get:

$$(ID + jG(\omega))\left(\left[\alpha(\omega)_R^C\right] + j\left[\alpha(\omega)_I^C\right]\right) = \left[\alpha(\omega)^N\right] \tag{11}$$

$$\left(\left[\alpha(\omega)_R^C\right] - G(\omega)\left[\alpha(\omega)_I^C\right]\right) + j\left(G(\omega)\left[\alpha(\omega)_R^C\right] + \left[\alpha(\omega)_I^C\right]\right) = \left[\alpha(\omega)^N\right] \tag{12}$$

since the right-hand side of the above equation is real so the imaginary part of the left-hand side must be equal to a zero matrix for all frequencies, hence:

$$G(\omega)\left[\alpha(\omega)_R^C\right] + \left[\alpha(\omega)_I^C\right] = 0 \tag{13}$$

$$G(\omega) = -\left[\alpha(\omega)_R^C\right]^{-1}\left[\alpha(\omega)_I^C\right] \tag{14}$$

using Eq. (6). The above equation becomes:

$$\left[\alpha(\omega)^N\right]([C]\omega + [D]) = -\left[\alpha(\omega)_R^C\right]^{-1}\left[\alpha(\omega)_I^C\right] \tag{15}$$

$$([C]\omega + [D]) = -\left[\alpha(\omega)_R^C\right]^{-1}\left[\alpha(\omega)_I^C\right]\left[\alpha(\omega)^N\right]^{-1} \tag{16}$$

In the matrix form Eq. (16) can be written as:

$$\begin{bmatrix}\omega & 1\end{bmatrix}\begin{bmatrix}[C] \\ [D]\end{bmatrix} = -\left[\alpha(\omega)_R^C\right]^{-1}\left[\alpha(\omega)_I^C\right]\left[\alpha(\omega)^N\right]^{-1} \tag{17}$$

since the normal dynamic stiffness matrix $[Z(\omega)^N]$ is inverse of normal FRF $\left[\alpha(\omega)^N\right]$ matrix $\left(\left[Z(\omega)^N\right] = \left[\alpha(\omega)^N\right]^{-1}\right)$ given in Eq. (3). Using Eq. (3), Eq. (17) becomes:

$$\begin{bmatrix} [C] \\ [D] \end{bmatrix} = -[\omega \quad 1]^{-1}\left[\alpha(\omega)_R^C\right]^{-1}\left[\alpha(\omega)_I^C\right]\left[Z(\omega)^N\right] \tag{18}$$

The above equation is the basic equation for the hybrid viscous-structural damping identification method. The identified viscous and structural damping matrices are both symmetric and positive definite. The Eq. (18) is applied to frequency range of interest. This makes Eq. (18) over determined. If nf is the number of frequency points used to identify the structural damping. The Eq. (15) becomes:

$$\begin{bmatrix} [C] \\ [D] \end{bmatrix}_{n\times n} = -\begin{bmatrix} \omega_1 & 1 \\ \omega_2 & 1 \\ \vdots & \vdots \\ \vdots & \vdots \\ \vdots & \vdots \\ \omega_{nf} & 1 \end{bmatrix}^+ \begin{bmatrix} \left[\alpha(\omega_1)_R^C\right] \\ \left[\alpha(\omega_2)_R^C\right] \\ \vdots \\ \vdots \\ \left[\alpha(\omega_{nf})_R^C\right] \end{bmatrix}_{(n\times nf)\times n}^+ \cdot \begin{bmatrix} \left[\alpha(\omega_1)_I^C\right]\left[Z(\omega_1)^N\right] \\ \left[\alpha(\omega_2)_I^C\right]\left[Z(\omega_2)^N\right] \\ \vdots \\ \vdots \\ \left[\alpha(\omega_{nf})_I^C\right]\left[Z(\omega_{nf})^N\right] \end{bmatrix}_{(n\times nf)\times n} \tag{19}$$

superscript '+' denotes pseudo-inverse of a matrix. Equation (19) is used to identify structural damping matrix [D] of the structure. $\left[\alpha(\omega)_R^C\right]$ and $\left[\alpha(\omega)_I^C\right]$ are the real and imaginary part of the FRF matrix, which are obtained from the experimentation. The accurate normal dynamic stiffness matrix (DSM) $\left[Z(\omega)^N\right]$ depends upon accurate knowledge of mass and stiffness matrices. For the proposed structural damping identification method mass and stiffness matrices are reduced to the degrees of freedom measured using iterated IRS method [25] considering the measurement points.

3 Case Study of Lumped Mass System

2 degree- of freedom lumped mass system shown in the Fig. 1 is described by lumped masses M_1, M_2 of 5 kg each, spring stiffnesses K_1, K_2 of 2×10^6 N/m each, the structural damping D_1, D_2 of 6×10^4, 8×10^4 N/m respectively and viscous damping C1, C2 of 200, 400 Ns/m respectively. Because this is a contrived example, the described result is known ahead of time. As the system has 2 DOF, 4 complex FRFs are calculated for each frequency forming a 2×2 complex FRF matrix of a function of frequency. The dynamic stiffness matrix is also calculated from the analytical mass and stiffness matrices for each frequency. As all the 2 natural frequencies are in the frequency range of 0–200 Hz.

Fig. 1 2 degree of freedom lumped mass system

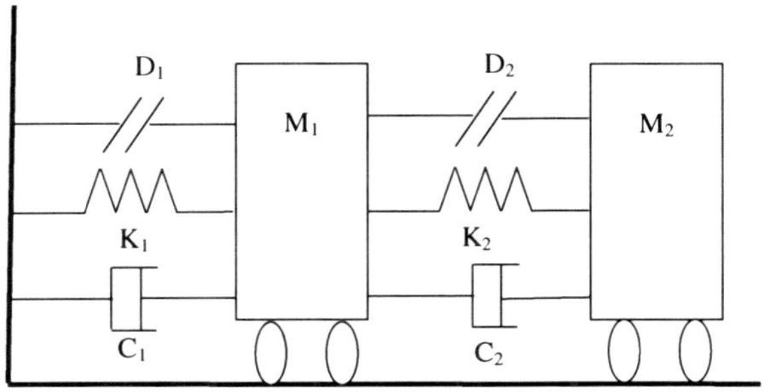

The frequency range from 0 to 200 is used to identify the viscous and structural damping using the proposed procedure. In this case study, it is assumed that accurate mass and stiffness matrices are known. Figure 2a shows the overlay of analytical FRF, which is undamped, and simulated experimental FRF obtained considering structural damping in the experimental data. It can be observed that the analytical FRF and simulated experimental FRF do not match with each other at resonance regions. Figure 2b shows the overlay of simulated experimental FRF and identified damped FRF. It can be observed from Fig. 2b that after damping identification, simulated experimental FRF and identified damped FRF match exactly and the two are virtual indistinguishable.

4 Case Study of Fixed-Fixed Beam Structure

A simulated study on a fixed-fixed beam is conducted to evaluate the effectiveness of the proposed method. The dimensions of the beam are $910 \times 50 \times 5$ mm. The beam is modeled using thirty as shown in Fig. 3. The displacements in y-direction

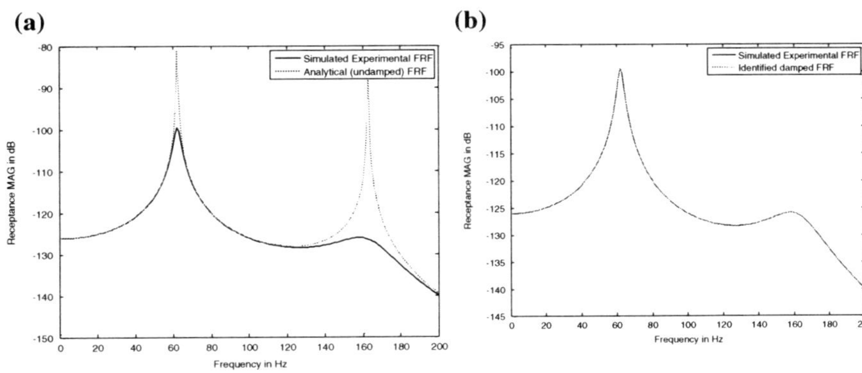

Fig. 2 Overlay of simulated experimental FRF and damped identified FRF of lumped mass system. **a** Before damping identification. **b** After damping identification

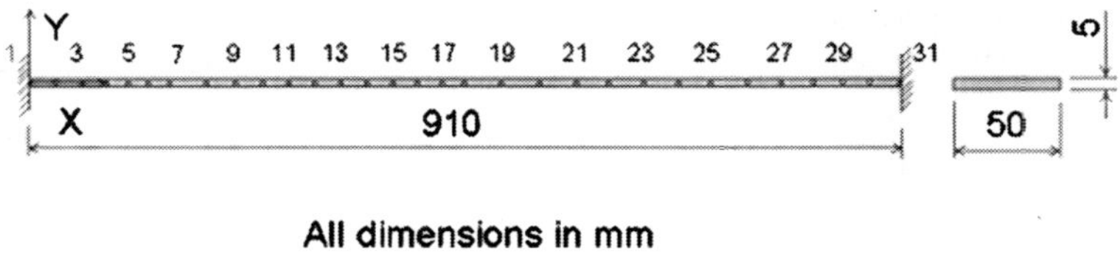

Fig. 3 Beam structure and its FE mesh

and the rotation about the z-axis are taken as two degrees of freedom at each node. It is assumed that prior knowledge of accurate and stiffness is known. Damping considered for all the cases is proportional damping being proportional to both mass $[M]$ and stiffness $[K]$ matrices [21]. The proposed method has been evaluated for the cases of different levels of damping in the numerical data. The real life measured data is always incomplete. Incompleteness is considered by assuming that only lateral degrees of freedom, at all the 29 nodes are measured.

The performance of the proposed method is judged on the basis of accuracy with which the identified damped FRFs obtained by proposed method with the simulated experimental FRFs. The case of simulated experimental data obtained using high level of structural damping and low level of viscous damping is considered. Figure 4a shows the overlay of analytical FRF, which is undamped, and simulated experimental FRF obtained considering high level of structural damping and low level of viscous damping in the experimental data. It can be observed that the analytical FRF and simulated experimental FRF do not match with each other at resonance and anti-resonance regions. After damping identification using proposed method, the identified and simulated FRFs match completely as shown in Fig. 4b.

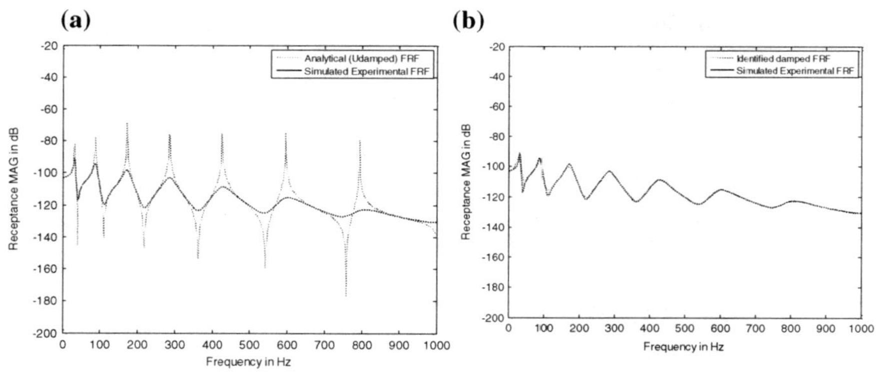

Fig. 4 Overlay of simulated experimental FRF and damped identified FRF for high structural and low viscous damping. **a** Before damping identification. **b** After damping identification

5 Conclusions

In this paper, a new method of hybrid viscous-structural damping identification method is proposed. The proposed structural damping identification method requires prior knowledge of accurate mass and stiffness matrices. The proposed method doesn't require initial damping estimates. The identified viscous structural damping matrices are both symmetric and positive definite. The proposed method addresses the difficulties for updating the damping matrices. The proposed method is working successfully for the cases of simulated numerical data. Simulated numerical case studies based on a lumped mass system and fixed-fixed beam structure with both viscous and structural damping models are carried out to assess the effectiveness of the proposed method. The success of these cases has proven the feasibility of the proposed method. The results have shown that the proposed method able to identify the structural damping matrices accurately in all levels of damping in the structure.

References

1. Rayleigh L (1897) Theory of sound (two volumes). Dover Publications, New York
2. Bert CW (1973) Material damping: An introduction review of mathematical models, measures and experimental techniques. J Sound Vib 29:129–153
3. Heckl M (1962) Measurements of absorption coefficients of plates. J Sound Acoust Soc Am 34:803–808
4. Oliveto G, Greco A (2002) Some observations on the characterization of structural damping. J Sound Vib 256:391–415
5. Maia NMM, Silva JMM, Ribeiro AMR (1998) On a general model for damping. J Sound Vib 218:749–767
6. Agrawal OP, Yuan L (2002) A numerical scheme for dynamic systems containing fractional derivatives. J Vib Acoust 124:321–324
7. Adhikari S, Woodhouse J (2003) Quantification of non-viscous damping in discrete linear systems. J Sound Vib 260:499–518
8. Chen SY, Ju MS, Tsuei YG (1996) Estimation of mass, stiffness and damping matrices from frequency response functions. J Vib Acoust 118:78–82
9. Minas C, Inman DJ (1991) Identification of viscous damping in structure from modal information. J Vib Acoust 113:219–224
10. Beliveau JG (1976) Identification of viscous damping in structures from modal information. J Appl Mech 43:335–338
11. Lancaster P (1961) Expression for damping matrices in linear vibration. J Aerosp Sci 28:256
12. Pilkey, D.F., 1998. Computation of damping matrix for finite element model updating. PhD thesis, Virginia Polytechnic Institute and State University, USA
13. Pilkey DF, Inman DJ, Friswell MI (1998) The direct updating of damping and stiffness matrices. AIAA J 36:491–493
14. Oho T, Okuma M, Shi Q (1990) Development of the experimental spatial matrix identification method. J Sound Vib 299:5–12
15. Lee JH, Kim J (2001) Development and validation of a new experimental method to identify damping matrices of a dynamic system. J Sound Vib 246:505–524

16. Adhikari S, Woodhouse J (2000) Identification of damping Part 1, Viscous Damping. J Sound Vib 243:43–61
17. Adhikari S, Woodhouse J (2000) Identification of damping Part 2, Non-Viscous Damping. J Sound Vib 243:63–88
18. Phani S, Woodhouse J (2007) Viscous damping identification in linear vibration. J Sound Vib 303:475–500
19. Imregun M, Visser WJ, Ewins DJ (1995) Finite element model updating using frequency response function data-1: Theory and initial investigation. Mech Syst Signal Process 9:187–202
20. Lin RM, Ewins DJ (1994) Analytical model improvement using frequency response functions. Mech Syst Signal Process 8:437–458
21. Arora V, Singh SP, Kundra TK (2009) Damped model updating using complex updating parameters. J Sound Vib 320:438–451
22. Arora V, Singh SP, Kundra TK (2009) Finite element model updating with damping identification. J Sound Vib 324:1111–1123
23. Pradhan S, Modak SV (2012) A method for damping matrix identification using frequency response data. Mech Syst Signal Process 33:69–82
24. Arora V (2014) Structural damping identification using normal FRFs. Int J Solids Struct 51:133–143
25. Friswell MI , Garvey SD , Penny JET (1998) The convergence of the iterated IRS technique. J Sound Vib 1211:123–132

Effectiveness of Adaptive Filter Algorithms and Spectral Kurtosis in Bearing Faults Detection in a Gearbox

Faris Elasha, David Mba and Cristobal Ruiz-Carcel

Abstract Bearing faults detection at the earliest stages is vital in avoiding future catastrophic failures. Many traditional techniques have been established and utilized in detecting bearing faults, though, these diagnostic techniques are not always successful when the bearing faults take place in gearboxes where the vibration signal is complex; under such circumstances it may be necessary to separate the bearing signal from the complex signal. The objective of this paper is to assess the effectiveness of an adaptive filter algorithms compared to a Spectral Kurtosis (SK) algorithm in diagnosing a bearing defects in a gearbox. Two adaptive filters have been used for the purpose of bearing signal separation, these algorithms were Least Mean Square (LMS) and Fast Block LMS (FBLMS) algorithms. These algorithms were applied to identify a bearing defects in a gearbox employed for an aircraft control system for which endurance tests were performed. The results show that the LMS algorithm is capable of detecting the bearing fault earlier in comparison to the other algorithms.

Keywords Vibration · Adaptive filter · Signal separation · Bearing diagnostics

1 Introduction

Monitoring of machine vibration for early fault detection is widely applied [1, 2]. The vibration signals from machines contain multiple sources which can be corrupted by noise from the transmission path. The diagnosis of bearing faults in

F. Elasha (✉) · D. Mba · C. Ruiz-Carcel
School of Engineering, Cranfield University, Bedford M43 0AL, UK
e-mail: F.elasha@cranfield.ac.uk

D. Mba
e-mail: D.mba@cranfield.ac.uk

C. Ruiz-Carcel
e-mail: c.ruizcarcel@cranfield.ac.uk

J.K. Sinha (ed.), *Vibration Engineering and Technology of Machinery*,
Mechanisms and Machine Science 23, DOI 10.1007/978-3-319-09918-7_19

gearboxes is not without its challenges [3, 4], therefore methods of enhancing the signal to noise ratio (SNR) are required [5, 6]. This is particularly the case in gearboxes where the gear mesh contribution to the overall vibration is of such significance as to mask bearing fault frequencies [7]. In practice, envelope analysis has been used to extract the bearing fault vibration signature in gearboxes [8], though in some cases envelope analysis has failed to reduce the gear mesh contribution to the total vibration signal. In such instances a narrow band-pass filter at high frequency has been applied to separate the high frequency component excited by bearing impacts [9].

Recently, signal separation techniques have been applied in the diagnosis of bearing faults within gearboxes. The separation is based on decomposing the signal into deterministic and random components. The deterministic part represents the gear component and the random part represents the bearings component of vibration. The bearing contribution to the signal is expected to be random due to slip effects [10]. Three algorithms were compared to assess their effectiveness in diagnosing a bearing defect in a gearbox; Least Mean Square (LMS), Spectral Kurtosis (SK) and Fast Block LMS (FBLMS).

These algorithms were applied to decompose the measured vibration signal into deterministic and random parts with the latter containing the bearing signal. In addition the result of the adaptive filter algorithms will be compared to the result of linear prediction and spectral kurtosis. This investigation assesses the merits of these techniques in identifying a natural degraded bearing under conditions of relatively large background noise. The gearbox considered in this study is part of a transmission system of an aircraft control system, which suffered premature bearing failure at an early stage of testing; therefore, these algorithms will be applied to examine their ability to identify the failure at the onset of degradation.

2 Theoretical Background

2.1 Adaptive Filter

An adaptive filter is used to model the relationship between two signals in an iterative manner; the adaption refers to the method used to iterate the filter coefficient. The adaptive filter solution is not unique, however the best solution is that which is closest to the desirable response signal [11]. FIR filters are more commonly used as adaptive filters in comparison of IRR filters [12].

The adaptive filter concept is based on Wold theorem which proposes that the vibration signal can be decomposed into two parts, deterministic $P(n)$ and random $r(n)$. This decomposition process can be represented by the following formula [13]:

$$x(n) = P(n) + r(n) \tag{1}$$

In the equation above the deterministic part can be predicted based on the history of the signal and the minimal prediction error, however the random part component cannot be predicted. The process of separation begins by applying adaptive noise cancellation (ANC), the fundamentals of this method have been detailed in [13, 14].

2.1.1 LMS Algorithm

The objective of the LMS algorithm is to optimize filter parameters and minimize prediction error, the prediction error ε_t is estimated according by [14]:

$$\varepsilon_t = d_t - h_i * x(t - i) \tag{2}$$

where, d_t denotes the desirable signal. The filter coefficient should be adjusted to minimize this error function. The error might be random in distribution and as such the expectation of the square error signal is used.

In the case of gearbox signal there is no reference signal, instead a delayed version of the input signal is used. An approximated adaptive LMS algorithm is proposed in [11], where the coefficients are updated by:

$$\mathrm{h_{t+1}} = \mathrm{h_t} + 2\mu\mathrm{x(t)}\varepsilon \tag{3}$$

In which, $\mathrm{h_{t+1}}$ denote the updated filter coefficient, and μ denotes the step size of the filter, this parameter should be selected carefully, the larger the step size the faster convergence, whilst on other hand a smaller step size leads to more accurate prediction, but the computation cost is high. The detail of step size selection can be found in [12]. In addition, the filter length should be selected carefully, the largest size filter decreases convergence speed and vice versa.

2.1.2 Fast Block LMS Algorithm

Applying of the standard LMS algorithm to adaptive filtering results in long processing time, this due to coefficients been updated sample by sample. This delay limits the use of the LMS algorithm for real time applications, therefore the Fast Block LMS (FBLMS) algorithm was proposed to reduce the process time [15].This algorithm is based on the transforming the time signal to the frequency domain and the filter coefficient is updated after transformation. In this algorithm the filter coefficient is updated for each segment, whereas the LMS algorithm updates the coefficients for each sample. The detail procedure of FBLMS is summarised in [16].

2.2 Spectral Kurtosis and Envelope Analysis

The basic principle of this method is to calculate the Kurtosis at different frequency bands in order to identify non-stationarities in the signal and determine where they are located in the frequency domain. Obviously the results obtained strongly depend on the width of the frequency bands Δf [17].

The Kurtogram [18] is basically a representation of the calculated values of the SK as a function of f and Δf. However, the exploration of the whole plane (f, Δf) is a complicated computation task, though Antoni [17] suggested a methodology for the fast computation of the SK. Antoni and Randall [19] demonstrated how the optimum filter which maximizes the signal to noise ratio is a narrowband filter at the maximum value of SK. Therefore the optimal central frequency fc and bandwidth Bf of the band-pass filter are found as the values of f and Δf which maximise the Kurtogram. In this investigation the SK computation and the subsequent signal filtration and envelope analysis were performed using the original Matlab code programmed by Jérôme Antoni.

3 Experimental Setup

The gearbox considered is used as part of a transmission driveline on the actuation mechanism of secondary control surfaces in civil aircrafts. The test rig was designed to simulate the actual operation conditions during the life cycle of the aircraft control system which implies the gearbox would experience a range of speed and torque conditions. A schematic of the testing is presented in Fig. 1. The gearbox consists of two spur bevel gears as shown in Fig. 2, each gear with 17 teeth producing a gear ratio of 1:1. Two angular contact bearings are used to support each gear.

During an endurance test of the flight system, a bearing within the gearbox failed after 860 h of testing, which is approximately 30 % of the expected bearing life. Vibration measurements were taken from the gearbox at different stages of the test. In addition, torque and angular velocity were also measured. The main rotational frequencies and bearing faults frequencies are summarized in Table 1.

Vibration data was acquired using an accelerometer fixed on the outer case of the gearbox as shown in Fig. 2. The operating frequency range of the accelerometers was 10–10000 Hz. A signal conditioner (Endevco 2775A) was employed and an NI DAQ system was used to acquire data at a sampling rate of 5 kHz. Data was acquired at different periods during the endurance test.

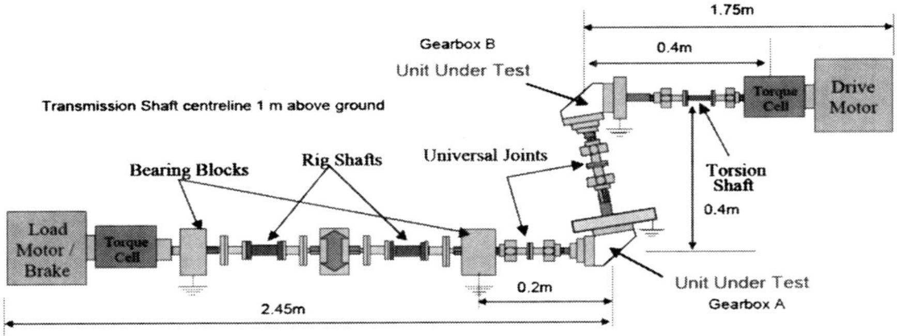

Fig. 1 Test rig layout

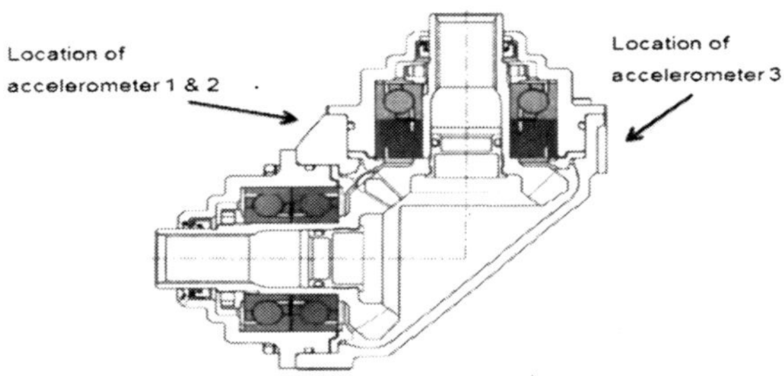

Fig. 2 Gearbox layout and sensors location

Table 1 Bearing faults frequencies

Parts	Frequency (Hz)
Shaft speed frequency (SS)	11.8
Gear mesh frequency (GM)	201.2
Inner race defect frequency (IRD)	83.2
Outer race defect frequency (ORD)	58.8
Cage defect frequency	4.9
Ball spin frequency	25.6

4 Result and Observations

4.1 Spectral Kurtosis

Analysis employing spectral kurtosis was undertaken on all data set, this yielded the frequency band and center frequency which were then used to undertake the envelope analysis. Defaults of the center frequencies are shown in Table 2. These frequencies were employed for envelop analysis in this section and for enveloping

Table 2 Maximum kurtosis location

Observation	Fc (Hz)	Δf (Hz)	Kmax	Frequency band (Hz)
1, 2, 3	2083.33	833.3	2.4	1666.7–2500

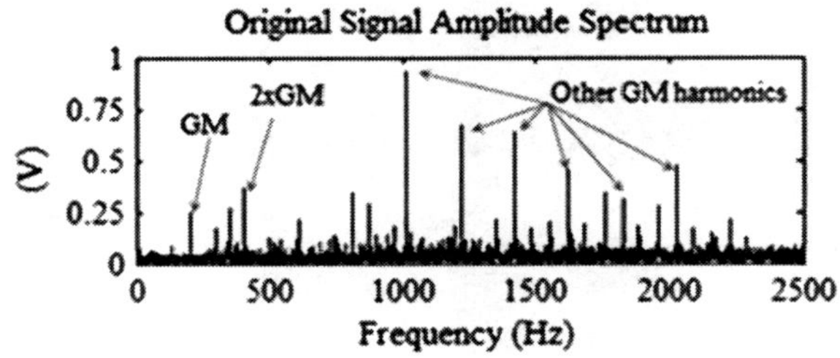

Fig. 3 Original signal amplitude spectrum

the separated signal obtained from linear prediction, LMS and FBLMS algorithms. The original signal collected at 24 percent is shown in Fig. 3.

Result from enveloped analysis using the filter parameters detailed in Table 2 is presented in Fig. 4. This shows clearly that the signal is dominated by the gear mesh frequency, but at this point (24 % of bearing life) it does not provide any information of an incipient fault in the system. In the second observation (27 % of bearing life) a new peak at the frequency of 58.4 Hz, indicating an incipient fault in the outer race of the bearing was evident. In addition, side-bands around the gear mesh frequency: 190.2 and 214 Hz were identified. The side-bands were spaced at running speed. In third observation (30 % of life) not only was bearing outer race frequency noted, but a second harmonic was also observed, see Fig. 4.

4.2 LMS and FBLMS Result

The Adaptive filter was applied to the measured bearing vibration signature at different stages of the endurance test, both LMS and FBLMS algorithms were applied to perform signal separation, the result of each algorithm will be compared later. After the bearing signal separation had been completed, envelop analysis was performed to identify any bearing defect frequencies. The envelop analysis was performed by applying a band-pass filter centered around 2,083 Hz with a bandwidth of 833.33 Hz. These parameters were selected based on the result from maximum Spectral Kurtosis as described in spectral kurtosis section.

The spectrum of enveloped signal associated with 24 % of bearing life is presented in Fig. 5 for the LMS algorithm. Observations showed the gear mesh (201.1 Hz) was dominant (see left plot in Fig. 5). The right plot in Fig. 5 represents a narrower frequency range in the original spectrum showing the presence of BPFO, shaft speed and harmonics of shaft speed. Figure 6 shows the result obtained by enveloping the FBLMS algorithms, which doesn't show the existence of any fault frequencies at this stage.

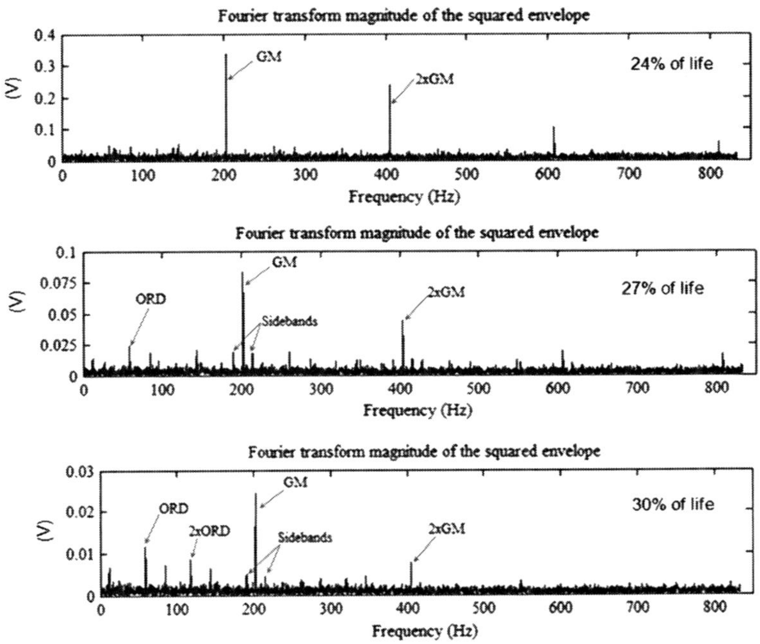

Fig. 4 Envelop spectrum using parameters from SK

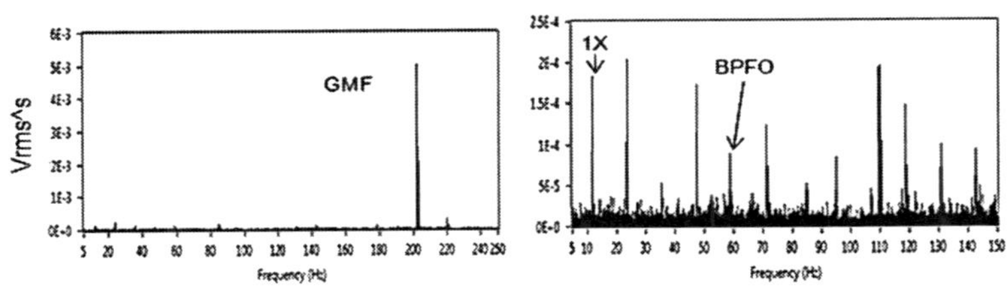

Fig. 5 Enveloped signal spectrum with LMS algorithm at 24 % of life

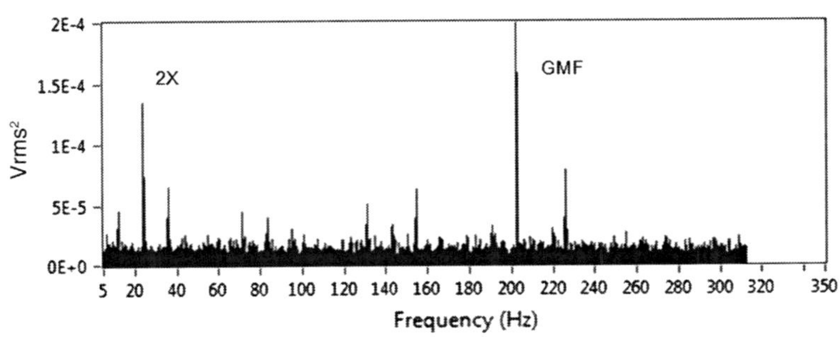

Fig. 6 Enveloped signal spectrum using a FBLMS algorithm at 24 %

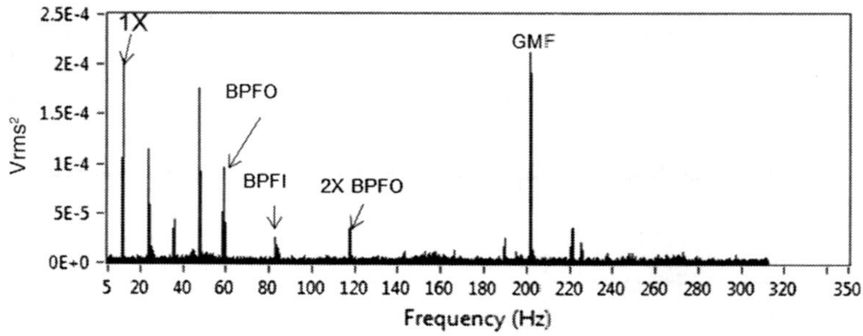

Fig. 7 Enveloped signal spectrum with LMS algorithm at 27 % of life

The spectrum of the enveloped signal from the second data (27 % of bearing life) using the LMS algorithm is shown in Fig. 7. The presence of the outer race defect frequency was evident; in addition a second harmonic of the BPFO was noted (118 Hz). This result supports the observation obtained from the first data set (see Fig. 5). Furthermore, the bearing inner race defect was also identified at 83.2 Hz as shown in Fig. 7.

Result from processing the data at 27 % using FBLMS algorithms showed the presence of both BPFI and BPFO frequencies at 58.85 and 83.2 Hz respectively, see Fig. 8. In addition, side-bands around gear mesh spaced by shaft frequency and shaft harmonics were identified by both algorithms as shown in Figs. 7 and 8.

For data set collected at 30 % of bearing life, the frequency spectrum obtained from the LMS algorithms showed BPFO, BPFI, and Ball Spin (BS) frequencies, see Fig. 9. In addition the second harmonic of the BPFI frequency (165 Hz) was evident. The frequency spectrum obtained from the FBLMS algorithms showed BPFI and BPFO frequencies as shown in Fig. 10. In both algorithms, the amplitude of BPFO decreased by 14 % from the previous data set; this is attributed to an increase of bearing clearances due to defects which result in a decrease of bearing vibration. In addition, this data was recorded during a loading cycle where the transmitted torque (50 Nm) was lower than in the previous measurements

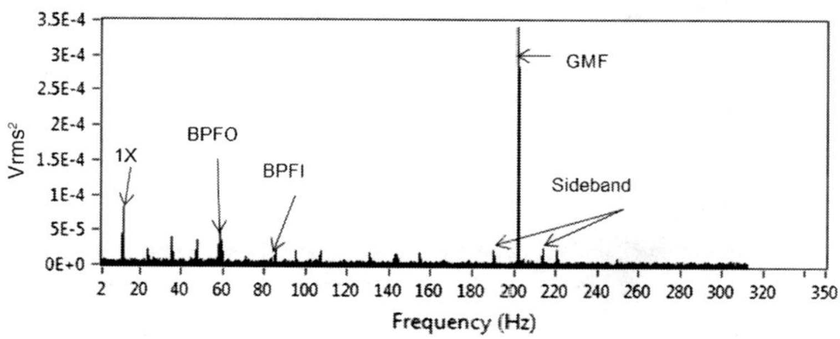

Fig. 8 Enveloped signal spectrum with FBLMS algorithm at 27 % of life

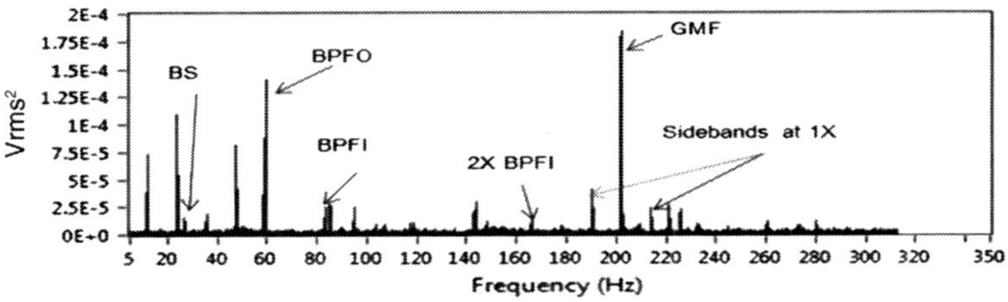

Fig. 9 Enveloped signal at 30 % of life using LMS algorithm

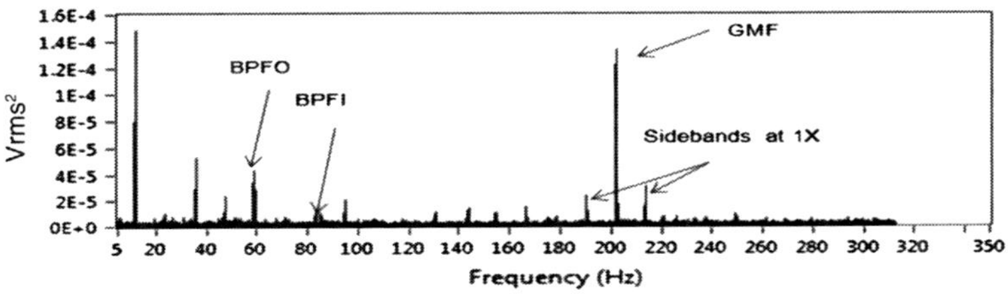

Fig. 10 Enveloped signal spectrum at 30 % of life using FBLMS algorithm

(126 Nm). Figures 9 and 10 show Sidebands around fundamental gear mesh spaced by shaft frequency and shaft harmonics were now more pronounced in the spectrum for the data set associated with 30 % bearing life, this is indicative of misalignment. This misalignment can affect the gear mesh and further accelerate bearing degradation, consequently the test was stopped at this stage and a visual inspection was performed to assess the damage.

At all three life stages the gearbox was disassembled for visual inspection, evidence of scratches in the bearing outer race at an early stage of 24 % was observed, see Fig. 11. At the end of the test the bearing ball and outer race were damaged as shown in Fig. 11.

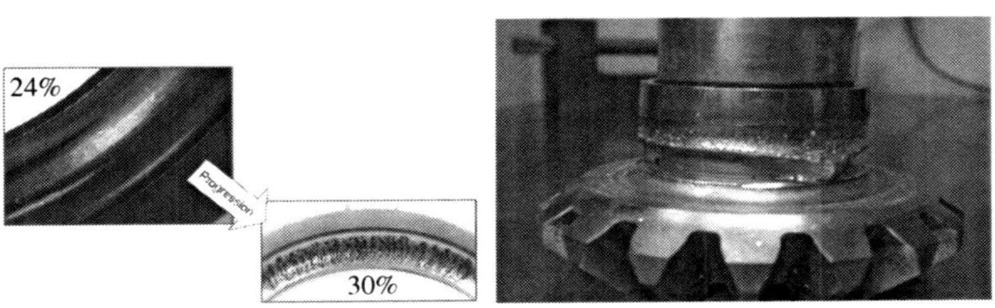

Fig. 11 bearing failure progress at 24 and 30 % of life time

5 Discussion

The techniques used in this paper are typically used for applications where strong background noise masks the defect signature of interest within the measured vibration signature. Of all the techniques presented, the LMS algorithm succeeded in detecting the bearing outer race fault earliest at 24 % of bearing life. FBLMS and SK techniques detected the bearing outer race fault at 27 % of bearing life. In addition, the LMS algorithm was the only technique that successfully identified both the outer race and ball spin faults. Comparing the LMS and FBLMS algorithms showed that LMS is able to detect the bearing fault earlier than FBLMS. However, the computational cost for LMS is high and therefore FBLMS is more suitable for online diagnostics where an immediate response is required. On the other hand, LMS can be used for offline diagnostics. However, with today's computer technology the delay time can be reduced significantly to a few minutes. In some applications such as wind and tidal turbines, vibration data is acquired once every hour depending on certain conditions. This makes the LMS algorithm the best candidate for diagnosing bearing faults in gearboxes. The main drawback of applying spectral kurtosis is the optimal selection of the window length. This was achieved estimating the maximum kurtosis for various window lengths. In addition other disadvantages of the spectral kurtosis technique are that it requires a high sampling rate and it is computationally complex. Also the performance of the technique diminishes significantly for signals with a non-Gaussian noise distribution.

6 Conclusion

The problem of bearing fault diagnosis in gearboxes has been investigated; Spectral Kurtosis, LMS and FBLMS algorithms were applied to separate the bearing signal. The LMS technique demonstrated the ability to identify the defect earlier than all other methods. This method is thus a very powerful tool for early fault detection in bearings, particularly for those applications where strong background noise from other sources in the machine masks the characteristics fault components in the frequency domain.

References

1. Randall RB, Antoni J (2011) Rolling element bearing diagnostics—a tutorial. Mech Syst Signal Process 25(2):485–520
2. Tan CK, Irving P, Mba D (2007) A comparative experimental study on the diagnostic and prognostic capabilities of acoustics emission, vibration and spectrometric oil analysis for spur gears. Mech Syst Signal Process 21(1):208–233

3. Li Z, Yan X, Tian Z, Yuan C, Peng Z, Li L (2013) Blind vibration component separation and nonlinear feature extraction applied to the nonstationary vibration signals for the gearbox multi-fault diagnosis. Measurement 46(1):259–271
4. Al-Ghamd AM, Mba D (2006) A comparative experimental study on the use of acoustic emission and vibration analysis for bearing defect identification and estimation of defect size. Mech Syst Signal Process 20(7):1537–1571
5. McFadden PD, Toozhy MM (2000) Application of synchronous averaging to vibration monitoring of rolling elements bearings. Mech Syst Signal Process 14(6):891–906
6. Fu-cheng Z (2010) Research on online monitoring and diagnosis of bearing fault of wind turbine gearbox based on undecimated wavelet transformation. In: *2010* IEEE youth conference on information computing and telecommunications (YC-ICT), p 251
7. Sawalhi N, Randall RB, Endo H (2007) The enhancement of fault detection and diagnosis in rolling element bearings using minimum entropy deconvolution combined with spectral kurtosis. Mech Syst Signal Process 21(6):2616–2633
8. Randall RB (2004) Detection and diagnosis of incipient bearing failure in helicopter gearboxes. Eng Fail Anal 11(2):177–190
9. McFadden PD, Smith JD (1984) Vibration monitoring of rolling element bearings by the high-frequency resonance technique—a review. Tribol Int 17(1):3–10
10. Barszcz T (2009) Decomposition of vibration signals into deterministic and nondeterministic components and its capabilities of fault detection and identification. Int J Appl Math Comput Sci 19(2):327–335
11. Douglas SC (1999) Introduction to adaptive filters. CRC Press, Boca Raton
12. Douglas SC, Rupp M (1999) Convergence issues in the LMS adaptive filter. In: Madisetti VK (ed) The digital signal processing handbook, 2nd edn. CRC Press, Atlanta
13. Widrow B, Glover JR Jr, McCool JM, Kaunitz J, Williams CS, Hearn RH, Zeidler JR, Eugene Dong J, Goodlin RC (1975) Adaptive noise cancelling: principles and applications. Proc IEEE 63(12):1692–1716
14. Widrow B, McCool J, Ball M (1975) The complex LMS algorithm. Proc IEEE 63(4):719–720
15. Dentino M, McCool J, Widrow B (1978) Adaptive filtering in the frequency domain. Proc IEEE 66(12):1658–1659
16. Ferrara ER (1980) Fast implementations of LMS adaptive filters. In: IEEE transactions on acoustics, speech and signal processing, vol 28, no 4, p 474–475
17. Antoni J (2007) Fast computation of the kurtogram for the detection of transient faults. Mech Syst Signal Process 21(1):108–124
18. Randall RB (2011) Vibration-based condition monitoring, 1st edn. Wiley, Chichester
19. Antoni J, Randall R (2006) The spectral kurtosis: application to the vibratory surveillance and diagnostics of rotating machines. Mech Syst Signal Process 20(2):308–331

Pitting Detection in Worm Gearboxes with Vibration Analysis

Faris Elasha, David Mba and Cristobal Ruiz-Carcel

Abstract Diagnostics of worm gearboxes are challenged by lack of research in this field; vibration analysis is rarely employed for the condition evaluation of worm gears due to the inherent challenges involved. However, these gears are commonly used in many applications such as escalators, mills and conveyors and significant cost may arise from their maintenance. This paper aims to apply various vibration analysis techniques to identify faults within worm gearboxes. The conditions of three different worm gearboxes were assessed. Various vibration signal analysis techniques including $FM4^*$, and spectral kurtosis and enveloping were employed to identify the presence of defects within the worm gearboxes. The analysis techniques applied were able to present fault features which indicated high levels of scratching-induced vibration, confirming profile damage (pitting) of Gearbox A worm wheel teeth. Among the employed techniques; $FM4^*$ shows high response to the worm gears pitting. In addition the results show sensitivity to the direction of vibration measurement.

1 Introduction

In a worm gearbox the worm meshes with a wheel to transmit motion between crossed shafts, usually perpendicular to each other. Worm gearboxes are generally used for applications requiring large gear reduction ratios; they have the advantage of taking up little space and are simpler in configuration compared to the other gears types of the same reduction ratio. Other advantages of worm gears include the

F. Elasha (✉) · D. Mba · C. Ruiz-Carcel
School of Engineering, Cranfield University, Bedford M43 0AL, UK
e-mail: F.elasha@cranfield.ac.uk

D. Mba
e-mail: D.mba@cranfield.ac.uk

C. Ruiz-Carcel
e-mail: c.ruizcarcel@cranfield.ac.uk

J.K. Sinha (ed.), *Vibration Engineering and Technology of Machinery*,
Mechanisms and Machine Science 23, DOI 10.1007/978-3-319-09918-7_20

self-locking effect, low backlash, damage tolerance as well as quiet operation. However, worm gearboxes have some disadvantages; they generate high friction compared to the other gears types due to their sliding action (as opposed to rolling) which results in heat generation and hence lower efficiency. Also the sliding motion gives rise to inevitable abrasive wear and scuffing, therefore worm gearbox wheels are usually made of a softer material (usually bronze) and worms are most commonly made of a harder steel [1]. Due to their unique properties, worm gearboxes have found their niche in a variety of industrial applications such as rolling and saw mills, mining machinery and escalators.

With a gradual shift towards condition-based maintenance for greater operational effectiveness, condition assessment via vibration analysis is common practice. However, it is seldom employed for the condition assessment of worm gears due to the inherent challenges involved in the analysis of worm gear vibration. Compared to other gears types where defects manifest as periodic impacts in the form of side-bands around the gear mesh frequencies [2, 3], such distinctive defect symptoms are not obvious for worm gear sets due to their continuous sliding interactions [4]. Consequently, the literature on this topic is relatively scarce.

Peng et al. [5, 6] combined use of oil and vibration analysis to establish detection methods in a worm gearbox; they concluded that wear analysis remains a useful tool for the detection of gear wear while the vibration analysis was more suited to identifying bearing failures. Vähäoja et al. [4] also reiterated the need to combine vibration and oil analysis to effectively diagnose defects in worm gearboxes. Acoustic emission (AE) has been used for fault detection in worm gearboxes, where parameters such as r.m.s and AE energy were found to be reliable and more capable of identifying defects compared to corresponding vibration parameters [7]. This paper aims to examine the effectiveness of different vibration analysis techniques in detection of faults in operational worm gearboxes employed for escalators.

2 Vibration Analysis Techniques

Whilst there are varied vibration diagnostic techniques available to the user this paper focuses on three techniques: Spectral Kurtosis and FM4*. The statistical measure FM4* is known to be effective in diagnosing gears pitting [8].

2.1 *Fm4**

FM4* is a relatively simple method used to detect changes in the vibration pattern resulting from damage on gear teeth. The FM4* parameter is the ratio between the kurtosis of the difference signal to the squared variance of a healthy gearbox's difference signal. The difference signal is computed by extracting gear mesh, shafts frequencies, their harmonics and associated side-bands from the vibration signature.

As a defect progresses on a tooth, vibration peaks will increase in the difference signal and as a consequence the kurtosis value will exceed 3, and this will lead to increase of FM4*. This metric is calculated as [9, 10]:

$$\mathrm{FM4}^* = \frac{\frac{1}{N}\sum_{i=1}^{N}(x_i - \mu')^4}{[Var(X)]^2} \tag{1}$$

where x is the difference signal, μ' is the mean value of the difference signal, N is the total number of the samples, and $Var(X)$ is the variance of the difference signal in good condition. This metric has been applied for vibration diagnosis of helicopter gearboxes even during torque variation [9, 11, 12].

2.2 *The Spectral Kurtosis (SK)*

The basic principle of this method is to determine the Kurtosis at different frequency bands in order to identify the energy distribution of the signal and determine where the high impact energy (transient events) are located in the frequency domain. Obviously the results obtained strongly depend on the width of the frequency bands Δf [13]. The kurtogram [14] is a representation of the calculated values of the SK as a function of f and Δf. However, exploration of the entire plane (f, Δf) is a complicated computational task, though Antoni [13] suggested a methodology for the fast computation of the SK.

The frequency band where the SK is maximized provides information which can be used to design a filter in order to extract the part of the vibration signal with the highest level of impulsiveness. Antoni and Randall [15] demonstrated how the optimum filter, which maximizes the signal to noise ratio, is a narrow-band filter at the maximum value of SK. Therefore the optimal central frequency fc and bandwidth Δf of the band-pass filter is found as the values of fc and Δf which maximise the kurtogram. The filtered signal can be finally used to perform an envelope analysis, which is a widely used technique for identification of modulating frequencies related to bearing faults. In this investigation the SK computation and the subsequent signal filtration and envelope analysis were performed using the original Matlab code programmed by Antoni [13].

3 On-Site Gearbox Measurements

Gearboxes vibration measurements were undertaken on three operational escalator worm gearboxes (A, B and C). Vibration measurement was taken in the vicinity where the worm meshes with the wheel and in all three directions, see Fig. 1. The worm gearbox was of the right-angle type with 85 teeth on the wheel gear and three

Fig. 1 Positions chosen for mounting accelerometers

threads on the worm. The gearbox had a nominal output speed of 870 rpm and a rated power of 38 kW.

The Y-axis of the tri-axial accelerometer arrangement was oriented in the vertical direction; the X-axis refers to the axis parallel to the worm shaft, and the Z-axis is parallel to the wheel shaft (axial) and in the direction of sliding motion. It was anticipated that vibration measurements in this direction would yield the most prominent indications of the presence of pitting.

For each vibration measurement 300-s of data was captured at a sampling frequency of 12 kHz using Monitran MTN/1100W accelerometers, configured in a tri-axial arrangement and mounted on a Monitran M8 (AA008) block with a magnetic attachment adaptor (Monitran MTN/MM001). The accelerometers had an operating frequency range of 0.8 Hz to 12 kHz. The acquisition system employed two National Instruments (NI) PXI-4472B cards (8-channel, 24-bit resolution). These cards were configured in a NI PXI 1031 chassis with a NI PXI 8360 MXI express remote controller that enabled interface to a Labview signal express software program.

Of the three gearboxes investigated it was known that gearbox B was in good condition following a recent overhaul, therefore this gearbox (B) was considered as the healthy reference condition. To aid diagnosis, all characteristic vibration frequencies were determined, see Table 1. These included the worm shaft speed with its harmonics and bearings defects frequencies. Each gearbox was supported by four bearings. The bearing defect frequencies for each gearbox were calculated and referenced to the rotating speed. The frequencies are listed in Table 1.

Table 1 Gearbox reference frequencies

Reference frequencies (Hz)	A	B	C
Gear mesh frequency GMF	43.5		
Wheel NDE Cage	0.205		
Wheel DE Cage	0.215		
Wheel NDE Roller Spin	1.32		
Wheel DE Roller Spin	1.73		
Wheel NDE Outer Race	3.28		
Wheel DE Outer Race	4.09		
Wheel NDE Inner Race	4.72		
Wheel DE Inner Race	5.41		
Worm DE Cage	6.04	6.07	6.16
Worm Shaft (1X)	14.14	14.2	14.42
Worm NDE Roller Spin	32.635	32.8	33.28
Worm DE Roller Spin	47	47.2	47.9
Worm NDE Outer Race	73.882	74.3	75.34
Worm NDE Inner Race	109.94	110	112.1
Worm DE Outer Race	121	121	123
Worm DE Inner Race	162	163	165

4 Results of Vibration Investigation

*4.1 Fm4**

Figure 2 shows the results of the FM4* analysis for each gearbox revealing comparatively high levels for Gearbox A. The value FM4* under healthy conditions (gearbox B) ranged from 2.8 to 5 for all vibration measurement directions, with the highest level noted in the z-direction. FM4* levels for Gearbox A were relatively high in Z and Y directions, 36 and 12 respectively, whilst the level in the X direction was relatively low at 3. The FM4* levels of gearbox C were comparative to levels noted on gearbox B except for the z-direction were a value of 11 was noted.

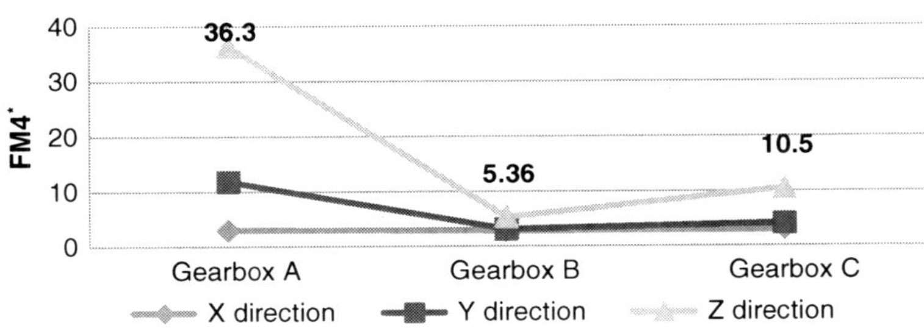

Fig. 2 FM4* analysis of gearboxes

4.2 The Spectral Kurtosis

The spectral kurtosis analysis was undertaken on data sets collected from all three gearboxes and this yielded the frequency bands and center frequencies which were then used to undertake envelope analysis, see Table 3. A typical kurtogram from measurement taken on gearbox A (Z direction) is shown in Fig. 3. Spectral plots of enveloped vibration signals following filtration, whose characteristics were determined with the aid of the kurtogram, are shown in Figs. 3, 4, and 5 for the data collected in all three directions (Table 2).

Observation from the spectra of the enveloped signal in the z-direction, see Fig. 3, showed interesting characteristics. Gearbox A showed worm shaft frequency and several harmonics. Such phenomenon is known to be associated with gear pitting [16]. This was not observed in the envelope spectra of the other gearboxes. Bearing frequencies observed included the wheel NDE outer race (3.5 Hz) and a harmonic, and wheel DE inner race (5.7 Hz) for Gearbox A. Result in y-direction also showed worm shaft harmonics and bearings frequencies for gearbox A (see Fig. 5) which reinforces the observations in the z-direction. Observations from gearboxes B and C in y- and x-directions were similar in pattern. Noted was the presence of a frequency component at 7.8 Hz, and its harmonics (Gearbox A and C) which corresponded to the 2nd harmonic of outer race wheel DE (see Fig. 6). The vibration amplitude of gearbox B was significantly low compared to the others gearboxes.

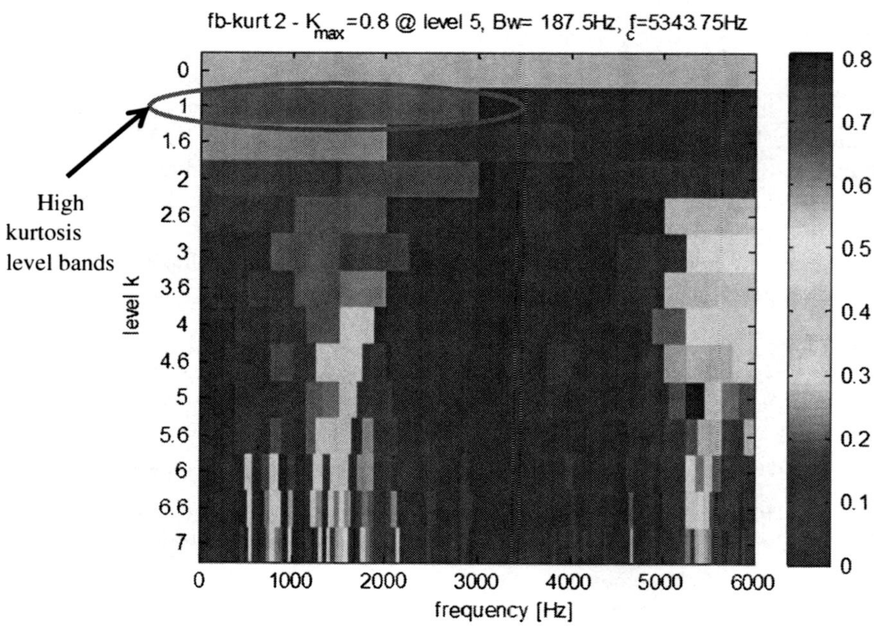

Fig. 3 Example of kurtogram for gearbox A (z-direction)

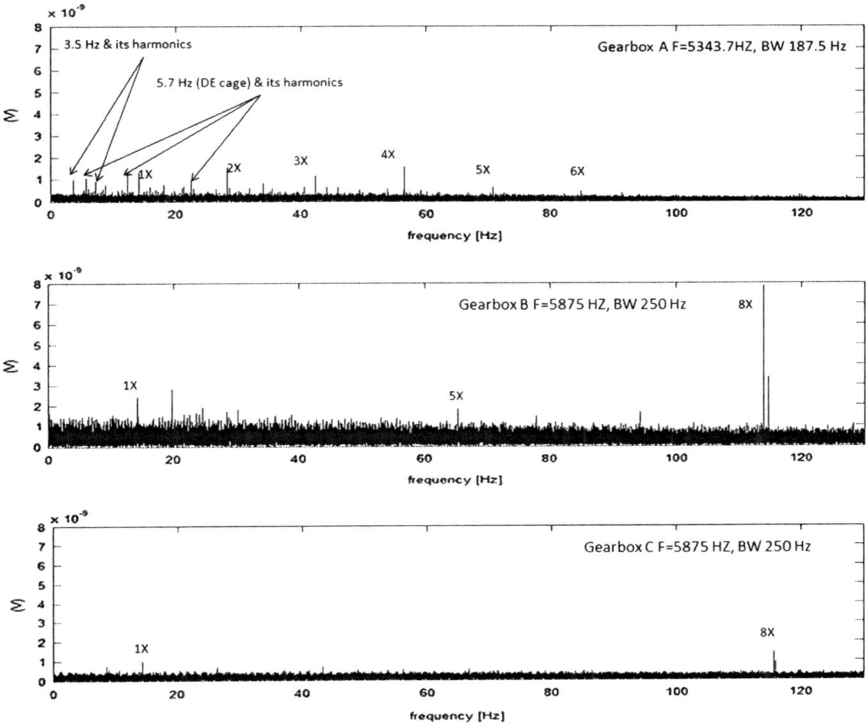

Fig. 4 Envelope spectrum for all gearboxes from spectral kurtosis in z-direction

5 Discussion

Two techniques were employed for the vibration analysis of the worm gearboxes. $FM4^*$ levels showed relatively higher values for gearbox A. The high levels associated with the $FM4^*$ of gearbox A pointed to the presence of gearbox damage, particularly the $FM4^*$ level (36) in the z-direction. It would appear that results in z-direction are more sensitive to the gear pitting fault conditions compared to other directions. This wasn't surprising as the direction corresponded to the sliding direction of the mating teeth.

Spectral kurtosis analysis results of gearbox B (healthy gearbox) in all directions showed no existence of defect frequencies whilst gearbox A was fraught with harmonics of the worm shaft frequency, a typical symptom associated with worn gears. The ability of SK analysis to identify the presence of worn gears is based on the identification of particular frequency regions with high impact energy; these impacts are due to presence of gears pitting which affect gears sliding motion. A summary of defect frequencies identified is presented in Table 3.

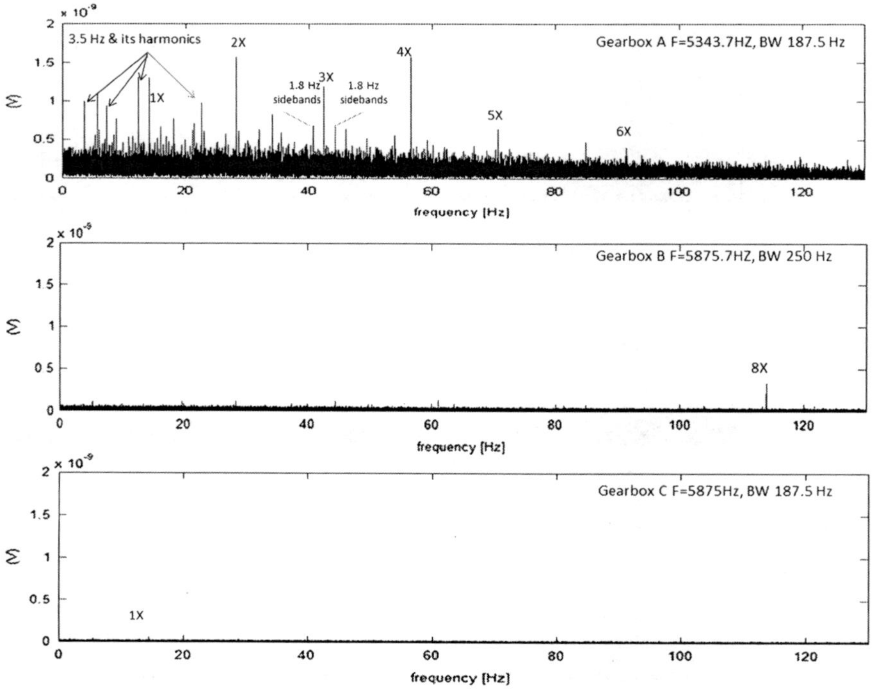

Fig. 5 Envelope spectrum from spectral kurtosis in y-direction

Table 2 Summary of spectral kurtosis frequencies

	Fc(Hz)	BW (Hz)	Value of kurtosis
Gearbox A (Z)	5343	187.5	0.75
Gearbox B (Z)	5875	250	0.73
Gearbox C (Z)	5875	250	0.72
Gearbox A (Y)	5343.75	187.5	0.68
Gearbox B (Y)	5875	250	0.68
Gearbox C (Y)	5875	250	0.68
Gearbox A (X)	726.6	46.9	0.72
Gearbox B (X)	5953	93.75	0.69
Gearbox C (X)	820	46.9	0.72

Given diagnosis of worm gearboxes with vibration analysis is dependent on the integrity of the relative sliding surfaces, any discontinuity in the gear profiles, due to pitting, will lead to impacts and such energies will be more sensitive across

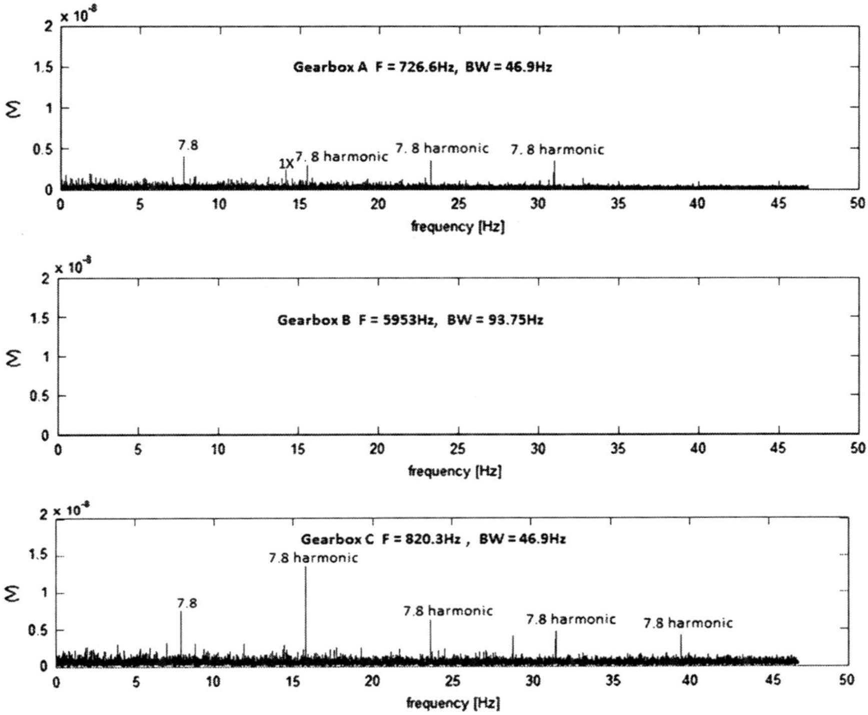

Fig. 6 Envelope spectrum from spectral kurtosis in x-direction

Table 3 Summary of defect frequencies observed from the SK spectrum

Gearbox	Spectral kurtosis observation
Gearbox A	Presence of numerous bearing defect frequencies (all directions) Shaft frequencies and its harmonics (all directions)
Gearbox B	–
Gearbox C	A bearing defect frequency (x-direction only)

particular frequency bands. For this reason the SK analysis is ideally suited for diagnosis of worm gearboxes.

Visual inspection revealed that gearbox A has visible pitting and bearing damage, while no obvious pitting damage was found on the wheel teeth of gearboxes B and C, see Fig. 7. These observations infer that statistical metric FM4*) can be used to set fault alarm levels with further diagnosis undertaken with spectral kurtosis.

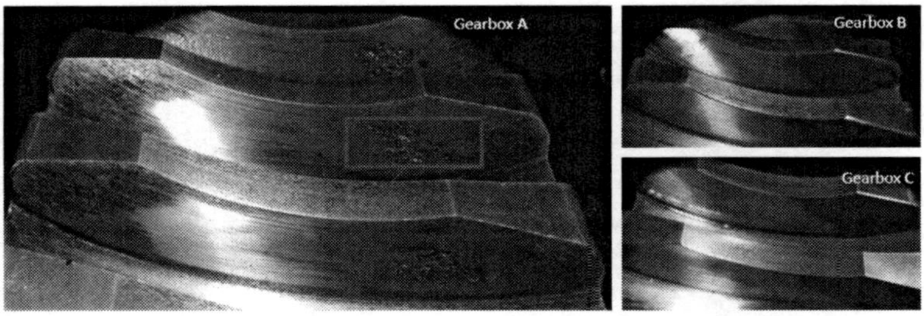

Fig. 7 Visual inspection of gearbox wheel teeth

6 Conclusion

Two vibration signal analysis tools were employed in this study. The techniques were able to present features which indicated pitting damage on one gearbox (gearbox A). The presence of a defective gear was confirmed by visual inspection, see Fig. 7. All techniques applied for diagnosis were sensitive to the direction of measurement with measurements taken horizontally and parallel to wheel shaft axis (z-direction) proving most sensitive, and more reliable results. The $FM4^*$ was deemed effective for the detection of pitting presence, the advantage of such a technique is its simplicity in implementation. SK followed by envelope analysis was valuable for the identification of both bearing and gears defects frequencies, though the SK analysis was less susceptible to the direction of measurement, making this technique relatively more robust. In conclusion, the conditions of three worm gearboxes were assessed; surface pitting has been identified in gearbox (A) wheel teeth, while no visible damage were identified for gearboxes B and C.

References

1. Black A (2014) The ins and outs of worm gears. http://www.machinerylubrication.com/Read/1080/worm-gears. Accessed Jan 16
2. Rao B (1996) Handbook of condition monitoring. Elsevier, Oxford
3. Alan D (1998) Handbook of the condition monitoring techniques and methodology, 1st edn edn. Chapman & Hall, London
4. Vahaoja P, Lahdelma S, Leinonen J (2006) On the condition monitoring of worm gears. In: Engineering Asset Management. Proceedings of the 1st World Congress on Engineering Asset Management (WCEAM), Mathew J, Kennedy J, Ma L et al (eds) Springer, London, pp 332–343
5. Peng Z, Kessissoglou N, Cox M (2005) A study of the effect of contaminant particles in lubricants using wear debris and vibration condition monitoring techniques. Wear 258 (11):1651–1662
6. Peng Z, Kessissoglou N (2003) An integrated approach to fault diagnosis of machinery using wear debris and vibration analysis. Wear 255(7):1221–1232

7. Elforjani M, Mba D, Muhammad A, Sire A (2012) Condition monitoring of worm gears. Appl Acoust 73(8):859–863
8. Zakrajsek JJ, Townsend DP, Decker HJ (1994) An analysis of gear fault detection methods as applied to pitting fatigue failure data. Syst Eng Approach Mech Fail Prev 16:199. Virginia Beach, Virginia, USA, NASA, USA
9. Decker HJ, Lewicki DG (2003) Spiral bevel pinion crack detection in a helicopter gearbox, NASA/TM-2003-212327—ARL-TR-2958. Glenn Research Center, NASA, Cleveland
10. Lebold M, McClintic K, Campbell R, Byington C, Maynard K (2000) Review of vibration analysis methods for gearbox diagnostics and prognostics. In: Proceedings of the 54th meeting of the society for machinery failure prevention technology, Virginia Beach, VA, p 623
11. Sait A, Sharaf-Eldeen Y (2011) A review of gearbox condition monitoring based on vibration analysis techniques diagnostics and prognostics. In Proulx T (ed) Rotating machinery, structural health monitoring, shock and vibration, vol 5. Springer, New York, p 307–324
12. Halim EB, Shah SL, Zuo MJ, Choudhury MAA (2006) Fault detection of gearbox from vibration signals using time-frequency domain averaging. In: American control conference, p 6
13. Antoni J (2007) Fast computation of the kurtogram for the detection of transient faults. Mech Syst Signal Process 21(1):108–124
14. Randall RB (2011) Vibration-based condition monitoring, 1st edn. Wiley, Chichester
15. Antoni J, Randall R (2006) The spectral kurtosis: application to the vibratory surveillance and diagnostics of rotating machines. Mech Syst Signal Process 20(2):308–331
16. Teng W, Wang F, Zhang K, Liu Y, Ding X (2014) Pitting fault detection of a wind turbine gearbox using empirical mode decomposition. Strojniski Vestnik/J Mech Eng 60(1):12–20

Impact Vibrations of Ultrasonic Multi-striker Hand Tool

Makhmut Ganiev, Ilshat Gafurov and Ildar Vagapov

Abstract The impact vibrations of the multi-striker ultrasonic hand tool are investigated theoretically and experimentally. The lightweight intermediate strikers are placed into the gap between the end-face of the high-power ultrasonic converter and the specimen being treated. The high-frequency impact force was measured by the calibrated piezoelectric sensor. The dynamical drift of the tool casing was measured by the linear displacement gauge. Both signals were simultaneously recorded with the use of the two-channel digital storage oscilloscope. The scan-pictures allow to determine the duration of the single impact pulse and its peak value, frequency of the pulse sequence, etc. It was observed, that slow oscillations of the tool casing are synchronously accompanied by the periodical changes of impact force. The dynamic model of the high-frequency impact treatment was constructed by the methods of theory of the vibro-impact systems. The induced impact stresses were calculated in the frames of the Hertz contact approximation with the use of experimental data. It is shown that the impact stress value reaches the yield point of material being treated. The ultrasonic impact processing can be interpreted as a cyclic alternation of setting-up and failure of the vibro-impact oscillation of the striker inside the variable gap. The changes in the mechanical properties and texture were analyzed by the metallographic methods. Metal hardening and creation of the profitable residual stresses on the surface of the specimen were observed.

Keywords Ultrasonics · Vibro-impact system · Residual stresses

M. Ganiev (✉) · I. Gafurov · I. Vagapov
Kazan Federal University, Kremlevskaja Str.,18, Kazan 420008, Russia
e-mail: vagapovik@yandex.ru

J.K. Sinha (ed.), *Vibration Engineering and Technology of Machinery*,
Mechanisms and Machine Science 23, DOI 10.1007/978-3-319-09918-7_21

1 Introduction

The high-frequency impact treatment is being applied successfully for relieving from the harmful tensile residual stresses and introducing the beneficial compressive residual stresses in the surface layer of the weldments [1]. This method employs a hand tool with the free needle-like intermediate strikers that vibrate in the gap between the ultrasonic transducer tip and the impact surface (Fig. 1). The energy generated by the ultrasonic transducer that is turned in resonance is imparted to the treated specimen through two impact joints. Depending on treatment conditions one of two vibro-impact modes can be observed: the continuous periodic impacts at the ultrasonic frequency or non-periodic impacts of the strikers upon the specimen surface [1].

The improved ultrasonic tool [2] provides better vibroprotection of the operator due to reduction of stiffness of the ultrasonic converter's connection with the tool body. The converter *3* is installed inside the sleeve *9* by means of the fastening flange, that is placed in the node of the longitudinal standing wave. The sleeve is able to reciprocally move along the guide rails *12*, which are attached to the casing *2* through the elastic gaskets *11* made of the vibroinsulation material.

Fig. 1 Ultrasonic tool scheme: *1*—pressing weight, *2*—tool casing, *3*—piezoelectric converter, *4*—intermediate striker, *5*—specimen, *6*—displacement gauge, *7*—piezoelectric impact sensor, *8*—oscilloscope, *9*—sleeve, *10*—spring, *11*—elastic gasket, *12*—guide rail

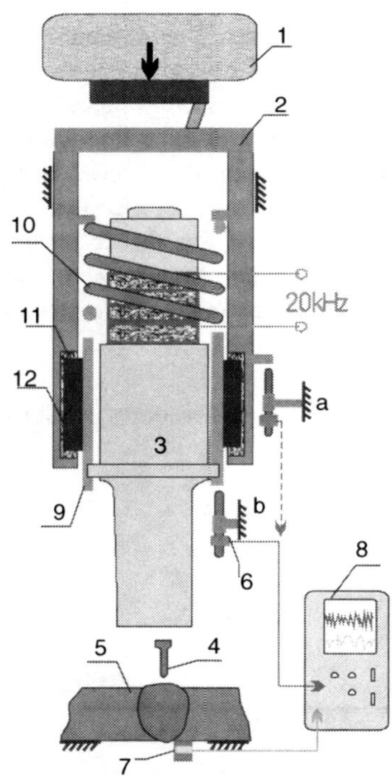

On the other hand, the reduction of tool stiffness can cause some undesirable consequences: weakening of impact force of the lightweight striker, widening of gap that can lead to interruption of impact treatment, until the spring will return the converter back toward the specimen surface.

In this paper both the vibration of tool casing and the force of the striker's impact upon the specimen are investigated. To facilitate the measurements, experiments were carried out for the tool with only one striker. The dynamic features of the ultrasonic impact tool are described in the frames of the theory of vibro-impact systems [3]. The values of impact stresses that are created in the material being treated were estimated.

2 Experiment

Ultrasonic tool was installed vertically into supporting frame (not shown on Fig. 1) and pressed to specimen by the weights *1* of 4, 8 and 10 kg. The displacement of the tool casing was measured by axial displacement gauge *6* when it is placed in position "a" on the frame. In position "b" the gauge *6* directly records the displacement of ultrasonic converter *3* during the processing. The high-frequency impact force was measured by calibrated piezoelectric sensor *7* and both signals were fed to the two-channel digital storage oscilloscope *8*.

The dynamical behavior of the ultrasonic impact tool is illustrated by the oscilloscope pictures with the different scanning rates. As it follows from Fig. 2a a motion of tool casing is a superposition of two almost harmonic oscillations with the frequencies in the neighborhoods of 20 and 180 Hz. More exact values of frequencies are indicated on the spectrum of the casing oscillation (Fig. 3b). The measurement of the period of motion for ultrasonic converter directly from Fig. 2b gives nearly the same value of frequency, viz. 178.57 Hz.

The analysis of the scan-pictures and specters under different experiment conditions as well as numerical estimations allows to consider the ultrasonic tool as two-mass oscillatory system consisting of the converter's mass and the total mass of the casing and the pressing weight. The masses are coupled by the elastic links (spring and vibroinsulation gaskets) and the friction unit (sleeve and guide rails). The above mentioned frequencies are the natural frequencies of the system with two degrees of freedom. The slow biharmonic oscillations are excited due to the multiple high-frequency collisions of vibrating tip of the converter with the rear end of the striker, which rebounds off the specimen. From the rapid scan-picture Fig. 2c the duration of the single impact upon the specimen can be estimated as 10 μs that is about one quarter of the ultrasound period.

The multiple impacts occur with the ultrasonic frequency that is also confirmed by the signal spectrum on Fig. 3a. The low-frequency oscillation of the ultrasonic converter are synchronously accompanied with the periodic change in the height of

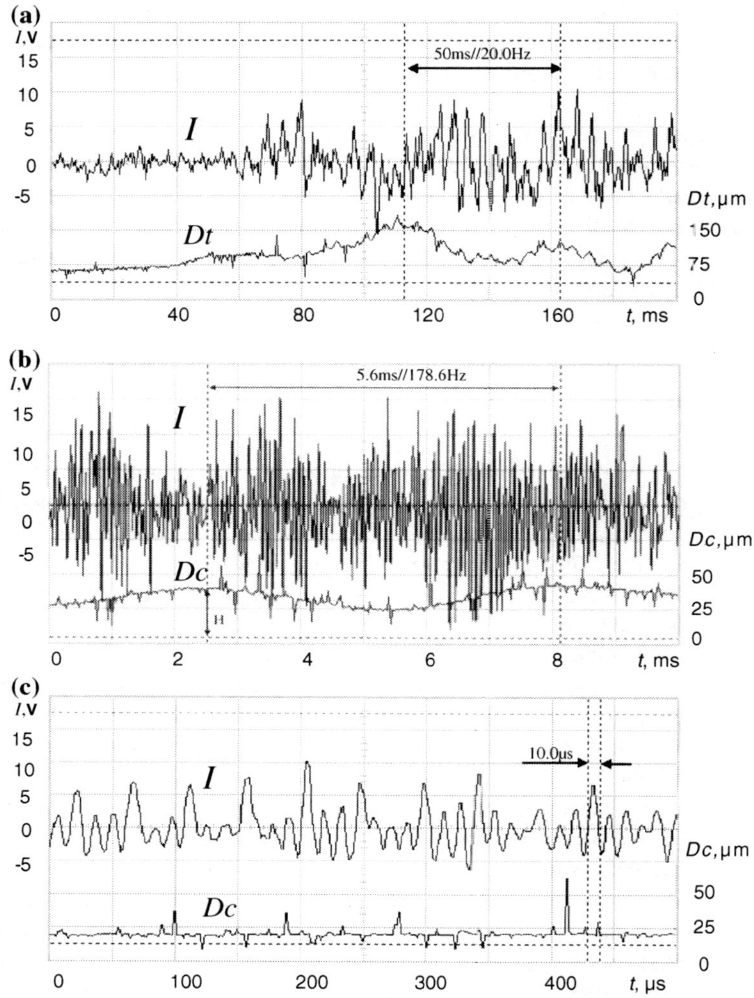

Fig. 2 Oscilloscope pictures of the impact force signal *I* and tool-casing drift *Dt* (**a**); impact force signal *I* and converter drift *Dc* (**b**, **c**). Pressure weight 8 kg

the impact peak. The powerful impacts upon the specimen are observed when the ultrasonic converter moves up from its the lowest position (see Fig. 2b). On the upper position the regular series of the pulses at ultrasonic frequency are transformed into a random set of the weak impact peaks alternating with the infrequent more strong peaks. The regular impacts at ultrasonic frequency are renewed within a few periods of ultrasound when the converter moves downwards.

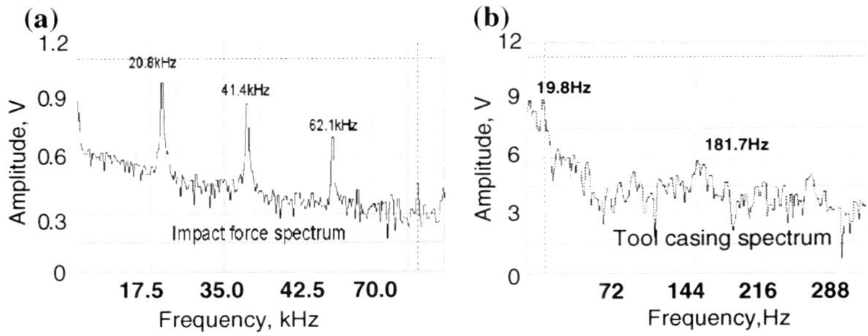

Fig. 3 Spectral representation of the impact force (**a**) and tool-casing motion (**b**)

3 Dynamic Model

3.1 Impact Interaction

The dynamic model constructed by the methods from the theory of the vibration-impact systems [3]. Interaction of the ultrasonic converter with the rear end of the striker is considered as a head-on collision of the rod tip with effective mass M_1 upon the body with concentrated mass M_2. The absolutely elastic approximation is acceptable because of high rigidity of the contact surfaces. Then the postimpact velocities of converter tip V_1^-, and the striker V_2^- are connected with preimpact velocities V_1^+ and V_2^+ by the relations [3]:

$$V_1^+ = V_1^- - \frac{M_2}{M_1 + M_2}(V_1^- - V_2^-), \tag{1}$$

$$V_2^+ = V_2^- - \frac{M_1}{M_1 + M_2}(V_1^- - V_2^-). \tag{2}$$

The striker impact upon the specimen surface is considered to be inelastic and is described by the velocity recovery factor R:

$$V_2^+ = -RV_2^- \tag{3}$$

It is worth considering steady-state of the vibro-impact mode with the continuous slow oscillations of the ultrasonic tool. In accordance with experimental observation, motion of the converter tip can be described as a superposition of the low-frequency oscillation of the converter as single whole and forced ultrasonic oscillations:

$$x(t) = (c_1 \cos 2\pi\Omega t + c_2 \sin 2\pi\Omega t) \cdot \exp(-\chi\Omega t) + a\cos(2\pi\nu t + \phi), \qquad (4)$$

where Ω is the natural frequency of the oscillation of the converter pressed by the spring *10* inside the tool casing, ν is the ultrasound frequency, χ is the coefficient of oscillation attenuation due to friction between sleeve *9* and rails *12* (Fig. 1). From the experiment data it follows that $\Omega \approx 180$ Hz and $\nu \approx 20$ kHz, i.e. $\Omega \ll \nu$.

The amplitude of the converter tip is kept at the level $a = 20 \pm 2$ μm by use of the automatic system for the adjustment of resonance frequency.

During the short time interval equal to period T of ultrasound the converter stays practically immobile in some position of dynamic equilibrium that is located higher than the static one (when ultrasound is switched off) by the height:

$$H = \Delta + h_2, \qquad (5)$$

where Δ is the dynamic gap that is opened as a result of the converter's shift under the influence of the multiple collisions with the striker rebounding off the specimen; h_2 is the space for reciprocal motion of the striker. During the direct and revers motion the intermediate striker covers the same distance:

$$h_2 = V_2^+ \tau_1 = -V_2^-(T - \tau_1 - \tau_y), \qquad (6)$$

where τ_y is impact duration, τ_1 is a time interval between the collision of striker with the horn tip and next impact upon the specimen (see Fig. 4).

The momentum conservation theorem can be applied to the ultrasonic converter at the dynamic equilibrium state:

$$GT = M_1(V_1^- - V_1^+), \qquad (7)$$

where G is the static pressing force. The system of five linear Equations (1)–(3), (6) and (7) for five unknown quantities $V_1^-, V_1^+, V_2^-, V_2^+$ and τ_1 has an unique solution. For example we adduce two of them:

$$V_1^- = \frac{GT}{M_2}\left(\frac{R}{1+R} + \frac{M_1 + M_2}{2M_1}\right), \quad V_2^+ = \frac{1+R}{R}\left(V_1^- - \frac{GT(M_1 + M_2)}{2M_1M_2}\right) \qquad (8)$$

After collision with the lightweight intermediate striker the converter tip moves in the same direction along the x axis, both V_1^- and V_2^+ mast be positive.

3.2 Steady-State Mode

When the steady-state mode is settled, the collisions of the converter tip with the rebounding back striker occur again and again in a certain permanent point with coordinate $x = \Delta$ between the dynamic equilibrium position and the specimen

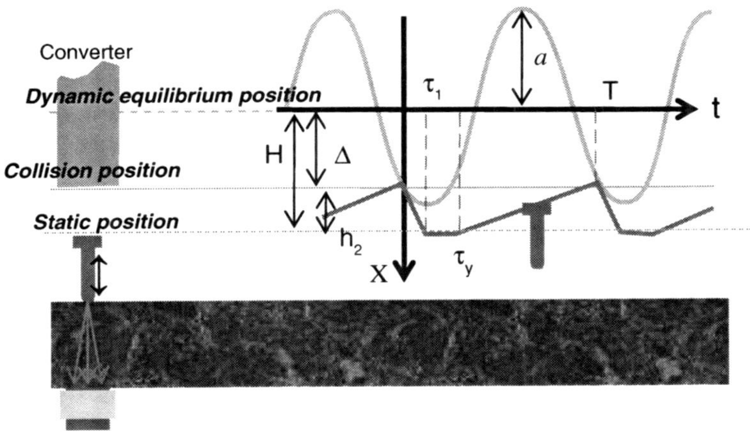

Fig. 4 Scheme of vibration of the converter tip and reciprocal motion of the striker

surface (Fig. 4). The requirement of periodicity for repetitive collisions should be written as [3]:

$$t = 0,\ x = \Delta,\ \dot{x} = V_1^+; \quad t = T,\ x = \Delta,\ \dot{x} = V_1^-. \tag{9}$$

Obeying the law of motion represented by Eq. (4) to conditions of Eq. (9) obtain the expression for dynamic gap Δ value:

$$\Delta = DV_1^-/v + \sqrt{a^2 - (BDV_1^-/v)^2}. \tag{10}$$

Coefficients B and D depend on design parameters of the ultrasonic tool [2] and processing parameters G, R and τ_y.

3.3 Range of Existence and Stability

The range of existence of the real solutions for the dynamic gap value is bounded by inequality following from Eqs. (8) and (10)

$$\frac{GT(M_1 + M_2)}{2M_1M_2} \leq V_1^- \leq \left|\frac{av}{BD}\right| \tag{11}$$

So, the steady-state mode with the impact frequency equal to ultrasonic frequency can be excited only if both the amplitude a of the converter tip and the collision velocity V_1^- exceed the threshold value, which explicitly depends on pressing force G.

Stability of the obtained solution of Eq. (4) under the requirement of Eq. (9) was investigated in the frames of the Schur criterion of asymptotic stability [3]. The analysis shows that all impact modes inside of the range of existence are stable with exception of the modes corresponding to the strict equalities in Eq. (11).

Equation (10) establishes the one-to-one correspondence between the dynamic gap Δ and collision velocity V_1^- for each steady-state vibrating-impact mode. When the velocity of collision continuously varies inside the range of existence and stability (described by the Inequality (11)), the frequency of the repetitive impacts stays changeless and equal to ultrasonic frequency. On the range boundaries the periodic vibrating-impact mode is broken because of loss of stability. So the periodic impacts being turned into the irregular rattle (see Fig. 2).

According to Eq. (8) the periodic change in the dynamic gap leads to synchronous variation of velocity V_2^+ of the striker's impact upon the specimen. Furthermore, the increase of impact velocity V_2^+ under the changeless value of impact frequency ν leads to widening of the spacing h_2 and the increase of total dynamic drift H of the converter (see Eqs. (5) and (6)).

In the frames of the first harmonic approximation [6] the reciprocal motion of the intermediate striker should be considered as the oscillation at the ultrasound frequency with the amplitude equal to $h_2/2$. So, the slow increase of spacing h_2 should be interpreted as the phenomenon of the pulling-up of the striker's oscillatory amplitude. That is the well-known feature of the vibro-impact systems [3].

The dependence of the dynamic drift parameters on the dimensionless collision velocity $v = V_1^- / a\nu$ within the range of existence and stability is shown in Fig. 5. The calculation have been made with the use of MATCAD 15 and substitution of experimentally measured impact parameters and design parameters of ultrasonic tool [1]. It is noted that at the lower boundary the displacement of converter has the minimal value equal to dynamic gap and both they are less than the amplitude of converter tip $H = \Delta \leq a$. From Eq. (8) it follows that $h_2 = 0$ i.e. striker stands on the specimen surface and the vibrating tip of converter touches the rear end of the striker.

3.4 Impact Stress

Let us assume that the impacting elements are deformed according to the Hertz law [3] and substitute the striker velocity V_2^+ (see Eq. 8) into the well-known expressions for the impact force F and the radius y of the contact spot:

$$F = q^{2/5}(M_2 V_2^{+2})^{3/5}, \quad y = \left(2\sqrt{2}F/q\right)^{1/3} \cdot \left(r_{st}^{-1} + r_{wd}^{-1}\right)^{-1/2}, \qquad (12)$$

where the factor q depends on the Young modulus and the Poisson ratio of the treated material; r_{st} and r_{wd} are the curvature radii of the contact surfaces for the

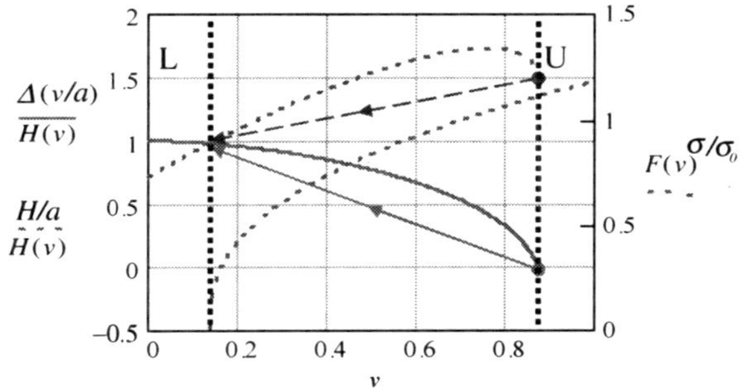

Fig. 5 The dependence of the dynamic parameters on the dimensionless collision velocity. Liters *L* and *U* denote the lower and upper boundaries of the range of existence and stability

striker and the welded joint. The value of impact stresses generated in a metal can be calculated as the ratio of the impact force to the contact spot area:

$$\sigma = F/\pi y^2. \tag{13}$$

The theoretical estimations corroborated with experimental measurements show that inside the local spot with the area of about 1–2 mm^2 the impact stress value reaches the yield point σ_0 of metal (see Fig. 5). So, redistribution and relaxation of internal residual stresses is caused by the micro-plastic deformation of the surface layers of metal.

4 Residual Stresses Distribution

Distribution of the residual stresses of the flat butt weld of chrome-nickel austenite steel with the yield point $\sigma_0 = 700$ MPa was investigated by the method of layer-by-layer metal etching. In Fig. 6 the results of treatment by three-striker tool for the

Fig. 6 Residual stresses distribution

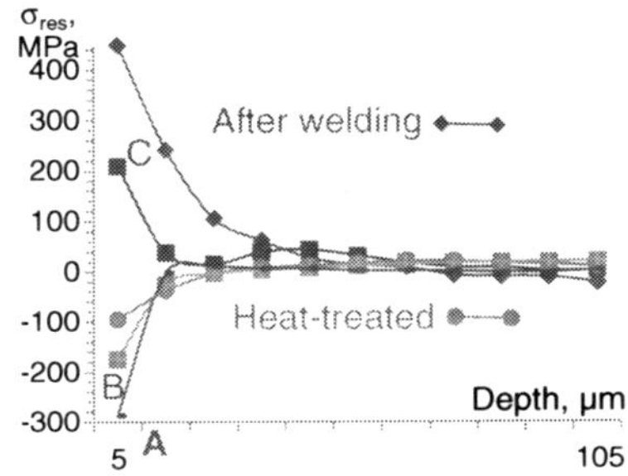

different regimes are presented. Average speed V of the tool movement along the weld seam and the pressure force G have the following values: $G = 100$ N $V = 20$ mm/s for regime “A”; $G = 80$ N $V = 20$ mm/s for regime “B” and $G = 80$ N $V = 40$ mm/s for regime “C”. It is seen that the compressive residual stresses can be created in the surface layer of metal as a result of the proper treatment regime.

5 Conclusion

On the base of the proposed model the dynamical behavior of the ultrasonic tool can be explained by the following way. The periodic change of dynamic gap due to the low-frequency oscillation of the tool casing causes the slow variation of the collision velocity within the limits of the range of existence and stability. The observed synchronous change of impact force is a result of its direct dependence on the impact velocity. If the displacement of the converter is increased so much that the collision velocity reaches the upper boundary of stability (Fig. 5), then the impact oscillations of striker are interrupted and it falls down on the specimen surface. Pressed by the spring the converter is returned back to the lowest position corresponding to the lower boundary of existence and stability. This transition is indicated by the arrows in Fig. 5. The vibrating converter tip comes into contact with the rear end of the striker and the periodic impact mode being resumed. The ultrasonic impact processing can be interpreted as a cyclic alternation of setting-up and failure of the vibro-impact oscillation of striker inside the variable gap.

The periodic change of the dynamic gap due to the low-frequency oscillation of tool casing causes the slow variation of the collision velocity. The impact frequency is kept and equal to ultrasonic frequency. The increase of gap width lead to pulling-up of oscillatory amplitude of striker that leads to the increase of impact force upon the treating sample.

References

1. Statnikov ES (2004) Physics and mechanism of ultrasonic impact treatment. IIW, Doc. XIII-2004–04
2. Shestakov SD, Ganiev MM (2005) Ultrasonic vibration-impact tool. Patent of Russian Federation, no. 2259912 IB
3. Babitsky VI (1998) Theory of vibro-impact systems and applications (foundations of engineering mechanics). Springer, Berlin, 318p

Use of the Roving Mass Technique and SWT to Identify and Locate Multiple Cracks in Simply-Supported Beams

Amirabbas Bahador and S. Olutunde Oyadiji

Abstract The roving mass technique is used to identify and locate multiple cracks in simply-supported beams. The technique involves the traversing of an auxiliary mass along the length of the beams. At each location of the roving mass, the stiffness of the beam is different which, in turn, causes the natural frequencies of the beam to change as the mass is traversed. A plot of the natural frequencies of the beam versus the location of the roving mass results in natural frequency curves (NFCs). If a crack is present in the beam, then the dynamics of the beam is affected because the crack causes a local reduction in the stiffness of the beam. This leads to very slight reductions in the natural frequencies of the beam. However, these slight reductions in the natural frequencies cannot be observed directly from the NFCs. Identification of crack information from the NFCs is achieved by the application of the stationary wavelet transform (SWT) to the NFCs. The presence of a crack is indicated as sharp peaks in the SWT detail coefficients plots. The effects of various roving mass magnitudes on the crack detection method are investigated. It is shown that the technique is efficient in detecting double symmetric cracks about the centre of the beam as well as detecting multiple cracks.

Keywords Crack detection · Simply-supported · Multiple cracks · Double symmetric cracks · Stationary roving mass · Stationary wavelet transform

1 Introduction

The dynamic characteristics of a structure, such as natural frequencies and mode shapes, are affected by cracks that are present in the structure. Cracks or local faults directly affect the natural frequencies by causing a local reduction in the stiffness of

A. Bahador · S.O. Oyadiji (✉)
Dynamics and Aeroelasticity Expert Group, School of Mechanical, Aerospace and Civil Engineering, The University of Manchester, Manchester M13 9PL, UK
e-mail: s.o.oyadiji@manchester.ac.uk

J.K. Sinha (ed.), *Vibration Engineering and Technology of Machinery*, Mechanisms and Machine Science 23, DOI 10.1007/978-3-319-09918-7_22

the structure. Numerous researches have been carried out on vibration based damage detection using natural frequencies. Narkis [1] identified cracks in simply-supported beams by measuring only the first two natural frequencies. They analytically defined the crack as a spring with a specific stiffness connecting the two segments of the beam. Messina et al. [2] identified single and multiple cracks in structures from changes in the natural frequencies of healthy and damaged structures. Morassi and Rollo [3] used the first three natural frequencies and the stiffness of a simply-supported beam to identify two cracks at different locations. Despite the ease of measuring natural frequencies of a structure, there are limitations in using natural frequencies in structural damage detection. The sensitivity of the natural frequencies to changes in the structural damages is insignificant. Moreover, the effects of multiple damages on the natural frequencies may be the same for a case with a single damage which results in ambiguities and limitations in damage detection. However, in recent years, the development of signal processing methods with high accuracy such as wavelet analysis, Fourier transforms and neural networks along with measurement devices and sensors with high accuracy such as scanning laser vibrometers has made significant improvements in the field of structural damage detection. Underwood et al. [4] used a 3-D scanning laser vibrometer to detect delaminations cracks in composite panels from changes in the magnitudes of frequency response functions (FRF). Kim and Melhem [5] detected fatigue cracks in a simply-supported pre-stressed concrete beam by decomposing the acceleration response of the beam using continuous wavelet transform (CWT). They detected the crack from decrease in the magnitudes of the high frequency components of the CWT.

Many real life mechanical systems consist of beams subjected to moving or stationary masses such as motors and engines mounted on cranes or moving trains and vehicles on bridges and roads made concrete slabs. Dynamics of a structure is affected by the presence of these moving or stationary loads and it has been used by researchers as a potential tool in the field of vibration damage detection. Zhang et al. [6] identified cracks in a beam by applying wavelet transform to the displacement response at a single point of the beam subjected to a moving mass. Lin and Chang [7] studied the effects of a concentrated moving mass on the vibration response of a cracked cantilevered beam for various moving mass speeds and crack depths. Nguyen and Tran [8] detected multiple cracks in beams by applying CWT to the displacements of the moving mass along the length of the beam. The velocity of the vehicle relative to the location of wavelet peaks indicated the location of the cracks. Zhong and Oyadiji [9] identified and located and a single crack in a simply-supported beam by applying the stationary wavelet transform (SWT) to the frequency plots obtained by traversing a mass with zero velocity on the beam.

This paper uses the auxiliary roving mass damage detection method used by Zhong and Oyadiji [9, 10] to investigate the capabilities of this crack detection method in detecting multiple cracks by applying the SWT to the NFCs. As it was discussed earlier, one of the main disadvantages of damage detection using natural frequencies is detecting multiple cracks with various depths. However, generating the NFCs using a roving mass allows for local detection of multiple cracks changes to the natural frequencies corresponding to the spatial location of the roving mass. The application of SWT to the NFCs reveals the small frequency fluctuations in the NFCs at the crack locations while allowing de-noising capabilities to reduce the amount of noise in the signals.

2 Predictions of Natural Frequency Curves of Cracked Simply-Supported Beams with Stationary Roving Mass Using FEA

2.1 Details of Finite Element Analysis

The stationary roving mass m is defined on ABAQUS FE as a point-mass subjected to a nodal point along the centre of the beam. The natural frequencies are computed with a sample distance of $x_s = 20$ mm for the roving mass thus, generating a natural frequency variation curve. The beams studied in this paper are 3D aluminium with the dimensions of 3,000 × 100 × 25 mm including single sided transverse cracks with a depth of H_c located at distance l_c from the left support. A uniform mesh is introduced along the length of the beam using a quadratic element denoted by ABAQUS FE as C3D20R. The defined material properties are Young's modulus $E = 70$ GPa, density $\rho = 2700$ kg/m^3 and Poisson's ratio $v = 0.3$.

2.2 Natural Frequency Curves of Simply-Supported Beams with Multiple Cracks

The efficiency of the crack detection method is investigated by studying the NFCs of a simply-supported beam for two cases of multiple cracks with various depths. In case 1, the cracks are located at $l_{c1} = 500$ mm and $l_{c2} = 2500$ mm with depths of $H_{c1}, H_{c2} = 2, 10$ mm respectively. In case 2, the cracks are located at $l_{c1}, l_{c2}, l_{c3}, l_{c4}, l_{c5} = 500, 1200, 1500, 1800, 2500$ mm with crack depths of $H_{c1}, H_{c2}, H_{c3}, H_{c4}, H_{c5} = 2, 5, 2, 1, 10$ mm respectively. The NFC of an intact beam is also included for comparisons. The effects of different roving mass magnitudes are also investigated for four different roving masses of 1, 2, 4 and 8 kg with mass ratios of $m/M_b = 4.9, 9.8, 19.7$ and $39.5\,\%$ respectively.

The first two transverse NFCs of the two crack cases for the roving mass of $m = 1$ kg are compared with the NFC of the intact beam in Figs. 1a, b and 2a, b. The figures show that the natural frequencies of the cracked beam are lower than the natural frequencies of the intact beam. Moreover, the natural frequencies decrease as the roving mass becomes closer to the anti-node of each mode where the flexibility of the beam is the highest.

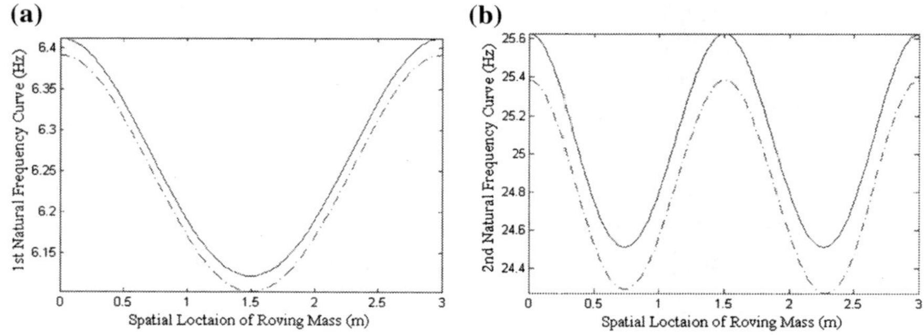

Fig. 1 The first two transverse NFCs of a simply-supported beam with double symmetric cracks: $x_s = 20$ mm, $m = 1$ kg, l_{c1}, $l_{c2} = 500$, 2500 mm, dash-dot line: $H_{c1,2} = 2$, 10 mm, solid line: Intact, **a** mode 1, **b** mode 2

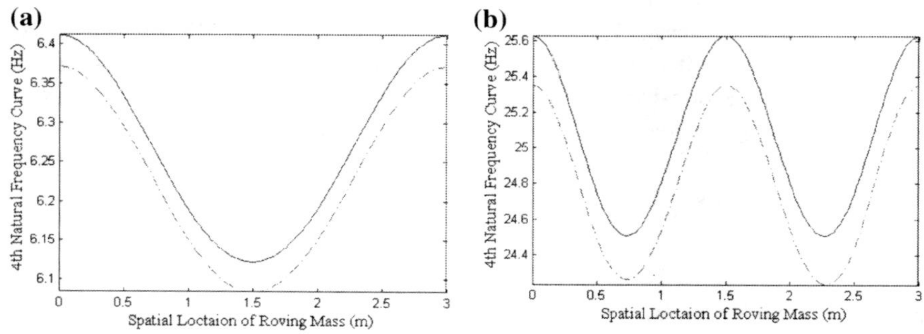

Fig. 2 The first two transverse NFCs of a simply-supported beam with multiple cracks: $x_s = 20$ mm, $m = 1$ kg, dash-dot line: H_{c1}, H_{c2}, H_{c3}, $H_{c4}, H_{c5} = 2$, 5, 2, 1, 10 mm, solid line: Intact, l_{c1}, l_{c2},$_{c3}$, $l_{c4}, l_{c5} = 500, 1,200, 1,500, 1,800, 2,500$ mm, **a** mode 1, **b** mode 2

2.3 Effects of Roving Mass Magnitude on Natural Frequency Curves

In this section, the effect of varying the magnitude of the roving mass on the predicted natural frequencies is investigated. The NFCs of the first two transverse modes of vibration of the cracked beams are compared to the NFCs of the intact beam for roving masses of 1, 2, 4 and 8 kg as shown in Figs. 3a, b and 4a, b. From these figures, it can be seen that the natural frequency values decrease significantly as the roving mass increases.

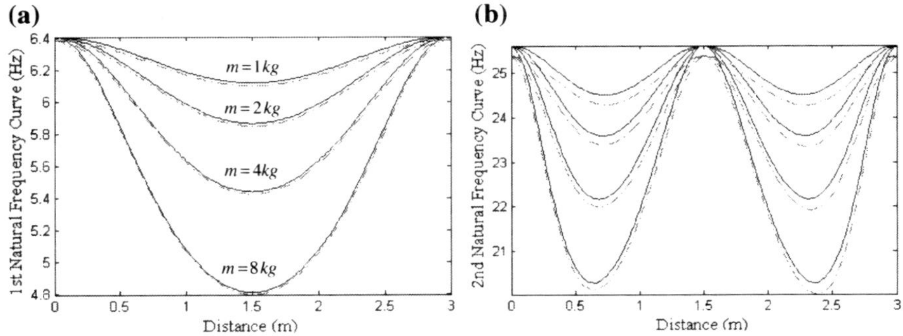

Fig. 3 Comparison of the first two transverse NFCs of a simply-supported beam with double symmetric cracks subjected to different roving masses: $m = 1, 2, 4, 8$ kg, $x_s = 20$ mm, $l_{c1}, l_{c2} = 500, 2500$ mm, dash-dot line: $H_{c1,2} = 2, 10$ mm, solid line: Intact, **a** mode 1, **b** mode 2

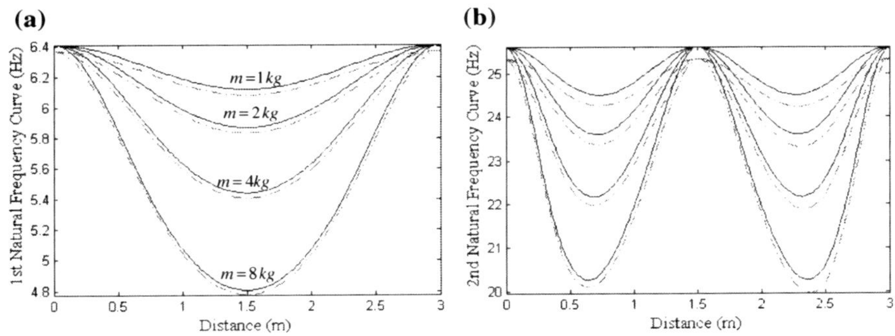

Fig. 4 Comparison of the first two transverse NFCs of a simply-supported beam with multiple cracks subjected to different roving masses: $m = 1, 2, 4, 8$ kg, $x_s = 20$ mm, dash-dot line: $H_{c1}, H_{c2}, H_{c3}, H_{c4}, H_{c5} = 2, 5, 2, 1, 10$ mm, $l_{c1}, l_{c2}, l_{c3}, l_{c4}, l_{c5} = 500, 1200, 1500, 1800, 2500$ mm, solid line: Intact, **a** mode 1, **b** mode 2

3 Crack Detection Using Stationary Wavelet Transforms of the NFCs

It was observed that the NFCs from are smooth curves without any visible local discontinuities due to the cracks. Therefore, in order to reveal the local frequency fluctuations due to the cracks and determine the severities of the frequency fluctuations, the SWT decomposition is performed on the NFCs.

3.1 SWT and SWT De-Noising

Wavelet analysis decomposes a signal by passing a wavelet through the signal which is then integrated over the entire length of the signal. The wavelet transform of a signal $f(x)$ is defined as [11];

$$W(a, b) = \int_{-\infty}^{+\infty} f(x)\psi_{a,b}(x)dx \tag{1}$$

The scaling and translating the mother wavelet $W(a, b)$ by parameters a and b respectively, generates the basis wavelet function $\psi_{a,b}$ given as [12];

$$\psi_{a,b} = \frac{1}{\sqrt{a}}\psi\left(\frac{x-b}{a}\right) \tag{2}$$

discretising the above equation by the dyadic values of a and b at resolution 2^j, the discrete approximation coefficient and the detail coefficient are given as [9];

$$c_{j,k} = \langle f(x), \phi_{j,k}(x)\rangle \tag{3}$$

$$d_{j,k} = \left\langle f(x), 2^{-j/2}\psi(2^{-j}x - k)\right\rangle \tag{4}$$

where $\phi_{j,k}$ is the dilation function and is defined as;

$$\phi_{j,k}(x) = 2^{-j/2}\phi(2^{-j}x - k) \tag{5}$$

The discrete approximation coefficient and the detail coefficient in Eqs. 3 and 4 satisfy the two-scale condition given by [9];

$$2^{-1/2}\phi\left(\frac{x}{2} - k\right) = \sum_{n=-\infty}^{+\infty} h(n-2k)\phi(x-n) \tag{6}$$

$$2^{-1/2}\psi\left(\frac{x}{2} - k\right) = \sum_{n=-\infty}^{+\infty} g(n-2k)\phi(x-n) \tag{7}$$

and they are obtained from Eqs. (3), (4), (6) and (7) [9];

$$c_{j+1,k} = \sum_{n=-\infty}^{+\infty} h(n-2k)c_{j,n} \tag{8}$$

$$d_{j+1,k} = \sum_{n=-\infty}^{+\infty} g(n-2k)c_{j,n} \tag{9}$$

Equations (8) and (9) represent the algorithm of discrete wavelet transform (DWT) which is not a time-invariant transform. Therefore, the approximation and detail coefficients of the SWT are obtained by up sampling the decomposed signal to its original size. The SWT detail and approximate coefficients are given as [11];

$$\tilde{c}_{j,k} = \sum_{n=-\infty}^{+\infty} h(l)\tilde{c}_{j,k+2^j l} \tag{10}$$

$$\tilde{d}_{j,k} = \sum_{n=-\infty}^{+\infty} g(l)\tilde{d}_{j,k+2^j l} \tag{11}$$

The detailed coefficients obtained from the predicted NFCs are contaminated by numerical noise. The effect of the numerical noise can be minimised by the application of filters based on the SWT de-noising approach. SWT de-noising includes two types: hard thresholding and soft thresholding. Hard thresholding returns the wavelet coefficient data values to zero for those values below the threshold [13]. Soft-thresholding at first returns the data values to zero, for those below the threshold, and then shrinks the nonzero coefficients by the threshold value [13]. The advantage of SWT de-noising is that it is completely independent of the wavelet system and only depends on the size of the signal [14]. In this paper, soft thresholding is applied to the SWT detail coefficients to reduce the noise and increase the accuracy of the crack detection.

Soft thresholding functions are defined as [9, 13];

$$\eta_{soft} = \begin{cases} sign(x)(|x| - th), & |x| \succ th, \\ 0, & |x| \leq th, \end{cases} \tag{12}$$

The threshold function is illustrated as [9, 15];

$$th = \sigma\sqrt{2\log N} \tag{13}$$

where x is the signal, th is the threshold, σ is the standard deviation of the noise and N is the length of the signal.

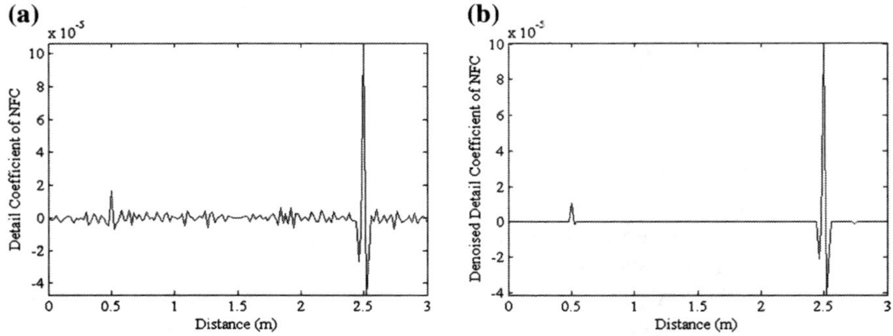

Fig. 5 SWT decomposition detail coefficient of the first NFC of a simply-supported beam with double symmetric cracks: l_{c1}, $l_{c2} = 500$, 2500 mm, $H_{c1,2} = 1$, 5 mm, $x_s = 20$ mm, $m = 2$ kg, **a** noisy, **b** De-noised by soft thresholding

Figure 5a and b show the SWT detail coefficients plots of the first NFC of the beam with double symmetric crack case of $H_{c1,2} = 2$, 10 mm before and after the SWT de-noising respectively. In Fig. 5, there are two sharp peaks at the crack locations at $l_{c1} = 500$ mm and $l_{c2} = 2500$ mm. In addition, the magnitudes of the SWT detail coefficients are higher for the deeper crack with a depth of $H_{c2} = 10$ mm at $l_{c2} = 2500$ mm but smaller for the crack with a depth of $H_{c1} = 2$ mm at $l_{c1} = 500$ mm. Comparing Fig. 5a with the de-noised plot in Fig. 5b, it is obvious that the SWT soft-thresholding significantly reduced the numerical noise. The sharp peaks identify the cracks at the correct locations.

4 Detection of Multiple Cracks in Simply-Supported Beams Using SWT

In this section, the identification of the location and severity of multiple cracks along with the effects of various roving mass magnitudes on the proposed crack identification method are investigated. For this reason, the SWT detail coefficients of the NFCs of the two cracked beam cases are compared for all four roving masses. Figure 6a shows the comparison of the SWT detail coefficients of the NFCs of the cracked beam case with crack depths of $H_{c1,2} = 2$, 10 mm for roving masses of 1, 2, 4 and 8 kg. Figure 6a, b show the zoom of Fig. 6a at the crack regions. Similarly, Fig. 7a shows the comparison of the SWT detail coefficients of the NFCs of the multiple crack case with crack depths of $H_{c1}, H_{c2}, H_{c3}, H_{c4}, H_{c5} = 2$, 5, 2, 1, 10 mm for 1, 2, 4 and 8 kg roving masses. Figure 7a1–a5 show the zoom of Fig. 7a at the crack locations. These figures show sharp discontinuities at the correct locations of the cracks. Also, as the roving mass magnitude increases, the

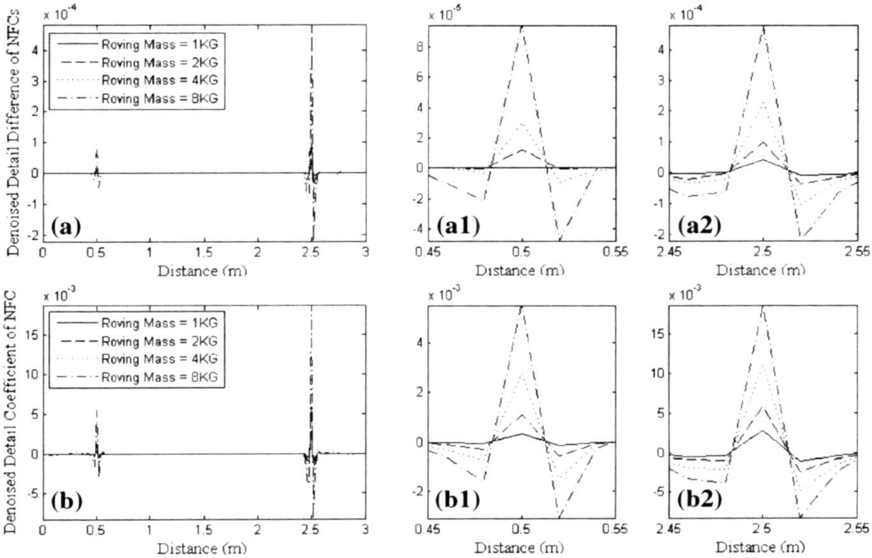

Fig. 6 Comparison of the SWT detail coefficients of the first two transverse NFCs of a simply-supported and intact beam with double symmetric cracks subjected to different roving masses: $x_s = 20$ mm, l_{c1}, $l_{c2} = 500$, 2500 mm, $H_{c1,2} = 2$, 10 mm, **a** mode 1, **b** mode 2

magnitude of SWT detail coefficients increases correspondingly. The peak values of the detailed SWT coefficients for mode one are greater than the corresponding magnitudes of the detailed SWT coefficients for mode two. Furthermore, it can be seen that the peak values of the detailed SWT coefficients increase relatively as the mode number increases.

The cracks with smaller depths relatively show lower peak values for each roving mass magnitude compared to the deeper cracks. From Fig. 6a, the smaller crack with a depth of $H_{c1} = 2$ mm at the location of $l_{c1} = 500$ mm, shows no peaks for the 1 kg roving mass which is due to the subtle effects of the roving mass on the crack and small beam deformations at that location for mode one. However, the method accurately identified the 2 mm crack for higher roving mass magnitudes of 2, 4 and 8 kg. It should be noted that the SWT detail coefficients of the NFCs of mode 2 become zero values for the crack in the middle of the beam at $l_{c3} = 1500$ mm showing no indication of any cracks at this region of the beam. This is due to the fact that the centre of the simply-supported beam is the nodal point for mode 2, thus when the roving mass is located at the centre of the beam, there are no changes in the dynamics of the beam and the resultant natural frequencies.

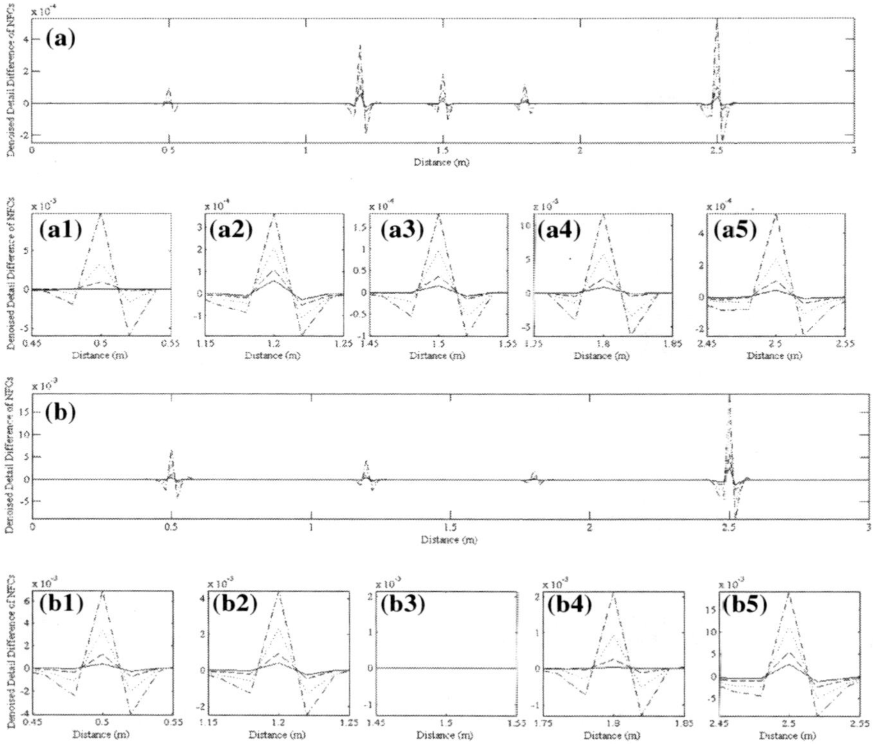

Fig. 7 Comparison of the SWT detail coefficients of the first two NFCs of a simply-supported beam with multiple cracks subjected to different roving masses: $x_s = 20$ mm, $l_{c1}, l_{c2}, l_{c3}, l_{c4}, l_{c5} = 500, 1200, 1500, 1800, 2500$ mm, $H_{c1}, H_{c2}, H_{c3}, H_{c4}, H_{c5} = 2, 5, 2, 1, 10$ mm, solid line: $m = 1$ kg, dashed line: $m = 2$ kg, dotted line: $m = 4$ kg, dash-dot line: $m = 8$ kg, **a** mode 1, **b** mode 2

5 Discussions and Conclusions

The use of a stationary roving mass and SWT for the identification of multiple cracks in simply-supported beams has been investigated in this paper. The natural frequencies of the beam change as the roving mass is traversed along the length of the beam. The natural frequencies are further affected as the roving mass comes into the vicinity of the crack, thus generating a small fluctuation in the NFCs.

These small natural frequency fluctuations are revealed as sharp discontinuities in the SWT detail coefficient plots. ABAQUS FEA has been used to model and simulate the modal responses of the cracked and intact simply-supported beams subject to stationary roving masses. The SWT detail coefficients detected the location of cracks with high precision. It was found that the magnitudes of the SWT detail coefficients at the crack locations corresponded to the crack depths. As the magnitude of the roving mass increases, the effects of the mass on the dynamics of the beam becomes more which, in turn, enhances the effects of the cracks on the natural frequencies.

Acknowledgments The award of a University of Manchester University Scholarship to Amirabbas Bahador is gratefully acknowledged.

References

1. Narkis Y (1994) Identification of crack location in vibrating simply supported beams. J Sound Vib 172(4):549–558
2. Messina A, Williams EJ, Contursi T (1998) Structural damage detection by a sensitivity and statistical-based method. J Sound Vib 216(5):791–808
3. Morassi A, Rollo M (2001) Identification of two cracks in a simply supported beam from minimal frequency measurements. J Vib Control 7(5):729–739
4. Underwood S, Koester D, Adams DE (2009) Damage detection in sandwich composite materials using laser vibrometry in conjunction with nonlinear system identification. pp 72952I–72952I
5. Kim H, Melhem H (2003) Fourier and wavelet analyses for fatigue assessment of concrete beams. Exp Mech 43(2):131–140
6. Wei Wei Zhang ZHW, Ma HW (2009) Studies on wavelet packet-based crack detection for a beam under the moving load. Key Eng Mater 413:285–290
7. Lin H-P, Chang S-C (2006) Forced responses of cracked cantilever beams subjected to a concentrated moving load. Int J Mech Sci 48(12):1456–1463
8. Nguyen KV, Tran HT (2010) Multi-cracks detection of a beam-like structure based on the on-vehicle vibration signal and wavelet analysis. J Sound Vib 329(21):4455–4465
9. Zhong S, Oyadiji SO (2008) Identification of cracks in beams with auxiliary mass spatial probing by stationary wavelet transform. J Vib Acoust 130(4):041001–041001
10. Zhong S, Oyadiji SO (2008) Analytical predictions of natural frequencies of cracked simply supported beams with a stationary roving mass. J Sound Vib 311(1–2):328–352
11. Seker S, Ayaz E (2003) Feature extraction related to bearing damage in electric motors by wavelet analysis. J Franklin Inst 340(2):125–134
12. Bagheri SK A (2013) Damage detection of structures under earthquake excitation using discrete wavelet analysis. Asian J Civil Eng (BHRC) 14(2):289–304
13. Mathworks (2010) Matlab Help manual
14. Berkner K, Wells RO Jr (1998) Wavelet transforms and denoising algorithms. In: Signals, systems and computers conference record of the thirty-second Asilomar conference on 1998
15. Zhong S, Oyadiji SO (2007) Crack detection in simply supported beams without baseline modal parameters by stationary wavelet transform. Mech Syst Signal Process 21 (4):1853–1884

Single and Multiple Crack Detection in Simply-Supported Beams Using SWT

Amirabbas Bahador and S. Olutunde Oyadiji

Abstract Any crack in a beam-like structure creates a local reduction in the stiffness which in turn, affects the dynamics of the structure. This local reduction in stiffness generates small discontinuities in the modal displacements of the beam at the crack locations. However, these discontinuities cannot be visually observed in the modal displacement plots. Therefore, the identification of cracks is performed by decomposing the modal displacements using the stationary wavelet transform (SWT). This paper investigates identification of multiple cracks in simply-supported beams using the SWT. The SWT decomposes a signal into two components that are known as the detail and the approximate coefficients. These coefficients can be viewed as the results of passing the signal through low-pass and high-pass filters. The approximate coefficient gives the low frequency components of the signal which in this case is the overall shape of the signal (mode shapes) and the detail coefficient gives the high frequency components of the signal which in this case contains any high frequency noise or effects of small structural discontinuities present in the signal. The detail coefficients of the SWT detect the existence of cracks by showing high values at the crack locations. The method presented can detect both single and multiple symmetric and non-symmetric cracks with high accuracy in beam-like structures. The modal displacement data of the cracked simply supported beams used are obtained from the finite element method. The results are validated experimentally.

A. Bahador · S.O. Oyadiji (✉)
School of Mechanical, Aerospace and Civil Engineering, The University of Manchester, Oxford Road, Manchester M13 9PL, UK
e-mail: s.o.oyadiji@manchester.ac.uk

A. Bahador
e-mail: amirabbas.bahador@manchester.ac.uk

J.K. Sinha (ed.), *Vibration Engineering and Technology of Machinery*, Mechanisms and Machine Science 23, DOI 10.1007/978-3-319-09918-7_23

1 Introduction

The aim of structural damage assessment is to identify, locate and determine the extent of any damage in a structure. The results of such assessments will be used to make key decisions on repair or retirement of the structure before any serious failures occur. Most structural faults are internal and are, therefore, not visible. In order to detect such invisible structural faults non-destructively, many methods of non-destructive testing (NDT) have been developed. These methods can be classified broadly into local and global methods. Because of their ease of use and cost-effectiveness, vibration-based methods, which fall into the global NDT classification, have increasingly become popular in non-destructive structural damage detection and structural health monitoring. The study of the effects of any structural damage on the dynamic behavior of the structure, such as natural frequencies, mode shapes and frequency response functions (FRF's), for damage detection has been the focus of many researchers. Natural frequencies of a structure can be obtained easily by only taking one or two measurements on the structure. Rizos et al. [1] identified cracks in cantilevered beams by measuring the vibration amplitudes of two points on the beam using analytical solutions. However, their method lacks accuracy and sensitivity in detecting small cracks. Nandwana and Maiti [2] identified cracks in cantilevered beams based on measurements of natural frequencies. Gudmundson [3] located cracks in cantilevered beams by studying the changes in natural frequencies. Using a mathematical model, Gudmundson showed that small changes in the resonance frequencies of a cantilevered beam are proportional to the square of the crack length, the square of the stress and inversely proportional to the length of the beam. Owolabi et al. [4] detected cracks in beams from changes in frequency response functions. They showed that the amplitudes of the FRFs decreased as the crack depth increased.

One of the major issues with vibration-based damage detection in structures is the need for baseline data of the healthy structures. Yoon et al. [5] detected the locations and sizes of narrow and wide notches in beams using the difference between the curvature of mode shapes. They used a baseline data of a healthy beam which was obtained using the gapped middle point by smoothing the curvature mode shapes near the damaged area. Schulz et al. [6] identified damages in cantilevered beams from changes in the FRF matrix of the damaged beam compared to that of the healthy one. The main advantage of natural frequencies is that they are easy to measure and can be used to evaluate changes in the dynamic characteristics of a structure. However, structural damages have small effects on the natural frequencies. Moreover, localized temperature and mass variations affects the natural frequencies. Mode shapes on the other hand, are more sensitive to any local damages. Therefore, some researchers have focused their work on damage detection using mode shapes. Cao et al. [7] studied the curvature sensitivity of a cantilevered beam for various deformations. Pandey et al. [8] identified cracks in beams from the absolute difference between the curvature mode shapes of the intact and damaged beams for the first five modes.

Mode shapes also encounter some limitations in the field of vibration-based damage detection. Principally, the changes in the displacement of the cracked region are very small and are undetectable in a mode shape compared to the effects of environmental noise. In order to overcome the problems associated with these methods, some researchers have concentrated on the application of more accurate signal processing algorithms and measurement methods such as laser vibrometers. Li et al. [9] detected cracks in cantilevered beams by applying wavelet transform to the strain-time responses of the beam. Douka et al. [10] identified transverse surface cracks in cantilevered beams by applying continuous wavelet transform to the mode shapes. Zhong and Oyadiji [11] detected small cracks in simply-supported beams using SWT of the mode shapes. In this paper, the modal displacement data is decomposed by the SWT and the detail coefficients are studied for crack detection. Sharp peaks in the detail coefficient plots of the SWT identify and locate the cracks. The magnitudes of the SWT detail coefficients are related to the crack depths. The proposed method demonstrates accurate crack detection for single cracks and for multiple cracks that are located symmetrically or non-symmetrically about the centre of the modal displacements.

2 FE Analysis

The beams modeled in the ABAQUS FE code are 3-D simply-supported beams with dimensions of 3,000 mm × 100 mm × 25 mm. The cracks on the beam are single-sided transverse cracks with a depth of H_c located at distance l_c from the left support. The chosen beam models are simply-supported aluminum beams using the 20 node 3-D quadratic brick element denoted in the ABAQUS FE code as C3D20R. A uniform mesh was used along the length of the beams with a fixed element length of 5 mm. The beam has a Young's modulus $= 70\,\text{GPa}$ $\text{E} = 70\,\text{GPa}$, density $\rho = 2{,}700\ \text{kg/m}^3$ and Poisson ratio $\nu = 0.3$.

3 SWT and Noise Reduction Method

A wavelet is a waveform with a finite length and zero average value. Decomposition of a signal by wavelet transform is such that the wavelet is passed through the signal and then integrated over the entire length of the signal. This allows for local analysis and revealing hidden aspects within the signal.

The wavelet decomposition of a signal $f(x)$ is defined as [12],

$$w(a,b) = \int_{-\infty}^{+\infty} f(x)\psi_{a,b}\mathrm{d}x, \tag{1}$$

The mother wavelet $w(a, b)$ is dilated and translated by the dyadic values of a and b at resolution 2^j which then gives;

$$\Psi_{j,k} = \left\langle f(x), 2^{-j/2}\,\psi(2^{-j}x - k)\right\rangle = \int_{-\infty}^{+\infty} 2^{-j/2} f(x)\psi(2^{-j}x - k)\mathrm{d}x, (j,k) \in Z \quad (2)$$

where, Z is a set of positive integers [13]. The wavelet transform Eq. (1) is derived from a type of analytical functions called scaling function $\phi(x)$ and it is given as [14],

$$\phi_{j,k}(x) = 2^{-j/2}\,\phi\left(2^{-j}x - k\right) \quad (3)$$

The mother wavelet $\psi(x)$ and the scaling functions $\phi(x)$ satisfy the so-called two-scale condition given by [12],

$$2^{-j/2}\,\phi\left(\frac{x}{2} - k\right) = \sum_{n=-\infty}^{+\infty} h(n - 2k)\phi(x - n) \quad (4)$$

$$2^{-j/2}\,\psi\left(\frac{x}{2} - k\right) = \sum_{n=-\infty}^{+\infty} g(n - 2k)\phi(x - n) \quad (5)$$

where the impulse responses of low-pass and high-pass filters are defined as $h(n)$ and $g(n)$.

The discrete approximation and detail coefficients $c_{j,k}$ and $d_{j,k}$ of signal $f(x)$ at resolution 2^j are given as [14],

$$c_{j,k} = \left\langle f(x), 2^{-j/2}\,\phi_{j,k}(x)\right\rangle = \int_{-\infty}^{+\infty} 2^{-j/2} f(x)\phi^*\left(2^{-j}x - k\right)\mathrm{d}x \quad (6)$$

$$d_{j,k} = \left\langle f(x), 2^{-j/2}\,\psi_{j,k}(x)\right\rangle = \int_{-\infty}^{+\infty} 2^{-j/2} f(x)\psi^*\left(2^{-j}x - k\right)\mathrm{d}x \quad (7)$$

where * represents the complex conjugation.

From Eqs. 4–7, the approximation and detail coefficients of the discrete wavelet transform (DWT) are obtained as [14],

$$\tilde{c}_{j,k} = \left\langle f(x), \frac{1}{2^{j/2}}\emptyset\left(\frac{x-k}{2^j}\right)\right\rangle \quad (8)$$

$$\tilde{d}_{j,k} = \left\langle f(x), \frac{1}{2^{j/2}} \psi\left(\frac{x-k}{2^j}\right) \right\rangle \tag{9}$$

DWT down samples the signal by half its length at each decomposition level. This is a drawback of the DWT as it is not a time-invariant transform. On the other hand, SWT up-samples the detail and approximation coefficients by a factor of two at each decomposition level thus the length of the decomposed signals are equal to that of the original signal. Therefore, the SWT detail and approximation coefficients are given as [14],

$$\tilde{c}_{j+1,k} = \sum_{l=-\infty}^{+\infty} h(l)\tilde{c}_{j,k+2^j l} \tag{10}$$

$$\tilde{d}_{j,k} = \sum_{l=-\infty}^{+\infty} g(l)\tilde{d}_{j,k+2^j l} \tag{11}$$

The SWT coefficients contain general response noise. The detail coefficients contain high frequency general response noise as well as local noise due to the crack. Therefore, in order to reduce the effect of general response noise and reveal the crack noise, an accurate de-noising technique should be applied to the detail coefficient. In this paper, the SWT soft-thresholding, due to its better mathematical properties, is used.

For a signal 'x', soft thresholding is defined as [15],

$$\eta_{soft} = \begin{cases} sign(x)(|x| - th), & | \quad |x| > th, \\ 0 & | \quad |x| \le th, \end{cases} \tag{12}$$

The threshold function used in this paper is computed as [14],

$$th = \sigma\sqrt{2logN} \tag{13}$$

where th is the threshold, σ is the standard deviation of the noise and N is the length of the signal which in this case is the length of the SWT decomposition detail coefficient.

4 Investigation of the Effects of Over Refinement of Mesh and SWT Sample Distances

In order to have a better understanding of the effects of mesh densities and the effects of the SWT sample distance on the crack detection results, a mesh study is performed on a series of intact beams with both uniform and localized refined mesh along the length. The SWT is then applied to the first transverse mode shapes and the detail coefficient plots are compared. Figure 1a shows the first transverse mode shape of an intact beam with 3,000 uniform elements with an element length of $l_e = 1$ mm. Figure 1b shows the SWT approximate coefficients of the mode shape in Fig. 1a. It can be seen that the approximate coefficient plot is similar to the mode shape. Figure 1c shows the SWT detail coefficients of the corresponding mode shape with a sampling distance of $x_s = 10$ mm. In another case, the number of elements of the mode shape is reduced to 120 elements with element length of $l_e = 1$ mm along the length of the beam. Figure 1d shows the SWT detail coefficients of this mode shape with a sampling distance of $x_s = 25$ mm. Comparing Fig. 1c with d, it can be seen that as the length of elements along the length of the beam and also the SWT sampling distance increase, the amount of noise in the detail coefficients decreases. However, it can be seen that the magnitudes of the detail coefficients of the noise are very small. The noise variations is due to the finite element method (FEM) being an approximate method.

The same procedure is applied to the intact beams with a localized refined mesh. Figure 2a and b illustrates SWT detail coefficients of the first mode shape of two intact beams with a localized refined mesh at $l = 750$ mm with the element lengths of 1 mm and 25 mm respectively. It should be noted that for the beams with localized denser mesh, the beams have uniform mesh densities with $l_e = 50$ mm in the unrefined region. Also, the SWT sampling distance includes all data points of the refined region such that the sampling distance is smaller at the localized refined mesh region. From Fig. 2a and b, it can be seen that there is a sharp discontinuity at the region where the SWT sampling distances are locally refined. From the results of this mesh study, it can be concluded that the mesh density and SWT sample distance in the modal analysis of beams obtained from ABAQUS FEA has profound effects on the results of damage detection method applied by the SWT. Therefore, the beams studied in this paper are modeled with a uniform mesh density and the SWT sampling distances are also kept constant.

5 Verification of the Efficiency of the Method for Detecting Single, Double Symmetric and Multiple Cracks

The crack detection method is applied to cracked beam cases with single, double symmetric and multiple cracks. Figure 3 shows the soft-thresholded SWT detail coefficients of the modal displacement of two beams one with a single and the other with double symmetric cracks of various depths for the first transverse mode shape with a sampling interval of $x_S = 5$ $x_s = 5$ mm. The cracks are located at distances of $l_{c1} = 500$ mm for the single crack and l_{c1}, $l_{c2} = 500$, 2,500 mm for the double symmetric crack cases from one end of the beam with the corresponding crack depths of $H_{c1} = 2$ mm and H_{c1}, $H_{c2} = 1$, 5 mm respectively. Similarly, Fig. 4 shows the soft-thresholded SWT detail coefficients of the modal displacement of the beam with multiple cracks of various depths for the first transverse mode shape with

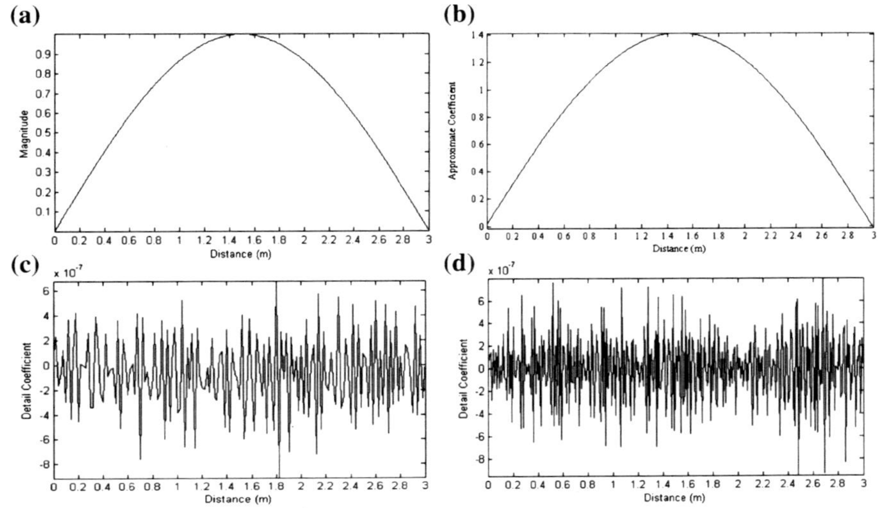

Fig. 1 SWT decomposition of the modal displacements of mode 1 of a simply supported intact beam with uniform mesh density, **a** mode shape, **b** Approximate coefficients, **c** SWT for $l_e = 10$ mm, IEEE transactions on SWT for $l_e = 25$ mm

a sampling interval of $x_s = 5$ mm. The cracks are located at $l_{c1}, l_{c2}, l_{c3}, l_{c4}, l_{c5} =$ 500, 1,200, 1,500, 1,800, 2,500 mm which correspond to the crack depths of $H_{c1}, H_{c2}, H_{c3}, H_{c4}, H_{c5} = 5, 1, 5, 10, 2$ mm respectively.

Figures 3 and 4 clearly indicate the existence of single, double symmetric and multiple cracks on the cracked beam cases. From Figs. 3b and 4, it can be seen that the peak values of the SWT detail coefficients increase as the crack depth increases.

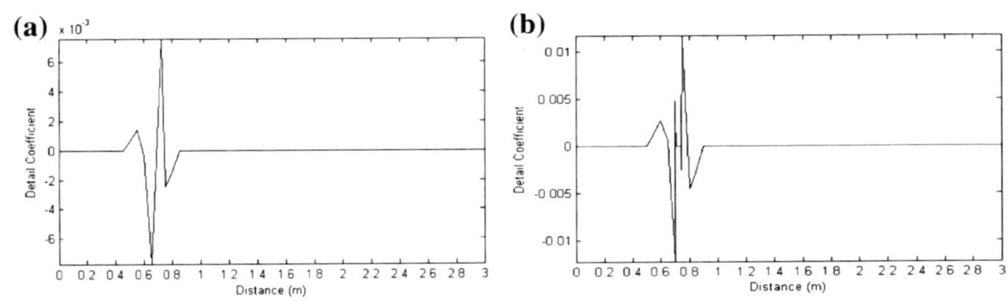

Fig. 2 SWT decomposition of the modal displacements of mode 1 of a simply supported intact beam with $x_s = 50$ mm (*unrefined region*) and a localized refined mesh, **a** SWT for $l_e = 1$ mm, **b** SWT for $l_e = 25$ mm

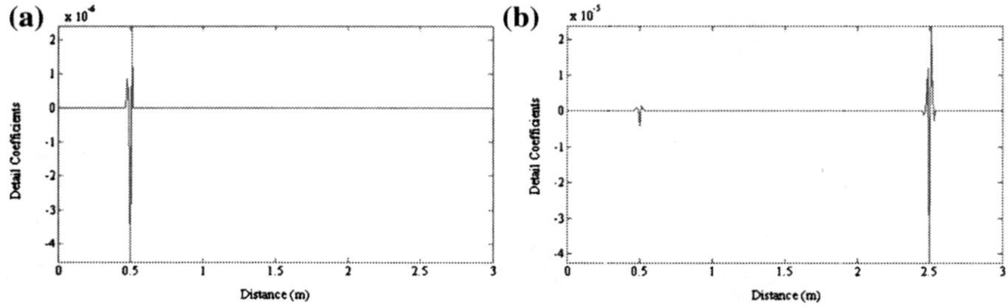

Fig. 3 The SWT detail coefficient of the first transverse modal displacement of simply supported signal and double cracked beams: $l_{c1} = 500$ mm, $l_{c1}, l_{c2} = 500,\ 2{,}500$ mm, $H_{c1} = 2$ mm, $H_{c1,2} = 1,\ 5$ mm, $x_s = 5$ mm, **a** single crack, **b** double symmetric crack

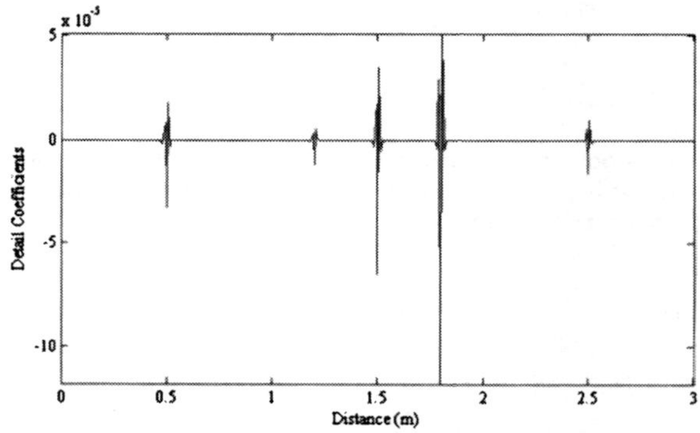

Fig. 4 The SWT detail coefficient of the first transverse modal displacement of simply supported multiple cracked beam, $H_{c1}, H_{c2}, H_{c3}, H_{c4}, H_{c5} = 5,\ 1,\ 5,\ 10,\ 2$ m, $l_{c1}, l_{c2}, {}_{c3}, l_{c4}, l_{c5} = 500,\ 1{,}200,\ 1{,}500,\ 1{,}800,\ 2{,}500$, $x_s = 5$ mm

6 Experimental Verification

In order to verify the numerical results an experiment is carried out on a simply-supported cracked aluminum beam with similar dimensions to the beams used in the numerical analysis in this paper. The experimental test is carried out at the University of Manchester School of MACE Dynamics and Vibrations Laboratory. The test beam includes a wire-cut crack located at l_c= 1,500 mm with a depth of $H_c = 12.5$ mm and a thickness of $t = 0.3$ mm.

A simply-supported boundary setup is designed from scratch and fabricated at the University of Manchester workshops. The simply-supported boundary configuration allows for the horizontal movement of the ends of the beam in x-axis while allowing for their rotation about the z-axis. The boundary setup uses a combination of linear slide and plummer bearings. The data acquisition systems used for this experiment is the LMS SCADAS III. A vibration chirp signal with a frequency

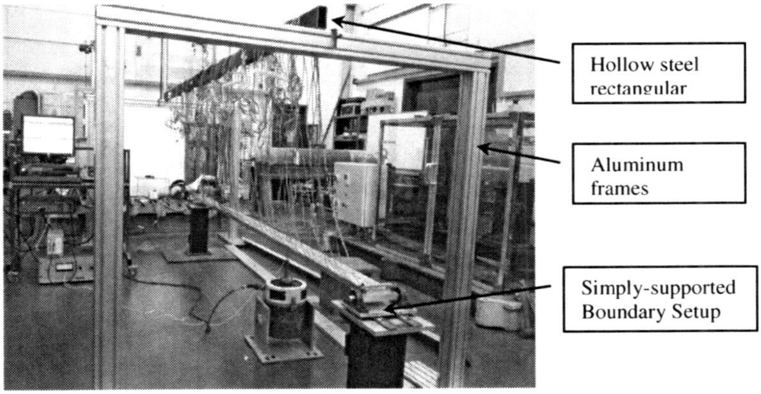

Fig. 5 Simply-supported experimental setup

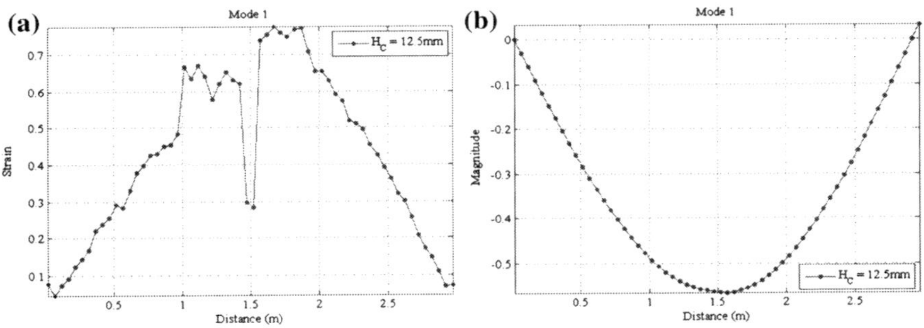

Fig. 6 First modal strain and displacement of the cracked simply-supported test beam with a crack depth $H_c = 12.5$ mm located at $l_c = 1{,}500$ mm, **a** modal strain of the first mode, **b** First transverse modal displacement

range of 0–250 Hz is generated by the LMS systems which is amplified by a power amplifier and exerted on the test beam via a shaker.

The response signals are measured using 60 strain gauges bonded to the beam on a line parallel to the centerline of the beam but with an offset distance of 10 mm and with equal intervals of 50 mm. Figure 5 shows the simply-supported experimental setup. It should be noted that in order to reduce the effects of the weights of the wires connected to the strain gauges on the dynamics of the beam, the wires are suspended from a steel hollow rectangular beam fixed on two aluminum frames.

The input signal is measured via a PCB force sensor placed between the shaker and the test beam. The response signals and the input signal are passed through a strain gauge amplifier and a signal conditioner respectively, before they are fed in the LMS data acquisition system. The modal strains are obtained from the resultant frequency response function FRF plots of 60 strain gauges.

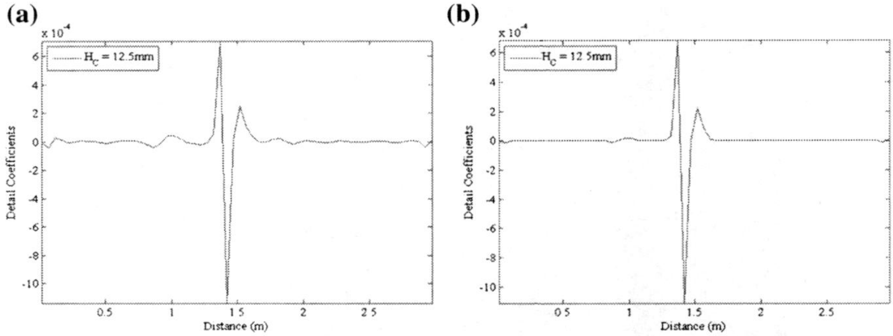

Fig. 7 SWT detail coefficient of the first transverse modal displacement of the cracked simply-supported test beam: x_s = 50 mm, l_c = 1,500 mm, H_c = 12.5 mm, **a** noisy, **b** De-noised by soft thresholding

Figure 6a shows the first modal strain of the cracked simply-supported beam. From this figure, it can be seen that there is a significant decrease in the strain values at the crack region which indicates the location and severity of the crack. In order to investigate the efficiency of the proposed crack detection method in detecting cracks form the modal displacements, the modal displacements are obtained by integrating the modal strains twice using MATLAB software and then the SWT is applied to the modal displacements.

Figure 6b shows the first transverse modal displacement of the cracked simply-supported test beam. From this figure, it can be seen that the modal displacement is a smooth curve with no clear indication of any discontinuities due to the crack. SWT is applied to the experimental modal displacements in Fig. 6b. Figure 7a shows the SWT detail coefficients of the first transverse experimental modal displacement of the test beam with a crack depth of H_c = 12.5 mm located at l_c= 1,500 mm and for the sampling distances of x_s = 50 mm. Figure 7b shows the soft-thresholded detail coefficients in Fig. 7b. From Fig. 7a and b, it can be seen that the SWT detail coefficients show sharp peaks at the correct crack location. Moreover, the magnitude of the detail coefficients peak can indicate the extend of the crack depth.

7 Conclusions

This paper investigates identification of multiple cracks in simply-supported beam-like structures using the SWT decomposition detail coefficients of the modal displacements. Based on the results from the FE mesh study, a uniform mesh is used along the length of the beam including around the crack region. Moreover, the SWT sample distances are also kept constant throughout the length of the beam. An experimental verification of the numerical results is performed by applying the crack detection method to the modal displacements of a cracked simply-supported beam. For this experiment, a roller and linear slide device, which simulates a simply-supported boundary condition, was designed and fabricated to match the test beam configurations and experimental requirements. The SWT of both the numerical and experimental modal displacements identify and locate the crack in the beam. In addition, the measured strain distribution directly identifies and locates the crack.

Acknowledgments The award of a University of Manchester University Scholarship to Amirabbas Bahador is gratefully acknowledged.

References

1. Rizos PF et al (1990) Identification of crack location and magnitude in a cantilever beam from the vibration modes. J Sound Vib 138:381–388
2. Nandwana BP, Maiti SK (1997) Detection of the location and size of a crack in stepped cantilever beams based on measurements of natural frequencies. J Sound Vib 203:435–446
3. Gudmundson P (1982) Eigenfrequency changes of structures due to cracks, notches or other geometrical changes. J Mech Phys Solids 30:339–353
4. Owolabi GM et al (2003) Crack detection in beams using changes in frequencies and amplitudes of frequency response functions. J Sound Vib 265:1–22
5. Yoon M-K et al (2009) Local damage detection with the global fitting method using mode shape data in notched beams. J Nondestruct Eval 28:63–74
6. Schulz MJ et al (1998) Locating structural damage using frequency response reference functions. J Intell Mater Syst Struct 9:899–905
7. Cao M et al (2011) Sensitivity of fundamental mode shape and static deflection for damage identification in cantilever beams. Mech Syst Signal Process 25:630–643
8. Pandey AK et al (1991) Damage detection from changes in curvature mode shapes. J Sound Vib 145:321–332
9. Li Z et al (2006) Damage detection of cracked beams based on wavelet transform. Int J Impact Eng 32:1190–1200
10. Douka E et al (2003) Crack identification in beams using wavelet analysis. Int J Solids Struct 40:3557–3569
11. Zhong S, Oyadiji SO (2011) Crack detection in simply supported beams using stationary wavelet transform of modal data. Struct Control Health Monit 18:169–190
12. Pesquet J et al (1996) Time-invariant orthonormal wavelet representations. IEEE Trans Signal Process 44:1964–1970
13. Hou Z et al (2000) Wavelet-based approach for structural damage detection. J Eng Mech 126:677–683
14. Zhong S, Oyadiji SO (2007) Crack detection in simply supported beams without baseline modal parameters by stationary wavelet transform. Mech Syst Signal Process 21:1853–1884
15. Mathworks (2010) Matlab Help manual (ed)

Modal Approach for Optimal Design of Two-Layer Piezoelectric Vibration Energy Harvesters

S. Olutunde Oyadiji and Xingyu Xiong

Abstract A modal approach for the development of optimal configurations of two-layer piezoelectric vibration energy harvesters is presented. The harvester comprises a primary cantilevered beam to which a secondary beam is attached via a mass that is located along the length of the primary beam. An additional mass is located along the secondary beam. The two masses are used to tune the natural frequencies of the composite system and one of them also serves as a spacer between the two beams. By varying the dimensions of the beams and masses, and the locations of the masses along the beams, the harvester can produce close resonance frequencies and significant power output. Thus, the frequency bandwidth of significant power generation by the harvester can be extended. To judge the performance of any harvester configuration requires a full analysis, using the coupled electromechanical equations of the piezoelectric harvester, to determine the electrical power output. However, the analysis is lengthy and time consuming. To hasten the process, a modal approach has been developed. The approach determines the modal performance by means of the mass ratio (which represents the influence of modal mechanical behaviour on the power density directly) and the modal electromechanical coupling coefficient. The modal parameters required by the approach are computed numerically by finite element analysis. The modal approach is used to select harvester configurations with optimal or near-optimal performance, which are harvester configurations with close resonances and moderate values of mass ratio. A full analysis is subsequently performed to determine the power outputs of these harvester configurations.

Keywords Energy harvesting · Piezoelectric energy harvester · Vibration energy harvester · Multiple resonances · Modal performance · Optimization

S.O. Oyadiji (✉) · X. Xiong
School of Mechanical, Aerospace and Civil Engineering, The University of Manchester, Oxford Road, Manchester M13 9PL, UK
e-mail: s.o.oyadiji@manchester.ac.uk

X. Xiong
e-mail: xingyu.xiong@postgrad.manchester.ac.uk

J.K. Sinha (ed.), *Vibration Engineering and Technology of Machinery*, Mechanisms and Machine Science 23, DOI 10.1007/978-3-319-09918-7_24

1 Introduction

Harvesting of small amounts of power from ambient sources, such as vibration, light and heat, has become a major field of research interest and activity within the last 10 years. This development has been driven by the need for self-powered, batteryless wireless sensors and sensor nodes for continuous monitoring of engineering assets some of which are located in remote areas. These small amounts of ambient energy can also be used to power mobile personal devices, such as smart phone, MP3 music player, tablets and laptop PCs, whose usage has become widespread within the same timeframe. Much of the development of energy harvesters has been focused on piezoelectric vibration energy harvesters (PVEHs) [1]. Like other vibration energy harvesters, PVEHs are designed to harvest vibration energy at the resonance frequency. Typically, a PVEH is a cantilevered steel beam to which a piezoelectric patch is bonded and has a tip mass that is used to tune the fundamental natural frequency of the PVEH. The higher resonance frequencies of the PVEH are much greater than the fundamental natural frequency but the levels of electrical power generated at those higher resonance frequencies are much lower. In addition, ambient or environmental vibration occurs at low frequency range and has random characteristics. Consequently, conventional vibration energy harvesters (VEHs), which are tuned to resonate at only one frequency, have limited performance; they become ineffective when the frequency of the ambient source of vibration is off-resonance. This drawback can be overcome by designing the harvester to possess broadband characteristics [2]. Approaches that have been developed to achieve this include resonance-tuning techniques, nonlinear system with magnets, and multi-resonance designs [3–6].

Several researchers have investigated multi-resonance harvester designs. Ferrari et al. [5] carried out experiments on an array of three cantilevered VEHs that were connected electrically. They tuned the resonance frequencies by varying the lengths of the beams and by attaching various tip masses. Qi et al. [6] developed, analysed and tested a multi-resonance VEH comprising a main clamped-clamped beam and side branches of cantilevered beams. Both the main and side beams generated the resonances but power was produced only by the main beam, which had piezoelectric material coverage. Ou et al. [7] and Zhou et al. [8] developed a two-mass cantilevered VEH consisting of a cantilevered beam with one mass attached to the middle of the beam, and the other mass attached to the free tip. But the model could only generate two close resonances only when the middle mass is much heavier than the tip mass. Chen et al. [9] developed a ten-mass VEH, which generates a wider frequency range of broadband power output. However, the use of nine masses degrades the power density significantly. Erturk et al. [10] developed an L-shaped VEH, with the second resonance frequency about twice the first resonance frequency. Karami and Inman [11] developed a zigzag cantilevered VEH that has up to 11 sub-branches. But the capability of the VEH to produce broadband power output is comprised by the wide spacing of the natural frequencies.

For the optimal design of a multi-resonance VEH, the objective of is to achieve close resonance frequencies with nearly the same power output level but also large values of the power output in each mode. This objective is achieved in this paper by using a modal approach to optimise a novel two-mass two-layer cantilevered PVEH. The first two modes are generated from the interaction between the base layer and the upper layer. The two masses are used to tune the natural frequencies of the PVEH and one of them also serves as a spacer between the two layers. The PVEH can easily generate close resonance frequencies and considerable power output in the two modes. The optimal or near-optimal locations of the masses along the lengths of the beams are determined by means of the modal approach.

2 Piezoelectric Vibration Energy Harvesting Model

2.1 Transverse Vibration of Cantilevered Beam

Figure 1 shows a cantilevered PVEH which is subjected to base excitation. The absolute transverse displacement of any point on the beam is given by,

$$u(x,t) = u_i(x,t) + u_o(x,t) \tag{1}$$

where $u_o(x, t)$ is the transverse displacement of any point on the beam relative to the clamped base of the beam, and $u_i(x, t)$ is the base excitation of the beam. For an internally-damped beam surrounded by air, the equation of motion for free transverse vibrations in terms of the absolute displacement $u(x, t)$ is given by the Bernoulli-Euler beam theory as [12, 13],

$$EI\frac{\partial^4 u}{\partial x^4} + c_s I\frac{\partial^5 u}{\partial x^4 \partial t} + c_a\frac{\partial u}{\partial t} + \rho A\frac{\partial^2 u}{\partial t^2} = 0 \tag{2}$$

in which $u = u(x, t)$, E is Young's modulus of elasticity, I is second moment of area, ρ is mass density, A is cross-sectional area, c_s is structural damping coefficient and c_a is air damping coefficient. Substituting Eq. (1) in Eq. (2) gives,

$$EI\frac{\partial^4 u_o}{\partial x^4} + c_s I\frac{\partial^5 u_o}{\partial x^4 \partial t} + c_a\frac{\partial u_o}{\partial t} + \rho A\frac{\partial^2 u_o}{\partial t^2} = -\rho A\frac{\partial^2 u_i}{\partial t^2} - c_a\frac{\partial u_i}{\partial t} \tag{3}$$

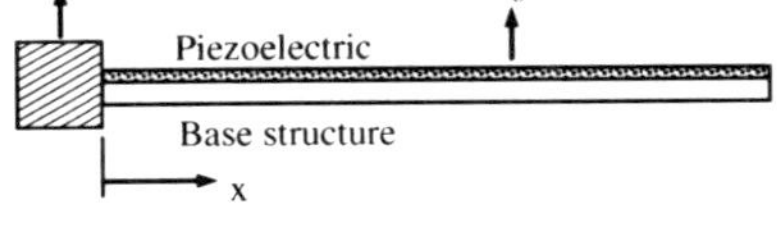

Fig. 1 Cantilevered PVEH subjected to base excitation

2.2 Distributed Parameter Electromechanical Model (DPM)

The distributed parameter electromechanical model (DPM) is based on the Bernoulli-Euler beam theory. For a base-excited cantilevered PVEH, the coupled electromechanical equations of motion are given by [12],

$$EI\frac{\partial^4 u_o}{\partial x^4} + c_s I\frac{\partial^5 u_o}{\partial x^4 \partial t} + c_a\frac{\partial u_o}{\partial t} + \rho A\frac{\partial^2 u_o}{\partial t^2} + \vartheta_1 v(t) = -\rho A\frac{\partial^2 u_i}{\partial t^2} - c_a\frac{\partial u_i}{\partial t} \tag{4}$$

$$C_p\frac{dv(t)}{dt} + \frac{v(t)}{\partial t^2} = -\int_0^L \vartheta_2\frac{\partial^3 u_i}{\partial x^2 \partial t}dx \tag{5}$$

where $v(t)$ is the voltage generated across the electrodes of the PVEH, C_p is the capacitance of the piezoelectric layer, and ϑ_1 and ϑ_2 are coupling coefficients. The relative transverse displacement response of the PVEH can be written as,

$$u_o(t) = \sum_{r=1}^{\infty}\phi_r\eta_r(t) \tag{6}$$

where ϕ_r and $\eta_r(t)$ are the mass-normalised eigenfuction and the modal coordinates of the rth mode, respectively. For a PVEH model with a resistive load, the governing equations of the vibration modal response and electrical behaviour then become [12],

$$\frac{d^2\eta_r(t)}{dt^2} + 2\zeta_r\omega_r\frac{d\eta_r(t)}{dt} + \omega_r^2\eta_r(t) + \chi_r v(t) = F_r(t) \tag{7}$$

$$\frac{v(t)}{R_l} + C_p\frac{dv(t)}{dt} - \sum_{r=1}^{\infty}\chi_r\frac{d\eta_r(t)}{dt} = 0 \tag{8}$$

where ζ_r, ω_r, χ_r and F_r are the damping ratio, angular resonance frequency, modal electromechanical coupling term and modal mechanical forcing term, respectively, and R_l is the resistive load. The steady state solution of Eq. (7) is,

$$\eta_r(t) = \frac{F_r - \chi_r v}{\omega_r^2 - \omega^2 + j2\zeta_r\omega_r\omega}e^{j\omega t} \tag{9}$$

For a PVEH that is essentially a uniform beam, ϕ_r, χ_r, and F_r can be determined analytically. In general, these parameters can be determined by Finite Element Analysis (FEA). Under short-circuit condition ($v = 0$) and applying harmonic base excitation $\ddot{u}_i = U_i\omega^2 e^{j\omega t}$, the forcing function F_r can be determined from Eq. (9) in the frequency domain as,

$$F_r = (\omega_r^2 - \omega^2 + j2\zeta_r\omega_r\omega)\eta_r(j\omega_r) = 2\zeta_r\ddot{u}_o\sqrt{m_{mr}}|_{\omega=\omega_r} \tag{10}$$

where m_m is the generalized modal mass associated with the mode and derived from the FEA natural frequency extraction analysis. For short-circuit condition with base excitation, the complex current i_r can be derived from Eq. (8) as,

$$i_r = j\omega\chi_r\eta_r(j\omega_r) = \sum_{r=1}^{\infty} \frac{j\omega F_r\chi_r}{\omega_r^2 - \omega^2 + j2\zeta_r\omega_r\omega}\Bigg|_{\omega=\omega_r} \tag{11}$$

and the voltage across R_l can be represented by [12],

$$v(t) = \frac{\sum_{r=1}^{\infty} \frac{j\omega F_r\chi_r}{\omega_r^2-\omega^2+j2\zeta_r\omega_r\omega}}{\sum_{r=1}^{\infty} \frac{j\omega\chi_r^2}{\omega_r^2-\omega^2+j2\zeta_r\omega_r\omega} + \frac{1}{R_l} + j\omega C_p} e^{j\omega t} \tag{12}$$

In the frequency domain, the complex power is given by

$$p(j\omega) = \frac{v(j\omega)v(j\omega)^*}{R_l} \tag{13}$$

3 Modal Approach for Performance Evaluation

3.1 A Modal Approach Based on Modified SDOF Model

The DPM model presented above can be used if the PVEH is essentially a uniform-section beam. For a PVEH with arbitrary cross-section or with side branches, the lumped parameter single degree-of-freedom (SDOF) model can be applied to each mode of vibration. For a base-excited SDOF model of a PVEH, the structural and electrical governing equations are:

$$\mathrm{m_m}\ddot{\mathrm{u}}_\mathrm{o} + \mathrm{c_m}\dot{u}_\mathrm{o} + \mathrm{k_m}\mathrm{u_o} - \alpha\mathrm{v} = \mathrm{F_m}; \tag{14}$$

$$\mathrm{C_p}\dot{v} + \frac{\mathrm{v}}{\mathrm{R_l}} - \alpha\dot{u}_\mathrm{o} = 0 \tag{15}$$

Erturk and Inman [14] derived a correction factor γ, which is the modal participation factor, for the SDOF model in order to obtain predictions that are comparable to the DPM model over a wide frequency band around the resonance. However, purely structural responses were predicted; the electrical outputs were not predicted. Guyomar et al. [15] used a SDOF model that included the electromechanical coupling to determine the normalized power for different resistive loads. They defined a modal electromechanical criterion k^2Q_m to evaluate the maximum average power output as follows,

$$P_{max} = \frac{F_m^2}{2c_m} \frac{\pi k^2 Q_m}{(\pi + k^2 Q_m)^2} \quad \text{when } k^2 Q_m \le \pi \tag{16}$$

$$P_{max} = \frac{F_m^2}{8c_m} \quad \text{when } k^2 Q_m \ge \pi \tag{17}$$

where k is modal electromechanical coupling coefficient (EMCC); $Q_m = 1/2\zeta$ is quality factor; F_m and c_m are lumped mechanical force and damping coefficient, respectively. However, Guyomar et al. did not consider the modal participation factor. The aim of this paper is to use a modal approach, based on the lumped SDOF model, to evaluate the performance at resonance and the maximum power output of two-mass two-layer PVEH configurations. The approach includes both the corrected structural response and the electromechanical coupling. A new modal structural criterion termed mass ratio N, which depends on the modal participation factor γ, is used. The mass ratio represents the influence of the modal mechanical/structural characteristics of the PVEH on the power density directly. The corrected lumped forcing function, the EMCC and lumped parameters are,

$$F_{mr} = -\gamma_r m_m \ddot{y}; \quad k^2 = \frac{\alpha_r^2}{k_m C_p + \alpha_r^2} = \frac{\omega_{oc}^2 - \omega_{sc}^2}{\omega_{oc}^2} \tag{18}$$

$$c_{mr} = 2\zeta_r \sqrt{k_{mr} m_{mr}}; \quad \omega_r = \sqrt{k_{mr}/m_{mr}} \tag{19}$$

and the corrected complex voltage developed on the resistor is,

$$v(j\omega) = \frac{\gamma_r \alpha_r R_l \dot{u}_i \omega^2 m_{mr}}{j\omega R_l \alpha_r^2 + (j\omega c_{mr} - \omega^2 m_{mr} + k_{mr})(1 + j\omega C_p R_l)} \tag{20}$$

where $\alpha_r = \chi_r \sqrt{m_{mr}}$ is lumped coupling factor; m_m is generalized modal mass; k_m is effective stiffness and $\omega_{sc} = \omega_r$ is the short-circuit undamped angular resonance frequency for the rth mode; ω_{oc} is the open-circuit angular resonance frequency. Substituting Eqs. (18) and (19) into Eq. (17) gives the maximum power as,

$$P_{max} = \frac{F_{mr}^2}{8c_{mr}} = \frac{\gamma_r^2 m_{mr}}{16\xi_r \omega_r} \ddot{u}_i^2 = \frac{m_{er}}{16\xi_r \omega_r} \ddot{u}_i^2 \tag{21}$$

where m_e is the effective mass which represents the mass participating in the forcing function and motion, and the sum of the effective masses for all modes is the total mass M of the whole model. A modal structural criterion, termed mass ratio, is defined as,

$$N_r = \frac{m_{er}}{M} = \frac{\gamma_r^2 m_{mr}}{M} \quad \text{where } m_{er} = \gamma_r^2 m_{mr}; M = \sum_{r=1}^{\infty} m_{er} \tag{22}$$

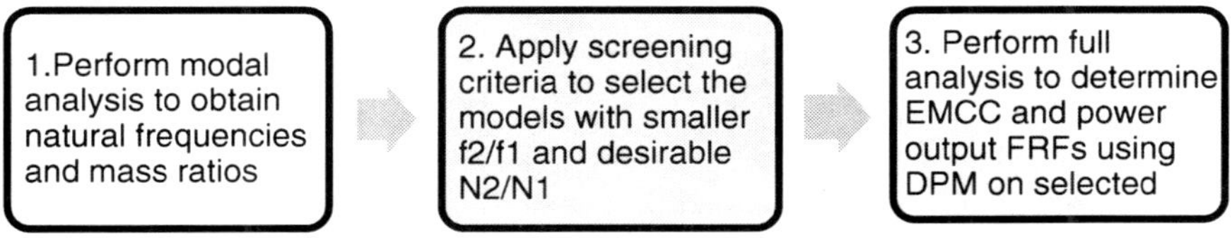

Fig. 2 Three key stages of the modal procedure

N_r denotes the mass ratio for the rth mode. The mass ratio indicates how much mass of each mode effectively participates in the overall motion and the contribution of each mode to the maximum power output. Because a PVEH with a bigger total mass is able to generate more power, it is necessary to normalise the output by the input and to use power density to evaluate the performance. Here, the mean squared acceleration weighted power density (MSAPD), which is the average power per 1 m/s^2 base excitation per cubic centimetre, is used. From Eq. (21), the maximum MSAPD can be written as:

$$\mathrm{MSAPD_{max}} = \frac{N_r \rho}{16 \xi_r \omega_r} * 10^{-6} \left(\mathrm{Ws^4/cm^3\,m^2}\right) \tag{23}$$

where ρ is the mass density. For a specified resonance frequency and damping ratio, the modal mechanical performance of the PVEH model in terms of the maximum power density can be evaluated using mass ratio directly.

3.2 Modal Procedure

A modal approach has been developed to evaluate the modal structural and electromechanical performance. The modal approach avoids the need to carry out full analysis using DPM at the first stage to compute the power output of a PVEH to assess its performance. Instead, it uses mass ratio and EMCC as two assessment criteria. The three key stages of the modal procedure developed for the evaluation of PVEH performance are shown in Fig. 2. Case studies using the modal approach and DPM to determine the performance of VEHs are given in the next section.

4 Optimising Two-Layer PVEHs by Modal Approach

4.1 Naming Convention for Two-Layer PVEH Configurations

A typical configuration of a two-layer (2L) two-mass (2M) PVEH is shown in Fig. 3a while Fig. 3b shows the general convention for naming the PVEH

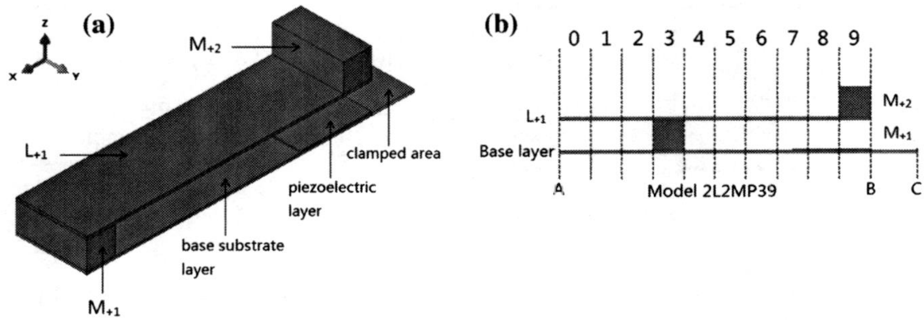

Fig. 3 Two-layer PVEHs; **a** two layers with two masses (*2L2MP09*); **b** numbering of mass positions; model *2L2MP39* has 2 layers and 2 masses, locations of masses are 3 and 9; A = tip, BC = clamped length of base layer

configurations. The beams are designated as L_0 (base layer) and L_{+1} (upper layer), while the masses are designated as M_{+1} (between L_0 and L_{+1}) and M_{+2} (above L_{+1}). The locations of the masses are variable from the free tip to the clamped end and are identified as P*mn* where *m* and *n* denote the positions of masses M_{+1} and M_{+2} given by $m \in [0,6]$ and $n \in [0,9]$. Variable *m* is restricted to the range of positions from 0 to 6 because the piezoelectric layer occupies positions 7 to 9. Thus, for example, Fig. 3a shows model 2L2MP09, which has two layers (2L) and two masses (2M) and the masses are located at positions 0 and 9 (P09).

4.2 Modal Screening for Optimal Two-Layer Two-Mass PVEHs

Many two-layer, two-mass (2L2M) PVEH models can be formed. In addition to the mass locations *m* and *n*, the thicknesses of the layers (t_{L0} and t_{L+1}) and the thicknesses of the masses (t_{M+1} and t_{M+2}) can be varied. Supposing that 5 thickness variations are considered in addition to the 7 × 9 mass locations, then a total of 43,750 2L2M PVEH models can be formed! But the performance of many of these models will be poor. Conventionally, this can only be ascertained and the strong models screened out by carrying out a full DPM analysis on all the 43, 750 models!! The modal approach considerably simplifies this process and makes it manageable as will be illustrated. The case study is based on models with different mass positions; the thicknesses of the layers and masses are assumed constant. Since the combinations of the mass positions are too many, the structural modal approach is used initially to screen the PVEHs with close resonance frequencies and the preferred values of mass ratios. The first step was the computation of the first two resonance frequencies of 2L2M models with M_{+1} and M_{+2} located at varying positions using FEA modal analysis in the ABAQUS FE code.

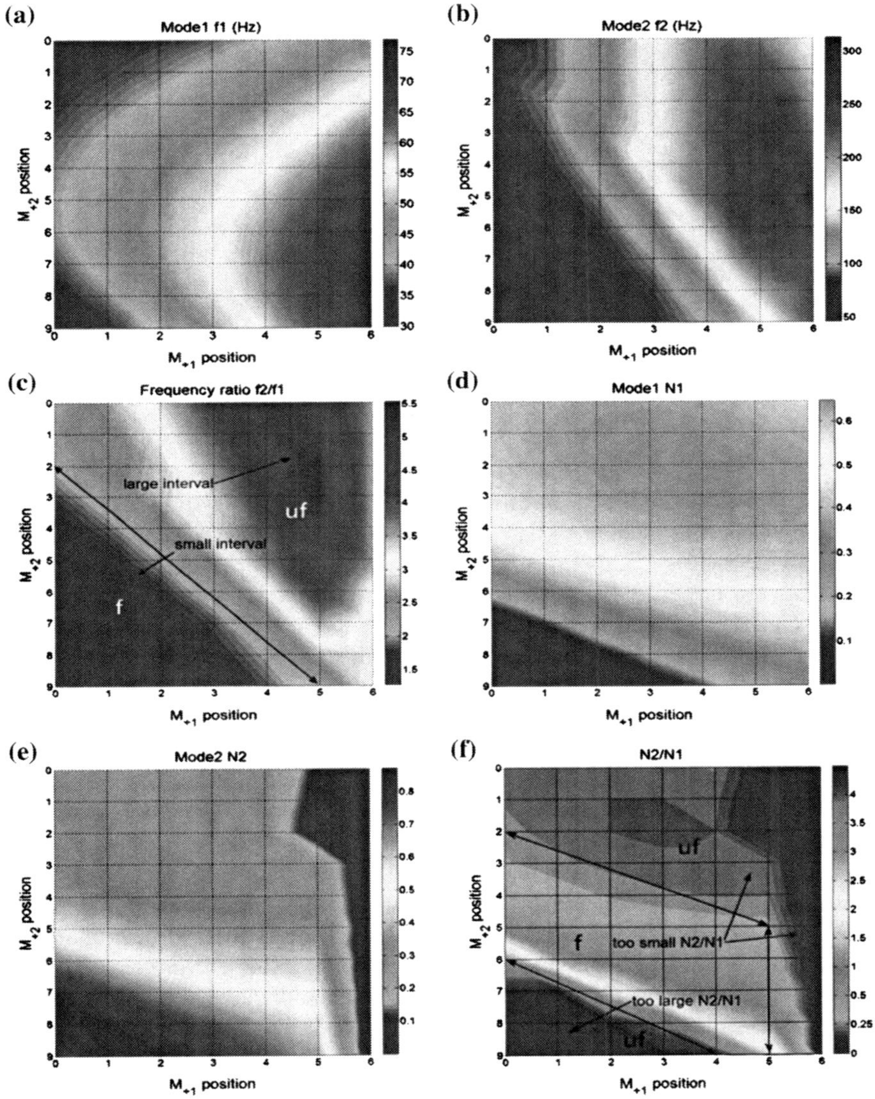

Fig. 4 Contour plots of first two resonance frequencies and mass ratios of the two-layer two-mass PVEH for different mass positions M_{+1} and M_{+2}. **a** f1; **b** f2; **c** f2/f1; **d** N1; **e** N2; **f** N2/N1; *f* favourable region, *uf* unfavourable region

The variations of the first two computed resonance frequencies with the mass positions M_{+1} and M_{+2} are shown by the contour plots of Fig. 4a and b while Fig. 4c shows the ratio of the resonance frequencies (f2/f1). The contour plots of Fig. 4d and e show the corresponding first two mass ratios, while Fig. 4f shows the

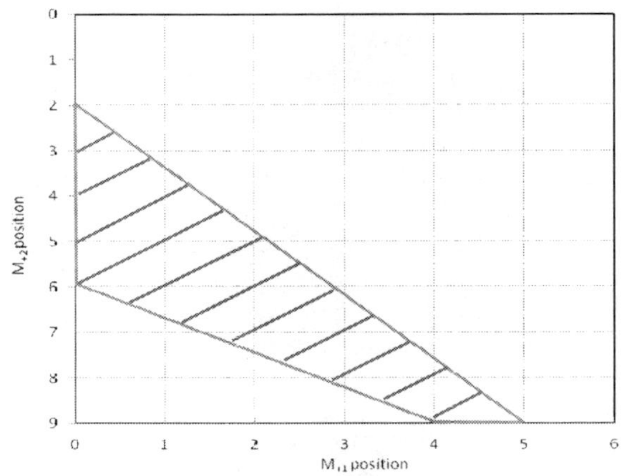

Fig. 5 Overlapped region

contour plot of the ratio of the mass ratios (N2/N1). In Fig. 4c and f, a small value of f2/f1 means that the interval between the two resonance frequencies is small and a favourable value of N2/N1 means that the mass ratio of each mode is neither too small nor too large. In this paper, an optimal, near-optimal or good design of the 2L2M PVEH is defined as that for which f2/f1 $\leq$ 2 and 0.25 $\leq$ N2/N1 $\leq$ 4. The 2L2M PVEH models that satisfy this criteria lie within the favourable (f) regions of the contour plots in Fig. 4c and f.

Combining Fig. 4c and f gives the favourable region shown in Fig. 5. The PVEH configurations in this region have small f2/f1 and favourable N2/N1, which produce superior mechanical performance. Most of the PVEH configurations outside the shaded region have poor modal mechanical performance. Therefore, instead of subjecting all the models to full electromechanical analyses, using a modal approach to screen the PVEHs initially is much more efficient.

To illustrate, some 2L2MPmn PVEHs that lie inside and outside the shaded region are subjected to full electromechanical analyses in order to determine their modal performance and MSAPD FRFs. The models inside the region have mass positions P03 to P06 while the models outside the region have mass positions P07 to P09. The results are presented in Fig. 6 and Table 1, which show that the position of M_{+1} is 0. Table 1 shows that when the position of M_{+2} is varied from 3 to 6, the value of f2/f1 lies within the range 1.5 $\pm$ 0.3 and the mass ratio is more evenly distributed. But when the position of M_{+2} is varied from 6 to 9, the values of N and k for the first mode decrease significantly. Then, the first mode becomes barely activated as can be seen in Fig. 6. The figure also shows that up to P06, the models have two clear resonances but from P07 to P09, the first resonance collapses and becomes subsumed in the second resonance. Therefore, the PVEH configurations with the optimal performance are those identified by the modal results shown in Table 1 with mass positions P03 to P06. Further results and discussions on the use of the modal approach for the development of PVEH can be found in [16–18].

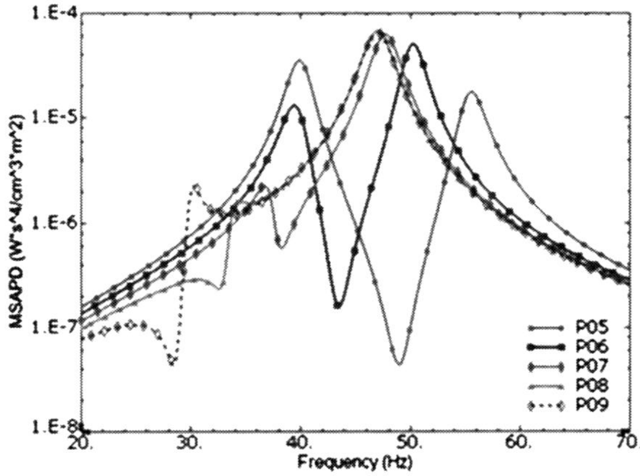

Fig. 6 MSAPD FRFs of models in and out of overlapped region

Table 1 Modal performance of model *2L2M* for various positions of M_{+2}

Models (2L2M)	Mode1			Mode2			f2/f1	N2/N1
	f1 (Hz)	N1 (%)	k1	f2 (Hz)	N2 (%)	k2		
P03	37.38	55.8	0.227	67.92	31.2	0.100	1.82	0.56
P04	38.79	50.3	0.226	61.58	36.7	0.100	1.59	0.73
P05	39.62	39.2	0.219	55.25	47.7	0.113	1.39	1.22
P06	39.35	18.9	0.220	50.09	67.8	0.103	1.27	3.59
P07	37.16	1.6	0.129	47.32	84.9	0.198	1.27	53.06
P08	33.60	0.9	0.092	46.54	85.4	0.219	1.39	90.05
P09	29.93	4.5	0.084	46.50	81.6	0.226	1.55	17.94

5 Conclusions

A modal screening procedure has been presented. The procedure uses the structural modal approach to filter the mass positions with favourable values of frequency ratio f2/f1 and ratio of mass ratios N2/N1. An optimal or near optimal mass position has a small interval between the two resonance frequencies and the mass ratio of each mode is neither too small nor too large. Since many configurations have poor modal mechanical performance, using a modal approach to screen the PVEHs obviates the need to carry out full electromechanical analyses on all possible configurations. Instead, only the screened configurations that have favourable mass positions are subjected to full electromechanical analyses.

References

1. Anton SR, Sodano HA (2007) A review of power harvesting using piezoelectric materials (2003-2006). Smart Mater Struct 16:1–21
2. Tang L, Yang Y, Soh CK (2010) Toward broadband vibration-based energy harvesting. J Intell Mater Syst Struct 21:1867–1897
3. Roundy S, Zhang Y (2005) Toward self-tuning adaptive vibration based micro-generators. Smart Struct Devices Syst 2:5649 (SPIE, pp 373–384)
4. Erturk A, Hoffmann J, Inman DJ (2009) A piezomagnetoelastic structure for broadband vibration energy harvesting. Appl Phys Lett 94:254102
5. Ferrari M, Ferrari V, Guizzetti M, Marioli D, Taroni A (2008) Piezoelectric multifrequency energy converter for power harvesting in autonomous microsystems. Sensors Actuators A 142:329–335
6. Qi SF, Shuttleworth R, Oyadiji SO, Wright J (2010) Design of a multiresonant beam for broadband piezoelectric energy harvesting. Smart Mater Struct 19:094009
7. Ou Q, Chen X, Gutschmidt S, Wood A, Leigh N (2012) An experimentally validated double-mass piezoelectric cantilever model for broadband vibration-based energy harvesting. J Intell Mater Syst Struct 23:117–125
8. Zhou W, Penamalli GR, Zuo L (2012) An efficient vibration energy harvester with a multi-mode dynamic magnifier. Smart Mater Struct 21:015014
9. Chen ZS, Yang YM, Lu ZM, Luo YT (2013) Broadband characteristics of vibration energy harvesting using one-dimensional phononic piezoelectric cantilever beams. Phys B 410 (2013):5–12
10. Erturk A, Hoffmann J, Inman DJ (2009) Modeling of piezoelectric energy harvesting from an L-shaped beam mass structure with an application to UAVs. J Intell Mater Syst Struct 20:529–544
11. Karami MA, Inman DJ (2011) Analytical modeling and experimental verification of the vibrations of the zigzag microstructure for energy harvesting. J Vib Acoust 133:011002-1
12. Erturk A, Inman DJ (2008) A distributed parameter electromechanical model for cantilevered piezoelectric energy harvesters. J Vib Acoust 130:041002
13. Oyadiji SO, Qi S, Shuttleworth R (2009) Development of multiple cantilevered piezo fibre composite beams vibration energy harvester for wireless sensors. In: Proceedings of WCEAM2009, Athens, Greece
14. Erturk A, Inman DJ (2008) On mechanical modeling of cantilevered piezoelectric vibration energy harvesters. J Intell Mater Syst Struct 19:1311
15. Guyomar D, Sebald G, Pruvost S, Lallart M, Khodayari A, Richard C (2009) Energy harvesting from ambient vibrations and heat. Intell Mater Syst Struct 20:609–624
16. Xiong X, Oyadiji SO (2013) Modal electromechanical optimization of cantilevered piezoelectric vibration energy harvesters by geometric variation. J Intell Mater Syst Struct 25(10):1177–1195
17. Xiong X, Oyadiji SO (2014) Optimal design of two-layer vibration energy harvesters using a modal approach. Smart Mater Struct 23(3):035005
18. Xiong X, Oyadiji SO (2014) A general modal approach for the development of optimal multi-layer stacked vibration energy harvesters. J Sound Vib 333(21):5386–5411

Synthesis of Vibration Gear Continuously Variable Transfer

Konstantin S. Ivanov, Roza K. Koilibaeva, Gachip Ualiev and Baurjan Tultaev

Abstract Drives of machines which are used now do not possess ability to be adapted for extreme working conditions. Such conditions are connected with possible difficulties of motion because of deviations from operation norms (for example, because of long inactivity, deterioration of conditions of lubricant, minor damages, a temperature difference, etc.). In the conditions of impossibility of elimination of minor failures on the move (for example in aeronautical engineering) insignificant discrepancy of operation of service mechanism can serve as a cause of accident. Recently the technological direction of use of adaptive drive of machines is advanced. The adaptive drive mechanism engages the engine and the self-regulated transmission mechanism. The adaptive gear transmission mechanism with two degrees of freedom has ability to actuate an executive working body with a speed back—proportional external load at constant engine power. It means that even in presence of handicap in motion of working body the drive mechanism breakage will not occur. The adaptive vibrating mechanism in the form of the closed gear differential is developed. The vibrating mechanism contains input carrier, output carrier and closed contour located between carriers. The closed contour contains input satellite, external block with two ring (epicycle) wheels, output satellite and internal block with two solar wheels. The wheels in each block are connected by elastic shaft. In an operating time elastic shafts transfer vibrating oscillations to the output carrier. Output carrier transfers vibrating action to working body. The vibrating mechanism provides reliable overcoming of operational

K.S. Ivanov (✉) · R.K. Koilibaeva · G. Ualiev · B. Tultaev
Institute of Mechanics and Machine Science MON RK, Almaty University of Power Engineering and Telecommunications AUPET, Baitursinov Street, 126, Almaty 050013, Kazakhstan
e-mail: Ivanovgreek@mail.ru

R.K. Koilibaeva
e-mail: r_k_k@mail.ru

G. Ualiev
e-mail: ualiev@mail.ru

B. Tultaev
e-mail: b.tultaev@mail.ru

J.K. Sinha (ed.), *Vibration Engineering and Technology of Machinery*,
Mechanisms and Machine Science 23, DOI 10.1007/978-3-319-09918-7_25

overloads. In offered work synthesis of the vibrating mechanism on the set vibrating action is executed. Work is executed on the basis of mechanics laws.

Keywords Adaptive mechanism · Vibrations · Synthesis

1 Introduction

Drives of machines which are using now do not possess ability to be accommodated for extreme working conditions. Such conditions are connected with possible difficulties of motion because of deviations from norms of maintenance (for example, because of long inactivity, a decline of conditions of greasing, insignificant faults, difference of temperatures, etc.). In the conditions of impossibility of elimination of small derangements on the move (for example, in an aeronautics) insignificant inaccuracy of maintenance of the service mechanism can serve as a cause of accident. The main deficiency of existing drives is a "rigid" constraint of the engine with the executive tool through the connecting gear with one degree of freedom. In this case the insignificant handicap on the move the tool or its wedging can call breakage of the drive.

Recently the scientific and technical direction of use of adaptive drives of machines develops [1–3]. The adaptive drive contains the engine and the self-controlled connecting gear. The adaptive gear connecting mechanism with two degree of freedom has ability to set in motion the executive tool with speed which is inverse—proportional to external loading at constant power of engine. It means that drive breakage will not occur even in the presence of a handicap on the tool motion. The engine can transfer motion to the connecting gear even at tool dead stop. Conditions of start-up of such mechanism admit substantial growth of a starting force at the expense of small starting speed of tool at constant power of engine. It is offered to use conditions of vibrating affecting on the tool to provide the beginning of motion of the tool in the presence of a handicap. For this purpose it is offered to include elastic links into the drive mechanism. In the present work synthesis of vibrating transfer on the set maximum vibrating effort is presented. Work is executed on the basis of mechanics laws.

2 Description of the Gear Adaptive Vibrating Mechanism

Working out of an adaptive vibrating gear variator which provides a variable speed of motion of the output tool depending on loading on him and creates vibrating affecting on the tool is observed. Vibrating affecting in a combination with power adaptation will provide reliable overcoming of starting resistance and overloading in operating conditions. Adaptive vibrating drive is shown on Fig. 1.

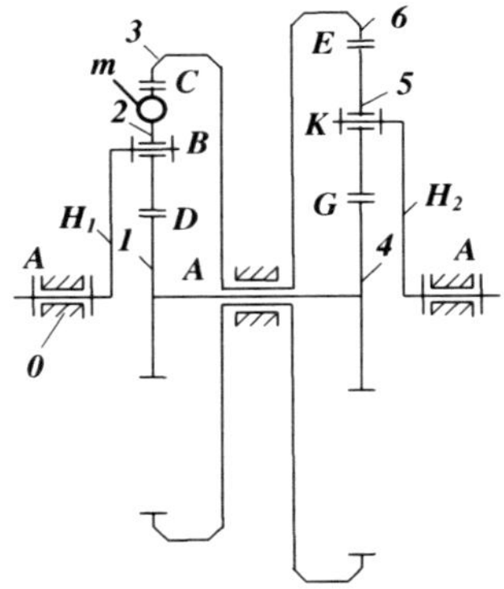

Fig. 1 Vibration gear continuously variable transfer

The adaptive vibrating drive represents the closed gear differential mechanism with two degrees of freedom or the continuously variable transmission. The mechanism contains frame 0, input carrier H_1, input satellite 2 with unbalanced masse m, block of central toothed wheels with external teeth (solar wheels) 1–4, block of central toothed wheels with internal teeth (ring wheels) 3–6, output satellite 5 and output carrier H_2. Toothed wheels 4–1, 2, 3–6, 5 are forming the closed contour. Wheels 1 and 4 are connected by elastic shaft. Wheels 3 and 6 are connected by elastic shaft also.

The external active moment M_{H_1} is acting on the input carrier H_1. External moment of resistance M_{H_2} is acting on the output carrier H_2.

The output carrier H_2 is moving quite definitely with a speed inversely proportional to output moment of resistance thanks to presence of two degrees of freedom and differential constraint in the closed contour [1, 2]:

$$\omega_{H_2} = M_{H_1}\omega_{H_1}/M_{H_2} \tag{1}$$

Formula (1) expresses effect of force adaptation of output link to variable load.

When the shafts connecting wheels 1–4 and 3–6 are rigid angular speeds of wheels are defined under formulas [1–3]:

$$\omega_3 = \frac{\omega_{H_2}(1 - u_{46}^{H_2}) - \omega_{H_1}(1 - u_{13}^{H_1})}{u_{13}^{H_1} - u_{46}^{H_2}}, \tag{2}$$

$$\omega_1 = u_{46}^{H_2}(\omega_3 - \omega_{H_2}) + \omega_{H_2}. \tag{3}$$

Here transfer ratios are next: $u_{13}^{(H_1)} = -Z_3/Z_1$, $u_{46}^{(H_2)} = -Z_6/Z_4$.

Presence of unbalanced masse m and elastic shafts connecting wheels 1, 4 and 3, 6 will lead to occurrence of dynamic pulses at transfer of motion on the output carrier.

Input carrier H_1 with the unbalanced satellite 2 is shown on Fig. 2.

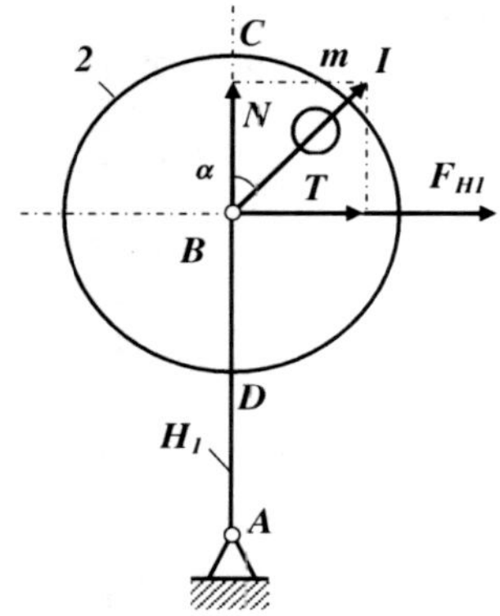

Fig. 2 Input carrier with unbalanced satellite

Carrier H_1 transfers input motive force F_{H_1} on satellite 2. Force F_{H_1} is transferred in points C and D to wheels 3 and 1 in the form of constantly operating forces $F_{H1}/2$. Unbalanced масса m creates on the satellite 2 centrifugal force of inertia

$$I = mr\omega_2^2, \tag{4}$$

where r—radius of an arrangement of the centre of masse m on satellite 2, ω_2—angular speed of satellite 2.

Let's transfer force I in the line of its action to point B and we will spread out it to components. We will designate α—angle of turn of radius r.

Component of inertia force $N = mr\omega_2^2 \cos\alpha$ acts on frame. Component of inertia force $T = mr\omega_2^2 \sin\alpha$ is transferred in points C and D to wheels 3 and 1 in the form of variable disturbing forces $T/2$.

Constant motive forces $F_{H_1}/2$ are transferred through blocks of wheels 1–4 and 3–6 on the output satellite 5 and further on the output carrier H_2. These forces are overcoming the resistance force R_{H_2} in point K. Variable disturbing forces $T/2$ are overcoming elastic resistance forces in blocks of wheels 1–4 and 3–6 (as the wheels in each block are connected by elastic shaft). Disturbing forces create oscillations on wheels 4 and 6 which are transferred to the output satellite 5 and further on the output carrier H_2 in the form of vibrations. These forces are creating the vibration action in point K which helps to overcome resistance force R_{H_2} at overloads.

At the greatest possible loading the output carrier will stop and the mechanism will continue movement with one degree of freedom. If thus on one of wheels of the closed contour to place a flywheel with mobile masses fixed by means of springs [4]. This flywheel will create a stock of potential energy of springs. Preservation of a stock of potential energy of a flywheel provides the constant input driving moment on input carrier. On the output satellite 5 the moment of resistance corresponding to the driving moment will take place. Dynamic impulses from wheels 1–4 and 3–6 will be transferred to the output satellite 5 and further on the motionless output carrier H_2. It will lead to resistance overcoming on motionless carrier both at emergency jamming and at start-up.

3 Dynamic Analysis of Motion of Mechanical System with Elastic Shafts

Let's consider the possibility to receive of definability of system with two degrees of freedom when elastic shafts connect wheels 1 and 4 and wheels 3 and 6.

Reactions $R_{23} = F_{H_1}/2 + T/2$, $R_{21} = F_{H_1}/2 + T/2$ are transferred on intermediate links 3 and 1 from input satellite 2. Reactions $R_{54} = R_{H_2}/2$, $R_{56} = R_{H_2}/2$ are transferred on intermediate links 4 and 6 from output satellite 5.

Here $F_{H_1} = \frac{M_{H_1}}{r_{H_1}}$, $R_{H_2} = \frac{M_{H_2}}{r_{H_2}}$,

M_{H_1}, M_{H_2}—moments on input and output carriers,

r_{H_1}, r_{H_2}—radiuses of input and output carriers,

r_i $(i = 1, 2\ldots6)$—radiuses of wheels.

Moments are transferred to wheels 1 and 3:

$$M_1 = R_{21} r_1, \quad M_3 = R_{23} r_3.$$

Moments are transferred to wheels 4 and 6:

$$M_4 = 0.5 M_{H_2} \frac{r_4}{r_{H_2}}, \quad M_6 = 0.5 M_{H_2} \frac{r_6}{r_{H_2}}$$

When elastic constraint between wheels 1–4 and 3–6 differential equations of motion of wheels 4 and 6 look like

$$J_4 \ddot{\varphi}_4 = M_1 - M_4 - c_4(\varphi_4 - \varphi_1), \tag{5}$$

$$J_6 \ddot{\varphi}_6 = M_3 - M_6 - c_6(\varphi_6 - \varphi_3). \tag{6}$$

Here J_4, J_6—moments of inertia of wheels 4 and 6, c_4, c_6—torsion rigidity of the shafts connecting wheels 1–4 and 3–6.

According to [5] on a method torsion rigidities at enough high power of the engine it is possible to consider angular speed ω_1 as a constant, then $\varphi_1 = \omega_1 t$.The angle φ_4 differs from φ_1 a little. Therefore it is convenient to take for generalized co-ordinate a difference $\varphi = \varphi_4 - \varphi_1$ instead φ_4. Then the differential Eq. (5) becomes:

$$J_4 \ddot{\varphi} + c_4 \varphi = M_1 - M_4. \tag{7}$$

The solution of Eq. (7) at zero entry conditions looks like:

$$\varphi = \frac{M_1 - M_4}{c_4} [1 - \cos(k_4 t)]. \tag{8}$$

Here $k_4 = \sqrt{\frac{c_4}{J_4}}$—oscillation frequency of wheel 4 concerning wheel 1 owing to elastics of the shaft connecting wheels 1–4.

Motion of a wheel 4 can be considered as the motion consisting of the basic motion with constant angular speed ω_1 and additional motion with speed $\dot{\varphi}$ having oscillatory character:

$$\dot{\varphi} = A_{\omega 4} \sin (k_4 t). \tag{9}$$

Here amplitude of angular speed $A_{\omega_4} = \frac{M_1 - M_4}{c_4} k_4 = \frac{M_1 - M_4}{\sqrt{c_4 J_4}}$.
Thus angular speed of wheel 4 in the presence of elastic constraint is

$$\omega_4 = \omega_1 + \frac{M_1 - M_4}{\sqrt{c_4 J_4}} \sin(k_4 t). \tag{10}$$

Angular speed of a wheel 6 is defined analogously:

$$\omega_6 = \omega_3 + \frac{M_3 - M_6}{\sqrt{c_6 J_6}} \sin(k_6 t). \tag{11}$$

Here $k_6 = \sqrt{\frac{c_6}{J_6}}$—oscillation frequency of wheel 6 concerning wheel 3 owing to elastics of the shaft connecting wheels 3 and 6.

Thus for transfer of oscillations on the satellite 5 and the output carrier H_2 without distortions (with frequency and amplitude of oscillation conservation) it is necessary to satisfy condition $k_4 = k_6$. From this condition which is condition for the selection of rigid and inertia parameters follows:

$$\frac{c_6}{c_4} = \frac{J_6}{J_4}. \tag{12}$$

The calculations have shown that at performance of Eq. (12) the equality of amplitudes of oscillation of wheels 4 and 6 takes place also.

Amplitudes and periods of oscillations are selected by the set of shafts torsion rigidities and moments of inertia.

On the found angular speeds of wheels 4 and 6 the angular speed of output carrier H_2 is defined:

$$\omega_{H_2} = \frac{\omega_4 - u_{46}^{H_2} \omega_6}{1 - u_{46}^{H_2}}. \tag{13}$$

As is shown in Fig. 3 the angular speed ω_{H_2} of the output carrier H_2 varies under the harmonious law about a mean $\omega_{H_{20}}$ defined by the formula (1).

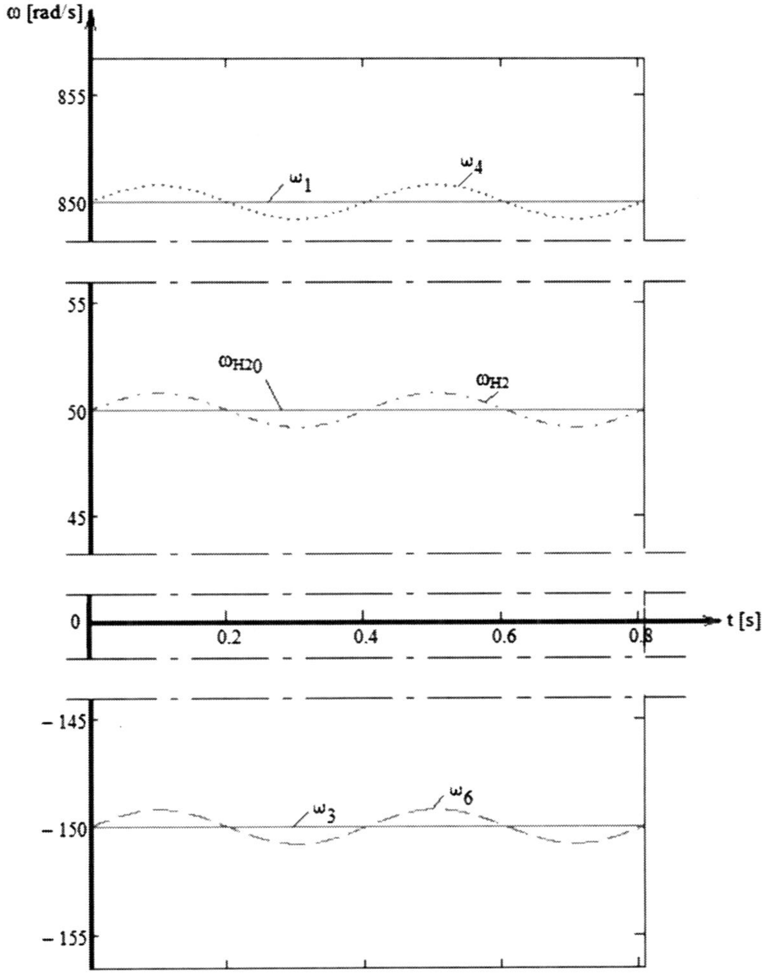

Fig. 3 Schedule of change of angular speeds

4 Synthesis of Vibration Adaptive Transfer

Synthesis consists in definition of parameters of the mechanism providing demanded vibrating influence, on following conditions:

(1) Overcoming of starting extreme resistance on the motionless output carrier, exceeding normal resistance in set number of times (10) in a starting mode of movement,

(2) Overcoming of extreme resistance on the moving output carrier exceeding normal resistance in set number of times (10) in an operational mode of movement.

The decision of a problem of synthesis of vibrating adaptive transfer is reduced to definition of parameters of a flywheel with mobile masses fixed by means of springs on a condition of overcoming of extreme resistance on the basis of equality potential and kinetic energy a flywheel with mobile masses fixed by means of springs.

5 Conclusion

The oscillation frequency in an elastic contour is high and has vibrating character. Vibrating affecting in combination with effect of power adaptation predetermines high reliability start and overcoming of emergency overloading.

The executed scientific researches allow creating simple and reliable adaptive vibrating drive of the service mechanism for the techniques working in extreme conditions (in aircraft, in space, etc.). The decision of a problem of synthesis of vibrating adaptive transfer is reduced to definition of parameters of a flywheel with mobile masses fixed by means of springs on a condition of overcoming of extreme resistance on the basis of equality potential and kinetic energy a flywheel with mobile masses fixed by means of springs.

References

1. Ivanov KS, Tultaev B (2012) Toothed continuously variable transmission (CVT)—industrial realization. New Trends Mech Mach Sci Theor Appl Eng 7:329–335. Springer, ISSN 2211-0984. In: Transactions of 4-th Euro conference on mechanisms. Springer, Santander, Spain
2. Ivanov KS, Koilybayeva RK, Ualiev GU (2013) Creation of vibration gear continuously variable transmission (CVT). In: 11th international conference on vibration problems (ICOVP 2013). Book of Abstracts, Lisbon, Portugal
3. Ivanov KS (2012) Paradox of mechanics—a basis of creation CVT. In: Transactions of 2-d IFToMM Asian conference on mechanisms and machines science, Tokyo, Japan, pp 245–264
4. Ivanov KS (1973) Flywheel with variable distribution of masses. The copyright certificate. The patent of the USSR. №365503, 8th Jan 1973
5. Levitsky NI (1979) Theory of mechanisms and machines. M: Science, Main edition of physical and mathematical literature, p 576

Crack Propagation in a 500 MW LP Turbine in Service

Wolfgang Hahn and David Futter

Abstract A 500 MW turbine generator set that operated through a rigorous two shift regime had developed a crack whilst in service. The shaft line was regularly subjected to multiple starts with oil whirl manifesting on load during certain operational conditions. First observations on generator no. 11 bearing showed an increase in the 1× component over a period of time, with changes through the critical speed during run-down triggering close monitoring. Further observations over time revealed that the Low Pressure turbine three, 1× and 2× components increased during steady state in conjunction with changes in phase angles and critical speed during transient conditions. Early indications showed that the likelihood of a crack in the main turbine shaft was high. Further observations over time revealed that the 1× and 2× steady state trends continued to increase up to the ISO limits set for the turbine generator set. The unit was shut down and qualitative inspections undertaken. The findings revealed that a crack had developed on the main rotor body in the pencil shaft.

Keywords Cracked rotor · Vibration · Low pressure turbine · Oil whirl

1 Introduction

In power plants, the low pressure (LP) steam turbines operating under two shift regimes are subjected to internal and externally influenced conditions. These turbines are often started up under high wetness fraction and limited warming

W. Hahn (✉)
EDF Energy, West Burton Power Station, Retford DN22 9BL, Nottinghamshire, UK
e-mail: wolfgang.hahn@edfenergy.com

D. Futter
E.ON New Build and Technology Limited Technology Centre, Ratcliffe on Soar, Nottingham NG11 0EE, UK
e-mail: David.Futter@eon.com

J.K. Sinha (ed.), *Vibration Engineering and Technology of Machinery*, Mechanisms and Machine Science 23, DOI 10.1007/978-3-319-09918-7_26

conditions due to the commercial competitive demands placed on the power industry. Two shifting can vary between hot, warm and cold starts and reach as many as 250 starts in total per year. Whilst under the rigorous start up regime, a combination of operational base load conditions could induce further damage to turbine LP rotors. The turbine under consideration covered more than 200,000 h of operation and in excess of 3,000 starts. The degradation mechanisms can be in the form of thermal influences, mechanical induced vibration from excitation, crack initiation and high cycle fatigue with rapid crack propagation.

Areas for consideration in LP turbines where cracks may potentially emanate and propagate under vibration and fatigue are likely to be pencil shaft collar (step changes in the stages for shrunk on discs), last stage discs, button drives and last stage blades. The Low Pressure turbine under consideration developed a crack in the pencil shaft, in the location between the collar and first stage disc stress relieving groove, the crack being of significant depth and circumferential coverage along the geometry.

2 Mechanisms Related to Turbine Rotor Vibration and Fatigue

The dynamic stiffness of a rotating mass is crucial to determining the system vibration behavior and hence the total support system stiffness can be derived from the stiffness curves using the critical speeds [1]. The theoretical critical speeds are not evident from the calculation during operation due to damping and these together with the operation deflection shapes were captured on load during transient conditions [2]. The results from the independent measurements were also compared with the on-line vibration monitoring system.

Sub synchronous vibrations are also known to occur in unstable systems at those frequencies that coincide with a natural frequency of the rotating mass. This occurs at slightly less than one half the running speed frequency due to oil film instability causing whirl to occur with high amplitudes at sub synchronous level [1]. Thermal sensitivity in turbines with surface developed cracks have the characteristic to close the crack during run up transients with surface heating and open the crack during cooling [3]. The heating transient delivers a smaller change in the 2× component vector than the cooling transient and heating transient delivers a larger change in the 1× component vector than the cooling transient in close proximity of the crack location.

Harmonic components in relation to the 1× and 2× components with the phase angles can behave differently in trends during steady state and transient modes of operation respectively. If phase angle of the change in response due to the crack is 180° (i.e. opposite) to the residual unbalance vibration response components, then 1× amplitude value may initially reduce. Similarly if the change in 2× response due to the crack is 180° to any pre-existing two-rev component due to inhomogeneity then the 2× amplitude value may initially reduce [2]. In time the propagation of the crack will lead to both these amplitudes increasing as the change exceeds double the amplitude of the original residual vectors.

High cycle fatigue in steam turbines is a result of and the response to forced excitation reaching sufficient amplitude to exceed the stress intensity threshold that eventually lead to failure after a given number of cycles. The strength and hardness of steam turbine rotor materials are limited by the susceptibility of stress corrosion cracking and corrosion fatigue when operating in warm wet environments. Crack propagation increases exponentially with high yield strength values [4].

3 Vibration Response of a Cracked in Service 500 MW LP Rotor

The shaft line arrangement on a 500 MW turbo generator set consists of one High Pressure turbine, one Intermediate Pressure turbine and three Low Pressure turbines with generator coupled along one shaft line as per Fig. 1. The Low Pressure turbine 3 was suspected of having a crack.

The effect of a shaft crack on the dynamics of a machine has been widely documented e.g. [5]; however in practice the data is rarely as clear-cut as the theory would suggest. The theory is based upon the asymmetry of stiffness introduced by a radial crack developing asymmetrically through the shaft. It will be seen later that the detected crack fits this pattern exactly.

Consider the vibration response at the bearing due to a perfectly symmetrical rotor with a small residual unbalance. The force generated by such a system can be expressed mathematically as:

$$F_t = M\ddot{x} + Kx + C\dot{x} \tag{1}$$

where,

F_t time dependant exciting force
M mass of system
K stiffness of system
C damping of system
x displacement of system
$\dot{x}$ velocity

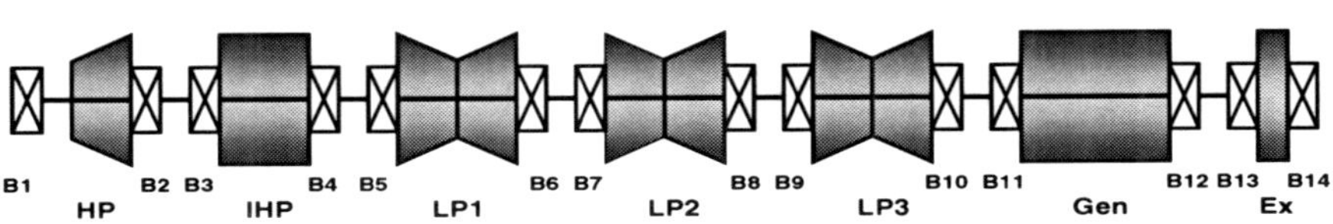

Fig. 1 Shaft line arrangement of the 500 MW turbine generator set

$\ddot{x}$ acceleration

For the ideal rotating system described above this will take the general form

$$F_t = F_0 \sin(2\pi f t) \tag{2}$$

where
F_0 amplitude of the force
f frequency of the rotation

It can be shown that the solution of the equation of motion when F_t is of the above form can be written as

$$x_t = A sin(2\pi f t + \emptyset) \tag{3}$$

where
A Amplitude of displacement of vibration
F Frequency of vibration
ϕ Phase of the vibration with respect to the force

Representing a vibration signal containing only one frequency component.

Next consider the stiffness component of the force when the stiffness changes depending upon the radial angle considered. This can be conceptualised simply with a standard plastic ruler: it is very easy to bend the ruler in its flat axis; yet almost impossible to bend it in the orthogonal plane, as illustrated in Fig. 2.

On large generator rotors, including the one in this machine train, due to the two-pole configuration and resulting configuration of winding slots, this phenomenon is already experienced. Careful engineering design changes in the form of compensating slots are often introduced to try to reduce the stiffness asymmetry, which

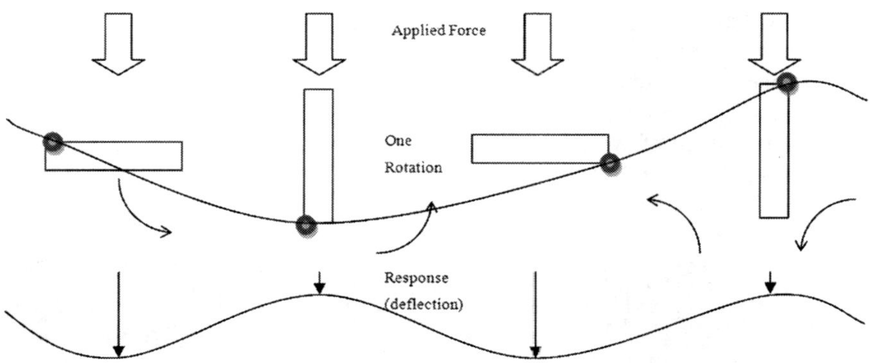

Fig. 2 Conceptualisation of the generation of a 2/revolution component from a stiffness asymmetry

generates a two per rev variation in stiffness, and therefore response. It remains normal for generator rotors to exhibit some 2/rev vibration component at operating speed.

Theoretically, it is clear that a radially asymmetric loss of stiffness in will generate a 2/rev response. In addition, since the overall stiffness of the system will reduce, the 1/rev response is also likely to change. The relative amounts of change between different components is governed by a host of considerations, such as the original stiffness of the system (rotor and supporting structure), the geometry of the crack and the mode shape of the shaft-line at operating speed. There have also been cases where thermal effects lead to a compressive force in the shaft such that the crack closes in certain situations. In addition these changes are superposed with the residual unbalance of the rotor; so may result in a reduction in vibration amplitude, at least initially.

In the case of this rotor train, the first indication of a problem was seen in the run-down traces in mid May 2012, Fig. 3. Although bearing 11 (generator front) showed the most apparent change, further investigation showed that the first significant indications occurred following a burst of sub-synchronous vibration and a spike in bearing oil temperatures in late April, with a suggestion of a trend in on-line vibration data in the period running up to that. Due to the sensitivity of these machines to condenser conditions, which relate to ambient conditions, it was not possible to immediately attribute the on-line change to any definite loss of integrity.

The design of the main bearings on the Low Pressure (LP) turbines are of the cylindrical fluid type and have been known to cause sub synchronous vibrations during periods where oil temperatures were low, associated with oil whirl around 19 Hz as can be seen in Fig. 4.

The increase in vibration from oil whirl was an intermittent problem. The turbine bearings are of tilting pad type which by virtue of their design solve most of the stiffness and damping characteristics in relation to synchronous and sub synchronous problems; however this particular turbine type remains very sensitive to periods of oil whirl, generally started by changes in oil conditions or shifts in alignment as the machine warms through. This machine in particular had suffered

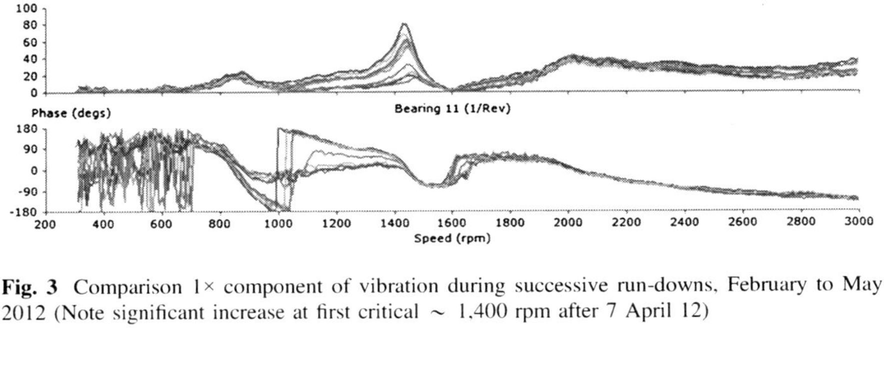

Fig. 3 Comparison 1× component of vibration during successive run-downs, February to May 2012 (Note significant increase at first critical ~ 1,400 rpm after 7 April 12)

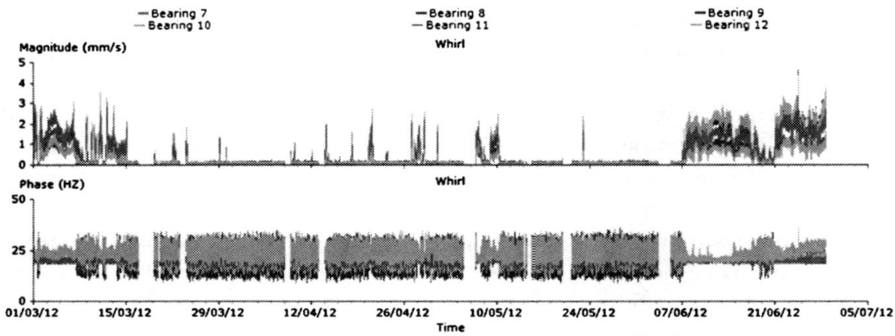

Fig. 4 Sub-synchronous vibration between bearings 7 and 12 due to oil whirl. Note how the whirl locks on to a frequency; the sub synchronous vibration when whirl is not present shows no favoured frequency

high levels of oil whirl on a number of occasions in the preceding months, which can lead to the potential for bearing failure through fatigue of the white metal. This behaviour was suppressed through changes to oil conditions in March, but returned in June as the failure developed, Fig. 4.

Whilst the problem of oil whirl was intermittently observed, the dynamic behaviour of LP rotor 3 showed signs of loss in stiffness and damping by virtue of the dynamic parameters observed from the 1× and 2× components, together with phase angles between bearings 9 and 10 respectively as per Fig. 5. The data showed slight changes in critical speed during the transient conditions. The rotor was initially monitored for a period of 1 month during the two shifting regime.

What is often overlooked when theory books discuss shaft cracking, is that the effect of losing the white metal from a bearing is to open the clearances, thus reducing the stiffness and the damping of the system. So bearing failure presents identical initial symptoms to a cracked shaft; however a bearing failure often gives other effects such as changes in oil system pressure, and a high temperature is experienced during the incident. It is often possible to continue running after a single bearing failure, especially where there are pairs of bearings between each rotor (as is the case here), as the neighbor simply takes up the load.

In this case there was no need to make any immediate intervention, so the rotor was maintained in service for a further period of two and a half months prior to removing the rotor from service, subject to close monitoring due to the known problem. During this time, a combination of whirl related problems and increased vibration levels were observed as the crack propagated in the main pencil shaft of the rotor. One curious observation is that the whirl frequency on load increased during the later stages of crack propagation, even though the critical speed during rundown decreased, as would be expected. It is postulated that the increasing frequency was driven by increased dynamic load experienced by the system as the overall vibration increased, leading to increased stiffness in the bearing oil film [6].

Fig. 5 LP 3 1× and 2× components during transients (19 May 12 compared to 18 March 12)

Figure 6 shows the typical change in the LP 3 rotor 2× component over 4 months, from prior to the point of detection to when the unit was removed from service. There was a small initial decrease, followed by a progressive increase. The 1× component showed a more sustained decrease in amplitude before increasing as the crack propagated and the associated phase changed accordingly. The first change in vibration was detected on the 1× component of the generator bearings during rundown, and in retrospect is discernible from early May.

The unit was shut down and re-started three times after the discovery of the vibration level changes of 1× components in bearing 11 of the generator. The operating unit was removed from service based on the limits placed on the turbine shaft line in accordance with ISO standards for Zone B to C and C to D respectively [7]. Another characteristic of a shaft crack is the increased sensitivity to thermal condition this gives. Figure 7 Illustrates (in polar format) the effect of increasing temperature whilst in service for two separate days, one before the crack was detected, and one shortly before the unit was finally shut down.

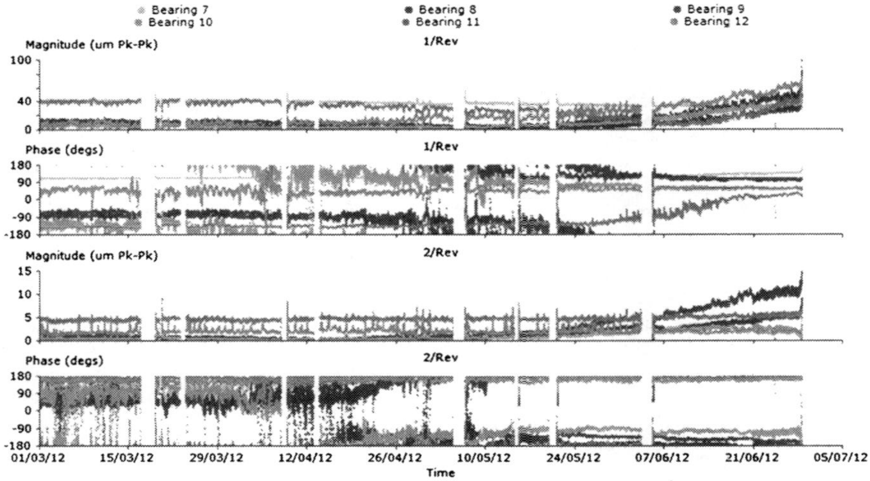

Fig. 6 Measured 1× and 2× components—at full speed of LP2, LP3 and generator bearings over 4 months

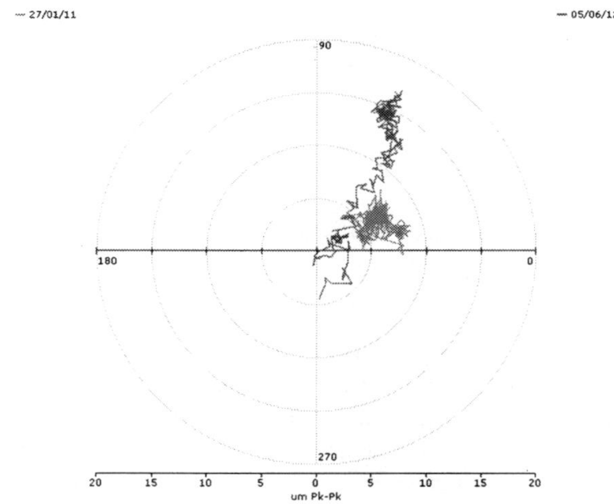

Fig. 7 Polar plot showing effect of load changes on the 1/rev vibration vector at bearing 10 (*red* from prior to crack detection; *blue* from 3 weeks before shutdown.)

4 Observations on LP3 Rotor After Removal from Service

Low pressure turbine 3 rotor was removed from service at the end of June 2012 after which it was inspected visually and a truth check performed. The maximum run out achieved on the rotor was 34 thousands of an inch during one full circumferential revolution. The maximum run out was obtained at locations V and W as per Fig. 8. This was the location in which the actual crack was located as well.

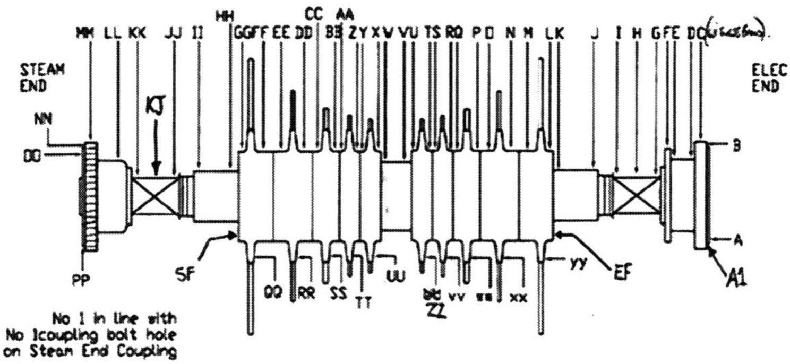

Fig. 8 Low pressure 3 rotor section—maximum run out at points V&W

The typical standard for run out during a truth check of large rotors is half a millimeter per foot length span [8]. In the case of the LP 3 rotor, the expected run out was around five thousands of a millimeter at points V and W. The readings reduced accordingly from a maximum of 34 thousands of an inch to five thousands of an inch closest to the bearing supports. A crack in the main pencil shaft or moved shrunk on discs was suspected at the time of the findings.

The extent of the bend in the rotor is much greater than could be tolerated by the machine at speed, however the presence of a crack in the vicinity of shrunk on disks leads to an exaggeration of the bend when the rotor is stationary. This is due to the Poisson effect. The shrink of the disk exerts a compressive force, which in normal circumstances will cause the rotor to elongate by an amount determined by the Poisson's ratio of the material. The crack in the rotor introduces a discontinuity in the material, thus interrupting the stress field and causing a significant bend to the shaft. The will straighten at speed, since the shrink will relax under centrifugal loading. (Large LP rotors shorten measurably between stationary and running speed due to the release of the shrink force, although the change is reversed by thermal expansion)

Further investigation into the rotor through non destructive examination methods revealed that the rotor had developed a crack to the magnitude of 200 mm deep and a circumferential span of approximately 270°. The crack appeared on the turbine end of the centre collar to stage 1 step change in the pencil shaft of the LP turbine rotor [9] (Fig. 9).

The statistics of the LP3 cracked rotor revealed that this spindle had covered more than 50,000 excess running hours (232,000 h) and more than double the total number of start ups (3,458 starts) to point of failure. Number of barring revolutions covered had not been accounted for. The turbine original design life was estimated at 150,000 operating hours and an equivalent of 1,500 starts. The turbine two shift regime may have influenced the fatigue life of the pencil shaft through the start up conditions, not withstanding the barring and lay-up conditions of the turbine during the through life period in operation.

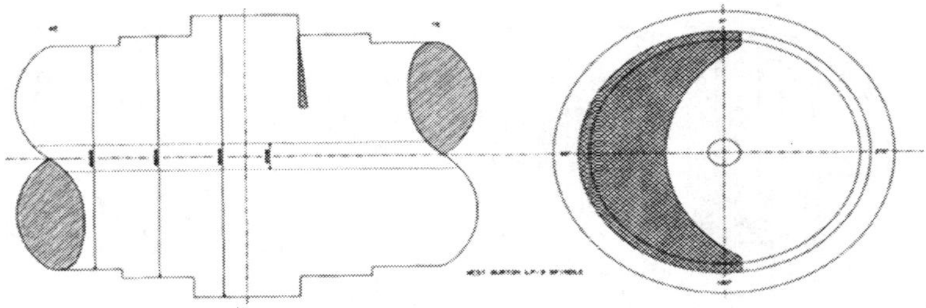

Fig. 9 Crack indication on the main shaft of LP3 turbine rotor (crack depicted by shaded area)

5 Conclusion

The Low Pressure turbine under consideration was subjected to a rigorous two shift regime whilst accumulating a significant number of running hours during the through life of the rotor. Vibration behaviour of the unit since the previous overhaul had been generally good, although intermittent periods of oil whirl were being experienced. The first indication of a progressive problem was detected through run-down analysis, and was clearest at the transition through the critical speed as measured at bearing 11.

Further observations indicated that the steady state 1× and 2× components of no.9 and 10 bearings respectively showed a progressive vector change in conjunction with continuing critical speed changes and phase changes during subsequent transient conditions. Upon shutdown and further inspections, a crack was revealed in the pencil shaft. The observations from the processed vibration information correspond well with historical descriptions of both oil whirl and the shaft crack propagation, allowing them to be singled out during data analysis and proven through the qualitative inspection and discovery of the crack in the main rotor body.

References

1. Bloch PH, Singh PM (2009) Steam turbines—design applications and re-rating, 2nd edn. McGraw-Hill Companies, New York
2. Sinha J, Hahn W, Elbhbah K et al (2012) Vibration investigation for low pressure turbine last stage blade failure in steam turbines of a power plant. In: Proceedings of the ASME Turbo Expo 2012, Copenhagen, Denmark
3. Bachschmid N, Pennachi P et al (2010) Cracked rotors—a survey on static and dynamic behaviour including modelling and diagnosis. Springer, Berlin
4. Jonas O, Mchemer L (2008) Steam turbine corrosion and deposits problems and solutions. In: Proceedings of the thirty-seventh turbo machinery symposium, Jonas Inc., Wilmington, Delaware
5. Davies WGR, Mayes IW (1984) The vibrational behaviour of a multi-shaft, multi-bearing system in the presence of a propagating transverse crack. J Vibr Acoust Stress Reliab Des 106:146—153

6. Futter D (2012) West Burton Power Station: Unit 4: vibration changes prior to forced outage June 2012. Eon New Build and Technology report ENT/12/TMR/CM/855/R
7. BS ISO 10816-2 (2009) Mechanical vibration—Evaluation of machine vibration by measurements on non rotating parts
8. Poursaeidi E, Kamalzedeh Y (2011) Causes of rotor distortions and applicable common straightening methods for turbine rotors and shafts. World Academy of Science and Technology 55 of 2011, pp 213—218
9. Mikkelsen T, Siemens, NDT03-2013, Automated UT examination of the west Burton Unit 4 LP3 Turbine and the retired LP1 spindle, 70PO03636

Effect of Ultrasonic Oscillations on the Interaction of al with Ga-In Eutectic Alloy

Alexey A. Novikov, Alexandr I. Nizovskii, Mikhail V. Trenikhin and Valerii I. Bukhtiyarov

Abstract The development of the compact sources of energy (~ 10 W) has been advancing during the recent years basing on the use of hydrogen fuel cells. One of the methods of obtaining pure hydrogen is the reaction of water with aluminum activated with Ga-In liquid eutectic alloy. The mechanism of aluminum activation by Ga-In (76 wt% Ga, t_m = 15.9 °C) eutectic alloy is based on the diffusion of the eutectic components along the grain boundaries of the original aluminum sample. In this paper, we attempted to use ultrasonic action to improve the efficiency of the diffusion process. It has been shown that the activation of aluminum in the ultrasonic field increased the rate of diffusion along the grain boundaries of the starting material by 2 orders of magnitude. The purpose of the work was to study the nature of the acoustic impact on the interaction between Ga-In eutectics and a massive aluminum sample. The heterogeneity of the structure causes corresponding changes in the velocity and direction of acoustic wave propagation in the sample and accordingly affects its resonance characteristics. The product obtained as a result of activation according to the described technique efficiently interacted with water to release hydrogen.

Keywords Ultrasonic · Gain liquid eutectic alloy · Aluminum activation · Pure hydrogen · Grain boundaries diffusion · Hydrogen · Fuel cell

A.A. Novikov (✉) · A.I. Nizovskii
Omsk State Technical University, Omsk, Russia
e-mail: yarus952@mail.ru

A.I. Nizovskii
e-mail: aleniz@inbox.ru

A.I. Nizovskii · V.I. Bukhtiyarov
Boreskov Institute of Catalysis SB RAS, pr. Lavrentieva, 5, Novosibirsk 630090, Russia
e-mail: vib@catalysis.ru

M.V. Trenikhin
Institute of Hydrocarbons Processing, SB RAS, 54 Neftezavodskaja st, 644040 Omsk, Russia
e-mail: tremv@yandex.ru

J.K. Sinha (ed.), *Vibration Engineering and Technology of Machinery*,
Mechanisms and Machine Science 23, DOI 10.1007/978-3-319-09918-7_27

1 Introduction

The last decade has seen increased interest in portable energy sources, in particular, with the growing number and diversity of various mobile devices (cellular phones, tablet and pocket computers, etc.). Today, one of the problems of search and rescue those in distress is the absence of compact sources of energy to ensure the work of portable communications. Existing battery is heavy enough and require regular charging, chemical power sources (batteries) and also have limited capacity and validity period.

Not in all situations is possible to use transported electric generators running on hydrocarbon fuels for charging accumulators. A possible solution in this situation is the use of compact fuel cells

One of directions of development such sources are compact hydrogen fuel cells elements. However, when using hydrogen there are problems with its low volume density, explosive, and also the high purity requirements for gas. In recent years there has been increasing on the research of the processes of hydrogen based on the interaction of metallic aluminum with water [1–3]. There are publications, which examined the possibility of reaction with water alloys of Al-Ga for obtaining pure hydrogen, which can be used in fuel cells without additional purification. The advantage of this method is that both reagents are stable and safe for storage, which does not require the provision of any special conditions for the provision of technical and ecological safety. High content of gallium (up to 30 at. %) in such alloys is a significant disadvantage because of the high cost of gallium. Our earlier results, showing the possibility of using industrially produced aluminum alloys that are widely used in machine building industries, for production of hydrogen. The high activity of aluminum to water appeared after special processing of the source material with a liquid metal eutectic alloys on the base of gallium [4, 5]. The question about the mechanism of activation of liquid aluminum alloys and the subsequent interaction with water still is not resolved. We have shown that activation of aluminum is connected with grain boundary diffusion components of eutectic Ga-In in activated metal. An important role will have a size of alloy, composition of crystal boundaries, composition and temperature of the environment. Activating according to our method, a relatively to a large sample at room temperature held several tens of hours since the diffusion processes have a high activation energy. To significantly increase the efficiency of activation was required heating of the sample. We discovered and described in the present work the effect of increasing the speed of aluminum activation by treatment with liquid eutectics based Ga-In ultrasonic field [6].

2 Experimental Set-Up

As the models used products from commercial duralumin alloy D16T (Russian analog alloy 2024T3). Activation is performed using the liquid eutectic alloy Ga-In with the MP = 15.9 °C, composition: Ga 76 % (mass.), In 24 % (mass.). The test bench is shown on Fig. 1. Scanning electron microscope JSM 6610 LV "JEOL"

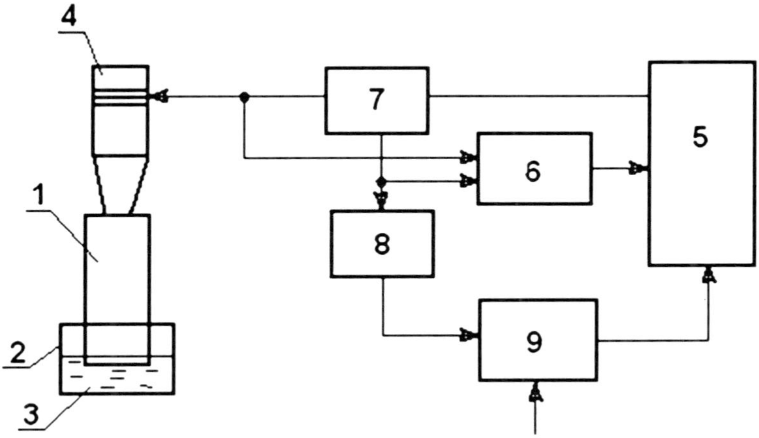

Fig. 1 Functional diagram of the test bench: *1* aluminum sample, *2* capacity, *3* galama, *4* ultrasonic emitter, *5* ultrasonic generator, *6* PLL, *7* sensor having to be removed power output, *8* comparator, *9* command device

(accelerating voltage –30 kV, resolution –3 nm) was used for investigation of activated aluminum in combination with energy dispersive X-ray spectrometer INCA "Oxford instruments".

3 Data Analysis and Results

The aim of this work was to study the acceleration of the process of the activation of aluminum Ga-In eutectics through the application of an additional wave acoustic factors. We previously found that the distribution of eutectic on the core material provides by diffusion along the grain boundaries and intergranular spaces [7]. It is assumed that the introduction of ultrasonic oscillations into the basic material (particularly resonant for this sample modes) must lead to the emergence of variables in time tensile and compressive stresses that cause, in turn, the fluctuations of grains in space.

Because these fluctuations have aimed wave character, in our opinion, they have influence on the process of diffusion components of eutectic, under grain between the edge of the sample to the opposite. Realization of acoustic influence to confirm or deny these assumptions can be made, at least, two different ways. First, the activation of aluminum can be made after immersing a sample in the eutectic and diffusion components of eutectic in the initial material to carry out ultrasonic influence on the eutectic. Secondly, at the stage of diffusion eutectic in the aluminum to produce the introduction of ultrasonic vibrations directly in an aluminum sample. In the second variant is possible resonant and nonresonant the input of ultrasonic fluctuations in the sample.

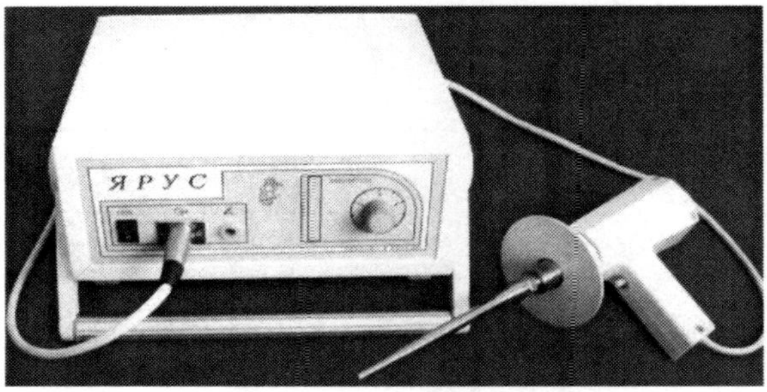

Fig. 2 Ultrasonic apparatus "Jarus"

To assess the impact of powerful high amplitude low frequency ultrasonic oscillations on the activation of aluminum at dipped in the eutectic Ga-In, we have developed and manufactured a test stand (see Fig. 1).

Since the greatest impact should be under the resonance input of ultrasonic fluctuations in the sample, and the implementation of the experiment is quite simple, this method was investigated first. As the basic equipment was used ultrasonic generator and radiator apparatus of the "Jarus" (see Fig. 2).

3.1 The Research of Acoustic Influence on Interaction of Eutectic with Aluminum Samples

According to the scheme shown in Fig. 3, at the working face of the radiator instead of the tool through the steel stud was fixed aluminum sample. To provide the efficient passing of ultrasonic oscillations through the sample with a minimum loss on heating was used solid duralumin alloy D16T. Because the device Jarus calculated on the frequency 42 kHz, as the sample was used, the column diameter of 20 mm and length 120 mm, which corresponds to about half wavelength of this frequency.

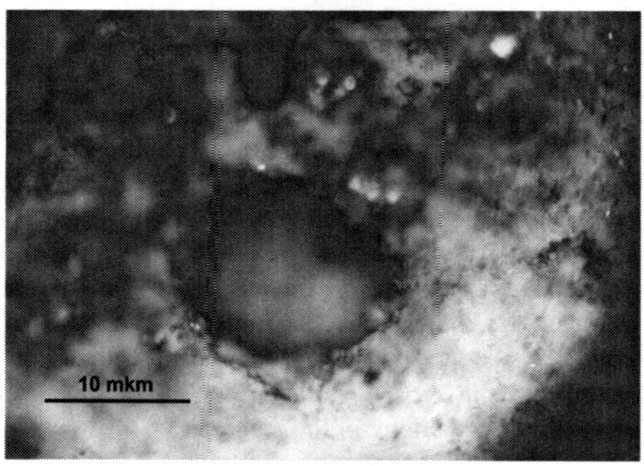

Fig. 3 Micrograph end of the sample after interaction with eutectic

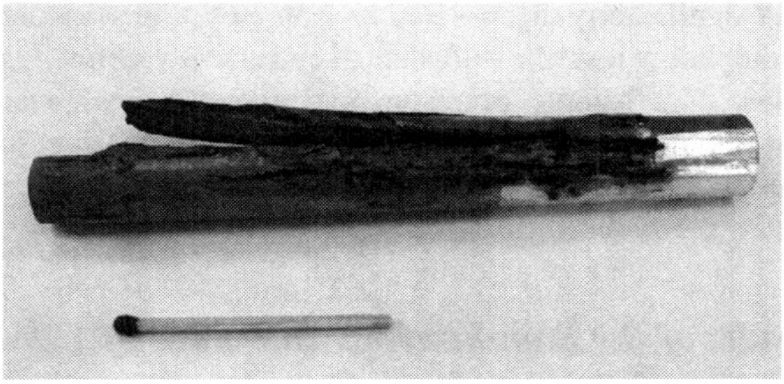

Fig. 4 Microphotograph of the sample after ultrasonic treatment and subsequent interaction with water

When the excitation power emitter 200 W the amplitude of working face of the emitter, and accordingly, and immersed in the eutectic end dural sample was 10–12 mm. One of the features used piezoelectric irradiator is its high-q of the resonant character, the corresponding serial oscillatory circuit. In the presence of a homogeneous structure of the sample with a minimum internal defects, sound power consumption, which when enabled ultrasound acts eutectic, is maximized for a given load.

Upon the spread of eutectic in the body of the sample, its resonance properties changed. Diffusion of eutectic in intergranular space of aluminum varied the density of the sample, since the density of the eutectic is different from the density of aluminum. This caused the corresponding change in the velocity of propagation of acoustic waves in the sample and, therefore, changes in its resonance frequency. An experiment conducted with the duralumin rod at a frequency of 42 kHz led to heating the sample to a temperature T = 62 °C and a complete loss of its resonance properties for 2 min. The loss of the resonance properties of the sample was accompanied by a decrease in the amplitude of excitation of the generator due to increasing resistance losses in the body of the sample. Microphotographs of the end face of the rod contacted with the eutectic after exposure to ultrasonic field are shown on Fig. 3. At the pictures, the cavitational cavities resulting from the interaction of aluminum with eutectic in an ultrasonic field are clearly visible.

Directly after ultrasonic treatment, the sample was placed in a glass beaker with water at room temperature. At the same time there was a violent chemical reaction with hydrogen evolution and the sample cracked (Fig. 4).

Observed in this experiment the effects can be explained as follows. Penetrating in intergranular space aluminum eutectic intensifies the process of interaction of oxygen with aluminum at grain boundaries and ultrasonic fluctuations of these boundaries even more his speed. The resulting alumina, with a greater molar volume than metal aluminium, "cracks" grain boundary, destroying the crystal structure of the sample. The appearance, formation and growth of these intergranular cracks lead to the fact that the energy of the ultrasonic fluctuations begins to stand out on the defect structure of the sample in the form of heat, which further accelerates the processes described above. The sample is heated and, as growth ultrasonic losses significantly deteriorated its resonant properties. By the time when the sample is completely lost, the diffusion of eutectic is practically cover the entire sample. The loss of a sample of resonance properties was accompanied by a decrease in the amplitude of excitation of the generator due to the increased resistance losses in the body of the sample.

3.2 The Results of the Simultaneous Influence of Ultrasound and Eutectic Alloy Aluminum Samples According to Scanning Electron Microscopy

The material obtained in the above experiment, was tested on reactivity with respect to water. Upon contact with water, there was a rapid reaction with hydrogen gas and intense heat. The sample quickly disintegrated into several relatively large pieces that actively responded with water. On Fig. 5 shows the photo (left) of the sample after contact with water for 3 min. In the photo are clearly visible characteristic of the grain structure of the material.

A study was conducted using scanning electron microscopy (SEM) the surface of the sample is removed after 10 min from reaction medium. On micrograph (Fig. 5) shows that the aluminum surface, partially oxidized, which is manifested in the form of bright spots due to charging of oxide particles (Fig. 6).

A more detailed analysis of the data SEM shows on some parts of the surface availability stereotonic particles near oxidized sample plots (Fig. 7). Using energy dispersive X-ray spectrometer for element analysis mode mapping the obtained maps of the distribution of elements on the surface. It is seen that Ga and In sufficiently distributed in aluminum. Analysis stereotonic particles demonstrates only the Ga and In. It is possible to state with confidence that they consist of components eutectic released their intergranular space in the interaction of activated material with water.

Fig. 5 Part of absorbent material after contact with water. Right: micrograph obtained using scanning electron microscopy (SEM)

Fig. 6 The area of the surface of activated sample after interaction with water for 10 min according to the SEM

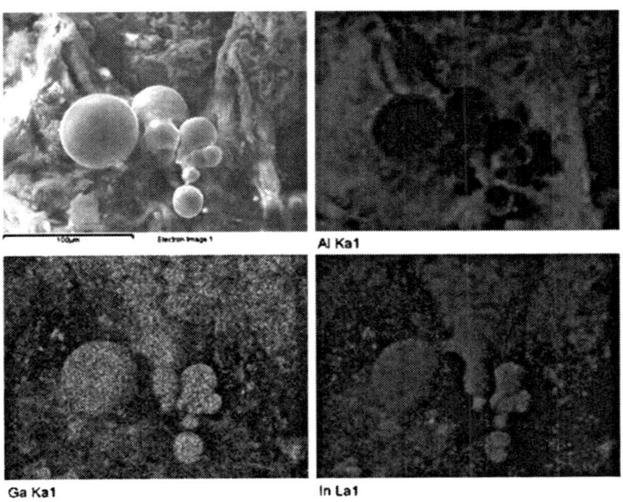

Fig. 7 Data scanning electron microscopy with simultaneous elemental analysis the sample surface (mapping)

4 Conclusion

Ultrasonic treatment is an effective factor of increase in the penetration of the components of the liquid metal eutectic alloys based Ga-In at grain boundaries of aluminium. High reactivity activated product is associated with a uniform distribution of the components of eutectic in intergranular space. Aluminum activated by the proposed method is a suitable reagent for obtaining hydrogen from water and can be used as material for hydrogen cartridge.

References

1. Deng Z-Y, Sakka Y (2008) Hydrogen-Generation Materials for Portable Applications. J Am Ceram Soc 91(12):3825–3834
2. Wang M, Ni HZ (2009) A review on hydrogen production using aluminum and aluminum alloys. Renew Sustain Energy Rev 13:845–853
3. Deng Z-Y, Zhong-Wen O (2010) Role of modification agent coverage in hydrogen generation by the reaction of al with water. J Am Ceram Soc 93(9):2534–2536
4. Trenikhin MV, Duplyakin VK (2006) The penetration of indium–gallium melt components into aluminum Russian. J Phys Chem 80(7):1110–1114
5. Trenikhin MV, Duplyakin VK (2006) Chemical Interaction of the in-ga eutectic with al and al-base alloys. Inorg Mater 42(3):256–260
6. Nizovskii AI, Novikov AA, Kirillov VA, Bukhtijarov VI (2011) Method of activating aluminium and apparatus for realizing said method. Patent (RU) 2 414 424 C2. Bull. No 8
7. Nizovskii AI, Vorobyov AA (2012) Peculiarities of aluminium interaction with Ga85In15 eutectics as evidenced By X-ray synchrotron diagnostics. Crystallogr Rep 57(5):693–699

Nonlinear Control of an Aeroelastic System with a Non-smooth Structural Nonlinearity

Shakir Jiffri and John E. Mottershead

Abstract Non-smooth nonlinearities such as freeplay, bilinear/piece-wise linear stiffness are among the various types of nonlinearity that have been encountered in aeroelastic systems. Freeplay, for example, may begin to appear as a result of ageing of components such as bolted joints and control surfaces, and has been known to be the cause of flutter-induced limit cycle oscillation (LCO). Therefore, it is evident that effectively controlling these nonlinearities is essential in avoiding the onset of LCO, or indeed any other type of nonlinear response. The present paper addresses the control of systems with non-smooth structural nonlinearities, through application of the feedback linearisation method. In systems with smooth nonlinearities, the required nonlinear feedback is also smooth, and therefore does not give rise to complexities associated with the feedback linearisation method. On the other hand, when controlling systems with non-smooth nonlinearities, the necessary control inputs are also non-smooth, and the applicability of feedback linearisation to such systems is of interest. This task is undertaken in the present work, through the use of numerical simulations on a 3 degree of freedom aeroservoelastic model. An example of a case where the parameterisation of the nonlinearity is uncertain is also addressed.

Keywords Nonlinear · Aeroelasticity · Active control · Feedback · Feedback linearisation · Adaptive feedback linearisation · Nonlinear control · Non-smooth · Freeplay

S. Jiffri (✉) · J.E. Mottershead
School of Engineering, Centre for Engineering Dynamics, The University of Liverpool, Brownlow Hill, Liverpool L69 3GH, UK
e-mail: s.jiffri@liv.ac.uk

J.E. Mottershead
e-mail: J.E.Mottershead@liv.ac.uk

J.K. Sinha (ed.), *Vibration Engineering and Technology of Machinery*, Mechanisms and Machine Science 23, DOI 10.1007/978-3-319-09918-7_28

1 Introduction

Vibration mitigation in aeroelastic systems continues to be an active area of research. Various methods, both passive and active, of achieving improved response under arduous operating conditions continue to be explored [1]. Much of this work includes the effects of various aerodynamic and/or structural nonlinearities, and presents appropriate control methods. A problem of particular importance is the suppression of flutter, which is the main motivation behind the present work.

There have been many publications in the literature dealing with the control of systems with smooth nonlinearities, including aeroelastic systems. The application of feedback linearisation on nonlinear aeroelastic systems with smooth structural nonlinearities, mainly of the hardening type, was investigated in [2–8]; both theoretical and experimental aspects have been addressed. Papers related to non-smooth systems are also available, albeit in less abundance and not dealing necessarily with aeroelastic systems. A method for adaptive control with feedback linearisation of systems containing a freeplay input was presented in [9], which was extended subsequently to include also a freeplay output [10]. The cases of partial feedback linearisation with and without relative degree were addressed subsequently in [11]. Other papers related to control of non-smooth nonlinear systems include [12, 13]. An important difference between [9–11] and the present work is that here, the non-smooth nonlinearity is modelled as part of the system dynamics, rather than as an input.

The present work applies partial input-output feedback linearisation on a 3-DOF aeroservoelastic numerical model with a freeplay nonlinearity in the pitch DOF, with the aim of stabilising the linearised response through pole-placement. The model employed is that developed by Edwards et al. [14], which includes actuator dynamics and approximates unsteady behaviour using two additional augmented aerodynamic states. Other work in which this model has been used include [15, 16]. In the present work, the parameters of the model are tuned to match the dynamics of the wind tunnel aerofoil rig (WTAR) at the University of Liverpool.

This paper begins by describing the nonlinear aeroelastic system, including equations of motion, followed by frequency and time-domain simulations based on the WTAR parameters. Expressions for input-output linearisation of the plunge DOF are derived, including those for the zero-dynamics. Numerical simulation results from the closed-loop system are then presented, demonstrating successful control of the system with non-smooth nonlinearity both when full knowledge of the nonlinearity is assumed, and when there is uncertainty associated with the nonlinearity.

2 Model Description

In this section, a detailed description of the aeroelastic model employed in this work is given. Thereupon, numerical simulation results performed using aeroelastic parameters pertaining to the WTAR at the University of Liverpool will be presented.

2.1 Equations of Motion

The aeroelastic model of Edwards et al. [14] featuring approximation of the unsteady aerodynamic loads through the use of augmented states is employed in the present work. This model consists of a total of 8 states in the first-order state-space representation. Six of these are structural states, namely plunge (normalised with respect to the semi-chord b), pitch, aileron flap (ξ, α, β respectively) and their time-derivatives ($\dot{\xi}$, $\dot{\alpha}$, $\dot{\beta}$ respectively). The remaining two are the augmented aerodynamic states mentioned above (x_{a_1}, x_{a_2}). Equations of motion for the model are given as

$$\dot{\mathbf{x}} = \mathbf{A}\mathbf{x} + \mathbf{B}u, \quad \text{where} \quad u = \beta_{com}, \mathbf{x} = \{ x_1 \quad x_2 \quad \cdots \quad x_8 \}^T = \{ \mathbf{q} \quad \mathbf{v} \quad \mathbf{q}_a \}^T,$$
$$\mathbf{q} = \{ \xi \quad \alpha \quad \beta \}^T, \qquad \mathbf{v} = \dot{\mathbf{q}} = \{ \dot{\xi} \quad \dot{\alpha} \quad \dot{\beta} \}^T, \qquad \mathbf{q}_a = \{ x_{a_1} \quad x_{a_2} \}^T,$$
$$\mathbf{A} = \begin{bmatrix} \mathbf{0}_{(3\times3)} & \mathbf{I}_{(3\times3)} & \mathbf{0}_{(3\times2)} \\ -\mathbf{M}_t^{-1}\mathbf{K}_t & -\mathbf{M}_t^{-1}\mathbf{C}_t & \mathbf{M}_t^{-1}\mathbf{D} \\ \mathbf{E}_1 & \mathbf{E}_2 & \mathbf{F} \end{bmatrix}, \qquad \mathbf{B} = \begin{bmatrix} \mathbf{0}_{(3\times1)} \\ \mathbf{M}_t^{-1}\mathbf{G} \\ \mathbf{0}_{(2\times1)} \end{bmatrix}, \qquad \mathbf{G} = \begin{Bmatrix} 0 \\ 0 \\ r_\beta^2 \omega_\beta^2 \end{Bmatrix} \tag{1}$$

and the definition of all quantities appearing within the above equation may be found in [14, 16]. The input in the above equation is the desired flap angle of the aileron. Figure 1 depicts the displacement co-ordinates and also various aeroelastic parameters that will be introduced later.

This particular model is chosen as it models the dynamics of the actuator, the means through which the input will be applied. As will be seen later, the existence of a non-smooth nonlinearity in the system will necessitate a non-smooth input during closed-loop control. Since such an input cannot be achieved in practice, modelling of the actuator dynamics is necessary to produce a numerical model that is representative of reality. This becomes an even more important issue when implementing feedback linearisation in practice, as using a control law that is based on a model without actuator dynamics will give rise to a discrepancy between the required non-smooth input and the actual smooth input provided by the actuator. It is likely that such a discrepancy will degrade controller performance.

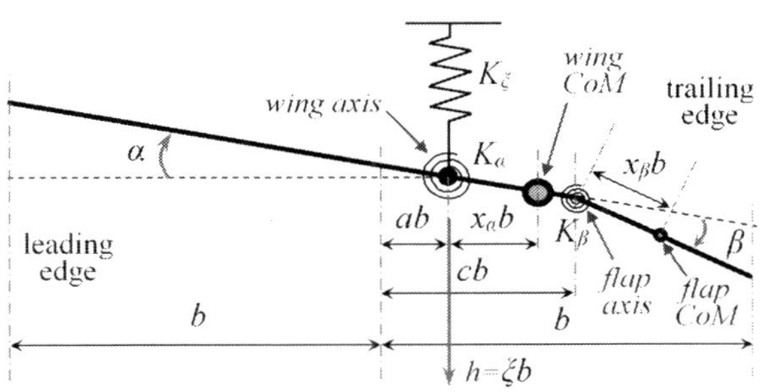

Fig. 1 Schematic of the WTAR showing displacements and parameters

The nonlinear case of the above model may be readily expressed in the affine form

$$\dot{\mathbf{x}} = \underline{\mathbf{f}}(\mathbf{x}) + \underline{\mathbf{g}}(\mathbf{x})\, u, \tag{2}$$

where, in the present case $\underline{\mathbf{g}}(\mathbf{x}) \equiv \mathbf{B}$ and the use of different symbols is aimed at maintaining conventionally accepted notation in the linear and affine nonlinear cases.

2.2 Aeroelastic Parameters of the WTAR

From previous experiments and related numerical simulations performed on the WTAR at the University of Liverpool [17], aeroelastic parameters that describe well the aerofoil behaviour were extracted. These are given in Table 1, along with estimates for the parameters describing flap dynamics (not found during the experiments).

This format of parameters is widely used in the literature, and is defined in [18] in addition to the papers referenced earlier.

Table 1 Parameters of the UoL wind tunnel aerofoil rig, used in the present numerical model

Parameter	Value
ω_α (rad/s)	35.354
r_α	0.4
x_α	0.09
ω_ξ (rad/s)	22.948
a	−0.33333
c	0.5428
b (m)	0.175
μ	69.0
ζ_ξ	0.002
ζ_α	0.015
ω_β (rad/s)	100
r_β	0.079057
x_β	0.0125
ζ_β	0.002

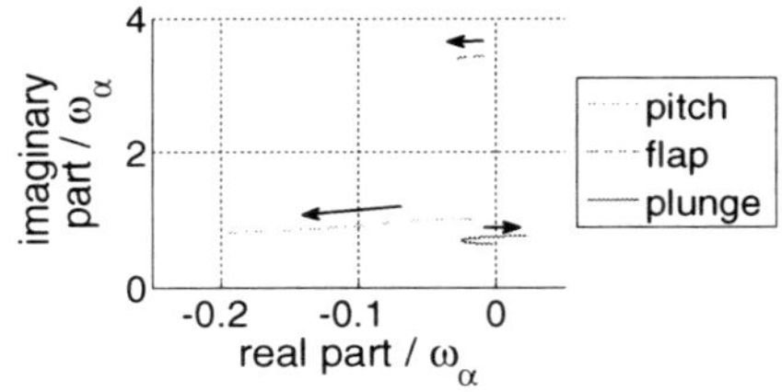

Fig. 2 Normalised eigenvalues of structural modes varying with airspeed

2.3 Frequency Domain Results for the Linear System

For the linear case of the aeroelastic system, one may plot the variation of the eigenvalues with respect to reduced air speed. For a speed range of $U^* = 0.1–3.0$ ($U^* = U/b\omega_\alpha$), the resulting plot is shown in Fig. 2.

The linear flutter speed (LFS) is located at the point where the normalised real part of an eigenvalue becomes positive. It is evident from Fig. 2 that this occurs with the plunge mode. The reduced LFS in the present system is found to be $U^* = 2.793$ (this translates to an absolute airspeed of 17.28 m/s).

2.4 Nonlinear Time-Domain Response with Pitch Freeplay

A symmetric freeplay nonlinearity is now introduced into the pitch DOF. The parameters describing the nonlinearity are given in Table 2.

The resulting pitch moment profile is depicted by the solid line in Fig. 3.

For the purpose of applying feedback linearisation (Sect. 3), it will be necessary to define a target linear system, i.e. the desired system once the nonlinearity has been eliminated. Naturally, this may be chosen as a system whose pitch stiffness is equal to the slope of the outer regions in the freeplay case. Such a choice would ensure that the linearised system will always have its eigenvalues located on the stable side of the s-plane, which would matter if the feedback linearisation cancels out only the nonlinear terms and not the entire open loop dynamics. The pitch moment profile in this case is shown by the dash-dot line in Fig. 3. It is now possible to define also the nonlinear moment, i.e. the moment which, when added to the linear moment produces the net, freeplay moment profile. This non-smooth nonlinear moment profile is shown by the dashed line in Fig. 3.

Table 2 Nonlinearity parameters for pitch freeplay

Parameter	Description	Value
g_α	Gap on either side of $\alpha = 0°$	$1°$
K_α	Stiffness in the outer regions ($\lvert\alpha\rvert > g_\alpha$), chosen to be equal to linear pitch stiffness K_α	$r_\alpha^2 \omega_\alpha^2$

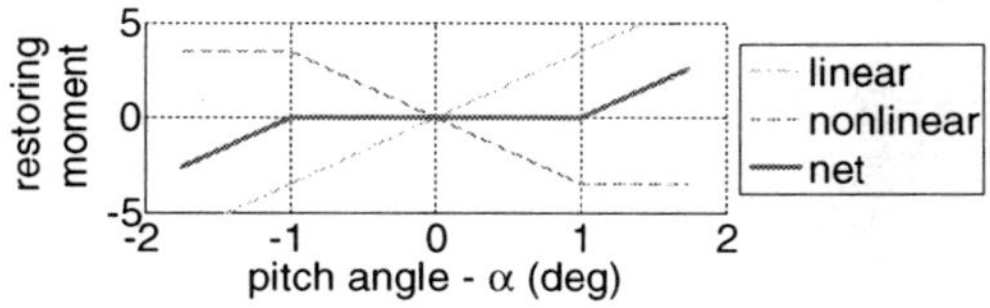

Fig. 3 Pitch moment profile in the presence of freeplay

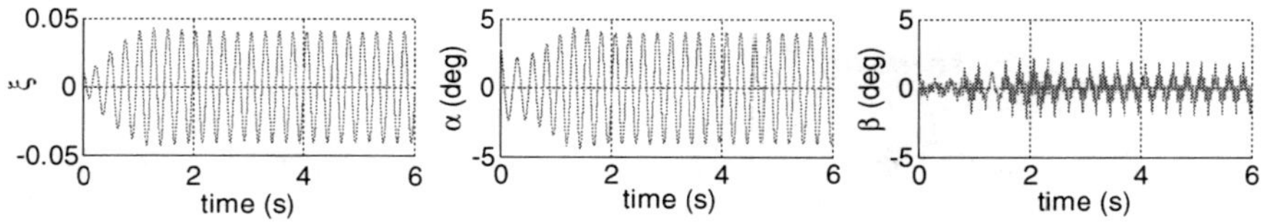

Fig. 4 Structural states of nonlinear system at $U^* = 2.0$

The nonlinear system is now simulated at a reduced velocity $U^* = 2.0$ with plunge and pitch initial values of $\xi = 0.01, \quad \alpha = 3^\circ$ respectively, and with all other states set to zero. The resulting structural responses are shown in Fig. 4.

It is evident that the response settles into an LCO, which occurs at an airspeed which is less than the LFS $U^* = 2.793$, which is expected for a system with freeplay [19].

3 Feedback Linearisation

This section presents the application of feedback linearisation on the nonlinear aeroelastic system described by Eq. (2). Feedback linearisation [20, 21] is a process whereby a nonlinear system is rendered linear through the application of nonlinear feedback and a co-ordinate transformation. The system in (2) is first expressed as

$$\mathbf{f}(\mathbf{x}) = \begin{Bmatrix} \mathbf{v} \\ \mathbf{\Psi q} + \mathbf{\Phi v} + \mathbf{\Lambda q}_a + \mathbf{\Omega f}_{nl} \\ \mathbf{E}_1\mathbf{q} + \mathbf{E}_2\mathbf{v} + \mathbf{F q}_a \end{Bmatrix}, \qquad \mathbf{g}(\mathbf{x}) = \begin{bmatrix} \mathbf{0} \\ \mathbf{\Xi} \\ \mathbf{0} \end{bmatrix}, \tag{3}$$

where

$$\begin{aligned} &\mathbf{\Psi} := -\mathbf{M}_t^{-1}\mathbf{K}_t\,, \quad \mathbf{\Phi} := -\mathbf{M}_t^{-1}\mathbf{C}_t\,, \quad \mathbf{\Omega} := -\mathbf{M}_t^{-1}\,, \quad \mathbf{\Xi} := \mathbf{M}_t^{-1}\mathbf{G}\,, \\ &\mathbf{\Lambda} := \mathbf{M}_t^{-1}\mathbf{D}\,, \end{aligned} \tag{4}$$

The nonlinear force vector arising from the freeplay, as illustrated in Fig. 3, is expressed as

$$\begin{aligned}\mathbf{f}_{nl} &= -\Delta\mathbf{K}_s\mathbf{q} \quad \text{for } |\alpha| \le g_\alpha \\ \mathbf{f}_{nl} &= -\Delta\mathbf{K}_s\mathbf{g}_\alpha \quad \text{for } |\alpha| > g_\alpha, \quad \alpha > 0 \\ \mathbf{f}_{nl} &= \Delta\mathbf{K}_s\mathbf{g}_\alpha \quad \text{for } |\alpha| > g_\alpha, \quad \alpha < 0\end{aligned} \tag{5}$$

where

$$\Delta\mathbf{K}_s = K_\alpha\mathbf{e}_2\mathbf{e}_2^T, \qquad \mathbf{g}_\alpha = \{0 \quad g_\alpha \quad 0\}^T \tag{6}$$

and where $\mathbf{e}_2$ is the second column of a 3 × 3 identity matrix. The feedback linearisation method requires that the outputs are continuously differentiable, and therefore smooth. The non-smooth nature of the freeplay nonlinearity would result in non-smooth—but continuous—forces/accelerations. However, the resulting changes in the system states (both displacement and velocity) will be smooth, as they are obtained as time-integrals of the accelerations (which are continuous, albeit non-smooth). Thus, all the states of the system are continuously differentiable, satisfying the conditions for feedback linearisation.

3.1 Plunge Output Linearisation

The classical input-output linearisation approach [20, 21] is now followed to apply feedback linearisation by controlling the plunge displacement. The co-ordinates of the linear system are obtained as

$$z_1 = y = x_1\ , \qquad z_2 = \dot{y} = \dot{x}_1 = x_4 \tag{7}$$

using Eq. (1). Here, the output y is chosen as the plunge displacement $\xi = x_1$. The partially linearised system may then be obtained as

$$\begin{Bmatrix}\dot{z}_1\\ \dot{z}_2\end{Bmatrix} = \begin{bmatrix}0 & 1\\ 0 & 0\end{bmatrix}\begin{Bmatrix}z_1\\ z_2\end{Bmatrix} + \begin{bmatrix}0\\ 1\end{bmatrix}v, \quad v = \mathbf{f}(\mathbf{x})_{(4)} + \xi_1 u, \quad \boldsymbol{\Xi} = \{\,\xi_1 \quad \xi_2 \quad \xi_3\,\}^T, \tag{8}$$

with v being an artificial input associated with the linearised system. Since there remains an un-linearised set of 6 states, it is necessary to examine the zero-dynamics to ensure their stability when designing a controller. Expressions for the remaining linear co-ordinates are first required to complete the transformation. These are chosen as

$$\begin{aligned}z_3 = x_2, \quad z_4 = x_3 + \xi_2 x_4 - \xi_1 x_5, \quad z_5 = \xi_3 x_4 - \xi_1 x_6, \\ z_6 = \xi_3 x_5 - \xi_2 x_6, \quad z_7 = x_7, \quad z_8 = x_8,\end{aligned} \tag{9}$$

completing the 8 × 8 transformation from nonlinear to linear co-ordinates as

$$\mathbf{z} = \mathbf{T}_{\mathbf{zx}}\mathbf{x}. \tag{10}$$

The resulting zero-dynamics are found as

$$\dot{\mathbf{z}}_{(3:8)_{ZD}} = \begin{bmatrix} \mathbf{P}_1 \\ \mathbf{T}_{\mathbf{zx}\,(4:6,4:6)}[\boldsymbol{\Psi} \quad \boldsymbol{\Phi} \quad \boldsymbol{\Lambda}](\mathbf{T}_{\mathbf{zx}}^{-1})_{(1:8,3:8)} + \mathbf{P}_2 \\ [\mathbf{E}_1 \quad \mathbf{E}_2 \quad \mathbf{F}_p](\mathbf{T}_{\mathbf{zx}}^{-1})_{(1:8,3:8)} \end{bmatrix} \mathbf{z}_{(3:8)_{ZD}} + \begin{bmatrix} \mathbf{0} \\ \mathbf{T}_{\mathbf{zx}\,(4:6,4:6)}\boldsymbol{\Omega}\mathbf{f}_{nl}(z_3) \\ \mathbf{0} \end{bmatrix} \tag{11}$$

where

$$\mathbf{P}_1 = \begin{bmatrix} 0 & 0 & -\frac{\xi_2}{\xi_1\xi_3} & \frac{1}{\xi_3} & 0 & 0 \end{bmatrix}, \quad \mathbf{P}_2 = \begin{bmatrix} 0 & 0 & -1/\xi_1 & 0 & 0 & 0 \\ 0 & 0 & 0 & 0 & 0 & 0 \\ 0 & 0 & 0 & 0 & 0 & 0 \end{bmatrix}. \tag{12}$$

The zero-dynamics are checked to verify stability of the internal dynamics of the partially linearised system. A stability investigation of the zero-dynamics yields that there exist 3 equilibrium points—one zero-equilibrium and two non-zero equilibria. The eigenvalues of the underlying linear system of the zero-dynamics (first part of Eq. (11)) are found to have negative real parts, viz.,

$$\begin{array}{lll} -3.02933 + 109.98286i & -3.30241 + 33.82588i & -35.57119 + 0.00000i \\ -3.02933 - 109.98286i & -3.30241 - 33.82588i & -3.80225 + 0.00000i \end{array}$$

Further investigation reveals that the zero equilibrium is unstable whereas the remaining two non-trivial equilibria are stable.

3.2 Linearised Response with Pole-Placement Implemented

A desired natural frequency ω_{n_ξ} and damping ratio ζ_{n_ξ} may be set for the controlled DOF ξ by choosing the artificial input as

$$v = -\omega_{n_\xi}^2 z_1 - 2\zeta_{n_\xi}\omega_{n_\xi} z_2. \tag{13}$$

For this simulation, target values are chosen as $\omega_{n_\xi} = 1\,\text{Hz}$, $\zeta_{n_\xi} = 0.1$. The resulting closed-loop response, for the same initial conditions as the open-loop case, is shown in Fig. 5.

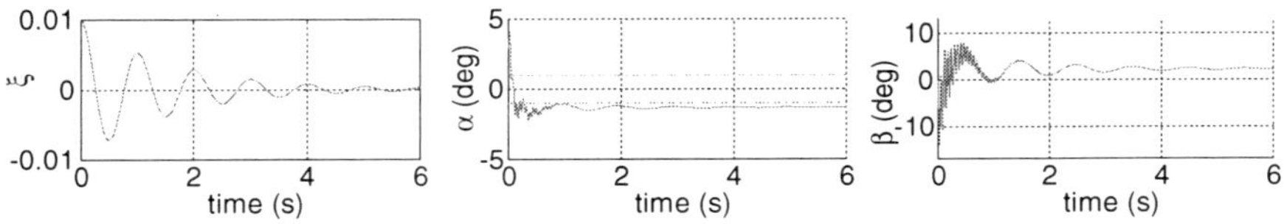

Fig. 5 Closed-loop response of system at $U^* = 2.0$

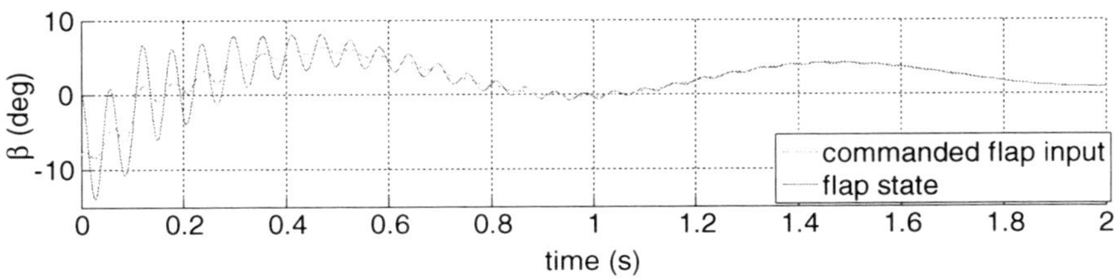

Fig. 6 Comparing commanded and actual flap angles $U^* = 2.0$

It is evident from the first subplot that the target natural frequency of 1 Hz is achieved in the plunge motion, as expected. The pitch motion, confined to the internal dynamics settles down to a non-zero equilibrium close to the lower bound of the freeplay gap $\pm 1°$ (indicated by the dashed horizontal lines), as seen in the middle plot (the pitch motion may settle down to either the upper or lower freeplay bound depending on the initial condition). The flap motion, given by the final subplot, is plotted alongside the commanded input in Fig. 6, where the difference between the two is highlighted.

Closer inspection of the input reveals non-smooth changes corresponding to the switching points in the pitch freeplay. This is expected, as the input is designed to cancel the system dynamics which include the non-smooth freeplay forcing terms. Since the dynamics of the actuator are accounted for in the model, the non-smooth nature of the input will not degrade feedback linearisation that is based on a nonlinear control law derived from the model, and exact pole placement will be achieved in the absence of nonlinearity parameter errors.

4 Adaptive Feedback Linearisation

In real situations, complete cancellation of the nonlinearity will not be achievable. This could be due to a variety of reasons such as inaccurate measurement of the nonlinearity, incorrect assumption of the form of the nonlinearity etc. Adaptive Feedback Linearisation is a method that may be used to guarantee asymptotic closed-loop stability in the presence of a discrepancy between the actual

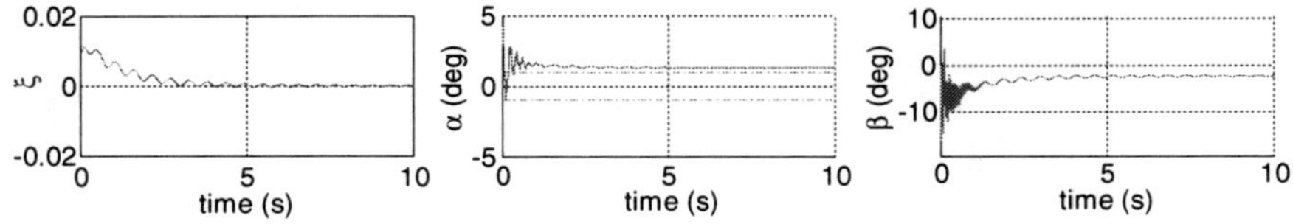

Fig. 7 Closed-loop response of system with adaptive feedback linearisation at $U^* = 2.0$

nonlinearity parameters and those assumed in the design of the controller. The assumed nonlinearity parameters are updated at every time step according to an adaptive law, which has the effect of driving the closed-loop controlled responses to zero.

The numerical simulation from Sect. 3.2 is continued. The inclusion of uncertainty/error in the description of freeplay would ideally require four nonlinearity parameters, namely the two slopes of the non-zero stiffness regimes and the gap on either side of the neutral position. Such an in-depth analysis is undertaken in [9], but in this work we assume symmetric freeplay and only the stiffness parameter K_α to be uncertain. Since the zero-dynamics have an asymptotically stable equilibrium, and the nonlinearity is linearly parameterisable, the conditions for Adaptive Feedback Linearisation are satisfied. A 40 % error in K_α is now assumed. Thus, $K'_\alpha = 1.4\ K_\alpha$. Commencing with a scalar quadratic Lyapunov function

$$V = \mathbf{z}^T_{(1,2)}\mathbf{P}\mathbf{z}_{(1,2)} + \tilde{K}^2_\alpha\ , \qquad \tilde{K}_\alpha = K_\alpha - K'_\alpha\ , \quad \mathbf{P} > 0\ , \tag{14}$$

it can be shown that a parameter update law

$$\dot{K}'_\alpha = \mathbf{r}^T \left(\boldsymbol{\Omega}_{(1,:)}\right)^T \underline{\mathbf{B}}_l^T \mathbf{P}\mathbf{z}_{(1,2)} \tag{15}$$

can be derived, which asymptotically drives the closed-loop controlled response to zero by ensuring that V is a decreasing function. Inclusion of this update law translates to an increase in the dimension of the state vector. In (15),

$$\begin{aligned} \mathbf{r}(\alpha) &= -\mathbf{e}_2\mathbf{e}_2^T\mathbf{q} \quad \text{for } |\alpha| \le g_\alpha \\ \mathbf{r}(\alpha) &= -\mathbf{e}_2\mathbf{e}_2^T\mathbf{g}_\alpha \quad \text{for } |\alpha| > g_\alpha, \quad \alpha > 0 \\ \mathbf{r}(\alpha) &= \mathbf{e}_2\mathbf{e}_2^T\mathbf{g}_\alpha \quad \text{for } |\alpha| > g_\alpha, \quad \alpha < 0 \end{aligned} \tag{16}$$

and $\mathbf{B}_l = [0 \quad 1]^T$ is the input matrix of the partially linearised system (Eq. (8)). The entries in the arbitrary matrix $\mathbf{P}$ are chosen judiciously so as to ensure rapid convergence of K'_α. For the same initial conditions as before and the same pole-placement requirement from the exact linearisation case above, the close-loop responses for the structural DOFs are given in Fig. 7.

It can be seen that the closed-loop plunge response is initially a little higher than before, although it eventually decays to zero. The pitch response settles to a stable equilibrium close to one of the two freeplay boundaries, as before. A noticeable difference between the controlled response in this case and in the case of exact feedback linearisation is that the pole-placement objective is not achieved here. This is expected, as the adaptive law does not take into account this objective, and merely guarantees the convergence of the response to the origin.

5 Conclusions

The application of partial feedback linearisation on an aeroelastic system with a freeplay nonlinearity has been investigated in this work. Although the nonlinear forces and the required inputs are non-smooth, the structural states themselves are smooth and continuously differentiable, thereby satisfying the requirements for feedback linearisation. Simulation results from the unsteady 3-degree of freedom aeroservoelastic model with pitch freeplay nonlinearity demonstrate successful pole-placement in the plunge degree of freedom, with the (uncontrolled) pitch response decaying to a stable equilibrium point close to one of the freeplay bounds. It is pointed out that modelling of the actuator dynamics in a non-smooth nonlinear system is essential, so that the feedback linearising control law that is designed will take into account the fact that in practice the actuator is only able to apply a smooth input. A simplified case of nonlinearity parameter uncertainty has also been investigated, and the utility of parameter adaptation with feedback linearisation in ensuring asymptotic stability of the closed-loop response has been successfully demonstrated.

Acknowledgments This research has been funded by EPSRC grant EP/J004987/1 under the project entitled "Nonlinear Active Vibration Suppression in Aeroelasticity".

References

1. Librescu L, Marzoeca P (2005) Advances in the linear/nonlinear control of aeroelastic structural systems. Acta Mech 178(3–4):147–186
2. Platanitis G, Strganac TW (2004) Control of a nonlinear wing section using leading- and trailing-edge surfaces. J Guid Control Dyn 27(1):52–58
3. Strganac T, Ko J, Thompson D (2000) Identification and control of limit cycle oscillations in aeroelastic systems. J Guid Control Dyn 23(6):1127–1133
4. Ko J, Strganac TW, Kurdila AJ (1999) Adaptive feedback linearization for the control of a typical wing section with structural nonlinearity. Nonlinear Dyn 18(3):289–301
5. Jiffri S, Paoletti P, Cooper JE, Mottershead JE (2014) Feedback linearisation for nonlinear vibration problems. Shock Vibr 2014:16. doi:10.1155/2014/106531. Article ID 106531
6. Jiffri S, Mottershead JE, Cooper JE (2013) Adaptive feedback linearisation and control of a flexible aircraft wing. In: Topics in modal analysis, vol 7, Proceedings of the 31st IMAC, a conference on structural dynamics 2013. Springer, Orange County CA, USA

7. Jiffri S, Mottershead JE, Cooper JE (2013) Active control of a nonlinear flexible aircraft wing, in international conference on structural engineering dynamics, Sesimbra, Portugal
8. Jiffri S, Mottershead JE, Cooper JE (2013) Nonlinear control of a flexible aeroelastic system, in international forum on aeroelasticity & structural dynamics. Bristol Marriott Royal Hotel, Bristol, UK
9. Recker DA, Kokotovic PV, Rhode D, Winkelman J (1991) Adaptive nonlinear control of systems containing a deadzone. In: Proceedings of the 30th IEEE conference on decision and control, vols 1–3. IEEE, New York, pp 2111–2115
10. Tao G, Kokotovic PV (1997) Adaptive control of systems with unknown non-smooth non-linearities. Int J Adapt Control Signal Process 11(1):81–100
11. Ma XL, Tao G (2000) Adaptive actuator compensation control with feedback linearization. IEEE Trans Autom Control 45(9):1705–1710
12. Zheng YF, Wen CY, Li ZG (2013) Robust adaptive asymptotic tracking control of uncertain nonlinear systems subject to nonsmooth actuator nonlinearities. Int J Adapt Control Signal Process 27(1–2):108–121
13. Tao G, Burkholder JO, Guo JX (2013) Adaptive output feedback actuator nonlinearity compensation for multivariable systems. In: 2013 American control conference. IEEE, New York, pp 3356–3361
14. Edwards JW, Ashley H, Breakwell JV (1979) Unsteady aerodynamic modeling for arbitrary motions. AIAA J 17(4):365–374
15. Conner MD, Tang DM, Dowell EH, Virgin LN (1997) Nonlinear behavior of a typical airfoil section with control surface freeplay: a numerical and experimental study. J Fluids Struct 11 (1):89–109
16. Li D, Guo S, Xiang J (2010) Aeroelastic dynamic response and control of an airfoil section with control surface nonlinearities. J Sound Vib 329(22):4756–4771
17. Papatheou E, Tantaroudas ND, Da Ronch A, Badcock K, Mottershead JE, Cooper JE (2013) Active control for flutter suppression: an experimental investigation, in international forum on aeroelasticity and structural dynamics (IFASD), 24–26 June 2013, Bristol, UK
18. Theodorsen T (1935) General theory of aerodynamic instability and the mechanism of flutter. NACA Report no. 496, pp 413–433
19. Wright JR, Cooper JE (2007) Introduction to aircraft aeroelasticity and loads. Wiley, Chichester
20. Isidori A (1995) Nonlinear control systems. Springer, Berlin
21. Khalil HK (2002) Nonlinear systems, 3rd ed. Prentice Hall, Upper Saddle River

A Block Decoupling Control Algorithm for Vibration Suppression of Linear Structures

Xiaojun Wei and John E Mottershead

Abstract In this article, a new block decoupling control algorithm is proposed to decouple undamped or damped structures into two independent substructures with desired closed-loop performance. The algorithm is featured by dealing directly with second-order dynamic structural systems. The block decoupling is implemented by imposing appropriate modal nodal constraints on the closed-loop right eigenvectors so that some modes are uncoupled with the others. The decoupling is synthesised into a process of eigenvalue assignment by the use of multi-input multi-output control, which causes no extra increase of constraints on the control gains and makes the block decoupling control problem very straightforward. A necessary solvability condition for the block decoupling problem is given. Two numerical examples are used to illustrate the working of the proposed approach in undamped and damped structural systems respectively.

Keywords Block decoupling · Modal nodal constraints · Eigenvalue assignment · Vibration suppression · Linear structures

1 Introduction

Multi-degree-of-freedom structural dynamic systems are characterised by the cross-couplings or interactions between their states, which makes it much more difficult to design and to implement control. Decoupling a large-scale structure into several subsystems by eliminating cross jointing effects using active control technologies is considered to be able to reduce the complexity of the control design. For instance, a

X. Wei (✉) · J.E. Mottershead
Centre for Engineering Dynamics, School of Engineering, University of Liverpool, Liverpool L69 3GH, UK
e-mail: xiaojun.wei@liv.ac.uk

J.E. Mottershead
e-mail: j.e.mottershead@liv.ac.uk

J.K. Sinha (ed.), *Vibration Engineering and Technology of Machinery*, Mechanisms and Machine Science 23, DOI 10.1007/978-3-319-09918-7_29

large-scale structure, a part of which is subjected to disturbances, may be decoupled into an undisturbed subsystem and a disturbed subsystem requiring control. The decoupling process, which generates several independent subsystems, is designated as block decoupling.

The attention of the current research is focused on how to block decouple a multi-degree-of-freedom linear structural dynamic system which is represented by second-order differential equations into two independent subsystems, and then control these two subsystems to have desired closed-loop performance. We begin this block decoupling control problem by writing the equations of motion of a n degree-of-freedom linear system as

$$\mathbf{M}\ddot{\mathbf{x}}+\mathbf{C}\dot{\mathbf{x}}+\mathbf{K}\mathbf{x}=\mathbf{0} \tag{1}$$

where **M**, C and **K** are $n \times n$ real symmetric matrices. **M** takes the lumped mass matrix. **M** is positive definite and **C** and **K** are positive semi-definite.

The mass, damping and stiffness matrices and the state vectors may be partitioned into the following forms

$$\begin{aligned}\mathbf{M}&=\begin{bmatrix}\mathbf{M}_1 & \mathbf{0}\\ \mathbf{0} & \mathbf{M}_2\end{bmatrix},\quad \mathbf{C}=\begin{bmatrix}\mathbf{C}_{11} & \mathbf{C}_{12}\\ \mathbf{C}_{21} & \mathbf{C}_{22}\end{bmatrix},\quad \mathbf{K}=\begin{bmatrix}\mathbf{K}_{11} & \mathbf{K}_{12}\\ \mathbf{K}_{21} & \mathbf{K}_{22}\end{bmatrix}\\ \mathbf{x}&=\begin{bmatrix}\mathbf{x}_1\\ \mathbf{x}_2\end{bmatrix},\quad \dot{\mathbf{x}}=\begin{bmatrix}\dot{\mathbf{x}}_1\\ \dot{\mathbf{x}}_2\end{bmatrix}\quad\text{and}\quad \ddot{\mathbf{x}}=\begin{bmatrix}\ddot{\mathbf{x}}_1\\ \ddot{\mathbf{x}}_2\end{bmatrix}\end{aligned} \tag{2}$$

where $\mathbf{M}_1, \mathbf{C}_{11}, \mathbf{K}_{11} \in \mathbb{R}^{m\times m}$; $\mathbf{M}_2, \mathbf{C}_{22}, \mathbf{K}_{22} \in \mathbb{R}^{(n-m)\times(n-m)}$; $\mathbf{C}_{12}, \mathbf{K}_{12} \in \mathbb{R}^{m\times(n-m)}$;$\mathbf{C}_{21}, \mathbf{K}_{21} \in \mathbb{R}^{(n-m)\times m}$;$\mathbf{x}_1$, $\dot{\mathbf{x}}_1$, $\ddot{\mathbf{x}}_1 \in \mathbb{R}^{(m)\times 1}$;
$\mathbf{x}_2$, $\dot{\mathbf{x}}_2$, $\ddot{\mathbf{x}}_2 \in \mathbb{R}^{(n-m)\times 1}$;$1 \le m < n$, $m \in \mathbb{Z}$; $\mathbb{R}$ and $\mathbb{Z}$ are designated as the fields of real and integer numbers respectively.

Now, q proportional and derivative feedback inputs are applied to decouple and control the system so that the closed-loop system may be written as

$$\mathbf{M}\ddot{\mathbf{x}}+\left(\mathbf{C}-\mathbf{B}\mathbf{F}^T\right)\dot{\mathbf{x}}+\left(\mathbf{K}-\mathbf{B}\mathbf{G}^T\right)\mathbf{x}=\mathbf{0} \tag{3}$$

where **B**, **G** and **F** are the force distribution and control gain matrices respectively; and

$$\mathbf{B}\mathbf{G}^T=\begin{bmatrix}\mathbf{B}_1\mathbf{G}_1^T & \mathbf{B}_1\mathbf{G}_2^T\\ \mathbf{B}_2\mathbf{G}_1^T & \mathbf{B}_2\mathbf{G}_2^T\end{bmatrix}\ \text{and}\ \mathbf{B}\mathbf{F}^T=\begin{bmatrix}\mathbf{B}_1\mathbf{F}_1^T & \mathbf{B}_1\mathbf{F}_2^T\\ \mathbf{B}_2\mathbf{F}_1^T & \mathbf{B}_2\mathbf{F}_2^T\end{bmatrix} \tag{4}$$

where $\mathbf{B}_1, \mathbf{G}_1, \mathbf{F}_1 \in \mathbb{R}^{(m)\times q}$; $\mathbf{B}_2, \mathbf{G}_2, \mathbf{F}_2 \in \mathbb{R}^{(n-m)\times q}$.

Then the closed-loop system becomes two independent subsystems of order $n-m$ and m respectively,

$$\mathbf{M}_1\ddot{\mathbf{x}}_1 + (\mathbf{C}_{11} - \mathbf{B}_1\mathbf{F}_1^T)\dot{\mathbf{x}}_1 + (\mathbf{K}_{11} - \mathbf{B}_1\mathbf{G}_1^T)\mathbf{x}_1 = \mathbf{0} \tag{5}$$

and

$$\mathbf{M}_2\ddot{\mathbf{x}}_2 + (\mathbf{C}_{22} - \mathbf{B}_2\mathbf{F}_2^T)\dot{\mathbf{x}}_2 + (\mathbf{K}_{22} - \mathbf{B}_2\mathbf{G}_2^T)\mathbf{x}_2 = \mathbf{0} \tag{6}$$

whenever

$$\mathbf{B}_1\mathbf{F}_2^T = \mathbf{C}_{12}; \quad \mathbf{B}_2\mathbf{F}_1^T = \mathbf{C}_{21}; \quad \mathbf{B}_1\mathbf{G}_2^T = \mathbf{K}_{12} \text{ and } \mathbf{B}_2\mathbf{G}_1^T = \mathbf{K}_{21} \tag{7}$$

where the zero matrices are of proper dimensions as required. It is understandable that, for a linear structure with second-order realisation, the receptance matrix is block decoupled if the dynamic stiffness matrix is block decoupled and vice versa.

The desired closed-loop performance may be obtained by the use of the pole placement approach described in [1]. In general, the force distribution matrix **B** is prescribed and the control gain matrices **F** and **G** are unknowns and require to be determined according to the desired performance. It can be seen that, for the decoupling approach by directly cancelling the off-diagonal blocks defined by (7) and the pole reassignment of the closed-loop system, the number of constraints on the control gains is $4m(n-m)+2nq$. However, the number of unknowns **F** and **G** is only $2nq$. Therefore, the number of the unknowns is less than that of the constraints, which will cause difficulties in precisely implementing the decoupling control. The decoupling control is demanding because the decoupling process may impose extra constraints on control gains.

The problem of decoupling linear time-invariant systems has received considerable attention for several decades. As one of the early researchers, Morgan [2] considered the input-output diagonal decoupling problem by using state feedback. Falb and Wolovich [3] gave necessary and sufficient conditions for decoupling with state feedback. Gilbert [4] extended the results of Falb and Wolovich and studied comprehensively the decoupling of multivariable systems by state feedback. Descusse et al. [5] firstly showed a necessary and sufficient condition for the diagonal problem without any restrictive assumption added neither to the system or control feedback matrices. Decoupling and pole placement in linear multivariable systems was considered in [6]. The block decoupling problem was investigated in [7, 8]. In these works, the system is represented by state-space model, which provides the basis for the feedback controller design. Alternatively, a transfer function matrix approach of decoupling control in linear systems, performed directly on the transfer matrix of the system under consideration, was initially proposed by Hautus and Heymann [9]. The decoupling or block decoupling problem with stability using a transfer function matrix approach was considered in [9, 10].

Decoupling control has been applied to structural vibration control. Zacharenakis [11, 12] investigated the decoupling problems of civil engineering structures via state/output feedback with the assumption that the number of inputs is equal to the number of outputs and also the order of the structure. By constructing state-space models of civil engineering structures, the algorithms developed in [13, 14] were used for decoupling. Li et al. [15] proposed decoupling control laws for vibration control of multi-story building modelled by a diagonal mass matrix and tri-diagonal damping and stiffness matrices. The control laws were designed based on the second-order differential equations of motion.

In the first-order state space approaches, the pole placement problem is intrinsically ill conditioned [16]. Another obvious drawback is that the first-order realisation requires significant numerical effort as the size of the problem doubles. The transfer function matrix approach, which requires much cumbersome algebraic manipulation in the field of rational functions, will become increasingly complicated with the increase of the dimension of the system, especially in the vibration control of industrial-scale structure. Therefore, decoupling control algorithms dealing directly with second order systems, which circumvent the difficulties related to the use of first-order realisations and also allow the second-order nature of the dynamic system to be maintained, are very desirable.

In this paper, the block decoupling control problem of second-order dynamic structural systems is firstly put forward in the Sect. 1. In Sect. 2, a block decoupling control algorithm is proposed to combat the block decoupling control problem in undamped systems. A theorem of block decoupling is given as well as a necessary solvability condition of decoupling control. This algorithm is extended to the block decoupling control problem of damped systems in Sect. 3. In the proposed block decoupling control law, the mass matrix is assumed to be diagonal. This is practical because in structural analysis a lumped-mass matrix can be employed. Also, it is assumed that all the states are measurable. This becomes possible and practical due to the progress in smart materials [17]. For the sake of clarification of several physical concepts, it should be noted here that the decoupling control algorithm aims to decouple dynamic structures into uncoupled substructures in physical coordinates by using active control techniques, which is different from the classical input-output decoupling, the modal decoupling by using coordinate transformation and substructure decoupling which is essentially the identification of a substructure. Eventually, two numerical examples are used to illustrate how the block decoupling control method works.

2 The Block Decoupling Control of Undamped Structures

To illustrate the idea of decoupling, we begin by the block decoupling control problem of undamped systems. The equations of motion of the closed-loop n degree-of-freedom undamped system (i.e., $\mathbf{C} = \mathbf{0}$ and $\mathbf{F} = \mathbf{0}$) may be represented as

$$\mathbf{M}\ddot{\mathbf{x}} + \left(\mathbf{K} - \mathbf{B}\mathbf{G}^T\right)\mathbf{x} = \mathbf{0} \tag{8}$$

And the closed-loop eigenvalue problem is

$$\left(\left(\mathbf{K}-\mathbf{B}\mathbf{G}^{T}\right)-\eta_{k}\mathbf{M}\right)\mathbf{w}_{k}=\mathbf{0},\quad k=1,2,\ldots,n \tag{9}$$

or

$$\left(\mathbf{K}-\mathbf{B}\mathbf{G}^{T}\right)\mathbf{W}-\mathbf{M}\mathbf{W}\Lambda=\mathbf{0} \tag{10}$$

By the application of the approach of imposing modal constraints on right eigenvectors proposed in [1], some nodal constraints may be imposed so that

$$\begin{aligned}\mathbf{w}_{j(i)}&=0,\quad j=m+1,m+2,\ldots,n,\quad i=1,2,\ldots,m\\ \mathbf{w}_{j(i)}&=0,\quad j=1,2,\ldots,m,\quad i=m+1,m+2,\ldots,n,\end{aligned} \tag{11}$$

or

$$\mathbf{W}=\begin{bmatrix}\mathbf{W}_{11} & \mathbf{0}_{m\times(n-m)}\\ \mathbf{0}_{(n-m)\times m} & \mathbf{W}_{22}\end{bmatrix} \tag{12}$$

where $\mathbf{w}_{j(i)}$, is the j-th entry of i-th right eigenvector of the closed-loop system; and $\mathbf{W}\in\mathbb{R}^{n\times n}$.

Theorem For undamped structures with lumped-mass matrix **M**, by the application of multi-input proportional feedback controller, if the closed-loop system is non-defective and has block diagonal eigenvector matrix **W**, then the closed-loop system is block decoupled.

Proof Since the lumped-mass matrix is employed for the system, then the mass matrix **M** is diagonal. Also, the closed-loop system is assumed to be non-defective and **W** is block diagonal. Then **W** is invertible and

$$\mathbf{W}^{-1}=\begin{bmatrix}\mathbf{W}_{11}^{-1} & \mathbf{0}_{m\times(n-m)}\\ \mathbf{0}_{(n-m)\times m} & \mathbf{W}_{22}^{-1}\end{bmatrix} \tag{13}$$

According to (10), the closed-loop stiffness matrix $\left(\mathbf{K}-\mathbf{B}\mathbf{G}^{T}\right)=\mathbf{M}\mathbf{W}\Lambda\mathbf{W}^{-1}$ is block diagonal with the same structure as **W**. Consequently, the closed-loop dynamic stiffness and receptance matrices are block diagonal and the system is block decoupled. □

The desired performance and/or system stability of the resultant two independent subsystems may be guaranteed by eigenvalue placement [1]. According to (11), there are m modal nodal constraints imposed on $(n-m)$ right eigenvectors and $n-m$ modal nodal constraints to the other m eigenvectors. According to [1], the number of the required control inputs should be no less than $1+\max\{m,n-m\}$, which constitutes a necessary solvability condition of the desired block decoupling problem.

If n inputs have been used, then $n-1$ modal nodal constraints may be imposed on every eigenvector $\mathbf{w}_i$, $i = 1, 2, \ldots n$ so that only the i-th entry of $\mathbf{w}_i$ is non-zero. Hence, the closed-loop eigenvector matrix $\mathbf{W}$ will be diagonal. Consequently, the closed-loop stiffness matrix will be diagonal and the closed-loop system is diagonally decoupled. Also, it can be seen that the decoupling control can be implemented exactly with predetermined force distribution matrix $\mathbf{B}$ because the block decoupling is obtained by the appropriate choice of arbitrary parameters $\alpha_{\mu_k,j}$ [1] rather than the imposition of constraints on control gains.

2.1 Numerical Example One

Consider a two degree-of-freedom mass-spring system with

$$\mathbf{M} = \begin{bmatrix} 1 & 0 \\ 0 & 1 \end{bmatrix} \text{ and } \mathbf{K} = \begin{bmatrix} 2 & -1 \\ -1 & 1 \end{bmatrix}.$$

The open-loop eigenvalues are $\lambda_1 = 0.3820$ and $\lambda_2 = 2.6180$, and the eigenvector matrix is

$$\mathbf{\Phi} = \begin{bmatrix} -0.5257 & -0.8507 \\ -0.8507 & 0.5257 \end{bmatrix}.$$

Now, a two-input proportional feedback controller is used to decouple the system into two uncoupled single-degree-of-freedom systems. The prescribed eigenvalues of the two uncoupled subsystems are $\mu_1 = 0.5$ and $\mu_2 = 3.0$.

According to the above analysis, modal nodal constraints are imposed to the closed-loop right eigenvectors so that the second entry of the first eigenvector and the first entry of the second eigenvector are zero. The force distribution matrix is chosen as

$$\mathbf{B} = [b_1 \quad b_2] = \begin{bmatrix} 2 & 2 \\ 2 & 3 \end{bmatrix}.$$

To impose the required modal node constraints, the parameters $\alpha_{\mu_k,j}$ defined in [1] are chosen as

$$\alpha_{1,1} = -\frac{\mathbf{e}_2^T \mathbf{r}_{1,2} \alpha_{1,2}}{\mathbf{e}_2^T \mathbf{r}_{1,1}}; \quad \alpha_{2,1} = -\frac{\mathbf{e}_1^T \mathbf{r}_{2,2} \alpha_{2,2}}{\mathbf{e}_1^T \mathbf{r}_{2,1}}$$

where

$$\alpha_{1,2} = \alpha_{2,2} = 1, \; \mathbf{e}_2^T = [0 \quad 1], \; \mathbf{e}_1^T = [1 \quad 0].$$

and

$$\mathbf{r}_{\mu_{k,j}} = (\mathbf{K} - \mu_k \mathbf{M})^{-1} \mathbf{b}_j, \quad k = 1, 2;\ j = 1, 2.$$

The control gain matrix is

$$\mathbf{G} = \begin{bmatrix} 3.25 & -2.5 \\ 0.5 & -1.0 \end{bmatrix}$$

The closed-loop eigenvalues are $\mu_1 = 0.5$ and $\mu_2 = 3.0$, and eigenvectors are $\mathbf{w}_1^T = \begin{bmatrix} 1 & 0 \end{bmatrix}$ and $\mathbf{w}_2^T = \begin{bmatrix} 0 & 1 \end{bmatrix}$. The closed-loop systems matrices are

$$\mathbf{M}_{CL} = \begin{bmatrix} 1 & 0 \\ 0 & 1 \end{bmatrix} \text{ and } \mathbf{K}_{CL} = \begin{bmatrix} 0.5 & 0 \\ 0 & 3.0 \end{bmatrix}.$$

Hence, the closed-loop system is found to be decoupled into two non-interactive single-degree-of-freedom systems with desired eigenvalues.

3 The Block Decoupling Control in Damped Structures

In this section, the idea of decoupling by the imposition of modal nodal constraint is used to tackle the block decoupling control problem of damped structures.

Similar to the decoupling laws of undamped systems, by the imposition of the modal nodal constraints given in (14), the non-defective closed-loop system with the lumped mass matrix will be block decoupled.

$$\begin{aligned} \mathbf{w}_{j(i)} &= \mathbf{w}^*_{j(i+n)} = 0, \quad j = m+1, m+2, \ldots, n,\ i = 1, 2, \ldots, m \\ \mathbf{w}_{j(i)} &= \mathbf{w}^*_{j(i+n)} = 0, \quad j = 1, 2, \ldots, m,\ i = m+1, m+2, \ldots, n. \end{aligned} \tag{14}$$

where $\mathbf{w}_{j(i)}$, is the j-th entry of i-th right eigenvector of the closed-loop system; $\mathbf{W}$ and $\mathbf{W}^* \in \mathbb{C}^{n \times n}$ are the partitioned sub-matrices of the right eigenvector matrix; $()^*$ is the conjugate of $()$; $\mathbb{C}$ is the field of complex number. The strict proof of this statement will be included in an upcoming journal paper.

Similarly, a necessary solvability condition of the block decoupling problem of damped structures is that the number of inputs is no less than $1 + \max\{m, n - m\}$, where m and $n - m$ are the number of modal nodal constraints on the closed-loop right eigenvectors. Compared with the approach of block decoupling by directly cancelling the off-diagonal blocks, it is evident that the method of the imposition of modal constraints is very straightforward because it synthesises the decoupling into the eigenvalue assignment process. Under the solvability condition, the decoupling control can be implemented exactly even though the force distribution matrix is predetermined.

3.1 Numerical Example Two

Consider the three degree-of-freedom system shown in Fig. 1.

The system matrices of the open-loop system are

$$\mathbf{M} = m\mathbf{I}, \quad \mathbf{C} = c\begin{bmatrix} 2 & -1 & 0 \\ -1 & 2 & -1 \\ 0 & -1 & 1 \end{bmatrix} \text{ and } \mathbf{K} = k\begin{bmatrix} 2 & -1 & 0 \\ -1 & 2 & -1 \\ 0 & -1 & 1 \end{bmatrix}.$$

If $m = 1$, $c = 1$ and $k = 5$, the open-loop eigenvalues are

$$\lambda_{1,4} = -0.0990 \pm 0.9902\mathrm{i};\ \lambda_{2,5} = -0.7775 \pm 2.6777\mathrm{i};\ \lambda_{3,6} = -1.6235 \pm 3.6877\mathrm{i}.$$

The prescribed closed-loop eigenvalues of closed-loop system are

$$\mu_{1,4} = -0.1 \pm 1.0\mathrm{i};\ \mu_{2,5} = -0.8 \pm 2.8\mathrm{i};\ \mu_{3,6} = -1.6 \pm 3.7\mathrm{i}.$$

Also, modal nodal constraints are imposed to the right eigenvectors of the closed-loop system so that the first two entries of the eigenvectors of the last mode and the last entry of the eigenvectors corresponding to the first two modes are zero. The three inputs are used and the force distribution matrix is chosen as

$$\mathbf{B} = [b_1 \quad b_2 \quad b_3] = \begin{bmatrix} 1 & 2 & 3 \\ 0 & 1 & 3 \\ 2 & 0 & 0 \end{bmatrix}.$$

The parameters $\alpha_{\mu_k,j}$ are chosen as

$$\alpha_{\mu_k,2} = 0.5,\ \alpha_{\mu_k,3} = 1, \quad k = 1,2,4,5; \quad \alpha_{\mu_k,3} = 1, \quad k = 3,6;$$
$$\alpha_{\mu_k,1} = \mathrm{inv}\left(\mathbf{r}_{\mu_k,1(3)}\right) \times \left(-\mathbf{r}_{\mu_k,2(3)}\alpha_{\mu_k,2} - \mathbf{r}_{\mu_k,3(3)}\alpha_{\mu_k,3}\right), \quad k = 1,2,4,5;$$

$$\begin{pmatrix} \alpha_{\mu_k,1} \\ \alpha_{\mu_k,2} \end{pmatrix} = \mathrm{inv}\left(\mathbf{R}_{\mu_k(1:2,1:2)}\right) \times \left(-\mathbf{r}_{\mu_k,2(1:2)}\alpha_{\mu_k,2} - \mathbf{r}_{\mu_k,3(1:2)}\alpha_{\mu_k,3}\right), \quad k = 3,6.$$

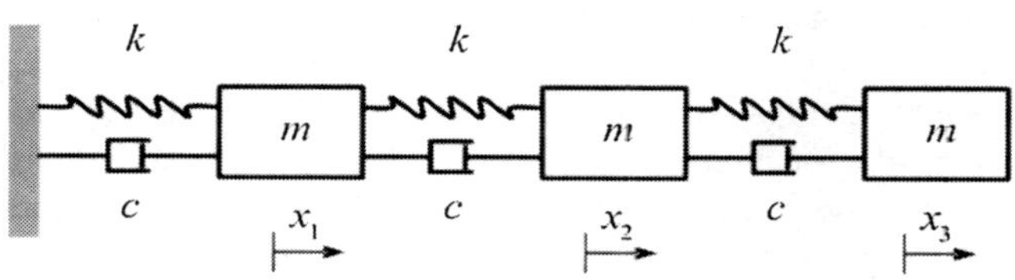

Fig. 1 The three degree-of-freedom system

where

$$\mathbf{r}_{\mu_{k,j}} = \left(\mathbf{M}\mu_k^2 + \mathbf{C}\mu_k + \mathbf{K}\right)^{-1}\mathbf{b}_j, \quad k = 1, 2, \ldots, 6;\ j = 1, 2, 3$$
$$\mathbf{R}_{\mu_{k,j}} = \left[\mathbf{r}_{\mu_{k,1}} \quad \mathbf{r}_{\mu_{k,2}} \quad \mathbf{r}_{\mu_{k,3}}\right].$$

The control gains are

$$\mathbf{G} = \begin{bmatrix} 0 & 3.0559 & 6.1117 \\ -2.5 & -1.9910 & -3.9820 \\ -5.6250 & 10.6250 & -5.2083 \end{bmatrix} \text{ and } \mathbf{F}$$
$$= \begin{bmatrix} 0 & 0.6617 & 1.3235 \\ -0.5 & -0.4420 & -0.8840 \\ -1.1000 & 2.1000 & -1.0333 \end{bmatrix}.$$

The closed-loop eigenvalues are

$$\mu_{1,4} = -0.1 \pm 1.0\mathrm{i};\ \mu_{2,5} = -0.8 \pm 2.8\mathrm{i};\ \mu_{3,6} = -1.6 \pm 3.7\mathrm{i}.$$

and eigenvectors are

$$\mathbf{w}_1 = \begin{pmatrix} 0.7122 - 0.0948\mathrm{i} \\ 0.8198 - 0.1091\mathrm{i} \\ 0 \end{pmatrix}, \mathbf{w}_2 = \begin{pmatrix} 0.0683 - 0.1660\mathrm{i} \\ 0.1245 - 0.2759\mathrm{i} \\ 0 \end{pmatrix}, \mathbf{w}_3 = \begin{pmatrix} 0 \\ 0 \\ -0.1062 + 0.2082\mathrm{i} \end{pmatrix},$$
$$\mathbf{w}_4 = \bar{\mathbf{w}}_1, \mathbf{w}_5 = \bar{\mathbf{w}}_2, \mathbf{w}_6 = \bar{\mathbf{w}}_3$$

The closed-loop system matrices are $\mathbf{M}_{CL} = \mathbf{I}$,

$$\mathbf{C}_{CL} = \begin{bmatrix} -3.2939 & 3.0359 & 0 \\ -5.6322 & 5.0939 & 0 \\ 0 & 0 & 3.2000 \end{bmatrix} \text{ and } \mathbf{K}_{CL}$$
$$= \begin{bmatrix} -14.4469 & 13.4281 & 0 \\ -26.3910 & 23.9371 & 0 \\ 0 & 0 & 16.2500 \end{bmatrix}.$$

Hence, the closed-loop system is found to be block decoupled with desired eigenvalues.

4 Conclusion

A block decoupling control algorithm is proposed to suppress the vibration of linear structures. The algorithm is designed based on the second-order differential equations of considered linear systems, preserving the sparseness, symmetry and definiteness in mass, damping and stiffness matrices. The block decoupling control algorithm, which synthesises the block decoupling into the process of eigenvalue assignment, is able to decouple undamped or damped structural systems into two uncoupled subsystems with desired closed-loop performance, which is demonstrated by two numerical examples. Also, the necessary assumptions and a necessary solvability condition of the block decoupling control are given.

Acknowledgment X Wei is supported by an award from the Chinese Scholarship Council and the University of Liverpool. The work forms part of a research project supported by EPSRC, grant number EP/J004987/1.

References

1. Ram YM, Mottershead JE (2013) Multiple-input active vibration control by partial pole placement using the method of receptances. Mech Syst Signal Process 40:727–735
2. Morgan B Jr (1964) The synthesis of linear multivariable systems by state-variable feedback. IEEE Trans Autom Control 9:405–411
3. Falb PL, Wolovich W (1967) Decoupling in the design and synthesis of multivariable control systems. IEEE Trans Autom Control 12:651–659
4. Gilbert E (1969) The decoupling of multivariable systems by state feedback. SIAM J Control 7:50–63
5. Descusse J, Lafay JF, Malabre M (1988) Solution to Morgan's problem. IEEE Trans Autom Control 33:732–739
6. Morse A, Wonham W (1970) Decoupling and pole assignment by dynamic compensation. SIAM J Control 8:317–337
7. Sato S, Lopresti PV (1971) On the generalization of state feedback decoupling theory. IEEE Trans Autom Control 16:133–139
8. Hikita H (1987) Block decoupling and arbitrary pole assignment for a linear right-invertible system by dynamic compensation. Int J Control 45:1641–1653
9. Hautus MLJ, Heymann M (1983) Linear feedback decoupling-transfer function analysis. IEEE Trans Autom Control AC-28:823–832
10. Wang Q-G, Yang Y (2002) Transfer function matrix approach to decoupling problem with stability. Syst Control Lett 47:103–110
11. Zacharenakis EC (1995) Input-output decoupling and disturbance rejection problems in structural analysis. Comput Struct 55:441–451
12. Zacharenakis EC (1996) On the input-output decoupling with simultaneous disturbance attenuation and h∞ optimization problem in structural analysis. Comput Struct 60:627–633
13. Paraskevopoulos PN, Koumboulis FN (1992) A new approach to the decoupling problem of linear time-invariant systems. J Franklin Inst 329:347–369
14. Tzafestas SG, Paraskevopoulos PN (1973) On the decoupling of multivariable control systems with time delays†. Int J Control 17:405–415
15. Li QS, Fang JQ, Jeary AP, Liu DK (2001) Decoupling control law for structural control implementation. Int J Solids Struct 38:6147–6162
16. He C, Laub AJ, Mehrmann V (1995) Placing plenty of poles is pretty preposterous. In: DFG-Forschergruppe scientific parallel computing, Preprint 95-17, Fak. f. Mathematik. TU Chemnitz-Zwickau, D-09107, Chemnitz, FRG
17. Inman DJ (2001) Active modal control for smart structures. Philos Trans R Soc A: Math Phys Eng Sci 359:205–219

Measurement and Simulation of the Vibroacoustic Performance of an Electric Motor

Alex McCloskey, Xabier Arrasate, Xabier Hernandez and Oscar Salgado

Abstract Noise and vibration of electrical machines is a major concern. Changes in the machine design to improve its efficiency can lead to unacceptable vibrations. Tools to predict its vibratory and acoustic performance at the design stage need to be developed. An improved finite element model has been developed to analyse the vibration behaviour of a permanent magnet synchronous motor (PMSM) using the finite element software ABAQUS. All components and subsets of the machine have been modelled and validated by experimental modal analysis (EMA) performed on them. Some modelling issues have been overcome so that an accurate enough model has been reached. The laminated stator, as it is formed by a pack of several steel sheets, has been treated as an orthotropic material and windings have been considered a solid orthotropic part as well. The rotor-shaft assembly has also been verified by EMA. The bearings that join the rotor to the assembly of the stator have been represented by radial springs. The electromagnetic forces are applied to the whole machine model in order to obtain the vibration response. These forces are obtained from the magnetic air-gap flux density which has been obtained with a 2D finite element model developed by FLUX. Finally, the vibration response has been used to calculate the radiated noise with an acoustic model developed in LMS

A. McCloskey (✉) · X. Arrasate
Department of Mechanical Engineering, Faculty of Engineering,
Mondragon Unibertsitatea, Loramendi 4, 20500 Arrasate/Mondragón, Spain
e-mail: amccloskey@mondragon.edu

X. Arrasate
e-mail: jarrasate@mondragon.edu

X. Hernandez
Department of Mechanical Engineering, Orona EIC, Polígono Lastaola,
20120 Hernani, Spain
e-mail: xhernandez@orona-group.com

O. Salgado
Department of Mechanical Engineering, Ik4 Ikerlan, Pº J.M. Arizmendiarrieta 2,
20500 Arrasate/Mondragón, Spain
e-mail: osalgado@ikerlan.es

J.K. Sinha (ed.), *Vibration Engineering and Technology of Machinery*,
Mechanisms and Machine Science 23, DOI 10.1007/978-3-319-09918-7_30

Virtual Lab. The results given by the acoustic numerical model are compared with sound power measurements.

Keywords Electric motors · Noise · Vibrations · Forced response · Electromagnetic noise

1 Introduction

The increasing use of inverters has introduced more excitation harmonics which originate vibration and noise problems. Noise and vibration have a direct influence on users perception of quality and comfort. Nowadays, in order to reduce costs, the efficiency of the machine needs to be improved and thus, the geometry of the stator needs to be optimised, being the vibration behaviour a restriction, as some geometry modifications can lead to unacceptable vibrations.

Noise and vibration in electric motors have different origins that can be classified into three groups:

- Electromagnetic noise: Originated by magnetic forces acting on the stator and rotor.
- Mechanical noise: Originated by bearings, shaft misalignment…
- Aerodynamic noise: Originated by the cooling fan.

The contribution of these sources changes depending on the working conditions. In the analysed case, due to the absence of a fan, the aerodynamic noise does not exist. According to Gieras et al. [2] and Timar and Lai [6], mechanical and aerodynamical noises increase with speed. However, Timar and Lai [6] affirm that in the very-low-speed domain mechanical and electromagnetic noises are present. Other authors as Kawasaki et al. [3] assess that—"noise below 1,000 Hz is due to radial electromagnetic excitation" and, according to Lakshmikanth et al. [4], the electromagnetic is the dominating source in low-medium power machines. As in the analysed machine the rotational speed is low, for a healthy machine mechanical noise should be low, so just electromagnetic noise will be calculated.

For this task a structural finite element model needs to be developed, which will be validated by experimental modal analysis (EMA).

In order to model the electric motor some modelling issues need to be overcome.

First, the laminated stator presents different stiffness values in radial and axial directions, according to Gieras et al. [2], Verdyck and Belmans [7] and Wang and Lai [8], as it is formed by a pack of several steel sheets. Thus, orthotropic material properties are applied to the solid part representing the laminated stator. It is hard to determine those properties the stiffness depends on the compression pressure acting in the axial direction (Wang and Williams [9] and Watanabe et al. [10]) during manufacturing and on the procedure used to tie the pack of steel sheets and, therefore, they are adjusted according to the experimental results.

Then, different modelling strategies are found in literature for the windings. Belmans et al. [1] and Wang and Lai [8] consider that the mass supplement due to windings is more important than the supplementary stiffness, so windings are treated as additional masses. However, it is hard to know if this strategy might be applicable for any stator geometry. So according to Gieras et al. [2], the volume of material representing the stator windings, which fills the slots of the stator, is also modelled as a solid part. Its orthotropic properties are again estimated according to the experimental modal analysis results.

When more parts are added, the junctions have to be modelled. Initially screwed parts are considered strongly joined so that contact surfaces of both parts are glued. In the case of the rotor, which is supported by bearings, the junction is modelled introducing radial springs where the bearings are placed.

2 Experimental Analysis

2.1 Modal Testing Set-Up and Procedure

The modal testing equipment consists on:

- Acquisition system: two Brüel and Kjaer Front-ends and laptop with Pulse Labshop software.
- Impact hammer.
- 10 triaxial accelerometers.

The system is elastically suspended by rubber bands to avoid the effect of constraints in the natural modes of the system.

For each analysed system a different spatial discretisation is used resulting in different number and position of measurement points. The analysed components are:

- Laminated stator.
- Laminated stator and windings.
- Laminated stator, windings and end-shields assembly.
- Rotor and shaft assembly.
- The previous two assemblies mounted with the corresponding bearings.

The last assembly (stator, rotor and end-shields) is used to analyse the behaviour of the bearings and their stiffness. Therefore, the modes of the rotor shaft system constrained by the stator need to be measured and thus, it is necessary to measure the acceleration in different points of the shaft. Because of this, other elements as the brake are not included in this system, as they block the way to the shaft, making impossible to place accelerometers on it.

2.2 Vibration Measurement Under Operating Conditions

The vibration measurements under operating conditions are carried out for the complete motor mounted rigidly on the floor using the same equipment as for the modal testing except the hammer.

The motor analysed under operating conditions includes the brake, which for the modal analysis was excluded. Accelerometers are placed on the outer surface of the machine regularly spaced. The machine operates in no load condition and at nominal speed.

2.3 Acoustic Power Measurement Under Operating Conditions

The acoustic power is measured in the same place where the vibration measurements were carried out and in the same operating conditions. Five microphones are placed at 0.5 m from the machine in an imaginary parallelepiped around it. From the sound pressure measured at these five points the acoustic power is obtained following the standard ISO 3746.

3 Structural Finite Element Model

The structural finite element model was developed verifying each of the main components with the modes obtained experimentally. For some components, due to the difficulty on estimating its properties, updating processes are carried out using Virtual.Lab optimisation module. A detailed analysis of the modes obtained for the different components, the properties assigned to the models and the relative errors given by the calculations, is presented at EURODYN 2014 [5].

The frequency range analysed by modal analysis is from 0 to 1,000 Hz as the noise generated by this machine is a low frequency noise (mostly below 1,000 Hz).

3.1 Stator Model

This component formed by a pack of several steel sheets is considered as a solid part and assigned orthotropic material properties. The resulting properties, similar to the steel in radial direction and around 3 % of it in axial direction, give relative errors for the simulated natural frequencies below 6 %.

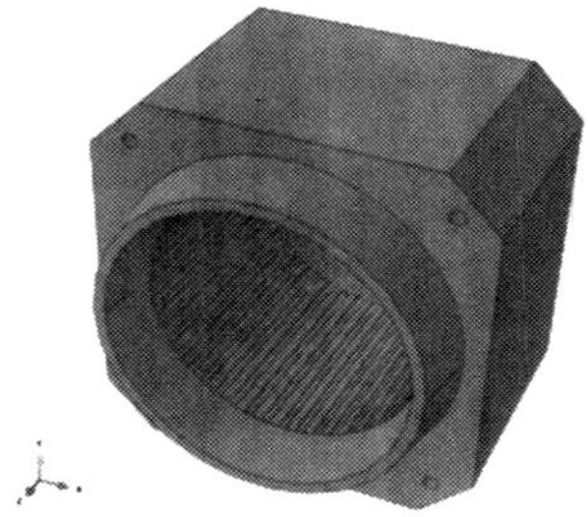

Fig. 1 Stator and windings model

3.2 Stator and Windings Model

The windings are composed of copper wires covered by an insulating material. Similarly to the laminated stator, the volume of the windings, which fills the stator slots (see Fig. 1) has been considered as a solid part and assigned orthotropic material properties, which are adjusted according to the experimental modes.

In the FE model this part is glued to the stator part (see assembly in Fig. 1) using the constraint "tie" provided by ABAQUS, which sticks the nodes of the slave surface to the elements of the master surface.

When the end-shields are assembled, axial deformations are constrained. Thus, the in-plane modes of the stator and windings system are the meaningful ones and the updating is focused on them. The relative error obtained for in-plane modes applied the updated material properties is below 3 %.

3.3 Stator, Windings and End-Shields

This model adds to the previous system the end-shields (see Fig. 2). One of these two parts is the cover of the motor at one of the sides of the stator and the other part holds the stator and rotor at one side and the brake at the other side and it is also the base of the entire motor, holding all its parts.

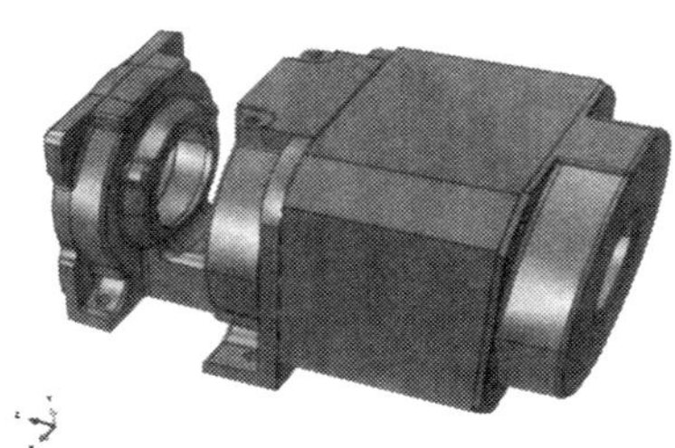

Fig. 2 Stator assembly: stator, windings and end-shields

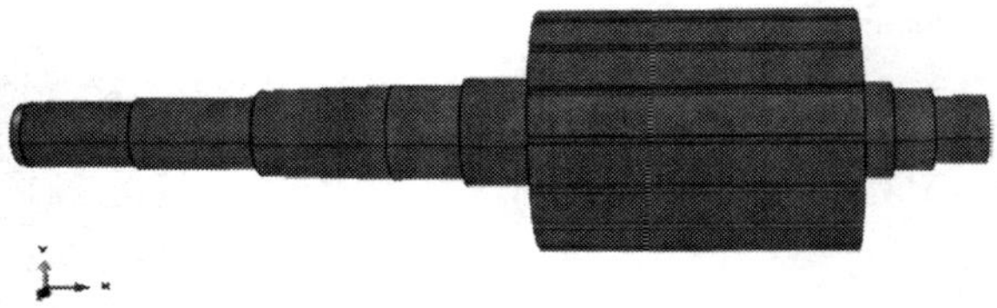

Fig. 3 Shaft and rotor assembly model

These two parts are screwed to the stator and it is considered that this join is strong enough to be modelled, as if the contact surface pair were welded. The material of these two parts is the grey cast iron EN-GJL-250 (GG-25) and, applied its mechanical properties, the biggest relative error for this model is around 5 %.

3.4 Rotor and Shaft Model

The shaft and the rotor (see Fig. 3) are assembled assuming that both parts are rigidly joined. The shaft is a steel solid part and the rotor is as the stator a pack of several steel sheets; therefore, similar orthotropic properties need to be applied.

3.5 Complete Motor System Excluding the Brake and the Pulley

The rotor and shaft assembly is introduced inside the stator, windings and end-shields assembly (see Fig. 4) and both assemblies are joined by springs. Each bearing is modelled by four springs placed with a 90° angle between them in the centre of the contact surface of each of the bearings. The stiffness of the springs is set to 50,000 N/mm according to the modes obtained by EMA and relative errors below 8 % are obtained. However, due to the few measurement points in the shaft, it is difficult to identify all the mode shapes.

Fig. 4 Stator and rotor assembly model

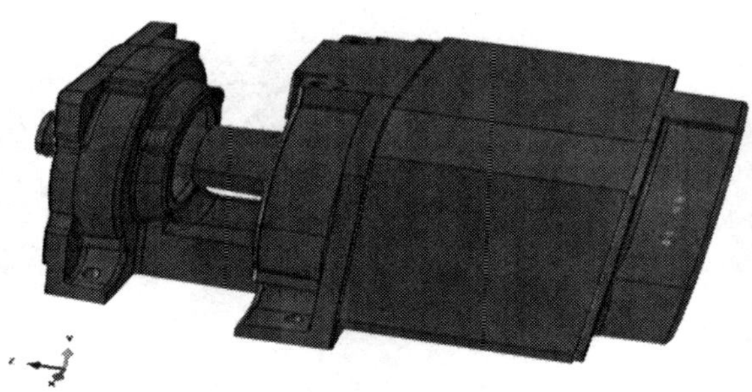

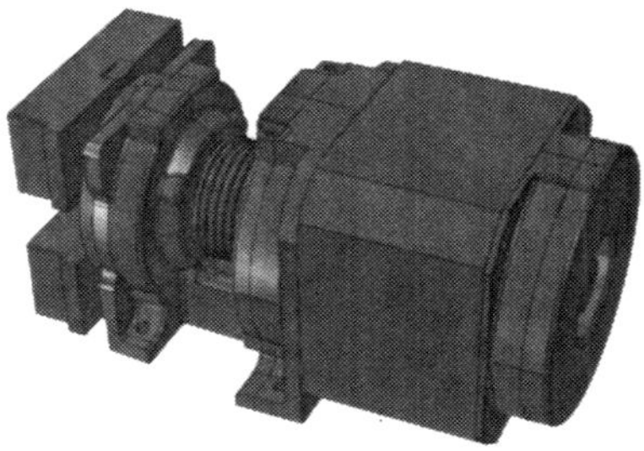

Fig. 5 Complete machine model

3.6 Complete Machine

The brake and the pulley are added to the motor model (see Fig. 5). These parts are assembled in the same way as in the previous assemblies, considering that they are joined to the other parts rigidly, so that the contact surfaces are glued.

4 Vibration Response Calculation

4.1 Electromagnetic Forces Calculation

The magnetic pressure distribution in the surface of a tooth and in the surface of a pole is calculated by FEM (FLUX 2D software) for the period of time corresponding to the fundamental frequency of the machine. In Fig. 6 the magnetic pressure obtained for the tooth surface is shown. Then, the magnetic pressure along time was decomposed obtaining the most important frequency components by FFT.

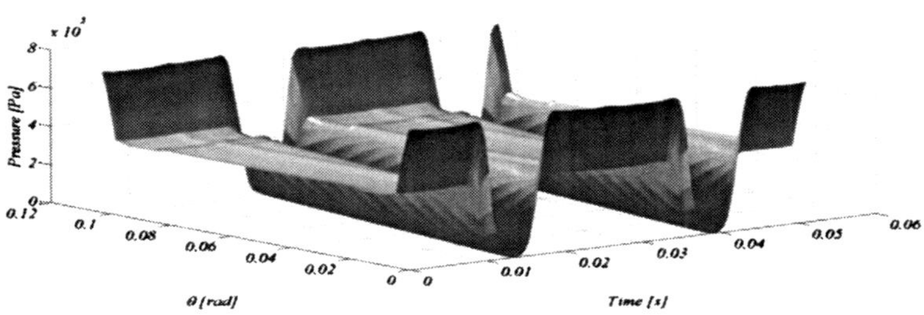

Fig. 6 Magnetic pressure on the tooth surface

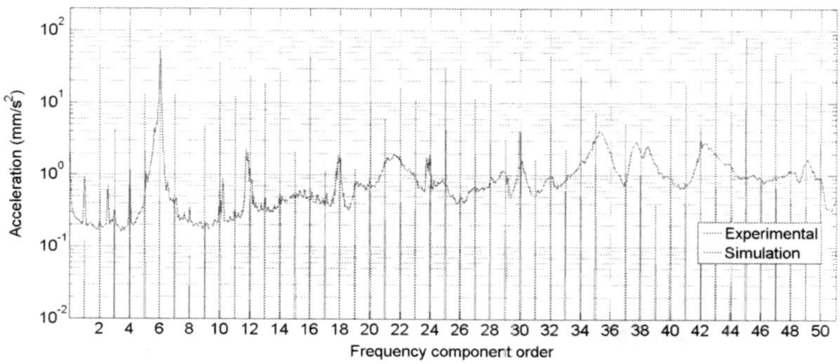

Fig. 7 Acceleration in mm/s^2 in the center of the stator, in vertical direction (normal to the surface) comparison, simulated versus experimental

4.2 Electromagnetic Forces Application to the Structural Model and Vibration Response Calculation

Introducing the electromagnetic pressure frequency components as pressure loads in the structural model, by modal superposition, the steady state dynamic response is calculated. The introduced loads are periodic and thus the obtained response spectrum is discrete. The damping value introduced to this model was obtained from the experimental vibration measurements.

The measurements under operating conditions are used to verify the calculated dynamic response of the machine. In Fig. 7, the discrete spectrum of the calculated vertical acceleration in the centre of the upper face of the stator, is shown with the corresponding continuous spectrum of the measured acceleration. As the energy of the measured acceleration is distributed in a continuous frequency band and the calculated is not, for a certain frequency the magnitude of the experimental one is generally below the calculated one.

5 Acoustic Response Calculation

5.1 Acoustic Finite Element Model

The motor finite element model is imported in Virtual.Lab and the air mesh is introduced between the skin of this model and a surrounding shell called "Convex mesh". The properties given to the air are the density (1.225 kg/m^3) and the speed of sound (340 m/s). Then, the "Automatically Match Layer (AML)" property is applied to the external surface of the air which applies the Sommerfeld condition that imposes to that surface to be a non-reflective boundary.

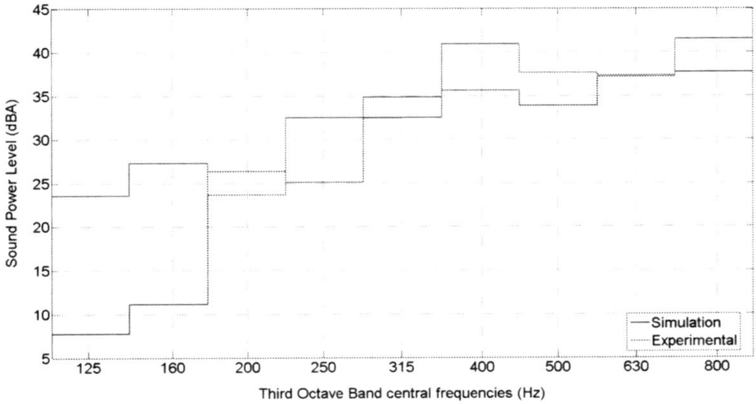

Fig. 8 Acoustic power comparison, simulated versus experimental

The previously calculated vibration accelerations, which could also be velocities or displacements, are applied as boundary condition to this acoustic model.

5.2 Acoustic Power Calculation and Experimental Validation

The calculated acoustic power is compared with the acoustic power obtained experimentally (see Fig. 8). Background noise acoustic power value for the 160 Hz third octave band is as big as the value obtained with the machine running. Therefore, according to the standard ISO 3746, the background noise value is plotted and it is pointed out that the machine acoustic power will be less than the plotted value but it is not possible to know how much it is.

Analysing the obtained results (see Fig. 8), it can be concluded that except for the first two frequency bands (125 and 160 Hz) the agreement between experimental and simulation results is good. In almost all the frequency bands the simulation is below. This makes sense as the calculated acoustic power is just the electromagnetic noise and the measured also contains some mechanical noise.

6 Conclusion

The first conclusion is that, as it was observed in the literature, the main modelling issue is to determine the mechanical properties of the laminated components as the stator and the rotor and specially the windings. In the case of the laminated stator and rotor the values of some material properties found at the literature may give good agreement with the experimental behaviour. However, there are many different windings topologies and sizes and, thus, it is hard to establish generally applicable mechanical properties for this part. To ensure accurate natural frequencies the strategy of modelling the windings as a solid orthotropic part should be followed adjusting its mechanical properties according to the experimental results.

The interaction between different parts is another modelling issue. Screwed parts, with the contact surfaces glued give good agreement with the experimental results. On the other side, the bearings are properly modelled by springs adjusting their stiffness according to the experimental results.

After the modes of the structural model have been validated, the main force excitation frequency components are applied and the vibration response is obtained. A satisfactory agreement between the calculated and real acoustic powers is obtained, and it is expected that a better matching between them would be obtained if the structural model were improved.

Acknowledgments The work presented in this paper has been carried out with the generous support of the company Orona EIC and the Basque Government (SIRUMA, S-PE12MU010, and HESIVAMO, UE2013-05).

References

1. Belmans RJM, Verdyck D, Geysen W, Findlay RD (1991) Electromechanical analysis of the audible noise of an inverter-fed squirrel-cage induction motor. IEEE Trans Ind Appl 27 (3):539–544
2. Gieras JF, Wang C, Lai JC (2005) Noise of polyphase electric motors. CRC press, New York
3. Kawasaki R, Hironaka Y, Nishimura M (2010) Noise and vibration analysis of elevator traction machine. In: INTER-NOISE and NOISE-CON congress and conference proceedings. Institute of Noise Control Engineering, pp 369–377
4. Lakshmikanth S (2012) Noise and vibration reduction in permanent magnet synchronous motors–a review. Int J Electr Comput Eng (IJECE) 2(3):405–416
5. McCloskey A, Arrasate X, Almandoz G, Hernandez X (2014) Vibro-acoustic finite element analysis of a permanent magnet synchronous machine. In: 9th international conference on structural dynamics, Eurodyn 2014. To be held at Porto, Portugal from the 30th of June to 2nd of July
6. Timar PL, Lai JCS (1994) Acoustic noise of electromagnetic origin in an ideal frequency-converter-driven induction motor. IEE Proc Electr Power Appl 141(6):341–346
7. Verdyck D, Belmans RJM (1994) An acoustic model for a permanent magnet machine: modal shapes and magnetic forces. IEEE Trans Ind Appl 30(6):1625–1631
8. Wang C, Lai JCS (1999) Vibration analysis of an induction motor. J Sound Vib 224 (4):733–756
9. Wang Huan, Williams Keith (1995) The vibrational analysis and experimental verification of a plane electrical machine stator model. Mech Syst Sig Proc 9(4):429–438
10. Watanabe S, Kenjo S, Ide K, Sato F, Yamamoto M (1983) Natural frequencies and vibration behaviour of motor stators. IEEE Trans Power Appar Syst 4:949–956

Forced Vibration of Eight Mistuned Bladed Disks on a Solid Shaft—Excitation of the First Compressor Bladed Disc

Romuald Rzadkowski and Artur Maurin

Abstract Considered here is the effect of multistage coupling on the dynamics of an aircraft engine rotor with eight mistuned bladed discs on a drum-disc shaft. Each disc had a different number of rotor blades. Free and forced vibrations were examined using finite element models of single rotating blades, bladed discs, and an entire rotor. Calculations of the global rotating mode shapes of flexible mistuned bladed discs-shaft assemblies took into account the excitation of the first compressor bladed disc with 0EO, 1EO and 2EO forces. The obtained maximal stress values of all of the rotor blades were carefully examined and compared with a tuned system to discover resonance conditions and coupling effects. Mistuning changes the stress distribution in individual rotor blades and the level of maximum stress increases or decreases as compared to bladed discs which are analyzed without the shaft.

Keywords Multistage coupling · Mistuning · Blades · Bladed disc

1 Introduction

The effects of multistage coupling on mistuned bladed disc dynamics were analysed by Bladh et al. [1], Shahab and Thomas [2], Sharma et al. [3], Sinha [4], Laxalde et al. [5]. Generally, these papers considered a small number of mistuned bladed discs, so the analysis of multistage coupling was insufficient.
Rzadkowski and Drewczynski [6] analyzed the forced vibration results for rotors with 8 tuned bladed discs, each with different numbers of blades. This showed that

R. Rzadkowski (✉) · A. Maurin
Department of Dynamics of Machines, Institute of Fluid Flow Machinery, Polish Academy of Sciences, Gdansk, Poland
e-mail: z3@imp.gda.pl

A. Maurin
e-mail: amaurin@imp.gda.pl

J.K. Sinha (ed.), *Vibration Engineering and Technology of Machinery*, Mechanisms and Machine Science 23, DOI 10.1007/978-3-319-09918-7_31

taking into account the differences in blade numbers is essential in the analysis of multistage coupling.

Rzadkowski and Maurin [7] analyzed the dynamic behavior of a rotor consisting of eight mistuned bladed discs on a shaft. Mistuning distorted the nodal diameters of mode shapes primarily in unshrouded bladed discs and caused multistage coupling between various nodal diameters. In the mistuned bladed system particular discs vibrated in modes with diverse nodal diameters. For example, there were modes where one bladed disc vibrated with two nodal diameters and at the same time another disc vibrated with one and zero nodal diameter.

The free vibration of mistuned aircraft engine rotor bladed discs were next analysed by Rzadkowski et al. [8].

In [9] we have considered how forced vibrations of a turbine bladed disc influenced the vibration of compressor stages in order to simulate how combustion chamber excitation might influence the level of compressor stage vibrations trough the shaft.

To better understand this problem, in this paper we present a forced vibration analysis on the same rotor, where the first bladed disc is excited by the fluid flow forces.

The responses of tuned bladed discs will be compared with those of mistuned bladed discs (henceforth referred to as case 1 and case 2, respectively).

More multistage coupling was revealed in the forced vibration analysis than in the free vibration analysis.

2 Model Description

A finite element model of aircraft engine bladed discs on a shaft is presented in Fig. 1. The main dimensions are as follows: the outer diameter of the largest—turbine disc is 0.512 m and the shaft length is 1.166 m. The number of blades on each disc corresponds to that of a real system. Therefore there were 28 rotor blades in compressor stage I, 41 in stage II, 41 in stage III, 47 in stage IV, 57 in stage V, 47 in stage VI, 49 in stage VII and 83 in the turbine stage. The final FE model had over 1.5 million DOFs.

Two axis-symmetric bearing types were modelled as springs with $k_{xx} = k_{yy} = k_{xy} = k_{yx} = 1{,}400$ N/μm stiffness for two roller bearings and $k_{xx} = k_{yy} = k_{xy} = = k_{yx} = k_{zz} = 1{,}000$ N/μm for the ball bearing. The material damping is assumed to be 0.005 %.

Two rotor models—tuned (case 1) and mistuned (case 2), were developed for the numerical analyses. The blade mistuning was modelled using the modified Young modulus for every blade in each stage, according to blade frequencies measured in real aircraft engine (see Table 1). Free vibration analyses were carried out in both models at 15,000 rpm.

Fig. 1 FEM of aircraft engine rotor

Table 1 Natural frequencies of the blades on the discs

Bladed disc (1F Mode)	Tuned (Hz)	Mistuning (Hz)	Mistuning (%)
1	499	498–515	3.2
2	670	619–660	6.11
3	1,210	1,187–1,341	12.7
4	1,130	1,099–1,247	13.1
5	1,387	1,228–1,366	9.9
6	1,597	1,585–1,623	2.4
7	1,635	1,604–1,645	2.5
T	1,278	1,101–1,245	11.3

3 Free Vibration Analysis

In both the tuned and mistuned models the natural frequencies for all the bladed discs were computed in [9] using the ANSYS code. The modes of each mistuned bladed disc were classified in a similar way to those of the tuned bladed disc but only with a small number of mode diameters. However, this was not possible in the case of larger nodal diameter modes as the mistuned blades distorted the nodal lines of the mode shapes. The natural frequencies of tuned and mistuned bladed discs on the shaft were divided into the modes dominated by the bladed discs and the modes dominated by the shaft with the discs. The number of frequencies of the tuned and mistuned systems is the same.

The natural frequencies of the mistuned cantilever blades, single mistuned bladed discs and the complete flexible shaft with eight mistuned discs were carefully examined to find resonance conditions and coupling effects.

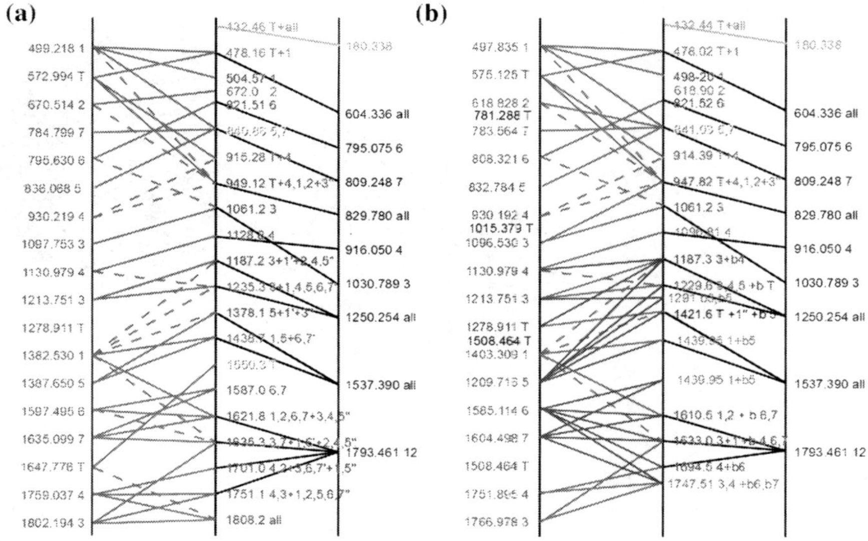

Fig. 2 Natural frequencies of **a** tuned and **b** mistuned bladed discs on the shaft corresponding to zero-nodal diameter modes

Figures 2, 3 and 4 present the zero-, one- and two-nodal diameter natural frequencies of tuned (left) and mistuned systems (right). The right axes indicate the natural frequencies of the eight discs (without blades) on the shaft, the middle axes show the natural frequencies of the mistuned bladed discs on the shaft, while the left axes present the uncoupled natural frequencies of single cantilever mistuned bladed discs corresponding to a given nodal diameter. The uncoupled modes of single blades and bladed discs were calculated separately. The frequencies of the turbine bladed disc are marked with the letter T, whereas those of the compressor are given the number (1–7) of their particular stage. The letter 'b' indicates single blade vibrations in a given stage.

The longitudinal modes are presented in black, the torsional modes in green, the bending bladed disc modes in red and the bending shaft dominated modes in blue. The frequencies with corresponding mode shapes are connected between axes with lines. Purple lines indicate couplings between single blades and entire stages. The lines connecting the natural frequencies, are divided into two types: continuous lines indicating strong coupling and dashed lines for weaker coupling (marked with single and double quotation marks next to the bladed disc symbol). Strong coupling occurs when the amplitudes of particular blades are very visible, whereas weak coupling occurs when the amplitude is relatively small. In the latter case we observe the vibrations of a shaft and disc without blades. The frequencies and mode shapes of mistuned bladed discs on a shaft were analyzed, starting from zero-nodal diameter modes.

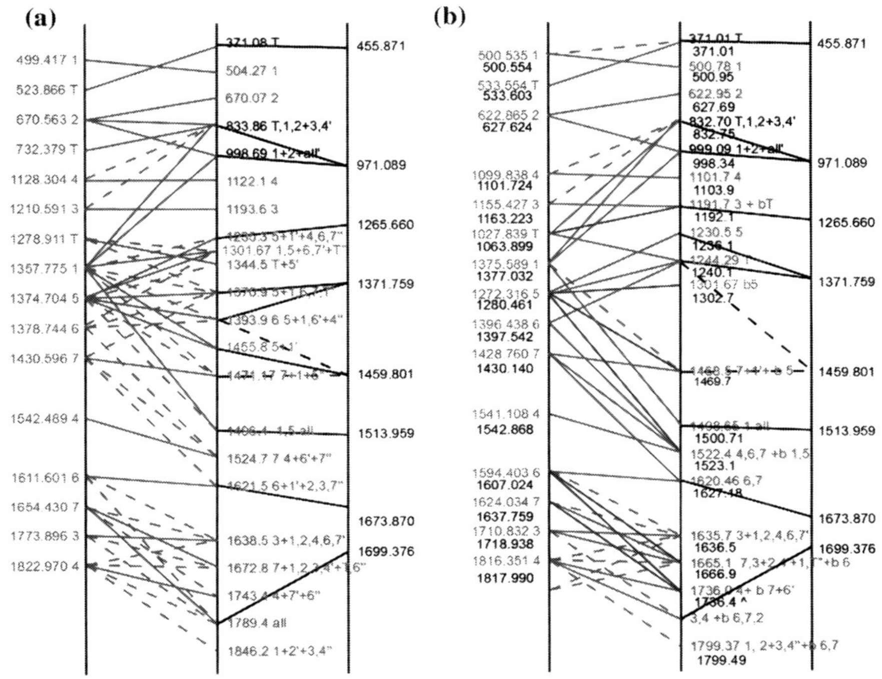

Fig. 3 Natural frequencies of (**a**) tuned and (**b**) mistuned bladed discs on the shaft corresponding to one-nodal diameter modes

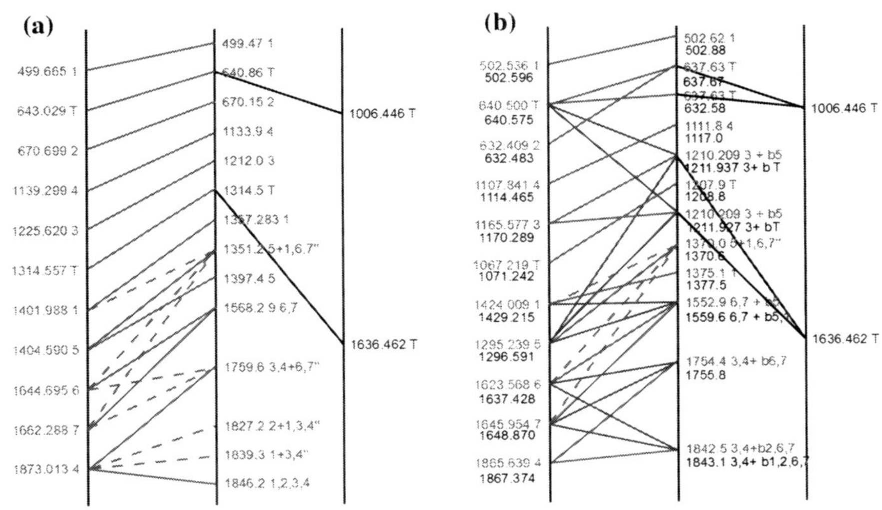

Fig. 4 Natural frequencies of (**a**) tuned and (**b**) mistuned bladed discs on a shaft with two-nodal diameter modes

3.1 Zero-Nodal Diameter Modes

As shown in Fig. 2, the first frequency mode corresponding to the zero-nodal diameter (132.46 Hz in case 1 and 132.44 Hz in case 2) was connected with a torsional rotor mode at 180.338 Hz, in both cases causing all the bladed discs to vibrate. The next frequency mode (478.16 Hz in case 1 and 478.02 Hz in case 2) produce coupling between the compressor first stage bladed disc and the turbine disc. These vibrations were also visible on the unbladed rotor at 604.336 Hz, thus the line connecting the two frequencies in Fig. 2.

In the frequency mode, 915.28 Hz (case 1) and 914.39 Hz (case 2), the turbine bladed disc and 4th compressor bladed disc vibrate predominately. At 949.12 Hz (case 1) and 947.82 Hz (case 2) the turbine bladed disc and 4th bladed disc are coupled along with smaller vibrations occurring in the 1st, 2nd and 3rd bladed disc.

Similar mode couplings and distortions also occurred in subsequent zero-nodal diameter modes, up to 1,187 Hz (see Fig. 2). It is too difficult to group frequencies on the basis of a free vibration analysis (see [9]), and therefore one has to refer a forced vibration analysis. Mistuning alters the vibration amplitudes in particular blades. For example each blade of the 1st bladed disc has a slightly different amplitude in an almost zero-nodal diameter mode.

3.2 One-Nodal Diameter Modes

Figure 3 presents the natural frequencies of tuned and mistuned bladed discs on the shaft, vibrating with a one-nodal diameter. The splitting of one-nodal diameter double frequencies occurs as an effect of blade mistuning. The bending shaft dominated modes are presented in blue. The 371.08 Hz mode and the 371.01 Hz mode are dominated by the bending motion of the shaft caused by the one-nodal diameter mode shape of the turbine disc at 455.871 Hz. The cantilevered turbine disc vibrates at 523.866 Hz (case 1) and at 533.554 and 533.603 Hz (case 2). Next, only the 1st bladed disc vibrates, with modes of 504.27 Hz in case 1, while in case 2 with 500.78 Hz. Similar mode couplings and distortions can be observed in subsequent one-nodal diameter modes, up to 1,370 Hz (Fig. 3).

One-nodal diameter vibrations of a turbine bladed disc at 1,344.5 Hz are coupled with the 5th bladed disc in case 1, whereas in case 2, the corresponding mode shape is approximately 100 Hz lower at 1,244.29 Hz, therefore vibrations only appear on the turbine disc. Here, the one-nodal diameter mode is greatly distorted.

3.3 Two-Nodal Diameter Modes

Figure 4 presents the natural frequencies of tuned and mistuned bladed discs on a shaft with two-nodal diameter mode shapes. The splitting into two two-nodal diameter frequencies appears in the mistuned system on account of the mistuning.

At 499.47 Hz only the 1st bladed disc of the tuned system vibrates with a two-nodal diameter mode. An equivalent mode shape appears in case 2 at 502.62 Hz. Next the turbine bladed disc vibrates at 640.86 Hz in the tuned system and at 632.58(637.63) Hz in the mistuned system.

Similar mode couplings can be observed in subsequent two-nodal diameter modes, up to 1,370 Hz (Fig. 4). The two-nodal diameter vibration mode only appears in the 5th bladed disc and to a lesser extent in the 1st, 6th and 7th bladed discs of the tuned system at 1,351.2 Hz, whereas in the mistuned system coupling with this mode is observed at 1,370.0 Hz. In the latter case, however, only some of the blades in the 5th stage vibrate, while in the 1st stage all the blades vibrate with an almost two-nodal diameter.

4 Excitation by 0EO

This paper aims to explain how the excitation of a first mistuned bladed disc influences mistuned turbine bladed disc (case 2) in comparison with a tuned compressor (case 1). For such an analysis modes had to be selected in which the first bladed disc and turbine bladed disc vibrate simultaneously. In case of 0EO excitation those modes are as follows (see Fig. 2): 478 Hz, 949 Hz—in the tuned system, and 478 Hz, 948 Hz, 1,421 Hz—in the mistuned system.

The first compressor bladed disc in case 1 and case 2 was excited by 0EO with frequencies ranging from 400 to 1,800 Hz (Fig. 5) and the maximal blades stress (von Misses stress) amplitudes measured in the blade roots were next analysed. The axial and circumferential amplitude forces obtained from CFD calculations were applied to the gravity centers of 10 profiles, evenly dividing blade length, for every blade of first compressor stage.

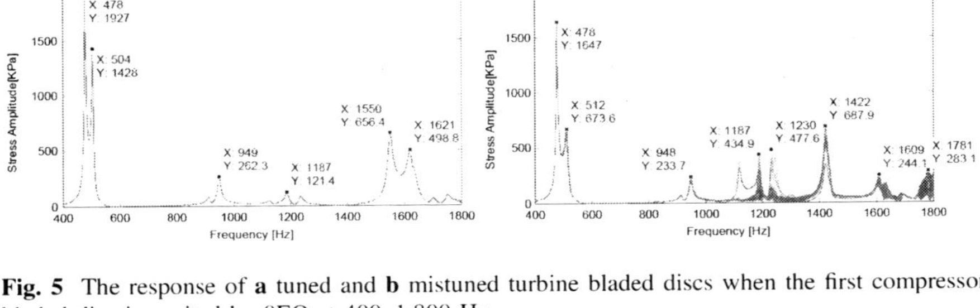

Fig. 5 The response of **a** tuned and **b** mistuned turbine bladed discs when the first compressor bladed disc is excited by 0EO at 400–1,800 Hz

4.1 Response of the First Bladed Disc to 0EO Excitation

0EO excitation (case 1) causes the first tuned bladed compressor disc to vibrate with a zero nodal diameter at 477 Hz and 505 Hz with the maximum von Misses stress values of 26 MPa and 173 MPa respectively (further in text those amplitude stress values will be given in parentheses next to resonance frequency value). The 477 Hz frequency corresponds to the mode in which the turbine bladed disc and first compressor bladed disc vibrate simultaneously (see Fig. 2). The 505 Hz frequency corresponds to the mode in which the 1st compressor bladed disc vibrates with a zero nodal diameter. In the free vibration analysis the vibration of the turbine bladed disc in this mode is not visible. Therefore 0EO of a first compressor bladed disc causes a zero nodal diameter mode in the turbine compressor bladed disc. The forced analysis provides more information than the free vibration analysis.

0EO excitation of the mistuned first compressor bladed disc (case 2) causes vibrations with a zero nodal diameter at 477 Hz (27 MPa) and 512 Hz (321 MPa).

While in the tuned system (case 1) 505 Hz corresponds to the mode in which the 1st compressor bladed disc vibrates with a zero nodal diameter, in the mistuned system a group of separate blade stress peaks appears in the region of 498–515 Hz (up to 321 MPa, which is almost two times higher than in case 1 (173 MPa)). These peaks originate from mistuned blade modes of the 1st compressor stage.

4.2 Response of the Turbine Bladed Disc to 0EO Excitation of the First Bladed Disk

In Fig. 5 we see that 0EO excitation of the first compressor bladed disc causes the turbine bladed disc to vibrate with a zero nodal diameter at 478 Hz with an amplitude of 1,927 kPa, 504 Hz (1,428 kPa), 949 Hz (262 kPa), 1,187 Hz (121 kPa), 1,550 Hz (656 kPa) and 1,621 Hz (498 kPa) in case 1 and at 478 Hz with an amplitude of 1,647 kPa, 512 Hz (673 kPa), 948 Hz (233 kPa), 1,187 Hz (434 kPa), 1,230 Hz (477 kPa) and 1,422 Hz (687 kPa) in case 2. The two remaining peak values: 1,609 and 1,781 Hz, are caused by strong blade coupling between the 1st and 7th mistuned bladed disks. Corresponding frequencies are presented in Fig. 2 on the middle axes for both diagrams (of the tuned and mistuned systems) with frequencies bearing the letter "T". In these modes the turbine bladed disc vibration is a result of coupling through the shaft. The new result for the mistuned system is that an additional group of modes with blade stress peaks appears in the region of 1,101–1,245 Hz (with maximum values of 434 and 477 kPa). These modes originate from mistuned blade modes of the turbine stage (1,101–1,245 Hz, see Table 1). We may conclude that overall stress level in case 2 region of 1,101–1,245 Hz is higher. The maximum blade stress level in the mistuned system reaches 1,647 kPa, whereas in the tuned system it is significantly higher: 1,927 kPa.

5 Excitation by 1EO

Having examined 0EO excitation, we shall now look at the influence of 1EO excitation of the first compressor bladed disc on the turbine bladed disc in an aircraft rotor. For such an analysis modes had to be selected in which the first bladed disc and turbine bladed disc vibrate simultaneously. In case of 1EO excitation those modes are as follows (see Fig. 3): 833 Hz, 999 Hz, 1,302 Hz, 1,371 Hz, 1,789 Hz—in the tuned system, and 833 Hz, 999 Hz—in the mistuned system.

The first compressor bladed disc in tuned (case 1) and mistuned systems (case 2) was excited by 1EO at frequencies ranging from 400 to 1,800 Hz (Fig. 3) and the stress response of turbine bladed disc was analysed. The axial and circumferential amplitude forces obtained from CFD calculations were applied to the gravity centers of 10 profiles, evenly dividing blade length, for every blade of first compressor stage.

5.1 Response of the First Bladed Disc to 1EO Excitation

In case 1, the 1EO excitation of the first compressor bladed disc causes it to vibrate with a one nodal diameter at: 504 Hz (see Fig. 3) (188 MPa). The 504 Hz frequency corresponds to the mode in which the 1st compressor bladed disc vibrates with a one nodal diameter.

The 1EO excitation of the mistuned first compressor bladed disc (case 2) causes the first mistuned bladed disc to vibrate. An additional group of modes with separate blade stress peaks appears in the region of 498–515 Hz (with a maximum blade stress of 323 MPa, 42 % higher than in case 1). These modes originate from mistuned blade modes of the 1st compressor stage.

5.2 Response of the Turbine Bladed Disc to 1EO Excitation of the First Bladed Disk

In case 1 (Fig. 6, left) 1EO excitation of the first compressor bladed disc caused the turbine bladed disc to vibrate with a one nodal diameter at 504 Hz (392 kPa), 834 Hz (315 kPa), 999 Hz (476 kPa), 1,302 Hz (828 kPa), 1,371 (484 kPa), and at 1,789 (627 kPa).

These frequencies are presented in Fig. 3 on the middle axes of both the case 1 and case 2 diagrams. Those of the turbine disc blades are marked ‘T’, whereas those of the compressor blades are given their relevant disc number. In these modes the turbine bladed disc and some of the compressor bladed discs vibrate as the result of coupling through the shaft as in the case of 0 nodal diameter modes (Fig. 2).

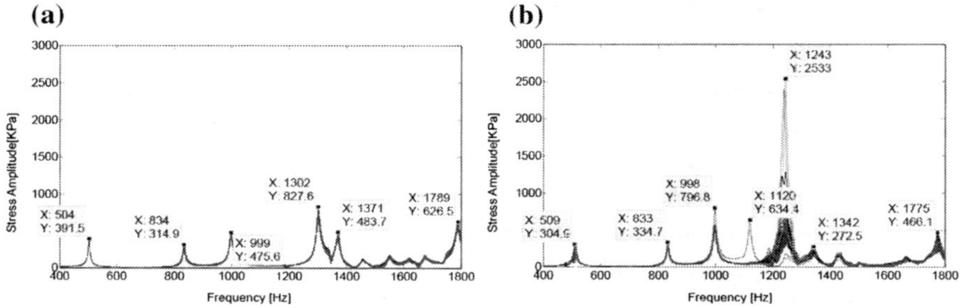

Fig. 6 The response of (**a**) tuned and (**b**) mistuned turbine bladed discs when first compressor bladed disc is excited by 1EO at 400–1,800 Hz

In case 2 (Fig. 6, right) 1EO excitation occurs at 509 Hz with an amplitude of 305 kPa, 833 Hz (335 kPa), 998 Hz (797 kPa) and at 1,775 (466 kPa) similar to case 1. The frequency of 1,371 Hz in case 1 is shifted down to 1,342 Hz (273 kPa).

In the mistuned system an additional group of modes with blade stress peaks up to 2,533 kPa appears in the region of 1,101–1,245 Hz, which is a new result. These peaks originate from mistuned blade modes (at 1,101–1,245 Hz) of the turbine stage (Table 1). The blade stress level in the mistuned system reaches 2,533 kPa, whereas in the tuned system it is no more than 828 kPa, which is over 3 times lower. Such difference occurs due to the fact that the frequency of one of the mistuned blades is covering the one nodal-diameter turbine disc mode at 1,244 Hz (refer to Fig. 3), causing the disc-blade resonance.

6 Excitation by 2EO

So far we have examined how 0EO and 1EO excitation of the first compressor bladed discs on a shaft influences the turbine bladed disc. Now we shall look at the effects of 2EO excitation. In free vibration analysis the two nodal-diameter modes that coupled the first compressor and turbine bladed disc were undetectable by the means of modal analysis for the tuned as well as for mistuned system.

6.1 Response of the First Compressor Disc to 2EO Excitation

In case 1 2EO excitation of the first compressor bladed disc causes it to vibrate with a one nodal diameter at: 499 Hz (see Fig. 4) (193 MPa). The 499 Hz frequency corresponds to the mode in which the 1st compressor bladed disc vibrates with a two nodal diameters.

Blade stress at 1,353 Hz have smaller maximal stress values than at 499 Hz. At 1,353 Hz 2EO excitation shows that the 1st bladed disc vibrates, whereas in free vibration analysis at this frequency the turbine disc and the compressor 5th bladed disc vibrate. Therefore the forced vibration analysis provides more information about multistage coupling.

2EO excitation of the mistuned first compressor bladed disc (case 2) causes the first mistuned bladed disc to vibrate. An additional group of modes with separate blade stress peaks appears in the region of 498–515 Hz (with a maximum blade stress of 288 MPa, 33 % higher than in case 1).

6.2 *Response of the Turbine Bladed Disc to 2EO Excitation of the First Bladed Disk*

The first compressor bladed disc in case 1 and case 2 was excited by 2EO with frequencies ranging from 400 to 1,800 Hz (refer Fig. 4) and the stress response of turbine bladed disc was analysed. The axial and circumferential amplitude forces obtained from CFD calculations were applied to the gravity centers of 10 profiles, evenly dividing blade length, for every blade of first compressor stage.

In case 1 (Fig. 7, left) 2EO excitation of the first compressor bladed disc causes the turbine bladed disc to vibrate with a two-nodal diameters at 504 Hz, 950 Hz, 1,186 Hz, 1,354 Hz, 1,551 Hz, 1,620 Hz and 1,753 Hz (refer to Fig. 4), all with stress amplitudes level under 200 kPa. In case 2 (Fig. 7, right) at 508 Hz the amplitude stress peak rises to 191 kPa—34 times higher than it case 1 it is still under level of 200 kPa. The overall response level of the turbine blades is low, due to the fact that the shaft construction between compressor and turbine disc disallows the two and higher nodal-diameters mode vibrations to transfer.

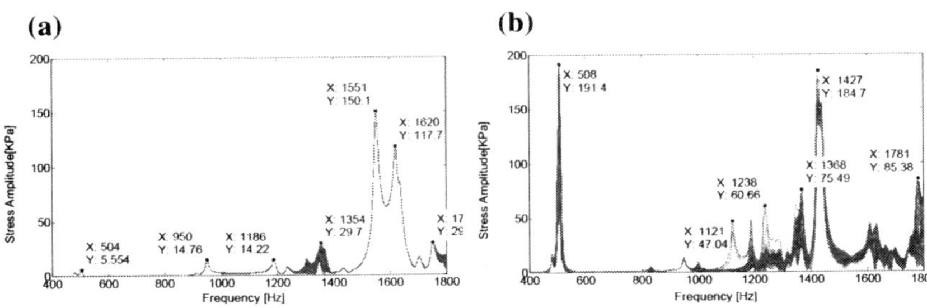

Fig. 7 The response of (**a**) tuned and (**b**) mistuned turbine bladed discs when first compressor bladed disc is excited by 2EO at 400–1,800 Hz

7 Conclusions

In this paper we consider the influence of the low engine order excitation of a tuned and mistuned first compressor bladed disc on the vibrations of tuned and mistuned turbine bladed disc.

The free vibration analysis is an insufficient tool to explain multistage coupling in mistuned system. The forced vibration analysis of an aircraft rotor showed that mistuning considerably increases multistage coupling when compared with a tuned rotor system. An additional modes may appear due to resonances caused by mistuned disc blade frequency shift.

The occurring resonances can cause a significant rise of blade stress levels and are potentially dangerous from the blade-life point of view.

The maximum blade stress response level measured in the mistuned system may be lower than in the tuned system. This occurs when the stress amplitudes of the exited stage are higher in the mistuned system than in the tuned system.

In an aircraft engine with a disc-drum shaft multistage coupling between the compressor stages and the turbine bladed disc primarily occurs in 0E0 and 1EO excitation of the turbine bladed disc. The effect appears in this particular system due to specific shaft connection between the turbine and compressor stages.

Acknowledgment The research work was supported by scientific project 2012/07/N/ST8/03390 "Coupled vibrations of mistuned bladed discs of steam and gas turbines".

References

1. Bladh R, Castanier M, Pierre C (2003) Effects of multistage coupling and disc flexibility on mistuned bladed disc dynamics. J Eng Gas Turbines Power 125:121–130
2. Shahab AAS, Thomas J (1987) Coupling effects of disc flexibility on the dynamics behavior of multi disc-shaft system. J Sound Vib 114(3):435–452
3. Sharma BK, Devadig HV, Singh AK (2005) Modal time history analysis of a steam turbine rotor to an earthquake excitation—A 3D approach. Advances in vibration engineering. Sci J Vib Inst India, 4(4):351–359
4. Sinha A (2007) Reduced-order model of mistuned multi-stage bladed rotor. In: Proceedings of ASME Turbo Expo 2007: power for land, sea and air, May 14-17, Montreal, Canada
5. Laxalde D, Lombard J-P, Thouverez F (2007) Dynamics of multi-stage bladed discs systems. In: Proceedings of ASME Turbo Expo 2007: power for land, sea and air, May 14-1, Montreal, Canada
6. Rzadkowski R, Drewczynski M (2010) Multistage coupling of eight bladed discs on a solid shaft. In: Proceedings of ASME Turbo Expo 2010: power for land, sea and air, June 14- 18, Glasgow, UK
7. Rzadkowski R, Maurin A (2012) Multistage coupling of eight mistuned bladed disc on a solid shaft, Part 1. Free vibration analysis, ASME TURBO-EXPO, Denmark, Copenhagen, June 11–15, 2012, ASME paper GT-2012-68391

8. Rzadkowski R, Maurin A, Drewczynski M (2013) Multistage coupling of mistuned aircraft engine bladed disc in a free vibration analysis. In: 11th international conference on vibration problems, Lisbon, Portugal
9. Rzadkowski R, Maurin A (2014) Multistage coupling of mistuned aircraft engine bladed disks in a forced vibration analysis. ASME Turbo Expo 2014, GT2014-26108

Dynamic Analysis of Parametrically Excited Piezoelectric Bimorph Beam for Energy Harvesting

S.K. Dwivedy, Anvesh K. Reddy and Anshul Garg

Abstract In this work, the dynamics of an energy harvester which is modeled as a base excited cantilever beam with tip mass having piezoelectric patches on both top and bottom surfaces has been investigated. The governing equation of motion of the system is developed using extended Hamilton's principle and it is solved by using method of multiple scales. Closed form solution has been developed to find out the voltage generated from this nonlinear energy harvester. While in most of the previously considered energy harvesters the system is considered without tip mass and axial load, here putting the additional tip mass and periodic axial load, different resonance conditions have been explored. A parametric study has been carried out to investigate the variation of generated voltage with different system parameters. The study will find application in the generation of voltage for a wide range of frequencies available in ambient vibration.

Keywords Energy harvester · Periodic axial load · Method of multiple scales · Piezoelectric patches

1 Introduction

Energy harvesting with piezoelectric material is a new field of interest among researchers which can find applications in the development of self powered and battery independent electronic and wireless devices with capability of surviving in remotely hostile environment. Mechanical vibration is one of the ambient energy which is found almost everywhere and can be utilized as source of energy for low

S.K. Dwivedy (✉) · A.K. Reddy · A. Garg
Mechanical Engineering Department, Indian Institute of Technology, Guwahati 781039, India
e-mail: dwivedy@iitg.ernet.in

A. Garg
e-mail: anshul.garg@iitg.ernet.in

J.K. Sinha (ed.), *Vibration Engineering and Technology of Machinery*, Mechanisms and Machine Science 23, DOI 10.1007/978-3-319-09918-7_32

power devices. Yildiz [1] gives an overview of different potential ambient energy-harvesting sources in terms of power density and it is observed that 200 $\mu W/cm^3$ energy can be generated by using vibration based piezoelectric harvester. Masana and Daqaq [2] carried out electromechanical modelling of a piezoelectric based energy harvester. Romani et al. [3] used a cantilever structure with a proof mass at the end as the most suitable structure for maximum energy harvesting design. Since geometry plays an important role in order to lower the frequency of vibratory system Karami and Inman [4] proposed a new zigzag bi-layered system to lower the frequency of the system. Mahmoodi and Jalili [5] developed a nonlinear model for the vibration analysis of a micro cantilever beam. Roundy and Wright [6] developed piezoelectric based generator for wireless electronics. Mak et al. [7] carried out experimental modelling and nonlinear analysis of an energy harvester incorporating a bump stop. In the present work a parametrically excited cantilever beam with bimorph configuration and tip mass is considered as an energy harvester and its nonlinear dynamics is studied to check its suitability as an energy harvester. The system is subjected to a periodic axial load with base excitation. The steady state frequency response and the generated voltage for two different resonance conditions are analyzed in this work.

2 Mathematical Modeling

Figure 1 shows the schematic diagram of an energy harvester modelled as a thin uniform flexible cantilever beam with piezoelectric patches on both the sides. The beam has a tip mass and is subjected to a periodic axial load $p(t) = p_0 + p\cos(\bar{\Omega}_2 t)$ at the free end and transverse base excitation $F(t)$ at the fixed end which produces acceleration g in the structure. Using Hamilton's principle, Euler-Bernoulli beam theory and linear constitutive equations of piezoelectricity, the governing equation of motion and boundary conditions are derived. The generalized Galerkin's method is applied to reduce the equation of motion to its temporal form. As the resulting equation is found to have weak nonlinear terms, it is solved using the method of multiple scales.

The cantilever beam is assumed to be inextensible. Both the longitudinal displacement $u(s,t)$ and transverse displacement $w(s,t)$ were considered for the beam. Here, s is the distance from the fixed end and t is the time. The potential energy U and kinetic energy T of the system can be written as

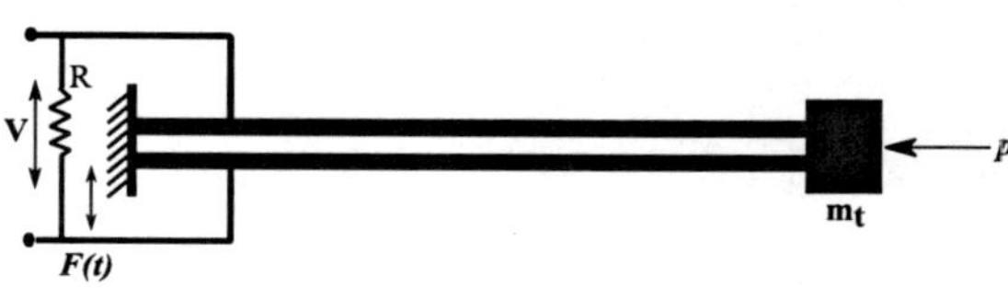

Fig. 1 Schematic diagram of an axially loaded energy harvester

$$U = \frac{1}{2}\int\limits_{v_s} \sigma_s \varepsilon_s dv_s + 2\frac{1}{2}\int\limits_{v_p} \sigma_p \varepsilon_p dv_p - 2\frac{1}{2}\int\limits_{v_p} E_v D_p dv_p \tag{1}$$

$$T = \int\limits_{v_s} \frac{1}{2}\rho_s(\dot{u}^2 + \dot{w}^2)dv_s + 2\int\limits_{v_p} \frac{1}{2}\rho_p(\dot{u}^2 + \dot{w}^2)dv_p + \frac{1}{2}m_t\dot{w}_L^2 \tag{2}$$

Here subscript s and p are used for substrate and piezoelectric layers, respectively. Also, σ, ε, E_v, ρ and m_t represent respectively, the stress, strain, electric field, density and the tip mass of the system. Using the linear constitutive equations of piezoelectricity, one can relate the stress (σ_p) and electric displacement (D_p) to the resulting strain according to the following equation.

$$\sigma_p = E_p\varepsilon_p(s, z, t) - e_{31}E_v \quad \text{and} \quad D_p = e_{31}\varepsilon_p(s, z, t) - \hat{\varepsilon}E_v \tag{3}$$

where E_v, D_p, e_{31}, $\hat{\varepsilon}$ and E_p are the electric field, electric displacement, piezoelectric constant, permittivity and the Young's modulus of piezoelectric element respectively. E_v can be related to the voltage $V(t)$ developed across a piezoelectric layer of thickness t_p according to the relation $E_v = V(t)/2t_p$. Similarly, stress in the structural element can be related to the resulting strain using the relation $\sigma_s = E_s\varepsilon_s(s, z, t)$. Also, one may use the area A and moment of inertia I of the substrate and piezoelectric layers in terms of their width b and thickness t_p, t_s using the following equations.

$$A_s = b_s t_s, \quad A_p = b_p t_p, \quad I_s = \frac{b_s t_s^3}{12}, \quad I_p = \frac{b_p t_p^3}{12} + A_p\left(\frac{t_p + t_s}{2}\right)^2 \tag{4}$$

The work done by the non conservative force can be expressed as

$$W = (\int\limits_0^{L_b} (\rho_s A_s + 2\rho_p A_p)wds)g + (p_0 + p\cos(\bar{\Omega}_2 t))u - \bar{V}(t)q \tag{5}$$

Here, q is the electric charge generated by the harvester. Using $w(s,t) = \sum_{i=1}^{\infty} r_g\phi_n(s)\bar{q}_n(t)$, where r_g is the scaling factor, $\phi_n(s)$ is the nth mode shape function and $\bar{q}_n(t)$ is the time modulation of the nth mode in Eqs. (1–5), and applying extended Hamilton's principle and generalized Galerkin's procedure, the temporal equation of motion of the system for the ith mode can be given by the following equations.

$$\ddot{q}_i + \mu_1\dot{q}_i + \omega_n^2 q_i + \Lambda_1 q_i^3 + \theta_{a1}V + \theta_{b1}q_i^2 V + F_2 q\cos(\Omega_2\tau) = F_1\cos(\Omega_1\tau) \tag{6}$$

$$-\dot{q}_i - \frac{r_g^2\theta_{c1}}{\theta_{a1}}q_i^2\dot{q}_i + \dot{V} + r_e V = 0 \tag{7}$$

where

$$\bar{q}_i = r_g q_i, \quad t = \frac{\tau}{\omega_1}, \quad \mu_1 = 2\mu, \quad \omega_n^2 = \frac{\bar{\omega}_n^2}{M\omega_1^2}, \quad \Lambda_1 = \frac{\bar{\Lambda}_1 r_g^2}{M\omega_1^2}, \quad F_{2s} = \frac{\bar{F}_2}{M\omega_1^2}$$

$$\theta_{a1} = \frac{\bar{\theta}_{a1}^2}{C_p M\omega_1^2}, F_1 = \frac{\bar{F}_1}{M\omega_1^2 r_g}, \quad r_e = \frac{1}{RC_p\omega_1}, \quad \Omega_1 = \frac{\bar{\Omega}_1}{\omega_1}, \quad \Omega_2 = \frac{\bar{\Omega}_2}{\omega_1}, \quad \theta_{b1} = \frac{\bar{\theta}_{b1}\bar{\theta}_{a1} r_g^2}{C_p M\omega_1^2}$$

$m_b = \rho_s A_s + 2\rho_p A_p$, $K_a = \frac{1}{2} b_p e_{31}(2t_s + t_p)(H(s) - H(s - L_b))$, $\omega_1 = \sqrt{\frac{K_1}{M}}$, $r_g = L$, $\bar{V} = \frac{\theta_{a1} r_g}{C_p} V$. Expressions for the other terms are given in the Appendix.

2.1 Solution of Temporal Equation

A uniform first order approximate solution of Eqs. (6) and (7) is obtained by using the method of multiple scales [8]. The time dependence can be expressed in multiple time scales as $T_n = \varepsilon^n \tau$ where ε is a book keeping parameter. The time derivative can be expressed as

$$\frac{d}{d\tau} = D_0 + \varepsilon D_1 + \varepsilon^2 D_2 + O(\varepsilon^3), \; \frac{d^2}{d\tau^2} = D_0 + 2\varepsilon D_0 D_1 + \varepsilon^2 D_1 + 2\varepsilon^2 D_0 D_2 + O(\varepsilon^3) \tag{8}$$

where $D_n = \partial/\partial T_n$. We expand q_i and V_i in the following forms:

$$q_i(\tau) = q_0(T_0, T_1, T_2) + \varepsilon q_1(T_0, T_1, T_2) + \varepsilon^2 q_2(T_0, T_1, T_2) + O(\varepsilon^3) \tag{9}$$

$$V(\tau) = V_0(T_0, T_1, T_2) + \varepsilon V_1(T_0, T_1, T_2) + \varepsilon^2 V_2(T_0, T_1, T_2) + O(\varepsilon^3) \tag{10}$$

Case 1: Simple resonance due to support motion ($\Omega_1 \approx 1$ and Ω_2 is away from 1) To express the nearness of the excitation frequency Ω_1 to the frequency of the harvester one may introduce detuning parameter $\sigma_1$and write $\Omega_1 = \omega_n + \varepsilon\sigma_1$. Substituting Eqs. (8–10), into Eqs. (6–7) and considering $\mu = \varepsilon\mu_s$, $\Lambda = \varepsilon\Lambda_s$, $\theta = \varepsilon\theta_{as}$, $F_1 = \varepsilon F_1$, $F_2 = \varepsilon F_2$ so that viscous damping appears at the same order of perturbation problem as the cubic nonlinearity, forcing and coupling and separating the terms of different order of ε one obtains the following equations.

$$O(\varepsilon^0): \quad D_0^2 q_0 + \omega_n^2 q_0 = 0 \; and \; D_0 V_0 + r_e V_0 = D_0 q_0 \tag{11}$$

$$\begin{aligned} O(\varepsilon^1): \quad D_0^2 q_1 + \omega_n^2 q_1 &= -2D_0 D_1 q_0 - \mu_1 D_0 q_0 - \Lambda_1 q_0^3 - \theta_{a1} V_0 \\ &\quad + F_1 \cos(\omega T_o + \sigma_1 T_1) - F_2 q_0 \cos(\Omega_2 T_0) \end{aligned} \tag{12}$$

$$D_0V_1 + r_eV_1 = D_0q_1 + D_1q_0 - D_1V_0 \tag{13}$$

Assuming the general solution of Eq. (11) as

$$q_0 = A(T_1)e^{i\omega_n T_0} + cc \textit{ and } V_0 = ZA(T_1)e^{i\omega_n T_0} + cc \tag{14}$$

where cc is the complex conjugate of the preceding term. Substituting q_0 into (12–13) and expressing $\cos(\omega_n T_0 + \sigma_1 T_1)$ in complex form, one obtains

$$-2i\omega_n D_1 A - i\omega_n \mu A - Z\theta_a A - 3\Lambda A^2\overline{A} + \frac{F_1}{2}e^{i\sigma_1 T_1} = 0 \tag{15}$$

To obtain the solution of the above equation, one may express the complex valued function A in the following polar form.

$$A(T_1) = \frac{1}{2}a(T_1)e^{i\beta(T_1)} \quad \overline{A}(T_1) = \frac{1}{2}a(T_1)e^{-i\beta(T_1)} \tag{16}$$

Substituting Eq. (16) in Eq. (15) and separating real and imaginary terms, one obtains the following reduced equations.

$$D_1 a = -\left(\frac{\mu}{2} + \mu_e\right)a + \frac{F_1}{2\omega_n}\sin\,\delta,\ aD_1\delta = \sigma_1 a - \gamma_s a - N_{eff}a^3 + \frac{F_1}{2\omega_n}\cos\,\delta \tag{17}$$

where, $\delta = \sigma_1 T_1 - \beta$, $\mu_e = \theta_a r_e/2(r_e^2 + \omega_n^2)$, $N_{eff} = \frac{3\Lambda}{8\omega_n}$, $\gamma_s = \theta_a\omega_n/2(r_e^2 + \omega_n^2)$

For steady state voltage and power, the time derivatives in the above equations can be set to zero and solving further the following equation is obtained:

$$\zeta_{eff}^2 a_0^2 + \left[(\sigma_1 - \gamma_s)a_0 - N_{eff}a_0^3\right]^2 = \frac{F_1^2}{4\omega_n^2} \tag{18}$$

$$zeta_eff = \mu_e + \mu/2$$

Voltage response of the system is given by the following equation.

$$\overline{V_{ss}}(t) = \frac{\theta_a r_g}{c}\left(\frac{\omega_n}{r_e^2 + \omega_n^2}\right)a_0\cos\left(\bar{\omega}_n t + \beta_0 + \psi\right), \quad \text{where} \quad \psi = \text{arc}\ \tan(r_e/\omega_n) \tag{19}$$

Case 2: Resonance due to periodic axial forcing ($\Omega_2 \approx 2$ and Ω_1 is away from 1) In this resonance condition, considering $\Omega_2 = 2\omega_n + 2\varepsilon\sigma_2$, one may obtain the following reduced equations.

$$D_1 a = -\mu_e a - \mu_s \frac{a}{2} - \frac{F_{2s}}{2\omega_n} \sin\,\delta,\; aD_1\delta = 2\sigma_2 a - 2\gamma_s a - 2N_{eff} a^3 - a\frac{F_{2s}}{\omega_n} \cos\,\delta \tag{20}$$

For steady state, the amplitude-detuning parameter relation becomes

$$\sigma_2 = \sqrt{\frac{F_{2s}^2}{4\omega_n^2} - (\mu_e + \frac{\mu_s}{4})^2} + N_{eff} a_0^2 + \gamma_s \tag{21}$$

3 Results and Discussion

For numerical results, the dimensions and material properties of the substrate and the piezoelectric bimorphs of the energy harvester are given in Table 1. For illustration purpose two examples are given below.

Case 1: Simple resonance due to support motion ($\Omega_1 \approx 1$ and Ω_2 is away from 1) Figure 2 shows the steady state frequency response and variation of voltage with variation in the frequency of the base excitation when the frequency of axial load is away from the natural frequencies. Here the amplitude of base acceleration g = 5 m/s^2, damping μ = 0.01, and the ratio of the total mass to the mass of the beam m_t/m_b = 2. In this case, it is shown that the amplitude of axial load has no significant effect on the voltage generation and one obtains a voltage of 5–30 μV by changing the frequency of base excitation from 280 to 310 Hz. Similarly, the voltage decreases from its peak 30–5 μV by increasing the frequency above the critical resonant frequency. Using this formulation and by changing system parameters one may obtain the required voltage for any application.

Case 2: Resonance due to periodic axial forcing ($\Omega_2 \approx 2$ and Ω_1 is away from 1) In this case the frequency of axial load is considered nearly equal to twice the natural frequency and the frequency of the base excitation is kept away from the natural frequency. This resonance condition is known as principal parametric

Table 1 Geometric and material properties of harvester

Material properties	Substrate	Piezo	Geometric properties (mm)	Substrate	Piezo
Young's modulus	190 GPa	66 GPa	Length	25.8	25.8
Mass density	8,700 kg m^{-3}	7,800 kg m^{-3}	Width	6.4	6.4
Dielectric constant	–	1,800	Thickness	0.14	0.19

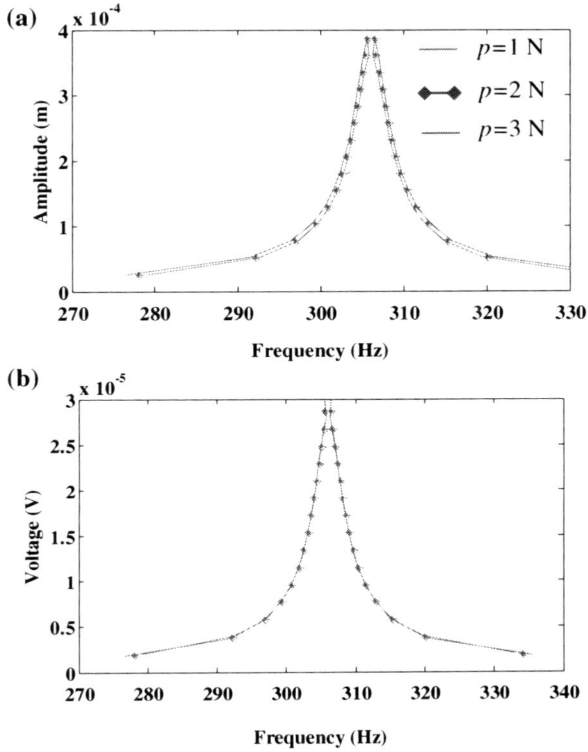

Fig. 2 Variation of **a** response amplitude **b** voltage with frequency of base excitation for axial load of 1, 2 and 3 N with $g = 5$ m/s^2 and $m_t/m_b = 2$

resonance condition and in this case, one obtains both trivial and nontrivial state. The trivial state become unstable by super and sub critical pitch-fork bifurcations. During this unstable trivial state, the system has stable nontrivial state. With increase in the amplitude of axial load, one can observe wide range of trivial unstable region and corresponding higher value of response amplitude. In this case the response amplitude is found to be very much larger than that observed in case of previous resonance case. As the voltage is linearly proportional to the response amplitude, the curve for variation in the voltage with frequency is similar to that shown in Fig. 3 but with different magnitude. In this case, one can obtain higher voltage in comparison to the previous case.

By changing the system parameters, it can be shown that one can generate voltage in the milli-volt range which can find many practical applications in the electronically control systems.

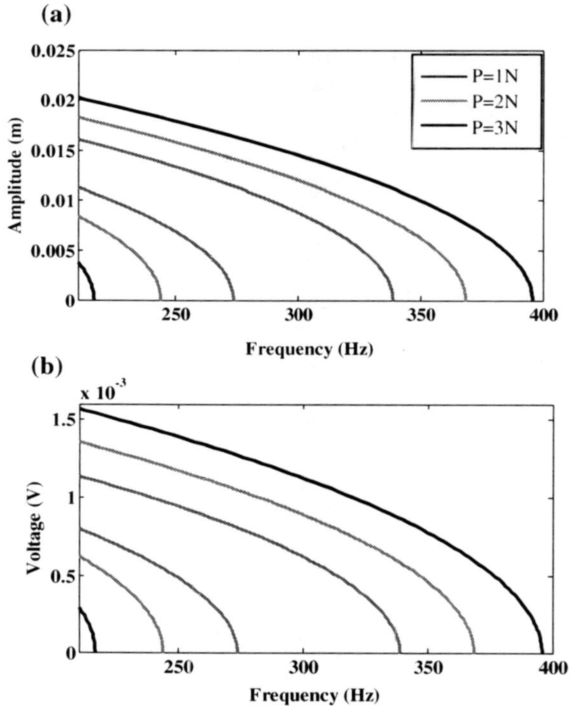

Fig. 3 Variation of **a** response amplitude **b** voltage with frequency of base excitation for axial load of 1, 2 and 3 N with $g = 5$ m/s^2 and $m_t/m_b = 2$

4 Conclusions

In this work a cantilever beam with piezoelectric bimorph and attached mass is considered for energy harvester where the fixed end is subjected to periodic acceleration and the free end is subjected to periodic axial load. The nonlinear governing equation of motion is derived and two resonance conditions have been studied. In the first case of simple resonance, the frequency of base excitation is kept nearer to the natural frequency of the system and the frequency of the axial load is kept away from the natural frequency. In the second case the frequency of axial load is kept nearer to twice the natural frequency. It is shown that by varying the frequency one can obtain a wide range of voltage.

Appendix

$\bar{\omega}_n = K_n - P_n$	$\theta_{an} = -\int_0^{L_b} \phi_n K_a'' ds$
$M = \int_0^{L_b} m_b \phi_n^2 ds$	$\theta_{bn} = -\int_0^{L_b} [\frac{1}{2}\phi_n (K_a \phi_n'^2)'' - \phi_n (K_a \phi_n' \phi_n'')']ds$
$K_n = (E_s I_s + 2E_p I_p) \int_0^{L_b} \phi_n (\phi_n'')'' ds$	$C_p = \frac{b_p \hat{\varepsilon} L_b}{2t_p}$
$P_n = p_0 \int_0^{L_b} \phi_n \phi_n'' ds$	$\theta_{cn} = \frac{3}{2}\int_0^{L_b} K_a \phi_n'^2 \phi_n ds$
$\bar{\Lambda}_n = (E_s I_s + 2E_p I_p) \int_0^{L_b} \phi_n [\phi_n' (\phi_n' \phi_n'')']' ds$	$\bar{F}_1 = m_b g \int_0^{L_b} \phi_n ds$
$\bar{F}_2 = p \cos(\bar{\Omega}_2 \bar{t})$	$z = (\omega_n^2 + i\omega_n r_e)/(\omega_n^2 + r_e^2)$

$$\phi_i(s) = -\frac{\sin \beta_n L_b + \sinh \beta_n L_b}{\cos \beta_n L_b + \cosh \beta_n L_b}(\cos \beta_n s - \cosh \beta_n s) + (\sin \beta_n s - \sinh \beta_n s)$$

β_n is obtained by solving the equation

$$1 + \cosh(\beta_n L_b)\cos(\beta_n L_b) + (\frac{m_t}{m_b})(\cos \beta_n L_b \sinh \beta_n L_b - \sin \beta_n L_b \cosh \beta_n L_b) = 0$$

References

1. Yildiz F (2007) Potential ambient energy-harvesting sources and techniques. J Technol Stud 35:40–48
2. Masana R, Daqaq MF (2011) Electromechanical modeling and nonlinear analysis of axially loaded energy harvesters. J Vib Acoust 133:011007-1
3. Romani A, Tartagni M, Sangiorgi E, Paganelli RP (2010) An energy autonomous switching converter for harvesting power from multiple piezoelectric transducers. In: IEEE Sensors Conference, pp 1173–1176. ISBN: 978-1-4244-8168-2/10
4. Karami MA, Inman DJ (2011) Analytical modeling and experimental verification of the vibrations of the zigzag microstructure for energy harvesting. J Vib Acoust 133:011002-1
5. Mahmoodi SN, Jalili N (2007) Non-linear vibrations and frequency response analysis of piezoelectrically driven micro cantilevers. Int J Non-Linear Mech 42:577–587
6. Roundy S, Wright PK (2004) A piezoelectric vibration based generator for wireless electronics. Smart Mater Struct 13(5):1131–1142
7. Mak KH, Popov AA, McWilliam S (2012) Experimental model validation for a nonlinear energy harvester incorporating a bump stop. J Sound Vib 331:2602–2623
8. Nayfeh AH (1979) Nonlinear Oscillations. Willey, New Jersey

About Multi-parametric Analysis of Drill String Vibrations

Almaz Sergaliyev, Julius Kaplunov and Lelya Khajiyeva

Abstract Coupled flexural-torsional vibrations of drill strings are subject to multi-parametric analysis. A drill string is modelled as a thin elastic beam taking into account the effect of primary static deformation induced by the prescribed compressional force and twisting moment. The appropriately normalized ratio of the two latter quantities along with dimensionless vibration frequency and beam relative thickness are assumed to be the main problem parameters. Three distinct dynamic regimes are determined. Two of them correspond to predominantly torsional or flexural vibrations. The third regime is not widely known. It is specific of the primary static deformation, caused by the aforementioned twisting moment. Simplified limiting equations are formulated for each of the regimes. Associated approximate dispersion relations are compared with the exact dispersion relation corresponding to the original problem. The developed methodology may be useful for qualitative interpretation of numerical and experimental data related to drill string modelling.

Keywords Drill string · Vibrations · Flexural-torsional vibrations · Elastic beam · Multi-parametric analysis

1 Introduction

Nowadays the main sources of energy are oil and gas. The demanded growth of volume of oil and gas production conducts to intensity of development of petroleum deposits. Therefore, oil and gas exploration represents scientific and practical

A. Sergaliyev (✉) · L. Khajiyeva
Al-Farabi Kazakh National University, Al-Farabi Avn.71, 050040 Almaty, Kazakhstan
e-mail: almaz.ss@gmail.com

L. Khajiyeva
e-mail: khadle@mail.ru

J. Kaplunov
School of Computing and Mathematics, Keele University, Staffordshire ST5 5BG, UK
e-mail: j.kaplunov@keele.ac.uk

J.K. Sinha (ed.), *Vibration Engineering and Technology of Machinery*,
Mechanisms and Machine Science 23, DOI 10.1007/978-3-319-09918-7_33

interest. The problem of security of reliability and durability of a borehole is one of the main problems of the oil and gas extraction branch.

Engineering and economical performance of drilling in the oil industry are dependent on many factors. These are the characteristics of drilling machines and tools, their operating modes, rock hardness, etc. Most boreholes are known to be rejected because of the curvature and drill rod breakage as a result of incorrect operating mode. Therefore, many works are devoted to the study of dynamics of boring columns. Among them are the works [1–9], where longitudinal, transverse and rotating vibrations of drilling columns under the effect of exterior forces and moments, inertial forces of torsion, inertial forces of flushing liquid, forces of friction of the drilling column against a slit and other internal and external actions, and also problems of their stability are considered. The author of the paper [1] considers the problems of investigation of the dynamic bending of elongated drill string tubes studied with allowance made for destabilizing effects of the longitudinal non-uniform preloading, action of torque, inertia forces of rotation and internal flows of washing liquid (mud). In the work [2] the stability of transverse vibrations of the nonlinear model of a compressed-torsioned drill-string is considered. For the analysis the Flock theory is used. For the analysis of transitions of a drilling column different approaches to model the drill string dynamics are used: one-mode approximation [3, 4], a beam model together with the finite element method [5–7], the Cosserat theory [8], and a discrete system with two degrees of freedom [9].

In the present paper, as well as in the paper [1], a drill string is modelled as a thin elastic beam taking into account the effect of primary static deformation induced by the prescribed compressional force and the twisting moment. Unlike the above specified methods of the analysis of models of drilling columns, a method of the multi parametrical analysis developed by the authors [10] is used here.

2 Statement of the Problem

The rotating drill string is considered under the effect of the constant longitudinal force T = const and constant torque M_z = const.

The equations of the bent axis for a rectilinear rod are represented as follows [1]:

$$\begin{aligned} &EJ\frac{\partial^4 u}{\partial z^4} - T\frac{\partial^2 u}{\partial z^2} - M_z\frac{\partial^3 v}{\partial z^3} - \rho F\omega^2 u + \rho F\frac{\partial^2 u}{\partial t^2} = 0, \\ &EJ\frac{\partial^4 v}{\partial z^4} - T\frac{\partial^2 v}{\partial z^2} + M_z\frac{\partial^3 u}{\partial z^3} - \rho F\omega^2 v + \rho F\frac{\partial^2 v}{\partial t^2} = 0, \end{aligned} \tag{1}$$

where F is a cross-section area; EJ is the bending stiffness of the rod; $u(z,t)$ and $v(z,t)$ are elastic displacements in the planes xOz, yOz, correspondingly (Fig. 1); ω is an angular velocity of rotation of the drill string.

Fig. 1 Design scheme of a drill string

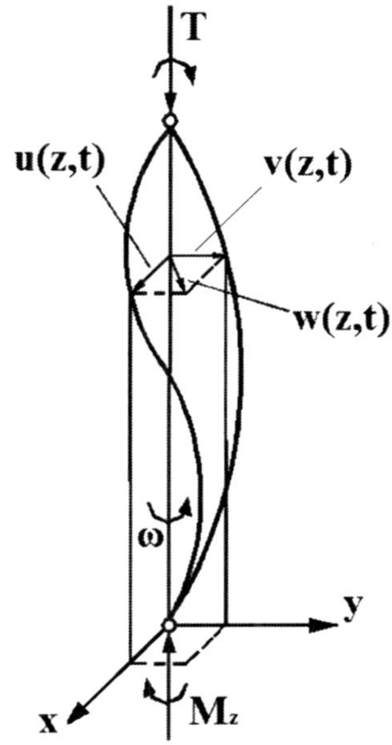

Further on we consider only simple harmonic motions, therefore the second derivative on time will be substituted everywhere for $(-\lambda^2)$, where λ is the circular frequency, and the temporal factor $\exp(i\lambda t)$ is lowered everywhere [10]. Let us introduce dimensionless variables:

$$z = \tilde{z}L, \quad u = \tilde{u}L, \quad v = \tilde{v}L, \tag{2}$$

where L is the length of the drill string.

The set of equations (l) in dimensionless quantities is represented as follows, thus the sign «$\sim$» is lowered further on:

$$\begin{array}{l} \frac{EJ}{TL^2}\frac{\partial^4 u}{\partial z^4} - \frac{\partial^2 u}{\partial z^2} - \frac{M_z}{TL}\frac{\partial^3 v}{\partial z^3} - \frac{\rho F \omega^2 L^2}{T}u - \frac{\rho F \lambda^2 L^2}{T}u = 0, \\ \frac{EJ}{TL^2}\frac{\partial^4 v}{\partial z^4} - \frac{\partial^2 v}{\partial z^2} + \frac{M_z}{TL}\frac{\partial^3 u}{\partial z^3} - \frac{\rho F \omega^2 L^2}{T}v - \frac{\rho F \lambda^2 L^2}{T}v = 0 \end{array}. \tag{3}$$

Let us introduce new notations:

$$\frac{EJ}{TL^2} = \eta^2, \ \frac{M_z}{TL} \sim \eta^\alpha, \ \alpha \geq 0$$
$$c = \sqrt{\frac{T}{\rho F}}, \ \Omega = c\omega L, \ \Lambda = c\lambda L$$

where c is the shear-wave velocity.

$$\frac{\partial}{\partial x} \sim \frac{\partial}{\partial \gamma}\eta^{-q},\ \frac{\partial}{\partial \gamma} \sim 1,\ q \geq 0$$
$$\Omega \sim \eta^{-a_1},\ \Omega = \Omega_* \eta^{-a_1},\ \Omega_* \sim 1,\ a_1 \geq 0$$
$$\Lambda \sim \eta^{-a_2},\ \Lambda = \Lambda_* \eta^{-a_2},\ \Lambda_* \sim 1,\ a_2 \geq 0$$

Substituting the received expressions in (3), after the simplifications we receive:

$$\begin{aligned} &\eta^{2-2q}\frac{\partial^4 u}{\partial \gamma^4} - \frac{\partial^2 u}{\partial \gamma^2} - \eta^{\alpha-q}\frac{\partial^3 v}{\partial \gamma^3} - (\Omega_*^2 \eta^{2q-2a_1} + \Lambda_*^2 \eta^{2q-2a_2})u = 0, \\ &\eta^{2-2q}\frac{\partial^4 v}{\partial \gamma^4} - \frac{\partial^2 v}{\partial \gamma^2} + \eta^{\alpha-q}\frac{\partial^3 u}{\partial \gamma^3} - (\Omega_*^2 \eta^{2q-2a_1} + \Lambda_*^2 \eta^{2q-2a_2})v = 0 \end{aligned} \tag{4}$$

The received set of Eqs. (4) is multiparametrical.

The analysis of the received multi-parametric system (4) will allow to allocate three distinct dynamic regimes. Two of them correspond to predominantly torsional or flexural vibrations. The third regime is not widely known. It is specific of the primary static deformation caused by the aforementioned twisting moment. These regimes are defined by way of detection of dominating parameters among α, q, a_1, a_2 and neglect of the "weak" components in the system (4). Simplified limiting equations are formulated for each of the regimes. Associated approximate dispersion relations are compared with the exact dispersion relation corresponding to the original problem.

3 Conclusion

The developed methodology may be useful for qualitative interpretation of numerical and experimental data related to drill string modelling. In this paper small transverse vibrations of the drill string are investigated. The results of their multi-parametric analysis can serve as the first approximation at the solution of problems of nonlinear oscillations of drill stings. Nonlinearity of the model of drill strings oscillations can be stipulated by removal of limitations on magnitudes of their deformations, that is their finiteness.

References

1. Gulyayev VI, Borshch OI (2011) Free vibrations of drill strings in hyper deep vertical bore-wells. J Petrol Sci Eng 78:759–764
2. Khajiyeva L, Sergaliyev A, Umbetkulova A (2013) Dynamic analysis of steel and dural drill rods. Adv Mater Res 705:91–96

3. Yigit A, Christoforou A (1996) Coupled axial and transverse vibrations of oil well drill strings. J Sound Vib 195:617–627
4. Christoforou AP, Yigit AS (1997) Dynamic modeling of rotating drill strings with borehole interactions. J Sound Vib 206:243–260
5. Khulief YA, AL-Naser H (2005) Finite element dynamic analysis of drill strings. Finite Elem Anal Des 41:1270–1288
6. Sampaio R, Piovan MT, Venero G (2007) Coupled axial/torsional vibrations of drill-strings by means of a nonlinear model. Mechanics Research Communication 34:497–502
7. Ritto TG, Soize C, Sampaio R (2009) Nonlinear dynamics of a drill-string with uncertain model of the bit–rock interaction. Int J Non-Linear Mech 44:865–876
8. Tucker RW, Wang C (1999) An integrated model for drill-string dynamics. J Sound Vib 224:123–165
9. Richard T, Germay C, Detournay E (2007) A simplified model to explore the root cause of stick–slip vibrations in drilling systems with drag bits. J Sound Vib 305:432–456
10. Kaplunov JD, Kossovich LYu, Wilde MV (2010) Boundary and interface resonance phenomena in elastic bodies. Physmathlit, Moscow. ISBN 978-5-9221-1280-2

Tuned Mass Damper Inverted Pendulum to Reduce Offshore Wind Turbine Vibrations

P.V.B. Guimarães, M.V.G. de Morais and S.M. Avila

Abstract Wind energy is in a fast development worldwide and is currently receiving great investments mostly for being a clean energy source. Advances in this area are resulting in increasingly high and slender wind turbines, intensifying vibrations in structures caused by its own operation and, also, by wind force. Searching a better performance leads to offshore devices, i.e. wind turbines installed at the ocean next to coast. These structures take advantage of wind forces generated by more intense and consistent wind with less turbulence in these regions. One of the conceptions for this kind of wind turbine is the floating device. A solution to the problem of excessive vibration, which has been studied by several researchers in recent years, is structural control. One of the structural control devices, already extensively studied and implemented in practice, is the Tuned Mass Damper (TMD). An alternative geometry for TMD is the pendulum absorber. However, previous studies have shown that pendulum TMD installed on floating wind turbines requires an excessive length, therefore a different approach is required. In this work it is proposed an inverted pendulum TMD to improve control performance in an offshore wind turbine modeled as an inverted pendulum. This device may be considered as a linear oscillator only for small vibration amplitudes. Numerical simulations are performed to define TMD parameters that improve the control device performance. Good performance is achieved. However, passive devices only work properly for the design frequency range, and wind forces are random type of excitations. Better results would be achieved if a robust control is developed. This study will serve as a basis for the proposition of a semi-active device.

Keywords Tuned mass damper · Vibration control · Offshore wind turbine · Inverted pendulum

P.V.B. Guimarães (✉) · M.V.G. de Morais · S.M. Avila
University of Brasilia, Brasilia, Brazil
e-mail: pedrobarca@hotmail.com

M.V.G. de Morais
e-mail: mvmorais@unb.br

S.M. Avila
e-mail: avilas@unb.br

J.K. Sinha (ed.), *Vibration Engineering and Technology of Machinery*, Mechanisms and Machine Science 23, DOI 10.1007/978-3-319-09918-7_34

1 Introduction

Wind energy generation is a renewable, clean energy with low environmental impact and available at various places around the world. This power source has been in fast development due to increasing requirements for better management of energy generating sources, requiring clean energy generation. One of the challenges of this field is the structural analysis of wind turbines.

Wind turbines capture wind kinetic energy through the blades, turning it into mechanical energy, which is transformed into electrical energy. A wind turbine comprises a set of equipment that consist of a rotor, which is coupled to the blades, a tower and the nacelle, where are the generator and other components.

Supporting towers due to its geometry and highness are slender and flexible, they can present excessive vibrations caused by its own functioning and also by wind action. A detailed analysis of the tower structural behavior is of great importance since its cost represents an amount of 30 % of system total cost [1].

When these devices are installed at the sea, near the coast, are called offshore wind turbines, their biggest advantage is that wind in these places are more consistent and intense with less turbulence [2]. One of the offshore wind turbine conceiving is the floating one. This structural system has an additional excitation source compared to onshore wind turbine, due to loads caused by the motion of sea waves at the base.

Reducing vibrations in structures excited by wind forces has increased its need and importance due to the tendency of more and more high and slender structures. These types of structures are more vulnerable to excessive vibrations caused by wind loading, that can affect both structural safety and people comfort. Wind turbine towers are also becoming more high and slender, they can reach more than 120 m of highness. One of the tallest towers is currently the Enercon E-126 which produces 7.5 MW and has a height of 135 m [1]. High amplitudes of vibration compromise safety of these towers as well as the good operation of wind turbines. Thus it justifies the study of alternatives to avoid these problems.

Structural control is a technology that aims to reduce excessive vibration installing external devices or applying external forces that promote changes in stiffness and damping of the system [3].

One of typical structural control devices is the tuned mass damper (TMD), it controls structural response transferring energy between main structure and an auxiliary mass. As the name suggests, this device is tuned to a given modal frequency, it is designed to control structures that vibrate mainly around a given mode vibration, usually the first one, such as high towers [4].

One conception of the TMD is the pendulum geometry [5]. This absorber has its vibration period depending on the pendulum length. It only can be considered as a linear oscillator when amplitude vibrations are small, otherwise the system turns to be nonlinear.

Recent studies have focused on vibration control in wind turbines using TMDs [6–10]. However, despite previous studies, obstacles need to be overcome. Among them are: cost reduction/maintenance and increased reliability, efficiency and robustness.

On a recent work [11] it was presented an inverted pendulum model to study the dynamic behavior and stability of a floating wind turbine. A passive control system, a pendulum TMD was connected to the structure aiming to reduce the angular amplitude of the main structure. To improve control system performance, a parametric study was carried out changing TMD damping and stiffness. However, results showed that a good control system efficiency could only be reached adopting high values of pendulum length, which is impracticable on constructive point of view. In this work, this study was continued, this time adopting an inverted pendulum geometry also for the TMD. Parametric studies were performed to find out best absorber parameters. Good performance is achieved using small pendulum length, which is a practice feasible solution.

2 Floating Wind Turbine Model: Inverted Pendulum

For modeling the dynamic behavior of an offshore wind turbine, this paper studies the behavior of a discrete model of an inverted pendulum [12] as shown in Fig. 1. It is a system with a spring and a damper at the base that simulate fluid resistance of the sea. The excitation of the system is a dynamic force applied to the mass m, simulating the wind force on the blades of the turbine.

On Fig. 1 m is the tip mass at the end of the bar, x_m and y_m are m coordinates, ρ is bar linear density, x_b and y_b are gravitational centre bar coordinates, θ is bar angular amplitude, l is bar length, K is torsional stiffness, g is gravity acceleration, m_c is car mass, u is car horizontal displacement, k_c is car stiffness, c is car damping coefficient.

Figure 2 shows the discrete model of an inverted pendulum with an absorber that is also an inverted pendulum. Where m_d is TMD pendulum mass, l_d is TMD pendulum length, d is distance of pendulum to bar base, r is TMD disc radius, c_d is TMD damping coefficient, k_d is TMD's stiffness and θ_d is TMD pendulum's angular amplitude related to the bar.

To consider the rod mass it is necessary to define coordinates x_{bd} and y_{bd}, s varies from 0 to l_d:

$$x_{bd} = d \cdot \sin\theta - s \cdot \sin(\theta_d - \theta) + u \cong d \cdot \theta - s \cdot (\theta_d - \theta) + u \quad (1)$$

$$y_{bd} = d \cdot \cos\theta + s \cdot \cos(\theta_d - \theta) \cong d + s \quad (2)$$

Considering rod density ρ_d, kinetic energy K_E, potential energy P_E and dissipated energy D_E can be calculated:

Fig. 1 Model of an inverted pendulum on a base

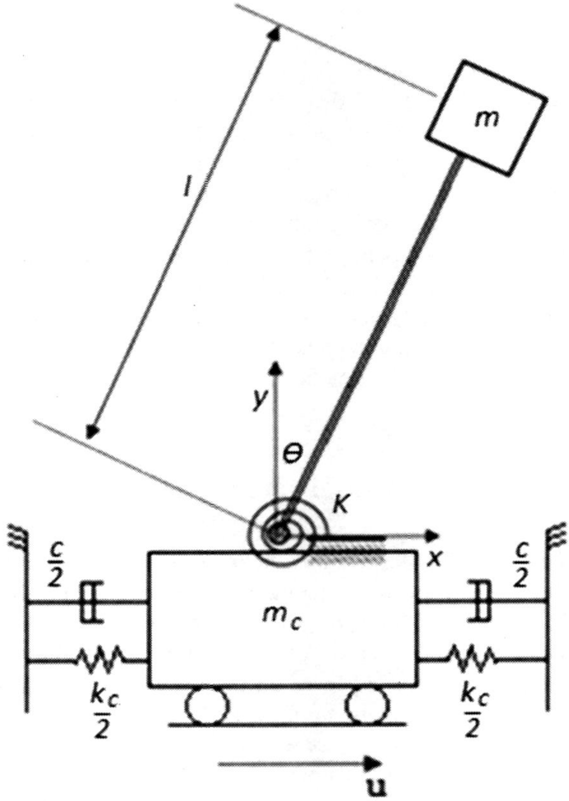

$$
\begin{aligned}
K_E = \frac{1}{2}\Big(& m_c \cdot \dot{u}^2 + m \cdot l^2 \cdot \dot{\theta}^2 + 2 \cdot m \cdot l \cdot \dot{\theta} \cdot \dot{u} + m \cdot \dot{u}^2 + m_d \cdot d^2 \cdot \dot{\theta}^2 - 2 \cdot m_d \cdot d \cdot l_d \cdot \dot{\theta}\left(\dot{\theta}_d - \dot{\theta}\right) \\
& + 2 \cdot m_d \cdot d \cdot \dot{\theta} \cdot \dot{u} + m_d d^2 \cdot \dot{\theta}_d^2 - 2 \cdot m_d d^2 \cdot \dot{\theta}_d \dot{\theta} + m_d d^2 \cdot \dot{\theta}^2 - 2 \cdot m_d \cdot l_d\left(\dot{\theta}_d - \dot{\theta}\right) \cdot \dot{u} \\
& + \frac{1}{3}\rho \cdot l^3 \cdot \dot{\theta}^2 + \rho \cdot l^2 \cdot \dot{\theta} \cdot \dot{u} + l \cdot \rho \cdot \dot{u}^2\Big) + \frac{\rho_d}{2} \cdot l_d \cdot d^2 \cdot \dot{\theta}^2 - \rho_d \cdot d \cdot \frac{l_d^2}{2} \cdot \dot{\theta}\left(\dot{\theta}_d - \dot{\theta}\right) \\
& + \rho_d \cdot l_d \cdot d \cdot \dot{\theta} \cdot \dot{u} + \frac{\rho_d}{2} \cdot \frac{l_d^3}{3}\left(\dot{\theta}_d - \dot{\theta}\right)^2 - \rho_d \cdot \frac{l_d^2}{2} \cdot \dot{u}\left(\dot{\theta}_d - \dot{\theta}\right) + \frac{\rho_d}{2} \cdot l_d \cdot \dot{u}^2
\end{aligned} \tag{3}
$$

$$
\begin{aligned}
P_E = \frac{1}{2}\big(& K \cdot \theta^2 + k_d \cdot r^2 \cdot \theta_d^2 + \left(2 \cdot m \cdot l + 2 \cdot m_d \cdot d + \rho \cdot l^2\right) g \cdot \cos\theta \\
& + 2 \cdot m_d \cdot l_d \cdot g \cdot \cos(\theta_d - \theta) + k_c \cdot u^2\big) + g \cdot \rho_d \cdot l_d \cdot \frac{l_d}{2} \cdot \cos\theta_d - \theta + g \cdot \rho_d \cdot l_d \cdot d \cdot \cos\theta
\end{aligned} \tag{4}
$$

Fig. 2 System with TMD

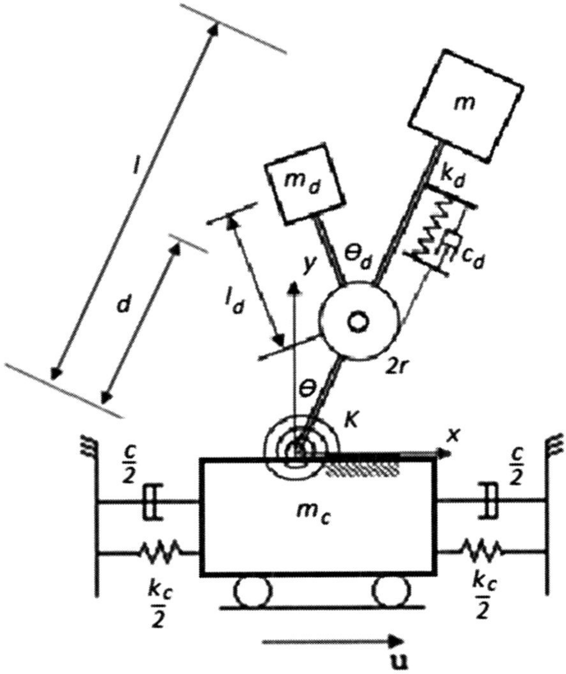

$$D_E = \frac{1}{2}\left(c_d \cdot \left(r \cdot \dot{\theta}_d\right)^2 + c \cdot \dot{u}^2\right) \tag{5}$$

Through the above energy expressions, using Lagrange equation, motion equations of the system can be written on a matricial form:

$$\begin{bmatrix} M_{1,1} & M_{1,2} & M_{1,3} \\ M_{2,1} & M_{2,2} & M_{2,3} \\ M_{3,1} & M_{3,2} & M_{3,3} \end{bmatrix} \begin{bmatrix} \ddot{\theta} \\ \ddot{\theta}_d \\ \ddot{u} \end{bmatrix} + \begin{bmatrix} 0 & 0 & 0 \\ 0 & c_d \cdot r^2 & 0 \\ 0 & 0 & c \end{bmatrix} \begin{bmatrix} \dot{\theta} \\ \dot{\theta}_d \\ \dot{u} \end{bmatrix}$$
$$+ \begin{bmatrix} K - g\left(m \cdot l + \frac{\rho^2}{2} + m_d(d + l_d) - \rho_d d\left(d + \frac{l_d}{2}\right)\right) & +m_d \cdot g \cdot l_d + \rho_d \cdot \frac{l_d^2}{2} \cdot g & 0 \\ +m_d \cdot g \cdot l_d + \rho_d \cdot \frac{l_d^2}{2} \cdot g & k_d \cdot r^2 - m_d \cdot g \cdot l_d - \rho_d \cdot \frac{l_d^2}{2} \cdot g & 0 \\ 0 & 0 & k_c \end{bmatrix} \begin{bmatrix} \theta \\ \theta_d \\ u \end{bmatrix}$$
$$= \begin{bmatrix} F(s) \cdot l \\ 0 \\ 0 \end{bmatrix} \tag{6}$$

where:

$$M_{1,1} = \frac{\rho \cdot l^3}{3} + m \cdot l^2 + m_d(d + l_d)^2 + \rho_d \cdot l_d(d + l_d) + \rho_d \cdot \frac{l_d^3}{3}$$
$$M_{1,2} = -m_d \cdot d \cdot l_d - m_d \cdot l_d^2 - \rho_d \cdot d \cdot \frac{l_d^2}{2} - \rho_d \cdot \frac{l_d^3}{3}$$
$$M_{1,3} = m \cdot l + \frac{\rho \cdot l^2}{2} + m_d(d + l_d) + \rho_d \cdot d \cdot l_d + \rho_d \cdot \frac{l_d^2}{2}$$
$$M_{2,1} = -m_d \cdot d \cdot l_d - m_d \cdot l_d^2 - \rho_d \cdot d \cdot \frac{l_d^2}{2} - \rho_d \cdot \frac{l_d^3}{3}$$
$$M_{2,2} = m_d \cdot l_d^2 + \rho_d \cdot \frac{l_d^3}{3}$$
$$M_{2,3} = -m_d \cdot l_d - \rho_d \cdot \frac{l_d^2}{2}$$
$$M_{3,1} = m \cdot l + \frac{\rho^2}{2} + m_d(d + l_d) + \rho_d \cdot d \cdot l_d + \rho_d \cdot \frac{l_d^2}{2}$$
$$M_{3,2} = -m_d \cdot l_d - \rho_d \cdot \frac{l_d^2}{2}$$
$$M_{3,3} = m_c + m + m_d + \rho \cdot l + \rho_d \cdot l_d$$

3 Numerical Results

Numerical simulations were carried out in order to analyze the effectiveness of an inverted pendulum TMD installed on a floating wind turbine discrete model. The system properties considered were those from a NREL 5 MW Wind turbine presented by [13]: Rotor, hub diameter: 126.3 [m]; hub height: 90 [m]; rotor mass 110,000 [kg]; tower mass 347,460 [kg].

Also the following values were considered for system parameters: $l = 90$ [m], $m_c = 240{,}000$ [kg], $g = 9.81$ [m/s^2], $K = 8.27 \times 10^8$ [N/rad], $m = 110{,}000$ [kg], $m_d = 11{,}000$ [kg], $d = 60$ [m], $l_d = 67.5$ [m], $k_d = 10^6$ [N/m], $c_d = 10^6$ [N s/m]. The wind dynamic loading was modeled as an applied load at the inverted pendulum mass. Two loading cases were considered: harmonic force ($F(t) = 33333 + 16666 * \sin(0.97 * t)$ [N]) and white noise force. The value of frequency of 0.97 rad/s was considered for being next to the system natural frequency.

Table 1 Best results for TMD parameters considering rod mass

d	l_d (m)	k_d ($\times 10^7$) (N/m)	c_d ($\times 10^6$) (N s/m)	RMS reduction (%)
77.5	9.5	1	1	32.44
77.5	9.5	1	10	29.4
76.0	11	1	0.1	28.26
76.0	11	1	1	33.74
76.0	11	1	10	29.9
74.5	12.5	1	10	29.32
58.0	29	10	100	28.35
56.5	30.5	10	100	28.94
55.0	32	10	100	29.46
53.5	33.5	10	10	28.65
53.5	33.5	10	100	29.88
52.0	35	10	10	29.78
52.0	35	10	100	30.23
50.5	36.5	10	10	31.05
50.5	36.5	10	100	30.49
49.0	38	10	10	32.48
49.0	38	10	100	30.67

In a way to study the TMD effectiveness a parametric study was carried out, considering the harmonic loading. The length l_d was varied from 2 to 38 meters, with a 1.5 m step, ensuring that the distance d varied too to maintain the rod top 3 m from the main tower top. The values of k_d and c_d were changed from 1 to 10^{10}. The best results found are presented on Table 1, that shows the *rms* angular displacement reduction for each case.

It can be observed that values of l_d, from 29 to 38 m doesn't lead to a considerable better performance, when compared to the shorter lengths. High values of l_d can lead to buckling phenomena. Parameter intervals where occur the better results are around $k_d = 10^7$ N/m, $c_d = 10^5$–10^7 N s/m, $l_d = 9$–12.5 m. Performing a new parametric analysis refining these intervals lead to the optimum values: $d = 79.6$ m; $l_d = 7.4$ m; $k_d = 5.9 \times 10^6$ N/m; $c_d = 4.4 \times 10^5$ N s/m.

Figures 3 and 4 show angular displacement time history, with and without control, considering the two loading cases: harmonic and white noise forces. Note that, in both cases, the control system has a very satisfactory performance, resulting in very small θ amplitudes, ensuring the linear system behavior.

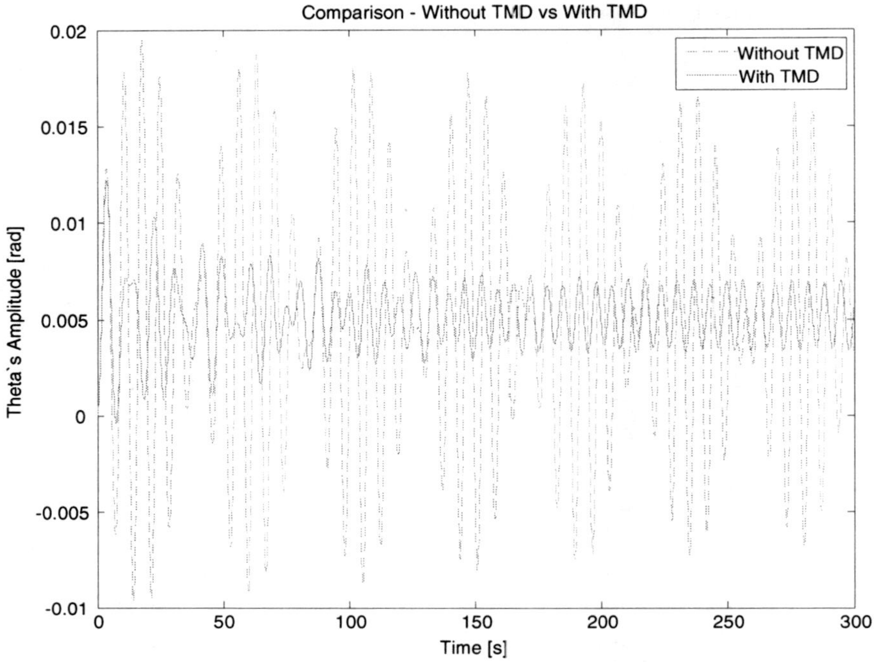

Fig. 3 Angular displacement θ time history: harmonic loading

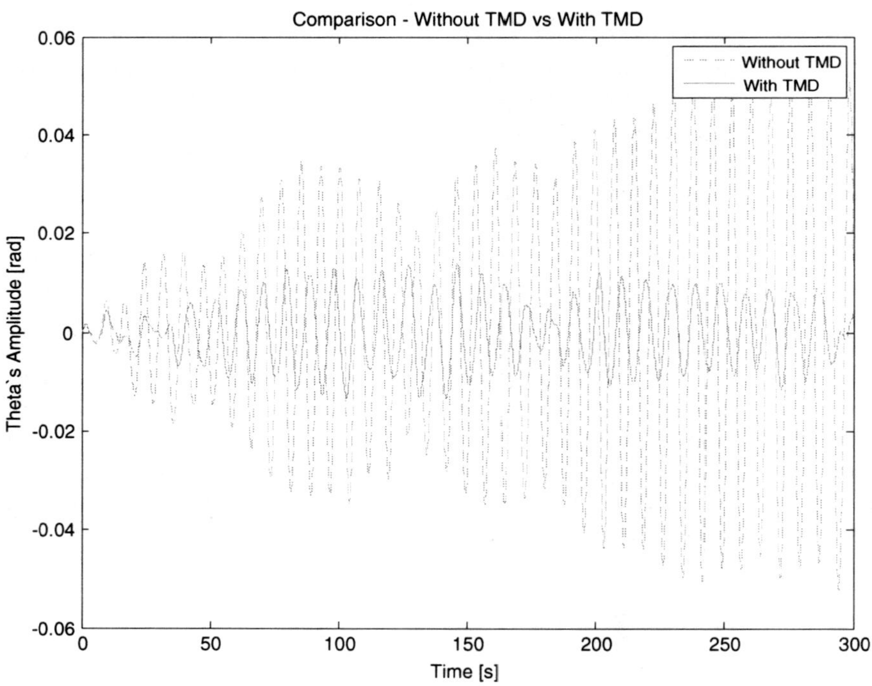

Fig. 4 Angular displacement θ time history: white noise

Table 2 Natural frequencies

Natural frequencies (rad/s)
0.000115835330287
0.660496037402893
1.135136775222053

A modal analysis gave the natural frequencies of the system (Table 2). These values correspond to the system with optimum parameter values. A comparison between frequency responses with and without control is showed on Fig. 5. It is observed that the TMD is able to reduce the resonance peak of the system without control about 10 times.

To ensure the hypothesis of linear behavior, the value of maximum angular amplitude θ_d was verified for both cases of loading, the values stayed within the given limit of 0.26 rad: $\theta_d = 0.176$ rad for harmonic loading and $\theta_d = 0.126$ rad for white noise loading. This response value is of great importance, from a practical point of view, in a way that the pendulum does not collide with the inner walls of the tower.

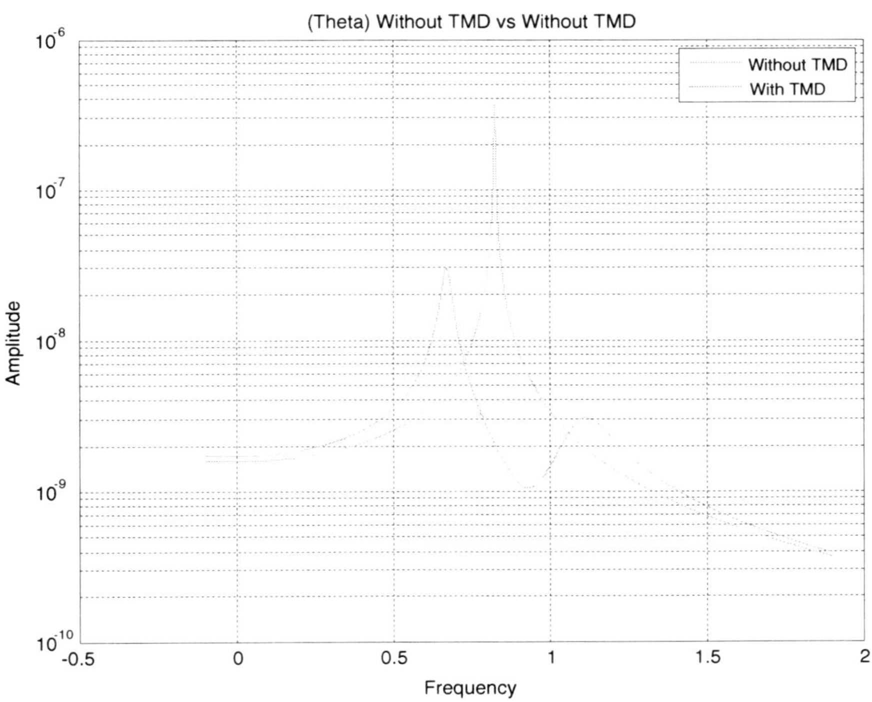

Fig. 5 System frequency response with and without control

4 Conclusion

This paper presents a study of the behavior of an offshore wind turbine by means of a simplified discrete model of an inverted pendulum. A proposal of a control system that minimizes excessive vibration and maintain the behavior of the system within the linear regime is performed. Using an inverted pendulum TMD, two cases of loading are considered to study the performance of the control system. A parametric study was conducted to achieve smooth operation. The control system showed to be very effective reducing significantly the amplitude vibration of the main system remaining within the boundary of linear behavior. This study is also serve to a semiactive device proposal.

Acknowledgment The authors gratefully acknowledge the financial support of the Brazilian agency CNPq for financial support of this study.

References

1. Quilligan A, O'Connor A, Pakrashi V (2012) Fragility analysis of steel and concrete wind towers. Eng Struct 36:270–282
2. Pao LY, Johnson KE (2009) A tutorial on the dynamics and control of wind turbines and wind farms, Proceedings of the 2009 American Control Conference: 2076–2089, St Louis, USA
3. Avila SM (2002) Controle Híbrido para Atenuação de Vibrações em Edifícios. PhD thesis, Pontifícia Universidade Católica do Rio de Janeiro, Rio de Janeiro
4. Soong TT, Dargush GF (1997) Passive energy dissipation in structural engineering. Wiley, Chichester
5. Lourenço R (2011) Design, construction and testing of an adaptive pendulum tuned mass damper. Master's dissertation, Waterloo, Ontario, Canada
6. Murtagh PJ, Ghosh A, Basu B, Broderick BM (2008) Passive control of wind turbine vibrations including blade/tower interaction and rotationally sampled turbulence. Wind Energy 11:305–317
7. Lackner MA, Rotea AR (2011) Passive structural control of offshore wind turbines. Wind Energy 14:373–388
8. Stewart G (2012) Load reduction of floating wind turbines using tuned mass dampers. Master's dissertation, University of Massachussets Amherst
9. Avila SM, Barcelos M, Morais MVG, Shzu MAM, Silva RC (2009) Vibration control of the set tower and wind turbine under the wind influence. In: 20th international congress of mechanical engineering COBEM, 2009, Gramado/RS. Proceedings of 20th International congress of mechanical engineering COBEM, 2009
10. Avila SM, Pereira WM (2011) Controle de vibrações em torres eólicas submetidas à ação de cargas harmônicas utilizando amortecedor de massa sintonizada na forma de pêndulo. In: 10ª Conferência Brasileira de Dinâmica Controle e Aplicações DINCON, 2011, Águas de Lindóia. Anais da 10ª Conferência Brasileira de Dinâmica Controle e Aplicações DINCON, 2011
11. Guimarães PVB, Avila SM, Shzu MAM, Prado ZJG, Morais MVG (2013) Vibration control of an offshore wind turbine modeled as an inverted pendulum, ICOVP 2013. Lisboa, Portugal
12. Drummond AC, Oliveira KC, Bauchspiess A (1999) Estudo do Controle de Pêndulo Inverso sobre Carro Utilizando Rede Neyral de Base Radial, IV Congresso Brasileiro de Redes Neurais
13. Stewart GM, Lackner MA (2011) The effect of actuator dynamics on active structural control of offshore wind turbines, Eng Struct 33:1807–1816

Vibration Control of Active Vehicle Suspension System Using Fuzzy Logic Controller

A. Shehata, H. Metered and Walid A.H. Oraby

Abstract Fuzzy logic control (FLC) algorithm grants a means of converting a linguistic control technique and it is widely used in vehicle applications. This paper demonstrates the application of fuzzy logic technique to design a controller for the active vehicle suspension system to improve the suspension system performance by altering the number and arrangement of the rules set and the universe of discourses. A mathematical model and equations of motion of quarter vehicle active suspension is derived and solved using MATLAB/Simulink software. The proposed fuzzy controllers using 9, 25 and 49 rules set with two different types of membership functions, trapezoidal and triangle, are implemented in a closed loop control system to demonstrate the influence of the numbers of rule set and the type of membership function on the performance of suspension system. Suspension performance criteria were assessed in both time and frequency domains. Performance comparisons between the passive suspension, as a reference, and the proposed controllers of the active suspension were achieved. The simulation results indicate that the proposed active fuzzy controllers can dissipate the energy due to road excitation effectively and improves suspension performance. Among the investigated systems, the 25 rules set with a trapezoidal membership function for the fuzzy controller gives the best performance.

Keywords Active suspension system · Fuzzy logic control · Ride comfort · Vehicle stability · Vehicle suspension control

A. Shehata (✉) · H. Metered · W.A.H. Oraby
Helwan University, Cairo, Egypt
e-mail: ahmedshehatagad@yahoo.com

H. Metered
e-mail: hassan.metered@yahoo.com

W.A.H. Oraby
e-mail: waho911@hotmail.com

J.K. Sinha (ed.), *Vibration Engineering and Technology of Machinery*,
Mechanisms and Machine Science 23, DOI 10.1007/978-3-319-09918-7_35

1 Introduction

Through the design of a suspension system, a number of conflicting requirements has to be solved. The suspension system has to provide a comfortable ride and vertical vehicle stability at the same time. Also, enough contact between tires and road surface is needed in various driving conditions in order to maximize vehicle stability. Instead of a passive suspension, an active suspension can be applied in order to provide better resolve the trade-off between these conflicting requirements. Designing a good suspension system with optimum vibration performance under different road conditions is a significant undertaking.

Fuzzy logic is used to hold the active suspension and the membership functions are optimized by using genetic algorithm operations [1]. A fuzzy controller is designed with Interval type-2 FLC to improve the disturbance rejection of the suspension system which can be sourced by road shocks [2]. The vehicle vibration and disturbance are reduced considerably with a fuzzy logic controller, to improve comfort in riding faced with uncertain road terrains [3]. The ride comfort is improved by means of the diminution of the body acceleration when road disturbances from smooth road and real road roughness [4]. Fuzzy logic is used to control active suspension of a half-car model [5]. The fuzzy logic controller is based on two inputs, namely, suspension velocity and body velocity. The output of the fuzzy controller is the damping coefficient of the variable damper [6]. Fuzzy logic is used to tune each parameter of PID Controller and input membership function of fuzzy controller optimized by Discrete Action Reinforcement Learning Automata (DARLA) technique [7].

Active suspension system of a quarter car model using adaptive fuzzy logic and active force control strategies [8]. A fuzzy logic controller is designed for the control of the seven degrees of freedom full vehicle model [9]. An active control based on fuzzy logic is constructed such as a result as of the vehicle body acceleration, suspension working space and dynamic tire load. The amplitude of vertical vehicle both, from the view of ride comfort of passengers, is minimized under the restraint of the suspension travel, relative velocity, body velocity, wheel speed and tire deflection. The development of a better-quality suspension system remains an important increase objective for the automotive industry. A vehicle suspension system should have the qualifications to reduce the displacement and acceleration of the vehicle body, to achieve its advantages not only in the ride comfort aspect, but also minimizes the dynamic binding of the tire to maintain better tire–terrain contact. In this article, six fuzzy controllers are proposed for active suspension system. Intervals of fuzzy rules are designed in order to improve system robustness to noise measurement and external disturbances araised from road irregularities to the systems studied (passive and active) suspension systems. The remainder of this article is organized as follows; a quarter vehicle model is presented in Sect. 2. Section 3 introduces a brief description of the Fuzzy logic controller approach. Finally, the results are discussed in Sect. 4 and the extracted concliosions are in Sect. 5.

2 Quarter Vehicle Model

Figure 1 shows the two-degree-of-freedom system that represents the quarter vehicle model. It consists of an upper or body mass, m_b, as well as a lower or wheel mass, m_w.

By applying the Newton's second law, the body and wheel accelerations can be written as follows:

$$\ddot{Z}_b = \frac{f_d - k_s(Z_b - Z_w) - c_s(\dot{Z}_b - \dot{Z}_w)}{M_b} \tag{1}$$

$$\ddot{Z}_w = \frac{-f_d + k_s(Z_b - Z_w) + c_s\left(\dot{Z}_b - \dot{Z}_w\right) - k_t(Z_w - Z_r)}{Mw} \tag{2}$$

$$x = [x_1, x_2, x_3, x_4] = \left[z_b, z_w, \dot{Z}_b, \dot{Z}_w\right] \tag{3}$$

where

Z_b Car body displacement
$\dot{Z}_b$ Car body velocity
Z_w Car wheel displacement
$\dot{Z}_w$ Car wheel velocity
f_d Actuator force
Z_r Excitation due to road disturbance

As a result, in the state space equation, the state variables are described in the A, B, C and D matrices as represented in the Eqs. (4) and (5)

$$\dot{x}(t) = Ax(t) + Bf_d(t) + DZ_r(t) \tag{4}$$

$$\begin{bmatrix} \dot{x}1 \\ \dot{x}2 \\ \dot{x}3 \\ \dot{x}4 \end{bmatrix} = \begin{bmatrix} 0 & 0 & 1 & 0 \\ 0 & 0 & 0 & 1 \\ -\frac{ks}{mb} & \frac{ks}{mb} & -\frac{cs}{mb} & \frac{cs}{mb} \\ \frac{ks}{mw} & -\frac{ks+kt}{mw} & \frac{cs}{mw} & -\frac{cs}{mw} \end{bmatrix} \begin{bmatrix} xb \\ xw \\ \dot{x}b \\ \dot{x}w \end{bmatrix} + \begin{bmatrix} 0 \\ 0 \\ -\frac{1}{mb} \\ \frac{1}{mw} \end{bmatrix} f_d + \begin{bmatrix} 0 \\ 0 \\ 0 \\ \frac{kt}{mw} \end{bmatrix} z_r \tag{5}$$

Let the measurements available for feedback be $y(t) = C\,x$ the output vector is assumed to be

$$y(t) = \begin{bmatrix} -\frac{ks}{mb} & \frac{ks}{mb} & -\frac{cs}{mb} & \frac{cs}{mb} \\ 1 & -1 & 0 & 0 \\ 0 & 1 & 0 & 0 \end{bmatrix} \begin{bmatrix} \ddot{z}b \\ sws \\ DTL \end{bmatrix} \tag{6}$$

The system parameters of the quarter car model are obtained from ref. [10].

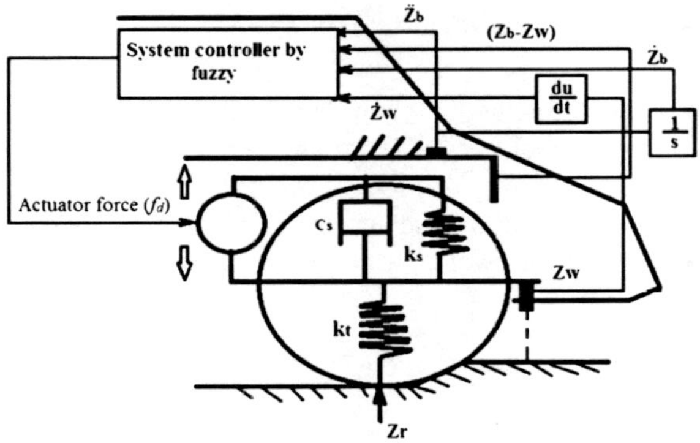

Fig. 1 Quarter-vehicle suspension model

3 Fuzzy Logic Controller Approach

This section offers a short description of the fuzzy controller that used to minimize the vibration levels to improve the suspension performance. Fuzzy rules are used to formulate the controller that can estimate expert reception and decision. The layout of the FLC is shown in Fig. 2.

The control rule Si, at any time i, is NB, ... , Z, ... , PB is represented by the linguistic descriptions (Mamdani), $S1 = (z_b - z_w)$, $S2 = (\dot{z}_b - \dot{z}_w)$, $S3 = \dot{z}_b$ and $S4 = \dot{z}_w$. There are four fuzzy inputs (S1, S2, S3, S4) and one output (f_d) with membership functions $\mu(z_b - z_w)$, $\mu(\dot{z}_b - \dot{z}_w)$, $\mu(\dot{z}_b)$ and $\mu(\dot{z}_w)$. FLC can be achieved by changing of universal of discourses and rule base through fuzzy to 9, 25 and 49 rules base. The fuzzy rule set is based on Table 1. Where, [NB (negative big), NM (negative medium), NS (negative small), ZE (zero), PS (positive small), PM (positive medium), PB (positive big)], e is the error and ce is the change of error with respect to the time. Fuzzy Control used "Mamdani" method, the reasoning is the "min-max" method. Two common membership functions are used; triangular and trapezoidal. The two membership function are chosen for the input and output variables, as shown in Fig. 3.

For instance, the linguistic control rules of the fuzzy logic controller obtained from fuzzy control with 49 rules set used in such case are as follows:

$$\begin{aligned}
&S1: \text{ IF } \{S1 = \text{ NB AND } S2 = \text{NB AND } S3 = \text{NB AND } S4 = \text{NB}\} \\
&\qquad \text{THEN } (f_d = \text{ NB}) \\
&S17: \text{ IF } \{S1 = \text{ NS AND } S2 = \text{ NS AND } S3 = \text{ NS AND } S4 = \text{ NS}\} \\
&\qquad \text{THEN } (f_d = \text{ NM}) \\
&S47: \text{ IF } \{S1 = \text{ PS OR PB AND } S2 = \text{ PS OR PB AND } S3 = \text{PS OR PB AND } S4 = \text{ PS OR PB}\} \\
&\qquad \text{THEN } (f_d = \text{ PB}) \\
&Si: \text{ IF } \{(S1, S2, S3, S4 = \text{Ei}) \text{ AND } ((S1, S2, S3, S4 = \text{Gi})\} \\
&\qquad \text{THEN } (f_d = \text{ Bi})
\end{aligned} \tag{6}$$

where Ei, Gi and Bi are labels of fuzzy sets representing the linguistic values $S1 = (z_b - z_w)$, $(\dot{z}_b - \dot{z}_w)$, $S3 = \dot{z}_b$, $S4 = \dot{z}_w$, e = error(S1), ce = change of error

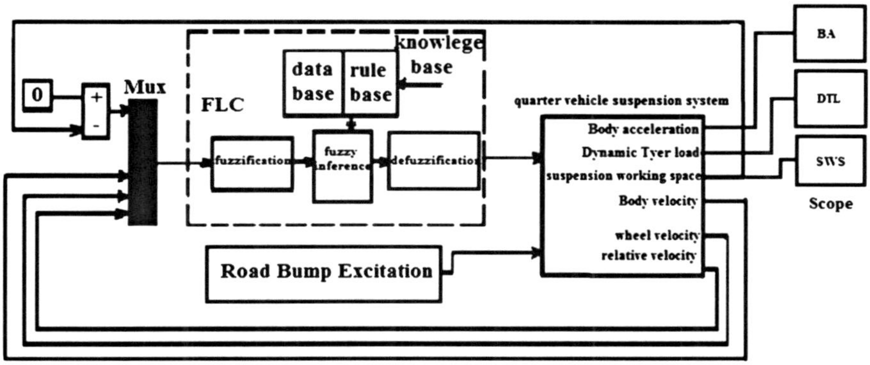

Fig. 2 Layout of the fuzzy logic control

Table 1 Fuzzy with 49 rules set

	e						
ce	NB	NM	NS	ZE	PS	PM	PB
NB	NB	NB	NB	NB	NM	NS	ZE
NM	NB	NB	NB	NM	NS	ZE	PS
NS	NB	NB	NM	NS	ZE	PS	PM
ZE	NB	NM	NS	ZE	PS	PM	PB
PS	NM	NS	ZE	PS	PM	PB	PB
PM	NS	ZE	PS	PM	PB	PB	PB
PB	ZE	PS	PM	PB	PB	PB	PB

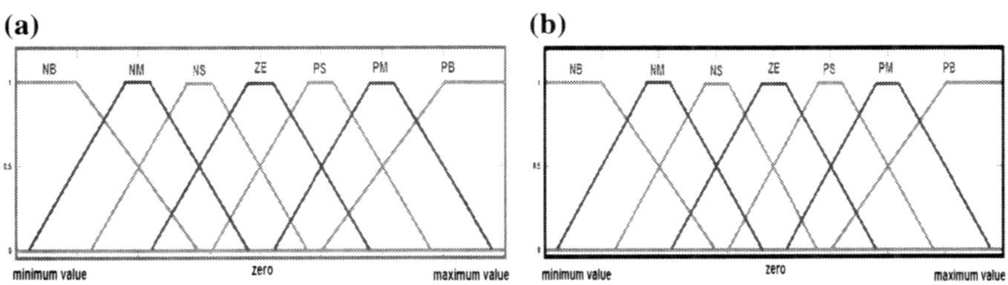

Fig. 3 Structure diagram for membership function trapmf type with 49 rules set. **a** e = S1, and ce = S2. **b** Actuator Force = f_d

(S2) and f_d = actuator force respectively, which are characterized by their membership functions. Also, the 9 and 25 rules set controller are formulated based on the same arrangement of the presented 49 rules shown in Fig. 3.

Defuzzification is the process for mapping from a set of secondary fuzzy control signals contained within a fuzzy output window to a non-fuzzy (crisp) control

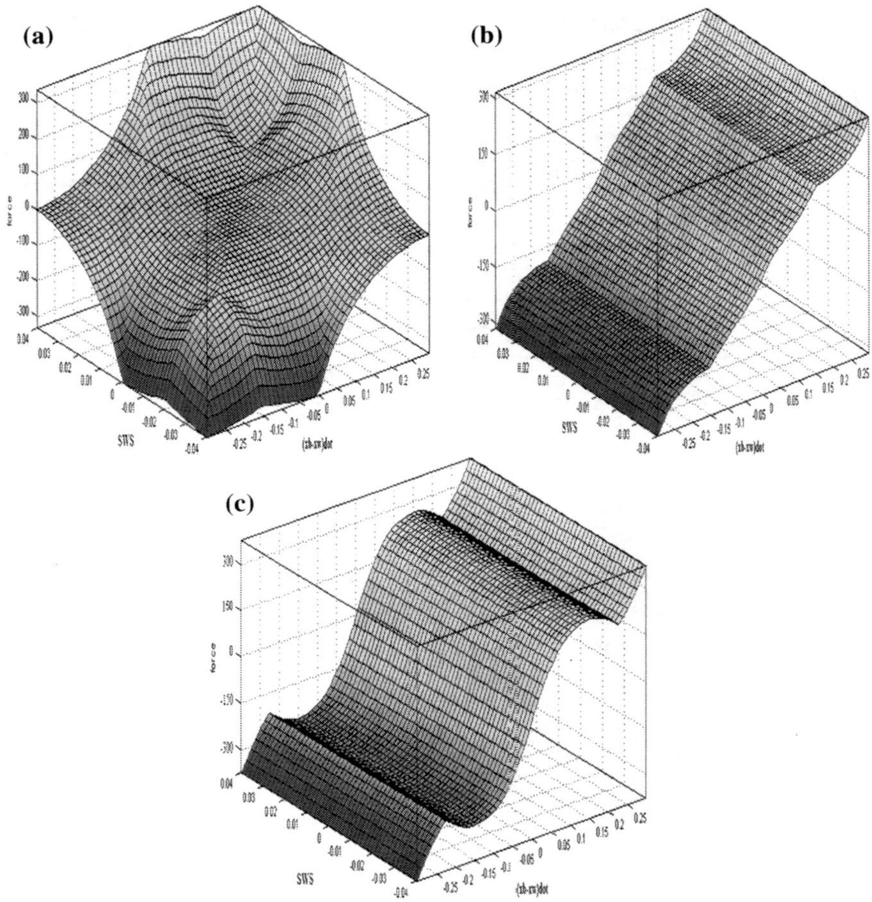

Fig. 4 Fuzzy surface with (9, 25 and 49) rules set, (**a**, **b** and **c**) respectively

signal. The center of region method is the most well identified defuzzification technique, which in linguistic conditions can be pronounced as:

$$f_d(\mathrm{t}) = \frac{\int f_d \times \mu(f_d)\mathrm{dt}}{\int \mu(f_d)\mathrm{dt}} \tag{7}$$

Figure 4 shows three fuzzy surfaces for the designed controllers using the trapezoidal membership function. The other surfaces using the triangle membership function are omitted.

4 Results and Discussion

The three main performance criteria in vehicle suspension design that manage the ride comfort and vehicle stability are suspension working space (SWS), vertical body acceleration (BA), and dynamic tire load (DTL). The BA is directly related to the ride comfort while tire's dynamic deformation ($Zw - Zr$) is required to achieve a good vehicle stability [11]. The suspension performance is improved when the SWS, BA, and DTL are minimized. There are four types of suspension systems investigated in this part:

(a) Conventional passive suspension system, with damping $c_s = 980$ Ns/m as in reference [12];
(b) Active suspension system, with 9 rules set fuzzy for both triangular and trapezoidal membership functions.
(c) Active suspension system, with 25 rules set fuzzy for both triangular and trapezoidal membership functions.
(d) Active suspension system, with 49 rules set fuzzy for both triangular and trapezoidal membership functions.

The above-mentioned performance criteria are used to calculate the performance of the control methods to select the best method of control to obtain the actuator force f_d for a given time history, as described in Sect. 3. Consequently, the damping of the conventional suspension system case is used as a reference for the comparisons with 3 active suspension systems proposed, noting that the value estimated for c_s is typical for automotive applications [12]. There are two types of road excitation, chosen to be very similar to the real-world road profiles, are considered in this study. First, time responses for the bumpy road input are being introduced. This style of road surface irregularity is one of the most encountered in actuality. Bump road input is formulated as presented in Eq. (8).

$$z_r = \begin{cases} a\{1 - \cos(\omega_r(t - 0.5))\}, & for\ 0.5 \leq t \leq 0.5 + \frac{d}{V} \\ 0, & otherwise \end{cases} \quad (8)$$

Where a is the half the bump amplitude, $\omega_r = 2\pi V/d$, d is the width of the bump, and V is the vehicle velocity. In this study the values for these parameters are, a = 0.035 m, d = 0.8 m, and V = 0.856 m/s [13]. The time history of the suspension system reaction under road bump excitation is shown in Fig. 5. The input of the fuzzy controller is composed of four input variables and one output variable. The relative displacement ($zb - zw$), relative velocity ($\dot{z}b - \dot{z}w$), body velocity $\dot{z}$b and wheel velocity $\dot{z}$w are defined as the input variables while the output variable is the controlling force based on the simulation software "Matlab/Simulink". The performance of the active control scheme is illustrated through a series of simulations. The time histories for SWS, BA, and DTL responses are shown in Fig. 5 and the Peak-To-Peak (PTP) values are shown in Fig. 6. The concluding figures show the comparison between the controlled active suspension system using

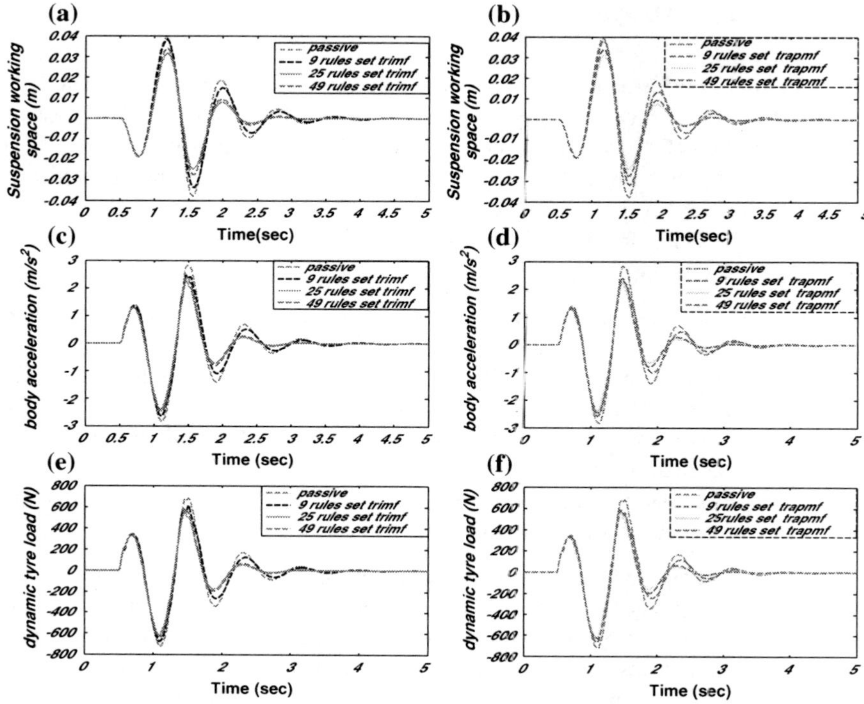

Fig. 5 The time domain of system responses under road bump excitation: **a**, **b** SWS; **c**, **d** BA; **e**, **f** DTL for both *trimf* and *trapmf* respectively

fuzzy with 9 rules set, fuzzy with 25 rules set and fuzzy with 49 rules set with both *trimf* and *trapmf*, and the conventional passive system. From these results it is seen that the fuzzy with 25 rules set controlled suspension can dissipate the energy due to bump excitation very well with both *trimf* and *trapmf*, reduce the settling time, and can improve not only ride comfort but also the vehicle stability.

The results indicate that, for the case of the fuzzy with 25 rules set control, the response is more continuous and the maximum value is lower than in the cases of fuzzy with 9 or 49 rules set controls for both *trimf* and *trapmf* and surely, better than the passive suspension system. The three controlled systems, both *trimf* and *trapmf* have the lowest peaks for the SWS, BA, and DTL, its values indicate an improved ride comfort and vehicle stability for the system response under road bump excitation. Figure 6 shows a complete comparison between all controllers for the peak-to-peak (PTP) values and their improvements percentages compared with passive suspension system. It's clearly shown that the 25 rules set with *trapmf* gives a superior performance. Moreover, the *trapmf* is counted as the best method of fuzzification and defuzzification process for SWS and BA while the *trimf* membership functions be in excess with DTL suspension performance.

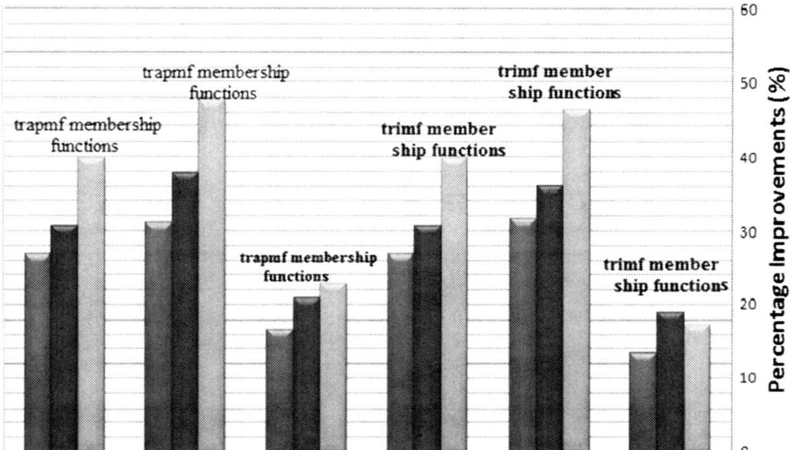

	49 rules set	25 rules set	9 rules set	49 rules set	25 rules set	9 rules set
SWS	39.7135	47.7865	22.7865	39.7135	46.3542	17.3177
BA	30.6169	37.7485	20.8075	30.6169	35.9934	19.0242
DTL	26.826	31.022	16.5138	26.826	31.5603	13.3917

Fig. 6 Percentage improvements in PTP values for the controlled systems compared to passive system for road disturbance excitation

The second type of road excitation was a random road profile selected from ref. [12] and described mathematically as;

$$\dot{x}_r + \rho V x_r = V \mathrm{w_n} \tag{9}$$

where, $\mathrm{w_n}$ is white noise with the intensity $2\sigma^2\rho V$, ρ is the road irregularity parameter, and σ^2 is the covariance of road irregularity. In random road excitation, the values of road surface irregularity (ρ = 0. 45 m^{-1} and σ^2 = 300 mm^2) were selected assuming that the vehicle moves on the paved road with the constant speed V = 20 m/s, as in reference [13]. Isolate the vehicle body from the road disturbances also to reduce the resonance peak of the body mass proximity to 1 Hz; it is important to improve the ride comfort, which is a sensitive frequency of the human body. Furthermore, to improve vehicle stability, it is important to maintain the tire in contact with the road surface and therefore reduce the resonance peak proximity at 10 Hz, which is the resonance frequency of the wheel [14]. In view of these considerations, the results obtained from the excitation described by Eq. (9) are presented in the frequency domain. The modulus of the Fast Fourier Transform (FFT) of the SWS, BA and DTL over the range 0–16 Hz are shown in Fig. 7. The FFT was suitably scaled and smoothed by curve fitting as done in [12–15]. The lowest resonance peaks in the body and the wheel can be achieved using the projected 25 rules set fuzzy control strategy. These figures indicate that the

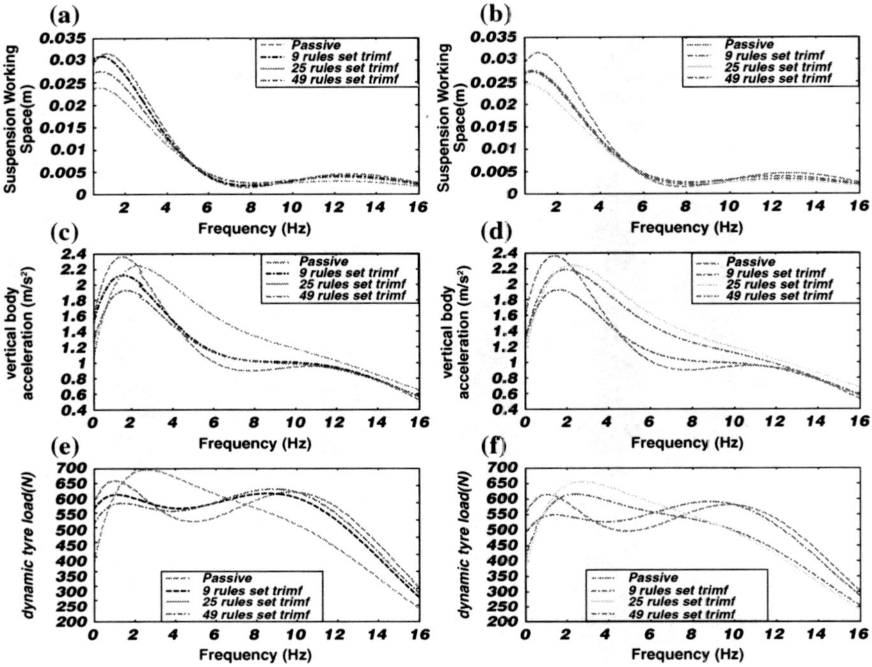

Fig. 7 The frequence domain responses under random road excitation: **a**, **b** SWS; **c**, **d** BA; **e**, **f** DTL for both *trimf* and *trapmf* respectively

technique of the fuzzy controlled system with 25 rules set, especially for *trimf*, dissipate the energy gained through the given road excitation in an excellent manner and thus, improves the SWS. On the other hand, the 49 rules set, for both *trimf* and *trapmf* achieves a significant improvement not only in terms of BA, which relates to the ride comfort, but also in DTL, which relates to the vehicle stability.

5 Conclusion

Control performance criteria were studied in the time domain and frequency domain to assessed the suspension effectiveness under bump and random road disturbance. The study implemented the 9, 25 and 49 rules set technique for the fuzzy controller of the active suspension system. From the time domain results, it had been concluded that these rules set can decrease the settling time and can improve ride comfort and vehicle stability over a conventional passive system. Comparing the number of rules set for the fuzzy controller for both *trimf*, *trapmf* implied that the 25 rules set improved SWS, BA and DTL over the 9 and even 49 rules set technique. Furthermore, frequency domain comparison between the three forms of fuzzy controllers, clarifies that the 49 rules set technique with *trimf* or *trapmf* achieved a significant improvement in BA and DTL. Moreover, the 25 rules set technique overcomes the 49 technique in SWS over the frequency domain. This is in addition to the improved dynamic criteria in time domain.

References

1. Pekgökgöz RK, Gürel MA, Bilgehan M, Kısa M (2010) Active suspension of cars using a fuzzy logic controller optimized by a genetic algorithm. Int J Eng Appl Sci 2(4):27–37
2. Ranjbar-Sahraie Bijan, Soltani M, Roopaie M (2011) Control of active suspension system: an interval type-2 fuzzy approach. World Appl Sci J 12(12):2218–2228
3. Rao MVC, Prahlad V (1997) A tunable fuzzy logic controller for vehicle-activesuspension Systems. Elsevier Fuzzy Sets Syst 85:11–21
4. Salem MMM, Aly AA (2010) Fuzzy control of a quarter-car suspension system. Int J Aerosp Mech Eng 4:235–240
5. Stříbrský A, Hyniová K, Honců J, Kruczek A (2003) Using fuzzy logic to control active suspension system of one-half-car model. Acta Montanistica Slovaca, Ročník, pp 223–227
6. Khajavi MN, Abdollahi V (2007) Comparison between optimized passive vehicle suspension system and semi active fuzzy logic controlled suspension system regarding ride and handling. Trans Eng Comput Technol 19:1305–5313
7. Ebrahimi N, Gharaveisi A (2012) Optimal fuzzy supervisor controller for an active suspension system. Int J Soft Comput Eng (IJSCE) V-2:36–39. ISSN: 2231–2307
8. Mailah M, Priyandoko G (2007) Simulation of suspension system with adaptive fuzzy active force control. Int J Simul Model 6:25–36
9. Tang C, Zhao G, Zhang Y, Ma Y (2012) The application of fuzzy control algorithm of vehicle with active suspensions. Res J Appl Sci Eng Technol 4(16):2744–2747
10. Metered H (2012) Application of nonparametric magnetorheological damper model in vehicle semi-active suspension system. SAE Int J Passing Cars: Mech Syst 5(1):715–726
11. Rajamani R (2006) Vehicle dynamics and control. Springer Science and Business Media, New York, p 308, 316
12. Metered H, Bonello P, Oyadiji SO (2010) An investigation into the use of neural networks for the semi-active control of a magnetorheologically damped vehicle suspension. In: Proceedings of the IMechE vol 224 Part D: J. Automobile Engineering, pp 1–20
13. Sims ND, Stanway R, Peel DJ, Bullough WA (2000) Controllable viscous damping an experimental study of an electrorheological long-stroke damper under proportional feed back control. Smart Mater Struct 8:601–615
14. Choi SB, Zhu WQ, Kim WK (2000) Vibration control of semi-active suspension featuring electrorheological fluid dampers. J Sound Vib 234:537–546
15. Sammier D, Sename O, Dugard L (2003) Skyhook and $H\infty$'control of semi-active suspensions some practical aspects. Veh Syst Dynamics 39:279–308

Two-Stage Data Driven Filtering for Local Damage Detection in Presence of Time Varying Signal to Noise Ratio

Jakub Obuchowski, Agnieszka Wylomanska and Radoslaw Zimroz

Abstract Local damage detection in rotating machinery can become a very difficult issue due to time-varying load or presence of another damage reflected in amplitude modulation of the raw vibration signal. In this paper a two-stage filtering method is presented to deal with this problem. The first stage is based on autoregressive (AR) modeling. It is incorporated to suppress high-energy components that mask an informative signal. High-energy amplitudes of mesh harmonics modulated by other damage or load variation can affect selectors of optimal frequency band as well, so they have to be suppressed. The second stage relies on filtering the AR-residual signal using a linear filter based on an informative frequency band selector. Here as a selector we propose to use the average horizontal distance on quantile-quantile plot. We compare the result of the second stage with the spectral kurtosis. The procedure is illustrated by real data analysis of a two-stage gearbox used in a belt conveyor drive system in an open-pit mine.

Keywords Local damage · Time varying conditions · Autoregression · Frequency band selector

J. Obuchowski (✉) · R. Zimroz
Diagnostics and Vibro-Acoustics Science Laboratory, Wroclaw University of Technology, Na Grobli 15, 50-421 Wroclaw, Poland
e-mail: jakub.obuchowski@pwr.edu.pl

R. Zimroz
e-mail: radoslaw.zimroz@pwr.edu.pl

A. Wylomanska
Hugo Steinhaus Center, Institute of Mathematics and Computer Science, Wroclaw University of Technology, Janiszewskiego 14a, 50-370 Wroclaw, Poland
e-mail: agnieszka.wylomanska@pwr.edu.pl

R. Zimroz
Research and Development Centre, KGHM Cuprum Ltd, Sikorskiego 2-8, 53-659 Wroclaw, Poland

J.K. Sinha (ed.), *Vibration Engineering and Technology of Machinery*, Mechanisms and Machine Science 23, DOI 10.1007/978-3-319-09918-7_36

1 Introduction

Local damage detection in rotating machinery operating in industrial conditions is a challenging problem. Inconveniences might be related to uncontrolled load and speed changes, unknown condition of a diagnosed machine, transmission of vibrations from other machines, acquisition of vibrations not related to operation of the machine, lack of data due to temporary failure of a signal acquisition system, etc. On the other hand, such problems lead to development of novel signal processing methods that substitute those which do not work properly in case of diagnostics in industrial conditions. In the paper we present a data driven method inspired by a vibration signal acquired during operation of a two stage gearbox operating in an open-pit mine. Both stages of the gearbox are damaged, thus the signal contains modulations related to two shaft frequencies. In this case, the signal of interest (SOI) related to the local damage of the first stage gear-pair is mixed with amplitude modulated signal due to the damage of the second stage. Thus, a time varying signal to noise ratio occurs which makes diagnostics more difficult. The approach proposed in this paper is an extension of the method presented in [1–3]. The method presented therein incorporates the idea of both decreasing power of narrowband spectral components by an autoregressive (AR) model and selection of an informative frequency band using the spectral kurtosis. We propose to extend this method by using another criterion of AR model goodness-of-fit and another measure of impulsiveness. It is worth mentioning that the AR model with constant coefficients might be used only in the case of constant speed of the machine. In other cases the model should be extended, e.g. for cyclic changes of rotational speed a periodic autoregressive model (PAR) might be used [4, 5]. If the speed is changing not in a cyclic way one can benefit from application of an adaptive AR filter (i.e. filter with coefficients varying in time) [6–8]. The second step of the method is designed especially for industrial conditions. Instead of classical kurtosis, we propose to use another measure which is as good as the kurtosis in the standard case and more effective in the case of a single excitation occurred during signal acquisition. Such excitation might not be related to operation of the machine. Since the kurtosis is very sensitive to such single artifacts, one can benefit from other measures of impulsiveness, especially those based on quantiles or empirical cumulative distribution function [9–12].

2 Methodology

The proposed two-stage procedure is based on signal filtering. Here we use both autoregressive modeling and optimal frequency band selection. Sometimes, the raw vibration signal contains a strong deterministic contamination which is highly amplitude modulated. In such case of time varying signal-to-noise ratio the signal of interest is invisible in both time series and envelope spectrum. Then, signal filtering based on measures of dispersion (e.g. the spectral kurtosis) may indicate a wrong frequency band as informative. We propose to filter out the deterministic signal using autoregressive filtering. The next step is based on linear filtering using frequency characteristics of the filter obtained by measures of impulsiveness. We compare filters driven by the spectral kurtosis [13] and one of informative frequency band selectors presented in [9–12].

As it was mentioned, we use the autoregressive model to filter out highly amplitude modulated mesh harmonics. The AR model of order p is defined as follows:

$$\sum_{i=0}^{p} \phi(i)X(t-i) = \varepsilon(t) \tag{1}$$

where $\phi(0) = 1$ and $\{\varepsilon(t)\}_{(t>0)}$ stands for noise.

It is known, that the AR time series model is able to model noisy sinusoidal pattern if its characteristic polynomial has complex roots. In the case of a large number of harmonics a high-order AR model is expected with at least two complex roots corresponding to one mesh harmonic. As the optimal order indicator we use the highest Kolmogorov-Smirnov criterion, i.e. AR(p) is said to be optimal if the Kolmogorov-Smirnov (KS) test statistic of residuals is the highest [14]. According to the fact that the residual signal in case of local damage should be impulsive, it is expected that the distance between empirical distribution and Gaussian one is high —the higher KS statistic, the more impulsive signal. Recall the KS statistic for signal X(t) is defined as follows [15]:

$$KS = \sup_x \left|\hat{F}(x) - F(x)\right|, \tag{2}$$

where $\hat{F}(\mathrm{x})$ is the empirical cumulative distribution function for given signal while F(x) is the cumulative distribution function of Gaussian distribution with parameters estimated form the signal.

Moreover, the results of AR filtering are also checked by comparing time-frequency maps of the residual signal with the raw signal. Parameters of the AR model are obtained by using Yule-Walker method [16].

Once mesh harmonics are suppressed during the previous step of the procedure, the residual signal might be still noisy, e.g. when the SOI is relatively narrowband. We propose to select the informative frequency band using the average horizontal distance on quantile-quantile plot (QQplot) [9–12]. Namely, we quantify the average distance between markers and reference line on the QQplot. The QQplot we use here is a graphical goodness-of-fit tool which compares quantiles of empirical distribution of the sample with the Gaussian distribution. The reference line connects first and third quartiles of both distributions. We compare it to the well-known spectral kurtosis. Both of them are based on analysis of narrowband slices of a time-frequency map. In this paper we design the filter using not only the characteristic given by a selector but we also enhance the characteristic by using

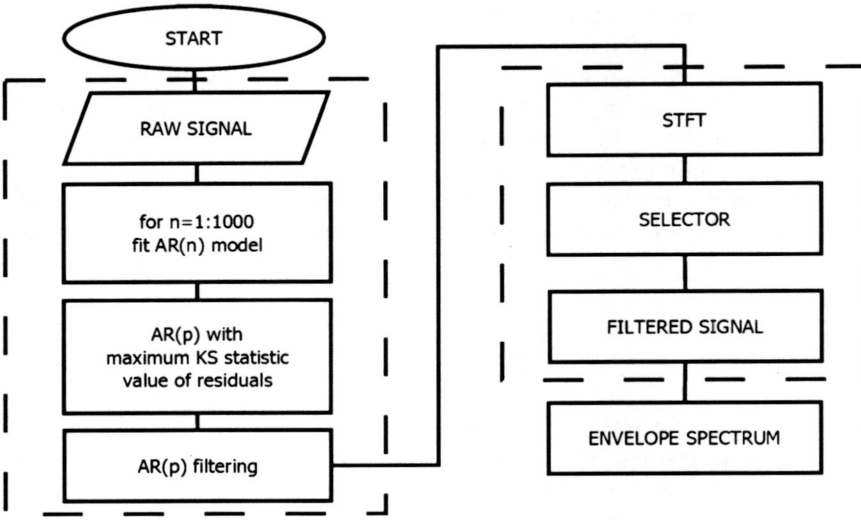

Fig. 1 Block diagram of the two-stage procedure

individual thresholds of selector's value for a given frequency bin. Namely, we put 0 in the frequency characteristic of the filter if the value of the selector is lower than the threshold for a given frequency bin. The thresholds are obtained using the Monte Carlo method and inverse pre-whitening [10, 11].

After designing the filter, we filter out the residual signal by computing Fourier transform of the signal, multiplying it by frequency response of the selector-based filter and, finally, using the inverse Fourier transform to return to the time domain. This method is an extension of the method presented in [13], where the filter is constructed using the spectral kurtosis.

A block diagram of our two-stage procedure is presented in Fig. 1.

3 Real Data Analysis

In this section we present analysis of the data that led us to establish the two-stage filtering procedure. The time series that we analyze is a raw vibration signal of a two-stage gearbox that operates in an open-pit mine and transfer torque from an engine to a belt conveyor pulley.

The gearbox reveals two kinds of damage. They are related to a local damage of first and rather distributed one for second shaft (fault frequencies 16.5 and 4.1 Hz, respectively). Length of the signal is 2.5 s and sampling frequency 16384 Hz.

Figure 2 presents the raw vibration signal (time series, envelope spectra and spectrogram). Time-frequency map of the signal shows that there are three interesting features of the signal. One is a strongly amplitude modulated high-energy

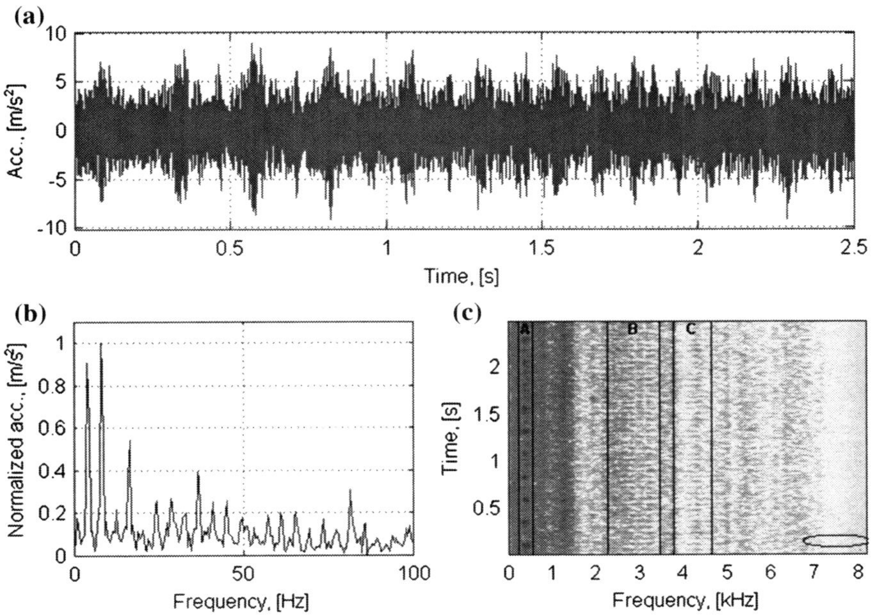

Fig. 2 Raw time series **a**, normalized envelope spectrum **b** and time-frequency map **c** of the raw vibration signal

component that affects the shape of the signal (denoted by A). The second is consisted of two informative frequency bands (B, C) with visible horizontal lines corresponding to cyclic wideband excitations related to damage. The third is an artifact (single, non-cyclic wideband excitation), denoted by an ellipse. Perturbations like this should not be associated with any kind of damage.

As it can be seen on the envelope spectrum, amplitude modulation frequency equals to 4.1 Hz and the first and second harmonic are clearly visible.

Single stage procedures based on informative frequency band selection do not work properly, due to high-energy amplitude modulated components (Fig. 3). Harmonics being multiplies of 4.1 Hz are dominating.

To suppress components related to 4.1 Hz amplitude modulation we fit an AR model to the raw time series. The highest Kolmogorov-Smirnov statistic criterion recommends AR(45) as the most suitable model (Fig. 4).

The residual signal obtained using AR filtering is presented in Fig. 5. Plot of time series presents the artifact and a barely visible cyclic pattern. The signal is too noisy to see a clear pulse train related to the local damage of the first shaft. The envelope spectrum shows that the frequency of excitations barely visible in time series equals to 16.5 Hz and the second harmonic is present. However, there is still quite large amount of noise that can be seen in the envelope spectrum. The time-frequency map presents two informative frequency bands (B and C, Fig. 5, panel (c)). These bands

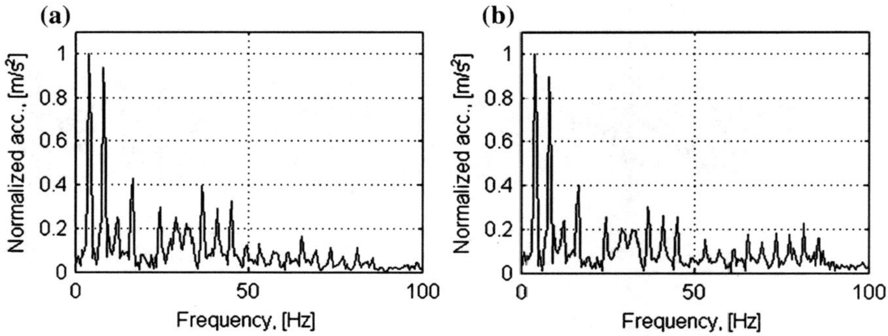

Fig. 3 Normalized envelope spectra of the raw vibration signal filtered using the spectral kurtosis (**a**) and the horizontal QQplot distance (**b**)

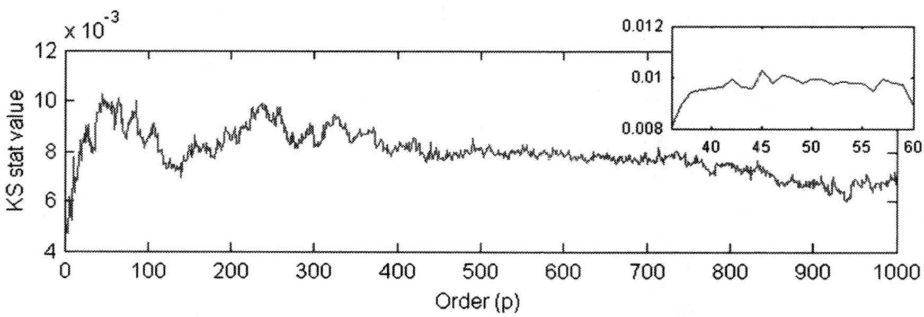

Fig. 4 Kolmogorov-Smirnov test statistic value versus order of autoregressive model. The highest value is obtained at $p = 45$

are relatively narrow which is consistent with large amount of noise in the time series plot. To extract only the SOI we decided to filter this signal using linear filters with frequency responses obtained by both the SK and QQplot average distance described in [9–12]. The thresholds are obtained using 1,000 Monte Carlo repetitions and quantiles of order 99 % are taken as individual thresholds for each narrowband frequency bin.

Figure 6 shows that the accidental impulse has great influence on the SK and the filtered signal is consisted mainly of high-frequency component which includes the artifact. As it can be seen on spectrogram (Fig. 6d), the SK also detected the informative frequency band, but values of frequency response at the highest bands are very large. Such structure of the signal results in noise in the envelope spectrum, Fig. 6c i.e. none of the harmonics related to the fault frequency are indicated. In the presence of artifact the result obtained by only AR filtering is much more efficient.

The filter driven by the selector that is a result of quantifying average horizontal distance on the quantile-quantile plot also indicated the high-frequency artifact, but its influence on the filtered signal is much lower (Fig. 7). Amplitude of the artifact is

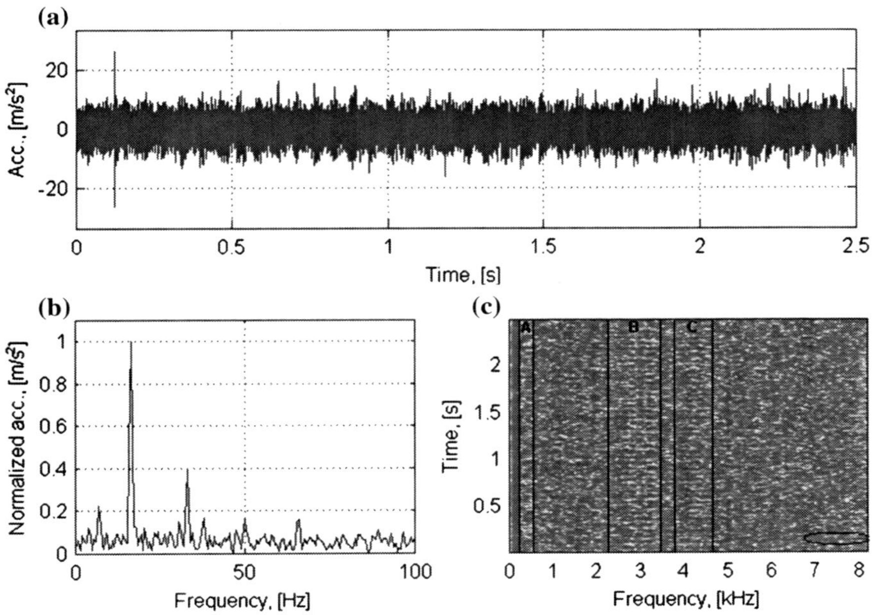

Fig. 5 Time series **a**, normalized envelope spectrum **b** and time-frequency map **c** of the signal after AR(45) filtering

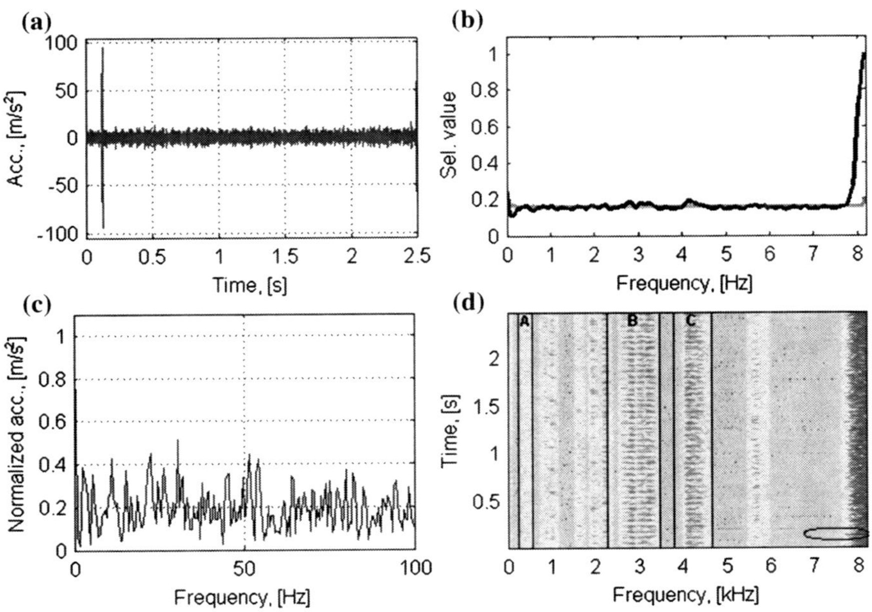

Fig. 6 Time series (**a**), frequency response of the filter based on the SK (**b**), normalized envelope spectrum (**c**) and time-frequency map (**d**) of the AR-residual signal after applying the filter based on the SK

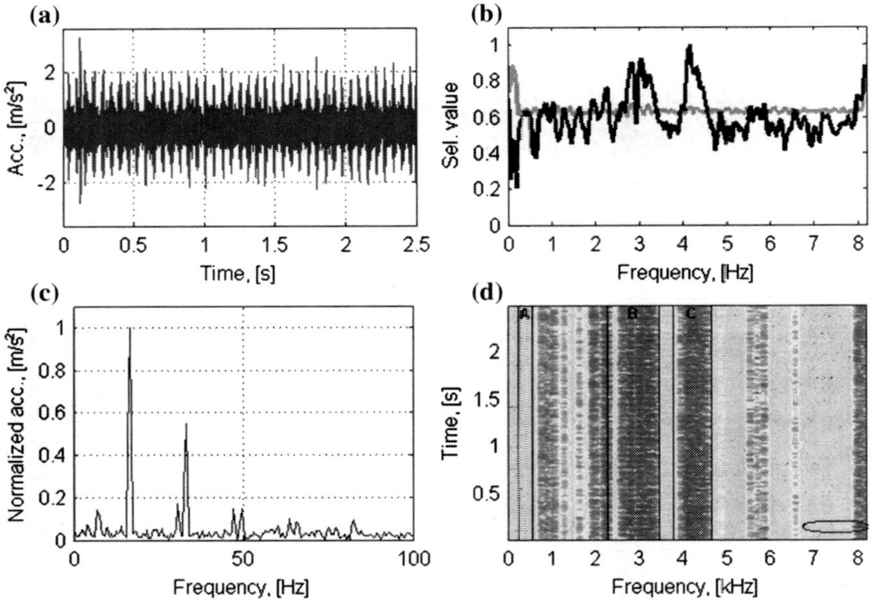

Fig. 7 Time series **a**, frequency response of the filter based on the average QQplot horizontal distance **b**, normalized envelope spectrum **c** and time-frequency map **d** of the AR-residual signal after applying the filter based on the average horizontal QQplot distance

just a little bit higher than average amplitude of the pulse train (Fig. 7a). Moreover, the pulse train is clearly indicated. The frequency characteristic of the filter shows that the selector based on average QQplot horizontal distance indicated frequency bands B and C as the most informative ones (Fig. 7b). Envelope spectrum of the filtered signal clearly shows that the artifact has no influence on the fault frequency recognition (Fig. 7c). This is the most valuable advantage of incorporating a different selector than the SK. Comparing the result of the two-stage filtering procedure to the signal filtered only by AR filter one can observe that our method results in a signal with significantly decreased amount of non-informative noise. Amplitude modulation of the signal of interest is clearly visible and several harmonics of the fault frequency are visible. The indicated frequency is related to the local damage of the second shaft, i.e. 16.5 Hz.

4 Summary

In this paper a new procedure of signal filtering for local damage detection is proposed. Presence of time-varying signal-to-noise ratio in the vibration signal leads to the two-stage procedure. It is particularly useful even if an accidental impulse (not related to damage) is present in the time domain. Such artifact affects frequency characteristic of a filter based on the spectral kurtosis, thus one can benefit from using another measure of impulsiveness, e.g. average horizontal distance on quantile-quantile plot. Such measure might indicate informative frequency band as well as kurtosis, but it is less sensitive to a single excitation that might occur during signal acquisition in industrial conditions.

Acknowledgement This work is partially supported by the statutory grant No. S30073 (J. Obuchowski).

References

1. Sawalhi N, Randall RB (2005) Spectral kurtosis enhancement using autoregressive models. In: 4th Australasian congress on applied mechanics, pp 231–236
2. Sawalhi N, Randall RB (2011) Signal Pre-whitening for fault detection enhancement and surveillance in rolling element bearings. In: Paper presented at the eighth international conference on condition monitoring and machinery failure prevention technologies, St David's Hotel, 20–22 June, Cardiff, UK
3. Sawalhi N, Randall RB (2011) Vibration response of spalled rolling element bearings: Observations, simulations and signal processing techniques to track the spall size. Mech Syst Signal Process 25(3):846–870
4. Wyłomańska A, Obuchowski J, Zimroz R, Hurd H (2014) Periodic autoregressive modeling of vibration time series from planetary gearbox used in bucket wheel excavator. In: Cyclostationarity: Theory and Methods, Lecture Notes in Mechanical Engineering, Chaari F et al (eds), pp 171–186
5. Broszkiewicz-Suwaj E, Makagon A, Weron R, Wyłomańska A (2004) On detecting and modeling periodic correlation in financial data. Phys A 336:196–205
6. Makowski R, Zimroz R (2014) New techniques of local damage detection in machinery based on stochastic modelling using adaptive Schur filter. Appl Acoust 77:130–137
7. Makowski RA, Zimroz R (2011) Adaptive bearings vibration modelling for diagnosis, lecture notes in computer science (including subseries lecture notes in artificial intelligence and lecture notes in bioinformatics) 6943 LNAI, pp 248–259
8. Makowski R, Zimroz R (2013) A procedure for weighted summation of the derivatives of reflection coefficients in adaptive Schur filter with application to fault detection in rolling element bearings. Mech Syst Signal Process 38(1):65–77
9. Obuchowski J, Wylomanska A, Zimroz R (2013) Stochastic modelling of time series with application to local damage detection in rotating machinery. Key Eng Mater 569–570:441–449
10. Obuchowski J, Wylomanska A, Zimroz R (2013) Novel enhancement techniques for noisy vibration signals applied to local damage detection in rotating machines. In: Proceedings of International conference surveillance 7, Oct 29–31, Chartres, France
11. Obuchowski J, Wylomanska A, Zimroz R (2014) Adaptation of Spectral Kurtosis based filtering procedure using novel criteria of impulsiveness, Manuscript submitted for publication
12. Obuchowski J, Wylomanska A, Zimroz R (2014) Selection of informative frequency band in local damage detection in rotating machinery. Mech Syst Signal Process, Advance online publication. doi:10.1016/j.ymssp.2014.03.011. 48(1–2):138–152
13. Combet F, Gelman L (2009) Optimal filtering of gear signals for early damage detection based on the spectral kurtosis. Mech Syst Signal Process 23(3):652–668

14. Zhan Y, Mechefske CK (2007) Robust detection of gearbox deterioration using compromised autoregressive modeling and kolmogorov-smirnov test statistic—Part I: compromised autoregressive modeling with the aid of hypothesis tests and simulation analysis. Mech Syst Signal Process 21(5):1953–1982
15. Justel A, Pen D, Zamar R (1997) A multivariate Kolmogorov-Smirnov test of goodness of fit. Stat Probab Lett 35(3):251–259
16. Brockwell PJ, Davis RA (2006) Time Series: theory and methods. Springer Series in Statistics, Springer, New York, ISBN: 978-1-4419-0320-4

Energy Harvesting from Near Periodic Structures

P.V. Malaji and S.F. Ali

Abstract In this research energy harvesting from near periodic structure is discussed. The near periodic system consists of two pendulums connected using a common linear spring. Mistuning in the simple coupled pendulum system is achieved by varying the length of one of the pendulums. Effect of this mistuning on amount of energy harvested is developed analytically and numerically. This will be discussed in this paper and at the same time effect of harvesting on mistuning will be presented. It is shown that with a proper electrical damping, optimal power can be obtained and effect of mistuning can be minimized. Same analysis is carried out with energy harvesting from both the pendulums. In case of harvesting from both the pendulums the harvesting bandwidth is increased and electrical damping required to minimize mistuning is more than that in case of harvesting with mistuned pendulum alone.

Keywords Energy harvesting · Near periodic structures · Electromagnetic · Mistuning

1 Introduction

With the advent of low powered electronics and miniature sensing devices, the requirement of power for such devices has reduced in one way and on the other hand their numbers have increased exponentially recently. This required a huge number of batteries to be replaced, which is not only costly but also poses a threat to the environment. Energy harvesting has recently become attractive in such situations as it meets the demands of these small devices without increasing their sizes to a huge extent. They replace regular power cells or can also recharge the power cells reducing the need for replacement.

P.V. Malaji (✉) · S.F. Ali
Department of Applied Mechanics, IIT Madras, Chennai 600036, India
e-mail: pradeepmalaji@gmail.com

S.F. Ali
e-mail: sfali@iitm.ac.in

J.K. Sinha (ed.), *Vibration Engineering and Technology of Machinery*,
Mechanisms and Machine Science 23, DOI 10.1007/978-3-319-09918-7_37

The key idea of energy harvesting is to scavenge energy from the ambient source that is available in various forms [1] like solar, wind, tidal etc. In this paper we focus on the mechanical vibration energy harvesting, where energy is harvested from a mechanical system that undergoes vibration due to some excitations. Previous results [1, 2] show that power scavenged by linear energy harvesters increases with the increase in the vibrations of the host structure. But the maximum power is harvested at the resonance frequency [2, 3]. In this paper we investigate the power harvesting capability of repetitive structures when they undergo mistuning.

Modern structures are built slender, multi-span and many a times same basic substructure is repeated, for example turbine blades, multi span bridges and repetitive space structures. In ideal case all substructures (like blades) are identical and have same amplitude for a given mode. But in real situation there will be slight difference between substructures because of geometry, material properties, tolerances in manufacturing, or in-service degradation. This makes the structure loose its repetitive nature, or we call mistuning. Mistuning may also happen if a sub component of a structure is not identical to others. For example in case of wind turbine blades, if geometry (say length) of one of the blade is slightly different than others, the structure is mistuned. Due to this mistuning at resonance some blades will vibrate with large amplitude and others with small amplitudes [4]. This behavior is known as vibration localization. Vibration localization may also occur due to mistuning of the structure, weak coupling between sub components and for low damping. For detailed study on vibration localization the readers are redirected to references [4, 5]. The objective of this manuscript is to study the effect on energy harvesting.

This paper aims at investigating power harvesting capacity of near periodic structures. As an example we have considered electromagnetic energy conversion using pendulum connected in series. Coupled pendulum [4] is one of the simplest model of repeated structure in which vibration localization is achieved by mistuning with low stiffness coupling and damping. The effect of mistuning and electrical damping on the harvested power is reported. Section 2 details the modeling of the near periodic structures; energy-harvesting equations are also given. Numerical results are shown in Sect. 3.

2 System Modeling

2.1 Model of Cyclic Structure

Repetitive structures can have variety of models [4, 5]. A simplified model consisting of pendulums attached in series with spring [4] is considered for the present study. Figure 1 shows the model, where n pendulums are attached through springs and the end springs at both sides are attached with fixed bases. Each of these pendulums represents a substructure in a periodic system. This system represents a cyclic structure with repeated subcomponents, which will have similar geometry, damping, stiffness etc. As shown in Fig. 1 the coupled pendulum system have same

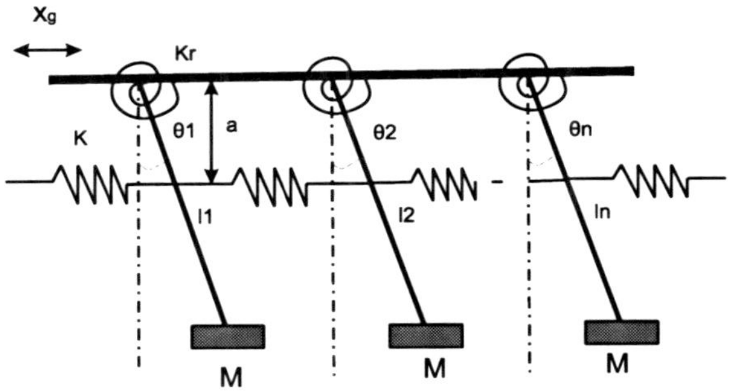

Fig. 1 General coupled pendulum system

mass m, torsional stiffness k_r, proportional damping constant c and translational spring k. Each pendulum is coupled by two translational springs. To introduce mistuning in the system one of the pendulums is assumed to have slightly different length than others (geometric discontinuity).

For simplification only two pendulums under support excitation (x_g) are considered. The equations of motion are given as follows;

$$ml_1^2\ddot{\theta}_1 + cl_1^2\dot{\theta}_1 + \left(k_r + 2\,ka^2\right)\theta_1 - ka^2\theta_2 = -ml_1\ddot{x}_g \tag{1}$$

$$ml_2^2\ddot{\theta}_2 + cl_2^2\dot{\theta}_2 + \left(k_r + 2\,ka^2\right)\theta_2 - ka^2\theta_1 = -ml_2\ddot{x}_g \tag{2}$$

The lengths of the two pendulums are considered different to account for mistuning. Converting above equation to frequency domain and introducing dimensionless parameters and variables as below.

$$\theta_j = \Theta_j\, e^{i\omega t}, x_g = X_g e^{i\omega t}$$

$$\propto = \frac{l_2}{l_1},\ \omega_0 = \sqrt{\frac{k_r}{ml_1^2}},\ \Gamma = \frac{c}{m\omega_0},\ \beta = \frac{ka^2}{k_r},\ f = \frac{X_g}{l_1},\ \Omega = \frac{\omega}{\omega_0} \tag{3}$$

Where, $\propto$—length ratio, Γ—damping factor, β—coupling factor, f—ratio of base excitation amplitude and length of first pendulum. Using above dimensionless parameters and variables Eqs. (1) and (2) in frequency domain can be written as;

$$\begin{bmatrix} -\Omega^2 + i\Gamma\Omega + 1 + 2\beta & -\beta \\ -\beta & -\alpha^2\Omega^2 + i\alpha^2\Gamma\Omega + 1 + 2\beta \end{bmatrix} \begin{bmatrix} \Theta_1 \\ \Theta_2 \end{bmatrix} = \begin{bmatrix} \Omega^2 f \\ \alpha\Omega^2 f \end{bmatrix} \tag{4}$$

Dimensionless amplitudes of pendulums are given by

$$\Theta_1 = \frac{((-\alpha^2\Omega^2 + i\alpha^2\Gamma\Omega + 1 + 2\beta)\Omega^2 + \beta\alpha\Omega^2)f}{\Delta} \tag{5}$$

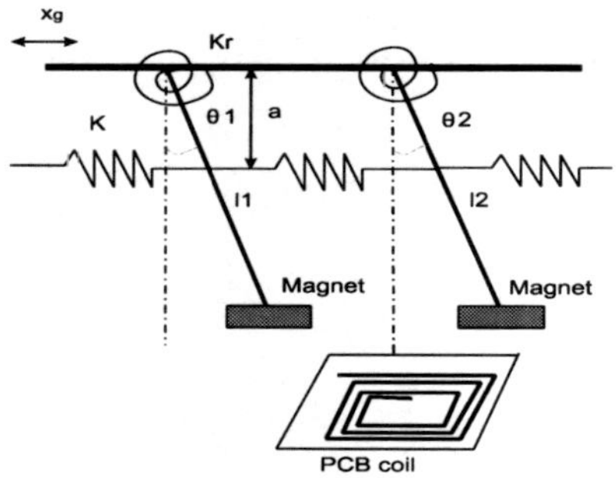

Fig. 2 Electromagnetic energy conversion system

$$\Theta_2 = \frac{(\beta + (-\Omega^2 + i\Gamma\Omega + 1 + 2\beta)\alpha\Omega^2)f}{\Delta} \tag{6}$$

Where,

$$\begin{aligned}
\Delta &= d_0 + d_1\Omega + d_2\Omega^2 + d_3\Omega^3 + d_4\Omega^4 \\
d_0 &= 3\beta^2 + 1 + 4\beta \\
d_1 &= i\Gamma + i\alpha\Gamma_e + 2i\beta\alpha\Gamma_e + 2i\Gamma\beta + 2i\beta\alpha^2\Gamma + i\alpha^2\Gamma \\
d_2 &= -2\beta - \alpha^2 - 2\beta\alpha^2 - \Gamma^2\alpha^2 - \Gamma\Gamma_e \\
d_3 &= -2i\alpha^2\Gamma - i\alpha\Gamma_e \\
d_4 &= \alpha^2
\end{aligned}$$

2.2 Electromagnetic Energy Harvesting

An electromagnetic harvester is conceived using mistuned double pendulum system. It is assumed that the pendulum system will scavenge vibration energy from the host structure. The host structure motion will excite the pendulum system at its base. As shown in Fig. 2, energy will be harvested from the motion of the pendulum by placing a PCB coil under the pendulum system. This will specifically be the mistuned pendulum as shown in Fig. 2. Permanent magnets will be placed as the pendulum masses with magnetic flux B. The power is delivered to the resistive load *R* connected across coil. Coil inductance is neglected, as it will be smaller than coil resistance [6].

Electromechanical equations for the system are given as

$$ml_1^2\ddot{\theta}_1 + cl_1^2\dot{\theta}_1 + \left(k_r + 2\,ka^2\right)\theta_1 - ka^2\theta_2 = -ml_1\ddot{x}_g \tag{7}$$

$$ml_2^2\ddot{\theta}_2 + cl_2^2\dot{\theta}_2 + \left(k_r + 2\,ka^2\right)\theta_2 - ka^2\theta_1 + BAI = -ml_2\ddot{x}_g \tag{8}$$

$$v = BLl_2\dot{\theta}_2 \tag{9}$$

Where, B—magnetic flux, A—surface area of coil, I—current trough coil = v/R L—length of coil, v—voltage across load, R—load resistance. Converting above Eqs. (7), (8) and (9) into frequency domain and using nondimensional parameters given in (3). From the relation between I and v the last term at LHS in Eq. (8) can be replaced with;

$$BAI = i\Omega\alpha\Gamma_e \tag{10}$$

$$\Gamma_e = \frac{B^2LA}{ml_2\omega_0R} \tag{11}$$

Where Γ_e is the electrical damping factor. Equations (7) and (8) can be written as;

$$\begin{bmatrix} -\Omega^2 + i^2\Gamma\Omega + 1 + 2\beta & -\beta \\ -\beta & -\alpha^2\Omega^2 + i\alpha^2\Gamma\Omega + 1 + 2\beta + i\alpha\Gamma_e\Omega \end{bmatrix} \begin{bmatrix} \Theta_1 \\ \Theta_2 \end{bmatrix} = \begin{bmatrix} \Omega^2 f \\ \alpha\Omega^2 f \end{bmatrix} \tag{12}$$

Electrical power harvested is given by $p = \frac{|V|^2}{R}$

$$p = \frac{|V|^2}{B^2LA} ml_2\omega_0\Gamma_e$$

From Eq. (9) and with $v = Ve^{i\omega t}$ we have

$$V = i\Omega\omega_0\alpha l_2BL\Theta_2$$

$$P = \frac{p}{ml_2^2\omega_0^3} = \Omega^2\alpha^2\Gamma_e\Theta^2 \tag{13}$$

Equation (13) gives the magnitude of power harvested.

3 Numerical Results and Discussion

Numerical simulation results are reported in this section. The dimensionless parameters considered α varied from 1 to 1.01. We considered α as the length ratio of second to first pendulum. The effect of mistuning on the coupled pendulum with low damping factor (Γ) and coupling factor (β) are studied initially. The amplitude responses of first and second pendulums are observed to find an optimal α. From these values the optimal parameters for maximum power harvested are obtained.

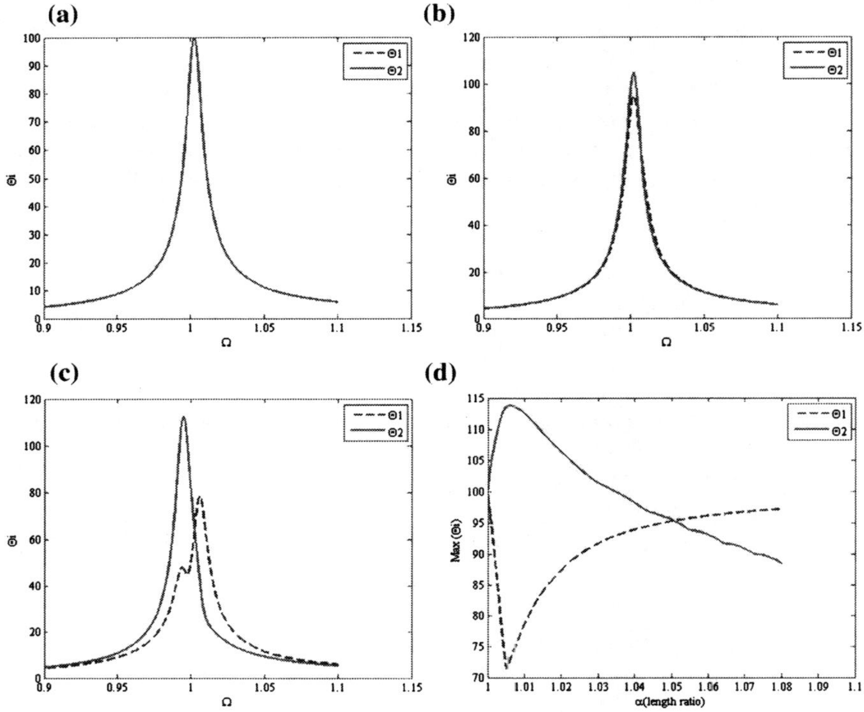

Fig. 3 **a** Dimensionless amplitudes for α = 1, Γ = 0.01, β = 0.005, f = 1. **b** Dimensionless amplitudes for α = 1.001, Γ = 0.01, β = 0.005, f = 1. **c** Dimensionless amplitudes for α = 1.01, Γ = 0.01, β = 0.005, f = 1. **d** Variation of dimensionless amplitudes due to variation in length ratio α

Figure 3a–c are obtained from Eqs. (5) and (6) showing the frequency responses with change in parameters of the pendulum system. Length of first pendulum is taken as nominal length. Mistuning achieved by varying length of pendulum 2. Figure 3a–c show dimensionless amplitudes of both pendulums with different values of α (1, 1.001 and 1.01) and Γ = 0.01, β = 0.005. These figures show that small non-uniformity of less than 1 % in subcomponents will cause significant difference in vibration response. Figure 3d shows the variation of maximum amplitudes with length ratio.

The difference between peak amplitudes of pendulum 1 and 2 is highest when length ratio (α) is 1.006. The difference increases till α = 1.006 and then it decreases. At this value the amplitude of pendulum 2 is 33.4 % more than pendulum 1. Comparison is done between maximum amplitudes even though they occur at slightly different frequencies [4]. If comparison is done at same frequency the difference between amplitudes keeps increasing.

Once the effect of mistuning on structure is known. The effect of mistuning on power harvested can be obtained from electromechanical relations given by

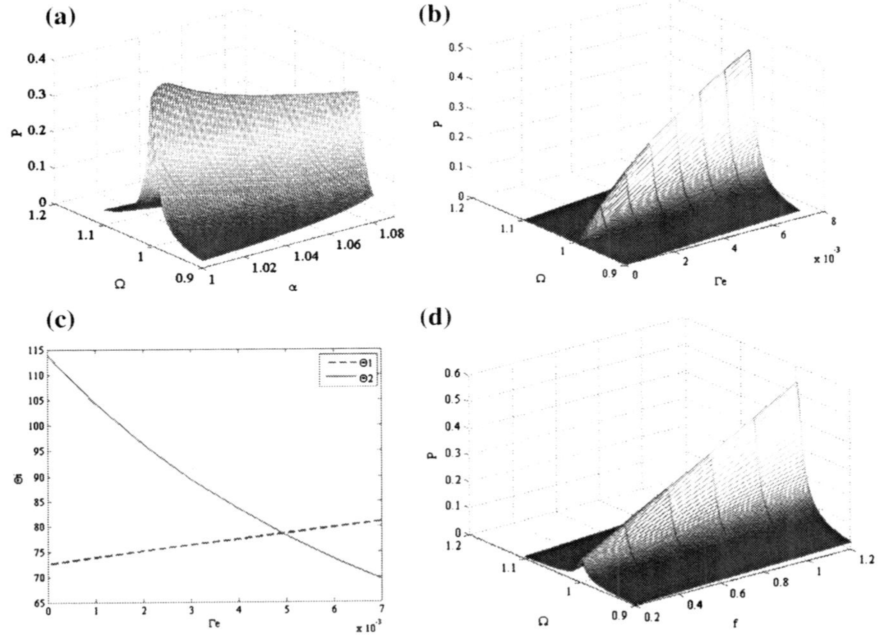

Fig. 4 **a** Power magnitude for variations in Ω and α, Γ = 0.01, β = 0.005, $\Gamma_e = 0.006$, f = 1. **b** Power magnitude P with Ω and Γ_e for α = 1.006, Γ = 0.01, β = 0.005, f = 1. **c** Dimensionless amplitudes with variation in electrical damping Γ_{e}, α = 1.006, Γ = 0.01, β = 0.05, f = 1. **d** Power magnitude P with Ω and f for α = 1.006, Γ = 0.01, β = 0.005, $\Gamma_e = 0.005$

Eqs. (12) and (13). Harvesting energy from pendulum will have a damping effect on the structure. The optimal value of this electrical damping has to be obtained which will maximize the power and minimize the difference between maximum amplitudes of pendulums.

Figure 4a shows the variation power magnitude with frequency ratio and mistuning (Ω and α) with $\Gamma_e = 0.006$. The maximum power is obtained near resonance (Ω = 0.99) and optimal mistuning (α = 1.006). It is observed that the power harvested of mistuned pendulum (α = 1.006) is 8.5 % more than that of tuned (α = 1) pendulum. Power harvested increases up to optimal mistuned value and then decreases. Figure 4b shows the variation of power magnitude with frequency ratio and electrical damping (Ω and Γ_e). The amount of power is increasing with increase in electrical damping Γ_e, which is inversely proportional to load resistance. Optimal value of electrical damping can be obtained by comparing Fig. 4b, c. Figure 4c shows the variation of maximum dimensionless amplitudes with electrical damping Γ_e. The effect of mistuning is completely vanished when Γ_e is 0.0045. Optimal value of Γ_e is chosen as 0.005 where power is maximum and mistuning effect is very less.

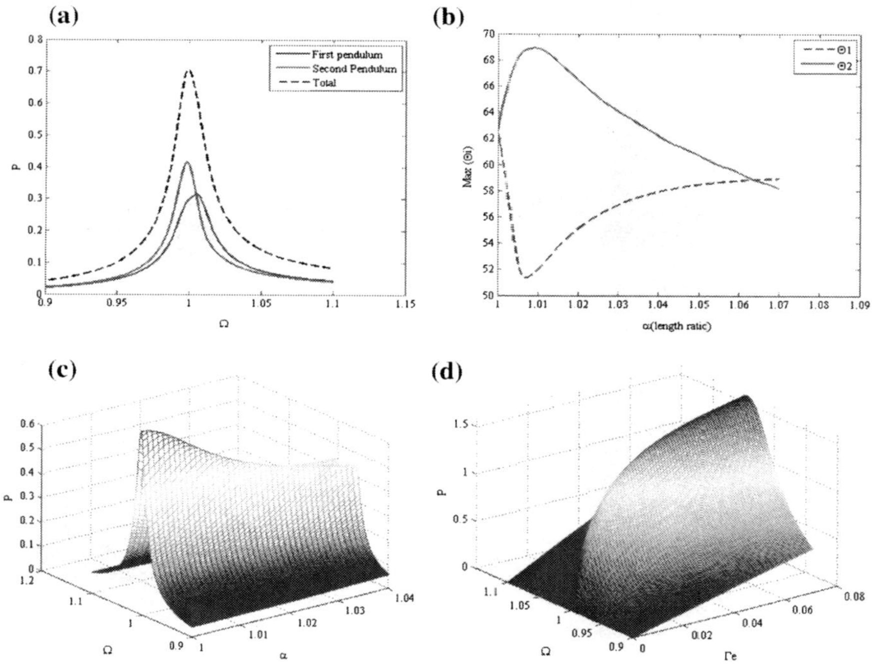

Fig. 5 **a** Power magnitude of both pendulums for $\alpha = 1.006$, $\Gamma = 0.01$, $\beta = 0.005$, $\Gamma_e = 0.006$, $f = 1$, **b** Variation of dimensionless amplitude with α_2. **c** Variation of total power magnitude with Ω and α, $\Gamma = 0.01$, $\beta = 0.005$, $\Gamma_e = 0.006$, $f = 1$. **d** Variation of total power magnitude with Ω and Γ_e, $\Gamma = 0.01$, $\beta = 0.005$, $\Gamma_e = 0.006$, $f = 1$

Figure 4d shows variation of power magnitude with frequency ratio and ratio of exited amplitude x_g and nominal length l (Ω and f) with $\Gamma_e = 0.005$. Magnitude of power P keeps increasing with increase in f.

In order to increases the amount of energy harvested, harvesting from both the pendulums is considered with mistuning. Figure 5a shows the power magnitude of first pendulum, second pendulum (mistuned) and total power magnitude. Because of mistuning there is difference in frequencies of the pendulums, which will increase the bandwidth of harvester. As harvesting is done from both the pendulums the amount of total power will be more than that harvested from single pendulum. Figure 5b is similar to Fig. 3d which shows the optimal value of α at which amplitude difference is maximum. Figure 5c shows the variation of total power magnitude with length and frequency ratio. In this case mistuning will not increases the total power harvested instead it will increase the bandwidth. From Fig. 5c it is clear that mistuning reduces the total power harvested, but that reduction is very small up to optimal value of length ratio ($\alpha = 1.006$) after which decay rate is fast. Figure 5d shows the variation of total power with electrical damping, which shows increase in total power with increase in electrical damping.

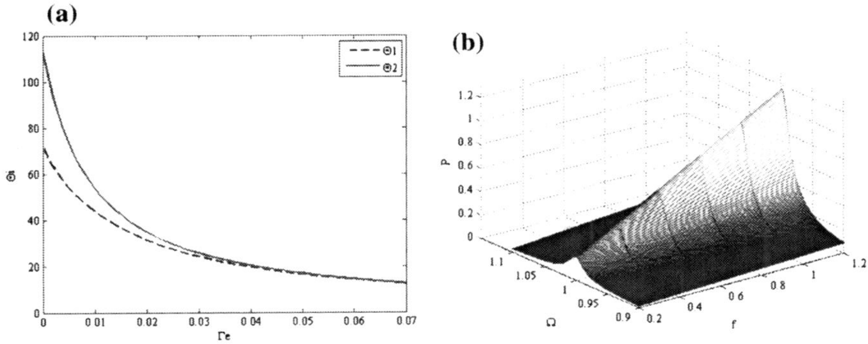

Fig. 6 **a** Variation of dimensionless amplitude with electrical damping Γ_e. **b** Variation of total power from with Ω and α $\Gamma_e = 0.02$

To get the optimal value of the electrical damping variation of maximum amplitudes of pendulums is plotted against electrical damping as shown in Fig. 6a. The effect of mistuning completely disappears when electrical damping is 0.04 with the huge reduction in amplitudes of pendulums (113 and 72 to 20 respectively for pendulum 1 and 2). Therefore electrical damping of 0.02–0.04 can be selected in order to increase power harvested and minimize mistuning effect. Figure 6b shows variation of power magnitude with frequency ratio and ratio of exited amplitude x_g and nominal length l (Ω and f) with $\Gamma_e = 0.005$. Magnitude of power P keeps increasing with increase in f.

4 Conclusions

In the present work the effect of mistuning on coupled pendulum is studied numerically with fixed low damping and coupling factor ($\Gamma = 0.01$, $\beta = 0.005$). It has been found that even a small variation of less than 1 % in length ratio ($\alpha_2 = 1.006$) will lead to large difference in maximum amplitudes of pendulums up to 33.4 %. Which will cause vibration localization of one of the pendulum and structure will behave as near periodic structure. Energy harvesting via electromagnetic conversion is considered from mistuned pendulum. Initial effect of mistuning on power harvested is studied with electric resistance $\Gamma_e = 0.006$ and α is varied from 1 to 1.1. It is observed that the power harvested increases by 8.5 % if mistuning of pendulum takes place ($\alpha = 1.006$). The optimum value of electrical damping Γ_e is obtained with mistuning present in system ($\Gamma = 0.01$, $\beta = 0.005$ and $\alpha = 1.006$). The optimum value of electrical damping is found to be $\Gamma_e = 0.005$, which will produce maximum power and minimizes mistuning effect.

If energy is harvested from both the pendulums with mistuning will increase the bandwidth, but the total power harvested will decrease. This decay rate is very small up to optimal value of mistuning. Also the amount damping required to minimize mistuning effect is very large ($\Gamma_e = 0.04$) compared to that of harvesting from single pendulum, which increases power output. These results conclude that not only useful power is harvested from near periodic structure but also the localization effect is reduced, which will make structure safer.

References

1. Roundy S, Wright PK, Rabaey J (2003) A study of low level vibrations as a power source for wireless sensor nodes. Comput Commun 26(11):1131–1144
2. Roundy S, Wright PK (2004) A piezoelectric vibration based generator for wireless electronics. Smart Mater Struct 13(5):1131–1142
3. Williams CB, Yates RB (1996) Analysis of a micro-electric generator for micro-systems. Sens Actuators 52:8–11
4. Yoo HH, Kim JY, Inman DJ (2003) Vibration localization of simplified mistuned cyclic structures undertaking external harmonic force. J Sound Vibr 261(5):859–870
5. Pierre C, Dowell EH (1987) Localization of vibrations by structural irregularity. J Sound Vibr 114(3):549–564
6. Kulah H, Najafi K (2004) An electromagnetic micro power generator for low-frequency environmental vibrations. In: Micro Electro Mechanical Systems—17th IEEE Conf. on MEMS Maastricht, The Netherlands, 25–29 January 2004, pp 237–240

Dynamic Analysis of Multilayered Sandwich Beam with MRE Core

S.K. Dwivedy and S.L.V.N. Avinash

Abstract In the present work dynamic analysis of multilayered sandwich beam with MRE core has been carried out. Finite element model has been developed for the multilayered sandwich beam with alternate stiff metallic and soft MRE core materials. Natural frequencies have been determined for 5 and 7 layered beams at different magnetic fields considering three different boundary conditions. Also the effect of increase in number of layers on modal frequencies, time and frequency responses have been studied which show that damping of multilayer beam occurs faster than the 3 layered sandwich beams with equivalent material.

1 Introduction

Sandwich beams are used for vibration and noise reduction of structures due to their light weight, high stiffness to weight ratio, good damping capability and high strength. Generally viscoelastic core is used in these structures. To make these structures adaptive research has been carried out to use magneto rheological elastomer core which changes its stiffness if magnetic field is applied. The application of multiple-layer treatment increases the damping characteristic of the structure significantly and thus helps to attenuate the vibration and noise for a broad range of frequency. The main advantage in using multilayer is that it allows the use of different core materials in different layers so that a wide range of the system natural frequency and damping can be tailored. Magneto rheological elastomers (MREs) are composed of tiny ferrous particles in chains and an elastomer such as synthetic rubber or natural rubber. When liquid state rubber combined with ferrous particles

S.K. Dwivedy (✉) · S.L.V.N. Avinash
Department of Mechanical Engineering, Indian Institute of Technology, Guwahati 781039, India
e-mail: dwivedy@iitg.ernet.in

S.L.V.N. Avinash
e-mail: sikhakolli@iitg.ernet.in

J.K. Sinha (ed.), *Vibration Engineering and Technology of Machinery*, Mechanisms and Machine Science 23, DOI 10.1007/978-3-319-09918-7_38

is exposed to a steady magnetic field, the ferrous particles will form chain-like structures arranged along the magnetic field (Liao et al. [1]). MRE is used for developing stiffness-controllable devices (Zhou and Wang [2]). Here a brief literature review related to sandwich beams is carried out.

Mead and Markus [3] obtained the sixth order differential equation of motion for a three layered sandwich beam with viscoelastic core. Arvin et al. [4] proposed a higher order theory for sandwich beam with composite faces and viscoelastic core using FEM. Sun et al. [5] studied the dynamic characteristics of MR material based adaptive sandwich beam. Zhou and Wang [2] investigated the magneto elastic loads subjected to conductive beam perpendicular to the magnetic field. Chen et al. [6] investigated experimentally the effect of carbon black on mechanical performance of MRE. Dwivedy et al. [7] studied the parametric instability regions of a three layered symmetric sandwich beam with conductive skins. Banerjee et al. [8] studied the free vibration of a three-layered sandwich beam using the dynamic stiffness method. The theory of flexural vibration of symmetric multilayered beam with viscoelastic core was analyzed by Agbasiere and Grootenhuis [9]. Amichi et al. [10] developed 3D Finite element model of sandwich beam with viscoelastic core. Many literature related to sandwich beams can be seen in work of Nayak [12].

Literature review reveals that majority of the studies are based on dynamic analysis of three-layer constrained sandwich beam using viscoelastic core and MRE core and very little work has been reported for dynamic analysis of multi-layered sandwich beam with viscoelastic core. No studies were reported on the work of multi-layered sandwich beam using MRE core. The present study is therefore aims to carry out dynamic analysis on multi-layered sandwich beam with MRE core.

2 Mathematical Modelling

Figure 1 shows a simply supported multilayered sandwich beam having $2n + 1$ layers with n number of viscoelastic MRE layers and $n + 1$ number of constraining isotropic layers. In numerical analysis 3, 5 and 7 layered sandwich beams are considered. This sandwich beam is subjected to magnetic field in transverse direction.

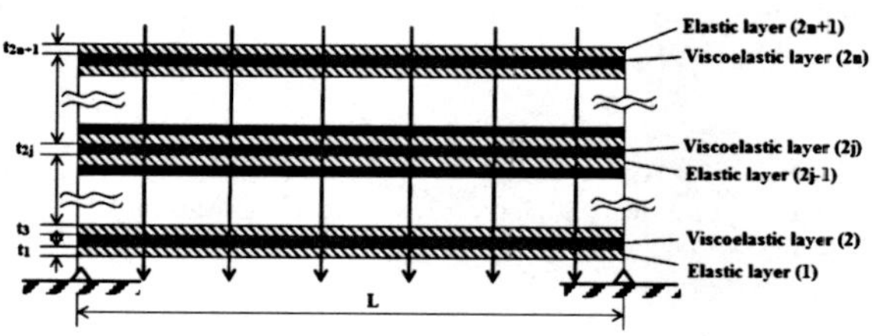

Fig. 1 MRE embedded multilayered sandwich beam subjected to magnetic field in the transverse direction

Let length, thickness, width, density, area, Young's modulus, shear modulus and second moment of area of ith *layer* be represented by L_i, t_i, b_i, ρ_i, A_i, E_i, G_i and I_i respectively. Here $i = 1$ represents the bottom layer. The following assumptions are considered for modeling the sandwich beam. The transverse displacement w of the beam is small and is same for all points in a given cross section. The rotary inertia and shear deformation in the constraining layers are neglected, that is the elastic layers obey the Euler–Bernoulli beam theory. Damping in the viscoelastic layer is primarily due to shear. Bending and extensional effects in the core are neglected. Young's modulus of the viscoelastic material is negligible in comparison to that of the elastic material.

The governing equation of motion is obtained by using Lagrange principle and finite element method as follows:

The potential energy of the sandwich beam can be given by,

$$U = \sum_{k=1}^{n+1} \frac{1}{2} \int_0^L E_{(2k-1)} I_{(2k-1)} \left(\frac{\partial^2 w}{\partial x^2} \right)^2 dx + \frac{1}{2} \int_0^L E_{(2k-1)} A_{(2k-1)} \left(\frac{\partial u_{(2k-1)}}{\partial x} \right)^2 dx$$
$$+ \sum_{l=1}^{n} \frac{1}{2} \int_0^L G^*_{v(2l)} A_{v(2l)} \gamma^2_{v(2l)} dx \tag{1}$$

where $G^*_{v(2l)} = G'_{v(2l)}(1 + i(\eta_v)_{(2l)}) =$ complex shear modulus of $(2l)$th MRE layer. $G'_{v(2l)}$ is the storage shear modulus and $(\eta_v)_{(2l)}$ is the associated core loss factor of $(2l)$th MRE layer, $i = \sqrt{-1}$, subscript 'v' is used for viscoelastic MRE core layers.

The total kinetic energy due to the metallic and core layers can be given as follows.

$$T = \sum_{k=1}^{n+1} \frac{1}{2} \int_0^L \rho_{(2k-1)} A_{(2k-1)} \left\{ \left(\frac{\partial w}{\partial t} \right)^2 + \left(\frac{\partial u_{(2k-1)}}{\partial t} \right)^2 \right\} dx$$
$$+ \sum_{l=1}^{n} \frac{1}{2} \int_0^L \rho_{v(2l)} A_{v(2l)} \left\{ \left(\frac{\partial w}{\partial t} \right)^2 + \left(\frac{\partial u_{v(2l)}}{\partial t} \right)^2 \right\} dx \tag{2}$$

For FEM analysis a standard two noded beam element (Fig. 2) with n $n + 3$ degrees of freedom (DOF) per node is considered, where n is number of MRE core layers. The DOF includes the transverse displacement w, axial displacement of constraining layer u_i, and slope φ of the beam. The elemental displacement vector is given below:

$$\{q^e\} = \{q^i \; q^j\}^{\mathrm{T}} = \left\{ w^i \; \varphi^i \; u_1^i \; u_3^i \; u_5^i \ldots\ldots u^i_{(2n+1)} w^j \; \varphi^j \; u_1^j \; u_3^j \; u_5^j \ldots\ldots u^j_{(2n+1)} \right\}^{\mathrm{T}} \tag{3}$$

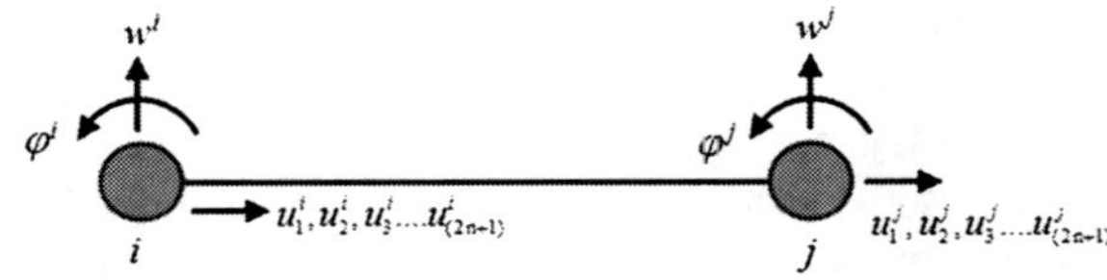

Fig. 2 Beam element with two end nodes and $n + 3$ 3 DOF for each node

Upon substituting the expressions for kinetic energy T and potential energy U into Lagrange equation $\left(\frac{d}{dt}\left(\frac{\partial L}{\partial\{\dot{q}_i\}}\right) - \frac{\partial L}{\partial\{q_i\}} = 0\right)$, where L = T − U is the Lagrangian, the governing equations of motion for MRE sandwich beam element in the finite element form are obtained as

$$[M^e]\{\ddot{q}^e\} + [K^e]\{q^e\} = \{F^e\}. \tag{4}$$

Here, $[M^e]$ and $[K^e]$ are the elemental mass and stiffness matrices, respectively, and F^e is the elemental force vector. Assembling the mass and the stiffness matrices and force vector for all the elements yields the global governing equation is as follows:

$$[M]\{\ddot{q}\} + [K]\{q\} = \{F\} \tag{5}$$

where, $[M]$, $[K]$ and F are the global mass and stiffness matrices and global force vector, respectively. It may be noted that the stiffness matrix is a function of shear modulus and core loss factor which changes with the applied magnetic field. A detailed derivation can be found in the work of Avinash [11].

In free vibration studies taking $F = 0$ in Eq. 5, the natural frequencies and loss factors are determined by finding the eigenvalues of $[M]^{-1}[K]$ matrix. While the square root of the real part of the eigenvalues gives the modal frequencies of the system, the loss factor is obtained from the ratio of the imaginary part to real part of the eigenvalues. To study the force vibration response, a harmonically varying point load $F = F_0 \cos(\omega t)$ is assumed to be applied at the nth node. Assuming the solution $\{q\} = \{q_0\} \cos(\omega t)$, Eq. 5 reduces to the following form.

$$\left([K] - \omega^2[M]\right)\{q_0\} = \left\{0\,0 \cdots F_0[N_w]|_{l_f}\,0\,0 \cdots\right\}^{\mathrm{T}} \tag{6}$$

For each frequency of excitation ω, the displacement amplitude q_0 is calculated by solving Eq. 6.

3 Results and Discussion

Based on the finite element formulations discussed in Sect. 2, a MATLAB code is developed for obtaining the free and forced vibration response of the multilayered sandwich beam. The developed code is validated by comparing the present results

Table 1 Comparison of the present modal frequencies with theoretically obtained modal frequencies (Zhou and Wang [2])

Boundary condition	Modal frequencies (Hz)			
	Mode	Zhou and Wang [2]	Present analysis	Percentage discrepancy (%)
Simply supported	1	68.816	68.24483	0.8369
	2	144.923	146.6002	1.1441
	3	222.145	224.9168	1.2324

with that of Zhou and Wang [2] where higher order theory has been used for 3 layered sandwich beam. The results (Table 1) are found to be in good agreement as the maximum discrepancy is found to be 1.23 %.

Here for numerical analysis 3, 5 and 7 layered sandwich beams with alternate metallic (Allulite) and MRE core having length 1 m, breadth 80 mm, metallic layer thickness 4 mm and MRE layer thickness 8 mm have been considered. The material properties of isotropic Allulite layer are $E = 72$ GPa, $\rho = 2700$ kg/m^3 and for MRE $\rho = 970$ kg/m^3. Following expressions for shear modulus (G'_v) and loss factor (η_v) have been taken from Nayak [12], developed by taking the experimental data of Chen et al. [6], where B is the applied magnetic field in Tesla. (1 Tesla = 1,000 G (Gauss)).

$$\begin{aligned} G'_v &= -6.9395B^6 - 9.1077B^5 + 71.797B^4 - 93.422B^3 + 38.778B^2 + 2.43B + 2.7006 \\ \eta_v &= 5.3485B^6 - 17.787B^5 + 22.148B^4 - 12.185B^3 + 2.3522B^2 + 0.1526B + 0.228 \end{aligned} \tag{7}$$

3.1 Effect of Increase in Number of Layers and Magnetic Field on Modal Frequency of Multilayered Sandwich Beam

Figure 3 shows sandwich beams with 3, 5 and 7 layers having same dimensions and mass in terms of both constraining layers and MRE core layers. For example the mass of MRE layer in Fig. 3a is same as the mass of 2 MRE layers in sandwich beam shown in Fig. 3b. The frequencies of all these beams for 3 different boundary and 2 different magnetic field conditions are presented in Tables 2, 3, and 4. It is observed that the natural frequency increases with increase in the applied magnetic field. This is because that the stiffness of the sandwich beams increases with the magnetic field strength due to the increase in complex shear modulus of MRE core (Eq. 7). It may be noted that the modal frequency values are maximum for fixed boundary condition and minimum for cantilever boundary conditions for a MRE embedded sandwich beam (Tables 2, 3, and 4). The results also show that there is a reduction in natural frequency when number of layers is increased. The percentage loss in frequency is more with increasing number of layers. It may be noted that the

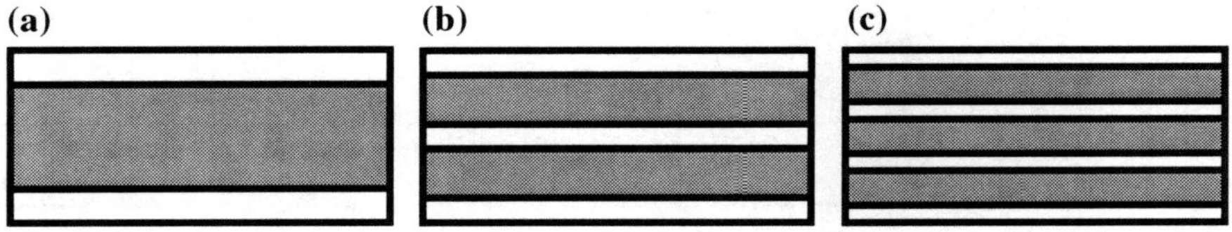

Fig. 3 Sandwich beam with **a** 3 layers ($L = 1$ m, $b = 80$ mm, $t_1 = t_3 = 6$ mm, $t_2 = 16$ mm), **b** 5 layers ($L = 1$ m, $b = 80$ mm, $t_1 = t_3 = t_5 = 4$ mm, $t_2 = t_4 = 8$ mm) and **c** 7 layers ($L = 1$ m, $b = 80$ mm, $t_1 = t_3 = t_5 = t_7 = 3$ mm, $t_2 = t_4 = t_6 = 16/3$ mm) having equivalent amount of materials

Table 2 Variation of modal frequency with magnetic field and number of layers for cantilevered boundary condition

Boundary condition	Modal frequencies (Hz)					
	Mode	3 layer beam	5 layer beam	7 layer beam	Percentage decrease (5 layer beam) (%)	Percentage decrease (7 layer beam) (%)
Cantilever (0 G)	1	11.750	11.404	11.227	2.94	4.45
	2	42.388	38.607	36.992	8.91	12.73
	3	91.576	77.000	71.218	15.91	22.23
Cantilever (600 G)	1	15.242	14.838	14.611	2.65	4.13
	2	53.482	50.638	49.508	5.31	7.42
	3	110.058	99.833	96.168	9.29	12.62

Table 3 Variation of modal frequency with magnetic field and number of layers for simply supported boundary condition

Boundary condition	Modal frequencies (Hz)					
	Mode	3 layer beam	5 layer beam	7 layer beam	Percentage decrease (5 layer beam) (%)	Percentage decrease (7 layer beam) (%)
Simply supported (0 G)	1	23.007	22.682	22.675	1.41	1.44
	2	61.690	54.036	51.546	12.40	16.44
	3	120.842	96.466	87.049	20.17	27.96
Simply supported (600 G)	1	31.860	31.405	31.287	1.42	1.79
	2	76.801	73.027	72.244	4.91	5.93
	3	139.567	122.500	117.088	12.22	16.10

difference in fundamental frequency is more in case of fixed-fixed beam. This difference may be attributed to the fact that for 3 layered sandwich beam the number of degrees of freedom is less than those used in 5 and 7 layered sandwich beam.

Table 4 Variation of modal frequency with magnetic field and number of layers for fixed-fixed boundary condition

Boundary condition	Modal frequencies (Hz)					
	Mode	3 layer beam	5 layer beam	7 layer beam	Percentage decrease (5 layer beam) (%)	Percentage decrease (7 layer beam) (%)
Fixed-fixed (0 G)	1	34.615	29.968	28.265	13.42	18.34
	2	84.566	67.757	61.126	19.87	27.71
	3	155.711	117.913	102.086	24.27	34.43
Fixed-fixed (600 G)	1	42.709	39.937	39.028	6.48	8.61
	2	97.872	85.712	81.457	12.42	16.77
	3	172.370	142.189	131.068	17.50	23.96

3.2 *Effect of Increase in Number of Layers and Magnetic Field on Time Response*

The effects of increase in number of layers and magnetic field on response amplitude for MRE embedded cantilever, simply supported and fixed-fixed sandwich beams having same volume in terms of both constraining layers and MRE core layers (Fig. 3) are shown. Figure 4a–c shows the time response comparison of the 3 layer (at 0 G and 600 G), 5 layer and 7 layer sandwich beams. It is observed that the response amplitude of 5 and 7 layered sandwich beam even at 0 G magnetic field is less than the amplitude of 3 layered beam at 600 G magnetic field of same volume i.e., response amplitude decreases with increase in number of layers for same volume. Also for multilayered sandwich beams the damping is higher with low settling time. This faster damping effect may be due to the evenly distribution of MRE layers in the multilayered sandwich beam.

3.3 *Effect of Increase in Number of Layers and Magnetic Field on Frequency Response*

The effects of increase in number of layers and magnetic field on frequency response for MRE embedded cantilever, simply supported and fixed-fixed boundary conditioned multilayer sandwich beam is presented in this section. The forced vibration analysis is done based on Eq. 6 with a harmonic force of 1 N amplitude applied at the free end of the cantilever beam and at the center for simply supported and fixed ended sandwich beam. The frequency response plots Fig. 5a–c shows that by increasing number of layers of the beam for the same volume in all 3, 5 and 7 layered beams (Fig. 3) the resonant peak is shifting towards left with increase in

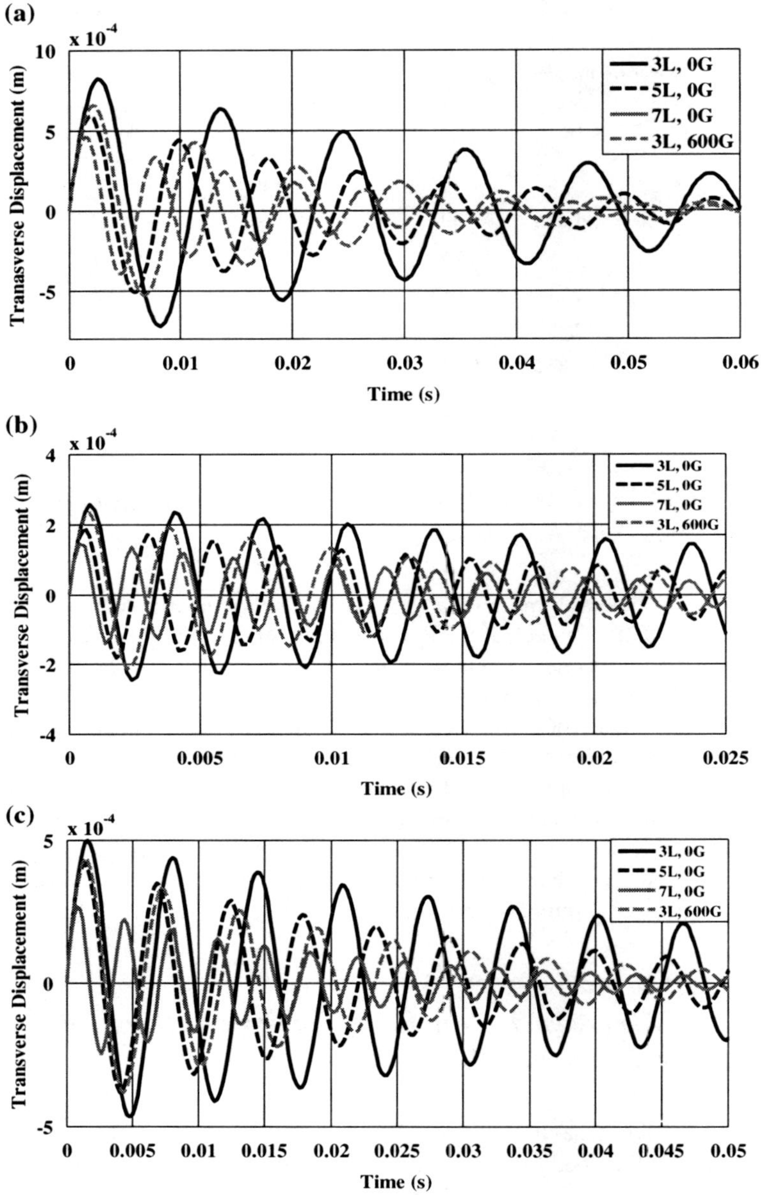

Fig. 4 Time response for 3, 5 and 7 layer **a** cantilevered, **b** simply supported, **c** fixed-fixed sandwich beam having same amount of material

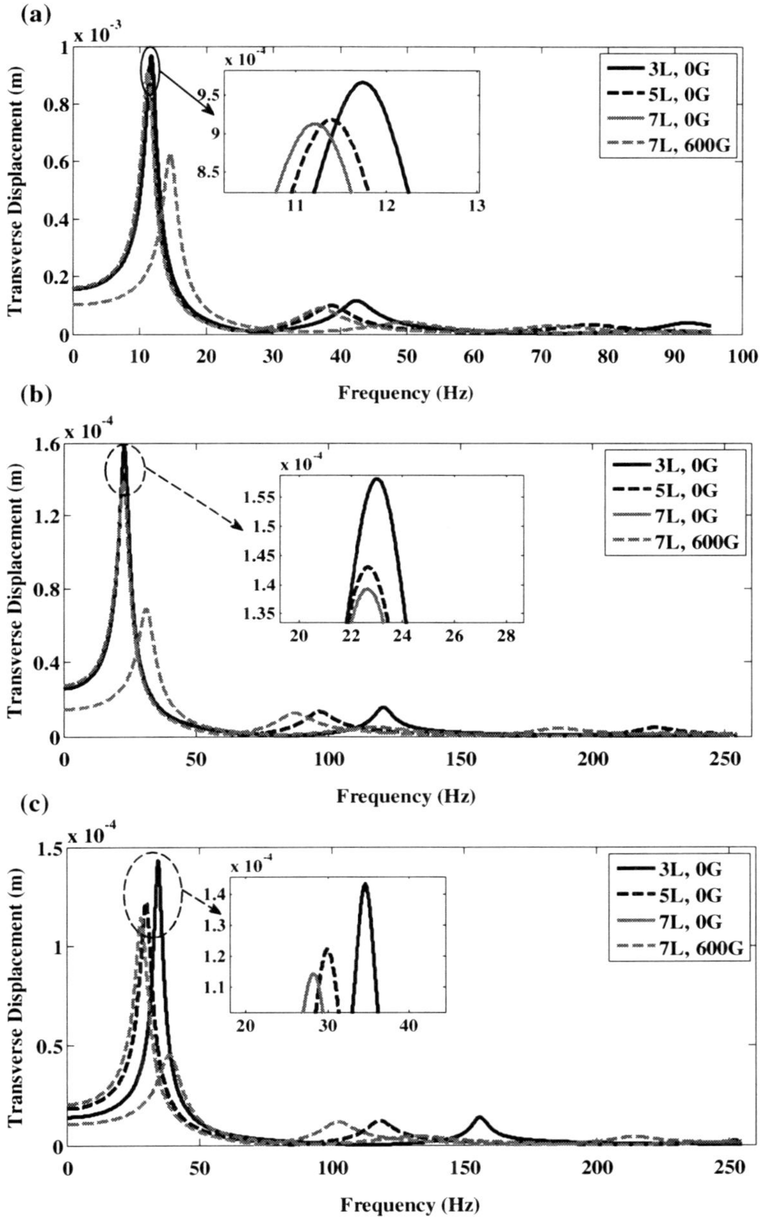

Fig. 5 Frequency response for 3, 5 and 7 layer **a** cantilevered **b** simply supported **c** fixed-fixed sandwich beam having same amount of material

number of layers, same as seen in Sect. 3.1 that increase in number of layers results in decrease in natural frequency. Also, the transverse displacement at primary frequency is decreased for the system with more number of layers for all boundary conditions. It is also shown that with increase in magnetic flux density, response amplitude decreases and the resonance condition is shifted to right side (for 7 layered sandwich beam, red lines).

4 Conclusions

In this work the finite element modeling of a multilayered sandwich beam with alternate metallic and MRE layers has been carried out for three different boundary conditions. The effects of magnetic field and the number of layers on modal frequencies, time and force responses of these systems have been studied. It is observed that the modal frequency of the beam increases with increase in applied magnetic field as this increase the stiffness of MRE core. For the same volume of the beam, with increase in number of layers the modal frequency decreases. In case of free vibration the damping of the beam increases with application of magnetic field or with increase in number of layers of the beam. In case of force vibration the response amplitude can be reduced either by increasing magnetic field or the number of layers. Both increase in number of layers and application of magnetic field have significant effect on reduction of response amplitude of the system. Depending on application, the designer can choose the optimum number of layers and magnetic field strength by using the developed method in this manuscript for effective vibration reduction.

References

1. Liao G, Gong XL, Xuan S, Guo C, Zong L (2012) Magnetic-field-induced normal force of magnetorheological elastomer under compression status. Ind Eng Chem Res 51:3322–3328
2. Zhou GY, Wang Q (2005) Magnetorheological elastomer-based smart sandwich beam with non conductive skins. Smart Mater Struct 14(5):1001–1009
3. Mead DJ, Markus S (1969) The forced vibration of a three-layer, damped sandwich beam with arbitrary boundary conditions. J Sound Vib 10(2):163–175
4. Arvin H, Sadigi M, Ohadi AR (2010) A numerical study of free and forced vibration of composite sandwich beam with viscoelastic core. Compos Struct 92(4):996–1008
5. Sun Q, Zhou JX, Zhang L (2003) An adaptive beam model and dynamic characteristics of magneto rheological materials. J Sound Vib 261(3):465–481
6. Chen L, Gong XL, Li WH (2008) Effect of carbon black on the mechanical performances of magneto rheological elastomers. Journal of Polymer Testing 27(3):340–345
7. Dwivedy SK, Mahendra N, Sahu KC (2009) Parametric instability regions of a soft and magneto rheological elastomer cored sandwich beam. J Sound Vib 325(4–5):686–704

8. Banerjee JR, Cheung CW, Morishima R, Perera M, Njuguna J (2007) Free vibration of a three layered sandwich beam using the dynamic stiffness method and experiment. Int J Solids Struct 44(2007):7543–7563
9. Agbasiere JA, Grootenhuis P (1968) Flexural vibration of symmetric multilayered beams with viscoelastic damping. J Mech Eng Sci 10(3):269–274
10. Amichi K, Atalla N (2009) A new 3D finite element for sandwich beams with a viscoelastic core. J Vib Acoust 131:1–9
11. Avinash SLVN (2014) Dynamic analysis of multilayered sandwich beam with MRE core. M. Tech. thesis, IIT Guwahati, India
12. Nayak B (2013) Dynamic stability of magnetorheological elastomer based sandwich beams. Ph.D. thesis, IIT Guwahati, India

Analysis of Coupled Transverse and Axial Vibrations of Euler Bernoulli and Timoshenko Beams with Longitudinal Crack for Its Detection

Jeslin Thalapil and S.K. Maiti

Abstract This paper presents a method to analyze the vibration of monolithic beams with longitudinal cracks for its detection. Both forward problem of determination of natural frequencies knowing the beam and crack geometry details as well as inverse problem of detection of crack with the knowledge of changes in the beam natural frequencies has been examined. Both long (Euler-Bernoulli) and short (Timoshenko) beams have been studied. For modeling a crack located at the free end of a cantilever, the beam is divided into three segments. For an internal crack located away from the free end of the beam, it is split into four segments. In both cases, two of the segments take care of beam portions above and below the crack. The cracked segments are constrained to have the same transverse displacements but different axial movements. The modeling shows good accuracy for both the forward and inverse problems. The formulation predicts the first five fundamental natural frequencies in the forward problems with a maximum difference of 5 % with reference to finite element solutions for short beams with edge or inner cracks. Further, in the case of inverse problems, edge crack with sizes varying 5 to 50 % of the beam length has been detected with errors less than 3 % in both short as well as long beams. In the case of inner crack located at the mid-span with sizes varying from 5 to 45 % of the beam length has been detected with errors less than 3 % in location and 6 % in size. The results thus show encouraging possibility of exploitation of the proposed method for crack detection in practice.

Keywords Structural health monitoring · Vibration of delaminated beams · Coupled vibrations · Longitudinal crack detection

J. Thalapil · S.K. Maiti (✉)
Department of Mechanical Engineering, Indian Institute of Technology, Mumbai, India
e-mail: skmaiti@iitb.ac.in

J. Thalapil
e-mail: jeslin.latheef@gmail.com

J.K. Sinha (ed.), *Vibration Engineering and Technology of Machinery*, Mechanisms and Machine Science 23, DOI 10.1007/978-3-319-09918-7_39

1 Introduction

Laminated beams are extensively used in aircraft, spacecraft and space structures because of high strength to weight ratio and stiffness to weight ratio. Ramkumar et al. [1] first proposed the model for vibration analysis of composite beams. The model was based on four Timoshenko beams. Wang et al. [2] examined the free vibrations of an isotropic beam with a through-width delamination by using four Euler–Bernoulli beams connected at the delamination boundaries. In their formulation they considered the coupling effect of longitudinal and flexural motions between delaminated layers. He assumed that the delaminated layers deformed ‘freely’ without touching each other. This is termed as ‘free model theory’, which was shown to be physically inadmissible. Mujumdar and Suryanarayan [3] then proposed a ‘constrained model theory’ where the delaminated layers are assumed to kiss each other, but are allowed to slide over each other. This model was physically admissible and the results showed good accuracy with the experimental results. Tracy and Pardoen [4] presented similar constrained model for a simply supported composite beam. Shu and Fan [5] extended this constrained model for the modeling of bi-material beams. Shen and Grady [6] found the separation of delaminated segments in their experiments. Luo and Hanagud [7] proposed an analytical model based on the Timoshenko beam theory, which uses piecewise-linear springs to simulate the ‘opening’ and ‘closing’ behavior between the delaminated surfaces. The model combined both free and constrained model, with the consideration that for free model the spring stiffness would be zero and infinity for constrained model. Saravanos and Hopkins [8] developed an analytical solution for predicting natural frequencies, mode shapes and modal damping of a delaminated composite beam based on a general laminate theory which involves kinematic assumptions representing the discontinuities in the in-plane and through-the-thickness displacements across each delamination crack. Shu et al. [9] presented an analytical solution to study a sandwich beam with double delaminations. His study emphasized the influence of the contact mode, ‘free’ and ‘constrained’, between the delaminated layers and the local deformation near the two fronts of the delamination. Shu and Mai [10] investigated the local deformation near the delamination fronts and identified the ‘rigid connector’ and the ‘soft connector’ conditions. A ‘rigid connector’ considers the differential stretching between the delaminated layers, while a ‘soft connector’ neglects the differential stretching. Lestari and Hanagud [11] studied a composite beam with multiple delaminations by using the Euler–Bernoulli beam theory with piecewise-linear springs to simulate the ‘opening’ and ‘closing’ behavior between the delaminated surfaces. Shu and Della [12] studied the composite beams with multiple delaminations by using the ‘free’ and ‘constrained’ models. Their study emphasized the influence of the multiple delaminations on the first and second frequencies and the corresponding mode shapes of the beams.

Large volumes of investigation have been reported involving crack(s) perpendicular to the longitudinal direction of a beam. Both forward and inverse problems have been addressed. The problem of longitudinal crack(s), particularly their detection, has not received that much of attention. The study of longitudinal cracks has been mostly driven by delamination in layered isotropic and composite beams. For the analysis of problems associated with such cracks, both free and constrained models have been proposed. Out of these two, the constrained model appears to give more accurate results. This paper examines the possibility of detecting longitudinal crack in monolithic long and short beams.

2 Mathematical Formulation for Long Beams

Beams with longitudinal crack show a coupling of axial and transverse vibration [1]. Figure 1 shows a typical beam with a longitudinal crack. The governing equation of transverse vibration, considering a Euler-Bernoulli beam, is given by

$$\frac{d^2}{dx^2}\left[EI\frac{d^2W}{dx^2}\right] - \omega^2\rho AW = 0, \quad \text{or} \quad \frac{d^4W}{d\xi^4} - \lambda^4W = 0 \tag{1}$$

where W is the transverse displacement, ω is the natural frequency of vibration, E is the modulus of elasticity, I is the second moment of inertia, ρ is the density of the beam, A is the cross sectional area, $\lambda^4 = {\rho AL^4\omega^2}/{EI}$ is the non-dimensionless frequency parameter and $\xi = {x}/{L}$ is the non-dimensionless length.

The governing equation of longitudinal vibration of the same beam is given by

$$\frac{d^2U}{dx^2} + \frac{\rho}{E}\omega^2U = 0, \quad \text{or} \quad \frac{d^2U}{d\xi^2} + p^2U = 0 \tag{2}$$

where

U is the axial displacement and $p^2 = \rho L^2\omega^2/E$.

The solutions to Eqs. 1 and 2 can be written as follows.

$$\begin{aligned} W &= C_1\cos(\lambda\xi) + C_2\cosh(\lambda\xi) + C_3\sin(\lambda\xi) + C_4\sinh(\lambda\xi) \\ U &= A_1\cos(p\xi) + A_2\sin(p\xi) \end{aligned} \tag{3}$$

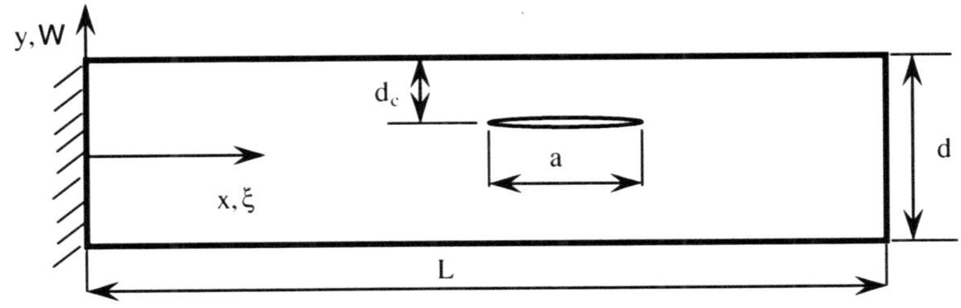

Fig. 1 Long beam with longitudinal crack

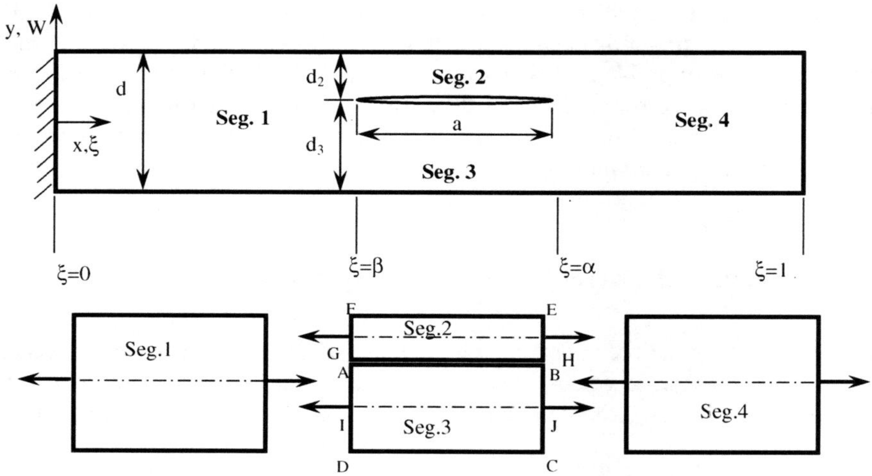

Fig. 2 Long beam with longitudinal beam

The beam with a longitudinal crack at depth d_c Fig.1 from top fibre can be divided into four segments as shown in Fig. 2. The vibration of each segment will be guided by equations of the type (1) and (2). However, there will be some constraints on their deformation as per the 'constrained model theory' [3].

The solutions to the equations of motion, Eqs. 1 and 2 for each segment can be written as follows.

For Segment 1,

$$\begin{aligned} W_1 &= C_1\cos(\lambda_1\xi) + C_2\cosh(\lambda_1\xi) + C_3\sin(\lambda_1\xi) + C_4\sinh(\lambda_1\xi) \\ U_1 &= C_{13}\cos(p\xi) + C_{14}\sin(p\xi) \\ \text{where} \quad \lambda_1^4 &= \frac{\rho A_1 L^4}{EI_1}\omega^2 \end{aligned} \tag{4}$$

The segments 2 (ABEF) and 3 (ABCD) kiss each other while undergoing transverse vibrations. Hence, they have the same vertical displacements. But, they have different axial displacements.

$$\begin{aligned} W_2 = W_3 &= C_5\cos(\lambda_2\xi) + C_6\cosh(\lambda_2\xi) + C_7\sin(\lambda_2\xi) + C_8\sinh(\lambda_2\xi) \\ U_2 &= C_{15}\cos(p\xi) + C_{16}\sin(p\xi) \\ U_3 &= C_{17}\cos(p\xi) + C_{18}\sin(p\xi) \\ \text{where} \quad \lambda_2^4 &= \frac{\rho(A_2 + A_3)L^4}{E(I_2 + I_3)}\omega^2 \end{aligned} \tag{5}$$

Segment 4 has the same form as that of segment 1.

$$
\begin{aligned}
W_4 &= C_9\cos(\lambda_1\xi) + C_{10}\cosh(\lambda_1\xi) + C_{11}\sin(\lambda_1\xi) + C_{12}\sinh(\lambda_1\xi) \\
U_4 &= C_{19}\cos(p\xi) + C_{20}\sin(p\xi)
\end{aligned} \tag{6}
$$

where A_1, A_2 and A_3 are the cross sectional areas of segment 1, 2 and 3 respectively and I_1, I_2 and I_3 are the area moments of inertia of the respective segments about an axis passing through the centers of the cross-sections and normal to the plane of the paper.

Using the boundary conditions at $\xi = 0$ and $\xi = 1$ for a cantilever beam,

$$W_1 = 0; W_1' = 0; U_1 = 0 \text{ for } \xi = 0 \tag{7}$$

$$W_4'' = 0; W_4''' = 0; U_4' = 0 \text{ for } \xi = 1 \tag{8}$$

Further at the left end A of the crack ($\xi = \beta$), there is a continuity of transverse displacement, transverse slope, shear forces, axial forces.

$$
\begin{aligned}
W_1 &= W_2;\; W_1' = W_2' \\
I_1 W_1''' &= (I_2 + I_3)W_2''';\; A_1 U_1' = A_2 U_2' + A_3 U_3'
\end{aligned} \tag{9}
$$

The jump conditions in axial displacement between A and G as well as A and I and balance of moments at the same location ($\xi = \beta$) can be expressed as follows.

$$
\begin{aligned}
U_2 &= U_1 - \frac{d_3}{2L} W_1'; U_3 = U_1 + \frac{d_2}{2L} W_1' \\
I_1 W_1'' &= (I_2 + I_3)W_2'' - \frac{A_2 d_3 L}{2} U_2' + \frac{A_3 d_2 L}{2} U_3'
\end{aligned} \tag{10}
$$

Similarly at the right end B of the crack ($\xi = \alpha$), the continuity conditions are

$$
\begin{aligned}
W_2 &= W_4; W_2' = W_4' \\
I_1 W_4''' &= (I_2 + I_3)W_2'''; A_1 U_4' = A_2 U_2' + A_3 U_3'
\end{aligned} \tag{11}
$$

At the same location, the jump conditions are

$$
\begin{aligned}
U_2 &= U_4 - \frac{d_3}{2L} W_4'; U_3 = U_4 + \frac{d_2}{2L} W_4' \\
I_1 W_4'' &= (I_2 + I_3)W_2'' - \frac{A_2 d_3 L}{2} U_2' + \frac{A_3 d_2 L}{2} U_3'
\end{aligned} \tag{12}
$$

Using Eqs. 7 to 12, the characteristic equation and the characteristic determinant $|Q|$ of size 20×20 is obtained. The characteristic equation is given by $|Q| = 0$.

2.1 Mathematical Formulation for Short Beams

The governing equation of transverse vibration of a short beam [13] is given by

$$EI\frac{\partial^2\psi}{\partial x^2}+k'\left(\frac{\partial y}{\partial x}-\psi\right)AG-\rho I\frac{\partial^2\psi}{\partial t^2}=0$$
$$\rho\frac{\partial^2 y}{\partial t^2}-k'\left(\frac{\partial^2 y}{\partial x^2}-\frac{\partial\psi}{\partial x}\right)G=0 \tag{13}$$

where y is the transverse deflection and ψ is the angle of rotation due to bending moment.

The solutions to these equations are [14]

$$Y=C_1\cosh(bp\xi)+C_2\sinh(bp\xi)+C_3\cos(bq\xi)+C_4\sin(bq\xi)$$
$$\Psi=C_1'\sinh(bp\xi)+C_2'\cosh(bp\xi)+C_3'\sin(bq\xi)+C_4'\cos(bq\xi) \tag{14}$$

where Y is amplitude of $y(x,t)$, Ψ is amplitude of ψ, ξ is the non-dimensional co-ordinate, and

$$\xi=\frac{x}{L},b^2=\frac{\rho AL^4\omega^2}{EI},r^2=\frac{I}{AL^2},s^2=\frac{EI}{k'AGL^2}$$
$$\left.\begin{matrix}p\\q\end{matrix}\right\}=\frac{1}{\sqrt{2}}\left[\mp(r^2+s^2)+\left[(r^2-s^2)^2+\frac{4}{b^2}\right]^{\frac{1}{2}}\right]^{\frac{1}{2}} \tag{15}$$

The governing equation of longitudinal vibration of a similar beam is given by

$$\frac{d^2U}{dx^2}+\frac{\rho}{E}\omega^2U=0\,\text{or}\,\frac{d^2U}{d\xi^2}+\lambda^2U=0 \tag{16}$$

where $\lambda^2=\rho L^2\omega^2/E$ and U is the axial displacement.

The solution to for axial vibration is given by

$$U=A_1\cos(\lambda\xi)+A_2\sin(\lambda\xi) \tag{17}$$

A beam with a crack is divided into 4 segments as shown in Fig. 2. The ‘constrained model theory’ [3] is again invoked. That is, segments 2 (ABEF) and 3 (ABCD) have the same transverse displacement but different axial displacements. The solutions to all the segments are given below.

Segment 1,

$$
\begin{aligned}
Y_1 &= C_1\cosh(b_1p_1\xi) + C_2\sinh(b_1p_1\xi) + C_3\cos(b_1q_1\xi) + C_4\sin(b_1q_1\xi) \\
\Psi_1 &= C_1'\sinh(b_1p_1\xi) + C_2'\cosh(b_1p_1\xi) + C_3'\sin(b_1q_1\xi) + C_4'\cos(b_1q_1\xi) \\
U_1 &= C_{13}\cos(\lambda\xi) + C_{14}\sin(\lambda\xi)
\end{aligned}
\tag{18}
$$

where

$$b_1^2 = \frac{\rho A_1 L^4\omega^2}{EI_1};\ r_1^2 = \frac{I_1}{A_1L^2};\ s_1^2 = \frac{EI_1}{k'A_1GL^2}$$

Segments 2 and 3 have identical transverse displacements but different axial displacements. Hence,

$$
\begin{aligned}
Y_3 = Y_2 &= C_5\cosh(b_2p_2\xi) + C_6\sinh(b_2p_2\xi) + C_7\cos(b_2q_2\xi) + C_8\sin(b_2q_2\xi) \\
\Psi_3 = \Psi_2 &= C_5'\sinh(b_2p_2\xi) + C_6'\cosh(b_2p_2\xi) + C_7'\sin(b_2q_2\xi) + C_8'\cos(b_2q_2\xi) \\
U_2 &= C_{15}\cos(\lambda\xi) + C_{16}\sin(\lambda\xi) \\
U_3 &= C_{17}\cos(\lambda\xi) + C_{18}\sin(\lambda\xi)
\end{aligned}
\tag{19}
$$

where

$$b_2^2 = \frac{\rho(A_2+A_3)L^4\omega^2}{E(I_2+I_3)};\ r_2^2 = \frac{(I_2+I_3)}{(A_2+A_3)L^2};\ s_2^2 = \frac{E(I_2+I_3)}{k'(A_2+A_3)GL^2}$$

Solutions for segment 4 are given by

$$
\begin{aligned}
Y_4 &= C_9\cosh(b_1p_1\xi) + C_{10}\sinh(b_1p_1\xi) + C_{11}\cos(b_1q_1\xi) + C_{12}\sin(b_1q_1\xi) \\
\Psi_4 &= C_9'\sinh(b_1p_1\xi) + C_{10}'\cosh(b_1p_1\xi) + C_{11}'\sin(b_1q_1\xi) + C_{12}'\cos(b_1q_1\xi) \\
U_4 &= C_{19}\cos(\lambda\xi) + C_{20}\sin(\lambda\xi)
\end{aligned}
\tag{20}
$$

The relations between C_i and C_i' in the above equations is given by

$$
\begin{aligned}
C_i' &= \frac{b}{L}\frac{p^2+s^2}{p}C_i, && \text{for } i = 1, 2, 5, 6, 9, 10 \\
C_i' &= -\frac{b}{L}\frac{q^2-s^2}{q}C_i, && \text{for } i = 3, 7, 11 \\
C_i' &= \frac{b}{L}\frac{q^2-s^2}{q}C_i, && \text{for } i = 4, 8, 12
\end{aligned}
\tag{21}
$$

Using the boundary conditions at $\xi = 0$ and $\xi = 1$ for a cantilever beam,

$$\begin{aligned} &Y_1 = 0; \Psi_1 = 0; U_1 = 0, \text{ for } \xi = 0 \\ &\Psi_4' = 0; \frac{1}{L}Y_4' - \Psi_4 = 0; U_4' = 0, \text{ for } \xi = 1 \end{aligned} \tag{22}$$

The second equation is connected with the shear force. Further at the left side A of the crack, $\xi = \beta$, there is a continuity of transverse displacement, slope, shear force, axial force. Therefore,

$$\begin{aligned} &Y_1 = Y_2;\ \Psi_1 = \Psi_2 \\ &A_1\left(\frac{1}{L}\frac{dY_1}{d\beta} - \Psi_1\right) = (A_2 + A_3)\left(\frac{1}{L}\frac{dY_2}{d\beta} - \Psi_2\right) \\ &A_1U_1' = A_2U_2' + A_3U_3' \end{aligned} \tag{23}$$

At the crack location, the beam segments 2 and 3 have different axial displacements. The jump conditions in axial displacements at A and G as well as A and I and balance of moments can be expressed as follows.

$$\begin{aligned} &U_2 = U_1 - \frac{d_3}{2}\Psi_1; U_3 = U_1 + \frac{d_2}{2}\Psi_1 \\ &I_1\Psi_1' = (I_2 + I_3)\Psi_2' - \frac{A_2d_3L}{2}U_2' + \frac{A_3d_2L}{2}U_3' \end{aligned} \tag{24}$$

Similarly at the right end B of the crack$\xi = \alpha$, the continuity conditions are given by

$$\begin{aligned} &Y_2 = Y_4; \Psi_2 = \Psi_4 \\ &(A_2 + A_3)\left(\frac{1}{L}\frac{dY_2}{d\beta} - \Psi_2\right) = A_1\left(\frac{1}{L}\frac{dY_4}{d\beta} - \Psi_4\right) \\ &A_2U_2' + A_3U_3' = A_1U_4' \end{aligned} \tag{25}$$

Similarly, the jump conditions at the right end B of the crack at the same location are as follows.

$$\begin{aligned} &U_2 = U_4 - \frac{d_3}{2}\Psi_4; U_3 = U_4 + \frac{d_2}{2}\Psi_4 \\ &I_1\Psi_4' = (I_2 + I_3)\Psi_2' - \frac{A_2d_3L}{2}U_2' + \frac{A_3d_2L}{2}U_3' \end{aligned} \tag{26}$$

Using the Eqs. 22 to 26, the characteristic equation of vibration of the cracked beam is obtained. The characteristic determinant $|Q|$ is of size 20×20. It can be solved to obtain the natural frequencies of a beam with a longitudinal crack.

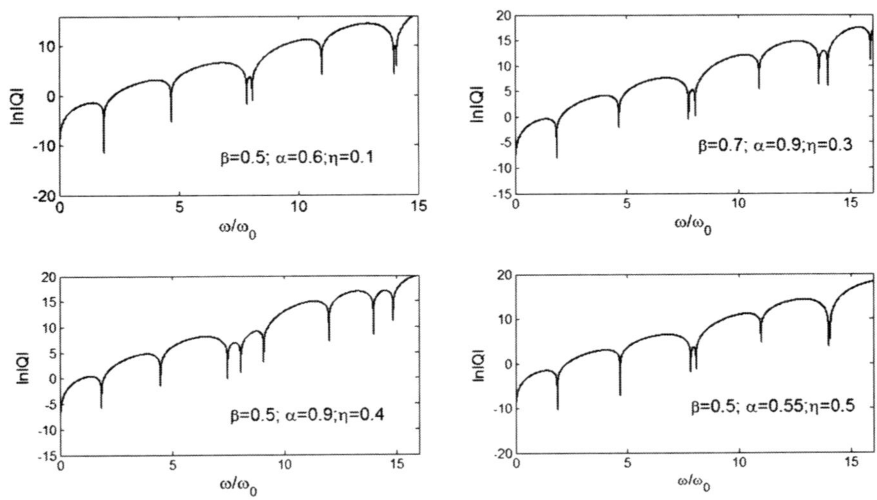

Fig. 3 Variation of $\ln|Q|$with $^{\omega}/_{\omega_0}$ for cantilever beams with longitudinal crack

2.2 Forward Problem

The characteristic equation $|Q| = 0$ is solved to obtain the natural frequencies of the cracked beam. Case studies were considered for long beam with the following dimensions and material properties: Length, L = 1200 mm, Width, B = 12.5 mm, Depth, D = 100 mm, E = 210 GPa, density, ρ = 7860 kg/m^3. A typical variation of $\ln|Q|$with$^{\omega}/_{\omega_0}$, where $\omega_0^2 = {}^{EI}/_{\rho AL^4}$ is shown in Fig. 3. The plot shows the variation for long beams with longitudinal cracks at various depths from the top fibre. The forward problem has also been done for short beams with the following geometric and material data: Length, L = 100 mm, Depth, D = 25 mm, Width, B = 12.5 mm, E = 210 GPa, density, ρ = 7860 kg/m^3.

2.3 Inverse Problem

The determination of crack location with the knowledge of natural frequencies has been done for long and short beams with longitudinal axial crack. The characteristic determinant Q of size 20 × 20 is a function of the crack location parameters and the natural frequency, i.e. Q = Q(β, α, ω). The characteristic determinant is plotted varying crack parameters β and α, each from 0 to 1. The contour plot technique of MATLAB is used to plot the value of Q for three or more measured natural frequencies and the intersection of three or more such plots gives the location and size of the crack. A typical contour plot is shown in Fig. 4. It is desirable to use four

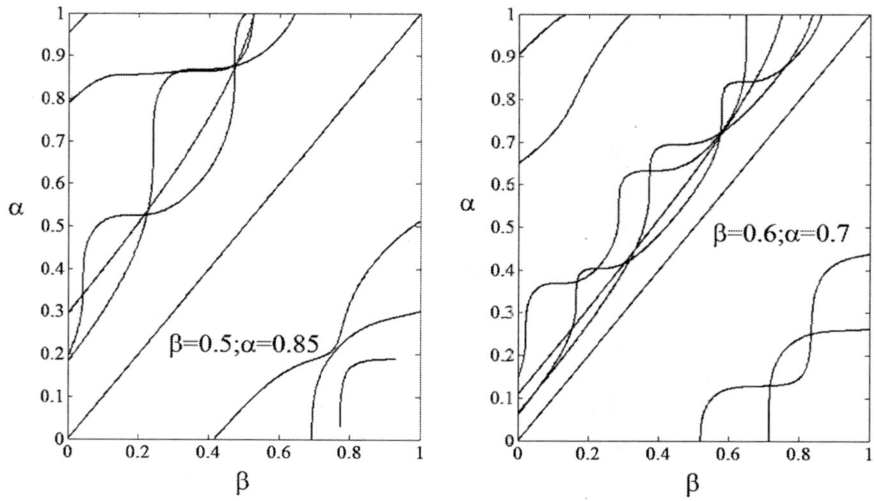

Fig. 4 Contour plot of $|Q| = 0$ with β and α for cantilever beam with longitudinal crack

or more modal frequencies for a better confirmation and accuracy of the location of crack and its size. In applications, the detection method therefore requires measurement of at least the first four transverse natural frequencies of the beam with a crack and the corresponding uncracked beam. In order to ensure a common intersection point of these plots, it is important that the modulus of elasticity 'E', which goes as input to the characteristic equation of beam with a crack for each mode, is calculated using the procedure given by Nandwana and Maiti [15]. According to this procedure two sets of frequency parameters are involved, the theoretical frequency parameter for the uncracked beam for a particular modeλ_{uct}, and the measured or computed frequency parameter of uncracked beam for the same mode λ_{ucm}.λ_{uct} for cantilever or simply supported beam is available in standard texts. λ_{ucm}, on the other hand, is determined from the measured or computed uncracked natural frequency (ω^{ucm}) through $\lambda_{ucm}^4 = (\omega_{ucm})^2\left(\rho AL^4/EI\right)$. These two parameters must be the same, but they are generally found to differ. This calls for a minor adjustment or zero setting. The modulus of elasticity (E) is so adjusted that $\lambda_{uct}^4 = (\omega_{ucm})^2\left(\rho AL^4/EI\right)$. This gives a value for E, which is given as input to the characteristic equation of beam with a crack. It may be noted that the value of E changes from one mode to another.

For case studies, the natural frequencies were computed using ANSYS 13.0 software. The beam was modelled using SOLID 183 elements. The zone around the crack tip is discretized using quarter point singular elements.

Table 1 Comparison of analytical and ANSYS frequencies for long cantilever beam with longitudinal crack

d_2/d	α	Size, a/L	First natural frequency ω1 (Hz)			Second natural frequency ω2 (Hz)			Third natural frequency ω3 (Hz)		
			Analytics	ANSYS	% Error	Analytics	ANSYS	% Error	Analytics	ANSYS	% Error
Uncracked beam			57.98	57.73	0.43	363.37	350.8	3.58	1017.5	938.9	8.37
0.1	0.6	0.1	57.98	57.72	0.45	363.36	350.7	3.60	1015.1	936.7	8.37
	0.8	0.3	57.89	57.63	0.46	361.78	349.2	3.58	1006.0	928.9	8.30
	0.9	0.4	57.82	57.55	0.47	358.61	346.2	3.56	1003.1	927.1	8.20
0.3	0.6	0.1	57.96	57.69	0.47	363.33	350.7	3.59	1008.3	928.3	8.63
	0.7	0.2	57.84	57.54	0.52	362.30	349.6	3.61	980.96	901.7	8.79
	0.9	0.4	57.25	56.90	0.61	343.66	331.3	3.73	955.0	882.9	8.17
0.4	0.7	0.2	57.76	57.46	0.53	361.82	349.2	3.61	962.77	885.6	8.70
	0.8	0.3	57.39	57.05	0.60	353.45	340.8	3.70	937.46	866.6	8.17
	0.9	0.4	56.90	56.53	0.65	335.06	323.2	3.64	924.49	857.7	7.78
0.5	0.6	0.1	57.97	57.73	0.42	363.03	350.7	3.50	1014.4	938.6	8.08
	0.7	0.2	57.97	57.71	0.46	359.8	347.8	3.43	970.94	899.2	7.97
	0.8	0.3	57.79	57.53	0.45	339.32	327.5	3.59	818.05	764.9	6.93

Geometric and Material Data: L = 1,200 mm, b = 12.5 mm, d = 100 mm, $\beta = 0.5$, E = 210 GPa, ρ = 7,860 kg/m^3

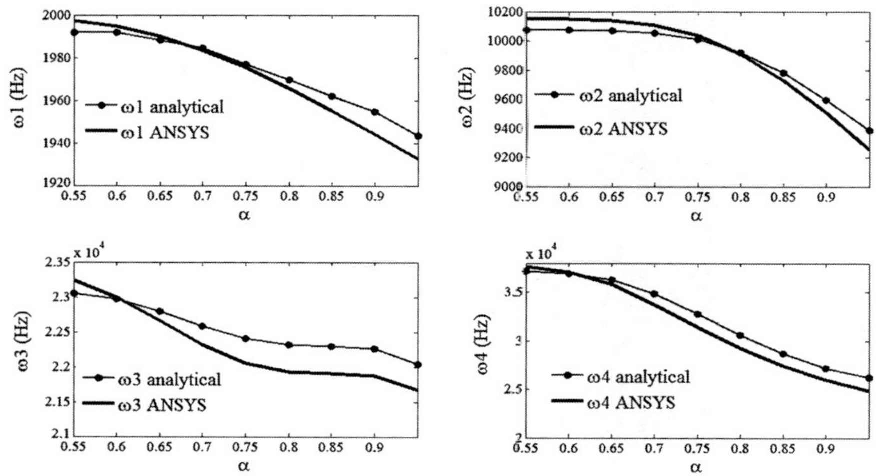

Fig. 5 Variation of natural frequencies for a short beam with longitudinal crack for β = 0.5

3 Results and Discussions

Table 1 presents the results of the forward problem for a long cantilever beam. The forward problem has been studied for crack with left end at β = 0.5. The location of the right end of the crack α varies from 0.6 to 0.9 and depth ratio d_2/d (Fig. 2) varies from 0.1 to 0.5. That is, non-dimensionless crack size a/L varies from 0.1 to 0.4.

Figure 5 shows the variation of the first four natural frequencies of a short beam with longitudinal axial crack with its left tip A at β = 0.5 and the crack size a/Lvaries from 0.05 to 0.45.

Table 2 presents the results of the forward problem for short beams with longitudinal crack. Results include cracks with left tip at β = 0.5, and the crack size a/L varying from 0.1 to 0.4 and depth ratio $\eta = d_1/d$ varies from 0.1 to 0.4. The results of the inverse problem of detection of longitudinal axial cracks in long and short beams are presented in Tables 3 and 4 respectively. The natural frequencies were again obtained using ANSYS 13.0.

In Table 3, results are presented for long beams with cracks at different locations and sizes. The error in prediction of crack location is 4 % and in size is 2 %.

In Table 4, results are presented for short beams with cracks at different locations and sizes. The error in prediction of crack location is 3 % and in size is 6 %.

Table 2 Comparison of analytical and ANSYS frequencies for short cantilever beams with longitudinal crack

Crack location			1st Mode, ω1 (Hz)			2nd Mode, ω2 (Hz)			3rd Mode, ω3 (Hz)		
η	α	a/L	Analytics	ANSYS	% Error	Analytics	ANSYS	% Error	Analytics	ANSYS	% Error
0.1	0.6	0.1	1991.6	1997.7	0.31	10075.5	10151.1	0.74	23056.3	23273.10	0.93
	0.7	0.2	1990.8	1996.2	0.27	10072.9	10144.8	0.71	23012.4	23182.70	0.73
	0.8	0.3	1989.2	1993.7	0.23	10056.5	10113.9	0.57	22990.8	23139.50	0.64
0.2	0.6	0.1	1991.41	1996.79	0.27	10075.51	10151.00	0.74	23037.44	23178.50	0.61
	0.7	0.2	1989.48	1992.47	0.15	10069.02	10133.30	0.63	22926.04	22912.20	0.06
	0.8	0.3	1985.45	1985.91	0.02	1002.91	10054.50	0.26	22866.87	22821.80	0.20
	0.9	0.4	1980.00	1978.06	0.10	9923.27	9889.33	0.34	22840.69	22771.90	0.30
0.3	0.6	0.1	1991.13	1995.79	0.23	10075.41	10150.30	0.74	23010.77	23082.00	0.31
	0.7	0.2	1987.58	1988.08	0.03	10063.41	10119.50	0.55	22800.34	22593.20	0.92
	0.8	0.3	1980.18	1976.58	0.18	9987.40	9986.46	0.01	22676.19	22411.20	1.18
	0.9	0.4	1970.21	1962.68	0.38	9797.10	9708.90	0.91	22635.51	22357.90	1.24
0.4	0.6	0.1	1990.82	1995.07	0.21	10075.31	10150.50	0.74	22980.92	23019.40	0.17
	0.7	0.2	1985.47	1985.02	0.02	10057.11	10112.30	0.55	22654.94	22393.50	1.17
	0.8	0.3	974.35	1968.81	0.28	9942.05	9931.98	0.10	22439.90	22060.60	1.72
	0.9	0.4	1959.43	1949.54	0.51	9659.61	9564.78	0.99	22389.73	22010.70	1.72

Geometric and Material Data: L = 100 mm, b = 12.5 mm, D = 25 mm, $\beta = 0.5$, E = 210 GPa, ρ = 7,860 kg/m^3

Table 3 Prediction of crack parameters for long cantilever beam with axial crack using contour plots

Actual parameters		Measured natural frequencies (Hz)				Predicted crack parameters			
Location β	Size a/L	ω1	ω2	ω3	ω4	Location β*	% Error	Size a/L*	% Error
0.5	0.05	57.7	350.8	935.7	1735.6	0.507	0.69	0.054	0.389
0.5	0.1	57.6	350.7	923.1	1716.4	0.503	0.3	0.108	0.838
0.5	0.15	57.5	350.4	901.8	1645.1	0.497	0.26	0.158	0.831
0.5	0.2	57.4	349.0	878.7	1525.5	0.498	0.21	0.209	0.869
0.5	0.25	57.2	345.6	862.2	1401.1	0.498	0.17	0.259	0.878
0.5	0.3	56.9	339.5	856.2	1300.1	0.498	0.15	0.309	0.878
0.5	0.35	56.6	330.5	855.5	1234.3	0.500	0.02	0.360	0.983
0.5	0.4	56.3	319.7	846.5	1207.1	0.531	3.06	0.394	0.555
0.5	0.45	56.0	308.7	824.0	1203.4	0.534	3.36	0.437	1.251
0.1	0.1	57.6	345.4	915.7	1706.4	0.115	1.49	0.116	1.576
0.2	0.1	57.6	346.9	934.9	1726.2	0.204	0.38	0.114	1.427
0.3	0.1	57.6	348.8	936.2	1683.5	0.292	0.78	0.112	1.193
0.4	0.1	57.6	350.3	923.3	1723.7	0.387	1.25	0.11	0.951
0.6	0.1	57.6	350.1	935.6	1677.7	0.595	0.76	0.108	0.837
0.7	0.1	57.7	349.5	937.4	1724.7	0.697	0.30	0.103	0.341
0.8	0.1	57.7	349.6	930.4	1723.1	0.808	0.79	0.097	0.254

Geometric and Material Data: L = 1,200 mm, b = 12.5 mm, D = 100 mm, E = 210 GPa, ρ = 7,860 kg/m^3

Table 4 Prediction of axial crack location in short cantilever beams using contour plots

Actual parameters		Measured natural frequency (Hz)				Predicted crack parameters			
Location β	Size a/L	ω1	ω2	ω3	ω4	Location β*	% Error	Size a/L*	% Error
0.5	0.05	1997	10150.3	23240.2	37615.2	0.487	1.264	0.091	4.06
0.5	0.1	1994.5	10149.2	22994.6	37102.9	0.478	2.16	0.145	4.47
0.5	0.2	1983.63	10108	22312.7	33683.1	0.473	2.724	0.250	4.99
0.5	0.25	1975.34	10036.1	22052.3	31366.4	0.473	2.728	0.301	5.07
0.5	0.3	1965.7	9910.8	21925	29208.5	0.473	2.728	0.351	5.09
0.5	0.4	1944.19	9510.31	21874.6	25937.4	0.473	2.737	0.452	5.23
0.2	0.1	1992.72	10007.4	23286.4	37429.3	0.176	2.391	0.145	4.46
0.3	0.1	1993.15	10082.1	23220.4	37068.8	0.271	2.926	0.145	4.51
0.4		1993.77							

(continued)

Table 4 (continued)

Actual parameters		Measured natural frequency (Hz)				Predicted crack parameters			
Location β	Size a/L	ω1	ω2	ω3	ω4	Location β^*	% Error	Size a/L^*	% Error
	0.1		10136.1	22948.5	37640.3	0.374	2.562	0.145	4.48
0.5	0.1	1994.59	10149.2	22994.6	37102.9	0.478	2.187	0.144	4.44
0.6	0.1	1995.56	10130.5	23264.1	36544.3	0.575	2.547	0.147	4.69
0.7	0.1	19955	10111.5	23299.6	37376.1	0.672	2.817	0.147	4.74
0.8	0.1	1997.42	10117.5	23182.1	37605.9	0.775	2.543	0.147	4.70

Geometric and material data: L = 100 mm, B = 12.5 mm, D = 25 mm, E = 210 GPa, ρ = 7,860 kg/m^3 , υ = 0.3

4 Conclusions

1. Beams with longitudinal crack show a coupling of transverse and axial vibrations. A method has been developed for the study of vibration characteristics of beams with longitudinal crack.
2. The accuracy of this method in prediction of natural frequencies is illustrated by case studies involving both long and short beams with crack. The natural frequencies show good match with the ANSYS results.
3. The model has been used to predict the parameters of a longitudinal axial internal crack in both long and short beams. The range of crack size considered is 5 to 45 %. In case of long beams, the maximum error in prediction of crack location is 4 % and the error in prediction of crack size is 2 %. The same errors in the case of short beams are 3 and 6 % respectively.

References

1. Ramkumar RL, Kulkarni SV, Pipes RB (1979) Free vibration frequencies of a delaminated beam. In: Reinforced Plastics/composites institute, 34th annual technical conference; 1979. The Society of the Plastics Industry Inc. 22-E, pp 1–5
2. Wang JTS, Liu YY, Gibby JA (1982) Vibration of split beams. J Sound Vib 1982 84 (4):491–502
3. Mujumdar PM, Suryanarayan S (1988) Flexural vibrations of beams with Delaminations. Journal of Sound and Vibration 125(3):441–461
4. Tracy JJ, Pardoen GC (1989) Effect of delamination on the natural frequencies of composite laminates. J Compos Mater 23(12):1200–1215
5. Shu D, Fan H (1996) Free vibration of abimaterial split beam. Compos: Part B 27(1):79–84
6. Shen MHH, Grady JE (1992) Free vibrations of delaminated beams. AIAA J 30(5):1361–1370
7. Luo H, Hanagud S (2000) Dynamics of delaminated beams. Int J Solids Struct 37 (10):1501–1519
8. Saravanos DA, Hopkins DA (1996) Effects of delaminations on the damped characteristics of composite laminates: analytical and experiments. J Sound Vib 192(5):977–993

9. Shu D (1995) Vibration of sandwich beams with double delaminations. Compos Sci Technol 54(1):101–109
10. Shu D, Mai YW (1993) Buckling of delaminated composites re-examined. Compos Sci Technol 47(1):35–41
11. Lestari W, Hanagud S (1999) Health monitoring of structures: multiple delamination dynamics in composite beams. In: 40th AIAA/ASME/ASCE/AHS/ASC Structures, structural dynamics and materials conference and adaptive structures forum
12. Shu D, Della CN (2004) Vibrations of multiple delaminated beams. Compos Struct 64 (3):467–477
13. Timoshenko SP (1928) Vibration problems in engineering. D. Van Nostrand Company INC, London
14. Huang TC (1961) The effect of rotatory inertia and of shear deformation on the frequency and normal mode equations of uniform beams with simple end conditions. J Appl Mech 28:579–584
15. Nandwana BP, Maiti SK (1997) Modelling of vibration of beam in presence of inclined edge or internal crack for its possible detection based on frequency measurements. Eng Fract Mech 58(3):193–205

Detection of Arbitrary Number of New Cracks Appearing in a Beam with Pre-existing Cracks

Dhanjee K. Mahto and Surjya K. Maiti

Abstract This paper is concerned with development of a scheme to detect fresh crack(s) in a beam with pre-existing crack(s). The cracks have been modeled by rotational springs. Relationship has been derived through energy method involving natural frequency, mode shape of uncracked beam, crack location(s) and size(s). Natural frequencies of the beam in the two states, before and after development of fresh crack(s), are the basis for the detection. Case studies have been done with uniform and stepped long beam(s). For the case studies, frequencies data have been generated through finite element method. Cantilever beams are examined with crack sizes in the range 15–50 % of section depth. Crack locations are varied within 92 % of the span from the fixed end. In the case of two pre-existing cracks and one fresh crack, the maximum errors in predicting location and size are 1.98 and 4 % respectively. In the case of one pre-existing crack and two fresh cracks the same errors are 6.99 and 8.33 % respectively. In the case of two-stepped cantilever beam with one initial crack and one fresh crack, the errors are 15 and 5.26 % respectively. The accuracy of prediction is thus very encouraging. One advantage of the proposed method is that it gives directly the location and size of the new crack(s).

Keywords Structural health monitoring · Cracked Euler Bernoulli beam · Crack detection in beam with initial crack

1 Introduction

Crack present in a structural or machine element can lead to its failure and disruption of service. It may cause catastrophe leading to loss of life and properties. It is quite necessary to detect such cracks before the accidents occur. The traditional methods to detect a crack are ultrasonic, X-rays, etc. These methods are time

D.K. Mahto · S.K. Maiti (✉)
Department of Mechanical Engineering, IIT Bombay, Bombay, India
e-mail: skmaiti@me.iitb.ac.in; skmaiti@iitb.ac.in

J.K. Sinha (ed.), *Vibration Engineering and Technology of Machinery*,
Mechanisms and Machine Science 23, DOI 10.1007/978-3-319-09918-7_40

consuming, costly and require full access of the component. Vibrations based methods have been developed to eliminate some of the shortcomings. The present paper is an effort in this direction.

An edge-normal crack in a structural member induces local flexibility at the crack location. This, in turn, causes a change in its stiffness, natural frequency and mode shape. Cracks can be detected using data on change in natural frequencies or mode shapes. Many researchers, e.g., Rizos et al. [1], Liang et al. [2], Nandwana and Maiti [3, 4], Nandwana [5] and Pandey et al. [6], have contributed in this direction.

A large number of investigators deal with detection of single crack in a beam like component. A few investigators (Hu and Liang [7]; Patil and Maiti [8, 9]) have considered the problem of detection of multiple cracks. Their strategy uses requisite number of frequencies of the beams with and without cracks to detect multiple cracks. There is a possibility that the reference configuration can be with certain number of crack(s). More new crack(s) can appear with further passage of time. Detection of these new cracks has not yet been addressed. The objective of this paper is to present a method that can be employed to detect new crack(s) in a beam with pre-existing crack(s). Both stepped and uniform beams have been examined.

2 Detection of Single Fresh Crack in Uniform Beam

2.1 Formulation

Figure 1a shows a rectangular beam with a crack of size a_1. In case a new crack of size a_2 (Fig. 1b) appears in the same beam, formulation given by Nandwana et al. [3] cannot be suitable for determining its location and size. The cracks appearing in the beam are considered to be open cracks in order to avoid any nonlinearity and facilitate solution to an inverse problem of crack detection without much complexity. The closing or breathing crack formulation involves nonlinearities and it cannot be cast to address the problems of crack detection. This is perhaps the reason why the closing crack formulation has not been applied to the detection of cracks. It has been shown by Ismail et al. [10] that the open crack formulation for a closing crack leads to an under-prediction of the crack size. For modelling, the beam with new and old cracks are represented by three beam segments connected by two rotational springs of stiffnesses K_{t1} and K_{t2} located at $x = \beta_1$ and $x = \beta_2$ (Fig. 1c) respectively.

The governing equation for mode shape of each segment is given by

$$\frac{d^2}{dx^2}\left[EI\frac{d^2Y_i}{dx^2}\right] - \omega^2\rho AU_i = 0 \tag{1}$$

where i can vary from 1 to 3 to cover the three segments, U_i is displacement of segment i. The solutions for the mode shape of the three segments are given by

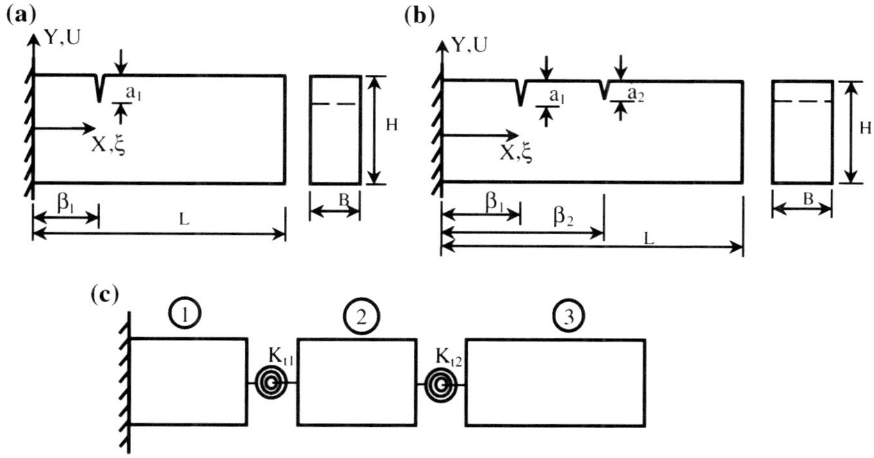

Fig. 1 **a** Cantilever beam with single crack **b** Cantilever beam with one initial and one new crack. **c** Representation of beam using two rotational springs

$$U_1 = A_1 \cos(\lambda\xi) + A_2 \cosh(\lambda\xi) + A_3 \sin(\lambda\xi) + A_4 \sinh(\lambda\xi) \tag{2a}$$

$$U_2 = A_5 \cos(\lambda\xi) + A_6 \cosh(\lambda\xi) + A_7 \sin(\lambda\xi) + A_8 \sinh(\lambda\xi) \tag{2b}$$

$$U_3 = A_9 \cos(\lambda\xi) + A_{10} \cosh(\lambda\xi) + A_{11} \sin(\lambda\xi) + A_{12} \sinh(\lambda\xi) \tag{2c}$$

where A_1 to A_{12} are arbitrary constants. These are determined using the boundary, continuity and compatibility conditions. The boundary conditions at the free and fixed ends respectively are

$$\text{bending moment, } U_3''|_{\xi=1} = 0; \quad \text{shear force, } U_3'''|_{\xi=1} = 0 \tag{3a}$$

$$\text{displacement, } U_1|_{\xi=0} = 0; \text{ slope, } U_3'|_{\xi=0} = 0 \tag{3b}$$

The continuity of displacement, moment, shear force and jump condition in slope at the crack locations $\xi = \beta_1$ and $\xi = \beta_2$ respectively can be written as follows.

$$U_1 = U_2, \quad U_1'' = U_2'', \quad U_1''' = U_2''', \quad U_2' = U_1' + \frac{U_1''}{K_1}, \tag{3c}$$

$$U_2 = U_3, \quad U_2'' = U_3'', \quad U_2''' = U_3''', \quad U_3' = U_2' + \frac{U_2''}{K_2}, \tag{3d}$$

where $K_1 = (K_{t1}L/EI)$ and $K_2 = (K_{t2}L/EI)$ are dimensionless stiffnesses of the two rotational springs. From the above twelve conditions the following characteristic equation is obtained.

$$A = det\begin{bmatrix} d_1 & d_2 & d_3 \\ d_4 & d_5 & d_6 \\ d_7 & d_8 & d_9 \end{bmatrix} = 0 \tag{4a}$$

where

$$d_1 = \begin{bmatrix} 1 & 1 & 0 & 0 \\ 0 & 0 & 1 & 1 \\ 0 & 0 & 0 & 0 \\ 0 & 0 & 0 & 0 \end{bmatrix}, \quad d_3 = \begin{bmatrix} 0 & 0 & 0 & 0 \\ 0 & 0 & 0 & 0 \\ -\cos(\lambda) & \cosh(\lambda) & -\sin(\lambda) & \sinh(\lambda) \\ \sin(\lambda) & \sinh(\lambda) & -\cos(\lambda) & \cosh(\lambda) \end{bmatrix},$$

$$D_{1i} = \begin{bmatrix} \cos(\alpha_i) & \cosh(\alpha_i) & \sin(\alpha_i) & \sinh(\alpha_i) \\ -\cos(\alpha_i) & \cosh(\alpha_i) & -\sin(\alpha_i) & \sinh(\alpha_i) \\ \sin(\alpha_i) & \sinh(\alpha_i) & -\cos(\alpha_i) & \cosh(\alpha_i) \\ \begin{Bmatrix} -(\frac{K_i}{\lambda})sin(\alpha_i) \\ -\cos(\alpha_i) \end{Bmatrix} & \begin{Bmatrix} (\frac{K_i}{\lambda})\sinh(\alpha_i) \\ +\cosh(\alpha_i) \end{Bmatrix} & \begin{Bmatrix} (\frac{K_i}{\lambda})\cos(\alpha_i) \\ -\sin(\alpha_i) \end{Bmatrix} & \begin{Bmatrix} (\frac{K_i}{\lambda})\cosh(\alpha_i) \\ +\sinh(\alpha_i) \end{Bmatrix} \end{bmatrix},$$

$$D_{2i} = \begin{bmatrix} -\cos(\alpha_i) & -\cosh(\alpha_i) & -\sin(\alpha_i) & -\sinh(\alpha_i) \\ \cos(\alpha_i) & -\cosh(\alpha_i) & \sin(\alpha_i) & -\sinh(\alpha_i) \\ -\sin(\alpha_i) & -\sinh(\alpha_i) & \cos(\alpha_i) & -\cosh(\alpha_i) \\ (\frac{K_i}{\lambda})\sin(\alpha_i) & (\frac{K_i}{\lambda})\sinh(\alpha_i) & (\frac{K_i}{\lambda})\cos(\alpha_i) & (\frac{K_i}{\lambda})\cosh(\alpha_i) \end{bmatrix},$$

$d_4 = D_{11}$, $d_8 = D_{12}$, $d_5 = D_{21}$, $d_9 = D_{22}$, d_2, d_6 and d_7 are null matrices of size (4×4), $\alpha_1 = \lambda\beta_1$ and $\alpha_2 = \lambda\beta_2$. λ is frequency parameter of the beam with two cracks. On simplification Eq. 4a takes the following form involving only variables β_2 and k_2.

$$f(\beta_2, k_2) = 0 \tag{4b}$$

Table 1 gives result on some case studies to illustrate the formulation.

2.2 *Methodology for Crack Detection and Numerical Studies*

The value of λ_n which goes as the input into Eq. 4b is calculated through the zero setting as follows.

$$\lambda_n^4 = \lambda_{ctn}^4 \left(\frac{\omega_n}{\omega_{cmn}}\right)^2 \tag{5}$$

Table 1 Accuracy of prediction of one new crack in cantilever beam with one pre-existing crack

Case no.	Actual data of new crack		ANSYS based natural frequencies (rad/s)			Predicted data of new crack			% Error	
	β_2	a_2/h	ω_1	ω_2	ω_3	β_2	K_2	a_2/h	β_2	a_2/h
1	0.3	0.2	272.68	1,741.58	4,526.94	0.3003	28.86	0.2005	0.03	0.05
2	0.4	0.3	270.21	1,668.26	4,517.36	0.3979	12.52	0.298	−0.21	−0.2
3	0.5	0.35	272.09	1,589.05	4,640.94	0.5025	8.83	0.348	0.25	−0.2
4	0.5	0.5	262.74	1,421.71	4,637.92	0.5025	3.49	0.495	0.25	−0.5
5	0.6	0.4	274.45	1,546.8	4,349.91	0.6002	6.4202	0.397	0.02	−0.3
6	0.7	0.3	277.96	1,680.72	4,319.45	0.6931	13.364	0.29	−0.69	−1
7	0.7	0.5	275.92	1,516.79	3,797.5	0.6998	3.466	0.496	−0.02	−0.4
8	0.8	0.4	278.4	1,703.72	4,193.44	0.7903	7.3182	0.376	−0.97	−2.4
9	0.85	0.3	278.68	1,743.36	4,522.59	0.8077	27.67	0.2048	−4.23	−9.52
10	0.9	0.5	278.7	1,743.92	4,491.41	0.8628	10.79	0.319	−3.72	−18.1

Beam geometric details: $L = 0.24$ m, $h = 0.02$ m, $B = 0.012$ m
Pre-existing crack details: $\beta_1 = 0.2$, $a_1/h = 0.2$
Frequency parameters of beam with initial crack: $\lambda_1 = 1.843$, $\lambda_2 = 4.693$, $\lambda_3 = 7.815$
Natural frequencies of beam with one crack obtained by FEM: $\omega_1 = 278.74$, $\omega_2 = 1{,}753.5$, $\omega_3 = 4{,}644.99$ (rad/s)

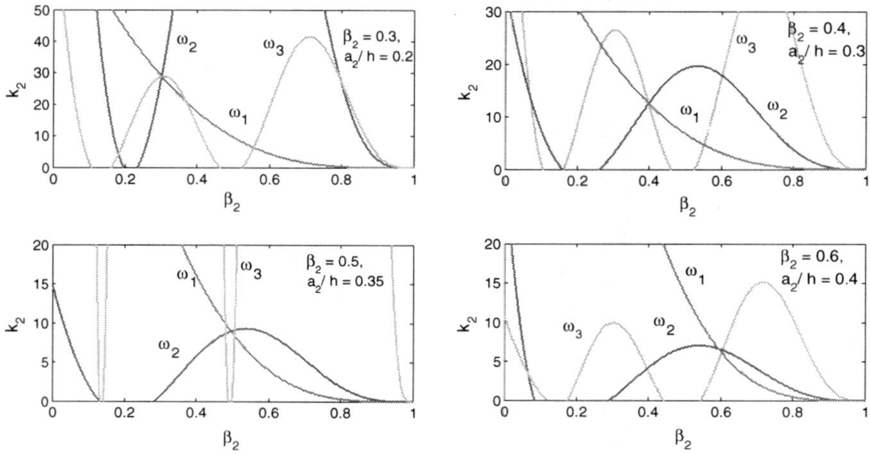

Fig. 2 Typical plots of variation of stiffness with crack location in cantilever beams

where λ_{ctn} is theoretical frequency parameter of beam with initial crack, ω_{cmn} is measured natural frequency of beam with initial crack and ω_n is the natural frequency of beam with two cracks in mode *n*. For case studies, the natural frequencies of the beam with one and two cracks simultaneously were computed using ANSYS (version 10.0) finite element software. Solid 8-noded 183 elements have been used for the modeling. In a particular discretization of the beam the number of elements is 460. The material used for all the case studies reported here is steel with E = 210 GPa, and $\rho = 7{,}860\ \text{kg/m}^3$. The theoretical frequency parameter (λ_{ctn}) is obtained by solving the forward problem of the beam with the initial crack. Using Eq. 4b, three different curves can be plotted between β_2 and K_2 for the first three modes and their common intersection point (Fig. 2) gives the crack location β_2 and K_2 corresponding to the unknown crack size.

In the case of a cantilever beam with two initial cracks and one new crack, a similar formulation is possible. A case study is presented in Table 2.

3 Detection of Multiple Fresh Cracks in Uniform Beam

When number of additional cracks is more than one, the formulation given in the previous section doesn't work. The energy method is utilized to derive a relationship involving location and size of new cracks and the beam natural frequencies.

Table 2 Accuracy of prediction of one new crack in cantilever beam with two pre-existing cracks

Case no.	Actual data of new crack		ANSYS based natural frequencies (rad/s)			Predicted data of new crack			% Error	
	β_3	a_3/h	ω_1	ω_2	ω_3	β_2	K_2	a_2/h	β_2	a_2/h
1	0.2	0.2	263.62	1,650.65	4,514.04	0.2165	22.97	0.224	1.65	2.4
2	0.25	0.35	250.79	1,647.04	4,283.58	0.2627	7.41	0.3742	1.27	2.42
3	0.3	0.25	263.21	1,638.1	4,402.34	0.3104	16.5	0.263	1.04	1.3
4	0.5	0.35	265.75	1,517.65	4,553.2	0.4982	8.87	0.347	−0.18	−0.3
5	0.65	0.3	270.7	1,574.13	4,264.02	0.6347	13.11	0.293	−1.53	−0.7
6	0.7	0.2	271.82	1,626.51	4,398.76	0.6872	32.65	0.189	−1.28	−1.1
7	0.7	0.4	270.54	1,541.73	3,957.25	0.6935	6.75	0.389	−0.65	−1.1
8	0.75	0.2	271.82	1,635.63	4,411.94	0.7302	36	0.179	−1.98	−2.1
9	0.8	0.4	271.66	1,612.65	4,094.2	0.7875	7.58	0.371	−1.25	−2.9
10	0.85	0.5	271.78	1,621.99	4,070.16	0.8376	4.3	0.46	−1.24	−4

Beam geometric details: $L = 0.24$ m, $h = 0.02$ m, $B = 0.012$ m
Pre-existing crack details
Crack 1: $\beta_1 = 0.1$, $a_1/h = 0.15$, Crack 2: $\beta_2 = 0.4$, $a_2/h = 0.3$
Frequency parameters of beam with two cracks: $\lambda_1 = 1.8199$, $\lambda_2 = 4.55$, $\lambda_3 = 7.7299$
Natural frequencies of beam with two cracks obtained by FEM: $\omega_1 = 271.98$, $\omega_2 = 1{,}651.5$, $\omega_3 = 4{,}560.23$ (rad/s)

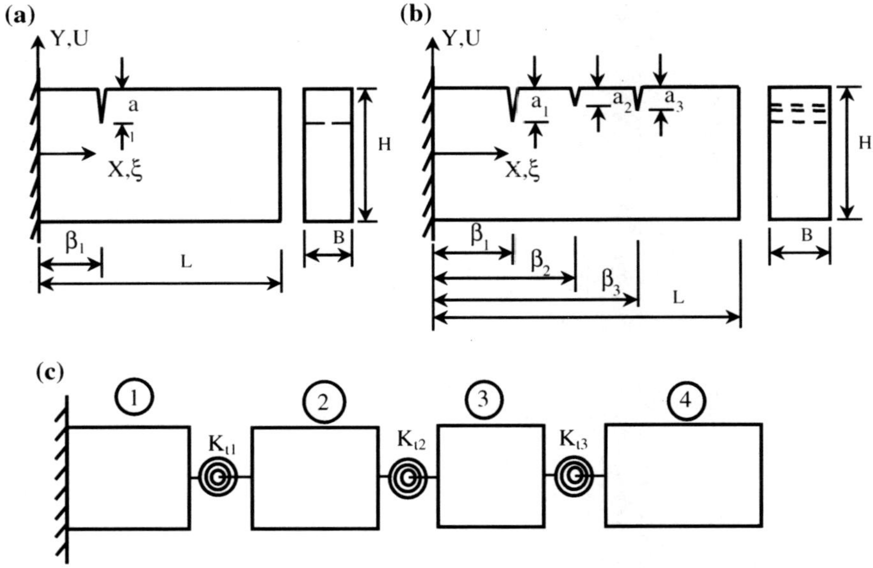

Fig. 3 **a** Cantilever beam with one crack. **b** Beam with one old and two new cracks. **c** Rotational spring based representation of beam

3.1 Formulation

A cantilever beam with one old crack (Fig. 3a) and two fresh cracks is shown (Fig. 3b). The formulation is done for the beam with three cracks. Each crack is represented by a rotational spring stiffness (Fig. 3c).

A crack in a beam reduces its strain energy (W). The reduction in strain energy is equivalent to amount of energy stored in the rotational spring. The strain energy of beam with a single crack is given by

$$W_{1cn} = W_n - \frac{M_{1n}^2}{2K_{t1}} \tag{6a}$$

where W_{1cn} is strain energy of cracked beam with one crack in mode n of natural vibration, M_{1n} is bending moment at the crack location in the same mode and K_{t1} is stiffness of rotational spring for the crack. When two fresh cracks appear in the same beam, strain energy is given by

$$W_{3cn} = W_n - \frac{M_{1n}^2}{2K_{t1}} - \frac{M_{2n}^2}{2K_{t2}} - \frac{M_{3n}^2}{2K_{t3}} \tag{6b}$$

where M_{in} stands for moment at location of crack *i* and K_{ti} is the spring stiffness for the same crack. The natural frequencies of the crack-free beam, and beam with one and three cracks respectively are given as follows.

$$\omega_n^2 = \frac{W_n}{V_n} \tag{7a}$$

$$\omega_{1cn}^2 = \frac{W_n - \frac{M_{1n}^2}{2K_{t1}}}{V_{1n}} \tag{7b}$$

$$\omega_{3cn}^2 = \frac{W_n - \frac{M_{1n}^2}{2K_{t1}} - \frac{M_{2n}^2}{2K_{t2}} - \frac{M_{3n}^2}{2K_{t3}}}{V_{3n}} \tag{7c}$$

where V_n, V_{1n} and V_{3n} are kinetic energy of uncracked (or crack-free) beam, beam with one crack and beam with three cracks respectively in mode *n* of natural vibration. ω_n, ω_{1cn} and ω_{3cn} are natural frequencies respectively of uncracked beam, beam with one crack and beam with three cracks in mode *n* of natural vibration. From Eqs. 7b, 7c, we can derive the following relation assuming that kinetic energy of cracked and uncracked beam is same and $\omega_{1cn} + \omega_{3cn} \approx 2\omega_{1cn}$.

$$\frac{2\Delta\omega_n}{\omega_{1cn}} = \frac{\frac{M_{2n}^2}{2K_{t2}} + \frac{M_{3n}^2}{2K_{t3}}}{W_n - \frac{M_{1n}^2}{2K_{t1}}} \tag{8}$$

where $\Delta\omega_n = \omega_{1cn} - \omega_{3cn}$. The beams under consideration are long beams (L/h > 10). Using the Euler Bernoulli beam theory, $M = EI\,(d^2U/dX^2)$. Equation 8 can be written in the form given below.

$$\frac{\Delta\omega_n}{\omega_{1cn}} = \frac{(EI)^2}{4\left(W_n - \frac{M_{1n}^2}{2K_{t1}}\right)}\left[\frac{\left[\left\{\frac{d^2U}{d\beta^2}\right\}_{\beta=\beta_2}\right]^2}{K_{t2}} + \frac{\left[\left\{\frac{d^2U}{d\beta^2}\right\}_{\beta=\beta_3}\right]^2}{K_{t3}}\right] \tag{9}$$

This equation consists of curvatures of the corresponding uncracked beam at locations β_2 and β_3. *U* stands for modal displacement.

3.2 Methodology for Crack Detection and Numerical Study

In Eq. 9, the unknown variables are β_2, K_{t2}, β_3 and K_{t3}. These variables can be evaluated using four independent equations that are formed by substituting four input natural frequencies in the equation. To solve these nonlinear equations, 'fsolve' command of MATLAB recommended by Maghsoodi et al. [11] is used.

The results for the detection of two additional cracks in steel beams lying on the right side of initial crack are given in the Table 3. The maximum error in predicting location of crack is 6.99 % and the error in prediction of size is 8.33 %.

4 Detection of Single Fresh Crack in Multi Step Beam

Detection of additional crack in a two-step beam is considered in this section (Fig. 4). The initial crack lies in segment 1 and position of the new crack can be considered in any of the three segments.

4.1 Formulation

The natural frequencies of beam with one and two cracks can be given in terms of strain (W) and kinetic (V) energies. The symbols used here have usual meanings as explained in the previous section.

$$\omega_n^2 = \frac{W_n}{V_n} \tag{10a}$$

$$\omega_{1cn}^2 = \frac{W_n - \frac{M_{1n}^2}{2K_{t1}}}{V_{1cn}} \tag{10b}$$

$$\omega_{2cn}^2 = \frac{W_n - \frac{M_{1n}^2}{2K_{t1}} - \frac{M_{2n}^2}{2K_{t2}}}{V_{2cn}} \tag{10c}$$

From Eqss. 10b, 10c, the following relation can be derived assuming that kinetic energy of cracked and uncracked beam is the same and $\omega_{1cn} + \omega_{2cn} \approx 2\omega_{1cn}$.

$$\frac{2\Delta\omega_n}{\omega_{1cn}} = \frac{\frac{M_{2n}^2}{2K_{t2}}}{W_n - \frac{M_{1n}^2}{2K_{t1}}} \tag{11}$$

where $\Delta\omega_n = \omega_{1cn} - \omega_{2cn}$. By using the Euler Bernoulli beam formulation, i.e. $M = EI\,(d^2U/dX^2)$, Eq. 11 can also be written in the form of

$$\frac{\Delta\omega_n}{\omega_{1cn}} = \frac{(EI_2)^2}{4\left(W_n - \frac{M_{1n}^2}{2K_{t1}}\right)}\left[\frac{\left[\left\{\frac{d^2U}{d\beta^2}\right\}_{\beta=\beta_2}\right]^2}{K_{t2}}\right] \tag{12}$$

Table 3 Accuracy of prediction of two additional cracks appearing on one side of the initial crack in cantilever beam

Case no.	Actual data		ANSYS based natural frequencies (rad/s)				Predicted data			% Error	
	β	a/h	ω_1	ω_2	ω_3	ω_4	β	K	a/h	β	a/h
1	0.3	0.3	262.353	1,648.56	4,379.5	8,073.95	0.2679	92,351	0.2916	−3.21	−0.84
	0.5	0.25					0.5699	99,309	0.282	6.99	3.2
2	0.35	0.3	263.978	1,517.35	4,132.8	8,369.04	0.4011	62,181	0.3469	5.11	4.69
	0.6	0.4					0.6566	67,539	0.3349	5.66	−6.51
3	0.35	0.4	253.051	1,355.35	3,480.95	8,401.45	0.3751	42,245	0.4061	2.51	0.61
	0.65	0.55					0.6638	29,668	0.463	1.38	−8.7
4	0.3	0.4	253.18	1,643.72	3,939.69	8,229.67	0.3105	50,849	0.3773	1.05	−2.27
	0.7	0.3					0.7014	90,762	0.2937	0.14	−0.63
5	0.4	0.5	249.847	1,442.02	3,709.74	7,786.6	0.3899	30,760	0.4571	−1.01	−4.29
	0.7	0.4					0.7372	48,446	0.3847	3.72	−1.53
6	0.6	0.5	270.78	1,407.42	4,077.53	7,519.83	0.5592	39,514	0.4167	−4.08	−8.33
	0.8	0.3					0.7349	60,701	0.3505	−6.51	5.05

Beam geometric details: L = 0.24 m, h = 0.02 m, B = 0.012 m
Pre-existing crack details: $\beta_1 = 0.2$, $a_1/h = 0.2$
Frequency parameters of beam with one crack: $\lambda_1 = 1.8425$, $\lambda_2 = 4.6933$, $\lambda_3 = 7.8149$, $\lambda_4 = 10.8534$
Natural frequencies of beam with one crack obtained by FEM: $\omega_1 = 278.742$, $\omega_2 = 1{,}753.5$, $\omega_3 = 4{,}644.99$, $\omega_4 = 8{,}506.3$ (rad/s)

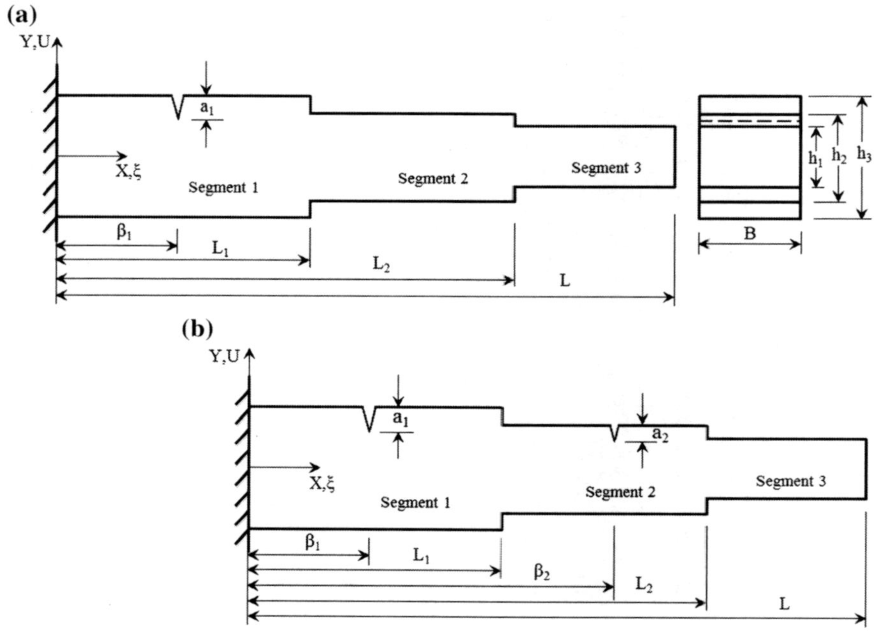

Fig. 4 **a** Cantilever beam with initial crack in first segment. **b** Beam with one additional crack in second segment

Table 4 Accuracy of detection of one new crack in two step cantilever beam with initial crack in first segment

Case no.	Actual data of new crack		ANSYS based natural frequencies (rad/s)		Predicted data of ew crack			% Error	
	β_2	a_2/h	ω_1	ω_2	β_2	K_{t2}	a_2/h	β_2	a_2/h
1	0.2	0.15	38.7	193.168	0.275	530,169	0.148	7.5	−0.2
2	0.2	0.3	37.927	192.905	0.275	131,203	0.293	7.5	−0.7
3	0.24	0.4	37.285	193.217	0.245	92,283	0.347	0.5	−5.3
4	0.36	0.3	38.223	191.832	0.297	67,116	0.336	−6.3	3.6
5	0.36	0.45	37.119	189.726	0.296	26,869	0.479	−6.4	2.9
6	0.4	0.25	38.551	191.682	0.327	101,863	0.279	−7.3	2.9
7	0.4	0.4	37.814	188.904	0.326	36,252	0.431	−7.4	3.1
8	0.45	0.25	38.651	190.975	0.362	106,328	0.273	−8.8	2.3
9	0.45	0.35	38.328	188.641	0.362	52,317	0.373	−8.8	2.3
10	0.5	0.3	38.634	189.25	0.397	79,883	0.311	−10.3	1.1
11	0.52	0.2	38.835	191.439	0.412	193,956	0.205	−10.8	0.5

(continued)

Table 4 (continued)

Case no.	Actual data of new crack		ANSYS based natural frequencies (rad/s)		Predicted data of ew crack			% Error	
	β_2	a_2/h	ω_1	ω_2	β_2	K_{t2}	a_2/h	β_2	a_2/h
12	0.55	0.42	38.464	184.236	0.425	41,605	0.409	−12.5	−1.1
13	0.62	0.36	38.77	187.528	0.47	76,104	0.318	−15	−4.2

Beam geometric details: $L_1 = 0.25$ m, $L_2 = 0.5$ m, $L = 0.75$ m, $h_1 = 0.024$ m, $h_2 = 0.02$ m, $h_3 = 0.016$ m, $B = 0.012$ m
Pre-existing crack details: $\beta_1 = 0.15$, $a_1/h = 0.25$
Frequency parameters of first segment with one crack: $\lambda_{11} = 1.9624$, $\lambda_{12} = 4.379$
Frequency parameters of second segment with one crack: $\lambda_{21} = 2.1497$, $\lambda_{22} = 4.7969$
Frequency parameters of third segment with one crack: $\lambda_{31} = 2.4034$, $\lambda_{32} = 5.3631$
Natural frequencies of beam with one crack obtained by FEM: $\omega_1 = 38.95$, $\omega_2 = 193.24$ (rad/s)

Using Eq. 12 and frequencies of the first two modes K_{t2} and β_2 can be determined. 'fsolve' command of MATLAB has been employed to solve for the two unknowns. The results for steel beams are presented in Table 4. The maximum error in predicting crack location is 15 % and the error in prediction of size is 5.3 %.

5 Conclusions

More studies have been done by varying both the number of pre-existing and new cracks. They are not reported here because of space restrictions. Considering all the investigations done, main conclusions are as follows.

1. In the case of detection of one new crack in uniform cantilever beam with one pre-existing crack, as the new crack shifts towards the free end the accuracy of prediction of crack details reduces.
2. As the number of new cracks increases, the accuracies of prediction of location and size reduce.
3. The additional crack appearing in a multi-step beam can be determined irrespective of the segment in which the additional crack occurs.
4. Single additional crack in multi-step beam too can be predicted with good accuracy using the first two natural frequencies only.

References

1. Rizos PF, Aspragathos N, Dimarogonas AD (1990) Identification of crack location and magnitude in a cantilever beam from the vibration modes. J Sound Vib 138(3):381–388
2. Liang RY, Choy FK, Hu J (1991) Detection of cracks in beam structures using measurements of natural frequencies. J Franklin Inst 328:505–518

3. Nandwana BP, Maiti SK (1991) Modeling of vibration of beam in presence of inclined edge or internal crack for its possible detection based on frequency measurements. Eng Fract Mech 58 (3):193–205
4. Nandwana BP, Maiti SK (1997) Detection of the location and size of a crack in stepped cantilever beams based on measurements of natural frequencies. J Sound Vib 203(3):435–446
5. Nandwana BP (1997) On the foundation for detection of crack based on measurement of natural frequencies. PhD thesis, Indian Institute of Technology, Bombay
6. Pandey AK, Biswas M, Samman MM (1991) Damage detection from change in curvature mode shapes. J Sound Vib 145(2):321–332
7. Hu J, Liang RY (1993) An integrated approach to detection of cracks using vibration characteristics. J Franklin Inst 330(5):841–853
8. Patil DP, Maiti SK (2003) Detection of multiple cracks using frequency measurements. Eng Fract Mech 70(12):1553–1572
9. Patil DP, Maiti SK (2005) Experimental verification of a method of detection of multiple cracks in beams based on frequency measurement. J Sound Vib 281:439–451
10. Ismail F, Ibrahim A, Martin HR (1990) Identification of fatigue cracks using vibration testing. J Sound Vib 140(2):305–317
11. Maghsoodi A, Ghadami A, Mirdamadi HR (2013) Multiple-crack damage detection in multi-step beams by a novel local flexibility-based damage index. J Sound Vib 332(2):294–305

Acoustic Technique for Diverse Safety Rod Drop Time Measurement in PFBR

V. Prakash, Hemant Prakash Agnihotri, P. Anup Kumar, R. Ramakrishna and K.K. Rajan

Abstract Diverse safety rods (DSRs) are used in Prototype Fast Breeder Reactor (PFBR) to shutdown the reactor during emergency conditions (SCRAM). These rods contain neutron absorber materials to reduce the reactivity for the safe shutdown of the reactor. During normal operation of the reactor, DSRs are held above the active core using electromagnets. During a SCRAM the electro-magnets are de-energized and DSR falls into the core. Since it is a safety related action, no active system is used for driving the rods into the core and the rod falls under gravity force. It is required to measure the fall time of DSR during each SCRAM in the reactor to ensure proper insertion of the DSR in the core. Experiments were carried out in various facilities to develop a measurement technique. This paper discusses the details of Diverse Safety Rod Drive Mechanism (DSRDM), acoustic technique used for fall time measurement, experiments carried out, analytical modeling details and its results and conclusion.

Keywords PFBR · Diverse safety rods · Drop time measurement · Acoustic technique · Wave propagation

V. Prakash (✉) · H.P. Agnihotri · P. Anup Kumar · R. Ramakrishna · K.K. Rajan
Fast Reactor Technology Group, Indira Gandhi Centre for Atomic Research, Kalpakkam 603102, India
e-mail: prakash@igcar.gov.in

H.P. Agnihotri
e-mail: agnihotri@igcar.gov.in

P. Anup Kumar
e-mail: panup@igcar.gov.in

R. Ramakrishna
e-mail: ramakrishna@igcar.gov.in

K.K. Rajan
e-mail: kkrajan@igcar.gov.in

J.K. Sinha (ed.), *Vibration Engineering and Technology of Machinery*, Mechanisms and Machine Science 23, DOI 10.1007/978-3-319-09918-7_41

1 Introduction

PFBR is equipped with two shutdown systems namely, Control and Safety Rod Drive Mechanism (CSRDM) and Diverse Safety Rod Drive Mechanism (DSRDM). DSRDM is used for the safe shutdown of the reactor. During SCRAM, DSR is released from its electromagnet and falls into the reactor core under gravity. It is a safety requirement to measure the fall time of DSR during each SCRAM. As no sensor can be attached to the moving part of DSR, non-contact type measurement techniques namely acoustic and eddy current methods are envisaged for the measurement of DSR fall time in PFBR.

Acoustic technique to assess the performance of shut off rods was attempted worldwide in several reactors. In PFR reactor (UK), 6 nos of acoustic wave-guides were installed in the primary vessel to gather data for acoustic boiling noise detection [1]. During the performance test of the shut off rods, acoustic signatures were recorded using these wave-guides. Measurement on rod release time and relative time for the acoustic events recorded using the wave-guides, confirmed the satisfactory insertion of the rods. Similar system was used in KNK-II reactor for the second shutdown system [2]. The termination of the SCRAM was determined by measuring the vibrations caused by the absorber hitting the bottom of the sheath. The Phenix reactor in France has six CRDMs and one complimentary shut off rod (SAC) [3] and the drop time of SAC is measured using acoustic method.

Studies for developing acoustic technique for DSR drop time measurement system for PFBR were initiated. Accelerometers mounted on upper part of DSRDM were used for the detection of acoustic events during the movement of DSR in the DSR subassembly [4]. Measurements were carried out in various water/sodium facilities and an on-line measurement system for PFBR has been developed [5]. The developed system was tested for its performance and results were compared with ultrasonic method to establish its measurement sensitivity [6]. Analytical modeling of the DSRDM has been performed using the finite element code ABAQUS. Excitation waves in the form of pressure were given at bottom end and output signal was obtained at the top end of the model. This paper discusses the details of the acoustic technique, experimental procedure, analytical modeling of the phenomenon, validation of the technique and its results and conclusion.

2 Description of DSRDM Mechanism

Three DSRDMs are provided in PFBR as a diverse shutdown mechanism. The mechanism is located in the control plug of PFBR and the total length is ~14 m. The top drive mechanism is extending outside the roof slab of the reactor vessel and the bottom part which holds the DSR is in high temperature sodium. DSR is kept inside the DSR subassembly using an electromagnet as shown in Fig. 1. Once the electromagnet is de-energized, DSR falls into the core under gravity and the total

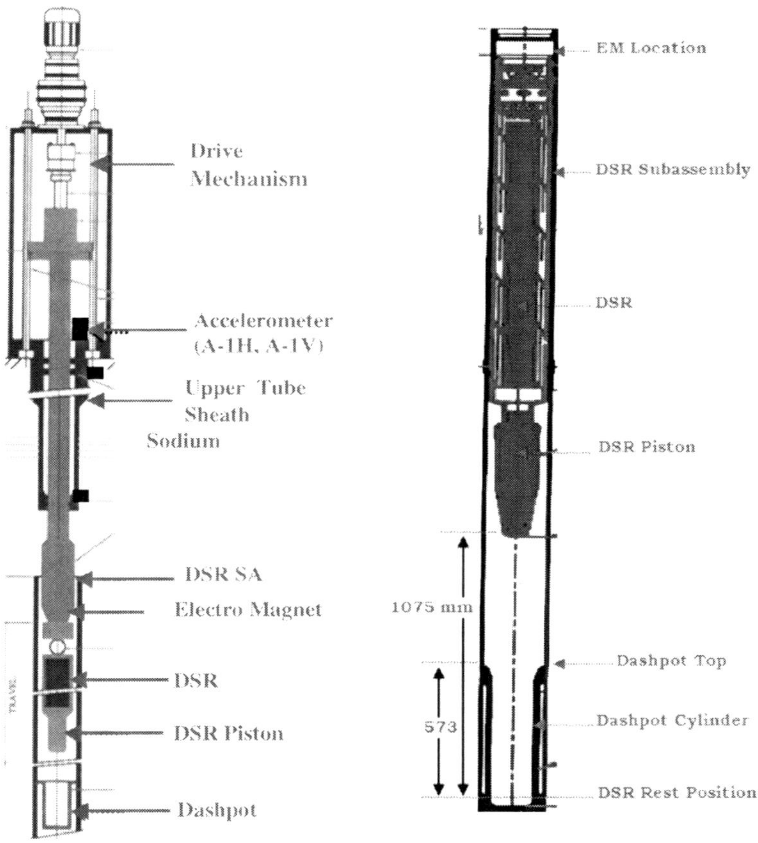

Fig. 1 Schematic of DSRDM and DSR subassembly

travel distance is 1,075 mm inside the DSR subassembly. A sodium dashpot is provided at the bottom of the subassembly for decelerating the DSR at the end of travel before bringing it into rest. The designed fall time of DSR is one second. To ensure proper insertion of the DSR into the core, the fall time of DSR needs to be measured.

Since no sensor can be attached to the mobile part (DSR) during its travel it is very difficult to measure the fall time of DSR. Moreover, DSR subassembly is located at the centre of the reactor vessel which is ~13 m in diameter, with hot sodium at 550 °C, contact type of measurement methods are practically impossible. So a non-contact type of measurement (acoustic technique) has been chosen to detect the DSR fall time.

3 Acoustic Technique

Acoustic technique is based on the detection of noise signals produced during the movement of DSR inside the subassembly. Once DSR is released from electromagnet (EM), at the end of free fall, DSR entry into the dashpot generates shock signals. Similarly at the end of braking (deceleration) time another shock signal is generated. By measuring the time delays between the electromagnet de-energizing signal and the above shock signals, free fall time, braking time and the total fall time can be estimated. Piezoelectric accelerometers are used as the primary sensing element for acoustic technique. The sensors are deployed on the top of the drive mechanism which is approximately 11 m away from the dashpot region as shown in Fig. 1. The free fall time is defined as the time elapsed between the instant at which the EM holding the DSR gets de-energizes and the instant at which the dashpot action is initiated. At the end of free fall, deceleration of DSR will take place in the dashpot region. The time taken for completion of deceleration of DSR is called braking time. The summation of free fall time and braking time will give the total drop time of DSR.

4 Measurements in Water

The acoustic events taking place in the dashpot region in water is successfully recovered at the top of the mechanism, where the sensors are installed. Several experiments were conducted in various test facilities and the measurement parameters were optimized. Figure 2 shows a typical time signal recorded using accelerometers at the dashpot region during experimental studies in water to characterize the signal at the source. The first peak in the time signal corresponds to the entry of DSR into the dashpot and the last peak corresponds to the final impact produced by the DSR at its final rest position. Other peaks observed in between the entry and the final impact correspond to the rubbing of DSR piston during its deceleration inside the dashpot due to DSR piston profile variation. This has been

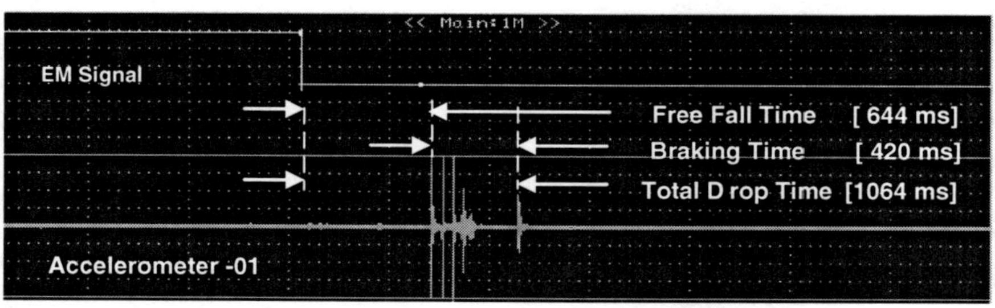

Fig. 2 Time signal for fall height of 1075 mm in water

verified by using an ultrasonic sensor, which was mounted below the dashpot to study the position of DSR during each instant after a SCRAM. The result was also verified by visual observation method, which clearly showed the rubbing marks on DSR Piston. The measured fall time is found to be 1,064 ms.

5 Studies on Acoustic Signal Transmission

The path of acoustic signal transmission is quite complex though DSRDM. Acoustic shock signals produced at the dashpot travels through the liquid medium and structural material (stainless steel) and reaches the sensors installed on top of the drive mechanism. To get an understanding of the phenomenon and signal transmission, further experimental studies were carried out in water.

5.1 Experimental Studies

Apart from the dashpot region, sensors were also installed on the DSRDM at different elevations and signals were recorded. Figure 3 shows the schematic of the mechanism and the position of sensors.

Accelerometers as well as acoustic emission sensors were used during the tests. Location of Sensor A-5 corresponds to the dashpot region where the shock signals are produced. The sensor location A-1 V corresponds to the top end of upper tube sheath (Fig. 4) of DSRDM which is approximately 11 m away from the dashpot region. During measurements, signal pattern at each sensor location is recorded.

Figure 5 shows the time signal recorded at three different locations on the DSRDM. As the signal propagates from the source to the drive mechanism top, the variation in time signal pattern due to dispersion can be clearly seen in the time signal.

5.2 Analytical Modeling

Analytical modeling is used for a better understanding of the system behavior. Modeling of DSRDM has been carried out in finite element code ABAQUS and is shown in Fig. 6a, b. Input load has been given as the pressure fluctuations at the bottom of the mechanism. This input was obtained from the waveforms recorded during experiments performed in air with water in the dashpot and DSR subassembly.

8 node linear brick, reduced integration, hourglass control (C3D8R) element is used for meshing the mechanism in the analysis. Dynamic explicit analysis was performed to obtain output wave in the form of stresses at the top of the mechanism.

Fig. 3 Test setup schematic

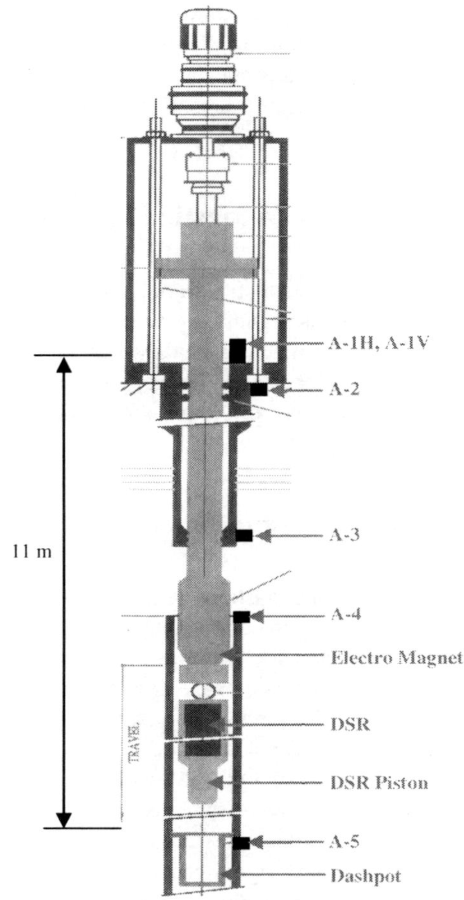

Element length is obtained by following 10 node per wavelength rule.

Frequency of the input wave (f)	2 kHz
Velocity of wave (v)	2,000 m/s (approx.)
Wavelength (λ)	v/f = 1 m

So the element length comes out to be 100 mm. However to refine the results further, maximum length of element is taken to be 15 mm. A typical analysis for running this code took around 180 min.

A pressure fluctuation proportional to acceleration signal measured at the dashpot region was given as the input in axial direction at the bottom surface of the DSRDM and is shown in Fig. 7. Output in the form of stress waves has been obtained at various nodes on the mechanisms. Figure 8 shows the output time signal

Fig. 4 Top end of upper tube sheath

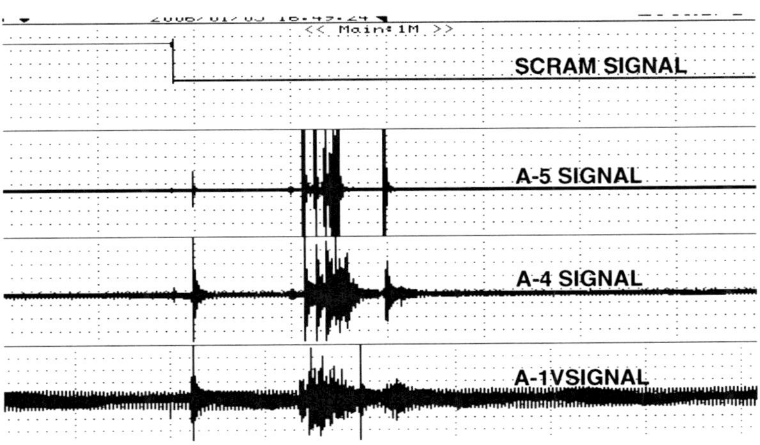

Fig. 5 Time signal at different locations

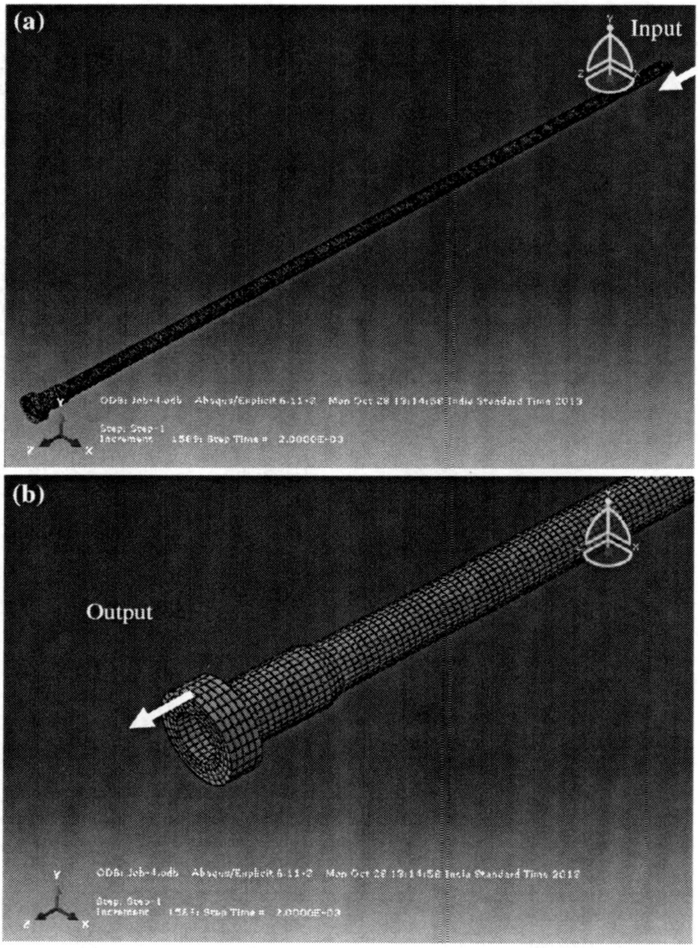

Fig. 6 **a** Full model of the mechanism, **b** Top portion of the mechanism

obtained at a node on the top of the model in axial direction. The signal pattern obtained at the DSRDM top location for various input signals were studied. During further tests in sodium, top sensor was mounted vertically and signals were recorded. Figure 9 shows the time wave form recorded from the top sensor for a free fall of 1,075 mm in sodium at 550 °C. The free fall time (552 ms) and total fall time (751 ms) was clearly distinguishable in the vertically mounted sensor at the top. The amplitude of final impact was also relatively high compared to horizontally mounted sensor (A-1H).

For PFBR an online system for DSR drop time measurement system was developed. Sensor (A-1 V) at the top of the mechanism with vertical orientation is used for measuring the fall time of DSR as the signal is more distinguishable with

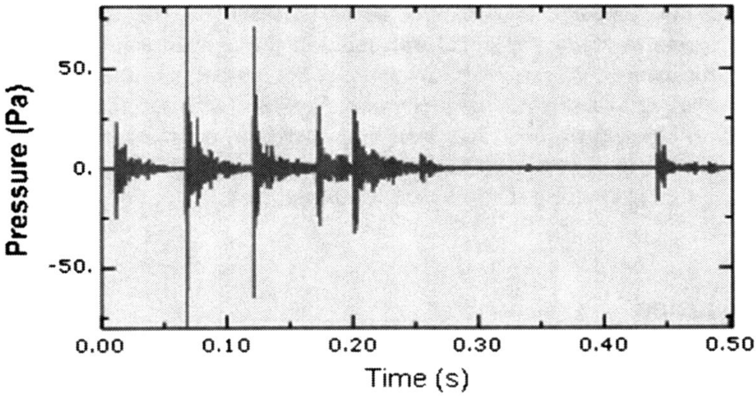

Fig. 7 A typical input signal

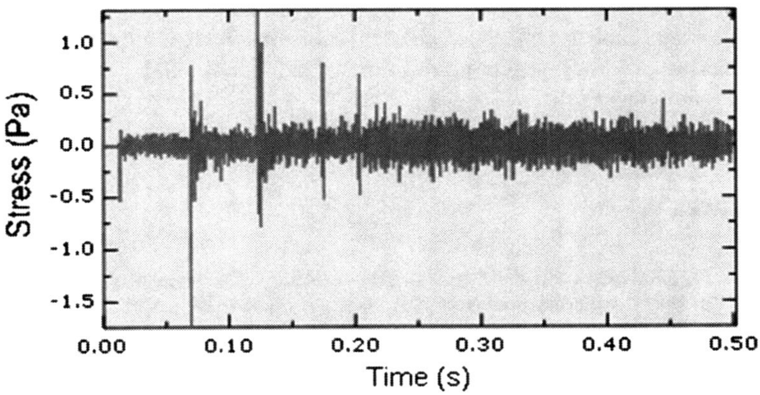

Fig. 8 Output stress wave pattern

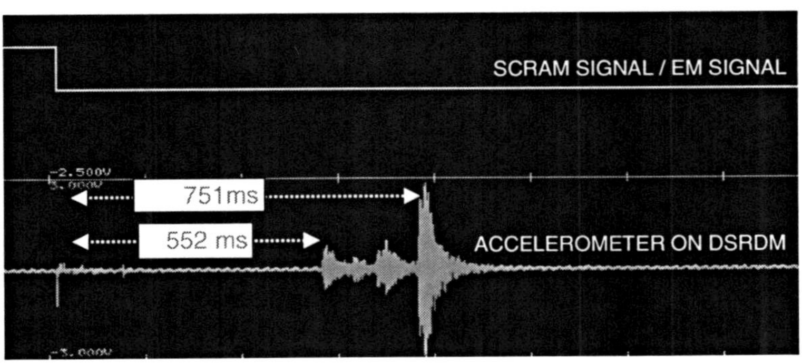

Fig. 9 Time signal recorded on the vertical sensor (test in sodium)

less dispersion compared to horizontal sensors (A-1H). Further modeling studies are in progress to study the signal transmission through different mediums. The DSRDM mechanism is seeing liquid sodium at the bottom part, inert gas hot argon at the middle region and air at the top region. Further studies are in progress to find the effect of these mediums and structural materials on measured time signal. Results obtained from analytical modeling are encouraging and is planned to extend the studies for interpreting the fall time when the three DSRs fall simultaneously.

6 Conclusion

Experiments were carried out in different test facilities for developing DSR drop time measurement system for PFBR. Characterization of the signal at the source was carried out and the dispersive nature of signal due to its transmission to the top region of DSRDM is studied. Analytical modeling is carried out to establish an input-output relationship and the preliminary results are encouraging. Studies are in progress to determine the effects of different mediums in signal transmission and to develop techniques for interpreting fall time of individual DSRs when all the three DSRs fall simultaneously.

References

1. Thomas PJ, Mc Knight JA, Rowley R (1991) Acoustic surveillance of fast reactor primary circuit experience with the dounreay PFR reactor SMORN VI , May 19-24, Gattinburg, Tennesse, USA
2. Glinsky HW (1978) International symposium on design, construction and operating experience of demonstration liquid metal fast breeder reactors; Bologna, Italy; 10-14 APr 1978, IAEA-SM–225/28
3. Sauvage J-F, Voix off (2004). Phénix: 30 years of history : the heart of a reactor. Bagnols-sur-Cèze: CEA, Valrhô, France
4. Prakash V, Anandaraj M, Thirumalai M, Kalyanasundaram P, Vaidyanathan G (2008) Application of acoustic technique for surveillance and anomaly detection in LMFBRs, Paper No: 48252, ICONE 16. Orlando, Florida, USA
5. Anandaraj M, Anup Kumar P, Ramakrishna R, Thirumalai M, Prakash V, Anandbabu C, Kalyanasundaram P, Vaidyanathan G (2009) online diverse safety rod drop time measurement system for prototype fast breeder reactor, national seminar and exhibitionon non-destructive evaluation, NDE 2009, Dec 10–12 2009, Tiruchirappalli, India
6. Prakash V, Nashine BK, Padmakumar G, Vijayashree R, Prashant Sharma, Sudheer Patri, Chandramouli S, Rajan KK (2013) Development of diverse methods for drop time measurement of PFBR shutdown mechanisms. In: International conference on fast reactors and related fuel cycles: safe technologies and sustainable scenarios (FR13), France, Mar 4–7 2013

Large Amplitude Free Vibration Analysis of Axially Functionally Graded Tapered Rotating Beam by Energy Method

Saurabh Kumar and Anirban Mitra

Abstract Geometrically nonlinear free vibration behaviour of axially functionally graded non-uniform rotating beams are investigated following variational form of energy method. Nonlinear strain displacement relations are employed to account for geometric nonlinearity present in the system. The static solution of the displacement field of the beam under centrifugal loading is obtained first and then the dynamic problem is formulated as an Eigenvalue problem based on the known static solution. The displacement fields are approximated by linear combination of orthogonally admissible functions and undetermined parameters. The method is validated successfully and results are presented in non-dimensional plane.

Keywords Geometric nonlinearity · Centrifugal stiffening · Axially functionally graded material (AFGM) · Loaded natural frequencies

1 Introduction

Functionally graded materials (FGMs) are the new generation of engineering materials with various advantages over traditional homogeneous and contemporary laminated composite materials due to the continuous transition of properties. Gradual variation of properties in FGMs may occur through depth-wise or length-wise directions. These advanced class of materials offer unique properties such as thermal resistance, high toughness, and low density and are currently used in

S. Kumar
Department of Mechanical Engineering, National Institute of Technology, Rourkela 769008, Odisha, India
e-mail: saurabhks88@gmail.com

A. Mitra (✉)
Department of Mechanical Engineering, Jadavpur University, Jadavpur, Kolkata 700032, West Bengal, India
e-mail: samik893@gmail.com

J.K. Sinha (ed.), *Vibration Engineering and Technology of Machinery*, Mechanisms and Machine Science 23, DOI 10.1007/978-3-319-09918-7_42

various engineering structures such as, turbine blades, helicopter rotor blades, ship propellers etc. The simplest representation of such a rotating elastic structure is a slender clamped-free beam under rotation with constant angular velocity. Hence, dynamic analysis of rotating functionally graded beams is an area of considerable research interest.

As already mentioned, the gradient variation in FG beams may be oriented along the transverse and/or axial direction. Majority of the work done in the corresponding field is concentrated on rotating FG beams with depth-wise property variation. Banerjee and Jackson [1] developed a dynamic stiffness method and applied the Wittrick-Williams algorithm to investigate the free vibration characteristics of rotating tapered Rayleigh beam. Piovan and coworkers [2, 3] studied the dynamic behaviour of such FG beams using finite element formulation and non-linear normal modes. Li et al. [4] investigated the free vibration characteristics of FG beams with dynamic stiffening effect using rigid–flexible coupled dynamics theory. Few literatures are also available for FG beams with axial gradation of properties. Shahba and coworkers [5, 6] analyzed the free vibration behaviour of centrifugally stiffened AFG beams by FEM approach and introduced new elements in terms of basic displacement functions (BDFs). Rajasekaran [7, 8] studied the free vibration of AFG Euler-Bernoulli and Timoshenko tapered beams with different boundary conditions using differential transformation and quadrature methods.

A review of the recent literature reveals that number of works on AFG rotating tapered beams are considerably fewer than those on depth-wise graded beams. Hence, the objective of the present work is to investigate the free vibration behaviour of rotating non-uniform AFG beam with clamped-free boundary condition.

2 Mathematical Formulation

An AFG non-uniform beam of length L with constant width (b) but varying thickness $t(x)$ is shown in Fig. 1. Three different taper profiles, namely, linear, parabolic and exponential, are considered for the present analysis and thickness variations along the longitudinal x-axis is given by the following expressions—Linear taper: $t(\xi) = t_0(1 - \alpha\xi)$; Parabolic taper: $t(\xi) = t_0(1 - \alpha\xi^k)$; Exponential taper: $t(\xi) = t_0\, exp(-\alpha\xi^k)$. Here t_0 is the thickness at the root of the beam, α and k are geometric parameters. Though the width is kept constant for present analysis the methodology is flexible enough to account for varying width as well. The model is made realistic by providing an offset between the axis of rotation and initiation of the beam at the fixed end. In Fig. 1 R is the offset distance of the root of the beam from the axis of rotation and Ω is the constant angular velocity of rotation. Material properties, namely, elastic modulus ($E(x)$) and density ($\rho(x)$), are also considered to be varying along the longitudinal x-axis as shown in Fig. 1.

The thickness of the beam being very small compared to its length, the effect of shear deformation and rotary inertia has been neglected. It is to be noted that the

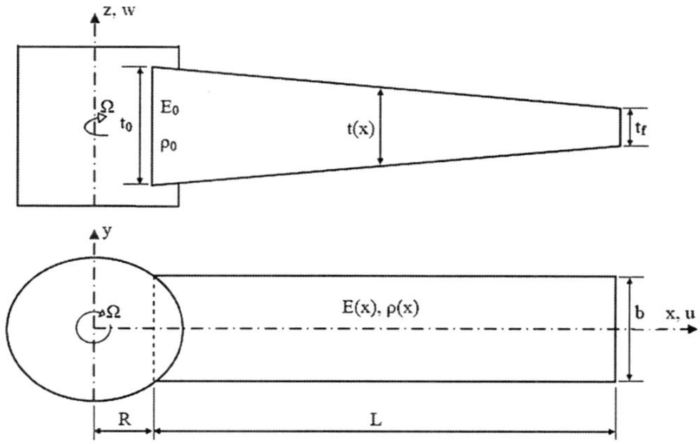

Fig. 1 Axially functionally graded tapered rotating beam

effect of non-uniformity at the root of the beam is neglected. The computations are carried out in normalized axial coordinate ξ which is given by $\xi = (x - R)/L$.

The large amplitude free vibration problem of axially functionally graded (AFG) tapered rotating beams is mathematically formulated following variational principle. The analysis is performed in two separate but interrelated steps. First the static analysis under centrifugal loading is carried out to obtain the deformed configuration. In the second phase, the dynamic problem, formulated as a standard Eigenvalue problem on the basis of known static displacement fields, is solved. Thus the calculated natural frequencies refer to the statically loaded configuration of the rotating beam arising from centrifugal forces.

2.1 Static Analysis

The mathematical formulation of the static analysis is based on the principle of minimum total potential energy, which is mathematically stated as,

$$\delta(\pi) = 0, \tag{1}$$

where, π (= $U + V$) is the total potential energy of the system. U is total strain energy stored in the system which consists of strain energies due to bending and stretching. Nonlinear strain-displacement relations are considered in order to incorporate stretching effect due to geometric nonlinearity in the system. The expression for strain energy is given by, $U = \frac{1}{2}\int_{vol} E(x)\,\varepsilon_x^2 dv$, where, ε_x is the axial strain of a fibre at a distance of z from the mid-plane and it is made up of both bending and stretching of mid-plane, as shown. $\varepsilon_x = -z\frac{d^2w}{dx^2} + \frac{du}{dx} + \frac{1}{2}\left(\frac{dw}{dx}\right)^2$.

Substituting the expression for ε_x, total strain energy in the normalized computational domain is obtained as,

$$U = (1/2L^3) \int_0^1 E(\xi)\, I(\xi) \left(\frac{d^2 w}{d\xi^2}\right)^2 d\,\xi$$
$$+ (1/2L)\, E(\xi)\, A(\xi) \int_0^1 \left[\left(\frac{du}{d\xi}\right)^2 + (1/4L^2) \left(\frac{dw}{d\xi}\right)^4 + (1/L) \left(\frac{du}{d\xi}\right) \left(\frac{dw}{d\xi}\right)^2 \right] d\,\xi \tag{2}$$

V is the potential of the external centrifugal force, calculated corresponding to constant angular speed of Ω and expressed as,

$$V = b\Omega^2 \int_0^1 \rho(\xi)\, t(\xi) u(\xi)\, \xi d\xi \tag{3}$$

Here, u and w denote the displacements along x- and z- directions, respectively. It is to be noted that u is stretching displacement whereas w is transverse deflection. These displacement fields, w and u, are expressed by linear combinations of unknown coefficients (c_i) and orthogonal admissible functions as follows,

$$w(\xi) = \sum_{i=1}^{nw} c_i \phi_i(\xi) \text{ and } u(\xi) = \sum_{i=nw+1}^{nw+nu} c_i \psi_i(\xi). \tag{4}$$

The start functions of these orthogonal sets of functions are selected by satisfying the flexural and membrane boundary conditions of the beam. Gram-Schmidt orthogonalization procedure is used to form appropriate sets of higher order coordinate functions. Substitution of the expressions of energy functionals and approximate displacement fields gives the governing set of equations as,

$$[K]\{c\} = \{f\} \tag{5}$$

where, $[K]$ is the stiffness matrix, $\{f\}$ is the force vector and $\{c\}$ is a vector of unknown coefficients. The set of governing equation of the static problem is nonlinear in nature and solved through an iterative direct substitution method, employing an appropriate relaxation technique.

2.2 Dynamic Analysis

The governing set of equations for the dynamic analysis is derived following Hamilton's principle, which is mathematically expressed as,

$$\delta\left(\int_{\tau_1}^{\tau_2}(T-U)\,d\tau\right)=0 \tag{6}$$

Here, T denotes the total kinetic energy of the system given by the expression,

$$T=\frac{1}{2}\int_0^1\left[\left(\frac{dw}{d\xi}\right)^2+\left(\frac{du}{d\xi}\right)^2\right]t(\xi)\rho(\xi)A(\xi)\,d\xi \tag{7}$$

The unknown dynamic displacement fields are assumed to be separable in time and space and the spatial part of the fields are approximated by finite linear combination of admissible orthogonal functions identical to those utilized for static analysis,

$$w(\xi,\tau)=\sum_{i=1}^{nw}d_i\phi_i(\xi)e^{j\omega\tau}\text{ and }u(\xi,\tau)=\sum_{i=nw+1}^{nw+nu}d_i\psi_i e^{j\omega\tau}(\xi) \tag{8}$$

where, ω is the natural frequency of the system and $j=\sqrt{-1}$. d_i represents a new set of unknown coefficients to be evaluated. nw and nu are the number of functions for w and u respectively. Substituting Eqs. (2) and (7) along with the dynamic displacement fields in Eq. (6) gives the governing set of equations for the beam in the following form,

$$-\omega^2[M]\{d\}+[K]\{d\}=0, \tag{9}$$

where, $[M]$ is mass matrix. In the above expression, $[K]$ is the stiffness matrix of the system at the deflected configuration. This implies that the unknown parameters obtained from the converged static solution are used to compute the starting values for solving the dynamic problem. Equation (9) is a standard eigenvalue problem which is solved numerically through a Matlab program. The square roots of the computed eigenvalues give the free vibration frequencies of the beam at the deflected configuration.

3 Results and Discussions

The present analysis is carried out for axially functionally graded non-uniform clamped-free beams under the action of centrifugal loading, while three different taper profiles (linear, parabolic and exponential) as well as three different material models have been considered. The elastic modulus and density are regarded as varying following various continuous functions as shown, Material 1: $E(\xi)=E_0$, $\rho(\xi)=\rho_0$; Material 2: $E(\xi)=E_0(1+\xi)$, $\rho(\xi)=\rho_0(1+\xi+\xi^2)$ and Material 3:

$E(\xi) = E_0 e^{\xi}$, $\rho(\xi) = \rho_0 e^{\xi}$, where E_0 is the modulus of elasticity and ρ_0 is the mass density at the root of the cantilever beam i.e. at $x = 0$ (Fig. 1). It should be noted that material 1 refers to a homogeneous material.

For clamped-free beam, start function for the definition of transverse deflection is chosen as $\phi_1 = \xi^2 (\xi^2 - 4\xi + 6)$. The number of functions for each of the displacements is taken as 8. The tolerance value of the error limit for the numerical

Table 1 Non-dimensional natural frequency of a linearly tapered cantilever homogeneous beam under different rotation speeds ($\alpha = 0.5$, $R/L = 0$)

Non-dimensional load, λ	Work done by	Non-dimensional frequency, μ		
		Mode 1	Mode 2	Mode 3
0	Present	3.8248	18.3217	47.2660
	Ref. [7]	3.8238	18.3173	47.2648
1	Present	3.9873	18.4781	47.4183
	Ref. [7]	3.9866	18.4740	47.4173
2	Present	4.4369	18.9399	47.8721
	Ref. [7]	4.4368	18.9366	47.8716
3	Present	5.0921	19.6859	48.6186
	Ref. [7]	5.0927	19.6839	48.6190
4	Present	5.8774	20.6855	49.6442
	Ref. [7]	5.8788	20.6852	49.6456
5	Present	6.7412	21.9038	50.9310
	Ref. [7]	6.7434	21.9053	50.9338
6	Present	7.6523	23.3057	52.4590
	Ref. [7]	7.6551	23.3093	52.4633
7	Present	8.5920	24.8590	54.2065
	Ref. [7]	8.5956	24.8647	54.2124
8	Present	9.5498	26.5358	56.1519
	Ref. [7]	9.5540	26.5437	56.1595
9	Present	10.5191	28.3127	58.2740
	Ref. [7]	10.5239	28.3227	58.2833
10	Present	11.4961	30.1706	60.5529
	Ref. [7]	11.5015	30.1827	60.5639
11	Present	12.4784	32.0943	62.9702
	Ref. [7]	12.4845	32.1085	62.9829
12	Present	13.4645	34.0714	65.5093
	Ref. [7]	13.4711	34.0877	65.5237

iteration scheme used to solve the static problem is taken as 0.50 % and the relaxation parameter is 0.50. Following geometrical dimensions and material properties are used for validation as well generation of results: L = 1.0 m, b = 0.02 m, t_0 = 0.01 m, E_0 = 210 GPa, ρ_0 = 7850 kg/m^3.

The present methodology and solution technique are validated by comparison with published results of Rajasekaran [7] for a non-uniform homogeneous rotating beam. The results are generated for a beam with uniform width and linearly varying thickness with $\alpha = 0.50$ and comparison of the non-dimensional natural frequency parameters ($\mu = \omega_i L^2 \sqrt{\rho_0 A_0 / E_0 I_0}$, i = 1, 2, 3...) corresponding to non-dimensional load ($\lambda = \Omega L^2 \sqrt{\rho_0 A_0 / E_0 I_0}$) is furnished in Table 1. The results were found to be in excellent agreement.

The variation of non-dimensional vibration frequencies (μ) with non-dimensional rotational speed (λ) for the first four modes are shown in Figs. 2 and 3 for different combinations of taper profile and material models. The figures are generated by keeping the *R/L* ratio and value of k constant at 0 and 0.5, respectively, while four different values of the taper parameter (α = 0.0, 0.2, 0.4 and 0.6) are considered in order to show its effect on the dynamic behaviour of tapered rotating beams.

The figures indicate that the natural frequencies increase with increase in rotational speed (hardening type nonlinearity) which is obviously due to the centrifugal stiffening effect. It is also noted from these set of figures that there is almost no effect of α on the first natural frequencies for all the cases, but the natural frequencies for other higher modes tend to decrease with increasing taper parameter. Moreover, the extent of decrease in the natural frequencies are greater for higher modes.

The effect of *R/L* ratio on the variation of non-dimensional out-of-plane vibration frequencies with non-dimensional speed of rotation for the first four modes is shown in Fig. 4a, b. The taper parameter is taken as 0.5 and kept constant while *R/L* ratio is varied from 0.0 to 0.3. The results show that for non-rotating beam the *R/L* ratios do not affect the natural frequencies but for rotating beam the natural frequencies increase with the increase in *R/L* ratios and the difference is larger for high rotating speeds.

To visualize the effect of centrifugal stiffening on amplitude of vibration, mode shape plots for first four modes are presented in Fig. 5. The figure is generated for Material-2 with linear taper (α = 0.4) and *R/L* = 0 for both $\lambda = 0$ and $\lambda = 12$. The difference in mode shape for non-rotating ($\lambda = 0$) and rotating ($\lambda = 12$) beam is clearly evident from the figure and it underlines the effect of centrifugal stiffening on the vibration amplitude.

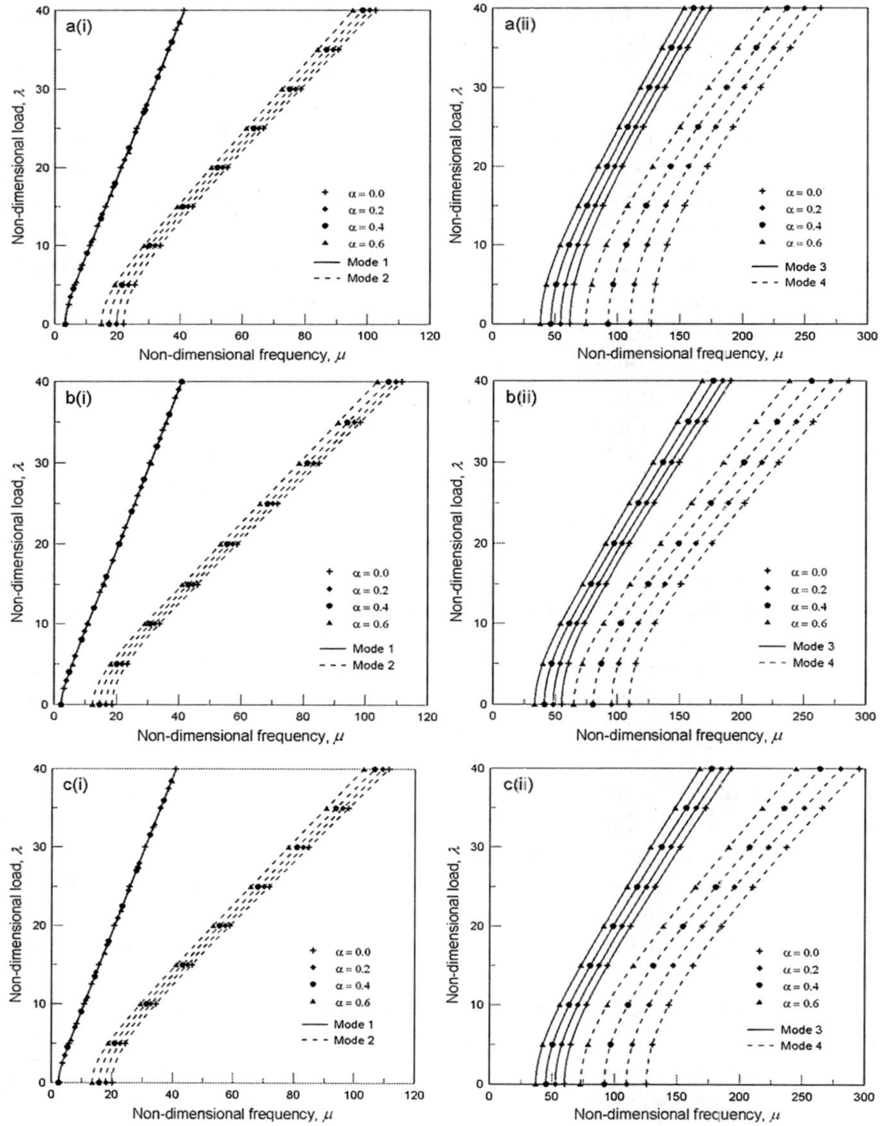

Fig. 2 Effect of taper parameters on dynamic behaviour of parabolic tapered AFG beam: **a** $E(\xi) = E_0$, $\rho(\xi) = \rho_0$. **b** $E(\xi) = E_0(1 + \xi)$, $\rho(\xi) = \rho_0(1 + \xi + \xi^2)$. **c** $E(\xi) = E_0 e^{\xi}$, $\rho(\xi) = \rho_0 e^{\xi}$

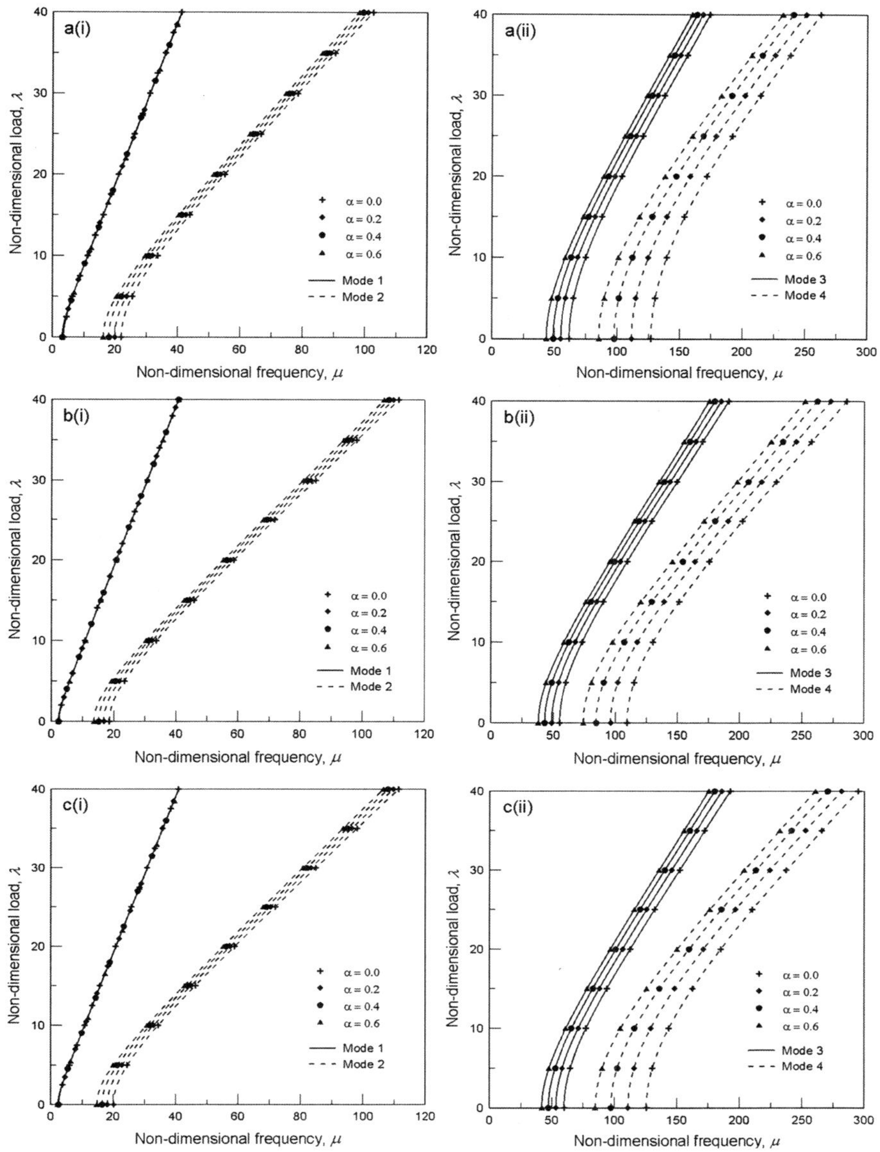

Fig. 3 Effect of taper parameters on dynamic behaviour of exponentially tapered AFG beam: **a** $E(\xi) = E_0$, $\rho(\xi) = \rho_0$. **b** $E(\xi) = E_0(1 + \xi)$, $\rho(\xi) = \rho_0(1 + \xi + \xi^2)$. **c** $E(\xi) = E_0 e^{\xi}$, $\rho(\xi) = \rho_0 e^{\xi}$

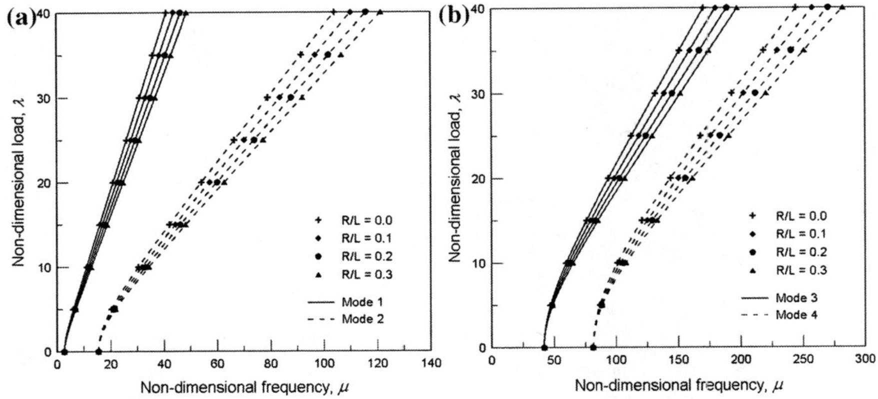

Fig. 4 Effect of *R/L* ratios on dynamic behaviour of linearly tapered AFG beam ($E(\xi) = E_0(1 + \xi)$, $\rho(\xi) = \rho_0(1 + \xi + \xi^2)$)

Fig. 5 First four mode shapes for rotating linearly tapered AFG beam ($E(\xi) = E_0(1 + \xi)$, $\rho(\xi) = \rho_0(1 + \xi + \xi^2)$)

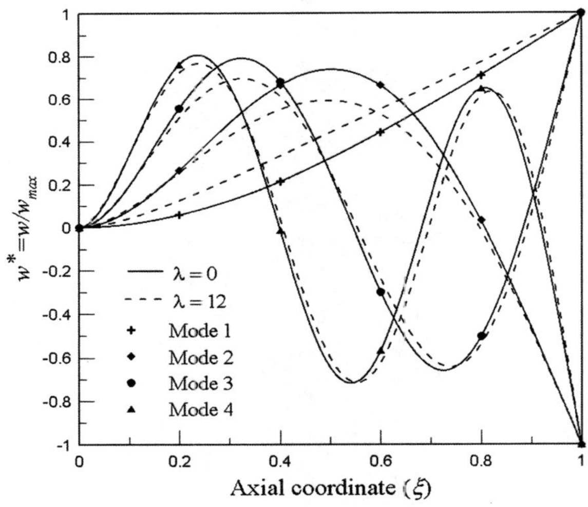

4 Conclusion

The present work deals with the large displacement free vibration of axially functionally graded non-uniform rotating beams by variational approach. Three different material property variations and three different types of taper profiles are considered for the analysis. Results for varying taper parameters and different *R/L* ratios are provided in a non-dimensional load-frequency plane and the effect of taper parameter and offset distance is discussed. Mode shape plots are also shown to elucidate the effect of centrifugal stiffening on amplitude of vibration. The results are validated with the benchmark results provided by other researchers. The present method can be extended to various other material property variations and taper patterns of rotating beam.

References

1. Banerjee JR, Jackson DR (2013) Free vibration of a rotating tapered Rayleigh beam: a dynamic stiffness method of solution. Comput Struct 124(2013):11–20
2. Piovan MT, Sampaio R (2009) A study on the dynamics of rotating beams with functionally graded properties. J Sound Vib 327(2009):134–143
3. Machado SP, Piovan MT (2013) Nonlinear dynamics of rotating box FGM beams using nonlinear normal modes. Thin-Walled Struct 62(2013):158–168
4. Li L, Zhang DG, Zhu WD (2014) Free vibration analysis of a rotating hub–functionally graded material beam system with the dynamic stiffening effect. J Sound Vib 333(5):1526–1541
5. Shahba A, Attarnejad R, Zarrinzadeh H (2013) Free vibration analysis of centrifugally stiffened tapered functionally graded beams. Mech Adv Mater Struct 20(2013):331–338
6. Zarrinzadeh H, Attarnejad R, Shahba A (2012) Free vibration of rotating axially functionally graded tapered beams. J Aerosp Eng 226(4):363–379
7. Rajasekaran S (2013) Differential transformation and differential quadrature methods for centrifugally stiffened axially functionally graded tapered beams. Int J Mech Sci 74 (2013):15–31
8. Rajasekaran S (2013) Free vibration of centrifugally stiffened axially functionally graded tapered Timoshenko beams using differential transformation and quadrature methods. Appl Math Model 37(2013):4440–4463

A Frequency-Based Criterion for Automatic Selection of the Optimal Intrinsic Mode Function in Diagnostics of Localized Gear Tooth Faults

Pawel Rzeszucinski, Michal Juraszek and James R. Ottewill

Abstract To date gearboxes remain one of the most important elements of virtually every power transmission system as far as a continuous operation of the shaft line is concerned. Any failure or breakdown may result in putting the whole production line, supply chain or even peoples life in jeopardy. Endeavours to detect an incipient fault within the system serve multiple purposes from increasing the safety of the people responsible for operating the machines, through to decreasing running and operational costs, allowing time to plan for the inevitable repairs and making sure that the downtime of the machine is kept to an absolute minimum. This, in turn, makes this branch of condition monitoring of rotating machinery one of the most intensively studied. The Empirical Mode Decomposition (EMD) is a relatively new method of signal decomposition, which breaks the original signal up into a number of so-called Intrinsic Mode Functions (IMFs). The decomposition represents a type of adaptive filtering which outputs a number of IMFs which, acquired according to two strict criteria, contain portions of the filtered version of the original signal and so can carry different information about the content of the signal. EMD has already been used in the field of condition monitoring of rotating machinery, but the selection of the optimal IMF for the task often requires the experience of a condition monitoring specialist. This paper proposes a frequency-based tool for automatic selection of the IMF that is best suited for the detection of localized gear tooth faults.

Keywords Empirical mode decomposition · Intrinsic mode function · Time synchronous average · Gearbox · Diagnostics

---

P. Rzeszucinski (✉) · J.R. Ottewill
ABB Corporate Research Center, Starowislna 13A, 31-038 Krakow, Poland
e-mail: pawel.rzeszucinski@pl.abb.com

J.R. Ottewill
e-mail: james.ottewill@pl.abb.com

M. Juraszek
AGH University of Science and Technology, 30 Mickiewicza Av., 30-962 Krakow, Poland
e-mail: juraszek@student.agh.edu.pl

J.K. Sinha (ed.), *Vibration Engineering and Technology of Machinery*,
Mechanisms and Machine Science 23, DOI 10.1007/978-3-319-09918-7_43

1 Introduction

Gearboxes are crucial elements of virtually all mechanical systems, used for transferring power and altering the speed and torque of driven shafts. Their important role in many different applications motivates early fault detection in order to prevent costly break-downs, stoppages of production lines and to avoid hazards for working personnel. Many different diagnostics techniques are being used for condition monitoring and diagnostics of gearboxes. The most widely used include vibration-based statistical techniques, time synchronous averaging, oil debris analysis, acoustic emission, shock pulse method or cepstrum analysis [1]. Most of these techniques, especially the classical statistics-based routines, assume the stationarity of the measured signals i.e. signals whose parameters do not change over time. Often however the operating parameters of the machines, e.g. shaft speed or loading, change with time. These non-stationary operating conditions call for the use of signal processing techniques capable of dealing with such waveforms.

This document outlines the relatively new techniques suited for the analysis of both the stationary as well as non-stationary signals i.e. Hilbert-Huang Transform (HHT) and Empirical Mode Decomposition (EMD). A concept of Intrinsic Mode Functions (IMF) generated as a result of the EMD is also introduced. Using these methods on Time Synchronously Averaged signals give a possibility to detect the onset of a localized gear tooth fault, when its severity is not significant. The efficiency of the described techniques is shown through comparison with the results obtained when only applying the standard TSA method.

2 Theoretical Background

Empirical Mode Decomposition is an adaptive signal decomposition technique that allows the analyzed signal to be decomposed into a number of empirical modes, referred to as Intrinsic Mode Functions (IMF). The sum of the IMFs leads to reconstruction of the originally decomposed signal. The decomposition of the signals into Intrinsic Mode Functions should fulfil two conditions [2]:

- The number of zero crossings and number of extremes in the waveform should be equal or differ by at most one,
- The mean value of the envelope defined by local maxima and the envelope defined by local minima at any point should be equal to zero.

The Hilbert-Huang transform constitutes a signal processing technique which analyses the signal in both the time and the frequency domain. In general time-frequency analysis techniques allow the changes in frequency content of a signal to be observed as a function of time. Over the years a number of time-frequency techniques have been proposed, but each contain their own characteristic limitations, the most commonly encountered being: counter-proportional resolution in each of the domains—the better the time resolution, the worse in frequency resolution and vice versa in Short Time Fourier Transform, uniform resolution which might be uniformly poor; results dependent on selection of the appropriate mother wavelet, different wavelets yielding different results in Wavelet Transform, cross-terms interference i.e. the appearance of components which are not present in the original signal, but result from the bilinear nature of the technique in Wigner–Ville Distribution [3].

HHT was firstly introduced by N.E. Huang et al. [4], and in principle it uses the properties of the analytic signal derived from the IMFs obtained through the use of EMD. This in turn allows for the instantaneous frequency to be obtained and projected as a function of time. This leads to the property of the HHT of being capable of analyzing non-stationary and non-linear signals without the obvious limitations of the above mentioned, classic time-frequency techniques [5].

3 Experimental Test Rig and Results

Data used to validate the results were generated on an experimental rig in the Diagnostics Laboratory of the ABB Corporate Research Center in Kraków, Poland. The system is comprised of a 1.5 kW induction motor operating at a speed of 1,400 RPM which was driving a two-stage, parallel shaft, helical reduction gearbox. The system was loaded by a 6 kW DC motor with nominal speed of 1,500 RPM which, in turn, was connected with a permanent magnet motor. This latter motor was used purely for the purpose of obtaining an encoder signal which was mounted to its shaft. The encoder signals were later used to derive the Time Synchronous Averages of the vibration signals. After 30 tests for a nominally healthy gearbox condition were conducted, the gearbox had a pit artificially seeded to one of the teeth on the 2nd stage output gear. The vibration generated by this gear was subject to the analysis in this text. Vibration signals were recorded for 26.88 s at the rate of 4444 samples per second with the use of SKF CMSS 2110 accelerometers. A much more detailed description of the test rig and the process of seeding the fault can be found in [6]. The gear of interest had 52 teeth and was rotating at the speed of 232 RPM resulting in the gear mesh frequency of 201.06 Hz.

3.1 TSA Vibrations

The output of the TSA routine is a waveform which may be analysed in the time domain (at least for the purpose of detecting the localized gear tooth faults). Time domain analysis provides a good basis for the characterization and differentiation of vibration signals generated under different condition of the meshing gears [7]. A localized gear tooth fault will generate an impulse-like transient event. In order to

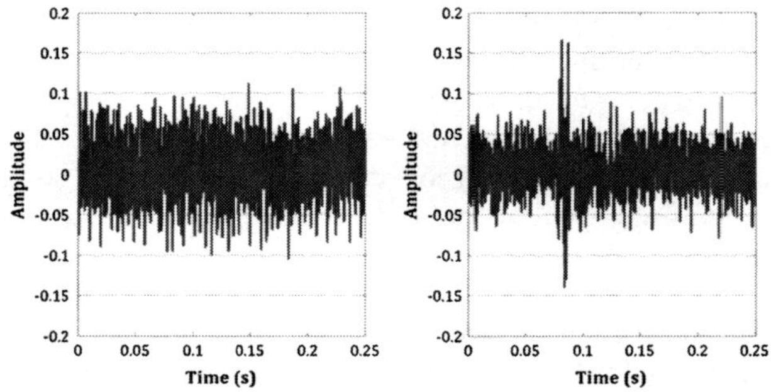

Fig. 1 Time synchronous averaged acceleration signals

automatically detect the presence of such an event a well-known statistical measure, kurtosis, is used.

It is the fourth normalized statistical moment of a signal defined as [8]:

$$\textbf{kurtosis} = \frac{N\sum_{i=1}^{N}(x_i - \bar{x})^4}{\left(\sum_{i=1}^{N}(x_i - \bar{x})^2\right)^2} \tag{1}$$

where N is the number of samples in signal x and $\bar{x}$ is the mean value of signal x.

The localized gear tooth fault results in high amplitude, impulse-like events synchronized with the rotational speed of the shaft [9]. The TSA was derived from the given dataset and the resultant waveform for both the healthy and the faulty conditions are shown in Fig. 1.

NOTE: Throughout the document the figures on the left hand side always corresponds to the results from healthy gearbox, whereas the figures on the right hand side are based on the data from the faulty state.

It can be seen that the averaging of the signal allowed to decrease the influence of noise in the healthy parts of the signal, whereas the impulse originating from the faulty gear (at around 0.8 s) was enhanced with reference to the remaining parts of the waveform.

3.2 Decomposition of Signals into IMFs

EMD, as a signal decomposition technique, is expected to divide the signal into a number of different waveforms of various scales, each having different frequency content. A question remains as to which IMF would be expected to contain the clearest indication of the incipient gear tooth fault, should such occur. Even though in theory an impulse excites the whole bandwidth of the frequency spectrum (including the structure resonance frequencies), the highest increase is expected to occur around the gear mesh frequency, its harmonics and shaft speed related sidebands of the analysed gearbox stage, where the majority of the gear operational energy is concentrated. The theoretical assumptions seem to be supported by the Hilbert-Huang spectra of IMFs generated from the faulty gearbox. Figure 2 shows spectra for IMF 1–4 color-scaled in the same range in order to present where the majority of energy is concentrated. One can see that, according to the expectations, the instantaneous amplitude of the highest energy is spread between the first and the second gear mesh frequency harmonics (in IMF 2 and 3). IMF 1 and 4 do not present any outstanding energy in the region where the fault was seeded. Therefore in this document a new criterion is proposed which analyses the amount of energy concentrated around the gear mesh frequency, its harmonics and shaft speed related sidebands of the analysed gearbox stage for each of the IMF signals and the one resulting in the greatest amount is chosen as optimal for further analysis.

For the sake of presentation, the first four IMFs derived for both the healthy and the faulty gear are presented in Fig. 3. By observing the presented IMFs one may see that the signatures of the fault can be observed in IMF 1–3, but only IMF 2 and 3 contain components that make the event clearly stand out in the waveform. The remaining IMFs, which contain lower frequency content, do not show any significant signatures of the damage. Table 1 shows the values of kurtosis of every IMF. It can be seen that, as has been highlighted previously, only the first three IMFs contain the signatures of gear tooth damage, however only IMF 2 and 3 present the fault in a clearly transparent way (kurtosis values above 4).

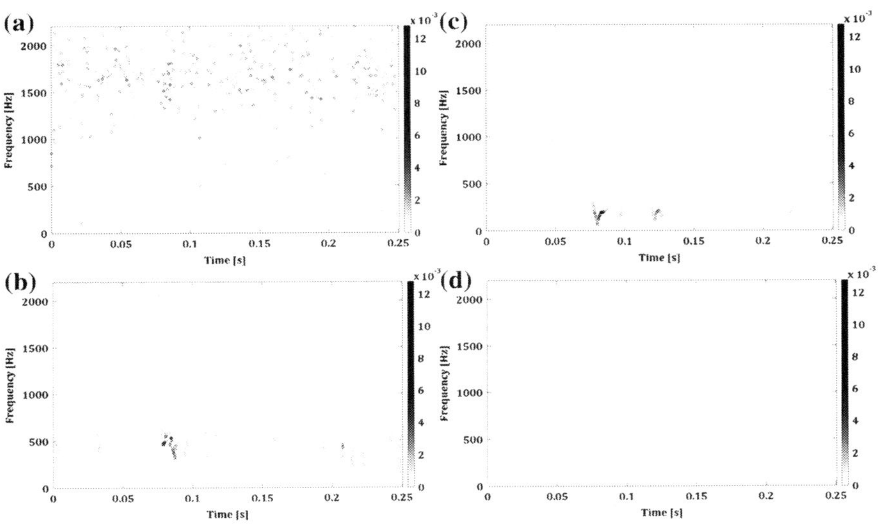

Fig. 2 HHT obtained for IMF 1–4 (**a–d** respectively) of a TSA signal from a faulty gearbox

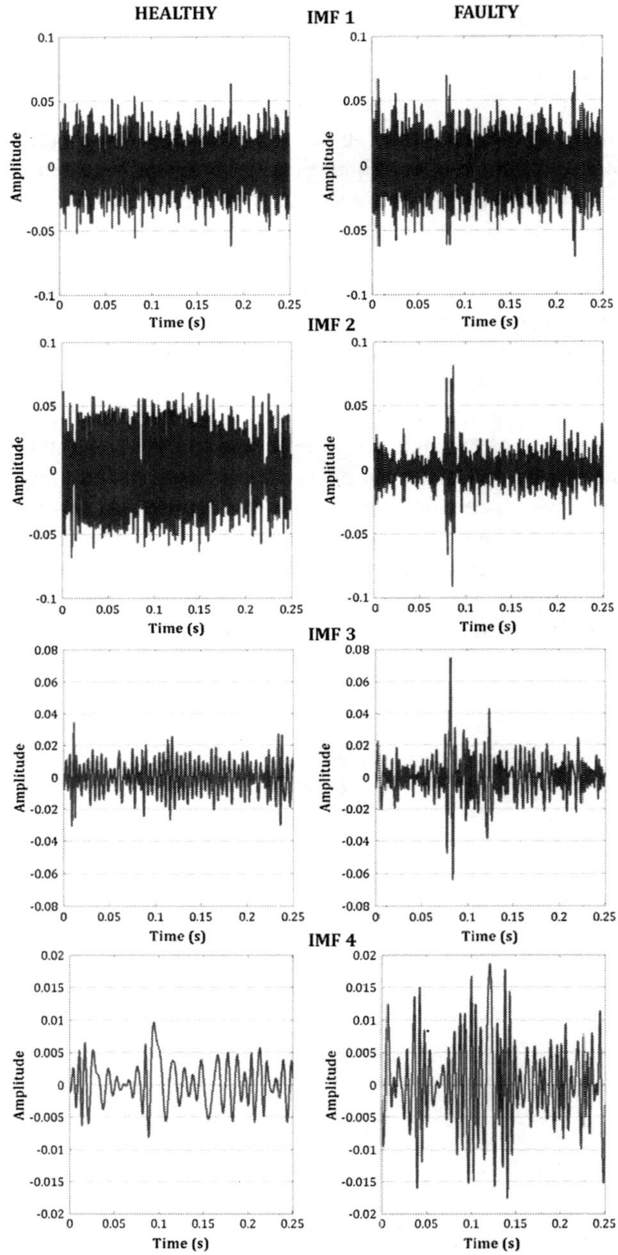

Fig. 3 IMF 1–4 of TSA signals from a healthy (*left column*) and faulty (*right column*) gearbox

3.3 Selection of Optimal IMFs

As mentioned in the previous chapter the proposed method of automatic selection of the optimal IMF for the detection of the localized gear tooth fault operates on the basis of determining the amount of energy concentrated around the gear mesh frequency of each IMF. The IMF which contains the greatest energy in the vicinity of the gear mesh frequency is selected for further evaluation. The selection of the optimal IMF does not only allow for the process to be fully automated, but also reduces the likelihood of false alarms that may arise as a result of applying the kurtosis metric on IMFs. As one could see in Table 1 the kurtosis value for IMF5 generated for a healthy signal resulted in a value of above 7, which would normally mean a warning condition. However, visual inspection revealed that the peak which caused the kurtosis value to increase was not located in the angular position of the shaft which corresponds to the failure of the gear and the dominant frequency of the IMF is too low to be truly representative of the damage.

Figure 4 shows the amount of energy contained in the first 4 IMFs when assessed in the frequency band ranging from the first to the second gear mesh

Table 1 Values of kurtosis for IMFs

	Kurtosis for healthy state	Kurtosis for faulty state
IMF_1	2.12	2.27
IMF_2	1.79	8.50
IMF_3	2.56	7.65
IMF_4	2.83	3.07
IMF_5	7.51	3.18
IMF_6	2.84	3.11
IMF_7	2.63	1.79
IMF_8	1.64	1.55
IMF_9	2.05	1.91

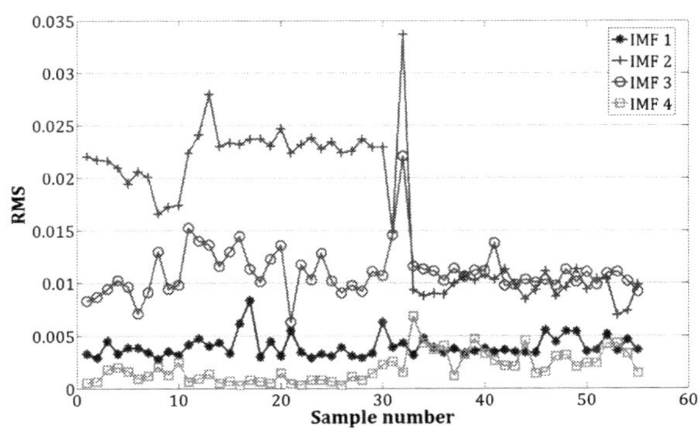

Fig. 4 The amount of energy in IMFs in the specified filtering band

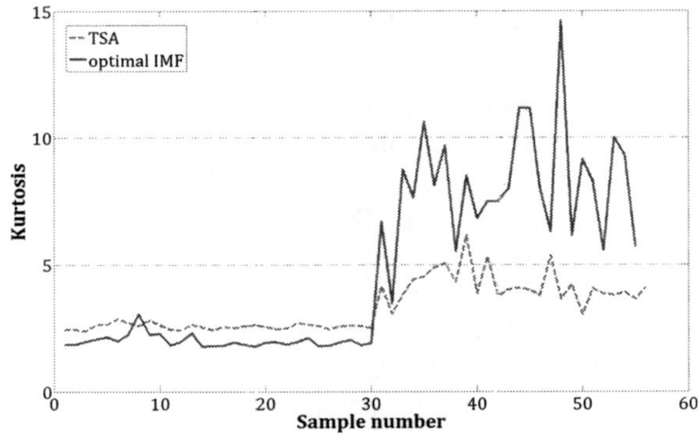

Fig. 5 Values of kurtosis applied on TSA and on the optimal IMF

frequency. The filtering was made with the use of a pass band 4th order Butterworth filter with the cut-off frequencies set at 150 and 450 Hz. It can be seen that IMF 2 contains the greatest energy up to the moment when the fault was seeded. This corresponds to the fact that for the healthy gearbox the spectrum was dominated by the 2nd GMF which might be attributed to some misalignment within the system. However, after stripping the gearbox, machining the fault and assembling the gearbox back together, the energy level became comparable with IMF 3 which might manifest much better alignment within the system. From the moment of seeding the fault up until the end of the test, the selection criterion navigates itself between IMF 2 and 3, choosing the optimal for detection of the gear tooth fault.

Figure 5 shows the values of kurtosis when applied on the TSA signal and on the optimal IMF selected on the basis of the proposed method. For the healthy part of the experiment, the kurtosis value of the IMF is lower than for the TSA. This is due to the decomposition process which results in signals resembling a sinusoidal shape hence the kurtosis value is closer to 1.5, when compared to the values for TSA oscillating around 3 which is characteristic for normally distributed signal. The faulty region of the plot shows a greater reaction of kurtosis to the onset of the fault when diagnosed on the IMF. The reactions might be even 300 % greater than for the TSA (samples 49). Overall it is expected that this difference might lead to the earlier detection of the fault, when utilized in real applications.

4 Conclusion

This document presented the possibility of enhancing the localized gear tooth fault diagnostic information contained within the Time Synchronously Averaged vibration signal by combining it with the Empirical Mode Decomposition technique. First the concept of the TSA was shortly presented, which was followed by a more detailed description of the two combined methods—Empirical Mode Decomposition (together with extraction of Intrinsic Mode Functions) and the

Hilbert-Huang transform. The resultant Hilbert-Huang spectra, showing the values of instantaneous amplitude and instantaneous frequency as a function of time, was used to observe the frequency distribution within a signal that contained an artificially seeded gear tooth fault. The results show that, as was assumed, the majority of the energy during the mesh of the faulty gear is concentrated around the gear mesh frequency, its second harmonic and the shaft speed related sidebands around these components. This information was used to set a criterion which allowed for the optimal Intrinsic Mode Function to be selected for the purpose of detecting impulsive content within the signal—a clear signature of localized gear tooth fault. Applying the EMD on a vibration signal subject to the TSA process showed that a clearer indication of the fault can be obtained from the waveform which resulted in higher output from the kurtosis metric. As a result it is expected that the EMD technique applied to a TSA signal might have the potential to indicate an incipient gear tooth fault earlier than the standard TSA approach alone.

References

1. Sait AS, Sharaf-Eldeen YI (2012) A review of gearbox condition monitoring based on vibration analysis techniques diagnostics and prognostics. In: Rotating machinery, structural health monitoring, shock and vibration vol 5. Springer, New York, pp 307–324
2. Dybała J, Zimroz R (2012) Application of empirical mode decomposition for impulsive signal extraction to detect bearing damage—industrial case study. In: Proceedings of the second international conference condition monitoring of machinery in non-stationary operations (CMMNO), Springer
3. Han L (2010) Gear fault detection and diagnosis based on Hilbert-Huang transform. In: 3rd International congress on image and signal processing
4. Huang NE, Shen Z, Long SR, Wu MC, Shih EH, Zheng Q, Tung CC, Liu HH (1998) The empirical mode decomposition method and the Hilbert spectrum for non-stationary time series analysis. Proc R Soc Lond 45(4A):903–995
5. Huang N, Shen S (2005) Hilbert-Huang transform and its applications. World Scientific, Singapore
6. Ottewill JR, Orkisz M (2013) Condition monitoring of gearboxes using synchronously averaged electric motor signals. Mech Syst Sig Process 38(2):482–498
7. Forrester BD (1996) Advanced vibration analysis techniques for fault detection and diagnosis in geared transmission systems. Ph.D. thesis, Swinburne University of Technology, Australia
8. Vecer P, Kreidl M, Smid R (2005) Condition indicators for gearbox condition monitoring systems. Acta Polytechnika 45(6):35–43
9. Rzeszucinski PJ, Sinha JK, Edwards R, Starr A, Allen B (2012) Amplitude of probability density function (APDF) of vibration response as a robust tool for gearbox diagnosis. Strain 48 (6):510–516

Vehicle Suspension Performance Analysis Based on Full Vehicle Model for Condition Monitoring Development

Moamar Hamed, Belachew Tesfa, Fengshou Gu and Andrew D. Ball

Abstract The objective of this research is to develop a mathematical model using a seven degree-of-freedom full car. The simulation analyses were conducted to predict the response of the vehicle when driven across speed bumps of different shapes and at range of speeds. Three bump sizes were considered in this study including bump 1 (500 mm × 50 mm), bump 2 (500 mm × 70 mm), and bump 3 (500 mm × 100 mm). These were run through the model at speeds of 8, 16, 24 and 32 km/hr. The model was validated using experimental data, which was collected by driving the vehicle across the bump 1 at a speed of 8 km/h. The performance of the suspension in terms of ride comfort, road handling and stability of the vehicle were analysed and presented. The vibration analysis for different speed levels of 8, 16, 24 and 32 km/hr indicated that, the effect of vehicle speeds on the vibration of the vehicle body increases at lower speeds up to a maximum value after which it began to decrease from the optimum point with increasing vehicle speeds. The model has been used for fault detection of under-inflation of vehicle tyre by 35 %, and also to predict possible future suspension faults.

Keywords Condition monitoring · Suspension modeling · Vibration measurement · Speed bump geometry · Vehicle speed

1 Introduction

Suspension systems and their components have significant influence on passenger safety, ride comfort, handling, and vehicle stability. According to the Ministry of Transport (M.O.T) data [1], between October 2010 and September 2011, approximately 24.2 million vehicle tests were carried out in UK. Figure 1 represents the

M. Hamed (✉) · B. Tesfa · F. Gu · A.D. Ball
Centre for Efficiency and Performance Engineering, University of Huddersfield, Huddersfield HD1 3DH, UK
e-mail: U0951001@hud.ac.uk

J.K. Sinha (ed.), *Vibration Engineering and Technology of Machinery*, Mechanisms and Machine Science 23, DOI 10.1007/978-3-319-09918-7_44

percentage failure by category for different car models. Lighting problems (19.79 %) attributed for a high number of re-tests followed by suspension faults (13.18 %), brake faults (11.47 %) and tyre faults (8.75 %). As shown in the pie chart (Fig. 1), suspension and tyres faults are the 2nd and the 4rt frequent faults in MOT tests respectively. To consider the performances of the suspension in terms of ride quality, handling and stability of the vehicle, some important parameters must be studied; these parameters are the wheel deflection, suspension travel and the vehicle body acceleration.

The aim is to achieve small amplitude value for these parameters [2]. Road handling is associated with the relative displacement between suspension and the road input (Zu – Zr). This is represented as wheel deflection as shown in the Fig. 2. Suspension travel is defined as the relative vertical displacement between the vehicle body and the wheel (Zs – Zu) as shown in the Fig. 2. This can be used for assessing the space required to accommodate the suspension spring. Ride comfort is related to vehicle body motion sensed by the passenger's comfort. This requires that the acceleration of the vehicle body (sprung mass) be relatively small. According to ISO: 2631-1-1997 [3] the proper road handling must be in the range of 0.0508 m whilst the standard value for suspension travel must be in the range of minimum of 0.127 m. Finally, the passenger feels highly comfortable if the RMS acceleration is below 0.315 m/s^2.

Fig. 1 Percentage of failure by category for different cars

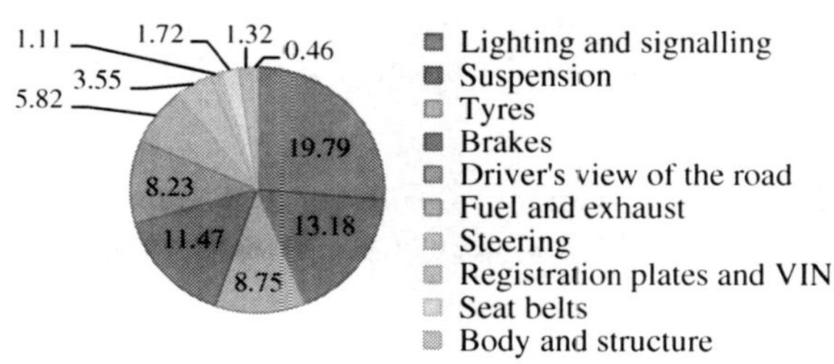

Fig. 2 Sketch of quarter car model

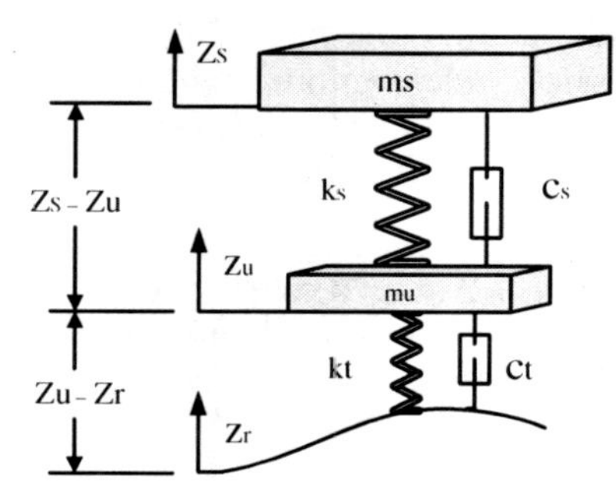

A number of researchers have investigated suspension performance using modelling/simulation [4]. A mathematical model for quarter car with 2-DOF and a half car with 4-DOF have been investigated by Faheem [5]. Also, a mathematical model of a 3-DOF quarter car with semi-active suspension system has been developed by Rao [6]. The model was used for testing of skyhook and other strategies of semi active suspension system. Modeling of one and two DOF for a quarter car design a semi-active twin-tube shock has been developed by Esslaminasa et al. [7]. Darus [8] adopted a state space approach in developing a mathematical model for both a quarter car and full car using MATLAB packages. In Metallidis [9], statistical system identification technique was applied for performing parametric identification and fault detection of nonlinear vehicle suspension system. A model-based fault detection applied on a vehicle control system has been presented by Kashi [10], which relies on mathematical descriptions of the system and which yields a robust fault detection and isolation of faults affecting the system. Agharkakli et al. [11] presents a mathematical model for passive and active of quarter car suspension system. A research study to improve road handling and rid comfort was presented by Ikenaga et al. [12]. Active suspension control system based on a full Vehicle model was presented, and the performance of suspension system was included. The effect of truck speed on the shock and vibration levels was discussed by Lu et al. [13]. They indicated that the effect of truck speed on root mean square acceleration of vibration was strong at a lower speed, but slight at a higher speed.

The objective of this research is to analyse the performances of suspension in terms of ride comfort, road handling and stability. The effect of road conditions, vehicle speed and any changing in suspension specifications include the effect of tyre pressure, have been investigated. For this a mathematical modelling of the full car has been conducted in this study.

To develop the vehicle model, it can be assumed that the vehicle is a rigid body and represented as sprung mass m_s, while suspension axles represented as unsprang mass m_u as shown in Fig. 3. The suspension between the vehicle body and wheels are modelled by simple linear spring and damper elements, and also each tyre is modelled as a single linear spring and damper. The equations of all motions are derived separately, and finally the equations of the body motions are achieved [8].

Equation of motion for bouncing of sprung mass:

$$\begin{aligned} m_s\ddot{z}_s = & k_f(z_{u1} - z_{s1}) + k_f(z_{u2} - z_{s2}) + k_r(z_{u3} - z_{s3}) + k_r(z_{u4} - z_{s4}) \\ & + c_f(\dot{z}_{u1} - \dot{z}_{s1}) + c_f(\dot{z}_{u2} - \dot{z}_{s2}) + c_r(\dot{z}_{u3} - \dot{z}_{s3}) + c_r(\dot{z}_{u4} - \dot{z}_{s4}) \end{aligned} \tag{1}$$

For pitching moment of inertia of sprung mass

$$\begin{aligned} I_p\ddot{\theta} = & k_f l_1(z_{u1} - z_{s1}) + k_f l_1(z_{u2} - z_{s2}) - k_r l_2(z_{u3} - z_{s3}) - k_r l_2(z_{u4} - z_{s4}) \\ & + c_f l_1(\dot{z}_{u1} - \dot{z}_{s1}) + c_f l_1(\dot{z}_{u2} - \dot{z}_{s2}) - c_r l_2(\dot{z}_{u3} - \dot{z}_{s3}) - c_r l_2(\dot{z}_{u4} - \dot{z}_{s4}) \end{aligned} \tag{2}$$

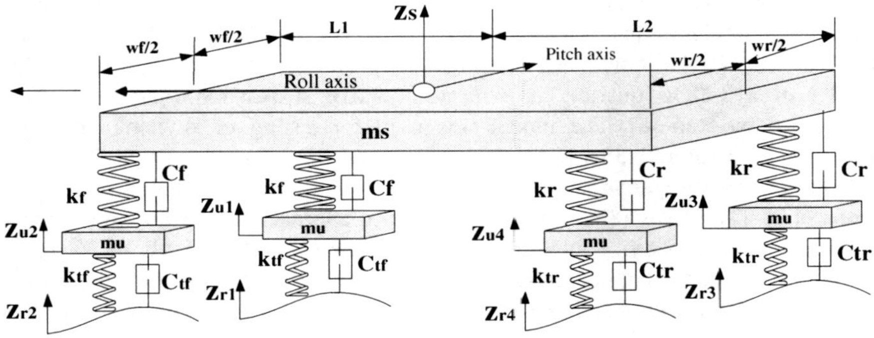

Fig. 3 Full vehicle models

For rolling motion of the sprung mass

$$I_r\ddot{\varphi} = \frac{k_f w_f}{2}(z_{u1} - z_{s1}) - \frac{k_f w_f}{2}(z_{u2} - z_{s2}) + \frac{k_r w_r}{2}(z_{u3} - z_{s3}) - \frac{k_r w_r}{2}(z_{u4} - z_{s4})$$
$$+ \frac{c_f w_f}{2}(\dot{z}_{u1} - \dot{z}_{s1}) - \frac{c_f w_f}{2}(\dot{z}_{u2} - \dot{z}_{s2}) + \frac{c_r w_r}{2}(\dot{z}_{u3} - \dot{z}_{s3}) - \frac{c_r w_r}{2}(\dot{z}_{u4} - \dot{z}_{s4}) \quad (3)$$

For each wheel motion in vertical direction

$$m_{uf}\ddot{z}_{u1} = -k_f(z_{u1} - z_{s1}) - c_f(\dot{z}_{u1} - \dot{z}_{s1}) + k_{tf}(z_{r1} - z_{u1}) + ctf(\dot{z}_{r1} - \dot{z}_{u1}) \quad (4)$$

$$m_{uf}\ddot{z}_{u2} = -k_f(z_{u2} - z_{s2}) - c_f(\dot{z}_{u2} - \dot{z}_{s2}) + k_{tf}(z_{r2} - z_{u2}) + ctf(\dot{z}_{r2} - \dot{z}_{u2}) \quad (5)$$

$$m_{ur}\ddot{z}_{u3} = -k_r(z_{u3} - z_{s3}) - c_r(\dot{z}_{u3} - \dot{z}_{s3}) + k_{tr}(z_{r3} - z_{u3}) + ctr(\dot{z}_{r3} - \dot{z}_{u3}) \quad (6)$$

$$m_{ur}\ddot{z}_{u4} = -k_r(z_{u4} - z_{s4}) - c_r(\dot{z}_{u4} - \dot{z}_{s4}) + k_{tr}(z_{r4} - z_{u4}) + ctr(\dot{z}_{r4} - \dot{z}_{u4}) \quad (7)$$

The parameters of suspension system and the definition of equation variables are summarized in Table 1 which was adopted from [7, 14], and some variables were emended to meet the specifications of the vehicle used in the experiment. The system can be represented in state space matrix forms as: $\dot{x} = A(x) + B(u)$ is the input state equation and $y = C(x) + D(u)$ output for the state equation. To simulate the state space matrixes MATLAB code has been developed.

The road profile is assumed to be a single bump with sin wave shape; it was calculated and created according to vehicle speeds, height and width of the bumps by the following equation:

$$u(p) = 1/2a * \sin(2 * \pi * Fp * t) \quad (8)$$

Table 1 Shows definition of equations variables and parameters of suspension

Variables	Definitions	Units
$m_s = 1290$	Sprung mass (mass of the body)	Kg
$m_f = 90$, $m_r = 85$	Unsprang mass	Kg
$I_r = 5340, I_p = 6430$	Roll and pitch of moment of inertia	Kg m^2
$l_1 = 2.13, l_2 = 2.0$	Distance from front and rear wheel to the car centre	m
$k_f = 36297, k_r = 19620$	Stiffness of front and rear spring	N/m
$k_{tf}, ktr = 182470$	Stiffness of front and rear tyre	N/m
$c_f = 3924, c_r = 2943$	Front and rear damper coefficient for	Nm/s
$w_f = 1.74, w_r = 1.70$	Front and rear vehicle width	m
z_s	Displacement of the vehicle body	m
θ, φ	Roll and pitch angles	rad

where (*a*) is the bump height, which was considered in this study as [50, 70 and 100 mm], *Fp* frequency of the bump which was calculated by considering length of the bump and vehicle speed, *t* is the time for crossing the vehicle on the bump. Three bump sizes were considered in this study such as: bump 1 (500 mm in width × 50 mm height), bump 2 (500 mm in width × 70 mm height), and bump 3 (500 mm in width × 100 mm height). The models were simulated for speeds of 8, 16, 24 and 32 km/hr for validations.

2 Experimental Set up and Test Procedures

To validate the theoretical model, a 2001 Vauxhall ZAFIRA car, with a front wheel drive and equipped with two different sensors was used. The sensors mounted on the car include: (1) a vibration sensor with sensitivity of (3.770 pc/ms-2) mounted on the upper mounting point of the front left shock absorber, and (2) a dynamic tyre pressure sensor (DTPS) with sensitivity of (11.43 Pc/0.1Mpa) connected to the valve stem of the front left wheel.

In order to insure significant installation for the sensors, two different adapters were designed and manufactured at the University of Huddersfield. In addition, a wireless measurement system was also designed and installed on the car to offer a complete remote measurement for vibration and pressure data. The two sensors were connected to the wireless sensor nodes (transmitters) assembled and situated in the centre rim of the front left wheel for the pressure sensor, and inside the car for the vibration sensor. The gateway (receiver) was equipped together with a laptop inside the car.

The main aim of this test was to obtain the acceleration (vibration) response of the suspension system in order to analyze the effect of velocity change and under-inflation of the tyre on the performance of the suspension system. The test was conducted with standard tyre pressure (2.3 bar) and vehicle speed of 8 km/h.

3 Results and Discussion

The model was validated using experimental data collected by driving the vehicle across the bump1 (located in the University of Huddersfield premises) with speed of 8 km/h.

The bump profile used was 5.80 m width, 0.50 m length and 0.050 m height. The MATLAB software was used to analyse the response of the vehicle. Figure 4 depicts the acceleration of vehicle body in time domain based on model simulation and experiments. It can be noted that the model fairly predicts the performance of the suspension in comparison with the experimental results. The plots of the road profile for the three bumps in time domain are shown in the Fig. 5a, for front and rear wheel of the vehicle.

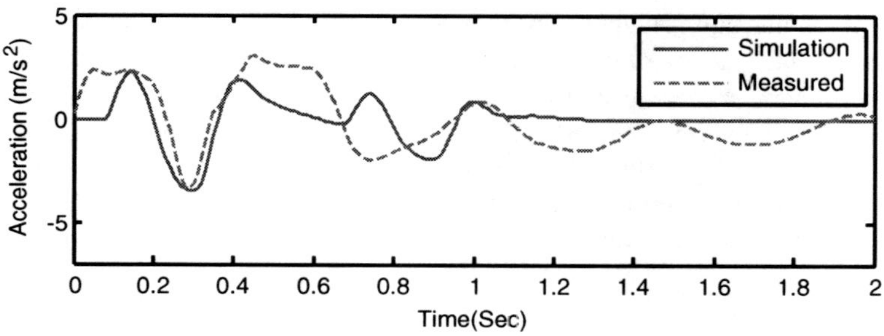

Fig. 4 Vibration (acceleration) of suspension simulation and experimental

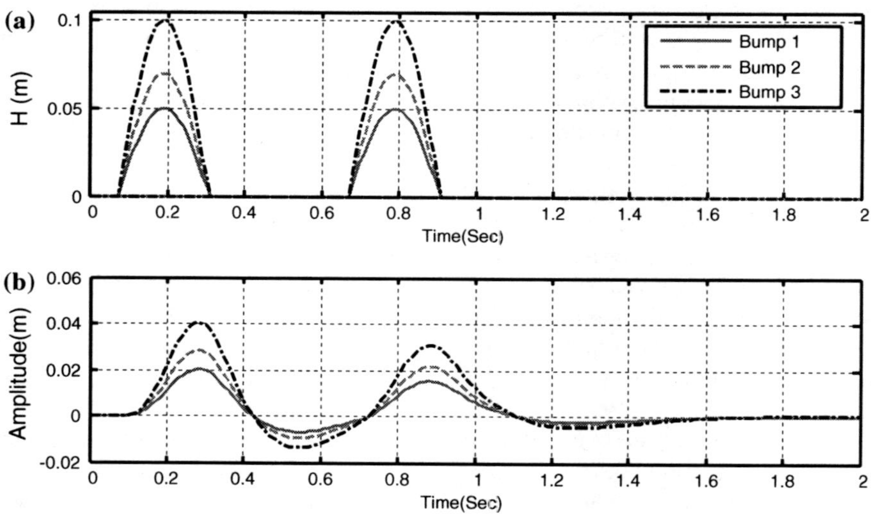

Fig. 5 **a** Road profile excitation and **b** displacement of vehicle body for different road bumps

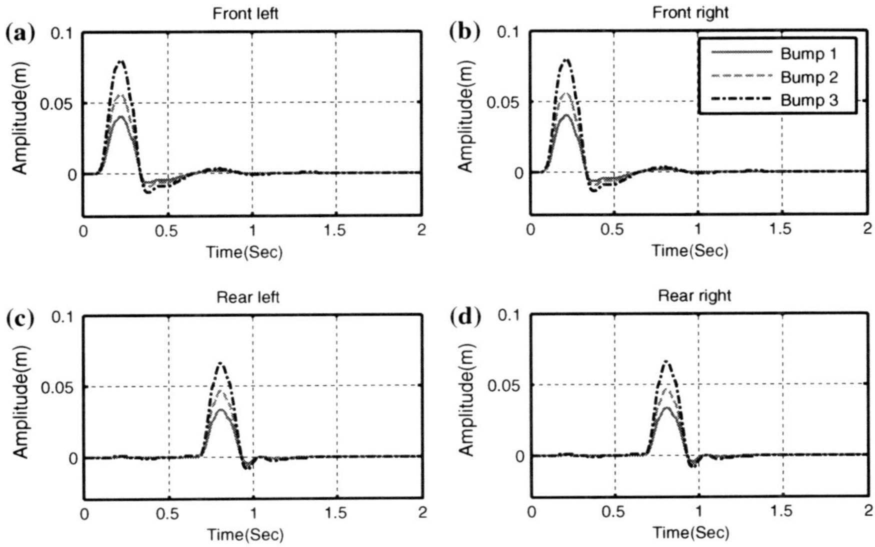

Fig. 6 Vehicle wheel's displacement with different road profile (bumps). **a** Front left, **b** Front right, **c** Rear left, **d** Rear right

For the simulation study, road disturbance is assumed as the input for the system. Figure 5b shows the displacement of the vehicle body (sprung mass) while Fig. 6 shows the displacement of four wheels (unsprung mass) in time domain. From these plots, it can be seen that the amplitude of the vehicle body and the wheels increase stably with increase in the bump height. This indicates that the performance of the suspension may be affected by this change in the geometry of bumps or road disturbances.

To analyse the performances of the suspension in terms of ride quality, handling and stability of the vehicle, road handling profile for vehicle is associated with the contact forces between the road surface and the vehicle tyre ($zu - zr$). The wheel deflection for this simulation was about 0.011 m, 0.015 m and 0.025 m for bump1, bump 2 and bump 3 respectively as presented in Fig. 7a. This range of wheel deflection seems to be acceptable compared with the proper road handling which must be in the range of 0.0508 m as per ISO: 2631-1-1997 [3].

The suspension travel can be defined as a relative displacement between the vehicle body and the wheel ($zs - zu$) as shown in Fig. 7b. The value is about 0.02, 0.028 and 0.04 m for passing the vehicle over bump1, bump 2 and bump 3 respectively. This range of suspension travel seems to be acceptable in comparison with the standard value which must have a minimum value of 0.127 m according to the ISO: 2631-1-1997 specification [3]. In addition, ISO: 2631-1-1997 [3] also states that, the passenger feels highly comfortable if the RMS acceleration is below

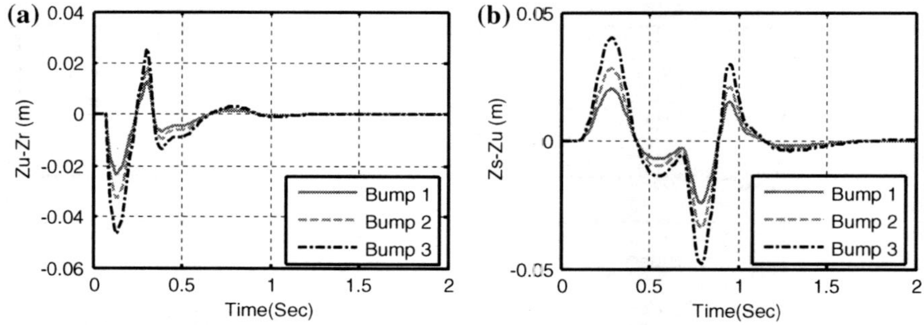

Fig. 7 **a** Wheel deflections and **b** suspension travel for different bumps

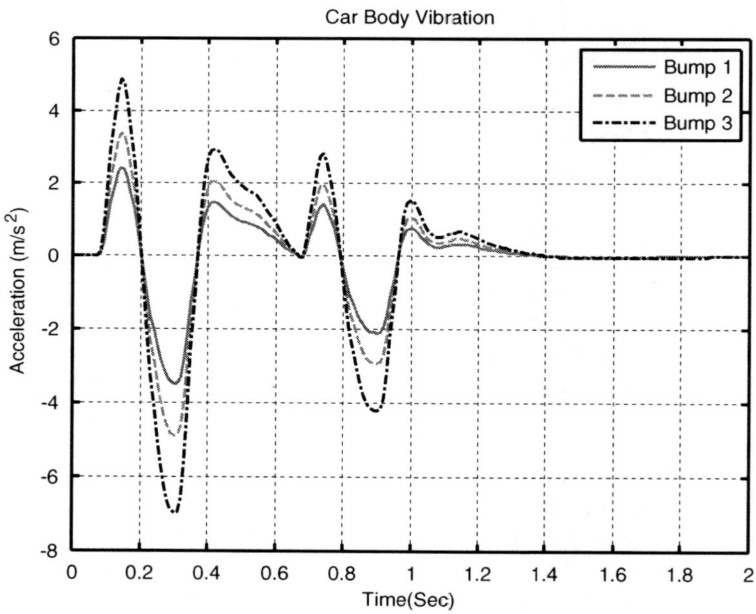

Fig. 8 Shows acceleration of the vehicle body with different bump sizes

0.315 m/s^2. In Fig. 8, the acceleration of the vehicle body in time domain is presented. The results agree with both the ISO specification and with results presented in previous researches [11]. Figure 9 shows a typical example of RMS value for acceleration of the vehicle body at different speed levels of 8, 11, 16, 24 and

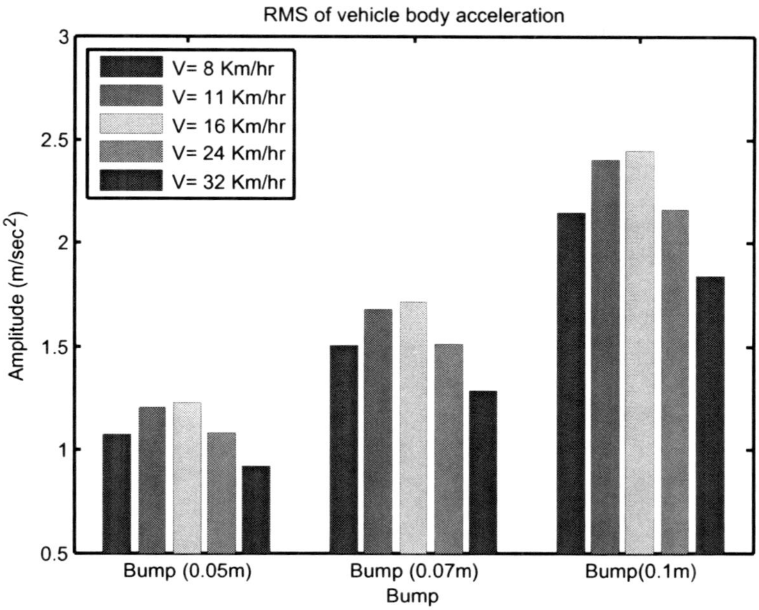

Fig. 9 Shows RMS for acceleration of the vehicle body with different speed and different bump sizes

32 km/hr, and with different bump sizes. These results show that the effect of vehicle speed on the acceleration of the vehicle body is strong at lower speeds and slight at high speeds. It can be clearly noted that, the change of the RMS value was high with changing speeds at values between 8 to 11 km/hr compared to changing the speed at high values between 11 to 16 km/hr. It should also be noted that, the highest acceleration occurs during the speed of 11–16 km/hr. These results have been compared with [13] and show some agreement.

The transfer function was used to detect the level of under-inflation of the tyre and also predict possible future suspension faults. In Fig. 10, the amplitude-frequency characteristic curves for changes of tyre stiffness in four different output cases (vehicle body vertical displacement, vehicle body velocity, displacement of front wheel, and displacement of rear wheel) are shown. From the plots, it can be seen that every vibration response value causes resonance phenomenon and generates peak values in the vicinity of about 1 and 9 Hz. The vibration response value becomes larger in resonance region as the decrease of tyre stiffness. In the high frequency resonance region, there is a change in the vibration response values which is related to the wheels however the vibration response values related to the vehicle's body shows also high change.

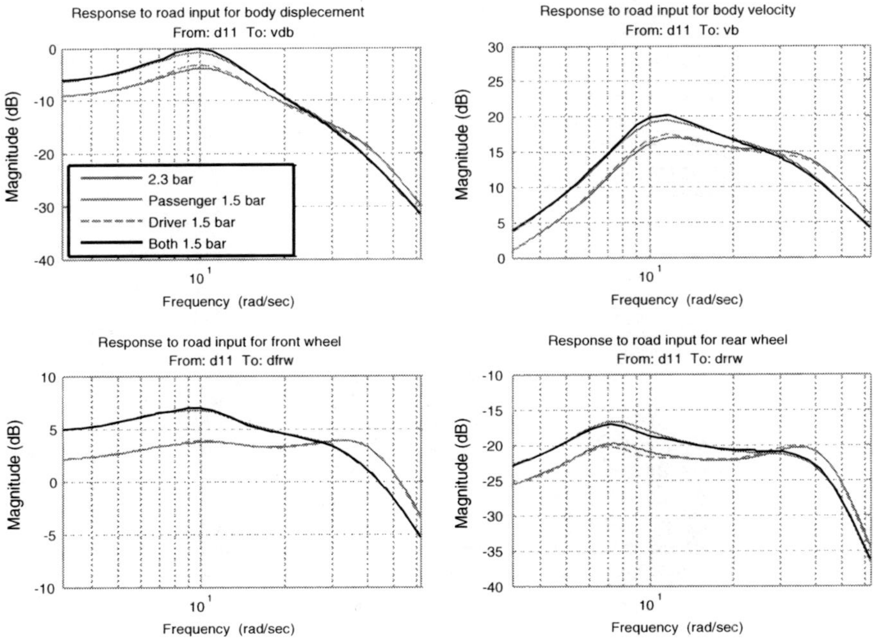

Fig. 10 Transfer function response for vehicle body and vehicle body velocity

4 Conclusion

A 7-DOF model for a full vehicle has been developed to analyse the time and the frequency response of the vehicle in MATLAB. The performances of the suspension in terms of ride comfort, road handling and stability of the vehicle were presented. The acceleration analysis of the vehicle body for different speed levels of 8, 16, 24 and 32 km/hr showed that the effect of vehicle speed on the acceleration of the vehicle body was higher at a low speed and reduced uniformly at higher speeds. Moreover, the influence of parameter variations on transfer functions as a method of fault detection of suspension has been introduced. This approach has been used for fault detection of under-inflation of tyre for three conditions.

References

1. John H Good Garages|Honest John (2012). http://good-garage-guide.honestjohn.co.uk/
2. Zohir BL (2009) Ride Comfort assessment in off road vehicles using passive and semi-active suspension. Asia pacific Conference on Defence & Security Technology (DSTC 2009)
3. Mitra A, Benerjee N, Khalane H, Sonawane M, JoshI D, Bagul G (2013) Simulation and analysis of full car model for various road profile on a analytically validated MATLAB/

SIMULINK model. IOSR J Mech Civ Eng IOSR-JMCE, ISSN(e) : 2278-1684, ISSN(p) : 2320–334X, 22–33
4. Verros G, Natsiavas S, Papadimitriou C (2005) Design optimization of quarter-car models with passive and semi-active suspensions under random road excitation. J Vib Control 11 (5):581–606
5. Alam F, Faheem A, Jazar R, Smith LV (2010) A study of vehicle ride performance using a quarter car model and half car model, pp 337–341
6. Rao R, Ram T, Rao k, Rao P (2010) Analysis of passive and semi active controlled suspension systems for ride comfort in an omnibus passing over a speed bump
7. Eslaminasab N, Biglarbegian M, Melek WW, Golnaraghi MF (2007) A neural network based fuzzy control approach to improve ride comfort and road handling of heavy vehicles using semi-active dampers. Int J Heavy Veh Syst 14(2):135–157
8. Darus R, Sam YM (2009) Modeling and control active suspension system for a full car model. In: 5th International colloquium on signal processing its applications, CSPA, pp 13–18
9. Metallidis P, Verros G, Natsiavas S, Papadimitriou C (2003) Fault detection and optimal sensor location in vehicle suspensions. J Vib Control 9(3–4):337–359
10. Kashi K, Nissing D, Kesselgruber D, Soffker D (2006) Diagnosis of active dynamic control systems using virtual sensors and observers. In: IEEE International conference on mechatronics, pp 113–118
11. Agharkakli A, Sabet G, Barouz A (2012) Simulation and analysis of passive and active suspension system using quarter car model for different road profile. Int J Eng Trends Technol 3(5):5
12. Ikenaga S, Lewis FL, Campos J, Davis L (2000) Active suspension control of ground vehicle based on a full-vehicle model. In: Proceedings of the 2000 American control conference, vol 6, pp 4019–4024
13. Lu F, Ishikawa Y, Kitazawa H, Satake T (2010) Effect of vehicle speed on shock and vibration levels in truck transport. Packag Technol Sci 23(2):101–109
14. Wong JY (2001) Theory of ground vehicles. Wiley, New York

Effects of Monograde and Multigrade Oils on the Friction Force in Four-Stroke Motor Engine: An Experimental and Analytical Approach

Anastasios Zavos and Pantelis G. Nikolakopoulos

Abstract In this work computational models with measured friction on a motor of four stroke engine are compared. Fluid Structure Interaction analysis is performed using the Navier Stokes equations in order to calculate the hydrodynamic friction for several engine operational conditions. The piston ring friction was experimentally measured under different conditions on a single-cylinder gasoline engine, using strain gauges along to the cylinder liner. The Hilbert-Huang Transform (HHT) is applied in the experimental signals. A differential deformation fluctuation signal of piston-piston rings and cylinder is adaptively decomposed into Intrinsic Mode Functions (IMFs) through the use of Empirical Mode Decomposition (EMD) methods. Further, the de noised deformation from strain gauges signal is obtained and the friction force can be determined. This method doesn't require any great engine modification, so the piston ring-cylinder friction reflects the reality. The experimental results for different grade type of oils (SAE 30, SAE 10w40), were compared with the numerical predictions, and presented.

Keywords Piston rings · Fluid structure interaction · Hilbert Huang transform · Friction · Hydrodynamic lubrication · Motor engine

1 Introduction

The piston assembly of an internal combustion engine has been recognized as a major source of mechanical friction. In most cases more than 40 % of the piston assembly friction is arising from the interfaces between the piston rings and cylinder

A. Zavos · P.G. Nikolakopoulos (✉)
Machine Design Laboratory, Mechanical and Aeronautical Engineering, University of Patras, Rio Patras, 26500 Patras, Greece
e-mail: pnikolak@mech.upatras.gr

A. Zavos
e-mail: zavosa@mech.upatras.gr

J.K. Sinha (ed.), *Vibration Engineering and Technology of Machinery*,
Mechanisms and Machine Science 23, DOI 10.1007/978-3-319-09918-7_45

liner. A number of the studies focus on an analytical and experimental study regarding the piston rings basic tribological parameters such us, the friction forces and oil film thickness. Their numerical models are mainly based on the Reynolds equation across the lubricating film [1–4]. A numerical model based on a Navier Stokes equations and power law models is proposed by Zavos and Nikolakopoulos [5]. Friction results regarding the Newtonian oils and Lubricants with additives compared in relation with similar operating condition. Later on, a friction force measurement system using the floating liner method to study the frictional behaviour of piston rings is developed [6]. However, in other studies presented different technique about the prediction of friction force in combustion engines, such as the ultrasound method [7] and the acoustic emission [8].

In all of the above treatments, the measurement and the calculation of friction force considered challenging for the modern engines. The HHT which can be adapted on non-linear and non-stationary signal and separate signals with difference phase, frequency, and amplitude to several Intrinsic Mode Function (IMF) by Empirical Mode Decomposition (EMD) was introduced by Huang in [9]. The current paper describes a piston ring-cylinder lubrication model based on Navier Stokes equations for different type of oils (monograde and multigrade). Additionally, the experimental results performed for a four stroke motor engine, where the friction was recorded using strain gauges. The denoising of deformation signals were achieved using the Hilbert-Huang Transform (HHT) and the friction force was predicted. Comparisons between the measured and calculated friction are presented and discussed.

2 Experimental, Strain Gauges and Signals Export Set-Up

Figure 1 shows the experimental set-up of a four stroke motor engine and some specifications. The engine was machined to accommodate the strain gauges along to the cylinder bore. Additional the strain gauge was also installed near to top dead center (TDC) to ensure safe operation and data reliability. The data was automatically logged using an interface program written in LabVIEW 11. In order to measure the strains prudence by the strokes in the cylinder, strain gauge of type N11MA512011 was chosen with length of 5 mm, nominal resistance of R_g = 120 ohm and preinstalled wire. Figure 2a, b shows the strain gauge set up and the position on the cylinder wall. The strain gauge has been placed near to the top dead center (TDC) where the combustion pressure is maximum. The deformation signals analyzed with the Labview 11. For this analysis, the strain gauge has the following parameters while the basic bridge configurations are also presented. The first quarter bridge configuration (Type I) is most commonly used in experimental stress analysis, where ambient temperature is relatively constant.

For the Type I configuration, the adjacent resistor in the lower arm is selected to have the same resistance as the strain gauge ($R_3 = R_g$). In this case, solving for the

Type	4-Stroke
Volume	107 cc
Bore & Stroke	52.4 × 49.5 mm
Max. HP	7.0 HP @ 7.500 rpm
Idle	1000 rpm
Compression ratio	10.0:1
Lubrication	Pressure/Splash

Fig. 1 Experimental set-up and the specifications for a four stroke motor engine

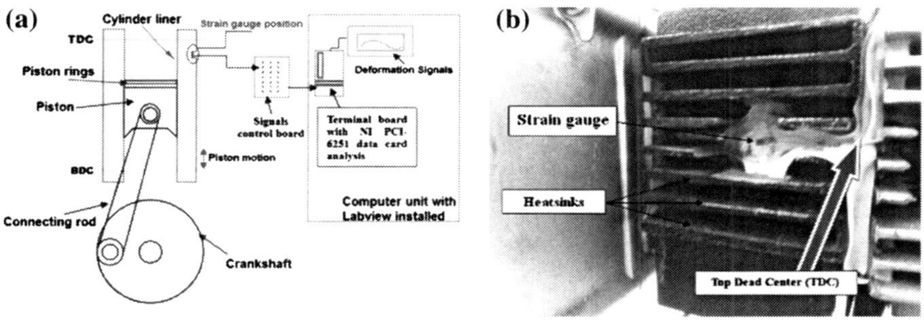

Fig. 2 **a** Strain gauge position and set-up; **b** Detail of strain gauge position in motor engine

resultant strain of the Quarter-Bridge Type I configuration will yield the following expression:

$$e = -4V_r \frac{1}{GF(1 + 2V_r)} \underbrace{(1 + \frac{Rl}{R_g})}_{\substack{Error \\ Term}} \tag{1}$$

where GF = 2 is the gauge factor of strain gauge, R_g is the nominal strain gauge resistance, and Rl is the lead-wire resistance. The above strain equation is in error by a factor of the ratio of the lead wire resistance to the nominal gage resistance. This factor is lead wire desensitization. Thus, the error minimized if the lead wire resistance is small ($Rl = 0$) and/or the nominal gage resistance is large. If ignoring the error term will cause an unacceptable error, then it should be added to the computer program.

2.1 Hilbert Huang Transform

Hilbert-Huang Transform (HHT) is a time-frequency analysis technique introduced by Huang to process non-stationary signals. It combines the Hilbert transform and the Empirical Mode Decomposition (EMD). For non-stationary signals this paper applied to the HHT method for reducing the noise of strain measurements. The main interest of the EMD is to consider the features of the analyzed signal, which are oscillations on determining the IMFs by using an iterative process. This explains that the time-scale of the decomposition will automatically be adapted to the dynamic of the analyzed signal. The below procedure for a signal $x(t)$ and their mean value $m_l(t)$ is followed:

$$h(t) = x(t) - m(t) \tag{2}$$

$$x(t) - c_1 = r \tag{3}$$

Setting r as new $x(t)$, repeat above steps and then obtain the 2nd IMF $2\ c$, the 3rd IMF $_3\ c$ …, until $_n\ c$ or r satisfies the stopping criterion. Then $x(t)$ can be decomposed into:

$$x(t) = \sum_{i=1}^{N} c_i + r \tag{4}$$

For each IMF, the Hilbert transform is defined as:

$$H[c(t)] = \frac{1}{\pi} \int_{-\infty}^{+\infty} \frac{c(\tau)}{1 - \tau} d\tau \tag{5}$$

Construct the analytic signal,

$$z(t) = c(t) + jH[c(t)] = a(t)e^{j\phi(t)} \tag{6}$$

The amplitude function is,

$$a(t) = \sqrt{c^2(t) + H^2[c(t)]} \tag{7}$$

Instantaneous frequency of the IMF component is defined,

$$f(t) = \frac{1}{2\pi} \frac{d\theta(t)}{dt} \tag{8}$$

Then after applying the Hilbert transform in each IMF, it obtained:

$$s(t) = \mathrm{Re}\sum_{i=1}^{N} a_i(t)e^{j\phi t} \tag{9}$$

and the Hilbert spectrum (10) and the marginal spectrum (11) are obtained as,

$$H(\omega,t) = \mathrm{Re}\sum_{i=1}^{N} a_i(t)e^{j\int \omega(t)dt}, \tag{10}$$

$$h(\omega) = \int_{0}^{T} H(\omega,t)dT \tag{11}$$

2.2 Monograde and Multigrade Test Oils

Two different type of oils are investigated and compared in this work. Monograde oil SAE 30 and Multigrade oil SAE 10w40 (Mobil Super™ 2,000 X1) are performed for constant temperature T = 26 °C for 10 s start up engine condition. Figure 3 shows the temperature field for this running time using the high performance thermo camera type FLIR SC660 with 640 × 480 resolution and ±1 % accuracy. The engine is arranged in secured base and cooled by an air condition. Hence, the external temperature along to the engine remains constant, so the strain gauge is not affected by the temperature variation. In practical terms, the viscosity of lubricants oils considered as, μ = 0.3 Pas for SAE 30 and μ = 0.095 Pas for SAE 10W40 for each crank angle respectively.

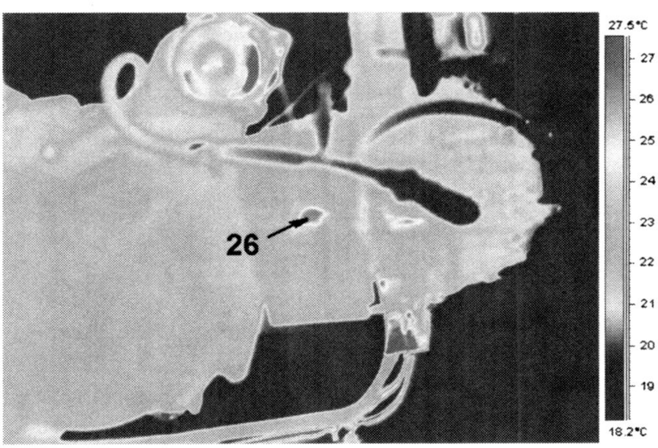

Fig. 3 Temperature field distribution near to the strain gauge area

3 Numerical Model and Boundary Conditions

A 2D simulation model is developed using the computational fluid dynamics. The Navier-Stokes equations, momentum Eq. (12) coupled with the continuity Eq. (13), are solved using the finite volume method. The flow is considered laminar, steady and two dimensional.

$$\frac{\partial}{\partial t}(\rho\vec{v}) + \nabla\cdot(\rho\vec{v}\vec{v}) = -\nabla p + \nabla\cdot(\bar{\bar{\tau}}) + \rho\vec{g} + \vec{F} \quad (12)$$

$$\frac{\partial \rho}{\partial t} + \nabla\cdot(\rho\vec{v}) = 0 \quad (13)$$

where $\rho\vec{g}$ and $\vec{F}$ are the gravitational body force and external body forces respectively.

In Fig. 4a the piston ring-cylinder model is presented. On the up stroke the compression ring is pushed to the lower surface of the groove, allowing oil float into the groove (Mode 1). On the down stroke, the ring switch side to the top, pumping oil into the groove to cylinder wall (Mode 2). Pressure build up is due to the converging geometry, as a result of the ring deformation, due to the chamber pressure. In addition, the piston ring diameter is, D_{ring} = 54.5 mm while the piston ring width and thickness are W = 3.35 mm and B = 0.5 mm respectively.

In pure hydrodynamic lubrication, there is no contact between the ring and liner, and the piston ring load is entirely support by hydrodynamic pressure in the oil film. In this case, the piston ring friction was obtained entirely from shear stress within the oil (see Fig. 4a). For this analysis, the maximum cylinder pressure at the top dead center (TDC) of motor engine calculated according to the following expression:

$$p_c = p_o CR^{\gamma} \quad (14)$$

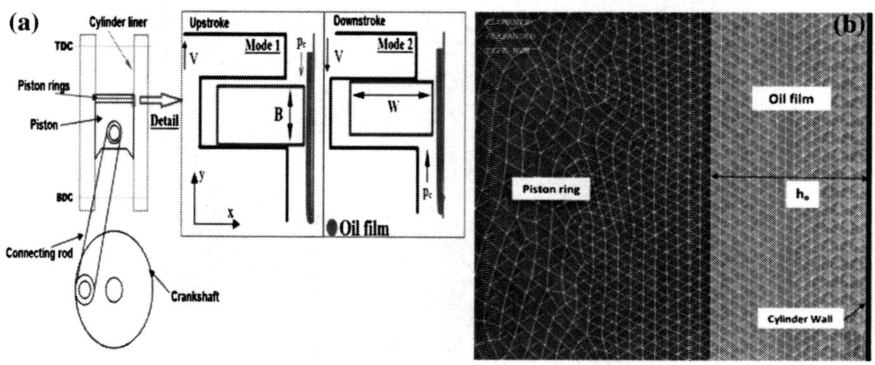

Fig. 4 **a** Piston ring-cylinder model and the piston ring modes; **b** Fluid structure meshing detail

where p_o is the cylinder pressure at bottom dead center (BDC) which is usually at 1 atm, *CR* is the compression ratio, and γ, is the specific heat ratio for the working fluid, which is about 1.4 for air.

Using the above equation the maximum pressure considered p_c = 2.51 MPa for this particular motor engine. Later, for the numerical solution the combustion gas pressure p_c applied on the piston ring profile for each crank angle. Moreover, the cylinder remains stationary, while the piston ring floats into the piston groove,as it is shown in Fig. 4. Additionally, the variation of the piston velocity is a function of the relative rotation angle of the crankshaft,and it is given by:

$$V = r\Omega(\sin\varphi + \frac{\lambda}{2}\sin 2\varphi) \tag{15}$$

where the crank radius is r = 25 mm and the control ratio is λ = 0.266.

The fluid field, is meshed using 79,343 tetrahedral. An initial distance between the ring and cylinder h_o = 12 μm is considered. The particular number of elements and nodes were obtained after extensively grid sensitivity tests for each examined case, while the pressure error considered as, $\text{Error}_{\text{pres}} \leq 10^{-6}$. For this analysis, the element size is in the order of 10^{-6} near to the piston ring area. The pressure error is achieved after 4,000 iterations for each crank angle. The Arbitrary Lagrangian Eulerian (ALE) formulation is also used in order, through the mesh movement, to move the fluid nodes in selected areas (non structural) to conform to the structural piston ring displacements [5].

3.1 Convergence Strategy and Design Quantities

A two-stage convergence process was applied. First, the generated pressures are deemed to have converged when,

$$Error_{pres} = \frac{\sum_{i=1}^{i=N_n^{CFD}} \left|p_i^k - p_i^{k-1}\right|}{\sum_{i=1}^{i=N_n^{CFD}} \left|p_i^k\right|} \leq 1 \times 10^{-6} \tag{16}$$

Second, the critical displacement on the ring in every time step is less/equal than five percent of the maximum displacement due to the chamber pressure, p_c.

$$DS_{cr} \leq 0.05 DS_{\max} \tag{17}$$

The time step defined for each crank angle for 4,000 iterations. The pressure distribution over the piston ring is obtained, after the numerical solution of the Eq. (12).

In fact, the numerical friction force is calculated by the integration of stress tensor $\overline{\tau}$ with ring surface which is interfaced with oil film, and it is expressed by:

$$F_f^{anal} = \int_A \overline{\tau} dA = \pi D_{ring} \int_{-B/2}^{B/2} (-p + 2\mu \frac{\vartheta v_y}{\vartheta y}) dy \qquad (18)$$

where v_y is the lubricant oil velocity in the y axis, D_{ring} is the piston ring diameter, μ is the dynamic viscosity and B is the piston ring thickness. About the experimental results, after the application of the Hilbert-Huang Transform to deformation signals, the experimental friction force predicted below according to the next equation:

$$F_f^{\exp} = \pi D_{ring} B \varepsilon \mathrm{E} \qquad (19)$$

where ε is the clear deformation after HHT and E is the modulus of elasticity. Steel piston ring considered with E = 200GPa for the specifications of motor engine.

4 Numerical Model Validation

Regarding the geometrical characteristics, the same geometry as Wakuri's work is used and thereafter, the boundary conditions applied on the piston ring profile as input to the program for each crank angle. Hydrodynamic numerical friction extracted for Newtonian lubricant SAE 30 with Ω = 2,400 rpm and presented in diagram at Fig. 5. There is a very good agreement between the values predicted by this analysis and [1]. From this configuration the deviation was 5 % from 0° to 180° and 10 % while crank angle varies between 180°–360°.

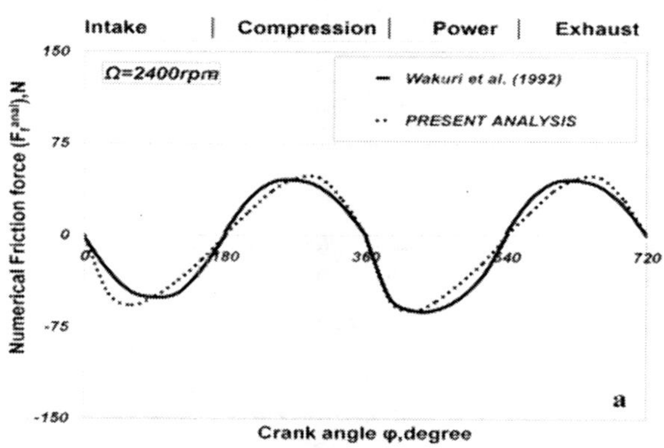

Fig. 5 Validation numerical friction results with Wakuri et al. (1992)

5 Results and Discussion

5.1 Deformation Extracted Signals

Figure 6 shows side by side the IMF components of the differential deformation fluctuation signal for rotational speed $\Omega = 1{,}000$ rpm using the type of oil SAE 30. It can be seen from Fig. 8 that the IMFs have different local frequencies. The EMD method picks out the highest frequency component present in the signal. Thus, the IMF 9 contains the appropriate frequency 16 Hz provided by the marginal spectrum.

Figure 7 presents the IMF components of the variation deformation signal for rotational speed $\Omega = 1{,}000$ rpm using the type of oil SAE 10w40. In detail, the IMF 9 provides the convenient frequency 16 Hz examined the marginal spectrum respectively.

5.2 Experimental and Numerical Friction Results

After the EMD method and the Marginal Spectrum, the experimental friction force predicted based on the IMF 9 for each case. Figure 8 illustrates the friction forces

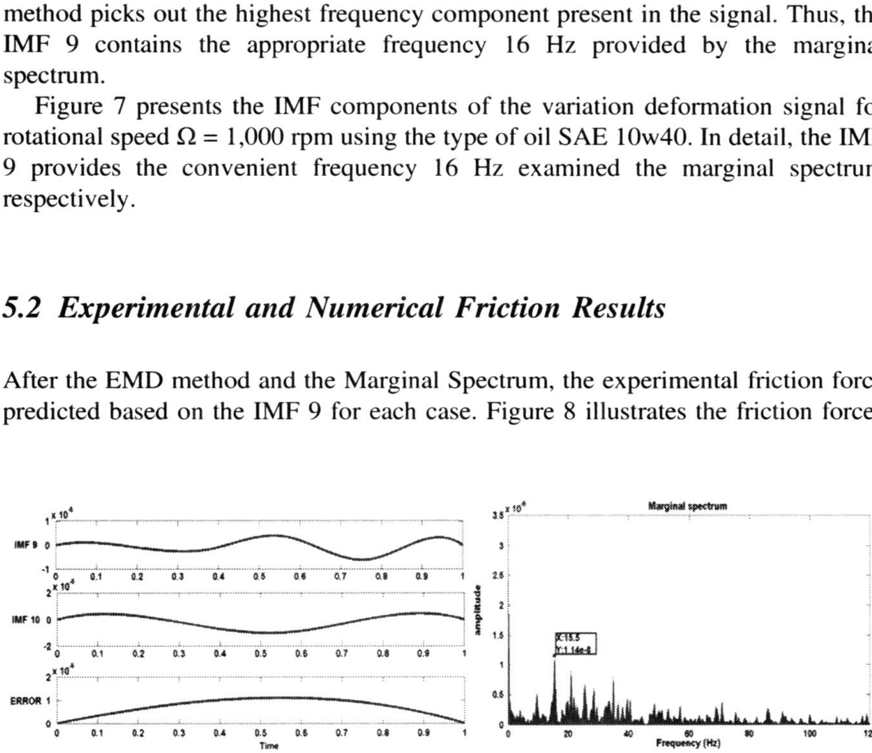

Fig. 6 IMF -9 analysis and the marginal spectrum for the monograde oil SAE 30

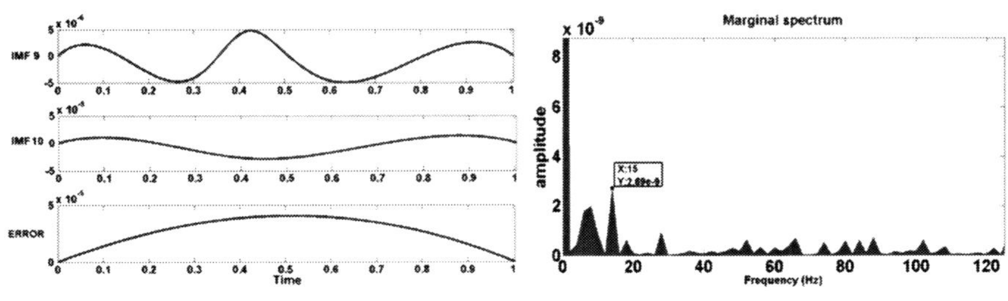

Fig. 7 IMF-9 analysis and the marginal spectrum for the multigrade oil SAE 10w40

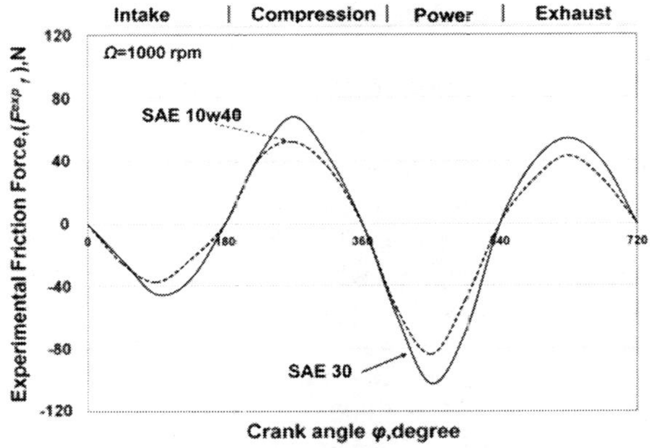

Fig. 8 Experimental friction forces versus crank angle

for Ω = 1,000 rpm having two different type of oils and constant temperature $T = 26$ °C. It is obvious that the friction reduced 18.87 % using the multigrade oil 10w40 in comparison with the monograde oil SAE 30 for the same operating conditions.

Numerical friction results compared with the experimental and presented in Fig. 9. The chamber pressure considered $p_c = 2.51$ MPa according to the given compression ratio for the above motor engine, as it is presented in paragraph 3. However, the complex shape of the clearance volume influences the value of compression ratio. Examined another case the pressure increased, $p_c' = 5.02$MPa. Figure 9a shows the numerical friction with experimental for SAE 30. The deviation of friction varies, 7–15 % in the intake and exhaust conditions. The results near to power and compression operations are with a relative large different 30–20 % in relation with the experimental results. Using the multigrade oil 10w40 the friction decreased. Figure 9b presents the variation of friction ranges between 5–12 % for intake, compression and exhaust conditions. In detail, in the power condition there is a difference of 50–55 % compared with the real engine operation.

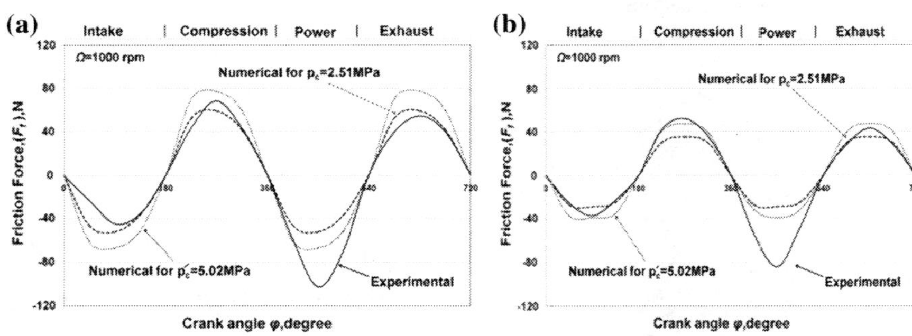

Fig. 9 Comparison numerical and experimental friction results **a** SAE 30, **b** SAE 10w40

6 Conclusion

This article presents a new and novel approach to measure the friction of piston rings using the strain gauge method. The HHT is used to denoise the stain signals. The simulation results show that is possible to improve the friction prediction in a modern motor engine using multigrade oils. Friction reduction of 18.87 % using a multigrade oil 10w40 in relation with the monograde oil SAE 30 for the same operation parameters was occurred.

Acknowledgment The authors would like to acknowledge the "KARATHEODORIS 2013" project, funded by the research council of the University of Patras, Greece.

References

1. Wakuri Y, Hamatake T, Soejima M, Kitahara T (1992) Piston ring friction in internal combustion engines. Tribol Int 25(5):299–308
2. Hamatake T, Wakuri Y, Soejima M, Kitahara T (2003) Effects of lubricant viscosity on the mixed lubrication of a piston ring pack in an internal combustion engine. Lubr Sci 15 (2):101–117
3. Hamatake T, Kitahara T, Wakuri Y, Soejima M (1993) Friction characteristics of piston rings in a reciprocating engine. Lubr Sci 6(1):21–40
4. Morris N, Rahmani R, Rahnejat H, King PD, Fitzsimons B (2013) The influence of piston ring geometry and topography on friction. Proc Inst Mech Eng Part J J Eng Tribol 227(2):141–153
5. Zavos A, Nikolakopoulos P (2013) The effects of Lubricants with additives on the friction force of smooth and artificially textured piston rings. In: Proceedings of world tribology congress, Italy
6. Cho S-W, Choi S-M, Bae C-S (2000) Frictional modes of barrel shaped piston rings under flooded lubrication. Tribol Int 33:545–551
7. Spencer A, Avan EY, Almqvist A, Dwyer-Joyce RS, Larsson R (2013) An experimental and numerical investigation of frictional losses and film thickness for four cylinder liner variants for a heavy duty diesel engine. Proc Inst Mech Eng Part J J Eng Tribol 227(12):1319–1333
8. Douglas RM, Steel JA, Reuben RL (2006) A study of the tribological behaviour of piston ring/ cylinder liner interaction in diesel engines using acoustic emission. Tribol Int 39 (12):1634–1642
9. Huang NE, Shen Z, Long SR, Wu MC, Shih HH, Zheng Q, Yen NC, Tung CC, Liu HH (1998) The empirical mode decomposition and the Hilbert spectrum for nonlinear and non-stationary time series analysis. Proc Roy Soc Lond Ser A 454(1971):903–995

Delamination Detection in a Laminated Composite Beam Based on Changes in Natural Frequencies

Kajal Khan and Surjya K. Maiti

Abstract This paper presents an analytical method for modelling delamination parallel to beam longitudinal axis that can help in solving both forward and inverse problems. For vibration analysis of a beam with internal delamination, it is divided into four segments. The forward problem of determination of frequencies due to a known delamination is solved analytically using the characteristics equation. Case studies are presented and the frequencies are compared with finite element frequencies which show a very good agreement between the two. To solve the inverse problem of locating damage from measured changes in natural frequencies, the characteristic equation is solved. In this paper a simple method to solve the inverse problem has been proposed. The method is fast and inexpensive and it has been validated using input frequency data generated by finite element method for both cantilever and simply supported beams. The predicted axial locations and sizes agree closely with the actual data. The maximum errors are 5 and 6 % in the prediction of location and size respectively for a known interface.

Keywords Structural health monitoring · Detection of delamination · Vibration of laminates with delamination · Laminated composites

1 Introduction

Composites are extensively employed in aerospace and marine applications for their excellent stiffness to weight ratio. But these materials have very poor interlaminar shear and transverse tensile strengths compared to in-plane strengths. So, very often they are prone to interlaminar delaminations due to residual stresses arising out of

K. Khan (✉) · S.K. Maiti
Department of Mechanical Engineering, IIT Bombay, Mumbai 400076, India
e-mail: kajalkhanme@gmail.com

S.K. Maiti
e-mail: skmaiti@iitb.ac.in

J.K. Sinha (ed.), *Vibration Engineering and Technology of Machinery*,
Mechanisms and Machine Science 23, DOI 10.1007/978-3-319-09918-7_46

fabrication, handling, aging, low velocity impact, etc. These delaminations are barely visible, but they significantly reduce the stiffness, strength and affect natural frequencies of the component. This reduction in the frequency can help in its possible detection. The vibration based detection techniques require modeling of delamination and an inverse algorithm to detect it. A considerable amount of analytical and numerical modeling techniques have been reported so far on the modeling of delaminations. Wang et al. [1] examined the free vibration of layered isotropic beam with through-width delamination by dividing the beam into four segments connected at the delamination boundaries (Fig. 1). They considered coupling between transverse and axial vibrations along with the assumption that the delaminated segment will move freely without touching each other. This is termed as "free model". Mujumdar and Suryanarayan [2] showed that the mode shapes obtained through the "free model" were physically inadmissible and they proposed a model based on the assumption that the delaminated layers are constrained to move together in the transverse direction and are free to slide on each other in the delaminated region. They referred this model as "constrained model". The "constrained model" was extended to modeling of delamination in composite slender beam by Tracy and Pardoen [3] and to composite Timoshenko beam by Shen and Grady [4] although they found delamination opening in their experiment. Shu and Fan [5] first applied the "constrained model" to delamination modeling in bimaterial split beam. They studied the effect of the modulus of the two materials on the frequency. Luo and Hanagud [6] developed a new model using piecewise linear spring in the delaminated segment that can simulate the observed experimental behavior between the delaminated segments involving considerable delamination opening. Lestari and Hanagud [7], Shu and Della [8] studied vibration of multiple delaminated composite beams. Kim et al. [9] presented a first order zigzag theory to model multiple delaminations.

Although considerable efforts have been put by several researchers to address the forward problem, i.e. to assess the effect of known damage/delamination on vibration characteristics, very few works have been published that consider the inverse problem of predicting location and size of damage knowing the changes in modal data. Valoor and Chandrashekhara [10] solved the inverse problem for midplane delamination in composite beams using feed forward back propagation neural network. Chakraborty [11] also presented a method of solving the inverse problem by using a back propagation artificial neural network (ANN). Zhang et al.

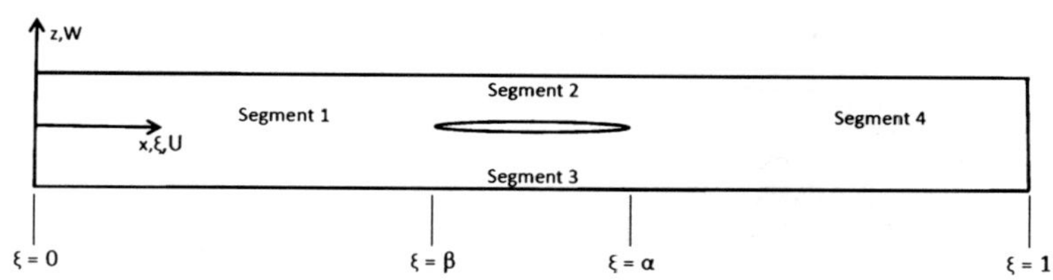

Fig. 1 Beam with internal delamination divided into four segments

[12] used the graphical techniques to determine crack location in the axial direction, its size and interface location. Ihesiulor et al. [13] used a method based on ANN (to provide surrogate models), and optimization algorithms (gradient based local search, non-dominated sorting genetic algorithm-II). Zhang et al. [14] compared different methods of delamination detection: Graphical techniques, ANN and surrogate assisted optimization using the experimental data. This paper presents an analytical model where numerical input data is used for the solution of inverse problem.

2 Mathematical Formulation

For solution of the forward problem, an analytical model consistent with the "constrained model" proposed by Mujumdar and Suryanarayan [2] is used here. The beam with an internal delamination is divided into four segments and free vibration analysis of the beam is done with the consideration that the delaminated segments will move together in the transverse direction although they have different axial vibrations. A coupling between the axial and transverse vibration has been considered. The interface designation and stacking sequence is shown in Fig. 2.

The transverse and axial vibration equations in a non-dimensional form are

$$\frac{d^4W}{d\xi^4} - \lambda^4 W = 0 \tag{2.1}$$

$$\frac{d^2U}{d\xi^2} + p^2 U = 0 \tag{2.2}$$

where,

$$\xi = \frac{x}{L}, \quad \lambda^4 = \left(\frac{mL^4}{D}\right)\omega^2, \quad p^2 = \left(\frac{mL^2}{A}\right)\omega^2 \tag{2.3a}$$

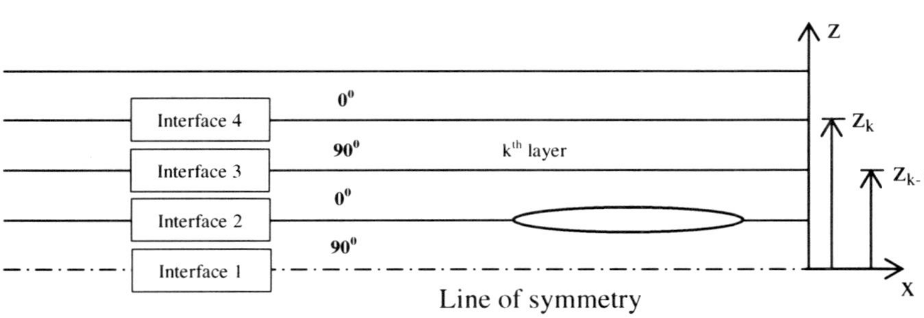

Fig. 2 Layer orientation and symmetric identification showing delamination at interface 2

$$m = b\sum_{k=1}^{n} \rho(Z_k - Z_{k-1}), \quad A = b\sum_{k=1}^{n} Q_{xx}(Z_k - Z_{k-1}),$$
$$B = \frac{b}{2}\sum_{k=1}^{n} Q_{xx}\left(Z_k^2 - Z_{k-1}^2\right) \tag{2.3b}$$

$$D = D_{11} - \frac{B^2}{A}, \quad D_{11} = \frac{b}{3}\sum_{k=1}^{n} Q_{xx}\left(Z_k^3 - Z_{k-1}^3\right) \tag{2.3c}$$

$$Q_{xx} = Q_{11}\cos^4\theta + Q_{22}\sin^4\theta + 2(Q_{12} + 2Q_{66})\cos^2\theta\sin^2\theta \tag{2.3d}$$

$$Q_{11} = \frac{E_1}{1 - \upsilon_{12}\upsilon_{21}}, \quad Q_{22} = \frac{E_2}{1 - \upsilon_{12}\upsilon_{21}}, \quad Q_{12} = \frac{\upsilon_{21}E_1}{1 - \upsilon_{12}\upsilon_{21}}, \quad Q_{66} = G_6 \tag{2.3e}$$

Here W and U are the transverse and axial modal displacements respectively, x is the distance from the left end of the beam, L is the beam length, ω is the natural frequency, m is the mass of the beam per unit length, A is the axial stiffness, B is the coupling stiffness, D is the bending stiffness, E_{11} and E_{22} are the longitudinal and transverse Young's moduli, respectively, ν_{12} and ν_{21} are the longitudinal and transverse Poisson's ratio respectively, θ is the angle of orientation of the kth lamina and z_k and z_{k-1} are the extreme locations of the kth lamina with respect to midplane of the beam.

The solution of Eqs. (2.1) and (2.2) is given by,

$$W = C_1\cos(\lambda\xi) + C_2\cosh(\lambda\xi) + C_3\sin(\lambda\xi) + C_4\sinh(\lambda\xi) \tag{2.4}$$

$$U = C_5\cos(p\xi) + C_6\sin(p\xi) \tag{2.5}$$

where C_1, C_2, C_3, C_4, C_5, C_6 are arbitrary constants.

Equations 2.4 and 2.5 will be applicable to each of the four segments shown in Fig. 1.

For beam segment 1:

$$W_1 = C_1\cos(\lambda_1\xi) + C_2\cosh(\lambda_1\xi) + C_3\sin(\lambda_1\xi) + C_4\sinh(\lambda_1\xi) \tag{2.6a}$$

$$U_1 = C_{13}\cos(p_1\xi) + C_{14}\sin(p_1\xi) \tag{2.6b}$$

where,

$$\lambda_1^4 = \left(\frac{m_1L^4}{D_1}\right)\omega^2, \quad p_1^2 = \left(\frac{m_1L^2}{A_1}\right)\omega^2$$

For beam segment 2 and 3:

$$W_2 = C_5 \cos(\lambda_2 \xi) + C_6 \cosh(\lambda_2 \xi) + C_7 \sin(\lambda_2 \xi) + C_8 \sinh(\lambda_2 \xi) \tag{2.7a}$$

$$U_2 = C_{15} \cos(p_2 \xi) + C_{16} \sin(p_2 \xi) \tag{2.7b}$$

$$U_3 = C_{17} \cos(p_3 \xi) + C_{18} \sin(p_3 \xi) \tag{2.7c}$$

where,

$$\lambda_2^4 = \left(\frac{m_2 + m_3}{D_2 + D_3}\right)\omega^2 L^4, \quad p_2^2 = \left(\frac{m_2 L^2}{A_2}\right)\omega^2, \quad p_3^2 = \left(\frac{m_3 L^2}{A_3}\right)\omega^2$$

For beam segment 4:

$$W_4 = C_9 \cos(\lambda_1 \xi) + C_{10} \cosh(\lambda_1 \xi) + C_{11} \sin(\lambda_1 \xi) + C_{12} \sinh(\lambda_1 \xi) \tag{2.8a}$$

$$U_4 = C_{19} \cos(p_1 \xi) + C_{20} \sin(p_1 \xi) \tag{2.8b}$$

where m_1, m_2, m_3 are the beam mass per unit length of segment 1, 2 and 3 respectively. D_1, D_2, D_3 are the bending stiffnesses and A_1, A_2, A_3 are the axial stiffnesses for segment 1, 2, 3 respectively. For segment 4, beam mass per unit length, axial and bending stiffness is the same as segment 1.

For cantilever beam, at the fixed end ($\xi = 0$), the transverse displacement, slope and axial displacement are zero

Therefore

$$W_1 = 0; \quad W_1' = 0; \quad U_1 = 0 \tag{2.9}$$

At the free end ($\xi = 1$), the bending moment, shear force and the axial forces are zero, i.e.

$$W_4'' = 0; \quad W_4''' = 0; \quad U_4' = 0 \tag{2.10}$$

At the left junction of the delamination ($\xi = \beta$) there is a continuity of transverse displacement, transverse slope, shear force and axial force and jump in the axial displacements and bending moment which can be expressed as:

Continuity conditions:

$$W_1 = W_2; \quad W_1' = W_2' \tag{2.11a}$$

$$D_1 W_1''' = (D_2 + D_3) W_2'''; \quad A_1 U_1' = A_2 U_2' + A_3 U_3' \tag{2.11b}$$

Jump Conditions:

$$U_2 = U_1 - \frac{d_3}{2L} W_1'; \;\; U_3 = U_1 - \frac{d_2}{2L} W_1' \tag{2.12a}$$

$$D_1 W_1'' = (D_2 + D_3) W_2'' - \frac{A_2 d_3 L}{2} U_2' + \frac{A_3 d_2 L}{2} U_3' \tag{2.12b}$$

where d_2 and d_3 are the height of upper and lower delaminated segments respectively. Similarly at the right junction of the delamination ($\xi = \alpha$) the conditions can be written as follows.

Continuity conditions:

$$W_2 = W_4; \;\; W_2' = W_4' \tag{2.13a}$$

$$(D_2 + D_3) W_2''' = D_1 W_4'''; \;\; A_2 U_2' + A_3 U_3' = A_1 U_4' \tag{2.13b}$$

Jump Conditions:

$$U_2 = U_4 - \frac{d_3}{2L} W_4'; \;\; U_3 = U_4 - \frac{d_2}{2L} W_4' \tag{2.14a}$$

$$D_1 W_4'' = (D_2 + D_3) W_2'' - \frac{A_2 d_3 L}{2} U_2' + \frac{A_3 d_2 L}{2} U_3' \tag{2.14b}$$

The above 20 equations (Eqs. 2.9–2.14b) give the characteristic determinant (|M|) of size 20 × 20. Solution to |M| = 0 gives the natural frequencies of the beam.

For simply supported beam the end conditions will only change. At both the end ($\xi = 0$ and $\xi = 1$), the transverse displacement, bending moment and axial force are zero. Therefore,

$$W_1 = 0; \;\; W_1'' = 0; \;\; U_1' = 0 \tag{2.15}$$

Similar characteristics equation can be obtained for simply supported beam using Eqs. (2.10)–(2.15).

3 Forward Problem

The characteristics equation is a non-linear equation. It can be solved using numerical techniques like Newton–Raphson method and bisection method to determine the frequencies. Here a graphical technique has been employed for getting the natural frequencies easily. A typical plot of log(det(M)) with ω/ω_0 is shown in Fig. 3 where $\omega_0 = \left({}^{D_1}\!/_{I_{11} L^4}\right)$. The frequencies corresponding to det (M) = 0, appear as discontinuities in the plot. These give the natural frequencies.

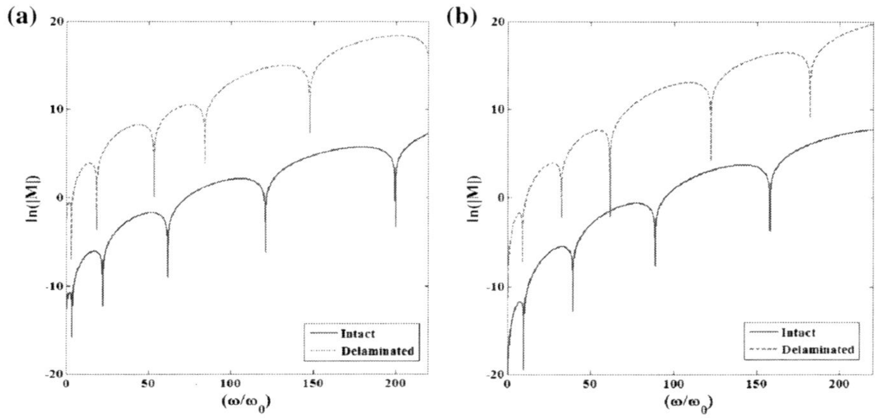

Fig. 3 Typical variation of ln|M| with respect to ω/ω_0 for intact and delaminated beam for **a** cantilever and **b** simply supported beams

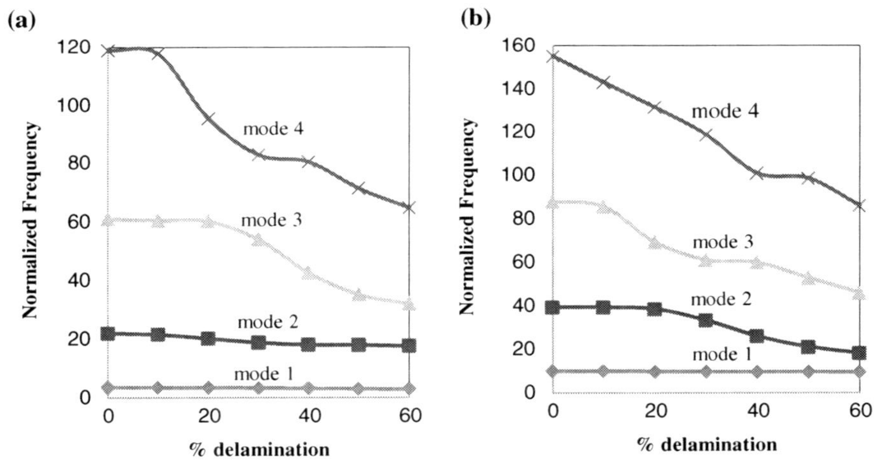

Fig. 4 Variation of normalized frequency with delamination size for **a** cantilever and **b** simply supported beams

An eight layer symmetric cross ply graphite-epoxy laminated beam with dimension 127 mm (length) × 12.7 mm (width) × 1.1049 mm (height) is considered for study here. The material properties are as follows: E_1 = 134.45 GPa, E_2 = 10.34 GPa, G_{12} = 5 GPa, G_{23} = 3.45 GPa, ν_{12} = 0.33, ρ = 1,477.5 kg/m^3. Figure 4 shows the variation of the normalized frequency (ω/ω_o) of the first four modes for various delamination lengths. As the size of delamination increases the 2nd, 3rd and 4th frequency decreases considerably whereas reduction in first frequency is not significant. Variation of the first natural frequency with the size of delamination for

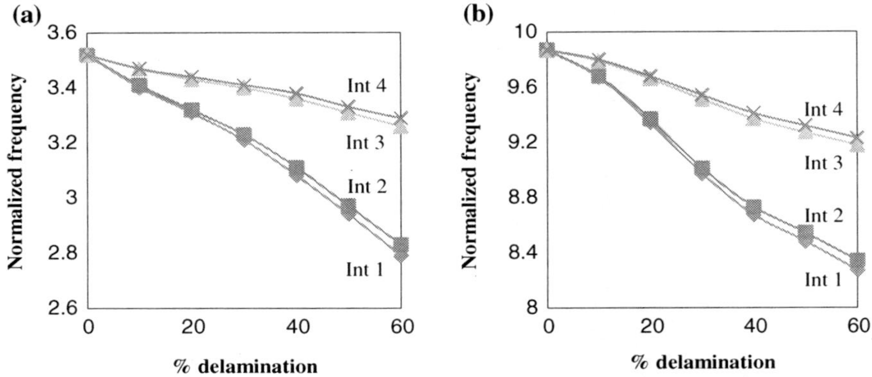

Fig. 5 Variation of first normalized frequency with delamination length for the four interfaces. **a** Cantilever and **b** simply supported beams

different interfaces has been plotted in Fig. 5. When the delamination occurs near to surface, i.e. at 3rd or 4th interfaces, the changes in frequency is very small. For delamination at 1st or 2nd interface there is a considerable reduction in the frequencies. The reduction pattern is same for all the modes.

4 Inverse Problem

The characteristics determinant |M| is a function of delamination parameters (its axial location, size and interface position) and the frequency. If the frequencies of a delaminated beam are measured experimentally or computed numerically, the delamination parameters become the only unknowns. For known interface location of a delamination if the variation of |M| with respect to α and β where α and β, are the two end locations of the delamination from the fixed end is plotted, a surface corresponding to each frequency is obtained. Thereby a number of surface plots (equal to the number of frequencies measured) can be obtained. For each given frequency of beam with delamination a contour plot can be obtained. All the contours corresponding to a number of frequencies intersect at a particular point, which gives the location details of the delamination. If they don't intersect at a particular point then the location given by the three paired intersections forming the smallest triangle is considered to determine the delamination location. The center of gravity is a good approximation for the location. If the interface is also unknown the above method is repeated for each interface, half of the total no of interfaces for symmetric laminates, and the standard deviation of the intersection triangle for each interface is calculated. The minimum standard deviation can be the basis to conclude on the interface location. Then the axial position details can be found out.

5 Finite Element Modeling

Finite element simulations have been done using ANSYS 14.0 software to determine the frequencies of delaminated beam and verify the analytical results. 8 noded shell elements (SHELL 281) with 6 degrees of freedom (translation in x, y, z axis and rotation about x, y and z axis) per node have been used to model the laminate. The section properties, i.e. layer height and orientation is provided in shell section commands. The number of integration points for evaluation of the stiffness matrix is taken as 3. The modeling is done in two segments, one for the upper segment and one for the lower segment of the delaminated beam and the corresponding identifications are assigned. After studying the effect of number of elements, the beam is divided into 100 elements in the longitudinal direction; in the depth direction one element is taken for each lamina. A total of 2,000 elements have been employed. The nodes for the undelaminated segments are merged together. To simulate the constrained mode, the coincident nodes corresponding to the delaminated segment are coupled in transverse direction so that they will have the same transverse displacement. After applying the boundary condition, modal analysis is performed with block Lanczos solver and the frequencies of only flexural modes are taken for comparison with the analytical results for the forward problem. For the solution of the inverse problems, the finite element based natural frequencies of both the damaged and intact beams were given as input to the analytical model.

6 Results and Discussions

The forward problem is solved analytically for different cases where the left end of delamination has been kept fixed at $\xi = 0.2$ and the size has been varied from 10 to 60 % of the beam length for all the interfaces. The analytical results are compared with the results obtained from finite element method and the percentage differences between the two are listed in Tables 1 and 2. The results show a very good accuracy of the analytical model.

The finite element frequencies are used as input for solution of the inverse problem. For intact beams, the finite element and analytical frequencies have differences. This calls for a minor adjustment in the stiffness values, which is termed as 'zero setting' [15]. According to this procedure two sets of frequency parameters are involved, say λ_{int}, the theoretical frequency parameter for the intact beam for a particular mode, and λ_{inm}, the measured frequency parameter of the intact beam for the same mode. λ_{int} is available in a standard textbook. λ_{inm}, on the other hand, is determined from the measured uncracked natural frequency (ω^{inm}) through $\lambda_{inm}^4 = (\omega_{inm})^2(m_1L^4/D_1)$. These two parameters must be the same, but they are generally found to differ. The stiffness (D_1) is so adjusted that $\lambda_{int}^4 = (\omega_{inm})^2(m_1L^4/D_1)$. This gives a value for D_1, which is given as input to the characteristic equation. D_2 is modified in such a way that

Table 1 Comparison of analytical and ANSYS frequencies for cantilever beam

Delamination parameters			% Difference between ANSYS and analytical frequencies				
I	β	Size	ω1	ω2	ω3	ω4	ω5
Undelaminated			0.28	0.46	0.98	1.68	2.62
1	0.2	10	−2.92	0.83	−0.96	−0.3	4.36
		20	−4.35	0.43	−2.85	2.96	1.97
		30	−4.66	−1.76	−1.3	1.32	3.56
		40	−4.24	−4.41	1.05	1.84	1.11
		50	−3.38	−6.46	1.27	1.14	2.21
		60	−2.41	−6.13	0.33	−0.72	2.56
2	0.2	10	−2.67	0.77	−0.92	−0.35	4.14
		20	−4.14	0.44	−2.74	2.91	1.96
		30	−4.34	−1.63	−1.35	1.37	3.48
		40	−3.9	−4.18	1.06	1.77	1.16
		50	−3.33	−6.14	1.23	1.27	2.14
		60	−2.31	−5.94	0.29	−0.55	2.53
3	0.2	10	−1.1	0.48	−0.05	0.47	2.6
		20	−1.95	0.29	−0.96	1.87	2.51
		30	−2.12	−0.61	−0.59	1.81	2.5
		40	−2.09	−1.74	0.89	1.39	2.58
		50	−1.98	−2.73	1.02	1.75	1.95
		60	−1.55	−2.99	0.83	1.35	2.64
4	0.2	10	−1.09	0.49	0.1	0.64	2.69
		20	−1.66	0.34	−0.77	1.89	2.64
		30	−1.91	−0.48	−0.45	1.88	2.54
		40	−1.72	−1.54	0.94	1.44	2.68
		50	−1.78	−2.4	1.07	1.79	2.03
		60	−1.25	−2.69	0.84	1.45	2.69

the ratio of modified D_1 and D_2 remains same as the ratio of actual D_1 and D_2. The same is done for D_3. A typical contour plot of the inverse problem is shown in Fig. 6 for both cantilever and simply supported beams. As the reduction in first frequency is almost negligible, it is not considered for solution of inverse problems. For most of the cases reported here, 2nd, 3rd and 4th frequencies are taken as input. Except in a few cases where change in 2nd natural frequency is very small, the 3rd, 4th and 5th frequencies are considered as input. The point which corresponds to the minimum standard deviation of the three intersections is chosen as the predicted location.

Table 2 Comparison of analytical and ANSYS frequencies for simply-supported beam

Delamination parameters			% Difference between ANSYS and analytical frequencies				
I	β	Size	ω1	ω2	ω3	ω4	ω5
Undelaminated			0.23	0.56	1.1	1.87	2.86
1	0.2	10	−1.5	−2.73	0.33	3.63	2.51
		20	−3.81	−4.15	3.65	1.29	3.88
		30	−6.41	−2.13	1.3	3.25	2.18
		40	−8.88	0.35	1.55	1.91	3.27
		50	−10.81	1.19	2.32	1.63	1.41
		60	−11.56	1.49	1.33	2.71	2.74
2	0.2	10	−1.43	−2.61	0.25	3.48	2.4
		20	−3.61	−3.96	3.53	1.31	3.65
		30	−6.2	−2.06	1.37	3.12	2.11
		40	−8.51	−0.47	1.41	2.02	3.13
		50	−10.4	1.05	2.27	1.58	1.41
		60	−11.16	1.44	1.38	2.66	2.69
3	0.2	10	−0.52	−0.95	0.44	2.12	2.18
		20	−1.57	−1.64	1.84	1.68	2.16
		30	−2.79	−1.1	1.61	1.72	2.36
		40	−4.06	0.47	1.06	2.36	2.26
		50	−5.06	0.95	1.52	1.49	1.97
		60	−5.57	1.18	1.77	2.06	1.97
4	0.2	10	−0.41	−0.78	0.54	2.15	2.35
		20	−1.38	−1.43	1.78	1.78	2.29
		30	−2.51	−0.95	1.68	1.74	2.47
		40	−3.68	0.52	1.07	2.39	2.33
		50	−4.57	0.94	147	1.55	2.05
		60	−5.1	1.16	1.75	2.05	2.04

For very small delamination the change in frequencies are too small to enable detection of delamination. Through the present case studies the delamination of size above 20 % has been detected for both simply supported and cantilever beams.

For simply supported beam two locations (as shown by point A and B in Fig. 6) are predicted for each case which is because of its symmetric geometry and vibration modes. The delamination size of 20–60 % of beam length has been considered keeping one end fixed and varying the position of the other end for both simply supported and cantilever beams. For simply supported beam, the delaminations were spread unequally over the two halves of the beam for delamination size of 40 and 50 % of beam length. The predicted values of location and size are listed

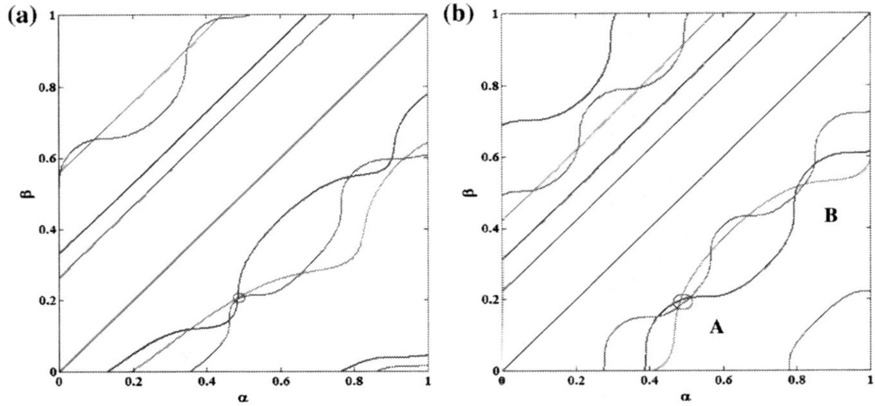

Fig. 6 Typical contour plot det(M) = 0 with **β** and **α** (Actual location: **β** = 0.2 and **α** = 0.5) for **a** cantilever and **b** simply supported beams

Table 3 Predicted delamination parameters for cantilever beam

Delamination parameters			Predicted parameters for known interface				Predicted parameters for unknown interface				
I	β	Size	β	% Error	Size	% Error	I	β	% Error	Size	% Error
1	0.2	20	0.205	0.49	20.1	0.13	1	0.205	0.49	20.1	0.13
		30	0.201	0.13	28.4	1.58	1	0.201	0.13	28.4	1.58
		40	0.223	2.34	39.1	0.91	1	0.223	2.34	39.1	0.91
		50	0.237	3.68	47.1	2.85	2	0.231	3.15	48.9	1.02
		60	0.244	4.37	54.6	5.39	1	0.244	4.37	54.6	5.39
2	0.2	20	0.205	0.51	20.1	0.11	1	0.209	0.86	19.5	0.53
		30	0.201	0.14	28.6	1.42	1	0.210	0.89	27.3	2.72
		40	0.224	2.38	39.1	0.87	2	0.224	2.38	39.1	0.87
		50	0.236	3.58	47.3	2.72	2	0.236	3.58	47.3	2.72
		60	0.243	4.34	54.8	5.13	2	0.243	4.34	54.9	5.13
3	0.2	20	0.205	0.50	19.6	0.43	1	0.256	5.58	11.2	8.82
		30	0.192	0.84	28.0	1.97	3	0.192	0.84	28.0	1.97
		40	0.203	0.31	39.8	0.23	3	0.203	0.31	39.8	0.23
		50	0.209	0.90	49.6	0.39	4	0.191	0.95	53.4	3.38
		60	0.194	0.64	59.9	0.09	3	0.194	0.64	59.9	0.09
4	0.2	20	0.204	0.42	19.6	0.42	1	0.257	5.68	10.7	9.31
		30	0.185	1.48	27.7	1.48	4	0.185	1.48	27.7	1.48
		40	0.202	0.23	39.9	0.23	4	0.202	0.23	39.9	0.23
		50	0.208	0.84	49.7	0.84	4	0.208	0.84	49.7	0.84
		60	0.193	0.65	60.0	0.65	3	0.174	2.64	57.6	2.37

Table 4 Predicted delamination parameters for simply supported beam

Delamination parameters			Predicted parameters for known interface				Predicted parameters for unknown interface				
I	β	Size	β	% Error	Size	% Error	I	β	% Error	Size	% Error
1	0.2	20	0.199	0.05	20.73	0.73	1	0.199	0.05	20.73	0.73
		30	0.190	1.00	29.95	0.05	2	0.184	1.60	30.88	0.88
		40	0.197	0.25	39.95	0.05	1	0.197	0.25	39.95	0.05
		50	0.203	0.31	50.51	0.51	2	0.196	0.38	52.27	2.27
		60	0.194	0.58	60.53	0.53	1	0.194	0.58	60.53	0.53
2	0.2	20	0.199	0.05	20.69	0.69	2	0.199	0.05	20.69	0.69
		30	0.189	1.04	29.91	0.09	2	0.189	1.04	29.91	0.09
		40	0.198	0.21	40.04	0.04	2	0.198	0.21	40.04	0.04
		50	0.203	0.33	50.51	0.51	2	0.203	0.33	50.51	0.51
		60	0.183	1.69	60.67	0.67	2	0.183	1.69	60.67	0.67
3	0.2	20	0.201	0.12	19.61	0.39	3	0.201	0.12	19.61	0.39
		30	0.196	0.43	29.99	0.01	4	0.190	0.99	31.87	1.87
		40	0.196	0.43	40.40	0.40	2	0.302	10.2	22.66	17.3
		50	0.205	0.50	50.59	0.59	3	0.205	0.50	50.59	0.59
		60	0.199	0.13	60.75	0.75	3	0.199	0.13	60.75	0.75
4	0.2	20	0.202	0.17	19.70	0.30	4	0.202	0.17	19.70	0.30
		30	0.197	0.31	30.19	0.19	4	0.197	0.31	30.19	0.19
		40	0.195	0.50	40.46	0.46	2	0.307	10.7	21.58	18.4
		50	0.205	0.47	50.62	0.62	4	0.205	0.47	50.62	0.62
		60	0.199	0.13	60.77	0.77	4	0.199	0.13	60.77	0.77

in Tables 3 and 4 for the cantilever and simply supported beams respectively. The maximum error in prediction of location and size in the case of cantilever beam is less than 5 and 6 % respectively whereas in the simply supported beam the error is less than 2 and 1 % respectively. Here the predicted location indicates the location of the left end of the delamination from the left support. When the interface (I) is also an unknown parameter the results showed that although the predicted location and size closely match with the actual values for most of the cases, the interface prediction is incorrect for few cases.

7 Conclusions

The modeling of free vibration of a laminated composite cantilever and simply-supported beams with a delamination has been presented. It is based on the coupled axial and transverse vibration through the "constrained model". This approach permits solution of both the forward and inverse problems. The following conclusions can be drawn from the case studies.

1. The analytical natural frequencies show very good agreement with the ANSYS result in the case of forward problem.
2. The changes in natural frequencies due to damage can be a basis of its possible detection.
3. The inverse problem has been solved for two classes of problems. In the first case the interface is taken as known; the error is less than 5 and 6 % in the prediction of location and size respectively for the cantilever beam. For the simply-supported beam the same errors are less than 2 and 1 % respectively.

In the second class of problem, the interface is unknown. In such situations, for most of the cases, the size and location is predicted correctly but the interface prediction is incorrect in some cases. This method can be applied to predict location and size of delamination with good accuracy.

References

1. Wang JTS, Liu YY, Gibby JA (1982) Vibration of split beams. J Sound Vib 84:491–502
2. Mujumdar PM, Suryanarayan S (1988) Flexural vibrations of beams with delaminations. J Sound Vib 125:441–461
3. Tracy JJ, Pardoen GC (1989) Effect of delamination on the natural frequencies of composite laminates. J Compos Mater 23:1200–1215
4. Shen MHH, Grady JE (1992) Free vibrations of delaminated beams. AIAA J 30:1361–1370
5. Shu D, Fan H (1996) Free vibration of bimaterial split beam. Compos B Eng 27:79–84
6. Luo H, Hanagud S (2000) Dynamics of delaminated beams. Int J Solids Struct 37:1501–1519
7. Lestari W, Hanagud S (1999) Health monitoring of structures: multiple delamination dynamics in composite beams. In: 40th structures, structural dynamics, and materials conference and exhibit, reston, virigina: American institute of aeronautics and astronautics, pp 1–10
8. Shu D, Della CN (2004) Vibrations of multiple delaminated beams. Compos Struct 64:467–477
9. Kim HS, Chattopadhyay A, Ghoshal A (2003) Dynamic analysis of composite laminates with multiple delamination using improved layer wise theory. AIAA J 41:1771–1779
10. Valoor MT, Chandrashekhara K (2000) A thick composite-beam model for delamination prediction by the use of neural networks. Compos Sci Technol 60:1773–1779
11. Chakraborty D (2005) Artificial neural network based delamination prediction in laminated composites. J Mater Des 26:1–7
12. Zhang Z, Shankar K, Tahtali M, Morozov EV (2011) Vibration monitoring of composite laminates with delamination damage. In: 18th international conference on composite materials, ICCM18, Korea, 2011

13. Ihesiulor OK, Shankar K, Zhang Z, Ray T (2013) Validation of algorithms for delamination detection in composite structures using experimental data. J Compos Mater 102:226–236
14. Zhang Z, Shankar K, Ray T, Morozov EV, Tahtali M (2013) Vibration-based inverse algorithms for detection of delamination in composites. Compos Struct 102:226–236
15. Nandwana BP, Maiti SK (1997) Modeling of vibration of beam in presence of inclined edge or internal crack for its possible detection based on frequency measurements. Eng Fract Mech 58:193–205

Investigation of the Effect of Biodiesel Blends on Fuel Injection Pumps Based on Vibration and Pressure Measurements

Ali Abruss, Yangchun Guo, Tie Wang, Fengshou Gu, Andrew D. Ball and D. Brown

Abstract Amongst alternative fuels for diesel-engine application, biodiesel is very attractive because it is biodegradable, an environmentally-friendly and sustainable source that can meet future energy demands. However, there are few published studies of the impact of biodiesel fuel and its blends on fuel injection pumps (FIPs). This study will investigate the influence of biodiesels derived from waste cooking oils with incremental blends of B10, B20, B30, B40 and B100. The FIP in this study is a rotary type attached to a four-cylinder, four-stroke direct injection, turbocharged diesel engine. Vibration and pressure measurements were made on the FIP. The results show the peak pressure close to the pump increases slightly the higher the proportion of biodiesel because of increased viscosity, density and bulk modulus of the fuel. Low frequency vibration increased as the proportion of biodiesel increased. These results demonstrate an increase in dynamic load on the pump components. However, high frequency vibration levels are lowest for the blends B10, B20 and B30, which may be helpful for improving the service life of the delivery.

Keywords Vibration · Fuel injection pump · Injection pressure · Biodiesel · Diesel engine

A. Abruss (✉) · Y. Guo · F. Gu · A.D. Ball · D. Brown
Centre for Efficiency and Performance Engineering, University of Huddersfield, Huddersfield, UK
e-mail: Ali.Aburass@hud.ac.uk

F. Gu
e-mail: F.gu@hud.ac.uk

T. Wang
Taiyuan University of Technology, Taiyuan, Shanxi, China

J.K. Sinha (ed.), *Vibration Engineering and Technology of Machinery*, Mechanisms and Machine Science 23, DOI 10.1007/978-3-319-09918-7_47

1 Introduction

Stringent emission laws, depletion of fossil fuels and public pressure have forced the motor industry to find alternatives to fossil fuels. Biodiesel fuel blends are one promising option currently being researched as a pathway to energy diversity and reduced petroleum dependence of the transportation sector. As shown in the latest studies [1, 2] the majority of research efforts are into improving engine combustion, engine performance and reducing emissions. However, various researchers have studied the characteristics of high pressure fuel injection systems and the timing of injection for fuels containing biodiesel. The fuel injection system when using biodiesel has an advancing injection timing and high injection pressure which can cause early combustion [2–6]. However, despite being a critical mechanism, FIPs for diesel engines have received little attention in terms of the dynamics of their operation. Obviously the physical properties such as density and viscosity of biodiesel are different from normal diesel fuel and these differences will affect not only the fuel supply characteristics but also the dynamics of the FIP. The latter being one of the decisive factors influencing FIP service life.

In this study, a vibration based approach is used to study the dynamics of FIPs used with diesel and different biodiesel blends. It commences by describing possible vibration sources in a FIP and then a systematic test study based on a diesel engine equipped with a rotary FIP. Finally, both the vibration signals and injection pressure signals are examined with different biodiesel blends under the same engine operation conditions so that the dynamic changes due to different fuel blends can be extracted.

2 Dynamics Rotary Fuel Injection Pump

A rotary distributor diesel FIP has a single plunger and barrel, where the plunger combines rotary and reciprocating movements by rotating the cam plate [7], as shown in Fig. 1. To achieve fuel injection, it must raise the pressure of the fuel sufficiently (several hundred bars) to rapidly open the delivery valve (micro-seconds). This allows the fuel to be forced into the injector nozzle at the correct time for fuel injection [9]. The valve impact may be reduced by the damping effect of the residual fuel between the seat and the valve body. The rotary and reciprocating movements will be accompanied by internal dynamic forces including that of fuel pressure in the barrel, the forces between the cam and roller, and also the recovery forces due to the spring. These forces will cause pump components and body to vibrate and possibly because of the high repetition rate may cause component resonance.

The impact of the delivery valve on the valve seat will cause the pump body to vibrate. These vibrations become stronger at higher engine speeds because of a higher rate of fuel delivery and could mean the character of the vibrations may

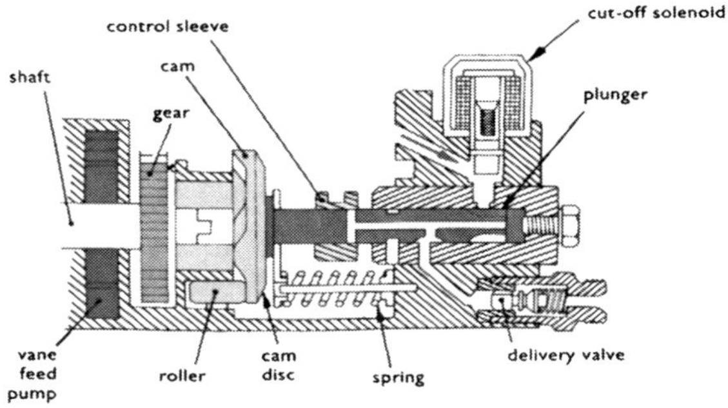

Fig. 1 Mechanical of a rotary distributor FIP of diesel engine [8]

change with different types of fuel and delivery rates. In particular, biodiesel has a high density and viscosity which will change the vibration excitation and system response [10–12]. Thus, it could be possible to use the vibration response to evaluate the changes of the internal dynamic effects due to different fuels.

3 Experimental Facilities and Test Procedures

3.1 Test Rig Setup

To identify the effects of biodiesel blends on the FIP, a four-cylinder, four-stroke, turbocharged, direct injection diesel engine was used for the experimental study. The test layout is shown in Fig. 2 and the specification of the diesel engine shown in Table 1.

The FIP on the engine is a rotary fuel injection pump. Its key specification is also provided in Table 1. The engine test rig is fully instrumented for in-cylinder pressure measurement, emission measurements and vibro-acoustics measurement. For this study, a dynamic pressure sensor is mounted on the fuel line close to the

Fig. 2 Test rig (diesel engine JCB 444 TCA)

Table 1 Specification of diesel engine and FIP

Diesel engine type	JCB 444 TCA74 kW
Number of cylinders	4
Compression ratio	18.3:1
Number of valves	16
Injection system	Direct injection
Displacement	4.399 l
Engine firing sequence	1–3–4–2
Maximum engine speed	@ 2,200 rpm
FIP type Delphi	DES /10130DEG
Number of injections	4
Pressure range of FIP	100–500 bar
Maximum speed of FIP	@ 1,100 rpm

pump, as shown in Fig. 3, for measuring the pressure changes of the FIP. An accelerometer is mounted on the body of the FIP body to obtain the desired vibration signal non-intrusively. For the experiments, the engine operated at two speeds: 1,200 and 1,600 rpm, each of which was under four increment loads: 100, 200, and 300 nm and to 400 nm. These settings were kept constant for different fuel blends to ensure accurate comparisons and hence obtain quantitative results of changes due to the use of different fuel blends.

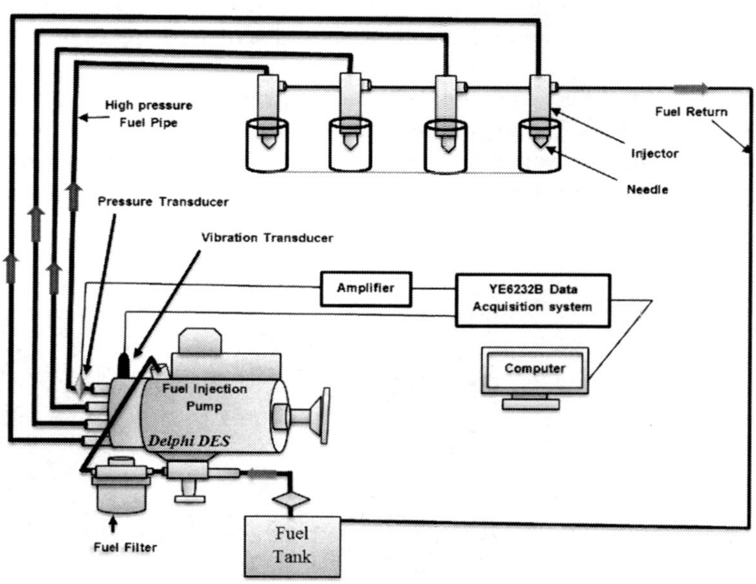

Fig. 3 Schematic diagram of the engine FIP

3.2 Biodiesel Fuel Physical–Chemical Properties

To perform the experimental study, a typical biodiesel from waste cooking oils was used as the base fuel and for creating different blends when mixed with standard diesel.

To obtain a better understanding of the effects of incremental increases in the relative volume of biodiesel blends B10, B20, B30 and B40 were also tested under the same engine operating conditions. Before the actual tests the density and viscosity of these blends are measured for more accurate comparison.

Figure 4 shows the measured densities increase with increase in proportion of biodiesel, which agrees with previously published literature [2, 12], however Fig. 5 shows that the percentage changes in viscosity are significantly greater, which may be a major factor influencing the fuel injection characteristics. Figure 5 also shows

Fig. 4 Density measurement results

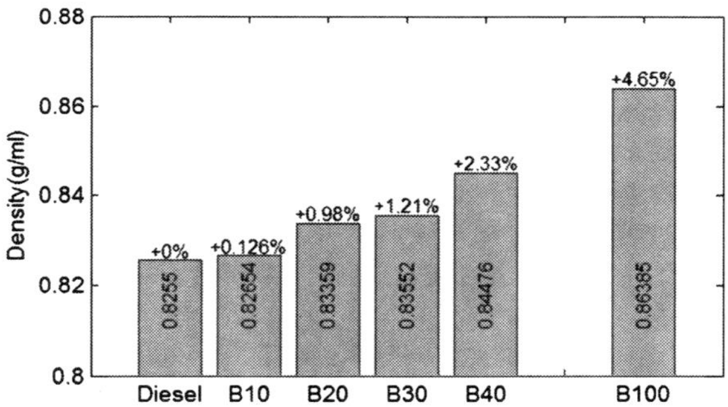

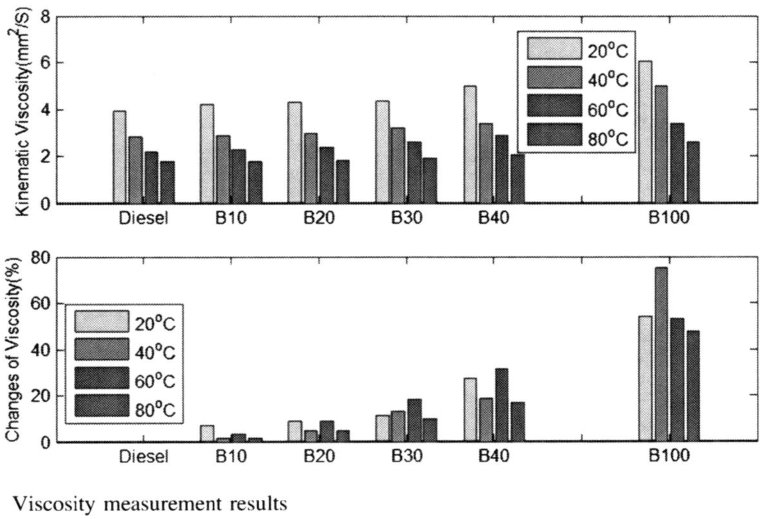

Fig. 5 Viscosity measurement results

that the relative changes are greater the higher the amount of biodiesel in the blend. These results agree with previously published literature [13, 14].

4 Results and Discussion

4.1 FIP Pressure Characteristics

Figure 6 shows the comparison of pressure waveforms obtained after 60 readings of pressure measured in the high pressure FIP under different operating conditions. It can be seen that the amplitude and duration of FIP pressure increases with both the load and the speed. However, when comparing biodiesel (B100) with standard diesel, the results show a clear increase in the amplitude and more advanced injection. In contrast, the changes for other blends cannot be observed as clearly.

In order to see the changes more accurately, three key fuel injection parameters maximum pressure, injection timing and injection duration are extracted from the waveform. Figure 7 presents the comparison of these parameters under different load and two speeds.

The results show that the peak pressure gets higher for all blends and is more significant at the lower speed due to the effect of the high viscosity under the interactions of prolonged time and at a low pressure. However, the increase is not proportional to the mixed proportions of biodiesel. This may be due to the nonlinear

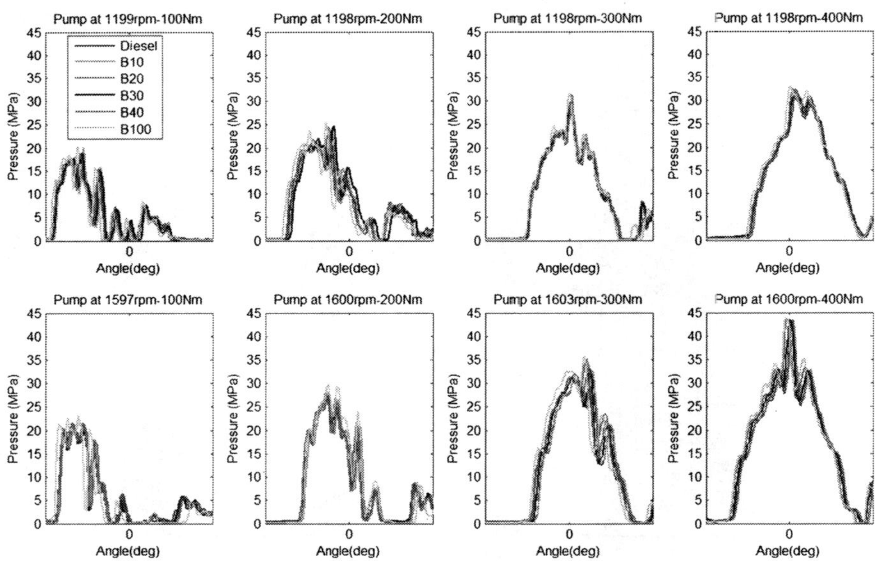

Fig. 6 Engine FIP pressures under different operating conditions

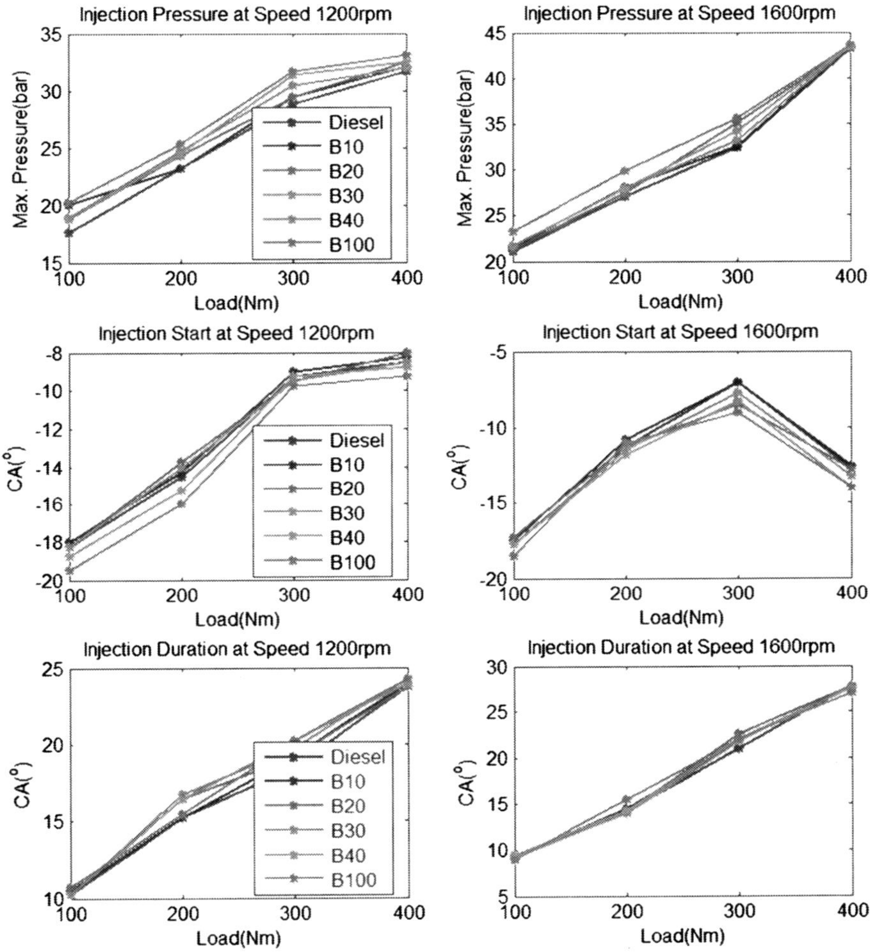

Fig. 7 Characteristics of key fuel injection parameters

effects of the fluid dynamics. For similar reasons, only higher proportions of biodiesel including B30, B40 and B100 show clearly the advanced injection whereas B10 and B20 show little change in the injection timing. In addition, the duration of the injection shows little change when compared to that of normal diesel. In general, it can be concluded that the higher proportion (B30, B40) of biodiesel alters the pressure supply characteristics, thereby influencing fuel injection quality and combustion. However, these changes are not as significant as shown in previous studies. Nevertheless, the increase in the peak pressure indicates a higher dynamic load applying to the pump and supplies lines, which may lead to the effects these components.

4.2 Engine FIP Vibration Characteristics

As mentioned in Sect. 2, the high-pressure impulses associated with the reciprocating forces of the plunger motion and the impacts due to opening and closing of the deliver valve may be the main causes of pump vibrations. These dynamic forces that generate the reciprocating motion also apply to the pump housing that generates vibration [8, 9]. Therefore, to see the influence of the fuel blends, the vibration signals are examined under different conditions as shown in Fig. 8.

Figure 8 shows vibration signals in association with the pump pressure. It can be seen that the vibration responses closely correlate with the pressure variations.

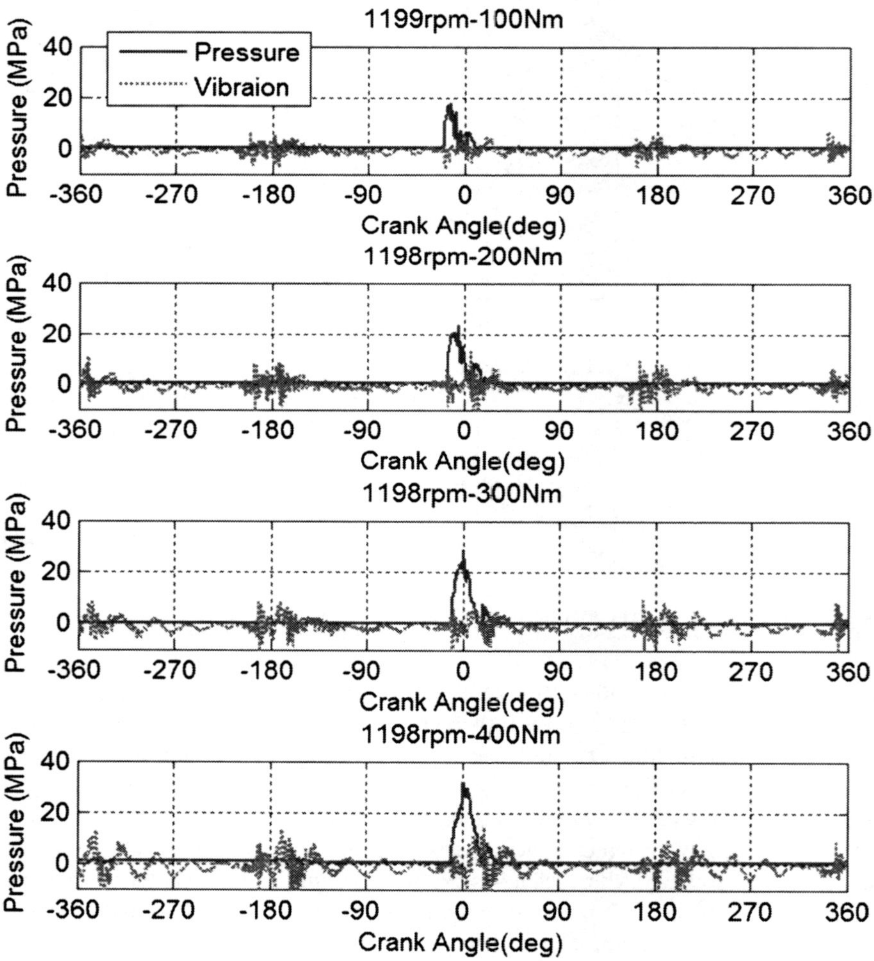

Fig. 8 FIP vibration signals under different load at 1200 rpm

Firstly, the high rate oscillations appear in the same period of the pressure pulse. In particular, there are two clusters of transient responses which correspond to the two impacts caused by the opening and closing of the delivery valve. Secondly, the amplitudes of the transient responses increase consistently with the pressure amplitude, showing that these vibrations could be an accurate indication of the amplitude of the pressure. Moreover, there are clear low frequency transient responses between two high frequency transients and their amplitude also increase with the pressure. These low frequency transients mainly come from the reciprocating forces due to the plunger, cam ring motions and associated transmissions. To gain a better understanding of the vibration, the signals are converted into the frequency domain. Figure 9a shows typical spectra for two different loads at the same speed. It shows that there are a string of harmonics in the spectrum due to the injection frequency which are calculated by

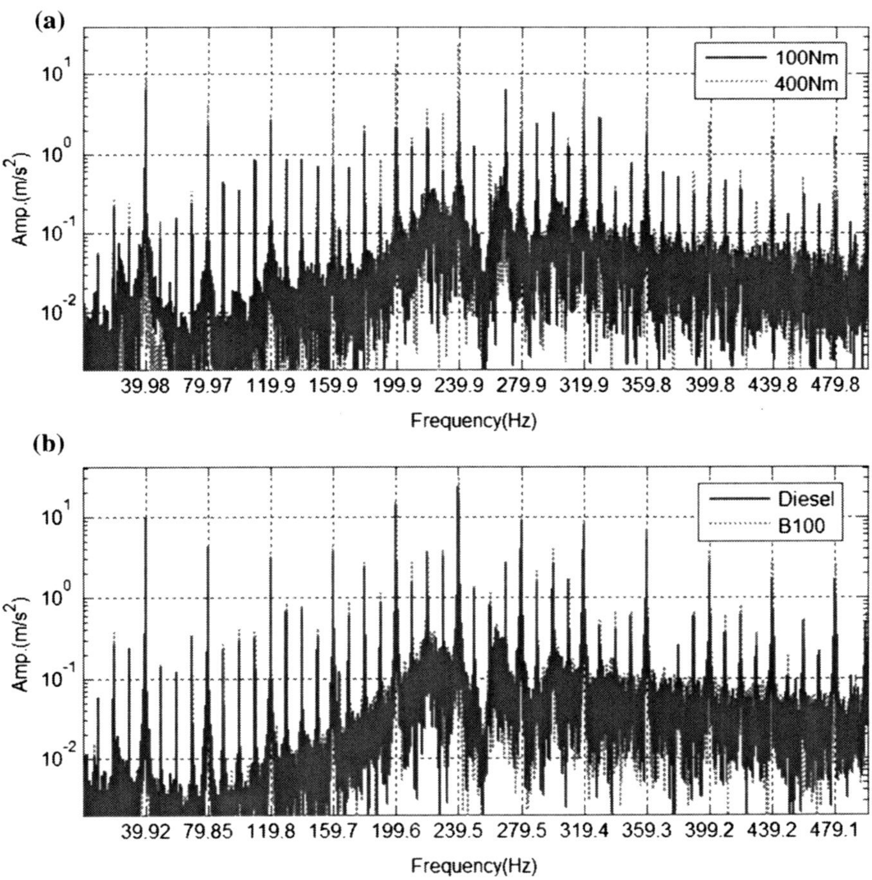

Fig. 9 Comparison of vibration spectrum between different cases. **a** Speed = 1199 rpm. **b** Load = 400 Nm and speed = 1198 rpm

$$f_i = \frac{n}{2 \times 60} \times 4 \tag{1}$$

where (n) is the engine speed in rpm. In addition, the amplitude increases clearly with the load. As these discrete components mainly relate to the reciprocating excision and its associated transmissions, their amplitude may be an effective indicator of the interaction between the mechanical motion and the fluid flow. Moreover, the spectrum also shows distinctive continuous contents in the high frequency range that also increase with load. As these contents relate more to the impulsive responses due to the impacts of the delivery valve, it can be a base for evaluating the valve motions. In addition, the spectrum comparison between the normal diesel and the biodiesel in Fig. 8b also shows clear differences. The spectral content of biodiesel at the injection frequency and its high order harmonics show visible increases due to high resistive loads caused by high density and high viscosity of biodiesel. The continuous contents also exhibit an increase for similar reasons. Although these changes are much smaller compared with that of the load effects in Fig. 9a, they show that the effect of density and viscosity on the FIP dynamics is similar to that of the load. Thus, the spectrum changes could be used to evaluate fuel quality.

As discussed above, there are two main sources of FIP vibration which exhibit in different frequency ranges. To show the effect of these two sources, the frequency contents below 400 Hz are averaged to obtain spectral amplitudes to represent the effects of the interaction due to reciprocating forces, whereas amplitudes are averaged between 400–2,000 Hz to illustrate the effects of the impact on the delivery valves.

Figure 10 shows the results of these two spectral values at two different speeds for the different fuel blends. It can be seen that for most load conditions the low frequency amplitudes for the biodiesel and its blends show a slight increase, compared with that of diesel. This shows that due to the additional load caused by the high viscosity and density of the biodiesel, more power is needed to drive the plunger and hence more dynamic loading to the system. For the high frequency amplitudes it shows that B20 and B30 have the lowest values. This means that high viscosity will reduce the impacts of the delivery valve and the vibration responses. However, as the viscosity becomes too high (such as B40 and B100), the effect of higher dynamic forces is more significant. This then results in a higher value of the high frequency amplitude. In general, it shows that lower proportions of biodiesel (B10, B20 and B30) will reduce the impact excitation and help maintain the contact surfaces of the delivery valve and potentially the nozzle valve in the injector.

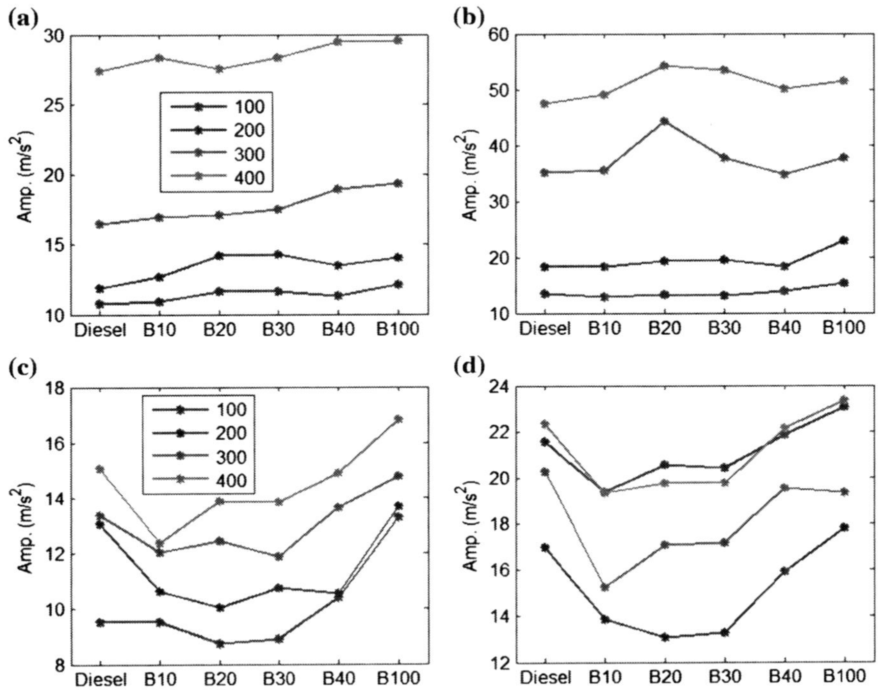

Fig. 10 Comparison of vibrations FIP between different biodiesel blends. **a** Low frequency at 1,200 rpm, **b** Low frequency at 1,600 rpm, **c** high frequency at 1,200 rpm, **d** high frequency at 1,600 rpm

5 Conclusion

This experimental study has confirmed that the injection pressure waveforms show higher peak value and advanced injection for higher proportions (B40 and B100) of biodiesels due to the effects of high density, high viscosity and bulk modulus. The high amplitude, especially, may indicate high dynamic loads upon the pump systems, which may in turn influence the service life of the pump. Furthermore, the vibration responses in the low frequency range also show higher amplitude for different biodiesel blends, which is consistent with the results from the pressure analysis. However, the high frequency vibrations due to the impacts of the delivery valves show a clear decrease in the smaller proportion biodiesel blends (B10, B20 and B30). This shows that these blends can be helpful for sustaining the contact surfaces of delivery valve with the high viscosity value.

Acknowledgment The authors would like to thank the Libyan culture attaché in London.

References

1. Xue J, Grift TE, Hansen AC (2011) Effect of biodiesel on engine performances and emissions. Renew Sustain Energy Rev 15(2):1098–1116
2. Tesfa B, Mishra R, Gu F, Powles N (2010) Prediction models for density and viscosity of biodiesel and their effects on fuel supply system in CI engines. Renew Energy 35:2752–2760
3. Yamane K, Ueta A, Shimamoto Y (2001) Influence of physical and chemical properties of biodiesel fuels on injection, combustion and exhaust emission characteristics in a direct injection compression ignition engine. Int J Engine Res 2(4):249–261
4. Radu R, Petru C, Edward R, Gheorghe M (2009) Fueling an DI agricultural diesel engine with waste oil biodiesel: effects over injection, combustion and engine characteristics. Energy Convers Manag 50(9):2158–2166
5. Ozsezen AN, Canakci M, Sayin C (2008) Effects of biodiesel from used frying palm oil on the performance, injection, and combustion characteristics of an indirect injection diesel engine. Energy Fuels 22(2):1297–1305
6. Kegl B, Hribernik A (2006) Experimental analysis of injection characteristics using biodiesel fuel. Energy Fuels 20(5):2239–2248
7. Eheim F (1982) Fuel injection pump for supercharged diesel internal combustion engines, in particular a distributor-type injection pump. U.S. patent no. 4,308,834. 5 Jan 1982
8. Edgar J Mechanical of A rotary distributor diesel fuel-injection pump type. http://www.autospeed.com/cms/article.html?&title=Our-Peugeot-Diesel-Part-4-Boost-and-the-Injection-System&A=109344
9. Tziourtzioumis D, Τζιουρτζιούμης Δ (2012) Experimental investigation of the steady state and transient operation of diesel engines fuelled by high percentage biodiesel blends. Ph.D. thesis
10. Crua C, Heikal M (2012) Investigation of the internal dynamics of diesel nozzles by time-resolved laser doppler vibrometry. In: ICLASS 2012. 12th triennial international conference on liquid atomization and spray systems, Heidelberg, Germany
11. Roosa VD (1978) Rotary distributor fuel injection pumps. U.S. patent no. 4,100,903
12. Ftoutou E, Chouchane M, Besbès N, Ouali R (2006) Injection fault detection of a diesel engine by vibration analysis
13. Alptekin E, Canakci M (2008) Determination of the density and the viscosities of biodiesel–diesel fuel blends. Renew Energy 33(12):2623–2630
14. Tat ME, Van Gerpen JH (2000) The specific gravity of biodiesel and its blends with diesel fuel. J Am Oil Chem Soc 77(2):115–119

Robustness Evaluation of Adhesively Bonded Ceramic Quad Flat Chips for Space Applications

Lassaad Ben Fekih, Georges Kouroussis, Christophe De Fruytier and Olivier Verlinden

Abstract Ceramic quad flat packs (CQFP) electronic components have long been used in power devices in the space industry. Soldering CQFPs to a printed circuit board (PCB) is not sufficient and thus the use of adhesives for reinforcement is compulsory. In fact, the risk of failure becomes very high during launch missions due to harsh constant and random accelerations. The paper investigates various distributions of two common thermally conductive and one epoxy adhesives. The intention is to compare the effects of these adhesive bonds on the robustness to vibrations of solder joints. Results are generated from a finite element (FE) model consisting of a CQFP component, an adhesive layer and a PCB. Key outputs such as the maximum deflection of the PCB, the stress in the solder joints and the fatigue lifetime of the assembly are discussed. Recourse to FE simulations allows competitiveness in terms of design cost and delivery time.

Keywords CQFP electronic component · Adhesive bonds · Space vibratory environment · Mechanical robustness

L. Ben Fekih (✉) · G. Kouroussis · O. Verlinden
Department of Theoretical Mechanics, Dynamics and Vibrations, Faculty of Engineering, University of Mons, 20-Place Du Parc, Mons 7000, Belgium
e-mail: lassaad.benfekih@umons.ac.be

G. Kouroussis
e-mail: georges.kouroussis@umons.ac.be

O. Verlinden
e-mail: olivier.verlinden@umons.ac

C. De Fruytier
Thales Alenia Space Belgium, 101, Rue Chapelle Beaussart, Mont-Sur-Marchienne 6032 Belgium
e-mail: christophe.defruytier@thalesaleniaspace.com

J.K. Sinha (ed.), *Vibration Engineering and Technology of Machinery*, Mechanisms and Machine Science 23, DOI 10.1007/978-3-319-09918-7_48

1 Introduction

The integration of surface mounted electronic components into space electronic boards drives the need to study their mechanical robustness and reliability. The industrial and academic teams are more concerned with mainstream electronics in use with computers, phones and cars [1]. The design of devices dedicated to space applications is subjected to stricter and more standardized specifications. This paper focuses on prospective enhancements of the mechanical integrity and durability of ceramic quad flat packs (CQFP) components. The latter are commonly used in flight power units. During their service life they are exposed to harsh vibrations with critical acceleration levels occurring mostly during the launch phase. The scope of this work differs also from studies such as [2], which explore the influence on the reliability of the architecture of the CQFP component, i.e. amongst other parameters its diagonal lead compliance, leads coplanarity and its dimensions. Here, a particular attention is given to second level interconnects (SLI) which stand for attachments of the electronic component to a printed circuit board (PCB). In simple words, our proposal is to evaluate the capacities of adhesive bonds in the reinforcement of solder joints in SLIs. Several forms of adhesive distributions can be found in the literature. The simplest solution in terms of the manufacturing process involves corner glue (CG) [3]. Pei et al. [4] studied the benefit of CG in alleviating the risk of thermal cycling for ball grid assembled (BGA) components. Meyyappan et al. [3] were also interested in BGAs. They concluded that CG mitigates the risk of failure in SLIs. Either by experience, using a loading pin acting on the PCB centre, or through finite element (FE) modelling, they spotted a prominent reduction of the PCB-substrate curvature together with the stress in solder joints using CGs. However, they cautioned, without giving any concrete evidence, against the loss of reliability in first level interconnects. Clementi et al. [5] were mainly interested in flip-chip encapsulation. They advocated the encapsulation of leads by stating that such a technique could reduce the failure resulting from solder wear out. Hongbin et al. [6] analyzed the drop test performance of pitch stackable very fine BGAs (PSvfBGAs) and figured out that UV-cured edge bonding, thermal-cured edge bonding and corner bonding are respectively about six, eight and seven times better than the PSvfBGAs without bonding. Yet the study of the impact of adhesive bonding on CQFP packages is still missing. In this paper, Sect. 2 is dedicated to a detailed description of a three dimensional FE model of a CQFP populated printed board. The next two sections are devoted to comparing the effects of three prospective adhesive bond strategies under quasi-static and random accelerations respectively. Before concluding, a final section is intended to guide the design engineer towards conveniently choosing his adhesives from a stiffness standpoint.

2 Finite Element Modelling of the Electronic Package

This work addresses the adhesive bonding of CQFP components to a PCB by different dispense modes of a silicone CV2946 adhesive, which is thermally conductive, and by a structural EC2216 epoxy glue. The CQFP component of interest counts 256 I/O leads disposed on four rows with a pitch (lead spacing) equal to 0.5 mm. The component area is square with an edge of 42 mm. The PCB has a 110×71 mm^2 rectangular shape and is meshed using ABAQUS S3R and S4R shell elements. CV2946 is intended to cover 80 % of the substrate bottom surface and EC2216 epoxy is to be injected in corners. Major difficulties arise from the choice of patterns for adhesives. The shape of the central adhesive was inspired from [7]. Its geometry and mesh details are illustrated in Figs. 1a, b. CGs are usually modelled by an L-shape pattern [8]. However, the CG appears as a dot on Fig. 2. Such a choice is made with the aim of preventing the glue from adhering to the leads due to the small vacant space near the corners. In this paper, we discuss a worst case made up of a rectangular PCB simply supported at its four edges. A CQFP component is intentionally placed at the PCB centre so that it experiences maximum bending moments. Material properties (Young's modulus E, Poisson ratio ν, density ρ) of different parts are reported in Table 1. We point out that the mechanical properties of the PCB and its damping ratio which is estimated to 2 % were obtained from a former experimental modal updating study [9]. Case studies to carry out are depicted in Fig. 3.

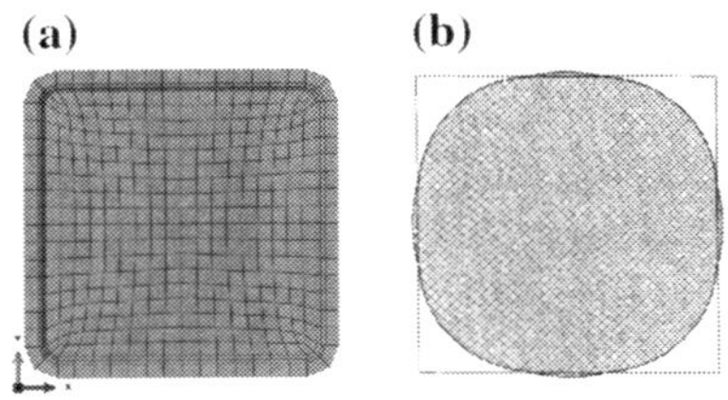

Fig. 1 Geometry and mesh of the central adhesive. **a** CV2946 adhesive (80 % coverage). **b** Typical distribution of adhesive beneath the substrate [7]

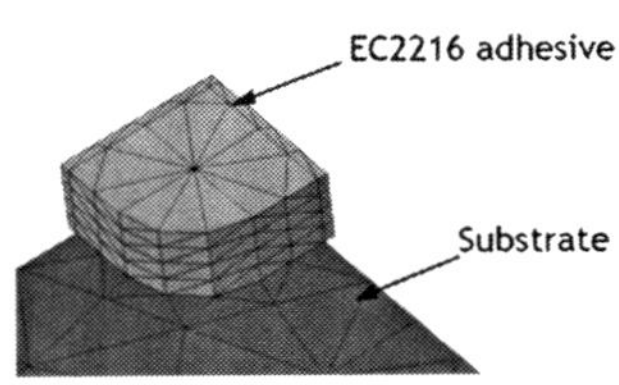

Fig. 2 Back view of a corner glue

Table 1 Material properties used

Components	Materials	E(GPa)	ρ (kg/m^3)	N
CQFP 256	Ceramic (A440)	310	3,800	0.23
	Kovar	138	8,360	0.317
	AgCu	110	8,930	0.35
Adhesives	CV2946	0.0207	1,530	0.49
	EC2216	0.999	1,300	0.46
Solder	Sn63Pb37	33.26	3,227	0.4
PCB	Polyimide copper	E_{xx} = 34.64 E_{yy} = 33.65	3,227	ν_{xy} = 0.49
		$G_{xy} = G_{xz} = G_{yz}$ = 7.55		

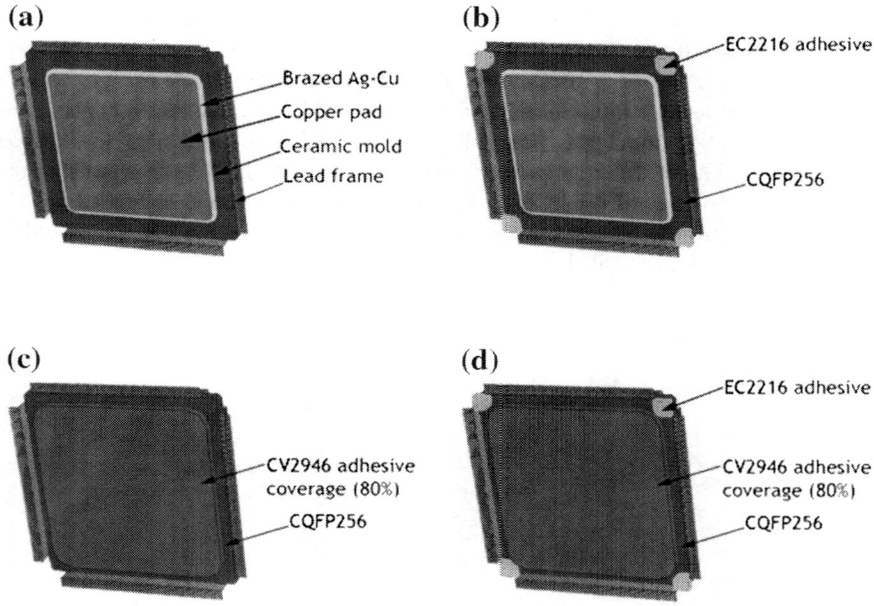

Fig. 3 Case studies of an adhesively bonded CQFP component. **a** Case study 1 (CS1): No adhesive. **b** Case study 2 (CS2): EC2216 CGs. **c** Case study 3 (CS3): CV2946 adhesive bond only. **d** Case study 4 (CS4): CV2946 adhesive bond+EC2216 CGs

3 Finite Element Analysis for Quasi-Static Accelerations

Our first interest was directed towards quasi-static accelerations. In such conditions an electronic board is strongly deflected along its out-of-plane direction due to acceleration levels usually greater than 100 g [10]. Besides, for space devices,

Table 2 Comparative study of different bonding strategies in terms of the mechanical robustness of the assembly

Case studies	Unit	CS1	CS2	CS3	CS4
VM_{max} in the lead	(MPa)	6.497	2.955	1.562	1.394
VM_{max} in the solder joint	(MPa)	0.860	0.347	0.556	0.250
VM_{max} in the PCB	(MPa)	0.411	0.593	0.510	0.675
Maximum PCB deflection	(μm)	6.518	5.509	4.616	3.387
VM_{max} in the EC2216 adhesive	(MPa)	n.a	0.2	n.a	0.148
VM_{max} in the CV2946 adhesive	(Pa)	n.a	n.a	15960	5479

a first-line mechanical design rule stipulates to not plastically deform materials and to stay within their elastic range. So for all FE simulations the assembly is subjected to a constant out-of-plane acceleration of 5 g, unlikely to cause the plastic work of any material. We attempt to locally retrieve indicators of the maximum equivalent Von Mises stress, VM_{max}, in main parts of the package as well as the maximum deflection of the PCB. In the space industry the Steinberg curvature criterion is a rule of thumb that allows validating or not an electronic board by only inspecting the ratio of its maximum deflection and its width [10]. Therefore, screening values of maximum deflections of the PCB makes sense. The list of critical materials includes in a decreasing severity the silicone adhesive characterized by a tensile strength of 1.3 MPa and 30% of maximum elongation followed by the epoxy glue and the Sn63Pb37 solder with 15 and 39.28 MPa yield strengths respectively. Thus, it can be directly observed that the solder joint is expected to be the first element to fail regardless of the case study. It appears in Table 2 that across CS2, CS3 and CS4, stress indicators were influenced by the use of adhesives and that the overall beneficial case is CS4. Except to stiffen the assembly, the deployment of only CV2946 seems of non-significant interest compared to CS2 where CGs dramatically contribute to alleviate stress levels in the vicinity of leads and solder rows. The relaxation of stress is obvious in Figs. 5b, d. It can also be noted that in CS4 the distribution of stress is less uniform around borders of the adhesive. This is due to the fact that the central adhesive hampers the flexure of the PCB. Therefore greater stress levels are transmitted to CGs and their beneath PCB areas.

Although all previous stress indicators are in the clear, an intriguing FE result is the possible bending of the leads against the substrate due to their short inter-distance (0.22 mm). This is due to the weak bending stiffness of the leads around the y-axis (Fig. 4). In the absence of an implemented contact law between the two parts in the FE model, it can only be seen from Fig. 4 a penetration of leads into the substrate. Within the scope of this study we are not able to comment on the potential penalty from this (Fig. 5).

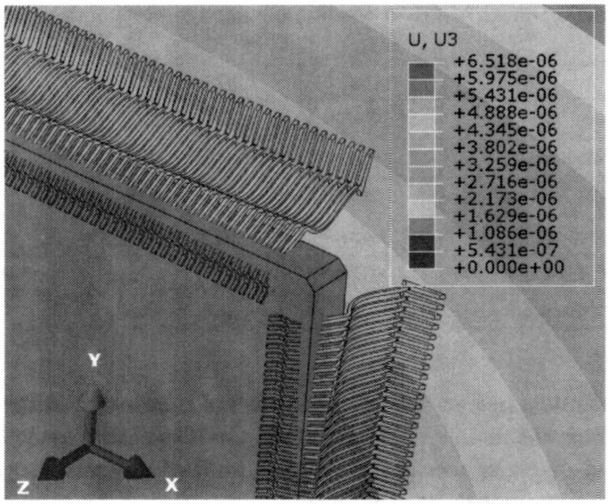

Fig. 4 Weakness of the leads under an out-of-plane acceleration of 5 g

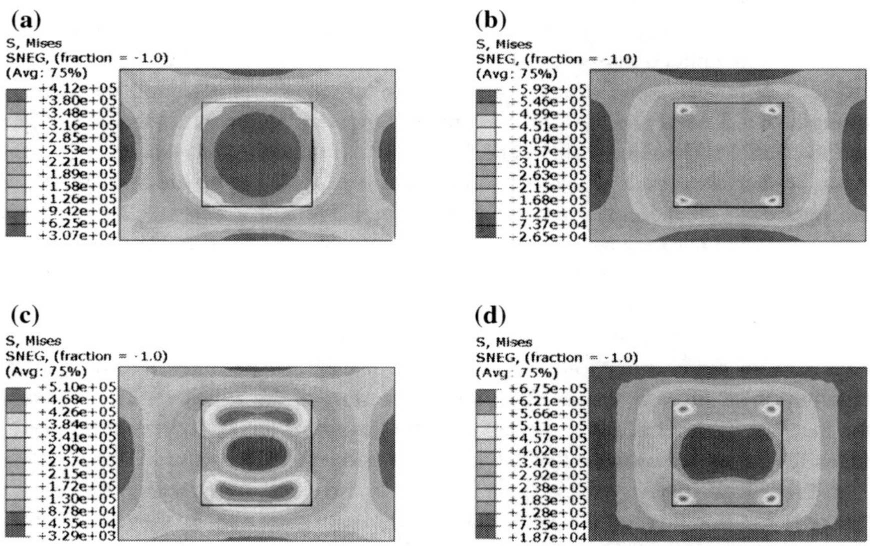

Fig. 5 Equivalent Von Mises stress distributions in the PCB. **a** CS1. **b** CS2. **c** CS3. **d** CS4

4 Finite Element Analysis for Random Accelerations

The random analysis was performed by applying the same excitation to all attachment nodes belonging to the PCB (base excitation). The input is defined in the frequency-domain by an acceleration power spectrum density (PSD) that emulates

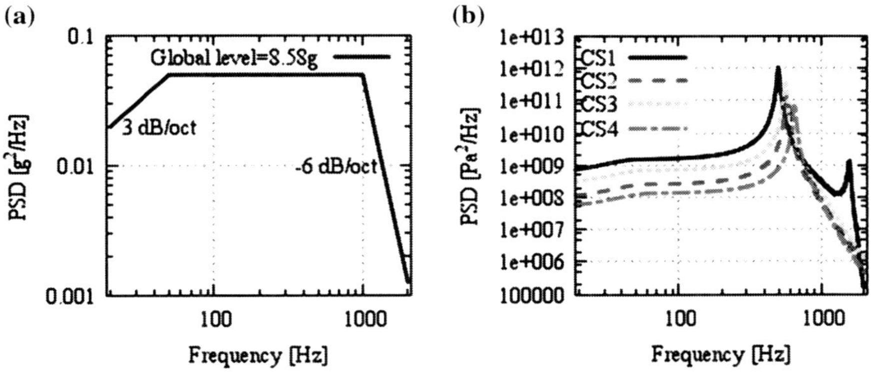

Fig. 6 Random vibration simulation of the CQFP package. **a** Input acceleration spectrum. **b** VM stress spectra in the most critical solder element

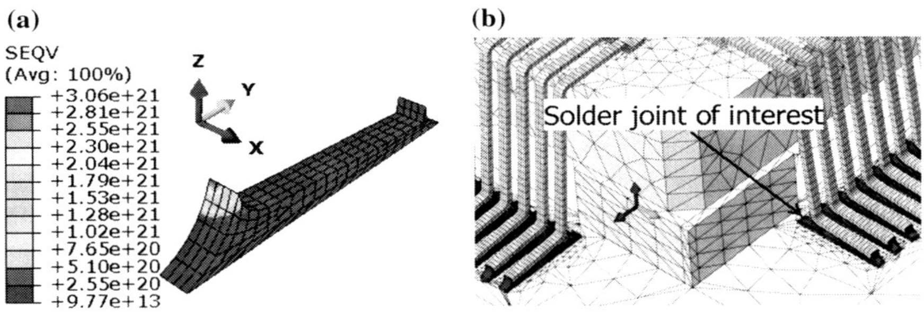

Fig. 7 VM stress in the critical solder joint at the resonance frequency. (SEQV = $2VM^2$). **a** Critical solder joint. **b** Location of solder joints in the package

the random vibratory environment of both satellites and launchers (Fig. 6a). The rectangular shape of the PCB induced that the most stressed solder rows are the ones parallel to the longest edge of the PCB (either top or bottom). However, the position of the critical element can be influenced by the adhesive bond strategy. Except in CS4, the critical element is found to be in the solder located in the top/bottom centre of the solder pattern. For all remaining cases, it is one amongst the nearest to the corners. In all cases, the maximum concentration of stress was always captured near the solder-lead interconnect (Fig. 7). Quantitatively, a standardized rainflow counting method is carried out [11]. The application of this method requires on the one hand a stress or strain history and on the other hand the knowledge of the high cycle fatigue (HCF) properties of the solder material. Specifically for the tin-lead alloy (Sn63Pb37), it was proven that the equivalent VM stress is a relevant estimator. Indeed, it leads to conservative results independently of the stress state which is actually multiaxial [12]. The HCF domain covers the linear part of the S-N (Wohler) curve which is comprised between 10^3 and 10^7 cycles. This segment is fitted in log-log scale by the Basquin's power law, $N_f \sigma_a^k = C$, where σ_a and N_f represent a given stress level and its corresponding number of cycles to failure respectively. Usual values of C and k for tin-lead solder equate 4.031 10^{33} Pa and 4 respectively [12]. The rainflow fatigue damage intensity $E_{RF}[D]$ is expressed by $E_{RF}[D] = \sum_{i=1}^{N} \sigma_a^{ik}/C$ where i denotes the ith rainflow extracted cycle. In practice, an IFFT algorithm is required to generate VM stress time sequences from PSDs (Fig. 6b). Note that the phase numbers are generated randomly. The fatigue lifetime is obtained by dividing the total exposure time set to 1 s by the damage intensity which is evaluated 1,000 times for different sets of phase vectors. Eventually we retain T_{RF} which is the mean of all computed lifetimes.

It emerges from Table 3 that without adhesive bonds the critical solder joint is mainly stressed under the action of two in-plane and one out-of-plane stresses which are (σ_{yz},σ_{zz}) and σ_{xx} respectively. So the most important issue is to track the evolution of these three components in function of the distribution of adhesives. It is obvious that the difference between CS1 and CS3 is minor as the percentage of enhancement never exceeds 24 %. This is consistent with previous findings of the quasi-static analysis that the CV2946 adhesive applied alone does not contribute to significant improvement. In contrast, the use of EC2216 CGs yields percentages of RMS stress reductions close to roughly 50 %. CS4, a combination of EC2216 CGs and 80 % coverage of CV2946, does not bring any large improvement compared to CS2 at this level. σ_{xx} should be taken into consideration as it reflects a tear out effort applied on the solder joint that could bring (short) adjacent leads closer. Fatigue lifetimes of the different cases are reported in Table 3.

Normally, the increase of the package stiffness through adhesive bonds leads to greater damage and consequently decreases fatigue lifetimes. However, Table 3 shows the opposite. The fatigue damage is more modulated by the aforementioned decrease of stress levels which outweighs the increase of the fundamental frequency of the package.

Table 3 Comparative study of the different bonding strategies in terms of the variance of stress components in the critical solder element and its fatigue lifetime

Case studies	CS1	CS2	CS3	CS4
σ_{xx}	0.25	0.13	0.19	0.23
σ_{xy}	0.031	0.012	0.028	0.029
σ_{xz}	0.122	0.0615	0.103	0.046
σ_{yy}	0.184	0.018	0.105	0.047
σ_{yz}	1.25	0.577	0.944	0.491
σ_{zz}	5.67	2.68	4.28	2.08
Lifetime (s)	1085	29734	4781	83731

5 Sensitivity Analysis: Effect of the Young's Moduli of Adhesives on the Fatigue Lifetime of Solder Joints

This section investigates possible improvements to the favourable case CS4. Licari et al. [13] reports Young's moduli of highly resilient adhesives ranging from 0.5 to 20 MPa. JEDEC J-STD-030 guideline [14] mentions typical moduli of underfill adhesives between 4 and 10 GPa. In linear dynamics simulations, the mesh size makes impossible to work with adhesives Young's moduli below 5 MPa. Considering these numbers, a design of experiments is performed by varying the moduli of central and corner adhesives starting from 5 MPa up to 10 GPa. We point out that the poisson's ratio and density of all adhesives are held equal to 0.46 and 1,500 kg/m^3 respectively. In total, 25 simulations under random accelerations are carried out. For a wide range of Young's moduli (2–10 GPa) (Fig. 8a), T_{RF} remains lower than 20 % of the maximum estimated lifetime, $T_{RF}^{\max}$, which is equal to 236177 s for a deployment of 1 GPa adhesives. Contrary to all expectations, the arrangement of two stiff adhesives is not an optimal solution. This result can be explained by the raise of the fundamental frequency of the package which counteracts the reduction of the stress level in the solder joint. Now, when turning to relatively resilient central adhesives (0.5–250 MPa), example of CV2946 silicone, the solder lifetime is found to be modulated mostly by the stiffness of the corners glue. In this situation the central adhesive is inefficient and CS4 comes down to CS2. Herein (Region 2 in Fig. 8a), most interest is for (4–5 GPa) corner adhesives where the lifetime is higher than 50 % of $T_{RF}^{\max}$. Note that the use of two highly resilient adhesives specifically lower than 250 MPa, provides values of lifetime below 50 % of $T_{RF}^{\max}$. Eventually, optimal solder lifetimes are reached with adhesives having similar values of Young's moduli and which are preferably included in the range (0.5–2 GPa) for both central and corner adhesives (Fig. 8b). In this way, the stress is uniformly distributed amongst the two adhesives behaving as parallel springs.

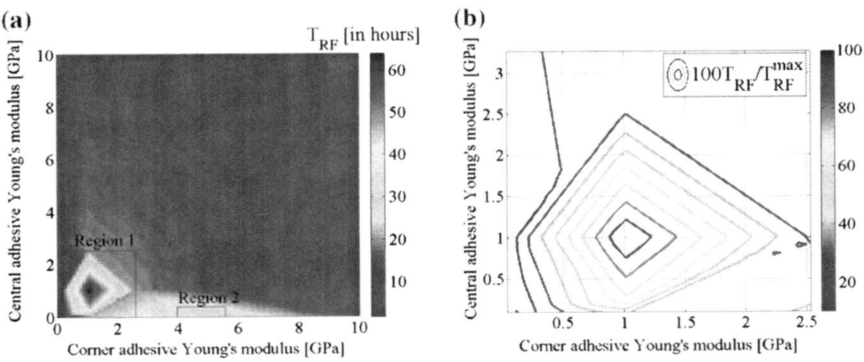

Fig. 8 Optimization of adhesives young's moduli in CS4. **a** Lifetime versus adhesives young's moduli. **b** Contour plot of the ratio of the lifetime per its maximum expressed in percentage (zoom in Region 1)

6 Conclusion

This paper demonstrated that adhesive bonds can play a positive role in enhancing the mechanical robustness of CQFP solder joints in a space vibratory environment. The percentage of improvement highly depends on the choice of the geometry and mechanical properties for the adhesive pattern. Under bending, it was assessed that optimal fatigue durability can be reached by the deployment of adhesives both into the corners and under 80 % of the area underneath the substrate. In such favourable distribution it is recommended to use adhesives of Young's moduli comprised between 0.5 and 2 GPa and to reduce the mismatch between the adhesives moduli. To decrease the lead-substrate contact, encapsulation of leads can be deemed a priori until a more elaborate solution is investigated. More work is to be performed for the experimental validation of our findings and for the study of other influential factors such as the geometry of the glue.

References

1. Gerke D, Kromann G, Srikantappa A (1998) Solder joint reliability of high I/O ceramic-ball-grid arrays and ceramic quad-flat packs in computer environments: the PowerPC 603 and PowerPC 604 Microprocessors. In: IEEE conference publications. doi: 10.1109/6144.814963
2. Kotlowitz RW (1990) Compliance metrics for surface mount component lead design. In: IEEE conference publications. doi: 10.1109/ECTC.1990.122316
3. Meyyappan K, McAllister A, Kochanowski M, Hsu I (2008) Effects of glue on the bend performance of flip chip packages. In: IEEE conference publications. doi: 10.1109/TCAPT.2008.922006
4. Pei M, Han R, Ge Y, Goyal S, Mukadam M (2013) The effect of corner glue on bga package temperature cycling performance: a modeling study. In: IEEE conference publications. doi: 10.1109/ECTC.2013.6575769
5. Clementi J, Carden T, Engle S (1994) Epoxy encapsulation on ceramic quad flat packs (CQFP's). IEEE conference publications. doi: 10.1109/ECTC.1994.367617
6. Hongbin S, Toshitsugu U (2011) Effects of edge and corner bond adhesives on the drop reliability of package-on-package bottom package assemblies. In: IEEE conference publications. doi:10.1109/ICCRD.2011.5764227
7. Tegehall PE, Dunn BD (2001) Evaluation of thermally conductive adhesives as staking compounds during the assembly of spacecraft Electronics. ESA STM-265
8. Hsieh G, McAllister A (2005) Flip chip ball grid array component testing under board flexure. In: IEEE conference publications. doi: 10.1109/ECTC.2005.1441385
9. Wattiaux D, Verlinden O, De Fruytier C (2009) Modelling of the dynamic behaviour of electronic boards used for spatial applications. In: 8th National Congress on Theoretical and Applied Mechanics, Belgium
10. Steinberg DS (2000) Vibration analysis for electronic equipment, 3rd edn. Wiley, New York, pp 45–56
11. Ben Fekih L, Kouroussis G, Verlinden O (2013) Spectral-based fatigue assessment of ball grid arrays under aerospace vibratory environment. Key engineering materials. doi:10.4028/www.scientific.net/KEM.569-570.425

12. Ben Fekih L, Kouroussis G, Verlinden O (2013) The applicability of multi-axial fatigue approaches using spectral analysis only. In: Proceedings of the 14th international conference on civil, structural and environmental engineering computing. Cagliary, Italy
13. Licari J, Swanson DW (2011) Adhesives technology for electronic applications: materials, processing, reliability, 2nd edn. William Andrew Elsevier, Massachusetts, p 133
14. JEDEC J-STD-030 (2000) Guideline for selection and application of underfill material for flip chip and other micropackages. IPC, Northbrook

Coupling Analysis for the Thermo-Acoustic-Vibration Response of a Thin-Walled Box

Yan Yang and Chen-Wu Wu

Abstract Monitor and control unit is essential to almost all of the modern aircrafts or spacecrafts, which always places in a relatively closed space and requires a compatible mechanical environment to ensure its normal function. The acoustic, vibration and thermal loads developed in the operation of the aircraft determine a complex environment for its monitor and control unit. In particular, there should be couple effect among the thermal, vibration and acoustic responses. This article presents numerical simulation studies on a thermo-acoustic-vibration response of a typical thin-walled box, to reveal the multi-physics environment of the monitor and control unit. The thermo-structural equations were proposed and solved numerically, following that the vibro-acoustic analysis was implemented. The coupled algorithm was developed to simulate the structure response and sound distribution, and predict the acoustic loss during transmitting through the box shell. The validation cases of the vibration and acoustic response of a plate were performed at first. The outcome provides the multi-physics environment prediction method for designing and optimizing the monitor and control unit.

Keywords Thermal · Vibration · Acoustic · Multi-physics environment

1 Introduction

Acoustic-vibration is an important coupled mechanical environment for modern aircrafts or spacecrafts, especially for the monitor and control unit, which comprises precision instruments and behaves sensitive to strong vibration and noise. In gen-

Y. Yang
LHD, Institute of Mechanics, Chinese Academy of Sciences, Beijing 100190
People's Republic of China
e-mail: yangyan@imech.ac.cn

C.-W. Wu (✉)
Institute of Mechanics, Chinese Academy of Sciences, No.15 Beisihuanxi Road, Haidian District, Beijing 100190, People's Republic of China
e-mail: chenwuwu@imech.ac.cn

J.K. Sinha (ed.), *Vibration Engineering and Technology of Machinery*, Mechanisms and Machine Science 23, DOI 10.1007/978-3-319-09918-7_49

eral, vibro-acoustic responses experiments should be performed in the design and development stages [1]. For decades, numerical simulation and prediction techniques have been developed increasingly to help the design and optimization of the mechanical environment of spacecrafts devices. As to vibro-acoustic analysis, coupled finite element method (FEM) and boundary element method (BEM) has been widely used, in order to reduce the computation cost.

As known, when the environmental temperature changes, the structure property will change, in particular the elastic modulus of the materials, which affects the material properties and vibro-acoustic behavior of the structure [2]. Not much work has been done in the thermo-vibro-acoustic analysis for multiple physics. Jeyaraj et al. [2] studied the vibration and acoustic response of a plate in a thermal environment subjected to a harmonic load, in which the thermal stress effect is included in a so called geometric stiffness matrix without considering the softening of the materials due to temperature elevation. Geng and Li [3] studied the vibration and acoustic radiation characters of an isotropic rectangular thin plate under thermal environments, in which the thermal structure stress is assumed to be the only contribution made by the thermal environment.

In this study, a thermo-vibro-acoustic analysis was implemented to investigate the multi-physics environment of a typical thin-walled box model, representing the monitoring and control unit of a spacecraft. Firstly, the thermo-structural equations were proposed and solved numerically with finite element method (FEM). Secondly, the vibro-acoustic analysis was made by using boundary element method (BEM).

2 Methods

The numerical simulation was performed, by using FEM and BEM through below procedure. In detail, the temperature and thermal stress are calculated by FEM for every thermal equilibrium state with considering the thermal softening of the materials. The vibration and acoustic response are computed by BEM for typical thermal equilibrium state.

2.1 *Thermo-Vibro-Acoustic Equation and Numerical Methods*

The equation of motion for a prestressed structure with the sound pressure load is as follows;

$$[\mathbf{M}]\{\ddot{\mathbf{U}}\} + [\mathbf{C}]\{\dot{\mathbf{U}}\} + [\mathbf{K} + \mathbf{K}_\sigma]\{\mathbf{U}\} = \{\mathbf{F}^\mathbf{t}\} + \{\mathbf{R}_s\mathbf{p}\} \tag{1}$$

where [**M**] is the inertial matrix, [**U**] is the displacement vector,[**C**] is the structural damping matrix, [**K**] is the temperature-dependent stiffness coefficient matrix, [$\mathbf{K}_\sigma$] is the increment stiffness coefficient matrix induced by thermal stress, [$\mathbf{F}^t$] is the applied load vector (supposed time harmonic), **p** is the acoustic pressure, and $\mathbf{R}_s$ is the coupled matrix describing acoustic pressure acting upon the structure.

For harmonic solution when $\mathbf{F}^t(t) = \mathbf{F}_s e^{j\omega t}$, assuming $\mathbf{U}(t) = \mathbf{U}e^{j\omega t}$, the Eq. (1) turns into,

$$([\mathbf{K} + \mathbf{K}_\sigma] + j\omega\mathbf{C} - \omega^2[\mathbf{M}])(\{\mathbf{U}\}) + \mathbf{L}_c\{\mathbf{p}\} = \{\mathbf{F}_s\} \tag{2}$$

where $\mathbf{L}_c$ is the vibro-acoustic coupled matrix, $\mathbf{F}_s$ is the harmonic amplitude of the force vector.

The basic governing equation of sound wave is Helmholtz equation (in frequency domain), assuming the sound wave is harmonic,

$$\nabla^2 p - k^2 p = -j\rho_0\omega q \tag{3}$$

where p is acoustic pressure, $k = \omega/c = 2\pi f/c$ is wave number, ρ_0 is the density of the fluid, and q is the sound source. The boundary element method for acoustic problem is written as,

$$\rho_0\omega^2[\mathbf{G}]\mathbf{u_f} + [\mathbf{H}]\mathbf{p} = \mathbf{F}_a \tag{4}$$

where **H** and **G** are boundary integral influence matrices, $\mathbf{F}_a$ is the sound source vector, **p** and $\mathbf{u_f}$ are the acoustic pressure and displacement of the boundary element nodes of the fluid. The coupled FEM/BEM is implemented to solve the structural-acoustic problem, using ANSYS and SYSNOISE respectively. First the pre stressed harmonic analysis is performed in ANSYS, the result file is imported into SYSNOISE to find the acoustic-structural response, where the indirect harmonic BEM is used.

2.2 The Model

The model was set up as a thin-walled box, as Fig. 1 shows, with geometry $0.5 \times 0.3 \times 0.2$ m^3, with thickness 2.5 mm. The material is specified as steel, which is herein modeled as isotropic and ideal elastic with Young's modulus $E = 200$ GPa, Poisson's ratio 0.3, mass density 7800 kg/m^3, thermal expansion coefficient 1.0×10^5 under ambient temperature condition [4]. The acoustic medium is modeled as air, with density 1.21 kg/m^3 and sound speed 340 m/s. The fluid material properties are supposed to be temperature-independent in this study. Without loss of generality, one normally incident harmonic plane sound wave, with sound pressure level 140 dB, is imposed upon the top of the box. In this study, there is no force excitation, in order to focus only on the acoustic environment.

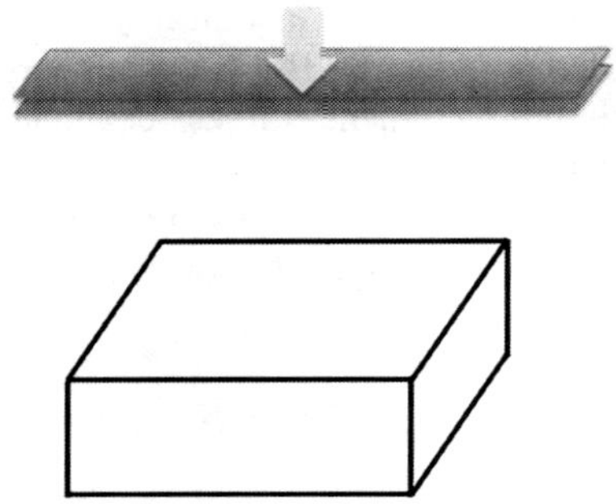

Fig. 1 The sketch of the model

Two cases, denoted by 'cool' and 'hot', where two values of Young's modulus determined by the material temperature are specified, that is, in the hot case the material is supposed to has 70 % Young's modulus than that of the cool one, i.e., 140 GPa in magnitude. Such stiffness reduction is always related to the temperature elevation of the materials, for which the linear approximation of elastic modulus dependent on temperature is assumed [4].

The volume element mesh of FEM model is matched with the shell element mesh of BEM, where the shell element size is set as 10 mm, and the thickness of volume element is equal to the thickness of the box, 2.5 mm. Such mesh size can resolve the acoustic analysis up to 5666 Hz.

3 Results and Discussion

The natural modes. The four corner points in the bottom wall plate of the box are fixed in computation. The first 25 modes were analysis with FEM. Table 1 shows the first 5 and the 25th natural frequencies of the box in cool and hot conditions. It is clear that the natural frequencies of the box reduce with increase in temperature. The first 2 mode shapes are shown in Fig. 2.

Vibro-acoustic response characteristics. The vibro-acoustic response may be represented by the acoustic field response at the center point of the box. Figure 3 is the sound pressure level at the center point (SPL_{center}), and the transmission loss (TL) here is defined as,

$$TL_{(1)} = SPL_{in} - SPL_{\mathrm{center}} \tag{5}$$

Table 1 Natural frequencies of the box under different ambient temperature

Natural frequencies (Hz)	Mode 1	Mode 2	Mode 3	Mode 4	Mode 5	Mode 25
Cool	179.3	299.0	391.5	394.8	435.9	1526.1
Hot	150.0	250.1	327.5	330.3	364.7	1276.8

Fig. 2 The mode shapes, Mode 1, 2, 3, 4, 5 and 25 (a–f), of the box under different ambient temperature

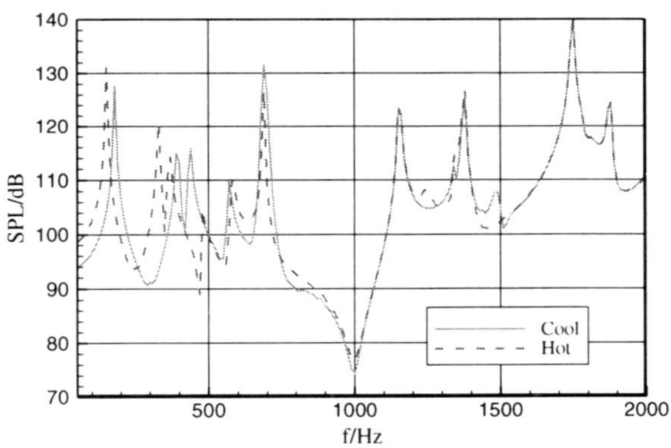

Fig. 3 The acoustic response, the sound pressure level ($p_{ref} = 2.0 \times 10^{-5}$ Pa) at the center point of the box

where SPL_{in} is the sound pressure level of the incident sound wave, which is 140 dB (Fig. 4).

The result shows that, around every natural frequency of the structure, the sound transmission would appear a corresponding peak. The transmission loss of the 'hot' box is slightly less than that of the 'cool' one. However, in higher frequency bands, above about 1500 Hz, such difference would disappear.

The acoustic response in the middle-'Z' section is given, as Fig. 5 shows. The sound wave transmits into the box;the acoustic field in the box appears well-distributed in the box, in the region near the wall, the sound pressure level decreases soon.

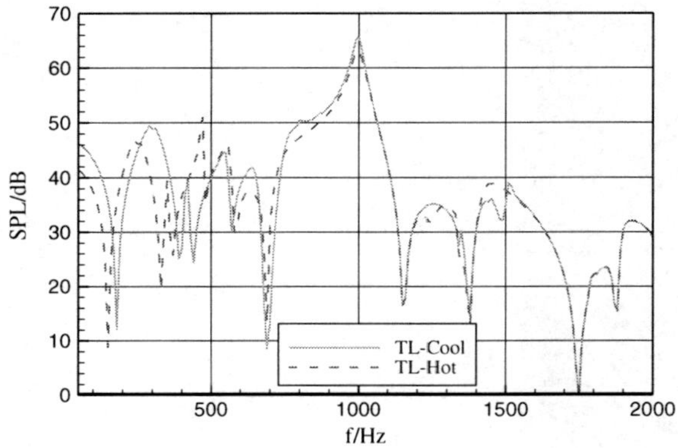

Fig. 4 The transmission loss, $TL_{(1)}$ of the box

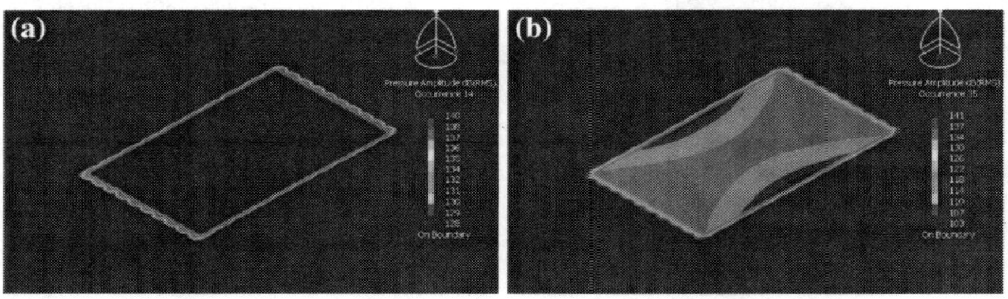

Fig. 5 The sound pressure level contour in the middle-'Z' section, the cool case at 180 Hz (**a**) and 430 Hz (**b**)

Furthermore, the vibration response can be found, shown in Fig. 6. It is seen that the displacement response shape is similar with the natural mode shape at the same frequency, by comparing Figs. 6 and 2. Thus, the resonance effect was numerically simulated.

The sound power on the middle-'Z' section, P_{mid}, is obtained, thus the transmission loss (TL) can be computed in another expression,

$$TL_{(2)} = 10\log(P_{\text{in}}/P_{\text{mid}}),\ \text{where } P_{\text{in}} = \frac{p_{\text{rms}}^2}{\rho_0 c} S \tag{6}$$

and S is the area of the top face of the box, and P_{in} is the sound power of the incident sound wave. Figure 7 gives the sound power level and the transmission loss, $TL_{(2)}$, according to formula (6). The result shows that, the difference caused by temperature is not so much. However, the qualitative property of the transmission loss is the same, comparing with the result of $TL_{(1)}$, that is, the transmission loss of the 'hot' box is a little less than that of the 'cool' one and the valley frequencies is

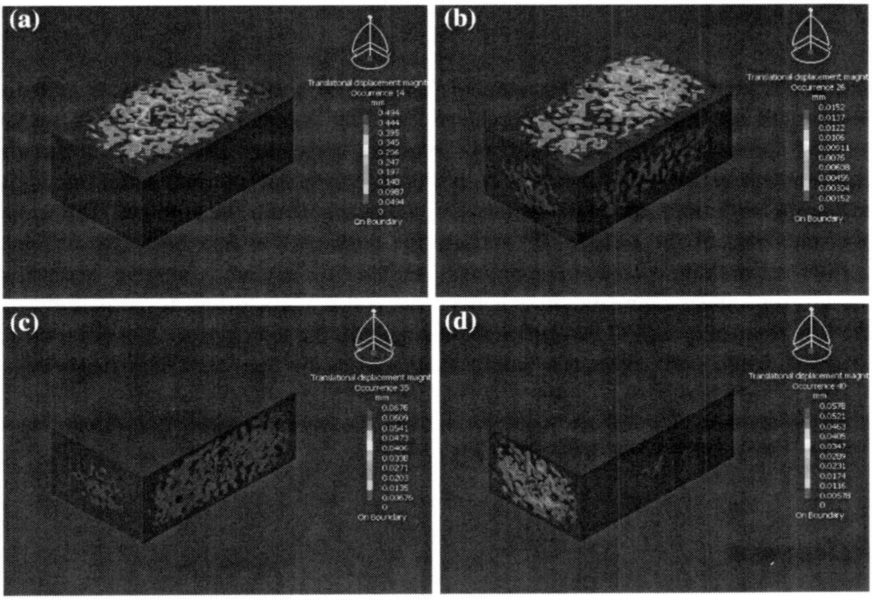

Fig. 6 The displacement response magnitude contour, on the inner surface of the 'cool' box, at 180 Hz (**a**), 300 Hz (**b**), 390 Hz (**c**) and 440 Hz (**d**)

Fig. 7 The transmission loss, $TL_{(2)}$ (defined by Eq. 6) of the box

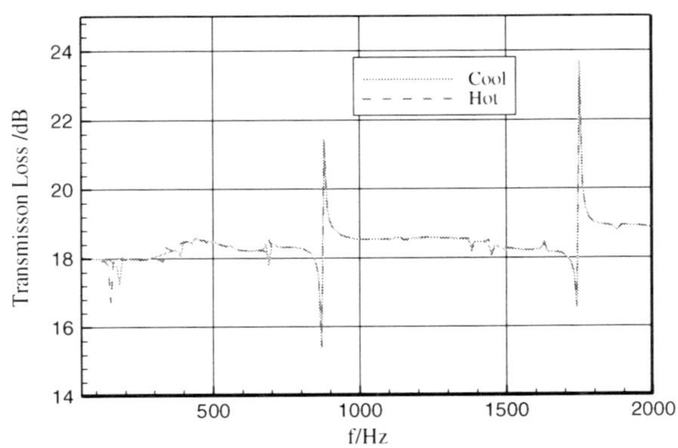

accord with the natural frequencies of the structure. Besides, in higher frequency bands, above 700 Hz around, the difference tends to disappear.

For the center point sound pressure level, as formula (5) defined, the transmission loss ($TL_{(1)}$) is much greater than that defined by (6) ($TL_{(2)}$), because the center point is a sensitive point, where the vibro-acoustic coupling is strong, than the points near the inner wall of the box and $TL_{(2)}$ reflects the total vibro-acoustic coupling effect.

4 Conclusion

The paper gives a thermo-vibro-acoustic analysis of a thin-walled box, in order to investigate the multi-physics characteristics. The coupled FEM/BEM is implemented to analyze the structural and acoustic response. The effect of thermal environment is studied to consider both thermal stress and thermal softening of the materials, with only one plane acoustic wave exerted from the ambient. The sound pressure level of the middle-'Z' section, the transmission loss with two different definitions, and the vibration response on the inner surface are computed, to observe the vibro-acoustic characteristics. It is revealed that the acoustic responses around the low frequency would be obviously changed by the temperature elevation of the material, while such difference would disappear in the region of high frequency.

Acknowledgement The authors would like to thank the financial support by National Natural Science Foundation of China under Grant No. 11332011.

References

1. Hughes WO, McNelis ME (2002) Recent advances in vibro acoustics. NASA/TM-2002-211697
2. Jeyaraj P, Padmanabhan C, Ganesan N (2008) Vibration and acoustic response of an isotropic plate in a thermal environment. J Vib Acoust 130(5):051005
3. Geng Q, Li Y (2012) Analysis of dynamic and acoustic radiation characters for a flat plate under thermal environments. Int J Appl Mech 4(3):1250028
4. Zhang ZX (ed) (1989) Handbook of enginery metal materials. Weapon Industry Press, Beijing (in Chinese)

Chatter Detection in CNC Milling Processes Based on Wiener-SVM Approach and Using Only Motor Current Signals

M. Lamraoui, M. El Badaoui and F. Guillet

Abstract The CNC milling is one of the most common processes in modern manufacturing, which characterized by highly nonlinear behavior and chatter problems. As other complex processes, Chatter detection in this situation is a crucial step for improving surface quality and reducing both noise and rapid wear of the cutting tool. This paper proposes a new methodology for chatter detection in computer numerical control milling machines. The originality of this method consists for using only motor current signals picked from electrical cabinet of machine and artificial intelligent. The methodology can be decomposed into four general tasks: (1) data acquisition, (2) signal processing, (3), features generation and selection, (4) classification. In signal processing task, electrical signals are resynchronized according to the electrical cycle (60 Hz) by exploiting the cyclo-stationarity of electrical signals through their cyclic statistics. After that, synchronous average is computed and subtracted from original signals in order to obtain the residual part. Wiener filter is then applied on residual signals by taking as reference the residual electrical signals acquired in spindle free rotation. This procedure allows estimating a signals corresponding to the electrical part and extracting the mechanical part, which linked to a chatter phenomenon and cutting mechanism. The signal processing task is primordial in order to decrease the dynamics of the electrical fundamental component and its harmonics and also to increase the contribution of mechanical parts. Extracted features computed from mechanical parts are then ranked based on theirs entropies in which only best features are selected and presented to the system for classification. At the classification step, the selected features are classified into two classes: stable and unstable utilizing a support vector

M. Lamraoui (✉) · M. El Badaoui · F. Guillet
Université de Lyon, Université de Saint Etienne, Jean Monnet, 42000 Saint-Etienne, France
e-mail: lamraoui.mourad@univ-st-etienne.fr

M. El Badaoui
e-mail: mohamed.elbadaoui@niv-st-etienne.fr

F. Guillet
e-mail: guillet@niv-st-etienne.fr

M. Lamraoui
LASPI, IUT de Roanne, 42334 Roanne, France

J.K. Sinha (ed.), *Vibration Engineering and Technology of Machinery*,
Mechanisms and Machine Science 23, DOI 10.1007/978-3-319-09918-7_50

machine (SVM). The intelligent chatter detection has accuracy above 96 % for the identification of cutting state after being trained by experimental data. The results show that it is possible to monitor chatter behavior in milling process by using motor current signal.

Keywords Chatter · Condition monitoring · Milling process · SVM · Current signals · Detection

1 Introduction

Chatter is a self-excited vibration that can arise in machining process at specific combinations of cutting parameters, depth of cut and spindle speed. This phenomenon affects the surface finish resultant from strong vibrations of the cutter. Furthermore, the machine, tool and bearings spindle wear out rapidly and a lot of noise is generated when chatter occurs. In machining field, chatter phenomenon is investigated extensively because manufacturing enterprises are turning to the automation system and the development of reliable and robust monitoring system for providing increased productivity, improved part quality and reduced costs [1]. Moreover, chatter monitoring is not an easy task for various reasons. Firstly, the non-linearity of machining processes and the time-varying of systems complicate this task. Secondly, the sensitivity and the dependency of acquired signals from sensors on different factors, such as machining condition, cutting tool geometry and workpiece material [2]. For these reasons, advanced automatic chatter detection and monitoring system is needed for optimizing and controlling machining processes and this becomes a topic of great interest. Advanced monitoring and detection methods are developed mostly relying on time, frequency and time-frequency analysis, etc. Insperger [3] analyzed system stability by modeling machine tool and developing differential equations. Smith and Tlusty [4] proposed the peak-to-peak force method to identify the limits of stability. Lamraoui et al. exploited the information of spindle encoder to compute the instantaneous angular speed signals and to detect the chatter [5]. Advanced techniques have also been used, such as multilayer perceptrons or back propagation neural networks [6], Information Theory [7], Radial basis function and features selection [8].

For chatter detection, several sensors are used to acquire data, such as accelerometer signal [9], dynamometers signals [10] and microphone [11]. Several methods dedicated to chatter monitoring, exploit the currents signal. Altintas has proposed an indirect method for sensing cutting force using a servo motor current [12]. Verl and Heisel and al. proposed a sensorless automate condition monitoring method utilizing position, speed and motor current [13]. Matsubara and Ibraki et al. controlled the steady cutting force using servo and spindle motor currents related to the cutting force in a milling process [14]. Li and Tso presented a method for on-line wear state monitoring and tool replacement decision-making using spindle motor and feed motor current signals in drilling [15]. Through these methods, the higher production costs and reduction in mechanical stiffness associated with additional sensors are removed.

In this paper a procedure based on current signal analysis and artificial intelligence for detecting chatter phenomenon is proposed. The approach is put forward to classify the current signals related to stable and unstable conditions. Signal processing strategy is mandatory after data acquisition and before generation and selection of features, due to the dynamics of the electrical fundamental components and theirs harmonics.

In this stage, a synchronous averaging and a wiener filter technic are proposed. The process is then followed by features generation from residual parts computed in the previous step. Extracted features are ranked based on their entropy in which only best features are selected and presented to the system for classification. At the classification step, the support vector machine (SVM) is tested. The system classifies the selected features into stable or unstable classes. The developed approach is applied for chatter detection in Huron K2X10 milling machine. The originality of this application is to study and detect the chatter phenomenon by using only internal information, which is the currents signal, and without any intrusive sensor. Additional, the detection is made regardless the operating condition of the machine (depth of cut, spindle speed).

The paper is organized as follows. In Sect. 2, data acquisition and taxonomy are presented. It covers the main stages involved in developing diagnosis tool for machinery system. Experimental data acquisition is then described and various parts of signal analysis are demonstrated. Finally, brief review of support vector machine classifier is presented. Classification results and discussion are presented in Sect. 3. Conclusion and perspectives are put forward in the final section.

2 Methodology and Taxonomy

The paper proposes a novel detection approach based on signal analysis and artificial intelligent that fulfills these tasks. This technique classifies sensitive features extracted from residual current signals to detect the chatter that is related to unstable conditions. The methodology consists in four major parts: (1) data acquisition, (2) signal processing, (3) features generation and selection, (4) classification. Each step is discussed in the following subsections. Figure 1 shows the developed methodology for chatter detection in milling machine process. The process is initiated by acquiring current signals. Three hall-effect current clamps are implemented on the electrical cabinet of machine to collect current signals in the three phases. At the processing step, currents signal is resynchronized relative to fundamental cycle (60 Hz). Synchronous averaging is then applied to estimate the average synchronous that will be removed from the current signal. Thereafter, wiener filter is applied to the residual signal obtained previously by taking as reference the current signal acquired in load conditions. Various statistical features

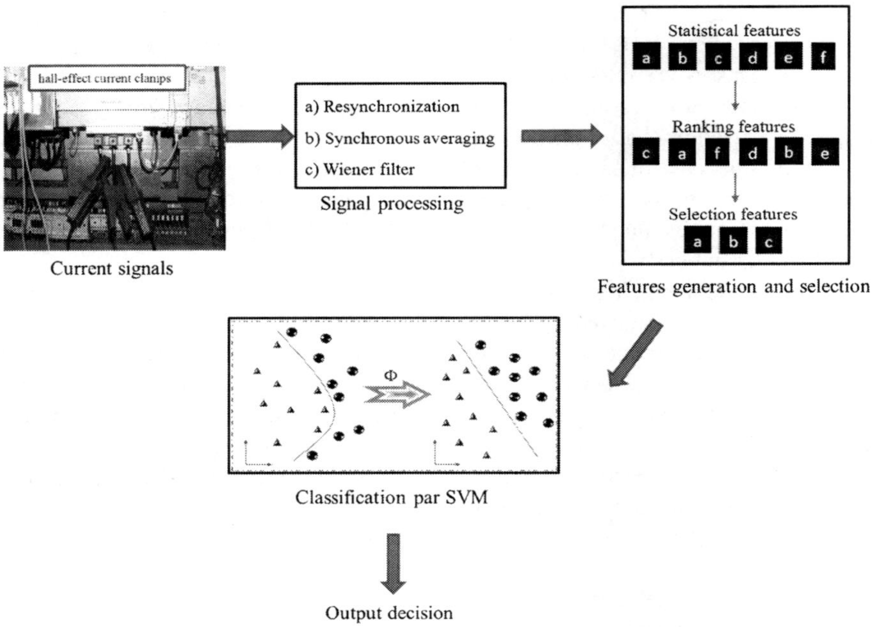

Fig. 1 Methodology proposed for chatter detection

in time domain are then extracted. They are ranked depending on their entropy factor and a group of them is selected for classification. At the classification step, the selected features are classified into two classes: stable and unstable utilizing Support Vector Machine approach. Finally an output decision is made for each tested sample. The different steps of the proposed methodology are discussed in detail in the following subsections.

2.1 Experimental Data Acquisition

The experimental milling tests are performed by dry cutting using vertical machine center Huron K2X10. Aluminum samples (7075-T6) of size [194, 133, 50 mm] are used and slot milling is created using 2 flutes micro grain solid carbide end mills characterized by a helix angle of 30°, maximum depth of cut of 38.1 mm and tool diameter of 25.4 mm. Three hall-effect current clamps (80i-500 s) are located inside the electric cabinet and are used to capture spindle motor current signals in the three phases. These electric sensors allow measurements of currents from 1 to 500 AC RMS and between 5 and 10 kHz. Data acquisition was conducted with a 24-bit accuracy. It is configured to obtain signals with a maximum sampling rate of 48 kHz. The acquisition is controlled by BETAVIB system. The parameters

Table 1 Cutting condition for chatter sets

Test #	Cutting depth (mm)	Cutting speed (m/min)	Rotational speed (rpm)	State
1	2	239.4	3,000	Stable
2	3	239.4	3,000	Stable
3	5	239.4	3,000	Stable
4	6	239.4	3,000	Unstable
5	3	398.9	5,000	Stable
6	5	398.9	5,000	Stable
7	6	398.9	5,000	Unstable
8	2	638.3	8,000	Stable
9	3	638.3	8,000	Stable
10	6	638.3	8,000	Unstable

corresponding to experimental data acquisition are synthesized in Table 1. In each test, the depth of cut is increased until chatter occurred. Tests are stopped at a cutting depth of 6 mm if no chatter is observed. The number of experimental tests is 14, ten of them correspond to stable conditions and the others four emerge chatter phenomenon. Data are recorded at different spindle speeds for stable and unstable conditions. Other tests were performed in spindle free rotation (without milling).

2.2 Processing Step

At the first processing step, we propose that collected data (in machining conditions and in spindle free rotation) are synchronized depending on their cyclic frequency (60 Hz). This operation allows to have a constant cyclic period and to avoid the effect of frequency fluctuation. It also allows to improve the synchronous averaging in the next step. The method of synchronization employed in this work is explained in [16].

In the second step, the synchronous average is estimated by synchronous averaging algorithm and subtracted from original signal. Residual signals are then filtered by using Wiener filter, taking as reference residual signals acquired in spindle free rotation [16, 17]. In fact, Mechanical information is linked to chatter phenomenon and milling process. This information contained in the electrical current is masked by the strong dynamic of the 60 Hz and the environmental electric noise. The contribution of synchronization to remove the synchronous average and the application of the wiener filter on residual current signals make it possible to separate the mechanical part of the electrical signals. It means that the 60 Hz peaks are strongly reduced and the electrical noises are decreased. The Wiener filter approach is described in Fig. 2. The first and the second step are applied for the three phases.

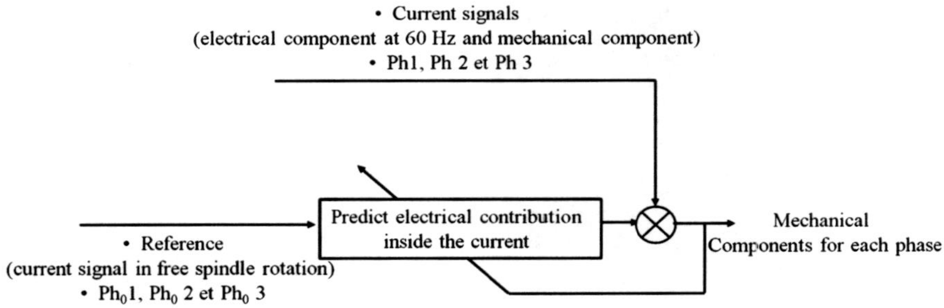

Fig. 2 Wiener filter approach

The third step is the extraction of statistical features from residual signals resulting from wiener filter approach. This extraction transforms signals into several descriptors that describe the signals. These indicators are correlated with milling states and influenced by chatter phenomenon. In this work, nine scalar descriptors were investigated: Variance, Skewness, Kurtosis, Peak value, Root Mean Square (RMS), Clearance factor, Crest factor, Shape factor, Impulse factor [10]. Table 2 summarizes these features.

After the features' extraction step, many descriptors are obtained. The features' selection step is crucial in order to find which features are the most significant and reliable for chatter detection. In fact, if features are selected with little discrimination power, the ulterior design of a classifier would conduct to poor performance. On the other hand, if features with rich information are selected, the classification accuracy increases and the computation time reduces. For this reason, a selection algorithm based on feature ranking is utilized. This algorithm is a salient approach among other selection techniques because it employs a metric to rank the features and excludes all ones that do not achieve appropriate accuracy. In this paper, relative entropy measure is proposed to rank features from best to worst and then, a set of only top listed features is presented for classification.

At the classification step, classical analytical techniques often cannot provide acceptable solutions to difficult diagnosis tasks. This explains why soft computing techniques such as neural networks become more and more popular in industrial

Table 2 Scalar description

Variance (Var)	$\frac{1}{N}\sum_{n=1}^{N}(s_n-\bar{s})^2$	Skewness	$\frac{\frac{1}{N}\sum_{n=1}^{N}(s_n-\bar{s})^3}{Var^{3/2}}$
Kurtosis	$\frac{\frac{1}{N}\sum_{n=1}^{N}(s_n-\bar{s})^4}{Var^2}$	Peak value (Pv)	$\frac{1}{2}[\max(s_n)-\min(s_n)]$
RMS	$\sqrt{\frac{1}{N}\sum_{n=1}^{N}s_n^2}$	Clearance factor (CLf)	$\frac{P_v}{\left(\frac{1}{N}\sum_{n=1}^{N}\lvert s_n\rvert\right)^2}$
Crest factor (C_f)	$\frac{P_v}{RMS}$	Shape factor (Sf)	$\frac{RMS}{\frac{1}{N}\sum_{n=1}^{N}\lvert s_n\rvert}$
Impulse factor (I_f)	$\frac{P_v}{\frac{1}{N}\sum_{n=1}^{N}\lvert s_n\rvert}$	Skewness	$\frac{\frac{1}{N}\sum_{n=1}^{N}(s_n-\bar{s})^3}{Var^{3/2}}$
Variance (Var)	$\frac{1}{N}\sum_{n=1}^{N}(s_n-\bar{s})^2$		

applications of fault diagnosis. The selected features are classified into two classes: stable and unstable utilizing a support vector machine (SVM). A brief theoretical review for this approach is presented in the following sections.

2.3 Support Vector Machine

SVMs have been recognized as powerful statistical learning theory, with good performance for classification and regress. SVM algorithms are concerned with learning a mapping $f : x_i \rightarrow t_i$ based on training sample of input-output pairs $\{(x_i, t_i), i = 1, \ldots, \tilde{N}\}$, and applying them for classification. $\tilde{N}$ is the number of samples and each sample belongs to a class by $t_i \in \{-1, 1\}$.

1. Separable case

The distance from the hyperplane to the closest positive (negative) example is known as the margin of the hyperplane. Maximizing the margin of the hyperplane is then equivalent to maximizing the distance between the class boundaries. The linear hyperplane boundary can be expressed as follows:

$$f(x_i) = w^T.x_i + \hat{b} = 0 \tag{1}$$

where w is a weight vector and $\hat{b}$ is a bias.

SVM guarantees to produce the optimal linear decision boundary in linearly separable case. The distance between the margin and the hyperplane is scaled to $\frac{1}{w}$, since support planes are symmetry with respect to hyperplane, the distance between the two margins becomes $\frac{2}{w}$. Find with maximal margins such that:

$$\begin{aligned} f(x_i) &= w^T.x_i + \hat{b} \geq 1 \text{ for } t_i = 1 \\ f(x_i) &= w^T.x_i + \hat{b} \leq -1 \text{ for } t_i = -1 \end{aligned} \tag{2}$$

Equation 2 can be combined into one set of inequalities:

$$t_i\left(w^T.x_i + \hat{b}\right) - 1 \geq 0, \forall i \in \tilde{N} \tag{3}$$

When the training samples are linearly separable, SVM yields the optimal hyperplane that separates two classes with no training error, and maximizes the minimum distance from the training samples to the hyperplane. Note that the supporting planes are parallel and that no training points fall between them. The parameter pair $(w; \hat{b})$ corresponding to the optimal hyperplane is the solution to the following optimization problem: "Minimize $\frac{1}{2}\|w\|^2$, subjected to $t_i\left(w^T.x_i + \hat{b}\right) \geq 1$, $i = 1, \ldots, \tilde{N}$".

Introducing non negative Lagrange multipliers $\hat{\alpha}_i$, $i = 1, \ldots, \tilde{N}$, the above optimization problem can be solved using the lagrangian function defined as a primary Lagrangian:

"Minimize $L = \sum_{i=1}^{\tilde{N}} \hat{\alpha}_i - \frac{1}{2}\sum_{i,j}^{\tilde{N}} \hat{\alpha}_i\hat{\alpha}_j t_i t_j \left(x_i \bullet x_j\right)$, subjected to: $\sum_{i=1}^{\tilde{N}} \hat{\alpha}_i t_i = 0$" where $(\bullet)$ is the dot product. The decision function can be obtained as follows:

$$f(x) = \text{sign}\left(\sum_{i=1}^{\tilde{N}} \hat{\alpha}_i t_i (x_i \bullet x) + \hat{b}\right) \tag{4}$$

So the decision function $f(x)$ represents the class assigned to data point x.

2. Non-separable case

In the case where a linear boundary is inappropriate, the transformation of vector x into a higher dimensional feature space is required so slack variables, ξ_i are introduced so that:

$$t_i\left(w^T.x_i + \hat{b}\right) \geq 1 - \xi_i, \forall i \in \tilde{N} \tag{5}$$

Each data vector can belong to one of the following three categories:

a. The vector is outside the margin band and correctly classified. $(\xi_i = 0)$
b. The vector is inside the margin band and correctly classified. $(0 < \xi_i \leq 1)$
c. The vector is misclassified. $(\xi_i > 1)$

The optimization problem becomes:
Minimize: $\frac{1}{2}\|w\|^2 + C_p \sum_{i=1}^{\tilde{N}} I(\xi_i)$
Subjected: $t_i\left(w^T.x_i + \hat{b}\right) \geq 1 - \xi_i$ and $\xi_i \geq 0$ where $i = 1, \ldots, \tilde{N}$
C_p is a controller parameter. Then the lagrangian function becomes:

$$L\left(w, \hat{b}, \xi, \hat{\alpha}, \hat{\mu}\right) = \frac{1}{2}\|w\|^2 + C_p \sum_{i=1}^{\tilde{N}} \xi_i - \sum_{i=1}^{\tilde{N}} \hat{\mu}_i \xi_i - \sum_{i=1}^{\tilde{N}} \hat{\alpha}_i \left[t_i\left(w^T x_i + \hat{b}\right) - 1 + \xi_i\right]$$

where $\hat{\alpha}$ and $\hat{\mu}$ are lagrangian multipliers.
The resulting SVM decision function is of the form:

$$f(x) = \sum_{i=1}^{\tilde{N}} \hat{\alpha}_i \hat{\mu}_i x_i \bullet x + \hat{b} \tag{6}$$

It is noticed that the way in which the data appears in the training problem is the form of dot products. First the data is mapped to some other Euclidean space H, using a non-linear mapping function $\Phi : R^d \to H, x \to \Phi(x)$

The training algorithm depends on the data through dot products. $\Phi(x_i) \bullet \Phi\left(x_j\right)$

A positive definite kernel is defined via the kernel trick which regards the kernel as generalized dot products. The transformation into higher dimensional feature space requires relative intensive computation. That's why it is possible to calculate a kernel which is simply the inner products of a feature space, in an original data space.

$$K(x_i, x_j) = \Phi(x_i) \bullet \Phi(x_j) \tag{7}$$

The idea of a kernel function is to enable operations to be performed in an input space, rather than in a high dimension feature space. Now if kernel function satisfies (7), there is a need to use K only in the training algorithm and it is not necessary to know an actual mapping function Φ, hence there is no need to evaluate an inner product in the feature. Therefore in this case, a dot product $(x_i \bullet x_j)$ can be replaced with $K(x_i, x_j)$ in this algorithm. SVM computes a value for $\hat{\alpha}_i$ that corresponds to the maximal margin hyperplane in the feature space, by replacing different kernel functions with $(x_i \bullet x_j)$. Several types of kernel functions that satisfy Mercer's condition can be used. Most common used kernel functions in SVM are polynomial kernel, hyperbolic tangent and Gaussian Radial Basis Function (RBF). The decision function for non-separable case is:

$$f(x) = \text{sign}\left\{ \sum_{i=1}^{\tilde{N}} \hat{\alpha}_i t_i K(x_i, x) + \hat{b} \right\} \tag{8}$$

The resulting classifier then becomes:

$$x \in \begin{cases} \text{classI}, f(x) > 0 \\ \text{classII}, f(x) < 0 \end{cases} \tag{9}$$

3 Application to Milling Machine for Chatter Detection

The proposed approach is applied on CNC milling machine. The principle goal is to detect the chatter phenomenon. During the manipulations of processing step, a total number of 23590 samples (one sample presents one spindle revolution) are collected at different depths and various rotational speeds. 17219 samples were corresponding to stable case and the others 6371 samples include chatter phenomenon. Current signals are passed through the different treatments of processing step. A good classification of data entered for training is important in order to get an efficient detection method. 40 % of data (known classes: stable and unstable samples) are used to train the network. The rest (60 % of data) are used for testing step. During the training step, all the extracted features are ranked from best to worst according to their capability to separate the two classes (stable and unstable), using relative entropy method. Then, the four best ranked features are selected and classified by SVM classifier. A radial basis function kernel is selected for SVM. To classify the parameters, a SVM was constructed using a Matlab toolbox. In order to achieve a good performance, SVM parameters were optimized by k-fold cross validation method, which is widely employed in SVM parameters design. Table 3 exhibits the features selected, with corresponding entropy values. The four best indicators (Variance, RMS, Clearance and Peak) were selected since their entropy is

Table 3 Best features selected with their entropy values

Order	Parameter	Entropy value
1	Variance	148,74
2	RMS	34,34
3	Peak	21,21
4	Clearance	9,33

greater than the others. The variance is revealed as the more sensitive, while the clearance was the worse sensitive features.

The classification rate errors of testing samples using SVM network classifier are presented in Table 4.

SE is the percentage number of unstable samples classified as stable and UE is the percentage number of stable samples classified as unstable. Table 3 demonstrates that the percentage of error of stable and unstable samples don't exceed 4 %. In order to confirm the accuracy of our methodology, classifications are realized with two classifiers that are the MLP (Multilayer Perceptrons) and the RBF (Radial Basis Function). The multi-Layer Perceptron (MLP) is a feed forward neural network which consists in units arranged in layers with only forward connections to units in subsequent layers. RBF is a feed-forward neural network trained using a supervised training algorithm. It is typically configured with a single hidden layer of units whose activation function is selected from a class of functions called basis functions.

Results are displayed in Fig. 3.

The results confirm that the SVM approach provides best chatter detection versus RBF and MLP approaches. MLP classifier allows having a better

Table 4 Classification rate errors using SVM approach

	Classification rate error (CRE) '%'
Stable error (SE)	0.88
Unstable error (UE)	3.87

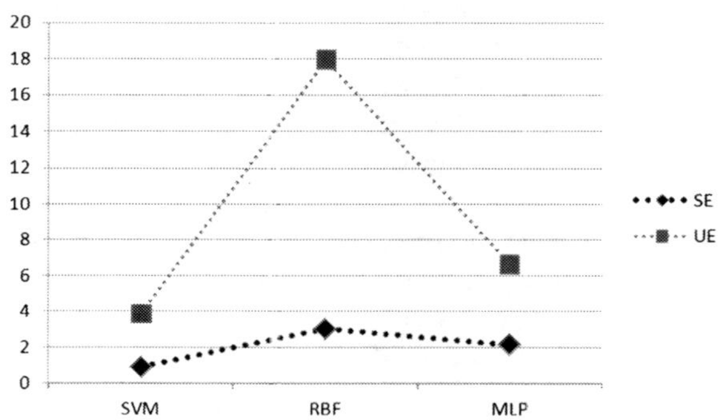

Fig. 3 Classification rate errors for different classifiers

classification by comparing to RBF. The results illustrate that by using the proposed approach; more than 96 % of testing samples (normal and chatter) are well classified.

4 Conclusion

This paper presents a new methodology for chatter detection in CNC milling machines based on current signal analysis and using neural networks. In signal processing task, electrical signals are resynchronized according to the electrical cycle (60 Hz) by exploiting the cyclostationarity of electrical signals through their cyclic statistics. After that, synchronous average is computed and subtracted from original signals in order to obtain the residual part. Wiener filter is then applied on residual signals by taking as reference the residual electrical signals acquired in spindle free rotation. This procedure allows estimating a signals corresponding to the electrical part and extracting the mechanical part, which linked to a chatter phenomenon and cutting mechanism. The signal processing task is primordial in order to decrease the dynamics of the electrical fundamental component and its harmonics and also to increase the contribution of mechanical parts. Extracted features computed from mechanical parts are then ranked based on theirs entropies in which only best features are selected and presented to the system for classification. At the classification step, the selected features are classified into two classes: stable and unstable utilizing a support vector machine (SVM). The intelligent chatter detection has accuracy above 96 % for the identification of cutting state after being trained by experimental data. SVM approach provides better results versus MLP and RBF approaches. The results show that it is possible to monitor chatter behavior in milling process by using motor current signal.

Acknowledgement Funding for this project was provided by a grant from la Région Rhone-Alpes.

References

1. Liang SY, Hecker RL, Landers RG (2004) Machining process monitoring and control: the state-of-the-art. ASME J Manufact Sci Eng 126:297–310
2. Vallejo AJ, Morales-Menéndez Rubén, Alique JR (2008) On-line cutting tool condition monitoring in machining processes using artificial intelligence. Robot Autom Control 1 (1):494–518
3. Insperger T (2002) Stability analysis of periodic delay differential equation modeling machine tool chatter. Ph.D. thesis, Budapest University of Technology and Economics
4. Smith S, Tlusty J (1990) Update on high-speed milling dynamics. J Eng Indus 112:142–149
5. Lamraoui M, Thomas M, El Badaoui M, Girardin F (2014) Indicators for monitoring chatter in milling based on instantaneous angular speeds. Mech Syst Signal Process 44(1–2):72–85

6. Tarng YS, Chen MC (1994) Intelligent sensor for detection of milling chatters. J Intell Manuf 5(3):193–200
7. Gradisek J, Govekar E, Grabec I (1998) Using coarse-grained entropy rate to detect chatter in cutting. J Sound Vib 214(5):941–952
8. Lamraoui M, Barakat M, Thomas M, El Badaoui M (2013) Chatter detection in milling machines by neural networks classification and feature selection. J Vibr Control
9. Li XQ, Wong YS, Nee AYC (1997) Tool wear and chatter detection using the coherence function of two crossed accelerations. Int J Mach Tool Manu 37(4):425–435
10. Ck Toh (2004) Vibrations analysis in high speed rough and finish milling hardened steel. J Sound Vib 278(1–2):101–115
11. Delio T, Tlusty J, Smith S (1992) Use of audio signals for chatter detection and control. ASME J Eng Indus 114(2):146–157
12. Altintus Y (1992) Prediction of cutting forces and tool breakage in milling from feed drive current measurements. Trans ASME 114:386–394
13. Verl A, Heisel U (2009) Sensorless automated condition monitoring for the control of the predictive maintenance of machine tools. Ann CIRP 58(1):375–378
14. Ibaraki S, Matsubara A, Murozumi M (2008) Efficiency comparison of cutting strategies for end milling processes under feedrate scheduling. Int J Autom Technol 2(5):377–383
15. Li Xiaoli, Tso SK (1999) Drill wear monitoring based on current signals. Wear 231 (2):172–178
16. Ibrahim A, Bonnardot F, El Badaoui M, Guillet F (2011) Detection of bearing damage using stator current, and voltage to cancel electrical noise. EURASIP J Adv Signal Process 2011
17. Zhou W, Lu B, Habetler TG, Harley RG (2009) Incipient bearing fault detection via motor stator current noise cancellation using wiener filter. IEEE Trans Indus Appl 45(4):1309–1317

Dynamics of the Unbalanced Support Centrifuge with Cavity Partially Filled with a Multiphase Liquid Medium

Almatbek Kydyrbekuly and Lelya Khajiyeva

Abstract The paper studies dynamics of the statically unbalanced support centrifuge with a cylindrical cavity partially filled with a multiphase liquid medium and mounted on an elastic foundation. The centrifuge is a rotor, installed cantilevered on the shaft. It is fixed at the lower hinged bearing and the upper elastic bearing and rotates with a constant angular velocity. It is easy to pass from the dynamic model of the support centrifuge to the model of the suspended centrifuge. A mathematical model of the system "centrifuge-multiphase liquid-foundation" is developed and solved analytically.

Keywords Rotor · Dynamics · Unbalanced support centrifuge · Liquid vibrations · Foundation vibrations

1 Introduction

It is necessary to develop such general universal dynamic and mathematical models that will allow to construct the rotor machine models by simple change of the system parameters. It will provide automatization of the rotor machines development on early stages of projecting as well as improvement and exploitation of the existing machines. Also it is necessary to develop such method of accounting of the system dynamic characteristics, which will enable to pass from one accounting scheme to another without any difficulties. It will allow realization of the numerical experiments with taking into account constructive particularities of the system Rotor-Frame-Foundation and such important factors as the elastic constraints, variability of the inertia parameters, interconnections of the rotor and the drive, etc. [1].

A. Kydyrbekuly · L. Khajiyeva (✉)
Faculty of Mechanics and Mathematics, al-Farabi Kazakh National University, al-Farabi av.71, 050040 Almaty, Kazakhstan
e-mail: khadle@mail.ru

A. Kydyrbekuly
e-mail: almatbek@list.ru

J.K. Sinha (ed.), *Vibration Engineering and Technology of Machinery*, Mechanisms and Machine Science 23, DOI 10.1007/978-3-319-09918-7_51

2 Equations of the System Motion

Let's consider an unbalanced support centrifuge (the dynamic model of the support centrifuge can be easily transformed to the model of the suspended centrifuge), representing a rotor with a cylindrical cavity partially filled with a multiphase fluid. Rotor is installed cantilevered on the shaft, fixed at the lower hinged bearing and the upper elastic bearing and rotates with a constant angular velocity Ω_0. Shaft is considered as an absolutely rigid. Such centrifuge can be considered as a symmetric gyroscope with a movable fulcrum. Rotor foundation (housing) with the mass M oscillates in the horizontal plane XO_1Y. Assume that the gravity center of the centrifuge foundation coincides with the point O. The centrifuge foundation is mounted on a fixed bearing by means of elastic isotropic devices with the stiffness coefficient c_1. The damping force of the foundation is also considered as the homogeneous with the damping coefficient n_1. Let us choose a fixed coordinate system O_1XYZ, whose origin coincides with the fulcrum O of the centrifuge in the equilibrium state of the system. The mobile coordinate system $OX_1Y_1Z_1$ is rigidly connected with the foundation (housing) of the centrifuge.

Let us also introduce a movable coordinate system $Oxyz$, rigidly connected with the centrifuge. The axis Oz is an axis of the centrifuge symmetry. By J and J_0 we denote the centrifuge's equatorial and polar moments of inertia with respect to the axes passing through the centrifuge center of gravity; $m,\ R,\ h$ are the mass, radius and height of the centrifuge; $c$is the stiffness coefficient of the upper elastic bearing; n is the damping coefficient of the centrifuge. Position of the rotating axis of the centrifuge is defined by the Resal angles α and β.

Coordinates of the gravity centers of the foundation and the centrifuge are denoted respectively as $(x_0,\ y_0)$ and $(x_s,\ y_s)$. Motion of the housing in the vertical direction, its rotations around the axes $O_1X,\quad O_1Y$ and O_1Z are neglected.

Taking into account that the generalized coordinates are $x_0,\ y_0,\ \alpha$ and β and using the Lagrange equations of the second kind, we obtain the motion equations of the foundation and centrifuge with fluid, which have the following complex form [1]:

$$\left.\begin{aligned} (m+M)\ddot{z}_0 + mL\ddot{\theta} + n\dot{z}_0 + c_1 z_0 &= me\Omega_0^2 e^{i\Omega_0 t} + F_r\,, \\ J_1\ddot{\theta} + mL\ddot{z}_0 + (n_1 - iJ_0\Omega_0)\dot{\theta} + \left(cl^2 - mgL\right)\theta &= me\left(L\Omega_0^2 + g\right)e^{i\Omega_0 t} + M_0. \end{aligned}\right\} \tag{1}$$

Here

$$z_0 = x_0 + iy_0, \quad \theta = \alpha + i\beta \tag{2}$$

determine the position of the gravity center of the housing and the rotating axis of the centrifuge on the complex plane; e is an imbalance of the centrifuge; l, L are the distance between the bearings and the distance from the lower bearing to the gravity center of the centrifuge.

It is assumed that the angular velocity $\Omega_0 = const$ is sufficiently high, so that the heterogeneous multiphase fluid at the cavity takes a cylindrical shape with the radius of the free surface r_0. The fluid gravity force is not considered, as $\Omega_0^2 R \gg g$ The fluid motion at the rotor cavity is determined relative to the movable cylindrical coordinate system (r, φ, z), rotating with the rotor. It is also assumed that under the action of the centrifugal forces a multiphase fluid medium (suspension, raw oil, etc.) separates strictly by the densities (phases), and the heterogeneous fluid is simulated by the n-layer immiscible fluids. After defining the translation acceleration and Coriolis acceleration of the fluid particle with unit mass of the j-th layer of fluid, the hydrodynamic Euler equation can be written in vector form:

$$\frac{\bar{V}_j}{t} - \nu_j \Delta \bar{V}_j + 2(\bar{\Omega} \times \bar{V}_j) + \dot{\bar{V}}_0 + \bar{\Omega} \times (\bar{\Omega} \times \bar{r}) + \dot{\bar{\Omega}} \times \bar{r} = -\frac{1}{\rho} \mathrm{grad} P_j, \quad (3)$$

where $\bar{V}_j$ is a vector of relative velocity of the particle of the j-th fluid, $\bar{r}$is radius-vector of this particle, $\dot{\bar{V}}_0$ is acceleration of the origin of the movable coordinate system (r, φ, z).

After projecting of the vector Eq. (3) on the axes of the cylindrical coordinate system and their linearization, we obtain the motion equations of the j-th layer of viscous fluid in the form [2]:

$$\frac{\partial U_j}{\partial t} - 2\Omega_0 V_j - \nu_j \left(\Delta U_j - \frac{2}{r^2}\frac{\partial V_j}{\partial \varphi} - \frac{U_j}{r^2}\right) = -\frac{1}{\rho_j}\frac{\partial P_j}{\partial r} - (\ddot{z}_0 + \ddot{\theta} z) e^{-i(\Omega_0 t + \varphi)}, \quad (4)$$

$$\frac{\partial V_j}{\partial t} + 2\Omega_0 U_j - \nu_j \left(\Delta V_j - \frac{2}{r^2}\frac{\partial U_j}{\partial \varphi} - \frac{V_j}{r^2}\right) = -\frac{1}{\rho_j r}\frac{\partial P_j}{\partial \varphi} + i(\ddot{z}_0 + \ddot{\theta} z) e^{-i(\Omega_0 t + \varphi)}, \quad (5)$$

$$\frac{\partial W_j}{\partial t} - \nu_j \Delta W_j = -\frac{1}{\rho_j}\frac{\partial P_j}{\partial z} + r(\ddot{\theta} - 2i\Omega_0 \dot{\theta}) e^{-i(\Omega_0 t + \varphi)}. \quad (6)$$

$$\frac{\partial (r U_j)}{\partial r} + \frac{\partial V_j}{\partial \varphi} + r\frac{\partial W_j}{\partial z} = 0, \quad (7)$$

where ν_j, ρ_j, P_j are the kinematic coefficients of viscosity, density and pressure of the j-th layer of the fluid; U_j, V_j, W_j are the velocity components of the particle of the j-th layer of the fluid; Δ is the Laplace operator in the cylindrical coordinate system; Eq. (7) is an equation of continuity by $\rho_j = const$.

Boundary conditions of the problem have the following form:
at the lateral boundary of the rotor by $r = R$

$$U_n = V_n = W_n = 0, \quad (8a)$$

at the upper and lower boundaries of rotor by $z = L + H, \ z = L - H$

$$U_j = V_j = W_j = 0 \quad j = 1, 2, \ldots, n; \tag{8b}$$

at the boundaries of the fluids sections by $r = r_j$

$$U_j = U_{j-1},\ V_j = V_{j-1},\ W_j = W_{j-1},\ \frac{\partial \xi_j}{\partial t} = U_j, \tag{9}$$

$$v_j \rho_j \left[\frac{1}{r} \frac{\partial U_j}{\partial \varphi} + \frac{\partial V_j}{\partial r} - \frac{V_j}{r} \right] = v_{j-1} \rho_{j-1} \left[\frac{1}{r} \frac{\partial U_{j-1}}{\partial \varphi} + \frac{\partial V_{j-1}}{\partial r} - \frac{V_{j-1}}{r} \right], \tag{10}$$

$$v_j \rho_j \left[\frac{1}{r} \frac{\partial W_j}{\partial \varphi} + \frac{\partial V_j}{\partial z} \right] = v_{j-1} \rho_{j-1} \left[\frac{1}{r} \frac{\partial W_{j-1}}{\partial \varphi} + \frac{\partial V_{j-1}}{\partial z} \right], \tag{11}$$

$$\begin{aligned} &\left[-P_j - \frac{1}{2} \rho_j \Omega_0^2 \left(r_j^2 - r_0^2 \right) + 2 v_j \rho_j \frac{\partial U_j}{\partial r} \right] \Big|_{r = r_j + \xi_j(\varphi, z, t)} \\ &= \left[-P_{j-1} - \frac{1}{2} \rho_{j-1} \Omega_0^2 \left(r_j^2 - r_0^2 \right) + 2 v_{j-1} \rho_{j-1} \frac{\partial U_{j-1}}{\partial r} \right] \Big|_{r = r_j + \xi_j(\varphi, z, t)}, \end{aligned} \tag{12}$$

at the free surface of the fluid by $r = r_0$

$$v_1 \rho_1 \left[\frac{1}{r} \frac{\partial U_1}{\partial \varphi} + \frac{\partial V_1}{\partial r} - \frac{V_1}{r} \right] = 0, \tag{13}$$

$$v_1 \rho_1 \left[\frac{1}{r} \frac{\partial W_1}{\partial \varphi} + \frac{\partial V_1}{\partial z} \right] = 0, \tag{14}$$

$$\frac{\partial \xi_0}{\partial t} = u_1, \tag{15}$$

$$\left[-P_1 - \frac{1}{2} \rho_1 \Omega_0^2 \left(r_1^2 - r_0^2 \right) + 2 v_1 \rho_1 \frac{\partial U_1}{\partial r} \right] \Big|_{r = r_0 + \xi_0(\varphi, z, t)} = 0. \tag{16}$$

3 Solution

Let us represent the complex deviation of the foundation and the centrifuge from the equilibrium position in the form:

$$z_0 = A e^{i \Omega_0 t} + B e^{i \omega t}, \tag{17}$$

$$\theta = C e^{i \Omega_0 t} + D e^{i \omega t}, \tag{18}$$

where the first terms in the formulas describe the forced vibrations of the system due to imbalance of the centrifuge, and the second terms determine the self-oscillation of the foundation and the centrifuge.

In order to solve the hydrodynamic Eqs. (4)–(6) and (7) with boundary conditions (8a, 8b)–(16), we introduce the stream functions Ψ_j and the velocity potential Φ_j^0 (the Lamba potentials). Then the velocity components of the fluid particle may be presented by the functions Ψ_j and Φ_j^0 in the following form:

$$U_j = \frac{\partial \Phi_j}{\partial r} + \frac{1}{r}\frac{\partial \Psi_j}{\partial \varphi}, \tag{19}$$

$$V_j = \frac{1}{r}\frac{\partial \Phi_j}{\partial \varphi} - \frac{\partial \Psi_j}{\partial r}, \tag{20}$$

$$W_j = \frac{\partial \Phi_j}{\partial z} + 2\omega\, r\, D\, e^{i(\sigma t - \varphi)}. \tag{21}$$

Substituting (19)–(21) into the equations of the fluid motion (4)–(6) and in the continuity Eq. (6), after some mathematical transformations with the introduction of additional functions F_j and G_j, we will obtain

$$\Delta \Phi_j = 0,\ \Delta * F_j = 0,\ \Delta * G_j = 0,\ \Delta * \Psi_j = 0,\ \Delta P_j^{/} = 0, \tag{22}$$

where

$$F_j = -\frac{\partial \Phi_j}{\partial t} - \frac{P_j^{/}}{\rho} - 2\Omega_0 \Psi_j, \tag{23}$$

$$G_j = -\frac{\partial \Psi_j}{\partial t} - 2\Omega_0 \Phi_j + v_j \Delta \Psi_j, \tag{24}$$

$$P_j^{/} = -\rho \frac{\partial \Phi_j}{\partial t} + \left(C_1^j r + \frac{1}{r} C_2^j\right) e^{i(\sigma t - \varphi)}, \tag{25}$$

Δ^* is the Laplace operator in the polar coordinate system.

Each from the functions Φ_j, F_j, G_j, Ψ_j is represented in the form:

$$/\Phi_j, F_j, G_j, \Psi_j/ = R_j(r) Z_j(z) \times exp\left[i(6t - \varphi)\right]. \tag{26}$$

Substituting (26) into (22) with taking into account of (23)–(25), and after separation of variables, we will obtain a Bessel equation relatively $R_j(r)$, and a linear second-order equation relatively $Z_j(z)$. Their solutions define the functions Φ_j, F_j, G_j, Ψ_j, and finally we may define the values of the components of the particle velocity and pressure at any point of the fluid j-th layer:

$$U_j = \sum_{k=1}^{¥} \left\{ \left[z'_{1j}(\alpha_{kj} r) - \frac{i}{r} R_{1kj} \right] \left[A_{1kj} e^{\alpha_{kj} z} + A_{2kj} e^{-\alpha_{kj} z} \right] + \left[\bar{z}_{1j}(\beta_{kj} r) - \frac{i}{r} R_{2kj} \right] \right.$$
$$\left. \times \left(B_{1kj} \cos \beta_{kj} z + B_{2kj} \sin \beta_{kj} z \right) + \left(C_{0j} - \frac{1}{r^2} D_{0j} - \frac{i}{r} R_{3kj} \right) \left(A_{0j} z + B_{0j} \right) \right\} e^{i(\sigma t - \varphi)}, \tag{27}$$

$$V_j = \sum_{k=1}^{¥} \left\{ \left[\frac{i}{r} z'_{1j}(\alpha_{kj} r) + R'_{1kj}(r) \right] \left[A_{1kj} e^{\alpha_{kj} z} + A_{2kj} e^{-\alpha_{kj} z} \right] + \left[\frac{i}{r} \bar{z}_{1j}(\beta_{kj} r) + R'_{2kj}(r) \right] \right.$$
$$\left. \times \left(B_{1kj} \cos \beta_{kj} z + B_{2kj} \sin \beta_{kj} z \right) + \left((C_{0j} - \frac{1}{r^2} D_{0j}) i + R'_{3kj}(r) \right) \left(A_{0j} z + B_{0j} \right) \right\} \tag{28}$$
$$\times e^{i(\sigma t - \varphi)},$$

$$W_j = \sum_{k=1}^{¥} \left\{ \left[z_{1j}(\alpha_{kj} r) \alpha_{kj} \left[A_{1kj} e^{\alpha_{kj} z} + A_{2kj} e^{-\alpha_{kj} z} \right] - \beta_{kj} \bar{z}_{1j}(\beta_{kj} r) \right. \right.$$
$$\left. \times \left(B_{1kj} \sin \beta_{kj} z - B_{2kj} \cos \beta_{kj} z \right) \right] + A_{0j} \left(C_{0j} r - \frac{1}{r} D_{0j} \right) + 2\omega r B_{34} \Big\} e^{i(-\varphi)}, \tag{29}$$

$$P_j = i\sigma\rho_j \sum_{k=1}^{¥} \left\{ z_{1j}(\alpha_{kj} r) \left[A_{1kj} e^{\alpha_{kj} z} - A_{2kj} e^{-\alpha_{kj} z} \right] \right\} - \bar{z}_{1j}(\beta_{kj} r)(B_{1kj} \cos \beta_{kj} z$$
$$+ B_{2kj} \sin \beta_{kj} z) + A_{0j} z (C_{0j} - \frac{1}{r} D_{0j}) + (C_{1j} r + \frac{1}{r} C_{2j}) e^{[i(\sigma t - \varphi)]} \tag{30}$$
$$+ \rho_j r \omega^2 (B - izD) e^{[i(\sigma t - \varphi)]} + \rho_j r \Omega_0^2 (A - izC) e^{(-i\varphi)}.$$

Using the complex expressions for the hydrodynamic force and its moment [3], certain to the second order because of the small coefficient of kinematic viscosity of the fluid, taking into account the difference between fixing of the support centrifuge and the introduced coordinate systems, taking into account the expression for the pressure in the form (30), we define F_r and M_θ. After substituting of F_r, M_θ, as well as (17) and (18), into the motion equations of the system (1), and further, after equating the expressions at the functions of time $e^{i\Omega_0 t}$, $e^{i\omega t}$ and the free terms, we obtain respectively a system of inhomogeneous algebraic equations concerning the amplitudes of forced vibrations of the centrifuge and the foundation, as well as the homogeneous system with respect to the coefficients B and D. From condition of equality to zero of the determinant of the homogeneous system, we obtain the characteristic equation with respect to ω, by nature of which roots the conditions of stability of the system motion are found.

The unknown integration constants contained in expressions of the components of the fluid particle velocity and pressure (27)–(30) are determined from the boundary conditions (8a, 8b)–(16).

4 Conclusions

The dynamic and mathematical models of the system "centrifuge-multiphase fluid-foundation" (CMPFF) was created and solved analytically for the first time.

The values of velocity components of the fluid particle and the values of pressure at any point of the fluid volume were determined in an analytical form.

Development of the methods for calculating the basic dynamic characteristics of the system and definition of the exact solutions of the linearized equations of motion of a multiphase fluid, centrifuge and foundation allow to study dynamics of the inter-related motions of other types of rotors with the cavities partially filled with fluid and his housing in the presence of elastic connections between them and the machine base, as well as to determine the influence of the wave properties of the fluid free surface on the motion of the entire system.

The results of the problem solving of the generalized mathematical model (CMPFF) allow to select appropriately those values of the parameters of centrifuge, foundation and fluid, by which the loads between the foundation and the housing, the housing and the centrifuge, the amplitude of the forced oscillations, as well as the width of the zone of the system instability are greatly reduced. Choice of the system parameters, which provide stable operation of the rotary machine, is of paramount importance.

References

1. Rahimov ER, Rahmatullaev AS, Kydyrbekuly AB (1987) Dynamics of unbalanced rotor in the interaction with the other physical fields. In: Transactions seventh world congress on the theory of machines and mechanisms, Sevilla, Spain, 1987, pp 532–539
2. Rahmatullaev AS, Kydyrbekuly AB (2003) Dynamics of rotor with cavity partially filled by weakly conducting ideal fluid. In: Proceedings of the IV international Kazakhstan–Russia scientific-practical conference "Mathematical simulation of scientific-technical and ecology problems of oil and gas industry", Almaty, 2003, pp 115–120
3. Kydyrbekuly AB (2012) Rotor-liquid-fundament system's oscillation. Advances in mechanisms design. Mechanisms and machine science, vol 8. Springer Science + Business, Media, Dordrecht, pp 223–229

Design and Test of a Piezoelectric Inertial Mass Actuator for Active Vibration Control

Sven Herold, Dirk Mayer, Tobias Melz and Tobias Röglin

Abstract Active control of vibrations can outperform passive systems in certain applications, e.g. when broadband damping is requested or when several orders of a periodic disturbance have to be cancelled. To generate dynamic forces for the active control system, inertial mass actuators are frequently used. Mainly, they comprise a force generating element driving a single-degree-of-freedom oscillator which is coupled to the structure to be controlled. Thus, those actuators can also be used for retrofitting of existing structures or for prototyping purposes. In this paper, a design for an inertial mass actuator utilizing piezoceramic actuators is studied. Since those actuators integrate both stiffness and force generation into one element, this enables more compact and mechanically robust designs. With respect to the future integration into industrial applications, standard multilayer piezoelectric actuators are considered to decrease the costs of the system and allow for a high reliability. Usually, an inertial mass actuator should possess a low resonance frequency in order to enable operation over a broad frequency range. Since the stiffness of piezoelectric multilayer actuators is rather high, a lever mechanism is designed which transforms the stiffness of the piezoelectric element into the desired range. An analytical model of the inertial mass actuator is derived and parameter studies are performed to investigate the characteristics of the design. A prototype is set up and the main parameters like resonance and block force are experimentally validated. Finally, the integration of the actuator into an active vibration control system for a lightweight structure is described.

Keywords Vibration control · Piezoelectric actuator · Inertial mass actuator

S. Herold (✉) · D. Mayer · T. Melz · T. Röglin (✉)
Fraunhofer Institute for Structural Durability and System Reliability LBF, Research Division Smart Structures, Darmstadt, Germany
e-mail: sven.herold@lbf.fraunhofer.de

T. Röglin
e-mail: tobias.roeglin@lbf.fraunhofer.de

D. Mayer
e-mail: dirk.mayer@lbf.fraunhofer.de

T. Melz
e-mail: tobias.melz@lbf.fraunhofer.de

J.K. Sinha (ed.), *Vibration Engineering and Technology of Machinery*,
Mechanisms and Machine Science 23, DOI 10.1007/978-3-319-09918-7_52

1 Introduction

Inertial mass actuators are a common component in active vibration control (AVC) systems [1]. The force of those actuators is generated by an inertial mass, which is excited by an active element. In order to mount the inertial mass to the structure, a soft spring is commonly used. Thus, an inertial mass actuator can be considered as a single-degree-of-freedom-oscillator which is driven above its resonance frequency.

Due to the need for a stiffness to mount the inertial mass and limitations of the maximum stroke delivered by the active element, the challenge of designing an inertial mass actuator is mainly to find a compromise between contradictory requirements, e.g. low resonance frequency for broadband operation and low static deflection and low inertial mass.

In many cases, electrodynamic actuation is used. For instance, inertial mass actuators of this type were successfully used for vibration control in cars [2], trains [3], or even for building floors [4]. The integration of piezoelectric actuators has also been investigated. Since those solid state actuators possess an inherent stiffness, they can carry the inertial mass while providing high dynamic forces. However, due to the high stiffness of the piezo elements the design of an inertial mass actuator with a low resonance frequency usually includes some lever mechanism for stroke amplification and stiffness reduction. In [5] several concepts are proposed, and a piezoelectric disk actuator is used in a prototype set up. Also bending beam configurations were studied, either with integrated stack actuators [6] or foils [7].

In this work, a design for a piezoelectric driven inertial mass actuator is proposed. In contrast to other designs, the basic idea is to develop a modular design which can be easily adapted to different requirements in terms of resonance frequency or block force.

In the following sections, first the basic characteristics for inertial mass actuators are summarized. The proposed design of the inertial mass actuator is introduced, and analytical formulations are derived for the main parameters block force and resonance frequency. A functional prototype is set up, and the characteristics are validated by experiments.

2 Inertial Mass Actuators

As shown in Fig. 1, an inertial mass actuator can be considered a spring-mass system with the stiffness k and the inertial mass m, enhanced by a damping d and an actuator force F_A [8]. Above the resonance frequency a rather constant force proportional to the actuator force is excited (Fig. 1). Therefore the resonance frequency of the actuator should be quite below its lowest working frequency. The relation between the block force F and F_A is given by the transfer function:

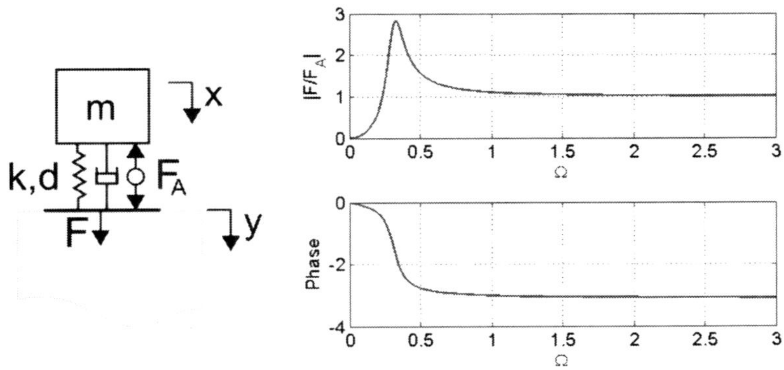

Fig. 1 Inertial mass actuator and its block force frequency response function

$$\frac{F}{F_A} = \frac{s^2}{s^2 + 2\theta\omega_0 s + \omega_0^2}, \tag{1}$$

where $\omega_0 = \sqrt{\frac{k}{m}}$ is the resonance frequency of the system and $\theta = \frac{d}{2m\omega_0}$ denotes the damping coefficient.

3 Design of the Piezoelectric Inertial Mass Actuator

The aforementioned considerations are applied to the design of an inertial mass actuator with integrated piezoelectric transducers.

Similar to the well-known bending beam with applied or integrated piezoelectric actuators [6] the elongation of the piezoelectric actuator is transformed into a bending moment, which deflects the tip of the beam. An attached mass excites a normal force and a moment in the root of the beam. The latter can be compensated for by the application of a symmetric set up with two actuators, which is not considered here, though.

3.1 Basic Equations

In this work, the integration of piezoelectric actuators and the supporting beam is simplified in order to enable a modular, reconfigurable actuator system (Fig. 2). Two piezoelectric actuators are mounted at a distance to the neutral axis of the beam. The inertial mass is intended to serve also as a lever for the actuation. By driving both actuators out of phase, a moment is generated at the tip of the beam,

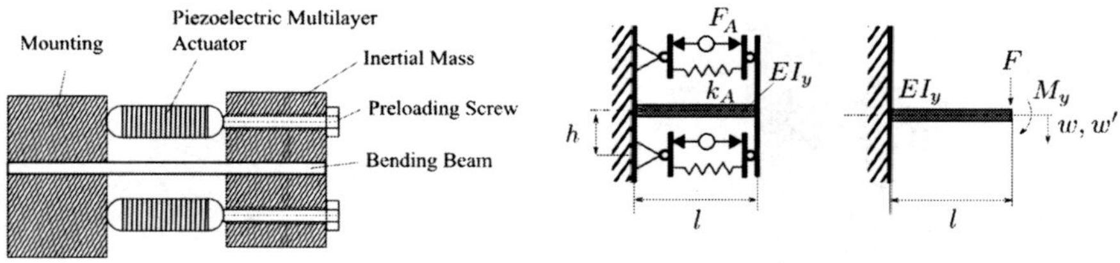

Fig. 2 Beam with coupled piezoelectric actuators, *left* design, middle: mechanical model, *right* simplified model

which causes a movement of the mass and in turn a reaction force at the mounting. For the analysis, generic piezoelectric actuators are assumed, which are described by their longitudinal stiffness k_A and block forceF_A. Both actuators are coupled to the tip of the beam by a lever of length h with respect to the neutral axis of the beam. The beam itself is characterized by its length$l$, Young's modulus E and the moment of inertiaI_y. Since the actuators are coupled mechanically by spherical end caps, only normal forces are assumed to be transmitted at the mounting points of the actuators. The effect of the preloading screw is studied in the next section.

This system can be regarded as a special case of an elastic stroke amplification or stiffness transformation. A simplified model is used to estimate the characteristics, i.e. the resulting overall stiffness in the bending direction of the beam with respect to the stiffness of the beam and the actuator, and the block force F generated in this direction with respect to the actuator force.

Since the inertial mass at the tip is expected to be much heavier than this actuation unit, the mass of the latter is neglected here in order to simplify the calculations using Euler's beam theory. The system can be represented by a bending beam with a tip force and moment M_y. The respective tip displacement w and angle w' can be found by superposition:

$$w(x = l) = \frac{l^3}{3EI_y}F + \frac{l^2}{2EI_y}M_y, \tag{2}$$

$$w'(x = l) = \frac{l^2}{2EI_y}F + \frac{l}{EI_y}M_y. \tag{3}$$

The moment M_y at the tip due to the actuator force F_A and the restoring force of the actuator stiffness k_A can be expressed by:

$$M_y = 2\left(hF_A - k_A h^2 w'(x = l)\right). \tag{4}$$

The free stroke w at the tip of the beam due to the actuator force F_A can be calculated by setting the force F to zero, eliminating w' by solving Eqs. 2 and 3 for the moment M_y and putting the result into Eq. 2:

$$w(x = l) = \frac{l^2 h}{EI_y + 2k_A l h^2} F_A \tag{5}$$

In a similar manner, the stiffness $k = \frac{F}{w}$ can be calculated by setting $F_A = 0$:

$$k = \frac{12EI_Y(EI_y + 2k_A h^2 l)}{l^3(4EI_Y + 2k_a h^2 l)}. \tag{6}$$

The block force of the actuator system can be calculated by:

$$F = kw. \tag{7}$$

3.2 Higher Order Effects

If higher active force amplitudes are required, the actuators have to be preloaded electrically and mechanically. The preloading compressive force on the actuator causes a tensile normal force on the bending beam. The integration of this normal force into the equations leads to a more complex differential equation (linearized) for the axially loaded Euler beam:

$$EIw'''' + F_N w'' = 0 \tag{8}$$

For compression (positive) forces F_N the stiffness of the beam decreases (a critical load F_N that produces zero beam stiffness defines the buckling problem for the beam). Tensile (negative) forces cause additional beam stiffness.

If a significant amount of mass is considered, the tip mass cannot be regarded as a point mass any more. The model thus has to be enhanced by incorporating the distance of the center-of-gravity of the actuator to the end of the beam.

Also, as for nearly every inertial mass actuator, the assumption of a single-degree-of-freedom oscillator is not valid over an infinitely large frequency range. In the design considered here, a higher order resonance is caused by the real tip mass not being a point mass, but also having a rotational moment of inertia J, resulting in a rotational degree-of-freedom for the tip mass. The natural frequency $\omega_{0,R}$ can be derived by

$$\omega_{0,R} = \sqrt{\frac{k_R}{J}} \tag{9}$$

with the overall rotational stiffness k_R that is mainly influenced by the actuator stiffness and the their distance from the neutral axis of the beam.

Fig. 3 Implemented prototype of the inertial mass actuator

3.3 Mechanical Design of the Inertial Mass Actuator

According to the model (Fig. 2) a prototype was built, that is shown in Fig. 3.

A thin beam is clamped between two steel blocks, which form the mounting of the inertial mass actuator. The width of the beam is large compared to its length to prevent torsion. The hole pattern of the mounting is suited for a connection to a shaker or a heavy breadboard. On the other side of the beam the inertial mass is attached. It consists of an equal number of steel blocks on each side of the beam which can be changed easily for experiments. The mass can be moved on the beam via slotted holes, which is necessary for the assembling of the piezoelectric actuators. These are monolithic multilayer actuators by CeramTec with a base area of 7.1×7.6 mm and a length of 30 mm. On both sides of the piezoelectric actuators spherical ceramic caps are glued to reduce damaging shear forces. To ensure the force transmission of the actuators and to avoid loosening and therefore nonlinear behavior the actuators have to be preloaded. This is done by screws in the two masses next to the beam with a distance of 7.5 mm to the beam's surface. The tips of the preloading screws are concave as counterparts of the actuator's ceramic caps. There are corresponding holes at the mounting at the same distance from the beam. This prevents shifting of the piezoelectric actuators in operation.

3.4 Test Set up

To analyze the dynamic characteristics of the inertial mass actuator, tests were performed. The whole actuator was mounted to a rigid, heavy base plate which can be treated as an infinitely small mechanical admittance in the considered frequency range (Fig. 4). The tip mass was instrumented with an accelerometer in order to estimate the block force from the acceleration of the mass by calculating $F_B = m\ddot{x}$.

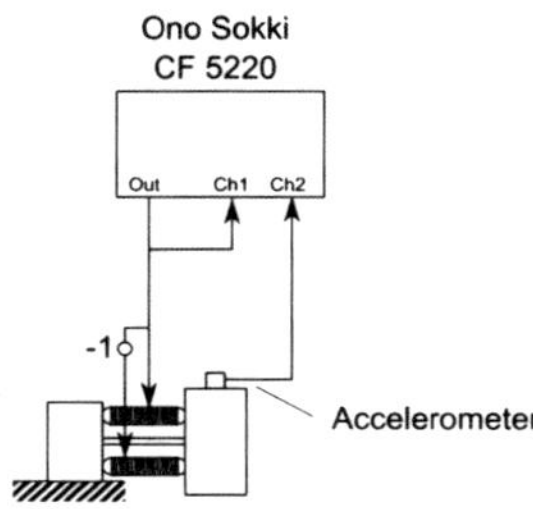

Fig. 4 Test set up for dynamic analysis of the inertial mass actuator

3.5 Experimental Validation

In order to avoid a pre-stressing of the beam, which would cause a stiffening effect and in turn a higher resonance frequency, the actuators were mounted just with a small preload. This reduces the usable amplitude range. However, this set up is still useful for experimental tests with small driving voltages.

To test the modularity of the concept, two different configurations were examined, each with a tip mass of approx. 1 kg, same type of piezoelectric actuators, but a different bending beam. The frequency response of the block force was calculated from Eq. 1, using the results from Eqs. 6 and 7. The design parameters can be found in Table 1.

The results from the FRF measurement and the calculation are compared in Fig. 5. The value for the damping had to be fitted to the measurement, which resulted in a damping coefficient of $\theta = 2\,\%$. Obviously, the calculation procures quite good estimates for the resonance frequency and the dynamic behavior over a relevant frequency range. Thus the relatively simple formulae from Eqs. 6 and 7 can be used to find proper design parameters for the inertial mass actuator.

Table 1 Configuration for the tests

Parameter		Configuration A	Configuration B
Beam			
Length	l	41.5 mm	
Width	b	60 mm	
Thickness	d	0.5 mm	0.95 mm
Young's modulus	E	220e9 N/m^2	190e9 N/m^2
Mass			
Mass	m	1.078 kg	1.087 kg
Actuator			
Stiffness	k_A	42e6 N/m	
Block force	F_A	2000 N	

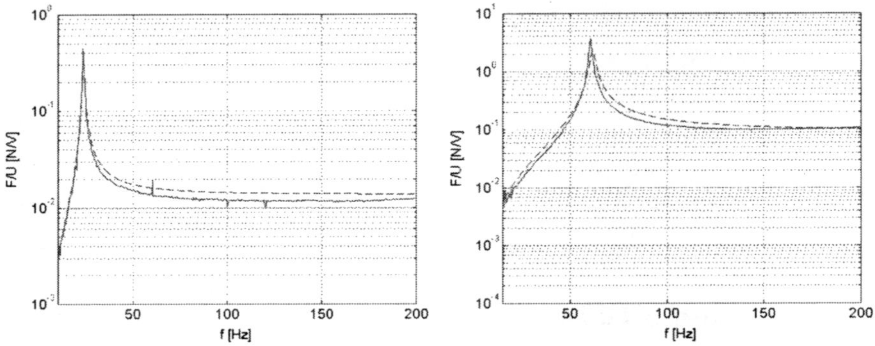

Fig. 5 Comparison of the measured (*solid line*) and calculated (*dashed line*) FRF for configuration A (*left*) and configuration B (*right*)

Afterwards, the FRF measurements were repeated with a preload on the actuators. The magnitude of the preloading force was estimated from the turning angle, flank lead of the preloading screw (0.75 mm), and the stiffness values for the beam and the actuators to:

$$F_N = \frac{k_B k_A}{k_B + k_A} \Delta S, \tag{10}$$

where $k_B = Ebd/l$ is the longitudinal stiffness of the beam and ΔS is the travel of the preloading screw. The comparison of the measured resonance frequencies and the calculated for the actuator configuration A is shown in Fig. 6. Obviously, the effect of preloading is significant and has to be taken into account. It has to be noted, that this holds also for the case of electrical preloading of the actuators,

Fig. 6 Variation of resonance frequency with respect to a preloading force on the actuators

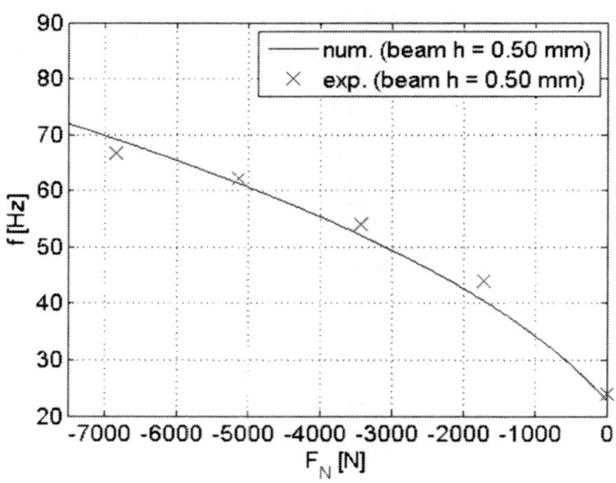

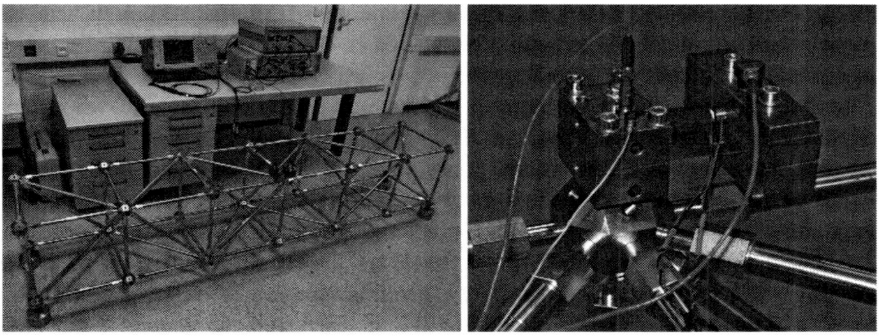

Fig. 7 Test set up for the active control experiment (*left*) and detail of the actuator instrumentation (*right*)

which results in a static displacement. However, this also allows for tuning of the actuator to desired frequencies, which will be used in the experiment described in the next section.

4 Application to an Active Vibration Control System

To evaluate the performance in an active vibration control system, the actuator was mounted to a lightweight truss structure (Fig. 7). The actuator was instrumented with accelerometers at its base and at the inertial mass.

In the first step, the actuator was tuned. Usually, an inertial mass actuator is tuned well below the resonances of the host structures. In the case considered here, a structural mode was in the vicinity of the resonance of the passive, mechanically preloaded inertial mass actuator, which resulted in a vibration absorption effect at 43 Hz (Fig. 8, left). This effect was enhanced by choosing an electrical preload of

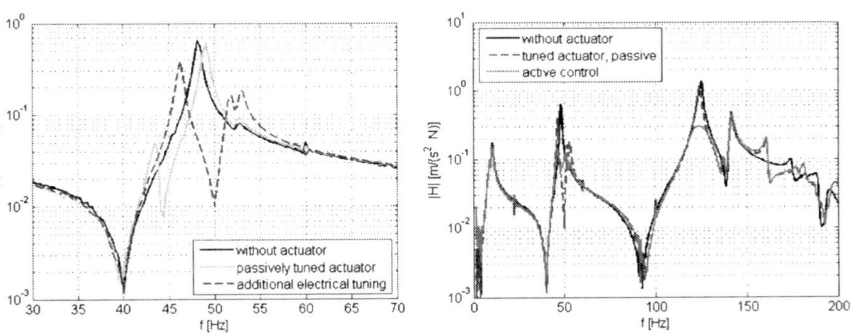

Fig. 8 Tuning of the actuator to the first elastic mode of the truss and control results

$50V_{DC}$, which resulted in an absorption frequency of 50 Hz. Although this frequency does not match the structural resonance at 48 Hz exactly, this configuration served well as a basis for adjusting the active control system.

In the next step, the active control system was set up and tuned. Two control loops were implemented. First, an active velocity feedback loop was used to enhance the damping of the inertial mass actuator, which caused a better performance of the absorption at the first mode at 48 Hz and a higher robustness against interaction between different control loops. Second, an acceleration feedback loop was tuned to the second mode of the truss at 125 Hz (Fig. 8, right). Since the actuator is not symmetric, it is exciting a bending moment at its base additional to the normal force. This results in lowering of a higher resonance frequency from 180 to 160 Hz and a deterioration of the vibration amplitudes there.

5 Conclusions

A concept for a piezoelectric driven inertial mass actuator, using a bending beam as an elastic lever, was developed and implemented. The modular design allows for tuning of the resonance by changing components. Due to the use of the bending beam also as a preloading spring, the actuator resonance is shifted during preloading. However, this effect can also be used to apply the actuator as a tuned vibration absorber at its resonance and as an active element at higher frequencies. Future developments can include the self-tuning of the absorber by automatic control of the electrical preload. Furthermore, the design should be improved towards a symmetric set up with two inertial masses in order to compensate for the bending moments and enhance the generated normal forces.

Acknowledgements This work was partly conducted within the framework of the LOEWE center AdRIA. The support by the state of Hesse is gratefully acknowledged.

References

1. Winberg M, Johansson S, Claesson I (2004) Inertial Mass Actuators—Understanding and Tuning. In: Proceedings of the ICSV 11, St. Petersburg
2. Belgacem W, Alain B, Masson P (2012) Active vibration control on a quarter-car for cancellation of road noise disturbance. J Sound Vib 331(14):3240–3254
3. Persson P, Lagö TL, Norberg A (1999) Active control of sleeper-induced sound in a high speed train. In: Proceedings of the sixth international congress on sound and vibration. pp 1747–1752
4. Díaz IM, Reynolds P (2009) Robust saturated control of human-induced floor vibrations via a proof-mass actuator. Smart Mater Struct 18(12):125024. doi:10.1088/0964-1726/18/12/125024
5. Lesieutre G, Rusovici R, Koopmann G, Dosch J (1995) Modeling and characterization of a piezoceramic inertial actuator. In: Proceedings of 36th Structures, Structural Dynamics and Materials Conference

6. Konstanzer P, Jänker P, Storm S (2007) A piezo inertial force generator optimized for high force and low frequency. Smart Mater Struct 16:1260–1264
7. Bös J, Janssen E, Kauba M, Mayer D (2008) Active vibration reduction applied to the compressor of an air-conditioning unit for trams. In: Proceedings of Acoustics
8. Preumont A (2011) Vibration control of active structures. Springer, Berlin

Heat Transfer and Thermoelastic Dynamics of a Rotating Flexible Disc in a Hard Disc Drive

Yong-Chen Pei and Huajiang Ouyang

Abstract Multi-field dynamic coupling takes place in computer hard disc drives, involving air flow, heat transfer and thermoelastic vibration of a rotating flexible disc in an enclosure filled with air. Air velocities and pressure induced by disc rotation in the enclosure are obtained by using penalty finite element method. Temperature increments in the rotating disc, driving shaft, enclosure and air flow are determined interactively by including the external heat sources from the shaft driving motor and enclosure circuit board, the internal heat source from aerodynamic heating due to viscous fluid dissipation, the heat convection in air flow, and the free convection heat loss at enclosure's outside surfaces. Natural frequencies of the rotating disc under the thermal stresses induced by the disc's temperature increment and centrifugal force are determined. Effects of air flow, heat convection and aerodynamic heating induced by disc rotation on heat balance in the enclosure and natural frequencies of the rotating flexible disc are investigated. This investigation is useful to hard drive design.

Keywords Rotating disc · Multi-physical interaction · Heat transfer · Air flow · Thermoelastic dynamics

1 Introduction

In computer hard disc drives, a high speed rotating disc works in the surrounding environment with air flow and heat. The air flow and heat transfer induced by disc rotation and their influences on the dynamics of rotating discs are the fundamental

Y.-C. Pei
Institute of Mechanical Science and Engineering, Jilin University, Nanling Campus, Changchun 130025, People's Republic of China
e-mail: yongchen_pei@hotmail.com

H. Ouyang (✉)
School of Engineering, University of Liverpool, The Quadrangle, Liverpool L69 3GH, UK
e-mail: h.ouyang@liverpool.ac.uk

J.K. Sinha (ed.), *Vibration Engineering and Technology of Machinery*, Mechanisms and Machine Science 23, DOI 10.1007/978-3-319-09918-7_53

and crucial issues to improve the performance of rotating discs [1]. Firstly, considerable vibration and even flutter instability of a rotating disc can be caused by the air flow induced by disc rotation. In view of acoustic and structural interactions, Jana and Raman [2] solved the wave equation governing the propagation of infinitesimal disturbances of air flow to analyse the vibration of a rotating flexible disc, and they found that coalescence between the acoustic and structural modes could lead to disc flutter instability at supercritical speeds. By treating the coupling dynamics of the air flow and rotating disc as fluid and structure interaction problems, Gad and Rhim [3] investigated dynamics of a flexible disc coupled to thin air film and rotating close to a rigid rotating wall. Cheng et al. [4] designed several enclosure covers of an optical disc drive to improve the flow-induced vibration of rotating discs. Secondly, heat transfer in the surrounding environment of rotating discs is affected considerably by the air flow [1]. Effects of disc rotation and heat transfer on the viscous fluid flow and temperature distribution between heated contracting rotating discs were analyzed by Nazir and Mahmood [5]. Jiji and Ganatos [6] analyzed the microscale steady axisymmetric flow and heat transfer in two infinite parallel discs separated by a gas-filled micro-gap.

There are three limitations in the above investigations on the air flow and heat transfer in rotating disc systems: (1) Some research has been done on the air flow induced vibration and flutter instability of a rotating disc [2–4], but there is no study on the problem of heat transfer and thermoelastic dynamics coupling with the fluid flow. (2) Aerodynamic heating effect due to fluid's viscous dissipation was neglected [5], whereas the corresponding results cannot be confirmed by the experimental observation by Kameya et al. [7] that the disc temperature increases around the disc's outer edge. (3) Heat transfer of the rotating disc was ignored [5, 6], although the dynamic characteristics of the rotating disc can be affected significantly by the thermal stresses induced by the disc temperature variation.

Steady state air flow, heat transfer and thermoelastic dynamics in the multi-field coupling system of a flexible disc rotating in an enclosure are studied in this paper for computer hard disc drives. The heat generated from the driving motor and enclosure circuit board is involved as the external heat source, and that from the aerodynamic heating of air flow is involved as the internal one. Temperature distribution in the rotating disc, driving shaft, enclosure and air is determined, and the natural frequencies of the rotating disc are computed under the thermal stresses induced by disc temperature distribution and centrifugal force.

2 System Modeling and Governing Equations

As shown in Fig. 1, a flexible annular disc rotates at a constant angular speed Ω in a cylindrical enclosure. The disc thickness h is assumed to be small compared with its outer radius a, and the driving shaft with radius b is assumed to be rigid to clamp and drive the disc. Due to electromagnetic induction and frictions in the driving motor, a heat flux q_S is generated and transferred to the driving shaft.

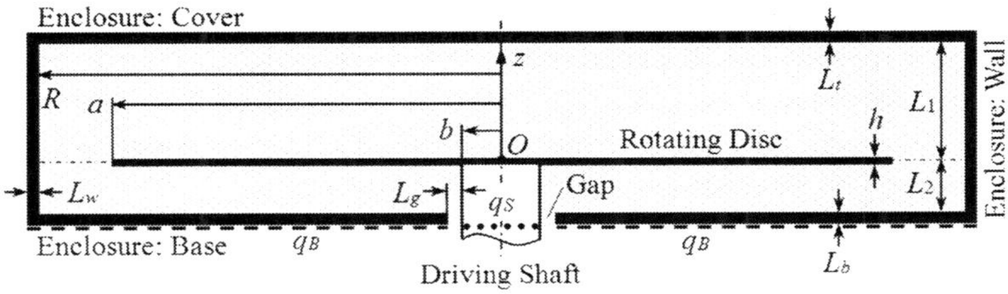

Fig. 1 Rotating disc and its driving shaft in cylindrical enclosure

The enclosure's base is always heated by circuit board below in hard disc drives, and a heat flux q_B is generated and transferred to the enclosure's base. In the case of hard disc drives, the read/write heads and suspension system are ignored in the current work to reduce the huge computational workload and to highlight heat balance and thermoelastic dynamics of the rotating flexible disc. Similarly, the same simplification also can be found in many references [2–4].

This paper studies the rotating flexible disc system in its normal steady working state. The axisymmetric steady-state air flow and temperature distribution should be much more prominent [6] than the transient and asymmetrical ones, and thus the latter can be neglected. However, as reported by Arafat et al. [10] and Pei et al. [11], the rotating flexible disc's transverse deflection due to non-axisymmetric modes (m, $n \neq 0$) is influenced significantly by disc's axisymmetric temperature distribution and thermal stresses. Therefore, non-axisymmetric vibration of the flexible disc is modeled and analyzed in the current work.

The axisymmetric Navier-Stokes and continuity equations [8, 9] for incompressible laminar flow can be written as

$$\begin{aligned}
\rho_a\left(u_r\frac{\partial u_r}{\partial r}+u_z\frac{\partial u_r}{\partial z}-\frac{u_\theta^2}{r}\right)+\frac{\partial p}{\partial r}&=\mu_a\left(\frac{\partial^2 u_r}{\partial r^2}+\frac{\partial u_r}{r\partial r}+\frac{\partial^2 u_r}{\partial z^2}-\frac{u_r}{r^2}\right)\\
\rho_a\left(u_r\frac{\partial u_\theta}{\partial r}+u_z\frac{\partial u_\theta}{\partial z}+\frac{u_r u_\theta}{r}\right)&=\mu_a\left(\frac{\partial^2 u_\theta}{\partial r^2}+\frac{\partial u_\theta}{r\partial r}+\frac{\partial^2 u_\theta}{\partial z^2}-\frac{u_\theta}{r^2}\right)\\
\rho_a\left(u_r\frac{\partial u_z}{\partial r}+u_z\frac{\partial u_z}{\partial z}\right)+\frac{\partial p}{\partial z}&=\mu_a\left(\frac{\partial^2 u_z}{\partial r^2}+\frac{\partial u_z}{r\partial r}+\frac{\partial^2 u_z}{\partial z^2}\right)
\end{aligned}\tag{1}$$

$$\partial u_r/\partial r+u_r/r+\partial u_z/\partial z=0\tag{2}$$

where $p(r, z)$ is air pressure relative to the ambient atmosphere; $u_r(r, z)$, $u_\theta(r, z)$, $u_z(r, z)$ are air velocity components in the radial r, circumferential θ and transversal z directions, respectively. The air velocity component u_θ at the disc and shaft surfaces equals to the local linear peripheral speed Ωr, while other velocity components on these surfaces are zeros. All velocity components vanish at the enclosure's inside surfaces. Within the driving shaft gap, the relative air pressure p vanishes and $\partial u_r/\partial r + u_r/r + \partial u_z/\partial z$ is zero in light of mass conservation [3].

The steady state pressure difference $P(r)$ between the upper and lower disc surfaces ($b \le r \le a$, $z = \pm h/2$) can be expressed as $P(r) = p(r,z)\big|_{z=-h/2} - p(r,z)\big|_{z=h/2}$.

Heat transfer takes place in the fluid medium of inside air flow, and between it and the solid media of shaft, disc and enclosure. The energy equation [9] for the fluid and solid can be expressed separately as

$$\rho_a c_a \left(u_r \frac{\partial \Gamma}{\partial r} + u_z \frac{\partial \Gamma}{\partial z} \right) = k_a \left(\frac{\partial^2 \Gamma}{\partial r^2} + \frac{\partial \Gamma}{r \partial r} + \frac{\partial^2 \Gamma}{\partial z^2} \right) + \mu_a \Phi(r,z) \tag{3}$$

$$k_s(\partial^2 \Gamma/\partial r^2 + \partial \Gamma/\partial r/r + \partial^2 \Gamma/\partial z^2) = 0 \tag{4}$$

where c_a is the specific heat of air, k_a (air) and k_s (shaft, disc and enclosure) is the thermal conductivity; $\Gamma(r, z)$ is the temperature increment relative to the atmospheric temperature T_a far away from the modelled domain; $q(r, z) = \mu_a \Phi$ represents the aerodynamic heating flux due to viscous air dissipation, Φ is the dissipation function [9]. At the enclosure's cover and walls, free convection boundary condition is considered with a convective heat transfer coefficient h_a. It is assumed that the heat generation power of the shaft base and enclosure base are P_S and P_B, and then the averaged heat flux $q_S = P_S/(\pi b^2)$ and $q_B = P_B/[\pi R^2 - \pi(b + L_g)^2]$. Heat fluxes q_S and q_B are assigned at the shaft base and enclosure base. The adiabatic condition is assumed at the driving shaft gap.

Subjected to the air pressure and thermal stresses and centrifugal force, the governing equation of the rotating flexible disc [10, 11] can be written as

$$\begin{aligned} &\rho h \left(\frac{\partial^2 w}{\partial t^2} + 2\Omega \frac{\partial^2 w}{\partial t \partial \theta} + \Omega^2 \frac{\partial^2 w}{\partial \theta^2} \right) - \rho h a^2 \Omega^2 \left[\frac{1}{r} \frac{\partial}{\partial r} \left(r \sigma_r \frac{\partial w}{\partial r} \right) + \frac{1}{r^2} \frac{\partial}{\partial \theta} \left(\sigma_\theta \frac{\partial w}{\partial \theta} \right) \right] \\ &+ \frac{Eh^3}{12(1-\nu^2)} \nabla^4 w + \frac{Eh^2 \alpha_T}{1-\nu} \nabla^2 \Theta_Q = P(r) + \frac{\partial^2 w}{\partial r^2} \left(\frac{\partial \Psi}{r \partial r} + \frac{\partial^2 \Psi}{r^2 \partial \theta^2} \right) \\ &+ \left(\frac{\partial w}{r \partial r} + \frac{\partial^2 w}{r^2 \partial \theta^2} \right) \frac{\partial^2 \Psi}{\partial r^2} - 2 \left(\frac{\partial^2 w}{r \partial r \partial \theta} - \frac{\partial w}{r^2 \partial \theta} \right) \left(\frac{\partial^2 \Psi}{r \partial r \partial \theta} - \frac{\partial \Psi}{r^2 \partial \theta} \right) \end{aligned} \tag{5}$$

$$\nabla^4 \Psi = -Eh\alpha_T \nabla^2 \Theta_M, \ \Theta_M = \int_{-h/2}^{h/2} \Gamma(r,z) dz/h, \ \Theta_Q = \int_{-h/2}^{h/2} \Gamma(r,z) z dz/h^2 \tag{6}$$

where $w(r, \theta, t)$ is disc's transverse deflection; ρ is mass per unit volume of the disc, E is Young's modulus, ν is Poisson's ratio, and α_T is coefficient of linear thermal expansion; σ_r and σ_θ are the radial and circumferential membrane stress resultants due to centrifugal force [12]. At the clamped edge $r = b$ and free edge $r = a$, the boundary conditions of the transverse deflection w and thermal stress function Ψ can be found in Deng and Ouyang [12].

3 Solution Procedures for Air Flow and Temperature

With the penalty finite element method by Elman et al. [13] and Reddy [14], a small penalty parameter ε ($\varepsilon \longrightarrow 0$) is introduced into the continuity equation Eq. (2),

$$\varepsilon p + \partial u_r/\partial r + u_r/r + \partial u_z/\partial z = 0 \tag{7}$$

By using the $Q_2 - P_{-1}$ element [13] to discretise the solution domain in the r-z plane, finite element equations of the system can be yielded from Eqs. (1–2) and Eqs. (3–4) respectively,

$$\mathbf{N}(\boldsymbol{X})\boldsymbol{X}/2 + (\Omega\mathbf{B} + \mathbf{D})\boldsymbol{X} + \boldsymbol{\Omega}(\boldsymbol{\Omega B}/2 + \boldsymbol{D}) = \boldsymbol{0} \tag{8}$$

$$\boldsymbol{x} = \boldsymbol{h}(\boldsymbol{X}) - [q_B\boldsymbol{Q}_B(\boldsymbol{X}) + q_S\boldsymbol{Q}_S(\boldsymbol{X})] \tag{9}$$

where the unknown variables for air velocities are collected as vector $\boldsymbol{X}$, those for temperature increment are collected as $\boldsymbol{x}$; Matrix $\mathbf{N}(\boldsymbol{X})$ and vector $\boldsymbol{h}(\boldsymbol{X})$ are derived from the nonlinear convection terms, $\mathbf{B}$ and $\boldsymbol{D}$ are coefficient matrices and vectors, $\boldsymbol{Q}_B(\boldsymbol{X})$ and $\boldsymbol{Q}_S(\boldsymbol{X})$ are related to the heat generation q_B and q_S. From Eq. (9), temperature increment $\Gamma(r, z)$ can be divided into two parts induced separately by the internal aerodynamic heating (q) denoted as $\Gamma_{\mathrm{Air}}(r, z)$ with solution $\boldsymbol{x} = \boldsymbol{h}(\boldsymbol{X})$, and by the external shaft and enclosure base heat generations (q_B, q_S) denoted as $\Gamma_{\mathrm{Base}}(r, z)$ with $\boldsymbol{x} = -[q_B\boldsymbol{Q}_B(\boldsymbol{X}) + q_S\boldsymbol{Q}_S(\boldsymbol{X})]$.

Equation (8) is a nonlinear quadratic vector equations, the iterative algorithm presented by Poloni [15] is used to solve $\boldsymbol{X}$ numerically.

The total heat generation power of aerodynamic heating can be determined by

$$P_A = 2\pi \iint_{V_a} \mu_a \Phi \; r dr dz \tag{10}$$

where V_a represents the solution domain of air media.

4 Solution Procedure for Disc's Natural Frequency

The transverse vibration of a rotating flexible disc [11] can be assumed as

$$w(r, \theta, t) = \sum_{n=-\infty}^{\infty} e^{in\theta} \sum_{m=0}^{\infty} \varphi_{m,n}(r) w_{m,n}(t) \tag{11}$$

where $\varphi_{m,n}(r)$ is the shape function of stationary disc mode (m,n).

Then an ordinary differential equation for the disc mode of n nodal diameters can be derived from Eqs. (11) and (5) with Galerkin's method [11],

$$\ddot{w}_n + i2n\Omega\dot{w}_n + (\mathbf{S}_n - \mathbf{H}_n + \Omega^2\mathbf{L}_n - n^2\Omega^2\mathbf{I})w_n = f_n \tag{12}$$

where $f_n = [\ldots f_{m,n}\ldots]^{\mathrm{T}}$, $f_{m,0} = \frac{1}{\rho h}\int_b^a \varphi_{m,0}(r)\left[P(r) - \frac{Eh^2\alpha_T}{1-v}\nabla^2\Theta_Q(r)\right] rdr$ and $f_{m,n} = 0$ for $n \neq 0$, since both $P(r)$ and $Eh^2\alpha_T\nabla^2\Theta_Q(r)/(1-v)$ are axisymmetric; $w_n = [\ldots w_{m,n}\ldots]^{\mathrm{T}}$ for $m = 0, 1, 2, \ldots$; matrices $\mathbf{S}_n$ and $\mathbf{L}_n$ can be found in Pei et al. [11], and the matrix induced by thermal stress can be written as

$$\mathbf{H}_n = \Theta_M(b)\mathbf{H}_n^b + \Theta'_M(b)\mathbf{H}_n^d + \sum_{m=0}^{\infty} J_m(\Theta_M)\mathbf{H}_{m,n}^r \tag{13}$$

From Eq. (13), dynamic property of the rotating disc is influenced by the mean temperature increment Θ_M in three ways: $\Theta_M(b)$, $\Theta'_M(b)$ and $J_m(\Theta_M)$. For the temperature distributions separately induced by the internal aerodynamic heating and by the external heating from the shaft and enclosure base, the three components $\Theta_M(b)$, $\Theta'_M(b)$ and $J_m(\Theta_M)$ of thermal membrane stress resultants are denoted as $\Theta_M^{Air}(b)$, $\Theta_M^{\prime Air}(b)$, $J_m(\Theta_M^{Air})$, and $\Theta_M^{Base}(b)$, $\Theta_M^{\prime Base}(b)$, $J_m(\Theta_M^{Base})$.

Substituting the homogenous solution $w_n = e^{(\lambda - in\Omega)t}A_n$ into the homogenous form of Eq. (12) yields an eigenvalue problem $[\lambda^2\mathbf{I} + (\mathbf{S}_n - \mathbf{H}_n + \Omega^2\mathbf{L}_n)]A_n = \mathbf{0}$ as presented in Pei et al. [11]. Then the dimensionless natural frequency of disc mode (m,n) is introduced as $\omega_{m,n} = \mathrm{Im}(\lambda)/\Omega_{0,0}$, where $\Omega_{0,0}$ is the natural frequency of disc mode (0,0) for a non-rotating disc [11] without any thermal stress.

5 Data Analysis and Results

Default geometric parameters [4] of the shaft-disc-enclosure system in Fig. 1 are selected as b = 15 mm, a = 60 mm, h = 1 mm, R = 62 mm, L_1 = 6 mm, L_2 = 4 mm, L_w = 2 mm, L_t = 1 mm, L_b = 1 mm and L_g = 1 mm. From Tan et al. [16] the heat generation powers of the shaft base and enclosure base are selected as P_S = 3 W and P_B = 2 W. In addition, some fundamental physical parameters of air, disc, shaft and enclosure are listed in Table 1.

Table 1 System fundamental physical parameters

Air media	Air fluid		$\rho_a = 1.205$ kg/m^3, $\mu_a = 1.8135 \times 10^{-5}$ kg/m/s, $v_a = 1.505 \times 10^{-5}$ m^2/s, $c_a = 1.012 \times 10^3$ J/kg/K, $k_a = 0.024$ W/m/K
Solid media	Rotor	Disc	$\rho = 2{,}700$ kg/m^3, $E = 70 \times 10^9$ Pa, $v = 0.35$, $k_s = 200$ W/m/K, $\alpha_T = 23 \times 10^{-6}$ K^{-1}
		Shaft	$k_s = 60$ W/m/K
	Enclosure	Cover	$k_s = 60$ W/m/K, $h_a = 8$ W/m^2/K
		Wall	$k_s = 230$ W/m/K, $h_a = 6$ W/m^2/K
		Base	$k_s = 230$ W/m/K

With Ω increasing from 0 rpm to 8,000 rpm, the total heat generation power of aerodynamic heating P_Λ in Eq. (10) is illustrated in Fig. 2. It can be seen that P_Λ increases very fast, and can get up to several watts, which are in the same order as heat generation power of the shaft base and the enclosure base at 8,000 rpm. It can be approximated by curve fitting to a polynomial in terms of Ω (krpm) as $P_\Lambda = -1.3041 \times 10^{-2}\Omega + 1.8334 \times 10^{-2}\Omega^2 + 3.0987 \times 10^{-3}\Omega^3$ with the maximum fitting error 6.5963×10^{-3} W, that is to say P_Λ increases with Ω cubed. From Eq. (10), P_Λ is the volume integral of aerodynamic heat flux $q(r, z) = \mu_a\Phi$, and Φ is related to the square of the gradients of air velocity components u_r, u_θ and u_z [9]. With the increase of Ω, velocity components u_r, u_θ and u_z increase, and their gradients also increase simultaneously.

With the increase of Ω and r/a, the mean temperature increments $\Theta_M^{Base}(r)$, $\Theta_M^{Air}(r)$, $\Theta_M(r)$ and $\Theta_Q(r)$ of the disc are illustrated in Fig. 3. Induced by q_S and q_B, $\Theta_M^{Base}(r)$ decreases along r/a and with Ω due to the convection induced by disc rotation as a "cooling fan", as shown in Fig. 3a. However, $\Theta_M^{Air}(r)$ increases considerably with Ω and r/a due to the aerodynamic heating in Fig. 3b. As sum of $\Theta_M^{Base}(r)$ and $\Theta_M^{Air}(r)$, the total ordinary mean temperature increment $\Theta_M(r)$ decreases first and then increases with Ω in Fig. 3c. $\Theta_Q(r)$ increases with Ω, and increases first and then decreases with r/a in Fig. 3d. It almost vanishes at both edges of the disc. Nevertheless, since the disc is very thin in the current investigation, $\Gamma(r, z)$ varies slightly across the disc thickness and thus the magnitude of Θ_Q is quite small compared with Θ_M as shown in Fig. 3. Moreover, $P(r)$ and the thermal moment $Eh^2\alpha_T\nabla^2\Theta_Q(r)/(1-\nu)$ act as the steady state forces on the disc in Eq. (12), and they do not affect natural frequency $\omega_{m,n}$, to be seen in the following numerical results and discussions.

With the increase of Ω, natural frequency $\omega_{m,n}$ is illustrated in Fig. 4a–f. Compared with the natural frequency without any thermal stress, disc natural frequencies $\omega_{m,n}$ for the disc modes (0,0), (0,1), (1,0) and (1,1) are decreased by the thermal stress from $\Theta'_M(b)$, whereas it is increased by the thermal stress from $\Theta_M(b)$ and $J_m(\Theta_M)$. However, the entirely opposite situation ($\omega_{m,n}$ being decreased or increased) takes place for the disc modes (0,2), (0,3). It can be concluded that the effect of the thermal stress from $\Theta'_M(b)$ on natural frequency is quite opposite to that from $\Theta_M(b)$ and $J_m(\Theta_M)$. Natural frequency $\omega_{m,n}$ with the thermal stress just

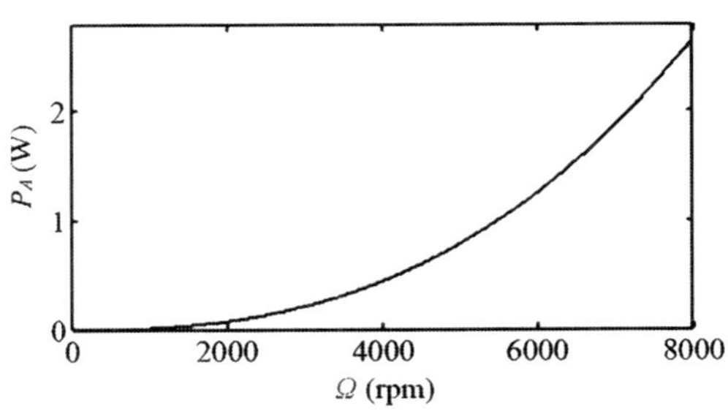

Fig. 2 Total heat generation power of aerodynamic heating varying with disc speed

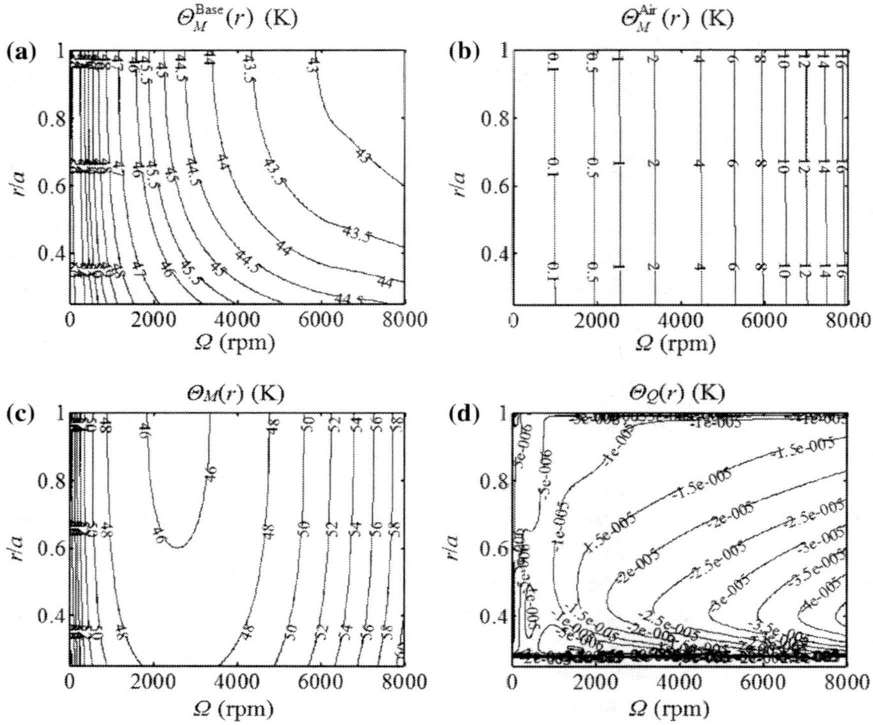

Fig. 3 Mean temperature increments varying with disc speed and disc radial position. (**a**) Ordinary mean temperature increment $\Theta_M^{Base}(r)$ induced by heat generations in the shaft and enclosure's base; (**b**) Ordinary mean temperature increment $\Theta_M^{Air}(r)$ induced by the aerodynamic heating; (**c**) Total ordinary mean temperature increment $\Theta_M(r) = \Theta_M^{Base}(r) + \Theta_M^{Air}(r)$; (**d**) Total quadratic mean temperature increment $\Theta_Q(r)$

from $\Theta'_M(b)$ and $J_m(\Theta_M)$ increases consistently with Ω from 0 rpm to 8,000 rpm, whereas the natural frequency with the thermal stress just from $\Theta_M(b)$ decreases first and then increases for disc modes (0,0), (0,1), (1,0) and (1,1), it increases for the disc mode (0,2), and it increases first and then decreases for disc modes (0,3). These can be explained by the effects of $\Theta_M(b)$, $\Theta'_M(b)$ and $J_m(\Theta_M)$ varying with Ω in Fig. 3. As shown in Fig. 4a–f, the effect of the thermal stress from $\Theta_M(b)$ on $\omega_{m,n}$ is very considerable, and it is more dominant than that from $\Theta'_M(b)$ and $J_m(\Theta_M)$. Thus it can be observed that the natural frequency with the thermal stress just from $\Theta_M(b)$ is very close to the natural frequency with all thermal stresses.

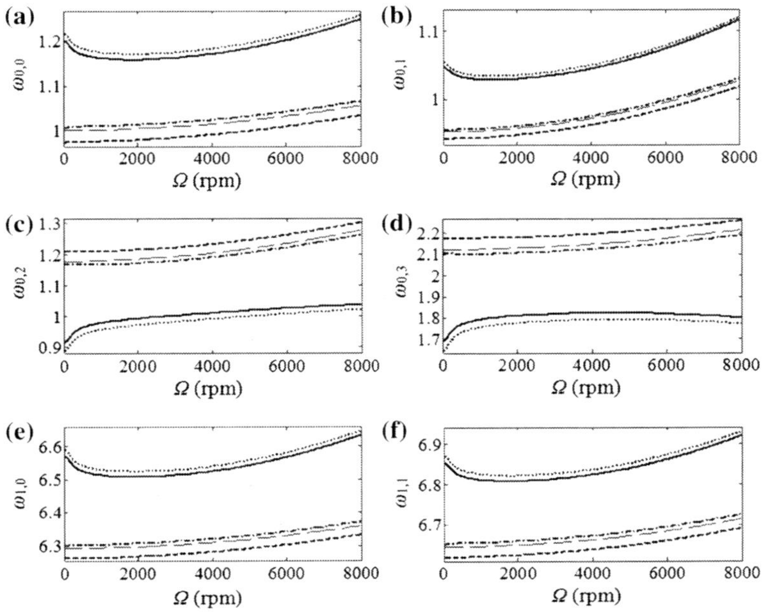

Fig. 4 System natural frequency varying with disc speed. – –: Without any thermal stress; ••••: With thermal stress just from $\Theta_M(b)$; —: With thermal stress just from $\Theta'_M(b)$; -•-: With thermal stress just from $J_m(\Theta_M)$; —: With all thermal stresses. (**a**) Disc mode (0,0); (**b**) Disc mode (0,1); (**c**) Disc mode (0,2); (**d**) Disc mode (0,3); (**e**) Disc mode (1,0); (**f**) Disc mode (1,1)

6 Conclusions

(1) Total aerodynamic heating power of viscous air dissipation increases very fast with disc speed, and can get up to several watts, which are in the same order as heat generation power of the shaft driving motor and the enclosure base circuit board.
(2) With the increase of disc speed, temperature increment decreases first due to the convection induced by disc rotation as a "cooling fan", and then increases due to the considerable aerodynamic heating.
(3) Natural frequency of the rotating flexible disc is influenced by temperature increment in three ways: the mean temperature increment and its radial derivative at disc clamped edge, and the distribution characteristics of mean temperature increment along disc radial direction, where the first way is more dominant than the others. The positive or negative effect of these three ways on the natural frequency varies with disc mode, and that of the second way is always quite opposite to the others.

References

1. Harmand S, Pellé J, Poncet S, Shevchuk IV (2013) Review of fluid flow and convective heat transfer within rotating disk cavities with impinging jet. Int J Therm Sci 67(2013):1–30
2. Jana A, Raman A (2006) Aeroelastic flutter of a disk rotating in an unbounded acoustic medium. J Sound Vib 289(3):612–631
3. Gad AMM, Rhim YC (2008) Study of a flexible disk rotating close to a rigid rotating wall considering fluid inertia effects. J Sound Vib 317(3–5):473–489
4. Cheng CC, Wu FT, Ho KL (2009) Reduction of flow-induced vibration and noise of an optical disk drive. J Sound Vib 320(1–2):43–59
5. Nazir A, Mahmood T (2011) Analysis of flow and heat transfer of viscous fluid between contracting rotating disks. Appl Math Model 35(7):3154–3165
6. Jiji LM, Ganatos P (2010) Microscale flow and heat transfer between rotating disks. Int J Heat Fluid Flow 31(4):702–710
7. Kameya T, Matsuda Y, Yamaguchi H, Egami Y, Niimi T (2012) Pressure-sensitive paint measurement on co-rotating disks in a hard disk drive. Opt Lasers Eng 50(1):82–86
8. Shevchuk IV (2009) Convective heat and mass transfer in rotating disk systems. Springer, Berlin. ISBN 978-3-642-00718-7
9. Bejan A, Kraus AD (2003) Heat transfer handbook. Wiley, New Jersey. ISBN 978-0-471-39015-2
10. Arafat HN, Nayfeh AH, Faris W (2004) Natural frequencies of heated annular and circular plates. Int J Solids Struct 41(11–12):3031–3051
11. Pei Y-C, He L, Wang J-X (2010) Rotating flexible disk under shaft temperature increment. J Sound Vib 329(17):3550–3564
12. Deng H, Ouyang H (2011) Vibration of spinning discs and powder formation in centrifugal atomization. Proc R Soc A: Math Phys Eng Sci 467(2011):361–380
13. Elman HC, Silvester DJ, Wathen AJ (2005) Finite elements and fast iterative solvers: with applications in incompressible fluid dynamics. Oxford University Press, New York. ISBN 978-0-19-852867-8
14. Reddy JN (2006) An introduction to the finite element method, 3rd edn. McGraw-Hill, New York. ISBN 0070607419
15. Poloni F (2013) Quadratic vector equations. Linear Algebra Appl 438(4):1627–1644
16. Tan CP, Yip TH, Tan DTD (2011) Thermal characteristics of enterprise-class hard disk drives. IEEE Trans Compon Packag Manuf Technol 1(6):868–872

A New Direct Method for Updating Mass and Stiffness Matrices with No Spillover

Jiafan Zhang, Huajiang Ouyang and Jun Yang

Abstract A direct method for model updating of mass and stiffness matrices of structures without spillover is presented, which requires the knowledge of only the few eigenpairs to be updated of the original undamped model. Upon a necessary and sufficient condition, proposed previously by authors, for the incremental mass and stiffness matrices that modify some eigenpairs while keeping other eigenpairs unchanged, the finite element model updating problem that preserves symmetry and avoids spillover is formulated as a semi-definite programming problem, which can be efficiently solved by existing semi-definite programming algorithms. Numerical examples are given to demonstrate the accuracy and effectiveness of the presented updating method.

Keywords Finite element model updating · Inverse eigenvalue problem · Direct method · Spillover · Spurious modes

1 Introduction

Updating a finite element model to match measured modal data is important in design, construction and maintenance of engineering structures and mechanical systems. Due to the limitations of available computational methods to handle distributed parameter systems, the finite element method is generally used to discretize such systems to form finite element models. However, accurate geometric and

J. Zhang · J. Yang
School of Mechanical Engineering, Wuhan Polytechnic University, Wuhan 430023, China
e-mail: jfz@whpu.edu.cn

J. Yang
e-mail: yangjun_1017@126.com

H. Ouyang (✉)
School of Engineering, University of Liverpool, Liverpool L69 3GH, UK
e-mail: h.ouyang@liverpool.ac.uk

J.K. Sinha (ed.), *Vibration Engineering and Technology of Machinery*, Mechanisms and Machine Science 23, DOI 10.1007/978-3-319-09918-7_54

material properties must be used in order for the finite element models to be predictable. Basically, finite element model updating (FEMU) is to incorporate the measured modal data into the finite element model to produce an improved finite element model with modal properties that closely match the experimental modal data. Then the updated model may be considered a better dynamic representation of the structure. Because of the importance of this problem, there now exist a relatively large number of works published in this field. Most of the works prior to 1995 were touched upon in the book by Friswell and Mottershead [1], who conducted a comprehensive review on papers published before 1995. Some of the more recent results can be found in review articles [2–4] and the references therein.

The existing methods of FEMU can be classified as working in the frequency domain or the time domain. They have also been classified by other researchers into the following three classes: (i) Direct matrix updating methods, (ii) Iterative methods using modal data, and (iii) Methods using frequency response data, and some of them have been widely used and successfully applied to FEMU for a variety of structures. Baruch, Berman and Nagy are the early advocates of the direct matrix updating methods [5–7]. They usually assumed that one matrix among the model mass matrix, the stiffness matrix, and the measured modal vector matrix was correct, and then used the Lagrange multiplier method and least-squares method to modify the other two matrices. Wei introduced an approach that could update the mass and stiffness matrices simultaneously using the measured eigenvector matrix [8]. The effects of the modifications were determined from the corrected mass and stiffness matrices. Additionally, Kuo et al. recently proposed a direct method which seemed more efficient and reliable [9]. The iterative updating methods use the sensitivity of the parameters to update the model. A number of chosen model parameters were systematically adjusted with respect to the measured modal data (natural frequencies and mode shapes) in order to minimize the objective function defined. There are a number of works that extensively discussed and demonstrated the results obtained from model updating using the optimization methods, for example, see [10–12]. In a class of model updating methods using measured frequency response functions (FRF), the FE models are updated based on the fully damped response along a frequency axis and not on an estimated set of modal data. Most of the FRF-based model updating techniques relied on expression of the system matrices as a polynomial of system parameters; for example, see [13–15]. The FEMU can also be performed with computational intelligence algorithms, such as neural networks, genetic algorithms, particle-swarm optimization, simulated annealing, and so on [16, 17].

In spite of these developments mentioned above, there are still some unsolved issues with the FEMU problem. For example, these existing methods may reproduce the given set of measured data while keeping symmetry of updated matrices; however, they cannot guarantee that extra, spurious modes are not introduced into the frequency range of interest after updating [1]. When updating a model, it is desirable to match only the measured modal data without tampering with the other unmeasured modal data in the original model. Such an updating method is known to have *no spillover*. In the last few years, Carvalho and Datta, and their collaborators, Mao and Dai respectively developed two novel direct methods to solve the problem with only the stiffness matrix updated [18], and with both the stiffness and mass matrices updated, for the undamped model [19]. However, the updating method proposed in [19] cannot reproduce the measured data accurately for a numerical example in their paper. Additionally, Chu et al. presented a complete theory on when model updating of damped systems with no spillover is possible [20].

The purpose of this paper is to propose a new direct method of updating mass and stiffness matrices simultaneously avoiding spillover and preserving symmetry, based on a necessary and sufficient condition for modification of mass and stiffness matrices with no spillover that was proposed by authors [21]. A matrix equation is formulated for updating mass and stiffness matrices, which requires the knowledge of only the few eigenpairs to be updated of the original undamped model. Setting the residual norm of this matrix equation as the objective function, the FEMU problem is posed as an optimization problem, which is solved by semi-definite programming (SDP) techniques. The presented updating method is shown to reproduce the measured data accurately for those numerical examples in [19], while avoiding spillover.

2 A Necessary and Sufficient Condition for No Spillover Updating

For an n-degree-of-freedom undamped vibration system, its natural frequency ω and the mode shape vector $\boldsymbol{x}$ must satisfy the following generalized eigenvalue equation:

$$(K_a - \lambda_i M_a)\boldsymbol{x}_i = 0,\ \ i = 1, 2, \ldots, n \tag{1}$$

where $M_a,\ K_a \in \boldsymbol{R}^{n\times n}$ are mass and stiffness matrices, respectively. In general, M_a is symmetric and positive definite, and K_a is symmetric and positive semi-definite, denoted by $M_a = M_a^{\mathrm{T}} > 0,\ K_a = K_a^{\mathrm{T}} \geq 0$, where $(\cdot)^{\mathrm{T}}$ is the transpose operation. $\lambda_i = \omega_i^2$ is the square of the ith natural frequency ω_i, called the ith eigenvalue, and $\boldsymbol{x}_i$ is the corresponding ith mode shape, called the ith eigenvector. For the sake of convenience, the system model is simply denoted by $\{M_a,\ K_a\}$. Equation (1) can be written in a compact representation as follows:

$$K_a X = M_a X \Lambda \tag{2}$$

where $\Lambda = \mathrm{diag}(\lambda_1, \ldots, \lambda_p, \lambda_{p+1}, \ldots, \lambda_n)$ and $X = (\boldsymbol{x}_1, \ldots, \boldsymbol{x}_p, \boldsymbol{x}_{p+1}, \ldots, \boldsymbol{x}_n)$ be eigenvalue and eigenvector matrices of the analytical model $\{M_a,\ K_a\}$. Let $(\mu_1, \ldots, \mu_p)$ and $(\boldsymbol{y}_1, \ldots, \boldsymbol{y}_p)$ be a set of p $(p \ll n)$ eigenvalues and eigenvectors measured from an experimental structure. Mathematically, the model updating problem may be formulated as follows:

Problem FEMU: Given an analytical model $\{M_a,\ K_a\}$ and a set of its associated eigenpairs $(\lambda_i,\ \boldsymbol{x}_i)$ $(i = 1, 2, \ldots, p)$ with $p \ll n$, and another set of measured eigenpairs $(\mu_i,\ \boldsymbol{y}_i)$ $(i = 1, 2, \ldots, p)$ from an experimental or a real-life structure have been obtained, update the analytical model $\{M_a,\ K_a\}$ to $\{M,\ K\}$ of the same structure such that:

(1)
$$M = M^{\mathrm{T}} = M_a + \Delta M > 0, K = K^{\mathrm{T}} = K_a + \Delta K \geqslant 0$$

(2) The subset $(\lambda_i,\ \boldsymbol{x}_i)$ $(i = 1, 2, \ldots, p)$ is replaced by $(\mu_i,\ \boldsymbol{y}_i)$ $(i = 1, 2, \ldots, p)$ as p eigenpairs of the updated model $\{M,\ K\}$.

(3) The remaining (unknown) *n-p* eigenpairs of the updated model $\{M,\ K\}$ stay the same as those of the original model $\{M_a,\ K_a\}$.

For convenience, the following partitions and notation are used:

$$\Lambda = \mathrm{diag}(\Lambda_1, \Lambda_2),\ X = (X_1, X_2), \Lambda_1 = \mathrm{diag}(\lambda_1, \ldots, \lambda_p),$$
$$\Lambda_2 = \mathrm{diag}(\lambda_{p+1}, \ldots, \lambda_n),\ X_1 = (\boldsymbol{x}_1, \ldots, \boldsymbol{x}_p),\ X_2 = (\boldsymbol{x}_{p+1}, \ldots, \boldsymbol{x}_n),$$
$$\text{and } \Sigma_1 = \mathrm{diag}(\mu_1, \ldots, \mu_p),\ Y_1 = (\boldsymbol{y}_1, \ldots, \boldsymbol{y}_p),$$

and assume that the eigenvector matrix X of the analytical model $\{M_a,\ K_a\}$ satisfies the normalization condition $X^{\mathrm{T}} M_a X = I_n$. Moreover, the following assumptions are made in this paper:

(A1) $(\lambda_1, \ldots, \lambda_p) \cap (\lambda_{p+1}, \ldots, \lambda_n) = \emptyset$ (an empty set);
(A2) $(\mu_1, \ldots, \mu_p) \cap (\lambda_{p+1}, \ldots, \lambda_n) = \emptyset$;
(A3) All the above-mentioned eigenvalues are distinct.

The FEMU problem, as stated above, concerns finding symmetric corrective matrices ΔM and ΔK such that the following eigen-matrix equations of the updated model $\{M,\ K\}$ hold simultaneously:

$$(\mathrm{M}_a + \Delta M) X_2 \Lambda_2 = (\mathrm{K}_a + \Delta K) X_2 \tag{3}$$

$$(\mathrm{M}_a + \Delta M) Y_1 \Sigma_1 = (\mathrm{K}_a + \Delta K) Y_1 \tag{4}$$

A necessary and sufficient condition was established for ΔM and ΔK that satisfy Eq. (3) in [21], which can be stated below.

Theorem 2.1 (No spillover updating). The real symmetric corrective matrices ΔM and ΔK satisfy Eq. (3), *if and only if* ΔM and ΔK satisfy the following matrix Eq. (5a) or (5b).

$$\Delta K(M_a^{-1} - X_1 X_1^{\mathrm{T}}) - \Delta M(M_a^{-1} K_a M_a^{-1} - X_1 \Lambda_1 X_1^{\mathrm{T}}) = 0 \tag{5a}$$

$$\Delta K(I_n - X_1 X_1^{\mathrm{T}} M_a) - \Delta M(M_a^{-1} K_a - X_1 \Lambda_1 X_1^{\mathrm{T}} M_a) = 0 \tag{5b}$$

Note that ΔM and ΔK satisfying Eq. (5a) or Eq. (5b) ensures that the eigenvalues $\lambda_{p+1}, \ldots, \lambda_n$ and corresponding eigenvectors $\boldsymbol{x}_{p+1}, \ldots, \boldsymbol{x}_n$ of the original analytical model $\{M_a, \ K_a\}$ are not affected by updating. Now, let ΔM and ΔK be in a specific form as follows:

$$\Delta M = M_a \Psi M_a, \ \Delta K = M_a \Phi \mathrm{M}_a \tag{6}$$

Where $\Psi, \ \Phi \in \boldsymbol{R}^{n \times n}$ are real symmetric matrices to be determined. Substituting (6) into Eqs. (5b) and (4), it is easy to show that the following two matrix equations can be obtained.

$$\Phi(M_a - M_a X_1 X_1^{\mathrm{T}} M_a) - \Psi(K_a - M_a X_1 \Lambda_1 X_1^{\mathrm{T}} M_a) = 0 \tag{7}$$

$$(M_a + M_a \Psi M_a) Y_1 \Sigma_1 - (K_a + M_a \Phi M_a) Y_1 = 0 \tag{8}$$

Here it is noteworthy that (i) determination of ΔM and ΔK that satisfy Eqs. (3) and (4) so as to solve the FEMU problem is now equivalently transformed to solve Eqs. (7) and (8) to determine Ψ and Φ; (ii) Eq. (7) works exclusively with only a small number of finite element frequencies and mode shapes that need to be updated in the analytical model $\{M_a, \ K_a\}$, in contrast to Eq. (3) that involves a large number of the unmeasured frequencies and mode shapes of $\{M_a, \ K_a\}$.

In what follows the procedure to determine Ψ and Φ, the FEMU problem is posed as an SDP problem based on Eqs. (7) and (8).

3 Formulating the FEMU as an SDP Problem

The SDP is an extension of linear programming with a linear objective function in a sense that in addition to linear constraints, the additional constraint that the matrix be positive semi-definite (note that this additional constraint cannot be expressed as a linear one) must be imposed [22, 23], that is,

$$\begin{gathered} Min \ tr(CX) \\ Subject \ to \ tr \ (A_i X) = b_i \ , \ i = 1, 2, \ldots, m \\ X \geq 0 \end{gathered} \tag{9}$$

where X is a variable matrix, $C, A_1, \ldots, A_m$ are given symmetric matrices, and $tr(\cdot)$ is the trace operator. The SDP unifies several convex optimization problems (e.g., linear and quadratic programming), and many primal-dual interior-point methods for linear and quadratic programming have been naturally extended to solve SDPs, where the number of arithmetic operations required by the algorithms is bounded by a polynomial of the problem size. Recently, Lin et al. [24] described some applications of SDP techniques to quadratic inverse eigenvalue problems (QIEPs). They claimed that unlike other numerical methods, the SDP approach presented a unified, efficient, and tractable scheme for solving QIEPs.

There are now many software packages for solving SDP problems of a certain size and a number of interfaces that facilitate the use of SDP software. One of them, called YALMIP [25], is used to implement numerical examples in this paper. YALMIP is a free MATLAB-based toolbox that serves as a convenient interface for multiple external optimization solvers. The YALMIP commands unify and facilitate the different formats in SDP software. When applied to QIEPs, it makes the description of various structural constraints, such as positive definiteness, nonnegativity, sparsity patterns, and prescribed entries of the coefficient matrices, extremely simple and offers via the well established SDP theory and algorithms a reliable and conclusive answer within the specified numerical tolerance [24].

The FEMU problem in this paper may be formulated as an SDP problem in two forms.

Formulation 1

$$\begin{aligned}\text{Min}J =&\alpha\left\|\Phi(M_a - M_aX_1X_1^{\mathrm{T}}M_a) - \Psi(K_a - M_aX_1\Lambda_1X_1^{\mathrm{T}}M_a)\right\|_F^2\\ &+ \beta\|(M_a + M_a)Y_1\Sigma_1 - (K_a + M_a\Phi M_a)Y_1\|_F^2\end{aligned}$$

where α and β are some preselected weight factors, and $\|\cdot\|_F$ is Frobenius norm. Note that J is a convex but nonlinear function in Ψ and Φ. This objective function can be easily rewritten as a second-order Lorentz cone programming problem via the YALMIP commands.

Formulation 2

$$\begin{aligned}&\text{Min } J = \left\|\Phi(M_a - M_aX_1X_1^{\mathrm{T}}M_a) - \Psi(K_a - M_aX_1\Lambda_1X_1^{\mathrm{T}}M_a)\right\|_F^2\\ &\text{Subject to } (M_a + M_a\Psi M_a)Y_1\Sigma_1 - (K_a + M_a\Phi M_a)Y_1 = 0\end{aligned}$$

Substituting Ψ and Φ, obtained by the SDP optimization computation above, into Eq. (6) for symmetric corrective matrices ΔM and ΔK, then one could optimally accomplish model updating with no spillover and symmetry preserved.

4 A Numerical Example

For the sake of saving space, a numerical example is given to show the application of the above-established method for solving the FEMU problem. All codes are run in MATLAB 7.11.

***Example* 4.1** (*Example* 4.2 in [19]). The matrices M_{a} and K_{a} of the original model are

$$M_a = \begin{pmatrix} 1.2940 & 0 & 0 & 0 & 0 \\ 0 & 1.2940 & 0 & 0 & 0 \\ 0 & 0 & 1.2940 & 0 & 0 \\ 0 & 0 & 0 & 1.2940 & 0 \\ 0 & 0 & 0 & 0 & 1.2940 \end{pmatrix}$$

$$K_a = \begin{pmatrix} 1188.5 & 196.6 & 0 & 0 & -642.4 \\ 196.6 & 626.3 & 0 & -555.6 & 0 \\ 0 & 0 & 1188.5 & -196.6 & -546.1 \\ 0 & -555.6 & -196.6 & 626.3 & 196.6 \\ -642.4 & 0 & -546.1 & 196.6 & 4019.1 \end{pmatrix}$$

□

Taking $p = 2$, the computed modal data are $\Lambda_1 = \mathrm{diag}(3297.1,\ 23.648)$, and

$$X_1 = \begin{pmatrix} -0.2101 & -0.2048 \\ -0.0216 & 1.0000 \\ -0.1817 & 0.1400 \\ 0.0671 & 0.9997 \\ 1.0000 & -0.0631 \end{pmatrix}$$

The 'measured' modal data are taken as $\Sigma_1 = \mathrm{diag}(3297.6,\ 23.148)$, and

$$Y_1 = \begin{pmatrix} -0.2184 & -0.1893 \\ 0.0166 & 1.0000 \\ -0.1768 & 0.1529 \\ 0.1055 & 0.9933 \\ 1.0000 & -0.1355 \end{pmatrix}$$

From *Formulation* 1 of the SDP optimization with weight factors $\alpha = \beta = 1.0$, the updated mass and stiffness matrices are given as follows:

$$M = \begin{pmatrix} 1.2939 & -0.0002 & -0.0002 & -0.0001 & 0.0010 \\ -0.0002 & 1.2957 & 0.0003 & 0.0017 & -0.0004 \\ -0.0002 & 0.0003 & 1.2939 & 0.0004 & 0.0007 \\ -0.0001 & 0.0017 & 0.0004 & 1.2955 & -0.0008 \\ 0.0010 & -0.0004 & 0.0007 & -0.0008 & 1.2896 \end{pmatrix}$$

$$K = \begin{pmatrix} 1200.4 & 165.0 & 0.10 & 33.8 & -665.5 \\ 165.0 & 625.3 & -26.7 & -543.4 & 148.5 \\ 0.10 & -26.7 & 1180.2 & -221.0 & -519.2 \\ 33.8 & -543.4 & -221.0 & 651.4 & 342.0 \\ -665.5 & 148.5 & -519.2 & 342.0 & 3975.6 \end{pmatrix}$$

It is easy to calculate $\|MY_1\Sigma_1 - KY_1\|_F = 1.0253\text{e} - 009$, $\|MX_2\Lambda_2 - KX_2\|_F = 1.0613\text{e} - 010$.

From *Formulation* 1 with weight factors $\alpha = 1.0e + 05$, $\beta = 1.0e + 07$, the updated mass and stiffness matrices are not listed for the sake of saving space. It is found, in this situation, that $\|MY_1\Sigma_1 - KY_1\|_F = 2.8121\text{e} - 011$, $\|MX_2\Lambda_2 - KX_2\|_F = 1.1405\text{e} - 010$, and the match with the measured modal data is improved further.

From *Formulation* 2, the updated mass and stiffness matrices are given as follows:

$$M = \begin{pmatrix} 1.2943 & 0.0005 & -0.0006 & -0.0002 & 0.0009 \\ 0.0005 & 1.2951 & 0.0006 & -0.0006 & 0.0001 \\ -0.0006 & 0.0006 & 1.2947 & -0.0005 & 0.0008 \\ -0.0002 & -0.0006 & -0.0005 & 1.2950 & -0.0008 \\ 0.0009 & 0.0001 & 0.0008 & -0.0008 & 1.2902 \end{pmatrix}$$

$$K = \begin{pmatrix} 1201.0 & 165.4 & -0.0003 & -34.1 & -665.9 \\ 165.4 & 626.1 & -26.2 & -544.3 & 148.7 \\ -0.0003 & -26.2 & 1181.1 & -221.7 & -519.4 \\ -34.1 & -544.3 & -221.7 & 652.2 & 342.0 \\ -665.9 & 148.7 & -519.4 & 342.0 & 3977.4 \end{pmatrix}$$

It is easy to calculate $\|MY_1\Sigma_1 - KY_1\|_F = 1.2383\text{e} - 009$, $\|MX_2\Lambda_2 - KX_2\|_F = 5.7886\text{e} - 013$.

The results show that the updated model matches the measured modal data accurately, and has no spillover using two SDP formulations. However, for the same example in [19], their method can update the model to avoid spillover, but cannot reproduce the measured data accurately.

5 Conclusions

A direct mass and stiffness matrices updating method is presented in this paper. Based on a matrix equation that the corrective mass and stiffness matrices should satisfy for no spillover updating, the finite element model updating problem is formulated as a semi-definite programming (SDP) optimization problem, which can be solved using the well established SDP algorithms and an efficient and tractable scheme. The updated mass and stiffness matrices are naturally symmetric, and are also positive definite or positive semi-definite, due to the characteristics of the SDP algorithm itself. Numerical experiments indicate that the presented method produces accurate results.

References

1. Friswell MI, Mottershead JE (1995) Finite element model updating in structural dynamics. Kluwer Academic, Dordrecht
2. Datta BN (2002) Finite-element model updating, eigenstructure assignment, and eigenvalue embedding techniques for vibrating systems. Mech Syst Sig Process 16(1):83–96
3. Dascotte E (2007) Model updating for structural dynamics: past, present and future. In: Proceedings of the international conference on engineering dynamics (ICED), Carvoeiro, Algarve, Portugal
4. Datta BN, Sokolov V (2009) Quadratic inverse eigenvalue problems, active vibration control and model updating. Comput Appl Math 8(2):170–191
5. Baruch M (1978) Optimization procedure to correct stiffness and flexibility matrices using vibration tests. AIAA J 16(11):1208–1210
6. Berman A (1979) Mass matrix correction using an incomplete set of measured modes. AIAA J 17(8):1147–1148
7. Berman A, Nagy EJ (1983) Improvement of a large analytical model using test data. AIAA J 21(8):1168–1173
8. Wei F-S (1990) Mass and stiffness interaction effects in analytical model modification. AIAA J 28(9):1686–1688
9. Kuo YC, Lin WW, Xu SF (2006) New methods for finite element model updating problems. AIAA J 44(6):1310–1316
10. Bakir PG, Reynders E, De Roeck G (2007) Sensitivity-based finite element model updating using constrained optimization with a trust region algorithm. J Sound Vib 305(1–2):211–225
11. Charbel F, Francois MH (1993) Updating finite element dynamic models using an element-by-element sensitivity methodology. AIAA J 31(9):1702–1711
12. Mottershead JE, Link M, Friswell MI (2011) The sensitivity method in finite element model updating: a tutorial. Mech Syst Sig Process 25(7):2275–2296
13. Friswell MI, Penny JET (1990) Updating model parameters from frequency domain data via reduced order models. Mech Syst Sig Process 4(5):377–391
14. Imregun M, Visser WJ, Ewins DJ (1995) Finite element model updating using frequency response function data – I: theory and initial investigation. Mech Syst Sig Process 9 (2):187–202
15. Esfandiari A, Bakhtiari-Nejad F, Sanayei M, Rahai A (2010) Structural finite element model updating using transfer function data. Comput Struct 88:54–64
16. Atalla MJ, Inman DJ (1998) On model updating using neural networks. Mech Syst Sig Process 12(1):135–161
17. Marwala T (2010) Finite-element-model updating using computational intelligence techniques, applications to structural dynamics. Springer-Verlag, London
18. Carvalho J, Datta BN, Gupta A, Lagadapati M (2007) A direct method for model updating with incomplete measured data and without spurious modes. Mech Syst Sig Process 21 (2):2715–2731
19. Mao X, Dai H (2012) Finite element model updating with positive definiteness and no spill-over. Mech Syst Sig Process 28(1):387–398
20. Chu MT, Datta BN, Lin WW, Xu SF (2008) Spillover phenomenon in quadratic model updating. AIAA J 46(2):420–428
21. Zhang JF, Ouyang H, Yang J (2014) Partial eigenstructure assignment for undamped vibration systems using acceleration and displacement feedback. J Sound Vib 333(1):1–12
22. Todd MJ (2001) Semidefinite optimization. Acta Numerica 10:515–560
23. Boyd S, Vandenberghe L (2004) Convex optimization. Cambridge University Press, Cambridge

24. Lin MM, Dong B, Chu MT (2010) Semi-definite programming techniques for structured quadratic inverse eigenvalue problems. Numer Algorithms 53:419–437
25. Lofberg J (2004) YALMIP: a toolbox for modeling and optimization in MATLAB. In: Proceedings of 2004 IEEE international symposium on computer aided control systems design, Taipei

Wave Propagation in a Honeycomb Composite Sandwich Structure in the Presence of High-Density Core Using Bonded PZT-Sensors

Shirsendu Sikdar and Sauvik Banerjee

Abstract "Honeycomb Composite Sandwich Structure" (HCSS), is one of the novel materials that has been adopted universally to form major structural components of aerospace, marine and automotive vehicles due to its high strength to weight ratio and high energy-absorption capabilities. In this research paper, the scope to study the general characteristics of guided wave (GW) propagation in an HCSS plate with the presence of a high-density (HD) core subjected to time-dependent transient surface excitations is presented numerically and experimentally. Significant effects are detected due to the presence of HD core in terms of decay in amplitude and reduction in group velocity of the output GW signals. A two-dimensional (2D) numerical model is made in ABAQUS. An effective non-reflecting boundary condition (NRBC) is modeled in ABAQUS in order to mitigate the undesirable boundary reflections during numerical simulation. Good agreements between the numerical and experimental results are noticed for all the cases studied.

Keywords Guided wave · Group velocity · Structural health monitoring · Honeycomb composite · High density core

1 Introduction

"Honeycomb Composite Sandwich Structures" often termed as "HCSS" are special class of composite structures in which thin composite skins are bonded with adhesives to the two faces of extremely lightweight and relatively thick composite

S. Sikdar (✉) · S. Banerjee
Civil Engineering Department, Indian Institute of Technology Bombay, Mumbai 400046, India
e-mail: shirsendu@civil.iitb.ac.in

S. Banerjee
e-mail: sauvik@civil.iitb.ac.in

J.K. Sinha (ed.), *Vibration Engineering and Technology of Machinery*, Mechanisms and Machine Science 23, DOI 10.1007/978-3-319-09918-7_55

honeycomb core. This is a novel material and adopted as a lightweight construction material in all severer conditions of aerospace, marine and automotive structures for specialized applications [1]. Due to its high strength to weight ratio, it is efficiently used to form major structural components such as wings, fuselage, blades, etc. High energy-absorption capability makes them very attractive for impact protection and mitigation related applications. Application of this material in various fields proves its novelty and hence, it is in great demand. Sandwich beams and panels with symmetric faces and cores of varying stiffness were investigated by Bozhevolnaya et al. [2]. A closed-form estimate of the stress-strain fields induced by local effects and its accuracy were verified experimentally for the case of a sandwich beam under three-point bending. Significant rise of the bending stresses in the faces at the vicinity of the core junctions are observed. Typical sandwich beam configurations with glass fiber-reinforced plastic face sheets and core junctions between polymer foams of different densities and rigid aluminum were tested under quasi-static and fatigue loading conditions by Johannes and Thomsen [3]. The Ultrasonic GW based techniques provide an efficient and accurate procedure for nondestructive evaluation (NDE) and structural health monitoring (SHM) of composite structures [4, 5]. The general features of GW propagation that can be transmitted in isotropic and anisotropic media have been well documented in classical textbooks [6, 7]. The current methodologies involve deployment of surface mounted broadband transducers or embedded piezoelectric wafer transducer (PWT) [8, 9]. Tuning of appropriate GW modes plays an important role for NDE and SHM of composite structures using piezoelectric transducers [10]. The dispersive properties of axially symmetric surface waves subjected to point loading on an isotropic elastic body with depth dependent elastic moduli and mass density was analyzed by Balogun and Achenbach [11] in the high frequency range [12]. The quantitative features of elasto-dynamic field produced in a homogeneous half-space and a unidirectional graphite/epoxy composite laminate due to surface loads are studied by Lih and Mal [13]. In order to reduce the computational efforts, the application of a non-reflecting boundary condition was studied by several authors [14–19]. A non-reflecting boundary condition (NRBC) was described for the scalar wave equation in [14]. Infinite elements in ABAQUS were used to design a non-reflecting boundary in [17], based on the concept by Lysmer and Kuhlemeyer [18].

With minimal research done in this field especially in ultrasonic guided wave (GW) based SHM, no theoretical model is available to carry out the analysis for a HCSS with jointed core of varying densities. The present research is motivated by the need to study the ultrasonic wave field produced by localized dynamic loads on the surface of HCSS with a soft-dense-soft core, to validate the experimental results with those obtained from numerical simulation and to develop an efficient NRBC in ABAQUS.

2 Experimental Set-Up and Data Acquisition

The photographic representation of the experimental NI set-up is shown in Fig. 1a, which is situated in the SHM Laboratory of the Indian Institute of Technology Bombay. The NI set-up is consists of an embedded arbitrary signal/function generator (FGEN) and a desktop monitor. A five-cycle sine pulse modulated by Hanning window is conveyed into the HCSS plate by means of FGEN soft-front panel, and a SCOPE soft-front panel is employed to amass Lamb wave signals, as shown in Fig. 1a.

The surface-bonded square (20 × 20 mm) piezoelectric wafer transducer (PWT) are mounted on top surface of the HCSS plate (600 × 450 × 13.5 mm) with a known HD-core region to actuate Lamb waves. The PWTs can serve as an 'Actuator' as well as a 'Receiver'. A detail schematic arrangement of PWTs is shown in Fig. 1b.

In order to perform the frequency modulation of the available sensors, two PWTs are attached at a distance of 180 mm on the top surface of the HCSS plate, as shown in the Fig. 2a.

Among the two PWTs one acts as a transmitter and other acts as a receiver. To identify the optimum frequency of excitation, the Lamb waves with various central frequencies ranging from 0.05 to 0.230 MHz were evaluated experimentally by applying a five-cycle sine pulse modulated with Hanning window. To know the

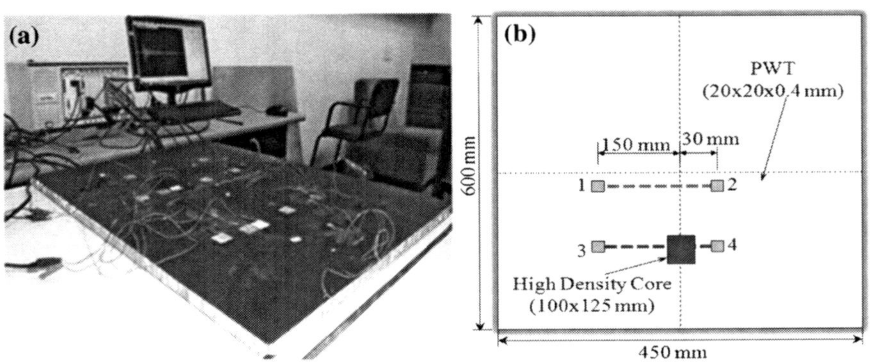

Fig. 1 Arrangement of PWTs; **a** complete experimental set-up, **b** schematic position of PWTs

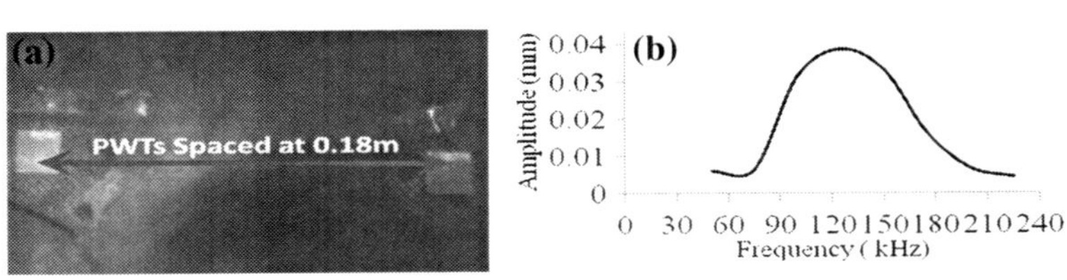

Fig. 2 Frequency modulation: **a** Experimental set-up, **b** calibration curve

efficiency of the PWTs at different frequencies, the frequency vs. amplitude graph was plotted as demonstrated in Fig. 2b.

3 Finite Element Modeling

The finite element method (FEM) is used as an alternative to numerically solve this class of problems. In this research work commercially available finite element software ABAQUS (explicit and implicit code) has been used for 2D simulation of ultrasonic GW propagation in HCSS. The 2 m long HCSS plate consists of two 0.74 mm thick Graphite/epoxy fiber reinforced composite (GFRC) skins, 0.02 mm thick adhesive layer (at the interface) and a 12 mm thick soft aluminum honeycomb core (Al5056) with an HD region (100 mm × 125 mm × 12 mm). The detailed material properties are shown in Table 1. Four numbers of PWTs (20 × 20 × 4 mm) are modeled on the top surface of the HCSS plate by using the ABAQUS implicit code. The PWT (SP-5H) material properties are used as same as in [20].

The numerical simulations are carried out with a 5-cycle 150 kHz sine pulse excitation modulated by Hanning window as shown in Fig. 3. In order to model the PWT excitations, the horizontal Symmetric surface source (S0) excitation is applied along the (+) ve X-direction at the top surface of the 2D plane strain model. The

Table 1 Elastic properties for the honeycomb-composite sandwich structure

Material type	E_1 (GPa)	E_2 (GPa)	G_{13} (GPa)	ν_{13}	ν_{23}	ρ (g/cc)
Graphite/epoxy	297.1	7.3	4.5	0.21	0.064	1.42
Adhesive	0.0487	0.0487	0.0174	0.4	0.4	1.25
Soft-core	0.0804	1.05	0.122	0.25	0.022	0.032
HD-core	0.734	1.584	0.4826	0.33	0.042	0.128

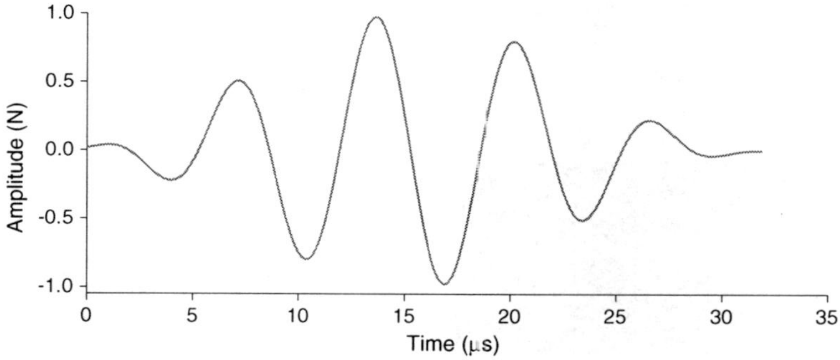

Fig. 3 Input signal of 150 kHz five-cycle sine pulse (modulated by Hanning window)

number of cycles in excitation signal is chosen to be 5 in order to avoid the broadband and narrowband excitation.

In ABAQUS the CPE4R elements (4-nodes bilinear plane strain quadrilateral, reduced integration, hourglass control) has been chosen for modeling of dimension (0.25 mm × 0.25 mm) for the honeycomb layer, (0.25 mm × 0.02 mm) for adhesive layer and (0.25 mm × 0.125 mm) for the composite skin. The CPE4E elements (4-nodes bilinear plane strain piezoelectric quadrilateral) are used for PWT (0.25 mm × 0.1 mm) and the CINPE4 are exclusively used by replacing the AC2D4R (4-noded linear 2-D acoustic quadrilateral, reduced integration, hourglass control) for the modeling of NRBC (Fig. 7).

3.1 Modeling of High-Density Core in HCSS

In order to perform the numerical simulation of HCSS plate with a (100 mm × 125 mm) HD-core in between two soft cores (AL5056), a 2 m long 2-D plane-strain model was made in ABAQUS. The properties of HD-core are given in Table 1. The numerical simulation was carried out with a 150-kHz loading. The schematic representation of the 2-D plane strain numerical model with high-density core is depicted in the Fig. 4.

In order to know the effects on Guided wave signal due to the presence of high density (HD) core and the local effects owing to the core junctions, initially, a 2D isotropic numerical modeling is done prior to the modeling for the HCSS. For the sake of simplicity, isotropic plane strain models of identical dimensions (1 m × 13.5 mm) for 'Aluminum' as well as for 'Aluminum-Steel-Aluminum (AL-Steel-AL)' plate has been created (Fig. 5). The elastic modulus taken for aluminum and steel are 70 and 200 GPa respectively.

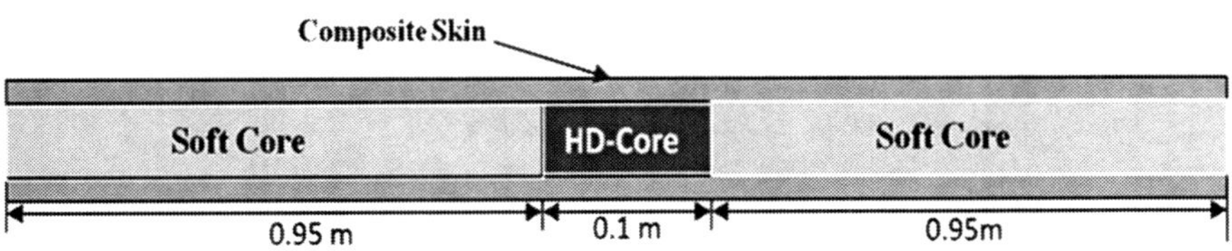

Fig. 4 Schematic representation 2D numerical HCSS model with HD-core

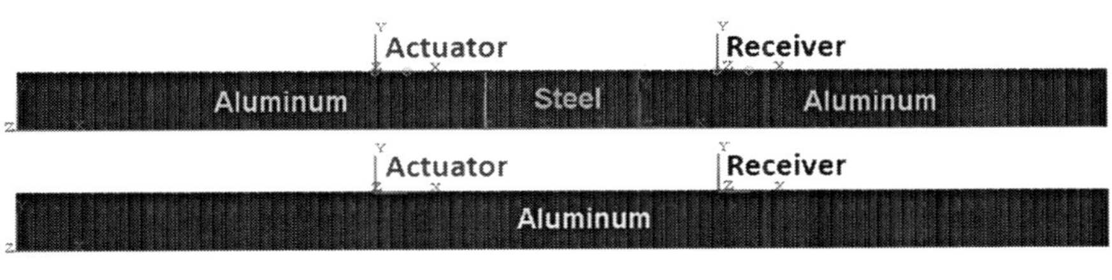

Fig. 5 Numerical model of Aluminum-Steel-Aluminum and Aluminum plate in ABAQUS

These two models can be considered as 'soft' and 'soft-dense-soft' models respectively, as the mass density of steel (7,850 kg/m^3) is considerably higher than aluminum (2,700 kg/m^3). The aluminum-steel-aluminum (A-S-A) plate model consists of a 100 mm steel plate in-between two 400 mm aluminum plates as partly shown in the Fig. 5. The bonding between the junctions was achieved by using the 'tie' option in ABAQUS. Similarly, 2D numerical models of HCSS plate are made for both the cases, 'without HD-core' and the 'with HD-core (Fig. 7)'.

4 Data Analysis and Results

A 150 kHz five-cycle Hanning pulse (Fig. 3) has been used as an input signal for all the numerical simulations. The PWT actuators and receivers are modeled at the top surface of the plates at a distance of 180 mm as shown in Figs. 1b, 5 and 7. For the isotropic simulation in ABAQUS (Fig. 5) it was observed that, there is a significant decay in amplitudes and a bit reduction in group velocity of the received signal for the case of soft-dense-soft (AL-Steel-AL) plate in comparison with the soft (Aluminum) plate model as shown in Fig. 6. In the figure, $Amp^* = Amp/Amp_{Max}$ is the normalized signal amplitude with *Amp* being the magnitude of the sensor signal and Amp_{Max} being the maximum value of all signal magnitudes plotted in the figure. The definition about *Amp** is also used in the following figures.

To investigate the effects on Guided wave signal due to the presence of the HD-core section, a 2D numerical HCSS plate with HD-core has been modeled in ABAQUS which is partly shown in Fig. 7. The NRBC is applied below 12 mm of the honeycomb cores by using the ABAQUS infinite elements (CINPE4) to mitigate the unwanted boundary reflections (Fig. 7). Whereas, a similar model having only soft-core is created in ABAQUS, in order to investigate the effect due to the presence of the HD-core.

Output response is collected at a distance of 180 mm, separately, for both without and with high-density core models. Significant change in amplitude has been observed (Fig. 8) which is quite similar to the earlier isotropic simulation of 'Aluminum' and 'AL-Steel- AL' plates (Fig. 6).

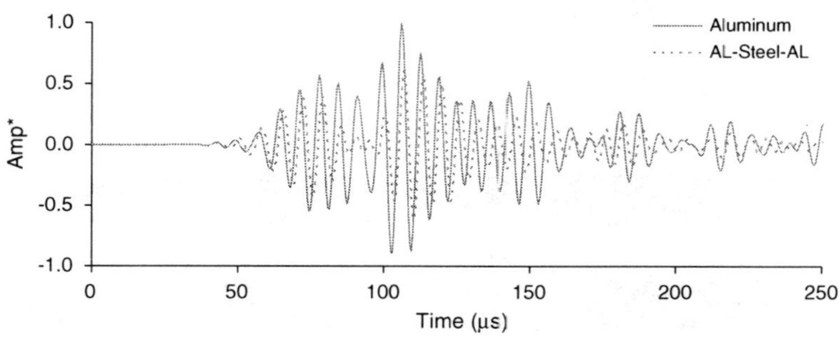

Fig. 6 Comparison of numerical output signals for Aluminum and AL-Steel-AL plate

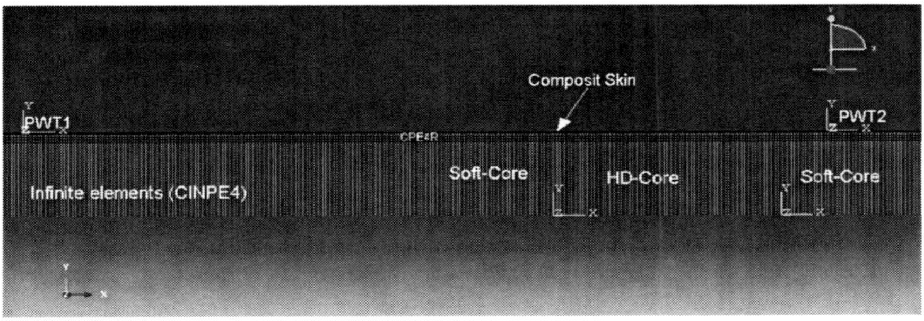

Fig. 7 Two-dimensional numerical model of HCSS with HD-core in ABAQUS

For the experimental work, two transducers (3–4) are mounted along high density core, as shown in Fig. 1. The FGEN soft-front panel in NI-setup is applied to produce a five-cycle sine pulse modulated with a Hanning window at a frequency of 150 kHz. In addition, the response along the same direction (0°–180°) without HD core Sects. 1–2 was measured to compare the results and analyze them to know the effect due to the presence of HD-core.

For the Sect. 1–2, which is sufficiently away from the HD-zone, the input signal is applied at PWT-1 and the output response collected at PWT-2. Similarly, for the Sects. 3–4, which is across the HD-zone, the input signal is applied at PWT-3 and the output response collected at PWT-4 as shown in Fig. 1b. The output signals collected at location 2 and 4 are compared and a significant change in amplitude and group velocity has been observed (Fig. 9), which is quite similar behavior as in the case of numerical simulation in Fig. 8.

In order to validate the results, the comparison of numerical and experimental output responses is needed. Therefore, the comparison of numerical and experimental output responses only for the soft-core region is done, which is shown in the

Figure 10 considerably good agreement is being observed with almost the same group velocity of primary symmetric (S0) modes.

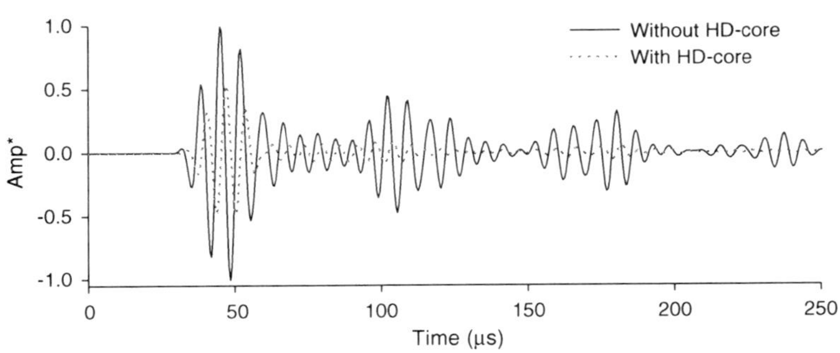

Fig. 8 Comparison of numerical output signals for with and without HD-zone

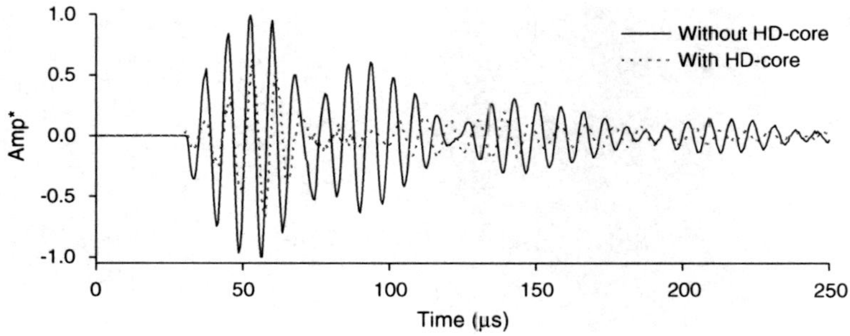

Fig. 9 Comparison of experimental output signals for with and without HD-core

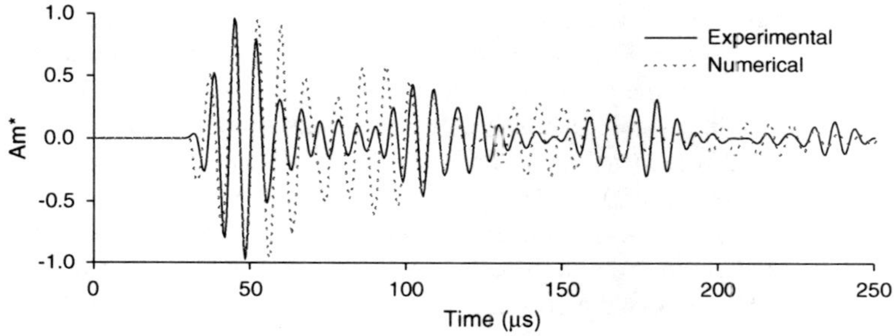

Fig. 10 Comparison of numerical and experimental response in absence of HD-core region

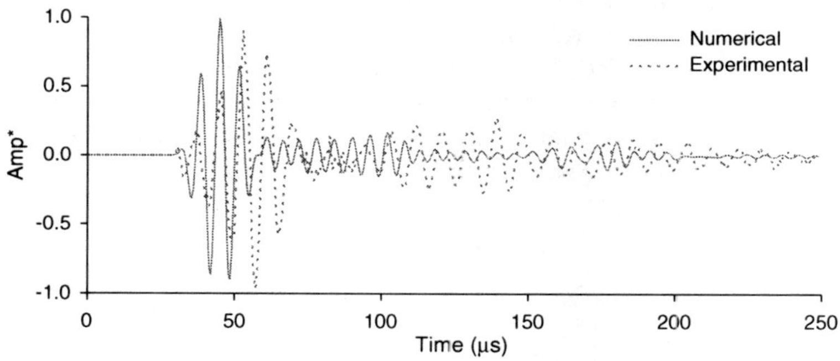

Fig. 11 Comparison of numerical and experimental response due to the presence of HD-core

The plots in Fig. 11 are depicting the numerical and experimental output signals due to the presence of HD-core. The comparison between two plots shows a fair match of the output response, with almost the same group velocity of primary symmetric modes (S0).

5 Conclusion

In this paper, the Guided wave propagation mechanism in Honeycomb Composite Sandwich structure is investigated both numerically and experimentally by using a surface-bonded piezoelectric actuator/sensor system. It is found that the wave mechanism is changed with the increase in core density. Good agreements between the 2-D numerical and experimental results have been noticed for all the cases. Significant effect due to the presence of high-density core in HCSS has been observed in terms of decay in amplitude and a slight reduction in group velocity of the output GW signal. Hence, the change in density of the medium leads to the change in amplitude in the propagating wave. The proposed non-reflecting boundary condition (NRBC) is found to be fairly effective for 2-D modeling in ABAQUS.

Acknowledgements The authors wish to thank the Indian Space Research Organization (ISRO) for supporting this work under grant 11ISROC001.

References

1. Peters ST (1998) Handbook of composites. Chapman and Hall, Boca Raton
2. Bozhevolnaya E, Thomsen OT, Kildegaard A, Skvortsov V (2003) Local effects across core junctions in sandwich panels. Compos B 34:509–517
3. Johannes M, Thomsen OT (2010) Examination of the failure of sandwich beams with core junctions subjected to transverse shear loading. J Sandwich Struct Mater 12(2):199–236
4. Raghavan A, Cesnik CES (2007) Review of guided-wave structural health monitoring. Shock Vib Dig 39(2):91–114
5. Su Z, Ye L, Lu Y (2006) Guided lamb waves for identification of damage in composite structures: a review. J Sound Vib 295:753–780
6. Achenbach JD (1978) Wave propagation in elastic solids. North-Holland, New York
7. Rose JL (1999) Ultrasonic waves in solid media. Cambridge University Press, Cambridge
8. Giurgiutiu V (2005) Tuned lamb wave excitation and detection with piezoelectric wafer active sensors for structural health monitoring. J Intell Mater Syst Struct 16:291–305
9. Mal AK, Ricci F, Banerjee S, Shih F (2005) A conceptual structural health monitor ing system based on vibration and wave propagation. Struct Health Monit 4:283–293
10. Maslov KI, Kundu T (1997) Selection of lamb modes for detecting internal defects in laminated composites. Ultrasonics 35:141–150
11. Balogun O, Achenbach JD (2012) Surface waves on a half space with depth- dependent properties. J Acoust Soc Am 132(3):1336–1345
12. Shuvalov AL (2008) The high-frequency dispersion coefficient for the Raleigh velocity in a vertically inhomogeneous anisotropic half-space. J Acoust Soc Am 13:2484–2487
13. Mal AK, Lih SS (1992) Elastodynamic response of a unidirectional composite laminate to concentrated surface loads, Parts I and II. ASME J Appl Mech 59:878–892
14. Alpert B, Greengard L, Hagstrom T (2002) Nonreflecting boundary conditions for the time-dependent wave equation. J Comput Phys 180:270–296
15. Berenger JP (1994) A perfectly matched layer for the absorption of electromagnetic waves. J Comput Phys 114:185–200

16. Drozdz M, Moreau L, Castaings M, Lowe MJS, Cawley P (2006) Efficient numerical modelling of absorbing regions for boundaries of guided waves problems. In: AIP Conf Proc 820:126
17. Liu GR, Jerry SQ (2003) A non-reflecting boundary for analyzing wave propagation using the finite element method. Finite Elem Anal Des 39:403–417
18. Lysmer J, Kuhlemeyer R (1969) Finite dynamic model for infinite media. J Eng Mech Div ASCE 95:859–877
19. Sim I (2010) Nonreflecting boundary conditions for time-dependent wave propagation. Ph.D. thesis, Faculty of Science, University of Basel, Switzerland
20. Bijudas CR, Mitra M, Mujumdar PM (2013) Coupling effect of piezoelectric wafer transducers in distortions of primary Lamb wave modes. Smart Mater Struct 22:065007

Health Monitoring of Stiffened Metallic Plates Using Nonlinear Wave Interaction and Embedded PZT Transducers

Deba Datta Mandal, Debashis Wadadar and Sauvik Banerjee

Abstract The ultimate strength of a structural component can be limited in the presence of microscopic imperfections, which serve as the origin of damage process. These micro imperfections are considerably shorter than the acoustic wavelength of the frequencies normally used in ultrasonic nondestructive technique. As a result, the sensitivity of the linear acoustic characteristics (attenuation, velocity etc.) is not sufficient enough to detect the microscopic imperfections of a structure. But, these micro imperfections produce local excess nonlinearity, which is higher than the intrinsic nonlinearity of the intact structure. On the other hand, weakly or incompletely bonded interfaces (e.g., contact-type defects, cracks, debondings, delaminations and loosening in bolted joints) also exhibit highly nonlinear behavior as the contact area changes frequently when elastic waves propagate through the interface. The various nonlinear techniques for damage detection developed in the recent years are: non-linear resonant ultrasound spectroscopy (NRUS) and non-linear wave modulation spectroscopy (NWMS).The later consists of two methods, namely harmonics generation and sidebands generation. In this work a simulation study has been carried out using ANSYS Finite Element (FE) software, where the NWMS methodology is adopted for the case of a stiffened aluminum plate for different degrees of damage in the form of different delamination lengths. A damage index (DI) is defined by taking into account the relative amplitude of the harmonics or the side-bands with respect to the carrier frequency amplitude. It is found that the DI increases with the increase in delamination length.

D.D. Mandal (✉) · D. Wadadar · S. Banerjee
Civil Engineering Department, Indian Institute of Technology Bombay, Mumbai 400076, India
e-mail: debu.ju.civil@gmail.com

D. Wadadar
e-mail: wadadardebashis89@gmail.com

S. Banerjee
e-mail: sauvik@civil.iitb.ac.in

J.K. Sinha (ed.), *Vibration Engineering and Technology of Machinery*,
Mechanisms and Machine Science 23, DOI 10.1007/978-3-319-09918-7_56

Keywords Nonlinear wave interaction · Contact-type defects · Non-linear wave modulation spectroscopy · Stiffened plate · Delamination · Structural health monitoring

1 Introduction

There are basically two types of approach in structural health monitoring (SHM) namely global approach and local approach. In global approach the changes in the vibrational properties of the structure caused by the damage are measured. This approach is only effective in dealing with large defects because the effects of small defects on the global vibrational properties are often below the noise level in large structures. On the other hand, in local approach the changes in the characteristics of ultrasonic waves propagating across the defects are measured and analyzed to detect and characterize them. This approach is effective in detecting small defects but it requires a dense network of sensors [1].

The existing conventional ultrasonic methods include the pulse-echo, the pitch-catch and through transmission. In the pulse-echo method, defects are detected in the form of additional echoes. In the pitch-catch method, wave dispersion and attenuation due to diffused damage in the material are used as flaw indicators. For thickness-wise inspection, the wave path traverses only a small portion of the material volume; hence, the transducer is mechanically moved along the surface to scan and interrogate the entire material volume. However, the previously mentioned methods are not so sensitive to detect micro cracks in material, delaminations in composite structure and unable to predict structural health of a joint or debonding of stiffeners, where contact surfaces play an important role. In the recent years nonlinear vibration and acoustic phenomena for SHM have emerged to be a promising technique for the detection of microcracks/flaws/delaminations/dislocations/disbondsetc in structural components. The nonlinear acoustical phenomena can be related to various imperfections of atomic lattices (intrinsic or material nonlinearity) and/or non-symmetric thermo-elastic behavior of interfaces (e.g. cracks, contacts, and rubbing surfaces). Intrinsic or material nonlinearity is global type nonlinearity while cracks, contacts, rubbing surfaces are local type nonlinearity. The later has demonstrated a great interest of theoretical and applied research for the last 15 years [2].

Previous experimental work on different materials (e.g., polymer, CL Lexan polycarbonate, nylon, adhesive layers and aluminum alloy) demonstrate that nonlinear phenomena are generally more sensitive to small damage severities than classical linear approaches (depending on velocity, attenuation, etc.) [3]. The features related to nonlinear effects have been investigated so far include: time-domain signal distortions, generation of higher harmonics, resonance frequency shifts, signal modulations, slow dynamics, and nonlinear dissipation. Despite many research works, there is still very little understandings of what physical mechanisms related to these nonlinearities are and the question still remains why material sensitivity of nonlinear effects is so strong if compared with linear responses. There are two major difficulties associated with these problems. The first one is the diversity of nonlinear mechanisms proposed. Similar nonlinear effects can be manifested by different mechanism and vice versa. For example, energy dissipation can be modeled using frictional, hysteretic or thermo-elastic mechanisms. Hysteresis in turn involves both elasticity and dissipation, and could be linear or nonlinear. The second one is, the various experimental evidence related to these nonlinear mechanisms have been observed. It is often very difficult, if not impossible, to separate all these mechanisms involved. It is also important to note that nonlinearities may result not only from cracks but also from other non-damage related effects such as: friction between elements at structural joints or boundaries, overloads, material connections between transducers and monitored surfaces, electronics and instrumentation measurement.

This article describes different types of nonlinearities present in damaged structures and different nonlinear wave interaction techniques developed over the last few years to estimate nonlinearity (damage) in structures in an effort to develop efficient techniques for health monitoring of stiffened plate using nonlinear elastic wave and embedded piezoelectric network.

2 Theoretical Background

Experimental evidence for the highly nonlinear behavior of micro cracked and damaged materials has existed for years from experiments of static stress–strain behavior and dynamic nonlinear wave interaction [4]. Those experiments imply that the dynamic elastic behavior of most solid materials cannot be described by a linear theory. Nonlinear elastic behavior may manifest itself in the generation of harmonics upon dynamic wave propagation, nonlinear attenuation, resonance frequency shift and slow time effects; while, hysteresis and discrete memory are commonly observed in static tests and persist even in low-strain dynamic experiments [5]. Micro-inhomogeneities such as cracks, voids, and contacts have a complex compliance and the local nonlinear forces may entirely dominate the relatively small atomic nonlinearity. Therefore, the theoretical description of nonlinear mesoscopic elastic materials contains terms that describe classical nonlinearity, as well as hysteresis, and discrete memory. The full theory of these effects has not been constructed yet. However, the approaches described below provide at least a qualitative explanation of much of the experimental evidence.

2.1 Classical Theory

The classical theory begins with the expansion of the elastic strain energy in powers of the strain tensor, ϵ_{ij}. The expansion coefficients designate the components of the second-order elastic tensor and the third-order elastic tensor. The equation of motion in Lagrangian coordinate x and time t is given as [6]:

$$\rho \frac{\partial^2}{\partial t^2}(u_i) = \frac{\partial \sigma_{ij}}{\partial x_j} \tag{1}$$

where, u_i is the displacement vector, ρ is the material density, σ_{ij} is the stress tensor. Now, for one-dimensional case and for non-linear elasticity the above equation of motion can be expressed as [7]

$$\rho \frac{\partial^2 u}{\partial t^2} = k_0 \left[1 + \beta \frac{\partial u}{\partial x} + \delta \left(\frac{\partial u}{\partial x} \right)^2 + \cdots\cdots \right] \frac{\partial^2 u}{\partial x^2} \tag{2}$$

where the stress field is expressed as

$$\begin{aligned} \sigma(x) &= k_0 \left[1 + \beta \frac{\partial u}{\partial x} + \delta \left(\frac{\partial u}{\partial x} \right)^2 + \cdots\cdots \right] \frac{\partial u}{\partial x} \\ &= k_0 \left[1 + \beta \epsilon + \delta \epsilon^2 + \cdots\cdots \right] \epsilon \end{aligned} \tag{3}$$

where

$\epsilon = \frac{\partial u}{\partial x}$ is the strain field, k_0 is the linear elastic modulus, and β and δ are non linear parameters are the measures of cubic and quartic anharmonicities [8]. These parameters are the linear combinations of the elastic constants [6].

Now, considering a time-harmonic plane (displacement) wave $A_1\cos(kX_1 - \omega t)$, where A_I is the amplitude, k is the wave number, and ω is the angular frequency. Assuming that the nonlinearity in the solid is small (for β only), then the solution to Eq. (2) for this time-harmonic wave is obtained by a perturbation analysis as [6]

$$\begin{aligned} u_1 &= -\frac{1}{8}\beta k^2 A_1^2 X_1 + A_1 \cos(kX_1 - \omega t) + \frac{1}{8}\beta k^2 A_1^2 X_1 \cos\{2(kX_1 - \omega t)\} + \cdots \\ &= A_0 + A_1 \cos(kX_1 - \omega t) + A_2 \cos\{2(kX_1 - \omega t)\} + \cdots \end{aligned} \tag{4}$$

$$\beta = \frac{8A_2}{k^2 A_1^2 X_1} \tag{5}$$

Therefore, it is noted that the amplitude of the second-harmonic displacement is proportional to the acoustic nonlinearity parameter and a subharmonic; that is, the static displacement is induced by the material nonlinearity. The acoustic nonlinearity parameter can be determined experimentally by measuring the absolute amplitudes of the fundamental (A_1) and the second-harmonic (A_2) displacement signals.

2.2 Hysteresis

The Classical nonlinear theory explained above cannot explain the nonlinear behavior generated by local nonlinear forces due to damage presence (such as cracks, voids and contacts). This evidence was found experimentally by several authors [4, 8, 9]. In order to include nonlinear behaviours like classical nonlinearity, as well as hysteresis and discrete memory, a stress dependence on strain time derivative (i.e., $\dot{\epsilon}$) can be added in the nonlinear classical stress–strain relationship. A theoretical description is given by the nonlinear mesoscopic elastic model as presented below.

$$\sigma = \int k(\epsilon, \dot{\epsilon}) d\epsilon \tag{6}$$

$$\begin{aligned} k(\epsilon, \dot{\epsilon}) &= k_0\left[1 + \left(\beta\epsilon + \delta\epsilon^2 + \cdots\right) + \propto \left\{\Delta\epsilon + \epsilon(t) sign(\dot{\epsilon}) + \cdots\right\}\right] \\ &= k_0[1 + \textit{nonlinear classical terms} + \textit{nonlinear nonclassical terms}] \end{aligned} \tag{7}$$

where, k is the combination of nonlinear classical and hysteretic (or non-classical) modulus wherein k_0 is the linear modulus, $\Delta\epsilon$ is the strain amplitude change over the previous period (i.e., ϵ is a function of time), $\dot{\epsilon} = d\epsilon/dt$ is the strain rate, $sign(\dot{\epsilon}) = 1$ if $\dot{\epsilon} > 0$ and $sign(\dot{\epsilon}) = -1$ if $\dot{\epsilon} < 0$, β and δ are classical nonlinear coefficients and $\propto$ is a material hysteresis measure and they can be negative. The increase in α reflects an increase in the nonlinear hysteretic behavior of the material Experimental and numerical studies proved that for purely hysteretic materials ($\beta, \delta = 0, \propto \neq 0$) and for mono frequency harmonic excitation, the third harmonic is generated which is quadratic with the fundamental amplitude (Eq. 8) [5, 10–12]. Hysteresis also causes the linear decrease in the resonance frequency (Eq. 9) and a linear increase in the modal damping ratio (Eq. 10) with the increase in strain levels. Hysteresis does not affect the level of even harmonics [5].

$$\Delta\epsilon_{3f} = c_1 \Delta\epsilon^2 \tag{8}$$

$$\frac{f_0 - f}{f_0} = c_2 \Delta\epsilon \tag{9}$$

$$\frac{\xi - \xi_0}{\xi_0} = c_3 \Delta\epsilon \tag{10}$$

where, f_0 = initial resonance frequency, f = resonance frequency at a particular strain level, ξ_0 = initial damping ratio, ξ = damping ratio at a particular strain level, c_1, c_2 and c_3 are the coefficients proportional to the hysteresis parameter, $\propto$.

2.3 Contacts

Experimental data on a cracked medium exhibiting high acoustic nonlinearity is a commonly observed phenomenon. This is due to the appearance of the nonlinearity parameters in the elastic modulus for the cracked medium for opening and closing of the cracks [13, 14]. Crack produces rough surfaces which are in contact. If observed on a microscopic scale, all machined surfaces are also rough, and the contact between surfaces is restricted to discrete areas at the tip of the surface asperities. The contact between the two nominally flat, but imperfect (or non-bonded), surfaces is no exception. In this respect, all the bolted joints also develop partial contact at their imperfect interfaces, unless the bolted joints are glued [15]. At these imperfect interfaces a true contact area is generated (a fraction of the nominal contact area). The true contact area is known to be smaller than the nominal contact area. This true contact area varies with the contact pressure. The behavior of the crack or the imperfect contact surfaces can be described using the model in which the interface is represented as an elastic contact of two rough surfaces, pressed one to the other under the action of internal stresses in the surrounding solid or due to the applied pressure [13, 14].

3 FE Simulation of the Stiffened Aluminum Plate

The models of the stiffened aluminum plate for different delamination lengths (100, 150, 200, 250 and 300 mm) are prepared using ANSYS (APDL) 14.0 FE software. The script files are written to do the entire modeling and analysis using APDL commands. A typical figure of the plate for 200 mm delamination is shown in Fig. 1, while the applied loading (20 cycles sine pulse modulated by Hanning window with central frequency of 2 kHz) in Fig. 2. The details of the material property and the geometry, contact surfaces, and the analysis parameters are given in Tables 1, 2, and 3 respectively.

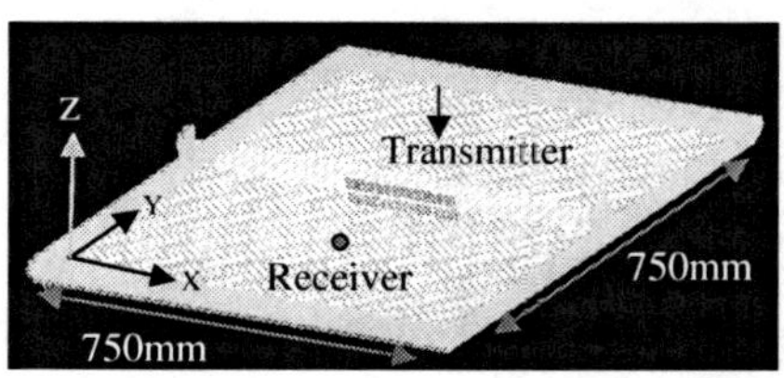

Fig. 1 Model of the stiffened aluminum plate in ANSYS (with delamination length of 200 mm)

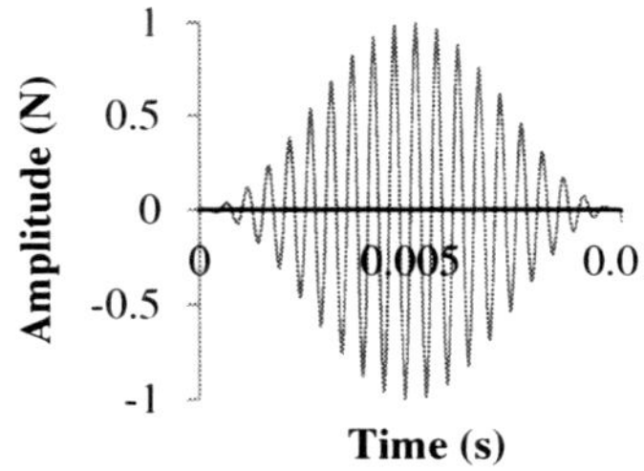

Fig. 2 Source signal (2 kHz 20-cycle Hanning pulse)

Table 1 Properties of stiffened aluminum plate and applied loading

Dimension of the plate	750 × 750 × 5 mm
Dimension of solid elements	5 × 5 × 5 mm
Material	Aluminum
Young's modulus (E)	70 Gpa
Poisson's ratio (ν)	0.33
Density (ρ)	2,700 kg/m^3
Element type	Solid element (SOLID45)
Boundary condition	Fixed at all 4 sides
Direction of applied load	Z-direction
Position of the stiffener	Middle of the plate
Thickness of the stiffener	5 mm
Width of the stiffener	100 mm
Delamination length	Variable (100,150,200,250,300 mm)
Delamination position	Middle of the interface
Excitation frequency	20 cycles sine pulse modulated by Hanning window with central frequency of 2 kHz

Table 2 Contact surface details

Contact type	Flexible to flexible
Contact element	CONTA173
Contact algorithm	Augmented Lagrangian
Target element	TAEGET170
Location of contact detection	Gauss points
Contact selection	Asymmetric
Gap/closure	No adjustment
Contact stabilization damping	Zero
Behavior of contact surface	Standard
Normal contact stiffness factor	1.0
Tangent penalty stiffness factor	Program controlled

Table 3 Analysis parameter details

Analysis type	Nonlinear transient
Type of nonlinearity	Contact
Analysis method	Full Newton-Raphson
Co-ordinate of the transmitter	(330, 565, 0 mm)
Co-ordinate of the receiver	(385, 250, 0 mm)

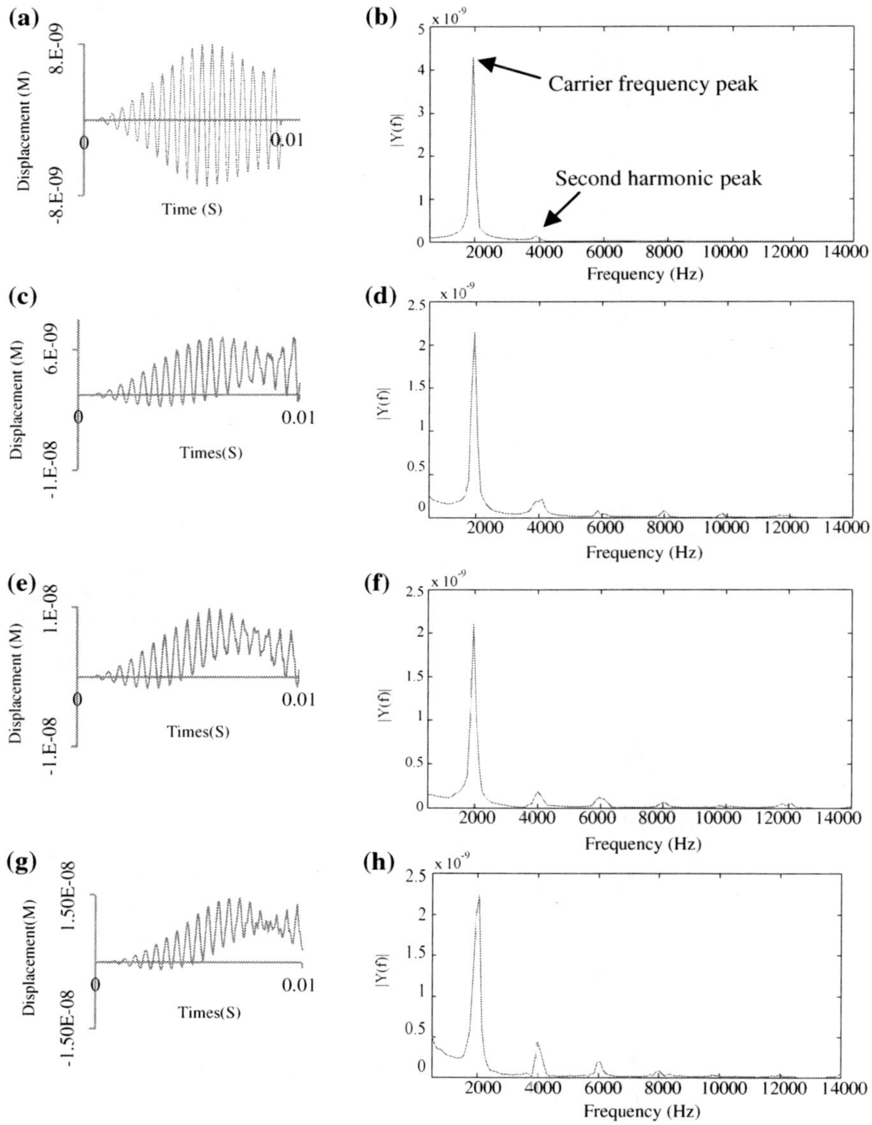

Fig. 3 Typical time domain signals for the delamination length of 100, 200, 250, 300 mm (**a**, **c**, **e**, **g**) and corresponding FFT of the signals (**b**, **d**, **f**, **h**)

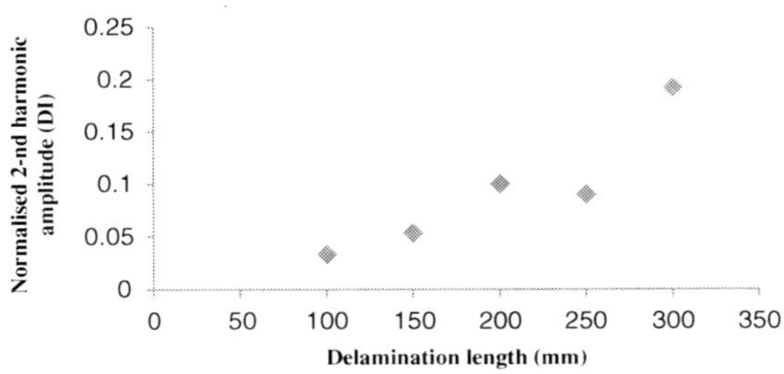

Fig. 4 Delamination length versus damage severity

3.1 Numerical Results and Discussion

The total health monitoring work comprises following steps: obtaining time domain data from the simulation, feature (i.e., appearance of harmonics) extraction from the data by performing FFT of the time domain signal and damage severity study by plotting damage index corresponding to the delamination length. *The damage index* (DI) is defined as the normalized second harmonic amplitude with respective to the carrier frequency amplitude. The typical time domain signals and the corresponding FFTs for the delamination lengths of 100, 200, 250 and 300 mm are shown in Fig. 3. In this figure, higher harmonics at 4, 6, 8, 10 and 12 kHz of the 2 kHz carrier frequency are shown. The final result in Fig. 4 shows the increase in the nonlinearity (DI) with the increase in the delamination length.

4 Conclusions

Since material strength is often reduced by micro defects, which are usually too small to detect by linear ultrasonic techniques (e.g., by measuring the change in attenuation, velocity, etc.), but can increase the acoustic nonlinearity significantly much earlier than the change in the linear parameters of the material. So, very small damage (e.g., micro cracks, contacts, etc.) can be detected by measuring nonlinear acoustic parameters. But to adopt the nonlinear techniques, it is very much important to check that degree of nonlinearity associated with the instruments including PZTs used for the experiments as they are also nonlinear in character. Also the boundary condition is of another concern for those techniques since boundary conditions may induce strong nonlinear effects.

The numerical investigation of the stiffened aluminum plate reveals that the modulation of the elastic wave occurring at the debonding interface between the plate and stiffener due to the wave interaction with the nonlinear contact stiffness. As a consequence, the higher harmonics of the input signal are generated. The corresponding graph for different delamination lengths show that the increase in the nonlinearity parameter (damage index) is due to the increase in the delamination length.

4.1 Scope for Future Research

The present work is the delamination detection of stiffened metallic structures can be extended for damage detection of composite plates as well as to monitor the health of riveted and bolted joint structures. So far only higher harmonics method is adopted for the stiffened plate simulation. The model can also be tested for other nonlinear methods to check how the other methods are sensitive to this type of contact nonlinearity. Experimental validation of the simulation results are required to rely on the methods.

References

1. Mal A, Ricci F, Banerjee S, Shih F (2005) A conceptual structural health monitoring system based on vibration and wave propagation. Str Heal Monit 4(3):0283–11
2. Klepka A, Staszewski WJ, Jenal RB, Szwedo M, Iwaniec J, Uhl T (2001) Nonlinear acoustics for fatigue crack detection—experimental investigations of vibro—acoustic wave modulations. Str Heal Monit 11(2):197–211
3. Nagy PB (1998) Fatigue damage assessment by nonlinear ultrasonic materials characterization. Ultrasonics 36(1998):375–381
4. Van Den Abeele KE-A, Johnson PA, Sutin A (2000) Nonlinear elastic wave spectroscopy (NEWS) techniques to discern material damage, Part I: nonlinear wave modulation spectroscopy (NWMS). Res Nondestr Eval 12:17–30
5. Van Den Abeele KE-A, Carmeliet J, Ten Cate JA, Johnson PA (2000) Nonlinear elastic wave spectroscopy (NEWS) techniques to discern material damage. Part II: Single mode nonlinear resonance acoustic spectroscopy. Res Nondestr Eval 12(1):31–42
6. Kim J-Y, Jacobs LJ, Qu J (2006) Experimental characterization of fatigue damage in a nickel-base superalloy using nonlinear ultrasonic waves. J Acoust Soc Am 120(3):1266
7. Klepka A, Staszewski WJ, Jenal RB, Szwedo M, Iwaniec J, Uhl T (2011) Nonlinear acoustics for fatigue crack detection–experimental investigations of vibro-acoustic wave modulations. Str Heal Monit 11(2):197–211
8. Guyer RA, TenCate J, Johnson P (1999) Hysteresis and the dynamic elasticity of consolidated granular materials. Am Phys Soc 82(16), April 19
9. Meo M, Polimeno U, Zumpano G (2008) Detecting damage in composite material using nonlinear elastic wave spectroscopy methods. Appl Compos Mater 15:115–126
10. Barbieri E, Meo M, Polimeno U (2009) Nonlinear wave propagation in damaged hysteretic materials using a frequency domain-based PM space formulation. Int J Solids Struct 46 (2009):165–180
11. Den Abeelea Van, Sutin A, Carmeliet J, Johnson PA (2001) Micro-damage diagnostics using nonlinear elastic wave spectroscopy (NEWS). NDT&E International 34(2001):239–248
12. Zumpano G, Meo M (2007) A new nonlinear elastic time reversal acoustic method for the identification and localisation of stress corrosion cracking in welded plate-like structures—A simulation study. Int J Solids Struct 44(2007):3666–3684
13. Buck O, Morris WL, Richardson JM (1978) Acoustic harmonic generation at unbounded interfaces and fatigue cracks. Appl Phys Lett 33(5):1
14. Ostrovsky LA, Johnson PA (2001) Dynamic nonlinear elasticity in geomaterials. Rivista Del Nuovo Cimento 24(7):2001
15. Yang J, Chang F-K (2006) Detection of bolt loosening in C-C composite thermal protection panels: I Diagnostic principle. Smart Mater Struct 15(2006):581–590

Using Roving Disc and Natural Frequency Curves for Crack Detection in Rotors

Zyad N. Haji and S.O. Oyadiji

Abstract In this work, a simple crack identification and localisation approach, which is based on a finite element model of a cracked rotor, is proposed. The model simulates the behaviour of a cracked rotor to provide data to be used to identify the location and depth ratios of a crack in stationary rotor systems. The cracked region, which is an open crack type, is modelled as time-varying stiffness. The rotor is considered as a simply-supported beam carrying a roving disc as an auxiliary mass which transverses along the shaft from one end to the other. A new technique, which is called frequency curve product (FCP) method, is presented to identify and locate a crack clearly in rotor systems. The proposed technique is based on the normalised natural frequency curves (NNFCs) of cracked and intact rotors using the principle of roving masses and natural frequency curves. This technique is developed in order to firstly, solve the disappearance of a crack effect from NNFCs when the crack is close to or exactly at a node of a mode shape. Secondly, FCP curves combine the first four NNFCs in a single plot to identify the crack location clearly. Different pairs of the normalised natural frequencies of different modes of vibration are multiplied together in order to enhance the identification and location of cracks. It is shown that this technique identifies the exact crack location through unifying all the first four natural frequency curves at the maximum positive value in the plot.

Keywords Rotor dynamics · Vibration analysis · Crack identification · Roving disc · Open crack · Finite element modelling

Z.N. Haji (✉) · S.O. Oyadiji
School of Mechanical, Aerospace and Civil Engineering, The University of Manchester, Oxford Road, Manchester M13 9PL, UK
e-mail: zyad.haji@postgrad.manchester.ac.uk

S.O. Oyadiji
e-mail: s.o.oyadiji@manchester.ac.uk

J.K. Sinha (ed.), *Vibration Engineering and Technology of Machinery*, Mechanisms and Machine Science 23, DOI 10.1007/978-3-319-09918-7_57

1 Introduction

The presence of cracks in shafts is inevitable and is due to flaws in the shaft material, manufacturing defects and cyclic loading. The growth of fatigue cracks in shafts may cause serious catastrophic failures and accidents if the cracks are not identified at the early stage of the crack initiation. It is known that a surface transverse crack in rotating systems results in reduction of local flexibilities and, hence, in changes to the vibrational behaviour of the rotor system [1, 2]. However, the presence of a crack in a rotor does not necessarily cause the rotor to promptly become unusable. The shaft is still usable and it gives an incipient signal that is embedded in the dynamic behaviour of the shaft. The shaft behaviour has to be monitored more carefully to avoid a sudden catastrophic failure. Therefore, the dynamic behaviour of a rotor with a surface transverse crack, which is a useful tool for the identification of a crack in rotating machinery, has been studied for a long time [3]. Over the past few decades, various methods and techniques of crack modelling and identification in rotating and stationary systems have been developed. Most of these methods are based on transverse vibration due to the fact that the fundamental frequency of the systems is due to transverse vibration. Localised methods, such as ultrasonics and X-ray techniques, have also been developed but they are not as versatile and do not offer high testing accuracy as transverse vibration method for crack and damage detection in huge and complex systems with inaccessible parts [3–5]. Patil and Maiti [3] have represented cracks in a beam by rotational springs which connect between the segments of the divided beam. They used a technique based on the transfer matrix method to identify cracks in beams. Lee [6] has proposed a simple method for the identification of multi-cracks in beams. Each crack was modelled as a massless rotational spring, and the Newton–Raphson and finite element methods were used to solve the inverse problem. Recently new crack modelling techniques, based on the finite element method, have been developed by Kulesza and Sawicki [7] and Liong and Proppe [8] and Al-Shudeifat and Butcher [9, 10].

Since the presence of cracks in structural systems reduces the local stiffness [4, 6], many investigations have been carried out using changes in natural frequencies because they can be measured more easily than other dynamic characteristics. Liang, Choy and Hu [11] have suggested that the intersection point of three frequency contour lines could be used for the identification and localisation of a crack in beams. Also, Nandwana and Maiti [12] have used the changes in natural frequencies for the identification of a crack when the crack was an inclined edge crack or an internal crack. The frequency contour plot method has been extended by Nikolakopoulos, Katsareas and Papadopoulos [13] to identify the location and size of a crack in beams and in plane frames based on the Timoshenko beam theory. In addition, the reduction of the natural frequencies due to the traversing of a roving auxiliary mass along cracked beams has been recently employed by Zhong and Oyadiji [14, 15] for the identification and location of cracks.

In this study, a simple technique, which is called frequency curve product (FCP) method, is presented to identify and locate a crack very clearly in stationary rotor systems. The proposed technique is based on the normalised natural frequency curves (NNFCs) of cracked and intact stationary rotors using the principle of roving masses and natural frequency curves which was introduced by Zhong and Oyadiji [14]. These NNFCs are obtained from the finite element modelling of the cracked stationary rotor with an open crack through using a roving disc. This technique is developed in order to firstly, solve the problem of the disappearance of a crack effect from NNFCs when the crack is close to or exactly at a node of a mode shape. This problem makes it difficult to decide the exact crack location in rotor systems. Secondly, FCP curves combine the first four NNFCs in a single plot to identify the crack location clearly. Different pairs of the normalised natural frequencies of different modes of vibration are multiplied together in order to enhance the identification and location of cracks. It is shown that this technique identifies the exact crack location through unifying all the first four natural frequency curves at the maximum positive value in the plot.

2 System Modelling

The illustrated rotor in Fig. 1 with dimensions in Table 1 has been used for the investigations of intact and cracked stationary rotors. The rotor, which is supported by rigid bearings as a simply-supported beam, consists of a shaft and a movable disc of 0.5 kg, is divided into N Bernoulli–Euler beam finite elements, where each element has two nodes with four degrees of freedom per node (Fig. 2). The finite element of the cracked and intact rotor has been programmed in the MATLAB environment.

2.1 Cracked Rotor Modelling

After assembling the N Bernoulli–Euler shaft elements, the rigid disc and the bearing supports, the free vibration equations of motion the intact rotor can be written as

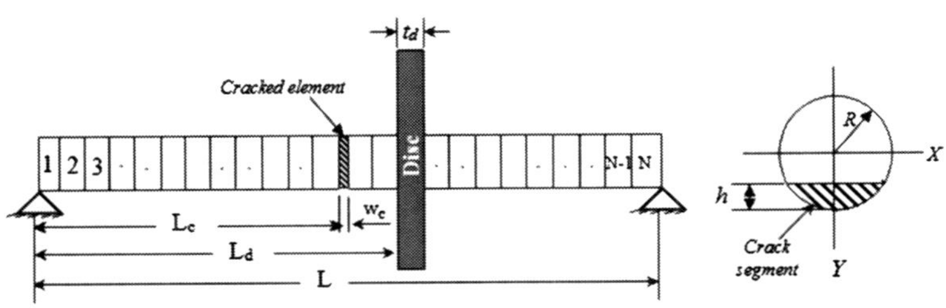

Fig. 1 Finite element model of an intact and cracked rotor

Table 1 Values of parameters in the numerical model

Parameters	Shaft	Disc
Material	Stainless steel	Aluminium
Young's Modulus, E_s (GPa)	200	70
Poisson's ratio, v_s	0.3	0.33
Density, r (kg/m^3)	7800	2700
Length, L (m)	1	0.01, 0.02, 0.3, …, 1
Outer diameter, D_o (m)	0.02	0.2
Inner Diameter, D_i (m)	–	0.02
Thickness, t (m)	–	0.04
Mass, m (kg)	2.45	0.5

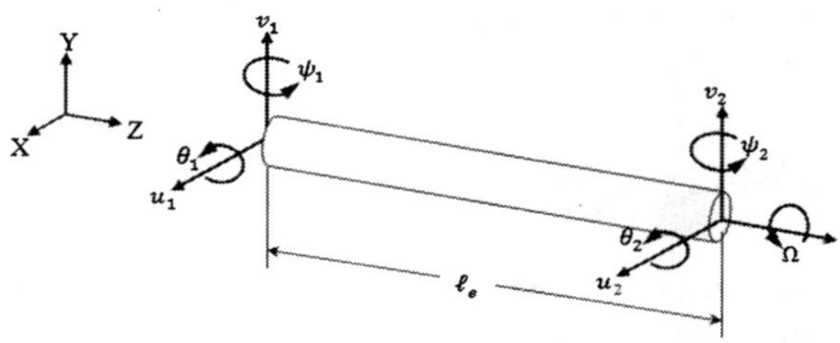

Fig. 2 Typical finite rotor element and coordinates for an intact and cracked rotor

$$\mathbf{M}\ddot{\mathbf{q}} + (\mathbf{C} + \Omega\mathbf{G})\dot{\mathbf{q}} + \mathbf{K}\mathbf{q} = 0 \tag{1}$$

where, $\mathbf{q}(t) = \left[\mathbf{q}_1^T \ldots \mathbf{q}_2^T \ldots \mathbf{q}_i^T \ldots \mathbf{q}_{N+1}^T\right]^T$ is the nodal displacement vector with dimension 4(N + 1) × 1 corresponding to the local coordinate vector $\boldsymbol{q}_e = [u_1 \quad v_1 \quad \theta_1 \quad \psi_1 \quad u_2 \quad v_2 \quad \theta_2 \quad \psi_2]^T$ for each element in Fig. 1, **M** is the global mass matrix which contains the mass matrices $\mathbf{M}_e$ for each element of the rotor and the mass matrix $\mathbf{M}_d$ of the disc corresponding to degrees of freedom $[u \quad v \quad \theta \quad \psi]^T$. **G** and **C** are the gyroscopic and damping effects, respectively.

In this paper, the cracked model given in Fig. 3, which was proposed by AL-Shudeifat and Butcher [9, 10], has been used in order to locally define the stiffness properties at the crack element. In this model, the stiffness reduction of the cracked element defined as

$$\mathbf{K}_{ce}^i = \mathbf{K}_1^i + \mathbf{K}_2^i \cos(2(\Omega t + \phi)) \tag{2}$$

Thus, the FEM equations of motion of the uncracked rotor bearing-system given in Eq. (1) are rewritten including the effect of the cracked element of an open crack model as

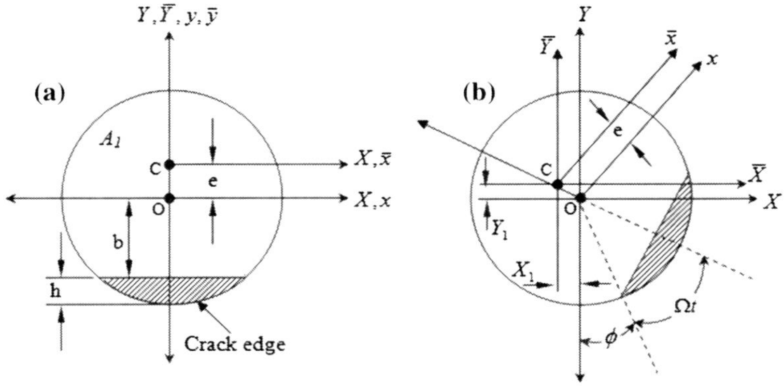

Fig. 3 A cracked element cross-section; **a** non-rotating, **b** rotating; the hatched part defines the area of the crack segment

$$\mathbf{M}\ddot{\mathbf{q}}(t) + (\mathbf{C} + \Omega\mathbf{G})\dot{\mathbf{q}}(t) + \left(\overline{\mathbf{K}} + \widehat{\mathbf{K}}\cos(2(\Omega t + \phi))\right)\mathbf{q}(t) = 0 \tag{3}$$

where $\overline{\mathbf{K}}$ is the global stiffness matrix derived from the global stiffness matrix $\mathbf{K}$ of the uncracked beams by replacing the uncracked element stiffness matrix $\mathbf{k}_1^i$ of element i by the cracked element stiffness matrix $\mathbf{k}_{1c}^i$. $\widehat{\mathbf{K}}$ is another global stiffness matrix; it has zero elements apart from those at the cracked element location where the elements are equal to $\mathbf{k}_{2c}^i$. In this study, both Ω and ϕ are equal to zero, according to the assumptions that the rotor is stationary and the crack is always fully open (Fig. 3a).

2.2 Crack Identification Approach

In this section, a simple approach, which is called frequency curve product (FCP) technique, is developed for the identification and location of a crack in rotor systems. This approach is implemented as follows:

Step 1: Implementation of Normalised Natural Frequency Approach

a. The natural frequencies f_{ci} of the cracked rotor are divided by the corresponding natural frequencies f_{oi} of the intact rotor for the same roving disc locations in order to obtain the natural frequency ratios ($\beta_i = f_{\mathrm{ci}}/f_{\mathrm{oi}}$) of each mode.
b. The natural frequency ratios β_i of each mode are divided by the maximum natural frequency ratio $\widehat{\beta}_i$ of each mode to determine the NNFCs λ_i (i.e. $\lambda_i = \beta_i/\widehat{\beta}_i$).
c. The variation of the NNFCs λ_i is plotted against the non-dimensional roving disc locations Γ.

Step 2: Implementation of Frequency Curve Product Technique

a. For each mode, and a specified crack depth ratio μ, let the normalised natural frequency value at the zero location ($\Gamma = 0$) of the roving disc act as a pivot, and subtract its value from the normalised natural frequencies corresponding to the roving disc locations in order to obtain the shifted NNFCs $\xi_{m,\mu}$ of a mode (m = 1, 2, 3 and 4) at an assigned crack depth ratio (μ) defined as,

$$\xi_{m,\mu} = \lambda_{i,\mu} + \lambda_{0,\mu} \quad (4)$$

where $\lambda_{0,\mu}$ is the normalised natural frequency at the pivot point at which $Ld/L = 0$ for each mode, $\lambda_{i,\mu}$ is the normalised natural frequencies corresponding to the roving disc locations, μ is the crack depth ratio and $i = L_d/L$ = 0, 0.01, 0.02, 0.03, 1.

b. The next sub-step is to form products of the NNFCs. This is carried out by multiplying the shifted normalised natural frequency curve $\xi_{1,\mu}$ of mode 1 by the shifted NNFCs $\xi_{2,\mu}$, $\xi_{3,\mu}$ *and* $\xi_{4,\mu}$ for modes 2, 3 and 4 respectively. Each result is normalised by dividing by its maximum value in order to obtain non-dimensional frequency curve product ($\sigma_{1,m}$) which are defined as

$$\sigma_{1,m} = \frac{(\xi_{1,\mu} * \xi_{m,\mu})}{Max(\xi_{1,\mu} * \xi_{m,\mu})} \quad \ldots\ldots \quad m = 2,\ 3,\ 4 \quad (5)$$

c. The threshold positive values of $\sigma_{1,m}$ (i.e. $\sigma_{1,m} > 0$) are plotted for each mode against the non-dimensional roving disc locations for each crack depth ratio.

d. This step unifies all the $\sigma_{1,m}$ (i.e. $\sigma_{1,2}$, $\sigma_{1,3}$ and $\sigma_{1,4}$) at all the crack depth ratios in a single plot under the designation σ_{1234}. This is determined by choosing the positive values of each the $\sigma_{1,2}$, $\sigma_{1,3}$ and $\sigma_{1,4}$ and multiplying them together according to the following equation.

$$\sigma_{(1234)i} = \sigma_{(1,2)i} * \sigma_{(1,3)i} * \sigma_{(1,4)i} \quad \text{for } \sigma_{(1,m)i} > 0 \quad (6)$$

where, i = 1, 2, 3, …, $N-1$.

2.3 Numerical Results and Discussions

The feasibility and applicability of the proposed technique for the identification and localisation of cracks in rotors have been investigated through two different numerical examples. The location and size of a crack of 0.5 mm width, w_c = 0.5 mm, for each example is defined in Table 2.

Table 2 Values for numerical examples

Example	Crack location ($\Gamma = Lc/L$)	Crack depth ratio (μ)	Disc location (L_d/L)
1	0.2	0.1, 0.3, 0.5, 0.7 and 1.0	0. 01, 0. 02, …, 1
2	0.5		

Example 1 In this example, numerical simulations have been carried out on the rotor with an open crack of various depth ratios (μ) at the location $\Gamma = 0.2$ from the left bearing. The results, which are natural frequencies of the cracked and intact rotors, are normalised to produce NNFCs (λ_i). Figure 4 shows the variation of the NNFCs against the non-dimensional roving disc locations for the rotors with cracks located at $\Gamma = 0.2$. The results in the figure clearly identify the presence, location and depth of the crack in the NNFCs λ_1 and λ_2 for modes 1 and 2, respectively, by the very large sharp discontinuity at the correct crack location. However, the NNFCs λ_3 and λ_4 for modes 3 and 4, respectively, do not clearly identify and locate the cracks. This is because of the

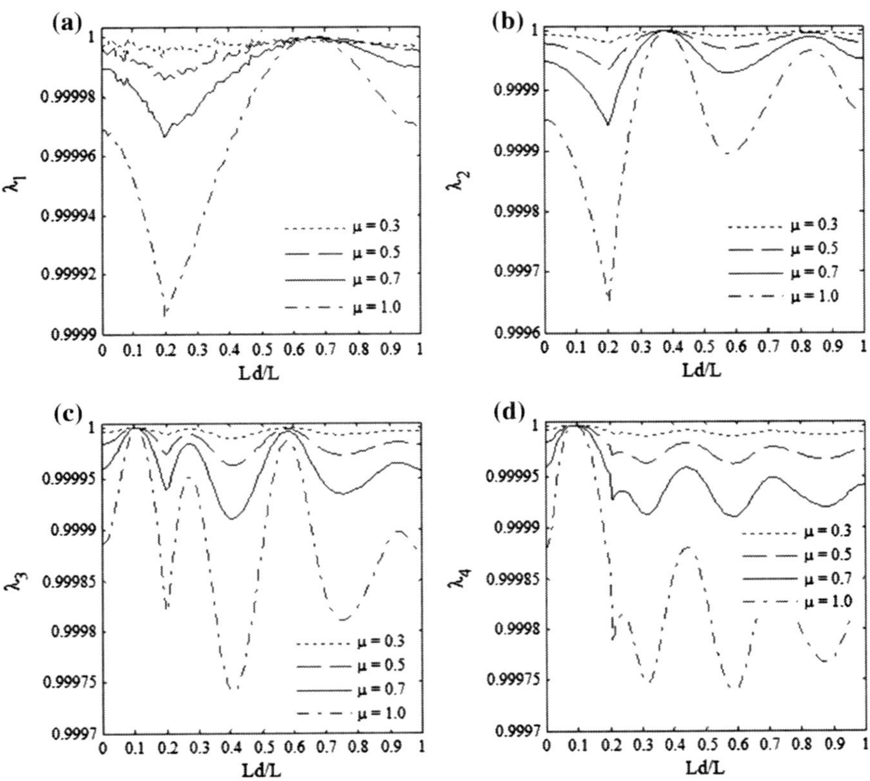

Fig. 4 Normalised natural frequency curves of the cracked rotor with cracks of different depth ratios at $\Gamma = 0.2$: **a** 1st mode, **b** 2nd mode, **c** 3rd mode, **d** 4th mode

closeness of the crack location to the nodes of the third and fourth modes. The figure shows that the curves turn sharply at the crack location but are rounded at the locations where there are no cracks. However, the lack of clear identification of the crack location in λ_3 and λ_4 generate some confusion and doubts about the exact crack location. For these modes, and for this crack location, the NNFCs do not uniquely identify the crack and its location. Also, the clarity of the identification decreases as the crack depth ratio μ decreases below 0.5 i.e. for $\mu \leq 0.3$. □

In order to improve the crack identification for $\mu \leq 0.3$ and for crack locations which are close to the nodes of some modes, for example $\Gamma = 0.2$ is close to the nodes of modes 3 and 4 (i.e. λ_3 and λ_4), the FCP method is applied to the frequency curves. The method uses all the four modes in order to produce results for a particular crack depth ratio as shown by Fig. 5a–c for μ = 0.5, 0.7 and 1.0, respectively, and for all crack depth ratios as shown in Fig. 5d. The crack is

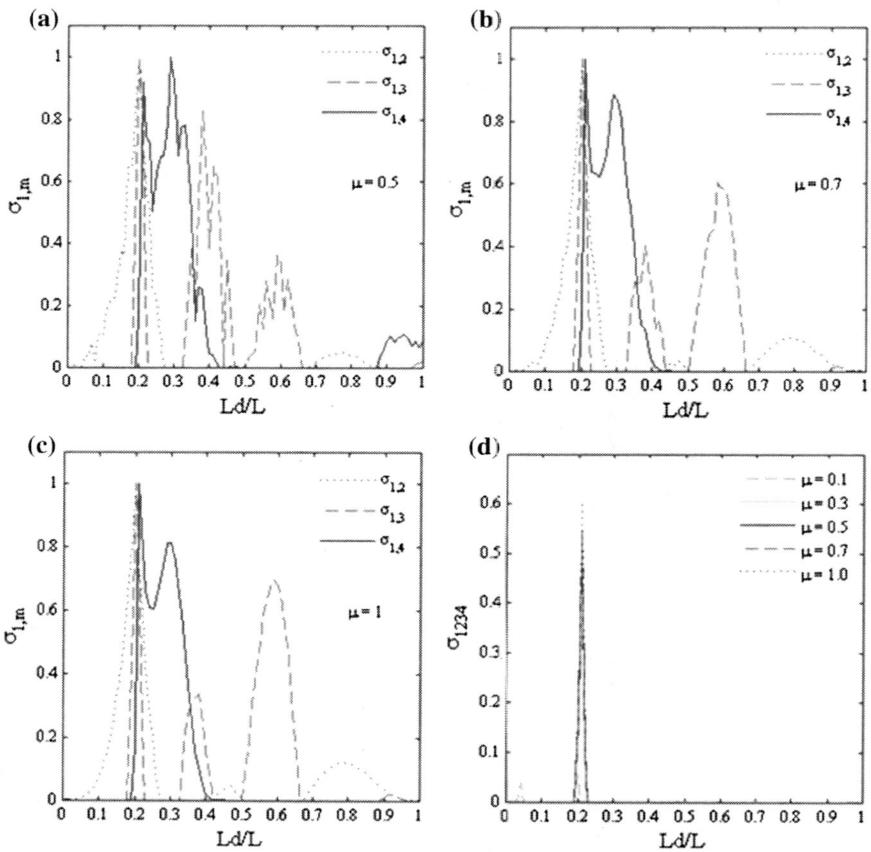

Fig. 5 Variation of frequency curves product (FCP) of the cracked rotor with a crack of various depth ratios at $\Gamma = 0.2$: **a**, **b** and **c** use Eq. (5); **d** uses Eq. (6)

identified at the location where the three frequency curve products $\sigma_{1,2}$, $\sigma_{1,3}$ and $\sigma_{1,4}$ show peak positive values at the same location $\Gamma = 0.2$ which is the correct crack location. However, the presence of other peaks in the results of the $\sigma_{1,2}$, $\sigma_{1,3}$ and $\sigma_{1,4}$ which are computed according to Eq. (5) means that extra effort is required to identify where the positive peaks of $\sigma_{1,2}$, $\sigma_{1,3}$ and $\sigma_{1,4}$ coincide which identifies the exact crack location. This extra effort is encoded as Eq. (6) unifies $\sigma_{1,2}$, $\sigma_{1,3}$ and $\sigma_{1,4}$ of a particular μ in a single plot under the designation σ_{1234} as shown in Fig. 5d. The figure shows that σ_{1234} identifies the crack location clearly without any ambiguity for all the crack depth ratios except for $\mu = 0.1$.

Example 2 Now in this example, open cracks of various depth ratios, which are induced in the rotor at $\Gamma = 0.5$ from the left bearing, have been investigated. It is known that the system natural frequencies are not affected, if a crack is located at modal nodes. That is, the identification of the crack location in this case is difficult to achieve. However, the application of the NNFCs technique on the results of the simulation has identified and localised the crack location with reasonable accuracy. These results, which are shown in Fig. 6, indicate that the location of the crack has

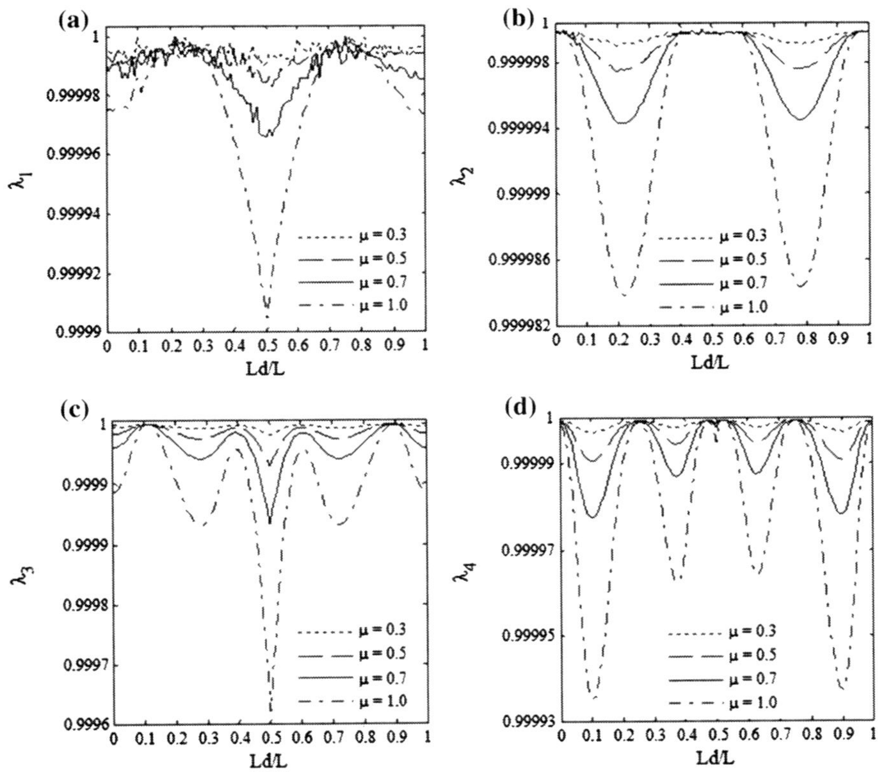

Fig. 6 Normalised natural frequency curves of the cracked rotor with cracks different depth ratios at $\Gamma = 0.5$: **a** 1st mode, **b** 2nd mode, **c** 3rd mode, **d** 4th mode

been very clearly identified in the λ_1 and λ_3 curves (Fig. 6a, c). Also, the figure shows that the discrepancy between the wedge- shaped characteristics of the curves at the cracked location, and the rounded shapes of the curves at the intact locations can be utilised for the identification of the crack. These curve characteristics give a reasonable indication of the crack location in the λ_4 curves (see Fig. 6d, whereas the crack location in the λ_2 curves is still not clearly identified (Fig. 6b) due to the location of the crack at the node of the mode. □

Figure 7 shows the results of applying the FCP method on the NNFCs for this case. It is seen that the peaks of $\sigma_{1,2}$, $\sigma_{1,3}$ and $\sigma_{1,4}$ somewhat coincide at the correct crack location. However, each of the frequency curves $\sigma_{1,2}$, $\sigma_{1,3}$ and $\sigma_{1,4}$ has many peaks, the overlay of these curves looks very chaotic, and identifying where the peaks of the three frequency curves coincide is very challenging. This chaotic behaviour is due to the location of the crack at exactly the middle of the rotor, which coincides with the nodes of modes 2 and 4 of bending vibration of the rotor.

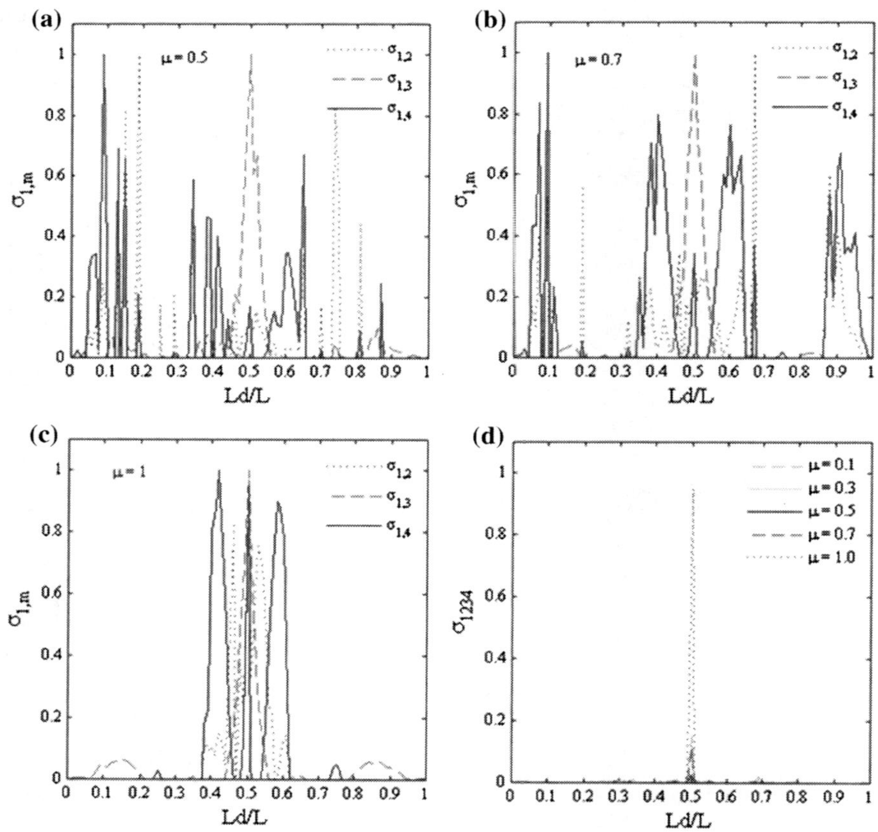

Fig. 7 Variation of frequency curves product (FCP) of the cracked rotor with a crack at $\Gamma = 0.5$ of various depth ratios: **a**, **b** and **c** use Eq. (5); **d** uses Eq. (6)

This chaotic behaviour makes it very difficult to identify the crack locations. In contrast, the results of the curves σ_{1234} in Fig. 7d identify and localize the cracks with high accuracy, clarity and without any chaotic behaviour irrespective of how small the crack depth ratio μ is. This unique feature of the composite frequency curve σ_{1234} facilitates the identification and location of cracks along the rotor without any ambiguity.

3 Practical Implementation of the Proposed Technique

In practice, there is the need to conveniently identify and locate cracks in rotors. A practical approach is to initially measure the natural frequencies of the rotor using very few ('coarse') locations of the roving disc. The results of this coarse measurement will give an approximate location for the structural fault. Further measurements can be made using 'refined' roving disc locations around the approximate location.To minimise the effect of noise, the mass of the roving disc should not be less than 5 % of the shaft mass. Also, to minimise the impact of the disc dynamics on the crack, the mass of the disc should be less than 50 % of the shaft mass.

4 Conclusions

The investigation results have shown that the application of the FCP approach enhances the identification and location of a crack that is located at or close the nodes of some modes. Despite the high chaotic behaviour generated in the FCP curves when the crack is located at the nodes of some modes and a small roving disc is used, the FCP curves identified and located the crack relatively better than the NNFCs. Also, the results showed that FCP method has the merit of unifying peaks of the products of frequency curves only at the crack locations.

References

1. Singh SK, Tiwari R (2013) Detection and localization of multiple cracks in a stepped shaft. Fatigue Fract Eng Mater Struct 36:85–91
2. Han Q, Chu F (2013) Dynamic response of cracked rotor-bearing system under time-dependent base movements. J Sound Vib 332:6847–6870
3. Patil DP, Maiti SK (2003) Detection of multiple cracks using frequency measurements. Eng Fract Mech 70:1553–1572
4. Behzad M, Ghadami A, Maghsoodi A, Michael Hale J (2013) Vibration based algorithm for crack detection in cantilever beam containing two different types of cracks. J Sound Vib 332:6312–6320

5. Cawley P, Ray R (1988) A comparison of the natural frequency changes produced by cracks and slots. J Vib Acoust Stress Reliab Design 110:366–370
6. Lee J (2009) Identification of multiple cracks in a beam using natural frequencies. J Sound Vib 320:482–490
7. Kulesza Z, Sawicki JT (2012) Rigid finite element model of a cracked rotor. J Sound Vib 331:4145–4169
8. Toni Liong R, Proppe C (2013) Application of the cohesive zone model for the evaluation of stiffness losses in a rotor with a transverse breathing crack. J Sound Vib 332:2098–2110
9. Al-Shudeifat MA, Butcher EA (2010) On the modeling of open and breathing cracks of a cracked rotor system. In: ASME Conference Proceedings, pp 919–928
10. Al-Shudeifat MA, Butcher EA (2011) New breathing functions for the transverse breathing crack of the cracked rotor system: approach for critical and subcritical harmonic analysis. J Sound Vib 330:526–544
11. Liang RY, Choy FK, Hu J (1991) Detection of cracks in beam structures using measurements of natural frequencies. J Franklin Inst 328:505–518
12. Nandwana BP, Maiti SK (1997) Modelling of vibration of beam in presence of inclined edge or internal crack for its possible detection based on frequency measurements. Eng Fract Mech 58:193–205
13. Nikolakopoulos PG, Katsareas DE, Papadopoulos CA (1997) Crack identification in frame structures. Comput Struct 64:389–406
14. Zhong S, Oyadiji SO (2008) Analytical predictions of natural frequencies of cracked simply supported beams with a stationary roving mass. J Sound Vib 311:328–352
15. Zhong S, Oyadiji SO (2008) Identification of cracks in beams with auxiliary mass spatial probing by stationary wavelet transform. J Vib Acoust 130:041001–041014

Vibration Based Diagnosis of Distributed Bearing Faults

Boštjan Dolenc, Pavle Boškoski, Jurij Pfajfar and Đani Juričić

Abstract Bearings are vital components in rotating machinery, being the only one that separate stationary and rotating elements of the machines. Bearings are subjected to degradation in various manners, e.g. scuffing, pitting, brinelling, electro erosion. In the early stage of degradation, faults emerge on the rolling surface locally. These tiny surface defects are hardly visible but do not cause problems in system operation. As time progresses, the process further deteriorates the rolling surface. The damage on the rolling surface widens in size and becomes distributed fault. The aim of this paper is to propose an approach for detection of distributed bearing faults. It is based on assumption that these faults excite particular vibrational pattern that can make them distinguishable from local faults. For this purpose, a lumped parameter model capable of simulating vibrational response of damaged bearing was adopted. The model was extended by incorporating fluctuations in ball diameter. By analyzing envelope spectra of simulated and measured vibrations we show that vibrational patterns produced by localized and distributed faults differ. The main contribution of the paper is a new feature which combines BPFO (ball pass frequency outer) and FTF (fundamental train frequency) component of the envelope spectrum. A simulation study dealing with broad range of damage propagation scenario is performed in order to assess performance of the proposed features. Finally, validation is performed on the set of damaged bearings.

B. Dolenc (✉) · P. Boškoski · Đ. Juričić
Department of System and Control, Jožef Stefan Institute, Ljubljana, Slovenia
e-mail: bostjan.dolenc@ijs.si

P. Boškoski
e-mail: pavle.boskoski@ijs.si

Đ. Juričić
e-mail: dani.juricic@ijs.si

J. Pfajfar
DOMEL, d.o.o, Železniki, Slovenia
e-mail: jure.pfajfar@domel.si

B. Dolenc
Jožef Stefan International Postgraduate School, Ljubljana, Slovenia

J.K. Sinha (ed.), *Vibration Engineering and Technology of Machinery*,
Mechanisms and Machine Science 23, DOI 10.1007/978-3-319-09918-7_58

Keywords Distributed bearing faults · Vibration analysis · Diagnostic features

1 Introduction

Bearing faults are one of the most common causes of failures in rotating machines [1]. Such failures are caused due to various reasons, for instance: pitting, spalling, electro erosion, wear etc., and are regarded as either localized or distributed faults. In this paper, an updated bearing model, capable of simulating both distributed and localized bearing faults is presented. By introducing imperfections in the bearing components it is shown how some of these faults manifest in vibrations and differ from each other.

The term distributed fault may appear somewhat confusing. In literature, Tandon and Choudhury [2, 3], Sopanen and Mikkola [4, 5] used the phrase to describe bearing ring imperfections. In these papers distributed bearing faults are regarded as irregularities in bearing geometry due to imperfect manufacturing, faulty mounting and misuse and not actual surface damage. Conversely, distributed faults are here regarded as bearing surface defects spreading over large area where two or more rolling elements can be located in damaged area at given moment.

The majority of published works addressing bearing diagnostics focus only on localized faults. Although these faults indicate the existing problem in bearing, it does not necessarily mean that the faulty bearing is at the end of its lifetime and should be replaced [6]. Gradually, localized faults evolve and spread over wider area, thus becoming distributed faults causing distinctive vibration pattern.

By monitoring the progress of the bearing surface deformation one could estimate the time until the bearing in question would not perform its function sufficiently well [7]. Even though better understanding of the impact of various irregularities such as distributed faults would improve prognostic capabilities [8], published work on analysis of vibrational pattern produced by bearing with distributed fault is scant.

One approach towards estimating the size of the fault is based on the time difference between the moments when the rolling element enters and exits the damaged area [9, 10]. Even though both methods successfully estimate the size of the fault, these papers do not study larger, distributed faults. Al-Ghamda and Mba [7] studied the effect of fault size on RMS and kurtosis values on acoustic emissions (AE) and vibration signals. Faults under consideration were somewhat larger than the ones used in previously presented work, yet significantly smaller than the ones used in this research. Eftekharnejad et al. [11] studied the effect of the growing defect on the envelope spectrum while monitoring vibrations and AE. However, the paper focuses on the comparison of AE and vibration methods with growing defect rather than on estimation of a fault size. Sawalhi and Randall [12] also studied envelope spectrum of the vibration produced by faulty bearing. Extended (few millimeters wide) faults were seeded on the outer race and simulated using their model.

Vibrations are compared by means of envelope spectrum inspection. Contrary to the above, de Castelbajac et al. [13] studied the effect of distributed fault for high spindle bearings introducing new criterion called Spindle Bearing Noise (SBN).

To sum up, previous work mainly focuses on analyzing vibrations generated by localized faults. Although most of the methods were successful in estimating the fault size, only a few of them consider naturally evolved faults. Defects were mostly seeded on the race surface, ignoring the possibility that damages can appear elsewhere.

The purpose of this study is to determine the effect of the distributed fault on the envelope spectrum of the vibrational signal. By upgrading the model developed by Sawalhi and Randall [14], one can show that distributed bearing faults can be distinguished from localized ones by introducing imperfections of the bearing elements into the model. The simulated results were validated utilizing vibration patterns generated by bearings with naturally born distributed faults spread over larger portions of the bearing circumference.

2 Modeling Vibration Response of Damaged Bearings

Ideally, when the bearing is in perfect condition, meaning bearing surface is intact and bearing components are impeccable, it performs flawlessly. However, in reality, even brand new bearings excite some vibrations. These vibrations originate from imperfect production. The shape of the bearing components somewhat differ from the desired. Moreover, during the mounting, components, such as bearing rings or rolling elements, can further deform in shape.

While in operation, bearings are subjected to various degradation processes (e.g. scuffng, pitting, brinelling, corrosion, electro erosion) which impede its performance. These processes manifest in surface defects which excite peculiar vibrational pattern through which they can be detected and localized.

Each time a rolling element hits the damaged area it excites impulse response $s(t)$, forming a train of impulses which can be expressed as [15]:

$$s(t) = \sum_{i=-\infty}^{i=\infty} A_i s(t - \tau_i) + n(t), \tag{1}$$

where A_i are random amplitudes, τ_i represents the time when random impacts occur and $n(t)$ is an additive random component that contains all non-modeled vibrations as well as environmental disturbances.

In theory, this model could be used to describe vibrations produced by any kind of fault. However, to properly characterize the distributions of random variables τ_i and A_i is formidable task. Naturally born faults may occur in several separate locations with different defect profiles, from small localized faults to large, distributed faults.

2.1 Bearing Vibration Model

To overcome the described problems, a bearing vibration model was proposed by Sawalhi and Randall [12, 14]. Although fairly simple, the model describes all the essential dynamic properties of a bearing mounted on housing.

The five DOF model, shown in Fig. 1, omits the mass and inertia of the rolling elements. Two DOF describe bearings inner race or shaft displacement (x_s, y_s), while (x_p, y_p) describe the motion of the motor housing where the outer race is mounted. The extra fifth DOF describes high frequency response of the bearing y_b. $k_p, c_p, k_s, c_s, k_r, c_r$ are stiffness and damping coefficients of the shaft, housing and bearing respectively. k_b is bearing load deflection factor.

The forces acting on the bearing rings, which, in turn, generate vibrations, are calculated using Hertzian contact theory. Elastic deformation of the *j*th rolling element δ_j is a function of the inner race displacement relative to the outer race, the rolling element radial position in time ϕ_j, radial clearance c (gap between rolling elements and races) and surface defect $C(\phi)$. The overall contact deformation of an individual rolling element reads [14]:

$$\delta_j = (x_s - x_p)\cos\phi_j + (y_s - y_p)\sin\phi_j - c - \beta_j C(\phi), \tag{2}$$

where β_j describes the position and the size of the modeled fault as:

$$\beta_j = \begin{cases} 1, & if\ \phi_d < \phi_j < \phi_d + \Delta\phi_d \\ 0, & otherwise. \end{cases} \tag{3}$$

The fault is positioned on the outer race between ϕ_d and $\phi_d + \Delta\phi_d$, as shown in Fig. 2. In the case of the inner race fault, the fault rotates with the shaft, which can be modeled as $\phi_d = \omega_n t + \phi_0$, where ϕ_0 is the initial position of the fault and ω_n is shaft speed.

Fig. 1 Model DOF

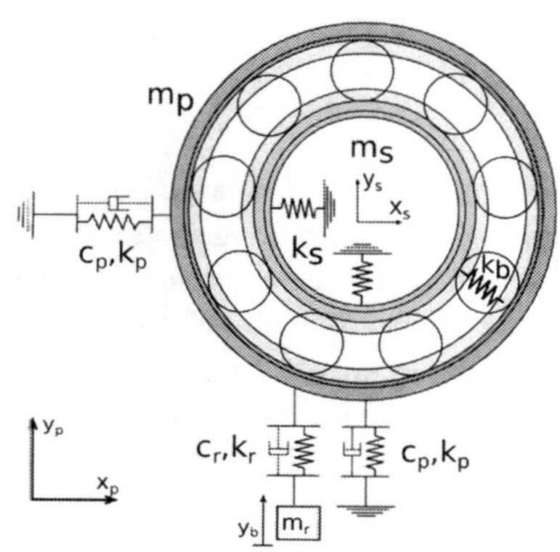

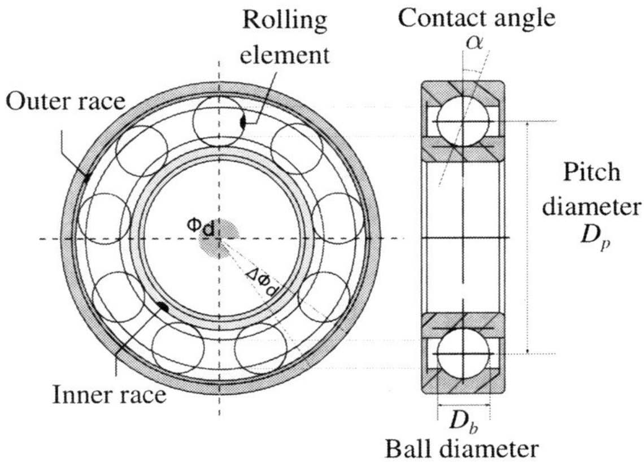

Fig. 2 Bearing dimensions and fault position

2.2 Modifications of a Model

In their paper, Sawalhi and Randall [12] described roughness of the surface defect $C(\phi)$ as low-pass Gaussian noise and studied the effect on envelope spectra of simulated vibrations.

However, to explain the vibration pattern produced by the distributed bearing fault, $C(\phi)$ needs to be modified. Considering imperfection in production mentioned earlier, rolling elements diameters slightly differ from the designed value. This detail turned out to be of great importance when describing vibration pattern produced by distributed bearing fault. The impact of these little differences on spectral content of the signal may be observed by modifying Eq. 2 as:

$$\delta_j = (x_s - x_p)\cos\phi_j + (y_s - y_p)\sin\phi_j - c - \beta_j\big(C(\phi) + A_j\big) + R(\phi), \quad (4)$$

where A_j models diameter deviation of the *j*th rolling element. $R(\phi)$ models geometrical deformations of rolling surfaces [5]. These occur when bearings are subjected to extreme forces.

$$R(\phi) = \sum_{m=1}^{n} A_{in}^{m}\cos\left[m(\phi_j - \theta_{in}) + \alpha_m^{in}\right] + \sum_{m=1}^{n} A_m^{out}\cos\left[m(\phi_j - \theta_{out}) + \alpha_m^{out}\right], \quad (5)$$

A_m^{in}, A_m^{out} are amplitudes and $\alpha_m^{in}, \alpha_m^{out}$ are the phase angles of the *m*th sine wave for inner and outer race respectively. ϕ_j, θ_{in} and θ_{out} are the angular positions of the *j*th rolling element, inner and outer race, respectively.

2.3 Characteristic Frequencies of the Distributed Fault

Localized faults produce fairly simple vibration pattern. Contrarily, vibrations excited by distributed fault are more complex. The faulty area excites vibrations composed of two major impacts, being entry and exit events [9] and series of vibrations produced by damaged area.

The frequency, by which the outer race fault can be diagnosed, is the frequency by which the rolling elements enter the damaged area and is denoted by BPFO (ball pass frequency outer). This frequency can be calculated from geometric properties of the bearing and is associated with the cage frequency FTF as $\mathrm{BPFO} = \mathrm{n_b} * \mathrm{BPFO}$, where $\mathrm{n_b}$ is the number of rolling elements. Keeping in mind the ball diameter differences, the vibrations excited by individual rolling element differ from one another. As each of the rolling elements enter the damaged area once per cage revolution, the envelope spectra, in addition to BPFO, contains peaks at FTF as well as its modulations to BPFO.

However, the impact of the ball diameter discrepancy differ from one bearing to another, even if the bearings are in perfect condition and only rolling elements differ in diameter [2]. As it will be shown in the following section, higher order characteristic frequencies can emerge in the envelope spectra of the measured vibrations. The possible reason of this peaks are differences in the profiles of the damaged surface areas, possible deformations of the rolling elements etc.

3 Vibration Based Diagnosis of Distributed Bearing Faults: Simulation Study and Measurement

In order to better interpret the experimental data the model briefly described in previous section was built using the same parameters as Sawalhi and Randall [14]. It should be noted that the model used is conceptual model of the realistic bearing at hand used in this study as the parameters are not obtained by any identification means from data. Consequently the content of the model response cannot be directly compared with the measured values, but rather the relative spectral contents should be analyzed and compared.

Presence of bearing faults is related to appearance of characteristic frequencies in the envelope spectra of vibration signals. Therefore the analysis is conducted only to show the emergence of the same spectral components under the same fault type, omitting the comparison of the amplitude values.

FAG 6205 bearings were mounted in motor to measure their vibration response. Two cases are studied: bearings with local fault, bearings with distributed fault. By comparing envelope spectra of the two faults we will show how their frequency contents differ.

In the first simulation a localized fault 0.5 mm deep is placed on the outer ring, while the angle $\Delta\phi_{\mathrm{d}}$ (Fig. 2) was set to 2°. By observing envelope spectrum of

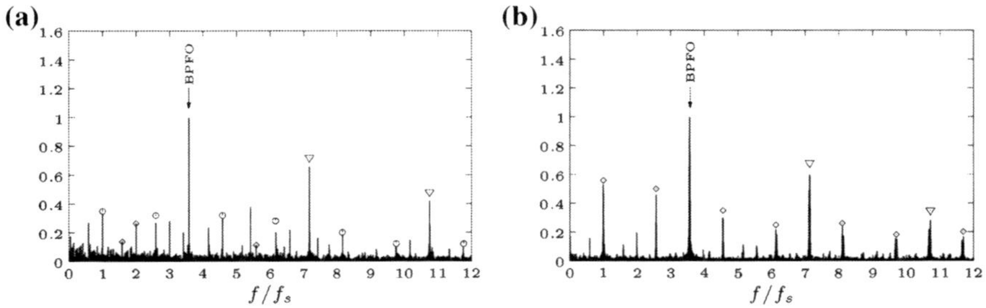

Fig. 3 Vibration envelope spectra under localized outer race fault: both spectra are clearly dominated by BPFO and its multiples indicated by triangles. RPM (x) and its multiples (*circles*) modulate as a result of ring waviness. **a** Simulation, **b** measurement

simulated vibrations in Fig. 3a one can see the prevailing characteristic frequency BPFO. However, closer inspection reveals additional peaks at shaft speed and its multiples and their modulations. In Fig. 3a (x) marks shaft speed (RPM) and its modulations to BPFO, while circle marks 2 (x) RPM. Remaining unmarked peaks were identified as multiples of RPM, but were left unmarked to avoid confusion. Modulations of the shaft frequency originate from rolling surface waviness.

The measurement was taken on the bearing with seeded outer race localized fault. Figure 3b shows the envelope spectrum of measured response. BPFO and its multiples clearly dominate the spectra. In addition, peaks at RPM and its modulation to BPFO can be observed. Most probable cause of these modulations is deformation in shapes of the race surface as simulated in previous example.

Simulated distributed bearing fault was placed on the outer race, while $\Delta\phi_{\mathrm{d}}$ (Fig. 2) was set to be 45°. The rough surface, as shown in Fig. 5, was modeled as low pass Gaussian noise. In both, simulated and measured cases, BPFO and FTF dominate spectra, justifying the model update.

By observing spectra of simulated vibrations in Figs. 3a and 4a along with spectra of measured vibrations in Figs. 3b and 4b one can see that spectra of localized and distributed fault differ in the presence of FTF related components.

Based on the analysis measured and simulated envelope spectra of a new feature for diagnosis of distributed bearing fault is proposed. Outer race distributed fault can be detected by observing amplitudes of two characteristic frequencies $(\mathrm{BPFO}, \mathrm{FTF})$.

To decide whether a fault is localized or distributed we propose a composite feature, being actually a vector of two features. From their position in a two-dimensional plane one can infer on the type of fault. New feature is designed to examine how much energy a particular frequency contributes to the envelope spectrum of the excited vibrations. Envelope signal is filtered around characteristic frequencies of BPFO, FTF, its second harmonic, and its modulations to BPFO. The output of filters is used to calculate the energy contribution of a particular frequency

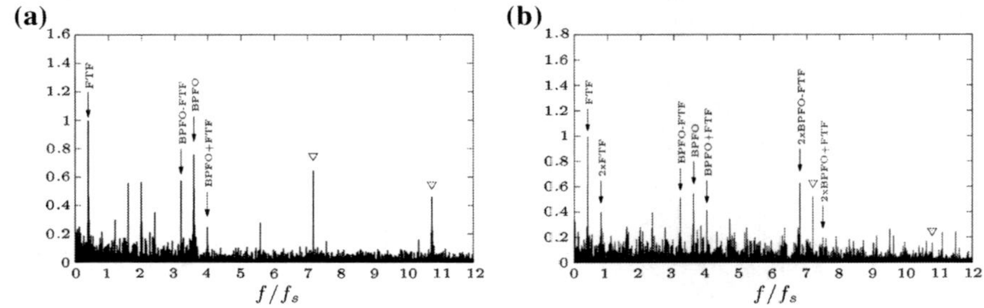

Fig. 4 Distributed outer race bearing fault: BPFO and its multiples are marked with *reverse triangles* while squares mark FTF and its modulations to BPFO. In both, simulated and measured case, BPFO and FTF dominate spectra, justifying the model update. **a** Simulation, **b** measurement

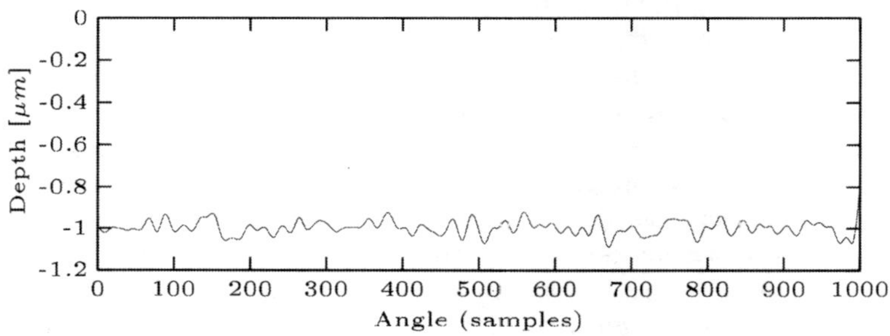

Fig. 5 Rough surface of the fault [12]

in a signal $E_f = \frac{1}{T}\int_0^T |e_f(t)|^2 dt$, where $e_f(t)$ is envelope signal filtered around characteristic frequency f. Features are then calculated as:

$$\Phi_1 = \frac{\sum_i^3 E_{i*BPFO}}{\sigma}, \quad \Phi_2 = \frac{\sum_i^2 E_{BPFO+i*FTF} + \sum_i^2 E_{i*FTF}}{E_{BPFO} + \sum_i^2 E_{BPFO+i*FTF} + \sum_i^2 E_{i*FTF}}, \tag{6}$$

where σ is energy of a band-passed signal around eigenfrequency of the motor determined by spectral kurtosis.

The proposed feature measures energy contribution of particular characteristic frequency in envelope signal. If detected fault is localized, than modulations of *FTF* are negligible. Therefore $\Phi_1 \gg \Phi_2$ holds. On the contrary, when detected fault is distributed sideband energy (i.e. *FTF*) has grater impact, thus Φ_1 and Φ_2 will have comparable values. Further, if the envelope signal does not contain any of these frequencies (i.e. bearing is fault-free), Φ_1 and Φ_2 will have near zero values.

4 Diagnostic Stability and Resolution

This section covers mainly the experimental results. In addition, simulation of faults of different sizes is performed to observe the effect of the growing defect on the proposed feature. There are envelope spectra of four different bearings with naturally born distributed fault shown on Fig. 6. Characteristic frequencies FTF and BPFO can be clearly recognized, suggesting the presence of distributed fault.

Further, diagnostic stability (DS) is tested. DS reflects the property of a feature whether it returns equal (or similar) values under repeated measurements. For this purpose, 20 measurements were taken on each of four bearings and are compared in Fig. 7. Measurements form obvious clusters. Each cluster corresponds to one bearing as marked in legend. Moreover, clusters are clearly separated from each other which suggests good diagnostic resolution. In addition to these, there are two more clusters marked with green and red which represent fault-free case and localized fault, respectively. As discussed in previous section, $\Phi_1, \Phi_2 \cong 0$ for fault-free bearing, $\Phi_1 \gg \Phi_2$ for localized and $\Phi_1,\ \Phi_2 > 0$ under distributed fault.

Resolution test is also performed on simulated vibrations. Faults were placed on the outer ring. Eight different cases were considered. The faults were extending over 2°, 10°, 30°, 45°, 50°, 55°, 60° and 65° of the bearing circumference ($\Delta\phi_d$ in Fig. 2). Figure 8 shows trend of the proposed feature. Simulation suggests fairly good resolution, although some uncertainty still remains when observing features marked green.

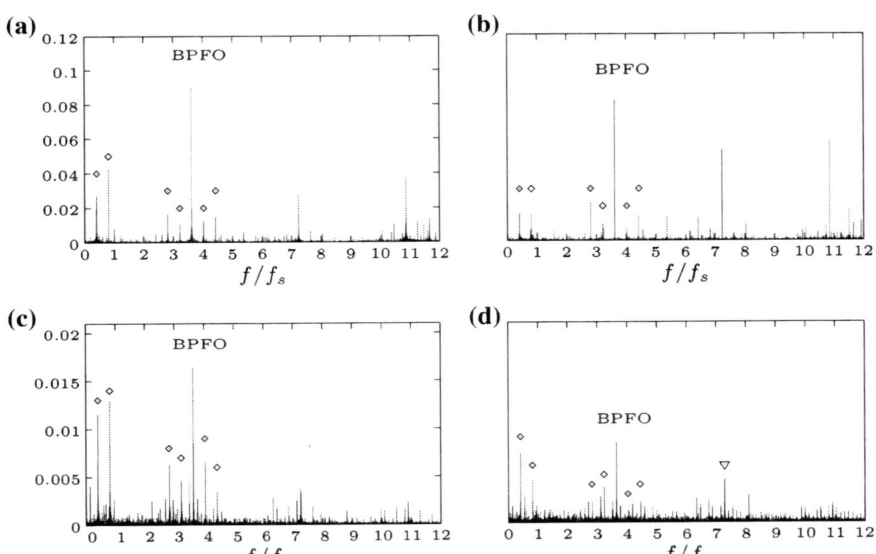

Fig. 6 Envelope spectra of four different bearings with outer race distributed fault: squares mark FTF, its harmonics and modulations to BPFO. **a** Bearing 1, **b** bearing 2, **c** bearing 3, **d** bearing 4

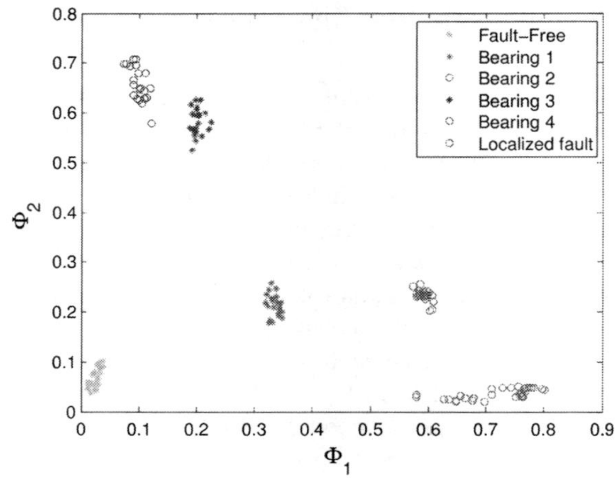

Fig. 7 Envelope spectra of four different bearings with outer race distributed fault: *squares* mark FTF, its harmonics and modulations to BPFO

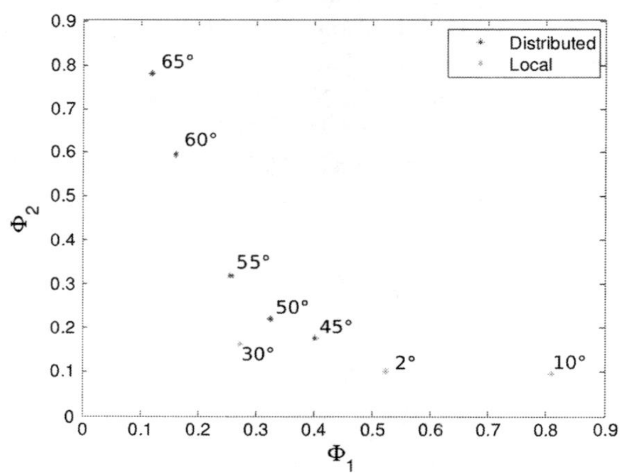

Fig. 8 Feature trend of simulated vibrations for different fault sizes

5 Conclusion

In this paper, a feature for diagnosis of the distributed fault was proposed. Bearing model was adopted and modified to identify the cause of the excited vibration pattern. Measuring vibrations of four different bearings were made to verify the presence of the same spectral components in the envelope spectra in the presence of distributed fault.

To thoroughly assess the performance of the feature, the vibration pattern should be monitored through the entire life of a bearing. The experiment would also provide better understanding of the connection between spatial distribution of the feature and fault severity.

Acknowledgment We like to acknowledge the support of the Slovenian Research Agency through the grant L2-4160 and MGRT project 3211-09-000130 (Prodismon).

References

1. Albrecht PF, Appiarius JC, McCoy RM, Owen E-L, Sharma DK (1986) Assessment of the reliability of motors in utility applications—updated. IEEE Trans Energy Conv EC-1:39–46
2. Tandon N, Choudhury A (1998) A theoretical model to predict vibration response of rolling bearings to distributed defects under radial load. J Vib Acoust 120:214–220
3. Tandon N, Choudhury A (1999) A theoretical model to predict the vibration response of rolling bearings in a rotor bearing system to distributed defects under radial load. J Tribol 122:609–615
4. Sopanen J, Mikkola A (2003) Dynamic model of a deep-groove ball bearing including localized and distributed defects. part 2: implementation and results. Proc Inst Mech Eng Part K J Multi-body Dyn 217:213–223
5. Sopanen J, Mikkola A (2003) Dynamic model of a deep-groove ball bearing including localized and distributed defects. part 1: theory. Proc Inst Mech Eng Part K J Multi-body Dyn 217:201–211
6. Kotzalas MN, Harris TA (2001) Fatigue failure progression in ball bearings. J Tribol 123:238–242
7. Al-Ghamda AM, Mba D (2006) A comparative experimental study on the use of acoustic emission and vibration analysis for bearing defect identification and estimation of defect size. Mech Syst Signal Process 20:1537–1571
8. Heng A, Zhang S, Tan ACC, Mathew J (2009) Rotating machinery prognostics: state of the art, challenges and opportunities. Mech Syst Signal Process 23:724–739
9. Sawalhi N, Randall RB (2011) Vibration response of spalled rolling element bearings: observations, simulations and signal processing techniques to track the spall size. Mech Syst Signal Process 25:846–870
10. Kumar R, Singh M (2013) Outer race defect width measurement in taper roller bearing using discrete wavelet transform of vibration signal. Measurement 46:537–545
11. Eftekharnejad B, Charnley B, Carrasco MR, Mba D (2011) The application of spectral kurtosis on acoustic emission and vibrations from a defective bearing. Mech Syst Signal Process 25:266–284
12. Sawalhi N, Randall RB (2008) Simulating gear and bearing interactions in the presence of faults: part II. Simulation of the vibrations produced by extended bearing faults. Mech Syst Signal Process 22:1952–1966
13. de Castelbajac C, Ritou M, Laporte S, Furet B (2014) Monitoring of distributed defects on hsm spindle bearings. Appl Acoust 77:159–168
14. Sawalhi N, Randall RB (2008) Simulating gear and bearing interactions in the presence of faults: part I. The combined gear bearing dynamic model and the simulation of localized bearing faults. Mech Syst Signal Process 22:1924–1951
15. Randall RB, Antoni J, Chobsaard S (2001) The relationship between spectral correlation and envelope analysis in the diagnostics of bearing faults and other cyclostationary machine signals. Mech Syst Signal Process 15:945–962

Freight Car Roller Bearing Fault Detection Using Artificial Neural Networks and Support Vector Machines

Daniel Maraini and C. Nataraj

Abstract This paper deals with fault detection and diagnostics of freight rail tapered roller bearings using statistical features and auto-regressive m.odel parameters extracted from bearing vibration signals. Two classifiers are used to distinguish cup, cone, and roller defects, namely, artificial neural networks (ANNs) and support vector machines (SVMs). Time domain vibration data is collected from normal bearings, and defective bearings with combinations of cup, cone, and roller defects. Features are used to train both ANNs and SVMs, and performance is compared for both classifiers. Mutual information between time domain features and target classes is computed in order to rank features with the highest relevance. Excellent performance was observed using laboratory data and the trained ANNs and SVMs, with testing accuracy as high as 100 % in some cases.

Keywords Condition monitoring · Artificial neural networks · Support vector machines · Mutual information · Feature selection · Bearing fault

1 Introduction

Knowledge of the condition of a freight car's roller bearings is valuable from both a maintenance and safety perspective. Although bearing reliability has increased considerably since the adoption of roller bearings, they are still subject to failure, which can lead to derailments and costly damage to infrastructure. In a study performed on 47,030 freight rail roller bearings, fatigue was identified as the number

D. Maraini (✉)
Amsted Rail Company, Inc, Chicago, IL 60606, USA
e-mail: daniel.maraini@villanova.edu

C. Nataraj
Department of Mechanical Engineering, Villanova University,
Villanova, PA 19085, USA
e-mail: nataraj@villanova.edu

J.K. Sinha (ed.), *Vibration Engineering and Technology of Machinery*,
Mechanisms and Machine Science 23, DOI 10.1007/978-3-319-09918-7_59

one cause of bearing failure, accounting for 74 % of damaged bearings [1]. Outer race fatigue accounted for 58 %, inner race fatigue 13 %, and rolling element fatigue for 3 % of all damaged bearings. From 2001 to 2010, 4,352 main-line derailments were reported by Class I railroads in the United States [2]. Of these derailments, 257 (1,739 cars derailed) were the result of bearing failure. Steel cleanliness has been identified as a key contributor to early bearing failure, where research has been devoted to detecting subsurface defects in bearing steel [3]. Detecting a defect as early as possible is imperative for both safety and scheduling of maintenance.

Railroad's currently use three main methods to identify defective roller bearings: (1) hot-box detection, (2) acoustic emission (AE), and (3) acoustic signal analysis. A hot-box detector (HBD) typically utilizes an infrared temperature transducer for measuring the bearing temperature as the train passes by. An alarm is issued for bearings with a temperature exceeding 105 °C (190 °F) above ambient conditions [4, 5]. This can be problematic, as the bearing temperature tends to rise only under severe faulty conditions. Acoustic bearing detectors use either AE or acoustic signal analysis to detect bearing faults. AE is a measure of transient elastic waves, and is capable of measuring both surface and subsurface defects. One drawback of AE is the difficulty in detecting inner race defects in a bearing [6]. Acoustic signal analysis is typically performed using both sound pressure and sound intensity. Doppler shift in the acquired acoustic signal and heavy noise are two of the main challenges in the wayside acoustic diagnosis of train bearings [7].

Overall, there are more than 6,000 wayside detectors in North America dedicated to bearing fault detection, with the majority utilizing temperature detection techniques. Detection sites can be sparsely distributed or unavailable in remote locations. In these cases, it may be advantageous to implement on-board monitoring equipment for bearing fault detection.

Very little literature is available in the area of on-board bearing fault detection for freight rail cars. Temperature based methods have been implemented, but lack early detection capability. Vibration based techniques have also been studied, with a focus on frequency domain analysis [8]. A considerable amount of literature is available in the general area of roller bearing fault detection using vibration. A summary of time, frequency, and time-frequency domain techniques is presented in [9]. Envelope analysis is discussed in [10], along with spectral kurtosis, a method for determining frequency bands with the highest impulsivity. A model based technique using cross-sample entropy was used to estimate the severity of defects in [11], and [12] presents an overview of auto-regressive (AR) modeling for general fault detection. Many of the available techniques rely on human interpretation and analysis of collected signals. Automated diagnostic systems that can provide a concise assessment of bearing health are imperative for remote monitoring.

Pattern recognition using extracted features from bearing vibration signals has gained popularity in bearing diagnostics. In general, time, frequency, and time-frequency domain features are extracted from short windows of vibration data. A classifier is trained to classify bearing faults based on the input features. Time-domain features with artificial neural networks (ANNs) and support vector machines (SVMs) classifiers have been used to classify multiple faults with excellent performance in [13]. The effect of the number of features on SVM classification accuracy was studied in [14], wherein it was determined that four statistical features were adequate for fault vs. no-fault classification. Feature selection using mutual information is presented in [15], wherein an optimal set of features with maximum relevancy and minimal redundancy was determined.

Selection of the optimal set of features is critical to minimize classification error. One of the most popular feature selection methods is to select features with the highest relevance or dependency on the target class [16]. This can be quantified in terms of the mutual information between a feature and a target class. In this max-relevance technique, the top m features in the descent ordering of mutual information are selected. This is very powerful in terms of bearing defect classification, as features can be selected that provide the highest information about specific bearing defects of interest.

In this paper, a bearing defect classification technique using time-domain features is presented. Bearing vibration data was collected in the laboratory using freight car tapered roller bearings with different defects. Statistical features and AR model parameters were extracted from bearing vibration data. Mutual information between features and classes was computed and used to aid in feature selection. These features were used for training both an artificial neural network (ANN) and support vector machine (SVM).

2 Algorithm

Figure 1 shows the proposed bearing fault detection algorithm. The purpose of the algorithm is to classify the bearing into one of four classes: (1) defect free (DF), (2) cup, cone, and roller defects (CCRD), (3) cup and cone defects (CCD), or (4) roller defect (RD). Bearing vibration data is first pre-processed, which consists of (1) demeaning and (2) down-sampling to 400 Hz. The pre-processed data is used to generate auto-regressive (AR) model parameters of the form

$$y(t) + a_0 y(t-1) + \cdots + a_n y(t-n) = e(t) \quad (1)$$

where a_0, a_1, … a_n are the model parameters, n is the model order, and $e(t)$ is a white noise disturbance [17]. The AR model parameters are used in place of typical FFT spectra for detection of bearing defect frequencies, which fall in the range of 4.2–120 Hz for the bearings tested. Smaller sample sizes can be used with comparable resolution to the FFT technique applied to longer data records.

The vibration data is also used to compute the following statistical features: (1) variance, (2) skewness, (3) kurtosis, (4) hyperskewness, (5) hyperflatness, and (6) crest factor. The mutual information is then estimated using the extracted features. This is used to rank and select features based on information content, or dependency on the target class. These selected features are used to train either a SVM or ANN. Once training is complete, the test data can be used to compute classification accuracy.

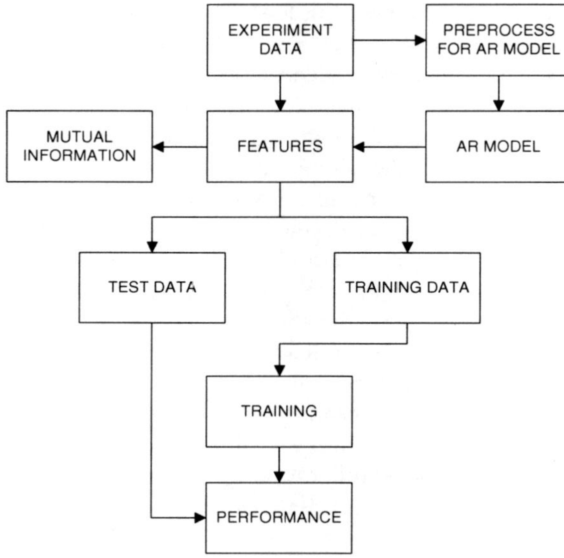

Fig. 1 Algorithm for bearing fault detection

3 Experimental Setup and Data Collection

All laboratory data was collected from a bearing life test rig, using 6 normal and 4 defective 6−1/2 × 12 Class F tapered roller bearings, with combinations of cup, cone, and roller defects. The bearings were run under 17 % load (6,079 Lbf) at a rotating speed of 560 rpm (9.33 Hz), corresponding to ball pass frequency on outer race (BPFO) of 96 Hz, ball pass frequency on inner race (BPFI) of 120 Hz, ball spin frequency (BSF) of 42 Hz, and fundamental train frequency (FTF) of 4.2 Hz. Accelerometers with a dynamic range of 200 g's were mounted on the bearing housing to capture radial acceleration and sampled at a rate of 10,240 Hz. Data was collected at each bearing for 20 s for a total of 204,800 samples. These samples were divided up into 20 bins of 1 s or 10,240 samples. A total of 220 bins of 1 s data were collected from the 10 bearings.

4 Support Vector Machines

A support vector machine (SVM) is a pattern classification technique that can be used for linearly separable patterns, or non-linearly separable patterns with the addition of kernel based methods [18]. Training data is used to determine the model parameters by solving a convex optimization problem, such that any local solution is also a global solution. The goal of the optimization problem is to find the hyperplane that yields the maximum separation distance between patterns in either the input or feature space. The output of the SVM is determined by the equation

$$y(\mathbf{x}) = \mathbf{w}^T \varphi(\mathbf{x}) + b \tag{2}$$

where $\varphi(\mathbf{x})$ denotes a feature space transformation, w is the weight vector, and b is the explicit bias parameter. The training set comprises N input vectors $x_1 \ldots x_N$ and target values $d_1 \ldots d_N$, where $d_i \in \{-1, 1\}$. The optimum hyperplane can be obtained as a solution to the optimization problem

$$Q(\mathbf{a}) = \sum_{i=1}^{N} a_i - \frac{1}{2} \sum_{i=1}^{N} \sum_{j=1}^{N} a_i a_j d_i d_j k(\mathbf{x}_i, \mathbf{x}_j), \quad i = 1, 2, \ldots, N \tag{3}$$

subject to the constraints $a_i \geq 0$ and $\sum_{i=1}^{N} a_i d_i = 0$, where $k(\mathbf{x},\mathbf{z}) = \varphi(\mathbf{x})^T \varphi(\mathbf{z})$ is the kernel function and $\mathbf{a}$ is a vector of Lagrange multipliers. For the present work, a Gaussian radial basis function (RBF) kernel was chosen of the form

$$k(\mathbf{x}, \mathbf{z}) = \exp\left[-\gamma \|\mathbf{x} - \mathbf{z}\|^2\right] \tag{4}$$

Once optimal values of a_i have been determined, new points can be classified by evaluating the sign of $y(\mathbf{x})$ using

$$y(\mathbf{x}) = \sum_{i=1}^{N} a_i d_i k(\mathbf{x}, \mathbf{x}_i) + b \tag{5}$$

where the bias is computed using

$$b = \frac{1}{N} \sum_{i=1}^{N} \left(d_i - \sum_{j=1}^{N} a_j d_j k(\mathbf{x}_i, \mathbf{x}_j) \right) \tag{6}$$

Equation (3) can be solved using many different optimization techniques.

5 Mutual Information

The mutual information is the amount of information one variable shares with another. For two variables x and c, it represents the reduction in uncertainty of x, given the new measurement c. For discrete variables x and c, the mutual information can be defined as

$$I(x; c) = \sum_{x,c} p(x, c) \log \frac{p(x, c)}{p(x)p(c)} \tag{7}$$

where $p(x)$, $p(c)$, and $p(x, c)$ are the marginal and joint probability mass functions. In order to calculate (7), estimates of the probability distributions are needed from the sampled data. For the present work, a kernel density estimate with a Gaussian kernel was chosen. Both $p(x)$ and $p(c)$ are easily estimated from the sampled data. The mutual information can thus be estimated using

$$\hat{I}(x;c) = \sum_{x} \hat{p}(x)\hat{p}(x|c)\log\left[\frac{\hat{p}(x|c)}{\hat{p}(c)}\right] \tag{8}$$

where the summation is over x, yielding the mutual information between a feature x and class c.

6 Bearing Vibration Data Observations

Figure 2 shows the time domain acceleration waveforms for both normal and defective bearings over four complete rotations (0.5 s). Amplitude modulated impulse responses are observed in the defective bearing signals. The highest magnitude accelerations are observed on the bearing with cup and cone defects. In general, the bearings with defects exhibit higher magnitude vibrations than normal bearings. Figure 3 shows the discrete Fourier transform (DFT) of the measured acceleration signal from both normal and defective bearings. For all bearings, the

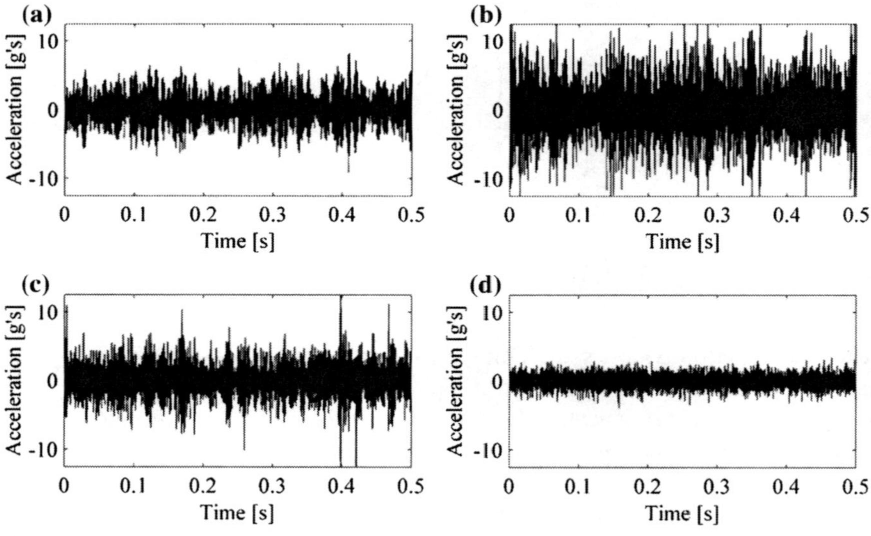

Fig. 2 Time domain waveforms for normal and defective bearings: **a** cup, cone, and roller defects, **b** cup and cone defects, **c** roller defects, and **d** defect free

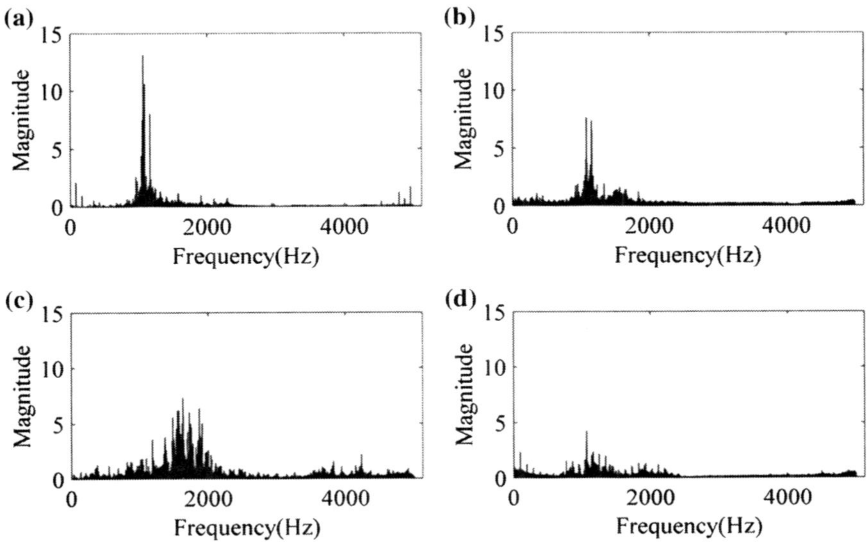

Fig. 3 Spectrum for normal and defective bearings: **a** cup, cone, and roller defects, **b** cup and cone defects, **c** roller defects, and **d** defect free

highest magnitude in the DFT is observed in the range of 1–2 kHz. Overall, the highest magnitude in the spectrum is observed for the bearing with cup, cone, and roller defects.

7 Feature Extraction

Figure 4 shows the normalized statistical features and first three AR parameters for normal and defective bearings. A model order of $n = 20$ was chosen for AR parameter estimation, using vibration data down-sampled to 400 Hz. From inspection, it is clear that no single feature can be used to classify a particular defect, and a combination of features is required for classification.

8 Mutual Information Estimation

Figure 5 shows the estimated mutual information between all features and classes. The highest relevance is found between the AR model parameters (features 7–26) and the 'normal' class. From the statistical features (1–6), the crest factor yields the highest relevance for all classes. The mutual information tends to decrease with increasing order of the central moments. For example, the lowest relevance is found for feature 5, the hyper-flatness.

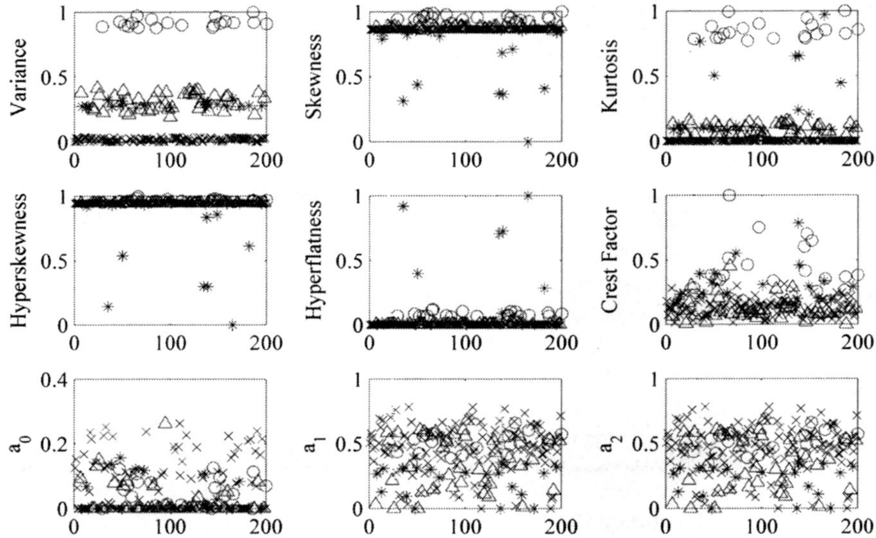

Fig. 4 Extracted features for bearings with CCRD (△), CCD (◯), RD (✳), and DF (×)

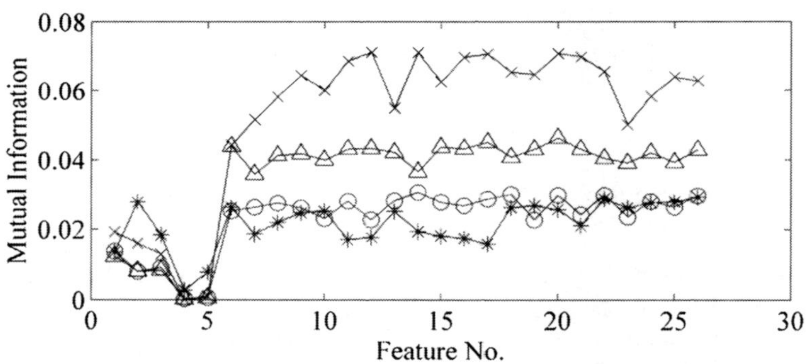

Fig. 5 Mutual information between feature and target class for central moments (features 1–5), crest factor (6), and AR model parameters (features 7–26) with CCRD (△), CCD (◯), RD (✳), and DF (×)

9 ANN and SVM Training

The feature sets extracted from the binned vibration data were divided with 70 % for training and 30 % for testing, giving a 26 × 140 training set, and a 26 × 60 testing set. The targets were assigned according to the four target classes and the type of classifier being used (SVM vs. ANN). Training and testing success was computed by averaging 20 training/testing runs for both classifiers.

Table 1 ANN and SVM testing performance

Features	ANN testing success (%)	SVM testing success (%)
1–26	96.5	100
1–6	91.5	88
1–3	94.5	88.8
4, 5	20	68.8
3–5	80.1	71.4
7–26	98.2	98.8
7–16	92.8	96.5
1–3, 6–26	100	100

All ANN training and testing was performed using Matlab's pattern recognition neural network function. An *N*–8–3 network was utilized, with *N* input features, 8 hidden neurons and 3 outputs. Updating of the weights and bias during training was carried out using the scaled conjugate gradient method. Backpropagation was used to calculate derivatives with respect to the weight and bias variables.

SVM training was completed using quadratic programming as the optimization method. The RBF kernel width was selected such that $\gamma = 0.5$. Classification was carried out using a one-versus-the-rest [18] approach for the four classes.

Table 1 shows the testing success from ANN and SVM testing. For both classifiers, the relationship between the feature set and accuracy are similar. The best performance for both classifiers was found using the combination of features 1–3 and 6–26. The worst performance was found using the ANN and features 4 and 5. This is expected, as features 4 and 5 have the lowest mutual information, or relevance to the target class.

10 Conclusion

A method for classifying freight rail roller bearing defects was presented using ANNs and SVMs. Statistical features and AR model parameters were extracted from bearing vibration data, and used as input features. The mutual information between each feature and target class was computed and used to aid in feature selection. For the present data, the AR parameters, crest factor, and variance yield the highest information. The hyperskewness and hyperflatness yield the lowest information, as was validated through ANN and SVM testing.

The selected features were used to train both ANNs and SVMs, with similar performance. 100 % testing accuracy was achieved with both classifiers, using the variance, skewness, kurtosis, crest factor, and AR parameters. Future work will aim to include model based features, as well as additional experiments with a wider range of defect sizes and rotating speeds.

References

1. Ferreira JLA, Balthazar JC, Araujo APN (2003) An investigation of rail bearing reliability under real conditions of use. Eng Fail Anal 10(6):745–758
2. Liu X, Saat MR, Barkan CP (2012) Analysis of causes of major train derailment and their effect on accident rates. Trans Res Rec: J Trans Res Board 2289(1):154–163
3. Koester L, Turner et al (2012) Near-race ultrasonic inspection of tapered roller bearing components for non-metallic defects. In: ASME 2012 rail transportation division fall technical conference (pp 107–110). ASME
4. Tarawneh CM, Cole KD et al (2008) Experiments and models for the thermal response of railroad tapered-roller bearings. Int J Heat Mass Transf 51(25):5794–5803
5. Tarawneh C, Wilson BM et al (2008) Dynamic bearing testing aimed at identifying the root cause of warm bearing temperature trending. In: Proceedings of the 2008 ASME RTD fall technical conference, RTDF2008-74036, Chicago, Illinois, 24–26 Sept 2008
6. Al-Dossary S, Hamzah RI, Mba D (2009) Observations of changes in acoustic emission waveform for varying seeded defect sizes in a rolling element bearing. Appl Acoust 70 (1):58–81
7. He Q, Wang J et al (2013) Wayside acoustic diagnosis of defective train bearings based on signal resampling and information enhancement. J Sound Vib 332(21):5635–5649
8. Tarawneh CM, Kypuros JA et al (2009) Vibration signatures of temperature trended bearings in field and laboratory testing. In: ASME
9. Tandon N, Choudhury A (1999) A review of vibration and acoustic measurement methods for the detection of defects in rolling element bearings. Tribol Int 32(8):469–480
10. Randall RB, Antoni J (2011) Rolling element bearing diagnostics—a tutorial. Mech Syst Signal Process 25(2):485–520
11. Kappaganthu K, Nataraj C, Samanta B (2009) Model based bearing fault detection using sample entropy features and support vector machines. (Preprint submitted to Elsevier)
12. Jardine AK, Lin D, Banjevic D (2006) A review on machinery diagnostics and prognostics implementing condition-based maintenance. Mech Syst Signal Process 20(7):1483–1510
13. Samanta B, Al-Balushi KR, Al-Araimi SA (2003) ANNs and SVMs with genetic algorithm for bearing fault detection. Eng Appl Artif Intell 16(7):657–665
14. Sugumaran V, Ramachandran KI (2011) Effect of number of features on classification of roller bearing faults using SVM and PSVM. Expert Syst Appl 38(4):4088–4096
15. Kappaganthu K, Nataraj C (2011) Feature selection for fault detection in rolling element bearings using mutual information. J Vib Acou 133(6):061001
16. Peng H, Long F, Ding C (2005) Feature selection based on mutual information criteria of max-dependency, max-relevance, and min-redundancy. IEEE Trans Pattern Anal Mach Intell 27 (8):1226–1238
17. Ljung L (1999) System identification. Wiley, New York
18. Bishop CM, Nasrabadi NM (2006) Pattern recognition and machine learning, vol 1. Springer, New York, p 740

Light Rotor-Stator Partial Rub Characterization Using Instantaneous Angular Speed Measurement

Sanjin Braut, Roberto Zigulic, Goranka Stimac and Ante Skoblar

Abstract Instantaneous angular speed (IAS) measurement is recently recognized as a promising technique for condition monitoring of various rotating machines and their subsystems. There are various approaches considering counting techniques and processing of signal. This paper considers application of analog signals from toothed wheel encoder or zebra tape encoder and acquisition at low to moderate sampling rates. Such a system can experience problems with non-uniformity of encoder segments geometry, thus correction of encoder segment non-uniformity was performed. The rubbing often occurs in rotating machinery at position with small clearances and can sometimes cause catastrophic breakdown of machine. So it is important to develop reliable tools for rub diagnosis. As the rubbing process is nonlinear this paper beside conventional Fourier transform also considered potential of sifting operation of Empirical mode decomposition (EMD) method. EMD was employed to produce intrinsic mode functions (IMFs) and to further improve the detection capabilities of Fourier transform. Developed procedure for IAS measurement was successfully tested on laboratory test rig for rotor to stator contact dynamics investigation at speed transition from no rub to light rotor to stator partial rub condition.

Keywords Rotor-stator partial rub · Rub characterization · Instantaneous angular speed · Analog encoder

S. Braut (✉) · R. Zigulic · G. Stimac · A. Skoblar
Faculty of Engineering, University of Rijeka, Vukovarska 58, 51000 Rijeka, Croatia
e-mail: sbraut@riteh.hr

R. Zigulic
e-mail: zigulic@riteh.hr

G. Stimac
e-mail: gstimac@riteh.hr

A. Skoblar
e-mail: askoblar@riteh.hr

J.K. Sinha (ed.), *Vibration Engineering and Technology of Machinery*,
Mechanisms and Machine Science 23, DOI 10.1007/978-3-319-09918-7_60

1 Introduction

Contact or rub between rotor and stator is one of the most intensive research subjects in rotor dynamics. Depending on rotor and stator structure configuration rotor-stator contact phenomena can be classified in rigid rotor disc-rigid stator contact [1–3], bladed disc-stator contacts [4] and rotor-stator contact in retainer bearings of the active magnetic bearings [5, 6].

From a data acquisition point of view [7–10], rotor vibration are usually measured by accelerometers at bearing pedestals or with non-contacting probes relatively from bearing pedestals to shaft in two radial directions (horizontal, vertical) and if needed in axial direction too.

Recently, many papers were published on measurement of Instantaneous angular speed (IAS) [11–13]. The idea is to use rotor encoder signal and try to determine new vibration condition patterns to improve overall diagnostic capabilities especially for those machines which experience change in speed during normal operation. If measurements have to be performed on toothed wheel encoder or zebra tape encoder with analog signals and acquired with a general purpose ADC system at low to moderate sampling rates, such a system can experience problems with non-uniformity of encoder segments geometry. In this case a correction procedure has to be performed using of normalized time passage ratio [14].

Empirical mode decomposition (EMD) is another new tool which has found application in vibration signal analysis [15–17]. It decomposes the data into a set of intrinsic mode functions (IMFs), not assuming linearity, stationarity, or any a priori bases for decomposition.

This paper considers IAS measurement with an optical encoder with a 30 segments at 20 kHz sampling frequency. To decrease error caused by low sampling frequency, raw signal is up-sampled by a factor 50 performing cubic spline interpolation. To resolve another source of error i.e. geometric non-uniformity, a correction procedure was performed using normalized time passage ratio of encoder segments. To test detection possibility of IAS measurement a partial rotor–stator rub failure is considered. Frequency spectra of IAS signal (FFT and STFT) clearly detect rub occurrence from normal operation. Rubbing situation is connected with presence of fractional sidebands of rotor speed in frequency spectra. Empirical mode decomposition was employed additionally to produce IMFs and to further improve the detection capabilities of Short time Fourier transform (STFT).

2 Instantaneous Angular Speed Measurement

In Li et al. [11] authors gave review of all kinds of angular speed measurement methods. They can be categorized into two groups; timer /counter methods and ADC based methods. The first one can be further divided on methods measuring an elapsed time between successive encoder pulses or counting encoder pulses during the prescribed time. The ADC based methods wasn't very attractive to researchers of condition monitoring until recent remarkable increase of speed and memory capacity of modern computers and general purpose data acquisition systems. Because ADC based method treat encoder signal as any other analog signal, there is elegant solution for software upgrade of measuring system originally built for lateral vibration measurements. This paper is focused on ADC method and discusses various techniques for signal conditioning.

While minimum measurable speed is affected by memory capacity of the acquisition system or length of the record for a given sampling frequency and encoder resolution, maximum measurable speed, theoretical speaking, is determined by the ratio of sampling frequency and encoder resolution.

Although resolution of the modern encoders can range from one to many thousands, for condition monitoring applications, a sensor with dozens or hundreds of pulses per revolution is sufficient for most machinery monitoring and diagnosis [11].

Required sampling frequency is given in [18] as

$$F_s = 4 \times \left[Nf_r + \left(f_x + \frac{N}{2\pi} \max|\delta\omega| \right) \right] \tag{1}$$

where $f_r = \omega/2\pi$ is the frequency of rotation, where f_x is the highest frequency of the varying part $\delta\omega$, N is the number of the teeth of the moving element or the resolution of the encoder. Based on authors' experience [14] sampling frequency could be even lower but the raw signal should be up-sampled/interpolated to virtually get approximately the same sampling frequency as can be obtained using Eq. (1).

2.1 Correction of Encoder Segment Non-Uniformity

In a case of toothed gear encoder as well as zebra striped disk, the geometric variations of encoder segments can't be neglected and are sufficient to induce significant corrupting effects in IAS vibration spectra. Therefore application of compensation technique is crucial for obtaining frequency spectra clear from order content. The whole procedure is based on determination of normalized time passage ratio of each encoder segment which is assumed to be characteristics of encoding device,

$$r_j(n,j) = \frac{t((j-1)N+n+1) - t((j-1)N+n)}{T(j)}, 1 \le n \le N, 1 \le j \le M. \tag{2}$$

where $T_j = t((j+1)N) - t(jN)$, is the time required for one revolution, N is the total number of encoder segments and j is the number of revolution of total M

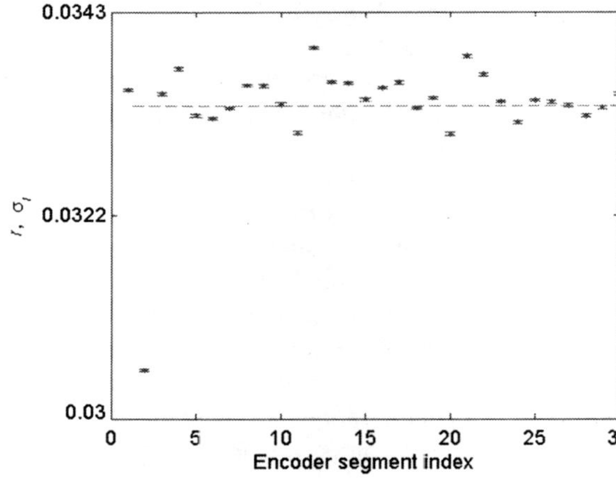

Fig. 1 Zebra tape encoder normalized time passage ratio

revolutions. Values of $r_j(n)$ are calculated for each encoder segment throughout a single revolution and then averaged over M revolutions:

$$r(n) = \frac{1}{M}\sum_{j=1}^{M} r_j(n,j), 1 \leq n \leq N, 1 \leq j \leq M. \tag{3}$$

Figure 1 shows the averaged normalized time passage ratios for a 30 black and white stripes zebra tape encoder used in this paper together with the time passage ratio for an equally spaced segments i.e. the reciprocal of the number of encoder segments (red dashed line).

2.2 *IAS Measurement with Geometric Non-Uniformity Correction*

The concept of IAS measurement is applicable for both constant and variable running speed so it is more flexible diagnostic tool then Time interval measurement of torsional vibration as explained in [14]. The thing what connects those two approaches is time array as well as normalized time passage ratio determination for encoders with pronounce of toothed wheel or zebra tape encoder Instantaneous angular speed (frequency) can be obtained considering the geometric non-uniformity according to:

$$\begin{aligned} f_{IAS}((j-1)N+n) = & r(n)/[t((j-1)N+n+1) - t((j-1)N+n)], \\ & 1 \leq n \leq N, 1 \leq j \leq M \end{aligned} \tag{7}$$

This instantaneous angular frequency can be further averaged if necessary.

3 Empirical Mode Decomposition (EMD)

The EMD is an empirically based adaptive data analysis method that was first proposed by Huang et al. [15] and has been utilized for nonlinear and non-stationary data analysis in various applications. The role of EMD, the first step of Hilbert Huang transform (HHT), is to decompose the complicated nonlinear and non-stationary signals into finite number of IMFs, i.e.:

$$x(t) = \sum_{k=1}^{m} c_k(t) + r_m(t), \tag{8}$$

where $c_k\,(t)$ is the k-th IMF of the signal $x(t)$ and r_m is the final residue which can be a constant of the signal mean trend. The EMD process is based on the assumption that any complicated signal is composed of IMFs and a trend, and each IMF represents the mono-component of oscillation with in a certain frequency band and satisfies the following conditions [15]:
(1) The number of extrema and the number of zero-crossings must be either equal or differ at most by one in the whole data set; (2) At any point, the mean value of the envelope defined by the local maxima and the envelope defined by the local minima is zeros. Obviously, each IMF only has one frequency value at any moment, therefore, it is feasible to compute the meaningful instantaneous frequency once the multi-component signal is reduced to a collection of mono-component functions through the EMD process.

4 Experimental Set-up

Experimental evaluation of the proposed algorithms for light rubbing detection at constant and variable speed condition is performed on a laboratory test rig shown in Fig. 2. The test rig, originally built for rotor—stator contact dynamics investigation, consists of a rotor supported by two roller bearings and connected via elastic coupling to induction motor with speed controller. The stator is made of annular plate elastically suspended on four circular beams.

The measurement system used for this purpose was based on National Instruments PCI card NI 4472 with four non-contacting displacement probes made by Schenck (Two for rotor radial displacement, one for stator radial displacement and one for toothed wheel encoder measurement) and two optical sensors also made by Schenck (one for phase measurement and one for zebra stripe encoder) all simultaneously sampled. While first encoding device had toothed wheel with 24 teeth/segments, the second one consisted of a disc with zebra stripes with alternating 30 black and white stripes.

Although acquisition card offered a higher sampling rate, selection of non-contacting eddy current displacement probe as a sensing device recommended

Fig. 2 Laboratory test rig; *a* toothed wheel encoder, *b* phase (1 pulse per revolution) sensor, *c* zebra stripe encoder, *d* elastically suspended stator plate, *e* two robust stator support, *f* drive

sampling rate up to 20 kS/s. Proposed procedure works as an off-line system because raw analog signal is first acquired with a simple LabVIEW application and then processed using a prepared Matlab routine.

Measurements and simulations [8] showed that first critical speed of the rotor is at 27.5 Hz, while the first natural frequency of the stator is at 90 Hz. Radial clearance between the rotor and stator was 0.4 mm. Previous analysis showed better measurements with optical encoder device so those measurement results will be presented below.

5 Experimental Analysis and Results

To test light rub detection capabilities of IAS measurement, dozen of tests are performed but due to limited space only one will be presented which combines several different conditions. During tests rotor had slight unbalance (5e–5 kg m). Different diagnostic tools are used i.e. FFT, Spectrogram (based on STFT) and HHT method and their results are compared. Test was performed starting at 25.7 Hz speed without establishing contact between rotor and stator. After that, speed was increased up to 27.6 Hz when light rotor—stator rubbing started. Results are presented in Fig. 3 (a) rotor disc orbit at $n = 25.73$ Hz (without rub—red dashed circle present clearance between rotor and stator), (b) rotor disc orbit at $n = 27.55$ Hz (light partial rub), (c) FFT of IAS signal at $n = 25.73$ Hz (without rub), (d) FFT of IAS signal at $n = 27.55$ Hz (light partial rub), (e) IAS as a function of time, (f) lateral vibration of rotor and stator in horizontal plane and g) Spectrogram (STFT)

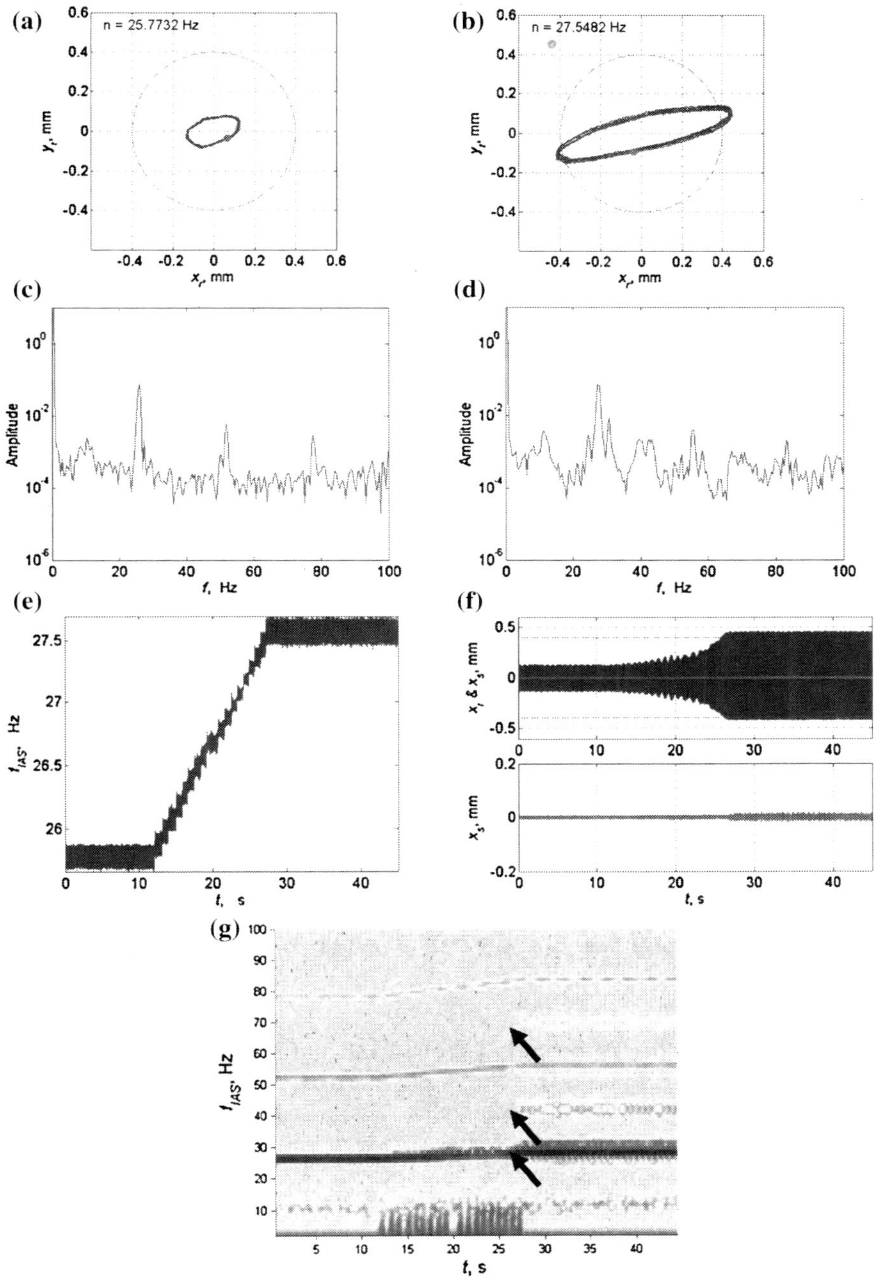

Fig. 3 Results of rotor-stator contact testing: as **a** rotor disc orbit at n = 25.73 Hz, **b** rotor disc orbit at n = 27.55 Hz, **c** FFT of IAS signal at n = 25.73 Hz, **d** FFT of IAS signal at n = 27.55 Hz, **e** IAS as a function of time, **f** lateral vibration of rotor and stator in horizontal plane and **g** Spectrogram (STFT) of IAS signal (*black arrows* show commencement of partial rubbing)

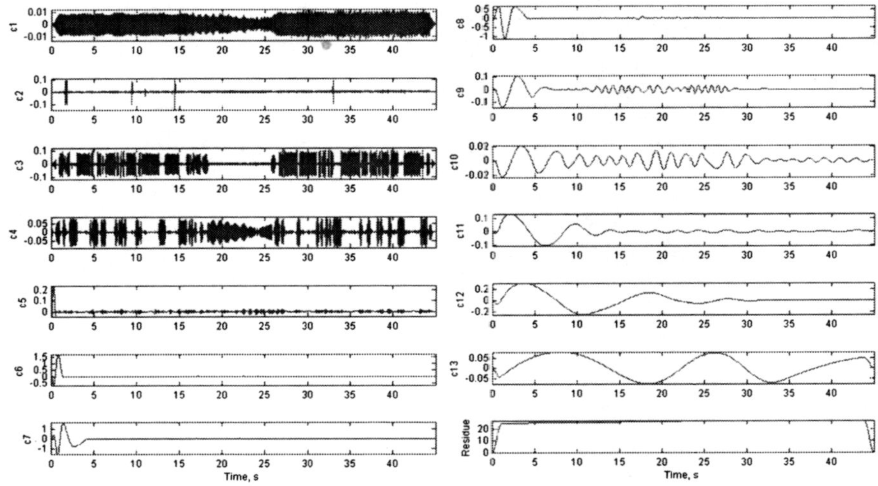

Fig. 4 IMFs of Instantaneous angular speed signal

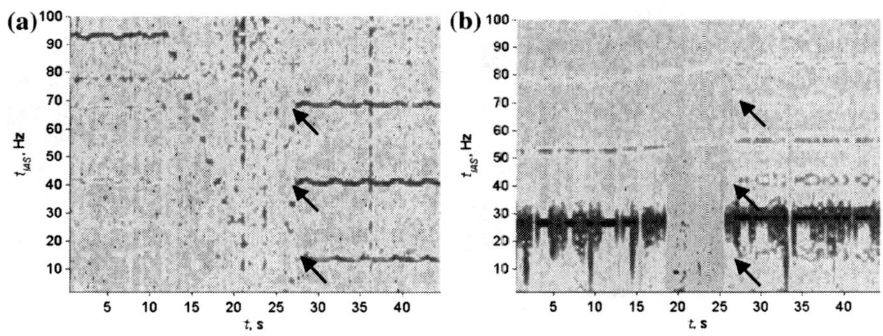

Fig. 5 Spectrogram of **a** IMF 1 and **b** IMF 3 of IAS signal (*black arrows* designated commencement of partial rubbing)

of IAS signal as well as in Fig. 4 in a form of IMFs of IAS signal and in Fig. 5 as Spectrogram of IMF 1 and IMF 3. From Fig. 3f, g can be concluded that rubbing started at $t = 26$ s from start of measurement. Vibration analyst could conclude that rubbing is going on, according to frequency spectrum and spectrogram shown in Fig. 5d, g, where fractional sidebands (at approx. 12, 41 and 70 Hz) of the first order of rotor speed (27.5 Hz) are present.

In Fig. 5 especially in Fig. 5 a this fractional sidebands as well as the time instant at which rubbing started (marked with black arrows) are even clearer than in Fig. 3g. In considered measurement of IAS signal these fractional sidebands are visible in IMFs 1 and 3 but this is not always the case. In some other measurements they are on IMF 5 etc. To obtain Intrinsic mode functions shown in Fig. 4 an EMD process had to be performed. In the future authors will investigate the possibility of the HHT method for diagnostics of different types of rotor failure. First part of HHT, EMD is already successfully implemented in this work, for sifting operation. It produced good enough IMFs of IAS for STFT to obtain clearer detection of partial rotor–stator rubbing than from the original IAS signal.

6 Conclusion

This paper verified potential of IAS measurement application as a diagnostic tool for partial rotor–stator rub failure. Measurement of IAS is performed with a optical sensor pointed on a zebra tape encoder with a 30 segments at 20 kHz sampling frequency. To decrease error caused by low sampling frequency, raw signal is up-sampled by an factor 50 performing cubic spline interpolation and finally obtaining 1 MHz of virtual sampling frequency. Another source of error is geometric non-uniformity. Therefore a correction procedure of encoder segment non-uniformity was performed using of normalized time passage ratio of encoder segments.

To test detection possibility of IAS measurement a partial rotor–stator rub failure is considered. Frequency spectra of IAS signal (FFT and STFT) clearly detect rub occurrence from normal operation. Rubbing situation is connected with presence of fractional sidebands of rotor speed in frequency spectra. Empirical mode decomposition was employed to produce IMFs and to further improve the detection capabilities of Fourier transform.

Future development of detection tools of rubbing phenomena will try to use HHT method.

Acknowledgments This study was supported by University of Rijeka (Grants No. 13.09.1.2.11 and No. 13.09.2.2.19.)

References

1. Choy FK, Padovan J (1987) Non-linear transient analysis of rotor-casing rub events. J Sound Vib 113(3):529–545
2. Bartha AR (2000) Dry friction backward whirl of rotors. Ph.D. Thesis, Swiss Federal Institute of Technology, Zurich
3. Von Groll G, Ewins DJ (2002) A mechanism of low subharmonic responce in rotor/stator contact—Measurement and simulation. J Vib Acoust 124 (2002):350–358
4. Ahrens J, Jiang J, Ulbrich H, Ahaus G (2001) Experimentelle Untersuchungen zum Schaufelanstreifen, Schwingungen in Rotierenden Maschinen—SIRM V Tagung, Wien, Austria, pp 97–108 (in German)
5. Fumagalli M, Schweitzer G (1994) Impact dynamics of high speed rotors in retainer bearings and measurement concepts. In: Fourth International symposium on magnetic bearings, ETH, Zurich, Switzerland

6. Orth M, Nordmann R (2002) ANEAS: a modeling tool for nonlinear analysis of active magnetic bearing systems. In: Second IFAC conference on mechatronic systems, Berkley, USA, pp 357–362
7. Sinha JK (2007) Higher order spectra for crack and misalignment identification in shaft of a rotating machine. Struct Health Monit 6(4):325–334
8. Braut S (2006) Analysis of the rotor-stator contact dynamics. Ph.D. thesis, University of Rijeka, Faculty of Engineering, Rijeka (in Croatian)
9. Braut S, Žigulić R, Butković M (2008) Numerical and experimental analysis of a shaft bow influence on a rotor to stator contact dynamics, Stroj Vest. J Mech Eng 54(10):693–706
10. Sinha JK, Elbhbah K (2012) A future possibility of vibration based condition monitoring of rotating machines. Mech Syst Signal Process 34(2013):231–240
11. Li Y, Gu F, Harris G, Ball A, Bennett N, Travis K (2005) The measurement of instantaneous angular speed. Mech Syst Signal Process 19:786–805
12. Renaudin L, Bonnardot F, Musy O, Doray JB, Remond D (2010) Natural roller bearing fault detection by angular measurement of true instantaneous angular speed. Mech Syst Signal Process 24:1998–2011
13. Sarma S, Agrawal VK, Udupa S, Parameswaran K (2008) Instantaneous angular and speed measurement using a DSP based resolver-to-digital converter. Measurement 41:788–796
14. Braut S, Žigulić R, Štimac G, Skoblar A (2012) Efficient approach for encoder geometry compensation in time interval measurement of torsional vibration. In: Proceedings of the 10th international conference on vibrations in rotating machinery, London, IMechE, London, pp 809–818
15. Huang NE, Shen Z, Long SR et al (1998) The empirical mode decomposition and the Hilbert spectrum for nonlinear and nonstationary time series analysis. Proc Roy Soc Lond 454:903–995
16. Qi K, He Z, Zi Y (2007) Cosine window-based boundary processing method for EMD and its application in rubbing fault diagnosis. Mech Syst Signal Process 21:2750–2760
17. Du Q, Yang S (2007) Application of the EMD method in the vibration analysis of ball bearings. Mech Syst Signal Process 21:2634–2644
18. Gu F, Yesilyurt I, Li Y, Harris G, Ball A (2006) An investigation of the effects of measurement noise in the use of instantaneous angular speed for machine diagnosis. Mech Syst Signal Process 20:1444–1460

Sequential Recurrence Analysis of Experimental Time Series of a Rotor Response with Bearing Outer Race Faults

C.A. Kitio Kwuimy, M. Samadani, K. Kappaganthu and C. Nataraj

Abstract Rolling elements bearings are one of the most common components used in expensive, high precision and critical machines such as gas turbines, rolling mills and gyroscopes. They can be subjected to various defects which could lead to catastrophic results. This includes inner and outer race defects and hence it is of interest to analyze the system response under such defects. A better understanding of the system performance under such defects can be beneficial when performing system diagnostics and system design. In this paper we are focused on the outer race defect and perform a comparative nonlinear time series analysis of a healthy system and a defective system. We consider various levels of outer race defects. The analysis is based on the recurrence properties of the system in its reconstructed state space. After determining the appropriate time lags through the average mutual information technique and the corresponding embedding dimensions through the false neighbor technique, we perform a sequential analysis of the system by subdividing the time series into bins and investigating the system response through recurrence quantification analysis parameters along with the entropy. This contributes to the enhancement of the science of diagnostics of outer race defects by analyzing the signature of various recurrence quantification analysis parameters as the system goes from a healthy state to a severely defective state.

Keywords Rolling bearings · Outer race defect · System diagnostics · Recurrence analysis · Entropy

C.A. Kitio Kwuimy (✉) · M. Samadani · C. Nataraj
Center for Nonlinear Dynamics and Control, Department of Mechanical Engineering, Villanova University, Villanova PA19085, USA
e-mail: cedrick.kwuimy@villanova.edu

M. Samadani
e-mail: msamadan@villanova.edu

C. Nataraj
e-mail: nataraj@villanova.edu

K. Kappaganthu
Cummins Engine Co, PO Box 3005, Columbus, IN 47202-3005, USA
e-mail: karthik.kappaganthu@gmail.com

J.K. Sinha (ed.), *Vibration Engineering and Technology of Machinery*, Mechanisms and Machine Science 23, DOI 10.1007/978-3-319-09918-7_61

1 Introduction

Rotating machines are one of the most widely components used in industry. It is important to constantly monitor the conditions of these machines during the operation and maintain them in proper working conditions. Rotating machines are composed of different sub-systems interacting with each other in a nonlinear fashion. Changes in any of these components can significantly affect the overall performance of the system. Bearings are the load carrying members and are the key elements for effective functioning of the machine, and are often the cause of failure. Hence, it is critical to be able to monitor the health of the bearings and to detect the defective ones before the system fails. Bearing failure can take place because of many reasons such as wear, contamination, and improper installation. One of the most common modes of failure in a rolling element bearing is a point defect on the inner or outer race of the bearing.

Bearings are essentially nonlinear and, in a faulty regime, prominence of nonlinear phenomena (for instance, multi resonance, chaos, bursting oscillations) are possible. This has motivated the use of nonlinear dynamic techniques to analyze the vibration measurements for the purpose of diagnostics and prognostics. These techniques include extracting features and analyzing them using pattern recognition algorithms. The features can be obtained using time [1], frequency [2–4], time-frequency domain techniques [4–6] and with dynamic invariants such as the correlation dimension, the Lyapunov exponent and the entropy [7–18].

Almost all research studies on diagnostics of rolling bearings are based on signal processing techniques. Rolling bearings are highly nonlinear components and many studies have reported nonlinear phenomena such as bifurcations, quasi-periodicity and chaos in rolling element bearings [19–23]. These methods have proven to be effective when the response of the system is mainly in the non-complex and linear domain. However, they lose their functionality when the response exhibits complex nonlinear phenomena such as quasi-periodicity and chaos. The goal of this research is to open a new horizon into the application of nonlinear dynamic analysis tools in the diagnostics of machinery. In this paper, we are using the method of recurrence plots (RP) for the first time in order to detect, isolate and predict the severity of defects in rolling elements bearings of a rotating machine.

The concept of recurrences was introduced by Henri Poincare for the first time; however, it took almost a century until his discovery was numerically exploited by Eckman et al [24]. They used the method of recurrence plots to visualize the recurrence of dynamic systems in a high dimensional phase space. In many cases, the phase space has dimensions higher than three which can only be visualized by a projection into the two or three-dimensional sub-spaces. However, recurrence plots enable us to visualize and investigate certain aspects of the phase space trajectory in a two dimensional representation. An RP is a visualization of a square matrix, in which the matrix elements correspond to those times at which a state of a dynamical system recurs. Technically, the RP reveals all the times when the phase space trajectory of the dynamical system visits roughly the same area in the phase space.

The method of RPs has been used recently by some researchers for data-based diagnostics to detect damages or dynamic changes along with the estimation of their severity [10, 25–29]. This is done by comparing some features, known as recurrence quantification analysis (RQA) parameters, for healthy and defective systems and by estimating the percentage of change or the presence of a defect. Along these lines, dynamic changes due to friction have been characterized by the use of RPs [10, 26], RPs have been used to identify changes in cracked aluminum plates, and the method has been considered by Iwaniec et al. [27] and Litak et al. [29] to detect dynamic changes during the composite milling process. The method has been used to analyze complexity in traffic flow under adverse weather condition by Vlahogianni and Karlaftis [30].

The limitation of this approach resides in the fact that the method of RPs does not always give detailed insight into the dynamics of the system and thus performs poorly when trying to classify the nature of the defect or dynamic changes [31, 32]. In our recent contribution, we considered an interesting idea that consists of using RPs to analyze a physics-based model for diagnostics and to obtain features that can be used for a hybrid approach for diagnostics: a physics-based model feature extraction is used to educate artificial intelligence systems, and similar features obtained from data collected from an actual system are used for diagnostics [31, 14].

We consider in this paper the idea of performing a sequential analysis of the system using the method of RPs. The main goal is to obtain a signature of various recurrence quantification analysis parameters for a healthy system and for a defective system with various levels of defects. We believe that, in a practical situation, knowing these signatures will contribute significantly to the classification of the system. Moreover, a singular feature of the method of RPs is its ability to extract features from a high dimensional system, and short, noisy and even non-stationary data as is the case in general for machinery.

The rest of the paper is organized as follows. In Sect. 2, we present the experimental setup along with the procedure used for data extraction. Section 3 recalls a general presentation of the method of RPs. In Sect. 4, we perform a sequential analysis of the system by subdividing the time series into bins and investigating the system response through recurrence quantification analysis parameters along with entropy. Section 5 presents a summary and conclusion.

2 Experimental Setup and Data Collection

2.1 The Experimental Setup

The machine fault simulator used as the experimental setup is shown in Fig. 1. It is a test rig with multiple parts that can be conveniently assembled and disassembled. It consists of a motor with a closed loop speed control connected to a shaft

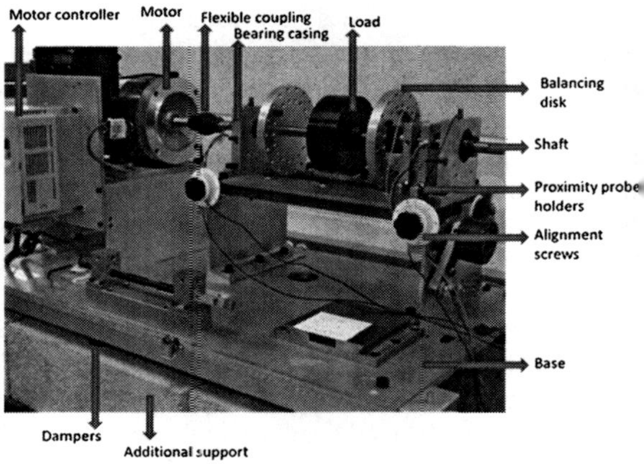

Fig. 1 Experimental setup [12]

supported on rolling element bearings. The span of the shaft is fixed but shafts of various diameters can be used. The bearings are placed in the bearing casing and can be easily replaced. The system can be loaded using a mass and balancing disks. A load mass of 5 kg is used in this research. The balancing disk has two rows of screw holes drilled near the edge into which different masses can be inserted to balance the shaft. The whole system is supported on a solid base with dampers so that ambient vibrations affecting the system are minimized. To provide more support, the setup is placed on heavy cement blocks. One of the primary sources of excitation in rotating systems other than unbalance is misalignment. The screws are used to position the bearing casings such that the misalignment is minimized and the flexible coupling dampens any excitation due to residual misalignment. The speed of the system is set using a computer attached to the motor controller. In this research, the data is collected at a constant speed of 50 Hz. During the data collection, the entire set of the operating condition is held stationary.

A rolling bearing has multiple moving parts as shown in Fig. 1 [12]. The rolling elements are enclosed in a cage that is placed between two races. The cage is press fitted so that all the components are held in place. Lubrication is provided between the rotating surfaces to reduce friction. The bearings used in this paper have spherical rolling elements and also have a seal to hold the lubricant in. Localized defects can occur in either the inner race or outer race of the rolling element. In this research only inner and outer race defects are considered as they are more common and often the cause of failure. The defects manifest themselves in the signals as amplitude modulations. The modulations occur at frequencies that are dependent on the type of the localized defect. These frequencies can be calculated using the bearings geometric parameters. For the purpose of fault severity identification, outer race defects are used in this paper. Two defect sizes are used: 30 and 90 mils. Figure 2 shows the bearings with outer race defects. The values of the parameters of the various components used in the experimental setup are listed in Table 1.

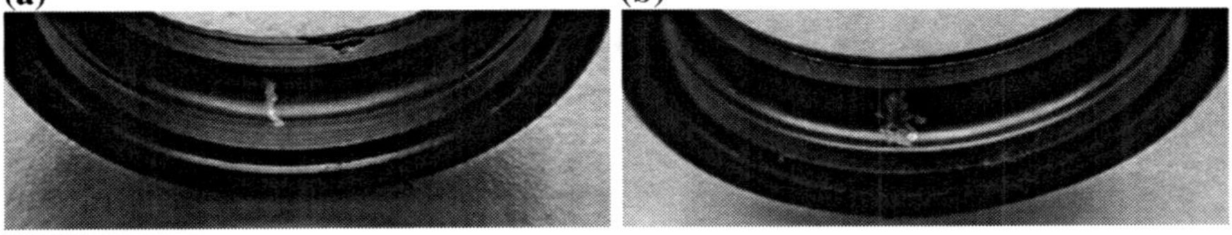

Fig. 2 Outer race defects. **a** Small defect. **b** Large defect [12]

Table 1 Parameters in the experimental setup [12]

Parameter	Values	Parameters	Values
Mass (shaft)	0.831 kg	Mass (disk)	0.66 kg
Diameter (shaft)	0.016 m	Diameter (disk)	0.152 m
Length (shaft)	0.35 m	Length (disk)	0.016 m
I (shaft)	3.1179×10^{-9} m^4	I_D (disk)	0.0019 kgm^2
I_p (shaft)	1.5588×10^{-9} m^4	I_p (disk)	$1.3101 \times 10 - 5$ m^4
E (shaft)	200 N/m^2	Diameter (rotation element)	0.079 m
k (shaft)	6.616×105 N/m	Mean diameter	0.335 m
Load mass	5 kg	Contact angle	90°
n (bearing constant)	1.5	Elasticity constant	1.8315×109 N/m
Inner race osculation	0.52	Outer race osculation	0.52
Inner race clearance	0	Outer race clearance	0

2.2 Data Collection

Data is captured with three different bearings at a constant speed of 50 Hz. Two of these bearings have an outer race defect and one bearing does not have any defect. The defect width in the smaller defect bearing is 30 mils and in the larger defect bearing is 90 mils. The system is loaded with a 5 kg mass. In this study the signals from the experimental setup are collected with accelerometers placed at the defective bearing. Data are captured at a sampling rate of 32,768 Hz. A total of four sets of 214 points are recorded and organized into 48 bins of 210 points. The first 10 bins are ignored because of the transient regime.

A sample of the data captured from the experimental setup is shown in Fig. 3 after applying a band pass filter. The horizontal axis represents time while the vertical axis represents the measured vibration of the system. In this figure, the first subplot is the vibration signal from a defect-free system; the second subplot is the signal from a system with the smaller outer race defect and the third subplot is the signal from a system with the larger outer race defect. Bearings with outer race defects have impulses excited by the rolling element entering the defect. Because of these impulses there is a significant difference between the signals from bearings with and without defects. It is relatively easy to identify defects in bearings using

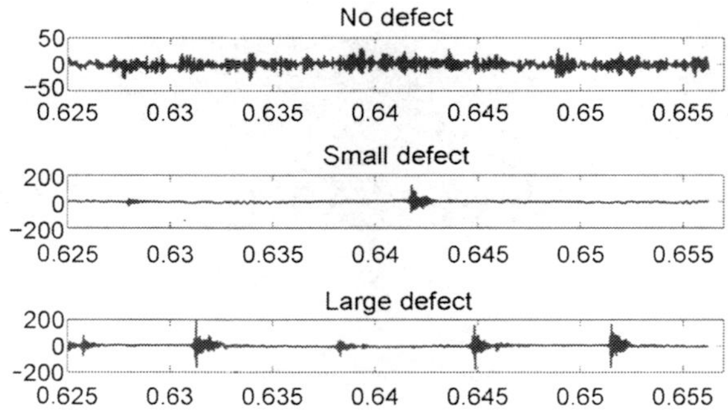

Fig. 3 Sample response of the system

many signal processing techniques like envelope spectrum, wavelet packet transforms, etc. [9, 12], which capture the presence of these impulses in time, frequency or time-frequency domains. It is more challenging to determine the severity of the defect, particularly in varying operating conditions as the relationship is not straightforward. Several studies have demonstrated that the correlation dimension can be used in an efficient way to detect and analyze faults in machineries and bearings [8, 11, 14–16, 18]. Our aim in this paper is to use the method of recurrence along with some statistical parameters to analyze the severity of defects in bearings. In a different contribution, we consider various dynamic invariants as tools to identify the severity of defects in bearings.

3 Recurrence Plots and Recurrence Quantification Analysis

The recurrence plots analysis for time series is based on the analysis of a matrix R whose elements are defined as

$$\mathbf{R}_{ij} = \begin{matrix} 1 : \mathbf{\Phi}_i \approx \mathbf{\Phi}_j \\ 0 : \mathbf{\Phi}_i \neq \mathbf{\Phi}_j \end{matrix}, \quad i,j = 1,\ldots,N$$

where $\mathbf{\Phi}_i = (\varphi_{1i}, \varphi_{2i}, \ldots, \varphi_{mi})$ is a state vector and m the dimension of the state space, N is the length of the time series, i and j are related respectively to the row and column of the matrix. $\mathbf{\Phi}_i \approx \mathbf{\Phi}_j$ means equality up to an error ϵ. The state vector $\mathbf{\Phi}$ is reconstructed from a single variable measurement s (in our case, the vibration in the x-direction). Based on the hypothesis that the time series itself contains information about the topology of the state space, Packard et al. [33] initiated work on rebuilding the state space using discrete data from dynamic systems. Based on the contributions of Withney and Takens [7, 10, 14, 17], the derivation of the state space is now well established. This is done in two steps: the first step consists of estimating the prescribed time lag and the second step consists on evaluation of the embedded dimension. In practice, the prescribed value of T corresponds to the first minimum of the average mutual information between the values of $s(n)$ and

$s(n+T)$, and the embedding dimension can be deduced from the method of false nearness neighbor [7, 14, 17, 34].

The elements of the matrix **R** arehus obtained by comparing the state of the system at time i and j with a threshold precision ϵ. Thus, formally, one has

$$\mathbf{R}_{ij} = \theta(\epsilon - \parallel \mathbf{\Phi}_i - \mathbf{\Phi}_j \parallel)$$

with $\|(.)\|$ been the Euclidian norm (L_2-norm) and $\theta(y)$ is the heaviside function defined as

$$\theta(y) = 1 \quad \text{for} \quad y > 0 \quad \text{and} \quad \theta(y) = 0 \quad \text{for} \quad y<0$$

An RP graph is obtained by plotting the $\mathbf{R}_{ij}$ elements with different colors; by definition, an RP graph is always symmetric ($\mathbf{R}_{ij} = \mathbf{R}_{ji}$) and always has a central diagonal.

In order to analyze the effects of the fault, we use major measurable quantities introduced by Webber and

Zbilut [35] and called recurrence quantification analysis (RQA) parameters [36]: Recurrence rate (*RR*), Determinism (*DET*), Laminarity (*LAM*), Average length of the diagonal lines (*L*) and Average length of the vertical lines (*TT*) respectively defined as:

$$RR = \frac{1}{N^2}\sum_{i,j=1}^{N} \mathbf{R}_{i,j}(\epsilon),\ DET = \frac{\sum_{l=l_{min}}^{N} lP(l)}{\sum_{l=1}^{N} lP(l)},\ LAM = \frac{\sum_{v=v_{min}}^{N} vP(v)}{\sum_{v=1}^{N} vP(v)},\ L = \frac{\sum_{l=l_{min}}^{N} lP(l)}{\sum_{l=l_{min}}^{N} P(l)}$$

$$\text{and } TT = \frac{\sum_{v=v_{min}}^{N} vP(v)}{\sum_{v=v_{min}}^{N} P(v)}$$

where, $P(x)$ is the histogram of x for a given threshold ϵ, l is the length of the diagonal, v is the length of the vertical line and $v_{min} = l_{min} = 2$.

To complement the analysis, we consider two statistical features and a dynamic invariant. Namely, we consider the Shannon entropy defined as

$$Entr = \sum_{l=l_{min}}^{N} P(l)\ln P(l)$$

4 Fault Analysis Using Recurrence Analysis

4.1 Fault Analysis Using the RQA

The first minimum of the average mutual information, plotted as a function of the time lag was obtained at $T = 2$ and, by application of the method of false nearest neighbor, we found that the minimum embedding dimension for the system is $d = 6$.

As shown in Fig. 3, there is no clear pattern to be deduced from the time representation of the system response. Fortunately, the specificity of the method of recurrence plots allows us to perform interesting visual analysis and extract patterns and quantitative features for the system response.

The graphs of Fig. 4 show the RQA parameters along with the entropy as a function of the bin number. The horizontal axis represents the bin number and the vertical axis is the RQA parameter. For the healthy system (Fig. 4a), the graphs show two bifurcations in the system response with slight fluctuations. For a bin number $N_{bin} < 20$, all the RQA parameters are almost constant before dropping for $20 < N_{bin} < 36$ and increase again for $N_{bin} > 36$. The slight fluctuation is certainly due to residual noise after the band-pass filter (see Fig. 3). The sudden drop can be due to an external event in the system response such as spin down motion. A more plausible explanation can be found in the fact that, both *DET* and *LAM* of the system have high values. The determinism describes the absence or presence of complexity and stochasticity in the system, that is, the predominance of periodic motion is described by large values of *DET*. The high value of *LAM* describes possible inner crises or intermittency in the system response [36]. This is correlated to a fine structure with parallel diagonals (and variable density) separated by a white shaded area. They basically indicate that some states are rare or have recurred in a certain region of the state space. That is, they evolve pretty much in parallel for a while and also resemble each other again after a certain interval of time [10, 14, 36]. This is also the signature of an underlying periodic behavior for the overall dynamics [36]. The RPs of Fig. 5b are potted for $N_{bin} > 20$ and appear to have the same structure as in Fig. 5b, except that the point is denser. More points are revisited in the state space. Our interpretation is that, at the bifurcation point, there is an event in the system that causes the amplitude or periodicity of the system to change (decrease). The entropy of the system is constant for all bins.

The RQA parameters of Fig. 4b are plotted for the system with a small outer race defect. In addition, and in comparison with the healthy system, one can observe the following. (i) The bifurcation time has shifted forward. (ii) Beyond the bifurcation, the RQA parameters tend to increase with an overall oscillation. (iii) All RQA parameters have increased. (iv) The entropy is an increasing function of the bin number and, beyond the bifurcation point, it becomes constant. Globally, this evolution shows that, in the presence of a small defect, either the tendency to periodicity has increased (high value of *RR* and *DET*) or after a certain time, windows of sporadic events cause the fluctuation of large values of *RR* and decreasing values of *DET*. One can definitely suspect the increase of intermittency (increasing of *LAM*); the intermittent appearance of bursting oscillations or the amplitude modulation appearing (see Fig. 3) at specific time to be contributing to increase in *LAM*, *LL* and *TT*. The increase in entropy reflects the increase in the complexity of the system until a certain point where it remains constant. Note that the bifurcation diagrams were not plotted adiabatically, thus, there is an "accumulation" or past history dependence. The value of a RQA at a specific bin depends on the state of the system at that bin and at the previous bin. This is well described

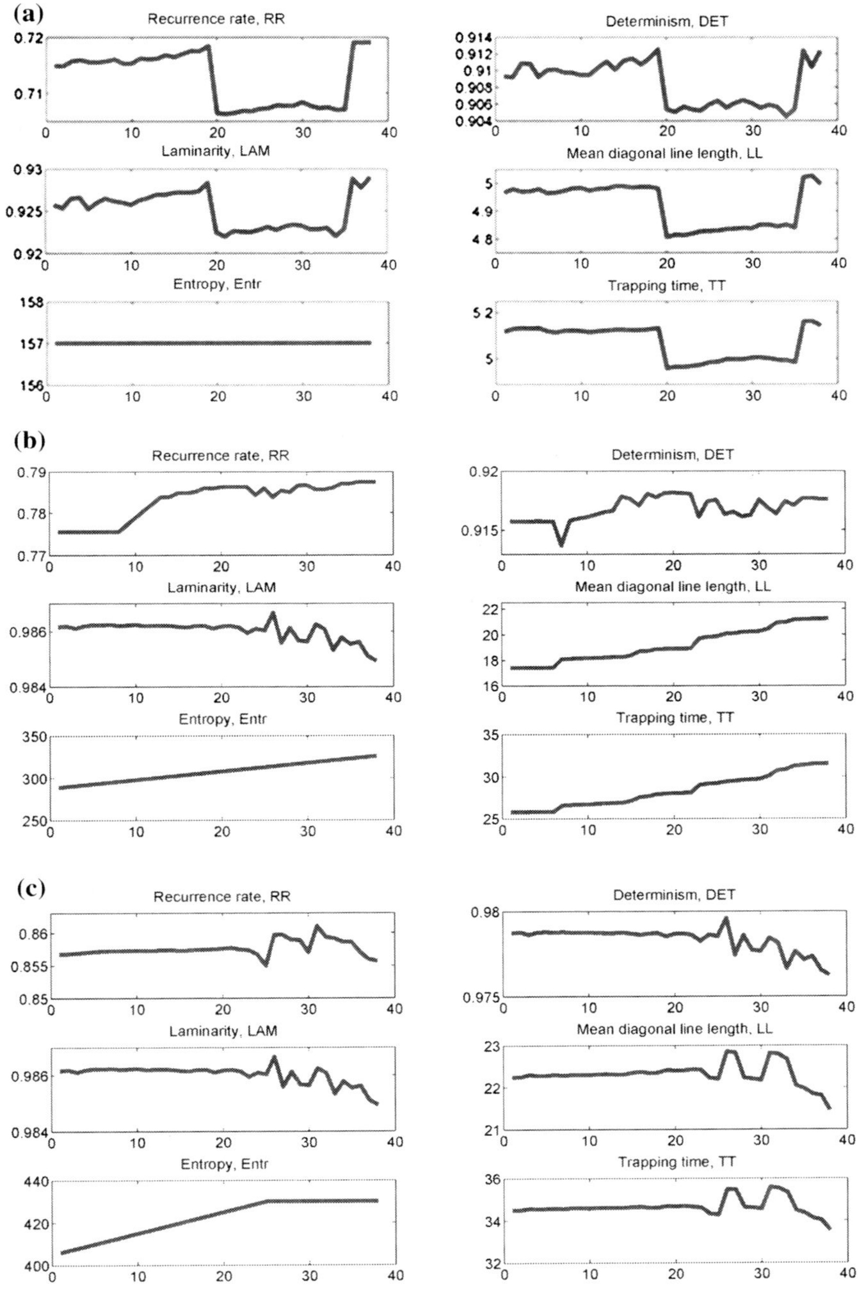

Fig. 4 RQA of the system. **a** System free of defect. **b** System with small defect. **c** System with large defect. The horizontal axis corresponds to the bin number while the vertical axis corresponds to the RQA parameter

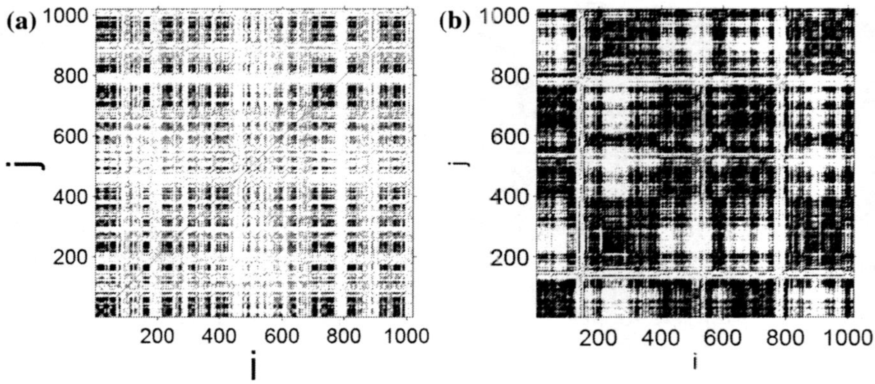

Fig. 5 RPs of the healthy system for $\epsilon = 2\sigma$. **a** RPs for bin $N_{bin} < 20$. **b** RPs for bin $N_{bin} > 20$

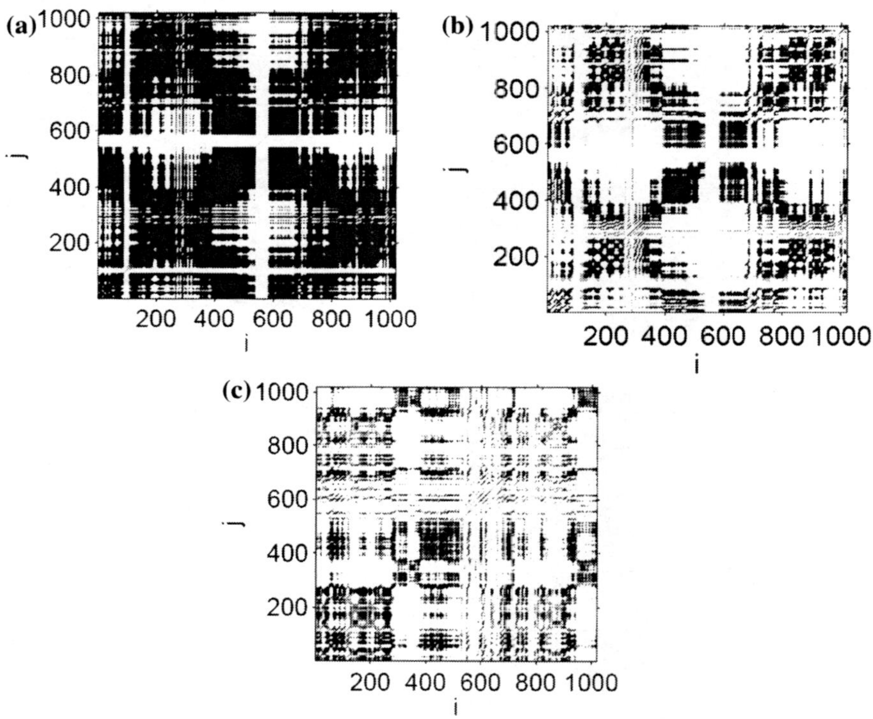

Fig. 6 RPs of the defective system. **a** RPs for bin $N_{bin} < 20$ and $\epsilon = 2\sigma$. **b** RPs for bin $N_{bin} < 20$ and $\epsilon = \sigma$. **c** RPs for bin $N_{bin} > 20$ and $\epsilon = 2\sigma$

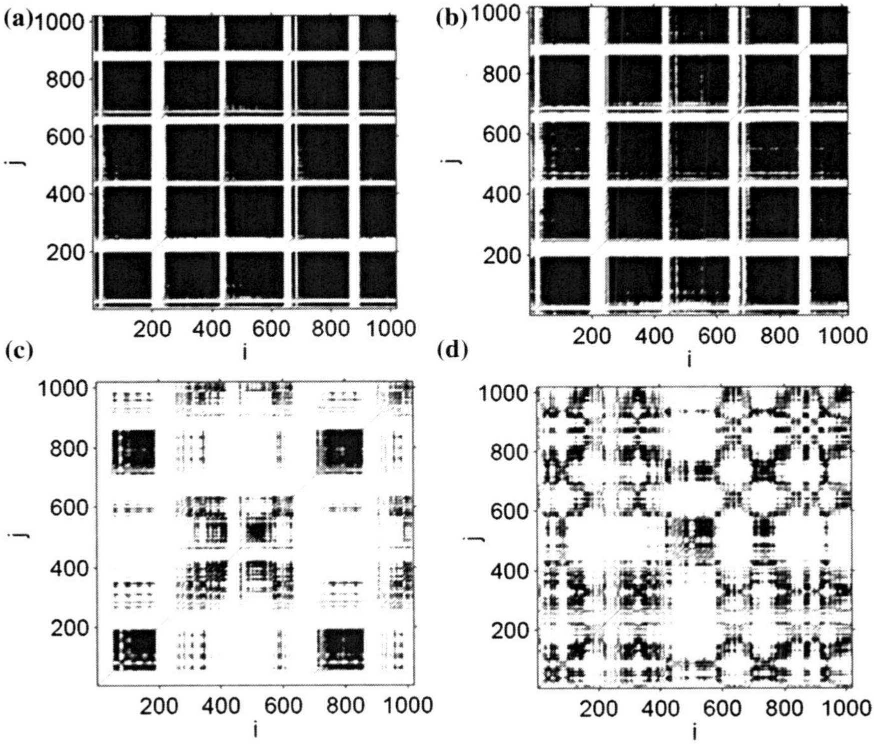

Fig. 7 RPs of the defective system. **a** RPs for bin $N_{bin} < 20$ and $\epsilon = 2\sigma$. **b** RPs for bin $N_{bin} < 20$ and $\epsilon = \sigma$. **c** RPs for bin $N_{bin} < 20$ and $\epsilon = 2\sigma$. **d** RPs for bin $N_{bin} > 20$ and $\epsilon = 2\sigma$

by the entropy: the system has ineluctably evolved toward complexity until the level of complexity remains "constant".

The RPs for the system with the small defect are plotted in Fig. 6 for the same threshold parameter as in Fig. 4. From Fig. 6a, one can identify fine structures with high density in comparison with the healthy system. This was predictable since the RR has increased. The information from this is that the threshold parameter has become too large to scan some details in the system response. A different view of the RPs is obtained (Fig. 6b) for half value of the threshold parameter. The structure shows more disrupted white bands, a signature of the fact that non-stationary transitions may have occurred (the bursting oscillation observed in the time response) [36]. Moreover, the vertically and horizontally spread black areas in the RPs corroborate the tendency to inner crises and intermittency. Figure 6c is obtained beyond the bifurcation point; the system response has the same structure, except that the white bands are more spread and visible. This suggests that the modulation phenomena or bursting oscillations are more visible.

The RQA parameters of Fig. 4c are plotted for the large outer race defect. In comparison with the case with no defect and the case with the small defect, one can deduce the following. (i) The bifurcation appears much earlier (backward shifted). (ii) *LAM* is not sensitive to the defect severity. (iii) All RQA parameters (except *LAM*) have decreased in comparison with the case with small defect, and they have increased in comparison with the case with no defect. (iv) The entropy is strictly increasing. The RPs of Fig. 7a–c shows more vertically and horizontally spread black areas as well as a clear white band.

Clearly, a visual inspection of Figs. 5a, 6b and 7c show that, as the severity of the fault increases, there is an increase of vertically and horizontally spread black areas as well as the appearance of a clear white band. This signifies more intermittency and a busting behavior. Also, from the above analysis of the RQA parameters, it follows that these parameters can be efficiently used to detect bearing defects; however, *LAM* is less robust in determining the severity of the defect.

5 Discussion and Conclusion

Failure in a large class of engineered systems is due to structural and/or parameter changes inducing soft and strong nonlinearities. These faults can be catastrophic for real engineering systems and there is a strong need to efficiently monitor the response of the system in order to understand/predict, and if possible, to control and suppress potential sources of failure. We believe that the development of the method of recurrence plots has given us new paradigms to detect complex nonlinear behavior and to perform system diagnostics. Caution is advised however, as although this method of RPs seems to be having success in various applications, detailed insight into the dynamics of the system needs to be teased out of the method as we have attempted to do in this paper.

The main goal of this paper was to propose an approach to extract signatures from the time series of bearings response with various levels of faults. This was done by performing a sequential analysis of the system using the recurrence quantification analysis parameters and the joint and cross recurrence plots. For all sequences, each RQA parameter was estimated. The following are some salient observations.

- The sequential approach is powerful in effectively distinguishing a healthy system, a system with small a defect and a system with a large defect.
- The transition time observed for the healthy system has shifted forward in the presence of small defect.
- All RQA parameters increase as the system goes from a healthy state to one with a large defect. However, *LAM* was found to be insensitive to the defect severity.
- The entropy of the healthy system is constant through all sequences, but increases with the sequence number in the presence of a defect. A large defect leads to a higher value of the entropy.

- The severity of the defect contributes to increased intermittent behavior. This is observed through the trapping time and mean diagonal length. This is valid also for the presence of vertical and horizontal structures in the RPs.
- A visual inspection of the RPs for a given sequence can help distinguish fault severity through the prominence of white and black areas.

Although the results of the present paper are significant, we found that defining the length and bound of each sequence can affect the results graphically and fundamentally. Our advice is that, one should initially plot the set of data as a function of time, and then identify various sequences. Or, a simple approach (the one used here) is to define the sequences such that all sequences have exactly the same integer multiples of the rotor frequency. Moreover, in practical situations, one would need to subdivide the data into training and testing sets and use the numerical values of the quantification parameters. This is currently under investigation; we are also exploring extension to fault localization.

Acknowledgments This work has been funded by the US Office of Naval research under the grant ONR N00014-08-1-0435. We wish to express our gratitude to the program manager, Mr. Anthony Seman III and also Dr. Frank Ferrese.

References

1. Tandon N (1994) A comparison of some vibration parameters for the condition monitoring of rolling element bearings. Measurement 12(3):285–289
2. Barkov A, Barkova N (1995) Condition assessment and life prediction of rolling element bearings. J Sound Vib 29(6):10–17
3. Randall R, Gao Y (1994) Extraction of modal parameters from the response power spectrum. J Sound vibr 176(2):179–193
4. Ypma A (2001) Learning methods for machine vibration analysis and health monitoring. University of Technology of Delft, Delft, Netherland
5. Cade IS, Keogh PS, Sahinkaya MN (2005) Fault identification in rotor/magnetic bearing systems using discrete time wavelet coefficients. IEEE/ASME Trans Mech 10(6):648–657
6. Mori K, Kasashima N, Yoshioka T, Ueno Y (1996) Prediction of spalling on a ball bearing by applying the discrete wavelet transform to vibration signals. Wear 195(1):162–168
7. Abarbanel HDI (1996) Analysis of observed chaotic data. Springer, New York
8. Chen Y, Du R, Qu L (1995) Fault features of large rotating machinery and diagnosis using sensor fusion. J Sound Vibr 188(2):227–242
9. Djebala A, Ouelaa N, Hamzaoui N (2008) Detection of rolling bearing defects using discrete wavelet analysis. Meccanica 43(3):339–348
10. Fontaine S, Dia S, Renner M (2011) Nonlinear friction dynamics on fibrous materials, application to the characterization of surface quality. Part II: local characterization of phase space by recurrence plots. Nonlinear Dynamics 66(4):647–665
11. Jiang J, Chen J, Qu L (1999) The application of correlation dimension in gearbox condition monitoring. J Sound Vibr 223(4):529–541
12. Kappaganthu K (2010) An integrative approach for machinery diagnostics. Ph.D. Thesis, Villanova University, Villanova, USA
13. Kappaganthu K, Nataraj C, (2011) Feature selection for fault detection in rolling element bearings using mutual information. ASME J Vibr Acoust 133(6):06100101–06100111

14. Kwuimy C, Samadani M, Nataraj C (2014) Bifurcation analysis of a nonlinear pendulum using recurrence and statistical methods: applications to fault diagnostics. Nonlinear Dyn 76 (2014):1963–1975
15. Kwuimy CAK, Samadani M, Nataraj C (2014) Preliminary diagnostics of dynamic systems from time series. In: Proceedings of the ASME International Design Engineering Technical Conference
16. Logan D, Mathew J (1996) Using the correlation dimension for vibration fault diagnosis of rolling element bearings. I: Basic concepts. Mech Syst Sig Proc 10(3):241–250
17. Takens F (1981) Dynamical Systems and Turbulence, Lecture Notes in Mathematics, chap. Detecting strange attractors in turbulence, pp 366–381
18. Wang W, Wu Z, Chen J (2001) Fault identification in rotating machinery using the correlation dimension and bispectra. Nonlinear Dyn 25(4):383–393
19. Ehrich F (1992) Observations of subcritical superharmonic and chaotic response in rotor dynamics. J Vibr acoust 114(1):93–100
20. Kim YB, Noah S (1996) Quasi-periodic response and stability analysis for a non-linear jeffcott rotor. J Sound Vib 190(2):239–253
21. Mevel B, Guyader J (1993) Routes to chaos in ball bearings. J Sound Vibr 162(3):471–487
22. Sankaravelu A, Noah ST, Burger CP (1994) Bifurcation and chaos in ball bearings. ASME Applied Mechanics Division-Publications 192:313–313
23. Tiwari M, Gupta K, Prakash O (2000) Effect of radial internal clearance of a ball bearing on the dynamics of a balanced horizontal rotor. J Sound vibr 238(5):723–756
24. Eckmann J, Kamphorst SO, Ruelle D (1987) Recurrence plots of dynamical systems. Europhys Lett 4(9):973–977
25. Elias J, Narayanan Namboothiri V (2013) Cross-recurrence plot quantification analysis of input and output signals for the detection of chatter in turning. Nonlinear Dyn 2013:1–7
26. Fontaine S, Dia S, Renner M (2011) Nonlinear friction dynamics on fibrous materials, application to the characterization of surface quality. Part I: global characterization of phase spaces. Nonlinear Dyn 66(4):625–646
27. Iwaniec J, Uhl T, Staszewski WJ, Klepka A (2012) Detection of changes in cracked aluminum plate determinism by recurrence analysis. Nonlinear Dyn 70(1):125–140
28. Litak G, Syta A, Gajewski J, Jonak J (2010) Detecting and identifying non-stationary courses in the ripping head power consumption by recurrence plots. Meccanica 45(4):603–608
29. Litak G, Syta A, Rusinek R (2011) Dynamical changes during composite milling: recurrence and multiscale entropy analysis. Int J Adv Manufact Technol 56(5–8):445–453
30. Vlahogianni EI, Karlaftis MD (2012) Comparing traffic flow time-series under fine and adverse weather conditions using recurrence-based complexity measures. Nonlinear Dyn 69 (4):1949–1963
31. Farrar CR, Hemez F, Park G, Robertson A, Sohn H, Williams TO (2003) A coupled approach to developing damage prognosis solutions. Key Eng Mater 245(2003):289–306
32. Jardine AK, Lin D, Banjevic D (2006) A review on machinery diagnostics and prognostics implementing condition-based maintenance. Mech Syst Sig Process 20(7):1483–1510
33. Packard N, Crutchfield J, Farmer J, Shaw R (1980) Geometry from a time series. Phys Rev Lett 45(9):712–715
34. Kantz H, Schreiber T (2004) Nonlinear time series analysis. Cambridge University Press, Cambridge
35. Zbilut JP, Thomasson N, Webber CL (2002) Recurrence quantification analysis as a tool for nonlinear exploration of nonstationary cardiac signals. Med Eng Phys 24(1):53–60
36. Marwan N, Romano MC, Thiel M, Kurths J (2007) Recurrence plots for the analysis of complex system. Phys Rep 438(5–6):237–329

Adhesive Layer Influence on Compressive and Tension Stiffness of Thin-Layer Rubber-Metal Elements

Vladimirs Gonca, Svetlana Polukoshko and Galina Hilkevica

Abstract The technique of "force–displacement" stiffness characteristics calculation for thin-layer rubber-metal damping devices is presented in this work. It is proposed to take into account adhesive layers and deformation of non-elastomeric layers working under tension and compression. A variational method of linear theory of elasticity for weakly compressible materials is used for calculations. The received dependences allow explaining nonlinearity of "force–displacement" stiffness characteristics and various behavior of glued rubber-metal device under tension and compression, observed in experimental investigations.

Keywords Elastomer · Compressive stiffness · Multilayer devices · Poisson's ratio · Variational procedure

1 Introduction

Rubber and rubberlike materials (elastomers) are unique family of materials which offer many engineering advantages because of their small volume compressibility and ability to maintain large elastic deformation [1–3]. Reinforced elastomeric structures consist of alternating thin layers of rubber and adhesive—bonded reinforcing layers of much more rigid material (usually metal). This allows to obtain the structures, which axial compression stiffness is in several orders greater than shear

V. Gonca
Institute of Mechanics, Riga Technical University, Ezermalu Str. 6, Riga 1006, Latvia
e-mail: vladimirs.gonca@rtu.lv

S. Polukoshko (✉) · G. Hilkevica
Ventspils University College, Engineering Research Institute "VSRC",
Inzenieru Str. 101, Ventspils 3601, Latvia
e-mail: pol.svet@inbox.lv

G. Hilkevica
e-mail: galina.hilkevica@venta.lv

J.K. Sinha (ed.), *Vibration Engineering and Technology of Machinery*,
Mechanisms and Machine Science 23, DOI 10.1007/978-3-319-09918-7_62

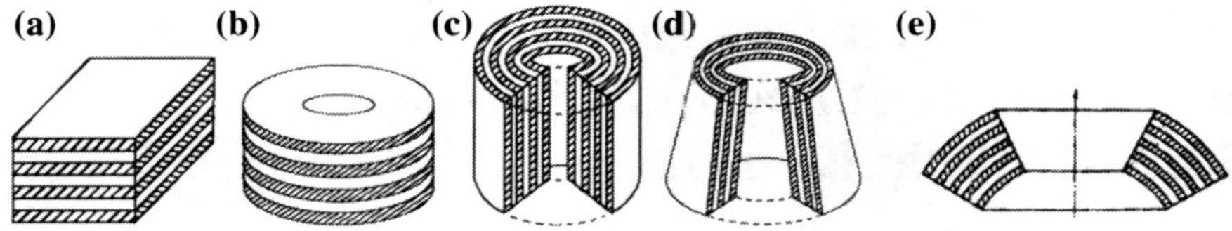

Fig. 1 Multilayer elastomeric structures examples: **a** flat of rectangular shape, **b** flat ring, **c** cylindrical, **d** conical, **e** spherical

stiffness. The connection of elastomeric with reinforcing layer is usually fulfilled by means of vulcanization or gluing. Packages of thin-layered rubber-metal elements (further TRME) are successfully used as bearing, joints, compensating devices, shock-absorbers etc. [3, 4]. In practice the TRME packages of different geometric shapes are used: flat, cylindrical, conical and others; number of layers may be different, at least three (Fig. 1). Elastomeric layer is considered as thin if its width/thickness ratio is much more than 10. For fastening of rubber layer to metal layer a rubber adhesive with vulcanization is used, during which the adhesive partially dissolves the elastomeric material on contact surface, forming bonding layer of some thickness. Mechanical properties of bonding layer may substantially differ from the mechanical properties of elastomeric layer. For very thin rubber elements thickness of elastomeric and bonding layer become comparable and the bonding layer influence on TRME stiffness becomes significant. Stiffness characteristic (relationship between imposed external forces and displacement) is very important for TRME practical applications and must be considered carefully.

In this paper analytical model taking into account effect of the adhesive layer and allowing describing different behavior of TRME packet under compression and tension is proposed for flat rectangular TRME. Variational method of linear theory of elasticity for weakly compressible materials was used.

2 Problem Definition

The authors performing tension and compression testing of the thin-layer ($\alpha = a/h$, $\beta = b/h \gg 10$, where a, b—geometrical dimensions in the plan, h—thickness of an elastomeric layer) multilayered rubber-metal elements, noted the next characteristic features of TRME: influence of elastomeric layers weak compressibility on tension—compression stiffness characteristic; noticeable nonlinearity of “force–displacement” dependence even in small deformations domain ($\varepsilon < 10$–15 %) under considerable specific loading; two-modularity of “force–displacement” dependence at tension and compression (“rigid” under compression and “soft” under tension) [5, 6]. For optimal design and efficient working of such TRME it is necessary to develop the analytical model taking into account the above listed features. Some authors proposed to describe the effect of high specific loading by taking into account the shear modulus G and bulk modulus K dependence on hydrostatic pressure [7–9]. Experiments show that at high specific loading these dependences are of "rigid" type for compression and "soft" for tension. The obtained results should be treat with a caution, since experiments were fulfilled not on a "clean" rubber elements, but on multilayer TRME packages which do not exclude the influence of the constructive nature (in particular, the deformation of nonelastomeric layers and work of adhesive layers) on the results of studies. In the works [7, 8] it is shown that "force–displacement" dependence is much more nonlinear for thin-layer rubber-metal elements, produced by vulcanizing, than for "pure" rubber elements under the same high levels of specific compressive load. Processing of the experimental data, using the known "force–displacement" dependence [5, 9] shows that disregarding of the deformation of nonelastomeric layers of thin TRME packages leads to the fact, that compressive stiffness characteristics may differ significantly from the experimental data, even at small levels of specific loading at which dependence of the share modulus G and volume modulus K on compressive loading value isn't yet observed. This confirms that the analytical model of TRME packages does not include all features of such structures.

When manufacturing of such elements rubber adhesive with the subsequent vulcanizing is used for fastening of a rubber layer to nonelastomeric rigid element. During vulcanization process the rubber element near contact surface is partially dissolving in the glue, forming a transitional layer of thickness h_c with mechanical properties different from the properties of the adhesive and elastomer.

Analysis of the rubber adhesives composition shows that the mechanical characteristics of the transitional layer (shear modulus G_c, bulk modulus K_c and Poisson's ratio μ_c) may significantly differ from the mechanical characteristics G, K and μ of elastomeric element [10]. For bulk modulus this difference may be more than an order of magnitude ($K_c \ll K$). Technologically, for the same curing techniques, regardless of the "fineness" ($\alpha = a/h$, $\beta = b/h$) of the elastomeric layer, the transition layer thickness h_c will be approximately constant. In this case, if TRME is not very thin ($h \gg h_c$), the contribution of the work of the transitional layer in response stiffness characteristic of TRME package will be negligible. In very thin TRME of thickness h and h_c may be of the same magnitude (but structurally, it is always $h > h_c$). In this case neglect of deformation of a transitional layer may lead to quantitative and qualitative errors at calculation of the compression stiffness characteristic of TRME packages. Calculation of thin elastomeric layers packages under compression show, that influence of volume deformation may be of one order (and possibly, more) with form change deformation even if Poisson's ratio of elastomer $\mu \to 0.5$ [7, 11, 12]. It a fortiori will be fair for a transitional layer, as soon as $\alpha_c = a/h_c > \alpha = a/h$, $\beta_c = b/h_c > \beta = b/h \gg 10$ and $K_c \ll K$.

In TRME packages metallic layers are generally used as intermediate non-elastomeric layer. Since the stiffness characteristic of the metallic layers are much greater than of the elastomeric layers, under compression of TRME packet the deformation of non-elastomeric layers influence may be only if these layers are very thin (such as a metal foil). Such layers undergo only a tensile strain in the plane of

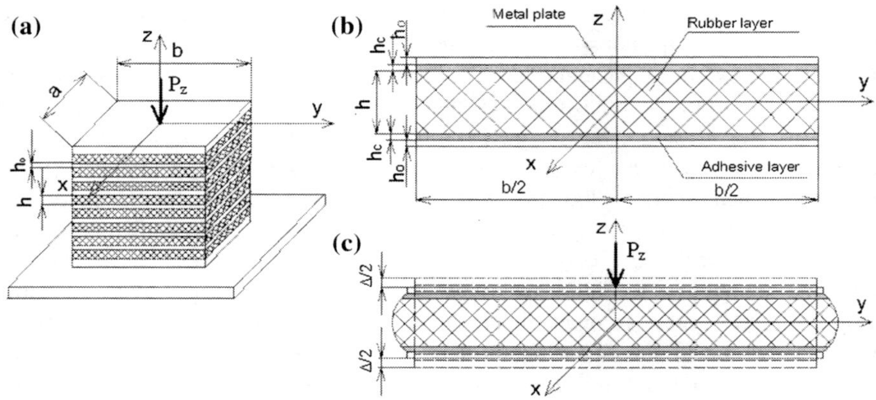

Fig. 2 Scheme of designed object: **a** flat TRME packet of rectangular form; **b** scheme of one TRME layer; **c** compressive deformation of rubber layer between two metal layers; h—thickness of elastomeric layer, h_c—thickness of transitional layers, h_o—thickness of metal layers

the layer under the influence of forces transmitted through the elastomeric layer tangential stresses in the cross section of their contact.

In this paper a variant of analytical model allowing to describe the "rigid" response in compression and "soft" in tension is proposed for glued TRME package. Qualitative and quantitative influence of transitional and nonelastomeric layers deformation on stiffness characteristics is considered by the example of the flat rectangular TRME package (Fig. 2).

3 Analytical Solution

Problem is solved by means of variational procedure, using principle of a minimum of total potential energy Π [7]:

$$\Pi = U + U_c + U_o - P\Delta_\Sigma, \tag{1}$$

where: U—potential energy of deformation of elastomeric layer (taking into account weak compressibility of elastomeric); U_c—potential energy of deformation of a transitional layer (taking into account the shear and volume deformations); U_o—potential energy of deformation of thin nonelastomeric layer, which thickness allows to consider only energy of tension of this layer; $\Delta_\Sigma = \Delta + 2\Delta_c$—one TRME element deformation, Δ—elastomeric layer deformation, Δ_c—transitional layer deformation.

$$U = G\int\limits_V \left[\frac{1}{2}\left(u_{i,j}\,u_{i,j} + u_{i,j}\,u_{j,i}\right) + \frac{3\mu}{1+\mu}s\,u_{i,i} - \frac{9(1-2\mu)}{4(1+\mu)^2}\,s^2\right]dV, \qquad (2)$$

$$U_c = G_c\int\limits_{V_c} \left[\frac{1}{2}\left(u^c_{i,j}\,u^c_{i,j} + u^c_{i,j}\,u^c_{j,i}\right) + \frac{3\mu_c}{1+\mu_c}s_c\,u^c_{i,i} - \frac{9(1-2\mu_c)}{4(1+\mu_c)^2}s_c^2\right]dV_c, \qquad (3)$$

$$U_o = 2(1+\mu_o)G_o\int\limits_{V_o}\left(u^{o\,2}_{x,x} + u^{o\,2}_{y,y} + 2\mu_0 u^o_{x,x}u^o_{y,y}\right)d\,V_o. \qquad (4)$$

In formulas (2)–(4) for elastomeric, transitional and nonelastomeric layers respectively: G, G_c, G_o—shear moduli; μ, μ_c, μ_o—Poisson's ratio; s, s_c, s_o—relative hydrostatic pressure; u_i, u_i^c, u_i^o—functions of movements; V, V_c, V_o—volumes of layers; i, j—coordinates x, y, z of rectangular coordinate system; on repeating subindexes summation is carried out, and the comma in subscripts indicate a partial derivative. Using functional (1) requires the mandatory fulfilment of the geometric boundary conditions and continuity conditions (joining conditions) of displacement on the borders between elastomeric, transitional and nonelastomeric layers. For the functional (1) boundary conditions for stresses and forces continuity conditions on the surfaces of layers joints are natural conditions for given loading scheme [7, 11].

For one TRME element the centre of its gravity is chosen as the origin of coordinates (Fig. 2c). Then the geometrical boundary conditions are written as:

$$u_z(x,y,\pm 0.5h) = \mp 0.5\varDelta,\ u^c_z(x,y,\pm 0.5(h+2h_c)) = \mp 0.5(\varDelta + 2\varDelta_c); \qquad (5)$$

conditions of displacement continuity on the surfaces of joining of layers:

$$\begin{aligned} &u_x(x,y,\pm 0.5h) = u^c_x(x,y,\pm 0.5h); \quad u_y(x,y,\pm 0.5h) = u^c_y(x,y,\pm 0.5h);\\ &u_z(x,y,\pm 0.5h) = u^c_z(x,y,\pm 0.5h);\\ &u^c_x(x,y,\pm(0.5h+h_c)) = K_{1_0}x; \quad u^c_y(x,y,\pm(0.5h+h_c)) = K_{2_0}y; \end{aligned} \qquad (6)$$

Considering the expected deformed state and boundary conditions (5), (6) and using a plane section hypothesis, we choose functions of displacement and hydrostatic pressure in the forms:for elastomeric layer: $s \approx C$;

$$u_x = C_1 x\left(4\frac{z^2}{h^2} - 1\right) + K_1 x;\ u_y = C_2 y\left(4\frac{z^2}{h^2} - 1\right) + K_2 y;\ u_z = 6\Delta\left(\frac{z^3}{3h^3} - \frac{z}{4h}\right); \qquad (7)$$

for transitional layer: $s \approx C$; $\quad u^c_x = (K_{1o} - K_1)x\left(4\frac{z^2}{h^2} - 1\right)\frac{1}{2\gamma+\gamma^2} + K_1 x;$

$$u_y^c = (K_{2o} - K_2)y\left(4\frac{z^2}{h^2} - 1\right)\frac{1}{2\gamma + \gamma^2} + K_2 y; \quad \gamma = 2\frac{h_c}{h};$$
$$u_z^c = -4\Delta\frac{z^3}{h^3} + 2B\frac{z}{h}\left(4\frac{z^2}{h^2} - 1\right); \quad B = \frac{0.5\Delta\left((1+\gamma)^3 - 1\right) - \Delta_c}{(1+\gamma)(2\gamma + \gamma^2)}; \tag{8}$$

for non-elastomeric layer:

$$u_{ox} = K_{1o}x; \; u_{oy} = K_{2o}y. \tag{9}$$

Force–displacement dependence is found from a condition of minimum for functional (1):

$$\frac{\partial \Pi(\Delta, \Delta_c, C, C_1, C_2, K_1, K_2, K_{1o}, K_{2o})}{\partial(\Delta, \Delta_c, C, C_1, C_2, K_1, K_2, K_{1o}, K_{2o})} = 0; \tag{10}$$

Influence of intermediate and nonelastomeric layers deformation on force–displacement dependence ($P - \Delta_\Sigma$) will be noticeable only for a very thin elastomeric layers ($\alpha = a/h \gg 10$, $\beta = b/h \gg 10$). For this geometry the system of the Eq. (10) may be simplified, leaving in multipliers and summands only terms proportional α^2, α_c^2, β^2 and β_c^2 and the system (10) falls into two poorly connected subsystems of the equations for deformation Δ and Δ_c definition. For TRME package consisting of n one-type layers having identical geometry and physical and mechanical properties, "force–displacement" dependence is written:

$$\Delta_\Sigma = \Sigma(\Delta + 2\Delta_c) = 0.4\frac{Pn}{F}\left(\frac{hB}{G} + 2\frac{h_c B_c}{G_c}\right);$$
$$B = \frac{1 + \frac{25}{48}\frac{D}{\chi}}{1 + \frac{2.5D}{6+5(1-2\mu)D}}; \; D = \frac{a^2 b^2}{h^2(a^2 + b^2)(1+\gamma)^2}; \chi = \frac{G_o h_o}{Gh(1-\gamma)}; \; \gamma = \frac{2h_c}{h}; \tag{11}$$
$$B_c = \frac{1}{1 + \frac{2.5D_c}{6+5(1-2\mu_c)D_c}}; D_c = \frac{a^2 b^2}{h_c^2(a^2 + b^2)}; \; F = ab.$$

If $h_c \ll h$ and $G_o h_o \gg Gh$, deformations of transitional and nonelastomeric layers in TRME may be neglected. From Eq. (11) for TRME displacement the equation is derived, which for very thin elastomeric layers ($\alpha > 10, \beta > 10$) practically coincides with results of work [11]. If $\alpha \gg 10$ and $\beta \gg 10$, thickness of an elastomeric layer h is commensurable with thickness of transitional layer h_c and Poisson's ratio of a transitional layer material is $\mu_c \approx 0.470$–0.485 [13], displacement of transitional layer Δ_c will be defined only by its volumetric deformation, and influence of distortion strain energy may be neglected:

$$\Delta_c = \frac{Ph_c}{FK_c}; \quad K_c = \frac{2(1+\mu_c)G_c}{1-2\mu_c}. \tag{12}$$

In this case for displacement of elastomeric layer Δ from (11) we receive:

$$\Delta = \frac{0.4PhB_1}{FG}; \quad B_1 = \frac{1+\frac{25}{48}\frac{D_1}{\chi_1}}{1+\frac{2.5D_1}{6+5(1-2\mu)D_1}}; \quad \chi_1 = \frac{G_o h_o}{Gh}; \quad D_1 = \frac{\alpha^2\beta^2}{\alpha^2+\beta^2}. \tag{13}$$

Displacement of TRME layer:

$$\Delta_\Sigma = \Delta + 2\Delta_c = 0.4\frac{P}{F}\left(\frac{hB_1}{G} + \frac{2h_c}{K_c}\right). \tag{14}$$

Dependences (11)–(13) are applicable if displacement of TRME does not exceed 10–15 %. From formulas (12) and (13) it follows that, depending on values G/G_c and K/K_c under condition of $h > h_c$ no more than an order of magnitude, in the elastomeric layer will be small deformations, and in a transitional layer—average deformations ($\Delta_c/h_c > 0{,}2$) and formula (12) is not applicable. In works [7, 8] it is shown that using a delta method, after simple transformations, on the basis of a formula (12) it is possible to receive dependence "force–displacement" for average deformations in a transitional layer:

$$\text{under compression } P = -K_c F \ln\left(1 - \frac{\Delta_c}{h_c}\right), \text{ under tension : } P = K_c F \ln\left(1 + \frac{\Delta_c}{h_c}\right).$$

Required dependences force displacement for glued TRME will be as follows:

$$\text{under compression : } \Delta_\Sigma^{comp} = \Delta + 2\Delta_c = 0.4\frac{PhB_1}{FG} + 2h_c\left(1 - e^{-\frac{P}{K_cF}}\right), \tag{15}$$

$$\text{under tension : } \Delta_\Sigma^{tens} = \Delta + 2\Delta_c = 0.4\frac{PhB_1}{FG} + 2h_c\left(-1 + e^{\frac{P}{K_cF}}\right). \tag{16}$$

The results of thin TRME compression experiments (up to 10–15 %), when compression load per unit surface can reach 200 MPa, show that the graphics "force–displacement" are essentially nonlinear; it is not possible to describe by the relation (11)–(13) obtained for small deformation [5, 6]. In experimental studies it is shown that shear modulus G and bulk modulus K of elastomeric layer depend on intensity of the specific loading if $s = P/F$ is more than 5 MPa [6, 14]. For taking into account of load intensity influence on "force–displacement" dependence it is proposed in [7] as an approximate solution to take the linear solution (11)–(13), in which instead of moduli G and K to substitute the values $G(s)$ and $K(s)$ which correspond to the average hydrostatic pressure values $s\ (x, y, z)$. For the thin flat elastomeric layers it can be assumed with sufficient accuracy that $s \approx P/F$ (F—the area of plane layer). This approximation will be the better, the less form changing energy of a thin elastomeric layer specific contribution into "force–displacement" dependence. This approach allows calculating of $G(s)$ and $K(s)$ from the volumetric tension—compression experiments, with accuracy of assumptions of smallness of the deformation.

Due to lack of experimental data it is proposed in [7, 13] at first approximation to assume that the dependence of $G(s)$ and $K(s)$ has the same type, and to assume:

$$G(s)/G \approx K(s)/K = 1 + \varphi \cdot s, \tag{17}$$

where factor φ is defined from experiment on pure volumetric compression.

4 Numerical Analysis and Experimental Results

Examples of proposed method application are presented in this part. A flat thin square shape TRME with typical for various engineering structures dimensions is considered. Dimensional data of element and mechanical properties of its materials are: $a = b = 80$ mm, $h = 1.0$ mm, $h_o = 0.2$ mm, $h_c = 0.2$ mm; $G = 1.0$ MPa, $\mu = 0.499$, $G_o = 8 \cdot 10^4$ MPa, $G_c = 1.5$ MPa, $\mu_c = 0.48$. Due to the lack of precision experimental data for calculation the shear and bulk modules in accordance with Eq. (17) factor $\varphi = 0.04$ MPa^{-1} is taken as constant [13].

Figure 3a shows plots of "force–displacement" dependence, calculated without considering physical nonlinearity of elastomeric for the cases: (1) in accordance with formula (13) with $\chi \rightarrow \infty$, without considering the transitional and nonelastomeric layers deformation; (2) in accordance with formulas (12) and (13) for $\chi \rightarrow \infty$, taking into account the transitional layers deformation; (3) taking into account the deformation of the transition and nonelastomeric layers using formulas (12) and (13); (4) in accordance with formulas (13) and (15) taking into account the average strain in the transition layer for compression; (5) in accordance with formulas (13) and (16) taking into account the average strain in the transition layer for tension. The curves are denoted in the same manner.

As it is clear from the plots, small deformation will be under force 150–180 kN action, i.e. 23–28 MPa.

Figure 3b shows plots of "force–displacement" dependence taken into account physical non-linearity of elastomeric. Since there are no experimental data for the transitional layer material moduli $G_c(s)$ and $K_c(s)$ dependence on loading intensity, deformation of the transitional layer do not consider physical nonlinearity. Graphs in Fig. 4 are plotted in accordance with: (1) formula (13), without considering nonelastomeric layers deformation and physical nonlinearity (given for comparison); (2) formula (13) without deformation of the nonelastomeric and transitional layers and taking into account the physical nonlinearity by Eq. (17); (3) considering

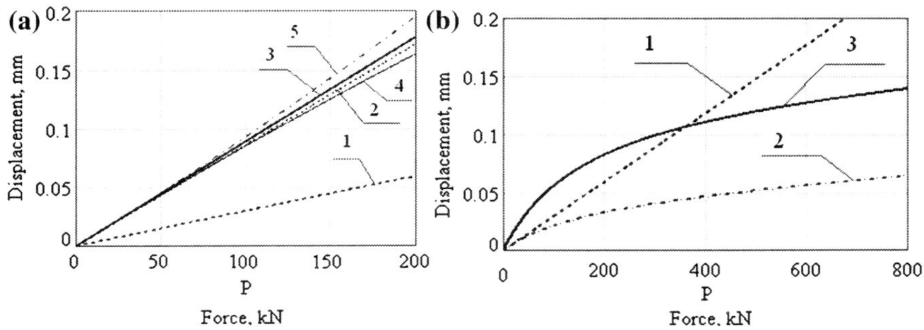

Fig. 3 Plots of "force–displacement" dependence for TRME with dimensions 80 × 80 × 1 mm: **a** without considering physical nonlinearity; **b** with physical nonlinearity accounting

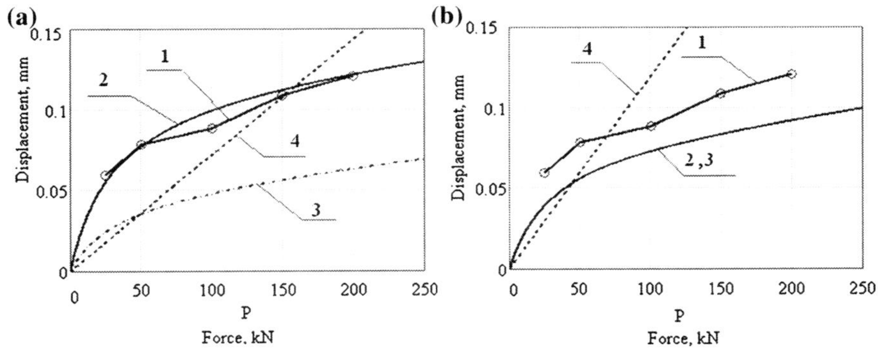

Fig. 4 Plots of "force–displacement" dependence for PRM-210 element: **a** taking into account glue layers; **b** without considering glue layer

the nonelastomeric and transitional layers deformation (14) and physical nonlinearity of the elastomer by (17).

The intensity of external loading is about 125 MPa and deformation is still small.

In Fig. 4 the results of compressive testing and analytical calculation of "force–displacement" dependence for TRME PRM-210 are presented. Experimental investigation was performed at the Moscow Institute of Thermal Technology.

Dimensions of PRM-210: $a = b = 35$ mm, total elastic layer and glue bonded layers thickness is 0.44 mm, $h_o = 0.1$ mm, $G = 0.45$ MPa, $\mu = 0.4981$, $G_o = 8 \cdot 10^4$ MPa (G, μ—measured data). It is assumed that thickness of glue bonding layer is $h_c = 0.08$ mm, then $h = 0.44$, $2h_c = 0.28$ mm, $G_c = 0.8$ MPa, $\mu_c = 0.48$.

Plots in Fig. 4a shows the results of calculation taking into account glue layers with $h_c = 0.08$ mm, $h = 0.28$ mm, plots in Fig. 4b—without considering glue layer ($h = 0.44$ mm, $h_c = 0$). In both cases physical nonlinearity of elastomer is considered. Denoted curves: 1—experimental data, 2—considering nonelastomeric and transitional layers deformation, 3—without considering deformation of transitional layers, 4—without deformation and nonlinearity (for comparison).

Analytical calculation with an adhesive layer and materials nonlinearity show good agreement with experiment. It proves the necessity of taking into account the work of the intermediate adhesive layers.

5 Conclusion

The improved analytical model of a thin rubber-metal element of flat shape is developed in this work based on variational principle of theory of elasticity for weakly compressible materials. The developed model and calculation method allows to take into account all the structural features of such elements (collaboration of elastomeric, transitional adhesive and rigid metal layers, physical nonlinearity of elastomeric material) when calculating the force–displacement dependence of element. In accordance with model "force–displacement" dependence is constructed, which is a necessary stiffness characteristic for optimal design of TRME products for all technical applications. This methodology permits to get a good results, the validity of the model is confirmed by the coincidence of calculation results with experimental data of other researchers.

References

1. Frolov NN, Moldovanov SJ, Lozovoy SB (2011) Mechanics of thin rubber elements, monograph. Kuban State Technological University, Publishing House "Jug", Krasnodar
2. Kelly JM, Konstantinidas DA (2011) Mechanics of rubber bearings for seismic and vibration isolation. Wiley, Chichester
3. Gent AN (2001) Engineering with rubber: how to design rubber components. Carl Hanser Verlag, Munich
4. Bauman JT (2008) Fatigue, stress and strain of rubber components: guide for design engineers. Carl Hanser Verlag, Munich
5. Gusyatinskaya NS (1978) Application of thin layer rubber-metal elements in machine-tools and other engines. Mashinostroenie, Moscow
6. Leykand NA, Lavendel EE, Lvov SV (1981) Experimental research of volume change of rubber under compression and tension. In: Lavendel (ed) Problems of dynamics and strength, Issue 38. Zinatne, Riga, Latvia, pp 49–54
7. Dimnikov SI, Lavendel EE, et al. (1980) Applied methods of design of products from highly elastic materials. Zinatne, Riga
8. Dimnikov SI (1991) Design of rubber elements of structures. Zinatne, Riga
9. Malkov VM (1998) Mechanics of multilayered elastomeric structures. St.-Petersburg University Press, St.-Petersburg
10. Kardashov DA (1982) Synthetic glues. Chemistry, Moscow
11. Lavendel EE (1976) Design of rubber products. Mashinostroenie, Moscow
12. Gonca VF (1970) influence of weak compressibility on solution of theory of elasticity problems for incompressible material. In: Lavendel (ed) Problems of dynamics and strength, Issue 20. Zinatne, Riga, pp 185–189

13. Leykand NA, Lavendel EE, Hrichikova VA (1981) Calculation of compression rigidity of thinlayer rubber-metal elements. In: Lavendel (ed) Problems of dynamics and strength, Issue 38. Zinatne, Riga, pp 57–63
14. Wood LA (1954) Values of physical constants of different rubbers. In: William E. Forsythe (ed) Smithsonian physical tables, 9-th rev. ed. Smithsonian Institution, Washington, USA

Multi Speed Model Updating of Rotor Systems

Manoj Chouksey, Jayanta K. Dutt and Subodh V. Modak

Abstract Accurate Finite Element (FE) models of rotor systems are required for predicting its dynamic behavior, in dynamic design and fault identification purposes. In inverse eigen-sensitivity method of finite element model updating, the limited number of measured eigenvalues available at any spin speed restricts the maximum number of parameters that can be updated. This paper proposes a multi speed model updating method based on inverse eigenvalue sensitivities to update parameters of a rotor system. The method uses eigenvalues obtained at more than one spin speed to update the model. Such an approach allows not only to update more number of parameters but also helps in obtaining a more consistent estimate of updating parameters.

Keywords Rotor systems · Finite element model updating · Inverse eigen sensitivity method

1 Introduction

Finite element model updating methods are used to update parameters of the model of structures [1–3]. Model updating method for rotor systems are not as developed as for non spinning structures [4–6]. A multi-speed formulation of inverse eigen-sensitivity method (IESM) for rotors is proposed with the objective of obtaining a consistent and a more reliable estimate of the updating parameters. The eigenvalues

M. Chouksey (✉)
Department of Mechanical Engineering, SGSITS, Indore 452003, India
e-mail: manoj_chouksey@yahoo.com

J.K. Dutt · S.V. Modak
Department of Mechanical Engineering, IIT Delhi, New Delhi 110016, India
e-mail: jkrdutt@yahoo.co.in

S.V. Modak
e-mail: svmodak@mech.iitd.ac.in

J.K. Sinha (ed.), *Vibration Engineering and Technology of Machinery*,
Mechanisms and Machine Science 23, DOI 10.1007/978-3-319-09918-7_63

of rotor systems change with spin speed. In view of this, it is expected that the model identified via updating is accurate in predicting dynamics of the system at not only one speed but at other speeds also. With a view to address this requirement, it is proposed in this paper to use eigenvalue data of the rotor corresponding to multiple speeds of operation in the updating process. With more amount of experimental data going into the identification process, the condition of the updating problem is expected to improve. This would lead to a more reliable estimate of the updating parameters. Since the proposed formulation uses eigenvalues at multiple speeds, it is called as multi-speed updating method for rotors. This particular approach is presented in this paper for updating those parameters that are spin-speed independent.

2 Multi Speed Updating Method for Rotors

The formulation of multi-speed updating method may be laid from the following equation:

$$\Delta \mathrm{q} \approx \mathbf{S} \times \mathbf{b} \tag{1}$$

In the above expression, '$\boldsymbol{\Delta q}$' is the vector formed by the differences between experimentally identified and corresponding finite element model eigenvalues, '$\mathbf{S}$' is the sensitivity matrix of all the 'm' eigenvalues considered with respect to 'nu' updating parameters the size of which is '$m \times nu$' and '$\mathbf{b}$' is the vector of fractional correction factors for 'nu' updating parameters. The eigenvalue sensitivities may be calculated by following the work of Plaut and Huseyin [7], Adhikari and Friswell [8], Garg [9] among others.

The above equation at a particular spin speed 'Ω' for updating 'nu' parameters by using 'm' eigenvalues, is written as:

$$\{\Delta \mathrm{q}_{\Omega}\}_{m\times 1} \approx [\mathbf{S}_{\Omega}]_{m\times nu} \times \{\mathbf{b}\}_{nu\times 1} \tag{2}$$

If 'n' spin speeds, $\Omega_1, \Omega_2, \cdots, \Omega_n$, are considered for measuring eigenvalues, then the above equation may be written as:

$$\begin{Bmatrix} \Delta \mathrm{q}_{\Omega 1} \\ \Delta \mathrm{q}_{\Omega 2} \\ \vdots \\ \Delta \mathrm{q}_{\Omega n} \end{Bmatrix}_{(n\times m)\times 1} \approx \begin{bmatrix} \mathbf{S}_{\Omega 1} \\ \mathbf{S}_{\Omega 2} \\ \vdots \\ \mathbf{S}_{\Omega n} \end{bmatrix}_{(n\times m)\times nu} \times \{\mathbf{b}\}_{nu\times 1} \tag{3}$$

The above matrix equation consists of complex numbers and therefore should be split into real and imaginary parts. Further the equation may be solved as per following equation:

The solution of Expression (3) for '**b**' is obtained as per the following Expression:

$$\mathbf{b} \approx \mathbf{S}_R^{\dagger} \times \Delta \mathbf{q}_R \tag{4}$$

In this work, the formulation and numerical validation of the multi-speed updating method has been shown by updating spin speed independent parameters only. However, the method may be extended to update both spin speed independent and spin speed dependent parameters. In such a case, the parameters which will depend on spin speed should be considered as new parameters for updating at each of the spin speed considered, whereas parameters which may be considered independent of spin speed of rotation should be taken as the same parameters at each speed of rotation. Thus, the proposed formulation provides a framework to identify, through a single formulation, a rotor system model, with both spin-speed independent and dependent updating parameters, that is applicable to predict the dynamics of the system over a whole range of spin speed.

The error indices corresponding to identification of stiffness and damping parameters may be calculated as per Eqs. (5) and (6). The average error in natural frequencies and modal damping factors can be calculated as per Eqs. (7) and (8):

$$\mathrm{AESC} = \frac{100}{n_s} \sum_{i=1}^{n_s} \left| \left(\frac{p_{RM} - p_{UM}}{p_{RM}} \right) \right|_i \tag{5}$$

$$\mathrm{AEDC} = \frac{100}{n_d} \sum_{i=1}^{n_d} \left| \left(\frac{p_{RM} - p_{UM}}{p_{RM}} \right) \right|_i \tag{6}$$

$$AENF = \frac{1}{n} \sum_{k=1}^{n} \frac{100}{m} \sum_{i=1}^{m} \left| \left(\frac{f_{RM} - f_{UM}}{f_{RM}} \right) \right|_i \tag{7}$$

$$AEMDF = \frac{1}{n} \sum_{k=1}^{n} \frac{100}{m} \sum_{i=1}^{m} \left| \left(\frac{\zeta_{RM} - \zeta_{UM}}{\zeta_{RM}} \right) \right|_i \tag{8}$$

where n is the number of spin speeds considered for finding the eigenvalues.

3 Numerical Example

For illustrating the updating process through simulation, a rotor system with one offset circular disc supported on rolling element bearings as shown in Fig. 1 is considered. The disc, mounted at 0.5 m from the left end support, is 0.05 m thick and has a diameter of 0.4 m. The shaft is 1.4 m long and has a diameter of 0.1 m. The modulus of elasticity of shaft material is assumed to be 200 GPa. For both shaft and disc, the mass density is assumed to be 7,800 kg/m^3.

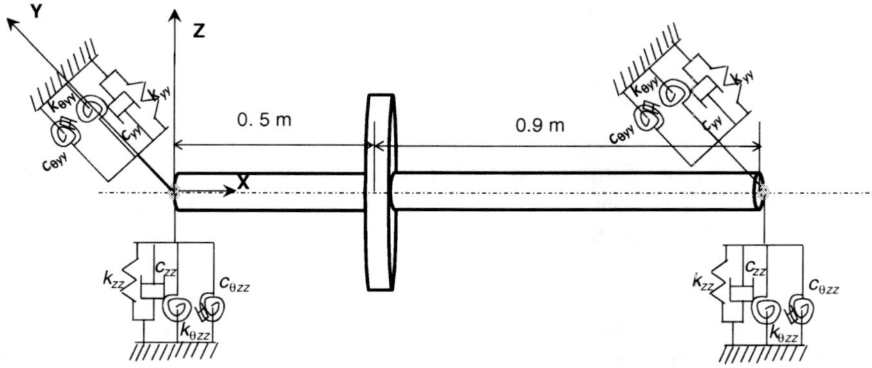

Fig. 1 Rotor-bearing system supported on rolling element bearings at the ends

To show the results, the parameters of the initial model (IM) are updated to match the eigenvalues of the reference model (RM). Two cases are considered for updating the parameters:

Case 1 When all the nine parameters are updated.
Case 2 When seven parameters are updated (excluding the two bearing radial damping parameters due to low sensitivity of eigenvalues to them).

The parametric values of bearing coefficients and shaft material damping coefficient for the reference model of Case 1 are taken same as given in Table 1. For Case 2, the parametric values of bearing radial stiffness, tilt stiffness, tilt damping parameters and coefficient of rotor shaft material damping are taken same as given in the Table 1, except the values of bearing radial damping coefficients (c_{yy}, c_{zz}), which are taken as 50 N-s/m.

To show the results of multi-speed model updating method to update the rotor finite element model, the eigenvalues are obtained for two rotor spin speeds, i.e. at $\Omega_1 = 2{,}000$ rpm and $\Omega_2 = 4{,}000$ rpm. Table 2 gives the results of model updating when initial 3 eigenvalues are considered in the updating frequency range for both of the cases. The table shows that the multi speed model updating method updates the natural frequencies, modal damping factors and the parametric values of the stiffness and damping coefficients nicely for case 1 (updating of nine parameters) as well as for case 2 (updating of seven parameters).

In further discussion, the results for case 2 are shown where all the parameters except the bearing radial damping parameters are updated by using initial three eigenvalues obtained at 2,000 rpm (Ω_1) and 4,000 rpm (Ω_2). The bearing radial damping parameters are excluded from the updating parameters due to their small eigenvalue sensitivities.

Table 1 Bearing and shaft material damping coefficients for IM and the corresponding deviations between the IM and RM

Radial coefficients								
	k_{yy}^L	k_{zz}^L	k_{yy}^R	k_{zz}^R	c_{yy}^L	c_{zz}^L	c_{yy}^R	c_{zz}^R
	In N/m				In N-s/m			
IM parameters	1×10^8	1×10^8	1×10^8	1×10^8	50	50	50	50
% deviations to build RM	−50	−40	−50	−40	200	175	200	175
RM parameters	5×10^7	6×10^7	5×10^7	6×10^7	150	137.5	150	137.5
Tilt coefficients								
	$k_{\theta yy}^L$	$k_{\theta zz}^L$	$k_{\theta yy}^R$	$k_{\theta zz}^R$	$c_{\theta yy}^L$	$c_{\theta zz}^L$	$c_{\theta yy}^R$	$c_{\theta zz}^R$
	In N-m/rad				In N-m s/rad			
IM parameters	1×10^6	1×10^6	1×10^6	1×10^6	50	50	50	50
% deviations to build RM	−70	−60	−70	−60	150	125	150	125
RM parameters	3×10^5	4×10^5	3×10^5	4×10^5	125	112.5	125	112.5
Rotating damping coefficient (η_v)								
$\eta_v = 0.0001$s for the IM; % deviations to build RM = 100; $\eta_v = 0.0002$s for the RM								

Figure 2 shows the variation of cumulative percentage corrections for the updating parameters for case 2 as obtained using the multi-speed model updating method. It shows that the cumulative correction factors converge to finite values showing the success of the model updating method.

Figures 3 and 4 show the variation of percentage error in natural frequencies and modal damping factors against number of iterations as obtained using the multi speed model updating method. It shows that percentage error in natural frequency and modal damping factors at both the speeds considered reduces nearly to zero, signifying the success of the method to update the model of the rotor system. Table 3 is given to show a comparison of the eigenvalues in the updating frequency range and out of updating frequency range and the percentage errors thereof before and after updating. The table shows that the percentage errors in the real and imaginary part of eigenvalues almost vanish after updating the system in both updating range and out of updating range modes.

4 Effect of Measurement Noise

Effect of measurement noise is also studied for the case of the multi-speed updating method. The multi speed model updating method performed well under the presence of measurement noise also and reduces the errors in both natural frequencies and modal damping factors that existed before updating the model. Table 4 shows

Table 2 Multi speed model updating results (Ω_1 = 2,000 rpm, Ω_2 = 4,000 rpm)

		AESC		AEDC		AENF		AEMDF	
	No. of modes	Before updating	After updating	Before updating	After updating	Before updating	After updating	Before updating	After updating
Case1	3	137.50	0.0291	57.64	1.0499	16.75	7.5×10^{-6}	39.34	1.6×10^{-4}
Case2	3	137.50	0.0082	53.88	0.0162	16.44	4.0110^{-5}	34.03	0.0018

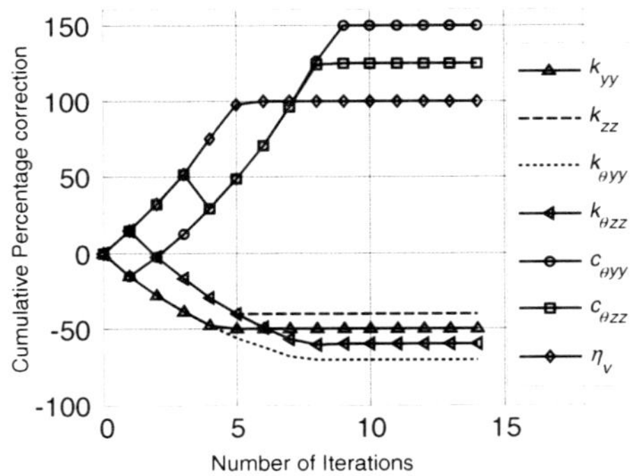

Fig. 2 Variation of cumulative percentage correction

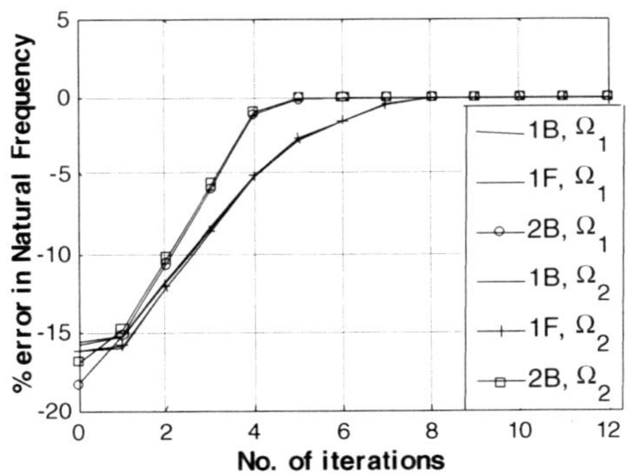

Fig. 3 % error in natural frequency without noise (Ω_1 = 2,000 rpm; Ω_2 = 4,000 rpm)

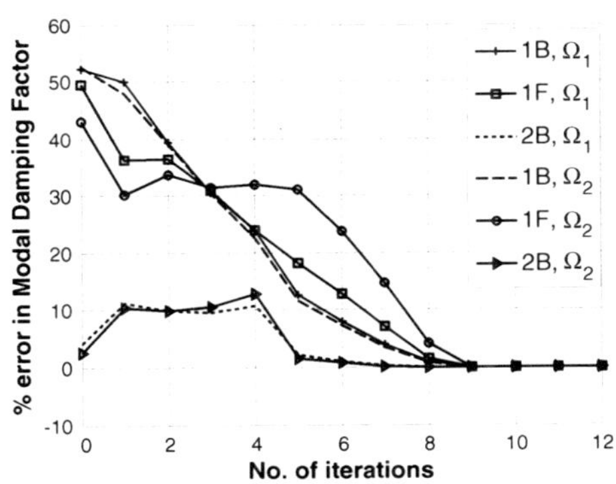

Fig. 4 % error in modal damping factor without noise (Ω_1 = 2,000 rpm; Ω_2 = 4,000 rpm)

Table 3 Comparison of eigenvalues without noise before and after updating in multi speed updating method

No.	Eigenvalues of IM (Hz)		Eigenvalues of RM (Hz)		Eigenvalues of UM (Hz)		% error before updating		% error after updating	
	Re (λ)	Im (λ)	Re (λ)	Im (λ)	Re (λ)	Im (λ)	Re (λ)	Im (λ)	Re (λ)	Im (λ)
1 (Ω_1)	−2.04	81.73	−3.70	70.52	−3.70	70.52	44.81	−15.91	−0.0002	0.00002
2 (Ω_1)	−1.01	82.04	−1.72	70.64	−1.72	70.64	41.39	−16.13	0.0004	−0.00001
3 (Ω_1)	−14.78	271.03	−13.04	229.15	−13.04	229.16	−13.37	−18.27	−0.0003	−0.00009
1 (Ω_2)	−2.56	81.59	−4.68	70.46	−4.68	70.46	45.32	−15.79	−0.0014	−0.00002
2 (Ω_2)	−0.49	82.19	−0.74	70.78	−0.74	70.78	33.79	−16.12	0.0086	0.00001
3 (Ω_2)	−16.40	268.56	−14.41	229.87	−14.41	229.87	−13.81	−16.83	0.0000	0.00009
4 (Ω_1)	−11.66	275.70	−15.89	241.47	−15.89	241.47	26.58	−14.18	−0.0039	−0.0013
4 (Ω_2)	−10.15	277.91	−14.59	240.82	−14.59	240.83	30.45	−15.40	−0.0045	−0.00148

Table 4 Multi speed model updating results ($\Omega_1 = 2{,}000$ rpm, $\Omega_2 = 4{,}000$ rpm) with ±1 % noise; case 2

No. of modes	AESC		AEDC		AENF		AEMDF	
	Before updating	After updating	Before updating	After updating	Before updating	After updating	Before updating	After updating
3	137.50	6.47	53.88	8.21	16.44	0.12	34.03	0.4105

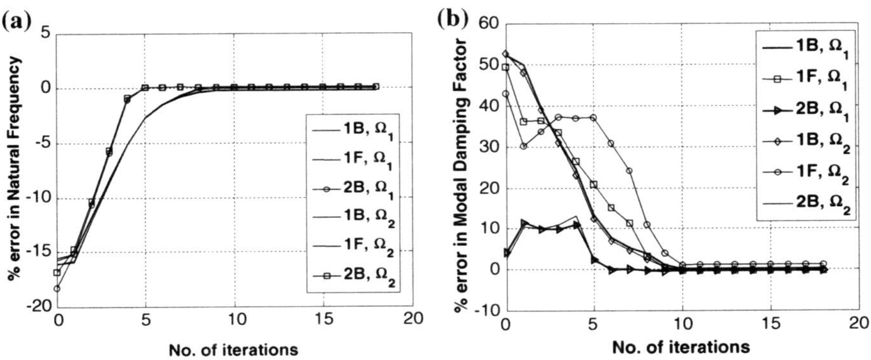

Fig. 5 Variation with ±1 % measurement noise ($\Omega_1 = 2{,}000$ rpm, $\Omega_2 = 4{,}000$ rpm) in **a** Natural frequency **b** Modal damping factors

various average error indices when the rotor finite element model of Case 2 is updated by using initial three eigenvalues obtained at two speeds, i.e. $\Omega_1 = 2{,}000$ rpm, $\Omega_2 = 4{,}000$ rpm. The table shows that the average errors in natural frequencies, modal damping factors, stiffness coefficients and damping coefficients reduce considerably in comparison to the errors which existed before updating. Figure 5a, b shows percentage errors in natural frequencies and modal damping factors respectively in presence of measurement noise (±1 %) obtained using the multi speed model updating method. It shows that percentage error in natural frequency and modal damping factors at both the speeds considered reduces substantially after updating even under the presence of measurement noise.

Table 5 shows that the errors in both natural frequencies and modal damping factors reduced to less than two percent and updated the eigenvalues not only within the updating frequency range but also outside the updating frequency range.

Table 5 Comparison of eigenvalues with ±1 % noise before and after updating (Ω_1 = 2,000 rpm, Ω_2 = 4,000 rpm)

No.	Eigenvalues of IM (Hz)		Eigenvalues of RM (Hz)		Eigenvalues of UM (Hz)		% error before updating		% error after updating	
	Re (λ)	Im (λ)	Re (λ)	Im (λ)	Re (λ)	Im (λ)	Re (λ)	Im (λ)	Re (λ)	Im (λ)
1 (Ω_1)	−2.04	81.73	−3.70	70.52	−3.68	70.42	44.81	−15.91	0.33	0.14
2 (Ω_1)	−1.01	82.04	−1.72	70.64	−1.73	70.82	41.39	−16.13	−0.78	−0.25
3 (Ω_1)	−14.78	271.03	−13.04	229.15	−13.08	228.95	−13.37	−18.27	−0.29	0.08
1 (Ω_2)	−2.56	81.59	−4.68	70.46	−4.69	70.44	45.32	−15.79	−0.16	0.04
2 (Ω_2)	−0.49	82.19	−0.74	70.78	−0.74	70.89	33.79	−16.12	0.87	−0.15
3 (Ω_2)	−16.40	268.56	−14.41	229.87	−14.42	229.73	−13.81	−16.83	−0.08	0.06
4 (Ω_1)	−11.66	275.70	−15.89	241.47	−16.13	241.28	26.58	−14.18	−1.54	0.08
4 (Ω_2)	−10.15	277.91	−14.59	240.82	−14.86	240.58	30.45	−15.40	−1.83	0.10

5 Conclusions

A multi-speed updating formulation based of inverse eigen-sensitivity method for rotors, that uses eigenvalues at more than one spin speed, is proposed with the objective of obtaining a consistent and a more reliable estimate of the updating parameters. The application of method is demonstrated with the help of numerical example of a rotor system. The method also worked well in the presence of measurement noise and reduced the errors in the natural frequencies and modal damping factors considerably. It is expected to be a useful tool to obtain a more consistent estimate of the updating parameters especially in an experimental setting.

References

1. Friswell MI, Mottershead JE (1995) Finite element model updating in structural dynamics. Kluwer Academic Publishers, Dordrecht
2. Mottershead JE, Friswell MI (1993) Model updating in structural dynamics: a survey. J Sound Vibr 167:347–375
3. Modak SV, Kundra TK, Nakra BC (2002) Comparative study of model updating methods using simulated experimental data. Comput Struct 80:437–447
4. Chouksey M, Dutt JK, Modak SV (2013) Updating bearing stiffness and damping coefficients of a rotor system. J Inst Eng (India) Series C 94:113–121
5. Chouksey M, Dutt JK, Modak SV (2013) Model updating of rotors supported on ball bearings and its application in response prediction and balancing. Measurement 46:4261–4273
6. Chouksey M, Dutt JK, Modak SV (2014) Model updating of rotors supported on journal bearings. Mech Mach Theory 71:52–63
7. Plaut RH, Huseyin K (1972) Derivatives of eigenvalues and eigenvectors in non-self-adjoint systems. AIAA J 11:250–251
8. Adhikari S, Friswell MI (2001) Eigen derivative analysis of asymmetric non-conservative systems. Int J Numer Methods Eng 51:709–733
9. Garg S (1973) Derivatives of eigen solutions for a general matrix. AIAA 11:1191–1194

Vibration Control of Semi-active MR Seat Suspension for Commercial Vehicles Using Genetic PID Controller

S. Gad, H. Metered, A. Bassuiny and A.M. Abdel Ghany

Abstract Nowadays, proportional integral derivative (PID) controller is the most popular control algorithm applied in engineering systems and has been generally accepted in industrial control. Recent developments in commercial vehicles have heightened the need for improving the ride comfort. The application of magneto-rheological (MR) dampers in a seat suspension has been shown to provide significant benefits in this area. In most research on seat MR dampers the control application was not quite suitable for the semi-active and nonlinear hysteretic nature of MR dampers. This paper introduces an investigation into the use of a controlled MR damper for a semi-active seat suspension for commercial vehicle, enabling more suitable control. The proposed control system comprises a system controller that computes the desired damping force using a PID controller tuned using genetic algorithm (GA), and a continuous state damper controller that provides a direct estimation of the command voltage that is required to track the desired damping force. A mathematical model of a six degree-of-freedom semi-active seat suspension with human body model using an MR damper is derived. The proposed semi-active seat suspension is compared to a passive seat suspension for prescribed base displacements. These inputs are representative of the vibration of the body (sprung) mass of a passive quarter–vehicle suspension under bump and random-profile road excitation. Control performance criteria such as seat travel distance and head acceleration are evaluated in both time and frequency domains, in order to quantify the effectiveness of the proposed semi-active control technique. The simulated results indicate that the proposed genetic PID of the semi-active MR seat suspension provides a significant improvement in ride comfort.

S. Gad (✉) · H. Metered · A. Bassuiny · A.M. Abdel Ghany
Helwan University, Cairo, Egypt
e-mail: sherifgad_301@yahoo.com

H. Metered
e-mail: hassan.metered@yahoo.com

A. Bassuiny
e-mail: bassiuny@yahoo.com

A.M. Abdel Ghany
e-mail: ghanymohamed@ieee.org

J.K. Sinha (ed.), *Vibration Engineering and Technology of Machinery*,
Mechanisms and Machine Science 23, DOI 10.1007/978-3-319-09918-7_64

Keywords Seat suspension · Vibration control · PID controller · Genetic algorithm (GA) · Heaviside step function · MR damper · Modified Bouc–Wen model

1 Introduction

Due to the exponential growth in different industry fields, the need for developing the commercial vehicles became very important to overcome many challenges for the drivers and to maintain safe and healthy working conditions. Previous literatures on the effects of long-term high-intensity whole-body vibration indicate an increased health risk to the lumbar spine and the connected nervous system of the segments affected [1]. This is due to the biodynamic behavior of the spine, horizontal displacement and torsion of the segments of the vertebral column. Also, Several vibration analysis have been made to compare the vibration dose between the normal car passengers and those who drive commercial trucks and it was found that commercial vehicles drivers are often exposed to vibrations 9–16 times higher than those who drive passenger cars [2]. Along with 10–20 h of driving per day for most commercial vehicle drivers. This work investigates vibration control for seat suspension in commercial vehicles. Semi-active suspension systems using MR and ER dampers have been very effective way to solve the obstacles and problems of active suspension systems. The benefits of semi-active suspension are the high performance with a low power consumption/complexity in comparison to active suspension systems [3].

Magnetorheological (MR) dampers are semi active devices that have recently become very popular in both seat and main suspension for vehicles. Magnetorheological (MR) fluid dampers are controlled through advanced algorithms and the damping force is adapted accordingly. In order to do so, two controllers are needed for MR fluid dampers [4]; system controller is used to calculate the required damping force under specific road conditions, damper controller is needed to estimate the command voltage applied to the MR fluid damper's electromagnetic coil.

2 Modeling of MR Seat Suspension with a Driver

In this subsection, the mathematical model for seated human body is represented, previous studies on vibration control have proposed single degree of freedom (SDOF) for human body but (SDOF) isn't a sufficient model and neglects a lot of dynamic properties of a human body which make the study not accurate enough. Different approaches have been taken by other researchers to describe the human body mathematically by multiple degree of freedom (MDOF) system [5–7], these approaches turned out to be more accurate representation for the human body.

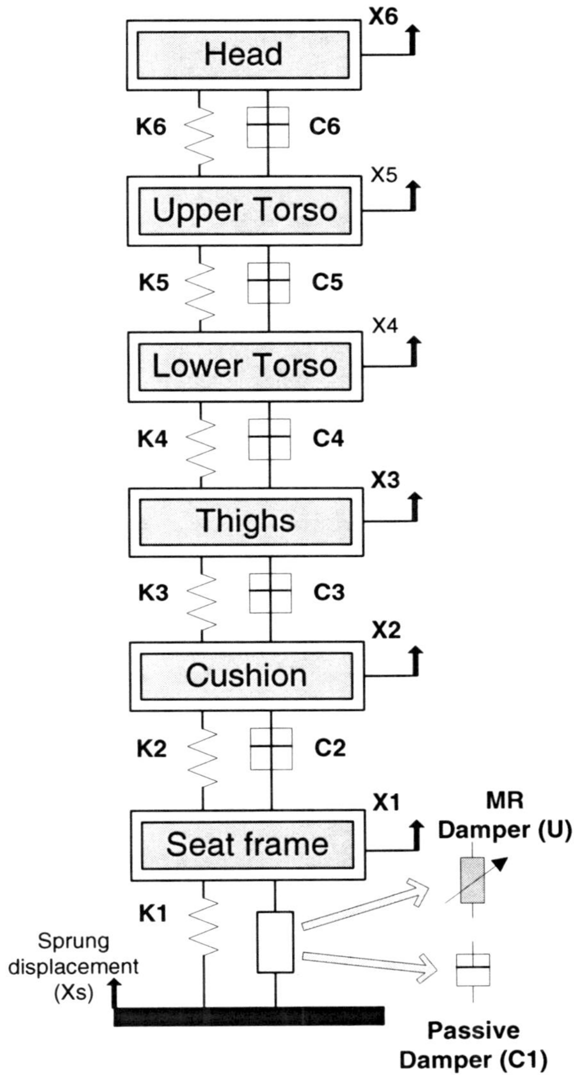

Fig. 1 Seat and human body model

The model used in this study is similar to the previous published model in Ref. [8]. The main advantage of this model is its relative simplicity while taking into account human body weight as a model parameter. The model details are shown in Fig. 1 and the model parameters are given in Table 1. The displacement, velocity, and acceleration for the model body parts in terms time and frequency domains are obtained by solving equations of motion represented by Eqs. (4–9) The effectiveness of the proposed technique is verified in simulation.

Table 1 Seat and human body parameters [8]

Mass (kg)		Damping coefficient (Ns/m)		Spring stiffness (N/m)	
m_1	15	c_1	830	k_1	31,000
m_2	1	c_2	200	k_2	18,000
m_3	12.78	c_3	2,064	k_3	90,000
m_4	8.62	c_4	4,585	k_4	162,800
m_5	28.49	c_5	4,750	k_5	183,000
m_6	5.31	c_6	400	k_6	310,000

The suggested human body consist of four degrees of freedom, It was built based on four different masses interconnected by four sets of springs and dampers, which includes head represented by m_6 connected to the upper torso through k_6 as a spring connection and c_6 as a damper connection, upper torso is represented by m_5 connected to the lower torso through k_5,c_5, lower torso and thighs are m_4, m_3. The arms and legs are combined with the upper torso and thigh, respectively.

The six degrees of freedom seat and human body model is separated from the vehicle suspension. The vehicle suspension is considered as a quarter car model and the parameters of it is similar to Ref. [9], Sprung mass (m_s) = 4,450 kg, Tyre mass (m_u) = 550 kg, sprung stiffness (k_s) = 622,180 (N/m), sprung damping (c_s) = 26,582 (N.s/m), Tyre stiffness (m_u) = 1,705,499 N/m.

In the above model, U is a representation for MR damper force, E_p is the connection force between cabinet floor and seat frame in passive suspension and MR suspension, the following equations for both cases respectively:

$$E_p = c_1\dot{x}_s + k_1x_s\,,\ \text{Passive suspension,} \tag{1}$$

$$E_p = U + k_1x_s\,,\ \text{MR damper suspension,} \tag{2}$$

For facilitating the study, it was assumed that only the vertical motion of the vehicle exists. The equations of motion are driven based of free body diagram analysis and the connection force between human body and seat cushion is obtained as follows:

$$F_h = c_3(\dot{x}_3 - \dot{x}_2) + k_3(x_3 - x_2), \tag{3}$$

where k_3 and c_3 are spring and damper connection for buttocks with thighs respectively. The equations of motion for a six (DOF) human body model with seat can be represented as follows:

Seat frame:

$$m_1\ddot{x}_1 = -c_1\dot{x}_1 - k_1x_1 + c_2(\dot{x}_2 - \dot{x}_1) + k_2(x_2 - x_1) - U + E_P, \tag{4}$$

Seat Cushion:

$$m_2\ddot{x}_2 = -c_2(\dot{x}_2 - \dot{x}_1) - k_2(x_2 - x_1) + F_h, \tag{5}$$

Thighs:

$$m_3\ddot{x}_3 = c_4(\dot{x}_4 - \dot{x}_3) + k_4(x_4 - x_3) - F_h, \tag{6}$$

Lower torso:

$$m_4\ddot{x}_4 = -c_4(\dot{x}_4 - \dot{x}_3) - k_4(x_4 - x_3) + c_5(\dot{x}_5 - \dot{x}_4) + k_5(x_5 - x_4), \tag{7}$$

Upper torso:

$$m_5\ddot{x}_5 = -c_5(\dot{x}_5 - \dot{x}_4) - k_5(x_5 - x_4) + c_6(\dot{x}_6 - \dot{x}_5) + k_6(x_6 - x_5), \tag{8}$$

Head:

$$m_6\ddot{x}_6 = -c_6(\dot{x}_6 - \dot{x}_5) - k_6(x_6 - x_5), \tag{9}$$

The state vector is defined as:

$$\begin{aligned} \mathrm{x} &= [X_1 \quad \dot{X}_2 \quad X_2\ \dot{X}_2 \quad X_3 \quad \dot{X}_3\ X_4 \quad \dot{X}_4 \quad X_5\ \dot{X}_5 \quad X_6 \quad \dot{X}_6] \\ \dot{x} &= AX + BU + DE_P, \end{aligned} \tag{10}$$

$$Y = CX, \tag{11}$$

System matrices are omitted here for the restriction of the paper length, $[A]$ matrix has a dimension of 12*12 and the state variables are listed at explained in matrix $[X]$, Output matrix $[C]$ is [2*12] and the control system defines the error signal as the subtraction of seat frame displacement and sprung mass displacement. $(X_I\text{-}X_s)$.

2.1 MR Fluid Damper

The most commonly used model for representing a hysteretic effect of the MR damper is the modified Bouc–Wen. The model was initially proposed by Bouc and generalized by Wen hence it has been called the Bouc–Wen model [10]. Many researchers adopted Bouc–Wen models [11] due to its extended ability to produce variety of hysteretic behavior. The mathematical model based on the modified Bouc–Wen model is described completely in Ref. [3] and illustrated by Fig. 2.

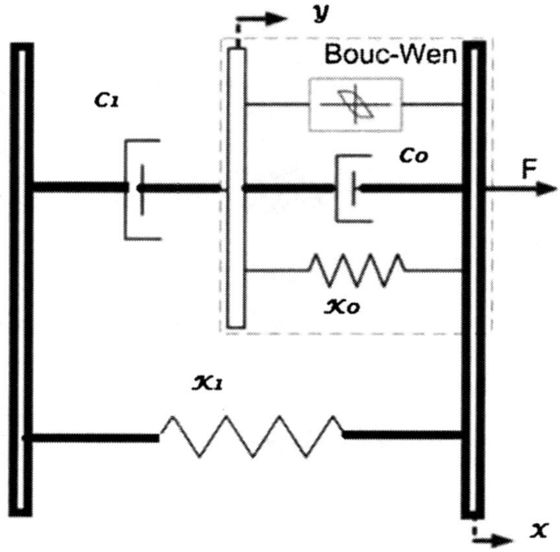

Fig. 2 Modified Bouc–Wen model

3 Controller Design

MR dampers are smart devices and produce the required damping force through two main controllers as explained in Fig. 3, System controller is used to evaluate the required damping force, and damper controller is the algorithm used to apply the DC voltage to the damper coil to change its damping force. In this section the most common methodologies used as system and damper controllers are summarized. System controller is an algorithm used to enable the MR damper to accurately track the desired control force, such as skyhook control [12–14], Linear–Quadratic–Gaussian (LQG) [15, 16] and relative displacement control [17], H infinity control (H∞) is also presented in previous publications [18, 19]. In this paper, a PID controller tuned using genetic algorithm (GA) is proposed to evaluate the ride comfort under different road profiles such as bump and random road excitations. PID controller is selected in this work due to its simplicity, efficiency and the extended abilities to change system response.

PID controller components create output based on measured error of the process being regulated. The suggested optimization technique is via GA. It is a stochastic algorithm based on principles of natural selection and genetics. GA is a global search method that mimics the process of natural evolution. The ride comfort for truck drivers mainly depend on the control algorithm, the main vehicle suspension system and the seat suspension system, the suggested objective function is to analyze the signal of the seat travel distance, minimizing the error between cabinet floor displacement and seat frame displacement, also, minimize the OS, and reduce settling time (t_s) as much as possible (less oscillations effect).

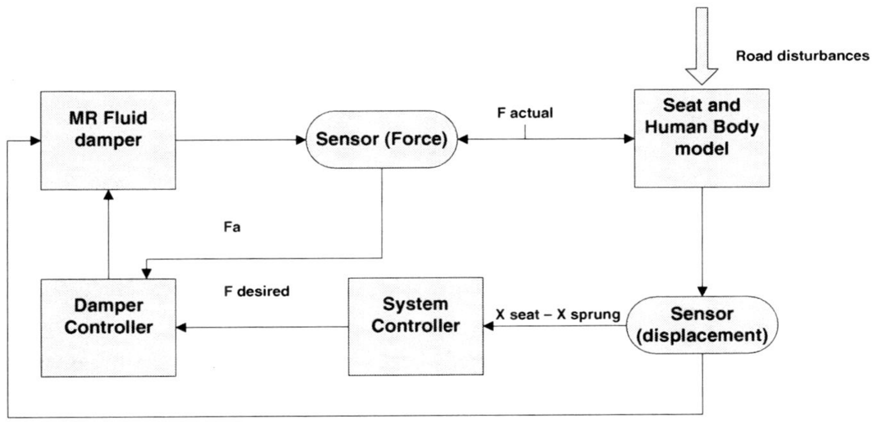

Fig. 3 Block diagram of semi-active MR seat suspension system

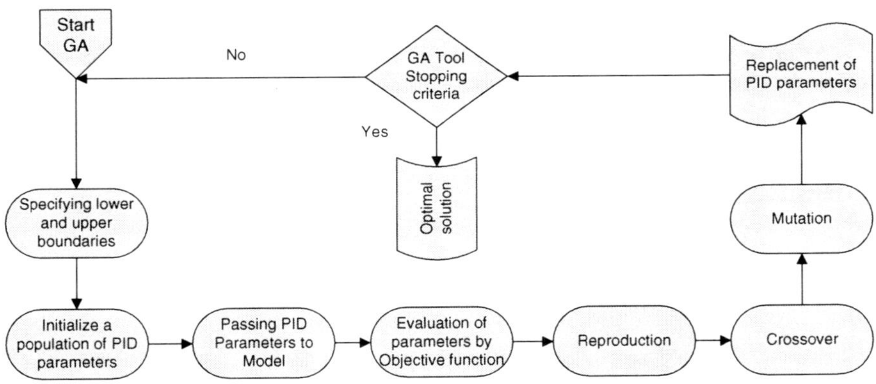

Fig. 4 Optimization via PID genetic algorithm

$$F_{val} = w_1(t_r - t_{rd}) + w_2(OS - OS_d) + w_3(t_S - t_{Sd}) + w_4(e) \tag{12}$$

F_{val} is the optimization target function, w_1, w_2, w_3 and w_4 are constants or weighting factors to specify the optimization priority for (GA) to search, in this study, equal values have been chosen for the weighing constants and it was set as 0.25 for each. Optimization stages via PID genetic algorithm is shown in Fig. 4, Table 2.

Damper controller is used to estimate the command voltage applied to the MR Damper's coil. Different types of damper controllers was proposed in previous work such as Heaviside step functions [20, 21], Signum Function method, and Continuous state controller (CSC) [7]. In this work (CSC) is used due to its structural simplicity and effectiveness, The Continuous state controller allow the

Table 2 Genetic algorithm parameters

Genetic algorithm parameters	Value
Population size	40
Generation	1,000
Fitness scaling	Rank
Crossover technique	Scattered
Probability of crossover	0.8
Mutation technique	Constraint dependent
Upper limit	[100,000 100,000 60,000]
Lower limit	[1,000 1,000 1,000]

command voltage v can vary continuously between minimum and maximum values 0, V_{max} according to the following algorithm.

$$V = \begin{cases} 0 & , \; G(F_d - HF_d)sgn(F_d) < 0 \\ G(F_d - HF_a)sgn(F_a) & , \; 0 \leq G(F_d - HF_d)sgn(F_d) \leq V_{max} \\ V_{max} & , \; G(F_d - HF_d)sgn(F_d) > V_{max} \end{cases}$$

The error signal is $(F_d - HF_a)$ where H is the feedback gain. The CS is only activated when the error and the actual damper force has the same sign.

4 Results and Discussion

In this study, two road profiles are selected [2] to evaluate the performance of the control strategy; the bump profile is introduced and it is represented as follows:

$$z_r = \begin{cases} a\{1 - \cos(\omega_r(t - 0.5))\}, for 0.5 \leq t \leq 0.5 + \frac{d}{V} \\ 0, \; otherwise \end{cases}, \tag{14}$$

where $w_r = 2\pi v_s/D$ and z_b is the half of the bump height, D is the width of the bump and V_S is the vehicle velocity, it is assumed that the vehicle's velocity is 3.08 km/h. The other road excitation used in this study is the random road profile, this type of road excitations is used for determining the frequency domain response of the seat, and the equation representing the random road input is as follows:

$$\dot{z} + \rho_r v_s z = v_s w_n, \tag{15}$$

w_n is the mathematical representation of the weight noise ρ_r is the road roughness parameter, and the vehicle speed assumed for this equation is 72 km/h.

The vibration control comparisons are held between four different suspension techniques: (a) passive oil damper seat suspension, (b) MR passive off damper seat suspension, (c) Classic PID controlled MR damper seat suspension system, (d) Genetic (GA) PID controlled seat suspension system. Peak to Peak (PTP) values are

Table 3 PID parameters for (Classic) and (GA) Matlab optimization tool

PID parameters	Classic PID controller	Optimization via GA
$[K_p\ K_i\ K_d]$	[9298.14 1201.6 300.12]	[12030 87247 55855]

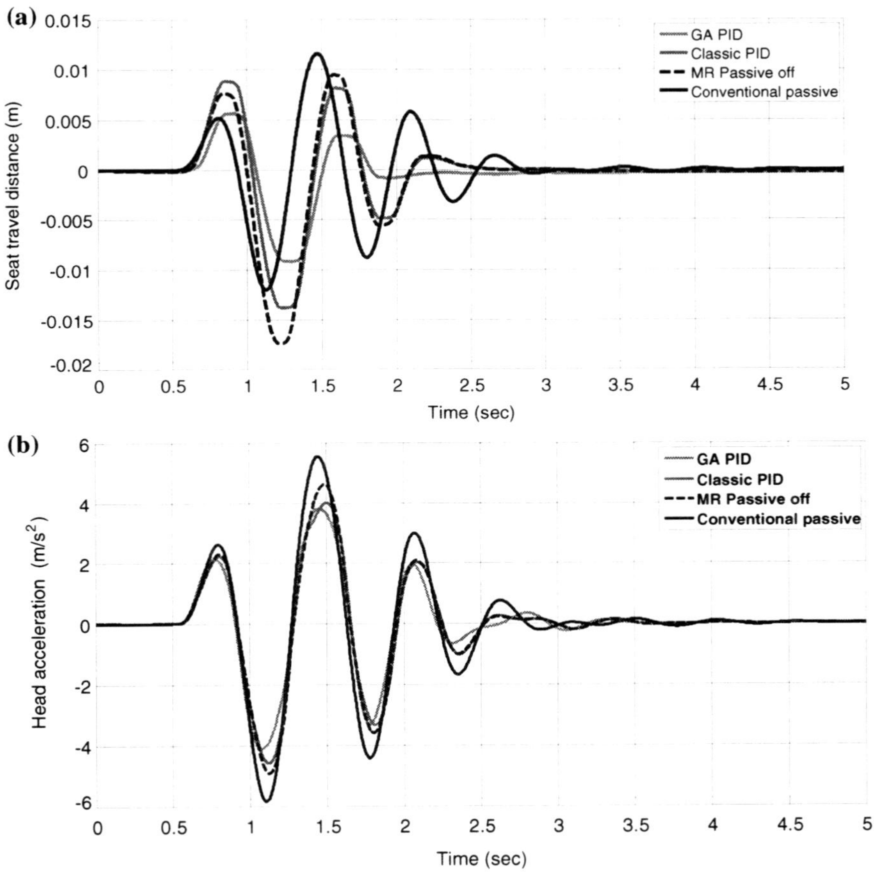

Fig. 5 The time history of system response under road bump excitation. **a** Seat travel distance. **b** Head acceleration

stated in Tables 4 and 5, which shows the effect of the control system on the peak to peak values, the values are compared with the passive suspension to obtain the percentage of improvement for ride comfort of commercial vehicles drivers (Table 3).

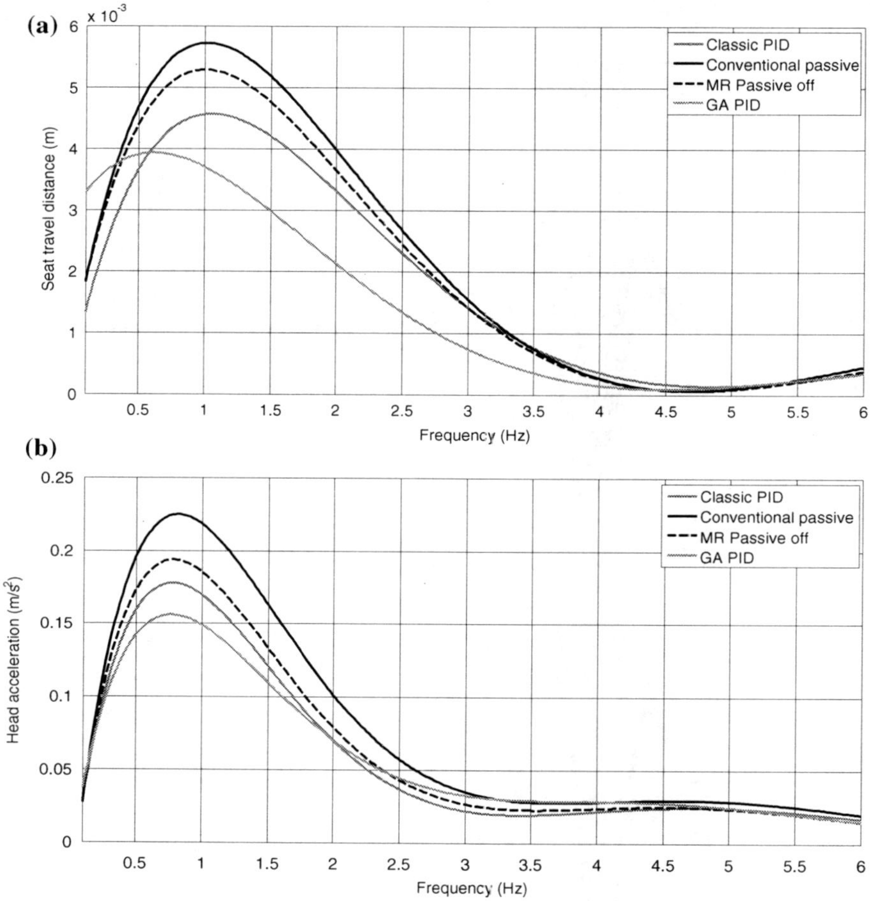

Fig. 6 The time history of system response under random road excitation. **a** Seat travel distance. **b** Head acceleration

Seat travel displacement and *Head acceleration* are two main performance criteria for driver's comfort. In order to achieve a high level of driving comfort, the body acceleration has to be minimized. Figures 5 and 6 give a complete comparison in both time and Frequency domain and indicated that the genetic algorithm PID (GA PID) gives a superior performance. Also, Tables 4 and 5 offer the improvement percentages compared to passive seat system for the peak-to-peak (PTP) and root-mean-square (RMS) for bump and random road profiles.

Table 4 Seat travel distance comparison under bump/random road conditions

Suspension system	Seat travel distance (PTP)	Improvement (%) respect to passive	Seat travel distance (RMS)	Improvement (%) respect to passive
	Bump road excitation		Random road excitation	
Passive	0.026	–	0.0398	–
MR passive OFF	0.025	4 %	0.0376	6.2 %
Classic PID	0.02	21.6 %	0.0361	10.3 %
GA PID	0.017	32 %	0.0341	15 %

Table 5 Head acceleration comparison under bump/random road conditions

Suspension system	Head acceleration (PTP)	Improvement (%) respect to passive	Head acceleration (RMS)	Improvement (%) respect to passive
	Bump road excitation		Random road excitation	
Passive	11.3942	–	0.3459	–
MR passive OFF	9.5472	16.210 %	0.2965	15.6 %
Classic PID	8.6006	24.518 %	0.2745	20.7 %
GA PID	7.9442	30.278 %	0.2697	22.1 %

5 Conclusions

In this paper, genetic algorithms are demonstrated to be an effective searching algorithm for optimum PID controller of semi active seat suspension system. The model used in the optimization process is a seat suspension using MR damper incorporating human body model. The system controller was formulated using Genetic PID (GA) and the continuous state damper controller is considered. The performance obtained by GA PID controller is compared with classic PID controller (Ziegler–Nichols), and MR passive off (zero applied voltage to the damper coil), and conventional passive damper. Control performance criteria such as seat traveled distance and Head acceleration were evaluated in time and frequency domains in order to quantify the suspension effectiveness under bump and random road excitations. This study adopted modified Bouc–Wen model for MR damper, the results indicated that (GA) PID control system is very effective and produces accurate tracking for the desired damping force.

References

1. Mechanical vibrations and shock evaluation of human exposure to whole-body-vibrations (ISO 2631-1, 2nd edn. 1997)
2. Choi SB, Choi JH, Lee YS, Han MS (2003) Vibration control of an ER seat suspension for a commercial vehicle. Int J Vehicle Design 33(1/2/3):2–16, 402–751, Korea
3. Metered H, Bonello P, Oyadiji SO (2009) Vibration control of a seat suspension system using magnetorheological damper, 2009, ASME international design engineering technical conferences & computers and information in engineering conference, San Diego, California, (paper no. DETC2009-86081), 30 Aug—2 Sept (2009)
4. Lam AH, Liao HW (2003) Semi-active control of automotive suspension systems with MR dampers. Int J Veh Des 33:50–75
5. Karkoub MA, Zribi M (2006) Active/semi-active suspension control using MR actuators. Int J Syst Sci 37(1):35–44
6. Du H, Sze KY, Lam J (2005) Semi-active H∞ control of vehicle suspension with MR dampers. J Sound Vib 283:981–996
7. Metered H (2010) Modelling and control of magnetorheological dampers for vehicle suspension systems, school of mechanical, Aerospace and Civil. Phd Thesis
8. Choi S-B, Han Y-M (2007) Vibration control of ER seat suspension with human-body model using sliding mode control, Sound Vib 303:391–404
9. Suna L, Caib X, Yanga J (2007) Genetic algorithm-based optimum vehicle suspension design using minimum dynamic pavement load as a design criterion. Sound Vib 301(1–2):18–27
10. Sahin TE, Cesmeci S (2010) Comparison of some existing parametric models for magnetorheological fluid dampers ‖. Smart Mater Struct 19:035012
11. Kwok NM, Ha QP, Nguyen TH, Li J, Samali B (2006) A novel hysteretic model for magnetorheological fluid dampers and parameter identification using particle swarm optimization‖. Sens Actuators A 132:441–451
12. Karnopp D, Crosby MJ, Harwood RA (1974) Vibration control using semi-active force generators 1974, ASME J Eng Ind 96:619–26
13. Lee HS, Choi BS (2000) Control and response characteristics of a MR fluid damper for passenger vehicles 2000, Intell Mater Syst Struct 11:80–87
14. Ahmadian M, Pare CA (2000) A quarter-car experimental analysis of alternative semi active control methods. Intell Mater Syst Struct 11(8):604–612
15. Liao WH, Wang DH (2003) Semi-active vibration control of train suspension systems via MR dampers 2003. Intell Mater Syst Struct 14:161–172
16. Lai CY, Liao HW (2002) Vibration control of a suspension system via a MR fluid damper 2002. J Vib Control 8(4):527–547
17. Choi BS, Lee HS, Park YP (2002) H∞ control performance of a full-vehicle suspension featuring MR dampers 2002, Veh Syst Dyn 38:341–360
18. Dyke SJ, Spencer BF, Sain MK, Carlson JD (1996) Modelling and control of magnetorheological dampers for seismic response reduction. J Smart Mater Struct 5:565–576
19. Wang DH, Liao WH (2005) Semi-active controllers for MR fluid dampers. J Intell Mater Syst Struct 16(11–12):983–993
20. Sims ND, Stanway R, Peel DJ, Bullough WA (1999) Controllable viscous damping: an experimental study of an ER long-stroke damper under proportional feedback control. J Smart Mater Struct 8:601–615
21. Lee Y, Jeon DA (2002) Study on the vibration attenuation of a driver seat using an MR fluid damper. J Intell Mater Syst Struct 13(7–8):437–441

Identification of Hysteretic Behavior of Magnetorheological Dampers Using NLARX Model

A. Shehata, H. Metered and W. Oraby

Abstract Magnetorheological (MR) dampers are the most promising devices for vibration control applications because it has many advantages such as mechanical simplicity, high dynamic range, low power requirements, large force capacity and robustness. In this paper, a new approach for studying the forward and inverse dynamical behavior of an MR damper using Non-Linear Autoregressive Models with Exogenous Inputs (NLARX) is presented. NLARX is a built-in function related to the identification toolbox, MATLAB/Simulink software, used to identify the nonlinear behavior of dynamic and engineering systems. The training and validation of the proposed model are done theoretically using the data generated from the modified Bouc–Wen model. Validation data sets representing a wide range of working conditions of the damper show that the use of the NLARX model to predict the forward and inverse dynamical behavior of MR dampers is reasonably accurate.

Keywords Identification · Magnetorheological (MR) damper · NLARX model

1 Introduction

A dramatic behavior of the MR fluid dampers change with a magnetic field applied on the dampers. Therefore, these fluids can reversibly and instantaneously switch from a free-flowing liquid to a semi-solid with controllable yield strength when exposed to a magnetic field [1]. There are two categories classified the Identification

A. Shehata (✉) · H. Metered · W. Oraby
Helwan University, Cairo, Egypt
e-mail: ahmedshehatagad@yahoo.com

H. Metered
e-mail: hassan.metered@yahoo.com

W. Oraby
e-mail: waho911@hotmail.com

J.K. Sinha (ed.), *Vibration Engineering and Technology of Machinery*,
Mechanisms and Machine Science 23, DOI 10.1007/978-3-319-09918-7_65

techniques for molding of the dynamic behavior of MR dampers parametric and non-parametric identification techniques. Firstly parametric models are based on mechanical idealization involving representation by an arrangement of springs and viscous dashpots [2–6]. The most popular parametric model is the modified Bouc–Wen model [2]. This is a semi-empirical relationship in which 14 parameters are determined for a given damper through curve fitting of experimental results. Secondly, non-parametric models do not make any presumptions, so the original input-output relationship of the system being modeled. Therefore an important measure of input/output data is utilized to identify the system, enabling the subsequent prediction of the system's response to arbitrary inputs. Proposed models in this category are based on identification toolbox. So far, the nonparametric techniques used for identifying MR dampers have involved neural networks [4–7]. In [7] a feed forward neural network and recurrent neural network were utilized. The preparation of the network depends on the input and output data sets obtained from experimental results or from a mathematical simulation of the MR fluid dampers. The modeling of MR fluid dampers with an adaptive neuro-fuzzy inference system (ANFIS) was proposed in [8]. The damper model developed in [9] took into account temperature variation and studied the effect of MR damper cylinder's surface temperature on the damping force through numerical and experimental studies [9]. CSC was introduced in [10] for an ER damper and was used in [11] for an MR damper, although no comparison was made in either [12] or [11] with alternative control strategies. An alternative method of commanding a continuous voltage signal is through a neural network of the inverse dynamics of the MR damper, as discussed above. This has the advantage over CSC of not requiring a force sensor. Such a strategy was introduced in [13] but it was only compared with simple "on-off" control. In the present paper, identification toolbox is applied for the first time in the parametric identification, modeling (Bouc–Wen) of MR fluid dampers both forward and inverse dynamic behavior model of its damper. The damping force will be articulated as a function of the displacement and input voltage. This functional demonstration could then be used to predict the damping force for unlike operating conditions, under any desired combination of voltage, amplitude, and frequency of the excitation signals, within the limits of the interpolation. The rest of this article is organized as follows: Sect. 2 introduces the modified Bouc–Wen model. The forward and inverse identifications and validations are done in Sects. 3 and 4, respectively.

2 The Modified Bouc–Wen Model

The mechanical idealization of the modified Bouc–Wen model is illustrated in Fig. 1. The phenomenological model is governed by the following equations:

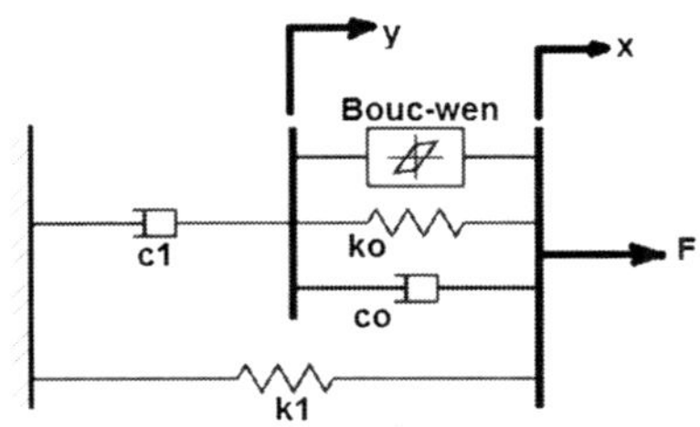

Fig. 1 Modified Bouc–Wen Model

$$F = c_1\dot{y} + k_1(x - xo) \tag{1}$$

$$\dot{z} = -\gamma|\dot{x} - \dot{y}||z|^{n-1}z - \beta(\dot{x} - \dot{y})|z|^n + A(\dot{x} - \dot{y}) \tag{2}$$

$$\dot{y} = \frac{1}{c_o + c_1}[\alpha z + c_o\dot{x} + k_o(x - y)] \tag{3}$$

$$\alpha = \alpha(u) = \alpha_a + \alpha_b u \tag{4}$$

$$c_1 = c_1(u) = c_1a + c_1b\ u \tag{5}$$

$$c_o = c_0(u) = c_oa + c_ob\ u \tag{6}$$

$$\dot{u} = -\eta(u - v) \tag{7}$$

where x and F are the displacement and the force generated by the MR fluid damper respectively; y is the internal displacement of the MR fluid damper; u is the output of a first-order filter and v is the command voltage sent to the current driver. In this model, the accumulator stiffness is represented by k_1 the viscous damping observed at large and low velocities are represented by c_o and c_1, respectively. k_0 is present to control the stiffness at large velocities; x_o is used to account for the effect of the accumulator α is the scaling value for the modified Bouc–Wen model. The scale and shape of the hysteresis loop can be adjusted by γ, β, A and n. The modified Bouc–Wen model is the most popular model for studying the dynamic behavior of MR dampers theoretically. It is used in various engineering systems with different control strategies to implement the MR damper and study system performance [7].

3 Modelling of MR Fluid Dampers Using NLARX

NLARX model can predict the behaviour of non linear system accurately and it has some advantages for application in engineering systems and their controllers. NLARX is considered as a nonlinear black box modelling of the dynamical behaviour of a magnetorheological fluid damper. There are three steps basically required to design the black box to identify the forward and inverse dynamics of the MR damper behaviour:

a. Identification of the inputs and output positions of the MR damper model.
b. Choosing the structural design of the specific black box.
c. Training the selected structural design with the theoretical or experimental data.

3.1 Forward NLARX MR Damper Model

The identification scheme for the direct model is depicted in Fig. 2. The displacement across the damper and the command voltage are considered as the two model inputs and the damper force is the output. The identification flow chart consists of two main operations illustrated as 'sequences 1' and 'sequences 2'. Firstly, sequences 1 gives a primary estimation of the model parameters to characterize the MR damper behaviour while the sequences 2 perform the validation process of the model by minimizing the error between target and predicted forces in the case of forward model and minimizing the error between target and predicted voltage signals in the case of inverse model.

In order to obtain a reliable NLARX model, appropriate training and validation data sets are required. The training data must cover the majority of situations of practical applications in order to ensure that the proposed NARX model trained using these samples can accurately represent the behaviour of the MR damper well. In this paper, training and validation data sets of the proposed NLARX model were obtained by input displacement and voltage signals to the MR damper using the modified Bouc–Wen model as listed in Table 1. Figure 3 represents the time histories of the training and validation data sets. The training data sets are the first 50 % of the total number of the signals elements and the validation data sets are the second 50 % of the total number of the signals elements.

The nonlinearity estimator block diagram of the forward model is illustrated in Fig. 4. The first step in estimating black box models is to choose model orders. The definition of orders depends on the type of the model. For NLARX models, the orders are represented by three numbers which define the number of past outputs, past inputs and input delays used for predicting the value of the output at a given time. The set of time-delayed for inputs and output variables defined by the orders are called "regressors". The model orders are chosen by trial and error. So, the determination of the best order for a NLARX model is an important issue to achieve a good fit. In this model, an IDNLARX model of order [2 2 1] and a sigmoid network as the type of nonlinearity are used together. So, X (t − 1), X (t − 2), V (t − 1), V (t − 2) and F (t − 1), F (t − 2) are selected as the standard regressors. To test the prediction quality of the proposed model, a comparison between the target and predicted damping force are done in Fig. 5 according to set 1 and a very good matching between them is achieved.

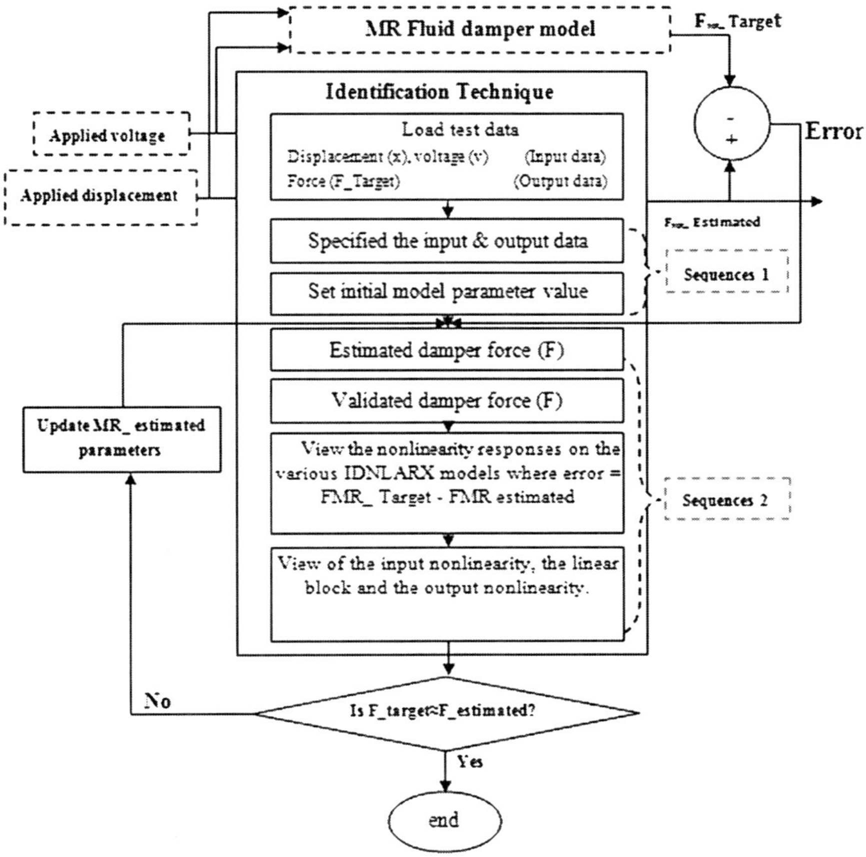

Fig. 2 Identification flow chart of the forward dynamics of MR damper using NLARX

Table 1 Training and validation data sets

No.	Voltage (v)	Displacement (m)	Force (N)
1	3	0.02 sin (4πt)	Generated by MR damper
2	2 + 2 sin (4πt)	0.008 sin (4πt)	Generated by MR damper

Figure 6 shows the effect of the fit value on the quality of the predicted damping force according to set 1 and set 2. It's clearly seen that the higher value of fit, the better matching can be obtained.

Figure 7 gives a complete analysis of the estimation of the nonlinearity of the forward model of the MR damper. It shows that a nonlinear relation between the displacement and the output force. Also, a nonlinear relation between the command

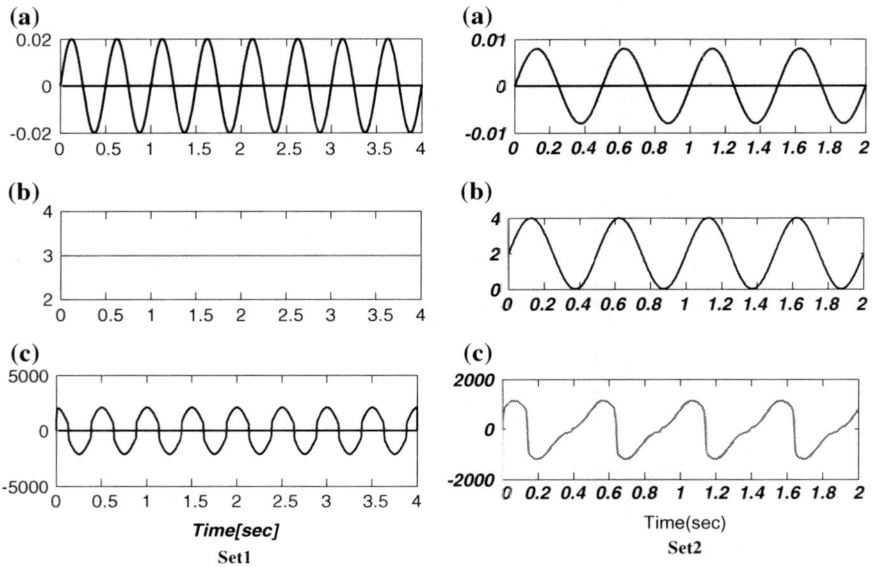

Fig. 3 Time histories of the training and validation data sets **a** displacement, **b** voltage, **c** force

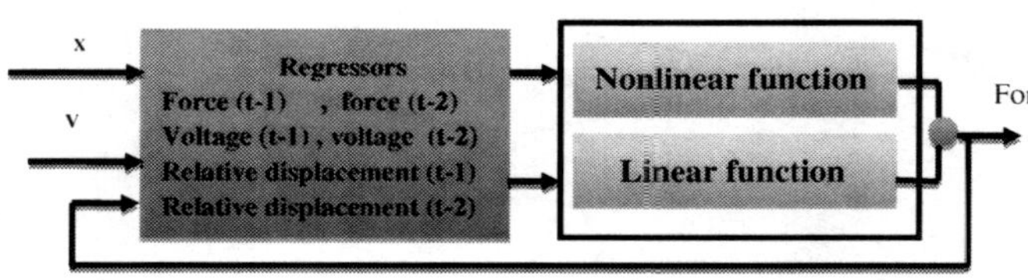

Fig. 4 The nonlinearity estimator block diagram of the forward model

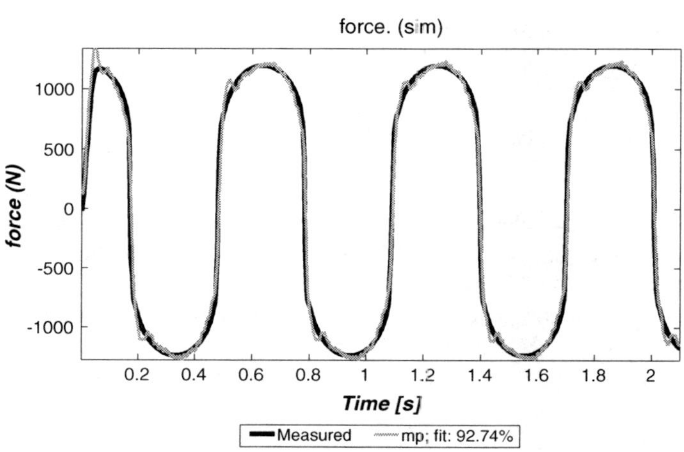

Fig. 5 Comparison between the target and predicted damping force according to set 1

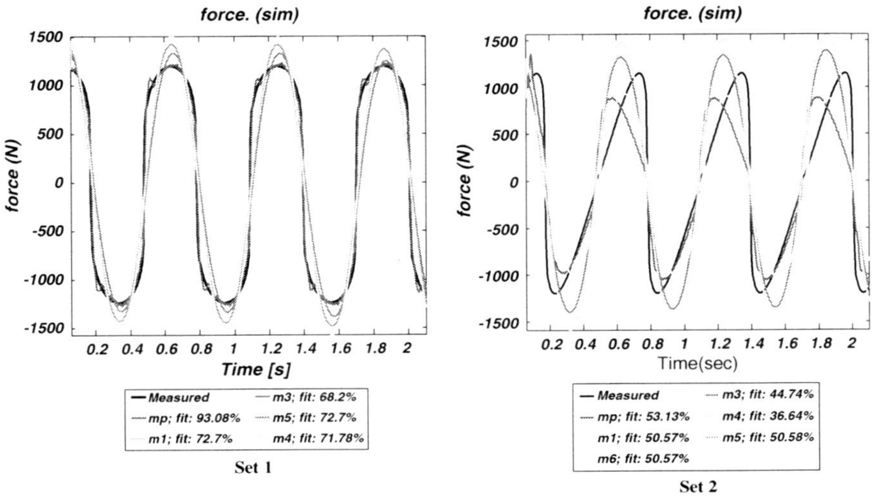

Fig. 6 The effect of fit value on the quality of the predicted damping force according to Table 1

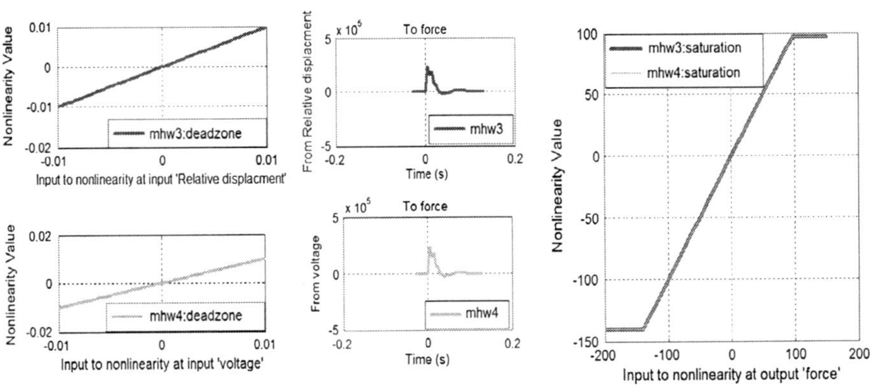

Fig. 7 Analysis of the estimation of the nonlinearity of the forward model

voltage and the output force is noticed. Finally, the NLARX forward model can predict the nonlinearity of the MR damper well.

3.2 Inverse NLARX MR Damper Model

The identification scheme for the inverse model is similar to Fig. 2 but the displacement across the damper and the damper force are considered as the two model

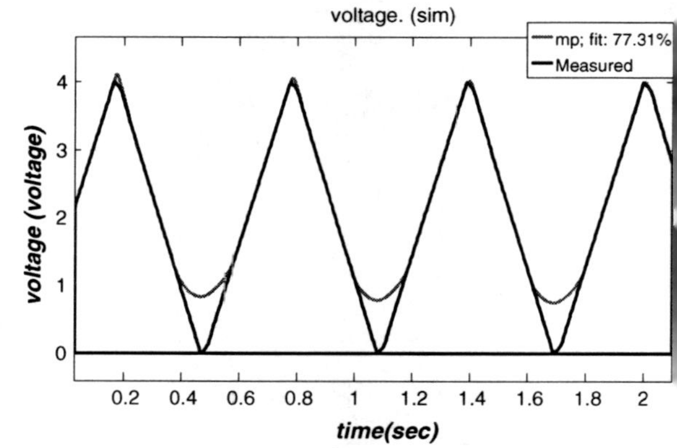

Fig. 8 Comparison between the target and predicted voltage according to set 2

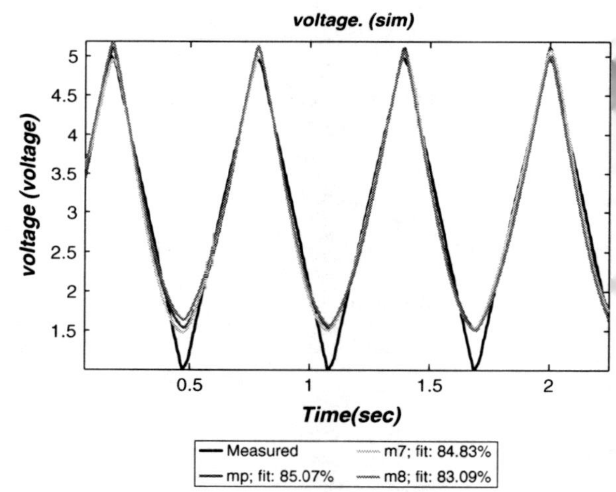

Fig. 9 The effect of fit value on the quality of the estimated voltage according to set 2

inputs and the command voltage is the output. Also, the nonlinearity estimator block diagram of the inverse model is similar to Fig. 2 but the command voltage is the output instead of the damping force. As done in the forward model, an IDN-LARX model of order [2 2 1] and a sigmoid network as the type of nonlinearity are used together again. So, X (t − 1), X (t − 2), V (t − 1), V (t − 2) and F (t − 1), F (t − 2) are chosen as the standard regressors. To test the prediction quality of the inverse model, a comparison between the target and predicted voltage are shown in Fig. 8 according to set 2 and a very good matching between them is achieved.

Figure 9 shows the effect of the fit value on the quality of the estimated voltage according to set 2. It's clearly seen that the higher value of fit, the better matching can be obtained.

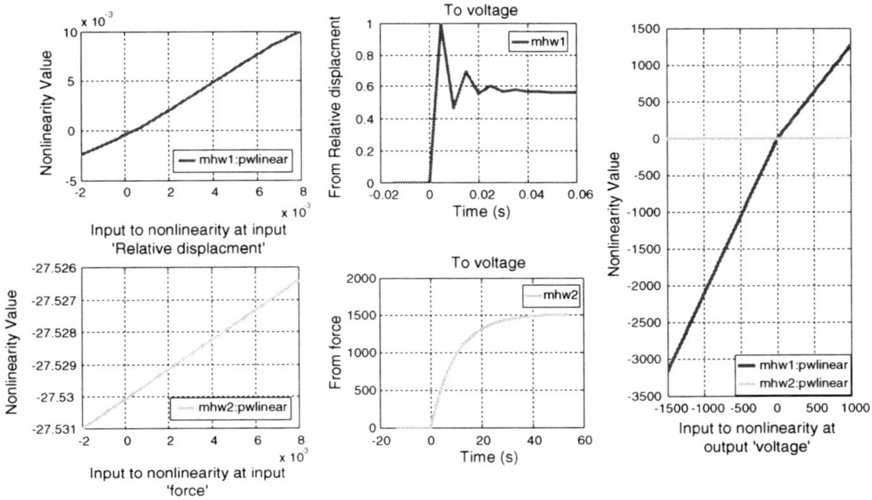

Fig. 10 Analysis of the estimation of the nonlinearity of the forward model

Figure 10 offers a complete analysis of the estimation of the nonlinearity of the inverse model of the MR damper. It shows that a nonlinear relation between the displacement and the command voltage. Also, a nonlinear relation between the damping force and command voltage is given. Finally, the NLARX inverse model can predict command voltage of the MR damper well.

4 Conclusion

In this paper, the NLARX identification of both direct and inverse dynamics of an MR fluid damper has been performed for the first time using based on a simulation behavior of the modified Bouc–Wen model. Both forward and inverse black boxes were considered and their architectures and the learning methods were described. The results obtained by the NLARX model of the MR damper were presented for different operating conditions. These results showed favorable agreement with those obtained by the modified Bouc–Wen model. The proposed technique has therefore been shown to be a fast and reliable system identification method, able to estimate the damping force under any desired combination of voltage and displacement. Moreover, the inverse NLARX model can predict the command voltage under any desired combination of displacement and force. This model will be useful as a damper controller.

References

1. Carlson JD, Weiss KD (1994) A growing attraction to magnetic fluids. Mach Des 66:61–66
2. SpencerJr BF, Dyke SJ, Sain MK, Carlson JD (1997) Phenomenological model for magnetorheological dampers. J Eng Mech: Am Soc Civil Eng 123(3):230–238
3. Wereley NM, Pang L, Kamath G (1998) Idealized hysteresis modeling of electrorheological and magnetorheological dampers. J Intell Mater Syst Struct 9(8):642–649
4. Li WH, Yao GZ, Chen G, Yeo SH, Yap FF (2000) Testing and steady state modeling of a linear MR damper under sinusoidal loading. J Smart Mater Struct 9:95–102
5. Lai CY, Liao WH (2002) Vibration control of a suspension system via a magnetorheological fluid damper. J Vib Control 8(4):527–547
6. Dominguez A, Sedaghati R, Stiharu I (2006) A new dynamic hysteresis model for magnetorheological dampers. J Smart Mater Struct 15:1179–1189
7. Metered H, Bonello P, Oyadiji S (2009) Nonparametric identification modeling of magnetorheological damper using Chebyshev polynomials fits. SAE Paper no. 2009-01-1222
8. Schurter KC, Roschke PN (2000) Fuzzy modeling of a magnetorheological damper using ANFIS. In: Proceedings of the IEEE international conference on fuzzy systems, pp 122–127
9. Batterbee D, Sims ND (2009) Temperature sensitive controller performance of MR dampers. J Intell Mater Syst Struct 20:297–309
10. Wang DH, Liao WH (2005) Semiactive controllers of magnetorheological fluid dampers. J Intell Mater Syst Struct 16:983–993
11. Lam AH, Liao HW (2003) Semiactive control of automotive suspension systems with magnetorheological dampers. Int J Veh Des 33:50–75
12. Sims ND, Stanway R, Peel DJ, Bullough WA (1999) Controllable viscous damping: an experimental study of an electrorheological long-stroke damper under proportional feedback control. J Smart Mater Struct 8:601–615
13. Liao WH, Wang DH (2005) Modeling and control of magnetorheological fluid dampers using neural networks. J Smart Mater Struct 14:111–126

Online Roundness Error Identification and Model-Based Monitoring for Rotors

Markus Rossner, Thomas Thuemmel and Heinz Ulbrich

Abstract Condition monitoring is vital for operating rotors safely. The application of model-based approaches allows a more detailed diagnosis than signal-based methods. Yet, model-based methods require identifying the properties of the undamaged rotor accurately. Measurements at single parts and modal analysis help to verify the rotor model. When monitoring rotor displacements with eddy-current sensors, also roundness errors must be considered. For roundness identification at operation speed, a method based on three non-orthogonal eddy-current sensors is presented. Several experiments show that the measurement data can be split into the contents rotor orbit and roundness error. The extracted orbit is evaluated by an academic model-based monitoring system developed for our test rig. Currently, it can evaluate unbalance, bow, misalignments and/or stator contact.

Keywords Rotating machinery · Condition monitoring · Model verification · Roundness error

1 Introduction

Due to the high rotation speeds of rotors, even small errors can have a large impact on the dynamics. Thus, an early identification is of great interest. In the recent years, several authors suggested model-based methods. Currently, there exist sophisticated methods and fault models [1]. Yet, many of the available model-based

M. Rossner (✉) · T. Thuemmel · H. Ulbrich
Institute of Applied Mechanics, Technische Universität München, Boltzmannstraße 15, 85748 Garching, Germany
e-mail: markus.rossner@tum.de

T. Thuemmel
e-mail: thuemmel@tum.de

H. Ulbrich
e-mail: ulbrich@tum.de

J.K. Sinha (ed.), *Vibration Engineering and Technology of Machinery*, Mechanisms and Machine Science 23, DOI 10.1007/978-3-319-09918-7_66

monitoring systems lack the ability to handle with multiple errors or to operate autonomously. These aspects lie within the focus of our research.

Commonly, the rotor orbit is observed by eddy-current sensors. However, the sensor data is also influenced by roundness errors or the so-called rotor runout[1] as the sensors measure the deflection of the rotor surface. For compensation, typically a record of the runout is taken at low operation speed and subtracted from the current measurement [2]. This procedure is simple, but it just can be performed offline.

For an online roundness monitoring of paper calenders, Kiviluoma designed a hand-held device with an accelerometer that is pressed on the calender [3]. Kim and Lee identified roundness errors at a rotor with magnetic bearings via adding several sample runout files to the controller input [4]. Jansen et al. added a setup of three eddy-current sensors at a lathe for identifying roundness errors directly when manufacturing the shaft [5].

This paper basically follows the latter approach and affirms it with several measurements. Before that, an overview of the test rig and the monitoring system is given in Sects. 2 and 3. Section 4 deals with model verification of the undamaged rotor which is essential for model-based monitoring. The roundness identification method is described in Sect. 5 and verified by the measurements in Sect. 6.

2 Rotor Test Rig

The model-based monitoring is developed for the academic rotor test rig shown in Fig. 1. The shaft has a diameter of 25 mm with a bearing distance of 590 mm. Including the disks (a) and the magnetic bearing rotor (near b, hidden), the rotor weights 11.1 kg. The rotor is supported by conventional bearings (c), the magnetic bearing (b) is used as an actuator. The motor (d) has a maximal speed of 3,000 rpm. Since the first bending mode lies at ca. 70 Hz, only subcritical operation is possible. The shaft displacement is measured with eddy current sensors (e).

Figure 1 also shows a sketch of the simulation model. Modelling the rotor with Euler-Bernoulli beam elements and elastic pedestals leads to following ODE [6, 7]:

$$\boldsymbol{M}\ddot{\mathbf{q}}+(\boldsymbol{D}+\boldsymbol{G})\dot{}+(\boldsymbol{K}+\boldsymbol{N})\boldsymbol{q}=\boldsymbol{h} \tag{1}$$

$\boldsymbol{q}$ are the generalized degrees of freedom. $\boldsymbol{M}$, $\boldsymbol{D}$, and $\boldsymbol{K}$ are the matrices of inertia, damping and stiffness, $\boldsymbol{G}$ and $\boldsymbol{N}$ the skew-symmetric gyroscopic matrices. The rotor errors can be modeled as an addend to the generalized force $\boldsymbol{h}$ [8].

[1] The term "runout" is used ambiguously. In geometric dimensioning and tolerancing, it comprehends roundness and concentricity errors. According to the standardized IFToMM terminology, runout does not include concentricity errors, cf. http://www.iftomm.3me.tudelft.nl/2057_1031/frames.html.

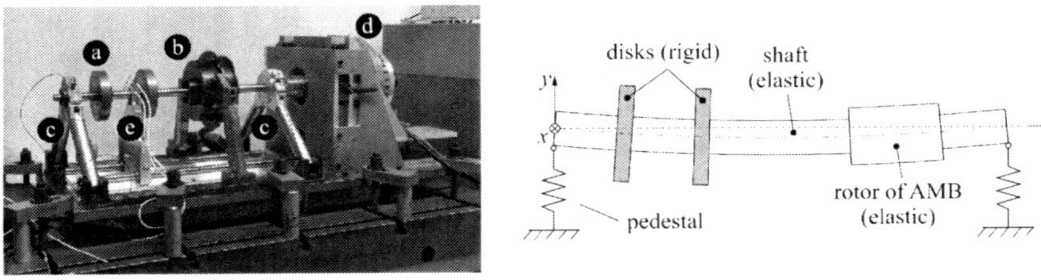

Fig. 1 Picture of the test rig (*left*) and sketch of the simulation model (*right*)

3 Overview About the Monitoring System

For autonomous monitoring, several phases during the monitoring process are needed [6]. Phase A is the initialization. Here, the rotor model and its parameter values are validated (see Sect. 4). This has to be done manually, while the following phases are executed autonomously.

Phase B regularly checks the signals or extracted features like orbit radius and compares them with the thresholds. If one threshold is exceeded, the further phases are activated. Phase C is signal-based. Here, more patterns of the signal are extracted in order to identify the type of error. Then, phase D includes all possible errors in the simulation. The unknown damage parameter are optimized thus that differences between simulation and measurement are minimized. This gives the most probable error. Phase E displays the results and offers post processing.

For monitoring of multiple faults, a damage library has to be built up. Within the library, typical error patterns for phase C have to be defined as well as adequate error models and objective functions for the optimization in phase D.

In the last years, our institute published several papers dealing with different errors. Ginzinger et al. present a monitoring for magnetic bearing rotor drops. Here, the contact forces should be estimated via simulation. For that, the unknown friction coefficient in the rotor-stator contact is identified via optimization [9].

Heckmann et al. show the monitoring of unbalance and misalignment [10]. Here, unbalance influences the 1X and thus the radius of the orbit, while misalignment changes the position of the orbit center.

In Rossner et al. [11], monitoring of bow is added. Bow means that the geometric centers of the cross sections do not confirm with the bearing centerline, e.g. due to plastic deformation. In the model, bow is included as equilibrium position. Both bow and unbalance excite the 1X of the rotor. As shown in this paper, rather small changes in operation speed are sufficient to separate bow and unbalance.

In past, our institute also investigated the monitoring of non-circular tram wheels. Hereto, Thuemmel et al. measured the acceleration of a prepared rail section while the tramcar passes over. In the acceleration spectrum, peaks at higher excitation orders like the 12X, 16X or 18X indicate damaged wheels [12].

4 Identification of Undamaged Test Rig

As the results of model-based monitoring rely on matching of simulation and measurement, any error in the model will corrupt the diagnosis. Thus, a good model of the undamaged rotor is essential. As rotors vary greatly in size, speed and operating conditions, no general procedure for model validation can be given. Yet, the steps performed at our test rig might give hints for validating other setups.

For slightly damped rotors, the main task of model validation is to identify mass and stiffness distribution. Hereto, the rotor was dismantled, so the rotor shaft and the disks can be weighted separately. For the shaft, the measured mass agrees with the calculated mass (using geometry and density). Adding the masses of the disks, this leads to a good mass model.

The geometry of the rotor is also used to calculate the stiffness. The two disks (a) in Fig. 1 are rather slim, so it is assumed that they do not stiffen the rotor. The magnetic bearing rotor is long compared to its diameter, so it is modeled as being elastic and it increases the overall stiffness. The stiffness of the pedestal is identified via applying static loads on the bearings and measuring the resulting deflection.

For verifying the overall model, the simulation model and test rig were investigated with modal analysis. At the test rig, first experimental modal analysis (EMA) using the impact hammer was performed. Here, the eigenfrequencies (EF) lie at 71.4/74.0 Hz for the first bending mode and 273/285 Hz for the second mode [6].

Yet, EMA can only be performed at the non-rotating rotor with the available equipment. For identifying the EF in operation, also operational modal analysis and waterfall analysis of a run-up were used. Here, the EF of the first bending mode lie at 67.6 Hz and 70.2 Hz, the second mode could not be identified due to limited speed. These measurements at rotation are more likely the correct values, so the simulation is fine-tuned to represent these values. Overall, the measured EF and eigenmodes are represented well by the simulation model, see Fig. 2.

5 Monitoring of Roundness Errors

5.1 Error Model

The motivation for monitoring roundness errors was given by the measured orbits in [10]: Due to roundness errors, they were rather square-shaped than circular. For getting better orbits, some sections of the shaft of our test rig were revised by grinding the surface. But for performing measurements at arbitrary positions, a roundness detection is necessary.

In the ideal state, the cross section of the shaft is circular, so every point of the surface has the same distance r_0 to the geometric center M, see Fig. 3. In case of roundness error, the distance to the geometric center is angle-dependent, so $r = r(\Psi)$. Ψ denotes the angle within the body-fixed coordinate system C. The

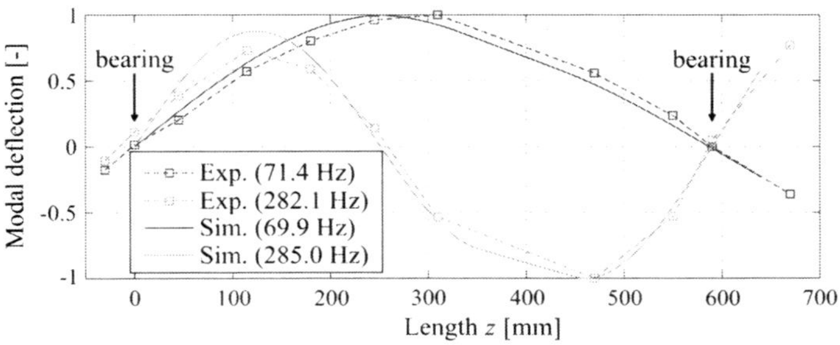

Fig. 2 Eigenfrequencies and mode shapes in simulation and measurement

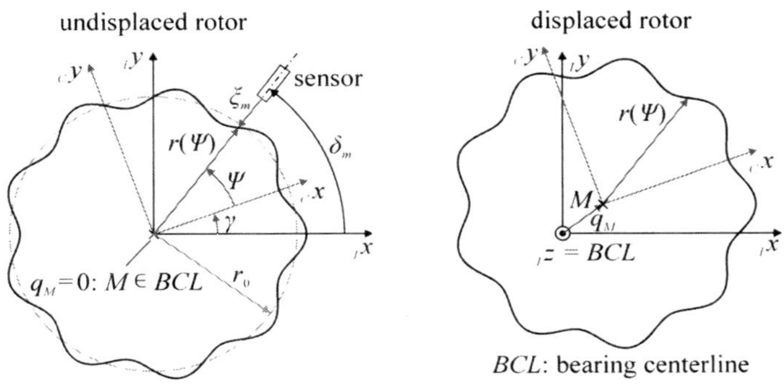

Fig. 3 Definition of quantities for modelling the roundness error

vector $\boldsymbol{q}_M = (x_M, y_M)^T$ describes the position of the geometric center. Note that in reality, $\boldsymbol{q}_M$ and the roundness error $\Delta r = r(\Psi) - r_0$ are very small compared to r_0.

From this definition of $r(\Psi)$, it can be deduced that r is 2π-periodic. So it is recommended to expand r in a Fourier series [4, 5]:

$$r(\Psi) = \sum_k \hat{r}_k \cdot \mathrm{e}^{jk\Psi} \text{ with } k \in \mathbb{Z} \text{ and } \hat{r}_k \in \mathbb{C} \tag{2}$$

The coefficient $\hat{r}_0 = r_0$ denotes the mean radius, so the deviation from the mean radius $\Delta r(\Psi) = r(\Psi) - r_0$ is better suited to describe roundness errors. Since $r(\Psi)$ has to be a real number, $\hat{r}_k$ is the conjugate-complex to $\hat{r}_{-k}$.

As $r(\Psi)$ is defined relative to the geometric center M, $r(\Psi)$ should not influence the position of M. The change of the position of M denoted as $(\Delta x_M, \Delta y_M)$ due to $r(\Psi)$ compared to the circular cross section with radius r_0 can be calculated thus:

$$\begin{pmatrix} \Delta x_M \\ \Delta y_M \end{pmatrix} = \frac{r_0^2}{A} \int_0^{2\pi} \begin{pmatrix} \cos\Psi \\ \sin\Psi \end{pmatrix} \cdot \sum_k \hat{r}_k \cdot \mathrm{e}^{jk\Psi} \mathrm{d}\Psi \overset{!}{=} 0 \tag{3}$$

The variable A is the area of cross section. Here, k = ±1 may lead to a non-zero integral, so it is required that $\hat{r}_{\pm 1} = 0$. This is consistent with the IFToMM nomenclature: A shift of the geometric center is defined as bow and so must not be mixed with roundness errors.

5.2 Separation of Rotor Orbit and Roundness Error

At operation speed, the eddy current signals contain orbit motion (= change of position of geometric center M) as well as roundness errors. For that, a separation routine is necessary. The following procedure has been published similarly by Jansen et al. [5], yet this work is not commonly known in rotor dynamics.

In this procedure, it is assumed that the orbit motion in the inertial frame I can be represented with a Fourier series:

$$_I\boldsymbol{q}_M = \begin{pmatrix} _I x_M \\ _I y_M \end{pmatrix} = \begin{pmatrix} \sum\limits_k \hat{x}_k e^{jk\gamma} \\ \sum\limits_k \hat{y}_k e^{jk\gamma} \end{pmatrix} \tag{4}$$

Here, $\gamma = \Omega t$ denotes the angle of the main rotor rotation. The rotation angle γ, the sensor angle δ_m and the body-fixed angle Ψ form the following equation (Fig. 3):

$$\delta_m = \Psi + \gamma \tag{5}$$

Equation (4) is now projected into sensor plane. As both orbit motion (4) and roundness error (2) are small compared to r_0, they can be superposed. Replacing Ψ by Eq. (5), the measured distance ξ_m between sensor and rotor surface is:

$$\xi_m = \sum_k \underbrace{\left(-\cos\delta_m \cdot \hat{x}_k - \sin\delta_m \cdot \hat{y}_k - e^{-jk\delta_m} \cdot \Delta\hat{r}_k^*\right)}_{\hat{\xi}_{mk}} \cdot e^{jk\gamma} \tag{6}$$

$\Delta\hat{r}_k^*$ means the conjugate-complex to $\Delta\hat{r}_k$. For identifying the parts, at least three sensors are required. When using exactly three sensors, the following set of linear equations can be generated:

$$\underbrace{\begin{pmatrix} -\cos\delta_1 & -\sin\delta_1 & -e^{-jk\delta_1} \\ -\cos\delta_2 & -\sin\delta_2 & -e^{-jk\delta_2} \\ -\cos\delta_3 & -\sin\delta_3 & -e^{-jk\delta_3} \end{pmatrix}}_{\mathbf{A}} \begin{pmatrix} \hat{x}_k \\ \hat{y}_k \\ \Delta\hat{r}_k^* \end{pmatrix} = \begin{pmatrix} \hat{\xi}_{1k} \\ \hat{\xi}_{2k} \\ \hat{\xi}_{3k} \end{pmatrix} \quad \text{for } |k| \geq 2 \tag{7}$$

The $\hat{\xi}_{mk}$ can be gained by fitting the measurement data as Fourier series. For a non-singular matrix $\boldsymbol{A}$, the $\hat{\xi}_{mk}$ can be decomposed into orbit motion $\mathbf{q}_M$ and roundness error $\Delta r(\Psi)$. As one measurement at operation speed is sufficient to identify the roundness error, this method can be performed online. Contrary, the classic runout identification is an offline identification as it requires operation at low speed.

For further investigations, two of the three sensors are mounted orthogonal, i.e. at $\delta_1 = 0°$ and $\delta_2 = 90°$. In this case, the determinant of the system is:

$$\det \boldsymbol{A} = \cos \delta_3 + \sin \delta_3 \cdot \mathrm{e}^{-jk \cdot \frac{\pi}{2}} - \mathrm{e}^{-jk\delta_3} \quad (8)$$

It can be shown that $\det \boldsymbol{A} \neq 0$ is guaranteed if the order k fulfills the condition $|k| \leq 4n_2 - 2$ when the third sensor is mounted at the angle $\delta_3 = n_1/n_2 \cdot 90°$. Here, n_1 and n_2 are coprime integers. At the test rig, $\delta_3 = 162° = 9/5 \cdot 90°$ is chosen, so the 19th order is the first one were $\boldsymbol{A}$ gets singular.

6 Examples

6.1 Example 1: Excitation Due to Resonance Speeds

In the first set, the roundness of a nearly circular cross section is examined. At 750 rpm, no resonance occurs, while at 1,000 rpm, the 4X of the rotor is in resonance with the first bending mode, see Fig. 4. Due to angle addition theorems, the 4X in combination with the 1X component leads to the pentagon shape of the *feigned orbit*.[2]

Using Eq. (7) and calculating the series (2) with the identified $\hat{r}_k$ for $|k| \geq 2$ leads to the roundness error displayed in Fig. 5. They agree for both speeds. As comparison, a classic runout measurement curve recorded at 10 rpm is also displayed (the 1X component was removed as this is the bow component). This curve shows the same characteristics as the online measured ones. The maximal error of 2.5 μm is good when considering the noisy raw data displayed in Figs. 4.

In the investigated revised sections of the shaft, the roundness error is a secondary effect as it is much smaller than typical orbits of the investigated rotor. Therefore, the real orbits are very similar to the feigned orbit in Fig. 4.

6.2 Example 2: Additional Excitation with Magnetic Bearing

The second set of measurements is taken at a cross section which was not ground after turning, so there is a large roundness error. It is dominated by the 3X component as the later experiments will show. Possibly, this is a result of manufacturing the rotor in a three-jaw chuck [13].

---

[2] In fact, the plot of the 90°-sensor data over the 0°-sensor is not the real orbit (= motion of geometric center) as the data is disturbed by roundness error.

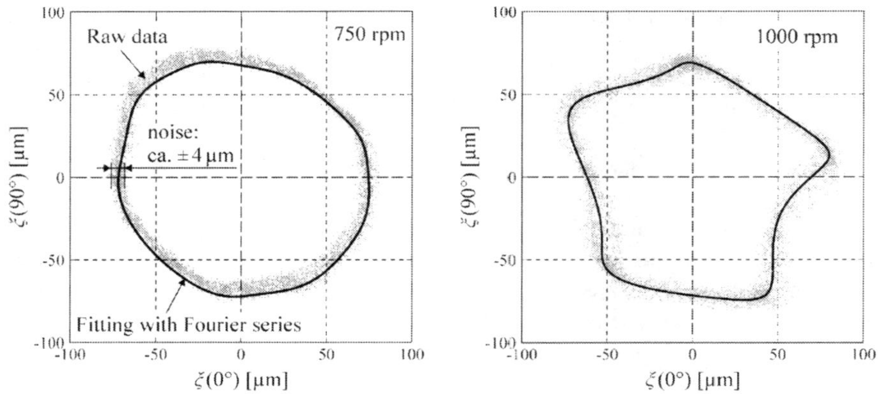

Fig. 4 The feigned orbits for the two speeds

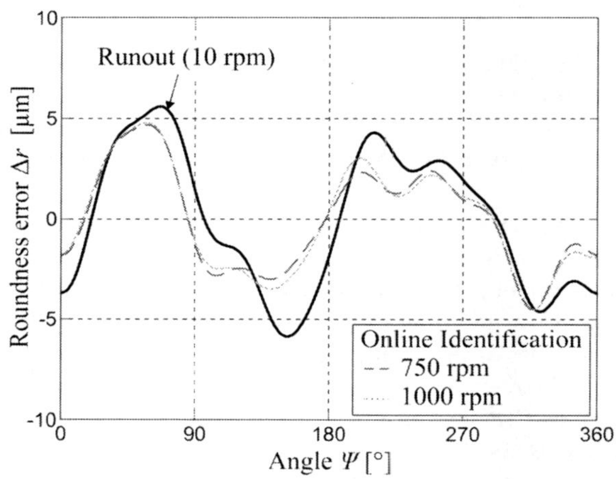

Fig. 5 Online identified roundness compared to classic runout measurement

For this measurement, the speed 910 rpm is chosen, here no resonance of higher excitation orders occurs. In this example, the magnetic bearing excites an additional backward 3X orbit motion. For that, a 3X square wave current was applied.

As shown in Fig. 6, the shape of the feigned orbit does not change basically with this specific excitation of the magnetic bearing, only its orientation has altered. Figure 7 shows the results of the suggested separation algorithm. The left picture shows the real orbit which clearly changes due to the excitation which was not quite obvious in the feigned orbit in Fig. 6. Here, also the curve without excitation differs significantly from the ideal circular shape due to the higher excitation orders. Yet, the higher orders are rather small with amplitudes of 4µm or less. Due to excitation,

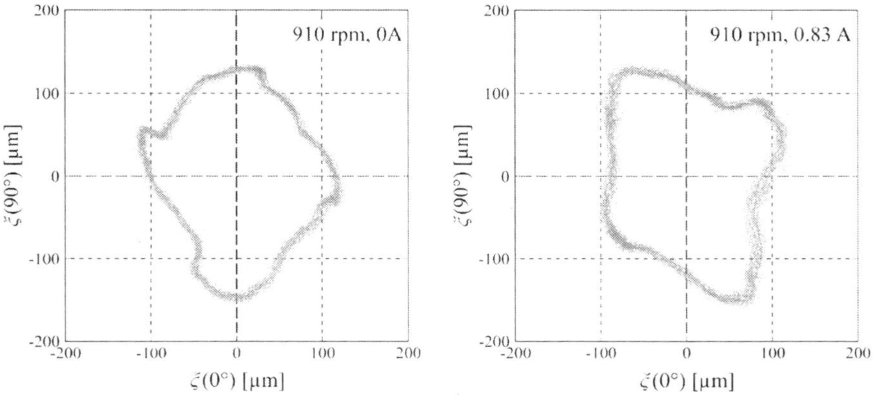

Fig. 6 Feigned orbits without and with additional excitation, cf. footnote 2 in Sect. 6.1

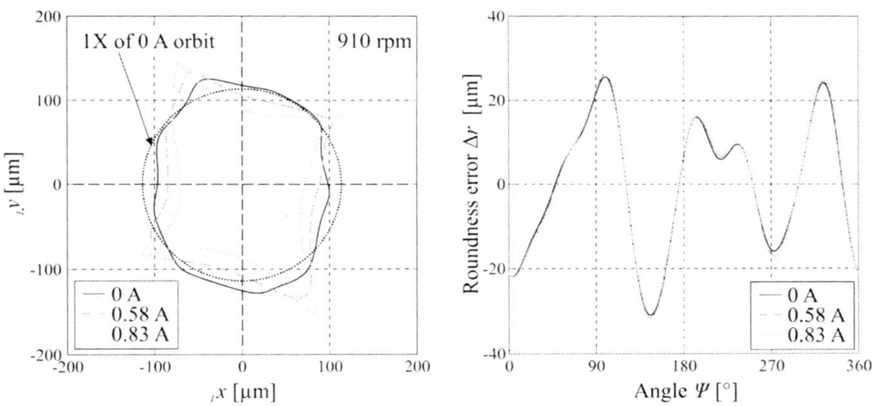

Fig. 7 Identified real orbits and roundness errors

mainly the 3X-component increases. In combination with the 1X due to bow, the real orbits get square-shaped as depicted in Fig. 7 left.

Figure 7 right shows that the identified roundness error is not influenced by modifying the current magnitude. As both roundness error and orbit have 3X components, this proves that the separation according to Eq. (7) works at a real rotor system.

7 Conclusion

As the effects of rotor errors are quite small, also secondary effects like roundness errors have to be considered in monitoring. For that error, a consistent model and a monitoring method were presented. Several experiments show that the excitation orders in the raw sensor data can be separated into orbit motion and roundness error. As the roundness error is identified at operation speed, this is an online method. With the isolated orbit motion, the monitoring can deliver more precise results. So, this work offers a useful expansion to the error library of our monitoring system.

Acknowledgments This research project is supported by the "Deutsche Forschungsgemeinschaft" under grant UL105/37-2.

References

1. Lees AW, Sinha JK, Friswell MI (2009) Model-based identification of rotating machines. Mech Syst Signal Process 23(6):1884–1893
2. Muszynska A (2005) Rotordynamics. Taylor & Francis Group, Boca Raton. ISBN: 978-0-8247-2399-6
3. Kiviluoma P (2009) Method and device for in situ runout measurement of calender thermo rolls. Ph.D. thesis, Helsinki University of Technology. TKK Dissertations 206. ISBN: 978-952-248-260-0
4. Kim C-S, Lee C-W (1997) In situ runout identification in active magnetic bearing system by extended influece coefficient method. IEEE/ASME Trans Mechatron 2 (1):51–57
5. Jansen MJ, Schellekens PHJ, De Veer B (2000) Advanced spindle runout-roundness separation method. Advanced Mathematical and Computational Tools in Metrology. pp 212–219. ISBN: 978-981-02-4494
6. Rossner M, Heckmann B, Thuemmel T, Ulbrich H (2012) Model-based monitoring of rotor systems—Identification of undamaged rotor and different failures. In: 14th International symposium on transport phenomena and dynamics of rotating machinery (ISROMAC). Honolulu, HI, USA
7. Ulbrich H (1986) Dynamik und Regelung von Rotorsystemen. Habilitation thesis, TU München. VDI, Düsseldorf. ISBN: 978-3-18-091568-5 (in German)
8. Bachschmid N, Pennacchi P, Vania A (2002) Identification of multiple faults in rotor systems. J Sound Vib 254(2):327–366
9. Ginzinger L, Sahinkaya MN, Heckmann B, Keogh P, Ulbrich H (2011) On the optimization problem of model-based monitoring. Sci China Technol Sci 54(5):1095–1106
10. Heckmann B, Ginzinger L, Ulbrich H (2010) Model-based identification of unbalance and angular misalignment in a rotor system. In: 13th International symposium on transport phenomena and dynamics of rotating machinery (ISROMAC). Honolulu, HI, USA
11. Rossner M, Heckmann B, Thuemmel T, Ulbrich H (2013) Modellbasiertes Monitoring von Schlag und Unwucht am Rotor. In: 10. Internationale Tagung Schwingungen in rotierenden Maschinen (SIRM), Berlin, Germany (in German)
12. Thümmel T, Mayr G, Waubke H, Breindl C (2000) Detektion unrunder Räder an Straßenbahnen mit einer Gleismeßstelle. VDI-Berichte 1568: Dynamik von Fahrzeug und Fahrweg. pp 367–386. ISBN: 978-3-18-091568-1 (in German)
13. Rahmann M, Ito Y (1985) Stability analysis of chatter vibration in turning process. J Sound Vib 102 (4):515–525
14. Ulbrich H (1988) New test techniques using magnetic bearings. In: Proceedings of the first international symposium on magnetic bearings, ETH Zürich, 6-8 June 1998

Study on Dynamic Load Sharing Characteristics of Double-Row Planetary Gear Train

Rupeng Zhu, Dongping Sheng, Guanghu Jin, Fengxia Lu, Heyun Bao and Miaomiao Li

Abstract The non-linear transverse-torsional coupled model of double-row planetary gear train is established, and planet's eccentricity error and time-varying meshing stiffness are taken into consideration. The solution of differential governing equation of motion is determined by applying the Fourier series method. The behavior of dynamic load sharing characteristics affected by the system parameters including gear eccentricity error, torsional stiffness of first stage carrier and input rate are investigated. Some theoretical results are summarized as guidelines for further research and design of double-row planetary gear train at last.

Keywords Double-row planetary gear · Load sharing · Time-varying stiffness · Eccentricity error · Torsional stiffness

R. Zhu (✉) · D. Sheng · G. Jin · F. Lu · H. Bao · M. Li
College of Mechanical and Electrical Engineering, Nanjing University of Aeronautics and Astronautics, Nanjing 210016, China
e-mail: rpzhu@nuaa.edu.cn

D. Sheng
e-mail: dong.ping.sheng@nuaa.edu.cn

G. Jin
e-mail: meeghjin@nuaa.edu.cn

F. Lu
e-mail: meefxlu@nuaa.edu.cn

H. Bao
e-mail: baoheyun@nuaa.edu.cn

M. Li
e-mail: limiaomiao@nuaa.edu.cn

J.K. Sinha (ed.), *Vibration Engineering and Technology of Machinery*,
Mechanisms and Machine Science 23, DOI 10.1007/978-3-319-09918-7_67

1 Introduction

Due to compact space, low noise and vibration, planetary gear train plays a very important role for power transmission and already applied widely in automotive, aerospace, marine and helicopter et al. Planetary gear sets contain a number of planets to split the input torque applied to one of their central members. This effectively reduces the force transmitted at each gear mesh, and allows the gear set to occupy less space.

During the past several decades, some significant studies and methods about solving the load sharing problem in planetary gear trains have been done and proposed. Seager [1] proposed a simplified model to show the importance of a floating sun gear for three to six planets planetary gear sets. Kahraman [2] constructed a dynamic mathematical model of a planetary gear train, which could be set to an arbitrary number of planets. In his 1999 paper, Kahraman [3] considered load sharing of planetary gear sets again, both in a mathematical model and in experimental work, and showed reasonable agreement between his experimental results and mathematical. Later, Bodas and Kahraman [4] utilized a finite element model of a planetary gear set to study the effects of various errors on planet load sharing behaviors. Ligata et al. [5] did further work along the lines of the study of Kahraman [3], but added the numbers of planets and torque as parametric variables in their experiment study. Ligata et al. [6] and Singh [7] present simplified models to predict the planet load sharing. The latest research done by Boguski, Kahraman and Nishino [8] raised a new method to measure planet load sharing and sun gear radial orbit of planetary gear sets.

Although a lot of researches related to single-row planetary gear train has been conducted and many new models were proposed as well, it should be noted, however, that there is very few relevant study on double-row planetary gear set which has been already applied extensively in helicopter main reducer and other high power transmission system. The main purpose of this study is to propose a new model, and analyze the influences and sensibility qualitatively caused by gear eccentricities error, first stage carrier's torsional stiffness and input rotation rate.

2 Modeling of Double-Row Planetary Gear Train

2.1 *Power Transmission and Generalized Coordinates Description*

Double-row planetary gear train consists of two rows of planetary gear set, which are combined together by the first stage's carrier and input shaft of the second stage's sun gear whose rotational stiffness is denoted by K_{c1s2}, as shown in Fig. 1a. In this paper, first stage's sun gear S_1 is assumed to be radially floating. The planets P_i and the ring gears r_i are considered to be flexibly supported. The whole system is

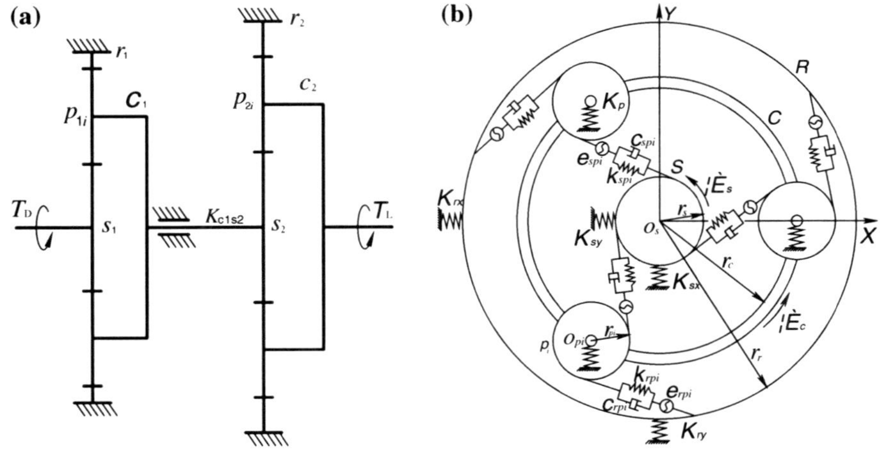

Fig. 1 **a** Power transmission schematic of double-row planetary gear train; **b** Dynamic model of the single-row planetary gear set

driven by prime mover T_D and loaded by external torque T_L, which are applied on the first stage's sun gear and second stage's carrier, respectively.

Given the common properties between the first and second stage planetary gear set, single-row planetary gear set is selected to describe dynamic model as shown in Fig. 1b. According to the practical application, flexibly supported ring gear has two-degree-of-freedom (2 DOFs), namely H_{ri} and V_{ri}, and its related supporting stiffness is denoted by K_{ri}. Sun gear has 3 DOFs, including θ_{si}, H_{si} and V_{si}. Planet gears are supported by flexible pin with the bearing stiffness of K_{pi} and each of them has 3 DOFs, namely θ_{pi}, ζ_{pi} and η_{pi}. The second stage's carrier only has one rotational degree of freedom in view of its restricted movement in the radial direction and given by θ_{c2}. The mesh damping is denoted by C_i. Based on the analysis above, the integrated double-row planetary gear system, with N planets in first stage and M planets in second stage, has $12 + 3(N + M)$ generalized coordinates.

2.2 Geometric Relations of Equivalent Gear Eccentricity Error and Translational Displacement on Action Line

(a) Equivalent gear eccentricity error along the line of action

With the aim to simplify the model, all the gears are considered as standard designed involute gear and without profile shifting. Except for the floating components, other gear's eccentricities and static transmission error are taken into account. Figure 2 shows the geometric relationship of the gear's eccentricity error and its equivalent on the line of action which is one of parameters in the meshing force calculation. Taking single-stage planetary gear train as an example for

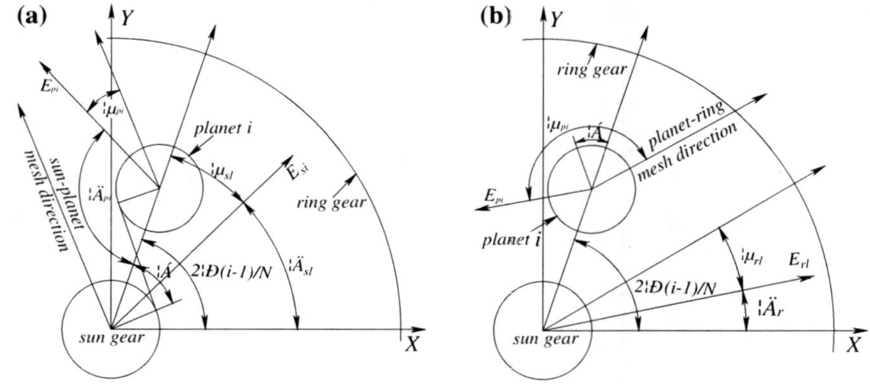

Fig. 2 **a** Planet's equivalent eccentricity error along sun-planet action line; **b** Planet's equivalent eccentricity error along planet-ring gear action line

analysis, Fig. 2a, without considering the errors of the floating sun gear, shows the planet's equivalent eccentricity error along sun-planet action line can be described as

$$\tilde{E}_{pi} = E_{pi}\cos\phi_{pi} = E_{pi}\sin(\omega_{pH}t + \beta_{pi} - \alpha) \quad (i = 1, 2, \ldots, N) \tag{1}$$

where $\tilde{E}_{pi}$ is the equivalent eccentricity error on sun-planet action line, E_{pi} is the value of the eccentricity, ω_{pH} is the eccentricity error's relative frequency, and equals to $(\omega_s - \omega_{\mathrm{H}})z_p/z_s$, ω_s is the rotation frequency of the sun gear, z_p and z_s are the tooth number of planet and sun, respectively. β_{pi} is the initial phase angle of the eccentricity, and α is the pressure angle of the gear pair.

By the same method, and known from the Fig. 2b, the planet's equivalent eccentricity error along planet-ring gear action line can be represented as

$$\hat{E}_{pi} = E_{pi}\cos\gamma_{pi} = E_{pi}\sin(\omega_{pH}t + \beta_{pi} + \alpha) \quad (i = 1, 2, \ldots, N) \tag{2}$$

(b) Equivalent translational displacement along the line of action

As the existence of the gear's translational displacement, which largely affects the meshing force, the analysis of equivalent displacement projected along the line of action becomes significantly important, and their geometric relations are shown in the Fig. 3. Likewise, taking the first stage planetary gear set for example, the projections of the translational displacements on the corresponding action line can be represented as

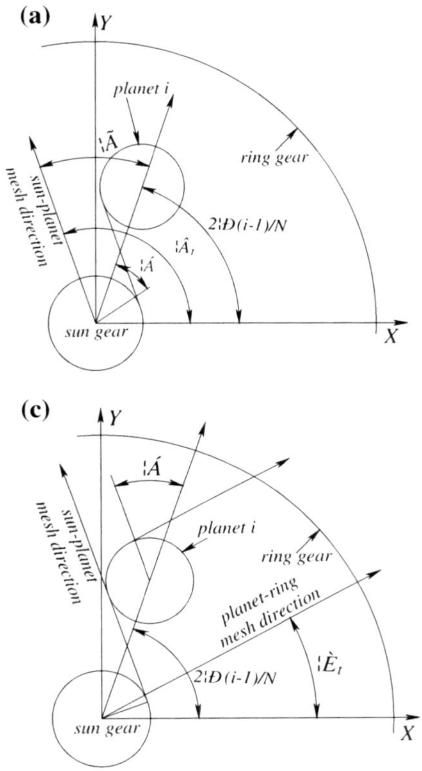

Fig. 3 **a** Sun gear's equivalent translational displacement along sun-planet action line; **b** Planet's equivalent translational displacement along sun-planet and planet-ring action line; **c** Ring gear's equivalent translational displacement along the planet-ring action line

$$\begin{cases} H_{s1}^{eq} = H_{s1}\sin[\alpha_1 - \frac{2\pi}{N}(i-1)],\ V_{s1}^{eq} = V_{s1}\cos[\alpha_1 - \frac{2\pi}{N}(i-1)] \\ \zeta_{1\_pi}^{eq} = \zeta_{1\_pi}\sin\alpha_1,\ \tilde{\zeta}_{1\_pi}^{eq} = \zeta_{1\_pi}\sin\alpha_1 \\ \eta_{1\_pi}^{eq} = \eta_{1\_pi}\cos\alpha_1,\ \tilde{\eta}_{1\_pi}^{eq} = -\eta_{1\_pi}\cos\alpha_1 \\ H_{r1}^{eq} = H_{r1}\sin[\alpha_1 + \frac{2\pi}{N}(i-1)],\ V_{r1}^{eq} = -V_{r1}\cos[\alpha_1 + \frac{2\pi}{N}(i-1)] \\ (i = 1, 2, \ldots, N) \end{cases} \tag{3}$$

where H_{s1}^{eq}, V_{s1}^{eq}, $\zeta_{1\_pi}^{eq}$ and $\eta_{1\_pi}^{eq}$ are the sun gear and planet's equivalent displacement projected on the line of sun-planet action line. $\tilde{\zeta}_{1\_pi}^{eq}$, $\tilde{\eta}_{1\_pi}^{eq}$, H_{r1}^{eq} and V_{r1}^{eq} are their equivalent displacements projected on the line of planet-ring action line.

2.3 Dynamic Meshing and Damping Force

Similarly, first stage's planetary gear set is taken to investigate the meshing and damping forces which are defined as follows

$$\begin{cases} P_{s1\_pi} = K_{s1\_pi}(t)x_{1\_spi} \\ P_{r1\_pi} = K_{r1\_pi}(t)x_{1\_rpi} \\ D_{s1\_pj} = C_{s1\_pi}(t)\dot{x}_{1\_spi} \\ D_{r1\_pj} = C_{r1\_pi}(t)\dot{x}_{1\_rpi} \end{cases} \quad (i = 1, 2, \ldots, N) \tag{4}$$

where P_{s1_pi} and P_{r1_pi} are the dynamic meshing force of sun-planet and planet-ring gear pairs. D_{s1_pj} and D_{r1_pj} are damping force of sun-planet and planet-ring gear pairs. $K_{s1\_pi}(t)$ and $K_{r1\_pi}(t)$ are the time-varying mesh stiffness which could be represented by a Fourier series of fundament frequency, and here only the fundamental harmonic is considered for the sake of simplifying. $C_{s1\_pi}(t)$ and $C_{r1\_pi}(t)$ are the damping coefficients of the engaged sun-planet and planet-ring gear pair. x_{1_spi} and x_{1_rpi} are the relative displacements along the related line of action, namely

$$\begin{cases} x_{1\_spi} = \hat{x}_{1\_spi} + H_{s1}^{eq} + V_{s1}^{eq} - \zeta_{1\_pi}^{eq} - \eta_{1\_pi}^{eq} - e_{1\_spi}(t) \\ x_{1\_rpi} = \hat{x}_{1\_rpi} + H_{r1}^{eq} + V_{r1}^{eq} - \tilde{\zeta}_{1\_pi}^{eq} - \tilde{\eta}_{1\_pi}^{eq} - e_{1\_rpi}(t) \\ \hat{x}_{1\_spi} = r_{s1}\theta_{s1} - r_{1\_pi}\theta_{1\_pi} \\ \hat{x}_{1\_rpi} = r_{1\_pi}\theta_{1\_pi} \\ e_{1\_spi}(t) = E_{s1pi}\sin(\omega_1 t + \beta_{s1pi}) - \tilde{E}_{1\_pi} \\ e_{1\_rpi}(t) = E_{r1pi}\sin(\omega_1 t + \beta_{r1pi}) - \hat{E}_{1\_pi} \end{cases} \quad (i = 1, 2, \ldots, N) \tag{5}$$

Here r_{s1} and r_{1_pi} are the radii of the base circle of the sun and planet, respectively. $\tilde{E}_{1\_pi}$ and $\hat{E}_{1\_pi}$ are already defined in Sect. 2.2. E_{s1pi} and E_{r1pi} are the amplitudes of engaged gear pair's static transmission error. β_{s1pi} and β_{r1pi} are the relevant initial phase angle. ω_1 is the meshing frequency of the internal and external engaged gear pair and equals to $\omega_s z_s z_r/(z_s + z_r)$.

2.4 Governing Differential Equation of Motion

According to the analysis above and d'Alembert principle, double-row planetary gear train's governing differential equation of motion can be derived as

$$\begin{cases}
J_{s1}\ddot{\theta}_{s1} + \sum_{i=1}^{N} (P_{1_spi} + D_{1_spi}) r_{s1} = T_D \\
M_{s1}\ddot{H}_{s1} + \sum_{i=1}^{N} (P_{1_spi} + D_{1_spi}) \sin[\alpha_1 - \frac{2\pi}{N}(i-1)] = 0 \\
M_{s1}\ddot{V}_{s1} + \sum_{i=1}^{N} (P_{1_spi} + D_{1_spi}) \cos[\alpha_1 - \frac{2\pi}{N}(i-1)] = 0 \\
J_{1_p}\ddot{\theta}_{1_pi} - (P_{1_spi} + D_{1_spi}) r_{1_pi} - (P_{1_rpi} + D_{1_rpi}) r_{1_pi} = 0 \\
M_{1_p}\ddot{\zeta}_{1_pi} - K_{1_p}(\zeta_{1_pi} - \theta_{c1} r_{c1}) - (P_{1_spi} + D_{1_spi}) \sin\alpha_1 - (P_{1_rpi} + D_{1_rpi}) \sin a_1 = 0 \\
M_{1_p}\ddot{\eta}_{1_pi} + K_{1_p}\eta_{1_pi} - (P_{1_spi} + D_{1_spi}) \cos\alpha_1 - (P_{1_rpi} + D_{1_rpi}) \cos a_1 = 0 \\
M_{r1}\ddot{H}_{r1} - \sum_{i=1}^{N} (P_{1_rpi} + D_{1_rpi}) \sin[\alpha_1 - \frac{2\pi}{N}(i-1)] + K_{r1} H_{r1} = 0 \\
M_{r1}\ddot{V}_{r1} - \sum_{i=1}^{N} (P_{1_rpi} + D_{1_rpi}) \cos[\alpha_1 - \frac{2\pi}{N}(i-1)] + K_{r1} V_{r1} = 0 \\
J_{c1}\ddot{\theta}_{c1} + \sum_{i=1}^{N} K_{1_p}(H_{1_pi} - \theta_{c1} r_{c1}) r_{c1} + K_{c1s2}(\theta_{c1} - \theta_{s2}) = 0 \\
J_{s2}\ddot{\theta}_{s2} + \sum_{j=1}^{M} (P_{2_spj} + D_{2_spj}) r_{s2} + K_{c1s2}(\theta_{s2} - \theta_{c1}) = 0 \\
M_{s2}\ddot{H}_{s2} + \sum_{j=1}^{M} (P_{2_spj} + D_{2_spj}) \sin[\alpha_2 - \frac{2\pi}{M}(j-1)] + K_{s2} H_{s2} = 0 \\
M_{s2}\ddot{V}_{s2} + \sum_{j=1}^{M} (P_{2_spj} + D_{2_spj}) \cos[\alpha_2 - \frac{2\pi}{M}(j-1)] + K_{s2} V_{s2} = 0 \\
J_{2_p}\ddot{\theta}_{2_pj} - (P_{2_spj} + D_{2_spj}) r_{2_pj} - (P_{2_rpj} + D_{2_rpj}) r_{2_pj} = 0 \\
M_{2_p}\ddot{\zeta}_{2_pj} - K_{2_p}(\zeta_{2_pj} - \theta_{c2} r_{c2}) - (P_{2_spj} + D_{2_spj}) \sin\alpha_2 - (P_{2_rpj} + D_{2_rpj}) \sin a_2 = 0 \\
M_{2_p}\ddot{\eta}_{2_pj} + K_{2_p}\eta_{2_pj} - (P_{2_spj} + D_{2_spj}) \cos\alpha_2 - (P_{2_rpj} + D_{2_rpj}) \cos a_2 = 0 \\
M_{r2}\ddot{H}_{r2} + \sum_{j=1}^{M} (P_{2_rpj} + D_{2_rpj}) \sin[\alpha_2 - \frac{2\pi}{M}(j-1)] + K_{r2} H_{r2} = 0 \\
M_{r2}\ddot{V}_{r2} + \sum_{j=1}^{M} (P_{2_rpj} + D_{2_rpj}) \cos[\alpha_2 - \frac{2\pi}{M}(j-1)] + K_{r2} V_{r2} = 0 \\
J_{c2}\ddot{\theta}_{c2} + \sum_{j=1}^{M} K_{2_p}(\zeta_{2_pj} - \theta_{c2} r_{c2}) r_{c2} = -T_L \quad (i = 1, 2, \ldots, N;\ j = 1, 2, \ldots, M)
\end{cases} \tag{6}$$

where J_{s1}, J_{s2}, J_{1_p}, J_{2_p}, J_{c1} and J_{c2} are the moment of inertias of the first and second stage's sun, planet and carrier. In addition, it should be noted that the subscripts like s, p and c represent sun, planet and ring gear, respectively. M_{s1}, M_{1_p}, M_{r1}, M_{s2}, M_{2_p} and M_{r2} are the masses of the related gears. K_{1_p} and K_{2_p} are the planet's bearing stiffness. K_{r1} and K_{r2} are the supporting stiffness of the ring gear. K_{c1s2} represents the torsional stiffness of the first stage carrier. Apart from r_{c1} and r_{c2} which represent the equivalent radii of the two stage's carrier, the symbol r with subscript stands for the radius of the relevant gear's base circle.

The derived equation of motion can be represented in a matrix-vector form which is useful in computing dynamic responses. The matrix-vector form of the Eq. 6 can be written as

$$[\boldsymbol{M}]\ddot{X} + [\boldsymbol{C}]\dot{\boldsymbol{X}} + [\boldsymbol{K}]\boldsymbol{X} = \boldsymbol{F} \tag{7}$$

where $\boldsymbol{X}$ is the generalized coordinate vector, $[\boldsymbol{M}]$ is the mass matrix, $[\boldsymbol{C}]$ is damping matrix, $[\boldsymbol{K}]$ is stiffness matrix, $\boldsymbol{F}$ is excitation vector.

3 Dynamic Load Sharing Characteristics Analysis

In order to analyze the influences to the load sharing behavior caused by system parameters like torsional stiffness and so on, a set of basic parameters are extracted from a real geared system, listed in Table 1. The fluctuation of torque is neglected as it is determined by the characteristics of the prime mover and load, and hard to eliminate or measure.

3.1 Influence Analysis Caused by Planet's Eccentricity Error

Under a certain set of parameters, the numerical solutions of the load sharing factor (LSF) varied with the increase of the planet's eccentricity error at low damping ratio ($\xi = 0.03$) are plotted in Fig. 4. According to Fig. 4(a), it can be observed that the first stage's LSF grows monotonically and significantly with the increase of the eccentricity error of the first stage's planet within the range $1 \leq E_p \leq 30$ μm, however, the second stage's LSF almost keep unvaried. By the same approach, with the increase of the eccentricity of the second stage planet, the LSF of the system is obtained as shown in Fig. 4b, it behaves similarly but has a larger slope which means the eccentricity error of second stage planet is more sensitive to second

Table 1 Parameters of double-row planetary gear set

	First Stage			Second Stage		
	Sun	Planet	Ring	Sun	Planet	Ring
Tooth number	69	31	131	94	36	167
Normal module (mm)	3.5	3.5	3.5	3.5	3.5	3.5
Pressure angle (deg)	20	20	20	20	20	20
Face width (mm)	45	45	45	60	60	60
Number of planet	3			4		
Input power (kW)	1,000					

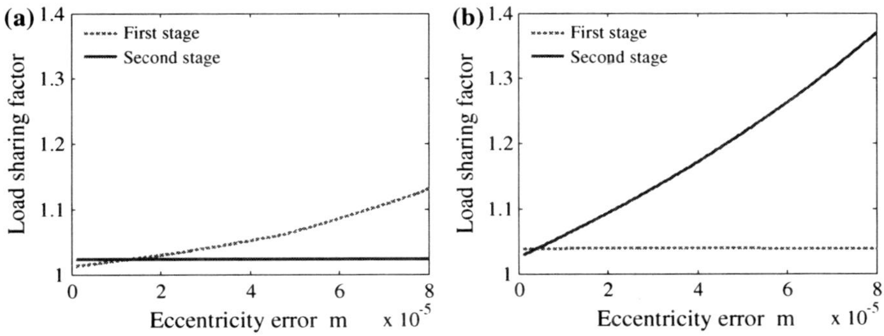

Fig. 4 **a** First and second stage's LSF varied with the increase of the first stage planet's eccentricity error; **b** First and second stage's LSF varied with the increase of the second stage planet's eccentricity error

stage's LSF than that in first second stage. Based on the analysis above, it could be concluded that very weak coupling effect is existed between the two stages and the increasing eccentricity error deteriorates the load sharing performance.

3.2 Infiuences Analysis Caused by Torsional Stiffness

The torsional stiffness of the component which combined the first stage's carrier and second stage's sun gear also act as an important parameter on load sharing characteristics shown in Fig. 5. As the torsional stiffness increase, it is noticeable that the first stage's LSF goes higher in an approximate linear way while the second stage's LSF lowers non-linearly and monotonically. This indicates that, as determining the torsional stiffness in design stage, a balance point is existed, which is valuable during the determination of two stage's load sharing performances.

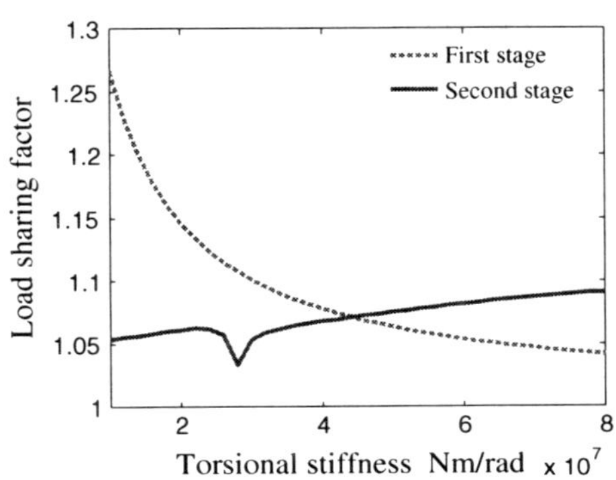

Fig. 5 First and second stage's LSF changed with the increase of the torsional stiffness

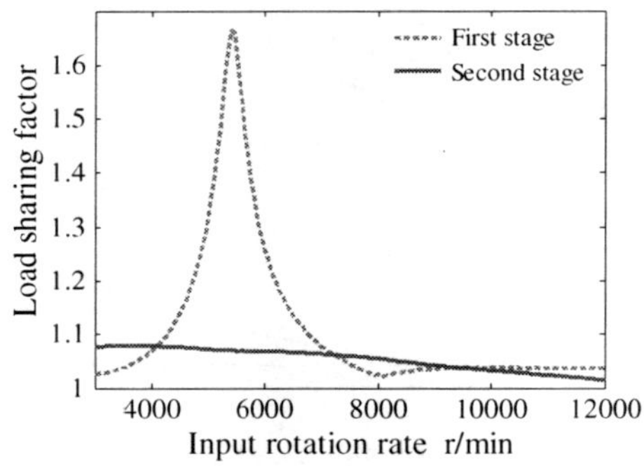

Fig. 6 First and second stage's LSF changed with the increase of the input rotation rate

3.3 Influence Analysis Caused by Input Rotation Rate

Based on the assumption of constant input power, rotation rate produce a strong effect to the first stage's LSF, but little effect to the second stage. As shown in Fig. 6, the input rotation rate versus LSF curves were plotted to illustrate the above point. As the rotation rate increases, a resonant peak of first stage's LSF is observed which happened in the speed range of 5,000 to 7,000 r/min, where the LSF changed intensely and the motion become unstable. However, this kind of phenomenon doesn't appear in the second stage whose LSF decrease smoothly at a very low slope with the increase of the speed. The reasonable interpretation is that the input speed of the second stage which has been reduced by first stage is outside of the second stage's main resonance region.

4 Conclusion

The analysis results enable us to draw the following conclusions

(1) The first stage planet's eccentricity error has large effect on the first stage's load sharing behavior, but produce negligible effect to second stage. The second stage's eccentricity error has the similar behavior.
(2) The planet's bearing stiffness of the first stage generates considerable effects on the first stage's load sharing factor, but has very weak effect to the second stage, and vice versa.
(3) The torsional stiffness of the first stage's carrier has significant effect on the first stage but very limited effect to second stage's load sharing. The balance point needs to be paid more attention during the designing.
(4) With the increase of the input rotation rate, a resonant peak is observed in first stage while the LSF decrease monotonously.

Acknowledgments This work is supported by National Natural Science Foundation of China (51,105,194); Doctoral Program Foundation of Institutions of Higher Education of China (20,113,218,110,017); Priority Academic Program Development of Jiangsu Higher Education Institutions (PAPD). Funding of Jiangsu Innovation Program for Graduate Education (CXZZ11_0199); "the Fundamental Research Funds for the Central Universities (NZ2013303, NZ2014201)".

References

1. Seager DL (1970) Load sharing among planet Gears. SAE Technical paper, pp 700178–700183
2. Kahraman A (1994) Load sharing characteristics of planetary transmission. Mech Mach Theory 29(8):1151–1165
3. Kahraman A (1999) Static load sharing characteristics of transmission planetary gear sets: model and experiment. SAE Trans 108(6):1954–1963
4. Bodas A, Kahraman A (2004) Influence of carrier and gear manufacturing errors on the static planet load sharing behavior of planetary Gear sets. JSME Int J Ser C 47(3):908–915
5. Ligata H, Kahraman A, Sing A (2008) An experimental study of the influence of manufacturing errors on planetary Gear stresses and planet load sharing. ASME J Mech Des 130 (4):041701–041709
6. Ligata H, Kahraman A, Singh A (2009) A closed-form planet load sharing formulation for planetary Gear sets using a translational analogy. ASME J Mech Des 131(2):021007–021012
7. Singh A (2010) load sharing behavior in epicyclic Gears: physical explanation and generalized formulation. Mech Mach Theory 45(3):511–530
8. Boguski B, Kahraman A, Nishino T (2012) A new method to measure planet load sharing and sun Gear radial orbit of planetary Gear sets. ASME J Mech Des 134(7):071002–071009

A Novel Method to Improve the Resolution of Envelope Spectrum for Bearing Fault Diagnosis Based on a Wireless Sensor Node

Guojin Feng, Dong Zhen, Xiange Tian, Fengshou Gu and Andrew D. Ball

Abstract In this paper, an accurate envelope analysis algorithm is developed for a wireless sensor node. Since envelope signals employed in condition monitoring often have narrow frequency bandwidth, the proposed algorithm down-samples and cascades the analyzed envelope signals to construct a relatively long one. Thus, a relatively higher frequency resolution can be obtained by calculating the spectrum of the cascaded signal. In addition, a 50 % overlapping scheme is applied to avoid the distortions caused by Hilbert transform based envelope calculation. The proposed method is implemented on a wireless sensor node and tested successfully for detecting an outer race fault of a rolling bearing. The results show that the frequency resolution of the envelope spectrum is improved by 8 times while the data transmission remains at a low rate.

Keywords Wireless sensor network · Envelope analysis · Bearing fault diagnosis · Overlap processing · Cascading

---

G. Feng (✉) · D. Zhen · X. Tian · F. Gu · A.D. Ball
Center for Efficiency and Performance Engineering, University of Huddersfield, Huddersfield HD1 3DH, UK
e-mail: u1273230@hud.ac.uk

D. Zhen
e-mail: d.zhen@hud.ac.uk

X. Tian
e-mail: u1178848@hud.ac.uk

F. Gu
e-mail: f.gu@hud.ac.uk

A.D. Ball
e-mail: a.ball@hud.ac.uk

J.K. Sinha (ed.), *Vibration Engineering and Technology of Machinery*,
Mechanisms and Machine Science 23, DOI 10.1007/978-3-319-09918-7_68

1 Introduction

As the equipment becomes increasingly complicated, the maintenance cost of industrial machines has been growing quickly. It is estimated that approximately half of operating cost can be attributed to maintenance in most processing and manufacturing operations [1]. Currently, wired online condition monitoring (CM) systems have been successfully employed in industry, but so far have been mainly restricted to large industrial machines or laboratories due to their high cost. One major cost for these wired systems is the investment in high quality cables and their installation costs in harsh industrial environment. For cases of remote condition monitoring, the installation cost may be even higher than the cost of the sensors [2].

With the rapid development of electronics and wireless technology, wireless sensor network (WSN) is gaining increasing popularity in CM fields. It brings several inherent advantages, such as low cost, convenient installation and easy to replace and upgrade [3]. Currently, using the WSN to monitor the static signals like temperature, pressure, humidity has become popular [4, 5]. However, the application of dynamic signals over the WSN is still limited, which is restricted by the conflicts between high data amount of dynamic signals and the low bandwidth of the WSN. Basically, there are two ways to monitor the dynamic signals over WSN. One way is to increase the bandwidth of the WSN, for example, employing Wi-Fi. However, it also brings new problems, such as higher power consumption in which case external power supply or frequent charging of the battery is needed. The other way is to employ on-sensor processing, in which, the raw data are firstly processed on the sensor node by the embedded algorithms and only the processing results are sent through the network. Thus, the transmission data amount can be greatly reduced. Another appealing reason for on-sensor processing is its potential to reduce power consumption of the node since the calculations on the microprocessor usually consumes less power than data transmitting [3, 6].

Bearing is widely used in the industrial plants and it is estimated that about 40 % of the induction motor failures are caused by bearing failures [2] and envelope analysis has been proved as an effective method to detect the early faults on the bearing. The author realized this algorithm on a wireless sensor node and its effectiveness for reducing the data amount have also been proved through experiments [7]. This paper extends previous works and an improved method is developed to increase the frequency resolution of the envelope spectrum. It is tested successfully on the same wireless sensor node for detecting the outer race fault, which makes the implementation of envelope analysis on the wireless sensor node more meaningful.

2 Theoretical Background

2.1 Envelope Analysis

As a classical diagnosis algorithm, envelope analysis is widely used for the early fault detection in gearboxes and rolling element bearings, etc. [8, 9]. It is suitable for diagnostics of machinery where faults have an amplitude-modulating (AM) effect on the characteristic frequencies of the machinery [8]. Figure 1 shows the envelope analysis to a simple modulation signal, in which, a 1,500 Hz carrier signal is modulated by an 80 Hz sine wave. In the spectrum of the modulation signal, there are three frequency components and the 80 Hz frequency component is carried to the high frequency band centered at the carrier frequency. After envelope analysis, the low frequency component is extracted out and the high frequency problem is transferred to a low frequency one. Comparatively, it is much easier to tell the modulating signal from the envelope spectrum especially when several modulating signals coexist.

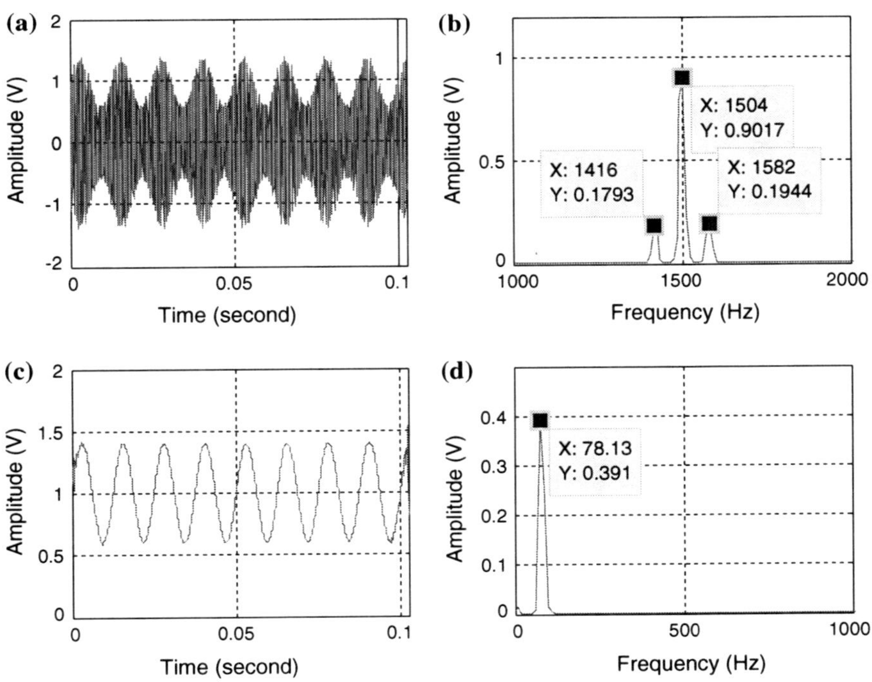

Fig. 1 Application of envelope analysis to a time waveform: **a** modulation signal, **b** spectrum of modulation signal, **c** analyzed envelope, **d** envelope spectrum

2.2 Down-Sampling, Cascading and Overlap Processing

For a signal sampled at F_s, its frequency resolution is determined by the acquisition time T, as shown in Eq. (1).

$$\Delta f = \frac{1}{T} = \frac{F_s}{N} \tag{1}$$

Obviously, increasing the number of samples N for calculating FFT so as to increase the acquisition time T is the only way to increase the frequency resolution. However, there is a maximum size of FFT calculations for a specific wireless sensor node. Take the processor used in [7] for example, a maximum size of 2,048 points FFT calculation can be realized when using the floating point format. Since wireless sensor nodes are usually designed with limited memory size and computing capability for low power consumption and large scale deployment considerations, this means the allowable size of FFT calculation for a wireless sensor node is usually not large enough and thus the frequency resolution is not sufficiently high for obtaining accurate diagnostic results.

However, for the implementation of envelope analysis, it is possible to improve the frequency resolution of envelope spectrum by using a scheme of down-sampling and cascading the computed envelope data. This is because that the interested frequency components of the envelope signal are shifted to the low frequency band after envelope analysis. As shown in Fig. 1, after envelope analysis, the effective

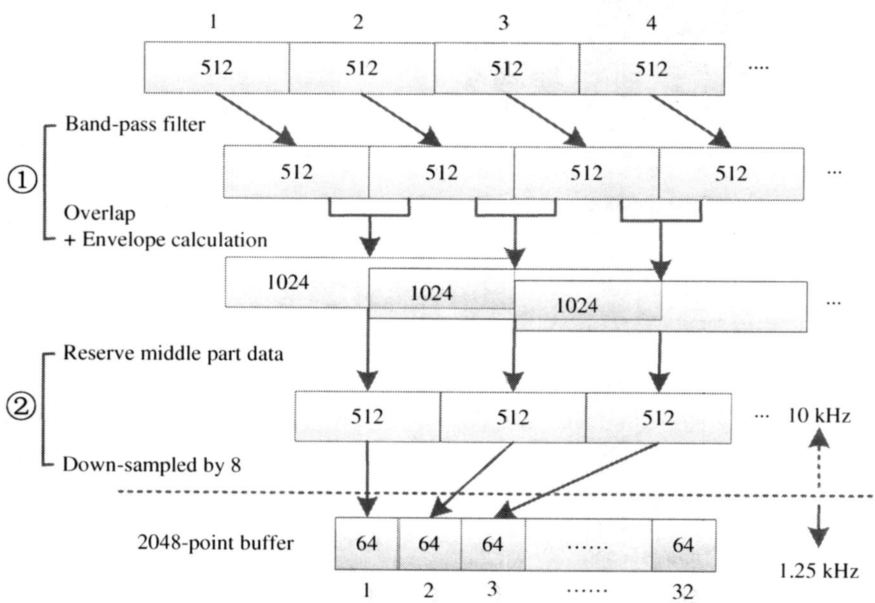

Fig. 2 Diagram for the overlap, down-sampling and cascading scheme

components are below 100 Hz. For cases dealt with in [7], the interested frequency band of the envelope signal was within 500 Hz.

Meanwhile, simply cascading the down-sampled signal may bring undesirable distortions to the spectrum. As shown in the analyzed envelope in Fig. 1c, obvious distortions appear at the front and end edge of the analyzed envelope signal. This is caused by the window effect when implementing Hilbert transform. In order to exclude these unwanted distortions, an overlap processing scheme can be employed according to [10].

A processing scheme which combines the overlap processing, down-sampling and cascading is given in Fig. 2. The collected vibration data are processed frame by frame with the size of 512 points and a 50 % overlap is employed in order to keep the çonsecutiveness of the analyzed envelope signal. As shown in the figure, the scheme consists of two main steps: envelope calculation and envelope down-sampling. In the first step, a data frame first passes through an FIR band-pass filter and then combines with last filtered data frame to form a new data frame of 1,024 data points. Then the envelope of this 1,024 data is computed using Hilbert transform, as descried in [7].

In the second step, only middle part of the computed envelope data is kept and down-sampled by 8 times, depending on the bandwidth of the band pass filter, to save computing memory and reduce the calculation amount. As shown in the fourth and the fifth row of the data flow diagram in Fig. 2, the original 512 points of data shrink to only 64 points.

The repeat of the above two steps will allow an envelope data of 2,048 points to be obtained. As the construction of the envelope data is based on the middle part of each frame, the side distortions are avoided. Moreover, because of the down-sampling operation, the envelope spectrum will have a high frequency resolution, for example, the resolution for the data described will be improved by 8 times.

3 Algorithm Implementation and Experimental Setup

3.1 Algorithm Implementation

In order to evaluate the algorithm, a wireless condition monitoring system employed is shown in Fig. 3. It consists of one sink node and one sensor node. Two Xbee wireless modules from Digi-international are employed to setup the Zigbee network. The sensor node is responsible for vibration signal collection and local processing while the sink node is used to route the processing results to the host computer for display and further analysis.

On the sensor node, vibration signal is measured using a Piezo-Electric (PE) accelerometer, which is selected because of its low power consumption compared to the Integrated Electronics Piezo-Electric (IEPE) one. A low-pass filter with 5 kHz cut-off frequency is added for anti-aliasing. Then, analog to digital conversion

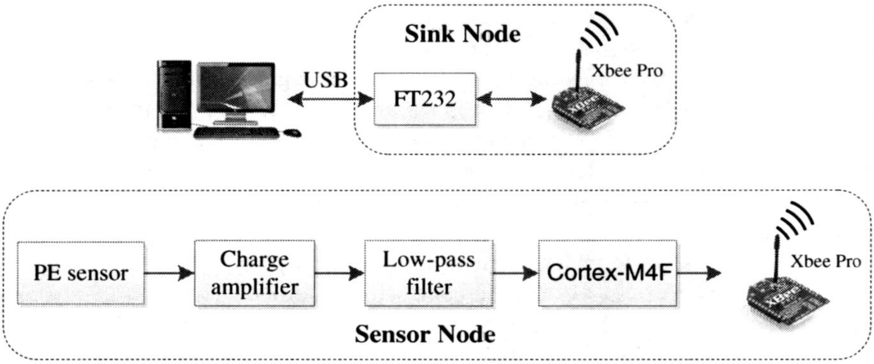

Fig. 3 Structure of sink node and sensor node

(ADC) and local processing algorithms are accomplished on a Cortex-M4F microcontroller TM4C1233H6PM, which has 256 k bytes flash, 32kB SRAM, 12-bit ADC with 1 MSPS sampling rate and various peripherals. According to the investigation in [11], Cortex-M architecture has been chosen as the new generation core of the WSN platform for low power and high performance considerations. With the integrated digital signal processing (DSP) unit and floating point unit (FPU), Cortex-M4F core is quite efficient for implementing complicated calculations.

The data processing flow inside the processor is given in Fig. 4. The vibration signal is firstly sampled by the on-chip ADC, which is triggered by setting the overflow rate of a timer at 10 kHz. Then, the conversion results are moved from the ADC register to the internal buffers automatically by a Direct Memory Access (DMA) unit. When the buffers are full, an interrupt will be generated to inform the central processing unit (CPU) to process the newly collected frame data (512 points). Here, data in the frame are converted from 16-bit fixed point format to 32-bit single floating point format for accurate calculations. Then, the overlap, down-sampling and cascading scheme described in Sect. 2.2 is implemented for computing the envelope spectrum. Finally, the calculated envelope spectrum is averaged by 4 times and transmitted over the Zigbee network.

As shown in Fig. 4, DMA unit together with the two buffers construct a double buffering structure, which enables the data acquisition and CPU calculations to be completed in parallel. As a result, signal processing can be implemented much efficient. The random access memory (RAM) consumed by the main buffers is as large as 24 kB, which is more than half size of the total RAM (32 kB). These buffers are used of high efficiency, especially for the main buffer *fBuf*, which is used for several calculations, such as envelope computation, down-sampling, cascading and FFT, etc.

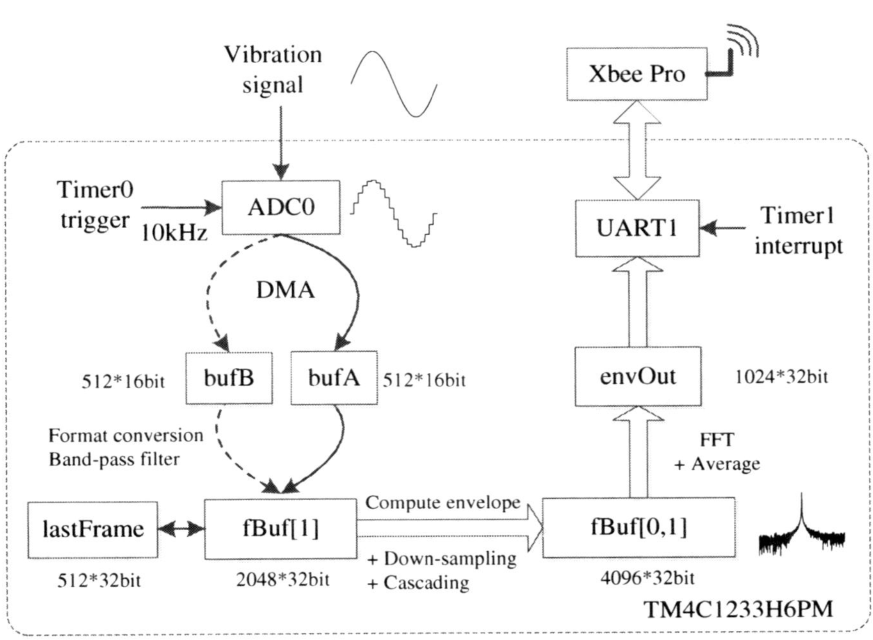

Fig. 4 Data processing flow inside the processor

3.2 Experimental Setup

In order to estimate the performance of the improved algorithm, the wireless sensor node was used to detect the fault on a rolling bearing test rig shown in Fig. 5. A faulty rolling bearing with a scratch on the outer race was installed inside the bearing house. When the bearing running, the scratch will be rolled over periodically, which results in a fault signal modulating with a carrier signal at the resonant

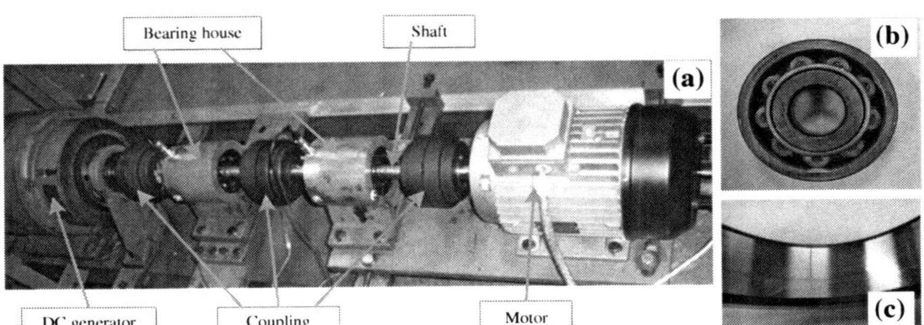

Fig. 5 Test rig: **a** bearing test rig, **b** rolling bearing, **c** defects on outer race

frequencies of the bearing house [8]. As explained in [7], an outer race faulty signal with the character frequency at 83.5 Hz is expected to be extracted using the envelope analysis algorithm. During the test, a PE accelerometer was mounted horizontally to the bearing house to measure the vibration signal and the sink node was placed about two meters away for receiving the processing results.

4 Results and Discussion

4.1 Processing Results on Sensor Node

The algorithm on the sensor node was developed using the Code Composer Studio (CCS) [12], which is a powerful integrated development environment provided by Texas Instrument. In order to fully examine the improved algorithm, several important internal processing results are firstly given by tracing into the program and visualizing the results via graph tools. The processing results viewed in CCS are illustrated in Fig. 6, including raw signal, calculated envelope, down-sampled envelope and cascaded envelope. As shown in Fig. 6, short period impact pulses can be noticed in the raw signal and high level noises are also observable. The high frequency components of the calculated envelope have been removed and the down-sampled envelope roughly matches calculated envelope while the points are different. As can be seen in Fig. 6c, the distortions at the front and end edge of the down-sampled envelope is not quite large.

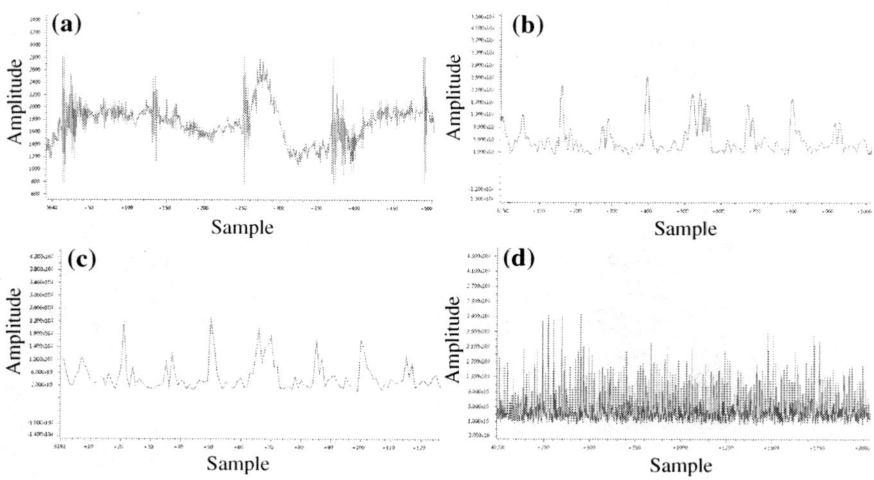

Fig. 6 Processing results on the sensor node: **a** raw data (512 points), **b** envelope (1,024 points), **c** down-sampled envelope (128 points), **d** Cascaded envelope (2,048 points)

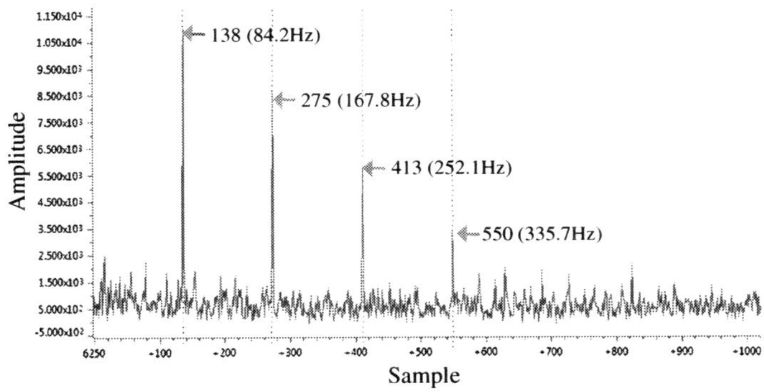

Fig. 7 Spectrum of the cascaded envelope

The spectrum of the cascaded envelope is presented in Fig. 7. In the test, the sampling rate of the AD converter is set at 10 kHz and after 8 times' down-sampling, the true sampling rate of the cascaded envelope is 1.25 kHz. Therefore, the frequency resolution after 2,048 points' FFT calculation is about 0.61 Hz. The index and corresponding frequency are illustrated in Fig. 7. As it shows, as high as the fourth harmonic of the outer race fault frequency can be clearly observed. This has indicated the existence of the outer race fault on the rolling bearing.

4.2 Remote Results and Comparison

In order to highlight the benefits of the improved algorithm, the envelope spectrum without cascading discussed in [7] is also used to analyze the signal from the same faulty bearing. The results of both methods are given in Fig. 8. Apparently, the outer race fault can be identified by both algorithms. However, the fault frequency

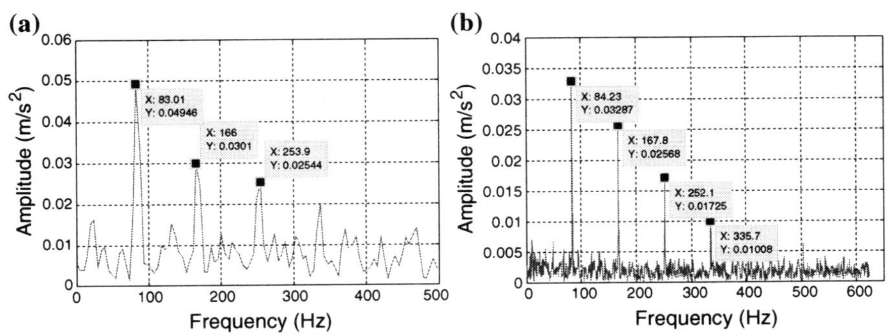

Fig. 8 **a** Envelope spectrum without cascading (2,048 points) **b** Envelope spectrum with cascading and overlap processing

Table 1 Effective data rate comparison

Data type	Data rate (kbps)
Raw data	160
FFT of raw data	80
Envelope analysis without cascading	8
Envelope analysis with cascading	9.7
Envelope analysis with cascading (within 500 Hz)	7.8

identified from the improved algorithm is 84.23 Hz which is much closer to the expected characteristic frequency 84.5 Hz, whereas that from normal envelope algorithm is 83.01 Hz, more deviation from the expected. This shows that the improved algorithm produces more accurate and reliable diagnostic results. In addition, it can be seen that the ratio from the spectrum of the improved algorithm between the amplitude at the fault frequency and the background noise level is more than 6 while that from the normal envelope spectrum is less than 5. This means that the improved algorithm provides a higher signal to noise ratio (SNR) for better detection and diagnosis performance.

The improved method not only increases the resolution but also performs well in the data amount reduction. Table 1 summarizes the effective data rate per second for different processing stages on the sensor node. All the data are supposed to be stored with 16-bit resolution. As it shows, the improved algorithm just transmits a bit more data than the envelope spectrum without cascading. If only data within 500 Hz are transmitted in the improved method, the data rate is even slightly lower. If frequency averaging is added, the transmitting rate can be further reduced.

5 Conclusions

An accurate envelope analysis algorithm is developed for a wireless sensor node for bearing fault diagnosis. The key processing steps of the algorithm include down-sampling, data frame overlapping and cascading. Through these steps, the resultant envelope spectrum can be obtained on the sensor node with sufficient frequency resolution and amplitude accuracy for reliable diagnostic results. Experimental evaluation shows that after down-sampling and cascading, the frequency resolution was increased by 8 times, compared with normal envelope algorithm, and the side distortions caused by Hilbert transform is avoided by a 50 % overlapping scheme. With these improvements, the tested bearing fault can be identified reliably in the envelope spectrum.

References

1. de Silva CW (2007) Vibration monitoring, testing, and instrumentation. Taylor & Francis Group, LLC, Boca Raton. ISBN 978-1-4200-5319-7
2. Lu B, Gungor VC (2009) Online and remote motor energy monitoring and fault diagnostics using wireless sensor networks. IEEE Trans Ind Electron 56(11):4651–4659
3. Hou L, Bergmann NW (2012) Novel industrial wireless sensor networks for machine condition monitoring and fault diagnosis. IEEE Trans Instrum Meas 61(10):2787–2798
4. Bin C, Xinchao J, Shaomin Y, Jianxu Y, Xibin Z, Guowei Z (2011) Application research on temperature WSN nodes in switchgear assemblies based on TinyOS and ZigBee. In: 2011 4th International conference on electric utility deregulation and restructuring and power technologies (DRPT), pp 535–538
5. Lu B, Wu L, Habetler TG, Harley RG, Gutierrez JA (2005) On the application of wireless sensor networks in condition monitoring and energy usage evaluation for electric machines. In: 31st Annual conference of IEEE industrial electronics society, pp 2674–2679
6. Lynch JP (2007) An overview of wireless structural health monitoring for civil structures. Philos Trans Roy Soc A 365(1851):345–372
7. Feng G, Mustafa A, Gu JX, Zhen D, Gu F, Ball AD (2013) The real-time implementation of envelope analysis for bearing fault diagnosis based on wireless sensor network. In: 19th International conference on automation and computing (ICAC), pp 1–6
8. McInerny SA, Dai Y (2003) Basic vibration signal processing for bearing fault detection. IEEE Trans Educ 46(1):149–156
9. Tandon N, Choudhury A (1999) A review of vibration and acoustic measurement methods for the detection of defects in rolling element bearings. Tribol Int 32(8):469–480
10. Basha SK, Pandey PC (2012) Real-time enhancement of electrolaryngeal speech by spectral subtraction. In: National conference on communications (NCC), pp 1–5
11. Feng G, Zhen D, Aliwan M, Gu F, Ball A (2013) Investigation of wireless protocols for remote condition monitoring. In: Proceedings of computing and engineering annual researchers' conference. University of Huddersfield, pp 19–24
12. Texas Instrument (2013) Code composer studio (CCStudio). Integrated Development Environment (IDE) v5—CCSTUDIO—TI Tool Folder. http://www.ti.com/tool/ccstudio#Technical Documents. Accessed 12 Dec 2013

Vibrational Characteristics of a Cracked Rotor Subjected to Periodic Auxiliary Axial Excitation

Heba H. El-Mongy, Younes K. Younes and Mohamed S. El-Morsy

Abstract Rotor failure due to cracks can be catastrophic with large costs in down time, damage to equipment and injury to personnel. Prevention of such problems is only attained by developing reliable methods for crack detection and diagnosis. In this paper, the vibratory response of a cracked rotor to an auxiliary axial excitation is investigated experimentally for the purpose of crack detection. An open crack (slot) is artificially created in the rotor-shaft of the experimental rig. The effects of running speed, crack depth and crack-unbalance angle variations on the vibratory response are investigated experimentally. It is shown that the auxiliary excitation method can be efficiently used in crack detection as an online condition monitoring technique. By properly choosing the frequency of the periodic axial excitation and the running speed, the presence of a crack in the excited shaft can be ruled out and the crack growth can be monitored. The crack-unbalance angle can influence the identification process negatively by masking the crack effect on the vibratory response. However, an unbalance trial mass test can help in finding out the unknown crack angle.

Keywords Condition monitoring · Crack detection · Vibration · Auxiliary excitation method

H.H. El-Mongy (✉) · Y.K. Younes · M.S. El-Morsy
Faculty of Engineering, Mechanical Design Department, Mattaria, Helwan University, P.O. BOX 11718 Cairo, Egypt
e-mail: heba_elmongy@yahoo.com

Y.K. Younes
e-mail: youneskhalil@yahoo.com

M.S. El-Morsy
e-mail: mohsabry1999@yahoo.com

J.K. Sinha (ed.), *Vibration Engineering and Technology of Machinery*, Mechanisms and Machine Science 23, DOI 10.1007/978-3-319-09918-7_69

1 Introduction

Cracks in rotating shafts are considered very serious and can lead to catastrophic accidents if not detected in time. There has been extensive research on the vibrational behaviour of cracked shafts and the use of response characteristics to detect cracks. The research in the past few decades on cracked structures and rotors is well documented in review papers [1–5].

Several researchers have identified that a coupling mechanism exists between different types of vibrations, i.e. axial, radial and torsional, in cracked shafts. Such coupling does not appear in absence of a crack. Papadopoulos and Dimarogonas [6–9] have extensively studied coupling of vibration due to crack. They showed that the presence of either of bending, longitudinal or torsional mode natural frequency in the vibration spectra of other modes can be considered a reliable crack indicator. However, they applied sweep excitation to a non-rotating shaft. Ostachowicz and Krawczuk [10] used an open crack model to demonstrate coupling of torsional and bending vibration in a rotating shaft.

Darpe et al. [11] used a non-linear response-dependent breathing crack model to study the coupling of bending and axial vibrations in a slow rotating cracked Jeffcott rotor using periodic axial impulse excitation. Predominant existence of lateral bending natural frequency, axial excitation frequency and side bands around it in lateral vibration spectrum have been shown as important crack indicators. Those analytical results were validated experimentally in [12]. In [13], Darpe et al. numerically studied the coupling between all types of vibrations; longitudinal, lateral and torsional vibrations for a rotating cracked shaft.

It should be noted that simulating a real fatigue crack in a rotating shaft is difficult to attain in a laboratory environment. This is performed as explained by Darpe et al. [14] by introducing an initial slit in the shaft under test then the shaft is placed on a three-point fatigue bending machine where the shaft is subjected to cyclic loading until the crack starts to propagate. In this case, the healthy readings are acquired from the rotor on solid condition, then the system is disassembled and the crack is generated. The system is re-assembled with the cracked shaft and the readings of the faulty condition are acquired. Aside that this procedure is expensive and time costing, it can be difficult to have a valid reference case of the healthy rotor due to misalignment and unbalance that can be introduced during assembly and disassembly [15]. Another disadvantage of this procedure is the possibility of introducing a bow in the rotor. These disadvantages may result in poor comparison between healthy and faulty conditions.

The effect of crack-unbalance angle is also a critical issue in the crack detection process. Early studies by Henry and Okah-Avae [16], Mayes and Davies [17] and Bachschmid et al. [18], indicated that a crack can only affect the synchronous unbalance response if the crack is able to open. This occurs when the unbalance is in phase with the crack where the shaft behaves as a slotted rotor. Similar results were reported for transient response of a cracked rotor as explained by Sawicki et al. [19]. In a detailed numerical and experimental study of the transient response of a cracked rotor, Darpe et al. [14] showed that the effect of unbalance angle variation depends upon the type of crack whether it is open or breathing which can be used as a diagnostic tool to distinguish both types. To the best of the authors' knowledge, the influence of crack-unbalance angle on the vibration response of a cracked rotor to an external excitation has not been investigated before.

In this paper, the coupling between longitudinal and lateral vibrations due to crack in a rotating shaft is experimentally investigated on a machinery fault demonstrator (PT 500) made by G.U.N.T.® An open crack (slot) is artificially created in the rotor-shaft of the test rig. A periodic axial excitation is applied to its rotor in both healthy and cracked conditions. A parametric study is conducted to investigate the effect of crack depth, running speed and crack-unbalance angle on the vibratory response to the applied external axial excitation. Effect of crack-unbalance angle is studied at two different running speeds and two different crack depths in an attempt to thoroughly investigate this interesting effect.

2 Experimental Set-Up

Figure 1 shows a photo and a schematic diagram of the G.U.N.T.® machinery fault demonstrator (PT 500) and the measurement and analysis setup located in the Vibration and Dynamics Laboratory of the Faculty of Engineering—Mattaria. The fault demonstrator consists of a steel shaft of diameter 0.02 m and length 0.5 m coupled to an electric motor (0.37 kW) via an elastic spacer ring coupling and supported by two ball bearings (SKF 6004-2ZR). The shaft carries one disc of mass 1.52 kg, diameter 0.15 m and thickness 0.01122 m at mid-span. Figure 1b shows a magnified view of the crack generated in the shaft.

The measurement and analysis setup includes a B&K permanent magnet exciter Type 4808, a B&K power amplifier Type 2712, a B&K exciter control Type 1047, two accelerometers (IMI 603C01) mounted via studs to the BRG#2 (Right bearing) in horizontal and vertical directions, a G.U.N.T.® measuring amplifier, a USB bmcm® data acquisition box and a laptop for data analysis and archiving.

The exciter control is used to generate a sinusoidal signal at the desired frequency. The excitation signal is then amplified by the power amplifier and transferred to the exciter which is positioned horizontally at the shaft end and carefully aligned with the shaft axis.

3 Experimental Approach and Data Analysis

For all the experiments conducted in this study, the vibration signals are acquired at a sampling rate of 8,000 Hz. Number of samples is taken as 64,000 and the time span is 8 s. A LabVIEW® virtual instrument is used for data acquisition. The frequency spectra are obtained using National Instruments Sound and Vibration

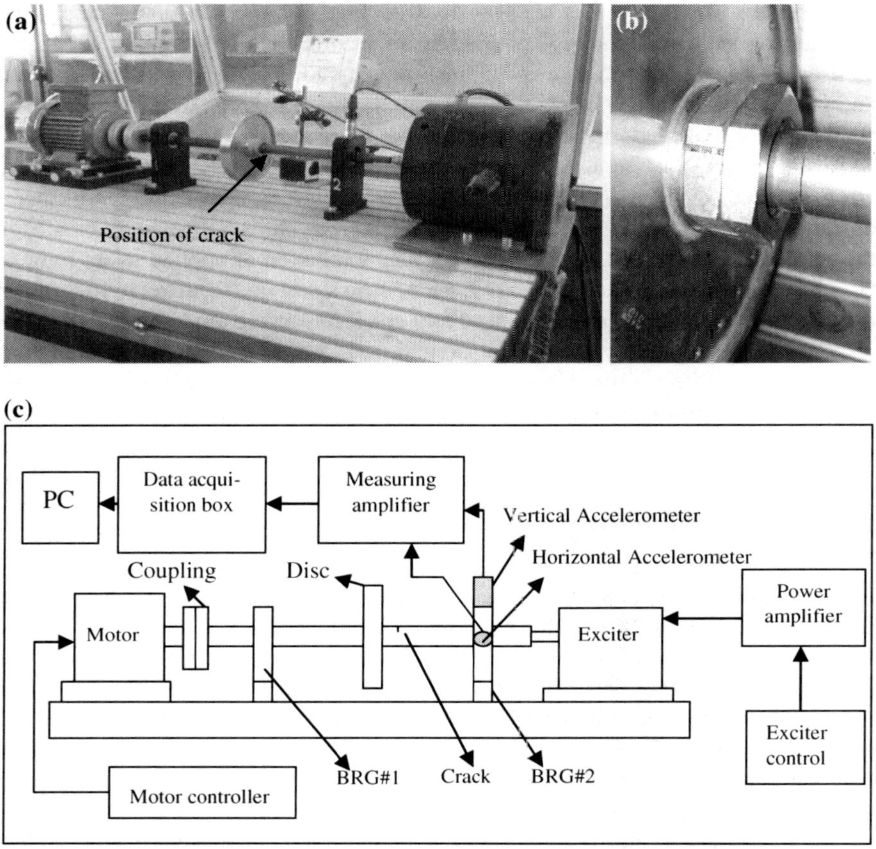

Fig. 1 G.U.N.T.® machinery fault demonstrator (PT 500) with measurement and analysis setup, **a** photograph, **b** crack shape, **c** schematic diagram

Assistant® software. Power spectrum is chosen for displaying frequency spectra and the vibration parameter is acceleration (RMS).

First, bump test [20] is performed for the shaft under test to find the fundamental bending natural frequency in the horizontal direction. The shaft with the disc situated centrally is suspended using a rope in a "free-free condition" and an accelerometer is fitted to it via magnet. The shaft is hit using an impact hammer and the vibration response is recorded and transformed to the frequency domain. The test is repeated several times to make sure that the results are repeatable and accurate. The fundamental bending natural frequency in the horizontal direction is found to be 313.3 Hz. This frequency is chosen to be the axial excitation frequency for all the experiments as it yields clear frequency spectra [11, 12].

The shaft is then mounted on the fault demonstrator. The machine is allowed to run in for about 4 h before acquiring baseline data for the shaft. Then the crack is created in the tested shaft at a distance of 5 mm from the taper lock of the disc using a jeweler's saw of width 0.35 mm so the crack is considered as a slot (open crack) as shown in Fig. 1b. The crack depth is increased without need to disassemble the system.

Tests are done at two different crack depths ($\bar{a} = 0.22$) and ($\bar{a} = 0.55$) where $\bar{a}$ defines the relative crack depth ($\bar{a} = a/R$), R is the shaft radius and a is the crack depth. Also, the tests are performed for two different running speeds to investigate the effect of the change of the running speed on the proposed method. The speeds are taken as 1/10th and 1/5th of the first critical speed, i.e., 1,878 and 3,756 RPM respectively.

To investigate the effect of the crack-unbalance angle, an unbalance mass of 3.947 g is placed on the disc at a radius of 60 mm giving eccentricity that equals 0.23682 kg mm. The unbalance mass is first aligned with the crack axis to establish the reference position ($\beta = 0$). The crack-unbalance angle is then varied 25 times in steps of 15° and the lateral vibration in both horizontal and vertical directions is measured for each test case taking into consideration the variation of the crack depth and running speed.

4 Results and Discussion

4.1 Coupling of Lateral and Axial Vibration Due to Crack

Figure 2 represents the frequency spectra of the balanced uncracked shaft without axial excitation where the running speed is taken as 1/10th of the first lateral critical speed of the rotor (1,878 RPM). The 1X component at 31.3 Hz can be seen in the vertical spectrum (Fig. 2b) but in small amplitude because the system is balanced. The bending natural frequency which matches with the 10X harmonic cannot be seen in both horizontal and vertical spectra. Then, axial excitation is applied to the shaft at a frequency equal to the first bending natural frequency (fi = 313.3 Hz). The frequency spectra of the balanced uncracked rotor with axial excitation are shown in Fig. 3. The horizontal spectrum (Fig. 3a) shows the bending natural frequency (fi) and its first harmonic (2fi), while only (2fi) appears in the vertical spectrum. Ideally, those frequencies should not have appeared because the lateral and axial vibrations are uncoupled in case of uncracked shaft. This could be due to inevitable slight misalignment between the exciter and the rotating shaft which results in coupling between axial and lateral vibrations, i.e. the excitation is not purely axial.

Figure 4 shows the frequency spectra of the cracked shaft ($\bar{a} = 0.22$) without excitation. The horizontal spectrum (Fig. 4a) shows higher harmonics of the 1X including the bending natural frequency (fi) which matches with the 10X harmonic component. Figure 5 shows the frequency spectra for the cracked shaft with excitation. The bending natural frequency (fi) and its first harmonic (2fi) are dominant in both spectra (Fig. 5a and b). Compared to Fig. 3a, the amplitude of (fi) increased from 0.00868 to 0.3708 g^2 and the amplitude of (2fi) increased from 0.00262 to 0.147 g^2. This increase is only attributed to the crack effect. For such a

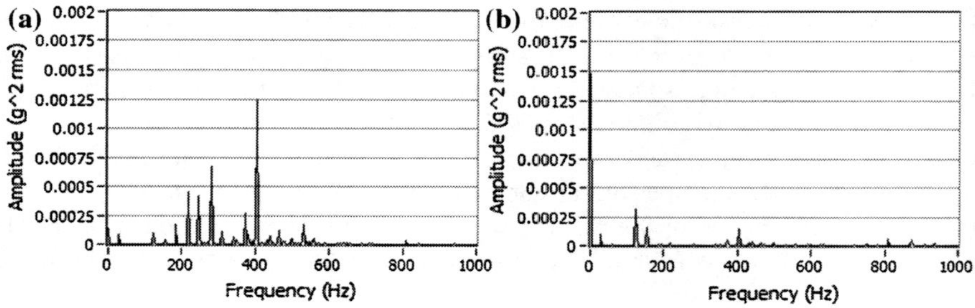

Fig. 2 Frequency spectra of the uncracked shaft without excitation, **a** horizontal, **b** vertical

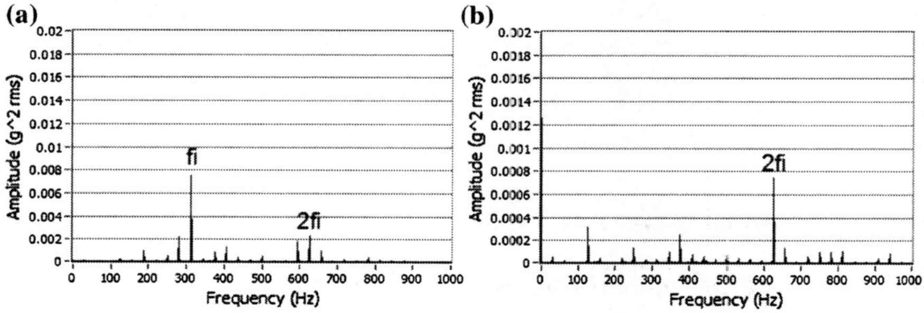

Fig. 3 Frequency spectra of the uncracked shaft with excitation, **a** horizontal, **b** vertical

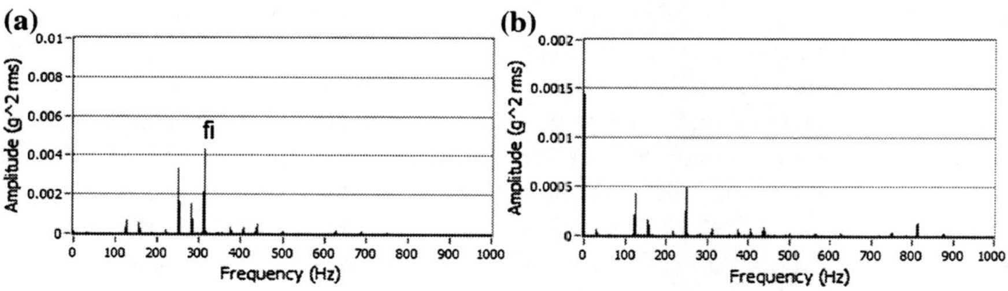

Fig. 4 Frequency spectra of the cracked shaft ($\bar{a} = 0.22$) without excitation, **a** horizontal, **b** vertical

small crack depth ($\bar{a} = 0.22$), coupling between lateral and axial vibrations is obviously established. It can be observed that the possible misalignment between the shaft and the exciter is shown to have lower effect on the vibration spectra when compared with the crack effect. Comparing Figs. 2a and 3a, the increase of the (fi) amplitude in the horizontal direction due to misalignment is about 9 times. Meanwhile, the increase of the (fi) amplitude in the horizontal direction due to both the crack effect and the possible misalignment is about 71 times as shown by comparing Figs. 4a and 5a.

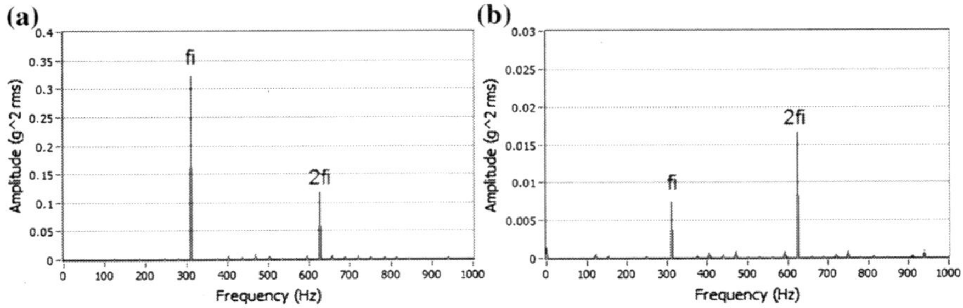

Fig. 5 Frequency spectra of the cracked shaft ($\bar{a} = 0.22$) with excitation, **a** horizontal, **b** vertical

4.2 Parametric Study

4.2.1 Effect of Crack Depth

Figure 6 shows the frequency spectra of the cracked rotor ($\bar{a} = 0.55$) running at speed equal to 1/10th of the first bending natural frequency and excited axially at a frequency equal to the first bending natural frequency (fi). The patterns of (fi) and (2fi) can be seen obviously in Fig. 6 as the dominant crack features in the horizontal and vertical spectra. However, the amplitudes of (fi) and (2fi) are much higher when comparing Figs. 5 and 6. It can be easily inferred from (fi) and (2fi) amplitude trending that the crack is growing in depth.

4.2.2 Effect of Running Speed

Figure 7 shows the frequency spectra of the unbalanced cracked rotor ($\bar{a} = 0.22$) running at speed equal to 1/5th of the critical speed (3,756 RPM) and subjected to axial excitation at a frequency equal to the first bending natural frequency (fi = 313.3 Hz). The unbalance is placed in phase with the crack. The horizontal

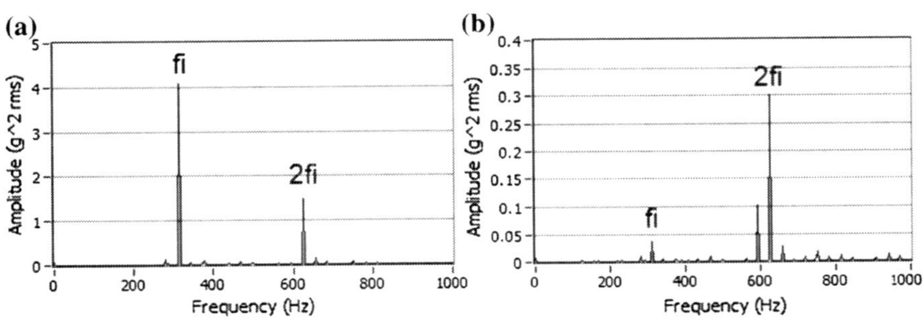

Fig. 6 Frequency spectra of the cracked shaft ($\bar{a} = 0.55$) with excitation, **a** horizontal, **b** vertical

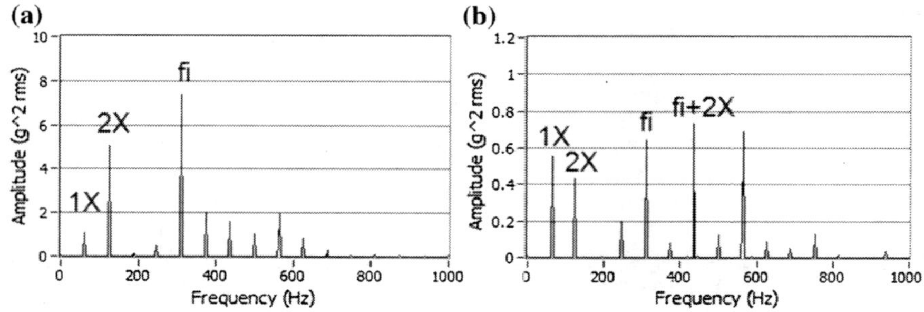

Fig. 7 Frequency spectra of the cracked shaft ($\bar{a} = 0.22$) running at 1/5th of first critical speed with axial excitation at frequency of first bending natural frequency, **a** horizontal, **b** vertical

spectrum (Fig. 7a) shows a high 2X which is the usual symptom of cracked rotors running at a subcritical speed [3, 15]. Also, the (fi) appears as the highest peak in the horizontal spectrum with sum and difference frequencies around it. But they are more obvious in the vertical spectrum (Fig. 7b) where the highest peak is seen at (fi + 2X). These results are in agreement with previous literature [11–13].

4.2.3 Effect of Crack-Unbalance Angle

The measured frequency spectra at different crack-unbalance angles showed similar pattern as the spectra shown in Fig. 5. However, the amplitudes of (fi) and (2fi) varied greatly with the crack-unbalance angle. The variation of the (fi) amplitude for the two crack depth cases is recorded and plotted in Fig. 8. It is obvious from Fig. 8 that the variation pattern of (fi) depends greatly on the crack depth and does not exhibit general behavior. Number of maxima is different in both cases and also the locations of the maxima are shifted. This can be attributed to the fact that increasing the crack depth will result in an increase in the crack angle. This will affect the locations of maxima in the (fi) variation as the regions of crack opening and crack closing will change. The (fi) amplitude for crack depth ($\bar{a} = 0.55$) is generally higher than that for crack depth ($\bar{a} = 0.22$) except in the region of $\beta = 120°{:}225°$.

For the case of running speed that equals 1/5th of the first critical speed, the same procedure is repeated and the 2X and (fi) are recorded and analyzed. The (fi) amplitude variation (not shown here due to space limitation) is not considered informative because it depends on the running speed and crack depth as shown before in Fig. 8. The 2X variation shown in Fig. 9a was much informative. Inspection of Fig. 9a shows that the 2X amplitude variation resembles a harmonic wave having two cycles with two maxima at $\beta = 0°$ and 105° and two minima at $\beta = 60°$ and 225°. However, in actual situations, unbalance position and crack angle are unknowns. The crack angle can be identified using a test that involves placing trial mass on the rotor and measuring the vibration. The crack angle will lie at one

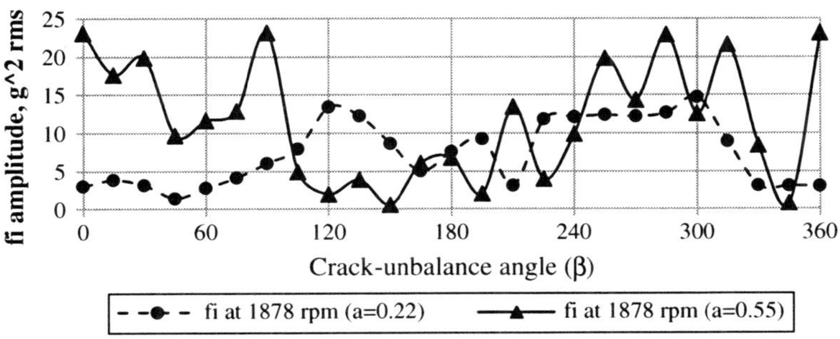

Fig. 8 Variation of amplitude of the horizontal component of first bending natural frequency (fi) with crack-unbalance angle (β) for different cases of crack depth

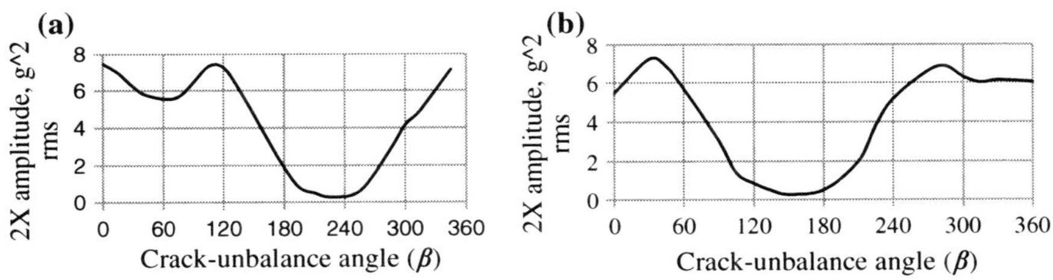

Fig. 9 Variation of amplitude of the horizontal 2X component with crack-unbalance angle (β) for cracked shaft ($\bar{a} = 0.22$) running at 1/5th of its first critical speed, **a** the crack axis is the reference for β, **b** the positive horizontal axis is the reference for β

of the two maxima in the trended results. To prove this observation, the previous test at the same conditions is repeated but the unbalance mass is placed on the disk starting from the positive horizontal axis. This situation simulates real condition where the crack angle is unknown. The result of this test, i.e. the 2X amplitude variation versus β is shown in Fig. 9b. The crack angle can be deduced from this figure as to be either at 30° or 285° which is in agreement with the real crack that lies at 285° from the specified reference position exactly.

5 Conclusion

Coupled lateral and axial vibrations due to crack in a rotating shaft have been investigated experimentally for the purpose of crack detection. Auxiliary axial excitation is applied to the rotating shaft of the experimental rig. Horizontal and vertical responses are measured and analyzed for the uncracked and cracked conditions. Effects of changing crack-unbalance angle, running speed and crack depth are investigated. The following conclusions are drawn:

1. For the experiments conducted in this paper, the axial excitation frequency (fi) and its harmonic (2fi) are dominant in both horizontal and vertical lateral spectra which indicates coupling between axial and lateral vibration due to crack. However, for all studied cases, the horizontal spectra exhibit higher values compared with the vertical spectra.
2. The measured lateral vibration responses to the axial excitation are found to be sensitive to crack depth change. When the crack depth is increased from ($\bar{a} = 0.22$) to ($\bar{a} = 0.55$), the amplitude of the main crack feature (fi) in horizontal direction has increased by 11 times. Therefore, crack growth can be effectively monitored using the auxiliary excitation method.
3. Increasing the running speed from 1/10th of the first critical speed of the rotor to 1/5th of the first critical speed helps in crack detection due to the appearance of the twice per revolution component (2X). In addition, the sum and difference frequencies appear around (fi) and (2fi) due to the interaction of the running frequency and its harmonics and the axial excitation frequency and its harmonics. Therefore, it is recommended to carry out the auxiliary excitation measurements at a running speed that equals one of the highest integer fractions of the first critical speed in the operating range of the machine as it will be the most informative. However, operating a cracked rotor that is subjected to an external force at very high speeds can be dangerous as it may cause crack propagation.
4. The crack-unbalance angle (β) greatly affects the horizontal and vertical responses to axial excitation. The pattern of (fi) amplitude variation versus crack-unbalance angle depends upon the crack depth and the running speed of the rotor. On the other hand, the pattern of 2X amplitude variation is more general and can be used to specify the crack angle using trial mass test.

The main advantages of auxiliary excitation method are that it is an online method that can be used while the rotor is running; it is suitable for early crack detection; and the associated tests can be carried out by measurements on bearings or shafts. It is recommended to extend this work to study the effect of other common faults that may coexist with the crack in order to establish a robust method of crack detection based upon auxiliary excitation and coupling between vibrations.

Acknowledgment The first author wishes to gratefully thank Dr. Tamer El-Sayed for his technical support in assembling the experimental rig to achieve minimum fault condition.

References

1. Wauer J (1990) On the dynamics of cracked rotors: a literature survey. Appl Mech Rev 43:13–17
2. Gasch R (1983) A survey of the dynamic behavior of a simple rotating shaft with a transverse crack. J Sound Vib 160(2):313–332

3. Dimarogonas AD (1996) Vibration of cracked structure: a state of the art review. Eng Fract Mech 55(5):831–857
4. Sabnavis G, Kirk RG, Kasarda M, Quinn D (2004) Cracked shaft detection and diagnostics: a literature review. Shock Vib Digest 36(4):287–296
5. Kumar C, Rastogi V (20090 A brief review on dynamics of a cracked rotor. Int J Rotating Mach 2009:6, Article ID 758108
6. Papadopoulos CA, Dimarogonas AD (1987) Coupling of bending and torsional vibration of a cracked Timoshenko shaft. Ing Arch 57:257–266
7. Papadopoulos CA, Dimarogonas AD (1988) Coupled longitudinal and bending vibrations of a cracked shaft. J Vib Acoust Stress Reliab Des 110:1–8
8. Papadopoulos CA, Dimarogonas AD (1987) Coupled longitudinal and bending vibrations of a rotating shaft with an open crack. J Sound Vib 117(1):81–93
9. Papadopoulos CA, Dimarogonas AD (1988) Stability of cracked rotors in the coupled vibrations. J Vib Acoust Stress Reliab Des 110:356–359
10. Ostachowicz WM, Krawczuk M (1992) Coupled torsional and bending vibrations of a rotor with an open crack. Arch Appl Mech 62:191–201
11. Darpe AK, Gupta K, Chawla A (2002) Analysis of the response of a cracked Jeffcott rotor to axial excitation. J Sound Vib 249:429–445
12. Darpe AK, Gupta K, Chawla A (2003) Experimental investigations of the response of a cracked rotor to axial excitation. J Sound Vib 260:265–286
13. Darpe AK, Gupta K, Chawla A (2004) Coupled bending, longitudinal and torsional vibrations of a cracked rotor. J Sound Vib 269(1–2):33–60
14. Darpe AK, Gupta K, Chawla A (2004) Transient response and breathing behaviour of a cracked Jeffcott rotor. J Sound Vib 272:207–243
15. Sinou JJ (2009) Experimental response and vibrational characteristics of a slotted rotor. Commun Nonlinear Sci Numer Simul 14(7):3179–3194
16. Henry TA, Okah-Avae BE (1976) Vibrations in cracked shafts. In: Vibrations in rotating machinery. The Institution of Mechanical Engineers, Suffolk, pp 15–19
17. Mayes IW, Davies WGR (1976) The vibrational behaviour of a rotating shaft system containing a transverse crack. In: Vibrations in rotating machinery. The Institution of Mechanical Engineers, Suffolk, pp 53–65
18. Bachschmid N, Diana G, Pizzigoni B (1984) The influence of unbalance on cracked rotors. In: Vibrations in rotating machinery. Institution of Mechanical Engineers, Suffolk, pp 193–198
19. Sawicki JT, Wu X, Baaklini GY, Gyekenyesi AL (2003) Vibration-based crack diagnosis in rotating shafts during acceleration through resonance. In NDE for Health Monitoring and Diagnostics. International Society for Optics and Photonics, pp 1–10
20. Scheffer C, Girdhar P (2004) Practical machinery vibration analysis and predictive maintenance. Elsevier, Amsterdam

Parametric Modeling of Main Excitation Sources on Board Vessels

Christoph Tamm, Georg Stoll and Sven Herold

Abstract Efficient analysis methods for predicting the vibro-acoustic behavior of vessels are important, especially during the early design stages. A prediction software tool can help to identify critical points in the ship design and to avoid costly rework subsequent to sea trials. Therefore, an approach based on numerical system-level simulation, to estimate noise, vibration and harshness performance during each step of the design process is proposed. This paper is focused on the development of numerical models of main excitation sources on board. Parametric excitation models for main engines and pumps as well as structural models for power trains and foundations are built up. The sources introduce excitation forces into the foundations or torsional moments into the power train. The overall simulation model is built up modular and contains structural submodels as well as excitation submodels. System-level simulations using *Matlab/Simulink* are performed and both stationary operational conditions and transient excitations, e.g. engine run-ups and misfiring, are simulated. The numerical results are presented and the accuracy of the model is evaluated by comparing numerical and experimental data.

Keywords Modeling · Numerical simulation · Vibro-acoustics · Prediction methods · Excitation sources

C. Tamm (✉) · G. Stoll · S. Herold
Department of Structure Dynamics and Vibration Technology, Division Smart Structures, Fraunhofer Institute for Structural Durability and System Reliability LBF, Bartningstraße 47, 64289 Darmstadt, Germany
e-mail: christoph.tamm@lbf.fraunhofer.de

G. Stoll
e-mail: georg.stoll@lbf.fraunhofer.de

S. Herold
e-mail: sven.herold@lbf.fraunhofer.de

J.K. Sinha (ed.), *Vibration Engineering and Technology of Machinery*, Mechanisms and Machine Science 23, DOI 10.1007/978-3-319-09918-7_70

1 Introduction

In recent years, the passenger comfort on board vessels became increasingly important. Owners and shipyards define low limit values for noise and vibration that must be observed during operation of the ship. On the other hand, light weight construction in combination with a large number of excitation sources lead to vibrations with negative effects on safety and comfort on board. Efficient prediction of the vibro-acoustic behavior of vessels can help to identify critical points in the ship design and to avoid costly rework subsequent to sea trials.

This paper is focused on a parametric modeling approach for the main excitation sources on board vessels. The developed models are described with a finite number of parameters and can be initialized very easily. This can be important in a pre-contractual phase of the ship design, where only a few latter built-in components are defined. Using parametric models can help to evaluate varying installation circumstances or varying components in order to optimize the ship design.

In this paper the developed parametric model is used to simulate excitation sources that cause periodic vibration phenomena, i.e. engines and ancillary machinery. Numerical models of a propulsion main engine and a cooling water pump are built up. The main engine introduces excitation forces into the foundation and torsional moments into the power train. The resulting torsional vibrations again induce structure-borne noise into the gear box, clutch and other parts of the power train.

A torsional vibration test stand is used to verify the proposed simulation approach. A numerical model is set up for a two-cylinder V engine and system-level simulations using *Matlab/Simulink* are performed. Both stationary operational conditions and transient excitations, e.g. engine run-ups and misfiring, are simulated. The numerical results are presented and the accuracy of the model is evaluated by comparing numerical and experimental data.

The numerical model presented in this paper is a part of a currently developed software toolbox to predict the vibro-acoustic behavior of a vessel efficiently. Figure 1 shows the examined vibro-acoustic process chain. One of the aims of this toolbox is the identification of acoustic excitation mechanisms in an early phase of design and also in parallel to the development process. Therefore a modular and hierarchic modeling scheme is proposed, a numerical model is implemented and the results are presented.

2 Numerical Model

The overall simulation model is built up modular and contains structural submodels of the resiliently mounted machinery, which are assumed to be rigid bodies as well as submodels of the foundation, the power train and an excitation submodel. Both the mechanical structures and the excitation submodels are modeled separately with

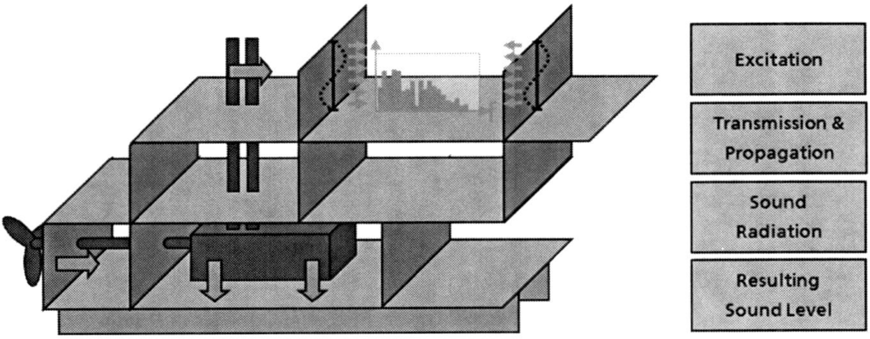

Fig. 1 Scheme of the vibro-acoustic process chain on board a vessel, consisting of excitation, transmission and propagation, sound radiation and a resulting sound level

regards to their compatibility within the overall model. Therefore the connection ports between the submodels must be defined accurately. The advantage of modular modeling is the possibility to easily substitute each submodel describing the same component. With this proposed modeling approach not only the description depth of a submodel can be enhanced but also the numerical formulation can be varied [1, 2]. In a pre-contractual phase an analytically determined model can be built up, which is easy to initialize and to parameterize. In the following design process each submodel can be replaced with a more accurate one, using more detailed modeling strategies or updating assumed parameters with experimentally determined.

2.1 Structural Submodels

2.1.1 Rigid Body Submodel

The housings and cases of the resiliently mounted machinery are modeled as rigid bodies with six degrees of freedom. The rigid body can perform linear translations in the three directions x, y and z and small rotations about the axes. Figure 2 shows a scheme of a rigid body with the global coordinate system located in the body's center of gravity.

The body is mounted to the foundation with a user-defined number of ideal three dimensional spring-damper systems. The mounts can be defined at a location α on the surface of the rigid body and the spring and damping coefficients c_α and b_α are time and frequency independent. External forces can be applied at any position in and on the rigid body and also external moments can be defined at the center of gravity. This leads to a linear time invariant system of equations (see [3, 4]).

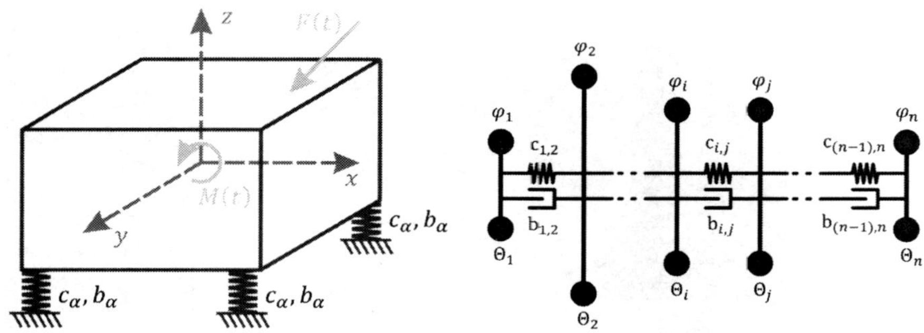

Fig. 2 Scheme of a rigid body (*left*), scheme of a torsional harmonic oscillator (*right*)

2.1.2 Power Train Submodel

The structural submodel of the power train is modeled as a torsional harmonic oscillator. It consists of a finite number of mass moments of inertia Θ_i. These mass moments are coupled by discrete torsional springs and torsional dampers. Figure 2 shows a scheme of a torsional oscillator with n degrees of freedom. For the complete torsional oscillator results a system of ordinary differential equations (see [4–6]). The systems of equation of motion for every installed component (e.g. combustion engine, gear box, clutch or propeller) are set up modularly. Finally, the complete power train is assembled by coupling the individual components. Thus it is easy to replace components or assembly groups to investigate their influence on the vibro-acoustic behavior.

2.1.3 Foundation Submodel

The submodels of the machinery foundation are described in two possible ways. Either the mobility of the foundation in every coupling point to the resiliently rigid body subsystems

$$Z_f(\omega) = \frac{\dot{X}(\omega)}{F(\omega)} \tag{1}$$

is determined experimentally [7], or a Finite Element (FE) model [8] of the foundation is available and the mobility is calculated. The experimental determination is performed by means of modal testing, where a force is supplied to the structure and both the force $F(t)$ and the resulting acceleration $\ddot{x}(t)$ are measured. Transformed into the frequency domain the velocity is calculated with Eq. (2) [7].

$$\dot{X}(\omega) = \frac{1}{\omega}\ddot{X}(\omega) \quad (2)$$

The foundation can also be described with a FE model. In order to determine the mobility transfer function a modal order reduced model, using the *Component Mode Synthesis* [9] implemented in the FE software *ANSYS*, is generated.

2.2 Excitation

The excitation submodels are based on the physical effects of the particular excitation source. For the main engines the excitation forces occur as a result of the combustion process. The considerable external forces and moments in pumps depend on several excitation mechanisms, such as the hydraulic imbalance, the rotating unbalanced masses and the vane passing [2, 5].

2.2.1 Combustion Engine

Both gas and inertia forces occur in consequence of the combustion process and excite vibrations in the mechanical structures. The calculation of the dynamic gas forces and moments are based on one cylinder's combustion pressure characteristics, determined at different mean effective pressures p_{mi}. The resulting cylinder pressure acting at one piston area can be calculated with Eq. (3) [4, 10].

$$p_{cyl} = \sum_k \left((\rho a_k + a_m) \cos\left(\frac{\pi}{T} k\varphi - \varphi_0\right) + \left(\rho b_k \sin\left(\frac{\pi}{T} k\varphi - \varphi_0\right)\right)\right) \quad (3)$$

The coefficients $a_k(p_{mi})$ and $b_k(p_{mi})$ of a Fourier series are calculated with the help of harmonic analyses. For each set of coefficients (e.g. $a_k(p_{mi,1})$ to $a_k(p_{mi,n})$) a continuous function is estimated using a polynomial interpolation. Thus, the cylinder pressure can be calculated for a user-defined number of revolutions $n(t)$ and an angle of revolution $\varphi(t) = \int 2\pi n(t)$ respectively. Setting a starting angle of revolution φ_0 for every cylinder, the ignition sequence of the engine is regarded.

With the coefficient ρ misfiring can be defined. In order to simulate the normal operation state, $\rho = 1$ must be set and the regular cylinder pressure is calculated. For values of $\rho < 1$ the cylinder pressure is decreased. Thus, it is possible to specify misfiring for each individual cylinder and simply vary the length and magnitude of the failure. For the integration of the oscillating inertia forces in Eq. (3) the coefficient

Table 1 Initial parameters of the submodels

Submodel	Assembly	Parameter
Structure submodel	Crank shaft	Moments of inertia, damping, stiffness
	Engine block	Inertia properties, mount positions, mass
Excitation submodel	Piston	Number, stroke, piston area, mass
	Connecting rod	Length, mass, strokes, V angle
	Propeller power curve	Rated speed (rpm), permitted rpm, power, firing order, cylinder pressure

$$a_m = \frac{m_{osc} r \omega^2}{A_k} A_i(\lambda) \tag{4}$$

is introduced, where m_{osc} is the oscillating mass of the piston, r is the radius of the crank shaft, A_k is the piston area and $A_i(\lambda)$ are approximated coefficients describing the piston stroke [3].

The developed analytical Eq. (3) was implemented using the *Matlab/Simulink* environment. With the resulting parametric excitation submodel the dynamic gas and inertia forces of a combustion engine can be simulated. The numerical results are dependent on several initial parameters (see Table 1).

All acting forces at the individual crank drives are calculated. Summed over the length of the crank shaft, considering the phase relationship and ignition sequence these forces are used as the excitation mechanism in the overall simulation model.

2.2.2 Cooling Water Pump

The occurring forces and moments in pumps depend on different excitation mechanisms. Hydraulic imbalances as well as rotating unbalanced masses lead to considerable periodic excitation forces. The vane passing is another physical effect, which has a significant influence on the response spectrum of the pump housing. Additionally stochastic excitations occur due to hydraulic pressure pulses and flow generated noise [5].

In addition to the expected vane passing frequencies numerous other rotational speed dependent harmonic oscillations occurred during the first experimental test cases. In order to take these additional effects into account, an inverse method to determine an excitation force spectrum on the basis of experimental data is proposed. Operating measurements were performed on a cooling water pump test stand to derive an excitation model. The data was recorded with eight triaxial accelerometers and under 26 different operating conditions. In the next step the measured accelerations in relative sensor coordinates were transformed into accelerations in global coordinates (center of gravity). Performing a Taylor series approximation and linearization of the sine and cosine terms, the acceleration of the center of gravity $\mathbf{x}$ can be calculated for n axis-parallel aligned sensors in relative coordinates $\mathbf{x_{rn}}$ with Eqs. (5)–(9).

$$\ddot{\tilde{\mathbf{x}}} = \mathbf{T}\ddot{x} \tag{5}$$

With:

$$\mathbf{x_{rn}} = (x_{rn}, y_{rn}, z_{rn}) \tag{6}$$

$$\ddot{\mathbf{X}} = (\ddot{x}, \ddot{y}, \ddot{z}, \ddot{\theta}, \ddot{\gamma}, \ddot{\phi}) \tag{7}$$

$$\ddot{\tilde{\mathbf{x}}} = (\ddot{x}_i, \ddot{y}_i, \ddot{z}_i; i = \{1..n\}) \tag{8}$$

$$\mathbf{T} = \begin{bmatrix} \frac{d\ddot{x}_1}{d\ddot{x}} & \cdots & \frac{d\ddot{x}_1}{d\ddot{\phi}} \\ \vdots & \ddots & \vdots \\ \frac{d\ddot{z}_n}{d\ddot{x}} & \cdots & \frac{d\ddot{z}_n}{d\ddot{\phi}} \end{bmatrix} \tag{9}$$

To determine the rigid body motion in global coordinates from the accelerations in relative coordinates, the inverse transformation must be solved. The inverse transformation problem has a reduced rank so that an approximation by minimizing the quadratic problem using the Moor-Penrose pseudo-inverse $\mathbf{T}^+$ was applied [11].

$$0 = T\ddot{\mathbf{X}} - \ddot{\tilde{\mathbf{x}}} \tag{10}$$

$$\ddot{\mathbf{x}} = \operatorname{argmin}_{\ddot{x}}\left(T\ddot{\mathbf{X}} - \ddot{\tilde{\mathbf{x}}}\right)^2 = \mathbf{T}^+\ddot{\tilde{\mathbf{x}}} \tag{11}$$

Using the inverse transformation (11) the measured time signals and the calculated complex frequency spectra can be transformed into those with respect to the assumed rigid body.

$$\begin{pmatrix} \ddot{\mathrm{x}}(\omega) \\ \ddot{\mathrm{y}}(\omega) \\ \ddot{\mathrm{z}}(\omega) \\ \ddot{\theta}(\omega) \\ \ddot{\gamma}(\omega) \\ \ddot{\phi}(\omega) \end{pmatrix} \approx \mathbf{T}^+ \begin{pmatrix} \ddot{\mathrm{x}}_1(\omega) \\ \ddot{\mathrm{y}}_1(\omega) \\ \\ \vdots \\ \\ \ddot{\mathrm{z}}_\mathrm{n}(\omega) \end{pmatrix} \tag{12}$$

3 Application and Results

The first application to verify the proposed simulation approach is the torsional vibration test stand depicted in Fig. 3. The test stand was built up in order to reproduce operational conditions in power trains of vehicles. The operational loads and vibrations can be experimentally simulated with reduced amplitudes. The test

Fig. 3 Torsional vibration test stand (*left*), combustion engine (*right*)

Fig. 4 Measured and simulated vertical accelerations at constant rotational speed of 3,000 rpm

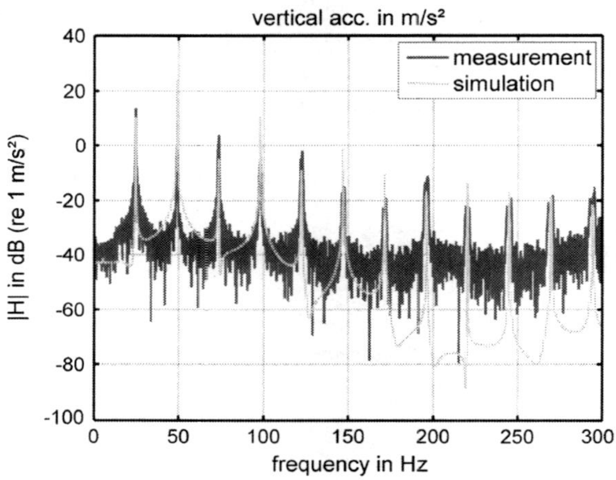

stand consists of a two-cylinder V engine, a rotor and an eddy current brake. The initial parameters of the submodels are shown in Table 1.

The results of the simulation are presented in this section. In the following Fig. 4 the measured and simulated accelerations in the vertical direction of the engine are compared. They refer to a defined position P on the surface of the base plate of the engine (see Fig. 3). In order to validate the simulation model, an experimental operational modal analysis was accomplished. The accelerations were recorded during a constant run and a run up, using a *LMS SCADAS mobile* system and PCB 100 mV/g triaxial accelerometers. The measured signals were interpreted with *LMS Test.Lab* software.

In Fig. 4, the measured and simulated results of a constant run at 3,000 rpm (revolutions per minute) are presented. The vertical accelerations are plotted in the

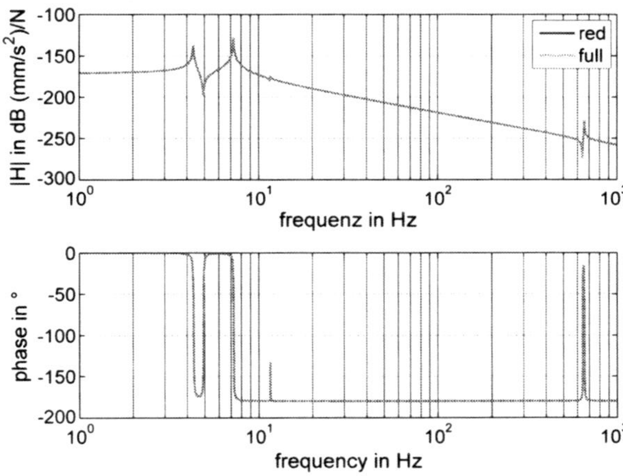

Fig. 5 Transfer function $H(\omega)$ of test stand foundation

frequency domain. The orange curve is the amplitude spectrum of the simulated acceleration, while the green curve shows the amplitude spectrum of measurement. The axis of ordinates is scaled logarithmically to a reference value of 1 m/s^2. The excited frequencies match very well and in the range from 0 to 200 Hz the amplitudes can be predicted accurately. Using the proposed models, both constant and user-defined time varying rotational speed (e.g. engine run-ups) can be specified and the numerical results can be simulated.

The second application to verify the simulation approach is a cooling water pump. The excitation force spectrum is calculated from experimental determined data, as described in Sect. 2.1. The accelerations were measured on a test stand and the dynamic behavior of the test stand was calculated. The geometry model of the test stand foundation was created; a FE model and a reduced order model using the *Component Mode Synthesis* were built up [8, 9]. Using this reduced model the transfer functions $\mathbf{H}(\omega)$ (see Fig. 5) from the pump's center of gravity to the coupling points were calculated. In combination with the measured accelerations $\ddot{\mathbf{x}}(\omega)$ the excitation force spectrum is derived.

$$\mathbf{F}(\omega) = \mathbf{H}^{-1}(\omega) \cdot \ddot{\mathbf{x}}(\omega) \tag{13}$$

In Fig. 6 the simulated excitation force spectrum is presented. For the numerical simulation only the magnitude and phase information of the 1st to 23rd order of the rotational speed are used. To consider also the stochastic parts of the spectrum, colored noise was added to the harmonic parts. The comparison on the right shows good accordance between the calculation from the measurement and the simulated results.

Fig. 6 Simulated (*green*) and from measurements derived (*orange*) excitation force spectrum

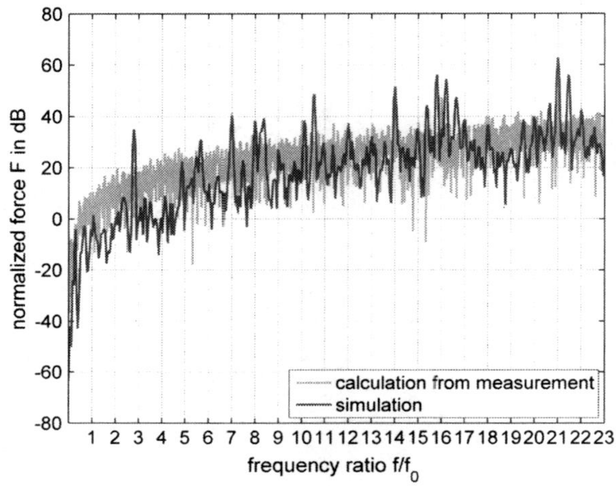

4 Conclusion

In this paper a parametric modeling approach for the prediction of main excitation sources on board vessels is presented. A modular and hierarchic numerical model has been set up, using the *Matlab/Simulink* environment. The structural submodels are based on analytical equations and dynamic forces $F(t)$ and accelerations $\ddot{z}(t)$ are simulated. Most of the initial parameters are taken from technical data sheets, installation drawing or supplier reports, so the models are easy to initialize.

Numerical simulations and measurements of stationary and transient operational conditions have been performed. The numerical results have been presented and the accuracy of the models has been evaluated by comparing numerical and experimental data. It is found that the proposed methods can be used to estimate noise, vibration and harshness performance in an early phase of the ship design process. Thus it might be possible to identify critical points in the ship design and to benchmark possible active and passive concepts for the reduction of vibrations in preliminary studies.

Acknowledgment The results presented in this contribution were developed within the collaborative EPES project. It was funded by the German Federal Ministry of Economics and Technology (project ref. no. 03SX305). This financial support is gratefully acknowledged.

References

1. Herold S, Jungblut T, Kurch M (2009) A systematic approach to simulate active mechanical structures. In: NAFEMS seminar: 'multi-disciplinary simulations—the future of virtual product development', Wiesbaden, Germany
2. Isermann R (2008) Mechatronische Systeme, Grundlagen. Springer, Berlin

3. Maass H, Klier H (1981) Kräfte, Momente und deren Ausgleich in der Verbrennungskraftmaschine, Die Verbrennungskraftmaschine Band 2. Springer, Wien
4. Tamm C, Kurch M (2013) Modeling and simulation of main engine excitation on board vessels, Marine 2013, Hamburg, Germany
5. Dresig H, Holzweißig F, Rockhausen L (2009) Maschinendynamik. Springer, Berlin
6. Hafner KE, Maass H (1985) Torsionsschwingungen in der Verbrennungskraftmaschine, Die Verbrennungskraftmaschine Band 3. Springer, Wien
7. Ewins DJ (2000) Modal testing: theory, practice and application. Research Studies Press, Baldock. ISBN:0-86380-218-4
8. Bathe KJ (1982) Finite element procedures in engineering analysis. Prentice-Hall Inc., New Jersey
9. Craig RR, Bampton MCC (1983) Coupling of substructures for dynamic analyses. AIAA J 6 (7):1313–1319
10. Tamm C, Atzrodt H, Kurch M (2012) Modellierung der Motoranregung zur vibroakustische Prognose im Schiffbau, 38. Jahrestagung für Akustik
11. Penrose R (1955) A generalized inverse for matrices. Proc Camb Philos Soc 51:406–413

Investigation of Motor Current Signature Analysis in Detecting Unbalanced Motor Windings of an Induction Motor with Sensorless Vector Control Drive

M. Lane, D. Ashari, F. Gu and A.D. Ball

Abstract Maintaining the efficiency of AC motors in site equipment is important, given the increasing cost of energy. Reduction of motor efficiency from baseline manufacturer data can go undetected until total failure of the equipment is experienced. This paper introduces motor current signature analysis methods used to detect the early onset of motor efficiency reduction in AC motors controlled by modern Sensorless-Vector Variable Speed Control inverters. A step increase in the resistance of one stator winding is simulated in stages. Off-line processing of motor current data signals using data analysis methods developed for the MATLAB platform is used to identify imbalances caused by subtle stator resistance increases. Initial results indicate that small increases in stator resistances can be observed in the motor current signals received after data processing techniques have been used on the measured signals. The test results are presented herein along with details on the research work to be continued.

Keywords Unbalanced · Efficiency · Single stator resistance increase · MCSA · PWM · Random switching pattern · IAS

1 Introduction

The model system of a standard Pulse Width Modulated (PWM) inverter, operating in Sensorless Vector mode and utilizing noise-reducing random PWM-pattern generation technology is used for this study. This research is intended to detect the

M. Lane (✉) · D. Ashari · F. Gu · A.D. Ball
Centre for Efficiency and Performance Engineering, University of Huddersfield, Huddersfield, UK
e-mail: mark.lane@hud.ac.uk

D. Ashari
e-mail: djoni.ashari@hud.ac.uk

J.K. Sinha (ed.), *Vibration Engineering and Technology of Machinery*, Mechanisms and Machine Science 23, DOI 10.1007/978-3-319-09918-7_71

incipient increases in the resistance of one single motor stator phase as a fault develops and before critical failure occurs.

Increases in single stator resistance of a three-phase induction motor can lead to voltage imbalances in the motor, causing a reduction of motor efficiency, increase in motor temperature and oscillatory running conditions. This in turn can lead to other electrical or mechanical failures occurring. Conditions such as corroded terminals, dissimilar metals, different conductor sizes, insulation breakdown of cables or connections have been reported as common causes of high resistance connections [1]. Installing a motor in these conditions can result in premature failure of the motor. However, this fault will remain undetected by the drive system because the motor is still functionally operational under these imbalanced conditions, but operating less efficiently.

A study on the detection of winding resistance faults in an AC squirrel cage induction motor fed directly from the mains supply is not a new research area. There have been a number of studies undertaken to detect critical stator faults in AC motors such as phase–phase or phase–neutral faults [2]. Modern inverter systems have fast fault detection systems in place to protect the Insulated Gate Bipolar Transistor (IGBT) device from failure and most drives indicate which phase of the IGBT device has experienced an overcurrent trip or other phase failure, allowing the engineer to trace the fault to one or more motor phase switching devices [3]. However, this is only effective in detecting critical faults such as open circuit or short-circuits. If these faults are not significant, the inverter drive will not detect an error and the drive system will continue to operate as normal.

The effects of unbalanced supply voltage feeds to an AC motor are well understood and result in the motor operating efficiency being reduced. The voltage imbalances are documented in the NEMA standards for AC induction motor performance [4] and can result in a loss of efficiency of up to 75 % for a voltage imbalance of 5 %.

The purpose of this paper is to examine the effects of small step changes in rotor resistance and develop a reliable means of detecting the change. The study of motor temperature using only the standard PT100 device and attempt to correlate this to an increase in motor temperature is included as are Instantaneous Angular Speed (IAS) and vibration measurements.

2 Test Facilities and Fault Simulation

A test rig consisting of a PWM inverter with 3 kHz switching frequency and a carrier frequency selected by a random pattern generator enabled by default was used. For the research to be valid with future drive systems, it is important that valid test results can be obtained from equipment that utilizes the very latest motor control technology and with these advanced switching techniques enabled. A standard inverter-rated 690 V AC motor was used, connected in Delta to the 415 V output inverter.

A shunt-wound DC motor is used to apply different loading to the AC motor drive system. The speed, load and test duration are all programmable and repeatable. The DC motor regenerates to the mains supply through a four-quadrant two-phase DC drive.

Previous research studies in this area have been limited to Volts/Hertz mode inverters—these being the most simple of variable speed control drives and traditionally the favoured workhorse. However, due to the reduced cost of sensorless vector drives, their use in industry is now more widespread [5] and their use is even applied to applications that traditionally a Volts/Hertz inverter would have sufficed, such as fan or pump applications. If this research was carried out using standard Volts/Hertz drives, the findings would have only been limited to a small, dwindling part of the AC drives market. Indeed, out of the Parker SSD Drives offering of eight drive variants ranging from 0.5 to 2000 kW, only one variant offers standard Volts/Hertz mode, the 650 model with a maximum size of 7.5 kW. By basing the research on latest drive technology, the results are able to be applied to existing and newer installations with confidence.

2.1 Healthy Operating Conditions

It is important to have a meaningful and consistent baseline data set from which to measure simulated faults. To this effect, healthy baseline data was measured in three separate tests, each with the same operating data applied. The design of the test rig is such that a Programmable Logic Controller provides a predictable load change to the AC driven motor system at precise, repeatable intervals.

A sampling rate of 96 kHz with a 40 s sample time was used for each test run. A total of 15 data sets were therefore obtained for the baseline data.

The baseline data measurement test parameters are shown in Table 1.

2.2 Fault Simulation

There are a number of phase resistance increments available on the test rig in 0.1 Ω steps obtained by changing the wiring tapping on the custom-built resistor unit.

It is proposed to simulate two types of phase resistance faults on the test rig. The first fault is one that develops in the supply connection between the inverter drive and AC motor shown in Fig. 1. On the test rig delta-connected motor, this increase in resistance affects two of the motor phase circuits.

The second fault is simulated to occur in one winding inside the AC motor thereby affecting only one motor phase. This is indicated in Fig. 2.

The same test as used for the baseline data tests was run for each fault condition.

Table 1 Test run speeds, duration of test and AC motor loading

Test	Speed (RPM)	Test duration (s)	Load (% of motor FLC)
1	1,470	40	0
2	1,470	40	25
3	1,470	40	50
4	1,470	40	75
5	1,470	40	100

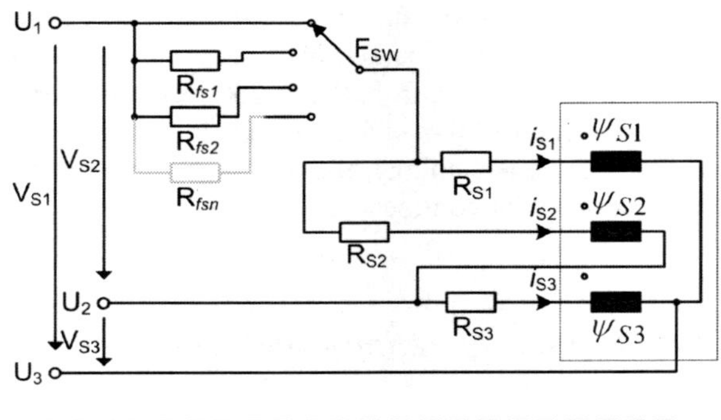

Fig. 1 Motor magnetizing circuit detailing the stator phase resistances used in one of the three phase connections between inverter and motor (DELTA circuit)

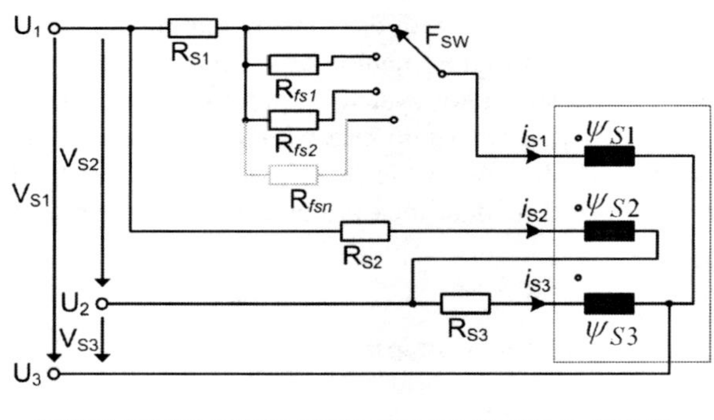

Fig. 2 Motor magnetizing circuit detailing the stator phase resistances used in one of the three phase connections internally in the motor on phase 1 (DELTA circuit)

2.3 Measuring Devices

A current and voltage measurement unit was designed and constructed to measure the supply connection between drive and motor. Closed-loop hall-effect current transformers from ABB with a 200 kHz bandwidth fed from a high-precision voltage source were used for each of the motor phases. So that motor power can be calculated, voltage measurement devices with a 15 kHz bandwidth also from ABB have been fitted for each phase.

Vibration, temperature and IAS measurements are also taken from the test rig.

3 Phase Winding Resistance Increments

After a series of baseline healthy tests are run, a stator resistance was applied to the stator winding. This was done in series with the winding and will be referred to as the resistance R_{fs}. The effect on the equivalent circuit of the stator is shown in Fig. 1 and the effect on one phase resistance increase on the stator equivalent circuit equation is as below:

$$V_{s1} = R_s is + = \frac{d\psi_{s1}}{dt} \rightarrow V_{s1} = \left(R_{fs} + R_s\right) is + = \frac{d\psi_{s1}}{dt}$$

This creates an imbalance in the stator circuit and the effect is to reduce the magnetic flux generated by one of the stator windings in proportion to the value of R_S . The effect of this is to be studied for motor instability and winding temperature increase compared to normal running conditions.

It is proposed to use the developed algorithms in MATLAB to indicate differences in the following motor parameters for each resistance increase.

- Voltage imbalance of three phases
- Current imbalance of three phases
- Baseline versus fault efficiency differences
- Effect on rotor asymmetry
- Effect on vibration
- Effect on IAS

4 Data Analysis and Results

Initial data results are included that demonstrate the difference between standard fixed-PWM frequency carrier drives and the more modern random pattern drives to illustrate the difficulties that could be present when attempting to harvest data from such drive systems.

4.1 Random Carrier Pattern Measurements

It was noted from initial motor current measurements that the random switching pattern generator introduces a noise floor to the data as can be seen in the measurements taken in Fig. 3. Compare this with Fig. 4, where the central carrier frequency and the associated sidebands can be readily identified and with post-signal processing potentially notch-filtered out.

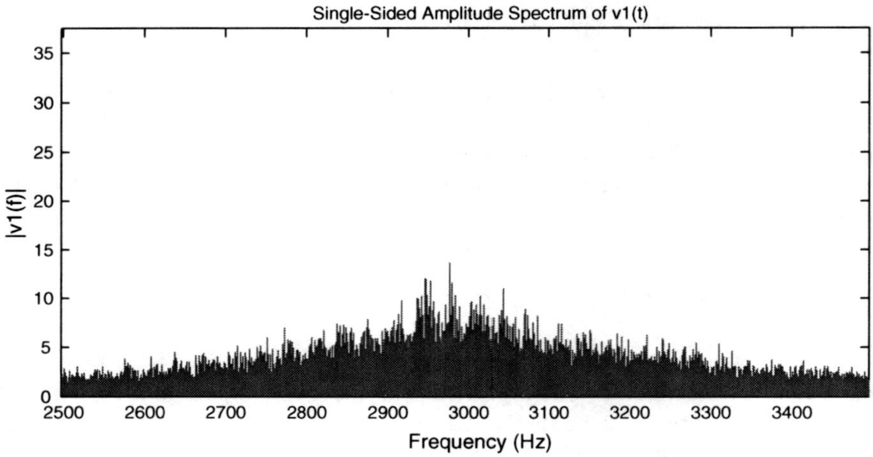

Fig. 3 Random pattern generation disabled; speed 1,470 RPM, load 100 %

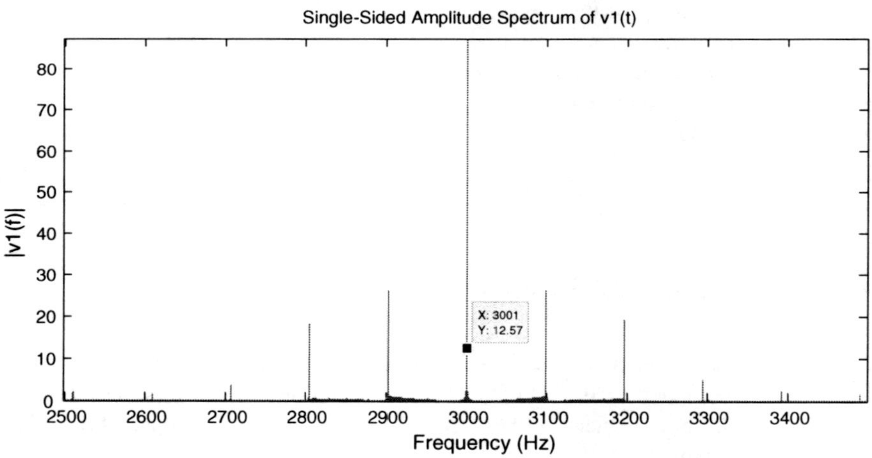

Fig. 4 Random pattern generation enabled; speed 1,470 RPM, load 100 %

The more random nature of measurements taken in the motor current signals may impact adversely on the ability to detect the small resistance changes to be applied. Nevertheless, the random pattern generation was enabled and test measurements taken under this condition. It is intended to run further tests to identify if the random pattern generation does affect any test results obtained.

4.2 Resistance Measurement Results

The test results taken from initial fault conditions generated show that at resistance increases 0.4, 0.6 and 0.8 Ω, there is a measurable increase in the voltage imbalance compared to the three baseline (BL) data sets. This is illustrated in Fig. 5. Current imbalance in each of the three phases is less obvious and this is given in Fig. 6.

The efficiency of the motor was also then examined and compared to the baseline data. The Efficiency by Air-Gap Torque is concerned with the losses as a product of the slip and air gap power calculated from the motor current, voltage measurement and IAS. No noticeable efficiency decrease was observed for the increase in stator resistance fault value shown in Fig. 7.

Figure 8 details the changes of efficiency measurements taken from the baseline data. It shows that the efficiency is increasing slightly at speeds lower than the motor rated base speed, indicating that the controller performs better at these

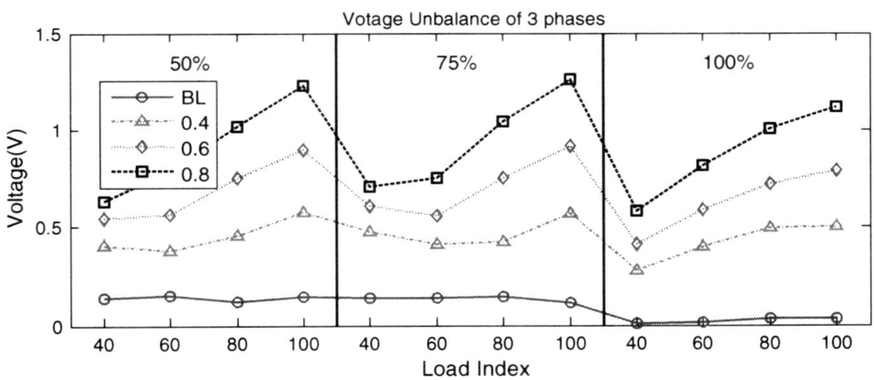

Fig. 5 Voltage imbalance (asymmetry) by means of NEMA definition

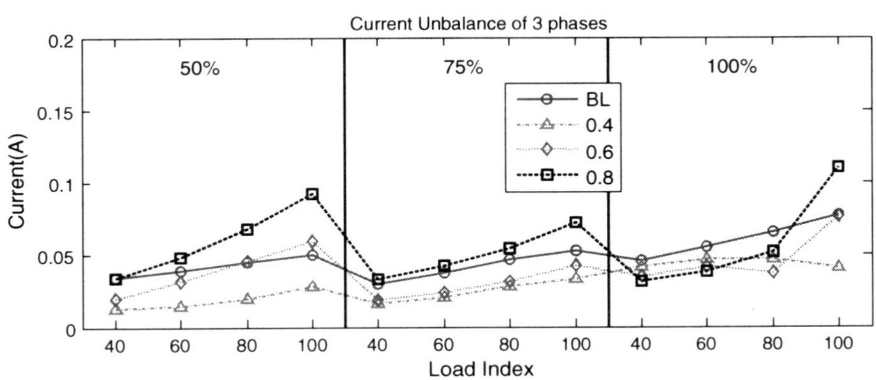

Fig. 6 Current imbalance (asymmetry) by means of NEMA definition

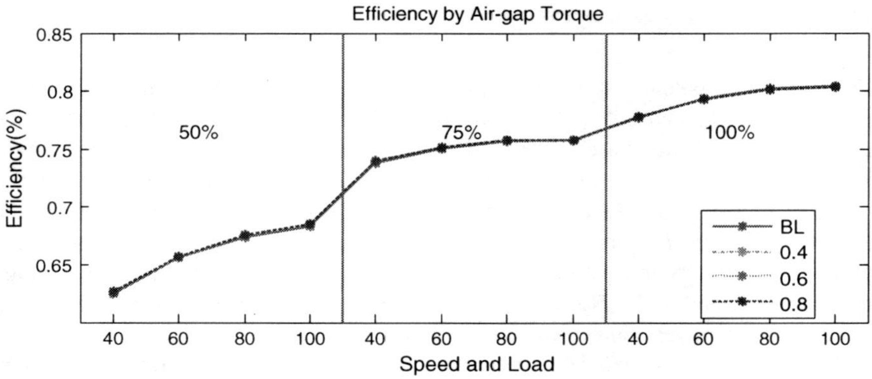

Fig. 7 Motor efficiency by air gap torque

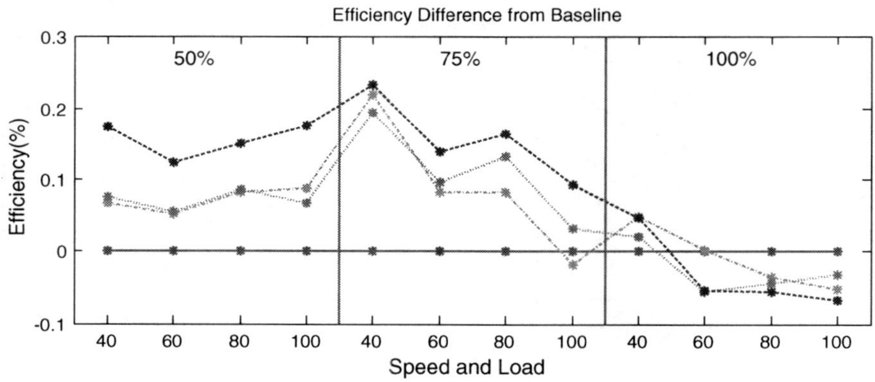

Fig. 8 Motor efficiency decrease from baseline data for each fault resistance increment

speeds. However, when the motor operates at the rated speed, the efficiency shows a slight decrease. This reduction may be due to the controller being unable to compensate for the resistance imbalance due to the increased resistance exceeding the motor specification for one phase.

Figures 9 and 10 show the results from vibration and IAS speed analysis on the motor. No noticeable differences were observed from each test data run. This is inconsistent with previous findings that the vibration increases with unbalanced supply conditions. The reasons behind may be due to the speed adjustment performance of the drive system.

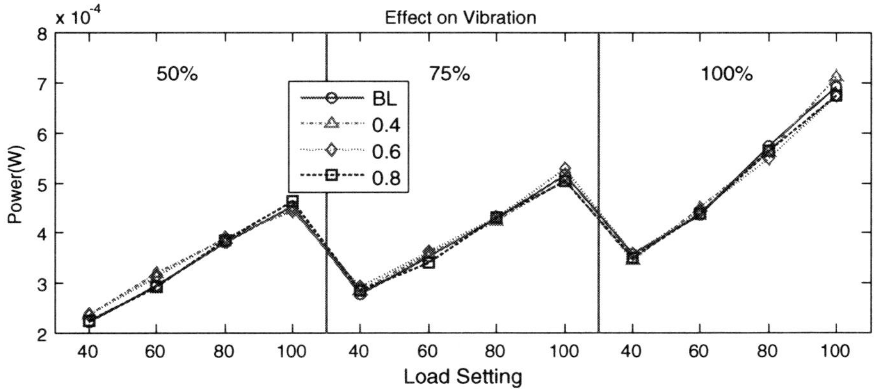

Fig. 9 Effect of resistance increase on rotor vibration

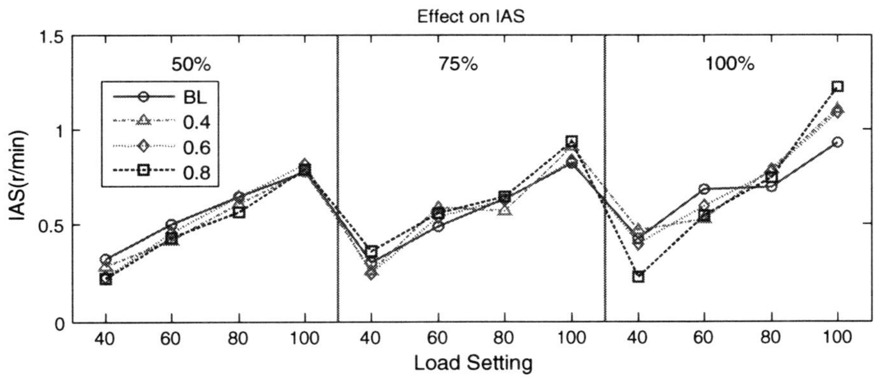

Fig. 10 Effect on rotor instantaneous angular speed

5 Conclusion

Initial results taken from the baseline and faulty data sets for the two resistance placements indicate that the small resistance increases in the stator windings of 0.2 Ω increments from 0.4 to 0.6 Ω can be readily observed by voltage and current imbalances in the supply. This change of resistance also leads to a slight increase in motor efficiency at the speed range below the motor rated speed, whereas there is a reduction in efficiency at the rated speed.

The small difference in IAS and vibration could be attributed to the inverter drive operating in sensorless vector control mode. This is likely to result in the rotor speed being maintained more accurately than for a Volts/Hertz unit. More tests are required under the two different inverter modes to confirm this behaviour.

At the time of writing, the motor temperature measurements based on internal (the motor thermistor included as standard) and external measurements have been inconclusive with respect to the efficiency decreases at full load but further tests will be carried out with a view to correlating this temperature data with efficiency reductions.

References

1. Noah B (2007) Identifying motor defects through fault zone analysis. http://www.reliableplant.com/Read/4288/motor-defects-fault-zone. Accessed 10 Feb 2014
2. Williamson S, Mirzoian K (1985) Analysis of cage induction motors with stator winding faults. IEEE Trans Power Apparatus Syst PAS-104(7):1838–1842
3. Siemens (2004) Simovert Masterdrives 2 Compendium, p 906
4. Austin HB (1999) The impact that voltage and frequency variations have on AC induction motor performance and life in accordance with NEMA MG-1 standards. In: Conference record of 1999 annual pulp and paper industry technical conference (Cat. No.99CH36338)
5. John S, Dan G (2007) New motors offer better performance. http://touchstoneenergy.coopwebbuilder2.com/sites/touchstoneenergy/files/images/BEA/Technology/NewMotorsOfferBetterPerformance.pdf. Accessed 10 Feb 2014

Analysis of Locally Nonlinear Two Dimensional Periodic Structures Using NOFRFs

C.M. Cheng, Z.K. Peng, W.M. Zhang and G. Meng

Abstract The analysis of locally nonlinear two dimensional (2D) periodic structures is studied using the concept of nonlinear output frequency response functions (NOFRFs). NOFRFs are a series of one-dimensional functions of frequency, which allow the analysis of nonlinear systems to be implemented in a manner similar to the analysis of linear system in the frequency domain. By inspecting the relationships between the NOFRFs of unit cell in the locally nonlinear two dimensional periodic structures, a series of properties about the NOFRFs of the nonlinear system are derived. The correctness of these properties is demonstrated by numerical studies. These properties obtained in this paper are of significance to the frequency response analysis for nonlinear 2D periodic structures. One important potential application of these properties is the location of damage in 2D periodic structures, which make system behave locally nonlinearity.

Keywords Volterra series · Nonlinear output frequency response function (NOFRF) · Two dimensional (2D) periodic structure · Damage detection

C.M. Cheng · Z.K. Peng (✉) · W.M. Zhang · G. Meng
State Key Laboratory of Mechanical System and Vibration, Shanghai Jiao Tong University, Shanghai 200240, People's Republic of China
e-mail: z.peng@sjtu.edu.cn

C.M. Cheng
e-mail: changmingcheng@126.com

W.M. Zhang
e-mail: wenmingz@sjtu.edu.cn

G. Meng
e-mail: gmeng@sjtu.edu.cn

J.K. Sinha (ed.), *Vibration Engineering and Technology of Machinery*, Mechanisms and Machine Science 23, DOI 10.1007/978-3-319-09918-7_72

1 Introduction

Periodic structures [1] are defined as structures consisting of identical substructures connected to each other in an identical manner. Analysis of the free and forced vibration and the mode analysis of linear periodic structures are of particular interests [2, 3]. Attentions have also been paid to the study of nonlinear periodic structures. Marathe and Chatterjee [4] have studied the wave attenuation in 1D nonlinear periodic structures using harmonic balance and multiple scale method. Narisetti et al. [5] investigated wave dispersion in two dimensional, weakly nonlinear periodic lattices using perturbation approach. In engineering practices, there are considerable periodic structures that behave nonlinearly just because one or a few components have nonlinear properties, and the nonlinear component is often the component where a fault or abnormal condition occurs. One of the well-known examples is beam structures [6] with breathing cracks, the global nonlinear behaviors of which are caused only by a few cracked elements. Therefore it is of great significance to effectively detect the position of the nonlinear components in a periodic structure. Based on a 1D periodic structure model, Sakellariou and Fassois [7], Fassois and Sakellariou [8], have used a stochastic output error vibration based methodology to detect the damage in structures where the damage elements were modeled as components of cubic stiffness.

The NOFRFs can be considered to be an alternative extension of the classical frequency response function (FRF) for linear systems to the nonlinear case. A series of important properties about the NOFRFs of locally nonlinear 1D periodic structure have been derived in Ref. [9]. In addition, based on these important properties of the NOFRFs, Peng and Lang [10] developed a novel method to detect the position of the nonlinear component in a 1D nonlinear periodic structure. In the present study, a series of important properties about the NOFRFs of 2D locally nonlinear periodic structure have been derived. In addition, based on these important properties of the NOFRFs, a novel method is developed to detect the position of the nonlinear components in a 2D locally nonlinear periodic structure. The simulation studies verify the effectiveness of the novel approach.

The outline of this paper is as follows. Section 2 briefly describes theoretical background of the Volterra series and the NOFRF concept. Some important properties of the NOFRFs for 2D locally nonlinear periodic structure are first derived in Sect. 3. Numerical simulations are used to verify the effectiveness of the proposed method in Sect. 4. Finally, the conclusions are given in Sect. 5.

2 Nonlinear Output Frequency Response Functions

The Volterra series is defined as

$$y(t) = \sum_{n=1}^{N} \int_{-\infty}^{+\infty} \int_{-\infty}^{+\infty} \cdots \int_{-\infty}^{+\infty} h_n(\tau_1, \tau_2, \ldots, \tau_n) \prod_{i=1}^{n} u(t - \tau_i) d\tau_1 d\tau_2 \cdots d\tau_n \quad (1)$$

where $y(t)$ and $u(t)$ are the output and input of the system, $h_n(\tau_1, \ldots, \tau_n), n = 1, 2, \ldots, N$ are the Volterra kernel functions.

The definition of the NOFRFs is based on the Volterra series theory of nonlinear systems. Lang and Billings [11] derived an expression for the output frequency response of this class of nonlinear systems to a general input. The result is

$$\begin{cases} Y(\omega) = \sum_{n=1}^{N} Y_n(\omega) \quad \forall \omega \\ Y_n(\omega) = \dfrac{1/\sqrt{n}}{(2\pi)^{n-1}} \displaystyle\int_{\omega_1 + \cdots + \omega_n = \omega} {}_{R^n} H_n(\omega_1, \ldots, \omega_n) \prod_{i=1}^{n} U(\omega_i) d\omega_{1 \to n} \end{cases} \quad (2)$$

where, $Y(\omega)$ is the spectrum of the system output, $U(\omega)$ is the spectrum of the system input, $Y_n(\omega)$ represents the nth order output frequency response of the system. In order to conveniently analyze nonlinear system in the frequency domain, Lang and Billings [12] recently proposed a new concept known as the nonlinear output frequency response function (NOFRFs). The NOFRFs are defined as,

$$G_n(\omega) = \frac{\int_{\omega_1 + \cdots + \omega_n = \omega} H_n(\omega_1, \ldots, \omega_n) \prod_{i=1}^{n} U(\omega_i) d\omega_{1 \to n}}{\int_{\omega_1 + \cdots + \omega_n = \omega} \prod_{i=1}^{n} U(\omega_i) d\omega_{1 \to n}} \quad (3)$$

under the condition that

$$\int_{\omega_1 + \cdots + \omega_n = \omega} \prod_{i=1}^{n} U(\omega_i) d\omega_{1 \to n} \neq 0 \quad (4)$$

By introducing the NOFRFs $G_n(\omega), n = 1, \ldots, N$, Eq. (2) can be rewritten as,

$$Y(\omega) = \sum_{n=1}^{N} Y_n(\omega) = \sum_{n=1}^{N} G_n(\omega) U_n(\omega) \quad (5)$$

3 GFRFs of 2D Locally Nonlinear Periodic Structures

Consider the two dimensional locally nonlinear periodic structures [13] where the component $\{N_1, N_2\}$ (shadow section in the middle of Fig. 1) is nonlinear, and the external force acting on mass (RF, CF) which is shown in Fig. 1. RN is the maximum number of row, and CN is the maximum number of column. Denote

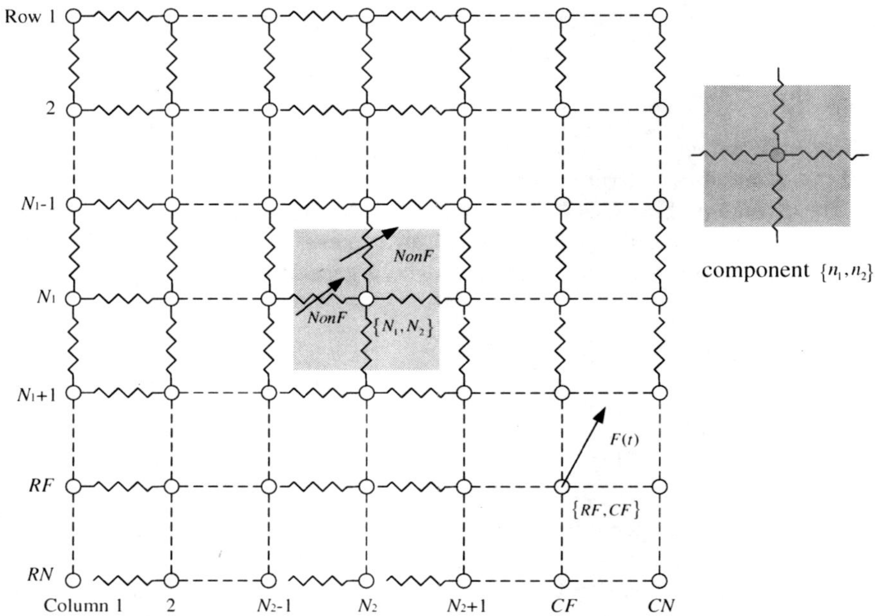

Fig. 1 Two dimensional locally nonlinear periodic structures

$$NonF\_row = \sum_{i=1}^{P} r_i\left(y_{n_1,n_2} - y_{n_1-1,n_2}\right)^i + \sum_{i=1}^{P} w_i\left(\dot{y}_{n_1,n_2} - \dot{y}_{n_1-1,n_2}\right)^i \tag{6}$$

$$NonF\_column = \sum_{i=1}^{P} r_i\left(y_{n_1,n_2} - y_{n_1,n_2-1}\right)^i + \sum_{i=1}^{P} w_i\left(\dot{y}_{n_1,n_2} - \dot{y}_{n_1,n_2-1}\right)^i \tag{7}$$

NonF_row and *NonF_column* represent the nonlinear restoring force produced by the nonlinear component in column direction and in row direction, respectively.

The equations of motion for the mass $\{n_1, n_2\}$ which comprises of the generalized coordinates y_{n_1,n_2} can be written in a general and compact form as [13],

$$m\frac{d^2 y_{n_1,n_2}}{dt^2} + \sum_{p,q=-1,0,1} c^{(p,q)} \frac{dy_{n_1+p,n_2+q}}{dt} + \sum_{p,q=-1,0,1} k^{(p,q)} y_{n_1+p,n_2+q} + FN = f^{ext}_{n_1,n_2}(t) \tag{8}$$

where, *FN* is the nonlinear restoring force produced by the nonlinear component, which can be described as

$$FN = \begin{cases} 0 & \text{mass}\{n_1, n_2\} \text{ not connected with nonlinear component}\{N_1, N_2\} \\ NonF\_row & \text{mass}\{n_1, n_2\} \text{ connected with top of nonlinear component}\{N_1, N_2\} \\ NonF\_column & \text{mass}\{n_1, n_2\} \text{ connected with left of nonlinear component}\{N_1, N_2\} \\ -NonF\_column - NonF\_row & \text{mass}\{N_1, N_2\} \end{cases} \quad (9)$$

To analyze the relationships between the *N*th order GFRFs, consider the input,

$$u(t) = e^{i\omega_1 t} + \cdots + e^{i\omega_N t} \quad (10)$$

Substituting this input into Eq. (1), yields

$$\begin{aligned} y_{n_1,n_2}(t) = H_{n_1,n_2,1}(\omega_1)e^{i\omega_1 t} + \cdots + H_{n_1,n_2,1}(\omega_N)e^{i\omega_N t} + \cdots \\ + N!H_{n_1,n_2,N}(\omega_1, \ldots, \omega_N)e^{i(\omega_1+\cdots+\omega_N)t} + \cdots \end{aligned} \quad (11)$$

where, $H_{n_1,n_2,p}(\omega_1, \ldots, \omega_N)$ $(p = 1, \ldots, N)$ is the *p*th order GFRF of mass $\{n_1, n_2\}$.
In order to reveal properties about the NOFRFs of 2D locally nonlinear periodic structure which can provide a more convenient way to detect the position of the nonlinear components, this paper divides all masses into the several cases according to whether the masses are connected to, below or above the nonlinear components, or suffered external force. Then, all these cases are analyzed respectively.

For the masses that are above the nonlinear spring and not connected to nonlinear spring and suffered no external force, in order to derive the relationships between the *N*th order GFRFs, substituting Eqs. (10)–(11) into Eq. (8), extracting the coefficients of $e^{i(\omega_1+\cdots+\omega_N)t}$, yields,

$$\begin{aligned} &- m(\omega_1 + \cdots + \omega_N)^2 H_{n_1,n_2,N}(\omega_1, \ldots, \omega_N) \\ &+ \sum\nolimits_{p,q=-1,0,1} c^{(p,q)}(i(\omega_1 + \cdots + \omega_N))H_{n_1+p,n_2+q,N}(\omega_1, \ldots, \omega_N) \\ &+ \sum\nolimits_{p,q=-1,0,1} k^{(p,q)} H_{n_1+p,n_2+q,N}(\omega_1, \ldots, \omega_N) = 0 \end{aligned} \quad (12)$$

In order to recursively obtain the relationships between the *N*th order GFRFs of the n_1 and the first row masses, adding all of the n_1th row unit cell $\{n_1, n_2\}$, gives

$$\sum_{n_2=1}^{CN} \left(\begin{aligned} &- m(\omega_1 + \cdots + \omega_N)^2 H_{n_1,n_2,N}(\omega_1, \ldots, \omega_N) \\ &+ \sum\nolimits_{p=-1,0,1,q=0} c^{(p,q)}(i(\omega_1 + \cdots + \omega_N))H_{n_1+p,n_2+q,N}(\omega_1, \ldots, \omega_N) \\ &+ \sum\nolimits_{p=-1,0,1,q=0} k^{(p,q)} H_{n_1+p,n_2+q,N}(\omega_1, \ldots, \omega_N) \end{aligned} \right) = 0 \quad (13)$$

All of the terms in Eq. (13) divide $\sum_{n_2=1}^{CN} H_{n_1,n_2,N}(\omega_1,\ldots,\omega_N)$, and define,

$$\lambda_N^{n_1-1,n_1}(\omega_1,\ldots,\omega_N) = \frac{\sum_{n_2=1}^{CN} H_{n_1-1,n_2,N}(\omega_1,\ldots,\omega_N)}{\sum_{n_2=1}^{CN} H_{n_1,n_2,N}(\omega_1,\ldots,\omega_N)} \quad (14)$$

in addition, for the sake of simplicity, denote

$$\begin{aligned} DS(\omega_1,\ldots,\omega_N) &= \left[-m(\omega_1+\cdots+\omega_N)^2 + c^{(0,0)}(i(\omega_1+\cdots+\omega_N)) + k^{(0,0)}\right] \\ CK^{(-1,0)}(\omega_1,\ldots,\omega_N) &= \left[c^{(-1,0)}(i(\omega_1+\cdots+\omega_N)) + k^{(-1,0)}\right] \\ CK^{(1,0)}(\omega_1,\ldots,\omega_N) &= \left[c^{(1,0)}(i(\omega_1+\cdots+\omega_N)) + k^{(1,0)}\right] \end{aligned} \quad (15)$$

where, $DS(\omega_1,\ldots,\omega_N)$, $CK^{(-1,0)}(\omega_1,\ldots,\omega_N)$ and $CK^{(1,0)}(\omega_1,\ldots,\omega_N)$ are functions that are only related to linear parameters of nonlinear system in the n dimensional hyper-plane $\omega_1+\cdots+\omega_n=\omega$, and are independent of nonlinear parameters of nonlinear system.

Then, Eq. (13) can be rewritten as

$$\lambda_N^{n_1+1,n_1}(\omega_1,\ldots,\omega_N) = \frac{-DS(\omega_1,\ldots,\omega_N) - CK^{(-1,0)}(\omega_1,\ldots,\omega_N)\lambda_N^{n_1-1,n_1}(\omega_1,\ldots,\omega_N)}{CK^{(1,0)}(\omega_1,\ldots,\omega_N)} \quad (16)$$

In order to recursively obtain the relationships between the Nth order GFRFs of the n_1 and the first row masses, the sequence of n_1+1, n_1 in $\lambda_N^{n_1+1,n_1}(\omega_1,\ldots,\omega_N)$ needs to be exchanged. Then, Eq. (16) can be rewritten as

$$\lambda_N^{n_1,n_1+1}(\omega_1,\ldots,\omega_N) = \frac{-CK^{(1,0)}(\omega_1,\ldots,\omega_N)}{DS(\omega_1,\ldots,\omega_N) + CK^{(-1,0)}(\omega_1,\ldots,\omega_N)\lambda_N^{n_1-1,n_1}(\omega_1,\ldots,\omega_N)} \quad (17)$$

Starting with Eq. (17), and iteratively using Eqs. (13)–(17) it can be deduced that,

$$\lambda_N^{n_1,n_1+1}(\omega_1,\ldots,\omega_N) = f\left(\lambda_N^{1,2}(\omega_1,\ldots,\omega_N), coef(\omega_1,\ldots,\omega_N)\right) \quad (18)$$

where $coef(\omega_1,\ldots,\omega_N)$ represents the coefficients which are uncorrelated to each order GFRFs. Similarly, $\lambda_N^{1,2}(\omega_1,\ldots,\omega_N)$ can be deduced by Eq. (12), which can be presented as

$$\lambda_N^{1,2}(\omega_1,\ldots,\omega_N)=\frac{-CK^{(1,0)}(\omega_1,\ldots,\omega_N)}{DS(\omega_1,\ldots,\omega_N)} \tag{19}$$

Similarly, for the mass $\{n_1-1, n_2\}$ that is connected to the up of nonlinear spring and suffer no external force, we can derived,

$$\begin{aligned}&DS(\omega_1,\ldots,\omega_N)\sum_{n_2=1}^{CN}H_{n_1-1,n_2,N}(\omega_1,\ldots,\omega_N)+CK^{(-1,0)}(\omega_1,\ldots,\omega_N)\sum_{n_2=1}^{CN}H_{n_1-2,n_2,N}(\omega_1,\ldots,\omega_N)\\&+CK^{(1,0)}(\omega_1,\ldots,\omega_N)\sum_{n_2=1}^{CN}H_{n_1,n_2,N}(\omega_1,\ldots,\omega_N)+\Lambda_N^{(n_1-1,n_2)((n_1,n_2))}(\omega_1,\ldots,\omega_N)=0\end{aligned} \tag{20}$$

For the mass $\{n_1, n_2\}$ that is connected to the bottom of the nonlinear spring and suffer no external force, similarly, the relationships between the Nth order GFRFs can be derived. It can be represented as

$$\begin{aligned}&DS(\omega_1,\ldots,\omega_N)\sum_{n_2=1}^{CN}H_{n_1,n_2,N}(\omega_1,\ldots,\omega_N)+CK^{(-1,0)}(\omega_1,\ldots,\omega_N)\sum_{n_2=1}^{CN}H_{n_1-1,n_2,N}(\omega_1,\ldots,\omega_N)\\&+CK^{(1,0)}(\omega_1,\ldots,\omega_N)\sum_{n_2=1}^{CN}H_{n_1+1,n_2,N}(\omega_1,\ldots,\omega_N)-\Lambda_N^{(n_1-1,n_2)((n_1,n_2))}(\omega_1,\ldots,\omega_N)=0\end{aligned} \tag{21}$$

For the mass $\{n_1-1, n_2\}$ that is connected to the up of nonlinear spring and suffer no external force, from Eqs. (14) and (20), it can be known that the following relationships hold for the GFRFs,

$$\begin{aligned}\lambda_N^{n_1-1,n_1}(\omega_1,\ldots,\omega_N)=&\frac{-CK^{(1,0)}(\omega_1,\ldots,\omega_N)}{DS(\omega_1,\ldots,\omega_N)+CK^{(-1,0)}(\omega_1,\ldots,\omega_N)\lambda_N^{n_1-2,n_1-1}(\omega_1,\ldots,\omega_N)}\\&\times\left\{1+\frac{\Lambda_N^{(n_1-1,n_2)((n_1,n_2))}(\omega_1,\ldots,\omega_N)}{CK^{(1,0)}(\omega_1,\ldots,\omega_N)\sum_{n_2=1}^{CN}H_{n_1,n_2,N}(\omega_1,\ldots,\omega_N)}\right\}\end{aligned} \tag{22}$$

For the mass $\{n_1, n_2\}$ that is connected to the bottom of nonlinear spring and suffer no external force, from Eqs. (14) and (21), it can be known that the following relationships hold for the GFRFs,

$$\lambda_N^{n_1,n_1+1}(\omega_1,\ldots,\omega_N) = \frac{-CK^{(1,0)}(\omega_1,\ldots,\omega_N)}{DS(\omega_1,\ldots,\omega_N) + CK^{(-1,0)}(\omega_1,\ldots,\omega_N)\lambda_N^{n_1-1,n_1}(\omega_1,\ldots,\omega_N)} \times \left\{ 1 - \frac{\Lambda_N^{(n_1-1,n_2)((n_1,n_2))}(\omega_1,\ldots,\omega_N)}{CK^{(1,0)}(\omega_1,\ldots,\omega_N)\sum_{n_2=1}^{CN} H_{n_1+1,n_2,N}(\omega_1,\ldots,\omega_N)} \right\} \tag{23}$$

For the mass $\{n_1, n_2\}$ that is suffered external force and not connected to the nonlinear components, similarly, the relationships between the Nth order GFRFs can be derived. It can be represented as follows.

If N equals 1, $\Lambda_1^{(n_1,n_2)}(\omega_1) = -1$, then it can be known that the following relationships hold for the GFRFs,

$$\lambda_1^{n_1,n_1+1}(\omega_1) = \frac{-CK^{(1,0)}(\omega_1)}{DS(\omega_1) + CK^{(-1,0)}(\omega_1)\lambda_1^{n_1-1,n_1}(\omega_1)} \left\{ 1 - \frac{1}{CK^{(1,0)}(\omega_1)\sum_{n_2=1}^{CN} H_{n_1+1,n_2,1}(\omega_1)} \right\} \tag{24}$$

If $N \geq 2$, $\Lambda_N^{(n_1,n_2)}(\omega_1,\ldots,\omega_N) = 0$, the following relationships hold for the GFRFs:

$$\lambda_N^{n_1,n_1+1}(\omega_1,\ldots,\omega_N) = \frac{-CK^{(1,0)}(\omega_1,\ldots,\omega_N)}{DS(\omega_1,\ldots,\omega_N) + CK^{(-1,0)}(\omega_1,\ldots,\omega_N)\lambda_N^{n_1-1,n_1}(\omega_1,\ldots,\omega_N)} \tag{25}$$

Similarly, for the masses that aren't connected to the nonlinear spring and suffered no external force and below the nonlinear components, similarly, the relationships between the Nth order GFRFs can be derived. It can be represented as

$$\lambda_N^{n_1,n_1-1}(\omega_1,\ldots,\omega_N) = \frac{-CK^{(-1,0)}(\omega_1,\ldots,\omega_N)}{DS(\omega_1,\ldots,\omega_N) + CK^{(1,0)}(\omega_1,\ldots,\omega_N)\lambda_N^{n_1+1,n_1}(\omega_1,\ldots,\omega_N)} \tag{26}$$

Similarly, it can be presented as

$$\lambda_N^{n_1,n_1-1}(\omega_1,\ldots,\omega_N) = f\left(\lambda_N^{RN,RN-1}(\omega_1,\ldots,\omega_N), coef(\omega_1,\ldots,\omega_N)\right) \tag{27}$$

where $coef(\omega_1,\ldots,\omega_N)$ represents the coefficients which are uncorrelated to each order GFRFs, and $\lambda_N^{RN,RN-1}(\omega_1,\ldots,\omega_N)$ can be deduced by Eq. (12), which can be presented as

$$\lambda_N^{RN,RN-1}(\omega_1,\ldots,\omega_N) = \frac{-CK^{(-1,0)}(\omega_1,\ldots,\omega_N)}{DS(\omega_1,\ldots,\omega_N)} \tag{28}$$

Moreover, following the same procedure, it can be easily derived that the relationships of GFRFs between column n_2 and the relationship of GFRFs between row n_1 are identical.

3.1 NOFRFs of 2D Locally Nonlinear Periodic Structure

According to the definition of NOFRFs in Eq. (3), the Nth order NOFRF $G_{n_1,n_2,N}(\omega)$ of the mass $\{n_1, n_2\}$ can be expressed as

$$G_{n_1,n_2,N}(\omega) = \frac{\int_{\omega_1+\cdots+\omega_N=\omega} H_{n_1,n_2,N}(\omega_1,\ldots,\omega_N)\prod_{i=1}^{N} U(\omega_i)d\sigma_{N\omega}}{\int_{\omega_1+\cdots+\omega_N=\omega}\prod_{i=1}^{N} U(\omega_i)d\sigma_{N\omega}} \tag{29}$$

According to Eq. (29) and the derivation in Sect. 3, the properties of NOFRFs of 2D locally nonlinear periodic structure can be obtained.

If the components $\{R(i), C(i)\}\ i = 1,\ldots,L$ are nonlinear without loss of generality, and assume $R(1) < \cdots < R(L)$, $C(1) < \cdots < C(L)$ the external force acting on mass (RF, CF), then the NOFRFs of 2D locally nonlinear periodic structure have the following properties in row direction.

(i) If $RF \le R(1)$, then for the masses $\{n_1, n_2\}(1 \le n_1 \le RF - 1$ and $R(L) \le n_1 < RN)$, the following relationships hold:

$$\lambda_1^{n_1,n_1+1}(\omega) = \cdots = \lambda_N^{n_1,n_1+1}(\omega) \quad (1 \le n_1 \le RF - 1 \text{ or } R(L) \le n_1 < RN) \tag{30}$$

For the masses $\{n_1, n_2\}(RF \le n_1 < R(1) - 1)$, the following relationships hold:

$$\lambda_1^{n_1,n_1+1}(\omega) \ne \lambda_2^{n_1,n_1+1}(\omega) = \cdots = \lambda_N^{n_1,n_1+1}(\omega) \quad (RF \le n_1 < R(1) - 1) \tag{31}$$

For the masses $\{n_1, n_2\}(R(1) \le n_1 < R(L))$, the following relationships hold:

$$\lambda_1^{n_1,n_1+1}(\omega) \ne \lambda_2^{n_1,n_1+1}(\omega) \ne \cdots \ne \lambda_N^{n_1,n_1+1}(\omega) \quad (R(1) \le n_1 < R(L)) \tag{32}$$

(ii) If $R(1) \leq RF \leq R(L)$, then for the masses $\{n_1, n_2\}$ $(1 \leq n_1 < R(1) - 1$ and $R(L) \leq n_1 < RN)$, the following relationships hold:

$$\lambda_1^{n_1,n_1+1}(\omega) = \cdots = \lambda_N^{n_1,n_1+1}(\omega) \quad (1 \leq n_1 < R(1) - 1 \text{ or } R(L) \leq n_1 < RN) \tag{33}$$

For the masses $\{n_1, n_2\}(R(1) \leq n_1 < R(L))$, the following relationships hold:

$$\lambda_1^{n_1,n_1+1}(\omega) \neq \lambda_2^{n_1,n_1+1}(\omega) \neq \cdots \neq \lambda_N^{n_1,n_1+1}(\omega) \quad (R(1) \leq n_1 < R(L)) \tag{34}$$

(iii) If $RF \geq R(L)$, then for the masses $\{n_1, n_2\}$ $(1 \leq n_1 < R(1) - 1$ and $RF \leq n_1 < RN)$, the following relationships hold

$$\lambda_1^{n_1,n_1+1}(\omega) = \cdots = \lambda_N^{n_1,n_1+1}(\omega) \quad (1 \leq n_1 < R(1) - 1 \text{ or } RF \leq n_1 < RN) \tag{35}$$

For the masses $\{n_1, n_2\}(R(L) \leq n_1 < RF)$, the following relationships hold:

$$\lambda_1^{n_1,n_1+1}(\omega) \neq \lambda_2^{n_1,n_1+1}(\omega) = \cdots = \lambda_N^{n_1,n_1+1}(\omega) \quad (R(L) \leq n_1 < RF) \tag{36}$$

For the masses $\{n_1, n_2\}(R(1) - 1 \leq n_1 < R(L))$, the following relationships hold:

$$\lambda_1^{n_1,n_1+1}(\omega) \neq \lambda_2^{n_1,n_1+1}(\omega) \neq \cdots \neq \lambda_N^{n_1,n_1+1}(\omega) \quad (R(1) \leq n_1 < R(L)) \tag{37}$$

Similarly, the properties of NOFRFs of 2D locally nonlinear periodic structure in column direction can be easily deduced, for the sake of simplicity, the properties of NOFRFs of 2D locally nonlinear periodic structure in column direction aren't listed in this paper.

Based on these important properties Eqs. (30)–(37) about the NOFRFs of 2D locally nonlinear periodic structure derived in this present study, the position of the nonlinear components in a 2D locally nonlinear periodic structure can be detected.

4 Numerical Simulation

In order to verify the effectiveness of the nonlinear components position detection method based on these important properties Eqs. (30)–(37) about the NOFRFs of 2D locally nonlinear periodic structure derived in this present study, one damped 5×5 2D locally nonlinear periodic structures are adopted in cases 1, as shown by Fig. 2.

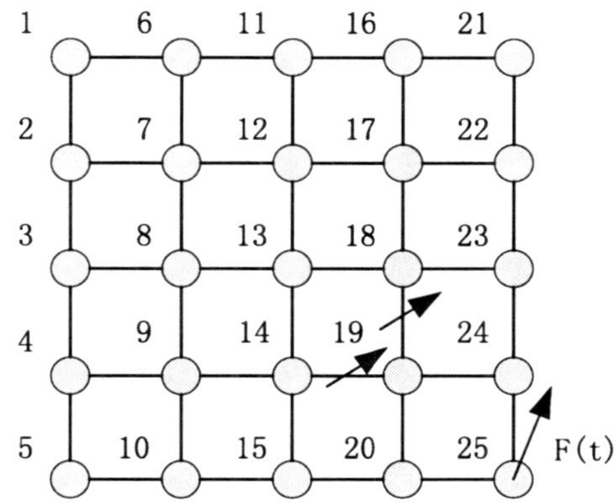

Fig. 2 5 × 5 2D locally nonlinear periodic structure, the nonlinear components are between mass 14–19 and mass 18–19

According to the frequency response of a nonlinear system subjected to a harmonic input [14], and only the NOFRFs up to the third order is considered, the frequency response outputs of the 5 × 5 2D locally nonlinear periodic structure,

$$\begin{aligned} Y_i(\omega) &= G_{(i,1)}(\omega)U_1(\omega) + G_{(i,3)}(\omega)U_3(\omega) \\ Y_i(2\omega) &= G_{(i,2)}(2\omega)U_2(2\omega) \\ Y_i(3\omega) &= G_{(i,3)}(3\omega)U_3(3\omega) \quad (i = 1, \ldots, 25) \end{aligned} \tag{38}$$

From Eq. (38), it can be seen that, using the method in Ref. [10], two different inputs with the same frequency but different amplitudes are sufficient to estimate the NOFRFs up to the third order. Therefore, in the numerical studies, two different inputs are $A = 0.8$ and 1.0, respectively.

Case 1. The nonlinear components are between mass 14–19 and mass 18–19, as shown by Fig. 2.

In this case, the springs between 14–19 and 18–19 are nonlinear. m is the unit cell mass, k_1 is the linear stiffness coefficient, k_2 and k_3 are the quadratic and cubic nonlinear stiffness coefficients, respectively. The damping is assumed to be proportional damping, e.g., $c = \mu k_1$. The values of the system parameters are listed in Table 1, and the input is a harmonic force $u(t) = A\sin(2\pi \times 20t)$, which is imposed on the 25th mass of this system.

In this case, according to property (i)–(iii) in the previous section, it can be known that the following relationships should be tenable in the row direction,

$$\begin{aligned} \lambda_1^{n_1,n_1+1}(\omega) &= \frac{\sum_{n_2=1}^{CN} G_{n_1,n_2,1}(\omega)}{\sum_{n_2=1}^{CN} G_{n_1+1,n_2,1}(\omega)} = \frac{\sum_{n_2=1}^{CN} G_{n_1,n_2,3}(\omega)}{\sum_{n_2=1}^{CN} G_{n_1+1,n_2,3}(\omega)} = \lambda_3^{n_1,n_1+1}(\omega) \quad (1 \le n_1 < 3) \\ \lambda_1^{n_1,n_1+1}(\omega) &= \frac{\sum_{n_2=1}^{CN} G_{n_1,n_2,1}(\omega)}{\sum_{n_2=1}^{CN} G_{n_1+1,n_2,1}(\omega)} \neq \frac{\sum_{n_2=1}^{CN} G_{n_1,n_2,3}(\omega)}{\sum_{n_2=1}^{CN} G_{n_1+1,n_2,3}(\omega)} = \lambda_3^{n_1,n_1+1}(\omega) \quad (3 \le n_1 < 4) \end{aligned} \tag{39}$$

and in the column direction,

Table 1 The values of the system parameters

Parameters	k_1	c	k_2	k_3	m
Value	3.6×10^4	$0.01k_1$	$4k_1^2$	$2k_1^3$	1

Table 2 The value of $\lambda_1^{i,i+1}$ and $\lambda_3^{i,i+1}$ in row direction

Row i	1	2	3	4
$\lambda_1^{i,i+1}$	0.5391 − 0.0630i	0.7407 − 0.1588i	**0.8219 − 0.2812i**	0.7993 − 0.3924i
$\lambda_3^{i,i+1}$	0.5392 − 0.0628i	0.7410 − 0.1583i	**−1.0995 − 1.0699i**	0.8300 + 0.2136i

Table 3 The value of $\lambda_1^{i,i+1}$and$\lambda_3^{i,i+1}$ in column direction

Column i	1	2	3	4
$\lambda_1^{i,i+1}$	1.0402 − 0.2300i	0.9223 − 0.4473i	**0.7212 − 0.4556i**	0.6677 − 0.3528i
$\lambda_3^{i,i+1}$	1.0179 − 0.2168i	0.9068 − 0.3923i	**−0.8136 − 0.9478i**	0.8017 + 0.1925i

$$\begin{aligned}
\lambda_1^{n_2,n_2+1}(\omega) &= \frac{\sum_{n_1=1}^{RN} G_{n_1,n_2,1}(\omega)}{\sum_{n_1=1}^{RN} G_{n_1,n_2+1,1}(\omega)} = \frac{\sum_{n_1=1}^{RN} G_{n_1,n_2,3}(\omega)}{\sum_{n_1=1}^{RN} G_{n_1,n_2+1,3}(\omega)} = \lambda_3^{n_2,n_2+1}(\omega) \quad (1 \le n_2 < 3) \\
\lambda_1^{n_2,n_2+1}(\omega) &= \frac{\sum_{n_1=1}^{RN} G_{n_1,n_2,1}(\omega)}{\sum_{n_1=1}^{RN} G_{n_1,n_2+1,1}(\omega)} \neq \frac{\sum_{n_1=1}^{RN} G_{n_1,n_2,3}(\omega)}{\sum_{n_1=1}^{RN} G_{n_1,n_2+1,3}(\omega)} = \lambda_3^{n_2,n_2+1}(\omega) \quad (3 \le n_2 < 4)
\end{aligned} \tag{40}$$

The evaluated results of $\lambda_1^{i,i+1}$ and $\lambda_3^{i,i+1}$ are given in Table 2 in row direction. Similarly, the evaluated results of $\lambda_1^{i,i+1}$ and $\lambda_3^{i,i+1}$ are given in Table 3 in column direction. It can be seen that the results shown in Tables 2 and 3 have a strict accordance with the relationships in Eqs. (39) and (40), respectively. Therefore, it can be known that there are nonlinear components between mass 14–19 and 18–19. The detecting results of the nonlinear component position have a strict accordance with the assumed location of the nonlinear components.

5 Conclusions

In this paper, significant relationships between the NOFRFs of 2D locally nonlinear periodic structure with nonlinear components have been derived. The potential of using these properties to locate the nonlinear component position is also studied. The effectiveness of this nonlinear components position detection method is demonstrated by numerical studies. As the position of a nonlinear component often corresponds to the location of defect in periodic structures, this new method is of great practical significance in fault diagnosis for mechanical and structural systems.

Acknowledgments The authors gratefully acknowledge that the work was supported by the National Science Fund for Distinguished Young Scholars (11125209), the National Natural Science Foundation of China (10902068, 51121063) and the Shanghai Pujiang Program (10PJ1406000) for this work.

References

1. Brillouin L (2003) Wave propagation in periodic structures: electric filters and crystal lattices. DoverPublications, Mineola
2. Duhamel D, Mace BR, Brennan MJ (2006) Finite element analysis of the vibrations of waveguides and periodic structures. J Sound Vib 294(2006):205–220
3. Mead DM (1996) Wave propagation in continuous periodic structures: research contributions from southampton, 1964–1995. J Sound Vib 190(1996):495–524
4. Marathe A, Chatterjee A (2006) Wave attenuation in nonlinear periodic structures using harmonic balance and multiple scales. J Sound Vib 289(2006):871–888
5. Narisetti RK, Ruzzene M, Leamy MJ (2011) A perturbation approach for analyzing dispersion and group velocities in two-dimensional nonlinear periodic lattices. J Vib Acoust Trans ASME 133(2011):061020
6. Neild SA, Mcfadden PD, Williams MS (2001) A discrete model of a vibrating beam using a time-stepping approach. J Sound Vib 239(2001):99–121
7. Sakellariou JS, Fassois SD (2006) Stochastic output error vibration-based damage detection and assessment in structures under earthquake excitation. J Sound Vib 297(2006):1048–1067
8. Fassois SD, Sakellariou JS (2007) Time-series methods for fault detection and identification in vibrating structures. Philos Trans R Soc A Math Phys Eng Sci 365(2007):411–448
9. Peng ZK, Lang ZQ, Billings SA (2007) Non-linear output frequency response functions of MDOF systems with multiple non-linear components. Int J Non-Linear Mech 42 (2007):941–958
10. Peng ZK, Lang ZQ (2007) Detecting the position of non-linear component in periodic structures from the system responses to dual sinusoidal excitations. Int J Non-Linear Mech 42 (2007):1074–1083
11. Lang ZQ, Billings SA (1996) Output frequency characteristics of nonlinear systems. Int J Control 64(1996):1049–1067
12. Lang ZQ, Billings SA (2005) Energy transfer properties of non-linear systems in the frequency domain. Int J Control 78(2005):345–362
13. Narisetti RK (2010) Wave propagation in nonlinear periodic structures. Ph.D. thesis, Georgia Institute of Technology, USA
14. Peng ZK, Lang ZQ, Billings SA (2007) Resonances and resonant frequencies for a class of nonlinear systems. J Sound Vib 300(2007):993–1014

Research on Multi-correlative Subsystems Analysis for Vibration Source Identification

S.M. Li, X.X. Jiang and C.W. Lim

Abstract The complex vibration source identification of structure in the vibration analysis and control is an important aspect that people pay more attention. First, the complex vibration system of structure is divided to several related subsystem each other in the paper. The model of subsystem-input/single-output system is established. The correlations between inputs and the model after average process in multi subsystem-input system are developed. The value changes of correlated input matrix and the affecting to the output in various relevant condition of subsystem-inputs are analyzed. According to coherence analysis, the method to estimate the contribution of inputs to the output is obtained. And the accuracies of method are verified by simulation. Finally, the vibration data are given by an automobile test. After the data analysis, the usefulness of new method provided in the paper is verified.

Keywords Source identification · Multi-correlative · Subsystem · Vibration

1 Introduction

To deal with the engineering vibratory problems in multi excitations/single response linear system, generally people get the contributions of excitations according to know the interaction of excitations [1]. A multi-input single-output model is used in the complex mechanical vibration system with different level of

S.M. Li (✉) · X.X. Jiang
College of Energy and Power Engineering, Nanjing University of Aeronautics and Astronautics, No.29 Yudao Street, Nanjing 210016, China
e-mail: smli@nuaa.edu.cn

C.W. Lim
Department of Building and Construction, City University of Hong Kong, No.83 Tat Chee Road, Kowloon Tong, Hong Kong, China
e-mail: bccwlim@cityu.edu.hk

J.K. Sinha (ed.), *Vibration Engineering and Technology of Machinery*, Mechanisms and Machine Science 23, DOI 10.1007/978-3-319-09918-7_73

inputs in current methods of vibration source identification [2–4]. All inputs are considered in the same level in the vibration system. Then the order of each source will be prioritized after analyzing this model. However there are many huge limitations in these methods for complex mechanical systems with multi-level inputs. If we don't take into account of the impact between different levels of inputs in subsystems and the output, the transmission path in major frequency points will be affected. The investigation of the sub-systems which has the hugest influence to the output is important [5, 6].

But when considering subsystem-excitation constituted of multi input, by means of knowing the interaction of subsystem-excitations and the impact of correlated excitations of each stage to response in each subsystem-excitation, the major excite stage of the subsystem to the response point cared about can be distinguished, hence the vibration measures can be taken with a clear goal [7].

Therefore, to solve the problem which contains different complex levels of subsystems, this paper establishes a multi-correlated subsystem model and using it to do coherence analysis from order into disorder by using the theories of collective average means. So the main vibration source of the structure will be identified by determining the contribution of every input of every subsystem to the output. And this procedure will be used in research of judging the main vibration source to a truck with actual data obtained in road test.

2 Model Establishment of Multi-subsystem-Inputs and Single-Output

2.1 Two-Subsystem-Inputs and Single-Output Model

Firstly, the model which contains two subsystem-inputs and single-output is established, which is shown in Fig. 1. Due to mechanical connection, the inputs of every subsystem are related to each other. This correlation can be preliminary analyzed through the characteristics of structure. For example, in a vehicle, the axle is connected to the frame by the leaf spring system, while the frame is connected to the cab through the suspension system. So the excitations, which contain the excitation from the wheel to axle and from engine or transmission to frame, turn to

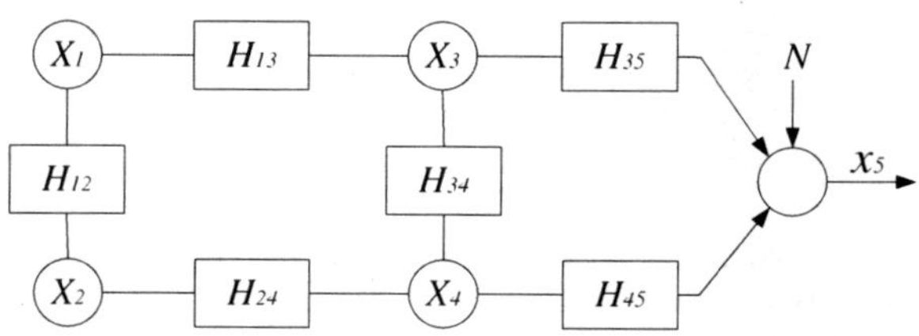

Fig. 1 Two correlated subsystem-inputs single-output system model

be correlated in this system. The same thing happens in the correlation among each inputs of every subsystems. Such as the random excitations due to road roughness become correlated because of the axle and frame which connect the four wheels. This connecting shafts are the energy transfer channel in physical explanation, while they can be replaced to transfer function in mathematical explanation [1].

Input X_1 transfer the energy to the output X_5 not only through the path of X_3 but also by the way of input X_2, X_3 and X_4. The energy that transfer from the input X_1 to X_4 by the path X_2 is generated from the correlated parts between X_1 and X_2 as well as between X_2 and X_4. And the energy that transfers from the input X_1 to X_4 by the path X_3 is generated from the correlated parts between X_1 and X_3 as well as between X_3 and X_4. The condition happen the same in the input X_2. So when we put the relevant parts into the two inputs in the same level respectively, the whole model will be turned into the two subsystems. Input X_1 and X_3 is defined as subsystem I, and Input X_2 and X_4 is defined as subsystem II. The transfer function H_{12} and H_{34} show the relationship between each input in subsystem I and II. Under normal circumstances, the transfer function H_{12} equals H_{21}, and H_{34} equals H_{43}, due to the non-direction between each inputs and the mechanical connection is mutual.

2.2 Multi-subsystem-Inputs and Single-Output Model

The model of multi-subsystem-input and single-output (MSISO) mechanical system is made based on the model of two-subsystem-inputs and single-output mechanical system which we discussed above. Similar to the two-subsystem-inputs and single-output model, the models with two level inputs that each level inputs contain n inputs, equals n-subsystem-inputs and single-output model. Figure 2 shows this MSISO model.

Fig. 2 Multi correlated subsystem-inputs single-output system model

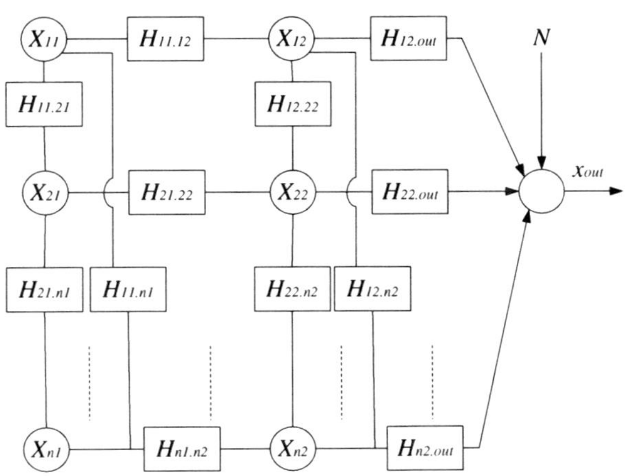

In the MSISO system, every level inputs of every subsystem are correlated, where the correlation are not only between the inputs of every level with of the adjacent subsystem, but also the inputs of among all other subsystems. The mechanical connection among the inputs is mutual, so the transfer function H_{ij} equals H_{ji} ($i \neq j$) due to the non-directional transformation between every pair of inputs.

2.3 Coherent Matrix Among Inputs

In order to define the correlation of this MSISO system, we put the regular coherence function [8] to describe the relevant level of each pair of inputs. So the matrix is used to represent the correlation between each level inputs. Equations (1) and (2) show us the matrix of subsystem I and II respectively.

$$ {}_1\gamma_{XX}^2 = \begin{pmatrix} 1 & \gamma_{11\cdot 21}^2 & \cdots & \gamma_{11\cdot n1}^2 \\ & 1 & \cdots & \gamma_{21\cdot n1}^2 \\ & & \ddots & \vdots \\ \mathit{symmetry} & & & 1 \end{pmatrix} \tag{1} $$

$$ {}_2\gamma_{XX}^2 = \begin{pmatrix} 1 & \gamma_{12\cdot 22}^2 & \cdots & \gamma_{12\cdot n2}^2 \\ & 1 & \cdots & \gamma_{22\cdot n2}^2 \\ & & \ddots & \vdots \\ \mathit{symmetry} & & & 1 \end{pmatrix} \tag{2} $$

In the equations, ${}_1\gamma_{XX}^2$ represents the correlation matrix of the first level inputs, and ${}_2\gamma_{XX}^2$ represents the correlation matrix of the second level inputs. Different value of γ_{XX}^2 show us the difference between each levels of inputs, which represents different physical connections. The correlation of each level inputs can be divided into the following situations.

(1) If the inputs of the two level multi-correlated-subsystem is fully correlated, the coherence function will be $\gamma_{ij}^2 = 1, i \neq j = 1, 2, \ldots, n$. And the n-subsystem-inputs and single-output system can be replaced as single-subsystem-input and single-output system. The matrix will be changed into a $n \times n$ square matrix, which all the elements are 1.

$$ {}_1\gamma_{XX}^2 = {}_2\gamma_{XX}^2 = \begin{pmatrix} 1 & 1 & \cdots & 1 \\ & 1 & \cdots & 1 \\ & & \ddots & \vdots \\ \mathit{symmetry} & & & 1 \end{pmatrix} \tag{3} $$

(2) If there is no physical connection between each inputs, then the inputs are completely unrelated to each other, and the coherence function will be $\gamma_{ij}^2 = 0$, $i \neq j = 1, 2, \ldots, n$. So the system can be turned into n independent subsystems inputs and single output system, and the matrix will become Eq. (4).

$$ {}_1\gamma_{XX}^2 = {}_2\gamma_{XX}^2 = \begin{pmatrix} 1 & & & \\ & 1 & & \\ & & \ddots & \\ symmetry & & & 1 \end{pmatrix} \quad (4) $$

(3) In ordinary conditions, the excitations in multi correlated subsystems are partially correlated, and the coherence function is expressed as $0 \leq \gamma_{ij}^2(f) \leq 1$, $i \neq j = 1, 2, \ldots, n$. The coherence function matrix of excitations of subsystems turns into a symmetrical matrix whose diagonal elements are all 1, shown as Eqs. (1) and (2).

3 Derivation of Collective Average Model

After establishing the MSISO model, the collective average means is used to deduce the correlated part and then make the average in order to build a new subsystem inputs collection. With this ideal, the collective average model is created which will make the inputs of MSISO system quite simply and turns coherence analysis from order into disorder. This model can be explained as Eq. (5).

$$ \begin{cases} X_{1\cdot in} = \frac{1}{n}(X_{11}, X_{11\cdot 21}, \ldots, X_{11\cdot n1}) + \frac{1}{n}(X_{12}, X_{12\cdot 22}, \ldots, X_{12\cdot n2}) \\ X_{2\cdot in} = \frac{1}{n}(X_{21\cdot 11}, X_{21}, \ldots, X_{21\cdot n1}) + \frac{1}{n}(X_{22\cdot 12}, X_{22}, \ldots, X_{22\cdot n2}) \\ \vdots \\ X_{n\cdot in} = \frac{1}{n}(X_{n1\cdot 11}, X_{n1\cdot 21}, \ldots, X_{n1}) + \frac{1}{n}(X_{n2\cdot 12}, X_{n2\cdot 22}, \ldots, X_{n2}) \end{cases} \quad (5) $$

In the equation, $X_{n\cdot in}$ represent the average input of subsystem n, $n = 1, 2, \ldots, n$, $X_{11\cdot n1}$ is the correlated inputs between the first level inputs of subsystem I and the first level inputs of other subsystem. $X_{12\cdot n2}$ is the correlated inputs between the second level inputs of subsystem I and the second level inputs of other subsystem. So does the nth subsystem, $n = 1, 2, \ldots, n$.

4 Coherent Analysis

When the external noise appears only in the output point, the total measured output spectrum $G_{yy}(f)$ will be divided into two parts as shown in Eq. (6), which contain the ideal linear output spectrum $G_{vv}(f)$ coming from the input $X(f)$ with transfer function $H(f)$ and the noise spectrum $G_{nn}(f)$.

$$G_{yy}(f) = G_{vv}(f) + G_{nn}(f) \tag{6}$$

$G_{xx}(f)$ is the auto-power spectrum density function of inputs; $G_{xy}(f)$ is the cross-power spectrum density function of inputs and output. It can be acquired that

$$G_{vv}(f) = \left|\frac{G_{xy}(f)}{G_{xx}(f)}\right|^2 G_{xx}(f) = \gamma_{xy}^2(f)G_{yy}(f) \tag{7}$$

In Eq. (7), the product of $\gamma_{xy}^2(f)$ and $G_{yy}(f)$ is called coherent output spectrum [9], which is often recorded as 'cop'. It indicates that how much is correlated to the spectrum $G_{yy}(f)$ in the measured power spectrum. So the total coherent output spectrum is the summation of all coherent power spectrums of all subsystems. By comparing the ratio of the coherent spectrum of one subsystem and the total coherent output spectrum, we can determine the impact between this subsystem and the output.

5 Simulation Analysis

In order to verify this method, the four subsystem-inputs and single-output model is established. That can be seeing in Fig. 3. This four subsystems can be expressed as that input 1 and 3 constitute subsystem I, 2 and 4 constitute subsystem II, 5 and 7 constitute subsystem III, 6 and 8 constitute subsystem IV. Firstly, the simulation input signal is at the frequency of 4.8 Hz. And the influence between the first level inputs and the output is greater than that between the second level inputs and the outputs. Then using the average method to the MSISO system, the coherence is analyzed in frequency domain. So the average inputs shown in Eq. (8) are obtained through putting the correlated parts into every input of every level on average. And the explanation of each variable can be seen in Eq. (5).

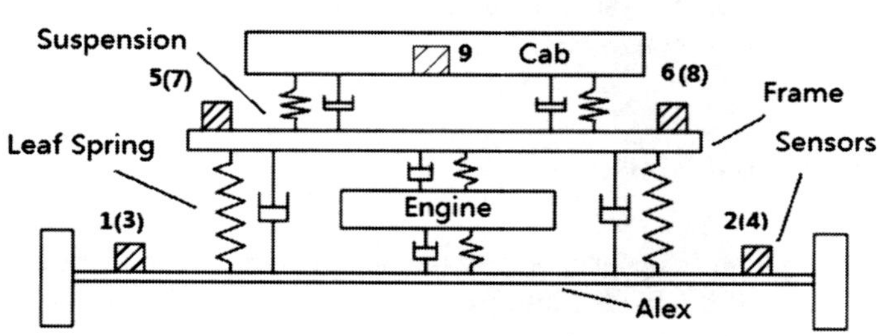

Fig. 3 Diagram of sensors mounting

Table 1 Results of coherent analysis between subsystem I and II

frequency f (Hz)	$\frac{cop1}{G_9}$ %	$\frac{cop5}{G_9}$ %	$\frac{cop1\cdot5}{G_9}$ %	$\frac{cop2}{G_9}$ %	$\frac{cop6}{G_9}$ %	$\frac{cop2\cdot6}{G_9}$ %
4.8	24.2	8.7	16.45	23.7	10.1	16.9

Table 2 Results of coherent analysis between subsystem III and IV

frequency f (Hz)	$\frac{cop3}{G_9}$ %	$\frac{cop7}{G_9}$ %	$\frac{cop3\cdot7}{G_9}$ %	$\frac{cop4}{G_9}$ %	$\frac{cop8}{G_9}$ %	$\frac{cop4\cdot8}{G_9}$ %
4.8	23.4	10.9	17.1	24.2	9.9	17.0

$$\begin{cases} X_{1\cdot in} = \frac{1}{4}(X_{11}, X_{11\cdot21}, X_{11\cdot31}, X_{11\cdot41}) + \frac{1}{4}(X_{12}, X_{12\cdot22}, X_{12\cdot32}, X_{12\cdot42}) \\ X_{2\cdot in} = \frac{1}{4}(X_{21\cdot11}, X_{21}, X_{21\cdot31}, X_{21\cdot41}) + \frac{1}{4}(X_{22\cdot12}, X_{22}, X_{22\cdot32}, X_{22\cdot42}) \\ X_{3\cdot in} = \frac{1}{4}(X_{31\cdot11}, X_{31\cdot21}, X_{31}, X_{31\cdot41}) + \frac{1}{4}(X_{32\cdot12}, X_{32\cdot22}, X_{32}, X_{32\cdot42}) \\ X_{4\cdot in} = \frac{1}{4}(X_{41\cdot11}, X_{41\cdot21}, X_{41\cdot31}, X_{41}) + \frac{1}{4}(X_{42\cdot12}, X_{42\cdot22}, X_{42\cdot32}, X_{42}) \end{cases} \tag{8}$$

At last, the coherent analysis is conducted to this four subsystems model and the results of ratios are shown in Tables 1 and 2. Where in this model $G_9 = cop1 \cdot 5 + cop2 \cdot 6 + cop3 \cdot 7 + cop4 \cdot 8$.

The coherent spectrum ratio between the average inputs in the first level is larger than that in the second level which can be obtained in Tables 1 and 2. So this answer indicates that the first level inputs have greater influence than the second level inputs. Besides these ratios show little difference which indicates that the contribution from every subsystems to the output is almost the same and there doesn't exist the subsystem that has the greater impact to this model.

The comparison of the contribution to output is based on the results in Tables 1 and 2. While Table 3 shows the coherent matrix among each level inputs in the characteristic frequency. We can tell from this table that the values in first coherent matrix is larger than that in the second matrix, which shows us that inputs in the first level is more relative.

So through the simulation, the practicability of this method is verified. By using the average MSISO model, the correlated parts are distributed to every inputs on

Table 3 Values of the coherent matrix in characteristic frequency

Frequency f (Hz)	$({}_1\gamma^2_{XX})$	$({}_2\gamma^2_{XX})$
4.8	$\begin{pmatrix} 1 & 0.94 & 0.90 & 0.92 \\ 0.94 & 1 & 0.93 & 0.92 \\ 0.90 & 0.93 & 1 & 0.94 \\ 0.92 & 0.92 & 0.94 & 1 \end{pmatrix}$	$\begin{pmatrix} 1 & 0.40 & 0.12 & 0.10 \\ 0.40 & 1 & 0.23 & 0.09 \\ 0.12 & 0.23 & 1 & 0.15 \\ 0.10 & 0.09 & 0.15 & 1 \end{pmatrix}$

Table 4 Value of the coherent matrix at the major frequencies

Frequency f (Hz)	$({}_1\gamma_{XX}^2)$	$({}_2\gamma_{XX}^2)$
3.36	$\begin{pmatrix} 1 & 0.916 & 0.952 & 0.914 \\ 0.916 & 1 & 0.902 & 0.856 \\ 0.952 & 0.902 & 1 & 0.958 \\ 0.914 & 0.856 & 0.958 & 1 \end{pmatrix}$	$\begin{pmatrix} 1 & 0.382 & 0.110 & 0.096 \\ 0.382 & 1 & 0.212 & 0.108 \\ 0.110 & 0.212 & 1 & 0.138 \\ 0.096 & 0.108 & 0.138 & 1 \end{pmatrix}$
3.66	$\begin{pmatrix} 1 & 0.946 & 0.936 & 0.902 \\ 0.946 & 1 & 0.882 & 0.864 \\ 0.936 & 0.882 & 1 & 0.968 \\ 0.902 & 0.864 & 0.968 & 1 \end{pmatrix}$	$\begin{pmatrix} 1 & 0.358 & 0.142 & 0.124 \\ 0.358 & 1 & 0.152 & 0.056 \\ 0.142 & 0.152 & 1 & 0.282 \\ 0.124 & 0.056 & 0.282 & 1 \end{pmatrix}$
3.97	$\begin{pmatrix} 1 & 0.956 & 0.934 & 0.862 \\ 0.956 & 1 & 0.944 & 0.908 \\ 0.934 & 0.944 & 1 & 0.958 \\ 0.862 & 0.908 & 0.958 & 1 \end{pmatrix}$	$\begin{pmatrix} 1 & 0.364 & 0.094 & 0.136 \\ 0.364 & 1 & 0.216 & 0.124 \\ 0.094 & 0.216 & 1 & 0.428 \\ 0.136 & 0.124 & 0.428 & 1 \end{pmatrix}$

average, and the first level inputs have the greatest impact to the output among all level inputs. There is no outstanding subsystem and every subsystem has almost the same influence to the output.

6 Experimental Analysis

The test vehicle has strong vertical vibration in the cab at the speed of 45–55 km/h when driving at the same road condition. For solving this problem, make use of the real vehicle road test, and the sensor is arranged as following Fig. 3. The experiment is conducted at the speed of 45, 50 and 55 km/h and the sensors are trying to obtain vehicle's vertical vibration acceleration signal.

Figure 3 shows the diagram of sensors installation. As shown in Fig. 3, four sensors are installed on the front and rear axes respectively, the other four sensors are mounted on the front and rear frames respectively where the location is almost vertical to the sensors mounted on the axes. The last sensor is installed on the cab baseboard. These sensors are respectively to measure the first stage excitations: 1, 2, 3, 4 and the second stage excitations 5, 6, 7, 8, and response 9, where the MISIO model is just like the model in simulation analysis. This vibration system is simplified into four correlated subsystems inputs and one output. The four wheel inputs and the corresponding frame consists the four correlated subsystems and the cab vibration is the output. The axles signal which include the wheel and road excitation are at the first level, and the frame signal which consist the engine and transmission excitation are at the second level.

Spectrum analysis is made to the measured data and the main frequency is 3.36, 3.66 and 3.97 Hz when the speed is 45, 50 and 55 km/h respectively. So the coherent matrixes at the main frequencies are shown in Table 4, where we know that the value in the first level matrix is still larger than the second level matrix.

Table 5 Results of coherent analysis

Frequency f (Hz)	$\frac{cop_{1\cdot 3}}{G_9}\%$	$\frac{cop_{2\cdot 4}}{G_9}\%$	$\frac{cop_{5\cdot 7}}{G_9}\%$	$\frac{cop_{6\cdot 8}}{G_9}\%$
3.36	16.75	16.76	16.47	15.81
3.66	16.90	16.43	16.73	16.20
3.97	16.55	17.16	17.41	16.86
Average	16.73	16.87	16.87	16.29

Then the collected average method is used to get the average inputs of each subsystem, and the coherent analysis is conducted to the whole system. The results by coherent analysis are shown in Table 5, where G_9 is just like the above one.

Considering data in Table 5, the first level inputs has greater influence than the second level inputs, so the excitation coming from engine and transmission is not the main vibration source to the truck. The road excitation is random excitation and doesn't match the frequency of the truck vibration. So after further consideration, the wheel's dynamic unbalance is the main force to the truck cab's vibration. The data of the four subsystems are almost the same, which indicates that there does not exist major vibration transmission path.

7 Conclusion

(1) For dealling with some complex vibration problem, the MSISO model has been made and related subsystems are created, which is helpful to obtain the main inputs that have the greatest influence to the output.

(2) The coherent matrix is defined in this paper which tells us the relationship of the inputs in the same level. By comparing the matrix, we can judge the correlation degree of different levels in the MSISO system.

(3) Using collective average method, the correlated part of excitation inputs of MSISO in different levels are divided averagely, so the MSISO system is simplified, and it can reflect the energy transmission correctly.

(4) By simulation and experiment, the practicability of this method is verified. To solve the truck's vibration problem, the truck is simplified to a four subsystem-inputs and single-output model, then use the collective average method and coherence function, which can be briefly conclude as MSISO method. This theoretical and experimental analysis can be used in the deferent other engineering problems, and have same expression.

Acknowledgment The research is supported by the National Natural Science Foundation of China (50675099) and Aviation science funds(2012ZD52054).

References

1. An M, Li S, Zhang Y, Liu J (2010) Coherence analysis of inputs using multiple correlated subsystem-inputs/single-output system modeling. Chin Mech Eng 21(21):2609–2613 unpublished
2. Chung JY, Crocker MJ, Hamilton JF (1975) Measurement of frequency response and multiple coherence function of the noise generation system of a diesel engine. J Acoust Soc Am N Y 58:635–542
3. Trethewey MW, Evensen HA (1981) Identification of noise source forge hammers during production and application of residual spectrum technique to transients. J Sound Vib 77:357–374
4. Park JS, Kim KJ (1992) Determination of priority among correlated inputs in source identification problems. Mech Syst Signal Process 6:491–502
5. Bae BK, Kim KJ (1998) A Hilbert transform approach in source identification via multiple-input single-output modeling for correlated inputs. Mech Syst Signal Process 12:501–513
6. Chen Moli, Li Shunming (2007) Coherence functions method for signal source identification. Chin Mech Eng Wuhan 18(1):95–100
7. Haiyan Hu (1998) Mechanical vibrations and impact. Aviation Industry Press, Beijing, p 2
8. Huang D, Wang Y (1989) Identifying main path of vibration transmission. J Vib Eng 2 (1):32–38
9. Bendat JS, Piersol AG (1993) Engineering applications of correlation and spectral analysis. Wiley, New York

A Nut-Type Ultrasonic Motor Driven with Single Phase Signal

Zhuzi Chen, Yu Chen, Tieying Zhou and Deyong Fu

Abstract Most travelling wave ultrasonic motors uses two phase signal with a 90° time phase difference to generate a traveling wave, this paper propose a novel nut-type travelling wave ultrasonic motor which uses only a single phase driving signal to excite two bending modes in $r - \theta$ plane of a nut-type stator simultaneously. Thus, both the motor structure and the driving circuit will be simplified. The working principle of the new single phase drive traveling wave nut-type ultrasonic motor was described. The working modes are observed with a finite element method software. A prototype nut-type ultrasonic motor was fabricated, and its electrical impedance was tested to observe the two working modes. The prototype motor worked at frequencies between two resonance frequencies under single phase drive as expected, and it also rotated bi-directional successfully.

Keywords Nut-type ultrasonic motor · Single phase drive · Mode frequency split

1 Introduction

For optical auto-focusing or zooming applications, thread-driven ultrasonic motors bear many advantages over other actuators for application in a limited space. A thread-driven ultrasonic motor consists of a stator and a rotor, whereas the stator is a metal tube which has piezoelectric plates bonded on outer surfaces. It is linear, large stroke, precise, low profile and easy to fabricate a mini version with low cost [1–4].

A nut-type linear ultrasonic motor was developed, which perfectly meets the linear and compact requirements of driving an optical zoom in mobile phone

Z. Chen · Y. Chen (✉) · T. Zhou · D. Fu
Physics Department, Tsinghua University, Beijing, China
e-mail: chenyu@tsinghua.edu.cn

Z. Chen
e-mail: chenzhuzi@qq.com

J.K. Sinha (ed.), *Vibration Engineering and Technology of Machinery*,
Mechanisms and Machine Science 23, DOI 10.1007/978-3-319-09918-7_74

cameras [1–4]. The excellent characteristics of this motor are: (1) Lens assemblies can be fixed in the screw rotor or nut stator to compose autofocus or optical zoom and realize the unique operating design of focus system; (2) The thread driving system directly drives the rotation of the rotor into the linear motion without using gears or cams, and thus can obviously reduce the number of components and thereby the cost.

To realize rotation, the existing nut-type ultrasonic motor must use both sine and cosine signals to excite two orthogonal bending vibration modes (with a phase difference of $\pi/2$) of the stator to generate circumferential travelling wave. In this paper, a nut-type travelling wave ultrasonic motor with simpler structure is presented, which uses only a single phase signal to excite two in-plane bending modes of the stator to generate a traveling wave. Bi-directional motion is achieved by simply switching driving voltage between two piezoelectric plates, keeping the other piezoelectric plate open circuit. Since only one piezoelectric plate is connected to driving voltage to excite two modes, the motor requires only one driving signal, simplifying the structure, fabrication, assembly, and driving circuit.

The theory for single phase drive exciting two in-plane modes of a ring was discussed in [5–7]. Ueha used a circular ring vibrator attached with a small mass as the theoretical model [5, 7], the degenerate vibration modes are perturbed and separated into two orthogonal vibration modes with different resonance frequencies. The two modes can be excited to vibrate at a phase difference of 90° by a single-phase source if the amount of the mass is appropriate.

Ceperley discovered that with perturbation node and antinode positions are fixed at the middle between the two perturbation locations [6]. For one of the modes the maximum amplitude happens at the middle while for the other mode the zero amplitude happens at the middle point.

In this paper, the motor proposed uses only two piezoelectric plates, so the structure, fabrication art and assembly streamline will be simplified since the original stator structure uses eight piezoelectric plates. The motor requires only one electrical signal to drive, so the driving circuit will be simplified since the original motor needs two phases signal to drive simultaneously with 90° phase difference in time domain.

2 Structure and Working Principle

2.1 The Original Design

The original nut-type ultrasonic motor structure mainly consists of three parts: stator, rotor and piezoelectric elements. The stator is a metal tube (Fig. 1) whose external walls are grounded into eight planes forming an octagon and the internal walls are fabricated into internal screw threads. The rotor is a cylinder or a tube with external screw threads which engage with the internal screw threads of the stator.

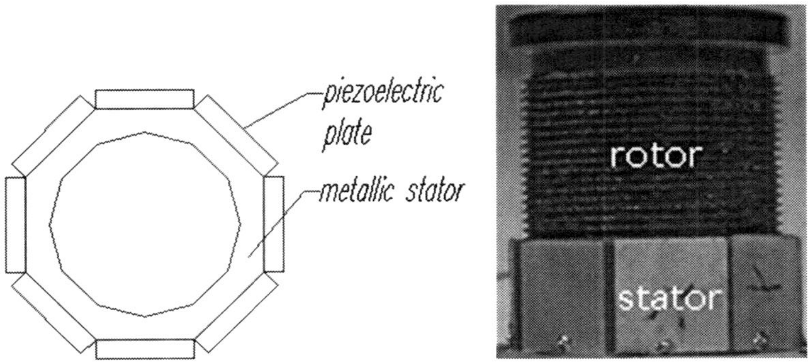

Fig. 1 Structure of the original nut-type ultrasonic motor

Fig. 2 The third order bending modes in r − θ plane

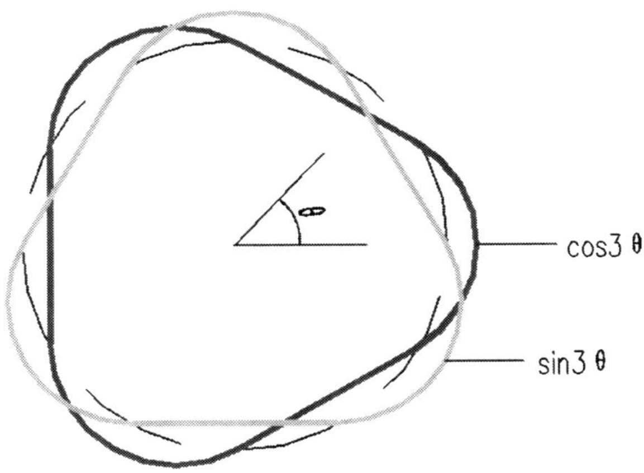

The rotor can rotate inside the tube of the stator and makes linear movement along axial direction. The piezoelectric elements are rectangular sheets with equal thickness whose polarization direction is the thickness direction. Moreover, they are bonded to the planes on external surfaces of the polygon metal tube.

As the piezoelectric elements are rather thin, the stator can be regarded as a thin wall tube. It has two degenerate third order bending vibration modes along the circumference (Fig. 2), which can be expressed as

$$\begin{pmatrix} u_{01} \\ u_{r1} \end{pmatrix} = A_1 \begin{pmatrix} \sin 3\theta \\ B_3 \cos 3\theta \end{pmatrix} \cos \omega t \tag{1a}$$

$$\begin{pmatrix} u_{02} \\ u_{r2} \end{pmatrix} = A_2 \begin{pmatrix} \cos 3\theta \\ -B_3 \sin 3\theta \end{pmatrix} \cos \omega t \tag{1b}$$

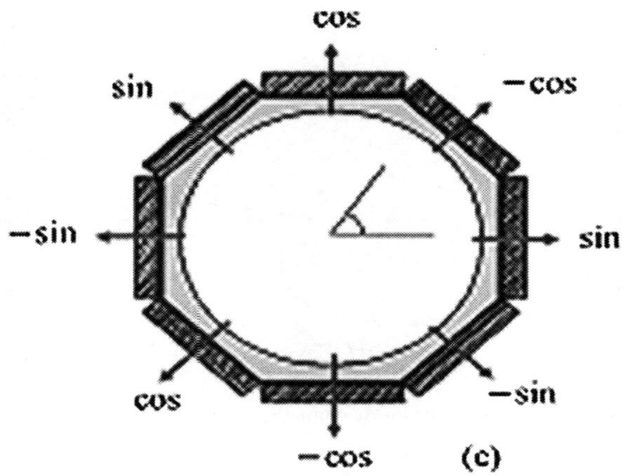

Fig. 3 The connected driving signal configuration for the original nut-type ultrasonic motor

For third order vibration (n = 3), when the signal connection pattern is as shown in Fig. 3, eight standing waves can be superposed into a traveling wave.

$$\begin{aligned}u_r &= A\Big\{\sin 3\theta \sin \omega t - \sin\Big[3\Big(\theta - \frac{\pi}{4}\Big)\Big]\cos \omega t + \sin\Big[3\Big(\theta - \frac{\pi}{2}\Big)\Big]\cos \omega t \\ &\quad + \sin\Big[3\Big(\theta - \frac{3\pi}{4}\Big)\Big]\sin \omega t - \sin[3(\theta - \pi)]\sin \omega t + \sin\Big[3\Big(\theta - \frac{5\pi}{4}\Big)\Big]\cos \omega t \\ &\quad - \sin\Big[3\Big(\theta - \frac{3\pi}{2}\Big)\Big]\cos \omega t - \sin\Big[3\Big(\theta - \frac{7\pi}{4}\Big)\Big]\sin \omega t\Big\} \\ &= A\Big[2\cos(\omega t - 3\theta) + 2\cos\Big(\omega t - 3\theta + \frac{9\pi}{4}\Big)\Big] \\ &= 2\sqrt{2+\sqrt{2}}A\cos\Big[\omega t - 3\theta + \cot^{-1}\Big(\sqrt{2}+1\Big)\Big]\end{aligned} \tag{2}$$

The displacement of a circular tube with travelling wave can be expressed as

$$\begin{pmatrix} u_0 \\ u_r \end{pmatrix} = A\begin{pmatrix} \cos(\omega t - 3\theta) \\ B_3 \sin(\omega t - 3\theta) \end{pmatrix} \tag{3}$$

Thus, the relationship between the r and θ displacement can be determined as

$$\frac{u_0^2}{1} + \frac{u_r^2}{B_3^2} = A^2 \tag{4}$$

The above equation indicates that the trajectories of stator points are elliptical of the same clock direction. It is the elliptic motion of points on inner wall of the stator that drives the rotor to rotate.

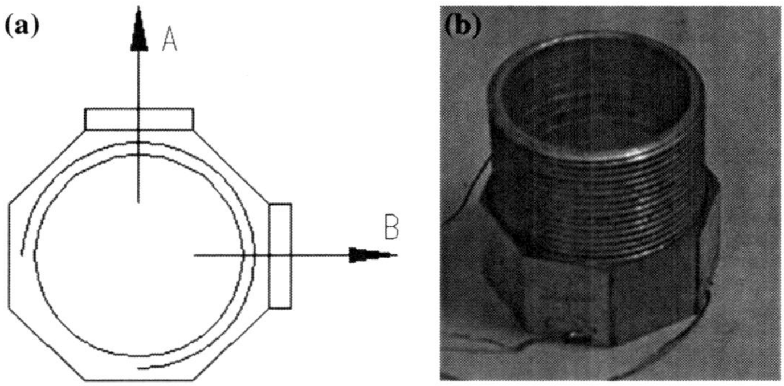

Fig. 4 Structure and prototype of the newly designed single phase driven nut-type traveling wave ultrasonic motor

2.2 The New Design

The newly proposed nut-type ultrasonic motor driven with single phase signal has two plates on its stator (Fig. 4a–b). The two plates are bonded to two side surfaces forming a 90° angle. As the piezoelectric elements are rather thin, effects of their thickness can be neglected and the stator can be regarded as a thin wall circular tube. The working modes are the two third order bending modes in the $r - \theta$ plane of the tube.

The number of piezoelectric plates used for the original stator is eight so the stator working modes remains degenerate due to octagonal symmetry. So the two working modes keep associated with the same resonant frequency. The newly designed stator uses only two plates, as a result the symmetry is broken and the degeneracy was removed. In this case, the degenerate modes are separated into two vibration modes associated with two different resonance frequencies with a small gap.

When the stator is excited with a single phase drive at a frequency between the two resonance frequencies, both modes will respond asynchronously with a phase difference due to the structural damping of the stator. The modes will combine to form a wave composed of both travelling and standing wave component. The travelling wave component will then drive the rotor to rotate consistently.

When AC voltage is applied on piezoelectric plate B, a clockwise transmitting traveling wave will be generated, as a result the rotor rotates counterclockwise under frictional actuation. When an AC voltage is applied on piezoelectric plate A, a counterclockwise traveling wave will be generated and the rotor rotates oppositely. Thus, bidirectional rotation can be achieved with a single driving source instead of the usually used two phase-shifted driving sources.

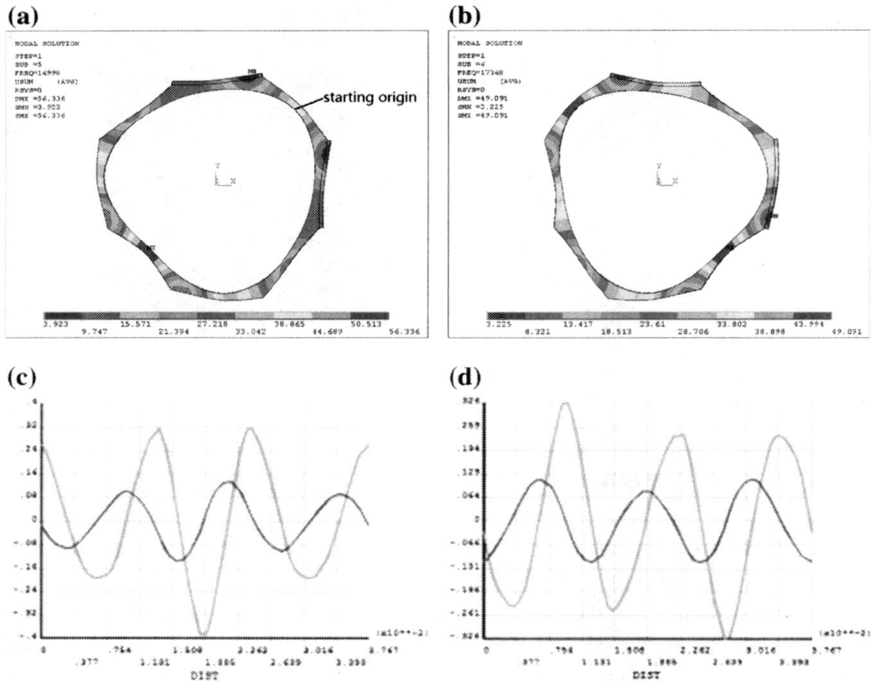

Fig. 5 The working modes' displacement contour plot and displacement curves against arc length of the new stator with two piezoelectric plates

3 Finite Element Analysis

With a finite element method software, modes of the new stator are analyzed. When the piezoelectric plates are reduced to two pieces resonance frequencies of the two third order modes depart (Fig. 5a, b), the resonance frequencies of the stator from modal analysis are 16.99 and 17.36 kHz.

The two third order working modes are orthogonal to each other, and for the given mode radial displacement Ur is orthogonal to circumferential displacement Uc (Fig. 5c, d). The mode shapes varied but still keep the shape similar to a sine function. These facts accord with the flexural vibration formation of rings and tubes given as Eq. (1a, b).

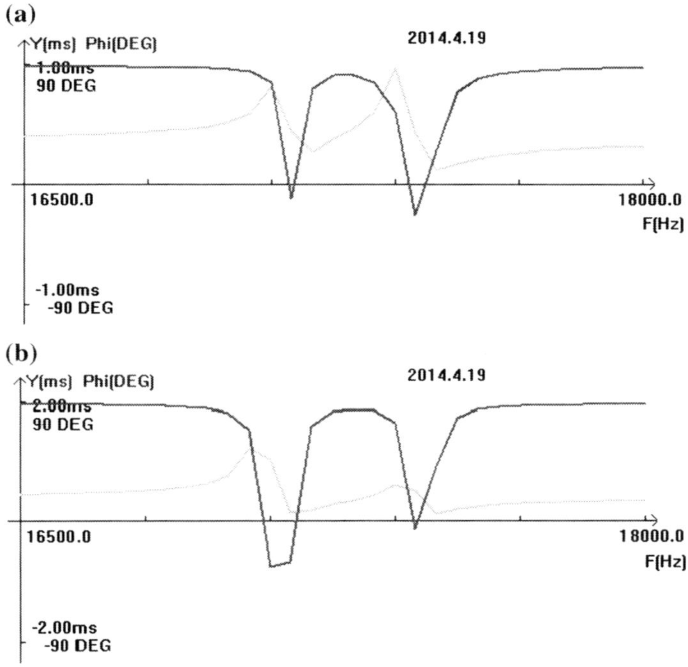

Fig. 6 **a** Electrical conductance of phase A, resonance frequencies are 17.15 and 17.40 kHz. **b** Electrical conductance of phase B; resonance frequencies are 17.10 and 17.45 kHz

4 Prototype and Its Experimental Test

A prototype nut-type ultrasonic motor of 12 mm thread diameter, 0.5 mm thread pitch, 5 mm side length, 6 mm height was fabricated (Fig. 4b). The thickness of the PZT plates used for actuation was 0.2 mm. Impedance of the prototype stator was tested. When only piezoelectric plate A was connected the resonance frequencies obtained are 17.15 and 17.40 kHz (Fig. 6a). When only piezoelectric plate B was connected, the resonance frequencies are 17.10 and 17.45 kHz (Fig. 6b). This is just the mode splitting phenomenon depicted in the Sect. 2.2.

Weight of the rotor is just the load force when the motor is erected, and speed–load curves were tested with rotors of different weight. A 40 V peak–peak, AC driving voltage of different frequencies was applied to the motor. The rotation speed of the motor was tested. The thread pitch of the stator and rotor is 0.5 mm, so linear motion speed can be calculated by multiplying rotation speed with 0.5 mm/round.

The motor operated smoothly at the frequencies between the working mode frequencies of 17.10 and 17.40 kHz (Fig. 7). This accord with theoretical analysis in Sect. 2.2. Bi-directional rotation was achieved by switching the connected piezoelectric plate.

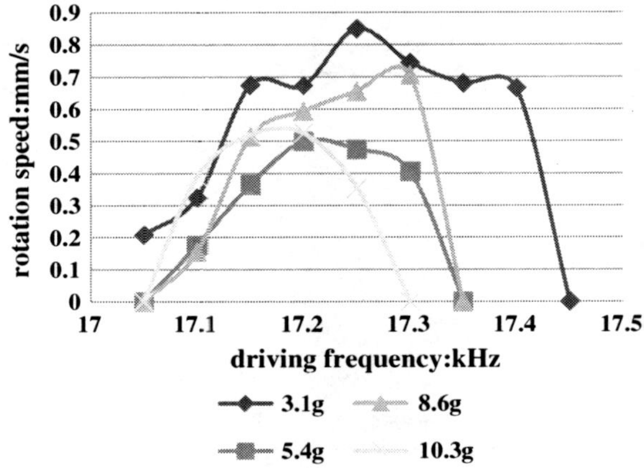

Fig. 7 Frequency dependence on rotation speed curve of different rotors

5 Conclusion

A novel nut-type ultrasonic motor is proposed. It uses two piezoelectric plates whereas the original structure uses eight plates, so the assembly art will be simplified and cost will be cut down. It can be driven with a single phase electric driving signal, so the driving circuit will be simplified.

The motor employs two third order bending modes in the $r - \theta$ plane of the stator, The stator mode frequencies split by a small gap of about 0.3–0.4 kHz for the newly designed stator with two piezoelectric plates because of symmetry breaking. When one of the two piezoelectric plates is connected to an AC voltage of a frequency between the two mode frequencies, both working modes are excited but vibrate with a small phase difference, so that a travelling wave component will be generated and the rotor will be driven to rotate by the travelling wave in the stator. Bi-directional rotation can be achieved by switching the connected piezoelectric plate.

The mode shape change of the new nut stator was observed with modal FEM analysis. A prototype was fabricated and the electrical impedance of the stator was tested, and two close working modes of 17.10 and 17.40 kHz was observed. The frequency dependence on the speed and mechanical power was tested. The motor operated smoothly at the frequencies between the working mode frequencies of 17.10 and 17.40 kHz. Bi-directional rotation was achieved by switching the connected piezoelectric plate.

References

1. Zhou TY, Zhang Y, Chen Y (2009) A nut-type ultrasonic motor and its application in focus system. Chin Sci Bull 54(20):3778–3783
2. Zhou TY, Chen Y, Lu CY (2009) Integrated lens auto-focus system driven by a nut-type ultrosonic motor USM. Sci Chin Ser E: Technol Sci 52(9):2591–2596

3. Zhou TY, Lu CY, Chen Y (2005) A thread screw driving polyhedral ultra-sonic motor. PRC Patent CN1767347
4. Zhou TY, Lu CY, Chen Y (2006) A thread driven polyhedron ultra-sonic motor. World Patent No. PCT/WO2006/003088
5. Ueha S, Tomikawa Y, Kurosawa M, Nakamura N (1993) Ultrasonic motors theory and applications. The Past Times Press, USA
6. Peter HC (1987) Split mode traveling wave ring-resonator. U.S. Patent 4,686,407
7. Minoru K, Sadayuki U (1991) Single-phase drive of a circular ultrasonic motor. J Acoust Soc Am 90(4):1723–1728

Theoretical and Experimental Investigation into 'Efficient Modal Control Strategies' as Applied on a Plate Structure

Manu Sharma and S.P. Singh

Abstract In this work some efficient modal control strategies for 'active vibration control' are theoretically simulated and experimentally implemented. Experimentation is done on a cantilevered test plate instrumented with one piezoelectric sensor and two piezoelectric actuators. Test system is modelled using finite element procedures. Finite element model is reduced to first three modes using modal reduction technique. Experiments are performed using modal space control (no weighting), modal space control (displacement weighting), modified independent modal space control, modal space control (energy weighting) and modal space control (fuzzy logic based weighting). It has been observed that, efficient modal control strategies conserve energy without sacrificing much on performance.

1 Introduction

Vibration control strategy is called active vibration control (AVC) if an external source of energy is employed to control the vibrations. An AVC scheme essentially requires sensor/sensors, processor and actuator/actuators. Vibration signal sensed by sensors is suitably processed via some control strategy by the processor and then fed back to the actuators. Variety of control strategies have been proposed for this purpose e.g. modal space control, optimal control, adaptive control, sliding mode control etc. Meirovitch [1] developed Independent Modal Space Control (IMSC) for controlling vibrations of a distributed mass body. In IMSC, equations of motion of the structure are decoupled using modal analysis and then modal control forces are determined by minimising a performance index. Baz and Poh [2] performed a

M. Sharma (✉)
University Institute of Engineering & Technology, Panjab University, Chandigarh, India
e-mail: manu@pu.ac.in

S.P. Singh
Department of Mechanical Engineering, Indian Institute of Technology, Delhi, India
e-mail: singhsp@mech.iitd.ernet.in

J.K. Sinha (ed.), *Vibration Engineering and Technology of Machinery*,
Mechanisms and Machine Science 23, DOI 10.1007/978-3-319-09918-7_75

numerical study to control the vibrations of a beam instrumented with piezoelectric patches, using IMSC. They also modified IMSC to control multiple modes and called it Modified Independent Modal Space Control (MIMSC). In MIMSC control effort shifts between modes to be controlled, depending upon the energy associated with the modes of vibration—mode with highest energy is controlled first. Baz and Poh [3] experimentally verified theoretical developments made by Baz and Poh [2]. Shijie et al. [4] have used genetic algorithms for controlling actuation of photostrictive actuator for active vibration control of a thin beam. Singh et al. [5] have suggested some efficient modal control strategies by suitably weighing the gains of IMSC. The essence of the method proposed by Singh et al. [5] is that a feedback force in different modes be applied according to the vibration amplitude in the respective modes i.e. mode having lesser vibration may receive lesser feedback. In the present work, strategies suggested by Singh et al. [5] have been experimentally implemented on a cantilevered test plate. The test plate is instrumented with piezoelectric sensors and actuators. Labview software is used to acquire sensor response on to the PC and to send control signal from PC to the piezoelectric actuators. Also, in this work, fuzzy logic based efficient modal control has been proposed and experimentally implemented.

2 Theoretical Study on 'Efficient Modal Control'

In vibration control strategies, control gains are computed without taking into consideration excitation force which disturbs the plant. In many cases, qualitative nature of excitation force is known. Say in a peculiar case, excitation force may be exciting fundamental mode predominantly and subsequent modes mildly. This excitation can be periodic also. This useful piece of information can be employed to modify the control gains. This study is directed towards employing nature of excitation for modifying control gains. While developing control law in IMSC/MIMSC/LQR etc. nature of external excitation is not taken into consideration. Singh et al. [5] developed some efficient modal control (EMC) strategies for active control of vibrations, using knowledge of the excitation and the uncontrolled response. They proposed efficient modal space control using displacement weighting and energy weighting. In the present work, theoretical conclusions of Singh et al. [5] are reproduced and experiments are conducted to establish the theoretical findings. Further, fuzzy logic is also employed to develop an efficient modal control strategy which is based upon energy of the mode.

2.1 EMC Using Displacement Weighting

According to EMC using displacement weighting, feedback in different modes is obtained from the following relations

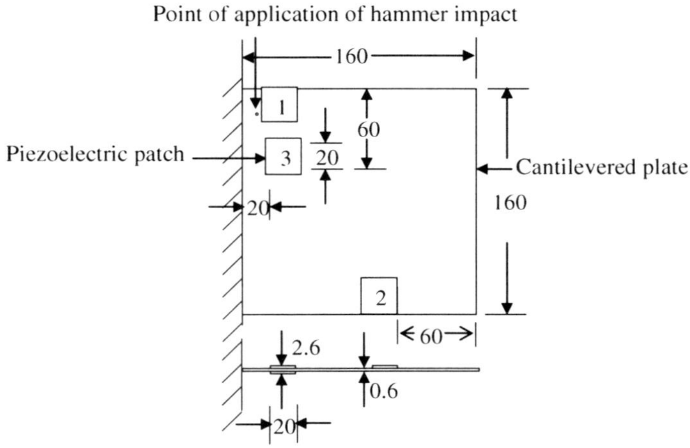

Fig. 1 Cantilevered smart plate

$$Feedback\,in\,mode\,i : Feedback\,in\,mode\,j : Feedback\,in\,mode\,k = 1 : \frac{displacement(j)}{displacement(i)} : \frac{displacement(k)}{displacement(i)} \quad (1)$$

Here displacement of a particular mode refers to height of the mode in frequency domain. Plate is disturbed by symmetrical triangular impulse.

As shown in Fig. 1, piezoelectric patch nearest to the 'point of application of impulse' is actuator 1, farthest to the point of application of impulse is actuator 2 and the third patch is the piezoelectric sensor. Figure 2a shows uncontrolled response of the plate and in Fig. 2b frequency spectrum of the sensor signal is given. First mode is excited by 2.9 V, second by 0.7 V and third by 0.05 V. First of all a multi-modal control strategy is adopted, where voltages to be applied on the actuators is obtained using following relation

$$\{V(t)\}_{2\times 1} = \{V_1(t)\}_{2\times 1} + \{V_2(t)\}_{2\times 1} + \{V_3(t)\}_{2\times 1} \quad (2)$$

where V_i is actuator voltage according to Independent modal space control of ith mode. First element of V_i is voltage to be applied on actuator '1' and second element of V_i is voltage to be applied on actuator '2'. According to independent modal space control (IMSC) [1], modal force to be applied on actuator can be obtained in closed form by

$$\mathrm{q_r} = \mathrm{g_{1r}}\eta_\mathrm{r} + \mathrm{g_{2r}}\eta_\mathrm{r}$$

where, $g_{1r} = -\lambda_r\lambda_r + \lambda_r\sqrt{(\lambda_r)^2 + \frac{1}{R_r}}$

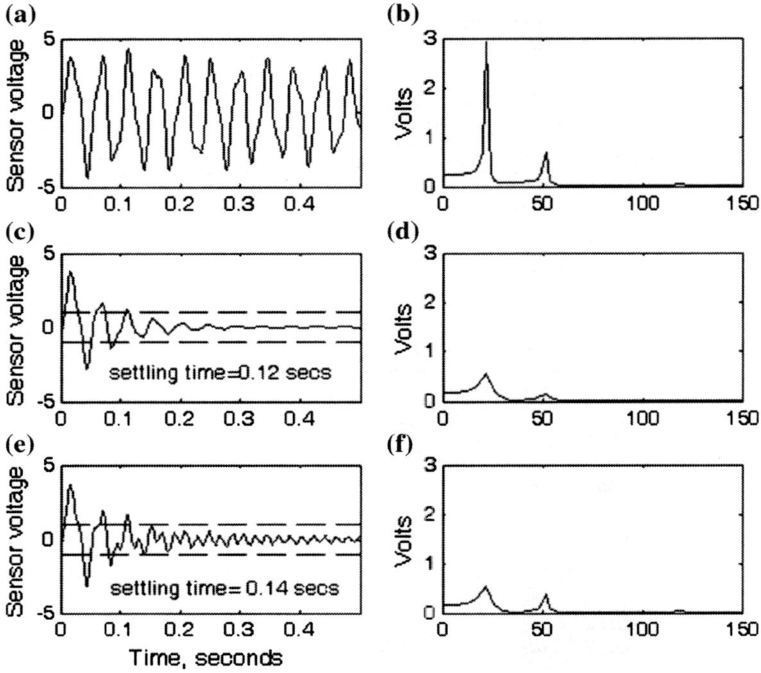

Fig. 2 Plate response uncontrolled (**a**, **b**), controlled without weighting (**c**, **d**) and controlled with displacement weighting (**e**, **f**)

$$g_{2r} = \sqrt{-2\lambda_r\lambda_r + \frac{1}{R_r} + 2\lambda_r\sqrt{(\lambda_r)^2 + \frac{1}{R_r}}}.$$

where λ_r is natural frequency of rth mode and parameter R_r for all the modes is taken as 0.003. In Fig. 2c, d plate is controlled by this strategy. Plate response settles down in 0.12 s. Plate response is defined to have settled down when response remains within ±1 V. From frequency response we find that first mode drops down to 0.56 V from 2.9 V, second mode drops to 0.14 V from 0.7 V and third mode drops to almost 0 V from 0.05 V. Actuator 1 applies maximum of 781 V (Fig. 3a) and actuator 2 applies maximum of 277 V (Fig. 3b). Next efficient modal control by displacement weighting is performed. Feedback in different modes is weighed as per Eq. (1)

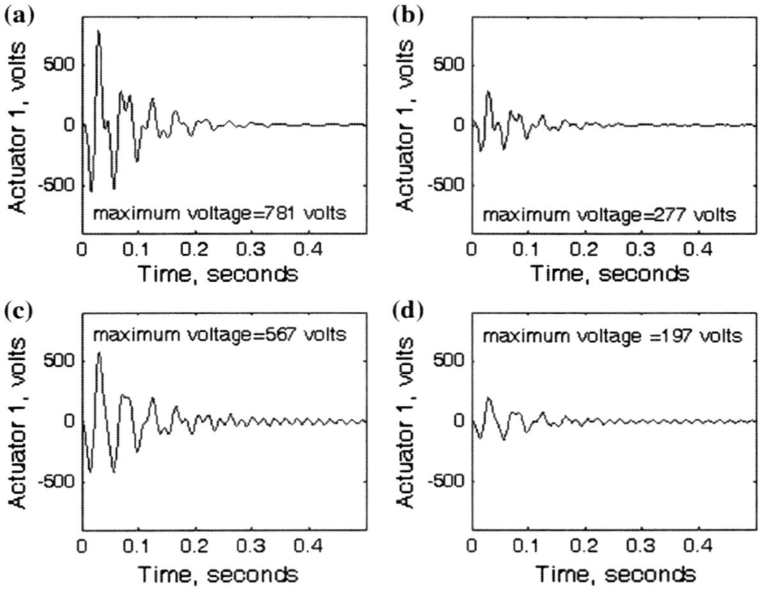

Fig. 3 Actuator voltages, without weighting (**a**, **b**), with displacement weighting (**c**, **d**)

$$\begin{aligned}&\textit{Feedback in } \mathrm{mod}\, e\, i : \textit{Feedback in } \mathrm{mod}\, e\, j : \textit{Feedback in } \mathrm{mod}\, e\, k\\ &= w_1 : w_2 : w_3\\ &= 1 : \frac{0.7}{2.9} : \frac{0.05}{2.9}\\ &= 1 : 0.24 : 0.017\end{aligned}$$

Thus equation gets modified as

$$V(t) = w_1 V_1(t) + w_2 V_2(t) + w_3 V_3(t) \tag{3}$$

Figure 2e gives sensor response with displacement weighting. Settling time is 0.14 s. From frequency response in Fig. 2f we find that first mode drops down to 0.55 V, second mode drops down to 0.4 V and third mode remains at 0.05 V. We find that settling time increases from 0.12 to 0.14 s i.e. 16 % increase. Actuator 1 applies maximum of 567 V (Fig. 3c) and actuator 2 applies maximum of 197 V (Fig. 3d). Maximum voltage applied by actuator 1 decreases by 27 % and maximum voltage applied by actuator 2 decreases by 29 %.

2.2 Fuzzy Logic Based EMC Using Energy Weightings

In MIMSC, at a particular instant of time, control effort is applied on mode of highest energy. So, control effort keeps on shifting from one mode to another. This can fatigue the actuator and can excite higher modes of the structure. In this light, multimodal control based upon Eq. (2) is free from this shifting. Singh et al. [5] have modified controller based upon Eq. (2) by modifying modal control gains based upon modal energies. Modal energy has been used as a criterion to weigh the modal gains. According to EMC using energy weighting, feedback in different modes is obtained from following relation

$$\begin{aligned} &\textit{Feedback in } \bmod e\, i : \textit{Feedback in } \bmod e\, j : \textit{Feedback in } \bmod e\, k \\ &= 1 : \frac{energy(j)}{energy(i)} \times \frac{frequency(i)}{frequency(j)} : \frac{energy(k)}{energy(i)} \times \frac{frequency(i)}{frequency(k)} \end{aligned} \tag{4}$$

Examination of Eq. (4) reveals that, EMC based upon energy weighting will fail if $energy(i)$ becomes zero and will weigh the mode too heavily if $energy(i)$ becomes very small. Keeping in mind these drawbacks of EMC with energy weighting, fuzzy logic based EMC is proposed in this section.

Sets *Low*, *Medium* and *High* are constructed over the range of Modal energy as shown in Fig. 4. *Low* signifies low values of modal energy, *Medium* signifies medium values of modal energy and *High* signifies high values of modal energy. Similarly, sets *Low*, *Medium* and *High* are constructed over the range of weight as

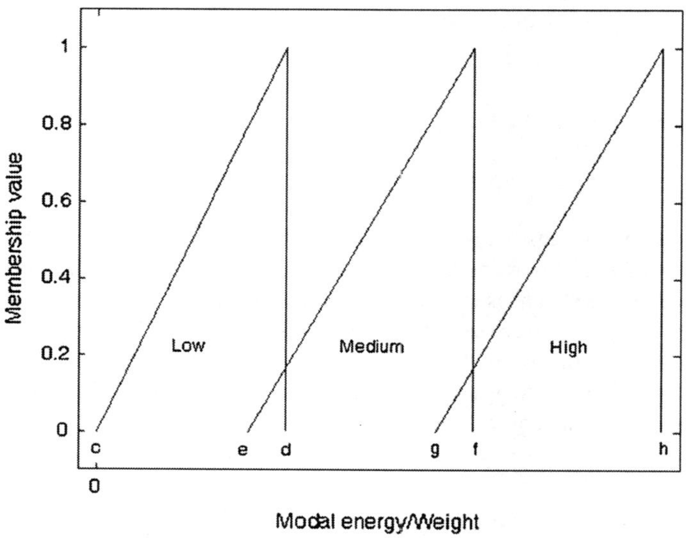

Fig. 4 Fuzzy sets for modal energy and weights to be applied on modal gains

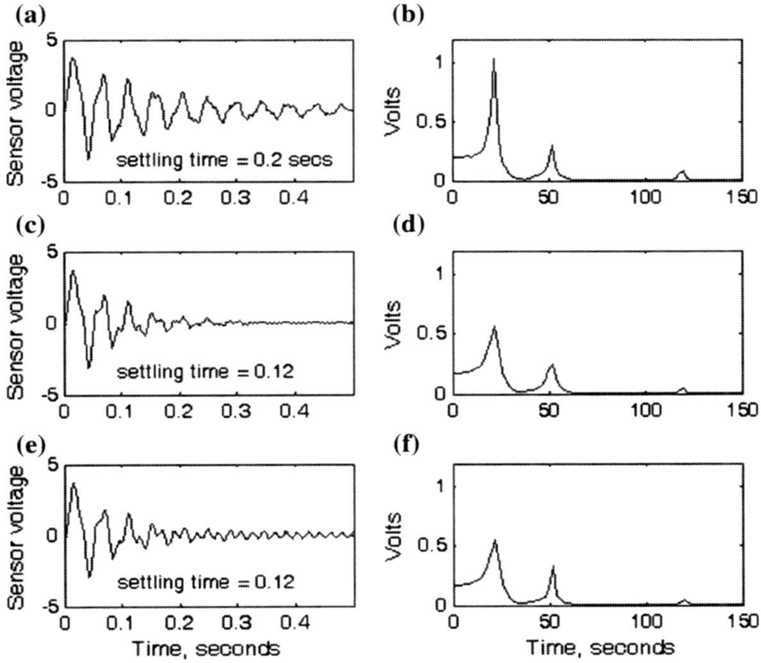

Fig. 5 Plate response, MIMSC (**a**, **b**), energy weighting (**c**, **d**) and fuzzy logic weighting (**e**, **f**)

shown in Fig. 4. Following rules are formulated to map fuzzy sets of modal energy with fuzzy sets of weight.

1. If modal energy is *Low* then weight to be applied on modal gain is *Low*.
2. If modal energy is *Medium* then weight to be applied on modal gain is *medium*.
3. If modal energy is *High* then weight to be applied on modal gain is *High*.

In fuzzy logic based EMC, weights are applied on second and third modal gains, according to the fuzzy logic procedure based on modal energies. In Fig. 5a plate is controlled by MIMSC. Plate response settles down in 0.2 s. First mode drops down to 1 V, second to 0.3 V and third rises to 0.08 V (Fig. 5b). MIMSC in this case, has increased the response of third mode by 0.03 V.

From Fig. 6a, b, it is observed that MIMSC applies maximum voltage of 567 V on actuator 1 and 190 V on actuator 2. It is observed that in MIMSC, a number of frequencies are present in actuator voltage response. In Fig. 5c plate is controlled by energy weighting. Plate response settles down in 0.12 s. Plate response drops down to 0.56 V in first mode, to 0.24 V in second mode and remains at 0.05 V in third mode. The corresponding actuator voltages are given in Fig. 6c, d.

From Fig. 6c, d it is observed that actuator 1 applies maximum of 547 V and actuator 2 applies maximum of 188 V. It is important to note here, that for same settling time, multi modal control with no weighting applies 42 % more peak

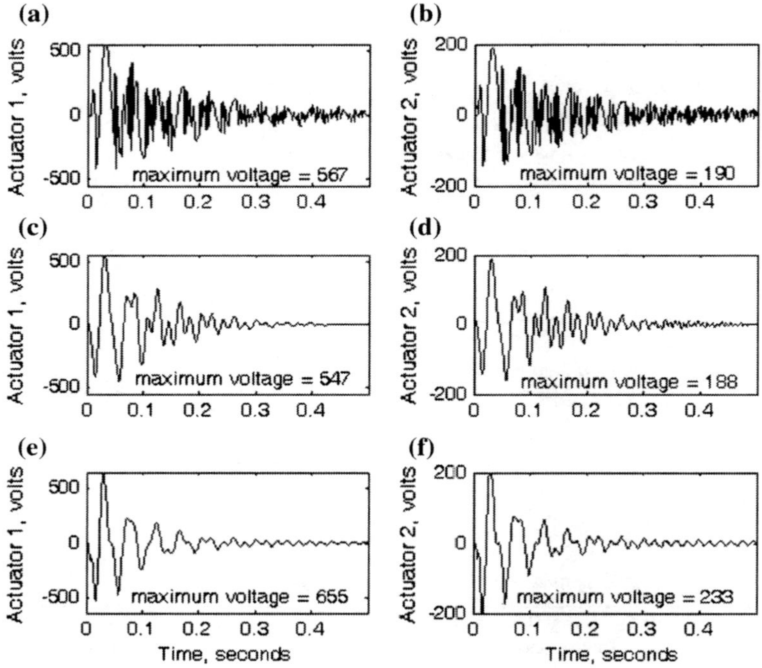

Fig. 6 Actuator voltages, MIMSC (**a**, **b**), energy weighting (**c**, **d**) and fuzzy logic weighting (**e**, **f**)

voltage on actuator 1 and 47 % more peak voltage on actuator 2 (refer Fig. 6a, b). In Fig. 6e plate is controlled by fuzzy logic based energy weighting. Plate response settles down in 0.12 s. Peak of first mode drops down to 0.56 V, second mode drops down to 0.34 V and third mode drops down to 0.04 V. Maximum voltage applied on actuator 1 is 655 V and on actuator 2 is 233 V (Fig. 6e, f). Again it is observed that, for same settling time, multimodal control with no weighting applies 16 % more peak voltage on actuator 1 and 16 % more peak voltage on actuator 2. So, it is concluded that suitable weighting of feedback gains will conserve energy, without much sacrifice of performance.

3 Experimental Results

A piezoelectric charge amplifier is used to amplify the piezoelectric sensor patch signal. Sensor response is acquired through a data acquisition card (PXI-6040E of National Instruments) on to a PXI based real time Real Time (RT) system (PXI-8170). RT system does control calculations on the sensor signal as per the computer program which is downloaded on it from the host PC via an ethernet cable. Plate is disturbed by hitting it at a specific point near the root by an instrumented hammer.

Table 1 Tabulation of experimental results on Multi-modal control of plates

Case	ST (s)	FMR (V)	SMR (V)	TMR (V)	V_1 and %D	V_2 and %D
Uncontrolled	3.76	3.4	1.5	0.4	–	–
Modal space control (no weights)	2.32	2.34	0.85	0.35	50 and –	60 and –
Displacement weighing	2.5	2.48	1.28	0.4	40 and 20	49 and 18
MIMSC	2.68	2.37	1.22	0.47	39 and 22	48 and 16
Energy weighing	2.64	2.38	1.35	0.37	34 and 32	49 and 18
Fuzzy logic weighing	2.65	2.13	1.4	0.44	33 and 34	52 and 13

Both hammer response as well as piezoelectric sensor response are stored in storage oscilloscope and then processed to get the frequency response function. Sensor patch (piezoelectric patch numbered 1) is placed in region where modal strains of first three modes is simultaneously high. Actuator 1 (piezoelectric patch numbered 1) is placed in region where modal strain of first and second modes is simultaneously high and actuator 2 (piezoelectric patch numbered 2) is placed in region where modal strain of third mode is high. Experiments were conducted using modal space control (no weights), displacement weighing, MIMSC, energy weighing and fuzzy logic weighing. Results are tabulated in Table 1.

Where ST is settling time, FMR is first mode response in frequency domain, SMR is second mode response in frequency domain, TMR is third mode response in frequency domain, DSTU is decrease in settling time compared to uncontrolled case, ISTM is increase in settling time compared to modal space control (no weighing), V_1 and %D is maximum voltage applied on actuator 1 and percentage decrease in this value compared to modal space control (no weighing), V_2 and %D is maximum voltage applied on actuator 2 and percentage decrease in this value compared to modal space control (no weighing). From Table 1 it is clear that efficient modal control strategies conserve energy without sacrificing much on performance.

4 Conclusions

Efficient modal control strategies presented by Singh et al. [5] have been experimentally verified for a plate structure. Extending the work of Singh et al. [5], fuzzy logic based control strategy is presented and experimentally verified. Weights applied on modal gains by presented fuzzy logic based strategy, are dynamic as they keep changing depending upon the modal energies of the plate. Proposed fuzzy logic based strategy is observed to conserve energy without sacrificing much on performance.

References

1. Meirovitch L (1989) Dynamics and control of structures. Wiley, New York
2. Baz A, Poh S (1988) Performance of an active control system with piezoelectric actuators. J Sound Vib 126(2):327–343
3. Baz A, Poh S (1990) Experimental implementation of the modified independent modal space control method. J Sound Vib 139(1):133–149
4. Shijie Z, Jingjing L, Hongtao W (2014) Genetic algorithm based wireless vibration control of multiple modal for a beam by using photostrictive actuators. Appl Math Model 38(2):437–450
5. Singh SP, Pruthi HS, Agarwal VP (2003) Efficient modal control strategies for active control of vibrations. J Sound Vib 262:563–575

Smart Machines with Flexible Rotors

Arthur W. Lees and Michael I. Friswell

Abstract The concept of smart machinery is of current interest. Several technologies are relevant in this quest including magnetic bearings, shape memory alloys (SMA) and piezoelectric activation. Recently a smart bearing pedestal was proposed based on SMAs and elastomeric O-rings. However, such a device is clearly relevant only for the control of rigid rotors; for flexible rotors there is a need for some modification on the rotor itself. In this paper, equations of motion are developed to describe a rotor with a force generator (for example, a piezoelectric element) mounted on it. It is shown that such a system may be used to compensate for imbalance by inducing a rotor bend. This leads to some questions as to the optimum control strategies, and the paper discusses some of the possibilities.

Keywords Rotordynamics · Control · Imbalance · Rotor bend · Piezoelectric

1 Introduction

Over recent years there have been significant developments in the study of smart structures, but the application of the same approaches to the automatic control of rotating machinery has received significantly less attention. It would appear, however, that this forms a very attractive field of study; if vibration levels were readily controllable, significant constraints on machine design could be eased and some machines could be lighter and more efficient. This lack of attention in this area is a little surprising since many of the requisite technologies are now common place but what is lacking is an integrated approach.

A.W. Lees (✉) · M.I. Friswell
College of Engineering, Swansea University, Singleton Park, Swansea SA2 8PP, UK
e-mail: a.w.lees@swansea.ac.uk

M.I. Friswell
e-mail: m.i.friswell@swansea.ac.uk

J.K. Sinha (ed.), *Vibration Engineering and Technology of Machinery*,
Mechanisms and Machine Science 23, DOI 10.1007/978-3-319-09918-7_76

Magnetic bearings [1] provide one possible component of a smart machine and are, of course, in regular use on certain types of machine. In particular they are readily applied to the control of rigid rotor machines, but their use with flexible rotors is somewhat more complicated. Lees et al. [2] suggested a rolling element bearing supported on a non-linear elastomer whose stiffness can be controlled be an external source. This also is primarily of interest for machines with rigid rotors.

For the control of machines with flexible rotors one is inevitably driven to some form of actuator mounted on the rotor and driven by slip-rings, telemetry or possibly some form of magnetic coupling. Horst and Wolfe [3] used piezoelectric (PZT) patches mounted on the shaft surface. The surface of the shaft was ground to an octagonal cross section to allow for the attachment of the patches and the actuators were connected via slip-rings. Their rotor comprised a single heavy disc overhung at the end of a shaft. The authors demonstrated the feasibility of this type of control and showed the effectiveness in terms of resonant response curves.

Sloetjes and de Boer [4, 5] used a combination of position feedback, integral control and scheduled balancing to give reduced vibration levels. They report on stability issues and show the response of a system during transient operation. Other relevant work in this area is that by Kunze et al. [6].

The current study seeks to extend the work of Lees [7] which analysed the rotordynamic behaviour of a rotor mounted with piezoelectric patches. The concept is that a voltage applied to the patch induces a shaft bend which compensates for the rotor imbalance. The analysis of Lees [7] treats the displacement and bend as two separate variables and derives the appropriate Lagrange equations. In the present study, the moment is applied directly to the rotor. Although the distinction between rotor bends and imbalance is often emphasized, given the ability to control the bend externally, the distinction ceases to be important.

Automatic balancing has been studied by a number of authors [e.g. 8–13]. The approaches described have been applied to specific types of machine but in the current study the aim is to investigate techniques which can be used more generally to compensate for a range of shaft forces, and imbalance is just one of these, albeit the most important.

2 Model of Shaft with Piezoelectric Patches

Piezoelectric patches are compact actuators used in a range of applications requiring precise positional control. The crucial component is a layer of piezoelectric material which undergoes a strain when a voltage is applied across the electrodes. The device is capacitive, a fact which must be considered within the overall system integration. The most convenient form of these patches is the Macro-Fiber Composite (MFC) which has a range of sizes and is conveniently packaged with external electrodes. As a voltage is applied to the electrodes, the strain will change by up to 0.2 %. If the piezoelectric composite is bonded to the shaft with appropriate epoxy cement, the overall effect will be to contract the composite, hence applying equal and opposite forces at either end of the device. If patches are mounted diametrically opposite to each other with appropriate bias and polarities the net effect will be equal and opposite bending moments applied to the shaft. These moments will induce a bend in the shaft which will be used to counteract any intrinsic imbalance.

At this point it is worth considering the attachment of the patches to the shaft. Given that the force is determined by the patch and voltage applied to it, the bending moment can be enhanced by increasing the offset from the shaft center line, perhaps by mounting between two discs. There is however a further difficulty here as the piezoelectric patches will not bend around a tight radius (although MFCs are much better than monolithic piezoelectric plates). This may be overcome by forming flats on an appropriate part of the shaft, an approach used by Sloetjes and de Boer [5]. Another possibility would be to abandon the composites and use PZT fibres directly, but at this investigative stage the MFCs are convenient.

Piezoelectric materials provide one way of implementing some form of rotor mounted actuation, but there are other possibilities which may be used. Piezoelectric materials have the advantage of good frequency response, but do require high voltages and this may be a disadvantage in some applications. An alternative activation could be provided by shape memory alloys. Such activation could operate at a few volts (as opposed to several hundred volts for the piezoelectric devices), and they can sustain higher strains. The disadvantage is the limited frequency response, particularly in the cooling phase, but this may not be too important for machines whose operating regime does not vary too quickly. An intriguing possibility is the use of Magnetic Shape Memory Alloys as this would render the use of slip-rings unnecessary. However, these possibilities are perceived as futuristic and for the present analysis, piezoelectric patches (for example MFCs) are taken as the actuation element.

There has been a significant effort in the modeling of piezoelectric materials as distributed transducers and many review papers have been published (see, for example [14–20]). Analyses range from simple devices such as uniform beams and plates in linear dynamics, to more complicated configurations such as composites under non-linear and non-uniform loading and dynamics (such as helicopter blades and aircraft wings). A majority of the research deals with the modeling of symmetric (bimorph) beams and plates, and a uniform strain model is assumed, where the through thickness variation of strain in the active piezoelectric device is uniform. Erturk and Inman [21] and Leo [20] give further details of the modeling of piezoelectric sensors and actuators integrated with beam structures.

Suppose that piezoelectric layers are added to a beam in a bimorph configuration. Then the moment about the beam neutral axis produced by a voltage V across the piezoelectric layers [17, 18] may be written as

$$M_A = \gamma_c V(t) \tag{1}$$

where the constant γ_c depends on the geometry, configuration and piezoelectric device. For a bimorph with piezoelectric layers in the 31 configuration, with thickness h_c, width b_c, and connected in parallel, then

$$\gamma_c = Ed_{31}b_c(h + h_c) \tag{2}$$

where h is the thickness of the beam, and d_{31} is the piezoelectric constant. These expressions assume a monolithic piezoceramic actuator perfectly bonded to the beam; Bilgen et al. [22] considered the effect of the structure of a Macro-Fiber Composite (MFC) on the coupling coefficient, and also the effect of the bond and Kapton layers. The mechanical stiffness and mass density of the piezoelectric layers may also be included in the shaft mass and stiffness.

3 Rotating Machine Models and Dynamics

Only a summary is provided here; for further details consult the book by Friswell et al. [23]. Suppose that the machine is modeled in the stationary frame and that the mass and stiffness properties of the shaft are symmetric. Two important points concern the response to unbalance and bends. First, the influence of rotor spin speed on the forces applied to the shaft is different for unbalance and bends; unbalance forces and moments increase with Ω^2 whereas the forces and moments due to a bent shaft are independent of the rotor spin speed Ω. Second, there is no reason why the spatial distribution of unbalance and the bend should be identical, and in general the total excitation force is never zero. One could choose the balance mass distribution for a rotor with an inherent bend; this would balance the bent shaft, but would do so only at a single rotor spin speed. This would also require adding mass at all nodes, which is impractical. However, the response is filtered using the dynamic stiffness matrix and hence adequate balance may be achieved by adding fewer balance masses. This could minimise some average response at all nodes, or achieve a zero response at particular degrees of freedom. A bend could also be used to counteract an inherent imbalance, and the issues are similar.

The key concept for the smart machine described in this paper is the application of moments to the shaft that essentially produce a static bend. Since the requirement is to produce a static bend, a variety of smart materials could be used, for example SMAs or piezoelectric patches. These strain actuators would be arranged in a bimorph configuration and hence may be modeled as a pair of opposing moments on the shaft at the ends of the actuators. They would have to be arranged in pairs of orthogonal bimorphs in order to apply moments to both shaft planes. Although the dynamics and hysteresis of the piezoelectric material can be important, here we assume that the applied moment is proportional to the demanded control input. These applied moments are fixed in the rotating frame, and are easily transformed to the stationary frame. Thus the motion is described by

$$\mathbf{M}\ddot{\mathbf{q}} + [\Omega\mathbf{G} + \mathbf{C}]\dot{\mathbf{q}} + \mathbf{K}\mathbf{q} = \Re\left(\left\{\Omega^2\mathbf{b}_0 + \mathbf{K}\mathbf{q}_{b0} + \mathbf{B}\mathbf{u}\right\}e^{j\Omega t}\right) \tag{3}$$

where $\Omega^2\mathbf{b}_0$ is the vector of forces and unbalance forces and moments in the rotating frame, and $\mathbf{q}_{b0}$ denotes the vector of deflections of the stationary rotor due to the permanent bend. $\mathbf{u}$ is the vector of control inputs and it is assumed that more than one set of actuators have been distributed on the shaft. The matrix $\mathbf{B}$ distributes the control inputs to the correct degrees of freedom. Note that the possibility of an inherent (permanent) shaft bend has also been included in this model. In fact the moment applied by the piezoelectric patches can interpreted as an induced bend in the shaft given by $\mathbf{q}_{b0} = -\mathbf{K}^{-1}\mathbf{B}\mathbf{u}$.

There are two ways of defining the control inputs for each pair of bimorphs. The physically based models require two real numbers that represent the voltage inputs to each bimorph. For steady state analysis in the frequency domain these may be combined as a complex number, and this provides a direct link to the common balancing approaches (for example the influence coefficient method).

4 Control Approaches

There are at least five possible approaches to use the forces arising in the piezoelectric MFC patches, which may be outlined as follows:

- Use proportional feedback in the rotating frame to directly change the natural frequencies. This approach is likely to be limited to small machines and does not directly counteract the inherent unbalance force.
- Apply an axial force to modify the rotor properties and therefore the natural frequencies of the machine. The axial force required is likely to be significant and hence this approach is impractical in most machines.
- Use derivative feedback to control damping rather than stiffness. This will improve the transient response but will not significantly affect the unbalance response, unless the machine is operated close to one of its critical speed.
- Use the patches to induce rotor bends, which can counteract the unbalance. This may be viewed as a standard balancing approach where the moments from the piezoelectric patches replace the forces from the balancing masses.
- Use integral feedback in the rotating frame. Since the unbalance response in the rotating frame can be viewed as a steady state error, applying integral feedback can reduce this error and hence balance the machine.

The first three approaches are unlikely to be practical and hence this paper will concentrate on the last two approaches.

Balancing by Bending the Shaft. Essentially the piezoelectric actuators apply moments to the shaft that may be easily controlled. The determination of the best moments to apply is identical to a balancing procedure, with the essential difference that the moments that may be applied take a particular form dictated by the position of the actuators on the rotor. The great advantage is that the moments applied may be continuously adjusted due to changes in the machine balance or changes in the rotor spin speed. Hence existing methods of balancing and automatic balancing may be used. Methods of automatic balancing based on changing the mass distribution of the rotor do have the advantage that a balanced machine will be balanced at all speeds, although in practice the range of spin speeds where the balance is satisfactory is limited.

Integral Feedback Control. Integral feedback is well established to reduce the steady state errors in a wide range of control systems. The error of interest in this case would be the response at the location of the sensors in the rotating frame. Note that this error would have components in both transverse directions and this is counteracted by two orthogonal bimorph actuators. Since the measurements will be taken in the stationary frame, and consist of proximity probe or accelerometer measurements, they must be transformed into the rotating frame, in the directions of the actuators. Most rotating machines will have a phase reference that gives both the rotor spin speed and also the phase correction; this reference enables the transformation matrix to be calculated. If the response in the rotating frame is aligned to the directions of the piezoelectric actuators, then each actuator voltage is simply proportional to the integral of the corresponding response.

Response measurements are assumed in this paper; if accelerations are measured then these must be integrated before transformation, and this is most readily done in the frequency domain by estimating the magnitude and phase of the unbalance response.

The measurements are translational displacements and the actuators apply moment pairs; even if the sensors and actuators are located at the same position on a machine they will not be collocated. Thus there is always a risk of spill-over where some modes, typically high frequency modes, are destabilised by the control system. This is made worse by the transient response excited during a run-up or rundown, and the need to balance the machine at a range of spin speeds where the response is significantly influenced by more than one critical speed.

5 Numerical Example

The numerical example will follow that given by Sloetjies and De Boer [5], and is motivated by a helicopter tail drive shaft. The shaft has length 1 m and is a square section aluminium tube with outside width of 10 mm and a thickness of 1 mm. Bearings are located at both ends of the shaft and are modeled as short bearings with stiffness 100 kN/m and damping 1 kNs/m in both transverse directions. The shaft is modeled using 20 beam finite elements [23]. The first critical speed for this machine is at 1771 rev/min. To reduce the calculation time for the simulations the model is projected onto the eight lowest frequency undamped modes of the stationary machine. The unbalance is modeled as a force at the mid-shaft position with

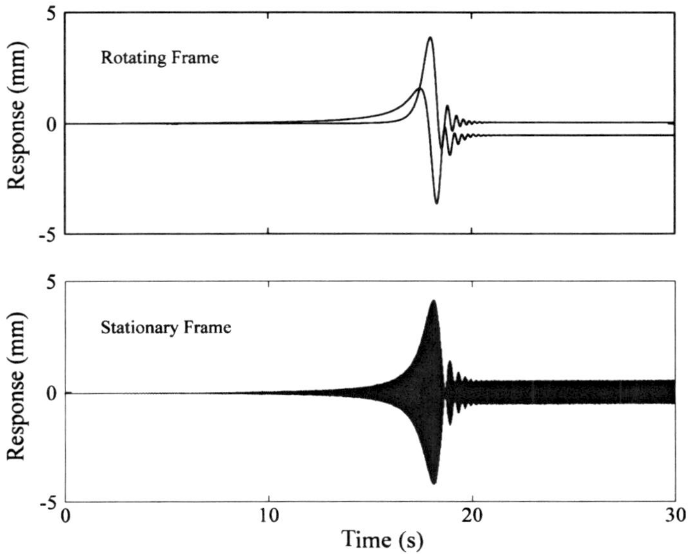

Fig. 1 Response at the mid-shaft position during the run-up

mass unbalance 5.7×10^{-6} kgm. The running speed of the machine is 2000 rev/min and the machine is run-up from zero speed in 20 s. Figure 1 shows the response of the machine in stationary and rotating coordinates for the first 30 s, and demonstrates the increased response as the machine runs through the first critical speed. Notice that in the rotating frame the response at the steady running speed is a static displacement.

Only one set of actuators (a pair of bimorph actuators) will be used in this paper, in contrast to the three sets of actuators used by Sloetjies and De Boer [5]. Although the piezoelectric properties may be used to directly estimate the coupling parameter, γ_c, here we assume the length of the piezoelectric actuators are 100 mm and adjust the coupling parameter to give a mid-shaft displacement of 1.1 mm/V, as given by Sloetjies and De Boer [5]. Thus $\gamma_c = 1.6 \times 10^{-3}$ Nm/V. The stiffness and mass properties of the piezoelectric material could be included in the shaft properties, but are neglected in this paper. Figure 2 shows the root locus when the integral parameter varies from 0 to 10^8 and shows that the control mainly affects the lowest frequency pole and causes some loss in stability. If we choose the integral control parameter as 5×10^5, then the closed loop response is shown in Fig. 3, in both the rotating and stationary frames of reference. Note that the controller is switched on at 30 s, after the machine has run-up and reached the steady state. The voltage applied to the piezoelectric patches is shown in Fig. 4. If the controller is switched on at the start of the run-up then the response as the machine runs through the critical speed may be reduced significantly, as shown in Fig. 5.

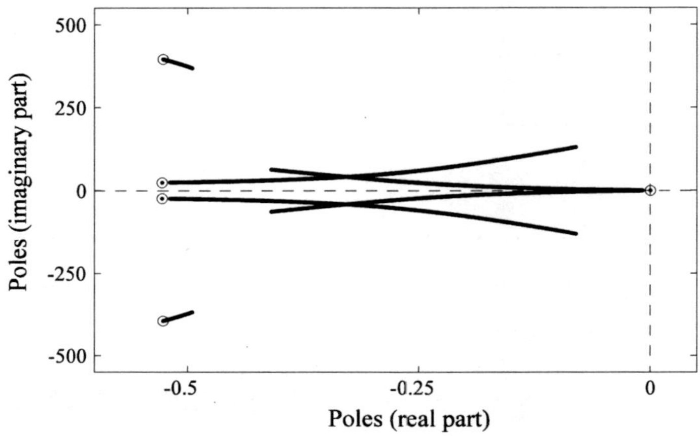

Fig. 2 The root locus for the integral controller. The *circles* denote the open loop poles

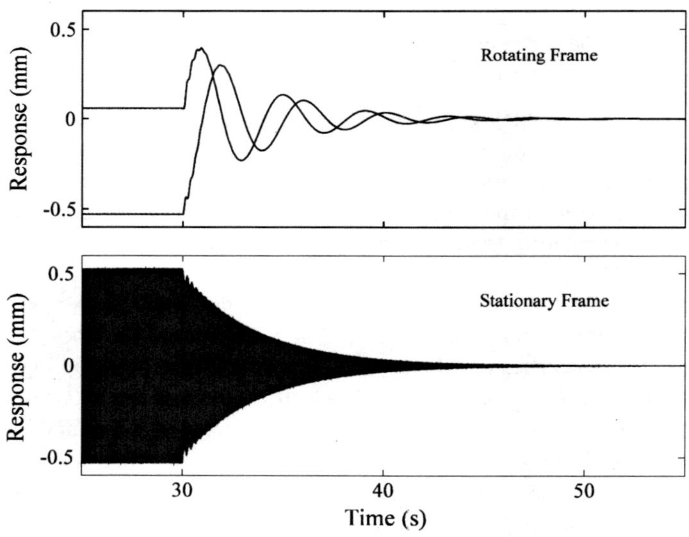

Fig. 3 Response at the mid-shaft position with the controller switched on at 30 s

The influence coefficient balancing approach is not simulated. In fact the PZT voltages required to balance the machine are identical to the steady state voltages given in Fig. 4 that clearly produce a zero steady state response at the sensor locations.

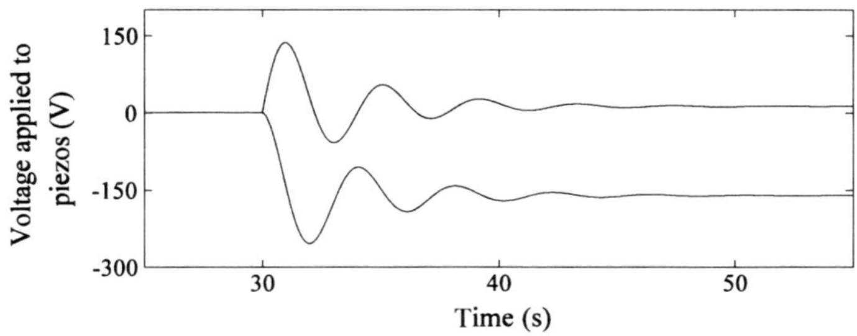

Fig. 4 Voltage applied to the piezoelectric bimorphs to control the unbalance response

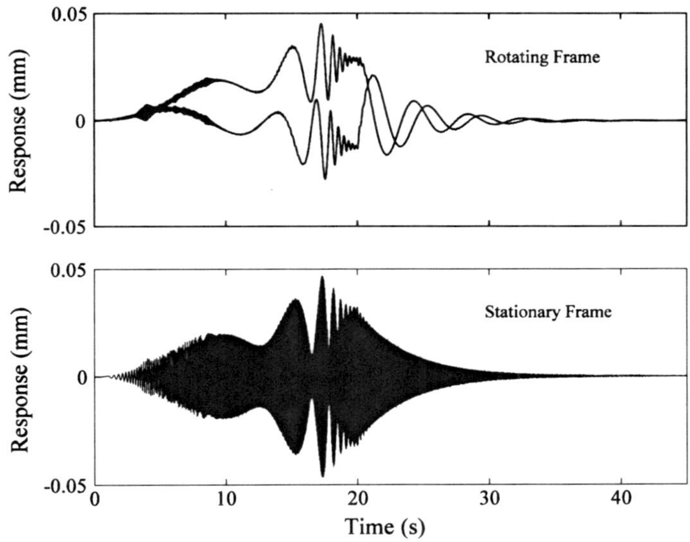

Fig. 5 Response at the mid-shaft position with the controller switched on at 0 s

6 Conclusion

This paper has demonstrated that rotor imbalance can be controlled by imposing a bend using shaft mounted actuation, and some possible concepts have been outlined. In particular, a simulated case study using piezoelectric actuation has shown that integral feedback control gives promising results. However, further work is needed to develop a robust system, considering issues such as the effective placement of actuators on the shaft, the effect of running through multiple critical speeds, the effect of asymmetries in the shaft and supports, and drift in the control system. Furthermore proportional and derivative control may be added to help improve stability and performance characteristics.

References

1. Schweitzer G (1998) Magnetic bearings as a component of smart rotating machinery. Proceedings of 5th international conference on rotordynamics, IFToMM, 7–10 Sept 1998
2. Lees AW, Jana S, Inman DJ, Cartmell MP (2007) The control of bearing stiffness using shape memory. Proceedings of ISCORMA, Calgary, Alberta, Aug. 2007
3. Horst H-G, Wolfel HP (2004) Active vibration control of a high speed rotor using PZT patches on the shaft surface. J Intell Mater Syst Struct 15:721–728
4. Sloetjes PJ, de Boer A (2006) Piezoceramics-based devices for active balancing of flexible shafts. Proceedings of ISMA, Leuven, Belgium, Sept 2006, pp 543–555
5. Sloetjes PJ, de Boer A (2008) Vibration reduction and power generation with piezo- ceramic sheets mounted to a flexible shaft. J Intell Mater Syst Struct 19(1):25–34
6. Kunze H, Riedel M, Schmidt K, Bianchini I (2003) Vibration reduction on auto- motive shafts using piezoceramics. Proc SPIE Smart Struct Mater 5054:382–386
7. Lees AW (2011) Smart machines with flexible rotors. Mech Syst Signal Process 25:373–382
8. Gosiewski Z (1985) Automatic balancing of flexible rotors, 1. theoretical background. J Sound Vib 100(4):551–567
9. Gosiewski Z (1987) Automatic balancing of flexible rotors, 2. synthesis of system. J Sound Vib 114(1):103–119
10. Vandevegte J (1981) Balancing of flexible rotors during operation. J Mech Eng Sci 23 (5):257–261
11. Vandevegte J, Lake RT (1978) Balancing of rotating systems during operation. J Sound Vib 57(2):225–235
12. Blanco-Ortega A, Beltran-Carbajal F, Favela-Contreras A, Silva-Navarro G (2008) Active disk for automatic balancing of rotor-bearing systems. Proceedings of the American control conference, 11–13 June 2008, pp 3023–3028
13. Zeng S, Wang XX (1998) The electromagnetic balancing regulator and the automatic balancing system. J Sound Vib 209(1):5–13
14. Chee CYK, Tong L, Steven GP (1998) A review on the modeling of piezoelectric sensors and actuators incorporated in intelligent structures. J Intell Mater Syst Struct 9(1):3–19
15. Chopra I (2002) Review of state of art of smart structures and integrated systems. AIAA Journal 40(11):2145–2187
16. Crawley EF (1994) Intelligent structures for aerospace: a technology overview and assessment. AIAA Journal 32(8):1689–1699
17. Crawley EF, Anderson EH (1990) Detailed models of piezoceramic actuation of beams. J Intell Mater Syst Struct 1(1):4–25
18. Crawley EF, de Luis J (1987) Use of piezoelectric actuators as elements of intelligent structures. AIAA Journal 25(10):1373–1385
19. Benjeddou A (2000) Advances in piezoelectric finite element modeling of adaptive structural elements: A survey. Comput Struct 76(4):347–363
20. Leo DJ (2007) Engineering analysis of smart material systems. Wiley, Hoboken
21. Erturk A, Inman DJ (2011) Piezoelectric energy harvesting. Wiley, Hoboken
22. Bilgen O, Erturk A, Inman DJ (2010) Analytical and experimental characterization of macro-fiber composite actuated thin clamped-free unimorph benders. J Vib Acoust 132(5):051005
23. Friswell MI, Penny JET, Garvey SD, Lees AW (2010) Dynamics of rotating machines, Cambridge University Press, Cambridge

Vibration Localization of Rotationally Periodic Structures

Arun Chandrashaker and Sondipon Adhikari

Abstract Structures with uniform periodic spacing and repeated geometry are found in complex engineering systems. Examples include turbine blades, ship hull, aircraft fuselage and oil pipelines with periodic supports. The vibration characteristics of these periodic structures are highly sensitive to its mass distribution, stiffness distribution and geometrical properties. Parametric uncertainties in structures which arise out of material defects, structural damage or variations in material properties can break the symmetry of periodic structures. These uncertainties can drastically change and localize the vibration modes. Identification of severe mode localization in the design process can help prevent failure due to high cycle fatigue in periodic structures such as turbine blades. A 2-DOF system is used to demonstrate the effect of parametric uncertainties on eigenvalue veering and mode localization phenomenon. The Finite Element Method (FEM) was developed to study the free vibrations of two linearly un-damped systems, namely, two cantilever beams coupled system and an idealized turbine blade. The effect of parametric uncertainties on eigenvalue veering and mode localization is demonstrated and a novel numerical method using the Modal Assurance Criterion (MAC) to quantify mode localization in complex systems is proposed.

Keywords Mode localization · Eigenvalue veering · Periodic structures · Quantifying localization · Modal assurance criterion · Parametric uncertainty

A. Chandrashaker (✉) · S. Adhikari
College of Engineering, Swansea University, Singleton Park, Swansea SA2 8PP, UK
e-mail: 740886@swansea.ac.uk

S. Adhikari
e-mail: S.Adhikari@swansea.ac.uk

J.K. Sinha (ed.), *Vibration Engineering and Technology of Machinery*,
Mechanisms and Machine Science 23, DOI 10.1007/978-3-319-09918-7_77

1 Introduction

Vibration modes of periodic structures are highly sensitive to parametric uncertainties. A slight variation to its mass or stiffness properties can isolate the vibrations to a specific region. A structure can be defined as periodic when its mass distribution, stiffness distribution and geometrical features repeats itself in space. Numerous examples of linearly and rotationally periodic structures can be found in many Aerospace, Civil and Mechanical engineering applications such as turbine blades, Aircraft fuselages and oil pipe lines with periodic supports. Excitation of the localized vibration modes can lead to High Cycle Fatigue (HCF) which contributes to premature failure of the structure [1, 2]. HCF in turbine blades has been identified as the major cause of Aircraft Engine failures [2].

Perfect periodic structures are idealized cases while in reality most structures are only nearly periodic due to parametric uncertainties such as material imperfections or structural damage. As structures are never periodic various authors have introduced intentional mistuning to reduce the effects of localization in turbine blades. Castanier and Pierre [1] discuss the importance of preventing HCF and summarize the design strategies used thus far to prevent extreme localization. An excellent review in [2] can be referred for more details about intentional mistuning and reduction of the forced response of bladed disks. Blair [3] proposed disk modifications to find the configuration that would give the best mistuning pattern using the Finite Element Method.

The design strategy in the case of intentional mistuning is to find the best and worst case mistuning pattern. It can be extremely difficult and tedious to individually observe all the vibration modes when making a number of design changes to the system. The veering phenomena [4, 5], where the eigenvalues are plotted against a varying design parameter, can be used to detect the localized modes [1, 6]. A method to numerically identify mode localization still needs further attention. This paper proposes such a method, which may help designers to easily obtain crucial information such as the degree of localization for all the vibration modes in the system. In Sect. 2 the veering phenomenon is described using an example of a two degree of freedom system. In Sect. 4, the Modal Assurance Criterion Localization Factor (MACLF) is proposed for the quantification of localization and illustrated through a complex coupled cantilever beam system and a bladed disk system. For the bladed disk system two separate numerical experiments were simulated by (a) varying the mass distribution (b) varying the stiffness distribution and in both cases the localization of vibration in bladed disks is shown in Sect. 5. The MACLF is applied for the two numerical experiments, it's effectiveness is validated with curve veering and visual inspection. To resemble realistic experiment numerical noise is also added to the modal data and the sensitivity of MACLF to noise is shown in Sect. 5.3. A set of conclusions drawn based on this study is given in Sect. 6.

2 Background of Mode Veering

The equations of motion governing the free vibration of a linear, undamped system with N degrees-of-freedom are

$$\mathbf{M}\,\ddot{\mathrm{u}}(t) + \mathbf{K}\mathrm{u}(\mathrm{t}) = 0 \tag{1}$$

where $\mathbf{M}$ and $\mathbf{K} \in \mathbb{R}^{\mathrm{N}\times\mathrm{N}}$ are the mass and stiffness matrices respectively. $\mathrm{u}(t) \in \mathbb{R}^N$ is the vector of generalized coordinates and $t \in \mathbb{R}^+$ denotes time. Parametric uncertainties in a system can be modeled as a change in the mass and stiffness matrix, where δ represents the change in the system matrices. The unperturbed mass and stiffness matrices are $\mathbf{M}_0$ and $\mathbf{K}_0$ respectively.

$$\mathbf{M} = \mathbf{M}_0 + \delta\mathbf{M} \quad \text{and} \quad \mathbf{K} = \mathbf{K}_0 + \delta\mathbf{K}. \tag{2}$$

A simple 2-DOF (Fig. 1) system has only two possible modes, namely, the first mode where m_1 and m_2 oscillate in the same direction and the second mode where the masses oscillate in the opposite direction. Pierre [7] showed that weak coupling can localize the vibration modes, vibrations can also be localized when the mass distribution in the system is disturbed. For example the equation of motion of a 2-DOF system are

$$\begin{bmatrix} m_1 & 0 \\ 0 & m_2 + \delta m \end{bmatrix} \ddot{\mathrm{u}}(t) + \begin{bmatrix} k_1 + k_2 & -k_2 \\ -k_2 & k_2 + k_3 \end{bmatrix} \mathrm{u}(t) = 0 \tag{3}$$

$$\begin{aligned} \mathbf{a}_1 = \begin{Bmatrix} -0.99 \\ -0.10 \end{Bmatrix} \mathbf{a}_2 = \begin{Bmatrix} -0.03 \\ 3.16 \end{Bmatrix} \mathbf{b}_1 = \begin{Bmatrix} -0.70 \\ -0.70 \end{Bmatrix} \\ \mathbf{b}_2 = \begin{Bmatrix} -0.70 \\ 0.70 \end{Bmatrix} \mathbf{c}_1 = \begin{Bmatrix} -0.13 \\ -0.71 \end{Bmatrix} \mathbf{c}_2 = \begin{Bmatrix} -0.99 \\ 0.09 \end{Bmatrix} \end{aligned} \tag{4}$$

Here $\mathbf{a}_i$, $\mathbf{b}_i$ and $\mathbf{c}_i$ are eigenvectors of the 2-DOF system where i represents the mode number. At the extreme end of the curve veering region (Fig. 2) ($\mathbf{a}_i$ and $\mathbf{c}_i$) the vibrations are confined to one region whereas the eigenvector $\mathbf{b}_i$ indicate evenly distributed vibrations. For a simple system the eigenvectors can be examined to detect localization but for realistic finite element models with 1,00,000 or more degrees of freedom it would be impossible to quantify localization. For a more complex system the changes in the mass and stiffness distribution can localize its modes with different levels of severity. In the case of periodic structures the perfect symmetry contributes

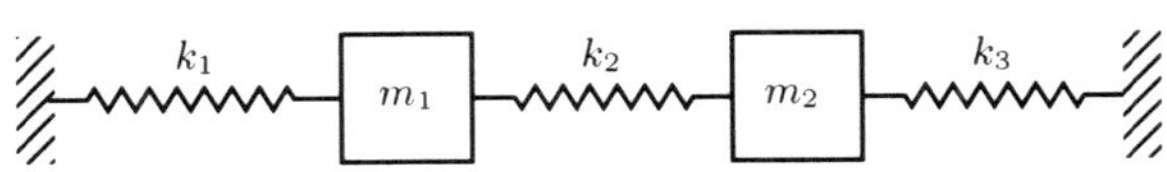

Fig. 1 Linear array of two spring-mass oscillators; $m_1 = m_2 = 1$, $k_1 = 100$, $k_2 = 10$, $k_3 = 100$

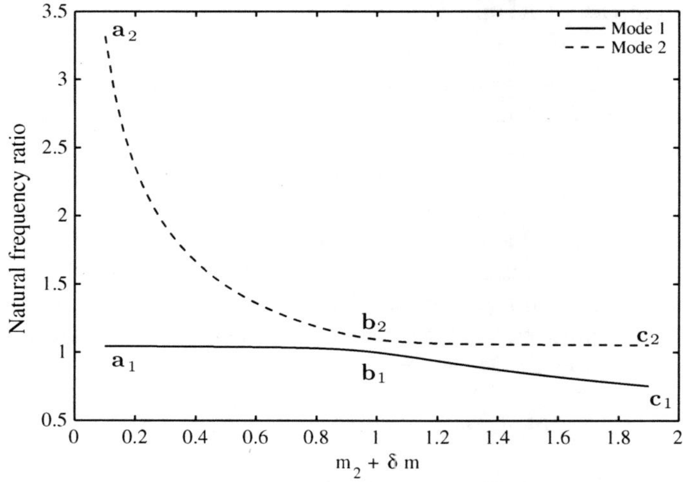

Fig. 2 Eigenvalue veering in a 2-DOF system; m_2 is varied ±90 % of m_1

to repeating eigenvalues, $\lambda_1 = \lambda_2,\ \lambda_3 = \lambda_4$ etc., where in the presence of parametric uncertainties the symmetry is broken and the repeating eigenvalues are split into distinct values [8–10]. Examining the derivative of the eigenvector expressed by [11] can provide an insight into the curve veering phenomena

$$\frac{\partial \mathbf{x}_j}{\partial p} = -\frac{1}{2}\left(\mathbf{x}_j^T \frac{\partial \mathbf{M}}{\partial p}\mathbf{x}_j\right)\mathbf{x}_j + \sum_{k=1\neq j}^{N} \frac{\mathbf{x}_k^T\left[\frac{\partial \mathbf{K}}{\partial p} - \lambda_j \frac{\partial \mathbf{M}}{\partial p}\right]\mathbf{x}_j}{\lambda_j - \lambda_k}\mathbf{x}_k \tag{5}$$

In Eq. (5) λ_k, λ_j and $\mathbf{x}_k$, $\mathbf{x}_j$ represent the eigenvalues and the eigenvectors of the system, where k and $j = 1, \ldots, N$ correspond to different mode numbers in the system. From Eq. (5) it can be observed when the values of λ_j and λ_k tends towards each other there is rapid change in the jth eigenvector [12]. This change in the eigenvector, due to some variation of the mass and stiffness matrices in the system, is what has to be quantified. In the next section we explain a possible way how this may be achieved.

3 Quantifying Localization Using the Modal Assurance Criterion

The mode shapes of the perfect system is the key to the quantification of localization. The change in eigenvector due to a some change in the mass and stiffness matrices can be quantified by using the MAC. The Modal Assurance Criterion is

Perfect Modes

Mistuned Modes	mode 1	mode 2	mode 3	⋯	mode k	
mode 1	MAC_{11}	MAC_{12}	MAC_{13}	⋯	MAC_{1k}	$\rightarrow MAX\left[MAC_{(1,k)}\right]$
mode 2	MAC_{21}	MAC_{22}	MAC_{23}	⋯	MAC_{2k}	$\rightarrow MAX\left[MAC_{(2,k)}\right]$
mode 3	MAC_{31}	MAC_{32}	MAC_{33}	⋯	MAC_{3k}	$\rightarrow MAX\left[MAC_{(3,k)}\right]$
mode j	MAC_{jk}	MAC_{jk}	MAC_{jk}	⋯	MAC_{jk}	$\rightarrow MAX\left[MAC_{(j,k)}\right]$

Fig. 3 Modal assurance crierion localization factor (MACLF)

normally used to compare experimental mode shapes to numerical mode shapes [13]. For the purpose of quantifying localization, the MAC is modified and used to compare two numerical finite element models with identical number of degrees of freedom. The Modal Assurance Criterion compares the kth mode of the perfect system to the jth mode of the mistuned system. As the mass and stiffness distribution is changed the mode shapes exchange their mode numbers, which does not allow mode number to number comparison.

In such instances where mode swapping takes place the maximum value of Modal Assurance Criterion Localization Factor (MACLF) (Fig. 3) is taken as the localization factor for every kth mode. Here $\mathbf{x}_{m_{(j)}}$ and $\mathbf{x}_{p_{(k)}}$ are the eigenvectors of the mistuned and perfect system respectively, where $\mathbf{x}_{m_{(j)}}$ and $\mathbf{x}_{p_{(k)}}$ k and $j = 1, 2, \ldots, N$ correspond to different mode numbers in the system.

$$\mathrm{MAC}_{\mathrm{Localization}} = \frac{\left|\mathbf{x}^{T}_{mistuned} \cdot \mathbf{x}_{perfect}\right|^{2}}{\left(\mathbf{x}^{T}_{mistuned} \cdot \mathbf{x}_{mistuned}\right)\left(\mathbf{x}^{T}_{perfect} \cdot \mathbf{x}_{perfect}\right)} \tag{6a}$$

$$\mathrm{MACLF}_{\mathrm{kthmode}} = MAX_{(jthmode)}\left[\frac{\left|\mathbf{x}^{T}_{m_{(j)}} \cdot \mathbf{x}_{p_{(k)}}\right|^{2}}{\left(\mathbf{x}^{T}_{m_{(j)}} \cdot \mathbf{x}_{m_{(j)}}\right)\left(\mathbf{x}^{T}_{p_{(k)}} \cdot \mathbf{x}_{p_{(k)}}\right)}\right] \tag{6b}$$

The MACLF is bounded by 1 and 0. When comparing the mode shapes of the perfect and mistuned system, the MACLF values can be used as a measure of the severity of localization. A MACLF value that is close to 1 indicates minimum or no localization and values that are close to zero will show strong localization. This will be explained using a coupled beam model shown in Sect. 4 and a bladed disk model in Sect. 5.

4 Localization in a Coupled Beam System

For this numerical experiment, a finite element model of the two coupled cantilever beam was developed. The uncertainty was introduced in the form of varying the density of only one beam and the rest of the system kept unchanged. The boundary

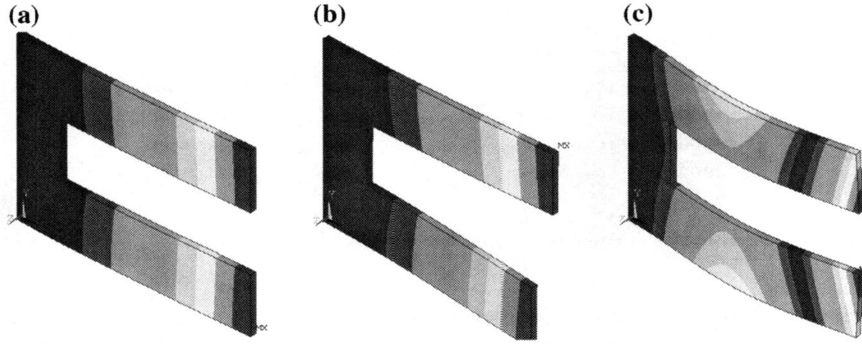

Fig. 4 Modes shapes of the system with no mistuning. **a** Mode 1. **b** Mode 2. **c** Mode 3

condition of the system is considered as fixed at one end throughout the numerical experiment. The first 15 modes of systems with ±25 % variation in density were considered for this study. In Fig. 4 we show the modes of the system with no mistuning. Figures 5 and 6 illustrate how the mode shapes of the system change to a −2.5 and −25 % variation in density. In Fig. 8 the curves of 15 modes in the coupled beam system is shown to veer when subject to a variation in density. A −25 % variation in density localizes the vibration modes severly as indicated by the veering regions. In Fig. 7 the MAC value and its applicability to quantify localization is demonstrated. It should be noted that the addition and deduction of density in a single blade do not produce identical results. Using Figs. 7 and 8 it is shown that the veering regions always correspond to a MACLF less than 1. The first three modes of the system with no mistuning (Fig. 4), −2.5 % mistuning (Fig. 5) and −25 % mistuning (Fig. 6) were selected to visually confirm the degree of mode localization.

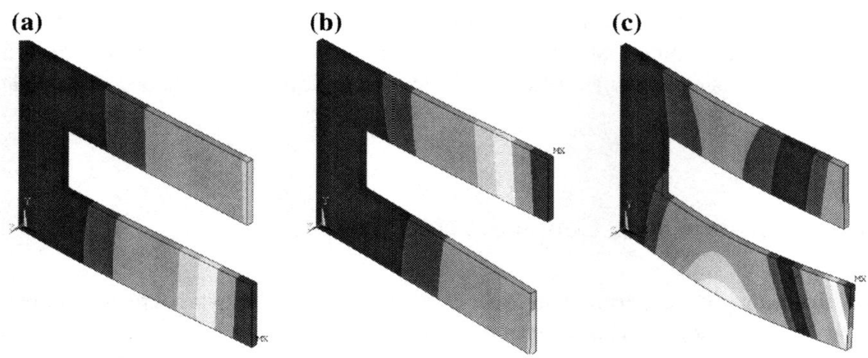

Fig. 5 Modes shapes of the system with −2.5 % mistuning. **a** Mode 1. **b** Mode 2. **c** Mode 3

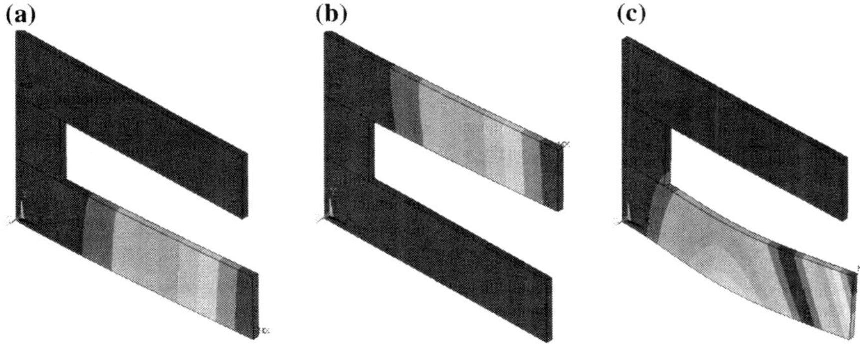

Fig. 6 Modes shapes of the system with −25 % mistuning. **a** Mode 1. **b** Mode 2. **c** Mode 3

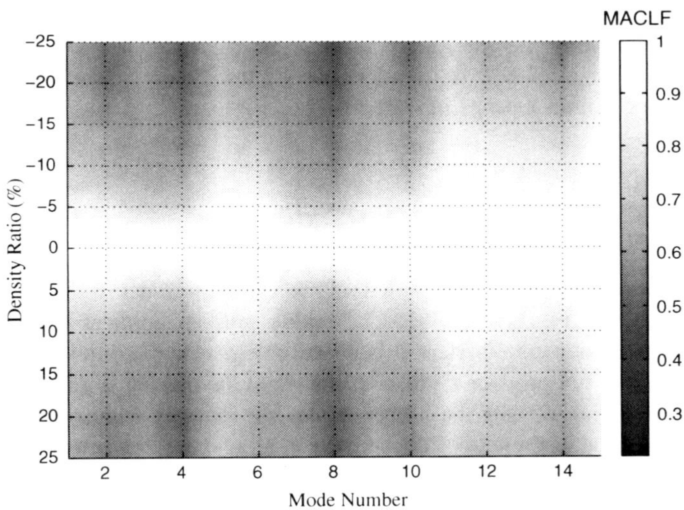

Fig. 7 MACLF applied to the coupled beam system for a variation of density

5 Dynamics of Bladed Disks

For this numerical experiment, a finite element model of the idealised bladed disk was developed. The bladed disk were modelled with 8-node solid brick elements, and has 12 sectors, 12 blades with 2,94,912 degrees of freedom. Throughout the the experiment the number of degrees of freedom was kept constant and no boundary conditions were applied to the system. Two numerical studies were performed on the bladed disk system namely (a) variation of the mass properties (b) variation of the stiffness properties. Experiment 1 involves the variation of the density only,

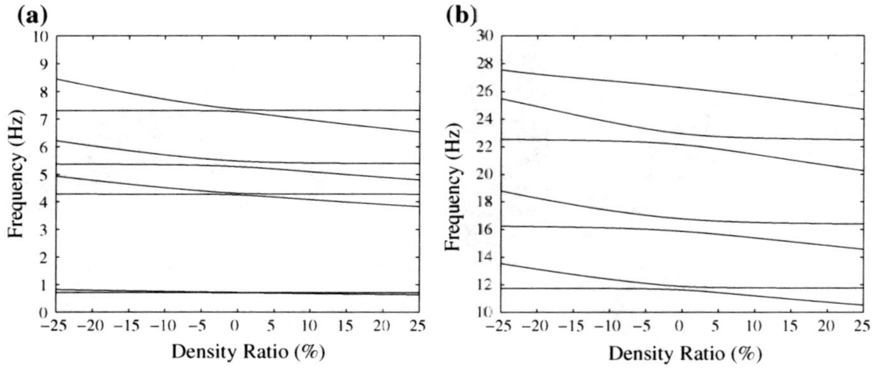

Fig. 8 Eigenvalue veering in a coupled beam system; system subjected to ±25 % variation of density

where the uncertainty was introduced in the form of varying the density of a single blade within a range of ±25 %. In experiment 2 the Young's Modulus of a single blade is varied within a range of ±10 %.

5.1 Variability in the Mass Distribution

The first 36 modes of the system with a ±25 % variation in density were considered for this study. In Fig. 10 the MACLF value and its applicability to quantify localization is demonstrated. Here it is observable that the addition of density in a single blade does not produce similar results when the density is deducted from the blade. When adding density the higher modes are strongly localized and it should be noted that some of the modes do not vary at all and are less sensitive to changes in the system. Using Fig. 9 the localization of modes is further validated, it is shown that the veering regions always correspond to a MACLF value less than 1. It can be seen that not all closely spaced eigenvalues veer in the presence of parametric uncertainties. It is observed that in the case of extreme curve veering, the MACLF value tends to be lower than 0.3. From Fig. 10 the degree of localization can be categorized into three different cases. Case 1 is when the MACLF values are between 0.6 and 1.0 indicating little or no localization, case 2 with MACLF values between 0.6 and 0.2 indicating moderate localization and finally case 3 with MACLF values between 0.2 and 0 indicating strong localization. Mode 15 (Fig. 11), Mode 21 (Fig. 12) and Mode 31 (Fig. 13) with varying levels of mistuning were selected to visually confirm the degree of mode localization.

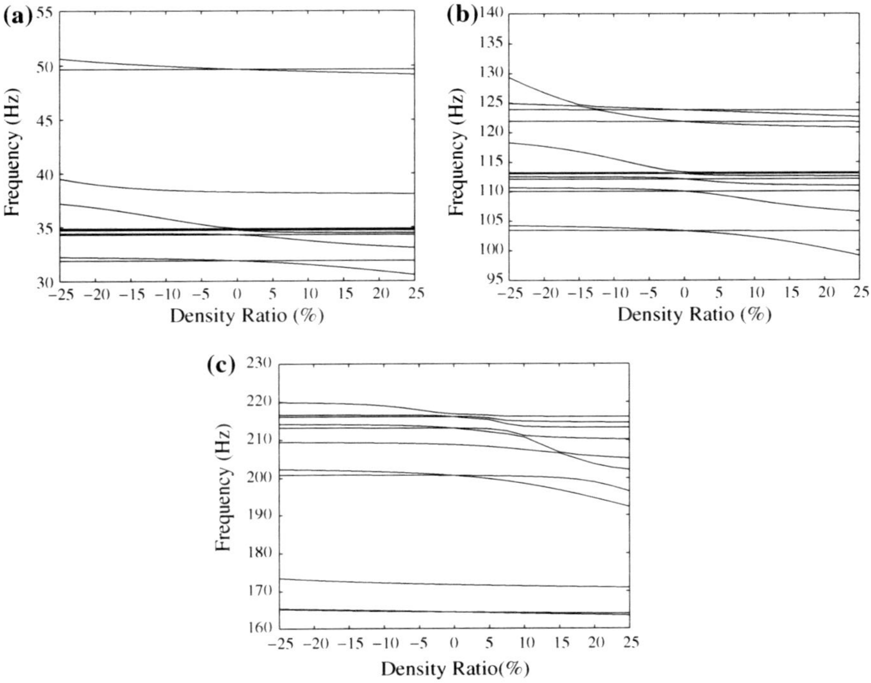

Fig. 9 Eigenvalue veering in a bladed disk system; system subjected to ±25 % variation of density in a single blade. **a** Modes 1:12. **b** Modes 13:25. **c** Modes 25:36

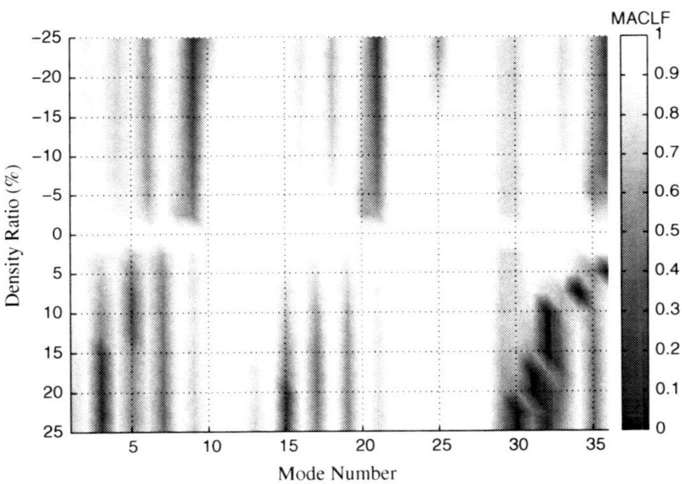

Fig. 10 MACLF for a bladed disk system for a variation in density

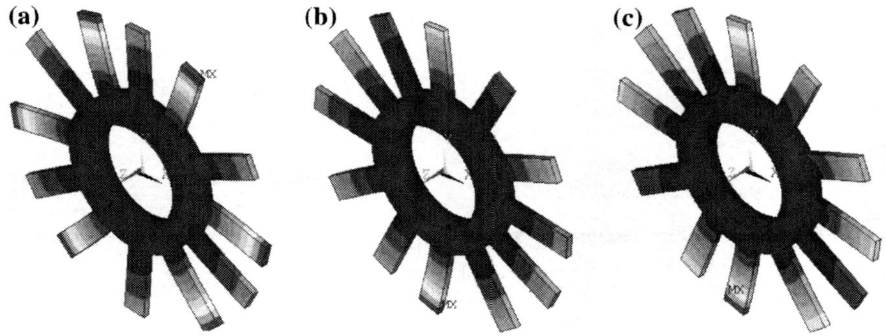

Fig. 11 Mode 15 of the system with 0, +10 and +20 % mistuning in a single blade. **a** No mistuning. **b** +10 % mistuning. **c** +20 %mistuning

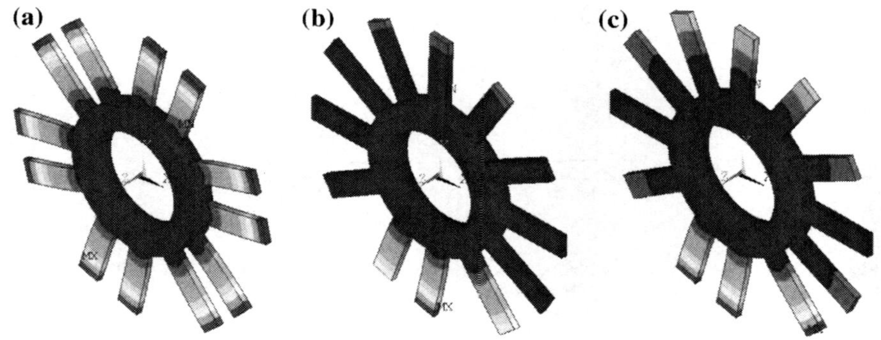

Fig. 12 Mode 21 of the system with 0, −10 and −17.5 % mistuning in a single blade. **a** No mistuning. **b** −10 % mistuning. **c** −17.5 % mistuning

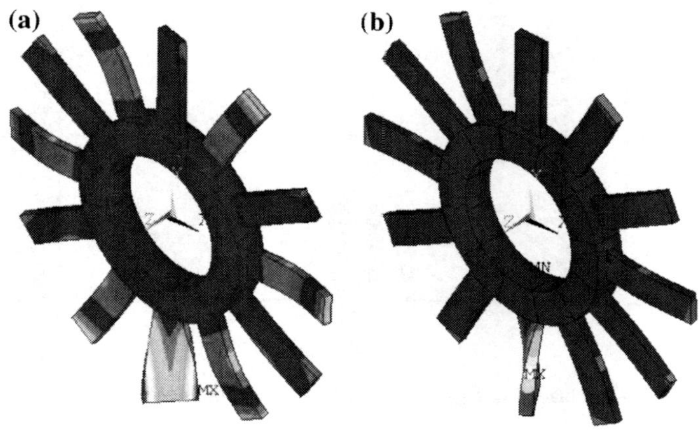

Fig. 13 **a** Mode 31: +20 % mistuning and **b** Mode 32: +15 % mistuning

5.2 Variability in the Stiffness Distribution

The first 36 modes of the system with a ±10 % variation in Young's Modulus were considered for this study. A MACLF plot for the 36 modes are shown in Fig. 14, the plot suggests that some of the modes are insensitive to localization due to parametric uncertainties. It can be observed from Figs. 10 and 14 that the non localized modes 14, 23, 26, 27 and 28 are not sensitive to changes to its mass or stiffness distribution. A similar pattern of MACLF generated by the previous study in Fig. 10 is also observed in Fig. 14. The degree of localization can be categorized into three different cases. Case 1 Fig. 15a the MACLF values are between 0.6 and 1.0 indicating minimum or no localization, case 2 Fig. 15b with MACLF values between 0.6 and 0.2 indicating moderate localization and case 3 Fig. 15c with MACLF values between 0.2 and 0 indicating strong localization (Fig. 15).

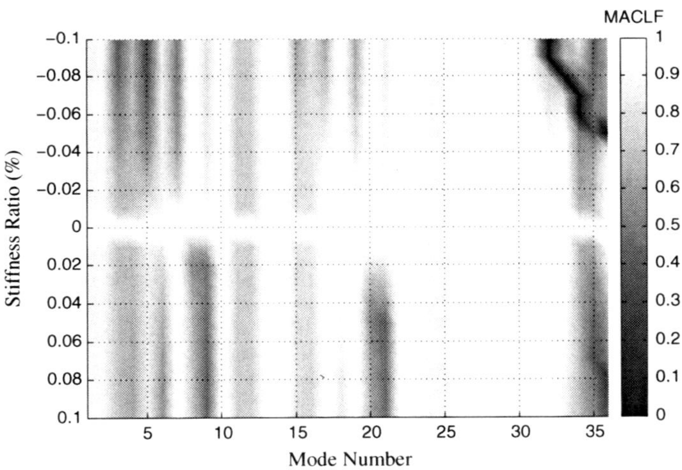

Fig. 14 MACLF for a bladed disk system for a variation in Stiffness in a single blade

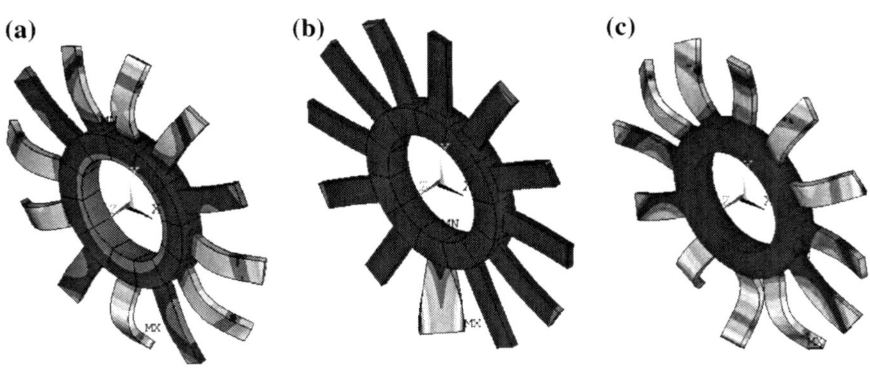

Fig. 15 Modes with −10 % mistuning in a single blade. **a** Mode 31. **b** Mode 32. **c** Mode 33

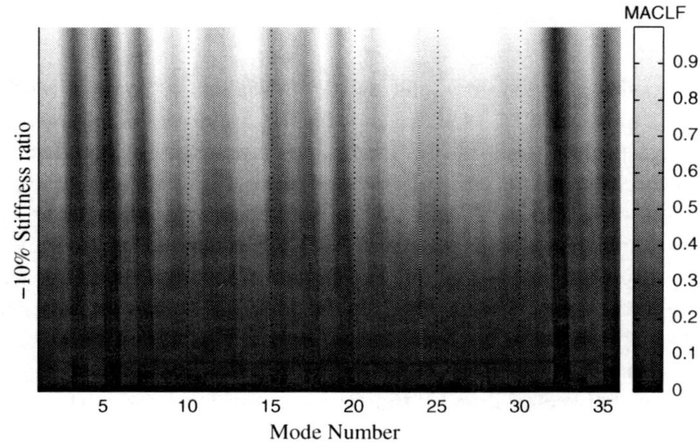

Fig. 16 MACLF of a system with −10 % stiffness ratio and no noise

5.3 Sensitivity with Respect to Noise

The MACLF can also be applied to experimental data to quantify localization, to simulate such a study to resemble a physical experiment the numerical modal data were corrupted with noise.

As an example the system with −10 % stiffness ratio were corrupted with 10 % (Fig. 17) and 20 % noise (Fig. 18).

$$\psi_j = \mathbf{x}_{m_j}\left(1 + \alpha\beta_{m_j}\right) \tag{7}$$

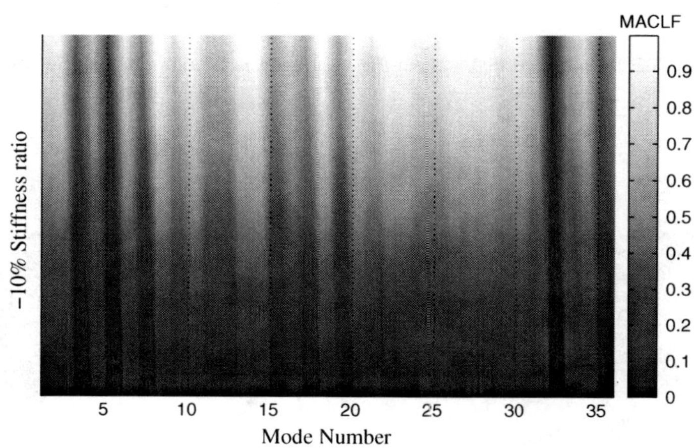

Fig. 17 MACLF of a system with −10 % stiffness ratio and 10 % noise

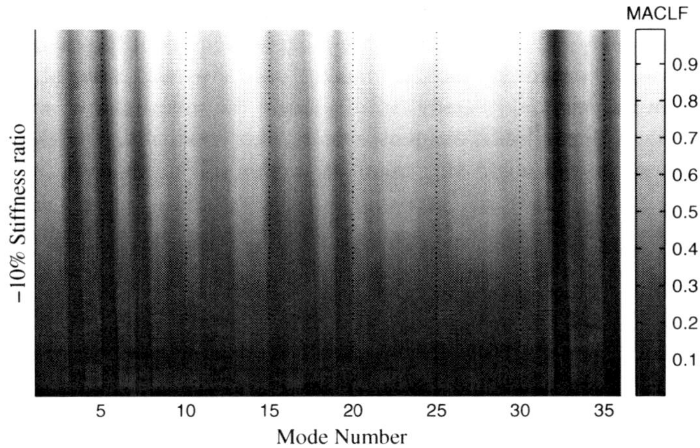

Fig. 18 MACLF of a system with −10 % stiffness ratio and 20 % noise

$$\psi_k = \mathbf{x}_{m_k}(1 + \alpha\beta_{m_k}) \tag{8}$$

Here the ψ_j (Eq. 7) and ψ_k (Eq. 8) are the eigenvectors of the corrupted mistuned and perfect system with the added noise. Here α is the percentage of added noise and β is the random vector generated from a Gaussian distribution. The addition of numerical noise does not significantly alter the MACLF. The MACLF in the presence of noise predicts the localized modes to a high degree of accuracy. In Fig. 16 the MACLF of the noiseless data is shown, Figs. 17 and 18 does not vary significantly.

6 Conclusions

A numerical method has been proposed to identify and quantify localization in complex systems. The method is based on the application of the modal assurance criterion to perfect and mistuned modes. For numerical quantification, we proposed the Modal Assurance Criterion Localization Factor (MACLF). MACLF generates a bounded value between 1 and 0 to indicate how much the vibration modes have localised with respect to the perfect vibration modes. The method is valid for the case when mode swapping takes place and the degree of localization of all modes in the system to a variation to the mass or stiffness distribution is easily obtained. The proposed method was applied to a two-cantilever coupled beam system and a bladed disk system then the results were verified by examining the MACLF values. MACLF values in the range of 1.0–0.6 indicates minimal localization, 0.6–0.2 and 0.2–0.0 indicates moderate and extreme localization respectively. To resemble a realistic experimental case, the modal data were perturbed with random noise. It was observed that the MACLF values still predicts the localized modes with a reasonable accuracy. The MACLF is very useful for quantifying localization in complex periodic structures such as bladed disk's where changes to vibration modes cannot be detected visually. In this paper the investigation is restricted to only un-damped linear finite element models, the applicability of MACLF to a realistic damped model possessing complex modes still needs to be verified.

References

1. Castanier MP, Pierre C (2006) Modelling and analysis of mistuned bladed disk vibration: status and emerging directions. J Propul Power 22:384–396
2. Nikolic M (2006) New Insights into the blade mistuning problem. Ph.D thesis, Imperial college (2006)
3. Blair AJ (1997) A design strategy for preventing high cycle fatigue by minimising sensitivity of bladed disks to mistuning. Master's thesis, Wright State University
4. du Bois JL, Adhikari S, Lieven NAJ (2009) Mode veering in stressed framed structures. J Sound Vibr 322:1117–1124
5. du Bois JL, Adhikari S, Lieven NAJ (2011) On the quantification of eigenvalue curve veering: a veering index. Transactions of ASME. J Appl Mech 78(041007):1–8
6. Liu XL (2002) Behaviour of derivatives of eigenvalues and eigenvectors in curve veering and mode localization and their relation to close eigenvalues. J Sound Vibr 256:551–564
7. Pierre C (1988) Mode localization and eigenvalue loci veering phenomena in disordered structures. J Sound Vibr 126:485–502
8. Rao JS (2006) Mistuning of bladed disk assemblies to mitigate resonance. Adv Vibr Eng 5 (1):17–24
9. Vijayan K, Woodhouse J (2014) Shock amplification, curve veering and the role of damping. J Sound Vibr 333:1379–1389
10. Vijayan K, Woodhouse J (2013) Shock transmission in a coupled beam system. J Sound Vibr 332:3681–3695
11. Fox RL, Kapoor MP (1968) Rates of change of eigenvalues and eigenvectors. AIAA J 6:2426–2429
12. Adhikari S (2013) Structural dynamic analysis with generalized damping models: identification. Wiley ISTE, UK, p 288 (in press)
13. Friswell M, Mottershead J (1999) Finite element model updating in structural dynamics. Kluwer Academic Publishers, UK

Forward and Backward Whirling of a Rotor with Gyroscopic Effect

Ali Fellah Jahromi, Rama B. Bhat and Wen-Fang Xie

Abstract The determination of whirling frequencies of high speed turbines is always challenging in rotor dynamics. The natural frequencies of a Jeffcott rotor are split in the presence of gyroscopic effect. It is quite well known that the lower branch corresponds to the backward whirl and the upper branch corresponds to forward whirl. The forward whirl mode of the rotor has been observed experimentally, however, the backward whirl has not been observed. In this study it is shown that the backward whirl can be observed when the rotor is coasting down to rest from above the critical speed corresponding to the backward whirl. In order to illustrate the forward and the backward critical speeds of a simple Jeffcott rotor, the natural frequencies are obtained analytically for the second natural frequency of the system because of the large gyroscopic effect present in that mode. An experimental set up was used to verify the presence of backward whirl while the rotor is coasting down to rest. The rotor is also simulated using finite element method by ANSYS, and Campbell diagram is plotted. The analytical, experimental and ANSYS simulations confirm the existence of the backward whirl when the rotor is coasting down.

Keywords Backward whirling · Gyroscopic effect · Finite element method · Campbell diagram

A.F. Jahromi (✉) · R.B. Bhat · W.-F. Xie
Department of Mechanical and Industrial Engineering, Concordia University, 1455 de Maisonneuve Blvd. W., Montreal H3G 1M8, Canada
e-mail: al_fel@encs.concordia.ca

R.B. Bhat
e-mail: rbhat@encs.concordia.ca

W.-F. Xie
e-mail: wfxie@encs.concordia.ca

J.K. Sinha (ed.), *Vibration Engineering and Technology of Machinery*, Mechanisms and Machine Science 23, DOI 10.1007/978-3-319-09918-7_78

1 Introduction

The analysis of the critical speeds of high speed rotors is necessary for defining design acceptance criteria of the industrial rotors. The forward whirl mode has been observed experimentally for rotors supported on isotropic and symmetric bearings [1]. However, a survey of the literature reveals that the backward whirling mode has not been observed experimentally. Difference in bearing stiffness in horizontal and vertical directions, considering negligible damping effect, exhibits backward whirling mode of rotor [2]. Subbiah et al. [3] observed backward whirling of a rotor supported by stiffness asymmetry of the bearings through experimental procedure. The backward whirl mode for the dynamic imbalance of rotor supported by fluid film damping and shaft structural damping is studied in the literature. Rao [4] studied the conditions for backward whirling for a flexible rotor supported by the hydrodynamic bearings. Moreover, the occurrence of the backward whirl is also studied for the cylindrical journal bearings [5, 6].

The start-up or coast-down results in a time variant frequency signal [7]. In the present paper, the occurrence of the backward whirling mode is studied considering the coast-down phenomenon for a rotor supported on isotropic ball bearings. An analytical model is developed to study the backward and forward critical speeds in presence of the gyroscopic effect. Moreover, the critical speeds of the Jeffcott rotor are calculated using Finite Element Method (FEM). Furthermore, an experiment is conducted to observe the backward whirling mode in a coast down phenomenon.

The paper is organized as follows: First, the analytical and FEM model are illustrated in Sect. 2. Then, the explanation of the experimental set-up is presented in Sect. 3. The results of the analytical model, the FEM and experimental study are presented in Sect. 4. Finally, the conclusions and remarks are given is Sect. 5.

2 Analysis and Simulation

In order to study the forward and backward critical speeds, a simple Jeffcott rotor is modeled considering the gyroscopic effect using analytical and FEM methods. The geometrical description and material properties of the Jeffcott rotor are presented in Fig. 1 and Table 1.

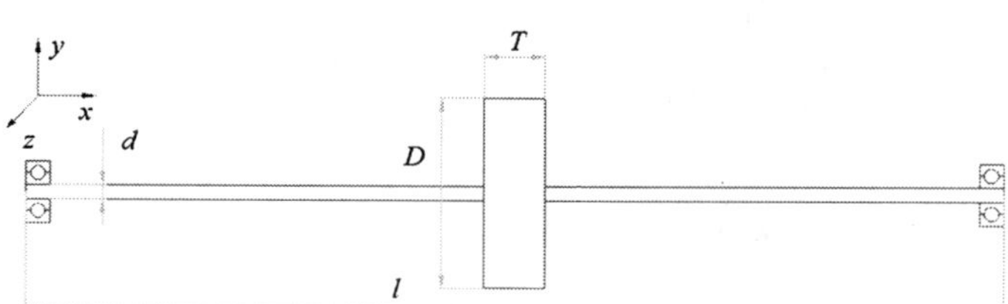

Fig. 1 Schematic of simple Jeffcott rotor

Table 1 Rotor parameters

Symbol	Description	Value
l	Shaft length	0.4 (m)
d	Shaft diameter	0.006 (m)
T	Disk thickness	0.025 (m)
D	Disk diameter	0.08 (m)
E	Modulus of elasticity	210 (GPa)
m	Disk mass	1 (kg)

2.1 Analytical Model

The disk is located in the middle of the shaft shown in Fig. 1. Therefore, the gyroscopic effect would be more prominent in the second mode compared to the first mode for the shaft with simply-supported boundary conditions. Consequently, based on the literature [8] the analytical model is derived to study the second bending mode of the system in Eq. (1). The straightening tendency of the shaft bending is modeled using torsional spring K_t which is shown in Fig. 2.

$$\ddot{\theta} - j\Omega\alpha\dot{\theta} + \omega_t^2\theta = Me^{j\Omega t} \tag{1}$$

where $\theta = \psi + j\varphi$ with ψ and φ being the rotations about the z and y axes, respectively, Ω is the angular velocity of shaft, ω_t and α are the natural frequency and ratio of mass moment of inertia respectively. M is the magnitude of the harmonic moment of excitation, and is given by $M = me\Omega^2$ where me is the unbalance. The e is considered 0.0001 (m) due to the static deflation of the shaft. Further:

$$K_t = J_T\omega_t^2 \tag{2}$$

$$\alpha = \frac{J_p}{J_T} \tag{3}$$

$$J_p = \frac{1}{8}mD^2 \tag{4}$$

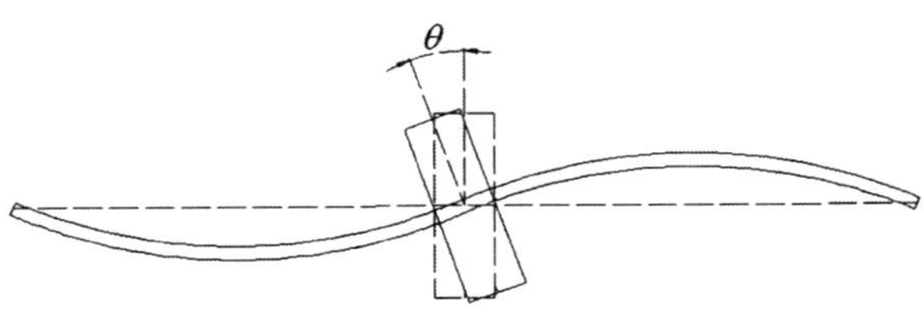

Fig. 2 Schematic of the rotor with torsional spring

$$J_T = \frac{m}{12}\left(\frac{3}{4}D^2 + l^2\right) \quad (5)$$

Equation (1) can be presented as:

$$\ddot{\psi} + j\ddot{\varphi} - j\Omega\alpha(\dot{\psi} + j\dot{\varphi}) + \omega_t^2(\dot{\psi} + j\dot{\varphi}) = Me^{j\Omega t} \quad (6)$$

By separating the real and imaginary terms in Eq. (6), it is written in the form of Eqs. (7) and (8) as:

$$\ddot{\psi} + \Omega\alpha\dot{\varphi} + \omega_t^2\psi = M\cos(\Omega t) \quad (7)$$

$$j\left[\ddot{\varphi} - \Omega\alpha\psi + \omega_t^2\varphi\right] = jM\sin(\Omega t) \quad (8)$$

In order to study the forward and backward whirling, let $\Omega = \beta t$ be the speed when the rotor starts from rest and increases towards the operating speed, and considering $\Omega = \Omega_0 - \beta t$ when the rotor is coasting down to rest. Based on the properties of the rotor studied, the β and Ω_0 are chosen as 300 (rad/s) and 4,500 (rad/s), respectively. Equations (9) and (10) represent the coupled differential equations for analyzing the forward critical speed and Eqs. (11) and (12) represent the mathematical model for the backward critical speed.

$$\ddot{\psi} + \beta\alpha\dot{\varphi}t + \omega_t^2\psi = M\cos(\beta t^2)\, 0 < t < 15 \text{ (s)} \quad (9)$$

$$\ddot{\varphi} - \beta\alpha\dot{\psi}t + \omega_t^2\varphi = M\sin(\beta t^2)\, 0 < t < 15 \text{ (s)} \quad (10)$$

$$\ddot{\psi} + (\Omega_0 - \beta t)\alpha\dot{\varphi} + \omega_t^2\psi = M\cos(\Omega_0 t - \beta t^2)\, 0 < t < 15 \text{ (s)} \quad (11)$$

$$\ddot{\varphi} - (\Omega_0 - \beta t)\alpha\dot{\psi} + \omega_t^2\varphi = M\sin(\Omega_0 t - \beta t^2)\, 0 < t < 15 \text{ (s)} \quad (12)$$

Equations (9)–(12) are solved in the time domain in Matlab using Runge-Kutta method for ordinary differential equations.

2.2 ANSYS Modeling

The model is defined using volume elements in ANSYS as shown in Fig. 3. In order to determine the gyroscopic effect, the model is meshed using the element SOLID 186 which is a 10-node tetrahedral element for homogenous structures which can be used for the calculation of Coriolis effect. The material model is assumed to be linear isotropic steel which is identified in Table 1. Simply-supported boundary conditions are applied at both ends of the shaft by fixing the center point at the ends of the shaft in y and z directions. In order to determine the forward and backward critical speeds

Fig. 3 Finite element model of the rotor in ANSYS

the angular velocity of the shaft is defined in range of 0 to 3,500 (rad/s) with steps of, 0 (rad/s), 400 (rad/s), 600 (rad/s), 1,000 (rad/s), 1,500 (rad/s), 2,000 (rad/s), 2,500 (rad/s), 3,000 (rad/s) and 3,500 (rad/s).

3 Description of the Experimental Set-Up

The experimental set-up consists of a rotor supported by the two identical isotropic ball bearings and is driven by a DC motor with variable speed range of 0 to 6,000 (RPM) as shown in Fig. 5. The rotor is connected to the motor using a flexible coupling. The deflection of the beams is measured by two displacement sensors, TM 620 and GUNT HAMBURG with 90° phase difference for monitoring the whirling orbit (Fig. 4). The DC motor is connected to the driver for controlling the angular velocity of the shaft. The proximity displacement sensors are connected to the signal conditioner (TM 151 GUNT HAMBURG) which sends signal to the oscilloscope. The data sampling time of the oscilloscope for recording the data is 1.6e-5 (s). The assembly of the experimental set-up is shown in Fig. 5.

4 Results and Discussions

The analytical model of a Jeffcott rotor which is presented in Eqs. (9)–(12) is solved using Matlab for 15 s with a time interval of 1e-4 s. In order to determine the forward and backward critical speeds the real part of the system response ψ in time

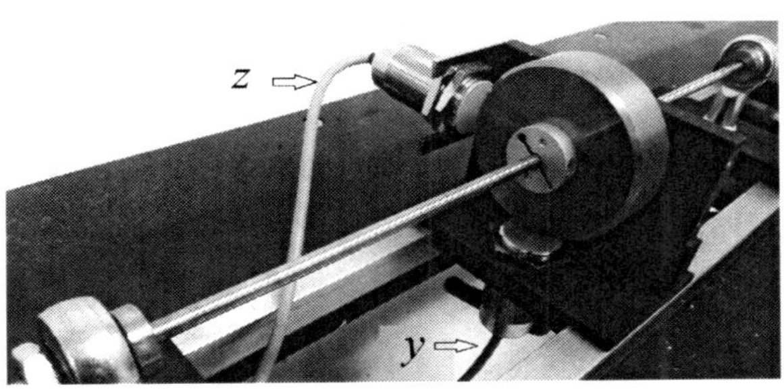

Fig. 4 Rotor and the proximity displacement sensors

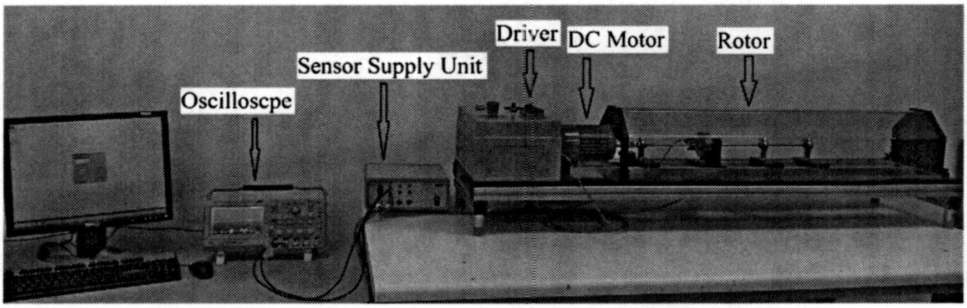

Fig. 5 Assembly of experimental set-up for whirl monitoring

domain is transformed to the frequency domain using Fast Fourier Transform (FFT). Figures 6 and 7 show the frequency responses for decreasing and increasing angular velocities of the shaft, respectively. Based on the calculated frequency response for the second natural frequency of the rotor, the backward critical speed is 105 (Hz) and forward critical speed is 445.6 (Hz). Figures 6 and 7 validate the splitting of the second natural frequency of the rotor by the proposed analytical model for the backward and forward critical speeds.

In order to calculate the variation of the first two natural frequencies of the rotor with changing rotational velocity of the shaft, the Campbell diagram of the rotor is plotted using ANSYS for the range of 0–3,500 (rad/s). Figure 8 shows that the gyroscopic effect is not significant in the first natural frequency of the rotor because the disk is located in the middle of the shaft and therefore, the disk does not rotate about the lateral axis in the first bending mode. Consequently, the gyroscopic effect is not significant in the first bending mode and the backward and forward critical speeds have same value which is 17.16 (Hz). The second natural frequency of the rotor is separated into forward and backward branches. Based on the FEM analysis, the backward critical speed is 97.2 (Hz) and forward critical speed is 458.4 (Hz) which are approximately equal to the analytical results.

The experimental set-up can reach up to rotational speed of 100 (Hz) which is in the range of the fundamental frequency of the system. Considering this limitation,

Fig. 6 Frequency response of the ψ for coasting down of the rotor from 4,500 (rad/s) to rest

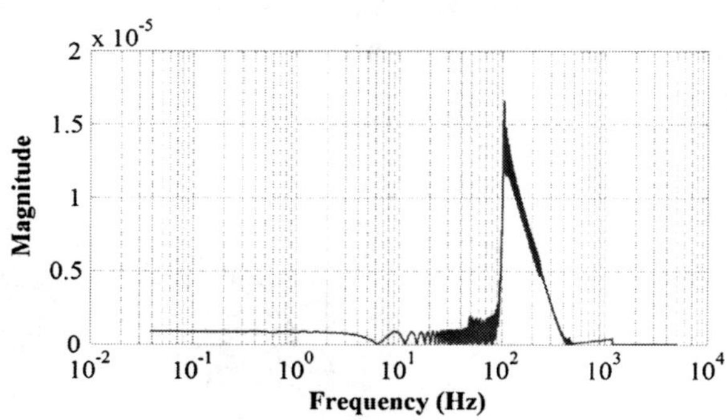

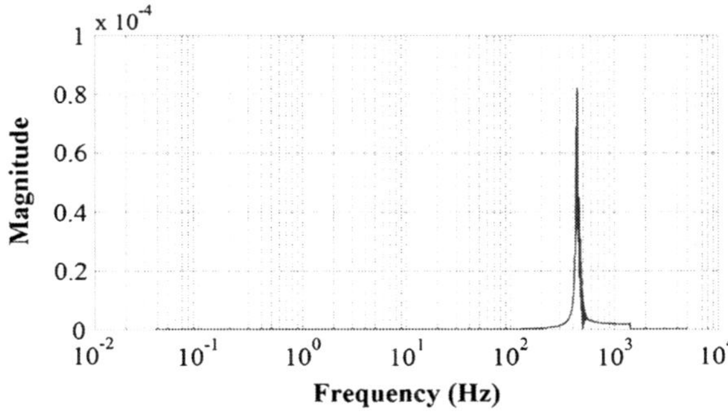

Fig. 7 Frequency response of the ψ for increasing angular velocity from 0 to 4,500 (rad/s)

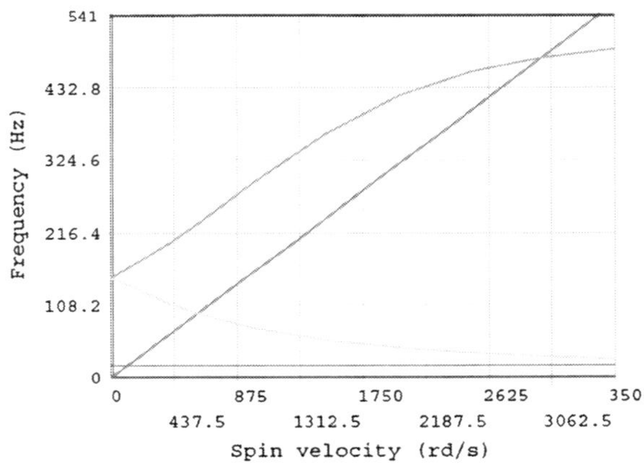

Fig. 8 Campbell diagram for first two bending modes of case study rotor

the experiment is designed for observing the whirling direction of disk in the first natural frequency of the system. The critical speeds for the forward and backward whirls have the same value due to the position of the disk. In order to observe the backward whirl the rotor is coasting down from 1,600 (RPM) to rest and the signals of the displacements sensors are recorded in 1.6e-5 (s) sampling time. Figure 9 presents the deflection of the beam in y and z directions, and the region shown inside the ellipse indicates the change in the whirl direction. Within the ellipse in Fig. 9 the backward whirl happens right after the critical speed. The phase change in the sensor signals confirms the existence of the disk backward whirling.

In order to illustrate the whirling direction of the disk, the deflection of the shaft in y and z directions is plotted against each other as a phase portrait for the portion of the data which are marked by ellipse in Fig. 9. The black arrows in Fig. 10 depict the forward whirling and red arrows present the backward whirling. Figure 10 confirms the changing whirl direction.

In order to quantitatively compare the determined frequency in the analytical study, FEM and experimental study, the difference between the results are presented as an error in Table 2. Table 2 presents the convergence of the results which are achieved from different methods for studying the forward and backward critical speeds.

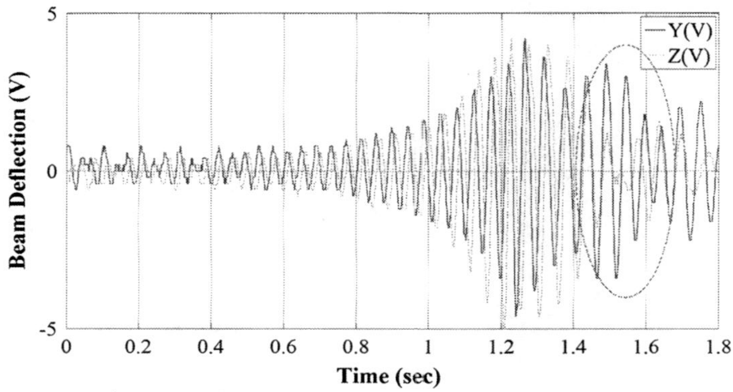

Fig. 9 Deflection of the beam in *y* and *z* directions

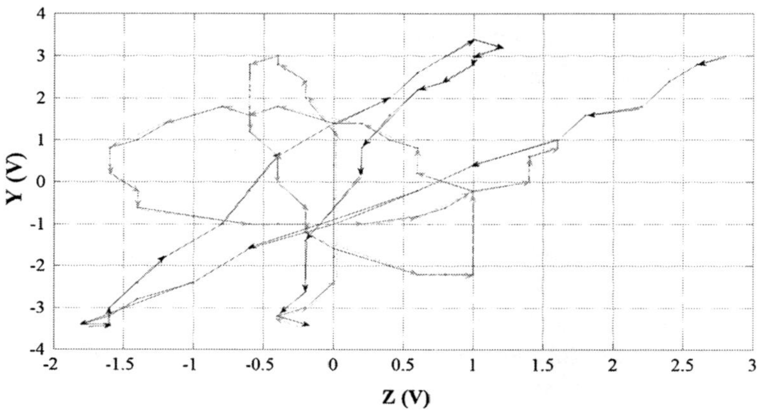

Fig. 10 Phase portrait of the shaft deflection in *y* and *z* direction

Table 2 Quantitative comparison among analytical, FEM and experimental results

	Forward		Backward	
First mode	Analytical	–	Analytical	–
	ANSYS	17.16 (Hz)	ANSYS	17.16 (Hz)
	Experiment	19.18 (Hz)	Experiment	19.18 (Hz)
	Error	10.53 %	Error	10.53 %
Second mode	Analytical	444 (Hz)	Analytical	105 (Hz)
	ANSYS	458.4 (Hz)	ANSYS	97.4 (Hz)
	Experiment	–	Experiment	–
	Error	3.15 %	Error	4.69 %

The analytical results differ slightly from those obtained using FEM analysis which may be because the analysis did not include shear deformation inertia effect.

5 Conclusions

An analytical model is derived to simulate the second mode of a Jeffcott rotor in start-up and run-down phases in the presence of the gyroscopic effect. The solution of differential equations for the case study rotor confirms the existence of backward whirl which occurs while coasting-down. The critical speed of the case study rotor is determined using ANSYS for the first two bending modes of the rotor. The comparison between analytical and FEM models shows approximately a difference of 4 % in critical speeds of the rotor. The proposed idea for the observation of the backward whirl which occurs during coast-down of the rotor to rest is experimentally demonstrated for a laboratory rotor supported on isotropic bearings. The backward whirl is experimentally observed after coasting down through the critical speed which is shown in Fig. 9. The comparison of the analytical, FEM and experimental results shown in Figs. 6, 7, 8, 9, and 10 and Table 2 confirm the existence of the backward whirling mode which manifests during coast-down.

Acknowledgement Authors wish to thank Mr. Gilles Huard and Mr. Dan Juras for their technical support and guidance in carrying out the experiments.

References

1. Yamamoto Y (1954) On the critical speeds of a shaft, vol 6. Memoirs of the Faculty of Engineering, Nagoya University, Japan, p 160–174
2. Downham E (1957) Theory of shaft whirling: a fundamental approach to shaft whirling. Engineer 204:518–522, 552–555, 660–665
3. Subbiah R, Bhat RB, Sankar TS, Rao JS (1985) Backward whirl in a simple rotor supported on hydrodynamic bearings. In: Proceedings of a symposium sponsored by Bently Rotor Dynamics Research Corporation and held in Carson City, June 1985, in NASA conference publication
4. Rao JS (1982) Conditions for backward synchronous whirl of a flexible rotor in hydrodynamic bearings. Mech Mach Theory 17(2):143–152
5. Rajalingham C, Ganesan N, Prabhu BS (1986) Conditions of backward whirling motion of a flexible rotor supported on hydrodynamic journal bearings. J Sound Vib 111:29–36
6. Rao C, Bhat RB, Xistris GD (1996) Experimental verification of simultaneous forward and backward whirling at different points of Jeffcott rotor supported on identical journal bearings. J Sound Vib 198(3):379–388
7. Sekhar AS (2004) Detection and monitoring of crack in a coast-down rotor supported on fluid film bearings. Tribol Int 37:279–287
8. Lee CW (1993) Vibration analysis of rotors. Springer, Dordrecht

Analysis of Nonlinear Dynamic Second-Order Effect of a Large-Scale Container Crane Under Seismic Excitations

Dong Wang, Gongxian Wang, Yeping Xiong and Jiquan Hu

Abstract Under seismic excitations, the second-order effect (P-Delta effect) exists in deformed structure under vertical loading even in linear-elastic system. For the vulnerability of the large-scale container crane, the effect is particularly remarkable in the slender structure, especially under large deformations. The current research on the rule of the static P-Delta effect has gained tremendous achievements, through systematic, reliable and simplified calculation methods. This paper however proposes the nonlinear dynamic second-order effect of the large-scale container crane. The aim of this paper is to analyze the influence of P-Delta effect on nonlinear dynamic seismic responses owing to different seismic conditions and seismic excitations. Meanwhile, by comparing with the methods of numerical FEM simulations, the simplified theoretical calculation conclusions of the dynamic second-order effect for the bending-type structure on portal frame can be verified. This research reveals clearly the regular pattern of P-Delta effect and influencing factors. The results have great significance on the seismic design of large-scale container crane. The main-findings are presented in the paper.

Keywords Seismic excitation · Large-scale container crane · Second-order effect · Nonlinear dynamics

---

D. Wang · G. Wang · J. Hu
School of Logistic Engineering, Wuhan University of Technology, Wuhan 430063, China
e-mail: wangdong.whut@163.com

G. Wang
e-mail: wanggongxian@gmail.com

J. Hu
e-mail: hjq580818@126.com

G. Wang · Y. Xiong (✉)
Fluid-Structure Interaction Research Group, Faculty of Engineering and the Environment, University of Southampton, Highfield, Southampton SO171BJ, UK
e-mail: Y.Xiong@soton.ac.uk

J.K. Sinha (ed.), *Vibration Engineering and Technology of Machinery*,
Mechanisms and Machine Science 23, DOI 10.1007/978-3-319-09918-7_79

1 Introduction

In recent years, the analysis of the structures of high and slender steel frame components has more and more taken the consideration of the influence of the action of axial, lateral forces and end moments on the deformed shape of beams. The axial loads and additional end moments further cause the increase of the lateral deformation, which is commonly referred to as P-Delta effect. The presence of P-Delta effect in a large-scale container crane possibly leads to the instability and failure of beams. The subject of structure design, including second-order effect design method, has been researched for many years.

The P-Delta effect is particularly remarkable in the flexible structures, especially under large deformations caused by major seismic loads. In the Hanshin Great Earthquake in western Japan (1995), there was a great damage of container cranes [1]. The beam of the portal frame in some container cranes produce the large deformation and buckling, and some had uplifted and derailed legs. For the vulnerability of the large-scale container crane, it is very important to study the seismic design of the large-scale container cranes with the consideration of the P-Delta effect.

Many researchers or designers have studied simplified methods of analyzing P-Delta effect of structures. Accounting for the aforementioned effect by using magnification factors applied to the results calculated by first-order analyses [2–4]. However, these methods are often unsatisfactory for cases especially under large deformations. Hence, some researchers have developed the method including nonlinear incremental stiffness due to the demand of more accurate second-order analysis [5]. In all these studies, the analysis of non-linear second-order effect accounting for dynamic and geometric nonlinearities is few and the influence of P-Delta effect on nonlinear dynamic seismic responses with the subject of large-scale container crane is rare, too.

The aim of this paper is to analyze the influence of P-Delta effect on nonlinear dynamic seismic responses of a large-scale container crane owing to different seismic conditions and seismic excitations. Meanwhile, by comparing with the methods of numerical simulations, the simplified theoretical calculation conclusion of the dynamic second-order effect for the bending-type structure on portal frame can be verified. This research reveals clearly the regular pattern of P-Delta effect and influential factors. The results can be used to improve the structural seismic design and response calculation.

2 Deformations and Simplified Model of a Large-Scale Container Crane

Recently, B.D. Kosbab and L.D. Jacobs et al. presented the analysis and testing of container cranes under earthquake loads [6–8]. As shown in Fig. 1, numerous responses, including lateral displacement and uplift were captured during their

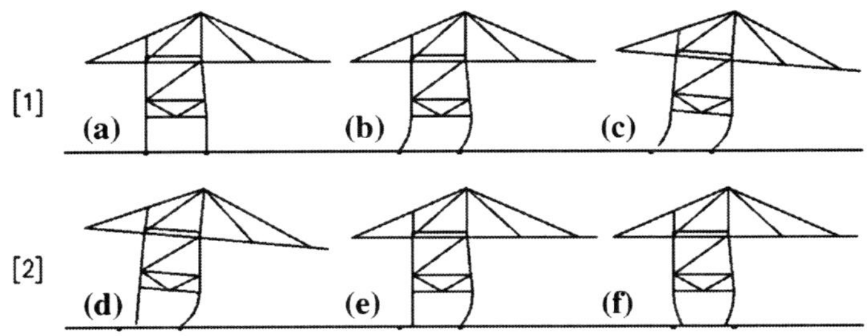

Fig. 1 Large-scale container crane [1](*a*) and portal beam subjected to bending deformations [1] (*b*)–[2](*f*) by seismic loads schematic

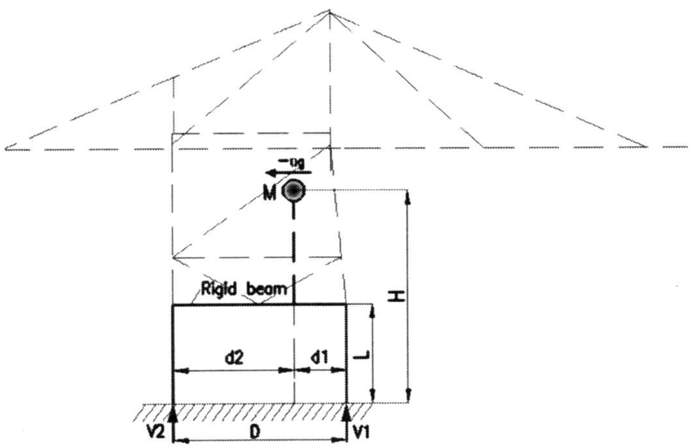

Fig. 2 Simplified model with concentrated mass in geometric gravity center

testing. Complimentary detailed nonlinear finite element time history analyses were also performed. It is found that the portal frame response dominates bending deformation and elastic behavior. Stronger excitation produced uplift, which coupled closely with elastic behavior. Nevertheless, in all of the aforementioned research efforts, the P-Delta effect is ignored.

In order to examine P-Delta effect in a simple way, consider the components of a large-scale container crane of the portal steel frame beams and a concentrated mass in the gravity center of total structure, as shown in Fig. 2. The stiffness of rectangular beams of seaside-legs and landside-legs is K_l with the length of L. The material of the beams, with elastic modulus E and second moment of cross section I, is adopted bilinear elastic-plastic material constitutive relation. By the deformations of the large-scale container crane under seismic loads shown in the studies, this paper can assumed that the cross-girder is a rigid beam connected with the upper substructure. The total mass of the large-scale container crane is M, with height of gravity center H, respectively.

Referring to simplified model in Fig. 2, regardless of the deformation of the structure subjected to horizontal seismic acceleration and impetus amplification, the moment equation at the seaside-leg is given as

$$D * V_2 + H * M * (-u_g) - d_1 * M * g = 0 \tag{1}$$

when the landside-leg starts to lift as shown in Fig. 1 [1](c), $V_2 = 0$. Lift acceleration can be described as

$$-u_g = \left({d_1}/{D}\right) * g \tag{2}$$

where $-u_g$ is the horizontal seismic acceleration, g is gravity acceleration, $V_i (i = 1, 2)$ is the axial reaction force, D is the wheel gauge.

It can be noted that the seismic acceleration $(-\alpha)$ expresses the lateral deformation degree of the structure. Therefore, this influence on P-Delta effect is remarkable. Moreover, the pertinent boundary condition of the problem becomes nonlinear, which is relevant to P-Delta effect, too. Simulation example with great practical interest is worked out by the end.

2.1 Numerical Model Considering P-Delta Effect

According to the aforementioned study on the structure of a large-scale container crane, the numerical model can be established as a single particle system container crane structure considering P-Delta effect, as shown in Fig. 3. Without loss of generality, it may be assumed that the model is subjected to bidirectional earthquake loads in the direction of x-axis and y-axis.

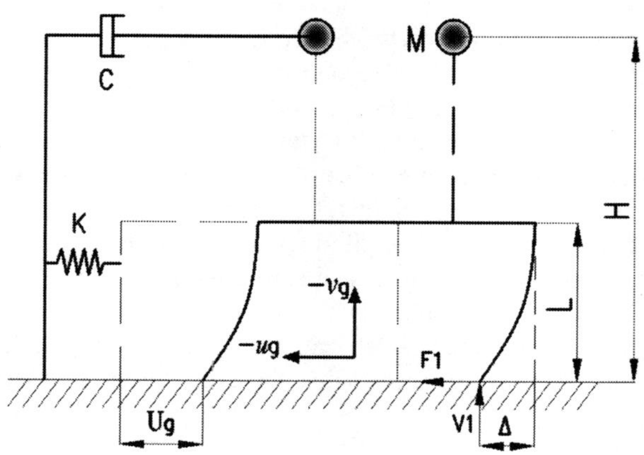

Fig. 3 P-Delta effect analysis numerical model

When considering P-Delta effect, the relative displacement of mass point generates the second-order effect with additional bending moment at the upper portal frame. The aforementioned moment is given as

$$M_0 = \left[Mg + M * \left(\ddot{v} + \ddot{v}_g\right)\right] * u \tag{3}$$

where $u = u(l)$ is the lateral deflection of beam; $\ddot{v}_g(t)$ is the time history of vertical seismic acceleration and $\ddot{v}(t)$ is the time history of relative vertical acceleration response of the mass point. While it is equivalent to an additional force in horizontal direction as

$$F_0 = \left[Mg + M * \left(\ddot{v} + \ddot{v}_g\right)\right] * \left({}^{u}\!/_{L}\right) \tag{4}$$

Due to the two components of the horizontal and vertical movement, the system has two degrees of freedom. Therefore, the equations of motion can be written as

$$M\ddot{u} + c_x\dot{u} + k_x u = -M\ddot{u}_g - \left[Mg + M * \left(\ddot{v} + \ddot{v}_g\right)\right] * \left({}^{u}\!/_{L}\right) \tag{5a}$$

$$M\ddot{v} + c_y\dot{v} + k_y v = -Mg - M\ddot{v}_g \tag{5b}$$

In the formulas, c_x and c_y are coefficients of horizontal and vertical damping; k_x and k_y are stiffness coefficients of the system, respectively. Thus it can be seen, $\left(\ddot{v} + \ddot{v}_g\right) * u$ appears in the formula (5a). It is apparent that the P-Delta effect is nonlinear problem. The solution of the motion equations can be obtained by solving the general nonlinear differential equations [9, 10]. After confirming the deformation configuration of the structure, reaction forces of the seaside-leg and landside-leg can be calculated for analyzing P-Delta effect.

2.2 Qualitative Analysis Method

Consider one side-leg of portal frame of a container crane with bending deformation under seismic loads as shown in Fig. 4. Under the action of horizontal support reaction force F, the lateral displacement at the bottom of leg is given as

$$\Delta_0 = {}^{FL}\!/_{3EI} \tag{6}$$

Due to the vertical support reaction force V, the aforementioned additional moment is given as

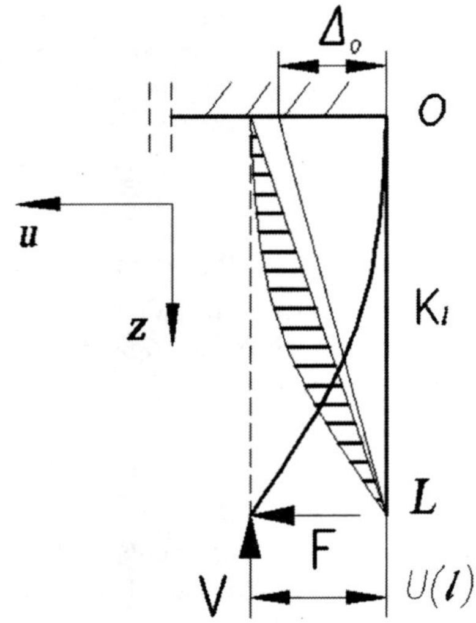

Fig. 4 Analysis sketch of additional bending moment distribution associated with P-Delta effect

$$M_0 = V\Delta_0 \tag{7}$$

while it is equivalent to an additional force in horizontal direction as

$$F_0 = {}^{V\Delta_0}/_{L} \tag{8}$$

The equivalent additional force F_0 will also generate additional lateral displacement as

$$\Delta_1 = {}^{F_0\Delta_0}/_{F} = {}^{VL^2\Delta_0}/_{3EI} = (\alpha L)^2 * {}^{\Delta_0}/_{3} \tag{9}$$

where $\alpha = \sqrt{{}^{V}/_{EI}}$, the total lateral displacement can be obtained as

$$\Delta = \Delta_0 + \Delta_1 = \Delta_0 * \left[1 + {}^{(\alpha L)^2}/_{3}\right] \tag{10}$$

The above formula is an approximate calculation method by considering P-Delta effect. The accurate formula is been deduced as follows by moment distribution as shown in the Fig. 4. The balance equation of the bending moment is

$$M(z) = F * (L - z) + V * [u(l) - u(z)] \tag{11}$$

while assuming that $u(l)$ is accurate lateral displacement, $u = u(z)$ is the displacement function, the partial differential equation of bending beam is given as

$$EIu''(z) = F * (L - z) + V * [u(l) - u(z)] \quad (12)$$

The general solution of the differential equation is

$$u(z) = c_1 \cos \alpha z + c_2 \sin \alpha z + {F * (L - z)}/{V} + u(l) \quad (13)$$

where α is aforementioned, $c_i (i = 1, 2)$ are given constants by the boundary conditions of equation as

$$u(z) = \begin{cases} u'(z) = 0, & z = 0 \\ u(l), & z = L \end{cases} \quad (14)$$

The accurate lateral displacement at the upper portal frame is obtained as

$$u(l) = {F * (\tan \alpha L - \alpha L)}/{V\alpha} = \Delta_0 \left[{3(\tan \alpha L - \alpha L)}/{(\alpha L)^3} \right] \quad (15)$$

Using Taylor formula, the solution of total lateral displacement is presented as

$$u(l) = \Delta_0 \left[1 + {2(\alpha L)^2}/{5} + {17(\alpha L)^4}/{105} + o(\alpha L)^6 \right] \quad (16)$$

The arising solution of equations is nonlinear due to the presence of the axial reaction forces V and horizontal reaction forces F at the bottom of seaside-leg and landside-leg. However, this qualitative analysis method can indicate that the lateral displacement of the structure increases along with the increase of the vertical seismic load which will be verified by the FEM simulations.

3 Simulations of FEM Numerical Example

FEM Analysis of structure is considered to be the most reliable and accurate method. The method can not only analyze structure qualitatively, but also give the quantified response of the structure under the seismic excitations by considering the P-Delta effect or not. To examine the influence of nonlinear dynamic second-order effect for the model presented here, on the basis of the analytical and numerical procedures presented in the previous Sect. 2.1, we carried out numerical simulations by structural time—history dynamic analysis [11, 12].

A typical large-scale crane was established by the FEM model as shown in Fig. 5. The basic design parameters of the container crane are shown as follows: base distance 15 m, wheel gauge 35 m, portal beam 18 m, the total weight of the container crane about 1,500t. The container crane wheels ride on the rails which attached to the shake table.

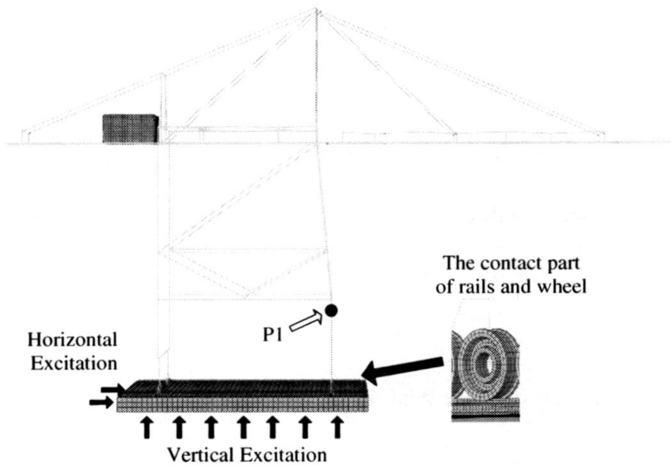

Fig. 5 Excitation sketch and the FEM model of a large-scale container crane

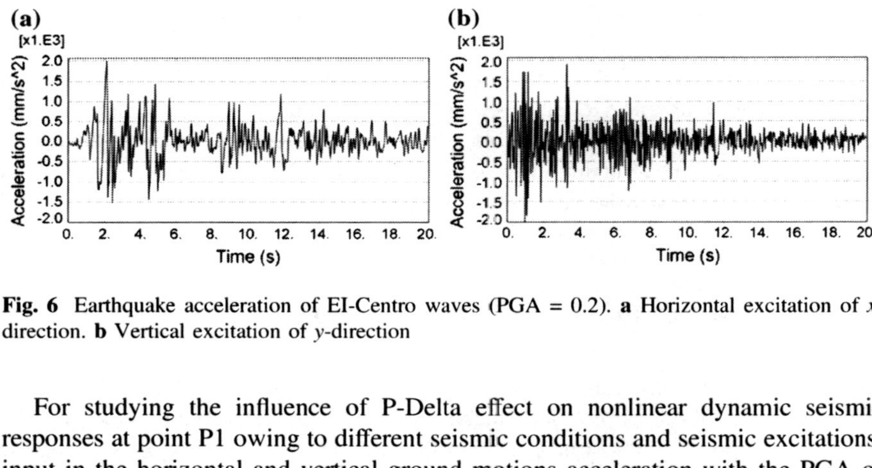

Fig. 6 Earthquake acceleration of EI-Centro waves (PGA = 0.2). **a** Horizontal excitation of *x*-direction. **b** Vertical excitation of *y*-direction

For studying the influence of P-Delta effect on nonlinear dynamic seismic responses at point P1 owing to different seismic conditions and seismic excitations, input in the horizontal and vertical ground motions acceleration with the PGA of 0.2 on the platform, which is according to fortification intensity of the XIII degree. The low amplitude excitations that do not cause uplift to isolate the elastic response. EI-Centro historical waves were selected with amplitude modulation as the seismic excitation acceleration input as shown in Fig. 6. Additionally, a low amplitude white noise was used for the determination of the various modes.

4 Responses Results and Comparison

As previously stated, the two-dimensional input motions considered are the horizontal component along the boom and the vertical component. To determine how the seismic inputs and seismic excitations amplitude can influence the nonlinear dynamic second-order effect on seismic responses, the analytical process is divided into four kinds of cases as shown in Table 1.

The qualitative responses of the model in this paper and B.D. Kosbab's studies are in good agreement. Four cases exhibit similar displacement trends and the dynamics of the structure were well captured. The comparing lateral displacement curves of response are shown in Figs. 7 and 8. Maximum displacements considering or ignoring P-Delta effect of nonlinear dynamic seismic responses are shown in the following Table 1.

According to the results in case1 and case2 without considering the P-Delta effect, when only input the ground motions in the horizontal direction, the maximum relative lateral displacements at P1 point are 40.1 mm and 39.5 mm in the forward and negative direction respectively. However, the values of responses increase to 44.5 mm and 44.4 mm after considering the P-Delta effect. The increasing ratio is approximately 10 %. In the two-dimensional motions analytical process, the response of the large-scale container crane is relatively greater. With ignoring the P-Delta effect, the maximum relative lateral displacements are 51.3 mm and 47.0 mm in the forward and negative direction respectively, While, the values of responses are increased to 59.8 mm and 54.6 mm after considering the P-Delta effect. The increasing ratio is about 15 %. The results are shown in the Table 2.

Table 1 Analysis cases of nonlinear dynamic seismic responses

PGA of seismic excitation	Case	Input direction	Second-order of effect
0.2 g	1	Horizontal	Consider
	2	Horizontal	Ignore
	3	Horizontal and vertical	Consider
	4	Horizontal and vertical	Ignore

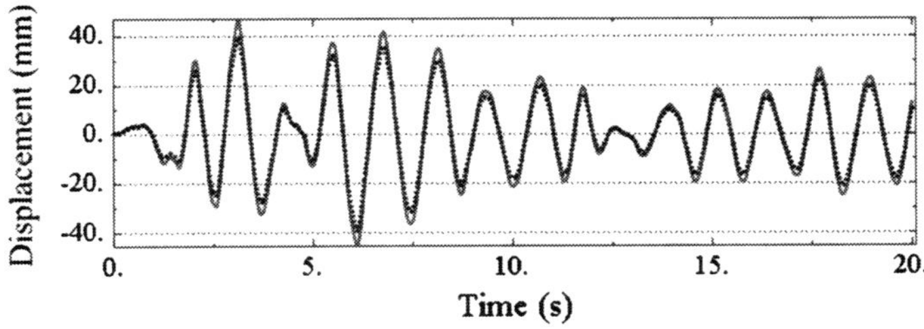

Fig. 7 The comparing lateral displacement curves of response between case1 () and case2 (···) at P1 location

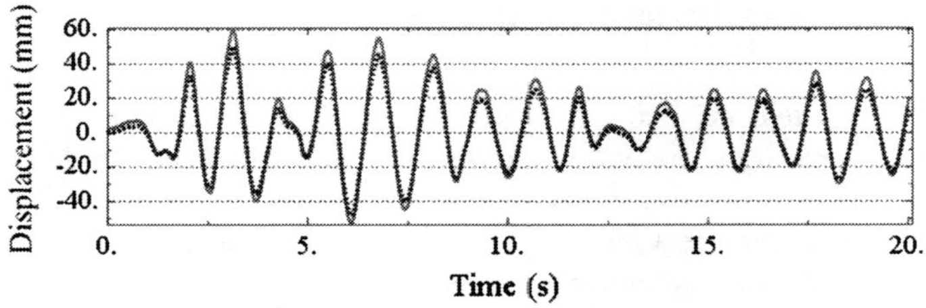

Fig. 8 The comparing lateral displacement curves of response between case3 (—) and case4 (···) at P1 location

Table 2 Maximum displacements (mm) in various cases of dynamic seismic responses

Case	Max. of displacement	Case	Max. of displacement	Discrepancy for P-delta effect
1	44.5	2	40.1	4.4
	−44.4		−39.5	4.9
3	59.8	4	51.3	8.5
	−54.6		−47.0	7.6

The lateral deformation increases even when considering the second-order of effect as comparing in case 1 and case 3, the vertical seismic excitation will significantly increase the structure's responses. One reason for this discrepancy is due to the influence of the vertical stiffness under the geometric nonlinear. P-Delta effect is prominent problem that needs research when vertical excitation exists.

5 Conclusion

In this paper, a simplified numerical model of a large-scale container considering of P-Delta effect is proposed for the second-order effect analysis. The portal frame of beams with a concentrated mass, taking into account P-Delta effect, is subjected to an additional force or moment, while variable axial and horizontal seismic loads are applied as ground motions. The analytical results of the numerical model indicate that the discrepancy of the results arising from the ignorance of second-order effect deformation is remarkable, especially when the presence of axial loads. The theoretical analysis of P-Delta effect is well suited for numerical simulations. The other main conclusions that can be drawn from the simulations are as the following.

(a) The vertical load can strengthen the influence of the horizontal earthquake, thus further increase the deformation of response and this effect is relevant to the original size of deformation.

(b) Due to the greater deformation of the portal frame, the bending moment in the upper position will increase arising from considering P-Delta effect.
(c) For the nonlinear response of the structure, the discrepancy between considering and ignoring P-Delta effect is more remarkable. It will increase the nonlinear deformation or permanent deformation of the structure.

Acknowledgement The authors would like to thank the National Natural Science Foundation of China for financially supporting this research under the contract No. 51275369.

References

1. Japan Society of Mechanical Engineers (1996) The Machines and industrial equipments damage due to the south hyogo-prefecture earthquake. In: JSME, the special Investigation committee on great Hanshin Awaji earthquake damage, pp 153–163 (in Japanese)
2. Goto Y, Chen WF (1987) Second-order elastic analysis for frame design. ASCE Struct Eng 113(7):1501–1529
3. Goto Y (1994) Second-order elastic analysis of frames. In: Chenand WF, Tom S (eds) Advanced analysis of steel frames. CRC Press Inc, Boca Raton, pp 47–90
4. Chen WF, Atsura T (1977) Theory of beam-column, vol I. McGraw-Hill Inc, New York
5. Chajes A, Churchill JE (1987) ĺonlinear frame analysis by finite element methods. ASCE, Struct Eng 113(6):1221–1235
6. Otani A, Nagashima K, Suzuki J (2002) Vertical seismic response of overhead crane. Nucl Eng Design, pp 211–220
7. Kosbab BD, Leon RT (2009) Seismic behavior considerations for jumbo container cranes. Structures 2009:1000–1009
8. Kosbab BD, Jacobs LD, DesRoches R, Leon RT (2009) Analysis and testing of container cranes under earthquake loads, lifeline earthquake engineering in a Multihazard environment ©2009 ASCE, pp 842–852
9. Hu Y (2006) Earthquake Engineering. Seismological Press, Beijing pp 163–173
10. Anil KC (2005) Dynamics of structures: theory and applications to earthquake engineering. Tsinghua University Press, Tsinghua pp 457–464
11. Vasilopoulos AA, Beskos DE (2009) Seismic design of space steel frames using advanced methods of analysis. Soil Dyn Earthq Eng 29:194–218
12. Yoshitaka T, Nobuyuki S (2005) Seismic response analysis system by means of multibody dynamics approach. In: Proceedings of IDETC/CIE 2005

Predictive Coefficient Method for Establishing Similarity Model of Jumbo Container Cranes

Zhe Li, Gongxian Wang and Jiquan Hu

Abstract Shaking table similar model test is an important and frequently-used research technique in dynamic response analysis of large structures. Similar conditions derived from dynamic similarity theory usually cannot be fully implemented on scale model, which causes model distorted. Moreover, the test results obtained from distorted similar model cannot be accurately inverted to predict the dynamic behaviors of the prototype structure. This research proposes Predictive Coefficient Method for establishing similarity model of Jumbo container cranes. In the case of container gantry cranes model test, Prediction Coefficient of dynamic response is obtained from distortion model and prototype in the method of Finite element prediction coefficient firstly; then Hammer modal test and seismic shaking table test of a container crane model and get its dynamic properties are carried out; thus the Predictive Coefficient can be amended and achieved by compare the numerical results and experimental results. The analysis results show that the distortion model test results according to the Prediction Coefficient Method can accurately predict the prototype structure dynamic characteristics and this method also can improve the reliability of similarity model test results.

Keywords Similarity theory · Scale modal test · Container crane · Shaking table test · Distortion model

1 Introduction

Earthquakes pose a significant threat to the safe use of large harbor machinery, and the increasing size of container cranes has resulted in vulnerability to seismic loads which has not previously been recognized. To explore the failure mode and mechanism of large container crane under earthquake loads and to improve the

Z. Li · G. Wang (✉) · J. Hu
School of Logistics Engineering, Wuhan University of Technology, Wuhan 430063, China
e-mail: wanggongxian@gmail.com; 172042756@qq.com

J.K. Sinha (ed.), *Vibration Engineering and Technology of Machinery*,
Mechanisms and Machine Science 23, DOI 10.1007/978-3-319-09918-7_80

seismic performance of the crane structures is one of the problems much needed to be solved and paid high attention in the field of safe use and design for large seaport machinery [1]. Model Test is an important part of large-scale vibration tests. Appropriate materials are adopted to make a model with geometric similarity and reduced scale with that of actual structure. Then such model was placed in an earthquake simulation shaking table test to measure data and be converted in accordance with appropriate scale ratio, leading to a result which can approximately reflect the dynamic characteristics of the actual structure during an earthquake. To design a model with reasonable scale according to dynamic similarity theory is significant for studies on seismic performance of container crane.

Generally in practical operation, the similar conditions cannot be satisfied completely. For example, the thickness of the beam cross-section cannot change in proportion to the length and width of the beam. So the designed model is not exactly the same as the original prototype and will become distortion model. The experimental results of the distortion model or the numerical results cannot be extended to the original prototype based on similar conditions. That means the dynamic characteristics of the prototype structure cannot be predicted through distortion model. In the existing design process of beam scale model, in order to deal with some certain parameters which can be changed in a similar ratio, for example, the similarity of radius of inertia was used to realize the similarity of bending stiffness and design the cross-section parameters of the curved beams, in order to minimize the radius of gyration error of cross-section [2, 3]. According to the design requirements of cross-section thickness and by combining the area moment of inertia formula, the parameters of the model of beam section were calculated. Finally, based on practical experience in engineering, the thickness, length, and width of the beam cross-section were obtained to design a 1:50 reduced scale model of container crane. Based on dynamic similarity theory, a container crane scale model with similar ratio of 1:20 was built [4, 5]. The quality compensation method was used in the model to handle the problem that the thickness of the beam section cannot change based on a similar ratio. Based on this, they conducted a series of seismic shaking table test and obtained good results.

Many scholars deduced similitude ratio based on the bending theory of Euler beam. That is, the thickness of the beam is set according to the design requirements, rather than scaling on a similar ratio. For example, the literature [2] considered the bending stiffness of the beam section, but did not consider the quality differences due to section thickness, as well as the effect of size on the displacement acceleration. Literature [4, 5] conducted quality compensation for the model according to the dynamic similarity principle, and put the compensating mass attached to the model. The value of the additional mass and the location were mainly based on experience and the calculated results of the qualitative study, without considering the increase of local stress of placing mass and stiffness problem. While none of the above studies established experimental or numerical model to calculate the relationship between the quantitative values of different thickness and the response of overall structure of the beam section under seismic loads. Literature [6] conducted experiment using the distortion model to calculate and analyze four scaled ship under water explosion, and summarized the model of section moment prediction bias. Literature [7] used the finite element prediction coefficient method to study the response of similar model of hull with single parameter distortion under explosive impact, and conducted numerical experiments using dynamic distortion model to predict the response of the distortion factor and using distortion model to predict the dynamic characteristics of the original model. None of the above studies conducted physical model tests, and just took the numerical analysis and study, without experimental verification. In this paper, finite element prediction coefficient method was adopted to conduct a numerical modeling calculation for a container crane and analyze the dynamic response of beam distortion coefficient for the beam model with section distortion. Shaking table was used to do distortion test for the cross-sectional thickness of the section entity, we launched a prototype reaction according to the similarity relation based on the model test results and compared it with the calculation value of prototype to validate the distortion prediction coefficients in the model test.

2 Similar Scale Analysis

2.1 *Similar Dimensional Analysis of Structural Dynamical Model*

In structural vibration problems, the mechanical properties of elastic-plastic structure is an important indicator to characterize structural deformation, displacement and destruction [8]. In similar model experiments, elastic modulus (E), geometry (l), density (ρ), time (t), dynamic stress (σ), displacement (u), velocity (v), acceleration (a), gravity (g), natural frequency (ω) can be chosen as the relevant parameters of the structure. The relationship among various parameters is determined by dimensional analysis, within the linear elastic range, the general physical function is [3]:

$$f(\sigma, l, E, \rho, t, u, v, a, g, \omega) = 0 \tag{1}$$

The density ρ, elastic modulus E, length l are taken as the basic dimension. Similar criterion is obtained according to π theorem:

$$\begin{aligned} &\pi_1 = \frac{\sigma}{\rho^0 El}, \ \pi_2 = \frac{t}{\rho^{0.5}E^{-0.5}l}, \ \pi_3 = \frac{a}{\rho^{-1}El^{-1}}, \ \pi_4 = \frac{g}{\rho^{-1}El^{-1}}, \\ &\pi_5 = \frac{u}{\rho^0 E^0 l}, \ \pi_6 = \frac{v}{\rho^{-0.5}E^{0.5}l^0}, \ \pi_7 = \frac{\omega}{\rho^{-0.5}E^{0.5}l^{-1}} \end{aligned} \tag{2}$$

Similarity ratio needs to meet the conditions for:

$$C_\sigma = C_E,\ C_t = C_\rho^{0.5} C_E^{-0.5} C_l,\ C_a = C_E C_\rho^{-1} C_l^{-1} = C_g,\ C_u = C_l,\\ C_v = C_E^{0.5} C_\rho^{-0.5},\ C_\omega = C_E^{0.5} C_\rho^{-0.5} C_l^{-1} \tag{3}$$

In Eq. (3), C represents the similarity ratio between prototype and model. Because the gravity acceleration of prototype and model are the same, namely $C_a = C_g = 1$, to satisfy similarity relation listed in (3) is quite difficult. Since the experiment mainly studies the response of container crane structure in the elastic stage, the design of the container crane scale model can be based on the Elasticity similitude law [8]. That is to ignore the similar requirements of $C_g = 1$, and add $C_m = C_\rho C_l^3$, as quality. Equation (3) is changed into:

$$C_\sigma = C_E,\ C_t = C_\rho^{0.5} C_E^{-0.5} C_l,\ C_a = C_E C_\rho^{-1} C_l^{-1},\ C_u = C_l,\ C_v = C_E^{0.5} C_\rho^{-0.5},\\ C_\omega = C_E^{0.5} C_\rho^{-0.5} C_l^{-1},\ C_m = C_\rho C_l^3 \tag{4}$$

Table $C_l = 15$ is chosen as the geometry scale, both the production model and the prototype is made by the same material and the geometric ratio, density ratio $C_\rho = 1$ and elastic modulus ratio $C_E = 1$ is taken as the basic scale. The main component of container crane is a beam and the research mainly focuses on beam bending. Other similar scales are derived according to Euler beam bending theory, such as the time scale ratio C_t, frequency ratio C_ω, mass ratio C_m, displacement ratio C_u, velocity ratio C_v, acceleration ratio C_a, strain ratio C_ε and stress ratio C_σ, etc. The condition of beam bending equilibrium is:

$$\rho A \frac{\partial^2 u}{\partial t^2} = EI \frac{\partial^4 u}{\partial l^4} \tag{5}$$

In formula (5), A and I are the cross-sectional acreage and moment of inertia of the beam cross section, respectively. Similar conditions can be obtained in accordance with dimensional analysis:

$$C_\rho C_A C_u C_t^{-2} = C_I C_E C_u C^{-4} \tag{6}$$

where C_A is section area ratio, C_I is the moment of inertia ratio:

$$C_t = C_\rho^{0.5} C_A^{0.5} C_l^2 C_I^{-0.5} C_E^{-0.5} \tag{7}$$

Frequency is the reciprocal of time, so we can get frequency scale C_ω:

$$C_\omega = C_\rho^{-0.5} C_A^{-0.5} C_l^{-2} C_I^{0.5} C_E^{0.5} \tag{8}$$

Quality ratio $C_m = C_\rho C_A C_l$, displacement ratio $C_u = C_l$, velocity and acceleration are the first and second order differential of displacement to time, so the speed ratio C_v and acceleration ratio C_a can be expressed as follows:

$$C_v = C_\rho^{-0.5} C_A^{-0.5} C_l^{-1} C_I^{0.5} C_E^{0.5},\ C_a = C_\rho^{-1} C_A^{-1} C_l^{-3} C_I^1 C_E^1 \tag{9}$$

According to beam bending strain formula (distance from neutral layer y is $C_y \approx C_l = 15$) and Hooke's law, we can get a similar relation of stress-strain in the elastic stage:

$$C_\varepsilon = C_m C_a C_l C_y C_I^{-1} C_E^{-1},\ C_\sigma = C_m C_a C_l C_y C_I^{-1} \tag{10}$$

2.2 Distortion Model

Container crane structure mainly consists of box girder. Take the column as an example, the width of the beam is L, height is H, and thickness is D, scale model beam section width is L_m, height is H_m and thickness is d_m. When L_m = L/15, H_m = H/15, d_m = d/15, beam section area ratio $C_A = 15^2$, moment of inertia ratio $C_I = 15^4$, the detailed scale that we derive between container crane prototype structure and 1:15 scale model is shown in Table 1.

In the modeling process, the length and the width of the beam cannot be changed by the same proportion with the section thickness, i.e. L_m = L/15, H_m = H/15, $d_m \neq$ d/15. Model distortion makes the model variables $\pi_d^m \neq \pi_d^b$ (π^b is distortion model and π^m is prototype value), and the dependent variable values are not equal. The characteristics of distortion model can not reflect the characteristics of original model. When $\pi_d^m \neq \pi_d^b$, define λ as the distortion coefficient (The meaning of distortion coefficient is the scale distortion between the actual value and the expected value in the model because the independent variable does not change as a whole similar ratio) make

$$\beta_d = \frac{\pi_d^b}{\pi_d^m} \tag{11}$$

Other similar conditions remain unchanged. Take the dependent variable as an example, define the acceleration prediction coefficients as $\delta_a = \pi_a^b / \pi_a^m$, and the equation of π_a^b and π_a^m are only different with π_d. So the acceleration coefficient δ_a is

Table 1 Similarity relation of the scale model

Similar parameters	Similar ratio	Similar parameter	Similar ratio
Geometry size	15	Mass	15^3
Density	1	Displacement	15
Modulus of elasticity	1	Speed	1
Cross section radius of inertia	15	Acceleration	1/15
Time	15	Stress	1
Frequency	1/15	Strain	1

a function of the predicted distortion coefficients λ, and it is not associated with non-distortion item. The variation between δ_a and λ can be obtained only through test or numerical calculation method to get the experience formula.

3 Distortion Prediction Coefficient

3.1 Numerical Calculation Model

In order to get the beam thickness distortion prediction coefficient, we modeled container a crane using finite element method. We established a full-scale numerical model according to the original prototype size. The structure of container crane and the finite element model are shown in Figs. 1 and 2. With a container crane (L × H = 66 m × 33 m, 520 t)as the research object, according to the size of vibration table (1.5 m × 1.5 m), and with pay load (2t) of the vibration table as test condition, we designed the model with 1:15 reduced scale. Container crane is mainly composed of portal rigid frame structure (including the land side column, sea side beam, lower beam and frame beam and side beam), front and rear beam, and front and rear rod.

A finite element modal analysis method was used to carry out numerical verification and analyze the similarity of the underwater structure of underwater structure model, coupled with vibration response and sound field radiated with good

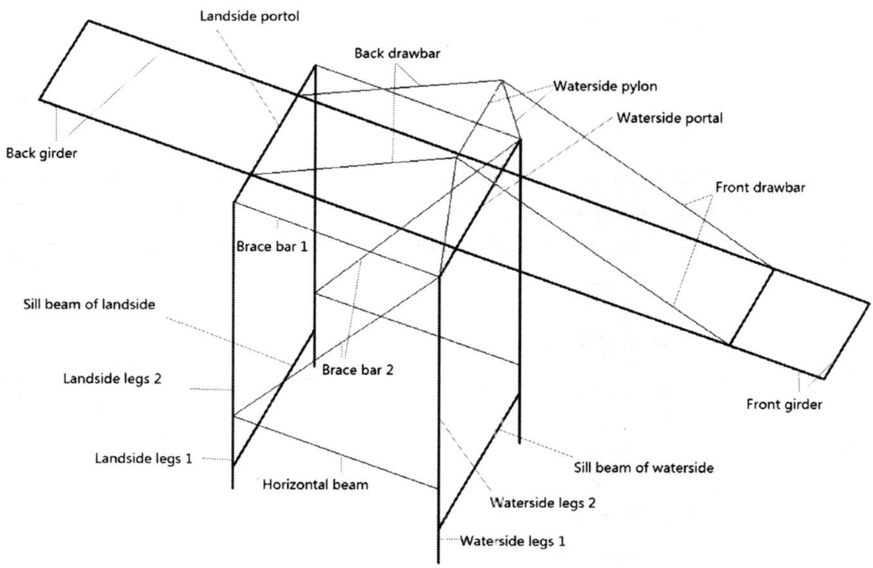

Fig. 1 Container crane structure model

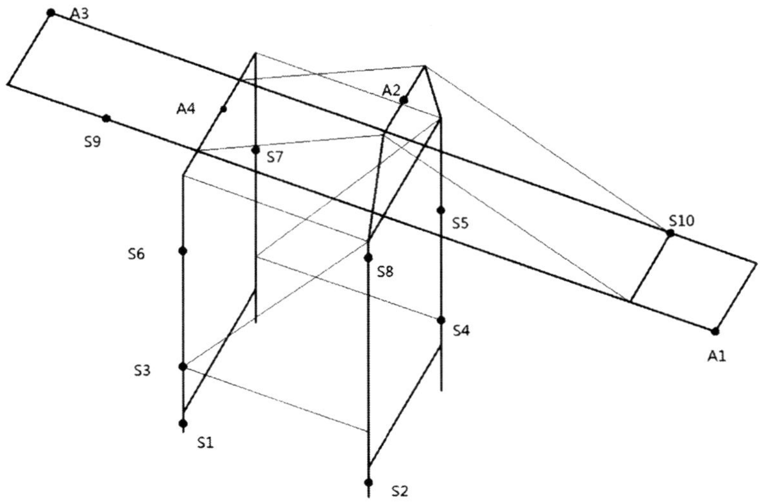

Fig. 2 Finite element model of container cranes

results obtained in literature [9]. A similar reference model with the reduced ratio of 1:15 was established (beam length, height width and thickness are all established according to the ratio 1:15). According to baseline model, a series of thickness distortion model were built. Take column as an example, the distortion data is shown in Table 2 (with only width listed in the table. The prototype width is 900 mm, length, width and height are all reduced in the ratio of 1:15. The thickness of some beam in the container crane structure is inconsistent in X, Z axis. The reason to select test column is that the beam thickness in X, Z is consistent with both as 10 mm, making the table more concise).

Table 2 Data of measured baseline model and distortion model

	Width/mm	Thickness/mm	Distortion coefficient λ
Baseline model	60	0.667	1
Model 1	60	1	1.5
Model 2	60	1.333	2
Model 3	60	1.667	2.5
Model 4	60	2	3
Model 5	60	2.333	3.5
Model 6	60	2.667	4
Model 7	60	3	4.5
Model 8	60	3.333	5

Table 3 Natural frequency prediction coefficients

	Thickness distortion coefficient λ	prediction coefficients δ_ω
Baseline model	1	1
Model 1	1.5	0.995
Model 2	2	0.99
Model 3	2.5	0.985
Model 4	3	0.981
Model 5	3.5	0.976
Model 6	4	0.972
Model 7	4.5	0.967
Model 8	5	0.963

3.2 Prediction Coefficients

We calculate the inherent frequency of models in Table 2, and obtain distortion model of natural frequency prediction coefficients δ_ω as shown in Table 3:

We use curve fitting method for data in Table 3 to get the function relationship between δ_ω and the thickness distortion coefficient λ,

$$\delta_\omega = 0.0002\lambda^2 - 0.0104\lambda + 1.0101 \tag{12}$$

As can be seen from Table 3, on the condition that the thickness is changing (the thickness is much smaller than cross-sectional width), the natural frequency of the model changes little, we use the frequency formula $\omega = \sqrt{k/m}$ to prove the correctness of the prediction coefficients. With beam quality as $m = [LH - (L - 2d)(H - 2d)]l\rho$, beam bending stiffness formula as $k = EI$ and section stiffness as $k = E[LH^3 - (L - 2d)(H - 2d)^3]$, assuming the thickness of section becomes 2 times as the original one, the distortion model mass $m^b = [LH - (L - 2nd)(H - 2nd)]l\rho$, stiffness $k^b = E\left[LH^3 - (L - 2nd)(H - 2nd)^3\right]$, $\delta = \omega^b/\omega = \sqrt{km^b/k^b m}$, and with L = 60 mm, H = 100 mm, d = 0.667 mm, n = 3, n = 4.5, we obtain $\delta = 0.9805$, $\delta = 0.9664$and we consider $\delta \approx \delta_\omega$. So the natural frequency prediction coefficients is correct.

Selecting A1–A4 as acceleration and displacement inspection point; S1–S10 as the stress and strain inspection point, as shown in Fig. 3. Selecting representative seismic record as the exciter input: EI-Centro north-south seismic acceleration record (20 s, Δt = 0.02 s), each curve are composed t by 1,000 output points according to finite element calculation. Take A1 acceleration time scale curve as an example. Take the feature point a of benchmark model curve (points with maximum peak are selected) and set coordinates of a benchmark model in point a as (t_a, a_a). Then take the corresponding feature point b with A1 acceleration time

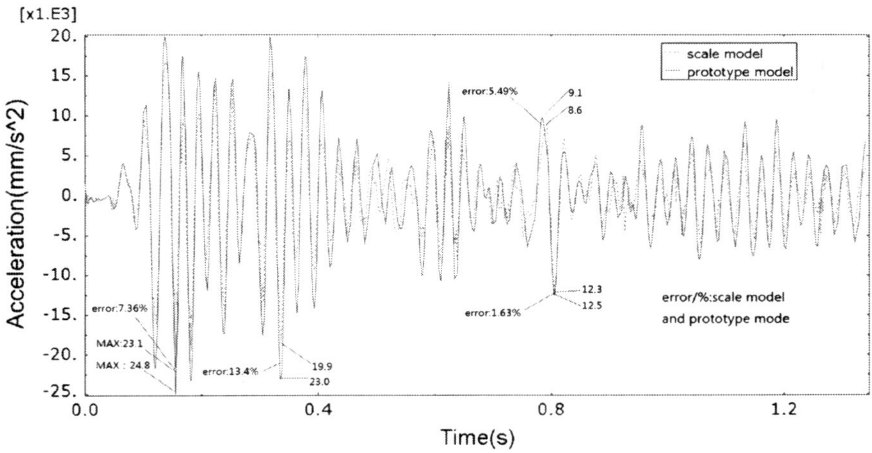

Fig. 3 measuring point of the model

history curve as an example. Take characteristic points b corresponding to distortion model and set the coordinates of distortion model in point B as (t_b, a_b). Then we can get the time prediction coefficient $\delta_t = t_b/t_a$, acceleration prediction coefficient $\delta_a = a_b/a_a$. We scale each abscissa point t in time-history curve of distortion model according to the time prediction coefficient δ_t. And at the same time, we scale the vertical point according to the prediction coefficients δ_a to get history curve. According to above paragraph, we can get 8 frequency prediction coefficients. Because the time prediction coefficient difference is very small, it can be neglected. The result obtained from above shows the function formula between acceleration prediction coefficient δ_a and the thickness of the distortion factor λ.

Factor formula of displacement prediction coefficient function δ_u and the thickness of the distortion factor λ:

$$\delta_u = -0.0257\lambda^3 + 0.2251\lambda^2 - 0.6057\lambda + 1.3996 \tag{13}$$

Factor formula of stress prediction coefficient δ_σ and the thickness of the distortion factor formula λ:

$$\delta_\sigma = -0.0197\lambda^3 + 0.1856\lambda^2 - 0.5582\lambda + 1.3907 \tag{14}$$

Figures 4 and 5 show A1 point acceleration, S5 point acceleration curve of stress of container crane prototype and distortion scale model ($\lambda = 3$, thickness d = 2 mm) under seismic excitation. (Note: the solid line represents the prototype and the dotted line represents the distortion model; time, acceleration, and stress have been converted according to similarity ratio ruler of Table 1; distortion model curve is converted according to the prediction coefficient). By comparing chart 5–6, it can be found the seismic response results of scale modal and prototype calculated by the

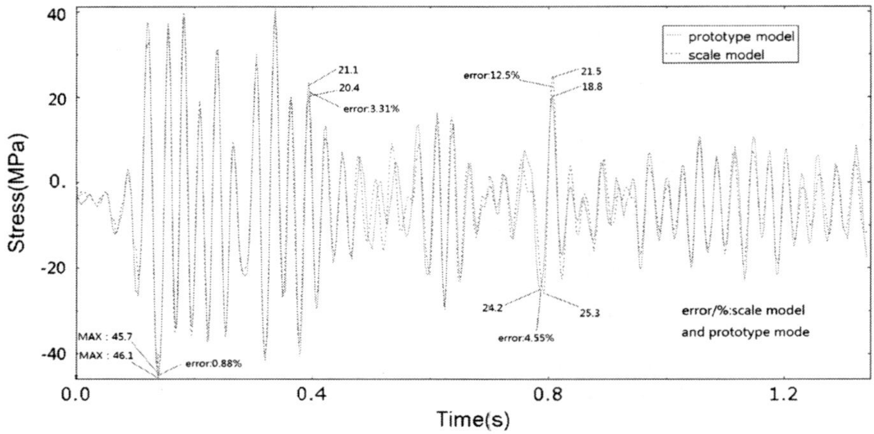

Fig. 4 The time history curve of acceleration of point A

Fig. 5 The time history curve of stress of point A

distortion scale modal are very close, indicating the distortion prediction coefficient was reasonable in numerical calculation.

4 Analysis of the Container Crane Scale Model Test

The laboratory scale model of a distortion is used to verify the validity of container crane structure distortion coefficients with the model shaking table test. The geometric similarity constant of the test model is 1/15, distortion coefficient $\lambda = 3$ and the thickness d = 2 mm. Please refer to Fig. 6.

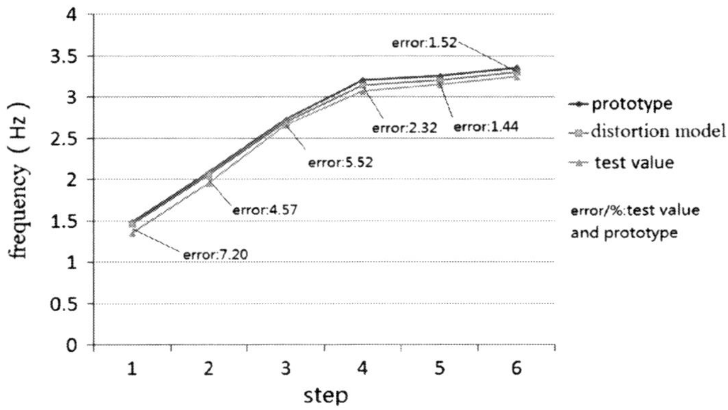

Fig. 6 The field site

4.1 Hammer Modal Test

Hammer mode method is used to obtain natural frequency of 1:15 scale modal of container crane. In order to get acceleration response, we emplace 47 One-way acceleration sensor in the front beam, the beam and gantry structures in the scaled model by using improved layout optimization method of independently effect sensor [10].

Table 4 shows the prototype modal finite element calculation results, and the first 6 natural frequencies of container crane distortion model modal test, and the corresponding error (error in the table is the one produced after the test data has been converted by the prediction coefficient and the values of the original prototype).

From Table 4, distortion scale model can accurately predict the dynamic characteristics of the prototype structure with smaller error. The accuracy of the frequency distortion prediction coefficient has been verified in this experiment.

Table 4 Natural frequency of prototype and distortion scale model

Step	Frequency/Hz			Error/%
	Prototype	Distortion model	Test value	Test and prototype
1	1.4868	21.973	20.299	7.21
2	2.0894	30.869	29.336	4.58
3	2.7259	40.384	39.887	5.59
4	3.2075	47.197	46.109	2.31
5	3.2593	48.167	47.286	1.43
6	3.3608	49.620	48.712	1.51

4.2 Shaking Table Test

The purpose of the test is to verify the validity of the distortion prediction coefficients, in addition to validating dynamic characteristics of the model in frequency domain, we also need to verify the changes of the property model in the time domain. We obtained the dynamic response of the model under earthquake excitation by a series of shaking table test, and compared the finite element calculation results with the prototype structure, so as to further verify the accuracy of distortion prediction coefficients.

The main parameters of the shaking table tests are: table size 1,500 × 1,500 mm, the maximum horizontal acceleration ±50 m/s^2, ± 30 m/s^2 vertical acceleration, the maximum horizontal displacement ±200 mm, maximum displacement in vertical direction ±100 mm, the maximum speed at horizontal direction and the vertical direction 0.8 m/s, the maximum bearing capacity 2T and the frequency range 0.1–100 Hz. In the experiment, the acceleration sensor and strain gauges are mounted on the door frame structure model in different position, as shown in Fig. 3. For the scale model, based on $C_t = 15$, $C_a = 1/15$, the time axis of seismic load is adjusted to 1/15 compared with that of original prototype and the peak acceleration is adjusted to 15 times of the prototype. After such processing, the period of seismic wave is very short and the test process is relatively fast. Therefore, in the process of the data collection, the sampling frequency of the data acquisition system can be adjusted larger.

We also provide acceleration time curve of points A1 and the strain curve of S5, as shown in Figs. 7 and 8. The distortion of 1:15 scale model can better predict the overall seismic behavior of the prototype structure. The curve in the figure includes

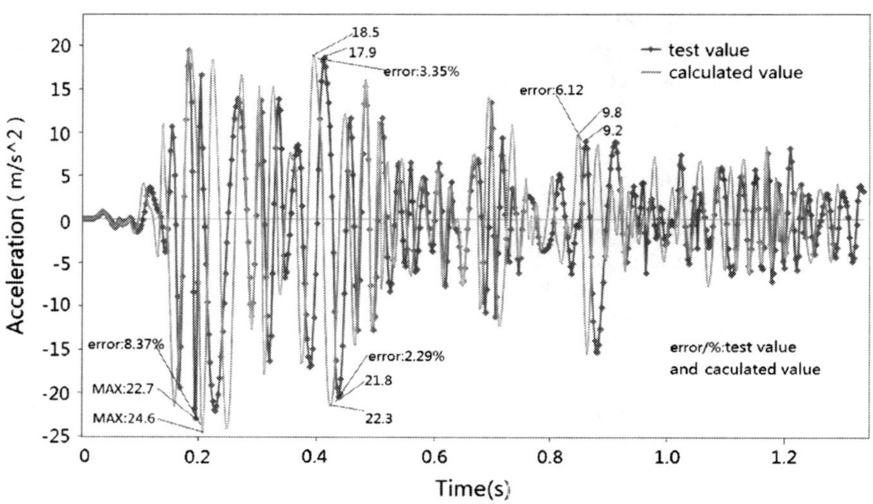

Fig. 7 Acceleration time curve of point A

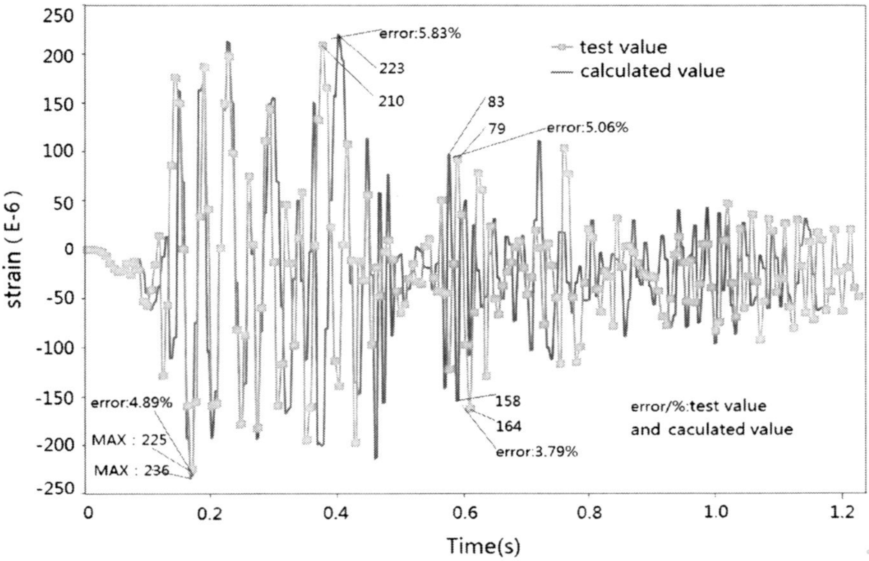

Fig. 8 Response time curve of point S5

prototype calculation and the scale model test, among which the test curve has quadrilateral markers. Map prototype value has undergone similarity transformation and the experimental value is tested with prediction coefficient test.

According to Figs. 7 and 8, Table 5, comparing numerical results and experimental result, it shows that the results have some errors with that of finite element calculations. Acceleration of the study point, extreme size of appeared strain and time are relatively similar to one another in the seismic load. The reasons for the difference of the results include: (1) the existence of simplified part in finite element modeling; (2) the error of sensor arrangement position and error of sensors themselves; (3) the presence of nonlinear problems of structural vibration test trials, the numerical model was based on linear elastic basis and similarity ratio was

Table 5 Maximum dynamic strain of different measured point

Measured point	Strain(E-6)		Error/%
	Prototype	Test value	Test and prototype
S1	91.6	82.9	9.49
S3	216.8	209.3	3.46
S5	186.7	173.5	7.07
S6	232.3	225.3	3.01
S9	71.3	75.4	5.75
S10	87.9	80.1	8.87

calculated in the elastic range; (4) external environmental effect. Relative error is within a certain allowable range and the overall test results are relatively in line with the finite element method, indicating the effectiveness of distortion prediction coefficients have been further verified.

5 Conclusion

With the Predictive Coefficient Method, numerical model of container crane was calculated and relation formula of distortion factor and prediction coefficients were obtained in this study. Furthermore, experiment was conducted on a distortion scale model. The results showed that:

(1) Distortion scale model can accurately predict the dynamic characteristics and seismic response of the prototype structure.
(2) The finite element prediction method is validated and the numerical results show that the prediction coefficients obtained by this method is correct.
(3) Distortion prediction coefficients are feasible in the scale model of the design, providing a reliable experimental model for subsequent researches.

References

1. Li Z, Wang G, Wang D et al (2013) Testing methods on dynamics behavior of jumbo container cranes under seismic loads[J]. J Wu Han Univ Technol 35(8):72–76 (in Chinese)
2. Jin YL, Li ZG (2012) Theoretical design and experimental verification of a 1/50 scale model of a quayside container crane [J]. Proc Inst Mech Eng, Part C: J Mech Eng Sci 226 (6):1644–1662
3. Jin Y-L, Wu T-X (2012) Dynamic similarity analysis and experimental verification on a quayside container crane [j]. J Shang Hai Jiao Tong Univ 46(10):1609–1615 (in Chinese)
4. Jacobs LD, Reginald DR, Roberto TL (2011) Seismic behavior of a jumbo container crane including uplift [J]. Earthq Spectra. 27(3):745–773
5. Jacobs LD, Reginald DR, Roberto TL (2010) Experimental study of the seismic response of container cranes [J]. Ports 2010: Building on the Past, Respecting the Future 210Don't Mess with Structural Engineers 2010 ASCE, PP 91–99
6. Zhang H-P, Yue Y-W, Su L-Q et al (2012) Model experimental scheme for longitudinal strength of a warship subjected to underwater explosion [J]. J Vib Shock 31(6):175–180 (in Chinese)
7. Wei Z, Yu W, Du J-Y (2010) Research on similarity of dynamic response of ship Grillage subjected to transient load [J]. Acta Armamentarii, 31(s1):172–178 (in Chinese)
8. He C, Wang Y, Fang Y et al (2012) Similarity model test of earth-pressure-balanced shield tunneling process [J]. China Civil Eng J 45(2):162–169 (in Chinese)
9. Wang Y-F, Zhou Qi-D, Ji G et al (2012) Numerial investigation on acoustic similarity of fluid-loaded stiffened cylinder [J]. J Vib Shock 31(16):66–71 (in Chinese)
10. Liu W, Gao W-C, Li H et al (2013) Improved optimal sensor placement methods based on effective independence [J]. J Vib Shock 32(6):54–62 (in Chinese)

Stability and Bifurcation Analysis of an Axially Accelerating Beam

Bamadev Sahoo, L.N. Panda and Goutam Pohit

Abstract The present work deals with nonlinear transverse vibration of a beam moving with a harmonically fluctuating velocity and subjected to parametric excitation at a frequency close to twice the natural frequency in presence of internal resonance. The geometric cubic nonlinearity in the equation of motion is due to stretching effect of the beam. The analysis is carried out using the method of multiple scales (MMS) by directly attacking the governing nonlinear integral-partial-differential equations and the associated boundary conditions. The resulting set of first- order ordinary differential equations governing the modulation of amplitude and phase of the first two modes are analyzed numerically to obtain steady state and dynamic bevaviour along with stability as well as bifurcation of the travelling system. The system exhibits trivial, single mode and two mode solutions with pitchfork, saddle-node and Hopf bifurcations. The sensitivity of the system towards the variation in frequency and amplitude of fluctuating velocity component, variation in damping, flexural stiffness and initial points are studied.

Keywords Internal resonance · Stability · Bifurcations · Perturbation technique · Principal parametric resonance

B. Sahoo (✉)
Department of Mechanical Engineering, IIIT, Bhubaneswar, India
e-mail: bamdev_sahoo@yahoo.com

L.N. Panda
Department of Mechanical Engineering, College of Engineering and Technology, Bhubaneswar, India
e-mail: lnpanda@yahoo.com

G. Pohit
Department of Mechanical Engineering, Jadavpur University, Kolkata, India
e-mail: gpohit@gmail.com

J.K. Sinha (ed.), *Vibration Engineering and Technology of Machinery*, Mechanisms and Machine Science 23, DOI 10.1007/978-3-319-09918-7_81

1 Introduction

Real life axially moving systems such as textile fibers, conveyor belts, paper sheets, band saw blades, magnetic tapes and robotic manipulators modeled as axially travelling beam move with variable axial speed which is assumed as a harmonic fluctuation over a mean travelling velocity. The amplitude and frequency of this fluctuating component of velocity causes parametric excitation to the system which leads to different parametric resonance cases, hence it is important enough to investigate. Wickert and Mote [1] reviewed the literature on transverse vibration of axially moving continua up to 1988. Oz and Pakdemirli [3] and Oz et al. [4] investigated principal and combination parametric resonances and stability analysis for an axially accelerating beam using the method of multiple scales. Ozkaya et al. [5] investigated non-linear transverse vibrations and 3:1 internal resonances of a beam with multiple supports and plotted frequency response curves for different support numbers. Chin and Nayfeh [6] and Panda and Kar [7, 8] investigated the principal parametric resonance and combination parametric resonance of hinged-clamped beams and pipe conveying pulsating fluid respectively. Pakdemirli and Oz [9] studied the transverse vibration of simply supported axially moving Euler-Bernoulli beam for infinite mode analysis and truncation to resonant modes. Ponomareva and Horssen [10] and Ghayesh [11] investigated nonlinear transversal vibrations of axially travelling continua involving variety of system models.

In this work, the influence of internal resonance is demonstrated by the two mode equilibrium solutions of both amplitude and frequency response analysis. Besides the two mode solutions there are single mode and trivial mode solutions. The system exhibits different types of bifurcations like pitchfork bifurcations, Hopf bifurcations and saddle node bifurcations for different control parameters. The influence of higher fluctuating velocity components, various damping parameters and internal frequency detuning parameter on the strength of modal interactions is studied.

2 Formulation of the Problem

Figure 1 shows a uniform horizontal beam simply supported at both ends and travelling with a harmonically variable velocity. The beam is assumed to be an Euler-Bernoulli type with transverse vibration and cubic nonlinearity due to mid-line stretching of the beam. The source of the cubic nonlinearity is the large amplitude of transverse vibration (assumed) and immovable support conditions. The non-dimensional equation of transverse motion of the beam considering viscous damping [12] and viscoelastic damping [13] is given by

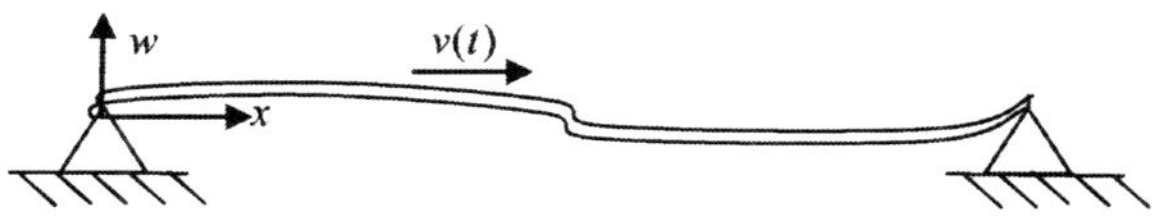

Fig. 1 Schematic diagram of an axially travelling beam with variable velocity

$$\ddot{w} + 2v\dot{w}' + \dot{v}w' + (v^2 - 1)w'' + v_f^2 w'''' + 2\varepsilon\alpha\dot{w}'''' + 2\varepsilon\mu\dot{w} = \frac{1}{2}v_l^2 w'' \int_0^1 w'^2 dx \quad (1)$$

The nondimensional scheme used here is

$$x = \frac{x^*}{L},\ t = t^*\sqrt{\frac{P}{\rho A L^2}},\ w = \frac{w^*}{L},\ v = \frac{v^*}{\sqrt{P/\rho A}},\ 2\varepsilon\alpha = \frac{E^*}{L^2}\left(\frac{I}{mE}\right)^{1/2},$$

$$2\varepsilon\mu = \frac{CL^2}{\sqrt{mEI}},\ v_l = \sqrt{\frac{EA}{P}},\ v_f = \sqrt{\frac{EI}{PL^2}}$$

where the variables with asterisk denote dimensional ones, dot and prime denote derivatives with respect to time and spatial variable x respectively. m is mass per unit length, ρ is density, A is cross sectional area, L is length, v_f is non dimensional flexural stiffness and v_l nondimensional longitudinal stiffness of beam. E^* is the coefficient of internal dissipation of the beam material (Kelvin-Voigt type). C is the external damping factor, α is nondimensional material damping and μ is nondimensional viscous damping. Reordering the transverse displacement with the relation $w = \sqrt{\varepsilon}w^{\#}$, where $\varepsilon < 1$, the system is converted into a weakly nonlinear one [4] and putting the variable velocity of beam, $v = v_0 + \varepsilon v_1 \sin\Omega t$, where v_0 is mean velocity, εv_1 is amplitude and Ω is the frequency of the harmonically varying component, Eq.(1) may be represented as (removing superscript "#")

$$\begin{aligned} &2\varepsilon\alpha\dot{w}'''' + v_f^2 w'''' + \left[v_0^2 + 2\varepsilon v_0 v_1 \sin\Omega t - 1\right]w'' + \varepsilon v_1 \Omega w' \cos\Omega t \\ &+ 2(v_0 + \varepsilon v_1 \sin\Omega t)\dot{w}' + 2\varepsilon\mu\dot{w} + \ddot{w} = \frac{1}{2}\varepsilon v_l^2 w'' \int_0^1 w'^2 dx \end{aligned} \quad (2)$$

With boundary conditions

$$w(0, t) = w(1, t) = w''(0, t) = w''(1, t) = 0 \quad (3)$$

3 Method of Analysis

The present adopted methodology is the direct perturbation Method of multiple scales (MMS) technique, where MMS [6–8] is directly applied to Eq. (2) and the problem is analyzed. The alternative method is carrying out Galerken discretization of the equation of motion (2) and then applying the MMS to analyzing the problem, which is not done here due to space constraint. An approximate solution to this weakly non-linear distributed parameter system in the form of a first order uniform expansion is aimed. The time scales and the time derivatives are

$$T_n = \varepsilon^n t,\ \frac{d}{dt} = D_0 + \varepsilon D_1 + \ldots\ldots,\ \frac{d^2}{dt^2} = D_0^2 + 2\varepsilon D_0 D_1 + \ldots\ldots, \\ D_n = \partial/\partial T_n,\ n = 0, 1, 2, 3\ldots \tag{4}$$

Assuming an expansion $w(x, t, \varepsilon) = w_0(x, T_0, T_1) + \varepsilon w_1(x, T_0, T_1) + \ldots\ldots$ (5)

and equating coefficients of like powers of ε on both sides, we get

$$O(\varepsilon^0) : D_0^2 w_0 + 2v_0 D_0 w_0' + (v_0^2 - 1)w_0'' + v_f^2 w_0'''' = 0 \\ w_0(0, t) = w_0(1, t) = w_0''(0, t) = w_0''(1, t) = 0 \tag{6}$$

$$O(\varepsilon^1) : D_0^2 w_1 + 2v_0 D_0 w_1' + v_f^2 w_1'''' + (v_0^2 - 1)w_1'' = -2v_0 D_1 w_0' - 2v_1 \sin\Omega t D_0\, w_0' \\ - 2D_0 D_1 w_0 - 2\alpha D_0 w_0'''' - 2\mu D_0 w_0 - 2v_0 v_1 \sin\Omega t\, w_0'' - v_1 \Omega \cos\Omega t\, w_0' \\ + \frac{1}{2} v_l^2 w_0'' \int_0^1 w_0'^2 dx = 0 \\ w_1(0, t) = w_1(1, t) = w_1''(0, t) = w_1''(1, t) = 0 \tag{7}$$

The solution of Eq. (6) may be written as

$$w_0(T_0, T_1, x) = \sum_{n=1}^{\infty} \phi_n(x) A_n(T_1) e^{i\omega_n T_0} + cc \tag{8}$$

where ϕ_n: mode shapes, ω_n: natural frequencies, and cc: complex conjugate. The mode shapes, dispersive relation and the support conditions are [3]

$$\phi_n(x) = C_{1n}\left\{ e^{i\beta_{1n}x} - \frac{\left(\beta_{4n}^2 - \beta_{1n}^2\right)\left(e^{i\beta_{3n}} - e^{i\beta_{1n}}\right)}{\left(\beta_{4n}^2 - \beta_{2n}^2\right)\left(e^{i\beta_{3n}} - e^{i\beta_{2n}}\right)} e^{i\beta_{2n}x} - \frac{\left(\beta_{4n}^2 - \beta_{1n}^2\right)\left(e^{i\beta_{2n}} - e^{i\beta_{1n}}\right)}{\left(\beta_{4n}^2 - \beta_{3n}^2\right)\left(e^{i\beta_{2n}} - e^{i\beta_{3n}}\right)} \right.$$
$$\left. e^{i\beta_{3n}x} + \left[-1 + \frac{\left(\beta_{4n}^2 - \beta_{1n}^2\right)\left(e^{i\beta_{3n}} - e^{i\beta_{1n}}\right)}{\left(\beta_{4n}^2 - \beta_{2n}^2\right)\left(e^{i\beta_{3n}} - e^{i\beta_{2n}}\right)} + \frac{\left(\beta_{4n}^2 - \beta_{1n}^2\right)\left(e^{i\beta_{2n}} - e^{i\beta_{1n}}\right)}{\left(\beta_{4n}^2 - \beta_{3n}^2\right)\left(e^{i\beta_{2n}} - e^{i\beta_{3n}}\right)} \right] e^{i\beta_{4n}x} \right\} \tag{9}$$

$$v_f^2 \beta_{in}^4 - \left(v_0^2 - 1\right)\beta_{in}^2 - 2v_0\omega_n\beta_{in} - \omega_n^2 = 0, \quad i = 1, 2, 3, 4 \tag{10}$$

$$\left(e^{i(\beta_{1n}+\beta_{2n})} + e^{i(\beta_{3n}+\beta_{4n})}\right)\left(\beta_{1n}^2 - \beta_{2n}^2\right)\left(\beta_{3n}^2 - \beta_{4n}^2\right) + \left(e^{i(\beta_{1n}+\beta_{3n})} + e^{i(\beta_{2n}+\beta_{4n})}\right)$$
$$\left(\beta_{2n}^2 - \beta_{4n}^2\right)\left(\beta_{3n}^2 - \beta_{1n}^2\right) + \left(e^{i(\beta_{2n}+\beta_{3n})} + e^{i(\beta_{1n}+\beta_{4n})}\right)\left(\beta_{1n}^2 - \beta_{4n}^2\right)\left(\beta_{2n}^2 - \beta_{3n}^2\right) = 0 \tag{11}$$

Since the higher modes will decay with time due to the presence of damping and coriolis terms present in the equation, the first two modes will contribute to the long term system response. Hence, Eq. (8) is replaced by

$$w_0(T_0, T_1, x) = A_1(T_1)\phi_1(x)e^{i\omega_1 T_0} + A_2(T_1)\phi_2(x)e^{i\omega_2 T_0} + cc \tag{12}$$

The case of principal parametric resonance of first mode and internal resonance is considered in [14] and in the present study the resonant frequency relations are

$$\omega_2 = 3\omega_1 + \varepsilon\sigma_1 \text{ and } \Omega = 2\omega_2 + \varepsilon\sigma_2 \tag{13}$$

Here σ_1 and σ_2 are detuning parameters. Putting Eqs. (12–13), into Eq. (7), one gets

$$D_0^2 w_1 + 2\,v_0 D_0 w_1' + \left(v_0^2 - 1\right) w_1'' + v_f^2 w_1'''' = \Gamma_1 e^{i\omega_1 T_0} + \Gamma_2 e^{i(\omega_1 T_0 + \sigma_1 T_1)}$$
$$+ \Gamma_5 e^{i\omega_2 T_0} + \Gamma_6 e^{i(\omega_2 T_0 - \sigma_1 T_1)} + \Gamma_8 e^{i(\omega_2 T_0 + \sigma_2 T_1)} + cc + NST \tag{14}$$

where the terms Γ_n are defined in the Appendix. *NST* stands for terms that do not produce secular or small divisor terms. As the homogeneous part of Eq. (14) with its associated boundary conditions has a nontrivial solution, the corresponding nonhomogeneous problem has a solution only if a solvability condition is satisfied [6–8].

$$2A_1' + 8S_1A_1^2\bar{A}_1 + 8S_2A_1A_2\bar{A}_2 + 8g_1\bar{A}_1^2A_2e^{i\sigma_1 T_1} + 2\mu\,C_1A_1 + 2\,\alpha\,e_1A_1 = 0 \tag{15}$$

$$2A_2' + 8S_4A_2^2\bar{A}_2 + 8S_3A_1\bar{A}_1A_2 + 8g_2A_1^3e^{-i\sigma_1 T_1} + 2\mu\,C_2A_2 + 2\,\alpha\,e_2A_2 + 2K_6\bar{A}_2e^{i\sigma_2 T_1} = 0 \tag{16}$$

where the prime denotes the differentiation with respect to slow time T_1 and S_i, g_i, K_i, C_i and e_i are nonlinear coefficients defined in Appendix. Over bar indicates complex conjugate.

4 Stability and Bifurcations

The evolutions of the equilibrium solutions, their stability and bifurcation analysis for principal parametric resonance are carried out from the modulation Eqs. (15–16) using

$$A_n = \frac{1}{2}[p_n(T_1) - i\,q_n(T_1)]\,e^{i\lambda_n(T_1)}, n = 1,2 \tag{17}$$

Putting this, in Eqs. (15–16), simplifying and separating the real and imaginary parts, we get the normalized reduced equations

$$\begin{aligned} p_1' = & -\vartheta_1 q_1 - S_{1R}(p_1^3 + p_1 q_1^2) - S_{1I}(p_1^2 q_1 + q_1^3) - S_{2R}(p_1 p_2^2 + p_1 q_2^2) \\ & - S_{2I}(q_1 p_2^2 + q_1 q_2^2) - g_{1R}(p_1^2 p_2 - p_2 q_1^2 + 2p_1 q_1 q_2) \\ & + g_{1I}(2p_1 q_1 p_2 - p_1^2 q_2 + q_1^2 q_2) - \mu C_{1R} p_1 - \mu C_{1I} q_1 - \alpha e_{1R} p_1 - \alpha e_{1I} q_1 \end{aligned} \tag{18}$$

$$\begin{aligned} q_1' = & \vartheta_1 p_1 + g_{1R}(2p_1 q_1 p_2 - p_1^2 q_2 + q_1^2 q_2) - g_{1I}(2p_1 q_1 q_2 + p_1^2 p_2 - p_2 q_1^2) \\ & + S_{1I}(p_1^3 + p_1 q_1^2) - S_{1R}(p_1^2 q_1 + q_1^3) - S_{2R}(q_1 p_2^2 + q_1 q_2^2) \\ & + S_{2I}(p_1 p_2^2 + p_1 q_2^2) - \mu C_{1R} q_1 + \mu C_{1I} p_1 - \alpha e_{1R} q_1 + \alpha e_{1I} p_1 \end{aligned} \tag{19}$$

$$\begin{aligned} p_2' = & -\vartheta_2 q_2 - S_{4R}(p_2^3 + p_2 q_2^2) - S_{4I}(q_2^3 + p_2^2 q_2) - S_{3R}(p_1^2 p_2 + p_2 q_1^2) \\ & - S_{3I}(p_1^2 q_2 + q_1^2 q_2) - g_{2R}(p_1^3 - 3p_1 q_1^2) + g_{2I}(q_1^3 - 3p_1^2 q_1) - \mu C_{2R} p_2 \\ & - \mu C_{2I} q_2 - \alpha e_{2R} p_2 - \alpha e_{2I} q_2 - K_{6R} p_2 + K_{6I} q_2 \end{aligned} \tag{20}$$

$$\begin{aligned} q_2' = & -\vartheta_2 p_2 - S_{4R}(q_2^3 + p_2^2 q_2) + S_{4I}(p_2^3 + p_2 q_2^2) - S_{3R}(p_1^2 q_2 + q_1^2 q_2) \\ & + S_{3I}(p_1^2 p_2 + p_2 q_1^2) + g_{2R}(q_1^3 - 3p_1^2 q_1) + g_{2I}(p_1^3 - 3p_1 q_1^2) - \mu C_{2R} q_2 \\ & + \mu C_{2I} p_2 - \alpha e_{2R} q_2 + \alpha e_{2I} p_2 + K_{6R} q_2 + K_{6I} p_2 \end{aligned} \tag{21}$$

$$\text{Where } \vartheta_1 = {}^1\!/_6(2\sigma_1 + \sigma_2) \text{ and } \vartheta_2 = 0.5\sigma_2 \tag{22}$$

Above equations are perturbed to evaluate the stability. The perturbed equation is

$$\{\Delta p_1' \; \Delta q_1' \; \Delta p_2' \; \Delta q_2'\}^T = [J_c]\{\Delta p_1 \; \Delta q_1 \; \Delta p_2 \; \Delta q_2\}^T \tag{23}$$

where T denotes transpose and $[J_c]$ is the Jacobian matrix whose eigenvalues determine the stability and bifurcation of the system.

$$a_1 = \left(p_1^2 + q_1^2\right)^{1/2} \quad \text{and} \quad a_2 = \left(p_2^2 + q_2^2\right)^{1/2} \tag{24}$$

where a_1 and a_2 are the amplitudes of first and second mode respectively.

5 Results and Discussion

The nonlinear steady state response of the system is obtained by solving the set of algebraic equations generated after setting $p_i' = q_i' = 0$ in the normalized reduced Eqs. (18–21). The stability and bifurcation of the equilibrium solutions are determined by using the continuation algorithm in which the eigenvalues of the Jacobian matrix at each point of the solution are evaluated. The frequency and amplitude response curves in the present analysis are symmetrical about σ_2 and v_1 axes respectively. For this reason, only positive sides of the response curves are shown. The normal continuous lines in the figure represent stable equilibrium solutions, the bold lines represent unstable foci, and the dotted lines denote saddles. Frequency response curves for first and second mode are shown in Fig. 2 for $\mu = 0.1$, $\alpha = 0$, $v_1 = 10$, $v_l = 40$ and $\sigma_1 = 92.39$, where the system response is plotted against frequency detuning parameter σ_2. The analysis in the form of frequency response plots depict the nonlinear phenomena viz. (i) bending of the back bone curve, (ii) rightward bending of the curves representing hardening type of nonlinearity, (iii) multi-branch solutions, (iv) jump phenomena and (v) a number of bifurcations.

In the case of principal parametric resonance of first mode and 3:1 internal resonance dealt in [14], the frequency response plots revealed two types of equilibrium solutions i.e. trivial and nontrivial solutions in both directly excited first mode and indirectly excited second mode. But in the present case, three distinct

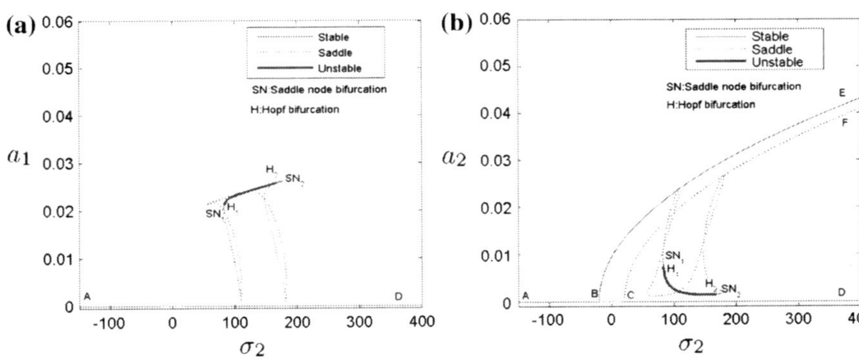

Fig. 2 Frequency response curves for the first and second modes when the second mode is parametrically excited for the system parameters $\mu = 0.1$, $\alpha = 0$, $v_1 = 10$, $v_l = 40$ and $\sigma_1 = 92.39$

equilibrium solutions i.e. trivial, single mode and two mode response curves are obtained. Unlike [14], the single mode solution is present in the directly excited second mode only while the two mode solution which is present in the directly excited second mode and indirectly excited first mode is isolated closed loop type and also limited between specific ranges of control parameter. The single mode solution has two branches, one is always stable (B-E) and shows hardening type nonlinearity while the other is unstable(C-F) at all time.

In addition to this, the stable (B-E) and unstable (C-F) branches show supercritical and subcritical pitchfork bifurcations from the trivial solution at parametric frequency detuning parameter $\sigma_2 = -20.069$ (at B) and $\sigma_2 = 20.069$ (at C) respectively. The isolated and closed loop two mode solutions for first mode (a_1) and second mode (a_2) which exist for specific ranges of parametric frequency detuning parameter have stable solutions between the limit point $SN_1(\sigma_2 = 82.727)$ and Hopf bifurcation point $H_1(\sigma_2 = 83.346)$ as well as between $SN_2(\sigma_2 = 177.839)$ and $H_2(\sigma_2 = 167.351)$.

The influence of increase of amplitude of travelling velocity fluctuating component is shown in Fig. 3 for $\mu = 0.1$, $\alpha = 0$, $v_1 = 50$, $v_l = 40$ and $\sigma_1 = 92.39$. The gap between the stable (B-E) and unstable (C-F) branches of single mode solution for second mode gets widened and the amplitude of response of this single second mode solution is increased. In addition, the isolated two mode solution also undergoes both qualitative and quantitative change; the isolated double loop solution transforms to isolated single loop solution in both the modes, the domain of these isolated two mode solutions shifts to higher values of frequency detuning parameter, and the amplitudes of response in both modes increase.

Figure 4 shows the influence of internal frequency detuning parameter (σ_1) on frequency response curves. When internal frequency detuning parameter (σ_1) decreases from 92.39 (Fig. 4a, b) to 44.00 (Fig. 4c and d), the amplitudes of both modes in the isolated closed loop two mode solutions decrease, but this decrease is

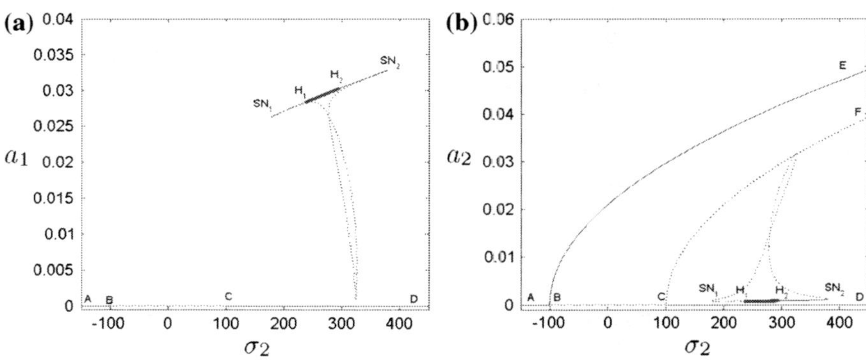

Fig. 3 Frequency response curves for the first and second mode when the second mode is parametrically excited for the system parameters $\mu = 0.1$, $\alpha = 0$, $v_1 = 50$, $v_l = 40$ and $\sigma_1 = 92.39$

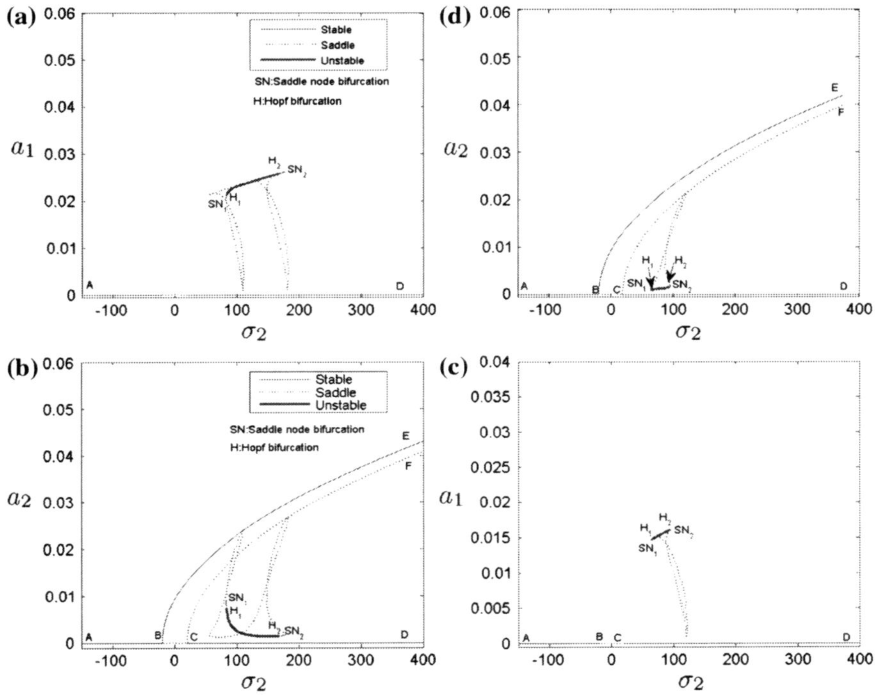

Fig. 4 Influence of internal detuning parameter on frequency response curves for first and second modes when second mode is parametrically excited for $\mu = 0.1$, $\alpha = 0$, $v_1 = 10$, $v_l = 40$ (a, b) $\sigma_1 = 92.39$ and $(c, d)\sigma_1 = 44.00$

more pronounced in case of the indirectly excited first mode. In addition to this, there is a qualitative change in the two mode solutions in the form of the double loop equilibrium branch transforming to single loop equilibrium branch and the frequency range of the closed loop solutions decreasing. In case of the single mode equilibrium solution branches occurring in second mode, the separation of non-trivial stable (B-E) and unstable (C-F) curves (a_2) decreases.

The influence of the type of damping on frequency response curves is compared in Fig. 5 for first mode and in Fig. 6 for second mode. When internal damping (α) in the system is considered in place of external damping, the size of closed loop two mode equilibrium solutions for both modes is substantially reduced which indicates weakening of nonlinear modal interaction.

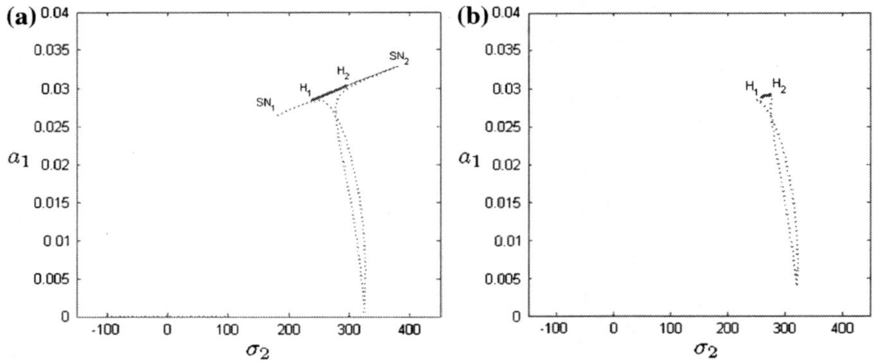

Fig. 5 Influence of damping on frequency response curves for first mode when second mode is parametrically excited for $\nu_1 = 50$, $\nu_l = 40$, $\sigma_1 = 92.39$ (a) $\mu = 0.1$, $\alpha = 0$ and (b) $\mu = 0$, $\alpha = 0.005$

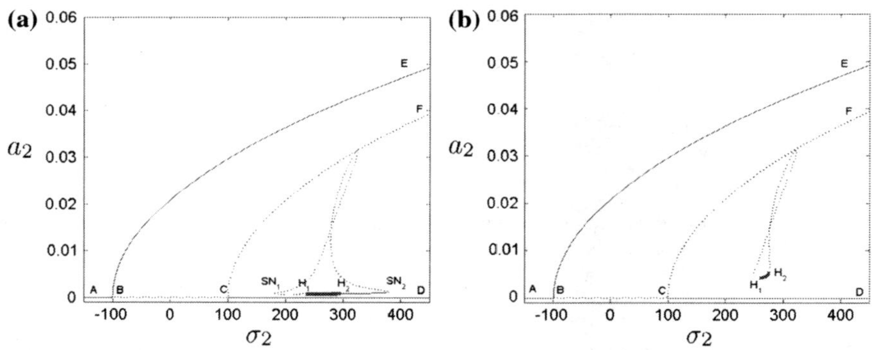

Fig. 6 Influence of damping on frequency response curves for second mode when second mode is parametrically excited for $\nu_1 = 50$, $\nu_l = 40$, $\sigma_1 = 92.39$ (a) $\mu = 0.1$, $\alpha = 0$ and (b) $\mu = 0$, $\alpha = 0.005$

6 Conclusion

The steady state response for principal parametric resonance of second mode comprises of three distinct solutions i.e. trivial, single mode and two mode solutions while that of first mode is having only trivial and two mode solutions [14]. The Single mode solution that exists only in directly excited second mode has two nontrivial branches. One is always stable and show hardening type nonlinearity while the other is unstable at all time. These two curves get separated from the trivial solutions through supercritical and subcritical pitchfork bifurcations respectively in frequency response plots and through only subcritical bifurcations for amplitude response plots. In addition to pitchfork bifurcations, the system experiences Hopf and saddle node bifurcations for variation of control parameters.

The isolated closed loop two mode solutions are because of the nonlinear modal interaction due to three to one internal resonance. They exist for specific ranges of control parameters. The strength of nonlinear modal interactions increases for higher values of harmonic component of travelling velocity and decreases in case of lower values of internal frequency detuning parameter and introduction of material damping.

Appendix

$$\begin{aligned}\Gamma_1 = & -2i\omega_1 A_1'\phi_1 - 2v_0 A_1'\phi_1' - 2i\mu\omega_1 A_1\phi_1 - 2i\alpha\omega_1 A_1\phi_1'''' + \\ & \frac{1}{2}v_l^2\left\{2A_1^2\bar{A}_1\phi_1''\int_0^1\phi_1'\bar{\phi}_1'\,dx + A_1^2\bar{A}_1\bar{\phi}_1''\int_0^1\phi_1'^2dx + 2A_1A_2\bar{A}_2\bar{\phi}_2''\int_0^1\phi_1'\phi_2'dx\right. \\ & \left. + 2A_1A_2\bar{A}_2\phi_1''\int_0^1\phi_2'\bar{\phi}_2'dx + 2A_1A_2\bar{A}_2\phi_2''\int_0^1\phi_1'\bar{\phi}_2'dx\right\}\end{aligned} \tag{A1}$$

$$\Gamma_2 = \frac{1}{2}v_l^2\left\{2\bar{A}_1^2A_2\bar{\phi}_1''\int_0^1\phi_2'\bar{\phi}_1'dx + \bar{A}_1^2A_2\phi_2''\int_0^1\bar{\phi}_1'^2\,dx\right\} \tag{A2}$$

$$\begin{aligned}\Gamma_5 = & -2i\omega_2A_2'\phi_2 - 2v_0A_2'\phi_2' - 2\mu i\omega_2A_2\phi_2 - 2\alpha i\omega_2A_2\phi_2'''' \\ & + \frac{1}{2}v_l^2\left\{A_2^2\bar{A}_2\bar{\phi}_2''\int_0^1\phi_2'^2\,dx + 2A_1\bar{A}_1A_2\phi_2''\int_0^1\phi_{11}'\bar{\phi}_1'dx + 2A_1\bar{A}_1A_2\bar{\phi}_1'\int_0^1\phi_1'\phi_2'dx\right. \\ & \left. + 2A_2^2\bar{A}_2\phi_2''\int_0^1\phi_2'\bar{\phi}_2'dx + 2A_1\bar{A}_1A_2\phi_1''\int_0^1\phi_2'\bar{\phi}_1'dx\right\}\end{aligned} \tag{A3}$$

$$\Gamma_6 = \frac{1}{2}v_l^2\left\{A_1^3\phi_1''\int_0^1\phi_1'^2dx\right\} \tag{A4}$$

$$\Gamma_8 = \bar{A}_2\left\{v_1\omega_2\bar{\phi}_2' - \frac{v_1\Omega}{2}\bar{\phi}_2' + i\,v_0v_1\bar{\phi}_2''\right\} \tag{A5}$$

$$S_1 = \frac{\frac{1}{16} v_l^2 \left\{ 2 \int_0^1 \phi_1'' \bar{\phi}_1 dx \int_0^1 \phi_1' \bar{\phi}_1' dx + \int_0^1 \bar{\phi}_1'' \bar{\phi}_1 dx \int_0^1 \phi_1'^2 dx \right\}}{-\left\{ i\omega_1 \int_0^1 \phi_1 \bar{\phi}_1 dx + v_0 \int_0^1 \phi_1' \bar{\phi}_1 dx \right\}} \tag{A6}$$

$$S_2 = \frac{\frac{1}{8} v_l^2 \left\{ \int_0^1 \bar{\phi}_2'' \bar{\phi}_1 dx \int_0^1 \phi_1' \phi_2' dx + \int_0^1 \phi_1'' \bar{\phi}_1 dx \int_0^1 \phi_2' \bar{\phi}_2' dx + \int_0^1 \phi_2'' \bar{\phi}_1 dx \int_0^1 \phi_1' \bar{\phi}_2' dx \right\}}{-\left\{ i\omega_1 \int_0^1 \phi_1 \bar{\phi}_1 dx + v_0 \int_0^1 \phi_1' \bar{\phi}_1 dx \right\}} \tag{A7}$$

$$S_3 = \frac{\frac{1}{8} v_l^2 \left\{ \int_0^1 \phi_2'' \bar{\phi}_2 dx \int_0^1 \phi_1' \bar{\phi}_1' dx + \int_0^1 \bar{\phi}_1'' \bar{\phi}_2 dx \int_0^1 \phi_1' \phi_2' dx + \int_0^1 \phi_1'' \bar{\phi}_2 dx \int_0^1 \phi_2' \bar{\phi}_2' dx \right\}}{-\left\{ i\omega_2 \int_0^1 \phi_2 \bar{\phi}_2 dx + v_0 \int_0^1 \phi_2' \bar{\phi}_2 dx \right\}} \tag{A8}$$

$$S_4 = \frac{\frac{1}{16} v_l^2 \left\{ 2 \int_0^1 \phi_2'' \bar{\phi}_2 dx \int_0^1 \phi_2' \bar{\phi}_2' dx + \int_0^1 \bar{\phi}_2'' \bar{\phi}_2 dx \int_0^1 \phi_2'^2 dx \right\}}{-\left\{ i\omega_2 \int_0^1 \phi_2 \bar{\phi}_2 dx + v_0 \int_0^1 \phi_2' \bar{\phi}_2 dx \right\}} \tag{A9}$$

$$C_1 = \frac{-i\omega_1 \int_0^1 \phi_1 \bar{\phi}_1 dx}{-\left\{ i\omega_1 \int_0^1 \phi_1 \bar{\phi}_1 dx + v_0 \int_0^1 \phi_1' \bar{\phi}_1 dx \right\}} \tag{A10}$$

$$C_2 = \frac{-i\omega_2 \int_0^1 \phi_2 \bar{\phi}_2 dx}{-\left\{ i\omega_2 \int_0^1 \phi_2 \bar{\phi}_2 dx + v_0 \int_0^1 \phi_2' \bar{\phi}_2 dx \right\}} \tag{A11}$$

$$e_1 = \frac{-i\omega_1 \int_0^1 \phi_1'''' \, \bar{\phi}_1 dx}{-\left\{ i\omega_1 \int_0^1 \phi_1 \bar{\phi}_1 dx + v_0 \int_0^1 \phi_1' \bar{\phi}_1 dx \right\}} \tag{A12}$$

$$e_2 = \frac{-i\omega_2 \int_0^1 \phi_2'''' \,\bar{\phi}_2 dx}{-\left\{ i\omega_2 \int_0^1 \phi_2 \bar{\phi}_2 dx + v_0 \int_0^1 \phi_2' \bar{\phi}_2 dx \right\}} \tag{A13}$$

$$g_1 = \frac{\frac{1}{16} v_l^2 \left\{ 2 \int_0^1 \bar{\phi}_1'' \bar{\phi}_1 dx \int_0^1 \phi_2' \bar{\phi}_1' dx + \int_0^1 \phi_2'' \bar{\phi}_1 dx \int_0^1 \bar{\phi}_1'^2 dx \right\}}{-\left\{ i\omega_1 \int_0^1 \phi_1 \bar{\phi}_1 dx + v_0 \int_0^1 \phi_1' \bar{\phi}_1 dx \right\}} \tag{A14}$$

$$g_2 = \frac{\frac{1}{16} v_l^2 \left\{ \int_0^1 \phi_1'' \bar{\phi}_2 dx \int_0^1 \phi_1'^2 dx \right\}}{-\left\{ i\omega_2 \int_0^1 \phi_2 \bar{\phi}_2 dx + v_0 \int_0^1 \phi_2' \bar{\phi}_2 dx \right\}} \tag{A15}$$

$$K_6 = \frac{\frac{1}{2} \left\{ v_1 \omega_2 \int_0^1 \bar{\phi}_2' \,\bar{\phi}_2 dx - \frac{v_1 \Omega}{2} \int_0^1 \bar{\phi}_2' \bar{\phi}_2 \, dx + i\, v_0 v_1 \int_0^1 \bar{\phi}_2'' \bar{\phi}_2 \, dx \right\}}{-\left\{ i\omega_2 \int_0^1 \phi_2 \bar{\phi}_2 dx + v_0 \int_0^1 \phi_2' \bar{\phi}_2 dx \right\}} \tag{A16}$$

References

1. Wickert JA, Mote CD Jr (1988) Current research on the vibration and stability of axially moving materials. Shock Vib Digest 20:3–13
2. Wickert JA, Mote CD Jr (1990) Classical vibration analysis of axially moving continua. J Appl Mech 57:738–744
3. Oz HR, Pakdemirli M (1999) Vibrations of an axially moving beam with time-dependent velocity. J Sound Vib 227(2):239–257
4. Oz HR, Pakdemirli M, Boyaci H (2001) Non-linear vibrations of an axially moving beam with time-dependent velocity. Int J Non-linear Mech 36, 227(2):107–115
5. Ozkaya E, Bagdatli SM, Oz HR (2008) Nonlinear transverse vibrations and 3:1 internal resonances of a beam with multiple supports. J Vib Acoust 130(2):1–11
6. Chin CM, Nayfeh AH (1999) Three-to-one internal resonance in parametrically excited Hinged-clamped beams. Nonlinear Dyn 20:131–158
7. Panda LN, Kar RC (2007) Nonlinear dynamics of a pipe conveying pulsating fluid with parametric and internal resonances. Nonlinear Dyn 49:9–30
8. Panda LN, Kar RC (2008) Nonlinear dynamics of a pipe conveying pulsating fluid with combination, principal parametric and internal resonances. J Sound Vib 309:375–406
9. Pakdemirli M, Oz HR (2008) Infinite mode analysis and truncation to resonant modes of axially accelerated beam vibrations. J Sound Vib 311:1052–1074

10. Ponomareva SV, van Horssen WT (2009) On the transversal vibration of an axially moving continuum with a time-varying velocity: transient from string to beam behavior. J Sound Vib 325:959–973
11. Ghayesh MH (2011) Nonlinear forced dynamics of an axially moving viscoelastic beam with an internal resonance. Int J Mech Sci 53:1022–1037
12. Chakraborty G, Mallick AK (1999) Stability of an accelerating beam. J Sound Vib 27 (2):309–320
13. Paidoussis MP (1975) Flutter of conservative systems of pipe conveying incompressible fluid. J Mech Eng Sci 17(1)
14. Sahoo B, Panda LN, Pohit G (2013) Parametric and internal resonances of an axially moving beam with time dependent velocity, modelling and simulation in engineering, volume 2013, Article ID 919517, p 18

Analysis of Indirect Measurement of Cutting Forces Turning Metal Cylindrical Shells

Kirill Kondratenko, Alexandre Gouskov, Mikhail Guskov, Philippe Lorong and Grigory Panovko

Abstract Cutting forces measurement is an important component of the machining processes development and control. The use of conventional direct measurement systems is often impossible as they interfere in the process's dynamics. This work proposes a method of cutting force indirect estimation during turning thin-walled cylindrical shells. Calculation of the flexibility matrix has enabled us to relate measured displacements of certain workpiece's points to the cutting force. An optimization approach for choosing the measurement points location has been proposed, based on the best conditioning of the flexibility matrix.

Keywords Cutting forces measurement · Thin-walled workpiece · Technological system · Optimization · Ill-conditioned systems

1 Introduction

In modern manufacturing, cutting forces measurement is a key element in understanding the operational conditions during machining workpieces. Usually, cutting forces are measured directly using different kinds of dynamometers. For instance, in

K. Kondratenko (✉) · A. Gouskov
Department RC5, Bauman Moscow State Technical University, Moscow, Russia
e-mail: kk92@ya.ru

A. Gouskov
National Research Centre "Kurtchatovsky Institute", Moscow, Russia
e-mail: gouskov_am@mail.ru

M. Guskov · P. Lorong
ENSAM-ParisTech, Paris, France
e-mail: mikhail.guskov@ensam.eu

P. Lorong
e-mail: philippe.lorong@ensam.eu

G. Panovko
IMASH RSA, Moscow, Russia
e-mail: gpanovko@yandex.ru

J.K. Sinha (ed.), *Vibration Engineering and Technology of Machinery*,
Mechanisms and Machine Science 23, DOI 10.1007/978-3-319-09918-7_82

[1], an experimental set-up is described where cutting forces are measured directly, with a Kistler 9257B three-component piezo-dynamometer. The direct measurement approach is also incorporated in [2]. Another approach, based on the use of currents drawn by a.c. feed-drive servo motors, is presented in [3], where the pulsating milling forces are measured indirectly within the bandwidth of the current feedback control loop of the feed-drive system.

Nevertheless, in some cases, the use of dynamometers can be problematic, for instance in case of thin-walled workpieces in presence of instabilities: due to the presence of resonances in the frequency response of the dynamometer can induce significant perturbations in the measured signals. This was the case for [4]: the addition of the dynamic system of the dynamometer has modified the conditions of the chatter onset.

In the present paper, we address the problem of quasi-static evaluation of the cutting force during turning cylindrical shells. The cutting force components are estimated indirectly, from the displacement measurements, with the help of the flexibility matrix, based on the elastic behavior of the structure.

Two arbitrary points belonging to the free end of the shell are chosen: point *A* and *B*. The points represent the location of two displacement sensors. The angular location of these points is determined by parameters Φ_1 and Φ_2, which is shown in Fig. 1. The radial displacements of these points are actually measured using non-contact displacement sensors. The relation between the two components F_C, F_R of the cutting force and the radial displacements w_A and w_B of points *A* and *B* at first approximation is linear and injective:

$$\mathbf{Af} = \mathbf{e} \tag{1}$$

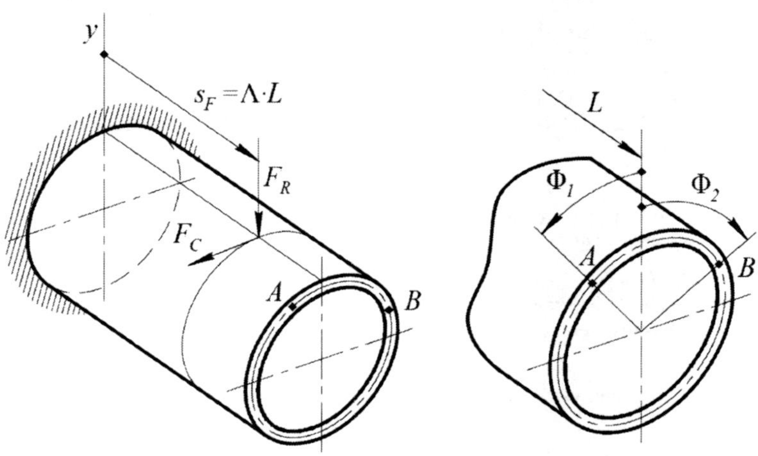

Fig. 1 The components of the cutting force (on the *left*). The free end of the shell (on the *right*) with angular location of the displacement sensors shown

where

$$\mathbf{A} = \begin{bmatrix} a_{11} & a_{12} \\ a_{21} & a_{22} \end{bmatrix}, \quad \mathbf{f} = \begin{bmatrix} F_C \\ F_R \end{bmatrix}, \quad \mathbf{e} = \begin{bmatrix} w_A \\ w_B \end{bmatrix} \tag{2}$$

Matrix **A** is called the *flexibility matrix*. Parameter Λ determines the axial position of the cutting force. In order for the solution $\mathbf{f} = \mathbf{A}^{-1}\mathbf{e}$ to be reliable, system (1) has to be *numerically stable*. The system stays numerically stable as long as its condition number is sufficiently close to 1. The question of the optimality of the sensors position is sought in terms of the condition number of the flexibility matrix.

Section 2 addresses the algorithm of calculation of the flexibility matrix, and Sect. 3 describes a particular way of avoiding numerical instabilities calculating this matrix.

2 Calculation of the Flexibility Matrix

In this section, we develop a quasi-static analysis of the force-displacement relation applied to the case of turning a thin-walled cylindrical shell. The cutting force is taken as concentrated.

The shell is shown in Fig. 2. The left-hand edge of the shell is rigidly fixed, and the right-hand edge is free. In other words, the shell is supported like a cantilever with "clamped-free" boundary conditions. Thickness of the shell is regarded as being much smaller than its diameter: $h \ll D = 2R$. We will be using cylindrical coordinates: axial s and angular φ. An arbitrary point M belonging to the middle surface of the shell is said to have coordinates (s, φ), as shown in Fig. 2.

The shell is subjected to pin-load, which is represented by the two components of the cutting force: radial F_R and circumferential F_C, as shown in Fig. 1. We neglect the axial component of the cutting force because it is always much smaller than the other two components [4], and because the stiffness in the z-direction is much higher than in the other two directions. Forces F_C and F_R act on the point with coordinates $(s_F, 0)$. We introduce Λ as a changeable dimensionless parameter so that $s_F = \Lambda \cdot L$, see (Fig. 1). We can now write down the expression for the shell's thickness:

$$h(s) = \begin{cases} h_{\text{fin}} & \text{if } s < s_F \\ h_{\text{ini}} & \text{if } s \geq s_F \end{cases} \tag{3}$$

where h_{ini} is the shell's thickness before cutting and h_{fin} is the shell's thickness after cutting.

According to [5], the general system of equations for a Kirchhoff-Love thin-walled cylindrical shell shown in Fig. 2 may be written [6] in matrix form

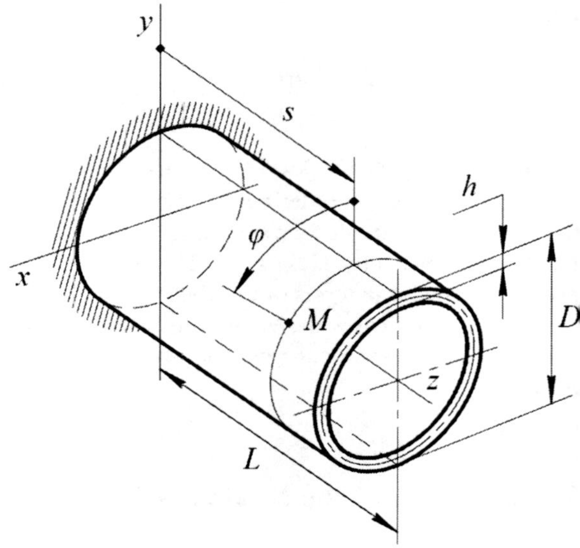

Fig. 2 The shell's dimensions and system of coordinates

$$\mathbf{L}\mathbf{y} = \mathbf{g} \tag{4}$$

where **L** is a linear partial differential operator represented by (8 × 8) matrix and

$$\mathbf{y} = \{\, u, \quad v, \quad w, \quad \vartheta_1, \quad RT_1, \quad RS_1^*, \quad RQ_1^*, \quad RM_1 \,\} \tag{5}$$

is the state vector and

$$\mathbf{g} = -R \cdot \{\, 0, \quad 0, \quad 0, \quad 0, \quad q_1, \quad q_2, \quad q_3, \quad 0 \,\}. \tag{6}$$

is the load vector. Here R is the shell's radius, u is the axial direction displacement, v is the circumferential direction displacement, w is the radial direction displacement, ϑ_1 is the surface normal's angular displacement, T_1, S_1^*, Q_1^*, M_1 are the internal forces, and q_1, q_2, q_3 represent distributed external loading. Equation (4) represents a linear system of partial differential equations. It is possible to separate s and φ with the aid of the Fourier method using complex Fourier series:

$$\mathbf{y} = \sum_{k=-\infty}^{+\infty} \mathbf{y}_{(k)} \exp(ik\varphi). \tag{7}$$

After separation of variables, system (4) is decomposed into the infinite amount of linear 8th order systems of ordinary differential equations (ODE). Each of these ODE systems can be written in matrix notation as follows

$$\frac{d}{ds}\mathbf{y}_{(k)} = \mathbf{F}_{(k)}\mathbf{y}_{(k)} + \mathbf{g}_{(k)} \tag{8}$$

where $\mathbf{F}_{(k)}$ is (8×8) constant square matrix:

$$\mathbf{F}_{(k)} = \begin{bmatrix} 0 & -\frac{ivk}{R} & -\frac{v}{R} & 0 & \frac{1-v^2}{EhR} & 0 & 0 & 0 \\ -\frac{ik}{R} & 0 & 0 & 0 & 0 & \frac{2(1+v)}{EhR} & 0 & 0 \\ 0 & 0 & 0 & -1 & 0 & 0 & 0 & 0 \\ 0 & -\frac{ivk}{R^2} & -\frac{vk^2}{R^2} & 0 & 0 & 0 & 0 & \frac{12\left(1-v^2\right)}{Eh^3R} \\ \frac{A}{R^2} & 0 & 0 & -\frac{A}{R} & 0 & -\frac{ik}{R} & 0 & 0 \\ 0 & \frac{Ehk^2}{R} & -B & 0 & -\frac{ivk}{R} & 0 & 0 & -\frac{ivk}{R^2} \\ 0 & B & C & 0 & \frac{v}{R} & 0 & 0 & \frac{vk^2}{R^2} \\ -\frac{A}{R} & 0 & 0 & A & 0 & 0 & 1 & 0 \end{bmatrix} \tag{9}$$

where E is the Young modulus, v is the Poisson's ratio, h is the shell's wall-thickness and

$$A = \frac{Eh^3k^2}{6R(1+v)}, \quad B = \frac{iEhk}{R}\left(1 + \frac{h^2k^2}{12R^2}\right), \quad C = \frac{Eh}{R}\left(1 + \frac{h^2k^4}{12R^2}\right). \tag{10}$$

Keeping in mind that $\dim \delta(x) = 1/\dim(x)$, it is quite obvious that

$$\{q_1, q_2, q_3\} = \{0, \quad F_C, \quad -F_R\}\frac{\delta(\varphi)\delta(s - s_F)}{R}. \tag{11}$$

Using the Fourier series of the Dirac delta function

$$\delta(\varphi) = (2\pi)^{-1}\sum\nolimits_{k=-\infty}^{+\infty} \exp(ik\varphi), \tag{12}$$

the load vector can be rewritten as

$$\mathbf{g}_{(k)} = \frac{1}{2\pi}\{0, \quad 0, \quad 0, \quad 0, \quad 0, \quad -F_C, \quad F_R, \quad 0\}\delta(s - s_F). \tag{13}$$

The left end of the shell $(s = 0)$ is rigidly fixed, and the right end $(s = L)$ is free. The followings are the boundary conditions:

$$\begin{aligned} &\{u, v, w, \vartheta_1\}_{(k)} = 0 && \text{at } s = 0 \\ &\left\{RT_1, RS_1^*, RQ_1^*, RM_1\right\}_{(k)} = 0 && \text{at } s = L \end{aligned} \tag{14}$$

Here are the continuity conditions at point $(s = s_F)$:

$$\mathbf{y}_{(k)}(s_F + \varepsilon) = \mathbf{y}_{(k)}(s_F - \varepsilon) + \frac{1}{2\pi}\{0,\ 0,\ 0,\ 0,\ 0, -F_C,\ F_R,\ 0\} \tag{15}$$

where ε is an infinitesimally small parameter. System (8) along with boundary conditions (14) represent a boundary value problem. We have used the initial parameters method [5] in order to solve this problem by means of numerical integration. The Godunov orthogonalization method [7] was incorporated to ensure numerical stability of the solution. Moreover, the method was further modified in order to eliminate the well-known Gram-Schmidt process's weakness [8]. The Gram-Schmidt process was replaced by the Householder transformation [9], which effectively performs the same thing—orthonormalizes a set of vectors in the Euclidean space $\mathbb{R}^n$. Harmonic $w_{(k)}(L)$ that corresponds to the radial displacements of points located on the free end of the shell, can be represented as a linear combination of the cutting force components:

$$w_{(k)}(L) = F_C \cdot \alpha_{(k)} + F_R \cdot \beta_{(k)} \tag{16}$$

where coefficients $\alpha_{(k)}$ and $\beta_{(k)}$ depend only on parameter Λ and have been obtained after numerical integration of the boundary value problem for different harmonics. According to expression (7),

$$w_{(k)}(L, \varphi) = \sum_{k=-\infty}^{+\infty} \left(F_C\alpha_{(k)} + F_R\beta_{(k)}\right)\exp(ik\varphi) \tag{17}$$

According to the definition (see Fig. 1), we can write that $w_A = w(L, \Phi_1)$ and $w_B = w(L, \Phi_2)$. Finally, according to formulas (1) and (2), matrix **A** can be represented as an infinite series

$$\mathbf{A} = \sum\nolimits_{k=-\infty}^{+\infty} \begin{bmatrix} \alpha_{(k)}\exp(ik\Phi_1) & \beta_{(k)}\exp(ik\Phi_1) \\ \alpha_{(k)}\exp(-ik\Phi_2) & \beta_{(k)}\exp(-ik\Phi_2) \end{bmatrix}. \tag{18}$$

It can be shown that the components of matrix **A** are always real numbers, which they must be, of course, since **A** is the flexibility matrix. The criteria of meeting the required accuracy has been

$$\|\mathbf{A}_{N+5} - \mathbf{A}_N\| \cdot \left\|\mathbf{A}_N^{-1}\right\| \le \varepsilon. \tag{19}$$

If the above inequality is satisfied, then we consider approximation $\mathbf{A}_N$ to be accurate enough. In our work, we have set the relative accuracy to $\varepsilon = 0.01 = 1\,\%$.

The magnitude of the cutting force is approximately 10^2–10^3 N, according to [4]. Using this data and our model, we have calculated that magnitudes of the displacements are within 10^{-5} m.

The components of matrix **A** are dependent on the three variable parameters that we have introduced earlier: $\mathbf{A} = \mathbf{A}(\Phi_1, \Phi_2, \Lambda)$.

3 Optimization of the Displacement Sensors Location

We need to determine the best set of parameters Φ_1 and Φ_2 (the displacement sensors angular location) that would make the system (1) as well-conditioned as possible. Measure of a square matrix's numerical stability is called its condition number μ. By definition [8],

$$\mu(\mathbf{A}) = \|\mathbf{A}\| \cdot \left\|\mathbf{A}^{-1}\right\|. \tag{20}$$

Condition number μ is always positive and cannot be less than one. The closer the condition number μ of the flexibility matrix **A** is to one, the better conditioned this matrix is. Therefore, to ensure the best numerical stability of the linear transformation (1) we must minimize the condition number μ of matrix **A**. Let us construct the target function

$$f(\Phi_1, \Phi_2) = \max_{\Lambda}[\mu(\mathbf{A}(\Phi_1, \Phi_2, \Lambda))]. \tag{21}$$

To accomplish our optimization goal we have had to minimize *f*:

$$f \to \min, \quad \left\{\Phi_1^{\text{opt}}, \Phi_2^{\text{opt}}\right\} = \arg\left[\min_{\Phi_1\in[0,2\pi] hfill \Phi_2\in[0,\pi]} f(\Phi_1, \Phi_2)\right]. \tag{22}$$

We have implemented the brute force approach minimizing function *f*. Angular increment has been set to 0.1°, and the increment for parameter Λ has been set to 0.05. Color plot of function $\mu^{-1}(\mathbf{A}(\Phi_1, \Phi_2, 1))$ is shown in Fig. 4. We have chosen to analyze function inverse to the condition number because this function is normalized: its range of values lies inside interval (0,1). The diagram reveals 20 *local maxima* of *f* (see Fig. 4 and Table 1). The best choice of parameters (i.e. the *global maxima*) turned out to be $\Phi_1^{\text{opt}} = \Phi_2^{\text{opt}} = 20.4°$, which corresponds to the target function value of 1.2. This displacement sensors' configuration is shown in Fig. 3.

Strictly speaking, this value is the global minimum of function $f(\Phi_1, \Phi_2)$. However, in practical terms, any of the displacement sensors configurations from Table 1 can be chosen because magnitude of *f* for any of those configurations does not even exceed the value of 10. It means that at the worst case scenario we could lose 1–2 significant digits [8] calculating the components of the cutting force using Eq. (1), which corresponds to relative accuracy $\varepsilon = 10^{-10}$. Such loss is not significant in comparison with other sources of error in our model.

Index 'S' in Table 1 represents *y*-symmetrical configurations, and indexes '1' and '2' represent identical configurations. Table 1 also features minimum value of norm of matrix **A**, which represents magnitude of displacements of measured points. Note that

Table 1 Optimal displacement sensors angular location configurations

#	Φ_1	Φ_2	f	$\|\mathbf{A}\| \cdot 10^6$	#	Φ_1	Φ_2	f	$\|\mathbf{A}\| \cdot 10^6$
AS	20.4°	20.4°	1.2	8.2	I1	208.3°	137.5°	1.8	0.59
BS	30.6°	30.6°	1.5	7.2	I2	222.5°	151.7°	1.8	0.59
C1	66.6°	24.3°	1.5	9.7	J1	241.1°	83.5°	2.0	2.1
C2	24.3°	66.6°	1.5	9.7	J2	276.5°	119.9°	2.0	2.1
DS	77.7°	77.7°	2.8	3.8	K1	268.5°	78.6°	2.0	2.7
ES	89.3°	89.3°	1.9	1.6	K2	281.5°	91.5°	2.0	2.7
F1	119.4°	83.5°	1.4	2.2	L1	293.5°	25.3°	1.4	7.1
F2	83.5°	119.4°	1.4	2.2	L2	334.7°	66.5°	1.4	7.1
GS	136.2°	136.2°	1.4	0.87	M1	329.3°	20.5°	1.2	7.6
HS	149.5°	149.5°	1.7	0.37	M2	339.5°	30.7°	1.2	7.6

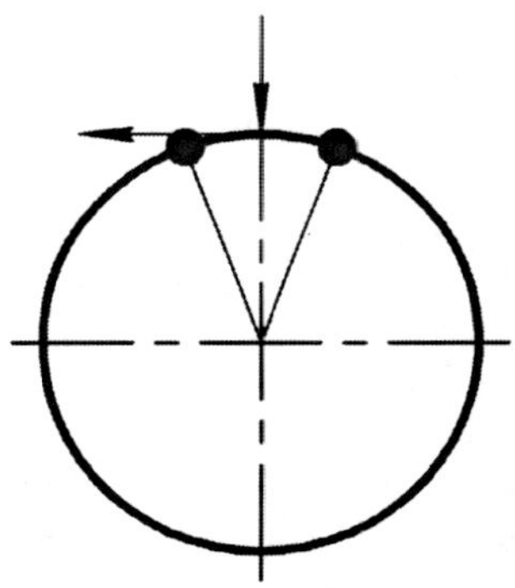

Fig. 3 Optimal displacement sensors location

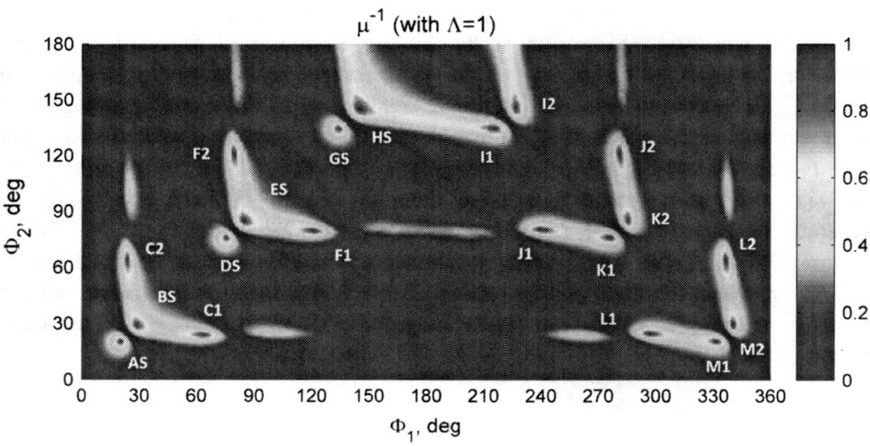

Fig. 4 Optimal displacement sensors' angular location configurations

$$\dim \mathbf{A} = \dim\|\mathbf{A}\| = \text{mm/N}. \tag{23}$$

We can see that configurations GS, HS, I1 and I2 are probably not preferable because the displacements would be significantly smaller than in the other configurations. From this point of view, configurations AS, BS, C1, C2, L1, L2, M1, M2 are optimal, and each of them can be readily picked by an experimenter.

4 Conclusion

In this work we have developed a mathematical model in order to be able to calculate the flexibility matrix that makes it possible to estimate the cutting force components based on displacement measurement in turning cylindrical shells. Analysis of the behavior of the flexibility matrix condition number has been performed. Based on this analysis, optimal configurations of the displacement sensors location have been suggested. These configurations make the flexibility matrix well-conditioned and the process of calculating the components of the cutting force numerically stable. Any of these configurations can be picked as they do not cause numerical instabilities when calculating the flexibility matrix.

References

1. Fang N, Wu Q (2009) A comparative study of the cutting forces in high speed machining of Ti-6Al-4 V and Inconel 718 with a round cutting edge tool. J Mater Process Technol 209 (9):4385–4389
2. Sun S, Brandt M, Dargusch MS (2009) Characteristics of cutting forces and chip formation in machining of titanium alloys. Int J Mach Tools Manuf 49(7–8):561–568
3. Kim T-Y, Kim J (1996) Adaptive cutting force control for a machining center by using indirect cutting force measurements. Int J Mach Tools Manuf 36(8):925–937
4. Lorong P, Larue A, Duarte AP (2011) Dynamic study of thin wall part turning. Adv Mater Res 223:591–599 doi:10.4028/www.scientific.net/AMR.223.591
5. Biderman VL (1997) Mekhanika tonkostennykh konstruktsiy. Moscow
6. Kondratenko KE, Gouskov AM, Guskov MA, Lorong. P, Panovko GY (2014) Analysis of indirect measurement of cutting forces in turning metal cylindrical shells. Vestnik MGTU imeni Baumana. doi:10.7463/0214.0687971
7. Godunov SK (1961) On the numerical solution of boundary value problems for systems of linear ordinary differential equations. Uspehi Mat Nauk 16:171–174
8. Higham NJ (2002) Accuracy and stability of numerical algorithms, 2nd edn. SIAM, Philadelphia
9. Householder AS (1958) Unitary triangularization of a nonsymmetric matrix. J ACM 5 (4):339–342. doi:10.1145/320941.320947
10. Gerasimenko AA, Gouskov AM, Guskov MA, Lorong P (2012) Spinning shell eigen modes calculation. Vestnik MGTU imeni Baumana

Hybrid Permanent Magnet and Foil Bearing System for Complete Passive Levitation of Rotor

Siddappa Iranna Bekinal, Tumkur Ramakrishna rao Anil, Sadanand Subhas Kulkarni and Soumendu Jana

Abstract This paper deals with the complete passive levitation for a typical Jeffcott rotor and rotation of the same at the speeds around 40,000 rpm. The passive levitation is achieved by supporting the rotor axially by a permanent magnet bearing and discrete bump foil bearings for the radial support. The permanent magnet bearing is made up of three pairs of ring magnets arranged in Halbach pattern. Bump foil bearings are designed for rotor weight to provide the radial support to the rotor system. The proposed rotor-bearing configuration is analysed using Finite Element Analysis (FEA) software (ANSYS) for rotor dynamic characteristics. The designed rotor bearing system is fabricated and tested up to the speeds of 40,000 rpm. The system response is acquired using advanced rotor-dynamic data acquisition system. The experimental results show that the rotor is completely airborne and stable at the desired speed.

Keywords Foil bearings · Permanent magnet bearings · Halbach structure · Passive levitation and rotor dynamics

S.I. Bekinal · T.R. rao Anil
Department of Mechanical Engineering, Gogte Institute of Technology, Belgaum 590008, Karnataka, India
e-mail: sibekinal@git.edu

T.R. rao Anil
e-mail: tranil@git.edu

S.S. Kulkarni · S. Jana (✉)
Propulsion Division, National Aerospace Laboratories, Bangalore 560017, Karnataka, India
e-mail: sjana@nal.res.in

S.S. Kulkarni
e-mail: sadanandsk@nal.res.in

J.K. Sinha (ed.), *Vibration Engineering and Technology of Machinery*,
Mechanisms and Machine Science 23, DOI 10.1007/978-3-319-09918-7_83

1 Introduction

High-speed applications such as energy storage flywheels, turbo molecular pumps, air cycle machines etc. require support system, which can withstand at rated speed with lesser complications and power losses. Ceramic bearings [1, 2] are one of the most sought options for high speed applications due to their ability to withstand higher loads at considerably wide operating conditions. Their application seizes at extreme temperatures and dusty atmosphere. Hence, there is a need to develop a support system, which can work in hostile environment with little or no lubrication. Permanent magnet bearings and foil bearings are emerging as best suited candidates for high-speed applications. Siebert et al. [2] have presented a passive magnetic bearing flywheel. In this, the rotor was supported by two sets of radial permanent magnet bearings operating in repulsive mode and jewel bearings on both ends of the rotor for the axial support. A prototype model of a repulsive-type magnetic bearing system using a novel arrangement of permanent magnets was fabricated by Mukhopadhaya et al. [3] for the application in the dairy industry. The magnetic bearing system is stable along the radial axis but is unstable along the vertical direction. A controlled electromagnet is used for controlling the rotor position along the vertical axis. A new permanent magnet repulsive type micro mass measurement system was developed by Hussien et al. [4]. The repulsive forces of the permanent magnet section are used effectively to keep the radial direction stable. A controlled electromagnet was used to stabilize the shaft in the axial direction. Sotelo et al. [5] have developed magnetic bearing sets for flywheel system. The bearing set consists of radial permanent magnet bearing and thrust superconducting magnetic bearing. In [6], Radial type superconducting magnetic bearings have been developed for a 10 kWh-class flywheel energy storage system by Ichihara et al. In the existing literature, jewel bearings, active or super conducting magnetic bearings are used in addition to permanent magnet bearings for complete levitation of the rotor. Use of jewel bearings introduces friction and limits the operating speed. The system will be complicated with electronic equipments and refrigeration system by the use of active and super conducting magnetic bearings. An alternate option has been proposed in the present paper. A new passive hybrid bearing set, composed of permanent magnet thrust bearing and radial foil bearings, is used to support the rotor. The rotor dynamic analysis of a passively levitated rotor configuration is performed using ANSYS. A test rig consisting of a typical rotor, permanent magnet thrust bearing, radial foil bearings and air turbines is designed and developed. The rotor is rotated up to 40,000 rpm with the help of air turbines and rotor dynamic characteristics are recorded using the ADRE (Automated Diagnostics of Rotor dynamic Equipments) data acquisition system.

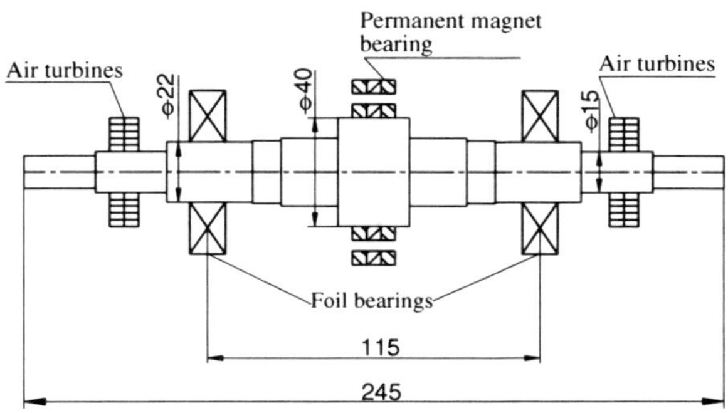

Fig. 1 The rotor-bearing configuration (all dimensions are in mm)

2 The Rotor Bearing Configuration

The main objective of the present work is to demonstrate the complete passive levitation of a typical rotor configuration using permanent magnet bearing and foil bearings. Schematic of the designed rotor bearing configuration is shown in Fig. 1. The permanent magnet thrust bearing made of three pairs of ring magnets arranged in Halbach pattern is provided at the centre to support the shaft axially and discrete bump foil bearings are provided at the ends to support the shaft radially.

2.1 The Permanent Magnet Thrust Bearing

Permanent magnet bearings are magneto-mechanical elements wherein the supporting property of bearing is realized by virtue of attractive or repulsive forces generated between the magnets. These are realized with axially, radially and perpendicularly magnetized ring magnets. The suitable characteristics (no lubrication, no maintenance, no power and no input energy etc.) of permanent magnet bearings made of ring magnets are utilized in many applications such as energy storage flywheels, turbo molecular pumps, hard disk drive spindle motors, moment gyro applications and conveyor systems [7–9]. The configuration of permanent magnet bearing considered in the present work is shown in Fig. 2. It consists of three ring pairs arranged in Halbach pattern and material of magnet rings is N35 (Nd-Fe-B) with the following properties: $B_r = 1.2$ T, $H_c = 868$ kA/m and $\mu_r = 1.1$. Simplified three-dimensional (3D) mathematical models to determine the force and stiffness in axial, radial and Halbach magnetized permanent magnet bearing configurations presented by Bekinal et al. [10–14] are used to design the selected configuration. The results of an axial force and axial stiffness as a function of axial displacement of the inner rings obtained using the mathematical models are shown in Fig. 3.

Results of the Fig. 3 show that the maximum axial force exerted by the outer rings on the inner rings is 271 N at 4.5 mm axial offset of the rotor and maximum axial stiffness of the bearing is 93,800 N/m at 0 mm axial offset. According to

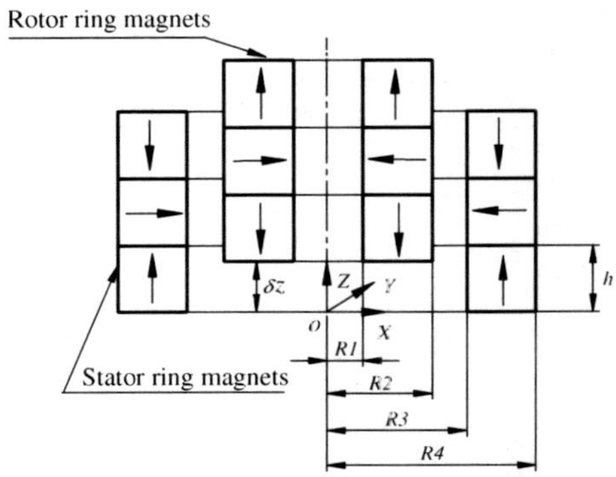

Fig. 2 Configuration of a permanent magnet bearing: $R1 = 0.02$ m, $R2 = 0.025$ m, $R3 = 0.029$ m, $R4 = 0.034$ m and $J = 1.2$ T

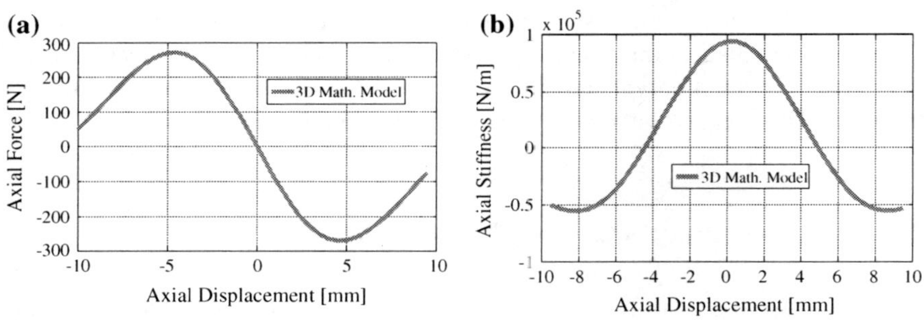

Fig. 3 Force and stiffness results of the permanent magnet bearing. **a** Axial force. **b** Axial stiffness

Earnshaw's theorem [15], the summation of bearing stiffnesses in Cartesian coordinate system must be equal to zero.

$$K_X + K_Y + K_Z = 0 \tag{1}$$

For circular rings, due to symmetry, $K_X = K_Y = K_r$, Eq. (1) leads to

$$K_Z = -2K_r \tag{2}$$

The axial permanent magnet bearing designed to support an axial load, will be unstable in the radial direction and vice versa. The negative radial stiffness of the configuration considered in the present work is 46,900 N/m.

2.2 Bump Foil Bearings

Foil-gas bearings are a class of hydrodynamic bearings that use the ambient gas as their working fluid and thus require no dedicated lubrication system [16]. Airfoil bearings consist of a compliantly mounted smooth inner top foil, which creates a wedge film between itself and the shaft during relative rotation, resulting in load carrying capacity. The top foil supported by compliantly mounted bumps foils, which deform elastically under the pressure field created by the aerodynamic film. The effective stiffness of the bearing is the summation of bump stiffness and air film stiffness acting in series. Foil bearing technology has been applied successfully to wide range of high-speed rotating machineries such as air cycle machines, auxiliary power units, and cryogenic turbo compressors [17–22].

The effective stiffness K_e of foil bearings is given by,

$$K_e = \frac{1}{\frac{1}{K_f} + \frac{1}{K_a}} \tag{3}$$

Where, K_f is the stiffness of bump foils and can be derived from solid mechanics relations as [20],

$$K_f = \frac{6G}{r_b^3} \tag{4}$$

Where, G is the flexure stiffness of one corrugation per unit length and calculated as,

$$G = \frac{Et_b^3}{12(1 - \upsilon^2)} \tag{5}$$

Where, E is the Young's Modulus and υ is Poisson's ratio. K_a is the stiffness of air film developed, which varies from 1 MN/m to 1,000 MN/m over the wide speed range. The maximum effective stiffness is obtained by assuming possible range of air film stiffness and calculated bump foil stiffness using Eq. (4).

The load carrying capacity *(W)* for the given geometry of the foil bearing and speed is obtained by using a thumb rule derived experimentally by Della Corte et al. [19] as,

$$W = D(Ld)(dkrpm) \tag{6}$$

Where, D is bearing load capacity coefficient (N/m^3 x *krpm*) which is determined experimentally for all three generation of foil bearings. From the consideration of load carrying capacity, bump stiffness and system dynamics, the configuration finalized for experimentation and performance analysis is given in Table 1. The calculated value of maximum effective stiffness of the designed bump foil bearings is 10^6 N/m and the load carrying capacity is 25 N at 50,000 rpm. The radial bump foil bearings developed for the experimental study are shown in Fig. 4.

Table 1 Dimensions of designed foil bearings

Shaft diameter	22 mm
Number of bump foils	6
Nominal clearance	0.050 mm
Bearing length	12 mm

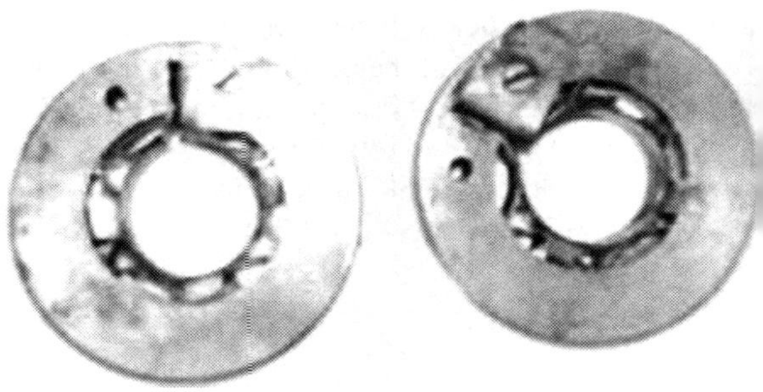

Fig. 4 Photograph of fabricated bump foil bearings

3 Dynamic Analysis of Rotor Bearing Configuration Using ANSYS

The rotor-bearing configuration as shown in Fig. 1 is modeled and critical speeds are determined using FEA software (ANSYS). The rotor is modeled with brick 8-noded solid 45 elements, the radial foil bearings are simulated using Combin 14 elements and permanent magnet bearings using Combin 214 element. The material properties used for the analysis are given in Table 2 and the meshed model of the rotor is shown in Fig. 5. The natural frequencies are presented in Table 2.

The results of finite element analysis show that the first natural frequency corresponds to axial mode, whereas second and third frequencies are conical cylindrical rigid body modes. The fourth natural frequency represents the flexural critical speed of the shaft.

Table 2 The rotor natural frequencies

Mode No.	Natural frequency (Hz)
1	50.4
2	226.36
3	233.4
4	2112.61

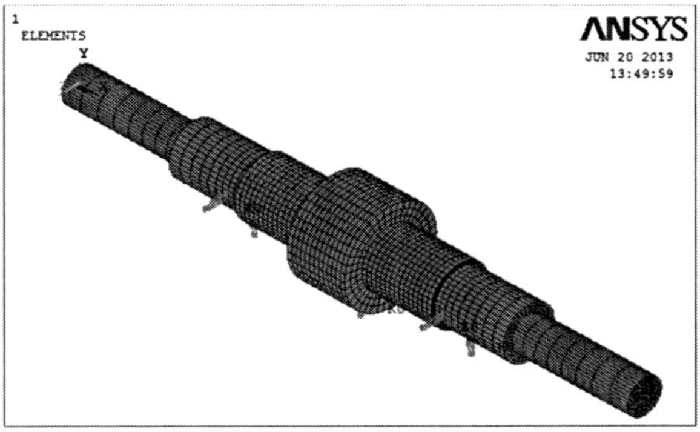

Fig. 5 Finite element model of the rotor bearing system in ANSYS

4 Experimentation

4.1 Test Rig Description

The designed rotor-bearing configuration is fabricated and assembled as shown in Fig. 6. It consists of a stepped rotor made of stainless steel (SS 304) and is supported axially by a permanent magnet bearing. The bearing consists of three rings on the stator (two axially polarized rings and one radial polarized ring comprising of eight radial polarized sectors each of 45°) and three rings on the rotor (two axially polarized rings and one radial polarized ring comprising of six radial polarized sectors each of 60°). The pre stressed stainless steel (SS 316) sleeves of 1 mm thickness are used to retain the magnets on both stator and rotor. The rotor is supported radially by 0.3 mm discrete bump foil bearings (refer Fig. 6). Five eddy current probes are used to measure the shaft vibrations. Two probes are mounted in horizontal directions and two in vertical directions between the permanent magnet bearing and radial foil bearings. Fifth probe is mounted axially at one end of the shaft to measure axial displacement. The shaft speed is recorded with the help of an optical key phasor. The eddy current probes and optical key phasor are connected to the automated data acquisition system, ADRE (Automated Diagnostics of Rotating Equipments) to determine the response of the system. Four air turbines are used to rotate the shaft at desired speed.

4.2 Experimental Results

The rotor is accelerated from rest to 40,000 rpm continuously by supplying compressed air to the turbines mounted on the shaft and maintained at maximum speed for small duration of time. The data is acquired using eddy probes and optical key phasor. Experiments are repeated to check the consistency of the results. The obtained results are represented in the form of plots such as Bode, Orbit, spectrum,

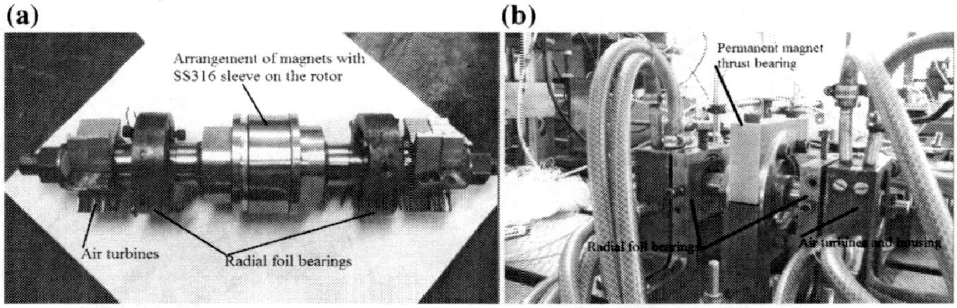

Fig. 6 The rotor bearing system. **a** Rotor with bearings and air turbines. **b** The test rig with rotor, bearings, air turbines and housings

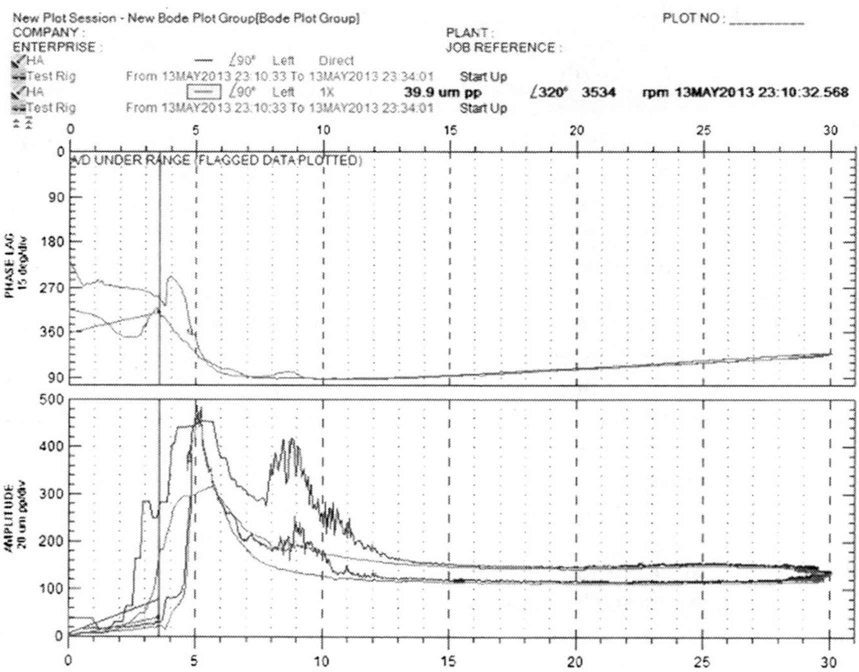

Fig. 7 Bode plot

time base etc., which are shown in Figs. 7, 8 and 9. ADRE is a automated tool for high speed rotor data acquisition and it selects the sampling rate based on the maximum RPM set for the given application. In the present investigation the limiting speed was set to 60,000 RPM. For this speed range, selected sampling rate is 32 samples per rotation. System takes data for ten rotations and gives the

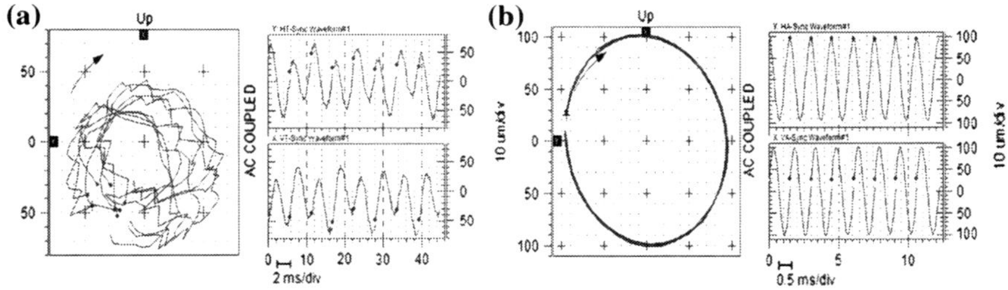

Fig. 8 Shaft Orbit plots at different speeds. **a** 10,654 rpm. **b** 39,383 rpm

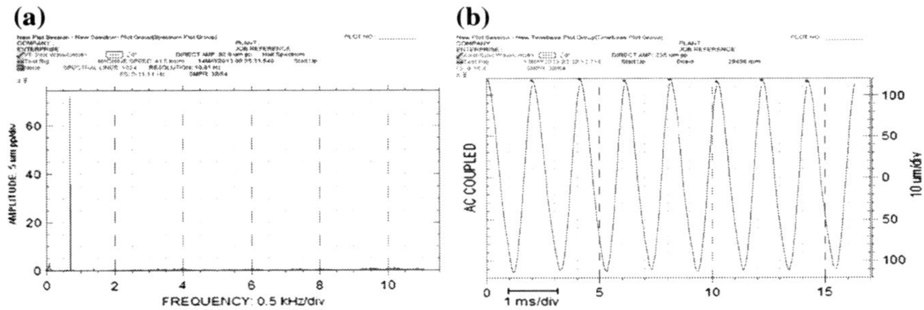

Fig. 9 **a** Frequency spectrum plot at 41,500 rpm. **b** Time base plot at 29,498 rpm

averaged values as a one set of data. The total data generated depends on the duration of the test. It has been observed from the Fig. 7 that the rotor levitates at a speed slightly above 7,500 rpm. In addition, the amplitude of shaft vibrations almost remains constant (~ 120 μm) once the shaft is airborne. The amplitude of vibration starts increasing from 3,000 rpm and peaks at around 5,000 rpm which is due to the presence of rigid body critical speed, 180 degrees phase angle shift justifies the inference. It also matches with the critical speeds determined using ANSYS and by theoretical calculation.

Initially the shaft orbit is not exactly circular as shown in Fig. 8a but once the shaft levitates the orbits (Fig. 8b) are almost circular, which is an indication of almost symmetrical stiffness of the bearing all around the circumference. The smooth and continuous shaft orbit depicts that the rotor is airborne without any rubbing or contact with the stator. The overall vibration of the system is mainly dominated by 1x components as shown in Fig. 9a. The small peaks observed at lower rpm can be attributed to the rigid body mode excitation due to initial rubbing. The passive support system restricted the amplitude of shaft vibration varies from 70 μm at lower speeds to 120 μm at maximum speed. The time base plot obtained from axially mounted eddy probe at different speeds (refer Fig. 9b). It is observed that the amplitude of vibration is almost constant around 110 μm once the shaft is airborne.

5 Conclusion

A test rig consisting of a stepped rotor supported by Halbach magnetized axial permanent magnet bearing and radial discrete bump foil bearings is designed and developed. The rotor is rotated up to 40,000 rpm with the help of air turbines and rotor dynamic characteristics are studied using the ADRE data acquisition system. A rotor bearing system with complete passive levitation is presented. The rotor levitates at a speed slightly above 3,500 rpm and the amplitude of shaft vibrations almost remains constant (~ 120 μm) once the shaft is airborne. It is observed that the rigid body critical speed is 3,000 rpm and which is found to match with rigid body speed obtained using ANSYS (3,029 rpm) and with theoretical calculations (2,671 rpm). The presented concept of hybrid bearing set can be used to support the rotors of high-speed rotating machinery such as air cycle machines, cryogenic turbo compressors, energy storage flywheels and turbo molecular pumps effectively.

Acknowledgment The authors would like to acknowledge the support of Propulsion Division, National Aerospace Laboratories, Bangalore and the KLS'S Gogte Institute of Technology, Belgaum in carrying out the research work.

References

1. Niizeki S (2000) Ceramic bearings for special environments. Motion Control 8:17–22
2. Sundberg DV (1974) Ceramic roller bearings for high speed and high temperature applications. SAE Transact 83(1974):1060–1066
3. Siebert M, Ebihara B, Jansen R (2002) A passive magnetic bearing flywheel. NASA/TM—2002-211159
4. Mukhopadhaya SC, Donaldson J, SenGupta G, Yamada S (2003) Fabrication of a repulsive-type magnetic bearing using a novel arrangement of permanent magnets for vertical-rotor suspension. IEEE Trans Magn 39(5):3220–3222
5. Hussien A, Yamada S, Iwahara M, Okada T (2005) Application of the repulsive-type magnetic bearing for manufacturing micromass measurement balance equipment. IEEE Trans Magn 41 (10):3802–3804
6. Sotelo G, Andrade R, Ferreira AC (2007) Magnetic bearing sets for a flywheel system. IEEE Trans Appl Super Cond (17)(2):2150–2153
7. Ichihara T et al (2005) Applications of super conducting magnetic bearings to a 10 Kwh -class flywheel energy storage systems. IEEE Trans Appl Supercond (15)(2):2245–2248
8. Guilherme GS, Andrade R, Ferreira AC (2007) Magnetic bearing sets for a flywheel system. IEEE Trans Appl Super Cond 17(2):2150–2153
9. Jinji S, Yuan R, Jiancheng F (2011) Passive axial magnetic bearing with Halbach magnetized array in magnetically suspended control moment gyro application. J Magn Magn Mater 323 (15):2103–2107
10. Ohji T, Ichiyama S, Amei K, Sakui M (2004) Conveyance test by oscillation and rotation to a permanent magnet repulsive-type conveyor. IEEE Trans Magn 40(4) 3057–3059
11. Bekinal SI, Anil TR, Jana S (2011) Force, moment and stiffness characteristics of permanent magnet bearings, proceedings of national symposium on rotor dynamics 2011, Indian Institute of Technology, Madras, India, 161–168

12. Bekinal SI, Anil TR, Jana S (2012) Analysis of axially magnetized permanent magnet bearing characteristics. Prog Electromagn Res B 44:327–343
13. Bekinal SI, Anil TR, Jana S (2013) Analysis of radial magnetized permanent magnet bearing characteristics. Prog Electromagn Res B 47:87–105
14. Bekinal SI, Anil TR, Jana S (2013) Analysis of radial magnetized permanent magnet bearing characteristics for five degrees of freedom. Prog Electromagn Res B 52:307–326
15. Bekinal SI, Anil TR, Jana S, Kulkarni SS, Sawant A, Patil N, Dhond S (2013) Permanent magnet thrust bearing: theoretical and experimental results. Prog Electromagn Res B 56:269–287
16. Earnshaw S (1842) On the nature of the molecular forces which regulate the constitution of the luminiferous ether. Trans Cambridge Philos Soc 7:97–112
17. Agrawal GL (1997) Foil air/gas bearing technology-an overview, ASME Publication 97GT-347
18. Heshmat H (2000) Operation of foil bearing beyond the bending critical mode. ASME J Tribol 122(1):192–198
19. Gray S (1975) Applying resilient foil air bearings to turbomachinery—techniques and challenges SAE 1975
20. Badgley RH, Reddecliff J (1975) Evaluation of a gas-lubricated foil bearing for control of gas turbine engine rotor critical speeds SAE -751072
21. Della Corte C, Zaldana AR, Radil KC (2004) A systems approach to the solid lubrication of foil air bearings for oil-free turbomachinery. ASME J Tribol 126(2004):200–207
22. Della Corte C (2011) Oil-free shaft support system rotor dynamics: past, present, and future challenges and opportunities (2011) NASA/TM – 217003

Amplitude Demodulation of Instantaneous Angular Speed for Fault Detection in Multistage Gearbox

Sankar K. Roy, A.R. Mohanty and C.S. Kumar

Abstract Traditional fault detection processes in gearbox are mostly based on vibration signal analysis. However, many a times extraction of fault features by the vibration signal analysis becomes difficult as the signal carries structure-borne noise. Moreover, position of the vibration sensor affects the signal to noise ratio of the signal. The presence of a fault in a gearbox changes the rotating dynamics of the system which varies the speed of the output shaft. Therefore, aim of this paper is to analyse the varying speed of the output shaft for detection of fault in a multistage automobile gearbox. Here, the varying speed is measured as instantaneous angular speed (IAS). Besides carrying information related to the fault, this IAS signal carries less structure borne noise due to direct response of rotating dynamics. Furthermore, in presence of fault, this varying speed signal or IAS signal is frequency modulated signal. Besides frequency modulation, the IAS signal becomes amplitude modulated under the application of load. Demodulation is a well known technique in the area of signal processing, which extracts the modulating part from a signal. Hence, the amplitude demodulation technique has been applied to the IAS signal which reveals various important frequencies related to the gearbox. Thus the technique helps to detect the fault in a multistage automobile gearbox under different load conditions.

Keywords Multistage gearbox · Fault detection · IAS signal · Amplitude demodulation

S.K. Roy · A.R. Mohanty (✉) · C.S. Kumar
Department of Mechanical Engineering, Indian Institute Technology Kharagpur, Kharagpur 721302, India
e-mail: amohanty@mech.iitkgp.ernet.in

S.K. Roy
e-mail: sayont007@gmail.com

C.S. Kumar
e-mail: kumar@mech.iitkgp.ernet.in

J.K. Sinha (ed.), *Vibration Engineering and Technology of Machinery*,
Mechanisms and Machine Science 23, DOI 10.1007/978-3-319-09918-7_84

1 Introduction

The gearbox is a most important component in power transmission system. Any defect presents in the system may lead to severe damage. Therefore continuous condition monitoring is a primary concern for all. Today, the monitoring system are developed based on vibration signal [1–5], acoustics signal [4], motor current signal [6, 7] etc. Among these, the vibration signal based system is mostly used. However, the vibration signal based condition monitoring system faces some problems. Firstly, the vibration signal carries structure borne noise which diminishes the actual signal's information. Secondly, strength of the signal depends on position of the vibration transducer. Hence, it requires experienced operator. In presence of defect in a gearbox, the dynamics of the gearbox changes and this definitely alters the rotational speeds of the shafts. Recently, the researchers are using this varying speed signal to monitor the rotating machines [8–18]. The speed signal has some advantages over the other signals. It carries less structure borne noise as it is a direct response of the rotating dynamics. The varying speed signal is measured as instantaneous angular speed (IAS). By using the IAS signal, the researchers have measured gear transmission error [8]. The IAS signal is measures by various speed measuring sensors such as encoder, Hall effect sensor etc. However, among these, the incremental rotary encoder is a cheaper sensor and provides highest resolution. The encoder generally provides pulse signal. Researchers have used Zero crossing technique to the pulse signal to estimate the IAS and used to detect the combustion related fault in diesel engine [9]. IAS signal is also estimated from the pulse signal by applying phase demodulation technique and further applied to detect the faulty rotor bar in the induction motor and cyclic and non-cyclic load conditions in helical gearbox respectively [10, 11]. Further, a detail review has been presented by the authors [12] to estimate accurate IAS from pulse signal. Gu et al. [13] have developed an IAS estimation technique based on frequency demodulation to monitor rotor misalignment. Based on IAS signal, researchers have monitored two stage reciprocating compressor [14], diesel engine [15], roller bearing [16], helical gearbox [17], feed axis gearbox [18] etc. So far we have discussed about the various signals to choose a best signal for condition monitoring process. Beside the best signal, the condition monitoring requires a perfect signal processing technique to extract out all the features buried in the signal. Fast Fourier transform (FFT) is a well known technique for signal analysis. However, the FFT is only suitable for analysis of a stationary signal. In real time the signal possesses the characteristic of non-stationarity. For analysis of non stationary signal the researchers have developed short time Fourier transform (STFT) [5], Wigner Ville distribution (WVD) [3], continuous wavelet transform (CWT) [4, 5] etc. In STFT, a signal is broken into different stationary segments by using windowing function and then FFT is applied on each segment. However, the technique faces problem because of poor time frequency resolution. WVD has developed to increase the time frequency resolution but the technique produces cross terms in time frequency domain. In CWT, time frequency is further improved as the technique is based on variable window.

However once the windowing function is selected then same wavelet will be applied for whole analysis. In presence of a defect, the speed varies. Therefore measured speed signal from the output shaft poses the characteristic of frequency modulation. Beside the frequency modulation, the IAS also possesses amplitude modulated characteristic. Hence amplitude demodulation technique has been applied to the IAS signal to detect the fault in multistage gearbox. Further a study has been carried out to understand the amplitude modulation behavior of IAS signal under different load conditions.

2 Theory

2.1 IAS Estimation

The speed measuring sensors measures the angular speed in terms rate of change of phase angle with time. In other word, the rate of change of phase angle with respect to time is also called instantaneous frequency (IF).

$$IF = \frac{d\phi}{dt} \tag{1}$$

If a machine is running at constant angular velocity (ω_0), then the output from the speed sensor can be written as

$$x(t) = A\cos[N_m\omega_0 t + \phi_0] \tag{2}$$

where A is the amplitude of the signal, N_m is resolution of the sensor, ϕ_0 is initial phase of the signal and t is time. However in real time, due to some unavoidable errors such as machining error, manufacturing error etc., even a newly installed machine cannot deliver a constant angular speed. Therefore, the varying speed will modulate the frequency of the sensor output. Hence it must be kept in mind that the modulation will be more in presence of additional defect. The varying speed (ω) can be modeled in terms two parts

$$\omega = \omega_0 + \delta\omega(t) \tag{3}$$

where $\delta\omega(t)$ is additional time varying component beside the constant one (ω_0). Hence the sensor output can be written as

$$\begin{aligned} x(t) &= A\cos\left[N_m\omega_0 t + N_m\int_0^t \delta\omega(t)dt + \phi_0\right] \\ &= A\cos[\omega_c t + \phi(t)] \end{aligned}$$

where

$$\omega_c = N_m\omega_0 \text{ and } \phi(t) = N_m \int_0^t \delta\omega(t)dt + \phi_0$$
$$= A\cos[\phi_s(t)] \text{ , where } \phi_s(t) = \omega_c t + \phi(t)$$

In case of encoder, the above output will be a frequency modulated pulse signal. Similar to FM signal, IF estimation from the above frequency modulated pulse signal is estimation of IAS. Hence, frequency domain technique has been adopted to estimate the IAS from the frequency modulated pulse signal. The frequency domain technique developed in the paper [13] has been introduced by a flow diagram as shown in Fig. 1.

Fig. 1 Flowchart for IAS estimation

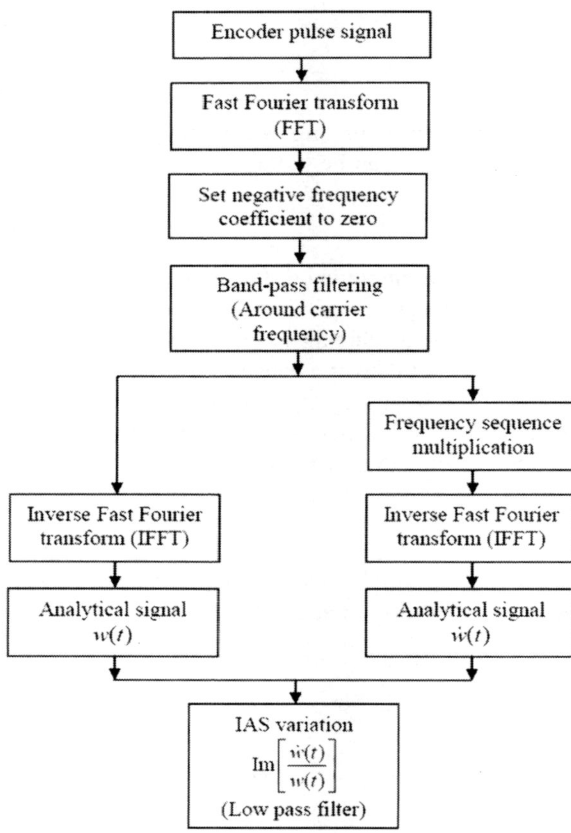

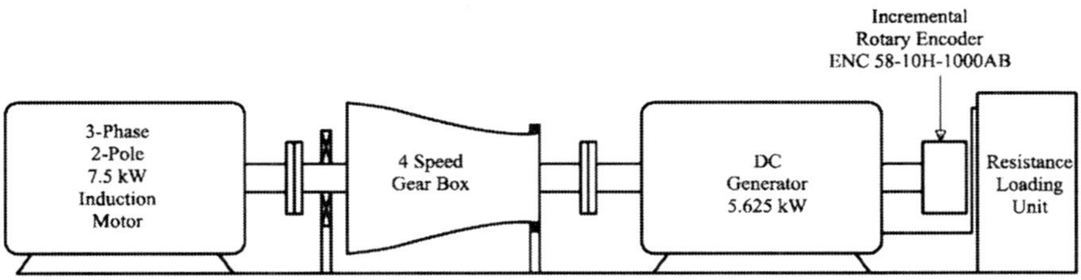

Fig. 2 Schematic diagram of experimental setup for multistage gearbox

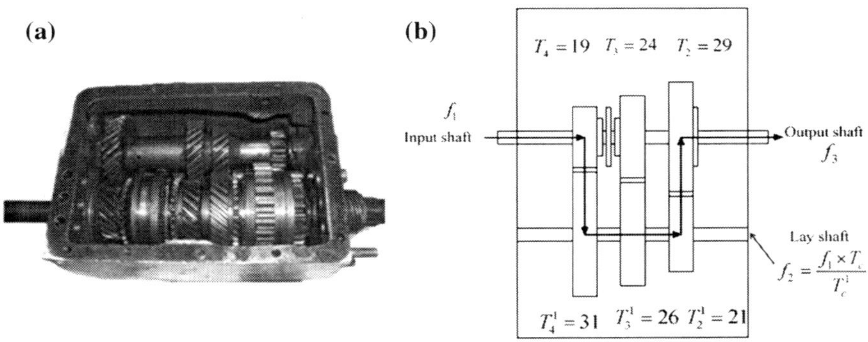

Fig. 3 **a** 4-speed synchromesh gearbox. **b** Line diagram of gear meshing

2.2 Amplitude Demodulation

The technique extracts the time varying amplitude from an amplitude modulated signal. The amplitude demodulation is done by using Hilbert transform. However, there are other techniques also. In Hilbert transform, a signal $s(t)$ is represented in analytical domain. The Hilbert transform of a function $s(t)$ is defined as [7]

$$H[s(t)] = \frac{1}{\pi} \int_{-\infty}^{\infty} \frac{s(t)}{t - \tau} d\tau \quad (5)$$

The technique is basically a convolution of $s(t)$ with the signal $1/\pi t$. Thus it shifts the phase of all frequency components by 90°. If $z(t)$ is an analytic signal of $s(t)$ then $z(t)$ can be defined by

$$z(t) = s(t) + iH[s(t)] = a(t)e^{i\varphi(t)} \quad i = \sqrt{-1} \quad (6)$$

Fig. 4 **a** Healthy 2nd gear. **b** One tooth broken in the 2nd gear. **c** Two teeth broken in the 2nd gear

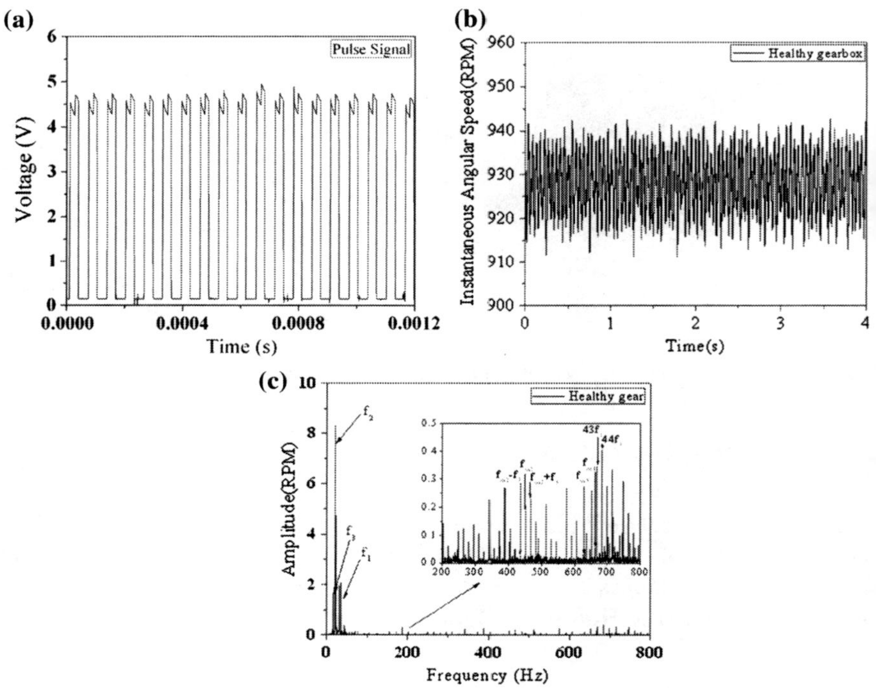

Fig. 5 IAS signal analysis for healthy gearbox at 2.625 kW load. **a** Pulse signal, **b** IAS signal and **c** frequency spectrum of IAS

Hence, amplitude of an analytical signal is

$$a(t) = \sqrt{[s(t)^2 + H[s(t)]^2} \tag{7}$$

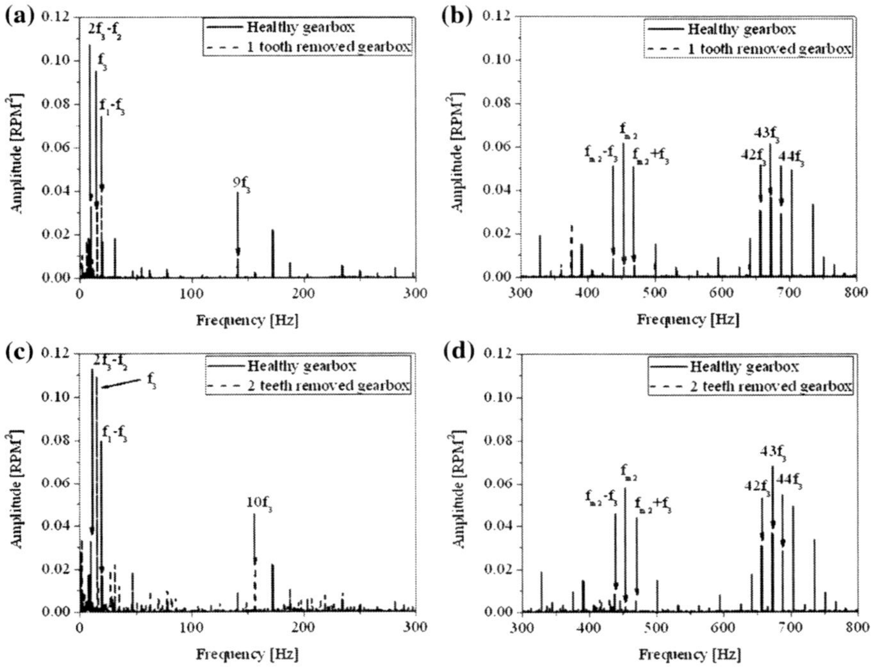

Fig. 6 Power spectra of amplitude demodulated IAS signal of Healthy gearbox versus one tooth defective gearbox from frequency range of **a** 0–300, **b** 300–800; Healthy gearbox versus two teeth defective gearbox from frequency range of **c** 0–300, **d** 300–800, at no load condition

3 Experimental Set-Up

The experimental setup (shown in Fig. 2) is equipped with 7.5 kW, two pole, three phase induction motor, a four stage automobile gearbox, a 5.625 kW DC motor, and a variable load resistor bank. The induction motor is used to drive the gearbox and an output shaft from the gearbox is connected with the DC motor without any speed reduction. The DC motor is acting here as generator. The load resistor bank is connected to the DC motor and used to load on the gearbox ranging from 0–5.625 kW load in step of 0.375 kW. The input speed of the induction motor is controlled by a variable frequency drive (VFD). Further, the speed of the output shaft from the DC motor is measured by an incremental rotary encoder ENC 58-10H-1000AB (Encoders India) of TTL (5 V) with resolution of 1,000 pulses. The pulse signal is further acquired by using a National Instruments (NI) 6110 A/D card of 16 bit resolution. The signal is saved in a personal computer and then processed by using Matlab.

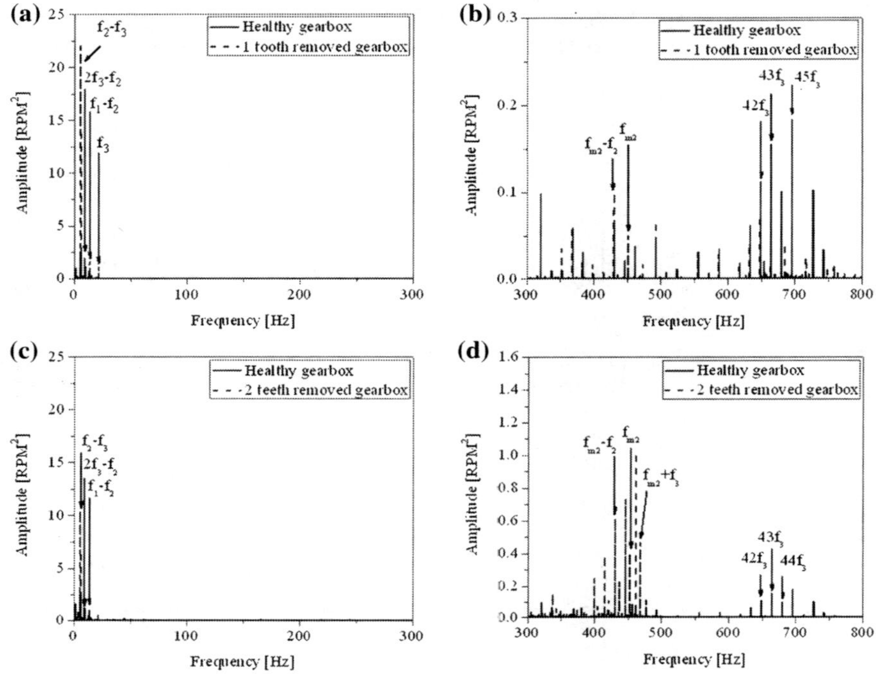

Fig. 7 Power spectra of amplitude demodulated IAS signal of Healthy gearbox versus one tooth defective gearbox from frequency range of **a** 0–300, **b** 300–800; Healthy gearbox versus two teeth defective gearbox from frequency range of **c** 0–300, **d** 300–800, at 2.625 kW load condition

The multistage gearbox is a four speed gearbox and 2nd, 3rd, 4th gears are synchro meshed. The gearbox and line diagram have shown in Fig. 3a, b respectively. Gearbox related important frequencies such as input frequency (f_1), lay shaft frequency (f_2), output frequency (f_3), 2nd, 3rd, 4th gear mesh frequencies (f_{m2}, f_{m3}, f_{m4}) are related as

$$f_2 = 0.6129 f_1,\ f_3^2 = \frac{f_1}{2.253},\ f_3^3 = \frac{f_1}{1.506},\ f_3^4 = f_1 \tag{8}$$

$$f_{m2} = 21 f_2,\ f_{m3} = 26 f_2,\ f_{m4} = 31 f_2 \tag{9}$$

Hence three gears (Healthy, one tooth removed and two teeth removed) are chosen for the experiment. The faulty gears are made by artificially removing one and two teeth from the 2nd main gear. The removal process has done by electro discharge machining (EDM) process. Three different gears are shown in Fig. 4.

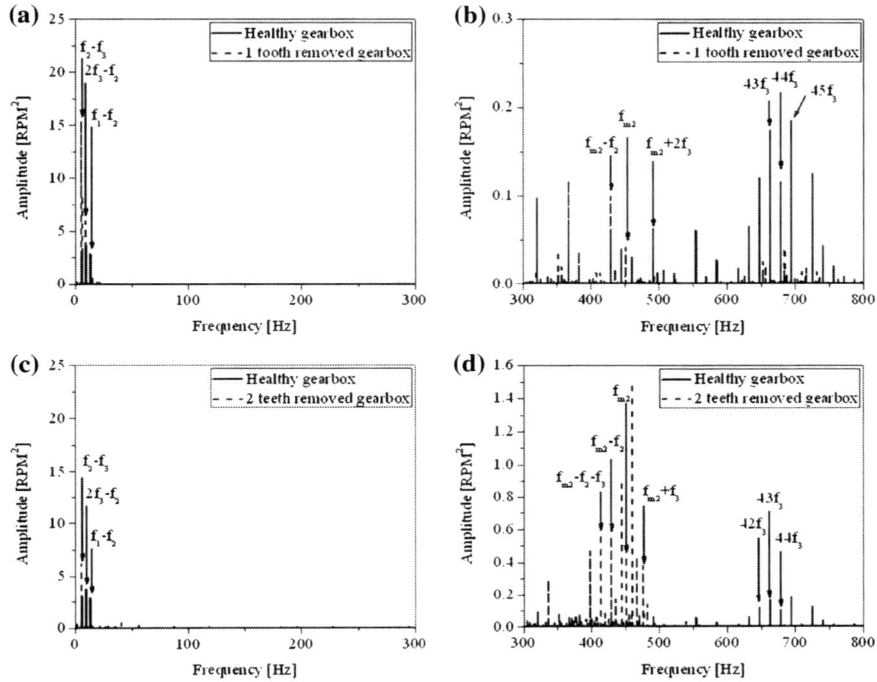

Fig. 8 Power spectra of amplitude demodulated IAS signal of Healthy gearbox versus one tooth defective gearbox from frequency range of **a** 0–300, **b** 300–800; Healthy gearbox versus two teeth defective gearbox from frequency range of **c** 0–300, **d** 300–800, at 5.625 kW load condition

4 Results and Discussions

The experiments were carried out in 2nd gear operation for three different gearboxes (Healthy, one tooth removed and two teeth removed) under three different loading conditions, while input speed is 37 Hz. The Fig. 5a, b show encoder pulse signal and estimated IAS signal from healthy gearbox under 2.625 kW load condition and input speed of 37 Hz. The Fig. 5c shows frequency spectrum of IAS signal. It reveals various important frequencies of input shaft, lay shaft, output shaft and as well as the gear meshing frequencies. However, the amplitude of 3rd and 4th gear meshing frequency is quite lower.

It is already mentioned that the IAS signal possesses amplitude modulation characteristics under load. Hence the IAS signal is amplitude demodulated by Hilbert transform and power spectrum of amplitude demodulated signal for three loading conditions (0 kW, 2.625 kW, 5.625 kW) have been shown in Figs. 6, 7 and 8. The Fig. 6 reveals that amplitude demodulation is unable to extract feasible information during no load condition. During loading condition, various important frequencies (f_{m2} and its side bands) are excited, which indicates presence of a fault in 2nd gear. It also reveals the excitation of lower frequency in case of faulty gearbox. The reason behind is that eccentricity of the shaft increases in presence of fault. The Fig. 7 reveals that amplitude of those frequencies increases while applied load is 2.625 kW. Further the amplitude decreases while applied load increases to 5.625 kW. This reveals that the higher load is acting as a damper. As the defect severity increases, the amplitude of f_{m2} and its side bands increases.

5 Conclusions

The paper studies the amplitude modulation characteristics of IAS signal from a multistage helical gearbox under different load and defect severity conditions. It is found that the no load condition is not convenient for fault detection of a gearbox. A certain amount load is always necessary to excite various important frequencies (gear meshing frequency and its side bands), which are very useful for fault detection in a gearbox and the load should not be so high that it reduces the amplitude. This amplitude variation increases as defect severity increases. Power spectra of amplitude demodulated IAS signal also reveal that low frequencies are getting excited in case of a faulty gearbox.

References

1. Randall RB (1982) A new method of modeling gear faults. J Mech Des 104(2):259–267
2. McFadden PD (1987) Examination of a technique for the early detection of failure in gears by signal processing of the time domain average of the meshing vibration. Mech Syst Signal Process 1(2):173–183
3. Staszewski WJ, Worden K, Tomlinson GR (1997) Time–frequency analysis in gearbox fault detection using the Wigner-Ville distribution and pattern recognition. Mech Syst Signal Process 11(5):673–692
4. Byder N, Ball A (2003) Detection of gear failures via vibration and acoustics signals using wavelet transform. Mech Syst Signal Process 17(4):787–804
5. Sait AS, Sharaf-Eldeen YI (2011) A review of gearbox condition monitoring based on vibration analysis techniques diagnostics and prognostics. Rotating machinery, structural health monitoring, shock and vibration 5:307–324
6. Mohanty AR, Kar C (2006) Fault detection in a multistage gearbox by demodulation of motor-current waveform. IEEE Trans Indus Elect 53(4):1285–1297
7. Kar C, Mohanty AR (2006) Monitoring gear vibrations through motor current signature analysis and wavelet transform. Mech Syst Signal Process 20(1):158–187
8. Sweeney PJ, Randall RB (1996) Gear transmission error measurement using phase demodulation. Proc Instit Mech Eng, Part C: J Mech Eng Sci 210(3):201–213
9. Yang J, Pu L, Wang Z, Zhou Y, Yan X (2001) Fault detection in a diesel engine by analysing the instantaneous angular speed. Mech Syst Signal Process 15(3):549–564

10. Sasi AB, Payne BS, Gu F, Ball A (2001) The exploitation of instantaneous angular speed for condition monitoring of electric motors. In: Proceedings of the 14th International Congress on Condition Monitoring and Diagnostic Engineering Management, 4–6 Sept 2001, Manchester, UK
11. Stander CJ, Heyns PS (2005) Instantaneous angular speed monitoring of gearboxes under non-cyclic stationary load conditions. Mech Syst Signal Process 19(4):817–835
12. Li Y, Gu F, Harris G, Ball A, Bennett N, Travis K (2005) The measurement of instantaneous angular speed. Mech Syst Signal Process 19(4):786–805
13. Gu F, Yesilyurt I, Li Y, Harris G, Ball A (2006) An investigation of the effects of measurement noise in the use of instantaneous angular speed for machine diagnosis. Mech Syst Signal Process 20(6):1444–1460
14. Elhaj M, Gu F, Ball AD, Albarbar A, Al-Qattan M, Naid A (2007) Numerical simulation and experimental study of a two stage reciprocating compressor for condition monitoring. Mech Syst Signal Process 22(2):374–389
15. Charles P, Sinha JK, Gu F, Lidstone L, Ball AD (2009) Detecting the crankshaft torsional vibration of diesel engines for combustion related diagnosis. J Sound Vib 321(3):1171–1185
16. Renaudin L, Bonnardot F, Musy O, Doray JB, Remond D (2010) Natural roller bearing fault detection by angular measurement of true instantaneous angular speed. Mech Syst Signal Process 24(7):1998–2011
17. Mohanty AR, Agarwal D (2010) Gear fault detection using instantaneous angular speed monitoring. In: Proceedings of the 1st international workshop and congress on E-maintenance, 22–24 June 2010, Luleå, Sweden
18. Zhou Y, Tao T, Mei X, Jiang G, Sun N (2011) Feed axis gearbox condition monitoring using built in position sensors and EEMD method. Robot Comput-Integr Manuf 27(4):785–793

Experimental Grey Box Model Identification and Control of an Active Gas Bearing

Lukas Roy Svane Theisen, Fabián G. Pierart, Henrik Niemann, Ilmar F. Santos and Mogens Blanke

Abstract Gas bearings have inherent dynamics that gives rise to low damping and potential instability at certain rotational speeds. Required damping and stabilization properties can be achieved by active flow control if bearing parameters are known. This paper deals with identification of parameters in a dynamic model of an active gas bearing and subsequent control loop design. A grey box model is determined based on experiments where piezo actuated valves are used to perturb the journal and hence excite the rotor-bearing system. Such modelling from actuator to output is shown to efficiently support controller design, in contrast to impact models that focus on resonance dynamics. The identified model is able to accurately reproduce the lateral dynamics of the rotor-bearing system in a desired operating range, in this case around the first two natural frequencies. The identified models are validated and used to design a model-based controller capable of improving the damping of the gas bearing. Experimental impact responses show an increase in damping by a factor nine for the investigated conditions.

Keywords System identification · Active gas bearings · Experimental techniques · Modelling · Rotor dynamics

L.R.S. Theisen (✉) · H. Niemann · M. Blanke
Department of Electrical Engineering, Technical University of Denmark, 2800 Kgs. Lyngby, Denmark
e-mail: lrst@elektro.dtu.dk

H. Niemann
e-mail: hhn@elektro.dtu.dk

M. Blanke
e-mail: mb@elektro.dtu.dk

F.G. Pierart · I.F. Santos
Department of Mechanical Engineering, Technical University of Denmark, 2800 Kgs. Lyngby, Denmark
e-mail: fpvas@mek.dtu.dk

I.F. Santos
e-mail: ifs@mek.dtu.dk

J.K. Sinha (ed.), *Vibration Engineering and Technology of Machinery*, Mechanisms and Machine Science 23, DOI 10.1007/978-3-319-09918-7_85

1 Introduction

Passive and active gas bearings are receiving growing attention for their high speed operation capabilities. Passive gas bearings offer advantages of high speed operation, low friction, and clean and abundant air as lubricant; however they suffer from low damping and vibration instabilities [1–3].

There are two main approaches to increase damping and stability; one is through foil bearings [4–9] that rely on friction between bumps and foil. Such a solution is relatively cheap, but the friction coefficient is a design challenge [4]. The other approach is via mechatronic solutions, which offer robustness and adaptability. Actively controlled gas bearings using piezo actuation [3] and active inherent restrictors [10, 11] is a promising mechatronic machine element approach to support clean and high speed solutions. An example of adjustable hybrid lubrication given in [12], proposes a rule based control strategy, where changing conditions caused changed critical speeds, which allowed them to cross the first critical speed. The design of model based controllers for gas bearings requires models that catch the dynamic behavior of the journal in the frequency range where control is needed. A model for control design should in a sufficiently simple manner describe the relation from piezo actuator electrical input to journal position.

The design of actively lubricated gas bearings relies on usage of the Reynolds equation, which models the behaviour of the compressible lubricant between the journal and the housing. The rotating journal in the gas bearing is then modelled as a mass subject to the force from the compressed air in the bearing. The Reynolds equation is solved to obtain a pressure distribution, which is integrated to obtain the aerodynamic forces. An input-output-model (IO model) was developed in [13] using a first order principles model to incorporate the behaviour of the piezo-electric valves into the Reynolds equation. This model catches well the behaviour of the bearing over a wide range of parameters but has too high complexity to be directly used for model based control. Another common approach is using a 2 DOF coupled mass-spring-damper system where parameters are experimentally identified from impacts. This model however does not properly describe the interaction of the pressurized air actively lubricating the journal, which is required for a model based controller design.

The present contribution adopts a systematic data based method for modelling of a gas bearing with external inputs and identifies low-order dynamical models that are shown to be adequate for model based controller design. The models are shown to describe the behaviour of the rotor-gas bearing system with high accuracy.

The paper is structured as follows; A brief description of the controllable gas bearing setup is first described. Section 3 presents the grey box model chosen and identification of its parameters is carried out based on experiments that cover the operational range of the bearing. The identified models are used for design of a model based controller to enhance damping properties. Special techniques to handle actuator hysteresis is discussed. Finally, an experimental characterization is done of the piezo actuated valves and the high closed loop damping is validated.

2 Experimental Setup of Gas Bearing

The experimental setup at hand is shown in Fig. 1: A flexible journal (a) is supported by both a ball bearing (d) and the controllable gas bearing (b). A disc (c) is mounted in one end to preload the journal. The horizontal and vertical journal positions (ϵ_x, ϵ_y) are measured at the disc location $(\cdot)_d$ using eddy current sensors (e). Disc deflections $(\cdot)_d$ are easily mapped to bearing deflections $(\cdot)_b$ under assumption of a rigid shaft, and the transformation then reads $\epsilon_{xb} = \kappa\epsilon_{xd}$, in which $\kappa = 0.625[\text{m/m}]$ is the fractional position of the gas bearing along the length of the journal relative to the disc. The angular position of the journal ϕ is measured by an optical encoder. A more thorough description of the setup is available in [3].

The gas bearing is controllable by injection of pressurized air through the opening and closing of four piezo actuated valves positioned as shown in Fig. 1. Each valve is controlled by a voltage $u_{p,i} \in [u_{p,\min}; u_{p,\max}]$, $i \in \{1, 2, 3, 4\}$, with limits $u_{p,\min} = 0\,V$, $u_{p,\max} = 10\,V$, where $u_{p,\min}$ corresponds to a fully opened valve and $u_{p,\max}$ fully closed. The position of the valves is measurable in the interval $y_{p,i} \in [y_{p,\min}; y_{p,\max}] = [0; 10]V$, $i \in \{1, 2, 3, 4\}$, and the expansion of the piezo actuator is proportional to the measured position with maximum expansion at $u_{p,\max}$ is $y_{\max} = 45$ µm which is approximately 10V, but it is not known exactly. Hence all piezo positions are listed in Volts. Piezo ceramics are subject to nonlinearities in form of hysteresis [14], which gives a nonlinear input output relationship. This nonlinearity is addressed in Sect. 4.1.

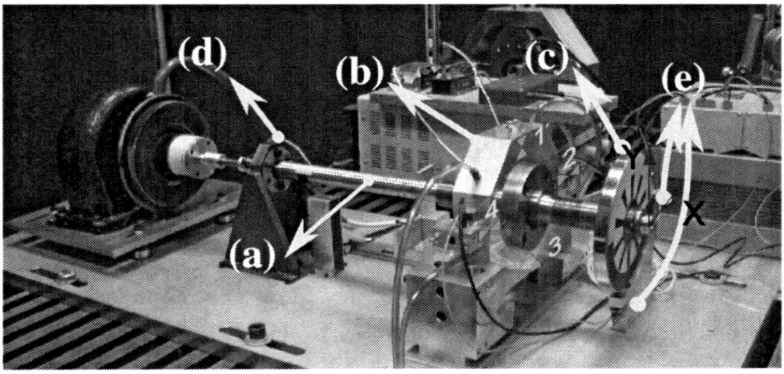

Fig. 1 The experimental bearing setup. A flexible shaft (**a**) is supported by both a ball bearing (**d**) and the controllable gas bearing (**b**) with four piezo actuated valves. A disc (**c**) is mounted in one end to preload the journal and displacement sensors (**e**) measure the lateral movement of the disc

3 Experimental Identification of Gas Bearing Parameters

Modelling of controllable gas bearings still represents an open challenge. The approach pursued in this work is to some extent considered model free, as the developed models are found experimentally based on excitation of the rotor-gas bearing system by changing valve positions thereby perturbing the fluid film. A model is assumed with known structure, though unknown parameters. Identification of the parameters is then sought through grey box modelling. Standard means of model validation (sign test, residual analysis) are then used to assess over/under parametrization of the model. The identified linear rotor-gas bearing models are found to have speed dependent parameters, hence polynomials parametrized in speed are fitted to allow a speed dependent model to be stated describing the gas bearing with the range of interest.

A common approach for modelling gas bearings is the linear mass-spring-damper-model. Using positions $p = [\epsilon_x, \epsilon_y]^{\mathrm{T}}$, with respective time derivatives $\frac{d}{dt}(\cdot) = (\dot{\cdot})$

$$\mathcal{M}\ddot{p}(t) + (\mathcal{D} + \Omega\mathcal{G})\dot{p}(t) + \mathcal{K}p(t) = f(t), \quad p = [\epsilon_x, \epsilon_y]^{\mathrm{T}} \tag{1}$$

where $f(t)$ is the force, Ω is the rotation speed and the matrices have structures as listed in Table 1. This models the gas bearing behaviour in a neighbourhood of the 1st and 2nd critical speeds and suggests a model structure.

The force $f(t)$ is the sum of external forces from disturbances, residual mass unbalance $f_u(t)$ and forces from the piezo actuators $f_p(t)$.

The modelling of equivalent applied force from the piezo valves is no trivial task. An approach presented in [13] uses a nonlinear state space model with states being position, and velocity of the piezo ceramic, and pressure in the chamber between the journal and bearing housing. The mentioned model however contains a coupling to the pressure of the lubricant, which in this case is not measured, making it challenging to use. Instead the approach pursued in the present work is taken from electromagnetically actuated bearings, in which opposing actuators are coupled pairwise (i.e. valves {1, 3},{2, 4} move oppositely). The gain from this approach is twofold: The coupled valves counteract possible nonlinear gains, and transforms the system from an overactuated (more control inputs than controlled outputs) to a fully actuated (equal number of control inputs and controlled outputs).

Due to piezo valves hysteresis which is shown later in Sect. 4.1 to be approximately 15 %, the valve input voltage is not suitable input for identification of linear models. Instead the piezo valve positions are used as model inputs for identification; more precisely the IO model input vector is chosen to be the difference of horizontal and vertical valve positions:

Table 1 Structure of system matrices

Mass matrix	Stiffness matrix		Damping matrix
$\mathcal{M} = \begin{bmatrix} m & 0 \\ 0 & m \end{bmatrix}$	$\mathcal{K} = \begin{bmatrix} k_{xx} & k_{xy} \\ k_{yx} & k_{yy} \end{bmatrix}$		$\mathcal{D} = \begin{bmatrix} d_{xx} & d_{xy} \\ d_{yx} & d_{yy} \end{bmatrix}$
	Input gain		Gyroscopic
	$\mathcal{B}_p = \mathcal{B}_d = \begin{bmatrix} b_{xx} & b_{xy} \\ b_{yx} & b_{yy} \end{bmatrix}$		$\mathcal{G} = \begin{bmatrix} 0 & g_{xy} \\ -g_{xy} & 0 \end{bmatrix}$

$$u(t) \triangleq [u_x(t), u_y(t)]^{\mathrm{T}} = [y_{p,2}(t) - y_{p,4}(t), y_{p,1}(t) - y_{p,3}(t)]^{\mathrm{T}} \tag{2}$$

The equivalent force acting on the journal is assumed linear i.e. $f_p(t) = \mathcal{B}_p u(t)$, where the structure of the input gain $\mathcal{B}_p$ is listed in Table 1.

Considering only the active forces $f(t) = f_p(t)$, and reformulating to a unit mass equation, the model reads:

$$\ddot{p}(t) + D\dot{p}(t) + Kp(t) = Bu(t) \tag{3}$$

where $D \triangleq \mathcal{M}^{-1}(\mathcal{D} + \Omega\mathcal{G}), K \triangleq \mathcal{M}^{-1}\mathcal{K}$, and $B \triangleq \mathcal{M}^{-1}\mathcal{B}_p$. Since a model structure is known (though the parameters are not), identification of the rotor bearing system parameters is sought through grey box modelling, which (using a suitable model structure and given reasonable initial parameter estimates) catches well both the stationary and dynamical behaviour of the bearing. The grey box modelling is eased by reformulation of the dynamics Eq. (3) to state space form, where a suitable choice of states is $x = [\epsilon_x, \epsilon_y, \dot{\epsilon}_x, \dot{\epsilon}_y]^{\mathrm{T}}$. The measurement noise and errors from simplified model are modelled as additive noise $d(t)$ entering both output and states with a disturbance input gain $\mathcal{B}_d$ and the chosen state space formulation of Eq. (1) then reads:

$$\begin{aligned} \dot{x}(t) &= \mathcal{A}x(t) + \mathcal{B}u(t) + \mathcal{B}_d d(t), \quad x(0) = x_0 \\ p(t) &= \mathcal{C}x(t) + d(t) \end{aligned} \tag{4}$$

where the system-, input gain-, and output matrix are

$$\mathcal{A} = \begin{bmatrix} 0 & I \\ -K & -D \end{bmatrix}, \quad \mathcal{B} = \begin{bmatrix} 0 \\ B \end{bmatrix}, \quad \mathcal{B}_d = \begin{bmatrix} 0 \\ B_d \end{bmatrix}, \quad \mathcal{C} = [I \quad 0], \tag{5}$$

where in the above the matrices to be identified are $\hat{K}, \hat{D}$, and $\hat{B}, \hat{B}_d\, \hat{x}_0$ with structure as listed in Table 1.

The parameters of (5) are identified by recasting the problem to a model parametrized in $\hat{\theta} \triangleq \{\hat{D}, \hat{K}, \hat{B}, \hat{x}_0, \hat{B}_d\}$ as $\mathfrak{M}(\theta)$, where the parameters to be identified are:

$$\hat{\mathcal{A}}(\theta) = \begin{bmatrix} 0 & 0 & 1 & 0 \\ 0 & 0 & 0 & 1 \\ -\hat{K}_{xx} & -\hat{K}_{xy} & -\hat{D}_{xx} & -\hat{D}_{xy} \\ -\hat{K}_{yx} & -\hat{K}_{yy} & -\hat{D}_{yx} & -\hat{D}_{yy} \end{bmatrix}, \quad \hat{\mathcal{B}}(\theta) = \begin{bmatrix} 0 & 0 \\ 0 & 0 \\ \hat{B}_{xx} & \hat{B}_{xy} \\ \hat{B}_{yx} & \hat{B}_{yy} \end{bmatrix}, \tag{6}$$

$$\hat{\mathcal{B}}_d(\theta) = \begin{bmatrix} 0 & 0 \\ 0 & 0 \\ \hat{B}_{d,xx} & \hat{B}_{d,xy} \\ \hat{B}_{d,yx} & \hat{B}_{d,yy} \end{bmatrix}, \quad \mathcal{C} = \begin{bmatrix} 1 & 0 & 0 & 0 \\ 0 & 1 & 0 & 0 \end{bmatrix}, \quad x(0) = \hat{x}_0(\theta) \tag{7}$$

The model then reads

$$\mathfrak{M}(\theta) : \begin{cases} \dot{x}(t) = \hat{\mathcal{A}}(\theta)x(t) + \hat{\mathcal{B}}(\theta)u(t) + \hat{\mathcal{B}}_d(\theta)d(t), & \hat{x}(0) = \hat{x}_0(\theta) \\ p(t) = \mathcal{C}x(t) + d(t) \end{cases} \tag{8}$$

3.1 Description of Experiments

The choice of input signal for system identification is an open challenge, but should certainly excite the system dynamics sufficiently in the frequency range of interest, i.e. up to 200 Hz covering the first two natural frequencies. For identification of the above model, control signals are designed manually to excite simultaneously both the horizontal and vertical lateral rotor dynamics. Such input sequences are applied to the bearing, during 15 data sequences of 10 s duration, with four different journal rotational velocities in the interval [2.9;6.0] krpm. During the experiments the gas injection pressure is kept constant at $P_{inj} = 8.0$ bar and sampling is made with $f_s = 5$ kHz. The horizontal and vertical valves are excited horizontally by a square input signal, while the vertical dynamics are excited by pseudo-random stepwise motion. Figure 2 shows measured input and output sequence of one of the 15 experiments.

3.2 Prefiltering

Before system identification, the measured journal positions are preprocessed (filtered) in two steps: First step is to remove the stationary run-out and mass unbalance response. For this purpose a 10 s data sequence is collected without any system excitation. This allows generation of a compensator function $\mathcal{F}(\phi)$ mapping positions as a function of encoder angle ϕ. The direct measurement $(\cdot)_r$ can then be converted to smoothed position $(\cdot)_s$ using the filtering $(\cdot)_s = (\cdot)_r - \mathcal{F}(\phi)$. The filtered positions can then be seen as deviations from the mass unbalance response. Second step consists of a low pass filtering is applied using a second order Butterworth filter, with cutoff frequency $f_0 = 1.0$ kHz being approximately five times faster than the bearing poles to be identified.

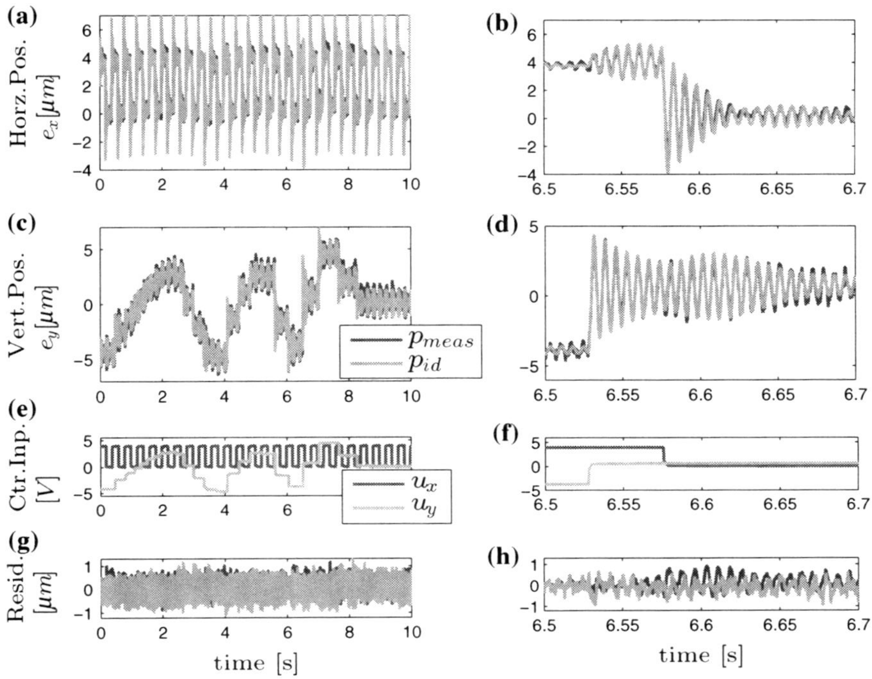

Fig. 2 **a**, **c** and **b**, **d** show an entire sequence and a time zoom of preprocessed measured and identified lateral journal positions $p_{meas} = [\epsilon_x, \epsilon_y], p_{id} = [\hat{\epsilon}_x, \hat{\epsilon}_y]$ used for system identification, excited by the valve positions u_x, u_y shown in **e** and **f**. **g** and **h** show the residual $e = p_{meas} - p_{id}$

3.3 Parameter Identification Using Prediction Error Method

The optimal model $\mathfrak{M}(\theta^{\star})$ is chosen as the minimum of the cost function $W(\theta) \triangleq \sum_{t=1}^{N} e(t)^{\mathrm{T}} e(t)$, where the prediction error e is defined as the difference between the measured and the estimated output $e(t) \triangleq y(t) - \hat{y}(t)$. The minimum is sought using the prediction error method (PEM) [15], such that the optimal parameter set reads:

$$\theta^{\star} = \min_{\theta} W(\theta) \tag{9}$$

The optimization is nonlinear and to ensure convergence to the optimal parameter set, good initial parameter estimates are required for $\hat{K}_{xx}, \hat{K}_{yy}, \hat{D}_{xx}, \hat{D}_{yy}, \hat{B}_{xx}, \hat{B}_{yy}$. These initial estimates are found as preidentified parameters describing the horizontal and vertical dynamics as 1-DOF mass-spring-damper systems individually. The estimates for all parameters (incl. the cross coupling gains) are then iteratively refined and identification is terminated when the improvement of

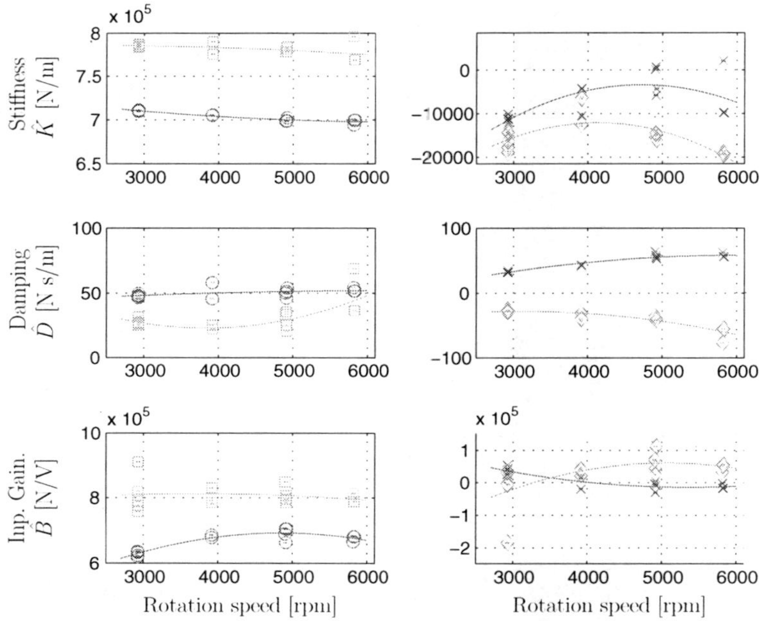

Fig. 3 Identified equivalent stiffness $\hat{K}$: $\{\hat{k}_{xx}, \hat{k}_{xy}, \hat{k}_{yx}, \hat{k}_{yy}\}$,damping $\hat{C}$: $\{\hat{c}_{xx}, \hat{c}_{xy}, \hat{c}_{yx}, \hat{c}_{yy}\}$, and input gain $\hat{B}$: $\{\hat{b}_{xx}, \hat{b}_{xy}, \hat{b}_{yx}, \hat{b}_{yy}\}$. $\circ\ \hat{(\cdot)}_{xx}\ \diamond\ \hat{(\cdot)}_{xy}\ \times\ \hat{(\cdot)}_{yx},\ \square\ \hat{(\cdot)}_{yy}$. Identified equivalents are shown for a unit mass. Speed varying quadratic polynomials (*solid lines*) have been fitted to measurements

error norm is less than bound $Q = 0.01$ indicating the optimization has reached a minimum. Figure 3 shows the identified equivalents of stiffness, damping and input gains for each of the 15 data sets along with a function fitted to the estimates. The identified parameters are very consistent for measurements made at the same rotational speed.

Figure 2 shows that the linear models, to a very high extent, catch the behaviour of the journal. A sign test is performed to assess over parametrization of the model. This test proves that identified parameters within 1 standard deviation σ of the estimate $\pm 1\sigma$ are sign consistent (i.e. 0 is not in the interval, and the parameter estimate is therefore sign consistent). Analysis of the residuals between the measurements and the identified model $\varepsilon \triangleq y - \hat{y}$ shown in Fig. 2 shows the residuals are not white noise as they should optimally be, and indicates presence of non-linearities or higher order dynamics not caught by the linear model. The residual however is bounded by $|\varepsilon| \leq 1.4$ μm, which indicates low magnitude of the neglected dynamics. Higher operation speeds will most likely require higher order models, though for the considered operation range the identified models are deemed sufficient.

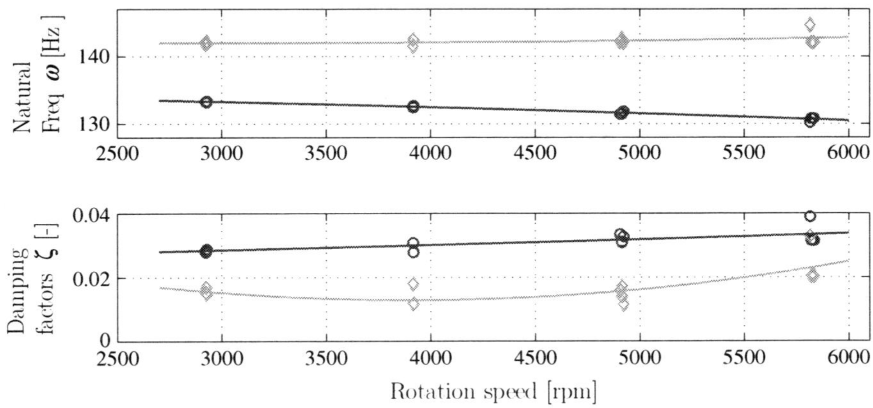

Fig. 4 Identified natural frequencies $\circ\,\omega_{1,2}$, $\diamond\,\omega_{3,4}$ (1st and 2nd critical speeds) and associated damping factors $\zeta_{1,2}, \zeta_{3,4}$. Speed varying quadratic polynomials (*solid lines*) have been fitted to measurements

The identified parameters $\hat{D}, \hat{K}, \hat{B}$ are used to fit speed varying quadratic polynomials to parameters $(\cdot)_{jk}$ of the form:

$$(\cdot)_{jk}(\Omega) = \alpha_{(\cdot),jk}\Omega^2 + \beta_{(\cdot),jk}\Omega + \gamma_{(\cdot),jk}, \tag{10}$$

with coefficients as listed in Table 3. The identified equivalent model then reads:

$$\begin{bmatrix} 1 & 0 \\ 0 & 1 \end{bmatrix} \begin{bmatrix} \ddot{\epsilon}_x(t) \\ \ddot{\epsilon}_y(t) \end{bmatrix} + \begin{bmatrix} \hat{D}_{xx}(\Omega) & \hat{D}_{xy}(\Omega) \\ \hat{D}_{yx}(\Omega) & \hat{D}_{yy}(\Omega) \end{bmatrix} \begin{bmatrix} \dot{\epsilon}_x(t) \\ \dot{\epsilon}_y(t) \end{bmatrix} + \begin{bmatrix} \hat{K}_{xx}(\Omega) & \hat{K}_{xy}(\Omega) \\ \hat{K}_{yx}(\Omega) & \hat{K}_{yy}(\Omega) \end{bmatrix} \begin{bmatrix} \epsilon_x(t) \\ \epsilon_y(t) \end{bmatrix} = \begin{bmatrix} \hat{B}_{xx}(\Omega) & \hat{B}_{xy}(\Omega) \\ \hat{B}_{yx}(\Omega) & \hat{B}_{yy}(\Omega) \end{bmatrix} \begin{bmatrix} u_x(t) \\ u_y(t) \end{bmatrix} \tag{11}$$

From each experiment, an estimate of the first and second critical speeds are directly calculable from the identified parameters as the system poles (the eigenvalues of $A(\theta)$). Figure 4 shows the obtained estimates of the two complex pole pairs $p_{1,2}, p_{3,4}$ listed in the form $p_{1,2} = -\omega_{1,2}\left(\zeta_{1,2} \pm j\sqrt{\zeta_{1,2}^2 - 1}\right)$, having natural frequencies ω and damping factors ζ. The standard deviation of estimate of natural frequencies are $\sigma_{\omega,1,2} = 1.22\,\text{Hz}$, $\sigma_{\omega,3,4} = 0.51\,\text{Hz}$, and of damping factors $\sigma_{\zeta,1} = 2.8 \cdot 10^{-3}[-]$, $\sigma_{\zeta,3} = 5.1 \cdot 10^{-3}[-]$. The identified parameters are shown for a unit mass bearing, but could be converted to real stiffness, damping and input gains on the assumption that the known mass of the journal m is equal to the equivalent mass.

4 Observer Based Controller Design

The undesirably low damping characteristics of the gas bearing can be addressed by model based control. This section shows design of a discrete time state feedback controller designed to increase closed loop damping of the gas bearing. The controller is to be implemented on a computer, hence the identified dynamics Eq. 4 from an experiment at 6.0 krpm are converted to discrete time using zero order hold with sampling time $T_s = 0.2$ ms to obtain the discrete time system, sampled at $x_k \triangleq x(kT_s)$:

$$\begin{aligned} x_{k+1} &= Fx_k + Gu_k + G_d d_k \\ y_k &= Cx_k + d_k, \end{aligned} \tag{12}$$

where the discrete time system matrices F, G are listed in Table 2. Increasing the damping is pursued using a pole placement controller, which requires state feedback. Alternatively an LQR or other optimal controller could have been used. Since only position is measured, a discrete time full order observer is designed to estimate the states $\hat{x}$ given the control input u_k and the position measurement p_k. Designing the state feedback gain K_{sb} and the observer gain L can be done independently. The latter is chosen to have sufficiently fast observer poles.

The state feedback gain K_{sb} is designed by pole placement [16] to increase the damping of the closed loop system by placing the discrete time closed loop poles $p_{d,cl}$ related to the discrete time open loop poles $p_{d,ol}$ according to $p_{d,cl} = Re(p_{d,ol})$ $+j1/2Im(p_{d,ol})$. This ensures a decent increase of damping. The control law then reads:

$$u_{f,k} = -K_{sb}\,\hat{x}_{k|k} \tag{13}$$

4.1 Experimental Piezo Actuator Characterization

The piezo ceramic hysteresis poses a challenge in controller design. A characterization of the hysteresis is developed by applying a stairwise increasing and then decreasing input voltage to a piezo valve. Figure 5 shows measured equilibrium positions as function of applied input voltage. Unfortunately the position exceeds the

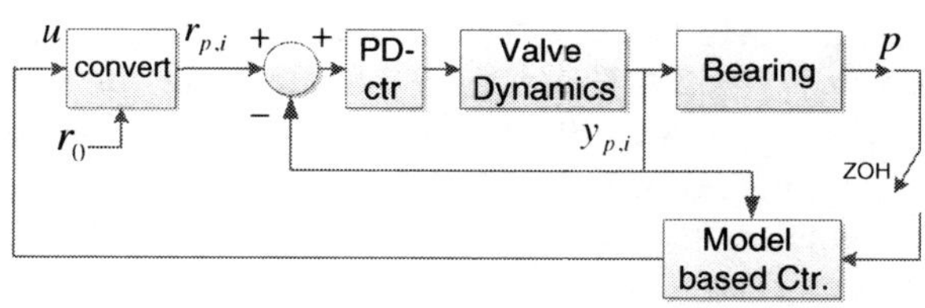

Fig. 5 Block diagram of the closed loop system

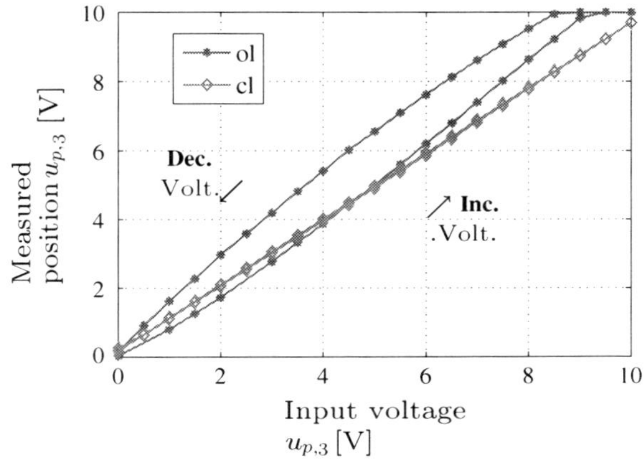

'ig. 6 Hysteresis plot of neas. valve pos. $y_{p,3}$ versus np. voltage in open loop (ol) $_{p,3}$, and PD-contr. closed oop reference $r_{p,3}$ (cl)

saturation limit of the measurement equipment (10 V), thus some of the hysteresis :urve cannot be recovered. It is evident though, that the equilibrium position can :hange $\approx$15 % due to the hysteresis. Seeking to counteract valve hysteresis and ncrease performance, local PD controllers are designed with discrete time gains $K_P = 18.2[V/V], K_D = -0.187[Vs/V]$. The valve positions are controlled towards 'eference positions $r_{p,i}$. Applying a similar staircase input as position reference to the :ontrolled valve, the equilibrium positions from Fig. 6 are obtained. The PD control oth counteracts the hysteresis and gives a desirable valve DC-gain close to unity. For control purposes, the PD controlled valves also make it reasonable to neglect valve dynamics. A block diagram of the closed loop system is shown in Fig. 6.

4.2 *Experimental Impact Response Using Damping Increasing Controller*

The designed controller is implemented and an impulse impact is applied both to :he controlled and uncontrolled gas bearing. Figure 7 shows the comparison of the open and closed loop system subject to manually induced impulse impacts with the journal rotating at $\Omega_t = 4.1$ krpm. The position measurements have been post-filtered to remove stationary run-out and mass unbalance response to allow visualization of the impact response, hence the control input variation. Quantitatively :he response magnitudes are not directly comparable as the impacts were forced manually, but qualitatively the increased damping is certainly clear. The logarithmic decrement method gives from $\epsilon_y(t)$ the damping to be increased from 0.0159 to).139—almost a factor nine.

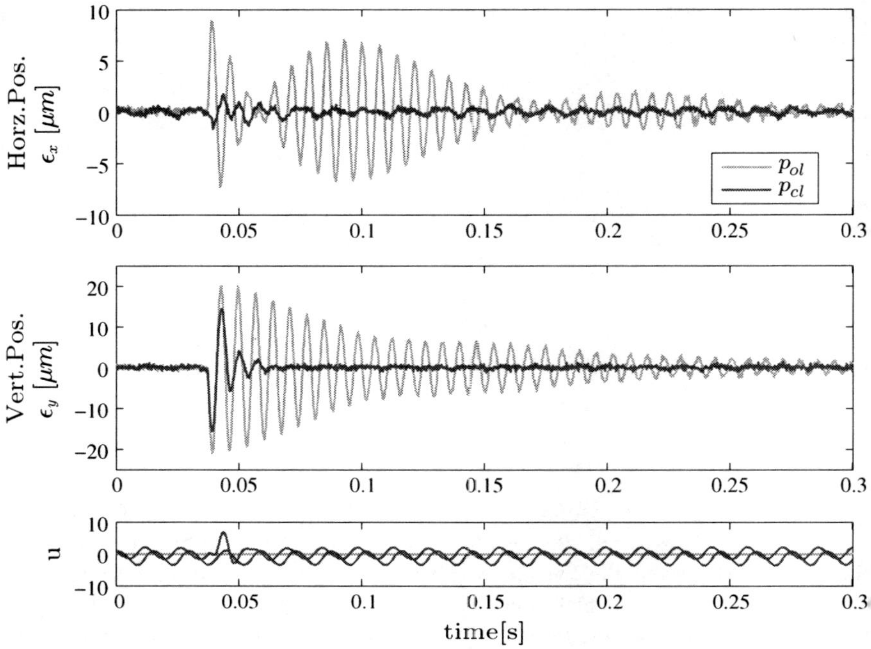

Fig. 7 Open and closed loop impact responses at rotation speed $\Omega_i = 4.1$ krpm. Shown positions are smoothed to highlight transient response. The controller increases the damping by almost a factor 9

5 Conclusion and Future Aspects

Grey box modelling was used successfully to develop accurate linear models of a rotor-gas bearing system based on experimental measurements. It was shown that it is possible to perturb the fluid film via active lubrication and extract relevant rotor-gas bearing model parameters based on input output relationships. The typical low damping properties of gas bearings could be significantly improved using the active lubrication principle. For a constant angular velocity of 4.1 krpm the damping was improved by a factor nine. Additional tests will be carried out in higher angular velocities in the near future. Future effort will focus on controller design addressing the speed dependent parameter variation.

A.1 Appendix

The appendix lists system parameters in Tables 2 and 3.

Table 2 Discrete time system, observer and controller parameters

Parameter	Parameter
$L = \begin{bmatrix} 0.0605 & -0.0006 \\ -0.0015 & 0.0720 \\ 39.9112 & 2.1157 \\ -3.9433 & 49.9684 \end{bmatrix}$	$K_{sb} = \begin{bmatrix} -0.7584 & -0.0131 \\ 0.0480 & -0.7212 \\ 0.0001 & -0.0001 \\ 0.0000 & 0.0001 \end{bmatrix}^T$
$F = \begin{bmatrix} 0.9863 & 0.0003 & 0.0002 & 0.0000 \\ 0.0003 & 0.9842 & -0.0000 & 0.0002 \\ -136.5 & 2.7635 & 0.966 & 0.0089 \\ 2.960 & -156.7 & -0.0184 & 0.9606 \end{bmatrix}$	$G = \begin{bmatrix} 0.0132 & 0.0007 \\ -0.0003 & 0.0161 \\ 130.96 & 7.4100 \\ -3.8668 & 159.96 \end{bmatrix}$

Table 3 Polynomial coefficients for fitted estimates According to Eq. (10)

$\hat{K}$	α	β	γ
$\hat{K}_{xx}(\Omega)$	9.78×10^{-4}	-12.9	7.4×10^5
$\hat{K}_{xy}(\Omega)$	-2.66×10^{-3}	21.97	-5.743×10^4
$\hat{K}_{yx}(\Omega)$	-2.511×10^{-3}	23.72	-5.94×10^4
$\hat{K}_{yy}(\Omega)$	-7.821×10^{-4}	3.607	7.814×10^5
$\hat{D}$	α	β	γ
$\hat{D}_{xx}(\Omega)$	-3.37×10^{-7}	4.3×10^{-3}	38.56
$\hat{D}_{xy}(\Omega)$	-4.26×10^{-6}	0.026	-68.84
$\hat{D}_{yx}(\Omega)$	-2.77×10^{-6}	0.03324	-41.09
$\hat{D}_{yy}(\Omega)$	5.418×10^{-6}	-0.04158	102.6
$\hat{B}$	α	β	γ
$\hat{B}_{xx}(\Omega)$	-0.017	165.6	2.91×10^5
$\hat{B}_{xy}(\Omega)$	-0.019	192.4	-4.3×10^5
$\hat{B}_{yx}(\Omega)$	8.5×10^{-3}	-91.46	2.31×10^5
$\hat{B}_{yy}(\Omega)$	-4.0×10^{-3}	29.98	7.57×10^5

Acknowledgments The authors would like to thank the Danish Ministry of Science, Innovation and Higher Education for the support to the FTP research project 12-127502.

References

1. Larsen JS, Santos IF (2013) Compliant foil journal bearings-investigation of dynamic properties. SIRM 2013
2. Powell JW (1970) A review of progress in gas lubrication. Rev Phy Technol 1(2):96–129
3. Morosi S, Santos IF (2012) Experimental investigations of active air bearings. In: ASME Turbo Expo 2012: turbine technical conference and exposition, American Society o Mechanical Engineers
4. Larsen JS, Varela AC, Santos IF (2014) Numerical and experimental investigation of bump foil mechanical behaviour. Tribol Int 74:46–56 (In print)
5. Heshmat H (1994) Advancements in the performance of aerodynamic foil journal bearings high speed and load capability. J Tribol 116(2):287–294
6. Heshmat H, Walowit JA, Pinkus O (1983) Analysis of gas lubricated compliant thrus bearings. J Lubr Technol 105:638–646
7. Heshmat H, Walowit JA, Pinkus O (1982) Analysis of gas-lubricated foil journal bearings J Lubr Technol 104:149–156
8. DellaCorte C, Lukaszewicz V, Valco MJ, Radil KC, Heshmat H (2000) Performance anc durability of high temperature foil air bearings for oil-free turbomachinery. NASA Technica Memorandum
9. Kim TH, San Andrés L (2008) Heavily loaded gas foil bearings: A model anchored to tes data. J Eng Gas Turbines Power 130(1):012504
10. Mizumoto H, Arii S, Kami Y, Goto K, Yamamoto T, Kawamoto M (1996) Active inheren restrictor for air-bearing spindles. Precis Eng 19:141–147
11. Mizumoto H, Arii S, Yabuta Y, Tazoe Y (2010) Vibration control of a high-speed air-bearing spindle using an active aerodynamic bearing
12. San Andrés L, Ryu K (2008) Hybrid gas bearings with controlled supply pressure to eliminate rotor vibrations while crossing system critical speeds. J Eng Gas Turbines Power 130 (4):042506
13. Morosi S, Santos IF (2011) Active lubrication applied to radial gas journal bearings. Part 1 modeling. Tribol Int 44(12):1949–1958
14. Dogan A, Uchino K, Newnham RE (1997) Composite piezoelectric transducer with truncated conical endcaps cymbal. IEEE Trans Ultrason Ferroelectr Freq Control 44(3):597–605
15. Ljung L (1999) System identification—theory for the user. Prentice-Hall, Upper Saddle River
16. Hendricks E, Jannerup O, Sørensen PH (2008) Linear systems control: deterministic anc stochastic methods. Springer, Berlin

Monitoring of the Damage in Rolling Element Bearings

Steven Chatterton, Pietro Borghesani, Paolo Pennacchi and Andrea Vania

Abstract Envelope Analysis of vibration signal is the most used and simplest method for the diagnostics of rolling element bearings. This method is based on the identification of bearing damage frequency components in the so-called Envelope Spectrum. If the recognition of the damage type is quite a simple task, the monitoring and the evaluation of the trend of a suitable damage index are complex tasks to be performed in an automatic way. The damage index must be robust against variations of system operating conditions and external vibration sources to avoid misleading results. In the paper, the case of a rolling element bearing in which the defect develops until a permanent failure is described as well as the algorithm implemented for alarm signaling.

Keywords Rolling element bearing · Envelope analysis · Square envelope spectrum · Bearing damage

1 Introduction

Rolling Element bearings (REBs) are often subjected to heavy operating conditions and aggressive environments and, therefore, they are prone to failure.

S. Chatterton (✉) · P. Pennacchi · A. Vania
Department of Mechanical Engineering, Politecnico di Milano, Via La Masa 1, 20156 Milan, Italy
e-mail: steven.chatterton@polimi.it

P. Pennacchi
e-mail: paolo.pennacchi@polimi.it

A. Vania
e-mail: andrea.vania@polimi.it

P. Borghesani
Queensland University of Technology, 2 George St, Brisbane, QLD 4000, Australia
e-mail: p.borghesani@qut.edu.au

J.K. Sinha (ed.), *Vibration Engineering and Technology of Machinery*,
Mechanisms and Machine Science 23, DOI 10.1007/978-3-319-09918-7_86

In the literature, most of the papers concerning bearing diagnostics, are focused on signal processing techniques applied to vibration measurements. Many different techniques have been developed for the detection of bearing damages [1]: from Envelope Analysis-EA up to recent developments of Square Envelope Spectrum-SES [2], 2nd order cyclostationary analysis [3, 4], Spectral Kurtosis-SK [5], Empirical Mode Decomposition-EMD [6] and Minimum Entropy Deconvolution-MED [7, 8]. Among these, the most used and simplest method for the assessment of the bearing health is the EA that is based on the identification of bearing damage frequency components in the so-called SES. The main critical point of this technique is the selection of a suitable frequency band for the demodulation of the vibration signal.

The selection of the frequency band, is based on the analysis of the peakedness of the analytical signal, by means of the Kurtosis index. The value of Kurtosis for a signal having a complex normal distribution is equal to 2.

Usually, the definition of the frequency band based on the band-Kurtosis (BK), that is the Kurtosis index of a filtered signal, is performed by mean of the Fast Kurtogram diagram, proposed for the first time by Antoni [9], or by more recent Protrugram by Barszcz and Jablonski [10], in which the value of BK is reported as function of the central frequency for a fixed bandwidth.

The constancy in the time of this optimal frequency band is also critical. In general, the constancy of this optimal band during the whole bearing life is particularly desirable. External phenomena, such as the temperature gradient of the bearings, can change the optimal bands suggested by the two diagrams.

The same BK index is assumed in the paper as damage indicator. Its simplicity allows the evaluation in the case of real-time applications. In the paper, the case of a rolling element bearing, in which the defect develops until a permanent failure, is described as well as the algorithm implemented for alarm signaling.

Vibration data are collected at a fixed time interval of 1 h. Current trend in industrial field is the adaptive monitoring in which the periodicity of the vibration measurement is varied accordingly to the trend of the damage indicator [11, 12].

2 Failure Description

In the paper, the damage occurred to one of the rolling element bearings of the frame supporting the output shaft of a train gearbox, shown in Fig. 1, is described.

This bearing was monitored by an accelerometer and a temperature dual-probe. After a period of about 39 millions of bearing cycles from the beginning of the experimental activity, high level of vibration had been detected by the sensor. The inspection revealed a severe damage of the bearing, mainly located in the inner ring with quite a number of brinelling marks, as shown in Fig. 2.

The damage was caused by an angular misalignment of the shaft with respect to the bearing house due to looseness of the frame support. In this operating condition, a local bearing overload occurred, leading to the premature failure.

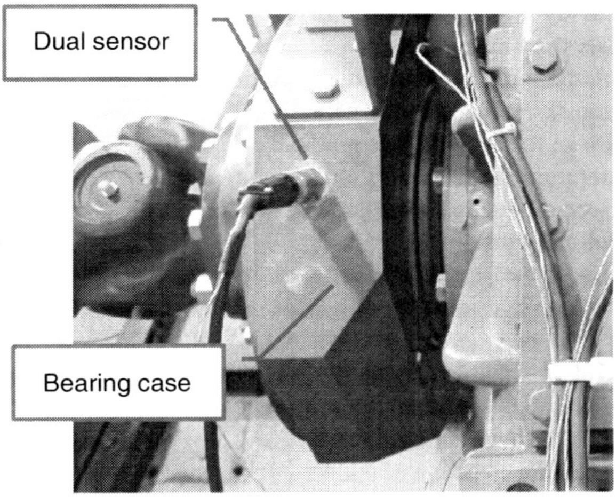

Fig. 1 Bearing case and position of dual sensor

Fig. 2 Inner ring of the damaged bearing after dismounting

Unfortunately, it was not possible to identify the time in which the misalignment of the shaft occurred. However, from the analysis of the vibration data acquired by the sensor it was possible to identify an increase of the damage indicator at about 5 millions of cycles in correspondence of a maintenance operation of the system. The bearing had been ran for about 100 days, for about 8 h in each day, at constant speed of about 970 rpm and constant radial load until the failure. After the failure, the bearing had been replaced.

3 Vibration Signal Analysis

The following algorithm has been used for the analysis of the vibration signal:

1. pre-evaluation of the signal on the basis of the gradient of the temperature. The first acquisition of each day is performed at cold temperature after about 10 min

from the start-up of the system. In this condition, the Kurtosis index is unstable. This suggests discarding vibration data when acquired in not a thermal steady state. The gradient of the temperature can be used as index of the thermal steady state. In the case analyzed in the paper, the system reached the steady state after about 2 h where the temperature gradient is less than 5 °C/h. The absolute value of the temperature is misleading due to the daily and seasonal variation of the room temperature, which oscillates between about 15 and 40 °C during the experimental activities.

2. band filtering of the signal in a suitable frequency band;
3. evaluation of damage level by means of the BK index of the analytic signal as function of time or cycles;
4. assessment of the damage type by the identification of damage bearing frequencies in the SES of the analytic filtered signal;
5. estimation of the Residual Useful Life (RUL) of the bearing on the basis of the Root Mean Square (RMS).

3.1 Band Filtering

The frequency band of the filter is characterized by two parameters: the bandwidth and the central frequency. The filter band has been selected on the basis of the maximum value of the BK. The trend of these two parameters in which the BK is maximum, is reported in Fig. 3 in a diagram called by the author "KurtoMap". It is

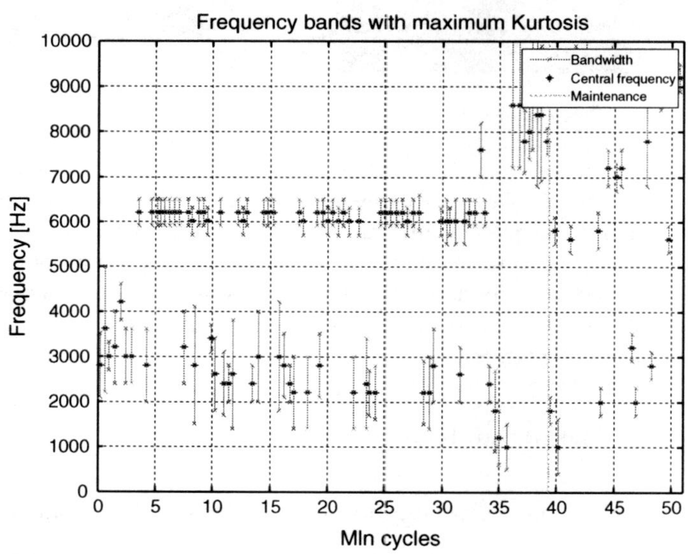

Fig. 3 KurtoMap for the frequency band selection

possible to observe two clusters: the first at a central frequency of 6,100 Hz and bandwidth of 1,000 Hz is assumed in the paper for tracking the damage index; the second at a central frequency of 2,500 Hz and bandwidth of 2,000 Hz.

3.2 Evaluation of Damage Level

The core of the algorithm is represented by the trend of BK index and RMS as shown in Fig. 4. In the figure, vibration data correspond to the 5th hour of test of each day when the system reached the thermal steady state. The optimal filter band is 5,625–6,875 Hz. The bearing has been replaced during the maintenance performed at about 39 millions of cycles.

By considering the BK and the RMS of the signal in Fig. 4, it is possible to identify two main stages of the damage evolution. The early stage lasts for about 34 millions of cycles. In this stage the BK deviates from the standard value of 2 to quite a constant value of 2.7, whereas the RMS value of the signal is very low.

The deviation from the standard value of 2 could be assumed as symptom of the damage growth. A threshold of 10 % with respect to the standard value of 2 could be used as alarm indicator. The transition between light and severe damage is indicated by simultaneous a drop of BK and an increase of RMS.

After the replacement of the damaged bearing with a new one (at about 39 millions bearing cycles), the BK index return within the threshold of 10 % with respect to the standard value of 2.

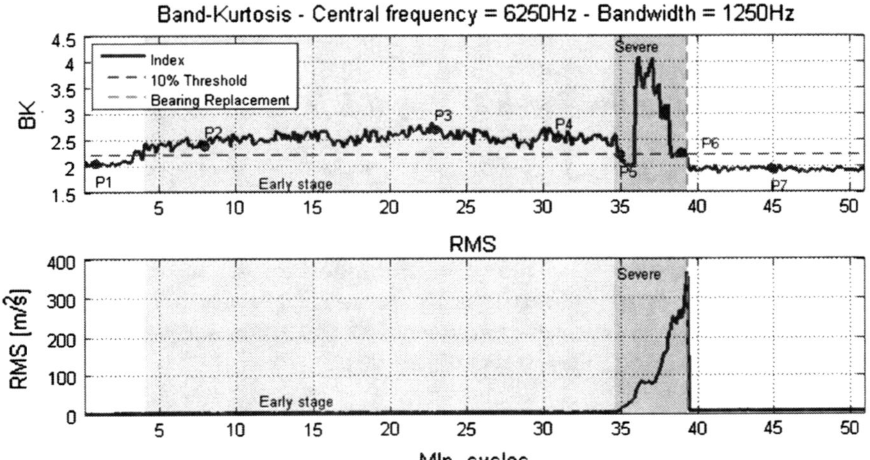

Fig. 4 Trend of BK and RMS values

The RMS of the filtered signal cannot give the same information provided by BK index. In the first part of the trend in Fig. 4, until about 34 millions of cycles the RMS is quite equal to zero.

3.3 Damage Type

The assessment of the damage type is based on the identification of fundamental defect frequencies of the bearing in the spectrum of the processed signal. The theoretical values of these frequencies depend only on the geometry of the bearing and represent the frequency of consecutive impacts of the defect on the rolling elements. Actual values of these frequencies are different from the theoretical ones. A frequency-shift, the "jitter", due to the actual kinematic contact among the elements of the bearing and depending mainly on the bearing load appears in real cases as described in [2, 13] and [14]. The following theoretical damage frequencies for the SKF NU1040M bearing running at 16.17 Hz are:

- Ball Pass Frequency Inner ring (BPFI) = 213.44 Hz;
- Ball Pass Frequency Outer ring (BPFO) = 174.64 Hz;
- Fundamental Train Frequency (FTF) = 7.28 Hz;
- Ball Spin Frequency (BSF) = 78.42 Hz.

Knowing the theoretical bearing defect frequencies, it is possible to identify the type of the bearing damage (inner ring, outer ring or rollers) by means of the SES [15–17].

The SESs of the analytic filtered signal are reported in Fig. 5 for different bearing cycles corresponding to several points indicated in Fig. 4. It is possible to observe in Fig. 5 the clear presence of cyclic frequency component close to the theoretical BPFI reported in dotted red line.

3.4 Estimation of the Residual Useful Life

As already said the RMS of the filtered signal cannot give the same information provided by BK index in the early stage of the damage, where its values is quite equal to zero till about 34 millions of cycles.

On the contrary, the RMS can give information for the second stage of the damage, where it is possible to forecast the residual useful life (RUL) of the bearing. In particular it is possible to interpolate RMS data samples by means of an exponential curve using different sets of samples:

$$y(c) = A \cdot e^{B \cdot (c - c_0)} \quad (1)$$

where c is in millions of cycles, and c_0 is the starting point of interpolation where the BK drops (at 34.9 millions of cycles).

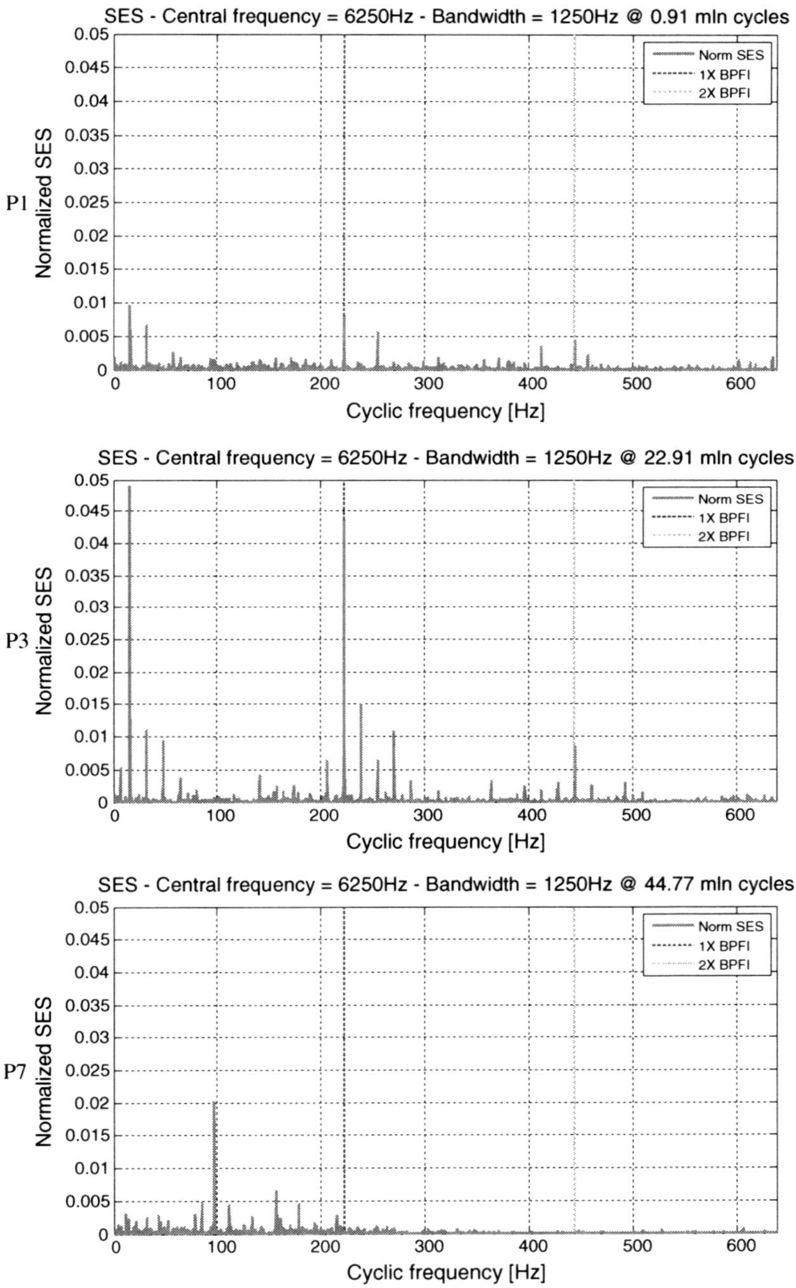

Fig. 5 Square Envelope Spectra for several bearing cycles

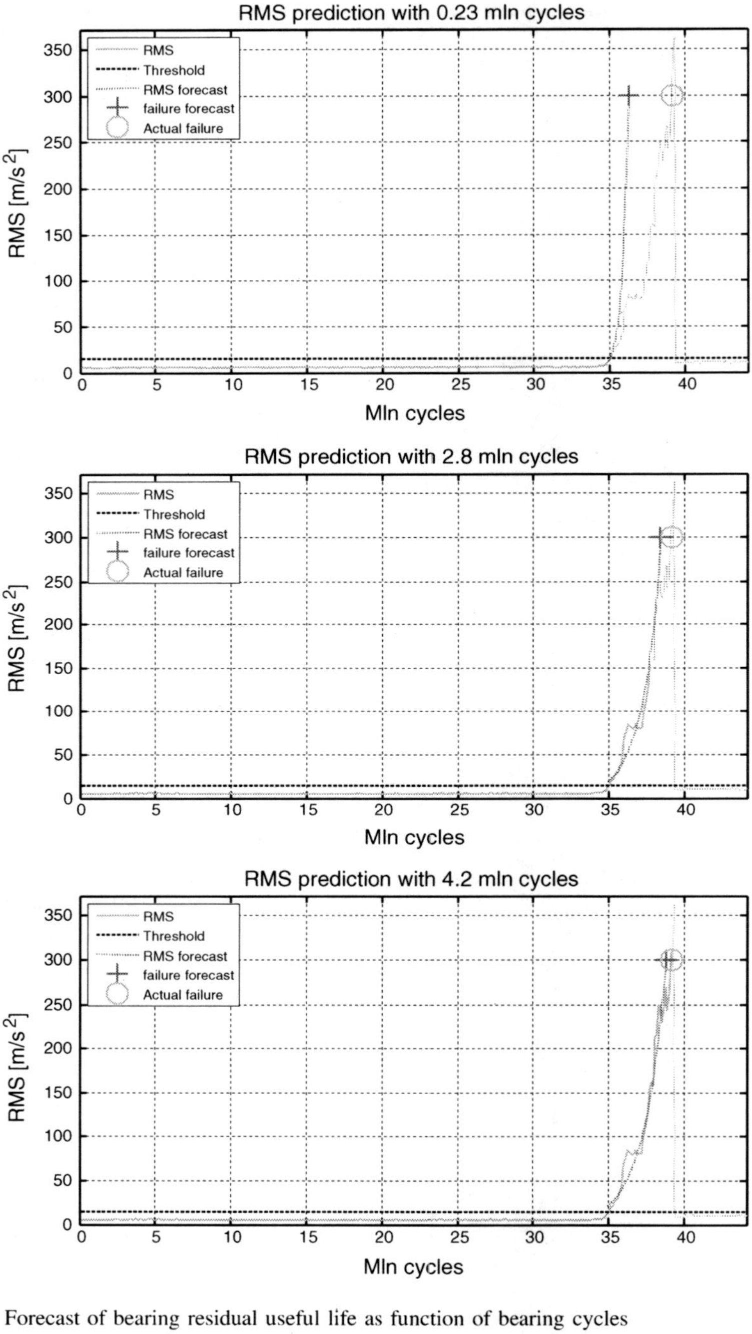

Fig. 6 Forecast of bearing residual useful life as function of bearing cycles

For instance, sets of 0.23, 2.8, and 4.2 millions of cycles are considered for the interpolating red dotted curve in Fig. 6, where a threshold of 300 m/s^2 is assumed as index of failure. The actual RMS curve is also reported in blue line. Obviously, the accuracy of estimation is better as the number of sample increases.

4 Conclusion

The problem tracking the level of damage in rolling element bearings has been discussed in the paper and an algorithm has been proposed as well.

The algorithm is based on an optimal use of standard tools as Envelope Analysis and Kurtosis index. The algorithm has been tuned and validated by means of experimental data of a failure occurred in an industrial rolling element bearing.

Eventually the estimation of the residual useful life on the basis of RMS value is also proposed.

References

1. Randall RB (2011) Vibration-based condition monitoring. Wiley, New York
2. Borghesani P, Ricci R, Chatterton S, Pennacchi P (2013) A new procedure for using envelope analysis for rolling element bearing diagnostics in variable operating conditions. Mech Syst Sig Process 38(1):23–35
3. Capdessus C, Sidahmed M, Lancome JL (2000) Cyclostationarity processes: application in gear fault early diagnosis. Mech Syst Sig Process 14:371–385
4. Randall RB, Antoni J, Chobsaard S (2001) The relationship between spectral correlation and envelope analysis in the diagnostics of bearing faults and other cyclostationary machine signals. Mech Syst Sig Process 15:945–962
5. Antoni J, Randall RB (2009) The spectral kurtosis: application to the vibratory surveillance and diagnostics of rotating machines. Mech Syst Signal Process 23:987–1036
6. Yu D, Cheng J, Yang Y (2005) Application of EMD method and Hilbert spectrum to the fault diagnosis of roller bearings. Mech Syst Signal Process 19:259–270
7. Randall RB, Sawalhi N (2011) Signal processing tools for tracking the size of a spall in a rolling element bearing. In: IUTAM symposium on emerging trends in rotor dynamics
8. Pennacchi P, Ricci R, Chatterton S, Borghesani P (2011) Effectiveness of MED for fault diagnosis in roller bearings. Vibration problems ICOVP 2011. Springer, Dordrecht. ISBN 978-94-007-2068-8. doi:10.1007/978-94-007-2069-5, pp 637–642
9. Antoni J (2007) Fast computation of the kurtogram for the detection of transient faults. Mech Syst Sig Process 21:108–124
10. Barszcz T, JabŁooski A (2011) A novel method for the optimal band selection for vibration signal demodulation and comparison with the Kurtogram. Mech Syst Sig Process 25 (1):431–451
11. Volkovas V (2011) Adaptable vibration monitoring. Vibration problems ICOVP 2011. Springer, Dordrecht. ISBN 978-94-007-2068-8. doi:10.1007/978-94-007-2069-5_80, pp 599–605
12. Volkovas V, Perednis A (2010) Adaptable vibration monitoring in rotor systems. J Vibro Eng 12(4):396–405

13. Pennacchi P, Chatterton S, Vania A, Ricci R, Borghesani P (2013) Experimental evidences in bearing diagnostics for traction system of high speed trains. Chem Eng Trans 33:739–744 doi:10.3303/CET1333124
14. Borghesani P, Pennacchi P, Chatterton S (2014) The relationship between kurtosis and envelope-based indexes for the diagnostic of rolling element bearings. Mech Syst Sig Proces 43:25–43
15. Chatterton S, Pennacchi P, Ricci R, Borghesani P, Vania A (2013) Development of a nev signal processing diagnostic tool for vibration signals acquired in transient conditions. Chen Eng Trans 33:61–66. doi:10.3303/CET1333011
16. Borghesani P, Pennacchi P, Chatterton S, Ricci R (2014) The velocity synchronous discret fourier transform for order tracking in the field of rotating machinery. Mech Syst Sig Proces 44(1–2):118–133
17. Borghesani P, Pennacchi P, Ricci R, Chatterton S (2013) Testing second orde cyclostationarity in the squared envelope spectrum of non-white vibration signals. Mec Syst Sig Process 40(1):38–55

Effect of Structural Dynamics on the Shaft Line Rotor Response in Turbomachines

E. Meli, G. Pallini, A. Rindi, F. Capanni and S. Rossin

Abstract The accurate model of the complicated dynamic phenomena characterizing rotors and support structure represents a critical issue in the rotor dynamic field. A correct prediction of the whole system behavior is fundamental to identify safe operating conditions and to avoid instability operating range that may lead to erroneous project solutions or possible unwanted consequences for the plant. Although a generic rotating machinery is mainly composed by four components (rotors, bearings, stator and supporting structure), many research activities are often more focused on the single components rather than on the whole system. The importance of a combined analysis of rotors and elastic supporting structures arises with the continuous development of turbomachinery applications, in particular in the Oil and Gas field where a wide variety of solutions, such as off-shore installations or modularized turbo-compression and turbo-generator trains, requires a more complete analysis not only limited to the rotor-bearing system. Complex elastic systems such as rotating machinery supporting structures and steel foundations might, in some situations, strongly dominate the entire shaft line rotor dynamic response (mode shapes, resonance frequencies and unbalance response). They give birth to transfer functions which will introduce coupling phenomena between machines bearings, becoming enablers of a new shaft line dynamic. Since FEM theory offers a number of different solutions to represent the rotor and the

E. Meli (✉) · G. Pallini · A. Rindi
Industrial Engineering, University of Florence, Florence, Italy
e-mail: enrico.meli@unifi.it

G. Pallini
e-mail: giovanni.pallini@unifi.it

A. Rindi
e-mail: andrea.rindi@unifi.it

F. Capanni · S. Rossin
Auxiliary Systems Engineering, GE Oil and Gas, Florence, Italy
e-mail: francesco.capanni@ge.com

S. Rossin
e-mail: stefano.rossin@ge.com

J.K. Sinha (ed.), *Vibration Engineering and Technology of Machinery*,
Mechanisms and Machine Science 23, DOI 10.1007/978-3-319-09918-7_87

rotating machine support system (beam models, solid models, transfer function etc.), in this paper a great emphasis is given to the results of an experimenta campaign done on a centrifugal compressor as validation of the new rotor dynamic approach.

Keywords Rotating machines · Shaft line · Structural dynamics · Dynamics stiffness · Bearings · Dynamic interaction

1 Introduction

This research activity aims to study how baseplates cause coupling transfer functions [1] between different bearings and the unavoidable need of the inclusion of this component in the analysis since project preliminary stages. There is clearly a limitation in the simplification of the supporting structure model and its effect on the dynamic of the system. As visible in Eq. (1), in the transfer function *Tf* the cross talking terms (K_{12}, K_{21}) link the DOFs of different bearings, as reported on Fig. 1 effecting the final dynamic response of the system.

$$Tf = \begin{bmatrix} K_{11} & K_{12} \\ K_{21} & K_{22} \end{bmatrix} \quad (1)$$

The appearance of cross talk terms deeply modify the response of the rotor [2, 5, 6], if compared to the classical approaches where $K_{12} \equiv K_{21} \equiv 0$ and the direct terms K_{11} and K_{22} act as a series of springs with the characteristic of bearings. In particular the influence of cross talk terms become more interesting in the second rocking or flexible mode of the rotor that is typical of the off-shore installation due to the isostatic anchoring system and to the deck flexibility.

The integration of the supporting structure dynamics may lead to the following:

- The system response highlights the modes of each individual component of the assembly;

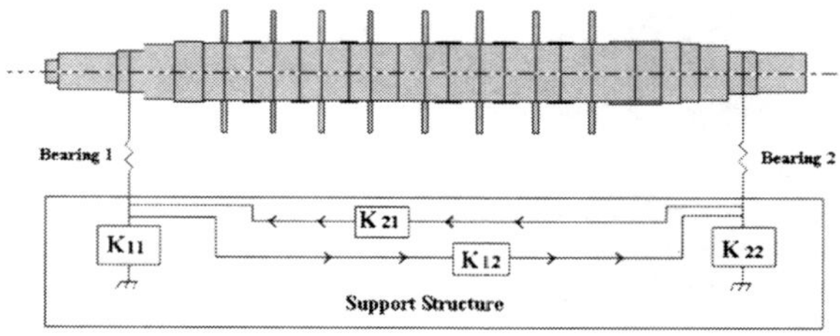

Fig. 1 Direct and cross talk terms

- The rotor modes shift in frequency without modification of the deformed shapes;
- The rotor keeps almost constant modes frequency but with different shapes.

As mentioned in [3], when the stiffness ratio $K_{support}/K_{brg} \leq 3.5$, the support flexibility begins to have a significant influence on the system's critical speeds and response characteristics. Such ratio is clearly a guideline to guarantee a good behaviour in the operating range; however the structural dynamic response is way from being constant in the machine speed range and a more deep characterization may be needed.

The general architecture (see Fig. 2) of this study highlights the mutual interaction between the three main components: rotor, bearings and support. The flow of local variables (forces, torques and kinematic variables) shows the primary importance of bearings as filter element interposed between the baseplate and the rotor. Bearings will be modelled as equivalent spring-damper, dependent on the running speed.

Two steps are required to accomplish the analysis: *the first step* considers the rotor mounted directly on foundations and there will not be any sort of model simplification; *the second step* leads to the reduction of calculation time instead, introducing assumption to represent the supporting structure in a simplified way. A scheme of the general architecture is visible in Fig. 2.

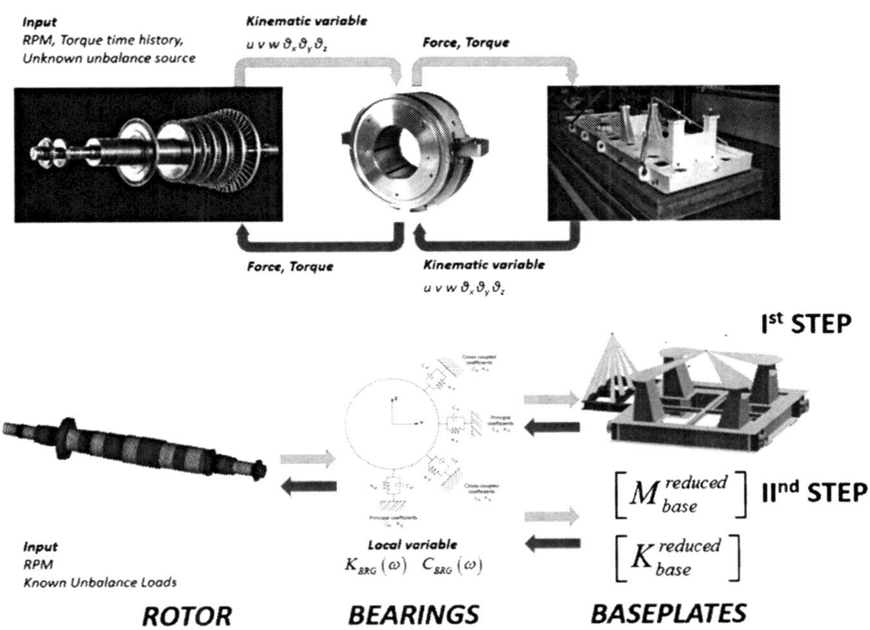

Fig. 2 General architecture

Fig. 3 FEM elastic support structure

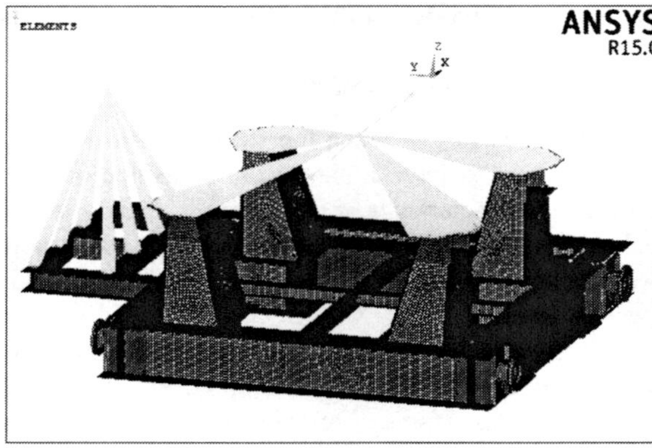

The calculation time reduction is strictly related to the use of Craig–Bampton Component Mode Synthesis technique [4] that generally can be considered an extension of the Guyan and modal reduction techniques.

The model in the first step is implemented using one of the most common FEM software, Ansys (Fig. 3), and then in the second phase is compared with $XLTRC^2$, a software fully dedicated to the rotordynamic field. To consider the dynamic contribution of the support structure, $XLTRC^2$ make use of the Mass and Stiffness reduced matrix as outcome of the Ansys CMS procedure.

This document highlights the correspondence between the rotor dynamic numerical investigation and experimental data acquisition on a machine start up high vibration issue experienced on a five stage MCL centrifugal compressor.

2 The Model

I Step: In the first step of the analysis the elements (rotor and support structure) are fully described considering six degrees of freedom and no simplification have been performed to reduce the dimension of the mathematical problem to solve.

The main component of the assembly is the rotor shaft. The rotor is the sum of many contributions and it must take into account the effect of the keyed and shrink fitted elements. The representation of the impeller simulate the increasing in mass due to the extra component and to properly describe the dynamics of the real part it includes concentrated mass and inertia features suitably distributed along the length of the shaft.

The resulting equation that describes the rotor is:

$$\begin{aligned}&\left\{\left[M_{shaft}\right]+\left[M_{sleeve}\right]+\left[M_{mi}\right]\right\}\left\{\ddot{q}_{rotor}\right\}\\&\quad+\left\{\left[C_{shaft}\right]+\Omega\left[G_{shaft}\right]+\Omega\left[G_{mi}\right]\right\}\left\{\dot{q}_{rotor}\right\}+\left[K_{shaft}\right]\left\{q_{rotor}\right\}=F\end{aligned} \tag{2}$$

where M_{shaft}, K_{shaft}, C_{shaft}, G_{shaft} respectively specifies the naked rotor mass, stiffness, damping and gyroscopic effect matrix, M_{sleeve} indicate the mass contribution of shrink fitted elements (like impeller), M_{mi} and G_{mi} explicit the dynamic contribute (concentrated properties m_x, m_y, m_z, I_p, I_t) of elements modelled as sleeves, then F represent external forces.

The F_y and F_z forces that the bearings exert on the rotor can be expressed in terms of linearized force coefficients for small perturbations about a stationary equilibrium at a given shaft speed. The forces are expanded as follows:

$$\{F_{BRG}\} = \begin{Bmatrix} F_y \\ F_z \end{Bmatrix} = \begin{bmatrix} -K_{yy} & -K_{yz} \\ -K_{zy} & -K_{zz} \end{bmatrix} \begin{Bmatrix} y \\ z \end{Bmatrix} + \begin{bmatrix} -C_{yy} & -C_{yz} \\ -C_{zy} & -C_{zz} \end{bmatrix} \begin{Bmatrix} \dot{y} \\ \dot{z} \end{Bmatrix} \tag{3}$$

where K_{yy}, K_{zz} and C_{yy}, C_{zz} are *Direct stiffness and damping coefficients*, while K_{yz}, K_{zy}, C_{yz}, C_{zy} represent the *Cross-coupling stiffness and damping coefficients.*

The entire baseplate, is bounded to the ground with spring element. The spring stiffness is extracted from dynamic stiffness simulations to satisfy the results of hammer test on the supporting pedestals.

Assembling the equations of the full system, Eq. (4), so introducing the effects of supports, the new set of coordinate $\{q\} = \{q_{rot}, q_{base}\}^T$ correspond to a rearrangement of the vector containing the coordinate of the node related to the rotor $\{q_{rot}\}$ and to the support structure $\{q_{base}\}$

$$\begin{aligned} &\{[M_{rotor}] + [M_{base}]\}\{\ddot{q}\} + \{[C_{rotor}] + [C_{base}] + \Omega[G_{rotor}]\}\{\dot{q}\} \\ &\quad + \{[K_{rotor}] + [K_{base}]\}\{q\} = F_{BRG} + F^{ext} \end{aligned} \tag{4}$$

II Step: The second step doesn't introduce any change to the rotor model.

The baseplate is the sum of different elements: *shell* to model the steel structure, *rigid connections* to represent joints and describe the casing, *concentrated mass and inertia properties* for the contribution of the centrifugal compressor casing to the assembly dynamic and *spring* to model the effect of the ground.

In the second step of the analysis a modal reduction of the baseplate (*fixed interface CMS*) has been performed to extract the reduced mass and stiffness matrices necessary to introduce the dynamic contribute of the supporting structure in XLTRC^2.

Master DOFs have been chosen on the boundary of the support structure and the result of this consideration tells that nodes that perfectly suit the circumstances are, as highlighted in Fig. 4 Master node location nodes on the extremity of elements that represent the linking between bearings and foundations. Acronyms DE and NDE, used to mark the left and right end, come from the original system where you can recognize the location of the torque input and actually DE means Drive End and NDE stand for Non Drive End.

As the goal of this analysis is to describe the flex-torsional behavior of the assembly each master node (located at the bearing interface) considers all the six DOFs (three rotations and three translations).

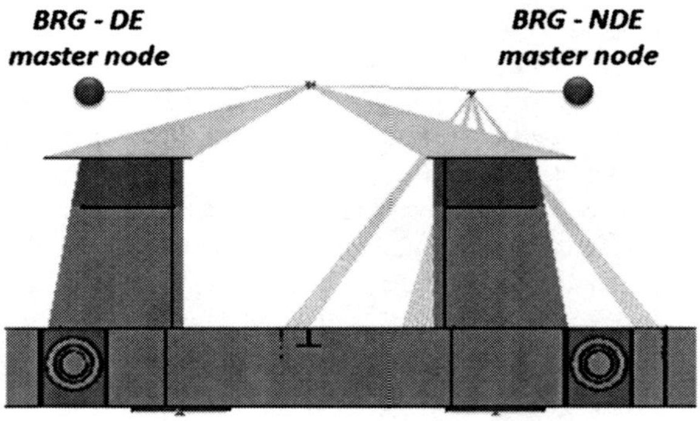

Fig. 4 Master node location

Considering the equation of the whole system, by this time the coordinate vecto contains the reduced coordinate of the baseplate $\{q_{red}\} = \{q_{rotor}, q_{base}^{red}\}^T$ and s the complete equation of the system turns into:

$$\begin{aligned} &\{[M_{rotor}] + [M_{base}^{red}]\}\{\ddot{q}_{red}\} + \{[C_{rotor}] + [C_{base}^{red}] + \Omega[G_{rotor}]\}\{\dot{q}_{red}\} \\ &\quad + \{[K_{rotor}] + [K_{base}^{red}]\}\{q_{red}\} = F_{brg} + F^{ext} \end{aligned} \quad (5$$

The biggest disadvantage of CMS approach is the increase of the problem size i compared to the matrices used with Static and Modal reduction. Furthermore CM provides dense matrices, neither diagonal, like modal reduction outputs, nor sparse as the original model. Despite all troubles, Craig–Bumpton's method guarantee excellent results with non-optimal master nodes distribution as well as in case of poor modal base, furthermore it gives great benefits for the reduction of calculatio time and matrices are always smaller than the original problem.

In general the fixed interface CMS considers only a subset of the constraine eigenmodes and the dimension of the matrix depends on a frequency band chose by the user $[0,\omega_{max}]$ that is a function of the frequency domain to investigate.

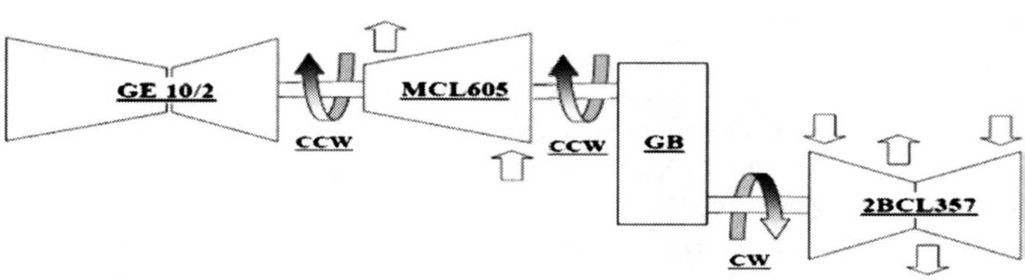

Fig. 5 Rotor train

3 Experimental Data: Turbo Compression Train

As result of a high shaft vibration issue on a centrifugal compressor experienced on field, a great number of vibration data have been gathered on both structure and proximity probes.

The compression system was made up of a gas turbine with two centrifugal compressors and one gearbox in one single shaft line (Fig. 5). Each rotating machine was installed on a separate baseplate steel structure and the individual rotors were connected by elastic coupling.

High vibration was detected at the proximity probes of both bearings at one of the two compressors, compromising anyway the proper train start up sequence. From the preliminary data acquisition on structure vibration reviled an unexpected response close to the operating speed but not to the extent to justify the train trip shut down.

Solution to the problem was the insertion of concrete underneath the baseplate to redefine a suitable anchoring system.

Several seismic probes were positioned at the compressor pedestals (Fig. 6 Seismic probes arrangement, steel baseplate structure and compressor casing, measuring vibration level along the horizontal and vertical directions (amplitude and phase)).

The structure mode shapes and frequencies were retrieved from a sequence of machine start up and shut down by means of a dedicated post processing procedure called Operating Modal Analysis (OMA). The extracted modes shape and frequencies were then compared to the one computed by dedicated FE models with the purpose of verification and subsequent model tuning.

Due to the complexity of the mechanical system under investigation and the strong focus on the effect of the structural dynamics on the overall shaft dynamics, a great accuracy of the baseplate FE model was fundamental for a robust experimental and numerical system validation.

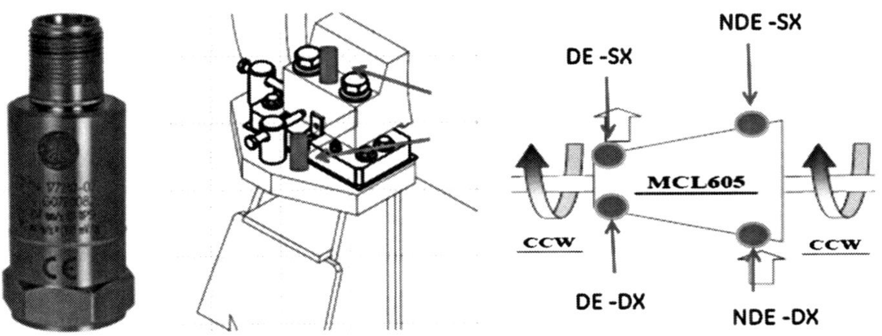

Fig. 6 Seismic probes arrangement

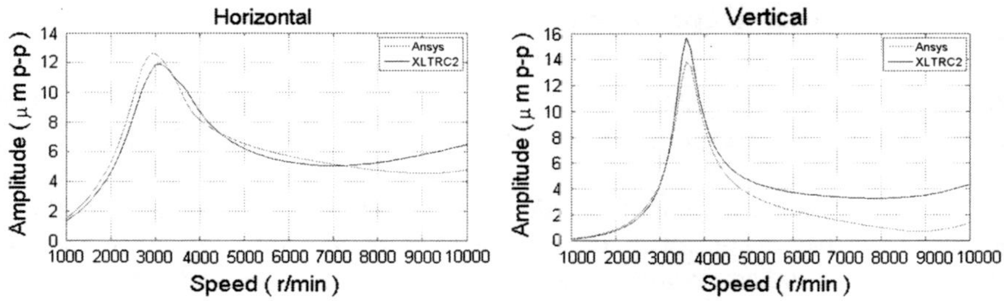

Fig. 7 Amplitude absolute displacements, first mode, only rotor, non drive end

4 Results and Major Findings

The amplitude of the synchronous unbalance U used for the system response is described in Eq. (6). According to the API standard the equation below represents the load to be applied at the location of maximum displacement, that to stimulate the first deformed shape corresponds to the bearings mid span

$$U = 4[6350(W/N)]\ [gmm] \quad (6)$$

where N represents the maximum continuous operating speed in rpm and W is the journal static weight load in kilograms.

Comparing the results obtained from both modelling software (Ansys and XLTRC^2) the response of the rotor without the support structure flexibility is revealing a strong match in the vicinity of the first shaft resonance mode (Fig. 7).

Even though one can appreciate the good match on the first rotor mode frequency (critical speed) there is still a substantial deviation for high frequencies as well as between vertical and horizontal directions.

Taking a close look to the dynamic stiffness of the supporting structure and comparing that with both the bearing stiffness and the limit identified from API617 as 3.5 times the bearings stiffness it is possible to see that indeed the structure on its own shows a degrading stiffness for shaft speed lower than the maximum operating speed. In Fig. 8 the areas below the bearing stiffness indicate a lack of stiffness along the vertical direction and so a potential "critical speed range".

When the support structure is included in the whole mechanical system, the overall picture of the shaft unbalance response drastically changes. The first shaft mode slightly shifts towards the high frequencies and the structural modes emphasize themselves as shaft response. Once again in the next picture (Fig. 9) the comparison of the models implemented in Ansys and XLTRC^2 underlines the good match between the two models after the introduction of the support system dynamics.

As the proximity probes measure a relative displacement between the compressor casing and the shaft, a proper post processing need to be done for a better

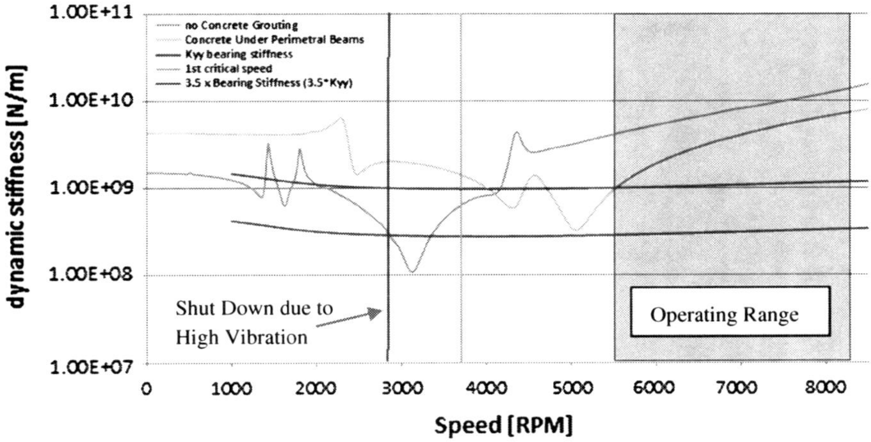

Fig. 8 Dynamic stiffness in vertical direction

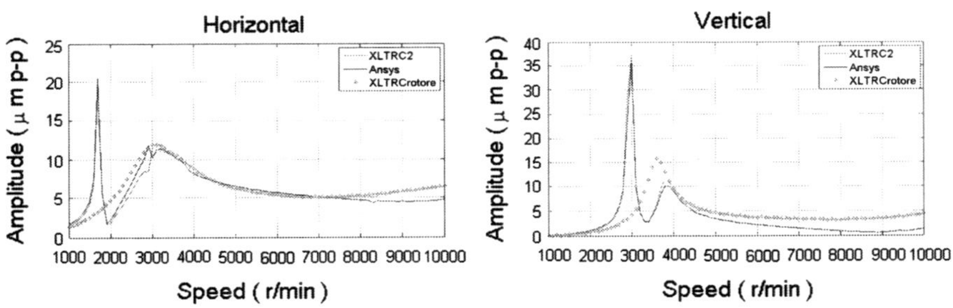

Fig. 9 Absolute displacements—rotor at bearing station

comparison with the experimental data (Fig. 11 relative displacements response of the system before and after concrete reinforcement).

The introduction of the concrete under the supporting structure produces a shift of the structural resonance (highlighted in the experimental test at 3,000 rpm) and this can be clearly seen on both the computational (Fig. 11) and experimental results (Fig. 10).

The response of the complete mechanical system with and without the additional concrete grouting (*grouting*—contraction *GR*) is overlaid to the dynamic response of the rotor alone (Fig. 11) understanding how much the study of the rotor in "stand alone" condition can deceive a rotordynamic study.

Looking at the grouting versus no grouting shaft response (Fig. 12 right column) the evidence of the strong benefit on the critical behavior at 3,000 rpm is clear. The proximity probe measurement retrieved after the additional concrete grouting (Fig. 12 left column) is in strong agreement with the final computational results even in sharp peak at pure structural response.

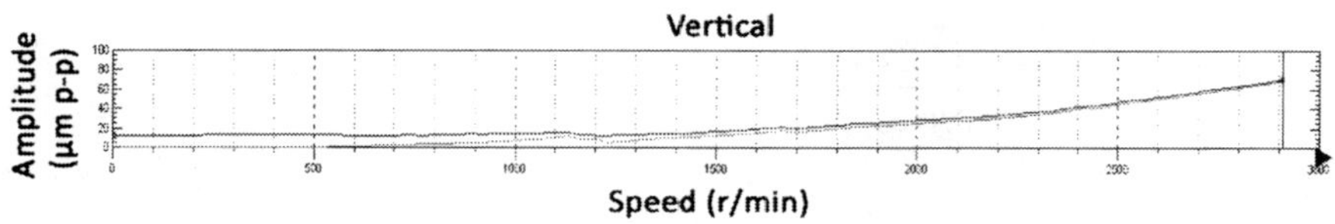

Fig. 10 Run up shut down condition: exceeded vibration level

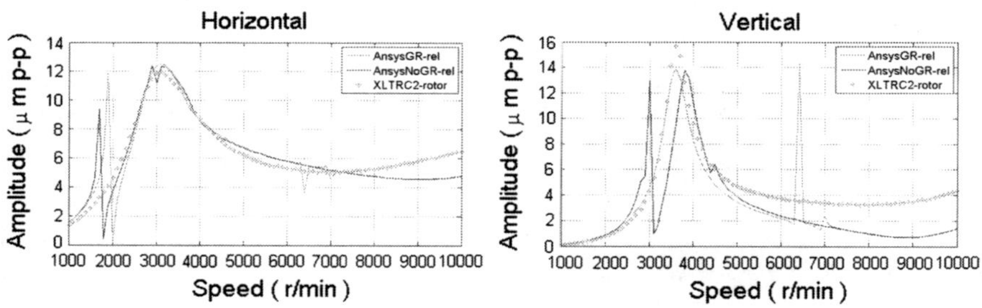

Fig. 11 Relative displacements response of the system before and after concrete reinforcement

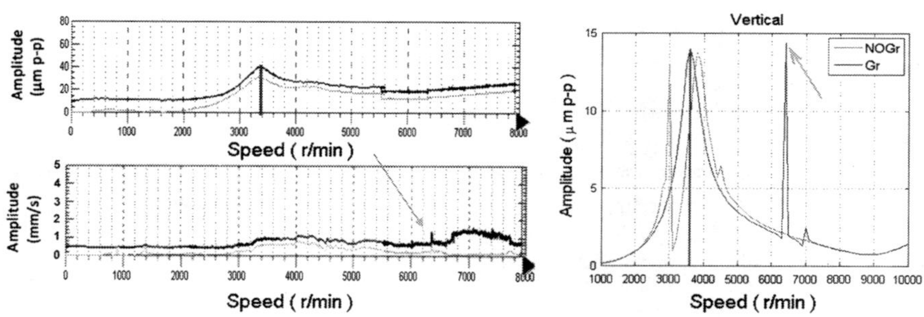

Fig. 12 Comparison between simulated (*right column*) relative displacements and experimental (*left column*) data

5 Conclusions

The behavior of a Rotor machine is not simple to predict and many tools and theories need to be compared at the same time to have a complete vision of the real behavior.

It is not sufficient to evaluate the different component of the rotating machine separately, it is important to have a deep knowledge of the entire assembly behavior (rotor, bearing and support structure) to predict the mutual influence of design parameter and allow different parallel working team to elaborate the problem with the right input.

The model, oriented in creating an accurate tool for the rotor dynamic of the whole machine, is at the moment a good improvement in terms of time and accuracy.

A future development is the representation of a complete train shaft line with experimental match out of dedicated experimental data.

Acknowledgment A special thanks to Valentina Peselli from Enginsoft s.r.l. for managing a thorough FE model debugging, finding the appropriate element selection within Ansys model, helping the team to find the pitfall of some element formulation which could not otherwise drive us to the result validation.

References

1. Cuppens K, Sas P, Hermans L (2000) Evaluation of the FRF based substructuring and modal sinthesis technique applied to vehicle FE data
2. Vazquez J, Barrett L (1999) Transfer function representation of flexible supports and casings of rotating machinary. In: IMAC-XVII-17th international modal analysis conference 1999
3. API Pubblication 684 (1996) Tutorial on the API standard paragraphs covering rotor dynamics and balancing: an introduction to lateral critical and train torsional analysis and rotor balancing
4. Bampton MCC, Craig RR (1968) Coupling of substructures for dynamic analysis. AIAA J 6 (7):1968
5. Kruger T, Liberatore S, Knopf E (2013) Complex substructures and their impact on rotor dynamic. In: SIRM 2013—10th international conference on vibrations in rotating machines, 2013
6. Ehehalt U, Luneburg B, Daniel C, Strackeljan J, Woschke E (2009) Methods to incorporate foundation elasticities in rotordynamic calculation. In: SIRM 2009—8th international conference on vibrations in rotating machines, 2009

Nonstationary Processes Studying Based on "Caterpillar"—SSA Method

Svetlana Polukoshko, Galina Hilkevica and Vladimirs Gonca

Abstract This paper deals with "Caterpillar"-SSA method—a new powerful model-free method of time series analysis and forecasting. It allows to decompose the nonstationary time series into trend, periodic components (sum of elementary harmonic), noise (random components) and to forecast subsequent behavior of system, if separability condition of time series and possibility of reverse synthesis of individual elements are fulfilled. This paper shows application of this method for random and nonlinear vibrations study on the example of building element vibration under seismic action.

Keywords Nonstationary time series · Singular spectrum analysis · Trend · Harmonic component · Noise · Forecasting

1 Introduction

Contemporary real-world systems (technical, economic, social, environmental, etc.) have very complex structure because of the complexity of their interconnection and influence of a large number of factors; it is not always possible taking into account all these action. The system can change the mode and the structure of its elements; such changing predetermines the new state of the system, qualitatively different from the previous ones, and unstationary development of all its processes. These

S. Polukoshko (✉) · G. Hilkevica
Engineering Research Institute "VSRC", Ventspils University College,
Inzenieru Str. 101, Ventspils 3601, Latvia
e-mail: pol.svet@inbox.lv

G. Hilkevica
e-mail: galina.hilkevica@venta.lv

V. Gonca
Institute of Mechanics,
Riga Technical University, Ezermalu Str. 6, Riga 1006, Latvia
e-mail: vladimirs.gonca@rtu.lv

J.K. Sinha (ed.), *Vibration Engineering and Technology of Machinery*,
Mechanisms and Machine Science 23, DOI 10.1007/978-3-319-09918-7_88

problems do not allow describing in detail the processes, using traditional approaches, particularly mathematical description in the form of the systems of differential equations. Describing the dynamics of a complex system and predicting of its future behavior may be fulfilled by studying of the real process sampling, i.e. nonstationary time series. Property of nonstationarity in complex systems is manifested by the origin of trends, characterized by fundamental dependencies in the system, and some jumps, the sources of which are random influence of individual factors. A model-free technique of time series analysis "Caterpillar"- SSA is usable tool for such problems solving. "Caterpillar" method is new powerful method of time series analysis and forecasting, which was independently developed in Russia and also in UK and USA (under the name of SSA-Singular Spectrum Analysis) [1, 3, 5, 10, 15]. The "Caterpillar"-SSA approach is used to study the time series in many various spheres wherever the trend or periodic behaviour can present, for instance, in meteorology, climatology, hydrology, geophysics, medicine, biology, sociology [2, 4, 6, 9, 13, 14]. It combines advantages of other methods, such as Fourier and regression analyses, with simplicity of visual control aids. A lot of problems can be solved by means of "Caterpillar"-SSA technique: finding trends of different resolution, smoothing, extraction of the seasonality components, change-point detection, simultaneous extraction of cycles with small and large period, simultaneous extraction of complex trends and periodicities; extraction of periodicities with varying amplitudes, finding structure in short time series, continuation of the time series [1, 4]. A modification of singular spectrum analysis for time series with missing data is developed and successfully tested. This method also can be used to low pass filter incomplete time series [11, 12]. Method is extensively developing, the variants of the "Caterpillar"-SSA method for analysis of multidimensional time series (MSSA), detection of structural changes are elaborated [7, 8, 11].

Last decade the methods for nonstationary and nonlinear time series based on Empirical Mode Decomposition (EMD) are developed. EMD method used to decompose series as sums of zero-mean components. The technique consists basically of computing an upper and lower spline envelope of the time-series and extracting the average of this envelope up to when the residuals are random noise or stopping criterion is met. SSA method is more mathematically grounded, while EMD is more empirical. The aim of SSA is to make a decomposition of the original series into the sum of independent and interpretable components: trend—sum of slowly varying components, oscillatory components—sum of elementary harmonic components and a structureless noise—the random component formed by a set of random factors. The "Caterpillar"—SSA method is useful tool for short and long, one-dimensional and multidimensional, stationary and nonstationary, almost deterministic and noisy time series analyzing. However, in mechanics application this method does not become widely used, but it may be successfully used for periodic motion detection, amplitude and frequency definition, as well as chaotic motion identification. The purpose of this work is a simple example to show the principle of the method, its ease of use and accessibility for mechanical applications.

2 SSA Algorithm

The basic version of method consists in transformation of one-dimensional series into multi-dimensional by one-parameter translation procedure, research of received multidimensional trajectory by means of principal components analysis (Singular Value Decomposition) and reconstruction of the series in accordance with the chosen principal components. Thus, the result of application of method is expansion of the time series into sum of simple components: slow trends, seasonal and other periodic or oscillating components, and also components of noises. The obtained decomposition may serve as basis of correct forecasting. Both of two complementary stages of SSA technique (decomposition and reconstruction) include two separate steps, hence the analysis is performed by four steps.

At the first step (called the embedding step) a one-dimensional series $F = (f_0, \ldots, f_{N-1}), f_n \in \mathbf{R}$, with length $N \geq 3$ is transferred to the L-dimensional series $X_i = (f_{i-1}, \ldots, f_{i+L-2})^T$, $i = 1, \ldots, K$, where $K = N - L + 1$ and L is called by window length, $L \in \mathbf{N}$, $1 < L < N$. This delay procedure gives the first name to the whole technique. Very important parameter of the embedding step is the window length L, it should be big enough but not greater than a half of series length. Vectors $\mathbf{X}_i$ form columns of the trajectory matrix $\mathbf{X} = [X_1 : \ldots : X_K]$, or:

$$\mathbf{X} = (x_{ij})_{i,j=1}^{L,K} = \begin{pmatrix} f_0 & f_1 & f_2 & \cdots & f_{K-1} \\ f_1 & f_2 & f_3 & \cdots & f_K \\ f_2 & f_3 & f_4 & \cdots & f_{K+1} \\ \vdots & \vdots & \vdots & \ddots & \vdots \\ f_{L-1} & f_L & f_{L+1} & \cdots & f_{N-1} \end{pmatrix} \quad (1)$$

Trajectory matrix $\mathbf{X}$ is a Hankel matrix, which mean that all the elements along the diagonal $i + j = const$ are equal.

The second step, which gives the second name of technique, SVD step, is the singular value decomposition of the trajectory matrix into a sum of rank-one bi-orthogonal elementary matrices:

$$\mathbf{X} = \mathbf{X}_1 + \ldots + \mathbf{X}_d. \quad (2)$$

Elementary matrix $\mathbf{X}_i$ is determined by the equality $X_i = \sqrt{\lambda_i}\mathbf{U}_i\mathbf{V}_i^T$, collection $(\sqrt{\lambda_i}, \mathbf{U}_i, \mathbf{V}_i^T)$ is called the i-th eigentriple of the matrix $\mathbf{X}$. Here $\sqrt{\lambda_i}$ is the square root of the i-th eigenvalue (i-th singular value) of the matrix $\mathbf{S} = \mathbf{X}\mathbf{X}^T$, $\mathbf{U}_1, \ldots, \mathbf{U}_L$- orthonormal set of eigenvectors of matrix $\mathbf{S}$, $d = \max\{i : \lambda_i > 0\}$, $\mathbf{V}_i = \mathbf{X}^T\mathbf{U}_i/\sqrt{\lambda_i}$. $\mathbf{U}_i$ and $\mathbf{V}_i$ are called as left and right singular vectors of the trajectory matrix. It is assumed that eigenvalues λ_i are arranged in the decreasing order of their magnitude ($\lambda_1 \geq \ldots \geq \lambda_L \geq 0$).

The third step (grouping step) corresponds to splitting the elementary matrices $\mathbf{X}_i$ into several groups and summing the matrices within each group. Let $I = 1, \ldots, d$, let us spilt of this set into the disjoint subsets $I_1, I_2, \ldots, I_m$, such that $I = \cup_{k=1}^{m} I_k$.

If $I_k = \{i_1, \ldots, i_p\}$ then $\mathbf{X}_{I_k} = \mathbf{X}_{i_1} + \ldots + \mathbf{X}_{i_p}$. The procedure of choosing the sets $I_1, \ldots, I_m$ is called the eigentriple grouping. For given group I_k the contribution of the component $\mathbf{X}_{Ik}$ into the expansion is measured by the share of the corresponding eigenvalues.

The last step (diagonal averaging) transfers each resultant matrix into a Hankel matrix, from which then the time series will be reconstructed (the additive components of the initial series F). How to create the Hankel matrix from any matrix $\mathbf{Y}$ and to receive the corresponded series is shown below. Let y_{ij} stands for an element of $L \times K$ matrix $\mathbf{Y}$, where $1 \le i \le L$, $1 \le j \le K$. Let denote $L^* = \min(L, K)$, $K^* = \max(L, K)$, $y^*_{ij} = y_{ij}$ if $L < K$, and $y^*_{ij} = y_{ji}$ otherwise. Diagonal averaging transforms the matrix $\mathbf{Y}$ into series $g_0, \ldots, g_{N-1}$, where $N = L + K - 1$, by formula:

$$g_k = \begin{cases} \frac{1}{k+1}\sum_{m=1}^{k+1} y^*_{m,k-m+2}, & \text{for } 0 \le k < L^* - 1, \\ \frac{1}{k+1}\sum_{m=1}^{k+1} y^*_{m,k-m+2}, & \text{for } L^* - 1 \le k < K^*, \\ \frac{1}{N-k}\sum_{m=k-K^*+2}^{N-K^*+1} y^*_{m,k-m+2}, & \text{for } K^* \le k < N. \end{cases} \quad (3)$$

Elements of Hankel matrix is obtained by averaging of y_{ij} over all i, j such that $i + j = k + 2$. In this way we obtain a decomposition of the initial series into several additive components. The result is the expansion of matrix $\mathbf{X}_{I_k}$ into simple components $F^{(k)} = (f_0^{(k)}, \ldots, f_{N-1}^{(k)})$ and initial series $F = (f_0, \ldots, f_{N-1})$ is decomposed into the sum of m series: $F = \sum_{k=1}^{m} F^{(k)}$, that is $f_n = \sum_{k=1}^{m} f_n^{(k)}$.

Forecasting time series, unlike the analysis, is only possible involving the mathematical model of the series. The model must either originate from the data itself, or be tested on the available data. When forecasting with SSA method, we consider a set of time series described by linear recurrent formulae (LRF). The class of time series governed by linear recurrent formulae is rather wide; it includes harmonics, polynomial and exponential time series. The time series governed by LRF generates recurrent extension as each its term is a linear combination of some of the previous ones. SSA forecasting method can be applied to the time series that approximately satisfy linear recurrent formulae.

From the original series $F_N = (f_0, \ldots, f_{N-1})$ after embedding procedure we receive the sequence of vectors $X_i^{(L)} = X_i(f_{i-1}, \ldots, f_{i+L-2}),^T i = 1,2, \ldots K$. Trajectory space of the series F_N is $\pounds^{(L)} = span(X_1, \ldots, X_K)$. Let $0 < d < L$; if dim $\pounds^{(L)} = d$, then series F_N has L -rank d: $rank_L(F_N) = d$ under condition of $d \le min(L,K)$. Number of $rank(F_N) = rank\boldsymbol{X} = rank\mathbf{XX}^T = rank\boldsymbol{X}^T\boldsymbol{X}$ is equal to the order of the singular value decomposition of the trajectory matrix $\boldsymbol{X}$.

Orthonormalized system of eigenvectors corresponding to the positive eigenvalues of matrix $\mathbf{XX}^T$ composes the orthonormalized basis of the space $\pounds^{(L)}$. The time series F_N has a dimension $fdim\,(F_N) \le d$ if $1 \le d \le N - 1$ and there $a_1, \ldots, a_d$ may be found as: $f_{i+d} = \sum_{k=1}^{d} a_k f_{i+d+k}$. For an eigenvector $\mathbf{U} \in \mathbf{R}^L$ we denote the vector of the first $L-1$ components of the vector U as $\mathbf{U}^{\nabla} \in \mathbf{R}^{L-1}$. Set $\Pi^2 = \pi_1^2 + \ldots.. + \pi_d^2 < 1$, where π_i is the last component of the eigenvector $\mathbf{U}_i$ $(i = 1, \ldots, d)$ (it is supposed that the space is not vertical). The last component $f_L$of any vector $\boldsymbol{F}$ is the linear combination of the first components($f_0, \ldots, f_{L-1}$); that is $f_L = a_1 f_{L-1} + \ldots + a_L f_0$, where the vector of coefficient $A = (a_1, \ldots, a_{L-1})$ may be expressed as $A = \sum_{i=1}^{d} \pi_i^2 U_i^{\nabla} / \sum_{i=1}^{d} \pi_i^2 U_i^{\nabla}(1 - \Pi^2).(1 - \Pi^2)$.The forecasts on the h steps $f_N, \ldots, f_{N-1+h}$ are obtained as: $f_i = f_i$ for $i = 0, \ldots, N-1$ and $f_i = \sum_{j-1}^{L-1} a_j f_{i-j}$ for $i = N, \ldots, N-1+h$. The main concept in studying SSA properties is separability, because decomposition of the series F_N may be successful only if the resulting additive components of the series are approximately separable from each other [1, 5]. If a time series F_N may be split as $F_N = F_N^{(1)} + F_N^{(2)}$, then the matrix terms of the SVD step may be split into $\mathbf{X}^{(1)}$and $\mathbf{X}^{(2)}$ respectively, i.e. $\mathbf{X} = \mathbf{X}^{(1)} + \mathbf{X}^{(2)}$. This would imply that each row of $\mathbf{X}^{(1)}$ is orthogonal to each row of $\mathbf{X}^{(2)}$. Since rows (and columns) of the trajectory matrix $\boldsymbol{X}$ are themselves subseries of the initial series, the orthogonality condition of the rows of $\mathbf{X}^{(1)}$and $\mathbf{X}^{(2)}$ is the condition of orthogonality of any subseries of length L and $K = N - L + 1$ of the series $F_N^{(1)}$ to any subseries of the same length, $F_N^{(2)}$. If this holds, then $F_N^{(1)}$ and $F_N^{(2)}$ are said to be weakly separable. In geometrical terms, $F_N^{(1)}$ and $F_N^{(2)}$ are separable if the subspace $\mathcal{L}^{(L,1)}$ spanned by the columns of $\mathbf{X}^{(1)}$ is orthogonal to the subspace $\mathcal{L}^{(L,2)}$ spanned by the columns of $\mathbf{X}^{(2)}$. One way to measure the separability between two series components, $F_N^{(1)}$ and $F_N^{(2)}$ (i.e. $F_N = F_N^{(1)} + F_N^{(2)}$) is to calculate the weighted correlation (w-correlations) using the formulas, where $i = 1, 2$:

$$\rho_{12}^{w} = \frac{\left\langle F_N^{(1)}, F_N^{(2)} \right\rangle_w}{\left\| F_N^{(1)} \right\|_w \left\| F_N^{(2)} \right\|_w}, \quad \left\| F_N^{(i)} \right\|_w = \sqrt{\left\langle F_N^{(i)}, F_N^{(i)} \right\rangle_w},$$

$$\left\langle F_N^{(1)}, F_N^{(2)} \right\rangle_w = \sum_{i=0}^{N-1} w_i f_i^{(1)} f_i^{(2)}.$$

The weights w_i is defined as follows: let $L^* = \min(L,K)$ and $K^* = \max(L,K)$, then $w_i = i + 1$ for $0 \le i < L^* - 1$, $w_i = L^*$ for $L^* \le i < K^*$, $w_i = N - 1$ for $K^* \le i < N - 1$.

A natural hint for grouping is the matrix of the absolute values of the w-correlations corresponding to a full decomposition. If the absolute value of the w-correlation is small then the corresponding series are almost w-orthogonal and is said to be weakly separable. The series $F^{(1)}$ and $F^{(2)}$ are w-orthogonal if $\langle F_N^{(1)}, F_N^{(2)} \rangle_w = 0$. Separability is analogous to independence of random variables whence the covariance and correlation between such random variables are zero.

Another way to enhance separability of the series is auxiliary information about the series to help in choosing the window length, for example, it is preferable to take the window length proportional to supposed period in order to achieve the better separability of the periodic components. Under the proper choice of window length L singular vectors in a sense 'repeat' the behavior of the corresponding tim series components: in particular, trend of the series corresponds to slowly varyin singular vectors, harmonic component produces a pair of left (and right) harmoni singular vectors with the same frequency, etc. The graph of the logarithms o eigenvalues is also very useful in eigentriples choosing: plateau in the eigenvalu spectra prompts ordinal numbers of the eigentriple and a slowly decreasin sequence of singular values corresponds to noise components. The problem o selection of "Caterpillar"- SSA parameters is thoroughly discussed (from theoretica and practical viewpoints) in [3, 5]. In work [5] the cooperation between SSA an Independent Component Analysis (ICA) has been considered in applications t blind signal separation. In this application Basic SSA is used for removal of nois and then ICA is applied for the extraction of independent components. The theor of ICA is developed for random variables and is not applicable in the deterministi case. The aim of ICA is finding statistically independent components from obser vations of their linear combinations. ICA has the important drawback: it does nc make ordering of the found components (vectors) like SVD does. By no means IC can be recommended as a full replacement of SVD. SVD provides the best matri approximations to the trajectory matrix; it often delivers the proper decompositio from the viewpoint of weak separability.

3 Numerical Example of "Caterpillar"-SSA Method Application

In recent years due to SSA method developing, corresponding computer programs ar elaborated for time series analysis: "Caterpillar"—SSA and Rssa in St-Petersbur University [1, 3], SSA-MTM Toolkit for Spectral Analysis, kSpectra Toolkit [15 MATLAB program is also widely used. In this work MathCAD program is applied a

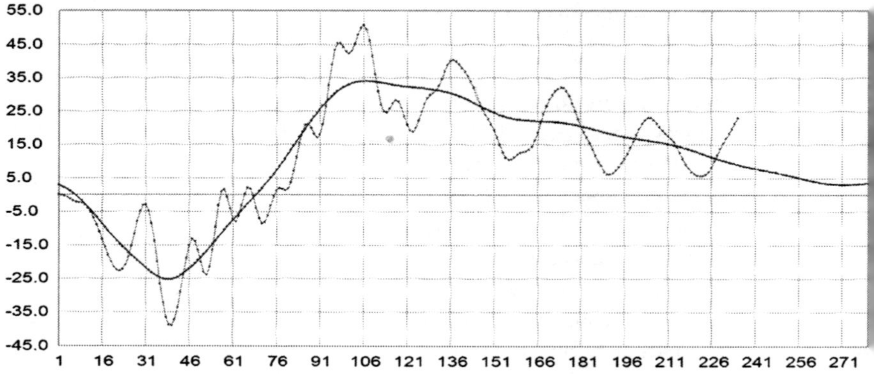

Fig. 1 Initial series (*line with dot*) and its trend (*tree eigentriples*)

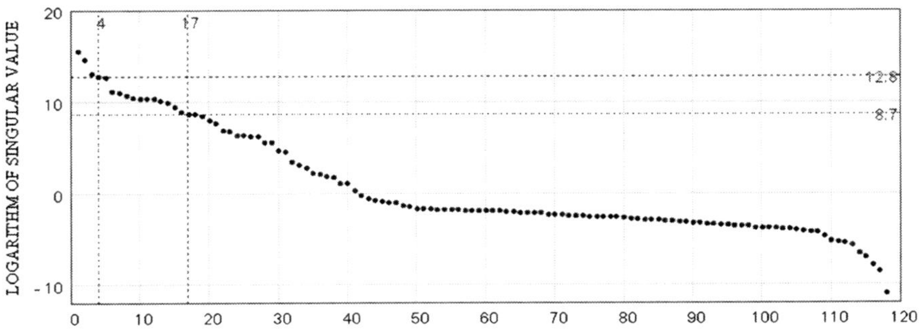

Fig. 2 Singular value spectrum: Plot of logarithms of 118 eigenvalues

easy-to-use environment for engineers and scientists research. MathCAD has a large number of the built-in functions, among them matrix functions, and programming tools. The analysis of horizontal displacement of the top of building column after seismic action is performed by means of "Caterpillar"—SSA. Data recording (mm) was made at regular interval 0.04 s, series length is 235 ($N = 235$). Initial time series and its trend (reconstructed series with the first tree eigentriples) are presented in Fig. 1, window length is set $L = 118$.

Plot in Fig. 2 shows the logarithm of eigenvalues in decreasing order; logarithm was introduced because of very large figures. It is clear that the first five eigenvalues have the largest value and are able to explain the main part of variance.

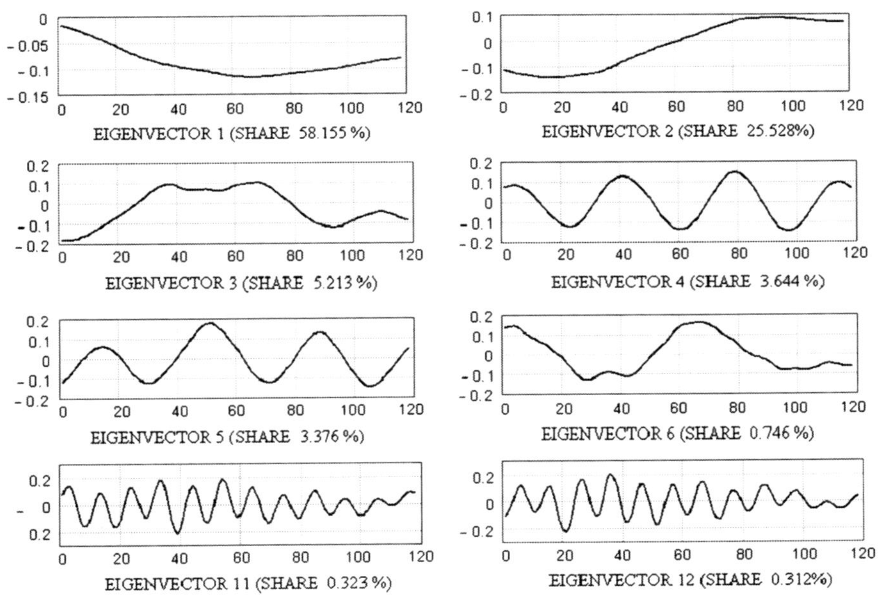

Fig. 3 One-dimensional plots of the first six, 11th and 12th eigenvectors

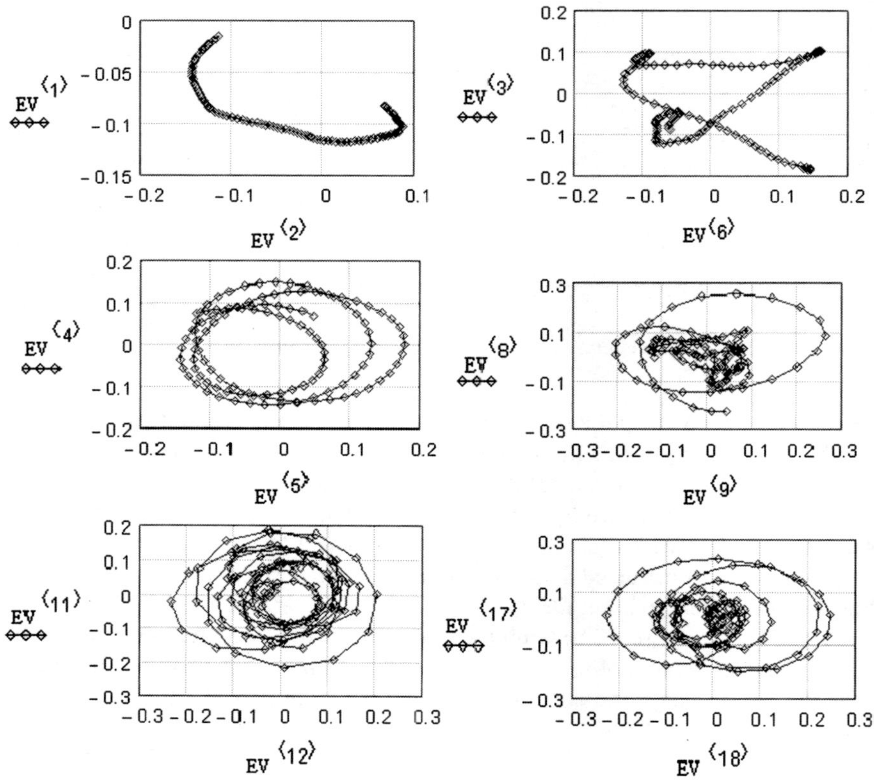

Fig. 4 Two-dimensional plots of eigenvectors

To determine the size of informative basis the next features of eigenvalue is used: the larger share of eqenvalue in the sum of elements of the eigenvalues vector the more background information includes a projection on the corresponding eigenvector. It may be supposed that there is noise after fortieth eigenvalue.

Plot in Fig. 2 may be used for periodical components finding. Since in ideal situation two main components with the same period ("sin" and "cos"), relevant to oneeigenvalue, correspond to one sinusoidal component, in real situation of these components correspond to close eigenvalues. Therefore, the "tread" may be observed in the plot: 4–5, 11–12, 17–18, 22–23, 24–25, 30–31, and 39–40.

In Fig. 3 one-dimensional plots of the first six, eleventh and twelfth eigenvectors are shown as the example. Analysis of eigenvectors shows that the first tree eigenvectors are a trend components (slowly changing component), sixth eigen-vector may relate to low frequency or to the slowly changing component, the fourth and fifth may be periodical components, as the same as 17th, 18th, 24th, 25th, 32th, 33th, 39th and 40th eigenvectors, that are not shown in Fig. 3.

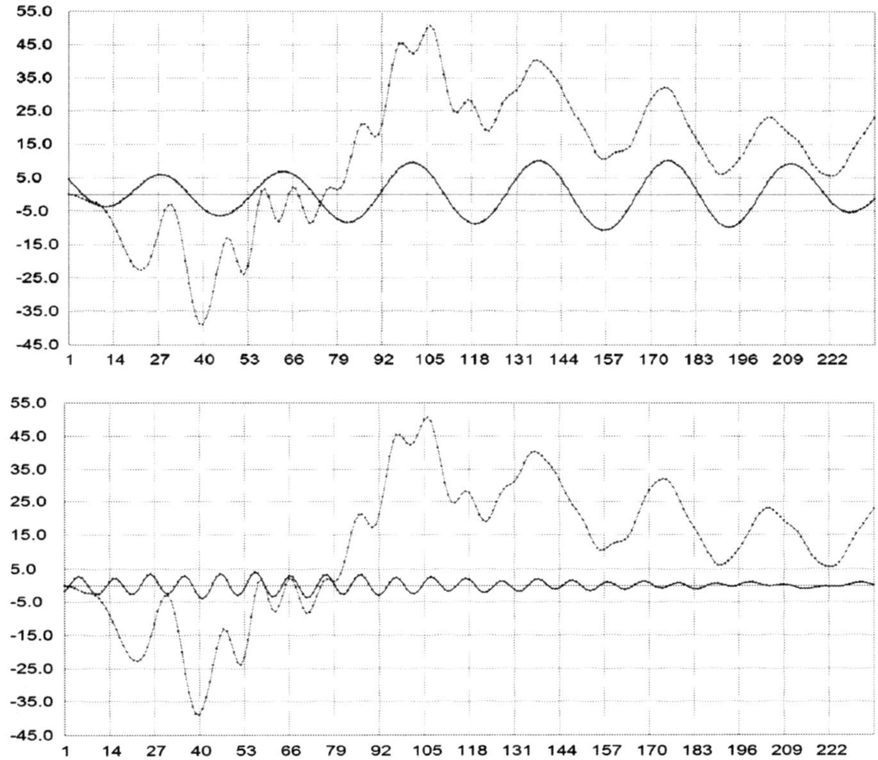

Fig. 5 Initial series and periodical components: Upper plot—fourth and fifth eigentriples, lower plot—eleventh and twelfth eigentriples

Figure 4 shows the examples of two-dimensional projection of the eigenvectors; vectors number is indicated in superscript of function and argument. Analysis of plots in Fig. 4 confirms the conclusion that the fourth and fifth eigenvectors (and also eleventh and twelfth) refer to the periodical components, sixth is not periodical.

By the same way the rest periodical or almost periodical components may be investigated: 17–18, 24–25, 32–33 and 39–40 eigenvalues. The amplitudes of this vibration are small.

In Fig. 5 the periodical components consisting of the fourth and fifth, eleventh and twelfth eigentriples are presented, it is clear that here oscillation are damped.

4 Conclusion

Example of application of the "Caterpillar"-SSA technique for analysis of one-dimensional time series is presented in this work: task of analysis and forecasting of nonlinear vibration of building element under seismic action. Simple solution shows the convenience of this method, its usability to nonstationary time-series investigation. In future it is necessary to fulfil the comparative analysis of the time series investigation and forecasting methods, based on SSA and EMD.

References

1. Alexandrov T, Golyandina N (2005) Automatic extraction and forecast of time series cyclic components within the framework of SSA. Proceedings of the 5th St. Petersburg, workshop on simulation, St. Petersburg State University, St. Petersburg, pp 45–50
2. Antoniou VV et al (2002) Principal component analysis of network traffic: the "Caterpillar"—SSA approach, VIII international workshop on advanced computing and analysis techniques in physics research, ACAT'2002, Book of abstracts, Moscow, p 176
3. Danilov D, Zhigljavsky A (1997) Principal components of time series: "Caterpillar" method. St. Petersburg University. http://www.gistatgroup.com/gus/
4. Ghil M et al (2002) Advanced spectral methods for climatic time series. Rev Geophys 40(1):1-1–1-41
5. Golyandina N, Zhigljavsky A (2013) Singular spectrum analysis for time series. Springer, Springer Briefs in Statistics
6. Golyandina N, Osipov E (2007) The "Caterpillar"-SSA method for analysis of time series with missing values. J Stat Plan Inf 137(8):2642–2653
7. Golyandina N, Usevich K, Florinsky I (2007) Filtering of digital terrain models by two-dimensional singular spectrum analysis. Int J Ecol Dev 8(F07):81–94
8. Golyandina NE, Nekrutkin VV, Stepanov DV Variants of the "Caterpillar"-SSA method for analysis of multidimensional time series. http://www.gistatgroup.com/cat/
9. Hassani H, Heravi S, ZhigljavskyA (2012) Forecasting UK industrial production with multivariate singular spectrum analysis, The 2012 international conference on the singular spectrum analysis and it's applications, May (2012), China. http://www.cefs.ac.cn/express/images/SSA%20China2012%20Hassani%20et%20al.pdf
10. Jolliffe IT (1986) Principal component analysis. Springer, New York
11. Keppenne CL, Lall U Complex singular spectrum analysis and multivariate adaptive regression splines applied to forecasting the southern oscillation. Experimental long-lead bulletin, Sect. D, pp 32–35. http://desastres.unanleon.edu.ni/pdf2/2005/noviembre/pdf/eng/doc9074/doc9074-d.pdf
12. Polukoshko S, Hofmanis J (2009) Use of "Caterpillar"—SSA method analysis and forecasting of industrial and economic indicators, environment, technology, resources. Proceedings of the 7th international scientific and practical conference, vol II, Rēzeknes Augstskola, Rēzekne, RA Izdevniecība (2009), pp 241–248. http://zdb.ru.lv/conferences/2/II_241-248_Polukosko2.pdf
13. Schoellhamer DH (2001) Singular spectrum analysis for time series with missing data. Geophys Res Lett 28(16):3187–3190
14. SSA –MTM toolkit for spectral analysis. http://www.atmos.ucla.edu/tcd/ssa/
15. Vautard R, Yiou P, Ghil M (1992) Singular spectrum analysis: a toolkit for short, noisy chaotic signals. Phys D 58:95–126

Vibration Measurements on PFBR Fuel Subassemblies

P. Anup Kumar, R. Vidhyalakshmi, H.P. Agnihotri, Sudheer Patri, S. Chandramouli, V. Prakash and K.K. Rajan

Abstract Prototype Fast Breeder Reactor (PFBR) is a 500 MWe liquid metal cooled fast breeder reactor, which is in the final stage of construction. The core of PFBR consists of 1,758 subassemblies supported at the bottom on the grid plate sleeves. Liquid sodium is used as the coolant and flow through the maximum rated fuel subassembly is 36 kg/s. The coolant flows axially from the bottom of the subassembly to top and it is in highly turbulent regime. This turbulent flow can excite flow-induced vibration of fuel subassembly which can cause failure of the fuel pin clad tubes from fatigue, wear and vibration induced fretting. Excessive vibration of fuel subassembly can also results in reactivity noise, fatigue or rattling. Flow induced vibration studies of dummy fuel subassemblies in water were conducted in subassembly test facility and the design was qualified for PFBR. However it is planned to measure the amplitude and frequency of vibration during

P. Anup Kumar (✉) · R. Vidhyalakshmi · H.P. Agnihotri · S. Patri · V. Prakash
Vibration Diagnostics Division, Fast Reactor Technology Group, Indira Gandhi Centre for Atomic Research, Kalpakkam 603102, India
e-mail: panup@igcar.gov.in

R. Vidhyalakshmi
e-mail: vidhyal@igcar.gov.in

H.P. Agnihotri
e-mail: agnihotri@igcar.gov.in

S. Patri
e-mail: patri@igcar.gov.in

V. Prakash
e-mail: prakash@igcar.gov.in

S. Chandramouli
Device Development and Rig Services Division, Fast Reactor Technology Group, Indira Gandhi Centre for Atomic Research, Kalpakkam 603102, India
e-mail: mouli@igcar.gov.in

K.K. Rajan
Fast Reactor Technology Group, Indira Gandhi Centre for Atomic Research, Kalpakkam 603102, India
e-mail: kkrajan@igcar.gov.in

J.K. Sinha (ed.), *Vibration Engineering and Technology of Machinery*, Mechanisms and Machine Science 23, DOI 10.1007/978-3-319-09918-7_89

pre-commissioning tests of PFBR. Measurements are planned during the isotherma run of PFBR at 200 °C with dummy subassemblies loaded in the core. Sinc measurement has to be carried out in high temperature sodium environment, con ventional contact type sensors such as accelerometers, strain gages etc. cannot b employed for vibration measurement. Non-contact measurement technique usin ultrasound waves was planned to be developed for vibration measuremen Extensive experiments were carried out in various test facilities and ultrasoni vibration measurement technique was established and demonstrated. Based on th experimental results, a device named SONAR was designed and developed fo PFBR. The SONAR device is equipped with ultrasonic sensors, which focuses o subassembly crown region, and is capable of movement in Z-axis (up and down and in Theta-axis (rotation). The movement of the subassembly is detected from th train of ultrasonic pulses and echoes from the target subassembly. Time signal an frequency spectra of vibration are extracted from the ultrasonic signals using signa processing technique implemented in LabVIEW platform. This paper discusses th details of the FIV measurements on PFBR fuel subassemblies, details of ultrasoni technique and SONAR device, its testing and results and conclusion.

Keywords PFBR · Subassembly · Vibration · Sonar · Ultrasonic technique

1 Introduction

Prototype Fast Breeder Reactor (PFBR) is a 500 MWe sodium cooled fast breede reactor under construction at Kalpakkam, India. It is a techno-commercial dem onstrator and the fore-runner of a series of FBRs planned in India. The core o PFBR consists of 1,758 subassemblies vertically supported on the grid plate. Out o the core subassemblies, 181 are fuel subassemblies which hold the fuel material i position [1]. Coolant sodium flows axially from the bottom of the subassembly t top and is in turbulent regime, in order to increase the heat transfer performance. A the subassemblies are free standing and supported only at the bottom, it is prone t flow induced vibrations (FIV), which can result in the failure of the fuel clad tube from fatigue, wear and vibration induced fretting [2]. Excessive vibration of fue subassemblies may also lead to reactivity fluctuations.

FIV studies were carried out on PFBR fuel subassembly on separate water tes rig to determine the amplitude and frequency of vibration. It is also planned t measure subassembly vibrations in PFBR during commissioning. Since it is ver difficult to carry out such measurement under sodium at high temperature, usin conventional contact type vibration sensors such as accelerometers, strain gage etc., other non-contact type techniques were investigated. It is planned to measur the subassembly vibration under sodium using ultrasonic technique. Extensiv experiments were carried out in various test facilities and a device (SONAR) i designed and manufactured. The prototype device was tested in water and i sodium for assessing its performance. This paper discusses the FIV measurement

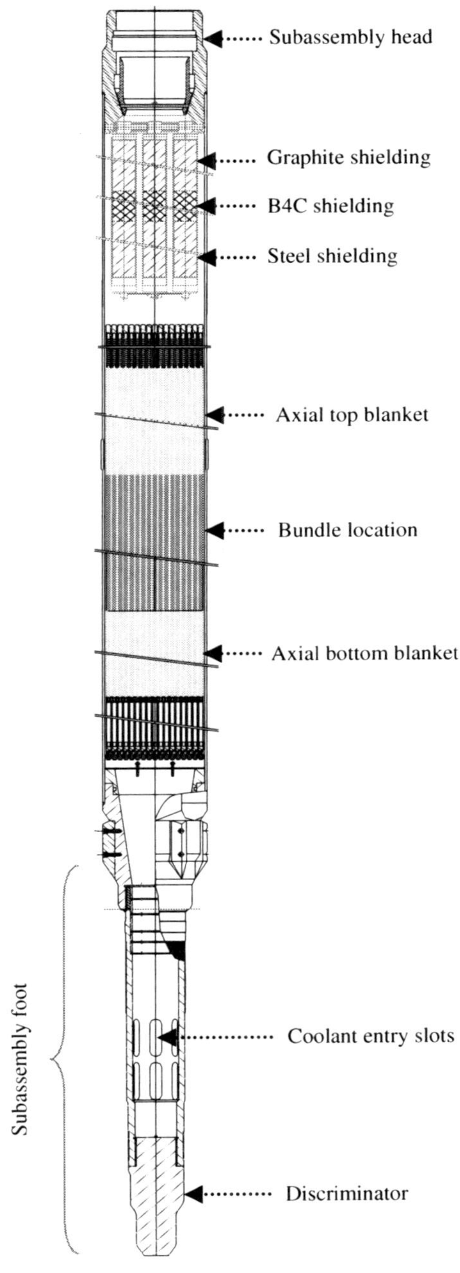

Fig. 1 PFBR fuel subassembly schematic

carried out on PFBR fuel subassemblies, details of ultrasonic technique anc SONAR device, its testing and results and conclusion.

2 Flow Induced Vibration Studies on PFBR Fuel Subassembly

Flow induced vibration studies were carried out on PFBR subassemblies as part o its design qualification for the reactor application [2]. The entire core subassemblie are supported in the sleeves of grid plate, which acts as the main coolant header Details of a typical fuel subassembly are shown in Fig. 1. The total height of the subassembly is 4.5 m out of which 3.9 m is above the grid plate. The total coolan flow from the two pumps in the reactor vessel is 7,020 kg/s out of which 6,425 kg/ (91.6 %) is utilized to transport heat generated in the active core. The flow throug the maximum rated fuel subassembly is 36 kg/s.

To understand the vibration characteristics of fuel subassembly, studies wer conducted on maximum rated experimental fuel subassembly in a water test rig [3] Prior to the FIV experiments modal analysis of fuel subassemblies were carried ou in air and water to extract the modal parameters. For FIV testing, a full scale dummy fuel subassembly was used and all the flow paths and support condition are simulated as in reactor. Perspex embedded SS window was assembled in the tes loop to view the movement of top end of subassembly during experiments.

The flow through the fuel subassembly is parallel flow and the main excitatio mechanism is turbulence. The simulation criterion from water to sodium is estab lished from the Burgreen correlation for the fuel subassembly vibration [4].

$$\left(\frac{\delta}{D_h}\right)^{1.3} = 0.83 \times 10^{-10} \times K \times T^{0.5} \times \Omega \quad (1$$

where δ is displacement of subassembly, D_h is the hydraulic diameter, K is the enc fixity factor, $T = \frac{\rho V^2 L^4}{EI}$ and $\Omega = \frac{\rho V^2}{\mu \omega}$. Where ρ is the fluid density (kg/m^3), V is the flow velocity (m/s), L is the span length, E is the modulus of elasticity (Pa), I is the moment of inertia (m^4), μ = Fluid viscosity (N-s/m^2) and $\omega = 2\pi f_n \propto \sqrt{\frac{EI}{M}}$. Where f_n is the modal frequency of subassembly (Hz) and M is the mass per unit length

The average temperature of flowing sodium in fuel subassembly is 472 °C. As per the Burgreen correlation for the same amplitude of vibration in sodium anc water, flow velocity ratio becomes,

$$\frac{V_W}{V_{Na}} = \frac{\rho_{Na}^{0.5} \times \mu_{Na}^{-0.33} \times E_{472}^{-0.33} \times M_U^{0.165}}{\rho_W^{0.5} \times \mu_W^{-0.33} \times E_t^{-0.33} \times M_{SS}^{0.165}}$$

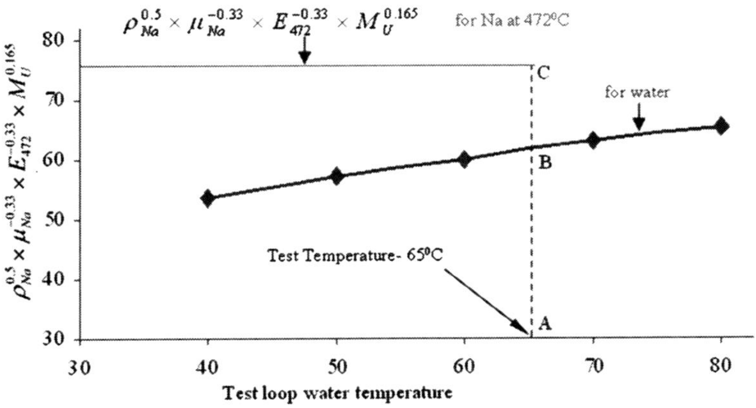

Fig. 2 Prototype versus model flow comparison

where E_t is the modulus of elasticity of Stainless Steel (SS) at the test temperature, E_{472} is the modulus of elasticity of SS at 472 °C, M_U is the mass of fuel subassembly at 472 °C and Mss is the mass of experimental fuel subassembly at test temperature.

The required test flow for FIV testing in water was calculated based on relation given in Fig. 2. Testing was carried out with 65 °C water. Hence test flow is fixed as,

$$\frac{V_W}{V_{Na}} = \frac{AC}{AB} = 1.22$$

Therefore the test flow,

$$Q_{Test} = 1.22 \times 154\,m^3/h = 188\ m^3/h.$$

where 154 m^3/h is the flow rate through the fuel subassembly in reactor. The Reynolds number in the water test at 122 % of rated flow is 0.75×10^5 and the Reynolds number in sodium at 472 °C is 1.0×10^5. The lower Reynolds number in water test is not of much significance, as the flow is in the fully developed turbulent region. The turbulence intensity does not change much over the Reynolds number range of 10^4–10^7. The excitation spectral density will depend upon mainly on the mean velocity of flow and depends weakly on the Reynolds number for the above range of Reynolds number [5].

Under-water accelerometers and strain gages were used for the FIV measurement. Accelerometers were also installed at different elevation of subassembly. The locations of the sensors are selected based on the analytically predicted mode shapes of the subassembly and the modal measurements carried out earlier.

Figure 3 shows the vibration spectra from accelerometers (0–200 Hz) for a flow of 188 m^3/h. It is found that the measured vibration frequencies of subassembly 3.6,

Fig. 3 Acceleration spectra

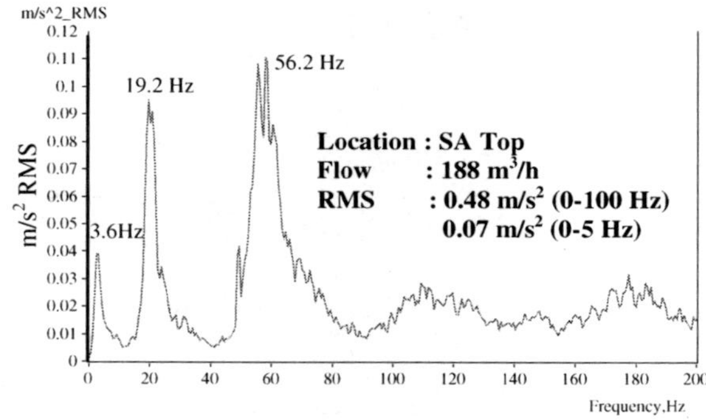

Fig. 4 Displacement spectra

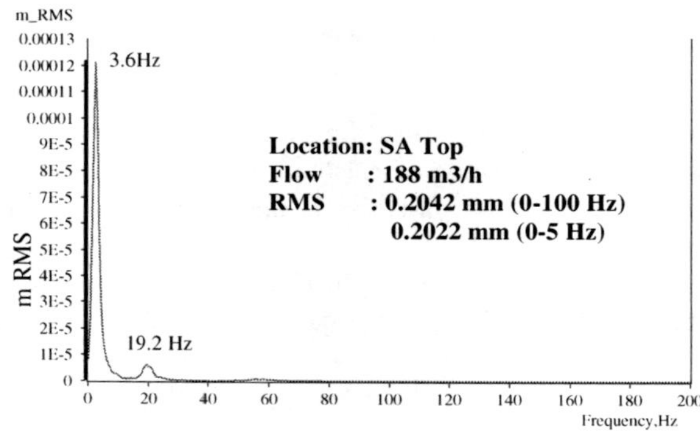

19.2 and 56.2 Hz are in close agreement with analytically predicted values. The overall amplitude of vibration displacement measured at the subassembly top is 0.2 mm as shown in Fig. 4. Higher modes are not excited significantly in the high frequency range and resonance types of vibrations are not observed. The maximum value of strain measured from the subassembly is 3.74 micro-Strain, which corresponds to the stress value of 0.68 MPa, which is well within the design limit of 20 MPa. Thus the design of subassembly is found to be safe against flow induced vibration.

3 Development of SONAR Device

It is planned to measure subassembly vibration during commissioning of PFBR. Measurements are planned at isothermal operation of PFBR at 200 °C with dummy subassemblies loaded in the core. Extensive experiments were carried out in various test facilities and the feasibility of using ultrasonic technique was established and demonstrated. A device named SONAR was designed and a prototype was manufactured to test in water and in sodium. After completing various measurement campaigns, SONAR device for PFBR was design and manufactured.

3.1 Ultrasonic Technique

Ultrasonic sensors in pulse echo mode is used for the detection of subassembly vibration. The time of flight technique is used to detect the position of the subassembly at an instant of time. The ultrasound waves emitted by the ultrasonic sensors are focused on the target (subassembly) and measures the time delay between each emitted and echo pulse, to determine the sensor-to-target distance. So continuous triggering of the sensor yields the time of flight corresponding to the subassembly movement. The movement of the subassembly will cause a variation in the time delay and thus the time delay between the transmitted pulse and received echo will be a representation of the subassembly position with respect to the sensor. By suitably extracting the time of flight data, subassembly movement/vibration is measured.

3.2 Description of SONAR Device

SONAR device consist of a drive mechanism to position the ultrasonic sensors for measurements. The lower part comprises of a guide tube and a spinner tube. Three numbers of ultrasonic sensors are deployed in the SONAR device. A sensor holder attached to the spinner tube is used to hold the ultrasonic sensors in position. Two numbers of ultrasonic sensors (including one spare sensor) are kept at an inclination of 60 degree with the horizontal so as to focus the crown region (Fig. 5) of the subassembly, which is the target to ultrasound waves. The third sensor is mounted

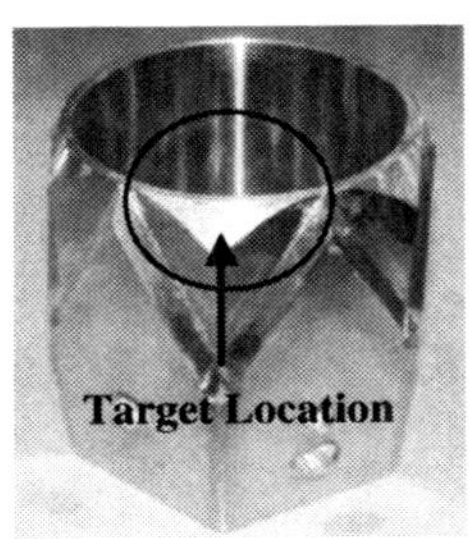

Fig. 5 Subassembly crown

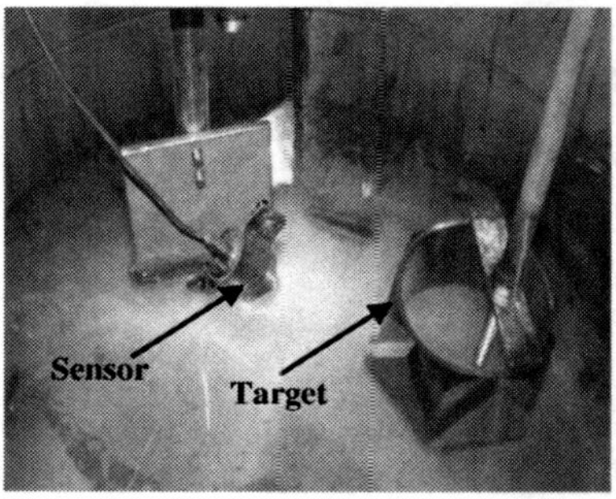

Fig. 6 Sensor focusing crown region

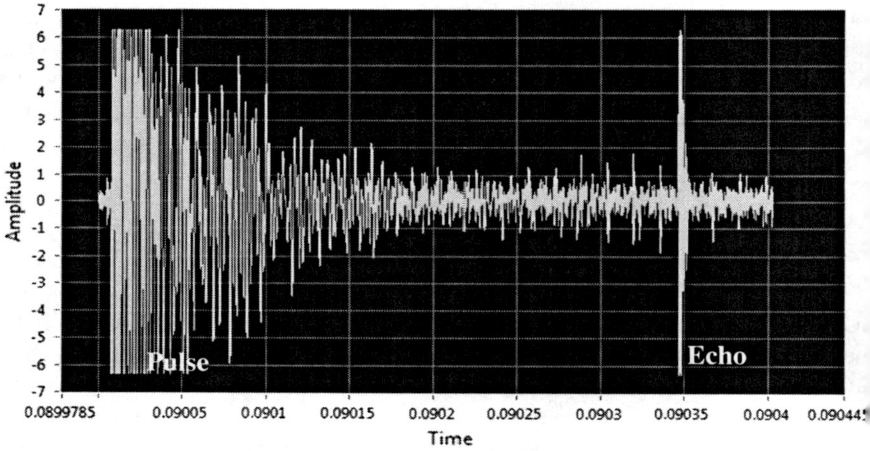

Fig. 7 Ultrasonic signal

on the spinner tube at a slightly higher elevation so as to focus to the control plu shroud tube. The data from this sensor is used for detecting the movement of th SONAR device itself in sodium flow.

3.3 Testing of Prototype Device

A prototype device was designed and fabricated for performance testing in wate and sodium. Stepper motors are used for vertical movement (Z-axis) and fo rotation (Theta-axis) of spinner tube, where the ultrasonic sensors are mounted. Th objective of the test was to assess the performance of the device in water and i

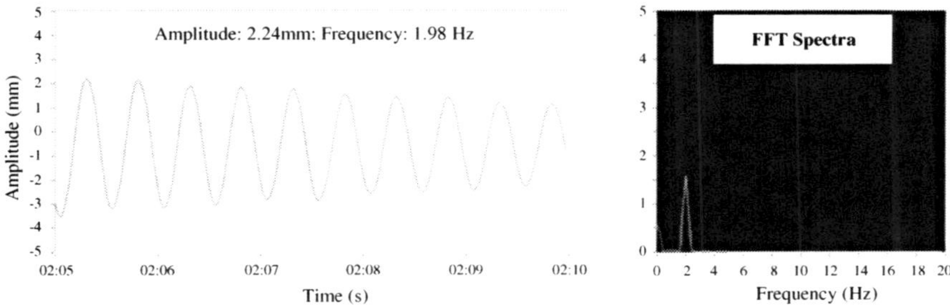

Fig. 8 Extracted vibration signal and spectra in water (frequency 2 Hz)

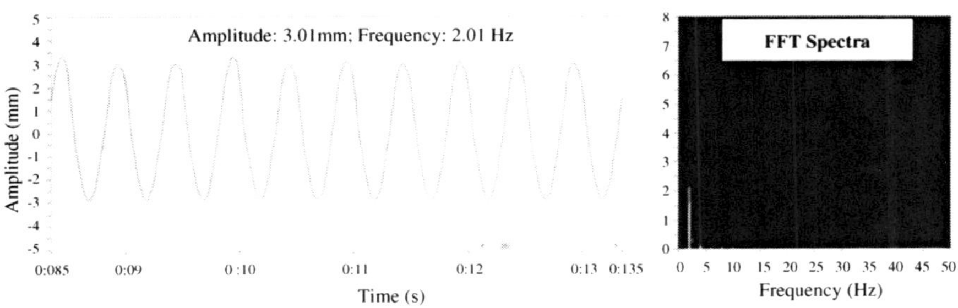

Fig. 9 Extracted vibration signal and spectra in sodium (frequency 2 Hz)

sodium, qualification of ultrasonic sensors and testing of ultrasonic electronics and data acquisition system.

For water testing, the prototype device was installed in a test vessel filled with water and the sensors are focused on the target in the water test setup. Crown region of subassembly head was used as the target simulating the subassembly and was suspended from the top. Figure 6 shows the sensor focusing on to the crown region. The movement of the target is achieved using an electro-dynamic exciter. Same set up was used for sodium testing. The movement of the transducer inside the vessel was done using the stepper motor attached to the spinner tube. Measurements in sodium were carried out in sodium at 200 °C. Known frequencies were fed to the exciter and movement of the subassembly was monitored and measured using the data acquisition system. The vibration signal is extracted from the ultrasonic signal using the software written in LabVIEW platform. Figure 7 shows a typical ultrasonic pulse.

Figures 8 and 9 show typical time signals and frequency spectra of vibration signal recorded during water and sodium testing respectively.

It was observed that the input vibration time signal is faithfully recovered from the train of ultrasonic transmitted pulse and received echoes using the developed software. The experiment was repeated for different input frequencies.

Fig. 10 PFBR SONAR upper part

3.4 Testing of PFBR SONAR Device

SONAR device for PFBR was manufactured and assembled. Figures 10 and 11 show the upper drive mechanism and the lower part of PFBR SONAR device.

A test vessel with subassembly heads positioned inside were used for performance testing.

During the preliminary testing, focusing trials on single subassembly head and multiple subassembly heads were carried out. Ultrasonic sensors are excited using the pulser-receiver units in the control panel, and the signal is monitored in the DAS. SONAR device, spinner tube is positioned in Z-axis and Theta-axis for obtaining echo from the subassembly crown region.

During the next phase of measurements, excitation of subassembly head was done using the electro-dynamic exciter and measurements were carried out. The vibration signal of the subassembly head is extracted from the ultrasonic signal and is plotted. Figure 12 shows a typical test result. Measurements were repeated for various input signal frequencies and amplitudes and in all the measurement trials, the SONAR was able to capture the echo and calculate the vibration signals.

Fig. 11 PFBR SONAR lower part

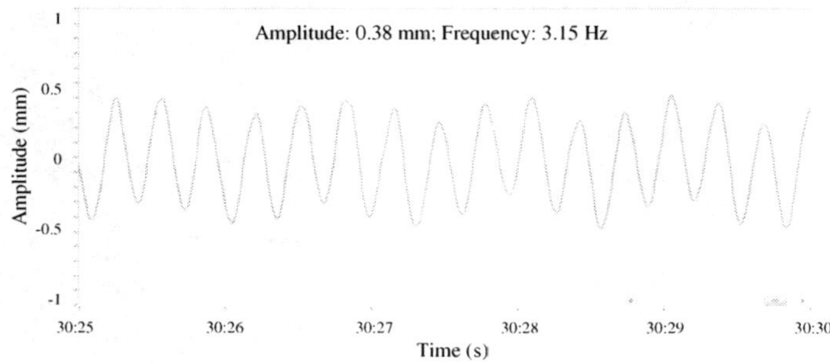

Fig. 12 Measured vibration signal (frequency 3.1 Hz)

4 Conclusion

Flow induced vibration measurements were carried out in water on PFBR fuel subassembly to determine the amplitude and frequency of vibration. Measured vibration amplitude and the corresponding stress developed is found to be well within the design limit. Subassembly vibration measurements are also planned in PFBR in sodium during commissioning and SONAR device employing ultrasonic technique is designed and manufactured. The device was tested for its performance and found to be satisfactory. This paper discusses the FIV measurements carried out on dummy fuel subassembly, details of ultrasonic technique and SONAR device, its testing and results and conclusion.

References

1. Chetal SC et al (2006) The design of prototype fast breeder reactor. Nucl Eng Des 236:852–860
2. Prakash V et al (2011) Experimental qualification of subassembly design for prototype fast breeder reactor. Nucl Eng Des 241(2011):3325–3332
3. Anandaraj M et al (2010) Flow induced vibration studies on PFBR fuel subassembly. In: Second international conference on Asian nuclear prospects 2010 (ANUP-2010), Chennai, India
4. Thirumalai M et al (2010) Subassembly hydraulic experiments. In: THEME meeting on challenges in thermal hydraulics of nuclear reactors, TM-CTHNR 2010, Kalpakkam, Tamilnadu, India, Feb 18–19
5. Beavers GS, Plunket R (1974) Modeling of flow induced vibrations in heat exchangers and nuclear reactors. Trans ASME, J Fluid Eng 358–364

Multiple Fault Classification Using Support Vector Machine in a Machinery Fault Simulator

S. Fatima, A.R. Mohanty and V.N.A. Naikan

Abstract The classification of various faults using a fault simulator and support vector machines (SVMs) has been studied. A database is created for number of faults by measuring vibration signals using seven accelerometers mounted on a machinery fault simulator (MFS). Statistical features are extracted in time domain from the vibration signals. Then, the sensitive features are selected using compensation distance evaluation technique. Multi-class SVMs ensemble algorithm is implemented for classification of the various faults by considering SVMs created by the possible combinations of sensitive features for each class of the fault. The effect of distance evaluation criterion for selection of sensitive features amongst the extracted twelve statistical features has been addressed. By using the developed algorithm, the effective location of accelerometer among seven accelerometers for better classification of the faults has been investigated. Measurements are done at five different rotational speeds. The robustness of the developed algorithm has been tested at different speeds.

Keywords Vibration · Rotational speed · Time domain · Compensation distance evaluation technique

S. Fatima · A.R. Mohanty (✉) · V.N.A. Naikan
Department of Mechanical Engineering, Indian Institute of Technology Kharagpur, Kharagpur 721302, India
e-mail: amohanty@mech.iitkgp.ernet.in

S. Fatima
e-mail: fatima.iitkgp@gmail.com

V.N.A. Naikan
e-mail: naikan@hijli.iitkgp.ernet.in

J.K. Sinha (ed.), *Vibration Engineering and Technology of Machinery*, Mechanisms and Machine Science 23, DOI 10.1007/978-3-319-09918-7_90

1 Introduction

Several studies have been carried out for fault detection using Motor current signature analysis (MCSA), vibration and acoustics signal. Now-a-days, soft computing based methods are increasingly being used for fault detection and prediction. Artificial neural networks (ANNs) have been utilized to detect faults in machinery in recent years and also statistical methods to preprocess the vibrating signal as input features. Support vector machine (SVM) is a machine learning method, which employs a small sample based on statistical learning theory and principle of risk minimization. Compared to the ANN, the SVMs are relatively new and their utility in condition monitoring remains unexplored [1]. Based on the idea of performing an excellent and easy maintenance program for machine fault diagnosis, AI systems such as artificial neural network, fuzzy experts system, and condition based reasoning were utilized in several studies but the application of support vector machines is still rare. Use of the SVMs in condition monitoring and fault diagnosis is tending to develop in future [2]. In order to compare these two methods, in an earlier study, time domain vibration signal of a rotating machine with normal and defective bearing were utilized for featured extraction to measure the fault detection [3]. It was noted that the time required to train the SVM was promisingly less than ANN [4]. The SVMs give better classification and high accuracy for detection of healthy and faulty conditions and the classification can be used for other rotating machineries [5]. Genetic algorithm was implemented with the ANN for identifying the required number of good features for more complex systems where some faults cannot be easily distinguished [6]. A novel fault diagnosis method for induction motor was proposed which is based on the classification of independent component analysis and multi-class SVM. Current and vibration data are considered in form of time domain features (mean, rms, shape factor, skewness, kurtosis, crest factor, entropy error, entropy estimation, histogram lower and upper) and frequency domain features [7]. Researchers have reported newer features of SVM that can be used for fault detection and fault classification; and showed their effectiveness compared to the traditional methods [8–10].

So here, in this study fault classification for five speeds in a wide speed range of 1,140–2,580 RPM is presented. Here the classification has been done considering the time domain data only, thus making it easy, to implement for on-line fault classification where a transformation to the frequency domain is not required. Based on the selected features, a multi-class SVMs algorithm has been used and then experimentally validated to classify the fault conditions of bearing and rotor unbalance.

2 Multiclass SVMs

In spite of the fact that SVMs are binary classifiers, multi-classification using one-against-all approach is implemented to classify the fault conditions mentioned in Table 2. In one-against-all approach [11], the N-SVM models are created where N is the number of conditions. For ith SVM, the model is trained by considering all the sample data for the ith condition as positive and the remaining condition's sample data as negative. Number of samples considered for classification in this approach are calculated as $N \times M$ where M is the number of the bins (each bin contains 4,096 samples) for each condition and N is the total number of conditions.

Here, the training data for N conditions is given by $\left\{\left(x_1, y_1\right), \left(x_2, y_2\right), \ldots, \left(x_N, y_N\right)\right\}$, where $x_i \in R^m, i = 1, 2, \ldots N$ and $y_i \in \{1, 2, \ldots, N\}$, the ith SVM model solves the following optimization problem

$$
\begin{aligned}
& \textit{Minimize} \ \tfrac{1}{2}\|\mathbf{w}^i\|^2 + C\sum_{i}^{N} \xi_j^i(\mathbf{w}^i)^{\mathbf{T}} \\
& \text{Subjected to} \quad (\mathbf{w}^i)^{\mathbf{T}}\phi(\mathbf{x}_j) + b^i \geq 1 - \xi_j^i, \ \text{ if } y = i \\
& \qquad (\mathbf{w}^i)^{\mathbf{T}}\phi(\mathbf{x}_j) + b^i \leq -1 + \xi_j^i, \ \text{ if } y \neq i \\
& \qquad \xi_j^i \geq 0, \ \ j = 1, 2, \ldots, N
\end{aligned}
$$

where C is the penalty term and data $\mathbf{x}_i$ is mapped into higher dimensional space using the function ϕ. The term $1/2\|\mathbf{w}^i\|^2$ minimizes the margin between two conditions and the penalty term $C\sum_{j=1}^{N}\xi_{j,j}$ reduces the training errors when the data is non-separable. The function $\operatorname{sign}\left((\mathbf{w}^i)^T\phi(x_j) + b^i\right)$ decides whether $\mathbf{x}_i$ is to be in the ith class or not. To select the condition of data $\mathbf{x}_i$, here the "Max Wins" strategy is used. This strategy predicts the condition of data $\mathbf{x}_i$ on the based largest vote [6].

In the present study, MATLAB SVM solver has been used for binary classification. The effectiveness of 2-norm soft margin SVM classification technique based on quadratic programming [QP] and least square based SVM [LS] has been investigated. Gaussian radial basis function (RBF) is considered as a kernel function to transfer the input data space to higher dimensional feature space.

The RBF kernel functions are very popular and are regarded as the best kernel functions for classifications. Insight regarding the RBF functions can be found elsewhere [12]. There are two parameters which are associated with these kernel function; C and σ. The upper bound parameter, C for penalty term and the kernel parameter, σ play important role in classification performance of SVMs. Improper selection of these parameters leads to under fitting and over fitting problems. The objective of cross-validation is to identify the optimum values of C and σ for which the SVM classifier is able to classify with the highest classification rate. In v-fold cross validation, the training set is divided into v subsets where $v - 1$ subsets are used to train the classifier and the remaining subsets are used for testing. This process is repeated for v times by considering that each of the subset is being used to test the classifier once. The cross validation accuracy is defined as the percentage of data correctly classified amongst the entire training data.

3 Experimental Setup

Experiments were conducted on the machine fault simulator (MFS) as shown in Fig. 1a. The transducers and their corresponding mount locations are represented in Fig. 1b. Vibrations of the drive end (DE) bearing and non-drive end bearing (NDE) were measured in three directions (x, y, z) with B&K 4,370 accelerometers and B&K 4,526 accelerometer was mounted on top of the motor to measure vibration in the vertical direction. Thus a total of seven channels of vibration data were acquired. All seven channels of data from the respective transducers were acquired and recorded simultaneously using B&K Pulse Time Data Recorder at 4,096 samples/s and a span of 1.6 kHz for duration of 5 min [13, 14].

4 Results and Discussions

In this paper, for combined fault condition classification having both bearing and unbalance defects, a total of 8 conditions (4 bearing, 3 unbalance and 1 normal) have been considered. In the first part of the parameter selection, a preliminary analysis has been carried out using the measured vibration signals. Further, feature extraction, feature selection and fault classification using multi-class SVMs has been discussed. Also, the effect of several parameters related to fault classification has been investigated in order to appropriately select them for the SVM process.

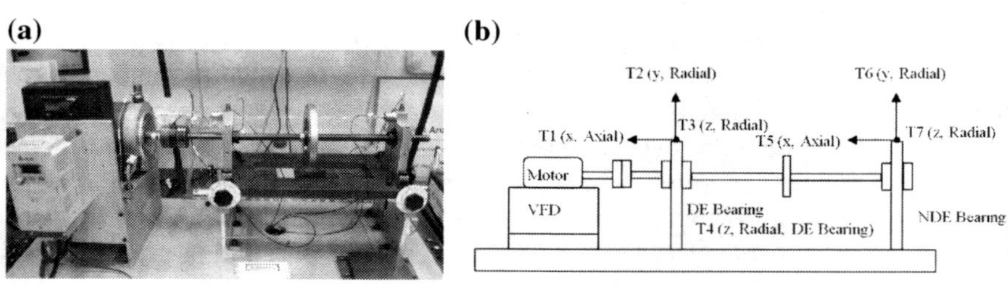

Fig. 1 **a** View of the machinery fault simulator used for the experiments. **b** Schematic representation of mounted transducers

4.1 Preliminary Analysis

Each data set has been divided into 50 bins for a total time of 1 s per bin consisting of 4,096 sampled data points. The vibration signals of a bin for all bearing and unbalance defect conditions in the rotor rig measured by the 1st transducer at 1,740 RPM are shown in Fig. 2. It is difficult to study these time signals for fault classification because of its high dimensionality. The dimensionality of these data has been reduced by extracting the twelve time domain features. So in all for 7 transducers there were a total of 84 features. The features are in group of seven as per the sequence given in the first column of Table 1. For example feature number 48 is [6 × 7] + 6 corresponding to the RMS of the signal and so on. One set of such time domain features for all conditions, for an operating speed of 1,740 RPM of the rotor-bearing rig for the vibration signal measured by 1st transducer is shown in Table 1. The normalized time features for all the conditions as measured by the vibration transducer at location 1 while operating at 1,740 RPM are shown in Fig. 3. It can be observed from Fig. 3 that the time domain features are significantly affected by the fault conditions. It is also difficult to classify the faults based on these features alone, since there is no noticeable pattern or effect of the defects on these features. Additionally, some of these features consist of irrelevant and redundant features. The feature selection using compensation distance evaluation technique has thus been used and distance evaluation criterion has been shown for

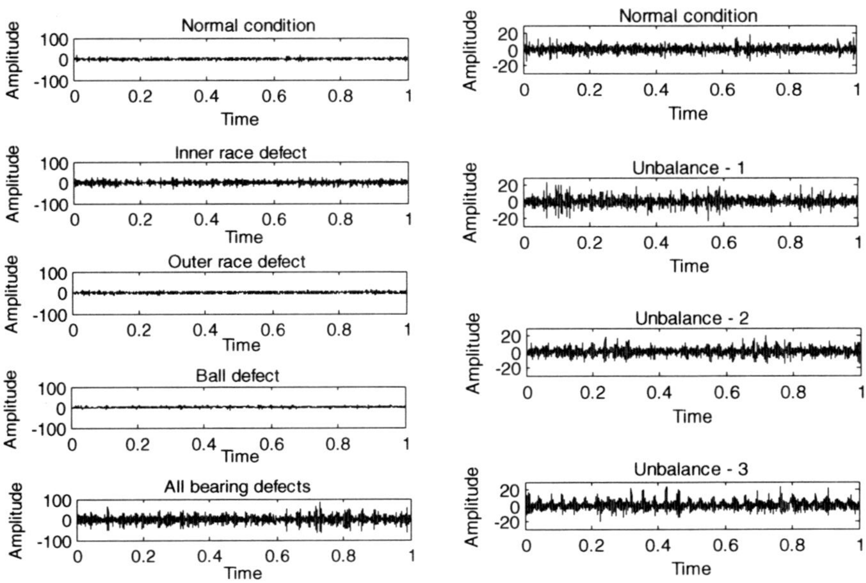

Fig. 2 Vibration signal in m/s^2 measured by 1st Transducer at 1,740 RPM for different conditions

Table 1 Time features extracted from the vibration signal measured by 1st transducer at 1,740 RPM

Time features	Normal	Inner race	Outer race	Ball bearing	All bearing defects	Unbalance-1 (6.12 g)	Unbalance-2 (12.25 g)	Unbalance-3 (23.88 g)
1. Mean (m/s^2)	−0.0095	0.0105	−0.0007	−0.0081	0.0074	−0.0072	−0.0051	−0.0045
2. Max (m/s^2)	19.96	27.63	11.78	11.23	84.06	23.30	19.62	24.65
3. Min (m/s^2)	−15.03	−29.59	−14.82	−13.44	−65.80	−24.07	−15.61	−20.34
4. Range (m/s^2)	34.99	57.22	26.60	24.67	149.90	47.37	35.23	44.98
5. Sum (m/s^2)	−38.95	42.82	−2.93	−33.01	30.51	−29.66	−21.08	−18.53
6. Median (m/s^2)	−0.027	−0.016	0.066	−0.040	0.217	−0.095	−0.115	−0.296
7. RMS (m/s^2)	3.63	7.67	3.47	2.81	15.53	4.79	4.17	4.50
8. Variance (m^2/s^4)	13.18	58.90	12.07	7.91	241.30	22.94	17.39	20.23
9. Kurtosis	4.60	3.73	3.14	3.85	4.23	5.12	4.24	5.08
10. Skewness (m/s^2)	0.22	−0.08	−0.04	0.05	−0.01	0.18	0.22	0.61
11. Crest factor	5.50	3.60	3.39	3.99	5.41	4.87	4.71	5.48
12. Shape factor	−381.65	733.95	−4861.00	−349.06	2085.00	−661.42	−810.06	−994.08

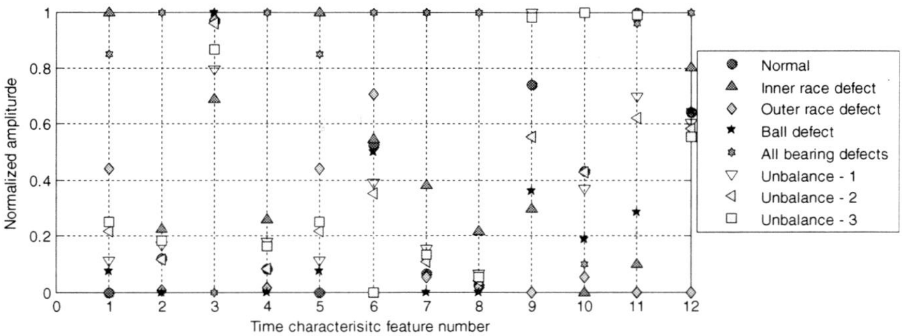

Fig. 3 Normalized time features of all defect conditions measured by 1st transducer at 1,740 RPM

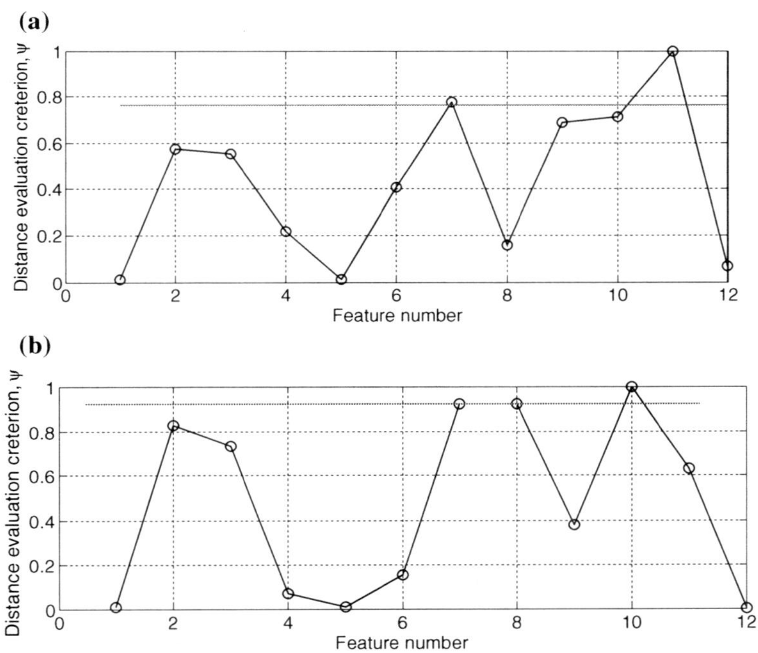

Fig. 4 Distance evaluation criterion for all 12 time features of signal measured by 1st transducer at 1,740 RPM. **a** Bearing, **b** unbalance

bearing and unbalance cases in Fig. 4. The feature which has the distance evaluation criterion value as 1 is considered as the most sensitive parameter for the classification of faults. From Fig. 4, it is seen that the 11th time feature (crest factor) is the most sensitive feature for bearing defect case and 10th feature (skewness) is the most sensitive feature for unbalance based on the vibration signal measured by 1st

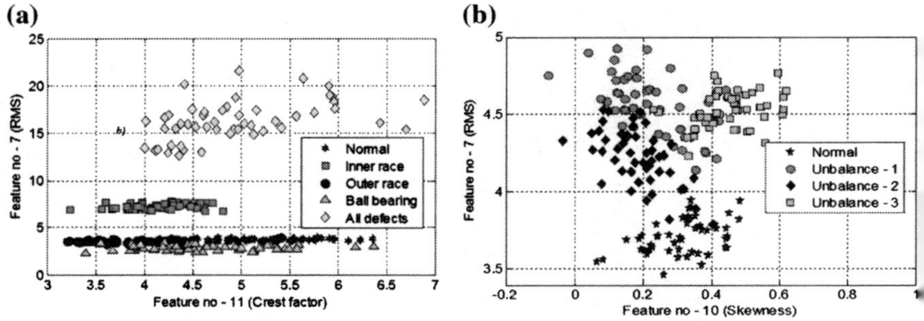

Fig. 5 Two sensitive feature distributions of signals measured by 1st transducer at 1,740 RPM **a** Bearing, **b** unbalance

transducer at 1,740 RPM. Distribution of the most two sensitive parameters hav been shown for all conditions of bearing and unbalance defects in Fig. 5. A simila analysis was done for all the other transducers and operating speeds.

Radial basis function is considered as the kernel function for SVMs classifie Firstly; preliminary analysis for fault classification using multi-class SVMs ha been carried out to choose appropriate SVMs parameters for better fault classification. During this analysis, effect of grid search for *C* and σ; the type of SVM solver; effect fivefold and 10-fold cross validation; repeatability and effect of distance evaluation criterion, ψ have been carried out.

4.2 Fault Classification of Combined Defects

To test the robustness of the fault classification algorithm, after the individua classifications for bearing and unbalance defects were done separately, then the dat set for all the 8 conditions were considered together for classification by SVM Table 2 summarizes the classification results for the combined defects at all speeds The most two sensitive features for each of the operating speed are selected base on the CDET criterion [15] as is shown in Fig. 4. A high classification rate i obtained in all the cases.

For a typical case, the most two sensitive feature distribution at 2,580 RPM fo all the conditions is shown in Fig. 6. This is further validated by the fact that 100 % classification is obtained at 2,580 RPM as shown in Table 2.

Table 2 Fault classification rate of bearing and unbalance defects

Rotational speed (RPM)	CDET criterion (ψ)	Two sensitive features	Classification rate							
			Normal	Inner race defect	Outer race defect	Ball bearing defect	All bearing defects	Unbalance-1 (6.12 g)	Unbalance-2 (12.25 g)	Unbalance-3 (23.88 g)
1140	0.80	46, 49	0.87	0.94	1.00	0.92	1.00	0.96	0.89	1.00
1380	0.80	45, 49	1.00	1.00	1.00	0.95	1.00	1.00	1.00	1.00
1740	0.60	44, 46	0.94	1.00	1.00	1.00	1.00	0.91	0.92	1.00
2220	0.42	44, 49	1.00	1.00	1.00	1.00	1.00	1.00	1.00	1.00
2580	0.60	48, 49	1.00	1.00	1.00	1.00	1.00	1.00	1.00	1.00

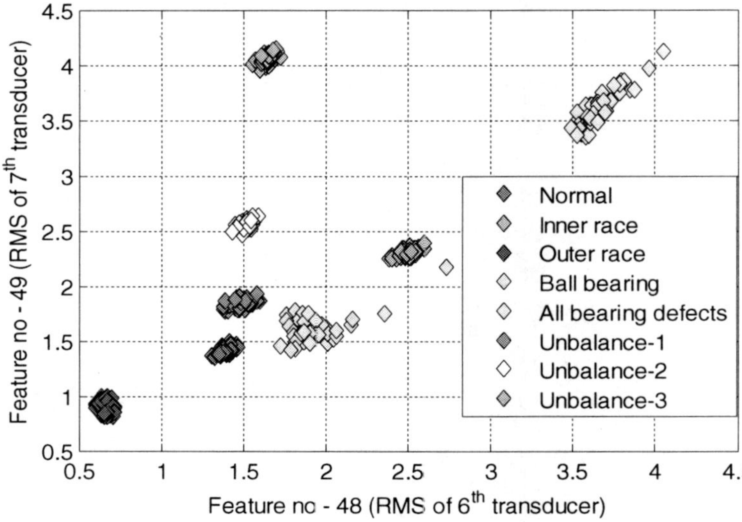

Fig. 6 Sensitive feature distribution of all defects of bearing and unbalance at 2,580 RPM

5 Conclusion

In this paper a methodology for fault classification of bearings and unbalance in a machine fault simulator has been developed. The classification technique was based on the compensation distance evaluation technique for feature selection. Support vector machines with a grid search approach were used to optimize the hyper-parameters for the fault classification process. The standard deviation of the classification rate is less than 1.5 % for all the cases. This methodology can be implemented for on-line condition monitoring of rotating machines.

References

1. Jack LB, Nandi AK (2002) Fault detection using support vector machines and artificial neural networks, augmented by genetic algorithms. Mech Syst Signal Process 16(2–3):373–390
2. Widodo A, Yang B (2007) Support vector machine in machine condition monitoring and fault diagnosis. Mech Syst Signal Process 21:2560–2574
3. Samanta B, Al-Balushi KR, Al-Araimi SA (2003) Artificial neural network and support vector machines with genetic algorithms for bearing fault detection. Eng Appl Artif Intel 16:657–665
4. Samanta B (2004) Gear fault detection using artificial neural networks and support vector machines with genetic algorithms. Mech Syst Signal Process 18:625–644
5. Yang B, Hwang W, Kim D, Tan AC (2005) Condition classification of small reciprocating compressor for refrigerators using artificial neural networks and support vector machines. Mech Syst Signal Process 19:371–390

6. Saxena A, Saad A (2007) Evolving an artificial neural network classifier for condition monitoring of rotating mechanical systems. Appl Soft Comput 7:441–454
7. Widodo A, Yang B, Han T (2007) Combination of independent component analysis and support vector machines for intelligent faults diagnosis of induction motors. Experts Syst Appl 32:299–312
8. Sugumaran V, Sabareesh GR, Ramachandran KI (2008) Fault diagnosis of roller bearing using kernel based neighborhood score multi-class support vector machine. Experts Syst Appl 34:3090–3098
9. Widodo A, Yang BS (2007) Support vector machine in machine condition monitoring and fault diagnosis. Mech Syst Signal Process 21:2560–2574
10. Lei Y, He Z, Yanyang Z, Chen X (2008) New clustering algorithm based fault diagnosis using compensation distance evaluation technique. Mech Syst Signal Process 22(2):419–435
11. Bottou L, Cortes C, Denker JS, Drucker H, Guyon I, Jackel LD, Cun YL, Muller UA, Sackinger E, Simard P, Vapnik, V (1994) Comparison of classifier methods: a case study in handwritten digit recognition. In: Proceedings of the 12th International Conference on Pattern Recognition and Neural Networks, IEEE, Jerusalem pp 77–82
12. Vapnik VN (1995) The nature of statistical learning theory. Springer, New York
13. Fatima S, Mohanty AR, Naikan VNA (2013) Most effective transducer locations for permanent health monitoring of a rotating machine. In: Proceedings of the 20th International Congress on Sound and Vibration Bangkok, Thailand
14. Fatima S, Mohanty AR, Dastidar SG, Naikan VNA (2013) Technique for optical placement of transducers for fault detection in rotating machines. J Risk Reliab 227(2):119–131
15. Lei Y, He Z, Yanyang Z, Chen X (2008) New clustering algorithm-based fault diagnosis using compensation distance evaluation technique. Mech Syst Signal Process 22(2):419–435

A Method for Vibro-Acoustic FE Model Updating of Cavities Using Frequency Response

D.V. Nehete, S.V. Modak and K. Gupta

Abstract Interior noise due to structure-borne sources has always been a cause of concern in the cavities encountered in aerospace, automotive and other applications. In the design of such cavities, the coupled vibro-acoustic model is the key to evaluate its structural and acoustic design. It can also be used as a diagnostic tool to identify the potential noise sources and to assess the effectiveness of the proposed design modifications. However effectiveness of this model greatly depends on the accuracy of predictions made by these models. It is however seen that often there are inaccuracies in the modeling of structural and acoustic domain of a cavity. Methods to deal with structural and acoustic modeling inaccuracies have been developed in the past through FE model updating for uncoupled FE model, while the same have not been addressed in the context of a coupled vibro-acoustic FE model. This issue forms the topic of this paper. This paper proposes a method for updating vibro-acoustic FE model using frequency response function with the aim of identifying the material property of flexible surface of cavity. The updating problem is formed as constrained optimization. The method requires measured vibro-acoustic frequency responses of the system. Updating parameters related to the material property and the stiffness of joint between the plate and rectangular cavity are used. The effectiveness of the proposed method is validated through numerical studies on a 3D rectangular box cavity. The robustness of the method under presence of noise is also investigated. The proposed method is found to work satisfactorily to correctly estimate the updating parameters and yield an updated model that correlates well with the measured frequency response.

D.V. Nehete (✉) · S.V. Modak · K. Gupta
Department of Mechanical Engineering, Indian Institute of Technology, Hauz Khas, New Delhi, India
e-mail: dvn1973@gmail.com

S.V. Modak
e-mail: svmodak@mech.iitd.ac.in

K. Gupta
e-mail: kgupta@mech.iitd.ac.in

J.K. Sinha (ed.), *Vibration Engineering and Technology of Machinery*, Mechanisms and Machine Science 23, DOI 10.1007/978-3-319-09918-7_91

Keywords Model updating · Vibro-acoustic · Interior noise · Vibro-acoustic finit element model updating · Constrained optimization

1 Introduction

When an acoustic cavity is enclosed by a flexible structure, the dynamic response c the structure and the cavity are coupled together. The coupled vibro-acoustic F model of a cavity is the result of coupling between the FE model of the structur and the FE model of the acoustic cavity. This model is extremely useful at th design stage as it can be used to fine-tune the cavity's structural and the acousti design for achieving specific NVH objectives. The availability of accurate F models for aerospace, automotive and other applications would make the process c dynamic design for NVH applications more economical, faster and more effective However, the effectiveness of this approach greatly depends on the accuracy of th predictions made using such a model. The accuracy of these models is compro mised due to structural dynamic modeling errors (related to material properties modeling of joints and boundary conditions, modeling of damping) and acousti modeling errors (mainly related to acoustic absorption properties of cavity surfaces that are often present in these models.

Updating of the structural dynamic finite element models has been an active are of research for last two decades and several approaches have been proposed [1 Model updating of a structure is carried out using modal test data [2]. A significan number of methods belonged to the direct category [3]. The modal data based o iterative method [4] is quite popular due to the freedom it allows in the choice of th updating parameters and the applicability of the method even with an incomplet data. Iterative methods have generally been based on either modal data or frequenc response function (FRF) data. FRF based iterative method is named as respons function method [5]. An alternative technique is proposed based on framing th updating problem as a constrained optimization problem [6]. The method was the used for updating joint parameters [7]. The method for finite element mode updating of coupled structural acoustic model, constituted as a problem of con strained optimization, is proposed [8].

The FE model updating has been investigated as well as studied a great deal fo purely structural dynamic systems. In updating of uncoupled acoustic FE model, method for updating acoustic models based on error in the constitutive relations i proposed [9], which is further extended and investigated to the updating of 3I acoustic models [10]. An updating technique is developed for the interior acousti problem with generalized boundary conditions based on constitutive relations erro [11]. A method is proposed based on constrained optimization problem for updatin acoustic cavity [12], except [13] in which a computational optimization has bee used to identify structural damping and acoustic absorption in a vibro-acousti passenger car cavity through model updating.

In the light of above observations, this paper proposes a method for updating vibro-acoustic FE model constituted as a problem of constrained optimization based on measured vibro-acoustic frequency responses of the system as reference data, with the aim of identifying the material property and the stiffness of joint between the plate and rectangular cavity. To evaluate the effectiveness of proposed procedure a numerical study on a 3D rectangular cavity with a flexible panel is carried out. The effect of 'measured' noise on updating is studied.

2 Coupled Vibro-Acoustic FE Model Updating Using Constrained Optimization

In the following section, a method is proposed coupled vibro-acoustic FE model updating problem as a constrained optimization problem.

The governing equation for a coupled undamped structural-acoustic system is,

$$M\ddot{u} + Ku = f \tag{1}$$

where,

$$
\begin{aligned}
M &= \begin{bmatrix} M_S & 0 \\ C_{AS} & M_A \end{bmatrix}, \\
K &= \begin{bmatrix} K_S & C_{SA} \\ 0 & K_A \end{bmatrix}, \\
u &= \begin{Bmatrix} u_s \\ p_a \end{Bmatrix}, \\
f &= \begin{Bmatrix} f_s \\ f_a \end{Bmatrix}
\end{aligned} \tag{2}
$$

In the above equations, M and K are the mass and stiffness matrices of coupled system, M_S and K_S are the mass and stiffness matrices of the structure, M_A and K_A are the mass and stiffness matrices of the acoustic field. C_{AS} and C_{SA} are coupling matrices. The subscripts s and A denote the structural and acoustic system respectively. u_s, p_a, f_s, f_a are the displacement vector of structure, acoustic pressure vector, excitation force vector acting on structure and the force vector which represents interior acoustic sources respectively. f is the excitation vector of coupled system, u is the nodal coordinate of coupled system and $\ddot{u}$ represents second differentiation of u with time.

To update an acoustic FE model, an objective function representing the error between the measured and the FE model frequency responses is minimized subjected to the bounds on the updating parameters. The updating parameters are the physical variables or parameters of the vibro-acoustic FE model chosen for correction during updating. The objective function is defined as,

$$J_{(u)} = \sum_{p=1}^{nm} \sum_{k=1}^{nf} \left[\left(\alpha_m^r - \alpha_a^r \right)^2 + \left(\alpha_m^i - \alpha_a^i \right)^2 \right]_{pk} \tag{3}$$

where α_m^r, α_m^i and α_a^r, α_a^i represent the real and imaginary parts of the 'measured' and the FE model pressure response respectively. α_a^r and α_a^i are the real and the imaginary parts of the pressure response computed from Eq. (1). The objective function is obtained by summing the error in the FE model pressure response over the 'nm' number of 'measured' responses at nf number of frequency points.

Upper and lower bounds on the vector of correction factors $\{ u \}$ are included in the form of following inequality constraints,

$$\{u\}_{\text{LB}} \leq \{u\} \leq \{u\}_{\text{UB}} \tag{4}$$

Thus the coupled vibro-acoustic model updating problem is defined as a constrained minimization of the objective function given by Eq. (3), subjected to the bounds given by Eq. (4). The minimization is done through a routine for constrained optimization available in MATLAB. This routine makes use of sequential quadratic programming (SQP) method.

The flow chart of coupled vibro-acoustic FE model updating based on constrained optimization using measured frequency response data is shown in Fig. 1. The 'measured' frequency response data are used as reference data. The objective function $J\{\text{u}\}$, defined in Eq. (3), is minimized in an iterative way subjected to constraints in Eq. 4 until the convergence is obtained.

3 Numerical Study

The proposed FE model updating method is illustrated through a case study.

3.1 Details of Case Study

A simulated case study of a 3D rectangular cavity of size 0.261 m × 0.3 m × 0.686 m along x, y and z axes respectively as shown in Fig. 2a, is considered here. The front side is a flexible surface while other five surfaces are rigid. The flexible surface of the cavity is modeled as an aluminum plate of size 0.261 m × 0.3 m. The modulus of elasticity and density of the plate are 7.1 e + 10 N/m^2 and 2,700 kg/m^3 respectively. The thickness of plate is 0.001 m.

The vibro-acoustic FE model to be updated is built as follows. The plate is modeled with 30 four noded isoparametric rectangular finite elements with three

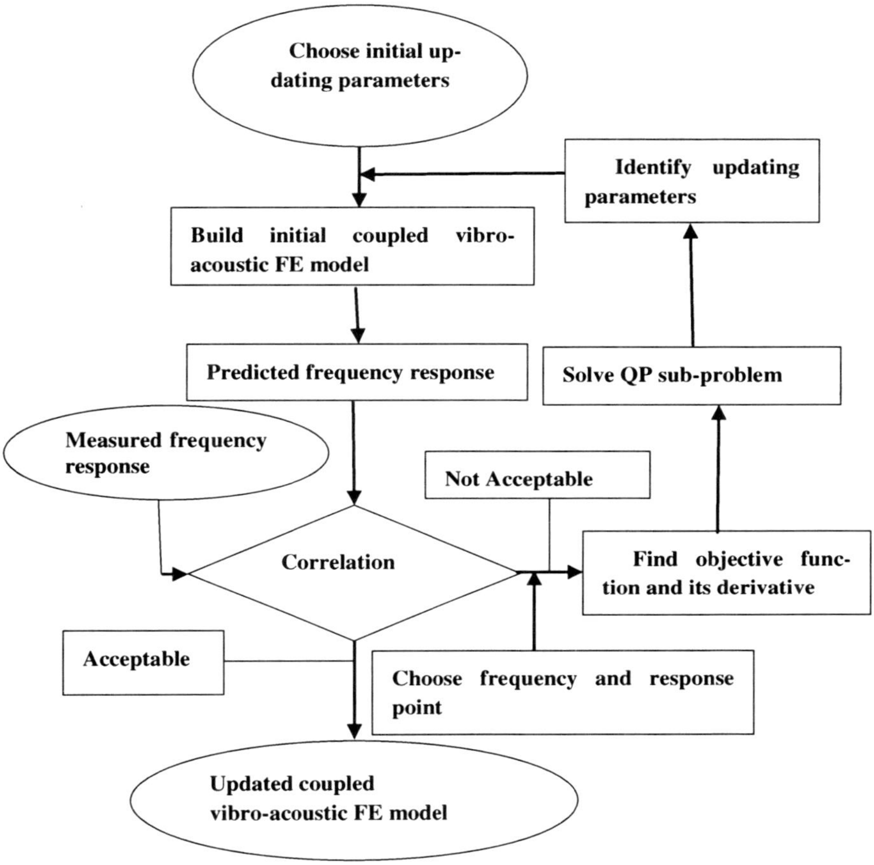

Fig. 1 Flow chart of the updating method

degrees of freedom at each node, one out of plane displacement and two rotational, at each node. The acoustic space of the cavity is modeled with 210 3D eight noded isoparametric acoustic finite elements with acoustic pressure as the degree of freedom at each node.

An eigen-value analysis of the coupled vibro-acoustic FE model is carried out, to get an idea of the coupled natural frequencies and the mode shapes. The natural frequencies are shown in Table 1. The table also shows the rigid-wall acoustic natural frequencies as well as the in-vacuo structural natural frequencies. The first coupled frequency, which is almost equal to zero, corresponds to the rigid body acoustic mode. The coupled frequencies dominated by the acoustic resonances and those dominated by the structural resonances are indicated with letters 'A' and 'S' respectively.

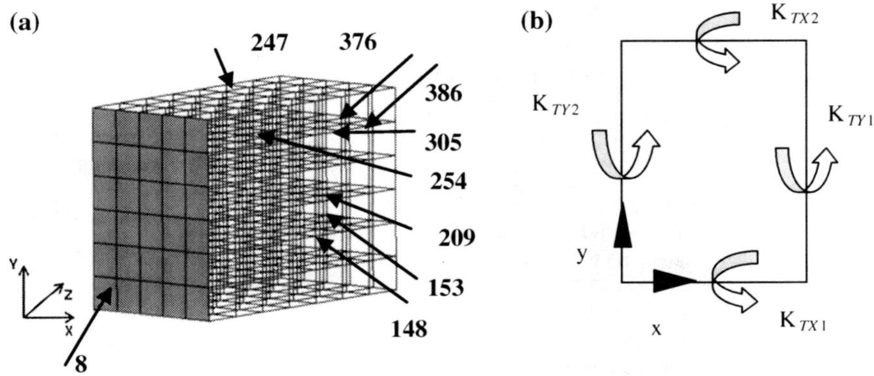

Fig. 2 **a** Simulated 3D rectangular cavity. **b** Torsional stiffness for rotational d.o.f

Table 1 Comparison of coupled, rigid-wall acoustic and in-vacuo structural natural frequencies

nm	Rigid wall acoustic natural freq. (Hz)	In-vacuo structural natural freq. (Hz)	Coupled vibro-acoustic natural freq. (Hz)
1	0.00	–	0.00
2	–	113.05	113.51(S)
3	–	207.83	205.48 (S)
4	–	248.70	246.23 (S)
5	252.10	–	255.77 (A)
6	–	321.97	319.84 (S)
7	–	362.99	361.66 (S)
8	–	453.82	451.60 (S)
9	–	458.76	455.78 (S)
10	–	513.22	510.33 (S)
11	516.90	–	521.29 (A)

3.2 Simulated/'Measured' Data

The 'measured' FRF of vibro-acoustic model is simulated by introducing certain known discrepancies in the vibro-acoustic FE model. These discrepancies are: (a) the modulus of elasticity of plate reduced by 10 % to simulate error in material property, (b) the rotational stiffness at the four edges of the fixed plate about the two axes is reduced by 95 % to simulate error in boundary condition. The second error is simulated by introducing at each node at the boundary a torsional spring for the rotational degree of freedom about each edge as shown in Fig. 2b. The sensitivity of the natural frequency to torsional spring corresponding to the rotational degree of freedom about each edge is seen to be higher than that of lateral stiffness. Therefore these stiffnesses are chosen as the updating parameters. These correction factors are represented by symbols K_{TX1}, K_{TX2}, K_{TY1} and K_{TY2}, as shown in Fig. 2b.

3.3 Updating Using Complete Data

The FE model is updated by updating 5 parameters. Forty numbers of frequency points are used to compute the objective function with frequency responses available at eight nodes 148, 153, 209, 247, 254, 305, 376 and 386 as shown in Fig. 2a. The 48 frequency points around acoustic resonances of FRF are used in updating. Figure 3a, b shows both the exact v/s identified values of updating parameters for complete and complete with noisy frequency response data respectively.

The five updating parameters that corresponds to the material property and the stiffness of joint between the plate and rectangular cavity of vibro-acoustic model. It is observed that, the method has identified the parameters exactly.

Figure 4a shows a comparison of the vibro-acoustic coupled FRF corresponding to simulated and initial FE models. This FRF is calculated with response on node number 59 in the acoustic domain to a unit harmonic force acting on node number 8 on the structure. Figure 4b shows a similar comparison for updated FE model. It is seen that this updated FE model FRF also matches exactly with 'measured' FRF.

3.4 Updating Using Complete Data with Measurement Noise

Accuracy of the identification of the updating parameters is affected by the noise in the measurement. Hence, the noise in the measurements is simulated by generating a random sequence r (k) of dimension equal to the number of frequency points in

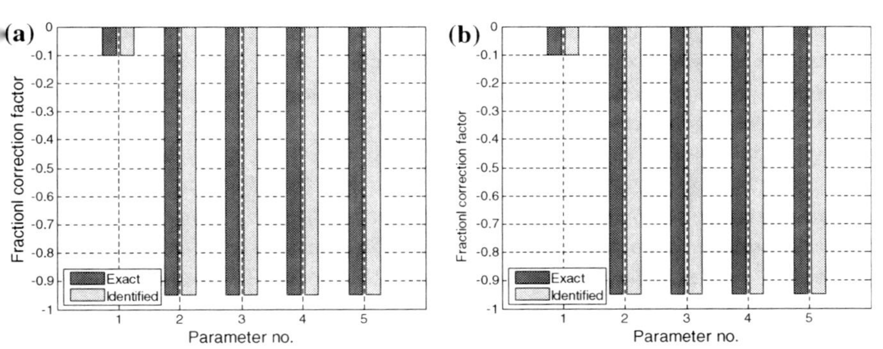

Fig. 3 Plot of exact as well as identified values of correction factors **a** case of complete data **b** case of complete data with measurement noise

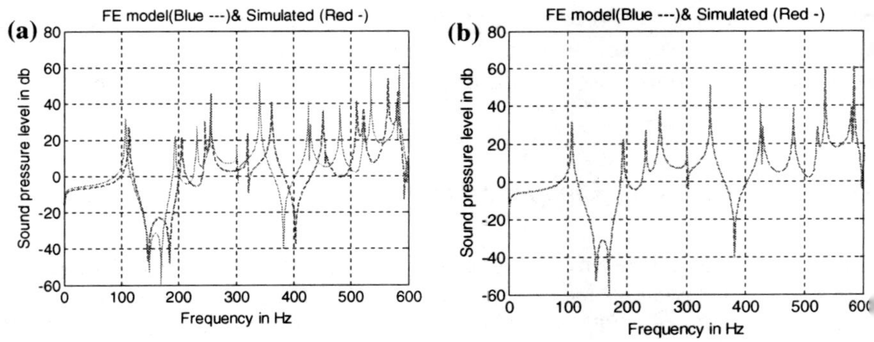

Fig. 4 Comparison of overlay of simulated and FE model coupled vibro-acoustic FRFs **a** befo and **b** after updating (case of complete data)

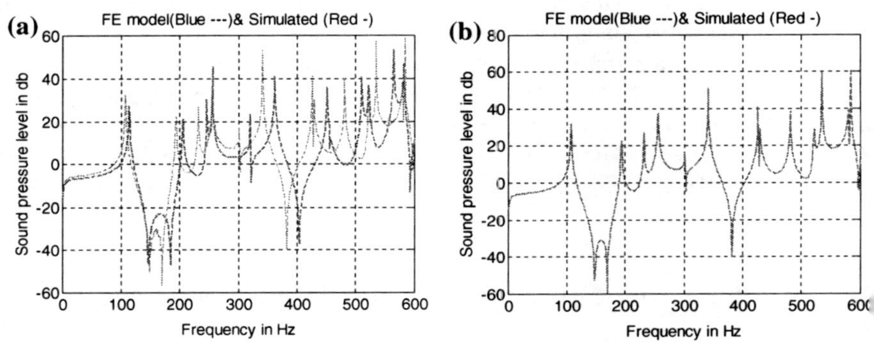

Fig. 5 Comparison of overlay of simulated and FE model coupled vibro-acoustic FRFs **a** befo and **b** after updating (case of complete data with measurement noise)

the measured responses. To simulate this condition, 5 % random noise with uniforr probability distribution is added in the real simulated frequency response an updating is carried out. Both the exact as well as identified values of updatin parameters are shown in Fig. 3b. In this case also it is seen that the method ha identified the parameters exactly.

Figure 5a shows a similar comparison of the vibro-acoustic coupled FRF cor responding to simulated and initial FE models. This FRF is calculated wit response on node number 59 in the acoustic domain to a unit harmonic force actin on node number 8 on the structure. Figure 5b shows a similar comparison fo updated FE model. It is seen that this updated FE model FRF also matches exactl with 'measured' FRF even in the presence of noisy data.

4 Conclusion

This paper proposes a method for FE model updating of coupled vibro-acoustic systems based on constrained optimization using frequency response function with the aim of identifying the material property and the stiffness of joint between the plate and rectangular cavity. The method minimizes the error between the measured and the FE model pressure responses. Inequality constraints in the form of upper and lower bounds on updating parameters are imposed. A numerical study is performed on a 3D rectangular box cavity backed by flexible plate. The robustness of the method in the presence of noise is studied. It is found that the proposed method effectively updates the coupled vibro-acoustic model to obtain a close match between FE model and corresponding 'measured' vibro-acoustic characteristics. The method also estimates the correction factors to the updating parameters with reasonable accuracy even in the presence of noise.

References

1. Mottershead JE, Friswell MI (1993) Model updating in structural dynamics: a survey. J Sound Vib 167(1993):347–375
2. Ewins DJ (2000) Modal testing: theory. Practice and Application, Research Studies Press, England
3. Nagy EJ, Berman A (1983) Improvement of a large analytical model using test data. AIAA Journal 21(1983):927–928
4. Collins JD, Hart GC, Hasselman TK, Kennedy B (1974) Statistical identification of structures. AIAA Journal 12(1974):185–190
5. Lin RM, Ewins DJ (1990) Model updating using FRF data. Proceedings of the 15th international seminar on modal analysis, Belgium, pp 141–162
6. Modak SV, Kundra TK, Nakra BC (2002) Model updating using constrained optimization. Mech Res Comm 27(5):543–551
7. Modak SV, Kundra TK, Nakra BC (2002) Use of an updated finite element model for dynamic design. Mech Syst Signal Proc 16(2):303–322
8. Nehete DV, Modak SV, Gupta K (2013) A method for FE model updating of coupled vibro-acoustic systems using constrained optimization ASME IMECE, San Diego, California
9. Decouvreur V, Bouillard P, Deraemaeker A, Ladeveze P (2004) Updating 2D acoustic models with the constitutive relation error method. J Sound Vib 278:4–5
10. Decouvreur V, Ladeveze P, Bouillard P (2008) Updating 3D acoustic models with the constitutive relation error method: a two-stage approach for absorbing material characterization. J Sound Vib 310(4–5):985–997
11. Progneaux A, ladeveze P, Deraemaeker A, Bouillard P (2010) Updating acoustical FE models: localization of unreliable physical data, proceeding of ISMA, pp 2689–2699
12. Dhandole SD, Modak SV (2012) A Constrained optimization based method for acoustic finite element model updating of cavities using pressure response. Appl Math Model 36 (2012):399–413
13. Schedlinski C, Wagner F, Bohnert K, Ksel M, Clasen D, Stein C, Glandier C, Kaufmann M, Tijs E (2008) Computational model updating of structural damping and acoustic absorption for coupled fluid-structure-analyses of passenger cars, proceeding of international conference on noise and vibration engineering. ISMA, Leuven, Belgium 2008:4229–4244

Experimental and Numerical Study of Friction-Induced Noise of Brake Pad Materials Having Grooved Surface

X.C. Wang, J.L. Mo, D.W. Wang, H. Ouyang, J. Zhao, G.X. Chen and M.H. Zhu

Abstract An experimental and numerical study was performed in this work to predict the squeal instability when pad specimens having different surfaces rotating on a brake disc specimen. The numerical results were consequently compared with the experimental results to investigate the effect of surface modification of pad materials on squeal generation. It was shown that surface modification of pad materials had a significant influence on the squeal instability. Diagonal grooved surface was found to be able to efficiently reduce squeal generation, which can be mainly attributed to its relatively smaller contact pressure distribution area compared to that of the smooth surface.

Keywords Squeal · Vibration · Surface modification · Pad materials · Finite element

1 Introduction

Automotive brake squeal is a noise problem caused by vibration induced by friction forces [1]. Usually, it occurs at frequency usually above 1 kHz and is described as sound pressure level above 78 dB or usually at least 20 dB above ambient noise level [2]. This type of noise has little effect on the brake performance. However, most customers would take this noise as a problem and demand its fixed, which results in significant yearly warranty costs [3].

In the past decades, extensive research efforts have been made to understand and eliminate brake squeal [3–5]. However, predicting and reducing brake squeal propensity still remains as challenging as ever, considering that the nature of brake

X.C. Wang · J.L. Mo (✉) · D.W. Wang · J. Zhao · G.X. Chen · M.H. Zhu
School of Mechanical Engineering, Southwest Jiaotong University, Chengdu 610031, China
e-mail: jlmo@swjtu.cn

H. Ouyang
School of Engineering, University of Liverpool, Liverpool L69 3GH, UK

J.K. Sinha (ed.), *Vibration Engineering and Technology of Machinery*, Mechanisms and Machine Science 23, DOI 10.1007/978-3-319-09918-7_92

squeal is fugitive, transient, and often non-repeatable [6]. Generally, there are a number of techniques and methods that have been performed to investigate the brake squeal. Among them, structural modification method is regarded as one of the most effective ways to reduce squeal [7–9]. The influence of structural modification on reducing brake squeal has been widely studied through the experimental methods in the last decades. It is found that the structure modification of brake pad has a significant effect on the squeal instability of brake system [4, 8, 9].

Recently, numerical studies have been widely performed to investigate the brake squeal [10–12], which can be mainly divided into two categories: complex eigenvalue analysis in the frequency domain and transient dynamic analysis in the time domain. The complex eigenvalue approach permits detection of the stability limit of the system by analyzing the eigenvalues and eigenvectors around the steady sliding state. However, the non-linear effects of contact are usually not negligible when instability occurs, and thus the transient dynamic analysis that takes into account the non-linear aspect of contact is preferred to study the evolution of the squeal instability. Therefore, complex eigenvalue analysis and dynamic transient analysis are complementary methods to study the generation of squeal [12–14].

In this work, a simplified setup of a disc brake, called pad-on-disc, was used to study the effect of grooved pad material on squeal generation both experimentally and numerically. Three different pad surfaces, named smooth surface, vertical grooved surface and diagonal grooved surface, were included. A numerical study was performed to simulate the experimental process by using the finite element software ABAQUS.

2 Experimental Procedure and Numerical Modeling

2.1 Experimental Procedure

The schematic view of the experimental setup is shown in Fig. 1. To start the test, the moving stage moves down to allow the upper holder carrying the pad specimen to contact the brake disc specimen with applying a constant normal load. Subsequently disc specimen is driven to rotate against pad specimen. During the test, the 3-D acceleration sensor mounted on the upper holder measures the vibration of the friction system, and the microphone located nearby the friction interface measures the noise signal.

Disc specimens (compacted graphite iron) were cut from the brake discs to the diameter of 25 mm and a thickness of 3 mm. The pad specimens (composite material) were manufactured to the size of 10 × 10 × 15 mm. Two kinds of grooves, i.e. vertical groove and diagonal groove were cut on the pad specimen surfaces. The width and depth of the groove was 1 and 0.3 mm, respectively. The tribological parameters are as follows: normal load of 100 N, rotate speed of 6.28 rad/s, testing time of 1800 s.

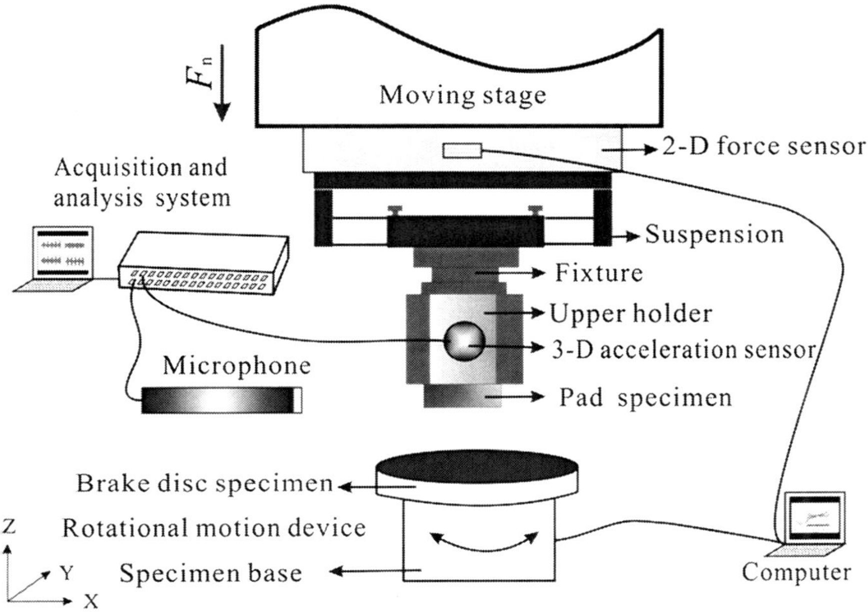

'ig. 1 Schematic view of the experimental set up

?.2 *Numerical Model*

'igure 2a shows the simplified finite element model of the experimental system presented in Fig. 1. The geometry of the numerical model was created according to he geometry of the experimental setup, which included a specimen base, a disc pecimen (active specimen), a pad specimen (passive specimen), an upper holder and a fixture. There is always frictional contact between the two specimens (passive and active). All the material parameters of the parts used in the numerical calcuation reflect those of the real experimental setup. The load and boundary conditions of the finite element model are shown in Fig. 2b: the bottom of the disc specimen is igidly constrained except the circumferential direction, which is used to apply the elocity boundary condition. For the pad specimen, the normal load is applied on he top of the up holder in the y-direction. All the constraint conditions were consistent with the real experimental setup except the part of threaded connection, which was simulated by a tie constraint.

In this study, surface modifications of pad specimen were conducted, by cutting centre and diagonal grooves, as shown in Fig. 3. In the numerical analysis, the element type was C3D8R, the contact formulation was finite sliding with the method of Lagrange multipliers. The friction formulation was penalty method. In he dynamic transient analysis, the observation point of the acceleration was located on the surface of the upper holder, which was the same as experimental study.

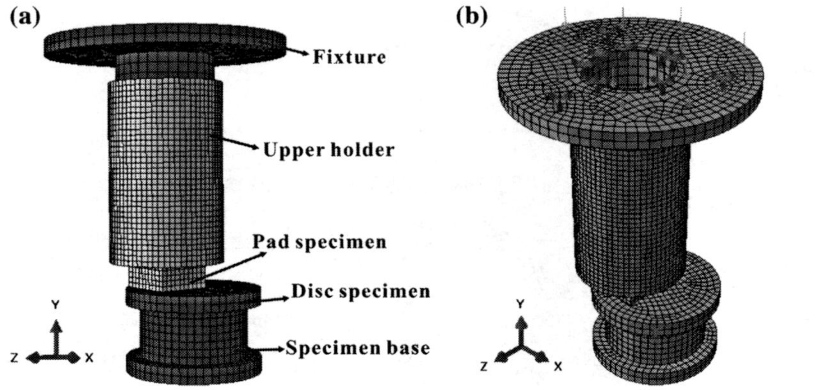

Fig. 2 Model of the experimental system: **a** finite element model; **b** load and boundary conditior of the finite element model

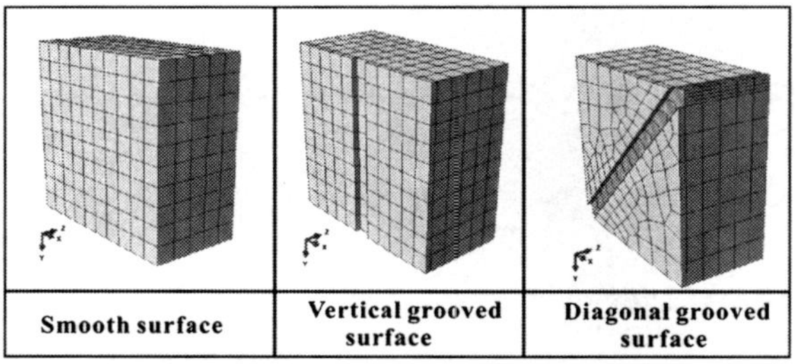

Fig. 3 Original smooth surface and two surface modifications of pad specimen

3 Experiment Results and Discussion

Equivalent continuous A-weighted sound pressure level during each 100 s tes duration was evaluated for the three pad surfaces, as shown in Fig. 4. It was foun that the sound pressure level generated by the diagonal grooved surface was lowe than that of other two surfaces throughout the test, indicating that cutting diagona groove on the pad specimen showed a great potential in reducing noise level.

Figure 5 shows the power spectrum density (PSD) of sound pressure throughou the test for the three surfaces. Four dominant frequencies of squeal can be observe for the smooth surfaces: 1,120, 2,225, 2,750 and 3,337 Hz. For the vertical groove surface, three dominant frequencies can be found: 1,150, 2,287 and 3,425 Hz which indicates that this modification of pad surface can suppress one of the squea frequencies in the range of 200 Hz to 3,000 Hz. In contrast, only one dominan frequency of 1,112 Hz was measured from the diagonal grooved surface during th whole test. Any noise signal having a dominant frequency of over 1 kHz was take

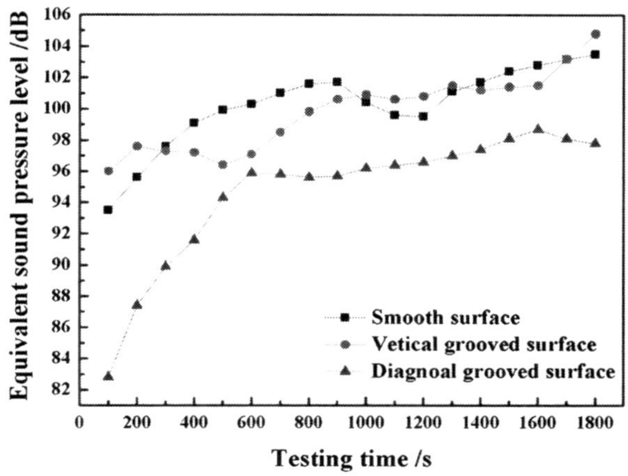

Fig. 4 Equivalent sound pressure level for these three pad surfaces

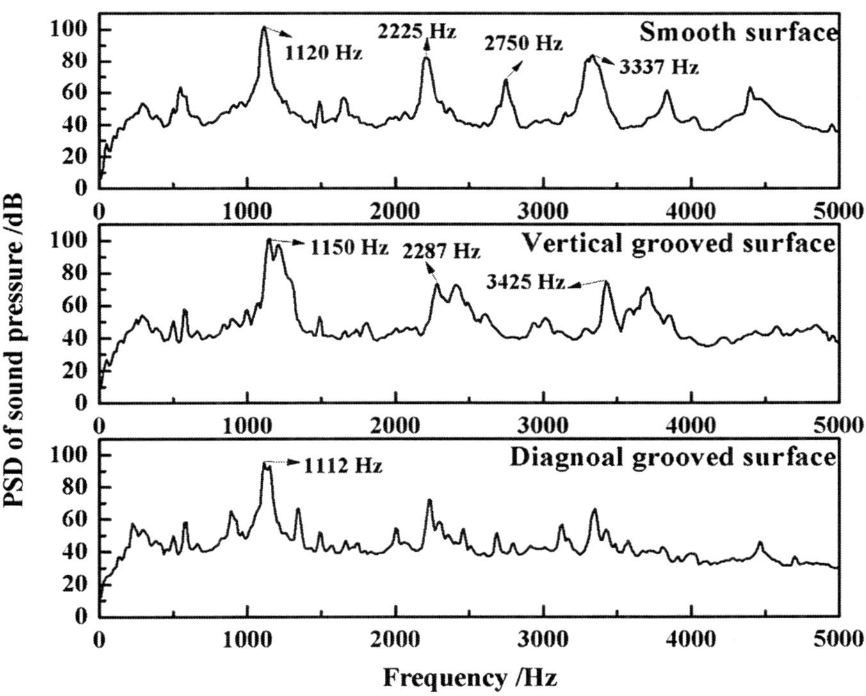

Fig. 5 Power spectrum density of sound pressure

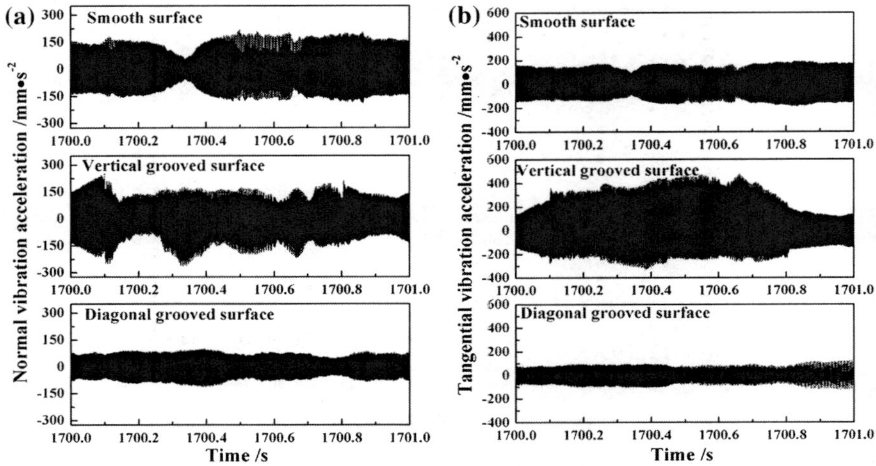

Fig. 6 Normal vibration acceleration (**a**) and tangential vibration acceleration (**b**) in the stable stage for the three pad surfaces

as a squeal event. Thus, modification of the contact surface by cutting diagonal groove on the pad surface had a significant effect on suppressing some high-frequency squeal.

The time history records of vibration accelerations in the stable stage were analyzed to further demonstrate that the diagonal grooved surface showed a better potential in suppressing squeal, as shown in Fig. 6. For the smooth surface and vertical grooved surface, both the normal and tangential vibration acceleration signals showed large amplitudes of vibration acceleration. In contrast, the amplitudes of vibration acceleration of the diagonal grooved surface were significantly smaller. Therefore, surface modification by cutting a diagonal groove on the pad surface showed a great potential in suppressing the generation of high-amplitude vibration, and consequently the generation of squeal.

4 Complex Eigenvalue Analysis

4.1 Introduction of Complex Eigenvalue Analysis

In the process of complex eigenvalue analysis, the frictional contact force has been introduced in the system at the contact interface. Thus, the general equation for the system [14], discretized by finite elements, is

$$\left(\lambda^2[M] + \lambda[C] + K_{NS}\right]\{\phi\} = 0 \quad (1)$$

This equation represents a non-standard eigenvalues problem, where $[M]$ is the mass matrix, $[C]$ is the damping matrix, $[K_{NS}]$ is the non-symmetric stiffness matrix of the system, λ is the eigenvalue and $\{\phi\}$ is the associated eigenvector. When the frictional contact is accounted for, $[K_{NS}]$ contains the supplementary terms due to contact forces at the contact nodes. The lack of symmetry leads to complex modes and complex eigenvalues. A general eigenvalue can be written as

$$\lambda = \alpha + j\omega \tag{2}$$

where the imaginary part ω is relative to the angular frequency of the corresponding system mode and the real part α is relative to the stability of the mode and thus the stability of the whole system. A positive real part of the eigenvalue means that the mode is unstable.

In the present paper, ABAQUS's complex eigenvalue analysis capability was applied to validate the finite element model, by comparing the dominant frequency of squeal between numerical study and experimental test. Moreover, the effect of these two pad surface modifications on the unstable model of friction systems can be observed by using complex eigenvalue analysis.

4.2 Complex Eigenvalue Analysis Results and Discussion

Figure 7 shows the complex eigenvalues of these three friction systems, versus the friction coefficient (μ). Two adjacent modes started to form an unstable complex mode in a certain value of friction coefficient. For all the three surfaces, a negative effective damping ratio appeared when the friction coefficient increased to 0.1, suggesting that the friction system had become unstable. The friction coefficient of 0.1 was found to be the critical value for the generation of unstable vibration at the dominant frequency of 1,383 Hz. Figure 7d shows the unstable mode of the friction system with smooth pad surface which mainly takes place on the upper holder in its flexural mode. This mode is related to the flexural motion of the upper holder carrying pad slides along the z-direction. In this work, modification of the pad surfaces did not show a significant influence on the unstable mode of the system by comparing the unstable modes of the three friction systems.

Comparing the complex eigenvalue analysis results with the squeal test results, the predicted frequency of 1,383 Hz was found to be close to one of the squeal frequencies of 1,120 Hz measured by experiments. The discrepancy between the numerical and experimental results was probably due to the required simplifications of the numerical model. Especially setting the threaded connection to a tie constraint would result in a relatively higher value of frequency.

It is noting that the complex eigenvalue analysis did not indicate all the experimental squeal frequencies above 2 kHz. Therefore, a numerical resolution of the complete nonlinear system needed to be performed additionally to estimate the

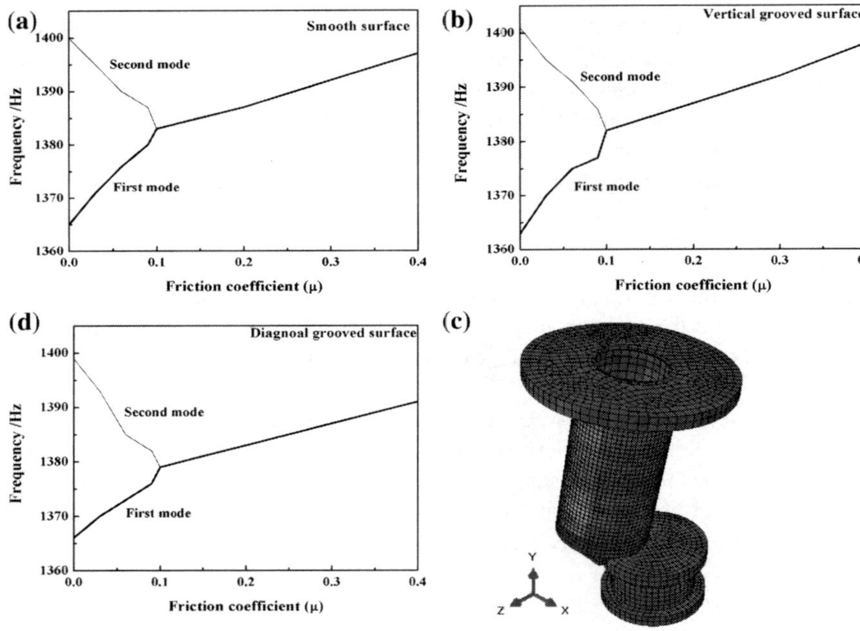

Fig. 7 Results of complex eigenvalues analysis for the smooth surface (**a**), vertical grooved surface (**b**), diagonal grooved surface (**c**), and the unstable mode of the smooth surface (**d**)

non-linear behavior of the solution far from the sliding equilibrium. The numerical model created above can be efficiently used to reflect the dynamic behavior of the experimental system, and will be used in the dynamic transient analysis thereafter, to reveal the effect of different pad surfaces on the friction vibration and noise.

5 Dynamic Transient Analysis

5.1 Introduction of Dynamic Transient Analysis

In the dynamic transient analysis, the equation of motion is as follows [14]:

$$[M]\ddot{x}_{(t)} = P_{(t)} - I_{(t)} \tag{3}$$

where $\ddot{x}$ is the acceleration vector, $[M]$ is the diagonal lumped mass matrix, $I_{(t)}$ is the internal force vector and $P_{(t)}$ is the applied external force vector including the calculated contact forces. The superscript t refers to the time increment. The velocity and displacement of the body are expressed in the following equations:

$$\dot{x}_{(t+\Delta t/2)} = \dot{x}_{(t-\Delta t/2)} + \frac{\left(\Delta t_{(t+\Delta t)} + \Delta t_{(t)}\right)}{2}\ddot{x}_{(t)} \tag{4}$$

$$x_{(t+\Delta t)} = x_{(t)} + \Delta t_{(t+\Delta t)}\dot{x}_{(t+\Delta t/2)} \tag{5}$$

where superscripts $t - \Delta t/2$ and $t + \Delta t/2$ refer to mid-increment values. Since the central difference operator is not self-starting because of the mid-increment of velocity, the initial values at time $t = 0$ for velocity and acceleration need to be defined. Both values are set to zero as the active specimen is stationary at time $t = 0$.

5.2 Time Domain Results and Discussion

For the transient dynamic analysis, the time history of the normal load and the rotational velocity of the disc specimen were used for describing real operating conditions, as shown in Fig. 8. A normal load was applied gradually until a prescribed value which is consequently kept constant. And then the disc started to rotate at t_0, and the rotational speed gradually increased up to t_1 which is consequently kept constant too. A constant friction coefficient of $\mu = 0.35$ was used, and the disc specimen was subjected to a normal load of 100 N and a rotational speed of 6.28 rad/s. The calculating time was set to 0.1 s.

Figure 9 shows the simulation results of vibration accelerations in both the normal and tangential directions in time domain. Visible limit cycles with the highest acceleration magnitude occurring at $t = 0.04$ s can be observed for both the smooth and vertical grooved surfaces. This type of dynamic behavior was consistent to that observed in experimental tests (see Fig. 6). This indicated that cutting a vertical groove on the pad surface could not prevent the generation of squeal. While for the diagonal grooved surface, no limit cycles were produced from the

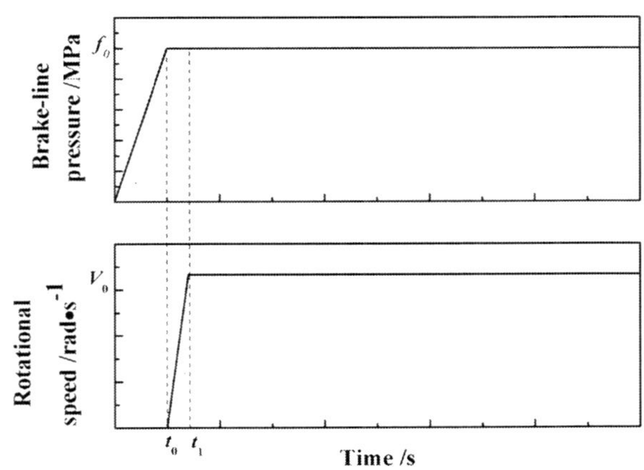

Fig. 8 Time history of the normal load and the rotational speed of the disc specimen in the dynamic transient analysis

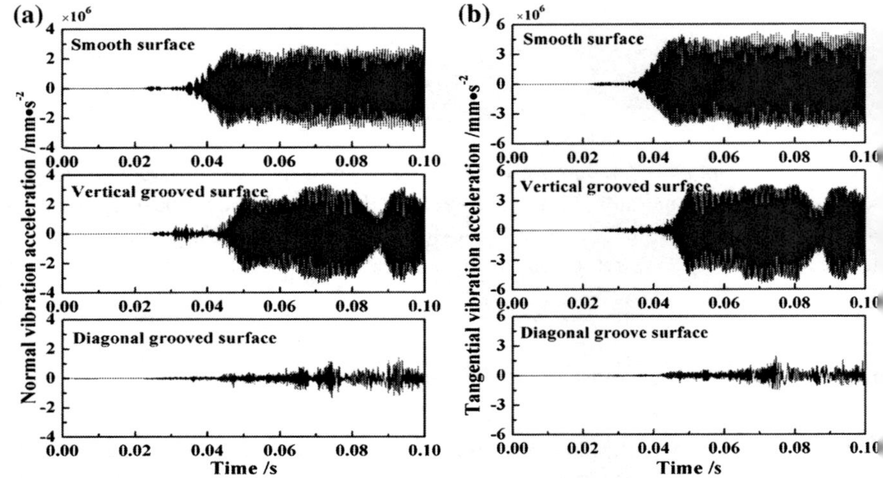

Fig. 9 Simulation results of vibration acceleration in the normal direction (**a**) and tangentia direction (**b**) in time domain

vibration response and the magnitude of predicted vibration acceleration was sig nificantly reduced. Friction system would not generate unstable vibration in thi condition. Therefore, surface modification by cutting a diagonal groove on the pa surface can be used to reduce the squeal instability.

To further investigate the phenomenon presented above, and propose a possibl physical explanation for why did the diagonal grooved pad show reduced vibratio amplitudes, the contact pressure distributions for both the smooth and diagona grooved surfaces at speed 6.28 rad/s were provided, as shown in Fig. 10. Th

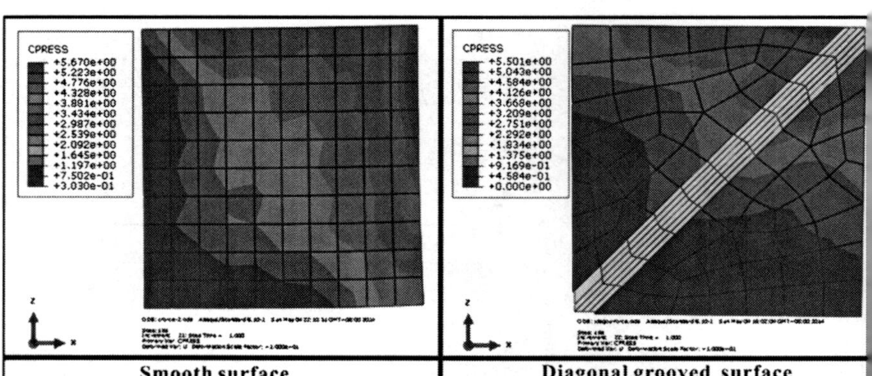

Fig. 10 Contact pressure distribution for smooth surface and diagonal grooved surface at speed o 6.28 rad/s

distribution area of contact pressure for the diagonal grooved surface was found to be much smaller than that of the smooth surface, which may mainly result in the significantly reduced vibration amplitude of the diagonal grooved surface.

6 Conclusion

This paper presented an investigation of brake squeal using a pad-on-disc experimental setup. Two modifications of pad surfaces were conducted, and the effect of grooved surface on squeal instability was investigated. From both the experimental and numerical results, conclusions could be made as follows:

(1) Both the experimental and simulation results showed that cutting a vertical groove on the pad surface had little effect on reducing the unstable vibration of the friction system; in contrast, cutting a diagonal groove on the pad surface showed a great potential in squeal reduction.

(2) The contact pressure distribution area for the diagonal grooved surface was much smaller as compared to the case of smooth surface. This may result in the reduction of vibration amplitude and consequently squeal suppression of the diagonal grooved surface.

References

1. Akay A (2002) Acoustics of friction. Acoust Soc America 114(4):1525–1548
2. Kinkaid NM, O'Reilly OM, Papadopoulos P (2003) Automotive disc brake squeal. J Sound Vib 267(2003):105–166
3. Júnior MT, Gerges SNY, Jordan R (2008) Analysis of brake squeal noise using the finite element method: A parametric study. Appl Acoust 69(2):147–162
4. Rabia AM, Ghazaly NM, Salem, MM, Abd-El-Tawwab AM (2013) Experimental studies of automotive disc brake noise and vibration: a review. Int J Modern Eng Res 3(1):199–203
5. Ghazaly NM, El-Sharkawy M, Ahmed I (2014) A review of automotive brake squeal mechanisms. J Mech Des Vib 1(1):5–9
6. Oberst S, Lai JCS (2008) A critical review on brake squeal and its treatment in practice. In: Internoise 2008, Shanghai
7. Shi TS, Dessouki O, Warzecha T, Chang WK, Jayasundera A (2001) Advances in complex eigenvalue analysis for brake noise. In: Proceedings of the 2011 noise and vibration conference, 30 April–3 May 2011, Traverse City, Michigan
8. Nouby M, Mathivanan D, Srinivasan K (2009) A combined approach of complex eigenvalue analysis and design of experiments (DOE) to study disc brake squeal. Int J Eng Sci Technol 1 (1):254–271
9. Oberst S, Lai JCS (2011) Statistical analysis of brake squeal noise. J Sound Vib 330 (2011):2978–2994
10. Liu P, Zheng H, Cai C, Wang YY, Lu C, Ang KH, Liu GR (2007) Analysis of disc brake squeal using the complex eigenvalue method. Appl Acoust 68(6):603–615
11. Soobbarayen K, Besset S, Sinou JJ (2013) Noise and vibration for a self-excited mechanical system with friction. Appl Acoust 74(10):1191–1204

12. Massi F, Baillet L, Giannini O, Sestieri A (2007) Brake squeal: linear and nonlinear numerical approaches. Mech Syst Signal Process 21(6):2374–2393
13. Meziane A, Baillet L, Laulagnet B (2010) Experimental and numerical investigation of friction-induced vibration of a beam-on-beam in contact with friction. Appl Acoust 71 (7):843–853
14. Ouyang H, Nack W, Yuan Y, Chen F (2005) Numerical analysis of automotive disc brake squeal: a review. Int J Veh Noise Vib 1(2005):207–231

Neural Network Modeling of Submarine Shell

Fei Wang, Yeping Xiong and Zhenping Weng

Abstract Submarine propeller operating in unsteady wake flow often results in dynamic fluctuating force which is transmitted through rotator-raft system and leads to shell structure vibration and structure-borne noise transmission. Even though methods like Flügge's equations of motion or FEM can be used to model cylinder shell, as for submarine equipped with devices the vibration behaviour of shell will not agree with that acquired from model developed with those methods. In this paper, Neural Network is used to analyse vibration of the submarine shell structure consisting of conical shell, hull and hemisphere shell. Firstly, a finite element model for the combined shell structure is developed for numerical simulations, whose aim is to acquire shell's vibration information so as to represent the actual output of sensors arranged on submarine shell. Then actuating force is applied on this model based on the prior knowledge of typical blade passing frequency and fluctuating force. The vibration characteristics and responses at the interested positions on the shell are analyzed numerically. These data are used in Neural Network as training and testing data respectively to further develop a non-linear submarine shell model. Vibration responses at the same points acquired from finite element model are compared with the Neural Network model respectively. Results show that Neural Network can be used to model the complex submarine shell structure for effective vibration analysis. Future research will be concentrated on applying Neural Network with the testing data that acquired from actual submarine to develop a controlled plant.

F. Wang · Z. Weng
Key Laboratory of Ship Vibration and Noise, China Ship Scientific Research Center, Wuxi, P.R. China
e-mail: myself3707@sina.com

Z. Weng
e-mail: wzp702@live.cn

Y. Xiong (✉)
Fluid-Structure Interaction Research Group, Faculty of Engineering and the Environment University of Southampton, SO171BJ Highfield, Southampton, UK
e-mail: Y.Xiong@soton.ac.uk

J.K. Sinha (ed.), *Vibration Engineering and Technology of Machinery*, Mechanisms and Machine Science 23, DOI 10.1007/978-3-319-09918-7_93

Keywords Submarine hull vibration · Shell combination · Neural network · Nonlinear model

1 Introduction

Submarine hull is a complex structure consisting of conical, cylindrical and hemisphere shells. Shell combination is of great interest in practical engineering applications. Since 1950s, studies on the vibration characteristics of shell structures have been well documented. A three-dimensional equation of motion of the vibrating cylindrical shell was first presented by Flügge [1]. Leissa [2] acquired the similar results through numerical analysis. Many researchers also studied vibration of various cylindrical shells, such as Forsberg [3, 4], Garnet et al. [5] and Thein Wah et al. [6]. Caresta et al. [7] used Flügge's equations of motion to model the fluid-loaded cylindrical shell by applying smeared theory in dynamic modeling which includes stiffeners, and an infinite shell model was used to solve fluid-structure interaction problem. Results showed a good approximation for finite shell. Merz et al. [8] studied a structure interacts with the fluid by a thin-walled axi-symmetric shell of finite elements based on Reissner-Mindlin theory, where transverse shear deformation is allowed, whereas the stress component normal to the shell is assumed zero throughout the shell thickness. To control acoustic radiation, for example, Cao et al. [9] used a pair of piezoelectric stack force actuation to control sound radiation from a cylindrical shell. In their work, only axial excitation is assumed; therefore only axial vibration is considered in the modeling.

However, most of the studies are on the elementary shell structures, such individual circular cylindrical, conical or spherical shells rather than the shell combinations. Recently, Qu et al. [10] analyzed the free and forced vibration of ring-stiffened conical-cylindrical shell combinations using a modified vibrational approach. And even though methods like that can solve problems of combined shell, when equipped with devices and working underwater, the submarine's vibration character will not agree with the results acquired by above methods.

In this paper, Neural Network is used to model and analyze vibration of the integrated submarine shell structure consisting of conical shell, hull and hemisphere shell combinations. A finite element model for the combined shell structure is developed for numerical simulations in order to acquire shell's vibration information so as to represent the actual output of sensors arranged on submarine shell. Our ultimate aim is to use practical measurement data to model the submarine so as to predict and control vibrations of the hull structure.

2 FEA Modeling of Submarine Structure

ABAQUS is a powerful engineering modeling software that is based on finite element method. It can solve problems from relative simple linear analysis to challenging nonlinear modeling. Here in this paper, we will develop a submarine shell model consisting of conical shell, hull and hemisphere shell. The detailed parameters in [11] are adopted for FEA modeling, which is listed in Table 1. The material is steel with density 7,800 kg/m^3, passion ratio 0.3, Young's module 2.1×10^{11}, respectively. The model developed in ABAQUS is shown in Fig. 1.

Submarine propeller operating in unsteady wake flow often results in dynamic fluctuating force which is transmitted through rotator-raft system and leads to shell structure vibration and structure-borne noise transmission. And the frequency of this force is equal to the speed multiplied by the number of blades. Furthermore, in engineering practices, the frequency of this force typically equals to 25 Hz, and the magnitude is approximately 2,000 N. Finally, all of the computations and analyzes will be conducted in frequency domain, and acceleration of the hull will be predicted as well.

Therefore, in the "Step" module of ABAQUS, the frequency is set to 25 Hz in "Steady-state dynamics, Direct". And in the "Load" module an axial force of 2,000 N is applied on the conical end to simulate propeller excitations while the other end, i.e. hemisphere end is free. The boundary condition is also free. Then adopt the mesh that ABAQUS suggested to define mesh, which has 516 elements, 515 nodes, and element type S3R S4R is chosen. Finally, define and submit this "Job", and export the acceleration as well as three dimensions special coordinator of nodes for next step.

3 Modeling Using Neural Network

In this paper Neural Network Fitting Tool GUI is used to develop submarine shell model. Firstly, the special coordinator of each node is used as input vector. For example, the coordinator of node-1 is [0, −1.2465, 0], then this means that the input layer of Neural Network not only has three neurons, but also each neuron's input is 0, −1.2465 and 0 respectively. Then the acceleration of y-direction (acceleration of other directions will be discussed later) is used as target, as for the node-1 the value is 2.0366.

While half of all the nodes will be used to develop the Neural Network model, the rest abundant nodes are used to validate the model.

Considering that Backpropagation Neural Network with biases, a sigmoid layer

$$f(x) = \frac{1}{1 + e^{-x}} \tag{1}$$

Table 1 The parameters of submarine shell model

Parameter	Value
Hull radius	3.5 m
Hull length	60.0 m
Hull thickness	0.025
Hemisphere shell thickness	0.014
Conical shell length	11.0 m
Conical shell thickness	0.014
Half-cone angle	18

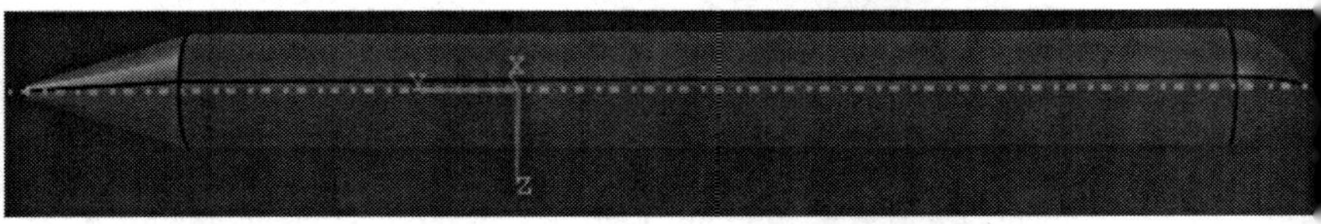

Fig. 1 Finite element model of submarine shell

and a linear output layer are capable of approximating any function with a finit number of discontinuities. So the choose of feed-forward Backpropagation Neur Network is appropriate. In the meantime the number of layer and neurons ar adapted to 5 and 20 respectively through trying. Besides, the input vectors an target vectors are randomly divided into three sets as follows:

- 60 % are used for training.
- 20 % are used to validate that the network is generalizing and to stop trainin before overfitting.
- The last 20 % are used as a completely independent test of networ generalization.
  The performance plot shown in Fig. 2 includes training errors, validation error and test errors. Consequently, it can be seen from the figure that the trainin stopped when the validation error increased for six iterations, which occurred iteration 22. In this case the result is reasonable because of the followin considerations:
- The final mean-square error is small,where the definition of mean-square is a follows:

$$e = \sqrt{\frac{\sum_{i=1}^{m} \sum_{j=1}^{n} \left(d_{ij} - y_{ij}\right)^2}{m \cdot n}} \quad (2$$

  d_{ij} is the target value, y_{ij} is the output value of network, m is number of inpu vector and n is the number of output, i.e. $n = 1$.
- The test set error and the validation set error have similar characteristics.

Fig. 2 The performance plot of network

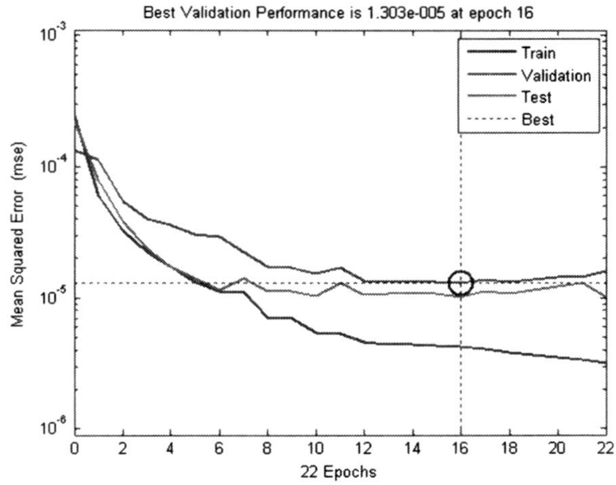

No significant overfitting has occurred by iteration 16 (where the best validation performance occurs). And the best validation performance is 1.303×10^{-5}.

The regression plot is shown in Fig. 3, which is composed by four drafts. Apparently, the output of Neural Network tracks the targets very well for training, testing, and validation, and the R-value is over 0.93 for the total response.

In this case, the network response is satisfactory, and we can now use the rest nodes to validate this network.

4 Validate the Network

Take the coordinators of the rest nodes as input vector of the network, and then compare the outputs with the results obtained from ABAQUS, as is shown in Fig. 4. It can be found that the difference between them is insignificant. In this case, the mean square error is 2.3337×10^{-5}. However, it is worth noting that to some nodes the error is so significant that one cannot just ignore them. One reason for this is that the nodes used to develop network can't represent the shell's whole characters well. In the following part, we will prove that when the number of the nodes increased from 515 to 3,251, the error accordingly becomes smaller.

Considering the fact that the prediction accuracy of network depends on the number of training vectors, so in order to improve the performance of network, the number of element in ABAQUS is enlarged to 3,281, and the number of node is enlarged to 3,251, the element type is still S3R S4R.

Adopt the same procedure as described above, the performance plot in this case is shown in Fig. 5. It can be seen that the best validation performance occurred at iteration 38, the best validation performance is 1.0583×10^{-4}. Compared with that obtained above, the MSE decreased about one level. Figure 6 is the regression plot,

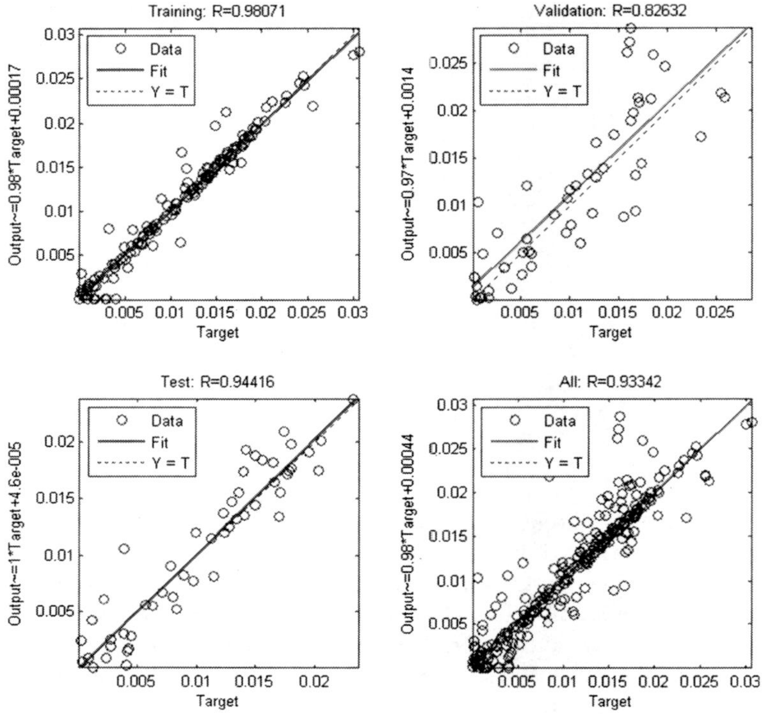

Fig. 3 The regression plot of network

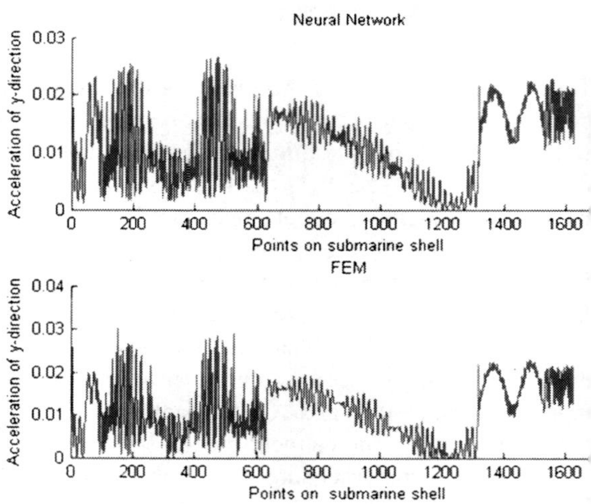

Fig. 4 As for the rest nodes, the results obtained from network and finite element model respectively

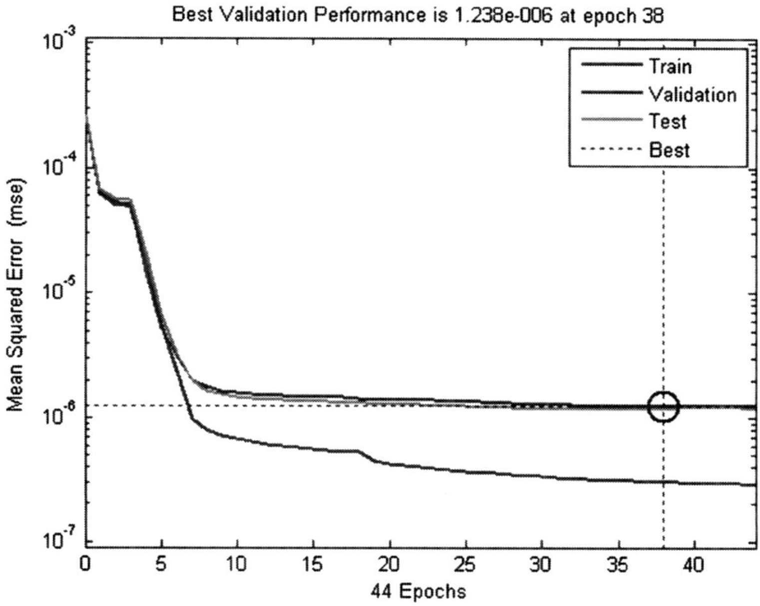

Fig. 5 The performance plot of network

which shows that the output of network can better track target for training, validation and test than that obtained above. In this case the R value is above 0.99.

It is still the case that the rest nodes are used to validate the network, and then the comparison between the results of network and finite element model is performed in Fig. 7. In this case the mean square error is 2.8856×10^{-5}, which is about 1/8 of that obtained above. And this can be regarded as a direct support that increasing the number of input vector can improve the performance of network, which will reduce the error between the results of network and finite element model as well. Similarly, same procedures can be taken to predict the acceleration of x- direction, and Fig. 8 shows the results acquired by Neural Network and FEM. In this case, the mean square error is 1.0583×10^{-4}.

5 Discussion

It can be concluded from the above that:

- In developing Neural Network model, the more input vector (which reflects in this paper is more nodes), the better performance the network will has.
- The information that input vector contains should be capable of representing the whole shell. In other words, we should avoid applying the local characters to

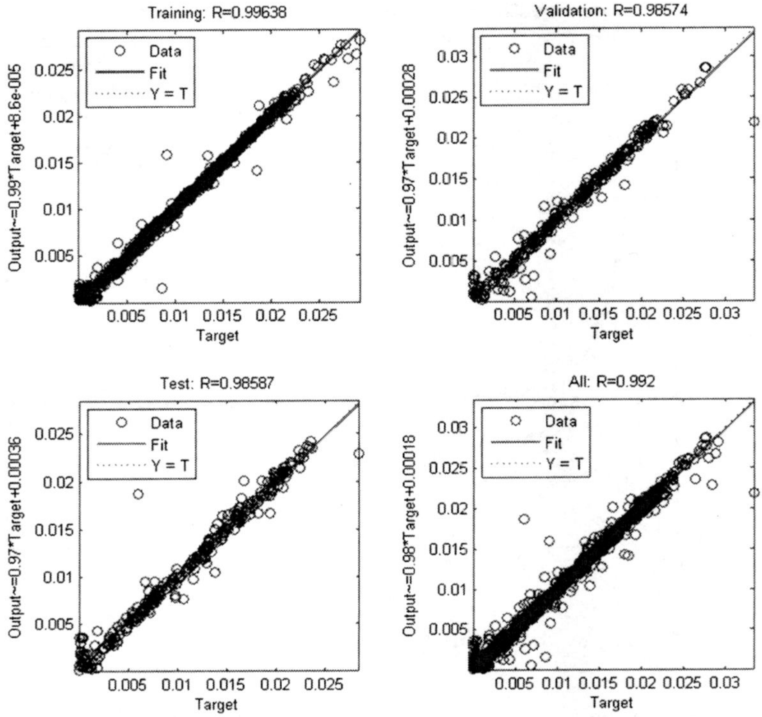

Fig. 6 The regression plot of network

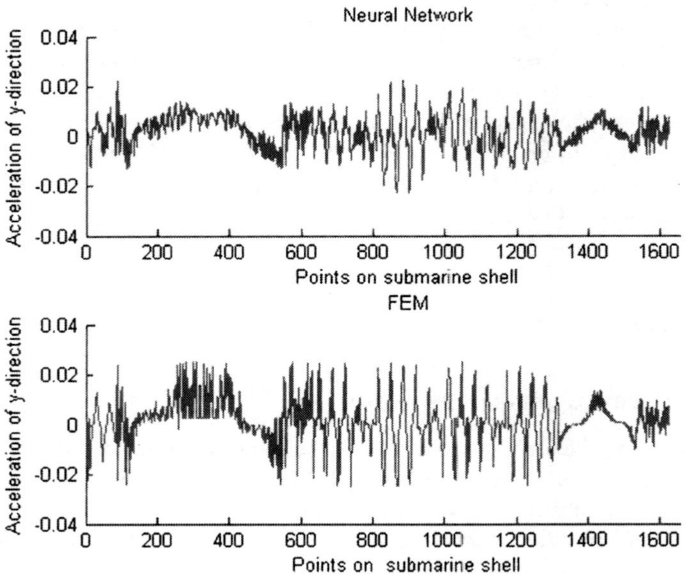

Fig. 7 As for the rest nodes, the results obtained from network and finite element model respectively

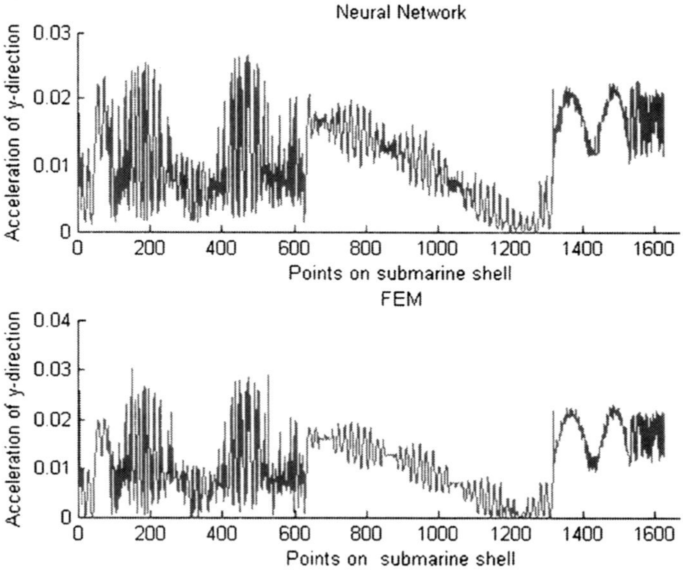

Fig. 8 As for the same nodes, the results predicted by network and simulated by finite element model respectively

represent the behavior of the whole coupled shell structure. Otherwise an accurate entire model cannot be developed even though the local information is abundant.

- Furthermore, the numbers of neuron and layer, training function, adaptation learning function as well as performance function all have different effect on the performance of network.

It is worth noting that, considering the limit of engineering practical, the number of actual sensors installed on the submarine shell should not be too large, which means that an optimization of the location and number of sensors should be performed before measuring, and this is one of the key points in our future research.

5 Conclusion

Submarine hull consisting of conical, cylindrical and hemisphere shells is a complicated shell structure. In this paper a feasibility study using Neural Network to develop a submarine shell model is presented for vibration analysis. Dynamic responses at the same points acquired from finite element model and Neural Network model respectively are compared. Results showed that Neural Network model developed allows obtaining an accurate prediction of the submarine shell vibration.

This would lay a solid foundation for making use of actual measuring data to mode submarine shell in the future, which would also construct a base for predicting an controlling the vibration of submarine shell. Future research will be concentrated o applying Neural Network with the sensing data collected from actual submarine t develop a controlled plant.

References

1. Flügge W (1973) Stresses in shells. Springer, Berlin, pp 204–259
2. Leissa AW (1973) Vibration of Shells. NASA SP-288, pp 1–184
3. Forsberg K (1964) Influence of boundary conditions on the modal characteristics of thi cylindrical shells. Am Inst Aeron Astron J 2(12):2150–2157
4. Forsberg K (1966) A review of analytical methods used to determine the modal characteristic of cylindrical shells. NASA CR-613
5. Garnet H, Goldberg MA (1962) Free vibration of ring-stiffened shells. Grumann Aircra Engrg. Cor Res Dedt Mech 21:156–106
6. Wah T, Hu WCL (1968) Vibration analysis of stiffened cylinders including inter-ring motio [J]. JASA 43(5):1005–1016
7. Caresta M, Kessissoglou NJ (2010) Acoustic signature of a submarine hull under harmoni excitation[J]. Appl Acoust 71(01):17–31
8. Merz S, Kinns R, NicoleKessissoglou N (2009) Structural and acoustic responses of submarine hull due to propeller forces[J]. J Sound Vib 325:266–286
9. Cao Y, Sun H, An F, Li X (2012) Active control of low frequency sound radiation b cylindrical shell with piezoelectric stack force actuators[J]. J Sound Vib 331(2012):2471–248
10. Qu YG, Chen Y, Long XH, Hua HX, Meng G (2013) A modified variational approach fo vibration analysis of ring-stiffened conical-cylindrical shell combinations. Eur J Mech A Solids 37:200–215
11. Pan X, Tso Y, Juniper R (2007) Active control of radiated pressure of a submarine hul J Sound Vib 311(1):24–242

Novel Approach for Structural Dynamic Topology Optimizations Based on Power Flow Mode Theory

Ye-Ping Xiong

Abstract The generalized damping based power flow mode theory [1] reveals natural power flow behaviours of a dynamic system based on the inherent characteristics of the system's damping distribution, which provides insight into energy dissipation mechanisms of the dynamic system. In this paper, a new Power Flow Mode dynamic Topology OPtimization (PFMTOP) approach is proposed based on the developed power flow mode theory to achieve topologically optimised systems' damping material distributions with enhanced vibration suppression capability. Conventional method of topology optimization focuses on minimizing the structural frequency response or dynamic compliance without considering structural damping. The new approach developed herein uses the trace of system's characteristic damping matrix as design objective to find an optimal damping material layout that maximize the energy dissipation for a given volume of the material to achieve minimum power flow response. Topology optimal design of damping distributions of two-phase structures subject to dynamic loading is studied. Example presented demonstrates the applicability and efficiency of the new PFMTOP approach. The obtained results reveal that the proposed approach can significantly enhance vibration energy dissipation and provides an effective method for optimised systems' damping material distributions with enhanced vibration suppression capability and reduced material usages. This new method can be readily extended to more complex structures to optimize the topology of damping material layer for improved vibrational power flow control.

Keywords Power flow mode theory · Dynamic topology optimizations · Trace of damping matrix · Damping distribution · Two phase structures

Y.-P. Xiong (✉)
Fluid-Structure Interaction Research Group, Faculty of Engineering and the Environment, Boldrewood Campus, University of Southampton, Southampton SO16 7QF, UK
e-mail: Y.Xiong@soton.ac.uk

J.K. Sinha (ed.), *Vibration Engineering and Technology of Machinery*, Mechanisms and Machine Science 23, DOI 10.1007/978-3-319-09918-7_94

1 Introduction

Engineering structures experience various dynamic loadings and the induce vibrations not only cause fatigue damage, but more importantly raise a grea environment concerns because of structure-borne, air-borne and waterborne noise There is an increasing demand in automobile, aero and maritime industries for low vibration, low noise products to satisfy performance standards and to reduce environmental noise pollutions. Meanwhile, with the global energy shortage becomes more critical, high energy performance efficiency is very demanding. Fo transportations in automobile, aero and maritime industries, there is high demand to reduce structure weight in order to reduce energy consumption, save material cost and increase transportation payload. Vibration problems of those lightweigh engineering structures under dynamic loading become more pronounced.

In recent years, structural topology optimization has been subjected to extensive study and developed to a great extent to handle different types of design objectives and constraints. In contrast to the traditional size and shape optimization, topology optimization provides the designers a design domain with more freedom in the initial stage of design. The book by Bendse and Sigmund [2] and a paper by Bendsøe et al. [3] in IUTAM symposium provide good references on topology design optimization of structures. However, in comparison to size and shape optimization problem, for dynamic topology optimization neither the theory nor the application is maturely developed. Most previous studies on topology optimization for vibrating structures have mainly used response or dynamic compliance [4], or fundamental frequency or frequency gap of the structure [5], or sound pressure for acoustic problem [6] as design objectives to minimise either structural frequency response, or dynamic compliance, or maximise fundamental or gaps between consecutive eigenfrequencies, or minimize sound radiations. Also structura damping is often ignored in previous studies on topology optimizations.

A power flow analysis provides a technique to describe the dynamic behaviour of complex structures and coupled systems at medium to high values of frequency where significant difficulties will be encountered in the higher frequency range when using FEA. Vibration power flow incorporates the influences of both force and velocity including their phase information and is valid over the whole fre- quency range. The approach focuses on the flow of vibrational energy rather than the detailed spatial pattern of the structural response. The fundamental concept of power flow is well discussed by Goyder and White [7] with comprehensive review in IUTAM symposium by Fahy and Price [8]. Various power flow analysis approaches have been developed by many researchers, such as, direct-dynamic stiffness method by Langley [9], mobility power flow approach by Cuschieri [10] four-pole parameter transfer matrix method [11], equivalent mobility method [12] energy flow density vector in a continuum [13]; power flow progressive approach by Xiong et al. [14], etc. Moreover, power flow is also used as a valuable parameter or cost function to assess passive, active or hybrid vibration control effectiveness by many researchers, [15–19] to name but a few. The author also developed a more generic power flow mathematical model consisting of multiple rigid/flexible structures of arbitrary configurations subjected to multiple excitations with arbitrary number of controllers considering structure-control interactions [20]. More recently, power flow research has been extended to nonlinear domains [21–24].

Using power flow analysis applied to topology optimization by Cho et al. [25] is an early work to study the structure topology optimization problems at high frequencies. In their paper, it was assumed that mechanism governing the flow of vibrational energy is similar to the flow of heat in thermal systems. An adjoint design sensitivity analysis (DSA) method is used to compute the variation of the structural response with respect to design variables. Deden and Hullber [26] performed topology optimization on sandwich beams for desirable mid-frequency response. A genetic algorithm optimization approach is used to determine candidate small cells, which can be used in the design of larger scale periodic sandwich structures. A spectral finite element analysis computation method is used to model mid-frequency wave propagation effects in structures. The results indicate that the optimization process is effective for designing new structure with decrease mid-frequency vibration response. Power flow response based dynamic topology optimization of bi-material plate structures is [27].

In this paper, an alternative approach based on power flow mode theory is proposed for topology optimization of structures. This allows not only the damping distribution can be optimized, but also the power flow of the structure can be controlled. The focus is placed on the feasibility and effectiveness of the new approach. Therefore, a commonly used square plate structure is considered for numerical simulations. To the best knowledge of the author, topology optimization based on power flow mode theory has not been addressed.

2 Power Flow Mode Theory-Brief Review

In this Section, the developed damping-based power flow mode theory [1] is briefly reviewed to aid the understanding for readers. It describes the essential power flow behaviour of a dynamic system. A characteristic damping matrix is introduced from physical damping components. The eigenvalues and eigenvectors of this characteristic damping matrix are defined as the characteristic damping factor and the power flow mode vectors of the system, respectively. These parameters are independent of external excitations. Power flow mode vectors are chosen as a set of base-vectors spanning the power flow space to describe completely the system's power flow behaviour. The generalized coordinate of the velocity vector decomposed in the power flow space is defined as the power flow response vector, which depends on external vibration sources. The time-averaged power flow is determined by the characteristic damping factor and the power flow response vector. For a unit power flow response vector, the time-averaged power flow value depends only on the system's characteristic damping factor.

2.1 New Power Flow Equation

For generality, consider the dynamics of a generalized linear system with N degrees-of-freedom described by the matrix equation

$$\mathbf{M}\ddot{\mathbf{X}} + \mathbf{C}\dot{\mathbf{X}} + \mathbf{K}\mathbf{X} = \tilde{\mathbf{f}}e^{\mathrm{i}\omega t} \tag{1}$$

where $\tilde{\mathbf{f}}$ denotes the complex amplitude of the excitation force vector, $\mathbf{M}$ is a real, symmetric and semi-positive definite mass matrix, $\mathbf{K}$ represents a symmetric and semi-positive definite stiffness matrix and $\mathbf{C}$ is a damping matrix. The solution of Eq. (1) takes the form $\dot{\mathbf{X}} = \tilde{\mathbf{V}}e^{\mathrm{i}\omega t}$. Here $\tilde{\mathbf{V}}$ is a velocity amplitude vector of the system. The conventional total time-averaged power input by all excitation forces applied to a system is given by

$$P = \frac{1}{2}\mathrm{Re}(\tilde{\mathbf{f}}^{\mathrm{H}}\tilde{\mathbf{V}}) = \frac{1}{2}\mathrm{Re}(\tilde{\mathbf{V}}^{\mathrm{H}}\tilde{\mathbf{f}}) \tag{2}$$

where the superscript H denotes a conjugate, transpose matrix. Xiong et al. [28] showed that after matrix manipulation the new time-averaged input power equation is derived as

$$P = \frac{1}{2}\mathrm{Re}\{\tilde{\mathbf{V}}^{\mathrm{H}}\bar{\mathbf{C}}\tilde{\mathbf{V}}\} \tag{3}$$

where

$$\bar{\mathbf{C}} = \frac{1}{2}(\mathbf{C} + \mathbf{C}^{\mathrm{T}}) \tag{4}$$

This result demonstrates that the time-averaged power of a system equals the power dissipated by damping. As is shown in [1], this characteristic damping matrix $\bar{\mathbf{C}}$ plays a significant role in determining the system's power flow.

The proposed theoretical mathematical model describes the time-averaged power flow in terms of system's damping and velocity response, which straightforwardly reveals vibration energy dissipation mechanisms of a dynamic system.

2.2 Power Flow Space

Because $\mathbf{C} + \mathbf{C}^{\mathrm{T}}$ is a real symmetric matrix, it follows that $\bar{\mathbf{C}}$ is a real symmetric matrix. In general, $\bar{\mathbf{C}}$ is a semi-positive definite matrix if there exists no negative active damping in the system. Based on matrix theory, the characteristic damping matrix $\bar{\mathbf{C}}$ can be decomposed into

$$\bar{\mathbf{C}} = \mathbf{\Phi}\mathbf{\Lambda}\mathbf{\Phi}^{\mathrm{T}}, \tag{5}$$

where

$$\mathbf{\Lambda} = \mathrm{diag}(\lambda_j), \tag{6}$$

$$\Phi = [\boldsymbol{\phi}_1, \boldsymbol{\phi}_2, \ldots, \boldsymbol{\phi}_N], \Phi^{\mathrm{T}}\Phi = \Phi\Phi^{\mathrm{T}} = \mathbf{I}. \tag{7}$$

Here λ_j, of dimension Nsm^{-1}, and non-dimensional real vectors $\boldsymbol{\phi}_j$, (j = 1, 2, …, N), denote the eigenvalues and the corresponding eigenvectors of the characteristic damping matrix $\bar{\mathbf{C}}$, respectively.

These power flow mode vectors are linearly independent of each other and they are chosen as a set of base vectors spanning the power flow space of the system.

2.3 Natural Power Flow Characteristics

The jth characteristic damping factor λ_j and jth power flow mode vector $\boldsymbol{\phi}_j$ of the characteristic-damping matrix $\bar{\mathbf{C}}$ are independent of external excitations. For a dynamic system, the characteristic-damping matrix $\bar{\mathbf{C}}$ is prescribed and accordingly its natural power flow characteristics, i.e. λ_j and ϕ_j, are determined. In the defined power flow space, the physical velocity vector $\tilde{\mathbf{V}}$ of the system decomposes into the form

$$\tilde{\mathbf{V}} = \Phi\tilde{\mathbf{q}}. \tag{8}$$

Since $\mathbf{\Phi}$ is the orthogonal matrix composed of the corresponding eigenvectors φ_j (j = 1, 2, …, N), thus

$$\tilde{\mathbf{q}} = \Phi^{\mathrm{T}}\tilde{\mathbf{V}}, \tilde{q}_j = \varphi_j^{\mathrm{T}}\tilde{\mathbf{V}} \tag{9}$$

which defines a complex valued velocity vector of dimension m s^{-1}. For a prescribed system, this power flow response vector $\tilde{q}$ defined in the power flow space depends on the external excitations and its jth element $\tilde{q}_j$ denotes the jth power flow response corresponding to the jth power flow mode φ_j.

2.4 Power Flow Mode

In power flow space the time-averaged power flow can be written in the form

$$P = \frac{1}{2}\mathrm{Re}\{\tilde{\mathbf{q}}^{\mathrm{H}}\Lambda\mathbf{q}\} = \frac{1}{2}\sum_{j=1}^{N}\lambda_j|\tilde{q}_j|^2 = \sum_{j=1}^{N}P_j, \tag{10}$$

where

$$P_j = \lambda_j |\tilde{q}_j|^2/2,\ P_j/|\tilde{q}_j|^2 = \lambda_j/2,\ (\ j\ = 1, 2, \ldots,\ N\ ) \quad (11$$

defines the *j*th mode dissipation power of the system. The power flow mode theor provides a new approach for vibration control applications through modifications o the system's damping distribution using passive and/or active control component Interested readers can refer to the full paper published in the royal society pro ceedings [1] for more details.

3 Power Flow Mode Based Topology Optimizations

In previous research on dynamic topology optimizations, dynamic compliance ha often been used as the objective function, where system damping existed in engi neering structures were neglected. As presented above, the generalized power flo mode theory [1] reveals natural power flow behaviours of a dynamic system base on the inherent characteristics of the system's damping distribution, which provide insight into energy dissipation mechanisms of the dynamic system.

In this paper, a new approach for structural dynamic topology optimization i proposed based on the developed power flow mode theory to address dampin based power flow dynamic topology optimizations. This approach presented herei aims to optimally design damping distributions of two phase structures subject t dynamic loading to achieve the maximum energy dissipation with reduced powe flow response and material usage.

3.1 Trace of System's Characteristic Damping Matrix

From Eq. (11), it reveals that for a dynamic system defined the time averaged powe dissipated per unit characteristic velocity $\tilde{q}_j$, (j = 1, 2, 3,…, N) is

$$P_{\mathrm{u}} = (\lambda_1 + \lambda_2 + \cdots + \lambda_N)/2. \quad (12$$

Based on matrix theory, the orthogonal transformation represented by Eq. (5 does not change the trace of the real symmetric matrix $\bar{\mathbf{C}}$ and therefore

$$P_{\mathrm{u}} = (\lambda_1 + \lambda_2 + \cdots + \lambda_N)/2 = \mathrm{trace}\ \bar{\mathbf{C}}/2. \quad (13$$

3.2 Problem Statement: Maximise the Trace of Characteristic Damping Matrix

Equation (13) demonstrates that the time averaged power dissipated per unit characteristic velocity is maximum if and only if the trace of its characteristic damping matrix is a maximum value. The main aim in vibration control is to maximize dissipation of the vibration energy. Applying the damping based power flow more theory [1], the trace of system's characteristic damping matrix is proposed as design objective to find an optimal material layout that maximize the trace of $\bar{\mathbf{C}}$ for a given volume of the material, i.e., the summation of eigenvalues of the global characteristic damping matrix $\bar{\mathbf{C}}$ of the system:

$$MAX\ P_u = \frac{1}{2}\text{trace}\{\bar{\mathbf{C}}\} \tag{14}$$

subject to

$$\mathbf{M}\ddot{\tilde{\mathbf{X}}} + \mathbf{C}\mathbf{X} + \mathbf{K}\mathbf{X} - \tilde{\mathbf{f}}e^{i\omega t} = \mathbf{0} \tag{15}$$

$$\sum_{e=1}^{N_e} \rho_e V_e - V^{m_S} \leq 0,\ V^{m_S} = \mu V_0 \tag{16}$$

$$0 < \rho_{\min} \leq \rho_e \leq 1,\ e = 1, 2, \ldots N_e$$

where N_e is the total number of finite elements, ρ_e denotes the relative density of the eth element and V_0 is the volume of admissible design domain. Here a bi-material design problem is considered, where ρ_e is the volumetric density of the stiffer material in element e and plays the role of the design variable in the problem. The parameter μ is the fraction of the given volume of the stiffer material m_S, which is the volume ratio of m_S to the admissible design domain V_0. The remaining part of the total volume is occupied by a weaker material noted by m_w.

3.3 Optimisation Procedure

The topology optimization procedure consists of several steps after sensitivity study. The first would be discretising the design domain, rectangular in the current case, into bi-linear elements. Then a proper material model has to be chosen which describes how the geometric property of an element decides the mass and stiffness properties. The two widely used models are the homogenization model and the solid isotropic material with penalization model (SIMP).

For the bi-material SIMP model considered in this paper, the finite elemen stiffness and mass matrices for the stiffer and weak materials be obtained usin mixture rule as follows:

$$\mathbf{K}_e(\rho_e) = \rho_e^p \mathbf{K}_e^{m_s} + (1 - \rho_e^p)\mathbf{K}_e^{m_w} \tag{17}$$

$$\mathbf{M}_e(\rho_e) = \rho_e^q \mathbf{M}_e^{m_s} + (1 - \rho_e^q)\mathbf{M}_e^{m_w} \tag{18}$$

where $p > 1$ and $q > 1$ are the penalty parameters used to enforce the concentrate distribution of material volume and normally $p = q = 3$ [6].

The Rayleigh damping model expressed by a linear combination of mass matri and stiffness matrix respectively is adopted, i.e. $\mathbf{C} = \alpha\mathbf{M} + \beta\mathbf{K}$ where α and β denote the mass damping and stiffness damping coefficients, respectively. The finit element damping matrix takes the following form:

$$\mathbf{C}(\rho_e) = \alpha\mathbf{M}_e(\rho_e) + \beta\mathbf{K}_e(\rho_e) \tag{19}$$

To demonstrate this new approach will lead to optimal vibration control effect, i the next section a numerical simulation is performed using the power flow topolog optimisation by designing materials' damping distributions to maximise the trace o damping matrix of the system.

4 Numerical Simulation and Analysis

The topology optimization problem of a squared plate with all edges clamped an subjected to uniformly distributed harmonic load of unit magnitude is considered The plate has dimension of 600 mm × 600 mm × 1 mm and is made from tw materials: stiffer material of Young's modulus Es = 210 GPa, mass density o $\rho_S = 7800$ Kg/m^3 and weak material of Young's modulus Es = 21 GPa with mas density of $\rho_w = 780$ Kg/m^3 are assumed. Poisson's ratios of both materials ar $v_S = v_w = 0.3$ and damping coefficients are chosen as $\alpha = 0.5$ and $\beta = 1$. Th domain is discretized by four-node S4R shell elements in ABAQUS for finit element modelling. The filter radius is $r_{\min} = 20$mm. The lower bound of th material density is set to be 0.001 which denotes weak element in the followin simulations. The objective volume of stiffer material is set to be 50 % of the desig domain, i.e., the volume fraction ratio $\mu = 0.5$ and the initial values of desig variable is set to be $\rho_e = 0.5$. Method of moving asymptotes (MMA) is employe for the optimization. Penalization with $p = q = 3$ is applied to stiffness and mas matrices in optimization. The objective is to design the bi-material structure wit minimized input power flow response and maximised trace of damping matrix o the bi-material structure with 50 % of material saving in the design domain.

Figure 1 shows the iteration histories of both the trace of system's damping matrix and power flow mode response of the two-phase structure. It clearl

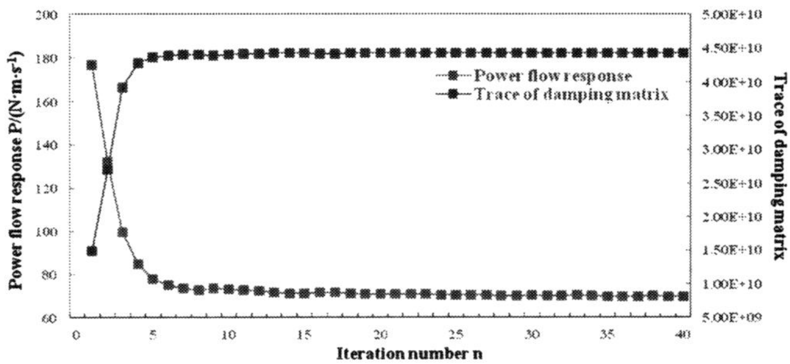

Fig. 1 Iteration histories of both the trace of system's damping matrix and power flow mode response for the two-phase Mindlin plate subjected to harmonic uniform pressure load

Fig. 2 Optimal topology design of two-phase material distribution (*blue* weak material, *red* strong material $Es = 10\ Ew$ and $\rho_S = 10\,\rho_w$)

evidenced that with optimisation progressed, the trace of damping matrix increases significantly from the initial value of 1.49×10^{10} to the final converged value of 4.43×10^{10} when the final optimum topology is obtained after the objective volume fraction decreases to 0.5 achieved at iteration $n = 40$. The trace of damping matrix is convergent to an almost constant value after 20 iterations.

More importantly, Fig. 1 also revels that from vibration power flow control viewpoint, the new approach not only increases the trace of system damping through redistributing system's two-phase material damping to achieve the optimal value, but also this leads to a significant reductions of power flow response under the same excitation to the structure.

Figure 2 displays the final optimal topology of two-phase material distributions using the proposed PFMTOP method for the squared plate under uniformly distributed loading at frequency of 100 Hz. The corresponding final optimised power

(a) (b)

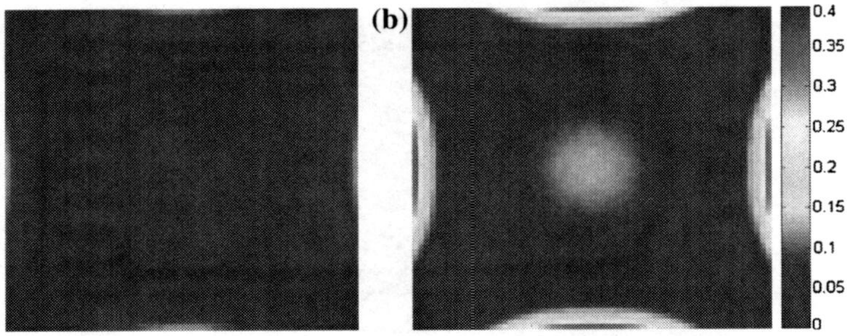

Fig. 3 The power flow mode responses of the plate: **a** Final optimized distribution; **b** Initial state

Table 1 Comparison of results of optimal design and effectiveness

Item	Initial value	Final value	Percentage (%)
Trace of damping matrix	1.49×10^{10}	4.43×10^{10}	+197.3
Power flow response ($N{\cdot}m{\cdot}s^{-1}$)	176.8	69.8	−60.5

flow mode response is shown in Fig. 3a and compared with the result at its initial design stage Fig. 3b.

Table 1 compares vibration control effectives for the initial design and optimal design regarding to trace of damping matrix and power flow mode response. When the final structure is convergent to a stable topology after 40 iterations, the power flow response reduced 60.5 % of its initial value.

5 Conclusion

In this paper, a new power flow mode based dynamic topology optimization (PFMTOP) approach is proposed applying the developed power flow mode theory [1]. The important benefit of using PFMTOP in dynamic topology design is the capability to achieve topologically optimised systems' damping material distributions with enhanced vibration suppression effectiveness from energy flow perspective. By optimising the material damping distribution of two phase structures subject to dynamic loading, the trace of system's damping matrix is maximised and power flow response minimized. This new method can be readily extended to more complex structures with multiple matrials. Further work is proposed to develop PFMTOP approaches for composite structures and damping layer on host structures to be reported in other papers.

Acknowledgments Financial support by EPSRC Shaping Capacity Award and Southampton Marine and Maritime Institute (SMMI) are gratefully acknowledged.

References

1. Xiong YP, Xing JT, Price WG (2005) A power flow mode theory based on a system's damping distribution and power flow design approaches. Proc R Soc A 461:3381–3411
2. Bendse MP, Sigmund O (2003) Topology optimization: theory, methods and applications. Springer, Berlin, Heidelberg, New York
3. Bendsøe MP, Olhoff N, Sigmund O (eds) (2006) Topological design optimization of structures, machines and materials—status and perspectives, proceedings of IUTAM symposium
4. Ma ZD, Kikuchi N, Cheng HC (1995) Topological design for vibrating structures. Comput Methods Appl Mech Eng 121:259–280
5. Yoon GH (2010) Maximizing the fundamental eigenfrequency of geometrically nonlinear structures using the element connectivity based topology optimization. Comput Struct 88:120–133
6. Du J, Olhoff N (2007) Minimization of sound radiation from vibrating bi-material structures using topology optimization. Struct Multidisc Optim 33:305–321
7. Goyder HGD, White RG (1980) Vibrational power flow from machines into built-up structures. J Sound Vib 68:59–117
8. Fahy FJ, Price WG (eds) (1999) IUTAM symposium on statistical energy analysis, Kluwer Academic, Dordrecht
9. Langley RS (1990) Analysis of power flow in beams and frameworks using the direct-dynamic stiffness method. J Sound Vib 136:439–452
10. Cuschieri JM (1990) Vibration transmission through periodic structures using a mobility power flow approach. J Sound Vib 143:65–74
11. Xiong YP, Song K, Ai X (1996) The influence of flexible foundation on isolators wave effects. Shock Vib 3(1):61–67
12. Xiong YP (1999) A new modelling method for machine foundation coupling system and its application to the control of power flow. Chinese J Mech Eng 12(2):123–129
13. Xing JT, Price WG (1999) A power-flow analysis based on continuum dynamics. Proc R Soc A 455:401–436
14. Xiong YP, Xing JT, Price WG (2001) Power flow analysis of complex coupled systems by progressive approaches. J Sound Vib 239:275–295
15. Miller DW, Hall SR, von Flotow AH (1990) Optimal control of power flow at structural junctions. J Sound Vib 140:475–497
16. Pan J, Pan JQ, Hansen CH (1992) Total power flow from a vibrating rigid body to a thin panel through multiple elastic mounts. J Acoust Soc Am 92:895–907
17. Fuller CR, Elliott SJ, Nelson PA (1997) Active control of vibration. Academic Press, London
18. Xiong YP, Song K, Wang C (1996) Optimum control of vibration and structure-borne noise for machinery on flexible foundation. Chinese J Mech Eng 9(1):13–20
19. Xiong YP, Xing JT, Price WG (2000) Hybrid active and passive control of vibratory power flow in flexible isolation systems. Shock Vib 7:139–148
20. Xiong YP, Xing JT, Price WG (2003) A general linear mathematical model of power flow analysis and control for integrated structure-control systems. J Sound Vib 267:301–334
21. Xiong YP, Xing JT, Price WG (2005) Interactive power flow characteristics of an integrated equipment—nonlinear isolator—travelling flexible ship excited by sea waves. J Sound Vib 287:245–276
22. Xiong YP, Cao Q (2011) Power flow characteristics of coupled linear and nonlinear oscillators with irrational nonlinear stiffness. In: Bernardini D, Rega G, Romeo F (eds) Proceedings of the 7th European nonlinear dynamics conference Rome, IT, Universita di Roma, pp 1–6
23. Yang J, Xiong YP, Xing JT (2012) Dynamics and power flow behaviour of a nonlinear vibration isolation system with a negative stiffness mechanism. J Sound Vib 332(1):167–183
24. Yang J, Xiong YP, Xing JT (2014) Nonlinear power flow analysis of the Duffing oscillator. Mech Syst Signal Process 45(2):563–578

25. Cho S, Park CY, Park YH, Hong SY (2006) Topology design optimization of structures a high frequencies using power flow analysis. J Sound Vib 298:206–220
26. Larsen AA, Laksafoss B, Jensen JS, Sigmund O (2009) Topological material layout in plate for vibration suppression and wave propagation control. Struct Multidisc Optim 37:585–59
27. Xue X, Xiong YP, Li GX, Gong JZ (2013) Power flow response based dynamic topolog optimization of bi-material plate structures. Chinese J Mech Eng 26(3):620–628
28. Xiong YP, Xing JT, Price WG (2004) A power flow mode theory based on inherer characteristics of damping distributions in systems and its applications. In: Gutkowski W Kowalewski TA (eds) Mechanics of the 21st century, proceedings 21th international congres Theoretical and applied mechanics ICTAM 2004, IUTAM, Warszawa, Poland. Paper N SM25S_11834

Accurate Calculation of the Phase Shift Angle of Sinusoidal Signals Using Trapezoidal Rule

Luis E. Monterrubio

Abstract In this work, the phase shift angle of a signal is obtained following a procedure that involves carrying out trapezoid rule numerical integration. The two required numerical integrations in this procedure define the tangent function of the phase shift angle. Accurate results (up to the 14th significant figures) are obtained using four equally spaced points (at 0°, 120°, 240° and 360°) on a periodic signal. But the procedure presented here has no limit in the amount of data points that can be used in the numerical integrations. Thus, this procedure can be used to estimate the phase shift angle when data includes noise, using as many points on a period as necessary.

Keywords Phase shift angle · Trapezoidal rule

1 Introduction

In the past, researchers have developed several methods to calculate the phase shift angle between two sinusoidal signals. A few of these methods are described in references [1–3]. Some of these methods are the N-bucket algorithm that includes the four-step and five-step algorithms, as well as the Carré method [1]. A characteristic common to all these methods is the use of the amplitude of the signal with a shift angle at equally distributed points. A difference of the method presented here with the methods mentioned above is that in the present method uses the same equation to calculate the phase shift angle regardless of the number of data points (four or more).

L.E. Monterrubio (✉)
Department of Engineering, Robert Morris University, 6001 University Blvd., Moon Township, PA 15108, USA
e-mail: monterrubio@rmu.edu

J.K. Sinha (ed.), *Vibration Engineering and Technology of Machinery*, Mechanisms and Machine Science 23, DOI 10.1007/978-3-319-09918-7_95

2 Computing the Phase Shift Angle

This work presents a procedure to compute the phase shift angle φ of sinusoidal signals using two reference signals. The first reference signal $y_{\sin}$ is defined with a sine function as

$$y_{\sin} = A\sin(\omega t), \tag{1a}$$

where t is time, while A and ω are the amplitude and circular frequency of the signal. To compute the phase shift angle φ with an arctan function it is necessary to have a second reference signal, equal to the signal given in (1a), but with a phase shift angle equal to 90° = π/2. Thus, the second reference signal $y_{\cos}$ is defined with a cosine function as

$$y_{\cos} = A\sin(\omega t + \pi/2) = A\cos(\omega t) \tag{1b}$$

Then, a signal with the same circular frequency but with an arbitrary phase shift is

$$y_2 = B\sin(\omega t + \varphi) + C, \tag{2}$$

where B and φ are the amplitude and phase shift angle of the signal, while C is the value from which the signal oscillates. This corresponds to a shift in the y axis of Fig. 1 showing the signals in (1a, 1b and 2) for $A = 8.45$, $B = 3.33$, $\varphi = 337°$ and $C = 0$.

The assumptions of this method are

- the sine and cosine reference signals oscillate from zero
- the size step is the same for all points and for all signals and
- four or more data points per cycle are available

The procedure in this work to compute the phase shift angle φ of sinusoidal signals is based on (3) and (4). Similar equations are given in [3]

$$\varphi = (T_\varphi/T)360° \tag{3}$$

Where = 1/f is the period of the two signals, $f = \omega/2\pi$ is the frequency in Hertz and T_φ is the time interval proportional to the phase shift angle that can be evaluated using the following integral:

$$T_\varphi = \int_0^T A\sin(\omega t)B\sin(\omega t + \varphi)dt \tag{4}$$

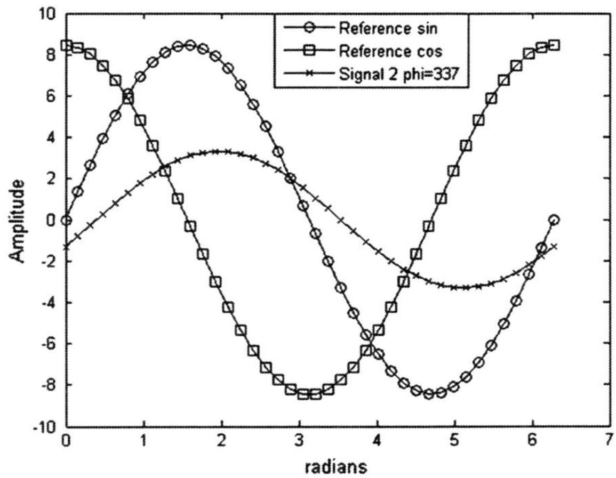

Fig. 1 Sine and cosine reference signals and signal with a phase shift angle $\varphi = 337°$

Solving the integral in (4) gives

$$T_{\varphi} = AB(\pi/\omega)\cos(\varphi) = AB\cos(\varphi)/2f, \tag{5}$$

Furthermore, (4) can also be solved using numerical integration. For periodic functions the trapezoidal rule is recommended [4]. Thus

$$\begin{aligned} T_{\varphi} &= \int_0^T A\sin(\omega t)B\sin(\omega t+\varphi)dt \\ &= AB\sum_{j=1}^{N-1}\frac{1}{2}\left[\sin(\omega t_j)\sin(\omega t_j+\varphi)+\sin(\omega t_{j+1})\sin(\omega t_{j+1}+\varphi)\right]\Delta t, \end{aligned} \tag{6}$$

where is the number of data points per period to be used from the signals and $\Delta t = T/N$ is the step size.

Equating the r.h.s. of (5) and (6) and rearranging

$$\cos(\varphi) = 2f\sum_{j=1}^{N-1}\frac{1}{2}\left[\sin(\omega t_j)\sin(\omega t_j+\varphi)+\sin(\omega t_{j+1})\sin(\omega t_{j+1}+\varphi)\right]\Delta t \tag{7}$$

A similar procedure for the second reference signal in (1b) gives

$$\sin(\varphi) = 2f\sum_{j=1}^{N-1}\frac{1}{2}\left[\cos(\omega t_j)\sin(\omega t_j+\varphi)+\cos(\omega t_{j+1})\sin(\omega t_{j+1}+\varphi)\right]\Delta t \tag{8}$$

Then the phase shift angle can be obtained from the arc tan function of the quotient of (8) divided by (7)

$$\varphi = \arctan \frac{\sum_{j=1}^{N-1} \frac{1}{2} \left[\cos(\omega t_j) \sin(\omega t_j + \varphi) + \cos(\omega t_{j+1}) \sin(\omega t_{j+1} + \varphi)\right]}{\sum_{j=1}^{N-1} \frac{1}{2} \left[\sin(\omega t_j) \sin(\omega t_j + \varphi) + \sin(\omega t_{j+1}) \sin(\omega t_{j+1} + \varphi)\right]} \tag{9}$$

Equation (9) is function only of the amplitude of the normalized arbitrary signal and the two normalized reference signals given in (1,2). Thus, all signals are normalized to vary from -1 to 1. The amplitudes of the signals A and B are not a source of error and thus can be included in (9). Furthermore, knowing the signs of the numerical integration ofsin φ and $\cos \varphi$ from (7 and 8), it is possible to determine the shift angle in the range [0,360].

It is worth noting that the result of the multiplication of the reference signals times the signal with a phase angle is another sinusoidal signal. This means that the numerical integration is carried out on a sinusoidal function for which the trapezoid rule is recommended [4]. Numerical experiments in MATLAB [5] showed that this procedure gives exact results (to the 14th significant figure) using only 4 points per cycle per signal. Furthermore, if the arbitrary signal oscillates from a value in y axis different to zero, the error increases proportional to the order of magnitude of the shift value C. For instance, for $C = 10^4$ results are accurate only till the 10th significant figure, providing all signals are centered and have maximum amplitude equal to one. For this reason, it is recommended to "center" the arbitrary signals. However, using more data points reduce this error. Errors of higher magnitude are also more likely to happen when the shift angle is smaller than 1°.

3 Conclusions

A very simple, flexible (in the number of data points) and accurate method to calculate the phase shift angle of sinusoidal signals using trapezoid rule numerical integration has been presented. As mentioned earlier trapezoid rule is very accurate to integrate periodic functions. Numerical experiments showed that the error of the numerical integrations cancel each other out. Thus, results for the phase shift angle using the present procedure are accurate to the 14th significant figure (using MATLAB), while the error of $\sin(2\pi)$ in MATLAB has an error in the 15th significant figure.

Acknowledgment The author wishes to acknowledge the financial support from the Marsden Fund through Professor S. Ilanko from The University of Waikato in New Zealand (http://www.royalsociety.org.nz/programmes/funds/marsden/).

References

1. Joenathan C (1994) Phase-measuring interferometry: new methods and error analysis. Appl Opt 33(19):4147–4155
2. Gajda J (2003) A multipoint algorithm to estimate the phase shift angle by the signal zero-crossing method. Syst Anal Model Simul 43(12):1615–1624
3. Aksamovic A, Konjicija S (2004) Numerical algorithm for measurement of angle phase shift for sines signal. Proceedings of the 2004 11th IEEE international conference on electronics, circuits and systems, pp. 451–454
4. Pozrikidis C (1998) Numerical computation in science and engineering. Oxford University Press, New York, pp 333–338
5. MATLAB version 7.10.0. Natick, Massachusetts: The MathWorks Inc., 2010

Effect of the Rail Support Stiffness on the Occurrence Propensity of Rail Corrugation Based on the Self-excited Vibration of a Wheel-Rail System

G.X. Chen, W.J. Qian, J.L. Mo and M.H. Zhu

Abstract From the viewpoint of friction-induced vibration, a self-excited vibration of a railway wheel-rail system is found to easily occur when the creep force between the wheel and rail is saturated. This self-excited vibration is probably responsible for rail corrugation. In this paper, a self-excited vibration model is established, which consists of a wheel, a rail and a series of sleepers. The model is used to study the effect of the rail support stiffness on the occurrence propensity of rail corrugation. Results show that the occurrence propensity of rail corrugation decreases with increasing rail support stiffness.

Keywords Rail corrugation · Self-excited vibration · Squealing vibration · Creep force

1 Introduction

Rail corrugation has been an elusive problem for more than 100 years [1]. Nowadays, metro lines have been built up in many large cities in China. It is found that almost all metro lines are subjected to rail corrugation to some extent. It is also found that rail corrugation almost occurs on the low rails of metro lines whose

G.X. Chen · W.J. Qian · J.L. Mo · M.H. Zhu
The State Key Laboratory of Traction Power, Tribology Research Institute, Southwest Jiaotong University, Chengdu 610031, China
e-mail: chen_guangx@163.com

W.J. Qian
e-mail: qwjst@163.com

J.L. Mo
e-mail: jlmo@swjtu.cn

M.H. Zhu
e-mail: zhuminhao@swjtu.cn

J.K. Sinha (ed.), *Vibration Engineering and Technology of Machinery*, Mechanisms and Machine Science 23, DOI 10.1007/978-3-319-09918-7_96

Fig. 1 Rail corrugation on the low rail at a curved radius of 350 m

curved radii are less than 350 m. In stark contrast, rail corrugation does not occu almost on the rails of metro lines whose curved radii are larger than 550 m.

The study history of rail corrugation can be traced back to 1930 [1–3]. Sinc then, numerous researches have been carried out. The existing theories associate with the generation mechanisms for rail corrugation may be divided grossly int two categories: rail corrugation due to the transient dynamic interaction between th wheel and rail, and rail corrugation due to stick-slip of the wheel-rail system. Th former theory is accepted generally [4–12]. However, the model established base on this theory is a non self-excited model. It needs a continuous irregularity input t sustain a vibration. A white noise input to the rail working surface can even excite the model to produce rail corrugation. This model cannot be used to satisfactoril explain why rail corrugation always firstly occurs on the low rail of sharp curve tracks and there is not rail corrugation on the track with large curved radii. Th latter theory is also supported by several investigators [13–17]. The model estab lished based on this theory is a self-excited model. It does not need a continuou irregularity input for a sustaining vibration. However, this model cannot be used t satisfactorily explain why rail corrugation always firstly occurs on the low rail o sharp curved tracks and there is not rail corrugation on the track with large curve radii. Therefore, these two existing theories have some drawbacks. Especially several cures originating from these two theories are found not to be able t effectively solve rail corrugation. In metro line 4 of Beijing, a large number o cologne egg sleepers were applied to suppress vibration of the track and ballast bed However, severe rail corrugation unexpectedly occurs as shown in Fig. 1. In thi case, the vibration of the track and ballast bed has become larger.

Most recently, the authors proposed the friction-induced self-excited vibration o a wheelset-track system as the mechanism for rail corrugation [18]. At the sam time, Kurzeck [19] also carried out similar studies. The self-excited vibration of wheelset-track system may be used to satisfactorily explain why rail corrugatio always firstly occurs on the low rail of sharp curved tracks and there is not rai corrugation on the track with large curved radii. It may be also used to explain why the friction modifier can cancel or suppress rail corrugation [18, 20]. In the present work, the friction-induced self-excited vibration of a wheel-rail system is used to study the effect of the rail support stiffness. Results show that the occurrence propensity of rail corrugation decreases with increasing rail support stiffness.

2 Self-excited Vibration Model of a Wheel-Rail System

2.1 Interaction Between the Wheel and Rail

Generally, a wheel is always kept to be in contact with the rail when a vehicle travels on a track. The dynamic simulation of vehicle curve negotiation and the field measurement all show that on a curved track the contact point between the inner wheel and low rail is roughly kept in the vicinity of the center of rail top for rail and the tread for wheel. The corresponding contact angle falls into the range of about 1–3°. In this case, the interaction forces between the inner wheel and low rail are shown in Fig. 2.

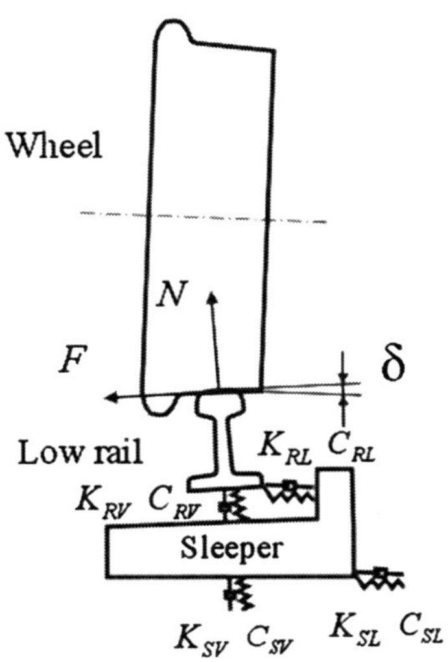

Fig. 2 The position of the contact point between the inner wheel and low rail and direction of the lateral creep force. In the figure, δ is the contact angle. N is the normal force. F is the lateral creep force. K_{RV} and K_{RL} are vertical stiffness and lateral stiffness of the rail support spring, respectively. C_{RV} and C_{RL} are vertical and lateral damping coefficients of the rail support dampers, respectively. K_{SV} and K_{SL} are vertical stiffness and lateral stiffness of the sleeper support springs, respectively. C_{SV} and C_{SL} are vertical and lateral damping coefficients of the sleeper support dampers, respectively

2.2 Finite Element Motion Equations of the Wheel-Rail System

In the present work, the creep force between the wheel and rail is assumed to be saturated and is approximately equal to the normal force multiplied by the coefficient of friction. In the presence of friction, modeling of the wheel-rail system is very complicated. The complexity focuses on modeling of the friction coupling between two elastic bodies. In the literature, two approaches available to model friction contact coupling between two elastic bodies are found [21]. The motion equations of the wheel-rail system can be rewritten into the following form [22]:

$$\mathbf{M}_r\ddot{x} + \mathbf{C}_r\dot{x} + \mathbf{K}_r x = 0 \tag{1}$$

where $\mathbf{M}_r$, $\mathbf{C}_r$ and $\mathbf{K}_r$ are $(n–m)$ by $(n–m)$ asymmetric matrices. The eigenvalue equation of Eq (1) is given as follows:

$$(\mathbf{M}_r\lambda^2 + \lambda\mathbf{C}_r + \mathbf{K}_r)\phi = 0 \tag{2}$$

The general solution of Eq (2) is as below:

$$x(t) = \sum \phi_i \exp(\lambda_i t) = \sum \phi_i \exp((\alpha_i + \mathrm{j}\omega_i)t) \tag{3}$$

Where ϕ_i is the ith eigenvector of Eq (2) and $\lambda_i = \alpha_i + \mathrm{j}\omega_i$ is the ith eigenvalue of Eq (2), j is the imaginary unit. From Eq (3), it is seen that when a real part of eigenvalues is larger than zero, the displacement will increase with time, that is, the vibration of the system becomes unstable.

Equations (1–3) constitute the analysis theory of self-excited vibration of frictional systems. The finite element stability analysis of the wheel-rail system in the presence of friction may be conducted using package ABAQUS.

Figure 3 shows the finite element model of the wheel-rail system. The rail is supported on a series of sleepers by a group of lateral springs and dampers, a group of vertical springs and dampers and a group of longitudinal springs and dampers at

Fig. 3 Model of the wheel-rail system

each position of sleepers. The number of each group of springs and dampers is the same as the number of the rail nodes on the contact area between the rail and each sleeper.

2.3 Nominal Parameters of the Wheel-Rail System

The density of the wheel and rail materials is $\rho = 7{,}800\ \mathrm{kg/m^3}$. The Youg's modulus of both the wheel and rail materials is $E = 2.1 \times 10^{11}$ N/m. The Poisson's ratio of both the wheel and rail materials $v = 0.3$. The contact angles $\delta = 1.96°$. The applied vertical force of the wheel $F_{\mathrm{V}} = 7{,}500$ N. The wheel of cars is 840 mm in nominal diameter. The rail is a type of 60 kg/m. The rail is 45,000 mm in length. The distance between two sleepers is 625 mm. The width of sleepers is 160 mm. The coefficient of friction between the wheel and rail is $\mu = 0.4$. In the present work, the stiffness value K_{RL} of the lateral rail support spring between the rail and each sleeper is assumed to be half of the stiffness value K_{RV} of the vertical rail support spring between the rail and each sleeper. The vertical and lateral damping coefficients of each rail support damper are $C_{\mathrm{RV}} = 5{,}000$ Ns/m, $C_{\mathrm{RL}} = 5{,}200$ Ns/m, respectively. The vertical and lateral spring stiffness values of each sleeper support are $K_{SV} = 8.9 \times 10^7$ N/m and $K_{SL} = 5.0 \times 10^7$ N/m, respectively. The vertical and lateral damping values of each sleeper support are $C_{SV} = 8.98 \times 10^4$ Ns/m and $C_{SL} = 4.0 \times 10^4$ Ns/m, respectively.

3 Results and Discussion

3.1 Self-excited Vibration Mode Shape of the Wheel-Rail System

The frequency of rail corrugation is generally considered to be in the range of 20–1,200 Hz. Therefore, the present results are limited to this frequency range. The effective damping ratio is defined as $\zeta = -2\mathrm{R_e}(\lambda)/|\mathrm{I_m}(\lambda)|$. The effective damping ratio is a parameter available to measure the propensity of self-excited vibration occurrence. The larger its absolute value, the more easily the corresponding self-vibration takes places. Figure 4 shows the mode shapes of unstable vibrations of the wheel-rail system. From this figure, it is seen that in two unstable vibrations the wheel vibrates on the rail in the lateral direction. It needs to be noted that the unstable vibration of 284.56 Hz is very close to the measured vibration frequency of corrugated rail, which is about 312.83 Hz.

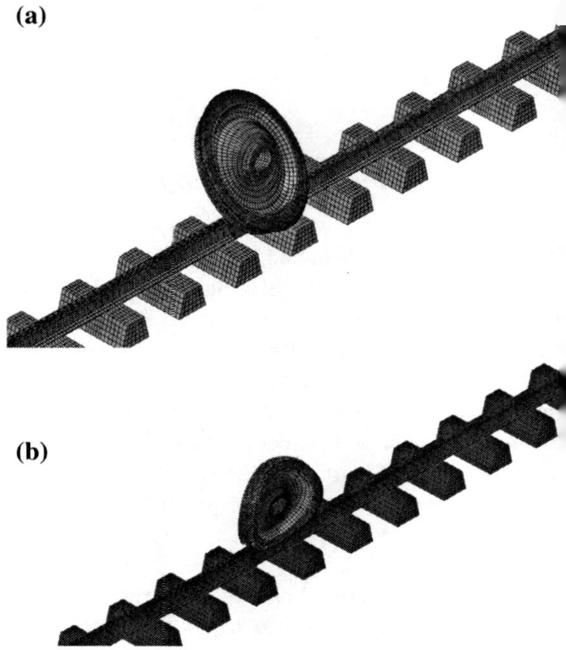

Fig. 4 Mode shapes of unstable vibrations of the wheel-rail system. **a** mode shape of unstable vibration of 284.56 Hz. **b** mode shape of unstable vibration of 522.28 Hz

3.2 Effect of Rail Support Stiffness on Self-excited Vibration of Wheel-Rail System

As mentioned above, rail corrugation very easily occurs on the wheel-rail-cologn egg sleepers system. The cologne egg sleeper is characterized by low suppo stiffness. Figure 5 shows the variation of the effective damping ratio ζ with the ra support stiffness. From Fig. 5, it is seen that the effective damping ratio ζ decrease with increasing rail support stiffness. This suggests that occurrence propensity of ra corrugation increases with decreasing rail support stiffness. This conclusion i consistent with the occurrence propensity of rail corrugation in the actual metro line

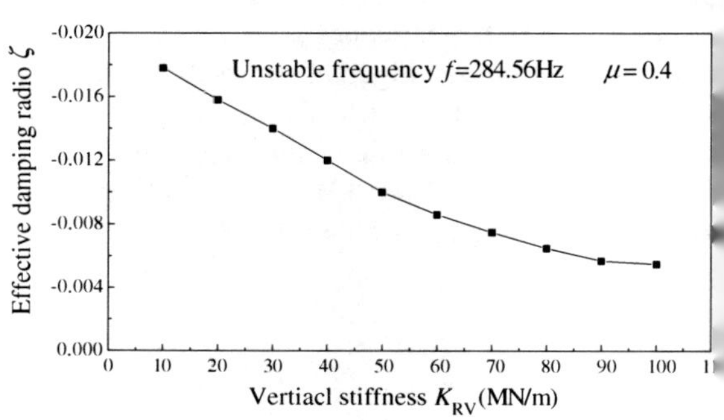

Fig. 5 Variation of the effective damping ratio with rail support stiffness

4 Conclusions

In the present work, the theory of friction-induced vibration is applied to study rail corrugation of the wheel-rail system. The following several conclusions can be obtained.

(1) When the creep force between the wheel-rail system is saturated, the wheel-rail system always has some unstable vibrations. This suggests that in this case rail corrugation is inevitable.
(2) The occurrence propensity of rail corrugation increases with decreasing rail support stiffness.

Acknowledgement This research received the financial supports from National Natural Science Foundation of China (No. 51275429), the innovation team development plan of the ministry of education (IRT1178) and the Fundamental Research funds for the Central University (SWJTU12ZT01).

References

1. Grassie SL, Kalousek J (1993) Rail corrugation: characteristics, causes and treatments. J Rail Rapid Transit 207:57–68
2. Sato Y, Matsumoto A, Knothe K (2002) Review on rail corrugation studies. Wear 253:130–139
3. Oostermeijer KH (2008) Review on short pitch rail corrugation studies. Wear 265:1231–1237
4. Hempelmann K, Knothe K (1996) An extended linear model for the prediction of short pitch corrugation. Wear 191:161–169
5. Matsumoto A, Sato Y et al (1996) Study on the formation mechanism of rail corrugation on curved track. Veh Syst Dyn 25:450–465
6. Vadillo EG, Tarrago JA, Zubiaurre GG, Duque CA (1998) Effect of sleeper distance on rail corrugation. Wear 217:140–146
7. Ilias H (1999) the influence of railpad stiffness on wheelset/track interaction and corrugation growth. J Sound Vib 227(5):935–948
8. Andersson C, Johansson A (2004) Prediction of rail corrugation generated by three-dimensional wheel-rail interaction. Wear 257:423–434
9. Wu TX, Thompson DJ (2005) An investigation into rail corrugation due to micro-slip under multiple wheel/rail interactions. Wear 258:1115–1125
10. Jin XS, Wen ZF, Wang KY, Zhou ZR, Liu QY, Li CH (2006) Three-dimensional train-track model for study of rail corrugation. J Sound Vib 293(3–5):830–855
11. Jin XS, Wen ZF (2008) Effect of discrete track support by sleepers on rail corrugation at a curved track. J Sound Vib 315:279–300
12. Daniel WJT, Cheng CY, Meehan PA (2008) Modelling the effects of friction modifiers on rail corrugation in cornering. Veh Syst Dyn 46:845–866
13. Clark RA (1984) Slip-stick vibrations may hold the key to corrugation puzzle. Railway Gaz Int 7:531–533
14. Clark RA, Scott GA, Poole W (1988) Short wave corrugation-an explanation based on stick-slip vibrations. In: Applied mechanics rail transportation symposium 96, ASME, pp 141–148
15. Ishida M, Moto T, Takikawa M (2002) The effect of lateral creepage force on rail corrugation on low rail at sharp curves. Wear 253:172–177

16. Suda Yoshihiro, Hanawa Mitsuo, Okumura Mikio, Iwasa Takashi (2002) Study on rail corrugation in sharp curves of commuter line. Wear 253:193–198
17. Sun YQ, Simson S (2008) Wagon-track modeling and parametric study on rail corrugation initiation due to wheel stick-slip process on curved track. Wear 265:1193–1201
18. Chen GX, Zhou ZR, Ouyang H, Jin XS, Zhu MH, Liu QY (2010) A finite element study on rail corrugation based on saturated creep force-induced self-excited vibration of a wheelset–track System. J Sound Vib 329:4643–4655
19. Kurzeck B (2011) Combined friction induced oscillations of wheelset and track during the curving of metros and their influence on corrugation. Wear 271(1–2):299–310
20. Eadie DT, Santoro M, Oldknow K, Oka Y (2008) Field studies of the effect of friction modifiers on short pitch corrugation generation in curves. Wear 265:1212–1221
21. Ouyang H, Nack W, Yuan Y, Chen F (2005) Numerical analysis of automotive disc brake squeal: a review. Int J Veh Noise Vib 1(3/4):207–231
22. Yuan Y (1996) An eigenvalue analysis approach to brake squeal problem. In: Proceedings of the 29th ISATA conference automotive braking systems, Florence, Italy

Modeling and Dynamic Force Simulation for Detection of Profile Error in Spur Gear Pair

Atul B. Andhare and Manish Kumar Verma

Abstract This paper describes modeling and simulation of spur gear pair in Adams–View. Initially, virtual model of an ideal gear pair is created in Adams and variation of the dynamic force is obtained by considering the tooth mesh stiffness and mesh damping coefficient. The magnitude of force variation thus obtained is confirmed with the design calculations. The dynamic force magnitudes thus obtained and those found in Adams are closely matching. Later on, tooth profile error is introduced in the gear pair by removing part of involute surface from one of the gears making one tooth surface as flat. Dynamic force variation is now obtained for the defective gear pair. After comparing the dynamic forces for defect free and defective gear pairs, it is found that change in dynamic force pattern is a good indicator of tooth profile error in spur gears.

Keywords Gear modeling · Spur gear · Gear dynamic load · Mesh stiffness · Mesh damping coefficient

1 Introduction

In mechanical systems power transfer is necessary from source to application. Gear systems are having several advantages compared with other transmission members such as constant transmission ratio, small overall dimensions and operating simplicity. Because of these advantages, gears have wide applications in mechanical engineering for transmission of power.

A.B. Andhare (✉) · M.K. Verma
Department of Mechanical Engineering, Visvesvaraya National Institute of Technology, Nagpur 440010, India
e-mail: abandhare@mec.vnit.ac.in

M.K. Verma
e-mail: manish.mech0@gmail.com

J.K. Sinha (ed.), *Vibration Engineering and Technology of Machinery*,
Mechanisms and Machine Science 23, DOI 10.1007/978-3-319-09918-7_97

When gears are run, they are subjected to all the forces applied to rotating bodies or equipments. Dynamic forces of gear are influenced by geometric deformations caused due to external process loads and manufacturing assembly loads. These deformations in action along with natural vibration of geared system cause sudden acceleration of gears followed by impact when the gear teeth come into mesh. The intermittent gear tooth action setup forced vibration in the pinion -gear pair and does not permit smooth continuous action as would generally be expected from a constant applied load.

Over the years, spur gear dynamics has been a topic of interest for many researchers as these are the simplest form of gears and are widely used. These research works are mainly focused on dynamics of healthy and faulty gears. Defects in gears are classified as [1] (1) geometric defects—including defects like run out, profile errors, misalignment, center distance errors, etc. caused due to errors in manufacturing and assembly, and (2) teeth errors like—cracks, spalling, chipping, pitting, etc. This paper is focused on the change in gear dynamics due to profile error which is a geometric error. Following paragraph describes some relevant works in this area.

Zoltan Korka [2] reviewed various mathematical models used in gear dynamics. According to the author the main aims of gear modeling are- stress analysis, reduction of surface pitting and scoring, improvement of transmission efficiency, reduction of radiated noise, analysis of loads on the other machine elements of the system especially on bearings and their stability regions, determination of natural frequencies of the system, analysis of vibration motion of the system and improvement in reliability and fatigue life. Some of the works which focus on local defects in gears are discussed first. Chen et al. [3] developed diagnosis theory for the gears by clarifying the variation of dynamic characteristic between healthy and faulty gear by using theories of signal processing, fuzzy and neural network. Chaari et al. [1] modeled the local damages in spur gear and effects on dynamic response under varying load conditions. Howard et al. [4] presented dynamic modeling of spur gear in mesh including friction and a crack. Apart from these works there are some works which focus on the manufacturing or assembly errors in gears. One such work by Shing et al. [5] suggested a model for the dynamics of spur gear systems with backlash consideration. Kamble and Saha [6] developed virtual prototype model of rack and pinion gear steering system with incorporated manufacturing error. They used ADAMS software for simulation. Shuing Li [7] showed the changes in the load capacity, load-sharing ratio and transmission error for pair of spur gears due to manufacturing error, assembly error and modification in tooth profile. Vedmar et al. [8] proposed a method to determine the dynamic load on spur gear teeth and bearing and it concluded that friction has no dynamic influence in the gear mesh line-of-action direction for gear contact force or on the bearing force. Similarly, gearbox simulation models with gear and bearing faults were also presented by Hiroaki and Nader [9]. Choy et al. [10] simulated gear vibrations and verified the results by performing experiments.

From the literature, it was found that there is less work describing gear dynamics for various manufacturing and assembly errors. Also, ADAMS is a powerful type for simulation of machine elements like gears. Therefore, it was planned to simulate gear dynamics under different manufacturing and assembly error conditions using ADAMS [11]. One such error considered was the tooth profile error. This paper describes the gear dynamics under the condition of error in tooth profile using ADAMS- View. The dynamic force on a defect free gear was obtained with virtual prototype model used in ADAMS-View. Effect of profile error for same working condition was simulated by incorporating the error in virtual model. Dynamic force was obtained by accounting for gear parameters like tooth mesh stiffness, mesh damping coefficient and contact ratio of the gears.

2 Virtual Model of Gears

For simulation purposes a virtual spur gear pair was modeled using CAD software. Initially the gears were modeled as ideal gears (without any defect). For modeling, both the gears in the pair were assumed to be of the same size. While creating virtual model it was necessary to assume some parameters of gear pair like involute curve equation, module, addendum, dedendum, base circle diameter and number of teeth. Parameters listed under Sect. 2.1 were used for simulation purpose. These parameters were selected to match with the gear dimensions of a physical set up which was available in the laboratory. This would facilitate the verification of simulation results with experimental results in the future. Figure 1 shows the spur gear pair virtual model used in this work along with the constraints used.

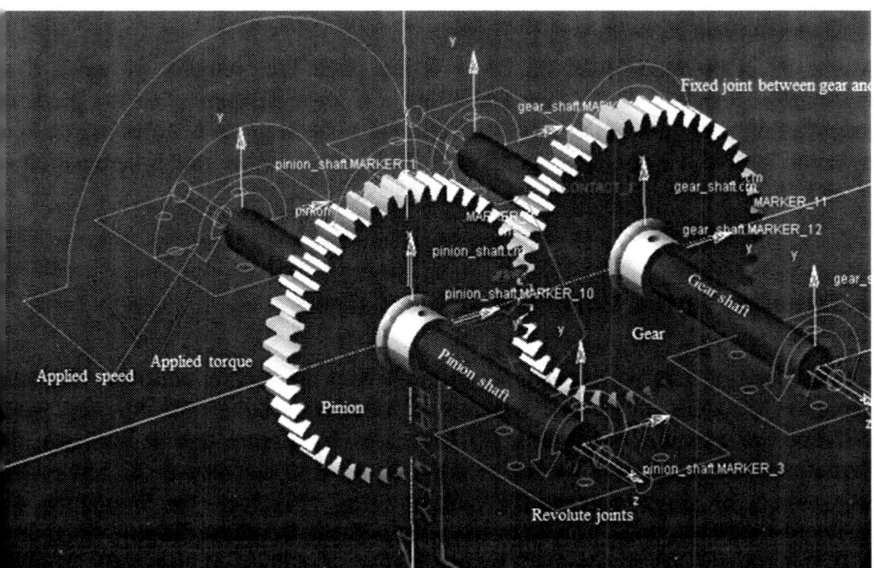

Fig. 1 Spur gear pair virtual model in ADAMS

2.1 Gear Parameters

No. of teeth = 42,
Module (m) = 3 mm,
Addendum = 1 m,
Dedendum = 1.25 m,
Pressure angle = 20^0
Face width (b) = 20.
Material = Steel (E = 207 GPa and Poisson ratio = 0.3)
Input power = 1 HP (0.75 kW)
Speed = 600 RPM

3 Dynamic Load Calculation for Ideal Gear

Using the parameters given in Sect. 2.1, the dynamic load of the ideal gear unde consideration was calculated using standard formulae available in hand books/tex books [12]. Lewis equation was used for static load calculation.

Tangential load $F_t = \frac{2T}{d}$

Radial load $F_r = F_t \cdot \tan(\theta)$

Resultant load $R = \sqrt{F_t^2 + F_r^2}$

where, T is applied torque, d is pitch circle diameter, and θ is pressure angle. Now incremental (dynamic) load is given as:

Dynamic load $F_d = F_t + \frac{21v(Ceb+F_t)}{21v+\sqrt{Ceb+F_t}}$

where, F_t is the static load (in N), v is the pitch line velocity (in m/s), C i deformation factor, b is the face width and e is error depending on the grade c manufacturing of gear. Here, gear was assumed to be of grade 6. After calculatior dynamic load (F_d) is found to be 2,956 N. The total force on gear is summation c F_t and F_d.

4 Dynamic Force Variation of Ideal Gear

Contact load of gear when gears are in mesh does not remain constant. The loa variation is mainly influenced because of following factors: (1) Varying mes stiffness and damping coefficient, (2) Impact load on tooth due to backlash, (3 Deviation from ideal tooth profile at the time of engagement and (4) Alternatin engagement of single and double pair of teeth. Therefore, for simulating th dynamic conditions it is essential to correctly estimate the above factors. Accordin to [13] mesh stiffness at gear tooth is combination of stiffness of gear body, bendin stiffness of teeth and stiffness of contact between teeth. For the gear pair used fc this work, contact ratio is 1.73. Mesh stiffness is calculated by considering the gears as two springs in series as suggested in the literature [13–16].

$$K_m = (K_p^{-1} + K_g^{-1})^{-1}$$

$$K_{p,duoble} = E \cdot b \cdot \left[\begin{array}{l} \left(f_B \cdot c_B \cdot \ln(\mathrm{r_d} - \mathrm{r_s})^{1.6} \cdot \mathrm{r_s^{1.6}}\right)^{-1} + \left(2 \cdot c_T \cdot m^2 \cdot z^{2.2}\right)^{-1} \\ \qquad + \left(c_C \cdot m^{1.85} \cdot z^2 \cdot T^{0.068}\right)^{-1} \end{array} \right]^{-1}$$

$$K_{g,double} = K_{p,double} \cdot \left(\frac{z_p}{z_g}\right)^2$$

where, variables are defined as given below.
E = Young's modulus of elasticity = 207 Gpa
f_B = Scale factor for single contact zone to double contact zone = 1.1
C_B = Coefficient for gear body = 0.0009555
C_T = Coefficient for bending of teeth = 0.000032
C_C = Coefficient for contact between teeth = 0.000079635
$Z_{p,g}$ = No. of teeth on pinion and gear = 42
m = Module in mm = 3
r_d = Root circle radius in mm = 62.5
r_S = Shaft radius in mm = 10.5
T = Applied torque in *N-m* = 11.93
Now, mesh damping coefficient is defined as [16]

$$D_m = 2 \cdot \xi \cdot \sqrt{K_m \cdot M}$$

$$M = (J_1 J_2)/(J_1 r_2^2 + J_2 r_1^2)$$

where, K_m is combined mesh stiffness, ζ is damping ratio, which is assumed as 5 % at pitch point, J_1, J_2 are the polar moment of inertias of pinion and gear shaft respectively, r_1 and r_2 are the base radii for pinion and gear, M is equivalent mass moment of inertia and it is calculated as 0.580. Using these values mesh damping coefficient D_m is calculated as 52.83. Using these parameters, the gear dynamics is simulated in ADAMS for the ideal gear condition. Results of this simulation are presented in the next paragraph.

Dynamic load variation for ideal gear obtained by simulation in ADAMS is shown below in Fig. 2a, b. It observed that maximum load is closer to the calculated value of 2,956 N as given in Sect. 3. Also, effect of contact ratio is clearly seen in Fig. 2b as two waves of different amplitudes appear—higher amplitude wave for less contact ratio and lower amplitude wave for higher contact ratio. In all these simulations, if static force is added, the nature of plot will remain same with shift in the mean value. From Fig. 2a, b, it is confirmed that gear model made in ADAMS is giving appropriate value of dynamic force and this model can be further used for simulating dynamic behavior of the gear with some defect on it.

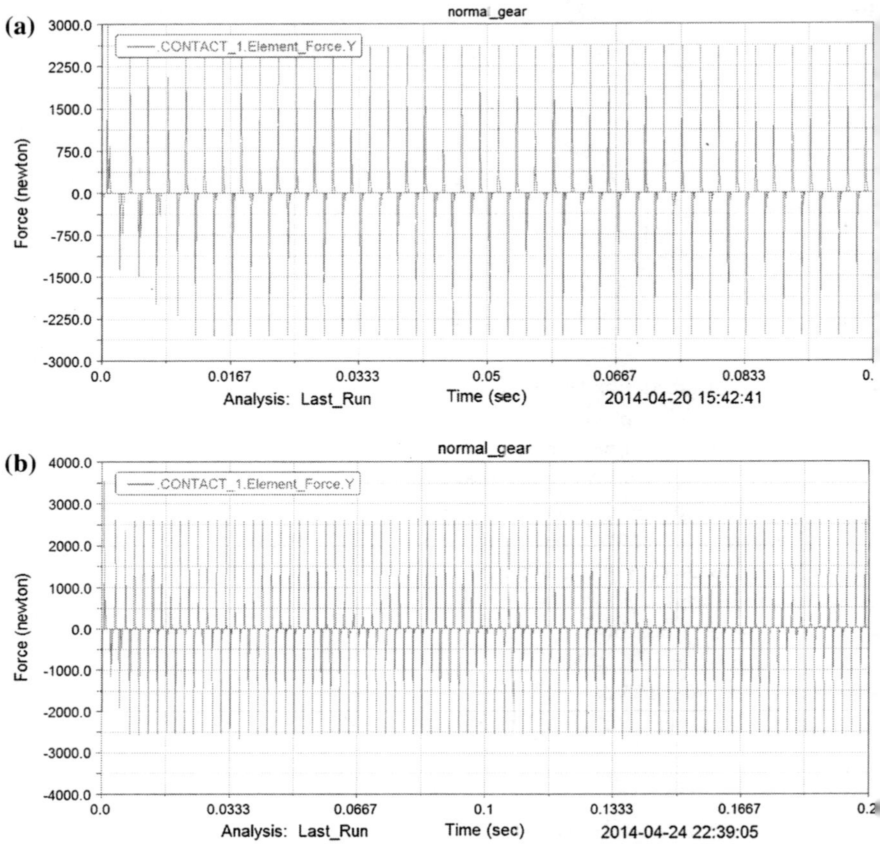

Fig. 2 **a** Dynamic load variation in single cycle. **b** Dynamic load variation for two cycles

5 Modeling of Gear with Profile Error

In this work, it is assumed that gear is having some profile error due to manufacturing errors. In the virtual model, profile error on gear tooth is modeled by removing part of involute surface from one of the gears making one tooth surface as flat. This is shown in Fig. 3.

6 Dynamic Force Variation for Gear with Profile Error

Dynamics of the gear with tooth profile error is simulated keeping same operating conditions as those of ideal gear and results are shown in Figs. 4, 5 and 6. It is observed that in case of faulty gear (Figs. 4b, 5b and 6b) the impulses do not remain

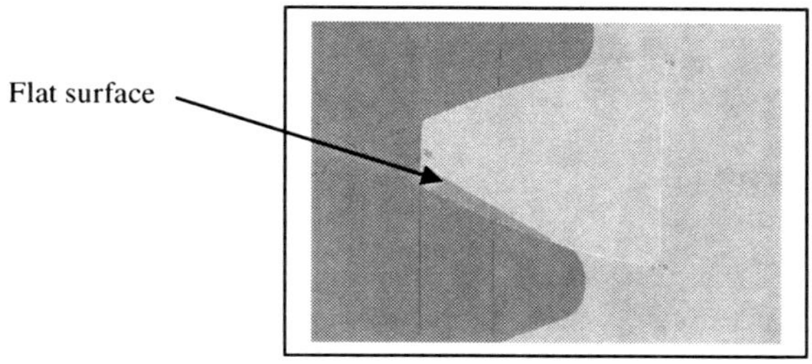

Fig. 3 Model of flattened tooth profile

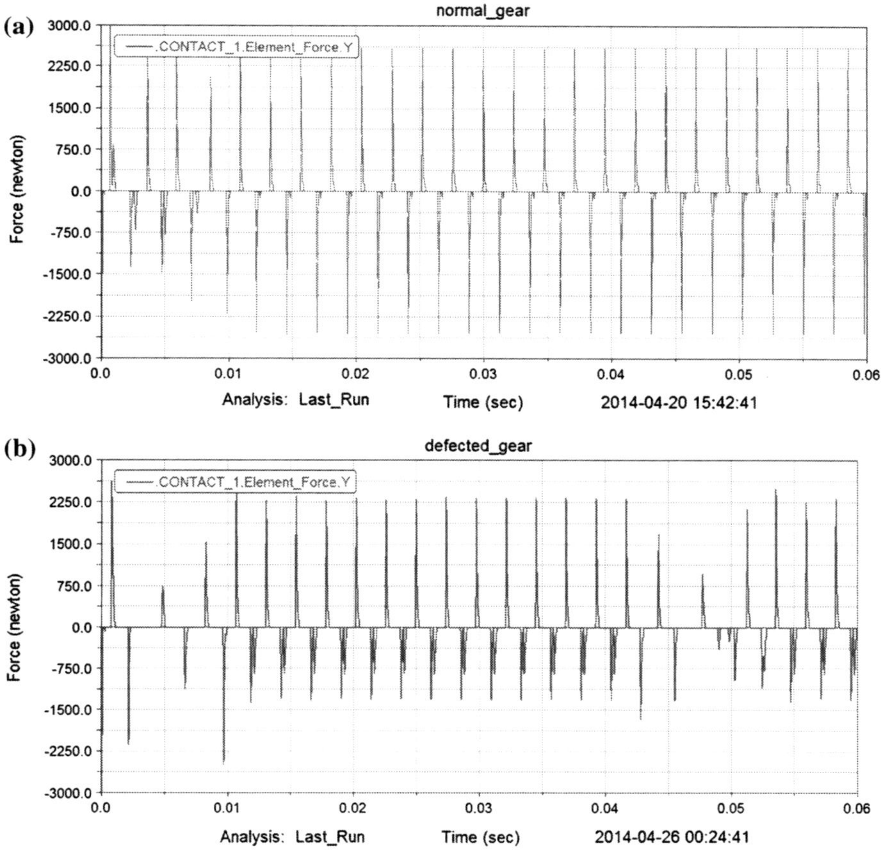

Fig. 4 Enlarged view of dynamic force variation for **a** Ideal gear **b** faulty gear

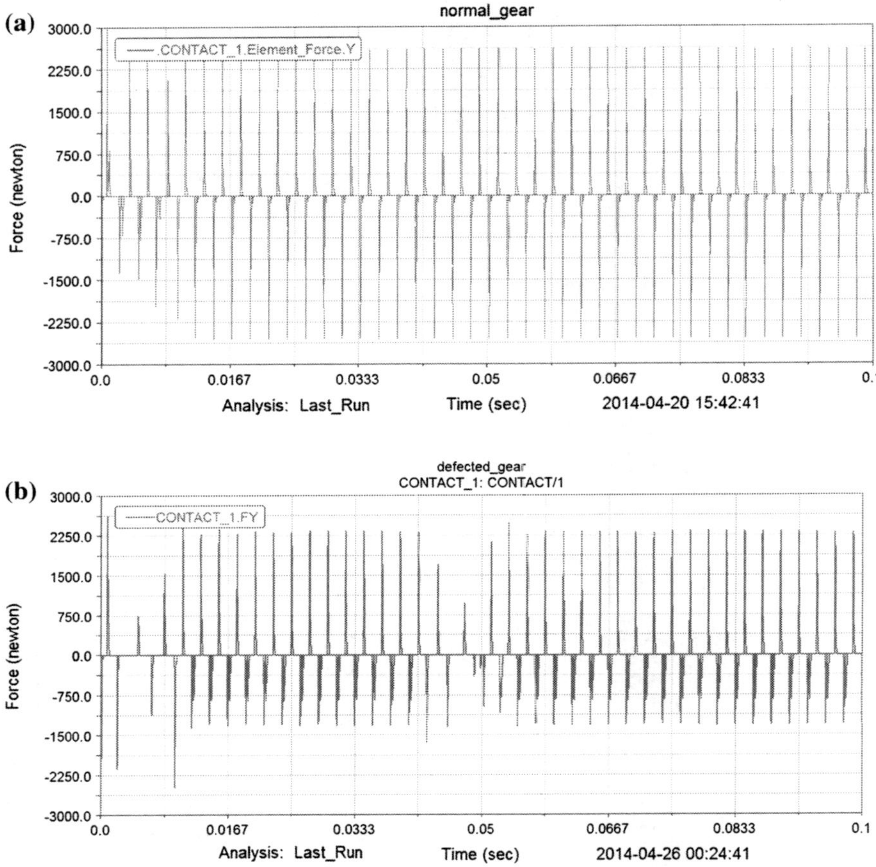

Fig. 5 Variation of dynamic force for one cycle for: **a** Ideal gear **b** faulty gear

as uniform as for ideal gear (Figs. 4a, 5a and 6a). Also, periodicity of the dynamic force pattern is getting disturbed when faulty tooth comes in mesh and the amplitude of impulses are lesser as load is shared by larger tooth area.

7 Conclusion

This paper has presented the simulation of dynamic force variation for a pair of spur gears under ideal and fault conditions. It has been shown that ADAMS—View can be used for gear dynamics simulation using virtual model of gear. Initially, virtual model of the ideal gear was made in CAD software and its dynamic force variation was obtained. This dynamic force magnitude was confirmed by using standard

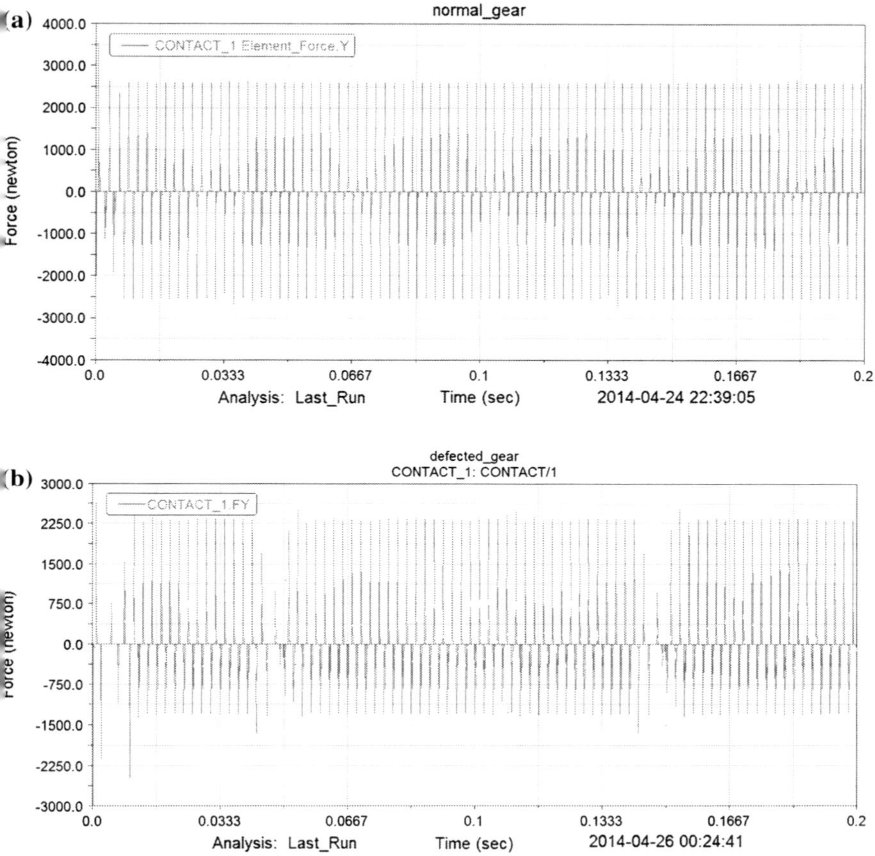

Fig. 6 Variation of dynamic force for two cycles for: **a** Ideal gear **b** faulty gear

design calculations. Later on, profile error was introduced on one of the gear tooth and the force variation was obtained. It was found that due to tooth profile error, the dynamic force pattern became irregular when the faulty tooth came into mesh. The force amplitude decreased as load was shared by more tooth area. Thus, the force amplitude varied once per rotation of the gear and produced an impact at gear mesh frequency. This feature can be used for identifying the profile error in condition based maintenance. The proposed method can be further extended to other kinds of gear defects like - manufacturing, assembly and tooth errors in gears. Thus, the proposed method can be used for analyzing signals from various gear defects without having physical models or experimental set up.

References

1. Chaari F, Zimroz R, Bartelmus W, Fakhfakh T, Haddar M (2011) Modeling of local damag in spur gears and effects on dynamics response in presence of varying load condition Proceedings of Surveillance 6, UTC Complegne, pp 1–19
2. Korka Z (2007) An overview of mathematical models used in gear dynamics. Rom J Acou Vib 4(1):43–50
3. Chen P, Feng F, Toyota T (2001) Dynamic analysis method of fault gear equipme COMADEM 2001, proceedings of the 14th international congress, Elsevier Science Lt pp 419–426
4. Howard I, Jia S, Wang J (2001) The dynamic modeling of spur gear in mesh including frictio and a crack. Mech Syst Signal Process 15(5):831–853
5. Shing TK, Tsai LW, Krishnaprasad PS An improved model for the dynamics of spur ge systems with backlash consideration. Technical research report, Institute for Syster Research, University of Maryland
6. Kamble N, Saha SK (2007) Developing a virtual prototype of a rack and pinion steeri system. Int J Veh Syst Model Test 2(1):61–79
7. Li S (2007) Effects of machining errors, assembly errors and tooth modifications on loadi capacity, load-sharing ratio and transmission error of a pair of spur gears. Mech Mach Theor 42:698–726
8. Vedmar L, Andersson A (2003) A method to determine dynamic load on spur gear teeth an on bearing. J Sound Vib 267(5):1065–1084
9. Hiroaki E, Nader S (2012) Gearbox simulation models with spur gear and bearing faults. I Gokcek M (ed.) Mechanical engineering, InTech
10. Choy FK, Ruan YF, ZakrajsekJJ, Oswald FB (1992) Modal simulation of gearbox vibratio with experimental correlation. 28th joint propulsion conference and exhibit cosponsored b the AIAA, SAE, ASME and ASEE, Nashville, Tennessee, July 1992. pp 1–15
11. MSC Inc., MSC ADAMS reference manual
12. Dudley DW (1992) Gear handbook: the design, manufacture and application of gear McGraw-Hill Book Company
13. Kiekbusch T, Howard I (2007) A common formula for the combined torsional mesh stiffne of spur gears. Proceedings of the Fifth Australasian congress on applied mechanics, ACAM 2007
14. Fernandez del Rincon A, Viadero F, Iglesias M, García P, de-Juan A, Sancibrian R (2013) model for the study of meshing stiffness in spur gear transmission. Mech Mach Theor 61:30–58
15. Khabou MT, Bouchaala N, Chaari F, Fakhfakh T, Haddar M (2011) Study of a spur ge dynamic behavior in transient regime. Mech Syst Signal Proc 25:3089–3101
16. Li S, Kahraman A (2011) A spur gear mesh interface damping model based o elastohydrodynamic contact behavior'. Int J Powertrains 1(1):4–21

Availability Improvement by Early Detection of Motor Bearing Failure Using Comprehensive Condition Monitoring Techniques at DTPS

Hemant M. Bari, Atul A. Deshpande and Suhas S. Patil

Abstract This paper shares a success story out of the Implementation of Co-ordinated Condition Monitoring techniques at DTPS, wherein imminent Motor bearing failure of one of the most critical HT auxiliary used during coal bunkering system i.e. Belt Conveyor-4B was diagnosed. On 15/01/2014, Motor NDE and DE bearing acceleration value deviated from 1.3 and 1.2 g to 2.2 and 4.9 g (Peak) respectively. Suspecting deterioration in Motor bearings and to confirm the same a Ultrasonic Monitoring carried out and observed that Ultrasonic level also increased from 4 to 8 LEDs. In view of Steep increasing trend of Motor bearings Acceleration value and Ultrasonic level, it was decided to replace the Motor with spare one. During inspection of motor, it was observed that Motor bearing clearances were found increased than the normal value and bearings rollers found damaged. Thorough inspection of Motor winding and Stator Rotor also done and found intact. No Scoring marks observed on both ends of Motor shaft. Vibration Analysis and Ultrasonic Monitoring helped in diagnosing the exact root cause of abnormity of deterioration of Motor bearings at early stage and thus avoided the catastrophic failure. The application saved the plant from huge losses in terms of power generation, down time, associated maintenance and spares costs which could have caused total cost of £58,854.3.

Keywords Vibration analysis · Ultrasonic monitoring

H.M. Bari (✉) · A.A. Deshpande · S.S. Patil
Department of Maintenance Planning, Condition Monitoring Cell, Reliance Infrastructure, Dahanu Thermal Power Station, 2x250 Mw, Dahanu, Mumbai, India
e-mail: hemant.bari@relianceada.com

A.A. Deshpande
e-mail: atul.deshpande@relianceada.com

S.S. Patil
e-mail: suhas.patil@relianceada.com

J.K. Sinha (ed.), *Vibration Engineering and Technology of Machinery*, Mechanisms and Machine Science 23, DOI 10.1007/978-3-319-09918-7_98

Fig. 1 2 × 250 MW Dahanu thermal power station

1 Introduction

Dahanu Thermal Power Station (DTPS) is one of the best power generation plant in the country, which commenced its commercial operations in January 1996 Recognized with innumerable awards, this power plant is known for its distinctive features that set it apart from others in terms of technological innovation, international performance and sustainability for a longer period. The plant utilizes a blending of Indian washed coal and imported coal as fuel. The general blending ratio is 80:20. The indigenous fuel is received from SECL (Korba) which is located about 1,400 Kms from the plant site. Imported coal is received from various countries like Indonesia, South Africa (Fig. 1).

Reliance Infrastructure DTPS has achieved remarkable results since it has consistently maintained all the major key performance parameters such as availability, plant load factor, auxiliary power consumption, specific oil consumption heat rate etc. The plant has strategically adopted Co-ordinated Condition Monitoring as one of the key maintenance management practices. The equipments are prioritized for close monitoring on the basis of criticality index to ensure high plant availability through various Co-ordinated Condition Monitoring techniques. These critical equipments are monitored more attentively to further improve overall performance.

2 Motor Bearing Failure of Belt Conveyor-4b

This 6.6 kV Conveyor-4B Motor rating is 224 KW with speed of 1,485 rpm and at full load, draws 25.8 Amps current. This 3Ø Induction Motor has Anti-friction bearing 6321 at NDE side and NU 324 at DE side. Conveyor-4B transfers coal from eithe

Fig. 2 Coal conveyor-4B

Conv 3A/B directly from rake and from Conv-9A/B (through coal yard) to Belt feeder 6 then subsequently to Conveyor 5A/B to transport coal finally at Boiler Bunker at speed of 2.7 m/s with capacity of 900 TPH. It is one of the most critical conveyors in Coal Handling Plant, which has longest inclination. It is very important conveyor for maintaining healthy bunker levels at all times and failure of this will directly affect the operational reliability of bunkering stream and coal supply to boiler will be interrupted which may lead to Power Generation loss (Fig. 2).

2.1 Pre-Maintenance History

1. Conveyor-4B vibration data collected on monthly basis. On 31.12. 2013, during routine monitoring, Offline data collected at Motor NDE, DE and at Gearbox input and output bearings with SKF Microlog CMVA 65 using cmss 2200 accelerometer sensor of sensitivity 100 mv/g and having frequency range of 1–5,000 HZ. At the same time, Motor and Reducer Gearbox Bearing Online vibrations were also found well within limits. And Motor DE and NDE bearing temperatures were found to be normal (Fig. 3).
2. But Motor NDE and DE acceleration value deviated from 1.3 and 1.2 g to 1.7 and 3.0 g (Peak) respectively (Figs. 4, 5).
3. Meanwhile, Ultrasonic monitoring done and observed that ultrasonic level also increased from 4 to 6 LEDs. Ultrasound instruments receive high-frequency sounds produced by operating equipment. These frequencies typically range from 20 to 100 kHz and are beyond the range of human hearing. The instruments electronically translate ultrasound frequencies down into the audible range where they are heard through headphones and observed as intensity on a display panel (Fig. 6).

Fig. 3 Online vibration of motor and reducer gearbox bearings

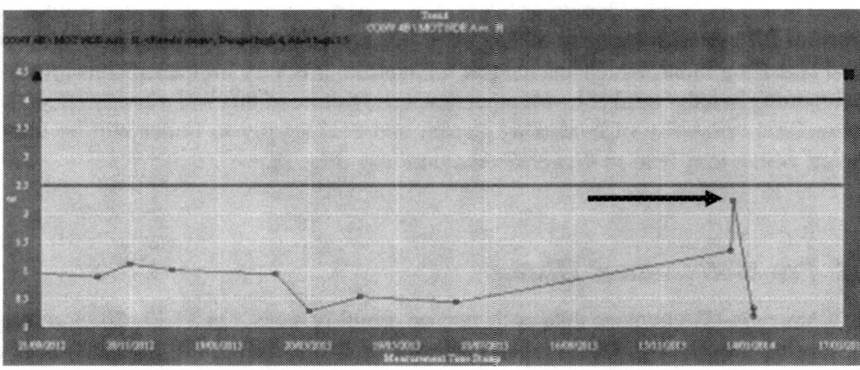

Fig. 4 Motor NDE bearing acceleration value trend

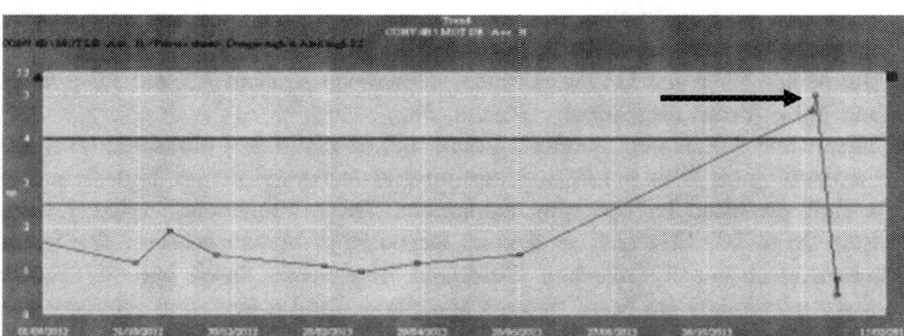

Fig. 5 Motor DE bearing acceleration value trend

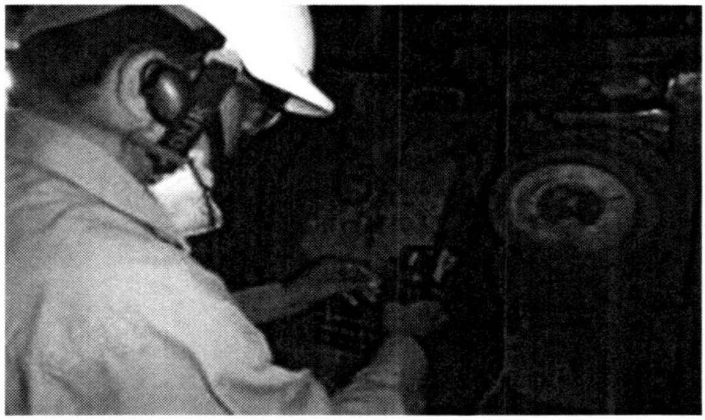

Fig. 6 Motor DE bearing ultrasonic monitoring

4. Acceleration value remained constant for next 10 days. On 15.01. 2014, Motor NDE and DE acceleration increased to 2.2 and 4.9 g. Ultrasonic level also increased from 6 to 8 LEDs. In view of Steep increasing trend of Motor bearings Acceleration value and Ultrasonic level, it was decided to replace the Motor with spare one.
5. Detailed analysis was carried out as follows [1, 2].

2.2 *Vibration Diagnosis*

The spectrum analysis for root cause is as follows (Figs. 7, 8):

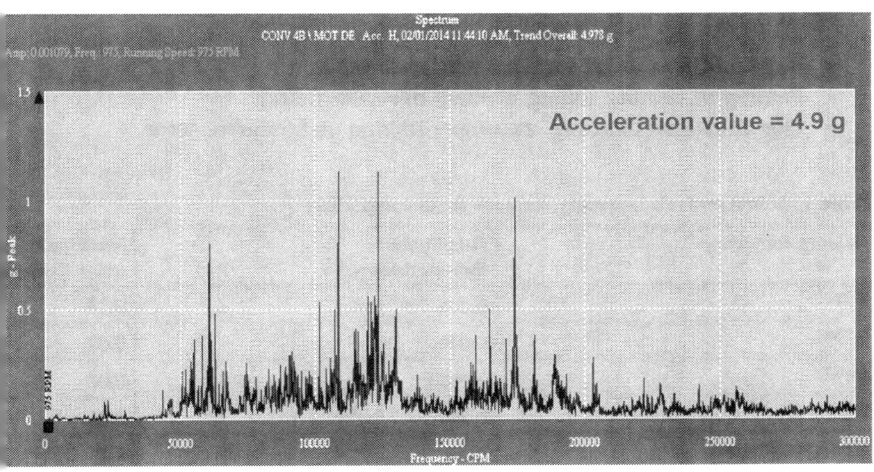

Fig. 7 Motor DE and NDE bearing acceleration spectrum

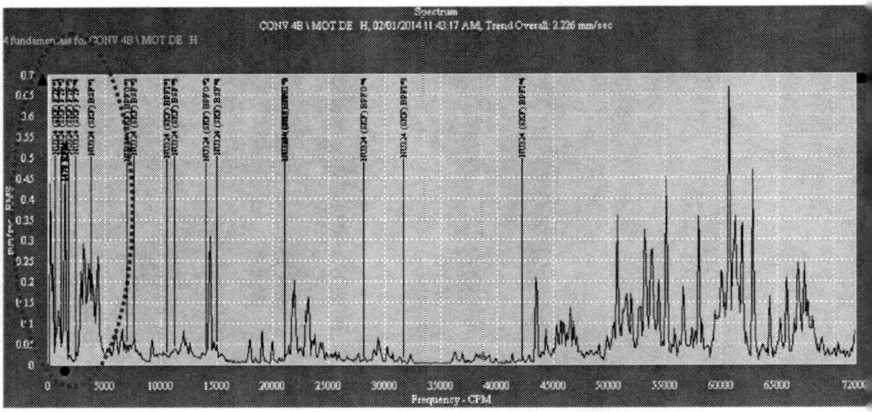

Fig. 8 Motor DE bearing vibration spectrum showing bearing defect frequencies

2.3 Frequency Analysis

1. On 15.01.2014, Motor NDE and DE acceleration increased from 1.3 and 1.2 to 2.2 and 4.9 g. Ultrasonic level also increased from 4 to 8 LEDs.
2. Acceleration spectrums showed Bearing Defect Frequencies at BSF, BPFO an BPFI which clearly pinpointed bearing defects (Table 1).
3. The above comparison show the contribution of Non- Synchronous Energy is o a higher side which is mainly due to bearing defects. When these frequenc stress waves' events occur at repeatable time intervals corresponding to bearing defect frequency, then it is likely that a bearing defect is present.
   Stress waves:

   Stress waves are high frequency vibrations, caused due to:

   - Impact of two metal surfaces with each other.
   - Passing of bearing rolling element over any defect.
   - Due to fatigue cracking, excessive friction and abrasive wear.

Table 1 Vibration fault frequency readings (before and after)

Bearing frequency	Amplitude (before damage)	Amplitude (after damage
BSF	0.004	0.15
BPFO	0.001	0.04
BPFI	0.005	0.02
FTF	0.004	0.06
Non synchronous	0.7	4.2

Stress waves are:

- Short in duration
- Micro seconds for fatigue cracking
- Milliseconds for Metal to Metal impact

4. Hence it was concluded that Acceleration and Ultrasonic value of Motor bearings increased due to deterioration of bearings
5. So it was decided to stop the Conveyor for Motor replacement.

2.4 *Observations During Maintenance Work*

1. Condition of cooling fan and Bearing housing found OK.
2. Bearing clearances were found on higher side than the original readings.
3. Scoring was observed on both DE and NDE bearings rollers (Fig. 9).
4. No Scoring marks observed over the Motor shaft
5. Thorough inspection of Motor Winding Stator rotor and was carried out and found OK. No Scoring marks observed (Fig. 10).

Fig. 9 Scoring on motor DE and NDE bearing

Fig. 10 No scoring on motor winding and stator rotor

Fig. 11 Motor alignment work in progress

2.5 Maintenance Work Carried Out

1. Fluid coupling from Motor DE side was removed.
2. Motor was replaced with the spare serviced Motor.
3. Motor trial was taken at No load decoupled condition and vibration and Ultrasonic Monitoring done. All readings found well within limit (Fig. 11).
4. Motor alignment done with rest of the drive.
5. Conveyor boxed-up and made available for operation and taken in service. At full load, all vibrations, acceleration and Ultrasonic level found in well within limit

3 Pre and Post Maintenance Vibration and Ultrasonic Readings

The detailed comparison of Pre and Post maintenance vibration and Ultrasonic readings is tabulated. It clearly shows that the parameters deviated due to damaged motor bearing came to normal level after replacement of the same (Table 2).

Table 2 Vibration and ultrasonic readings (before and after)

Measurements		Vibration velocity (mm/s) RMS		
Point	Position	Initial readings	After replacing motor	
		15/01/2014	16/01/2014	16/01/2014
		@ Full load	Decoupled trial	@ Full load
Motor (NDE)	H	1.5 (2.2 g)	2.2 (0.1 g)	0.9 (0.3 g)
	V	1.7	0.9	0.7
	A	1.4	0.8	0.6
Motor (DE)	H	2.2 (4.9 g)	1.3 (0.4 g)	1.0 (0.8 g)
	V	1.1	1.4	0.8
	A	1.9	0.7	0.5
Ultrasonic Level		8 LEDs	2 LEDs	3 LEDs

4 Results after Motor Replacement

After replacement of Motor Vibration spectrum does not show any abnormality such as Looseness and Bearing defect frequencies Also, Ultrasonic Monitoring done. No abnormal sound noticed.

4.1 Reduction in Motor DE Acceleration Value from 4.9 to 0.8 G

See Fig. 12.

4.2 Reduction in Motor Ultrasonic Value from 8 LEDs to 3 LEDs

See Fig. 13.

4.3 Financial Benefits

Vibration analysis and Ultrasonic Monitoring helped in diagnosing the exact root cause of abnormity at early stage. The early detection of Motor bearing damage of one of the most critical HT auxiliary COAL CONVEYOR-4B avoided costly catastrophic failure of Motor.

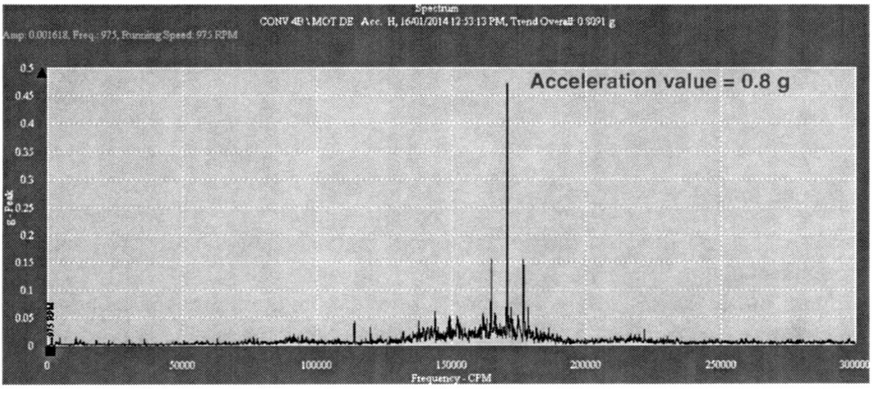

Fig. 12 Motor DE bearing acceleration spectrum

Fig. 13 Reduced motor ultrasonic level

- Generation loss Calculation

$$
\begin{aligned}
0.09\ \text{MUs/hr}\ (1\text{MU} &= 1{,}000{,}000\ \text{Units}) \\
\text{Total saving for 12hr} &= 0.09 \times 12 \times 1{,}000{,}000 \\
&= 1.080{,}000\ \text{Units} \\
\text{Cost per unit} &= £0.05 \\
\text{Total Saving} &= 0.05 \times 1.080{,}000 \\
&= £54{,}000
\end{aligned}
$$

- Cost of Motor = £4,854.3
- Total Cost saving = Generation loss + Cost of Motor

$$
\begin{aligned}
&= 54{,}000 + 4854.3 \\
&= £58{,}854.3
\end{aligned}
$$

5 Conclusion

- Co-ordinated Condition Monitoring helped in diagnosing the exact root cause of abnormality at a very early stage of Coal Conveyor-4B
- Early diagnosis has resulted in timely action savings in cost and avoidance of unplanned shutdown. Avoided costly catastrophic failure of Motor

- To avoid this kind of failure of bearings does not happen in future, pre-cautionary measures taken by adding more checks in preventive maintenance and modifying the vibration alarm level from 7.1 to 4.5 mm/s with close real time monitoring.

References

1. Rao JS (2000) Vibratory condition monitoring of machines, Narosa Publishing House
2. Harris TA, Kotzalas MN (2007) Essential concepts of bearing technology, 5th edn. Taylor & Francis

Base Isolator of Vertical Seismic Vibration Using a Negative Stiffness Mechanism

Yusuke Mochida, Naohiro Kida and Sinniah Ilanko

Abstract A model of a base isolator for vertical seismic vibration is considered. It consists of a mass and two springs, one of which is a linear type and the other is of varying stiffness. It is assumed that the varying stiffness spring can be changed from positive to negative stiffness. The varying stiffness is accomplished by the horizontally placed springs which are also connected the vertical spring by rigid links providing geometric stiffness which is negative due to compression in the spring and the axial stiffness from the inclined links which decrease as their orientation approaches the horizontal direction. The numerical investigation shows that the force induced in the vertical spring becomes constant regardless of the displacement around half of the maximum contraction. The experiment also confirmed this force-displacement relationship. In the range of displacement, the force transmitted to an object supported by the isolator due to the ground motion can be limited. This mechanism offers a possible way to prevent an object from being damaged by excessive force during earthquakes.

Keywords Base isolator · Vertical seismic vibration · Negative stiffness · Earthquake

Y. Mochida (✉) · S. Ilanko
School of Engineering, Science and Engineering, The University of Waikato, Hamilton, New Zealand
e-mail: yusuke@waikato.ac.nz

S. Ilanko
e-mail: ilanko@waikato.ac.nz

N. Kida
Graduate School of Mechanical Engineering, Hokkaido University, Sapporo, Japan
e-mail: naohiro41@frontier.hokudai.ac.jp

J.K. Sinha (ed.), *Vibration Engineering and Technology of Machinery*, Mechanisms and Machine Science 23, DOI 10.1007/978-3-319-09918-7_99

1 Introduction

Methods to prevent buildings from damage during earthquakes have continued to receive considerable attention from researchers including isolation and control systems. The control methods may be categorised into mainly two types of mechanisms: active and passive control systems. Active control system requires an external power supply, which is usually large, and the system itself also requires a substantial amount of space to store the controller and sensors [1]. In contrast passive control systems do not require an external power source and are therefore generally preferred [2]. Base isolation, an approach to decouple the superstructure and the ground motion of an earthquake, can be considered as one of the passive systems. Most of the isolation and control techniques currently in use generally work well for horizontal excitation. However, they cannot minimise structural damage caused by large vertical ground motion, which can occur close to the epicenter [3]

There are only a few systems designed to support structures during vertical motion. In 2008, Asai et al. [4] developed a vertical seismic isolation device using constant-load springs for limiting response to vertical acceleration. Lee, et al proposed a method to improve the vehicle driver vibration isolation using negative stiffness springs [5]. The horizontal beams which are axially pre-loaded with compressive stress are connected to the vertical spring. The beams have a negative stiffness effect on the vertical spring and therefore, the effective stiffness is reduced This helps to improve the transmissibility in vertical vibration especially in low frequency region [6]. Euler buckled beam that is placed horizontally can be used a negative stiffness corrector in similar manner [7]. The Euler springs are also used in vertical vibration isolators, however, their stroke is quite small [8, 9]. Duke, et al developed a scissor type tractor seat with horizontally placed springs and a dampe to reduce the seat acceleration during an operation. The damper prevents an end stop impact [10, 11].

In this paper, we theoretically investigate a vertical base isolation system using combination of a linear spring and a varying spring that can change its stiffnes from positive to negative values. The stiffness of the varying spring is changed effectively by the connected horizontal springs instead of the horizontal beams (the Euler buckled beam). This may increase the working range. The experiment is also carried out using the system with the varying springs to verify the theoretical model

2 Numerical Model

A simple one degree of freedom model which consists of a mass and two springs, a shown in Fig. 1, is considered. One is a linear spring k_1 that has a constant stiffnes $\bar{k}$, the other is a varying spring k_2, which has a stiffness changing according to the displacement of the mass. A schematic model of the system will be given in Sect. 4.

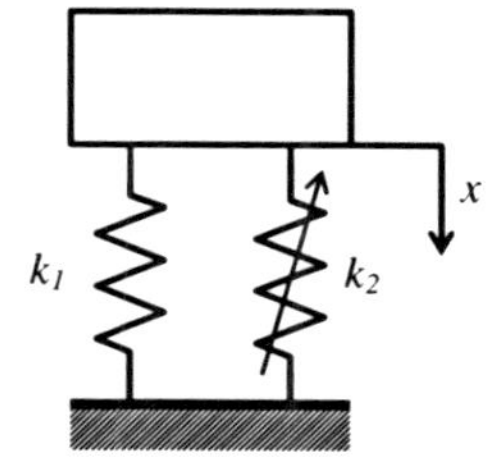

Fig. 1 One degree of freedom model of the isolator

We investigate two cases where the restoring forces induced by k_2 are expressed by a sinusoidal function (case 1) or a sum of polynomials (case 2). The equations are given as follows,

$$F_2 = a\bar{k}\sin(2\pi x/H) \quad (0 \le x \le H) \tag{1}$$

or

$$F_2 = \bar{k}\sum_{i=1}^{n} a_i x^{i-1} \quad (0 \le x \le H) \tag{2}$$

where a's are adjustable coefficients and H is the maximum allowable contraction of the spring. The spring with the varying stiffness will have a negative stiffness in a certain range of displacement.

3 Numerical Results

The force induced by the two springs was calculated for the given two cases. The relationships between the spring force and the displacements are given in Figs. 2 and 3 for Case 1 and Case 2 respectively. The horizontal axis is the displacement of the mass and the vertical axis is the forces induced by the springs. However, they are non-dimensionalised by the maximum contraction H and the spring constant of the linear spring $\bar{k}$. The dashed line and the long dash dot line show forces induced by the liner spring and the varying spring respectively. The solid line depicts the combined force of these two springs.

As can be seen from Fig. 2, the induced force becomes constant regardless of the displacement around the half of its maximum contraction. The coefficient in the above equation (a) was chosen so that the linear spring has the spring constant that is the positive value of the spring constant of the varying spring at the half of its maximum displacement in order to maximize the region of the constant force.

Figure 3 shows the relationships between the spring forces and displacement for Case 2. Similar to Case 1, the resultant force becomes flat around the middle point.

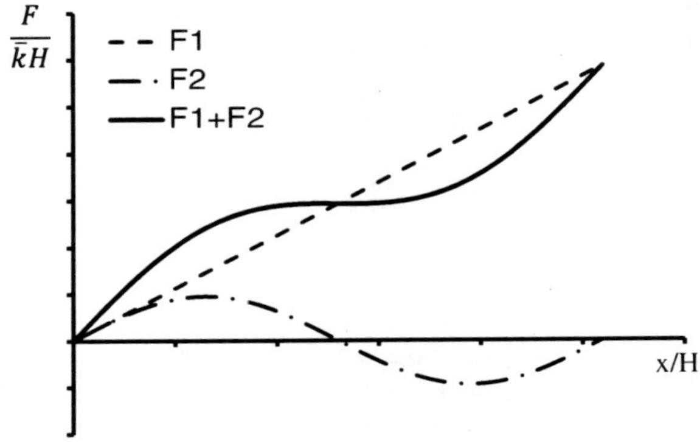

Fig. 2 Forces induced by the springs (case 1)

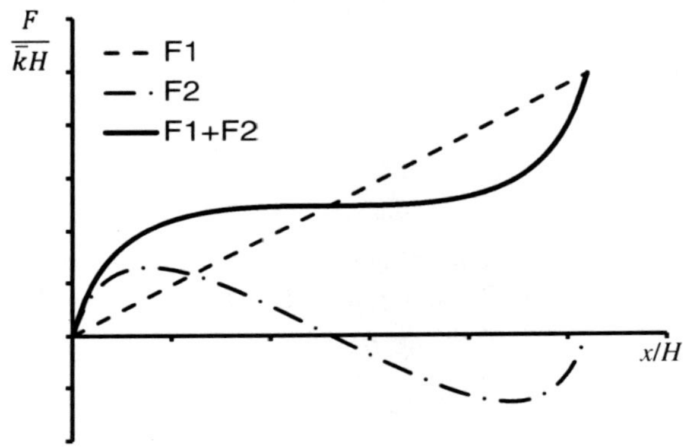

Fig. 3 Forces induced by the springs (Case 2)

The region is, however, wider than that of Case 1. This is observed because the distance between the positive and negative peaks of force induced by the varying spring is larger and the force between the peaks changes more linearly.

If an object is supported by these two springs, the force transmitted to the object due to the ground motion can be limited at the certain range of displacement and it can be prevented from damage during earthquakes.

4 Experiment

A experiment has been carried out to verify the numerical model presented in the previous section statically. Figure 4 shows a schematic diagram of the apparatus used. It consists of three springs, one is set in vertical position and the other two are set in horizontal position. These three springs are connected by links as shown in the figure.

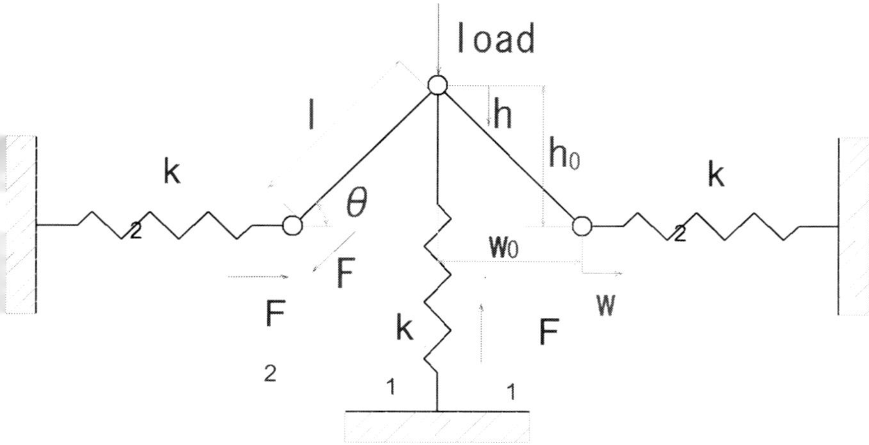

Fig. 4 A schematic diagram of the experiment apparatus

The force at the top of the apparatus, which is induced by the springs due to their displacement, is given by Equations (Xa) and (Xb).

$$F_{total} = F_1 + F_2 = k_1 h + k_2 w \tan\theta \tag{3a}$$

and

$$\tan\theta = \frac{h_0 - h}{w_0 + w} \tag{3b}$$

where h and w are the displacements of the springs in vertical and horizontal direction respectively, and h_0 is the initial hight from the horizontal axis and w_0 is the initial horizontal distance from the centre axis. By choosing the appropriate values for k_1, k_2, h_0 and w_0, it is possible to produce the range of displacement where the force is constant. The typical force-displacement relationship can be shown as Figs. 2 or 3, and the force induced by the spring with k_2 can be expressed approximately by Eqs. (1) or (2).

An equipment based on the diagram given in Fig. 4 is made and tested using a loading machine (Figs. 5 and 6). The actual equipment used in the experiment is shown in Fig. 5. The dimensions of the equipment are given as l = 140 mm, h_0 = 99.5 mm (H/2) and w_0 = 98.5 mm. Four identical sprigs, which spring constant is 0,952 N/m, are used and two of them are put in series vertically and in parallel horizontally. Hence the spring constant in the vertical position (k_1) and in the horizontal position (k_2) effectively become half and twice of the original spring constant respectively. The natual length of the springs is 63 mm.

Figure 7 shows the comparison between the experimental values and the numerical values. Due to the limitation of the test apparatus, only half of the full

Fig. 5 The actual apparatus used in the experiment

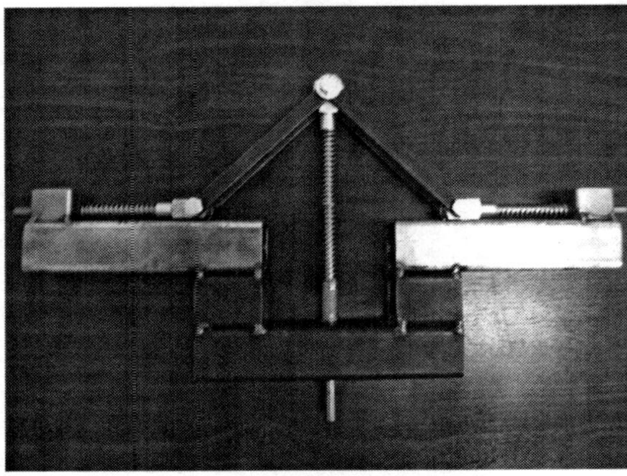

Fig. 6 The loading machine which is applying force on the equipment

Fig. 7 Comparision between the numerical prediction and the experiment result

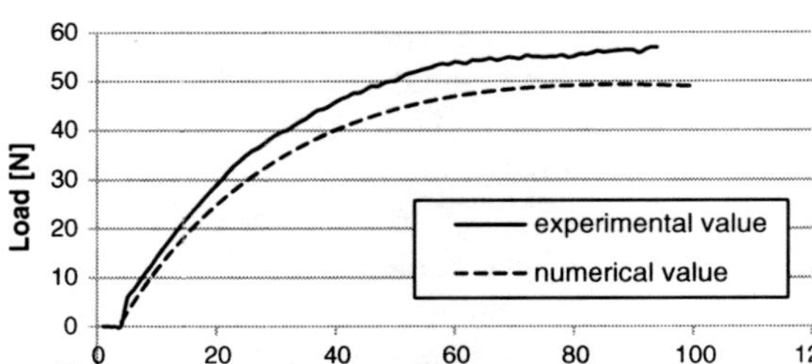

displacement range was tested. The solid line shows the experiment result and th dot line shows the result obtained numerically. It can been that this is a goo agreement between the results and it is confirmed that the induced force is constar round 50–100 mm of displacement, which is round a half of the maximum displacement. There is a discrepancy between the results, which may caused by friction on the bearings that sport the links. However, its effect is small.

5 Conclusion

A base isolator that consists of two springs, one linear spring and the other non-linear (varying stiffness), has been considered. It is assumed that the stiffness of the non-linear spring can be changed from being positive to negative. The numerical modelling shows that the forces induced by the springs become constant regardless of the displacement around half of the maximum contraction. An experiment carried out to verify the numerical model statically, produced results which are in good agreement with predicted numerical results. The friction at the bearing may have caused the small discrepancy, however its effect is not significant. In the range of displacement an object supported by the isolator can avoid damage during earthquakes because the force transmitted to the object due to the ground motion can be reduced significantly.

Acknowledgments The authors wish to acknowledge the financial support from the Marsden Fund (http://www.royalsociety.org.nz/programmes/funds/marsden/).

References

1. Fisco NR, Adeli H (2011) Smart structures: part i—active and semi-active control. Scientia Iranica 18:275–284
2. Parulekar YM, Reddy GR (2009) Passive response control systems for seismic response reduction: a state-of-the-art review. Int J Struct Stab Dyn 9:151–177
3. Carr AJ (2011) Inelastic response spectra for the christchurch earthquake records. In: Technical report to the canterbury earthquakes royal commission
4. Asai T, Yoshida N, Masui T, Araki Y (2008) Vertical seismic isolation device using constant load supporting mechanisms. J Struct Constr Eng 73:1511–1518
5. Lee CM, Goverdovskiy VN, Temnikov AI (2007) Design of springs with "negative" stiffness to improve vehicle driver vibration isolation. J Sound Vib 302(4–5):865–874
6. ORLIN Technologies Ltd (2014) Negative-stiffness mechanisms. http://www.orlin.co.uk/negative_stiffness.htm. Accessed May 2014
7. Liu X, Huang X, Hua H (2013) On the characteristics of a quasi-zero stiffness isolator using Euler buckled beam as negative stiffness corrector. J Sound Vib 332:3359–3376
8. Chin EJ, Lee KT, Winterflood J, Ju L, Blair DG (2005) Low frequency vertical geometric anti-spring vibration isolators. Phys Lett A 336:97–105
9. Winterflood J, Barber TA, Blair DG (2002) Mathematical analysis of an Euler spring vibration isolator. Phys Lett A 300:131–139
10. Duke M, Goss G (2007) Investigation of tractor driver seat performance with non-linear stiffness and on-off damper. Biosyst Eng 96(4):477–486
11. Duke M, Fow A (2012) Comparison of simulation and experimental results of a tractor seat with nonlinear stiffness and dead-band damping. J Vib Acoust 134(5):051006

A Novel Piezoelectric Rotational Speed Meter

Weiqing Huang, Meng-xin Sun, Jian-hui Zhang, Qing-jun Ding and Yin Wang

Abstract A novel piezoelectric rotational speed meter which is within the family of solid-state wave gyroscope based on direct and inverse piezoelectric effect is proposed in the paper. The hard core of the speed meter is a vibratory ring with circular inner surface and octagonal outer surface. First, the paper presents the principle introduction and structure design of the rotational speed meter. Second, simulation results and mode test results show appropriate working modal frequency and modal shape. We found that the working frequencies of active and sense mode are nearly equal. In addition, coriolis test are done to validate the working principle. Finally, the data from the experiment are analyzed to obtain the equation between output signals and angular velocity. We conclude two equations which can be used to work out the angular velocity input and the scale factors of the equations are 16.4 mV/min/r and the correlation coefficient is 0.998 and 0.995 respectively. These results are more accurate when the angular velocities range from 5 to 700 RPM.

Keywords Piezoelectric structure · Solid-state gyroscope · Angular velocity

1 Introduction

Solid vibration gyroscope can be used as the angular velocity sensor. In the middle of the 19th century, Foucault found that the vibration plane of a vibration rod keep the same when it rotates. And Brian found this principle can be used to measure the

W. Huang (✉) · M.-x. Sun · J.-h. Zhang · Q.-j. Ding · Y. Wang
State Key Laboratory of Mechanics and Control of Mechanical Structures, Nanjing University of Aeronautics and Astronautics, 29 Yudao Road, Nanjing 210016, China
e-mail: mehwq@nuaa.edu.cn

M.-x. Sun
e-mail: sunmengxin891106@163.com

Y. Wang
e-mail: wangyin@nuaa.edu.cn

J.K. Sinha (ed.), *Vibration Engineering and Technology of Machinery*,
Mechanisms and Machine Science 23, DOI 10.1007/978-3-319-09918-7_100

angular velocity and the linear acceleration [1]. Solid vibration gyroscope is divided into many types such as vibration beams, tuning forks, hemispheres, rings, cylinders, Microelectromechanical system and so on. Of these types of the gyroscope, hemispheres, rings and cylinders are within the family of solid-state wave gyroscope which is often used to achieve the measurement of angular velocity based on inverse piezoelectric effect to excite and direct piezoelectric effect to sense. It has advantages such as long life, high sensitivity, low power consumption, and low cost, which make it play an important role on aerospace, automotive, defense,and bio-medical engineering [2].

In recent years, many novel types of solid-state wave gyroscope had been designed and manufactured. Delco has been developing HRG for several decades. All kinds of new inertial gyro sensors, which are under development by Delco Systems Operations that utilize the HRG basic principle, promise low cost, small size, and high reliability [1]. In 2001, Farrokh Ayazi presents the design, fabrication, and testing of a high aspect-ratio polysilicon ring gyroscope (PRG) [3]. In 2008, Sushma Kotru propose a new design for the gyroscope which should increase the accuracy by increasing the mass of the gyro, that is, produce a meso-scale gyro using MEMS technology and increase signal amplitude by using piezoelectric actuators and sensors [2]. In 2011, Tao Yi and the others presents the modeling, simulation, fabrication, and performance characterization of a novel ring vibratory gyroscope based on piezoelectric effect. The resonator of this gyroscope, which has the advantages of high Q-factor and high sensitivity, has a simple structure, comprising of a substrate, eight thin beams and a vibrating ring. The active and sense piezoelectric elements are attached to the outer surface of the beams [4].

To conclude all types of the solid-state gyroscope, we find the resonator of it plays a crucial role in the precision and sensitivity of the gyroscope. Therefore, the design and processing method of resonator are particularly crucial. Nevertheless, existing gyros has some defects. Firstly, the resonator requires high demands on the processing accuracy due to its structure characteristics. Secondly, we cannot paste the piezoelectric ceramic on the surface of the resonator because of its circular shape. Pasting the piezoelectric ceramic on the external support structure will inevitably have an impact on the input signal. To solve this problem, we designed a novel resonator with octagon outer surface while circle inner surface. The piezoelectric ceramic pieces are pasted on the eight outer surfaces.

2 Structure Design and Operation Principle

2.1 Modeling and Analysis

In the article, we proposed a novel piezoelectric rotational speed meter which is actually a type of solid-state wave gyroscope. Its resonator is circle structure which is shown in Fig. 1. The outer surface of the structure is octagon shape and the inner surface is circular shape. The piezoelectric ceramic pieces are pasted on the eight

'ig. 1 Schematic of the esonator

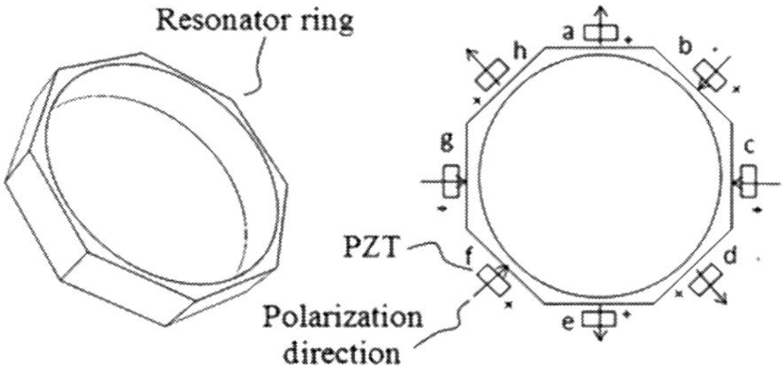

uter surfaces a to f. Since its outer surface is octagon shape and its inner surface is ircular shape, it will be easily and conveniently processed. The piezoelectric eramic pieces are pasted on the eight outer surfaces of the ring-shaped resonator lirectly (as shown in Fig. 1), which avoid transferring vibration by support beam onnected to the ring structure, so that it will improve the vibration amplitude and neasurement accuracy. To use eight piezoelectric ceramic pieces will obviously nake it easier to detect the output signal of the structure. In addition, the different olarization direction is used to produce the vibration mode that we need.

The working principle of the speed meter is like the other solid-state wave yroscope. Firstly, we should simplify the resonator ring to an eight-beam structure. Dne of the beam and the piezoelectric ceramic piece pasted on its outer surface is nodeling as shown in Fig. 2.

We use the following equation to describe the stress and strain of piezoelectric lements caused by piezoelectric effect.

$$\begin{bmatrix} T \\ D \end{bmatrix} = \begin{bmatrix} C^E & -e \\ e & \varepsilon \end{bmatrix} \begin{bmatrix} S \\ E \end{bmatrix} \tag{1}$$

vhere, S is strain, D is electric displacement, C^E is stiffness coefficient, e is pie-oelectric coefficient, ε is dielectric constant, T is stress, E is field strength. Spe-ially, for the transverse effect piezoelectric element which is made of PZT-5 lesigned in the paper.

In active mode, we apply a sinusoidal signal on the piezoelectric ceramic pieces n the surface a, c, e, g. As the stresses of x and z direction (T_1 and T_3) in iezoelectric element have little influence on the beam we analyzed, we mainly sed d31 effect to make the deformation of structure.

'ig. 2 Modeling of the esonator

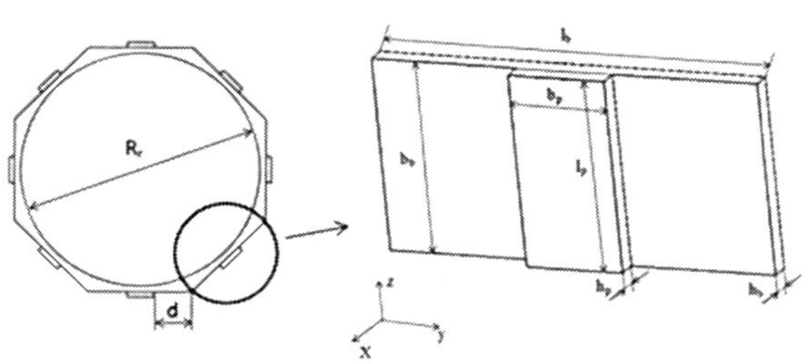

The inner stress T_2 in the piezoelectric element can make the beam to contract o extend, which applies sinusoidal moment acting on the beam. This moment can b written approximately as follow.

$$M_p = \frac{1}{2} T_2 l_p h_b (h_p + h_b) \quad (2$$

On the effect of the moment M_p, the force situation of the beam is as shown i Fig. 3 The external force of the beam is only F_a.

Apparently, F_a shown in Fig. 3 is the driving force of active mode.

When the input signal is a sinusoidal signal, F_a will change as sinusoidal forn which can be written as,

$$F_a = F_0 cospt \quad (3$$

where F_0 is the maximum force. Using energy method to the structure from time to t, we obtain that,

$$8 \int_0^t F_a v_a \Delta t = 4 \times \frac{1}{2} m v_a^2 + 4 \times \frac{1}{2} I_{zo} \omega_o^2 \quad (4$$

where, m is the mass of every beam, v_a is the velocity of beam a, c, e, g., I_{zo} is th rotational inertia of one beam with instant center, and ω_O is the angular velocity of th beam with O. Therefore, velocity equation of beam can be expressed as following.

$$v_a = \frac{12F_0}{5mp} sinpt \quad (5$$

where p is circular frequency of the sinusoidal signal.

When the structure is rotating around z axis with the angular velocity ω, it wi produce coriolis force as shown in Fig. 4.

To analyze the sense mode, we simplify the coriolis force mentioned above a shown in Fig. 4 The force F_s can be expressed as following.

$$F_s = F_c \cos \alpha = \frac{\sqrt{2}}{2} F_c = \frac{\sqrt{2}}{2} m \omega v_a \quad (6$$

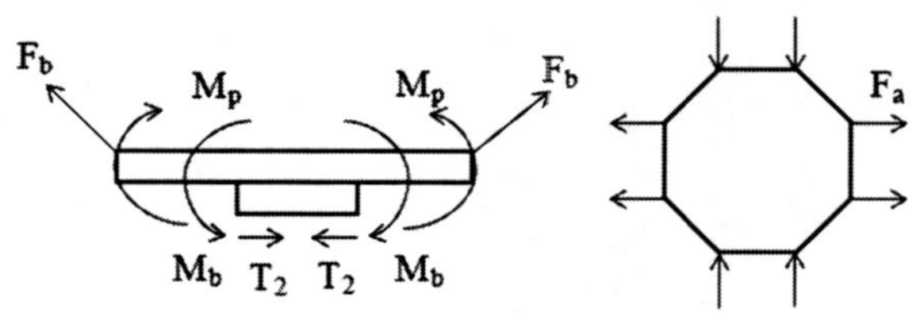

Fig. 3 Force analysis and simplified of active mode

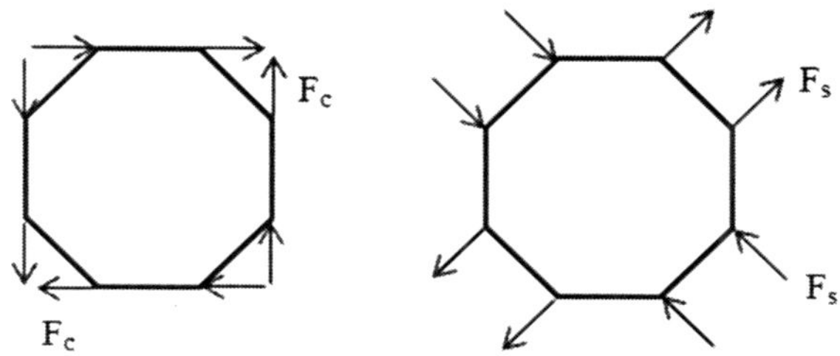

Fig. 4 Force analysis and simplified of sense mode

where F_c is the coriolis force. As driving force of the sense mode, this force F_s will produce strain S which will sensed by piezoelectric ceramic pieces pasted on surfaces b, d, f, h of resonator. According to Eq. (1), we can obtain the electric displacement as following.

$$D = \varepsilon S = \frac{2\varepsilon \int(\int \frac{2F_s}{m} dt)dt}{l_b} b_p \tag{7}$$

Then we could obtain the equation

$$V_s = \frac{Q}{C} = \frac{\int D dA}{C} = \frac{\int \frac{2\varepsilon \int(\int \frac{2F_s}{m} dt)dt}{l_b} b_p dA}{C} \tag{8}$$

where, V_s is the output voltage of sense mode, Q is the quantity of electric charge of piezoelectric ceramic pieces pasted on surfaces b, d, f, h and C is the capacitance of piezoelectric ceramic piece. We found the piezoelectric ceramic piece will create a voltage which can be used to detection the coriolis force caused by rotating resonator. In this way, we could finally find the angular velocity by processing the electrical signal.

2.2 FEM Analysis

FEM software ANSYS is used to help us to get the appropriate modal frequency and modal shape. In order to develop the sensitivity and reliability of the structure, the resonator ring uses copper and the piezoelectric ceramics use PZT-5A. Here we select SOLID 186 as the element type of the copper ring and select SOLID 5 as the element type of the PZTs.

Then, the first four order modes of mode analysis are as shown in Fig. 5. From this mode analysis, we found that the first-order and second is the active mode and sense mode respectively. The active mode resonating frequency is 1355.2 Hz while the sense mode is 1355.6 Hz. The third-order mode frequency is 3083.8 Hz, and the fourth is 3135.6 Hz. To conclude the mode analysis, the frequencies difference

Fig. 5 Mode analysis by ANSYS 14.0 of **a** active mode (1355.2 Hz), **b** sense mode (1355.6 Hz), **c** third-order mode (3083.8 Hz), **d** fourth-order mode (3135.6 Hz)

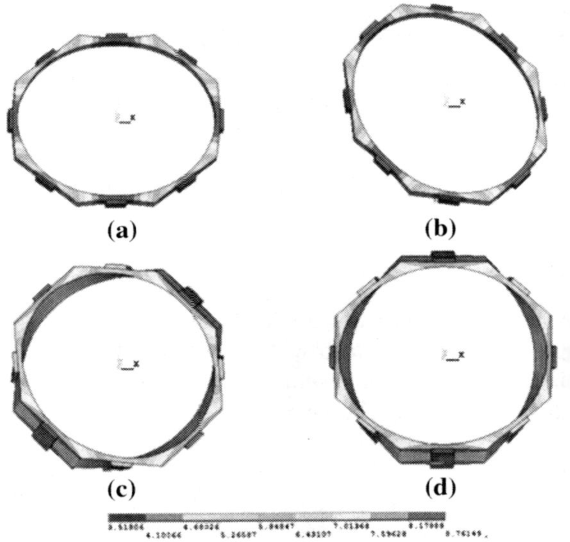

mainly caused by truncation error of numerical computation of active and sense mode is small. Furthermore, the frequencies of the third-order mode and the fourth-order mode that may be caused in z axis are far from the active mode frequency, which shows the working mode of the resonator is an in-plane (x-axis and y-axis) vibration.

According to the results above, we take a harmonic response analysis. To simulate the operation principle of the resonator, we apply sinusoidal voltage on the PZTs on surfaces a, c, e, g, provide angular velocity in z axis and activate the coriolis effect of ANSYS. Therefore, we found the graded distribution of electric charge of PZTs on the surfaces b, d, f, h created in sense mode, as shown in Fig. 6. This response, which can be used to obtain the angular velocity input, validates the operation principle of the resonator.

Fig. 6 Graded distribution of electric charge in sense mode

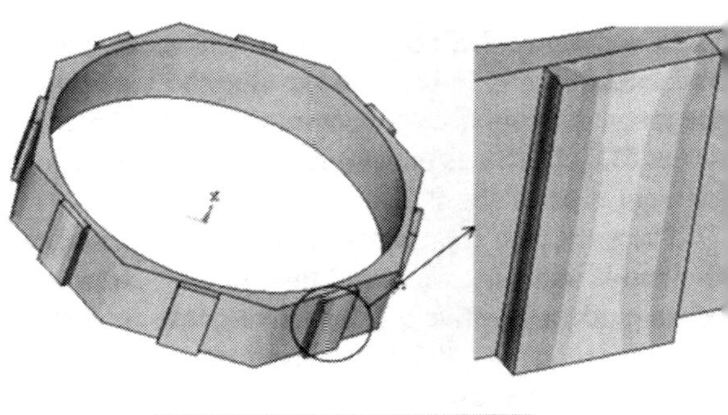

3 Experimental Investigation

In order to make the resonator work normally accord with the working principle, we design and manufacture the whole speed meter as shown in Fig. 7. It includes substrate, spindle, the piezoelectric vibration ring structure, and the brush.

Firstly, we take a mode test to validate the vibration of the structure's work mode using a laser Doppler vibrometer. In driving mode test, we applied the alternating voltage on the piezoelectric ceramic pieces of outer surfaces a and e. Then we pick the response up from outer surfaces c and g. The resonating frequency is 1,368 Hz as shown in Fig. 8a. In detection mode test, we applied the alternating voltage on the piezoelectric ceramic pieces of outer surfaces b and f. Then we pick the response up from outer surfaces d and h. The resonating frequency is 1,369 Hz as shown in Fig. 8b. From the results above, the frequency split between two working mode is small which will ensure the accuracy of the following tests.

Then we take a coriolis force test which will validate measurement ability of the structure. We applied a sinusoidal signal on piezoelectric ceramic pieces on surfaces a, c, e, g by signal generator and power amplifier. The output signal is scanned by oscilloscope which is connected to piezoelectric ceramic pieces on surfaces b, d, f,

Fig. 7 Prototype of the rotational speed meter

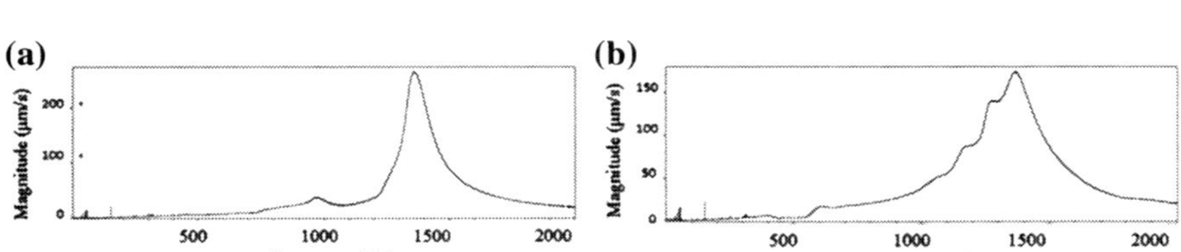

Fig. 8 **a** Resonating response of active mode. **b** Resonating response of sense mode

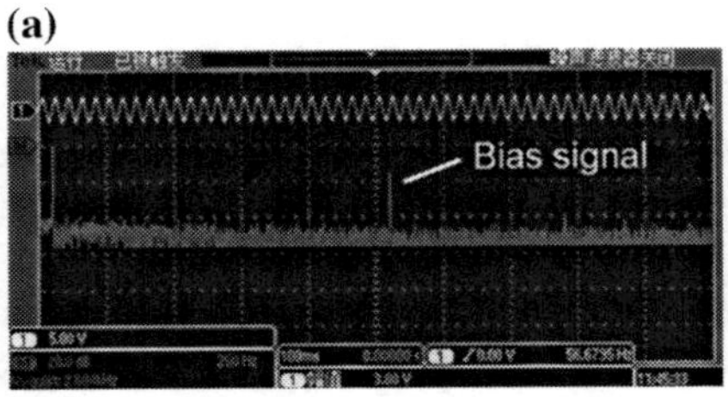

Fig. 9 **a** The output signal without angular velocity input. **b** The output signal with angular velocity input

h. The frequency spectrum when it doesn't rotate is shown in Fig. 9a. The frequency spectrum when it rotates is shown in Fig. 9b. The coriolis force signals are produced around resonator frequency signal from the comparison of the results.

In order to obtain the angular velocity from the output signal, we should build relationships between output signal and the angular velocity. So we continue to observe frequency spectrums as shown in Fig. 10.

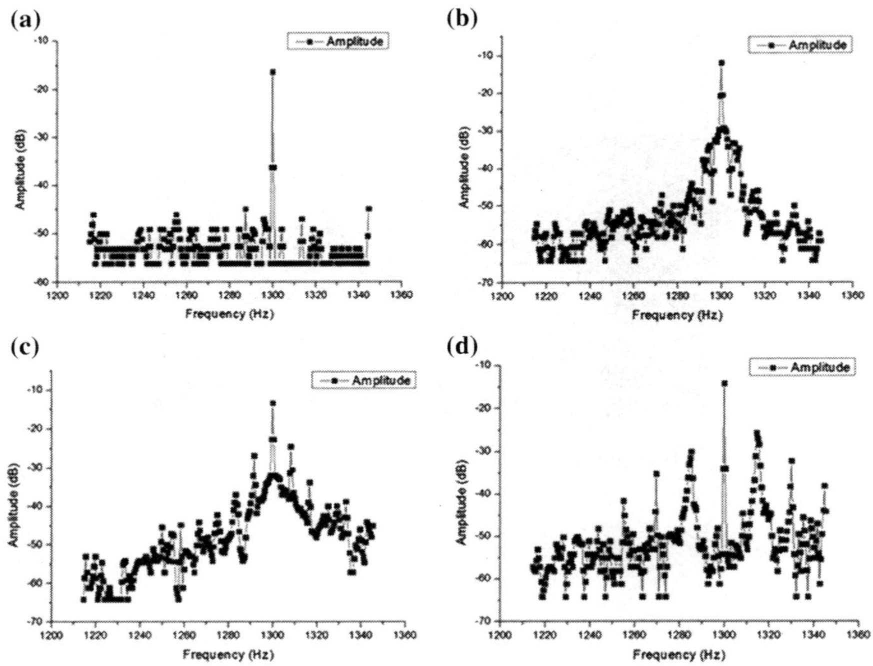

Fig. 10 Frequency spectrums with angular velocity of **a** 0 rpm, **b** 100 rpm, **c** 500 rpm, **d** 900 rpm

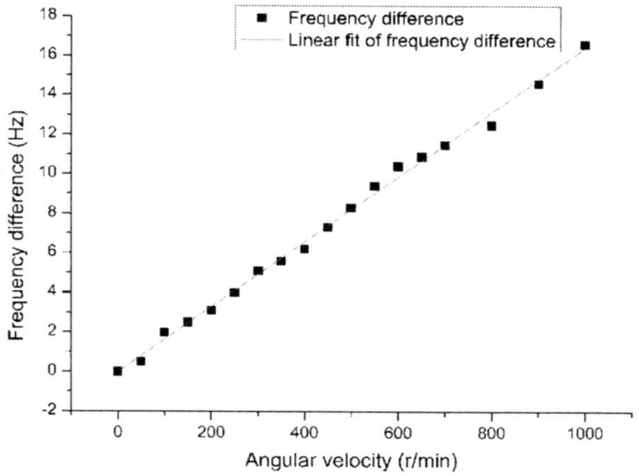

Fig. 11 The relationship of frequency difference and angular velocity

After the further analysis of the results in Fig. 10, we could conclude that the amplitude and frequency of the coriolis signal are changed when the input signal is different.

To describe the input angular velocity with a proper way, we firstly draw a graph of the relationship between frequency difference (of bias signal and coriolis signal) and angular velocity as shown in Fig. 11. Obviously, there is a linear relation. So we fit the data as a line when the correlation coefficient r is 0.998. The linear relation can be deduced as,

$$f_s = f_0 + k_f\omega = -0.0118 + 0.0164\omega \tag{9}$$

where f_s is the amplitude of output signal, ω is angular velocity and the measured scale factor k_f is 16.4 mV/min/r.

Secondly we draw a graph to find the relationships between amplitudes and angular velocity. We fit the data as a line, as shown in Fig. 12, when the correlation coefficient r is 0.995. The linear relation can be deduced as,

$$V_S = V_0 + k\omega = 1.5297 + 0.0164\omega \tag{10}$$

where V_S is the amplitude of output signal, ω is angular velocity and the measured scale factor k is 16.4 mV/min/r.

Both the Eqs. (9) and (10) can be used to obtain the input angular velocity from the output signal. Experimental error is mainly caused by the generated heat of the structure and slight fluctuations of the input angular velocity. As the structure has a certain mass, input speed is not stable when it is too slow ($r < 5$ rpm). Also, when the speed is too fast ($r > 1000$ rpm), the experimental device generate heat continuously, thus the output signal is not accurate. So when the angular velocity ranges from 5 to 700 rpm, these results is more accurate. To compare with the

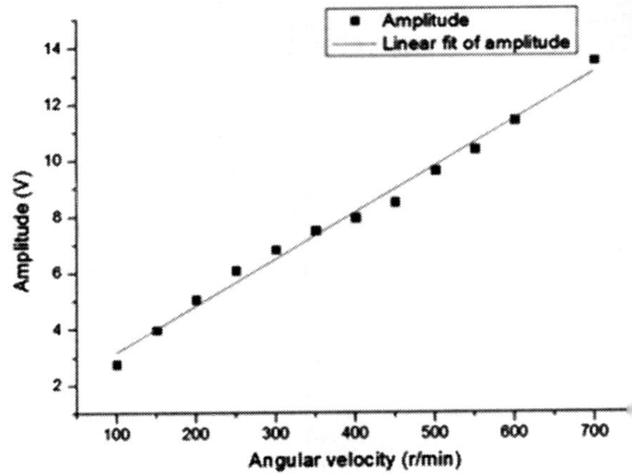

Fig. 12 The relationship of amplitude and angular velocity

similar sensors, which could always measure the angular velocity range from 0 to 150°/s (25 rpm), the measurement range of the proposed sensor is wider. Also, the accuracy of this sensor can be seen from the linear correlation coefficient as mentioned above.

4 Conclusion

This paper presented design, analysis, simulation and experiment of a novel piezoelectric rotational speed meter based on piezoelectric effect. First, we introduce the structure design and operation principle of the novel rotational speed meter. Based on simulation results, we obtain the proper modal frequency and modal shape. After processing the whole structure, the mode tests are done to validate the working principle of the structure. In addition, the coriolis tests show that we can obtain the coriolis force signal when there is angular velocity input from the frequency spectrum. Finally, the data from the experiment are analyzed to obtain the relationship of output signals and angular velocity. We get two kinds of measure of input angular velocity from output amplitude (voltage) and frequency difference respectively. The results showed that the piezoelectric rotational speed meter has high accuracy and reliability when the angular velocities range from 5 to 700 RPM.

Acknowledgments The work was supported by the scientific research base innovation foundation of NUAA (NJ20120001). Basic Research Operating expenses Program of NUAA (No. NP2012201)

References

1. Matthews A, Rybak FJ (1992) Comparison of hemispherical resonator gyro and optical gyros. IEEE Aeros Elect Syst Mag 5:40–46
2. Kotru S, Zhong J, Highsmith A, Jackson JE (2008) Design and fabrication of a meso-scale gyroscope, microelectronics and electron devices. WMED 2008. IEEE Workshop on
3. Ayazi F, Najafi K (2001) A HARPSS polysilicon vibrating ring gyroscope. J Microelectromech Syst 2:169–179
4. Tao Y, Wu X, Xiao D, Wu Y, Cui H, Xi X, Zhu B (2011) Design, analysis and experiment of a novel ring vibratory gyroscope. Sens Actuators A: Phys 2:286–299

A New Linear Motor with Two Piezoelectric Stacks Based on Differential Friction Effects

Weiqing Huang, Xifu Chen and Jianhui Zhang

Abstract The existing resonant linear piezoelectric motors must operate with high working voltage in resonant condition, resulting that their operating frequency range is narrow and their running stability is poor. So a kind of linear piezoelectric stepping motor with broad frequency domain was proposed based on friction difference. First, the friction difference driven principle of motor is proposed by analyzing the continuous stepping motion in a period. Then the pizoelectric actuator unit is designed with flexible structure and the motor was constructed with two piezoelectric actuator units in two perpendicular direction. further, both the displacement characteristics of mover and the amplitude charateristics of the actuator were tested with keyence laser displacement sensor (LDS). Finally, the stepping charateristics of the motor were reserched on by analyzing the displacement testing data with mathematical software. The experimental results shows that the motor can operate stably within the scope of 350–750 Hz. When excitation voltage is 30 V and pre-load is 3 or 10 N, the lateral amplitude of the drive foot is approximately 4 μm and the stable average steplength ranges from 3.1 to 3.2 μm with the error rate about 5–7.5 %.

Keywords Linear motors · Piezoelectric stacks · Stepping motors · Friction difference

W. Huang (✉) · X. Chen · J. Zhang
State Key Laboratory of Mechanics and Control of Mechanical Structures, Nanjing University of Aeronautics and Astronautics, 29 Yudao Road, Nanjing, China
e-mail: mehwq@nuaa.edu.cn

X. Chen
e-mail: cxf@ycit.cn

J. Zhang
e-mail: zhangjh@nuaa.edu.cn

J.K. Sinha (ed.), *Vibration Engineering and Technology of Machinery*,
Mechanisms and Machine Science 23, DOI 10.1007/978-3-319-09918-7_101

1 Introduction

Linear piezoelectric motors have showed wide application prospect in precision positioning, instrumentation, biomedical and many other fields and caused the attention of the researchers at home and abroad [1–3]. For resonance linear piezoelectric motors have the advantages of compact structure, fast response, low noise and no electromagnetic interference, outstanding researches have been done in this field by domestic and foreign scholars [4, 5]. While resonance linear piezoelectric motors are largely affected by the temperature and the structure design due to they work in resonance condition [6]. In recent years, therefore, there is a boom in the development of non-resonant piezoelectric motors [7].

In recent decades, using the perfect characteristics of the piezoelectric stack under non-resonant conditions, domestic researchers develop piezoelectric stepping motors with high positioning accuracy [8, 9], among which the inchworm principle and inertial impact non-resonant type linear motors are well-developed [10]. However, the operation frequency of above two motors is low. At the same time, high requirements were put forward in drive signal and structure acurracy. In recent years, professor Huang from Nanjing University of Aeronautics and Astronautics proposed non-resonant friction drive motors based on elliptical orbit [11]. The investigation of friction drive mechanism shows that it is necessary for the stator and the mover to separate effectively in the second half of the elliptical motion cycle. Thus the motor can operate stably only under high frequency and the piezoelectric ceramics driver need larger power. In addition, effective separation of motor also put forward high demand on contact condition between stator and mover, resulting that preloading device should be designed properly and the assembly accuracy should be guaranteed.

In this article, the sawtooth wave and square wave signals will be used to produce two vertical linear motion trajectories at drive foot so as to realize the friction drive between the stator and the mover. Due to the step characteristics of square wave signals, it is easier to implement vertical separation between the stator and mover so that the motor can run in a wider and lower frequency domain.

2 Working Mechanism of Motor

This article expected to put forward a piezoelectric actuators based on two groups orthogonal structure (as shown in Fig. 1) includes lateral actuator (parallel to the motion direction of the mover) and vertical actuator (perpendicular to the motion direction of the mover). Two groups of actuators are permanent connected to the drive foot of stator. Lateral actuators provide drive foot of stator reciprocating displacement in motion direction, and the elongation and shortening of another actuator can realize the contact and separation for drive foot and mover. Thus the separation between the stator and the mover would just be controlled by the vertical actuator.

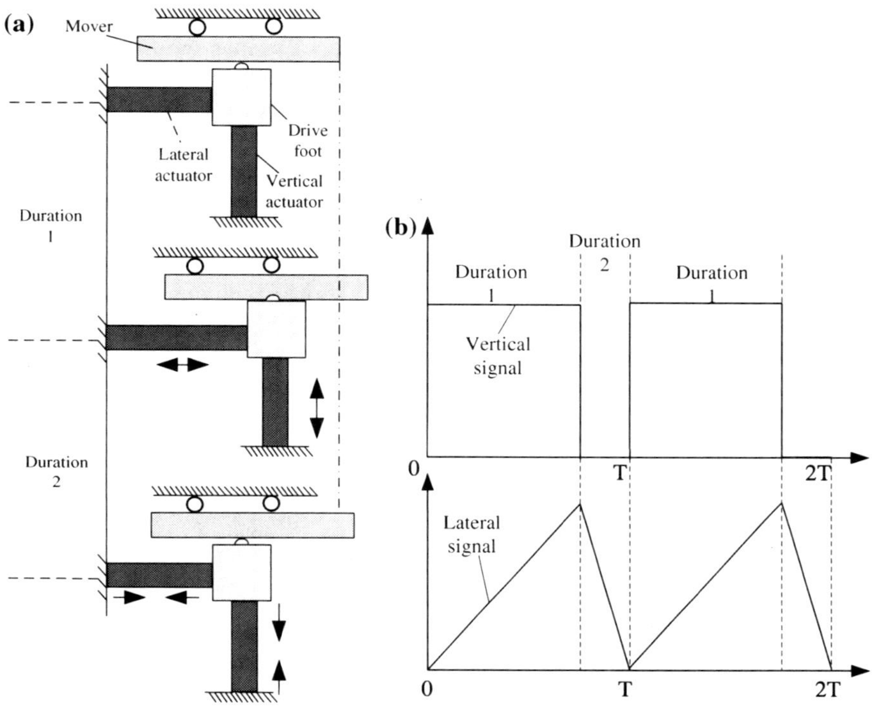

Fig. 1 The operation principle of the motor **a** The operational cycle of motor **b** The exciting signal of piezoelectric actuators

As shown in Fig. 1, in the initial state, vertical and lateral actuators are dormant, so the contact force between stator and mover is the initial pre-pressure. When we apply cycle square wave voltage on the vertical actuator, the duty ration of the square wave voltage should be greater than 0.5, this article 0.75.

In duration 1, excitation voltages are applied to both actuators at the same time. Vertical actuator is excited by the square wave voltage. At the beginning, vertical actuator elongates rapidly and the contact force of stator and mover is the most stressful. Lateral actuator is excited by the triangle wave voltage. At this time, triangle wave is at up slope stage, the lateral actuator continues to elongate. Therefore, the mover will be driven rightward by the friction of contact interface between stator and mover.

In duration 2, the vertical actuator is excited by square wave voltage which is zero and the vertical actuator shrink to the initial state rapidly. At this moment, the pressure between stator and mover is the minimum. At the same time, we apply lateral actuators with the triangle wave voltage. Triangle wave is located in the down slope stage, so the lateral actuator continues to contract. At this time, the mover will return under the effect of friction force due to the initial pressure between the stator and mover.

However, because the duty ratio of square wave voltage is greater than 0.5 and the friction between stator and move in duration 2 is less than the friction force in duration 1, the movement distance in duration 2 is smaller than that of duration 1. It can realize the dynamic motion to the right within the whole motion cycle. At the same time, the drive foot will restore the initial state.

From above analysis,the vertical actuator and lateral actuator are excited by the excitation signal as shown in Fig. 1b. At the beginning of duration 1, the drive foot keeps compressing with the elongation of vertical actuator. In duration 1, the drive foot move to the right with the rising of voltage signal. At the end of duration 1 vertical actuator shrinks rapidly, so the drive foot shrinks with the vertical actuator. Entering into duration 2, lateral actuators also shrinks rapidly and the drive foot moves along with the shrinkage of lateral actuator. At the end of Sect. 2, the drive foot returns to the initial position. Therefore, the trajectory of the drive foot is a rectangular path in the whole cycle.

3 The Structure of Motor

The motor is composed of two piezoelectric actuators(lateral actuator and vertical actuator), mover (linear guide), flexible support, preload guide rail and preloading pin. One end of the stator group is fixed with preload guide rail, and the other end is fixed to base. With restoring force produced by flexible support, the stator stay in touch with the mover to make the piezoelectric linear motor self-lock on the power-off. The structure of the motor is shown in Fig. 2.

The vertical and lateral actuators have the same structure as shown in Fig. 3. They both include the support block (a1, a2), the tension spring (a3, a4), pin (a5 and a6), circlip (a7, a8), the laminated piezoelectric ceramic (a9), fixed wedge (aa), adjusting wedge (ab) and adjusting block (ac). Support blocks connect into an organic whole through the tension spring and pin. Laminated piezoelectric ceramic group (a9) and fixed wedge (aa), adjusting wedge (ab) and adjusting block (ac) concatenated axially between support blocks (a1, a2).

The stator of motor can be viewed as single-degree-freedom systems both in vertical direction and in lateral direction. In both direction, the equivalent stiffness is about 9.6 N/μm and the equivalent mass is about 0.1 kg. Therefore, the resonant frequency is about 9.8 kHz.

4 The Experimental Researches

The prototype was designed and fabricated according to Fig. 3. Keyence XL-80 Laser displacement sensor (LDS) was adopted to measure displacement of mover and the amplitude of the actuator. The experiment set-up of motor is shown in Fig. 4.

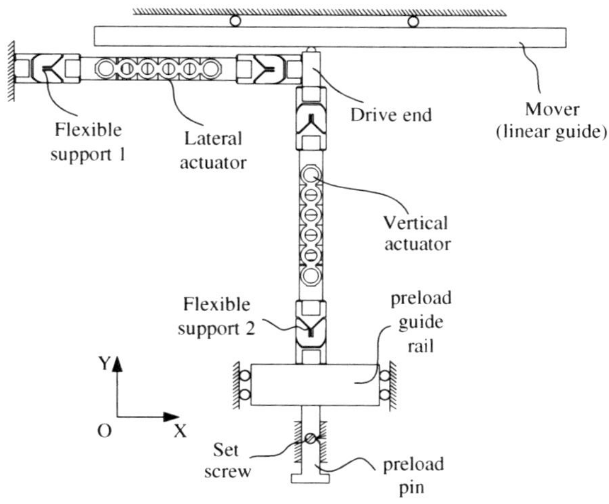

Fig. 2 The schematic of motor

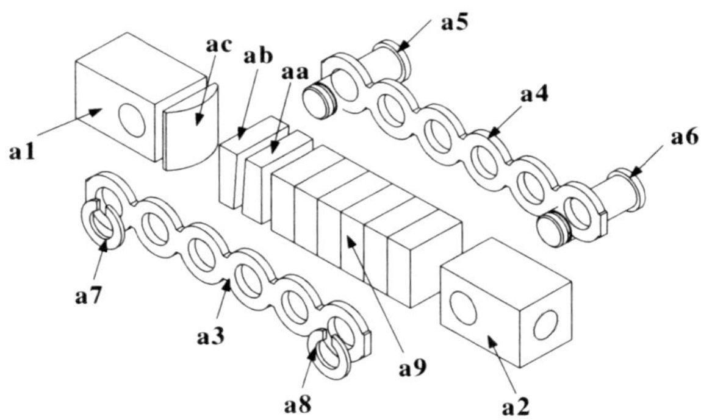

Fig. 3 The diagram of Piezoelectric actuator unit

It can be seen from Fig. 5 that the lateral amplitude of the drive foot of stator excited by 450 Hz sinusoids. When the preload is 3 or 10 N, vertical amplitude are almost the same, and both are stable during a long time. In Fig. 5, the duration 1 is the drive process and the duration 2 is the return process. In duration 1, there is oscillation while the drive foot try to push the mover because of the friction coupling effect between the stator and the mover.

Figure 6 shows large stroke displacement curve of motor when frequency is 450 Hz. It can be seen that the motor remains good characteristics of speed stability in twide frequency range.

Figure 7 shows the magnified displacement curves in Fig. 6, It can be seen that the stepping motor has high accuracy and stability under the same pressure.

Figure 8 shows the average pace curve of motor under different frequency.

From Fig. 8, it can be seen that the steplength ranges from 3.08 to 3.33 μm, the average pace is 3.20 μm and the average error is within 4 %, under conditions that the preload is 3 N and frequency ranges from 350 to 600 Hz. When preload is 10 N

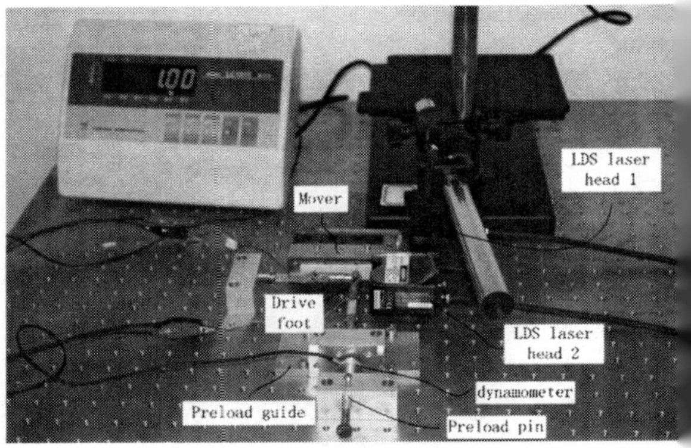

Fig. 4 The experiment device of motor

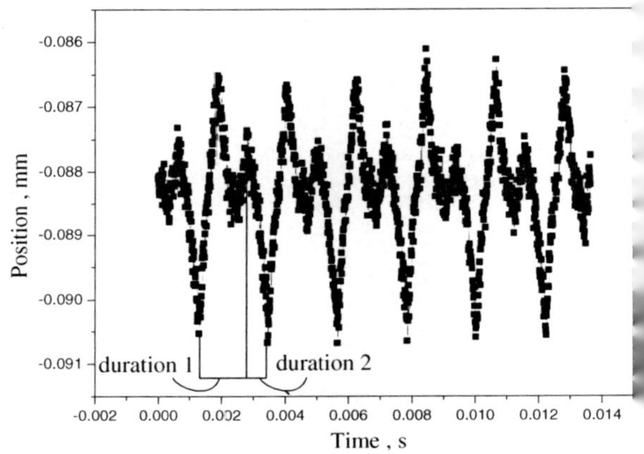

Fig. 5 Vertical displacement curve of drive foot

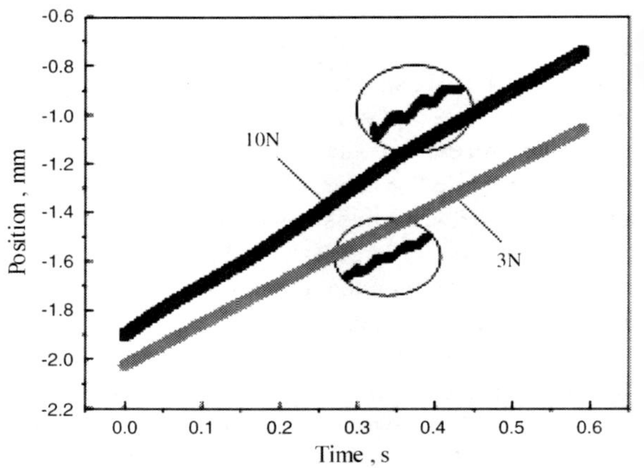

Fig. 6 The displacement curves of motor

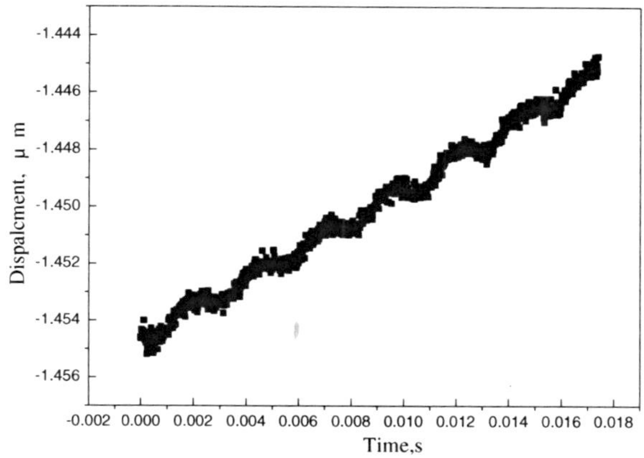

Fig. 7 The displacement curves of motor

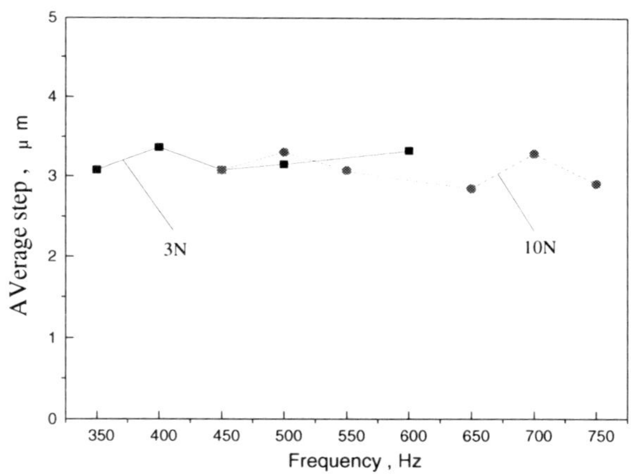

Fig. 8 The average pace curves of motor

and frequency ranges from 450 to750 Hz, the steplength ranges from 2.85 to 3.31 μm, the average interval is 3.08 μm and the error is within 7.5 %. The stepping characteristics experiments indicate that the motor has uniform steplenth in 350–750 Hz frequency range.

5 Conclusion

The theoretical study and experimental research show that the friction drive and non-resonant piezoelectric linear motor can achieve stable operation in a low and wide frequency range. We can obtain the following conclusions from the experimental results.

(1) The horizontal amplitude of the stator drive foot is stable in a wide frequency range, which provides a precondition for high positioning accuracy.
(2) The displacement experiments indicate that motor presents good linearity in a large stroke, which implies that the motor possesses good speed stability.
(3) The motor have good stepping character. The experimental results show that the average step of the motor varies in the scope of 3.1–3.2 μm in the frequency range of 350–750 Hz.

Further, it is found that when the driving frequency is beyond the range of 350–750 Hz, the operating speed of the motor decreases sharply. This is because there only exits static friction forces between motor stator and mover when the frequency is lower than 350 Hz. Thus, the stator drive the mover move back and forth in whole operation cycle. When the frequency is lower than 350 Hz, the act force between stator and mover is mainly the impact force, so the displacement in drive stage is nearly equal to that in return stage and the mover make reciprocating motion in a whole period. That is to say, when the frequency is too low or too high, the relative slip rate between the stator and the mover is almost unchanged in the extending stroke and return stroke, which is against the high speed driving of the motor. In addition, the stepping resolution of motor could be improved by the structure optimization design of the motor, the reduction of the working voltage, and the optimizing of operation factors such as working frequency and preload.

Compared with the friction drive type piezoelectric stack linear motors, the displacement resolution of this motor is slightly lower in 350–750 Hz. The friction drive type piezoelectric stack linear motor can reach the displacement resolution of 0.1 microns. However, the displacement resolution of the proposed motor can be improved by controlling the amplitude and frequency of the excitation signals and by the structural optimization design and the choice of friction materials between the stator and mover.

Acknowledgement we wishes to thank the National Natural Science Foundation of China (51375224), Basic Research Operating expenses Program of NUAA (No. NP2012201), Project Funded by the Priority Academic Program Development of Jiangsu Higher Education Institutions.

References

1. Yin W, weiqing H (2012) Linear ultra-sonic motor using longitudinal vibration [J]. Trans Nanjing Univ Aeronaut Astronaut 01:40–45
2. Suzuki M, Hosaka H, Morita T (2012) Resonant-type smooth impact drive mechanism actuator with two langevin transducers[J]. Adv Robot 26(3–4):277–290
3. Deshpande M, Saggere L (2007) An analytical model and working equations for static deflections of a circular multi-layered diaphragm-type piezoelectric actuator. Sens Actuators, A:Phys 136(2):673–689
4. Jiantao Z, Weiqing H, Zhu hua et al (2009) Lead screw linear ultrasonic motor using bending vibration modes. Trans Nanjing Univ Aeronaut Astronaut 02:89–94

5. Liu W, Zhu H, Zhou S et al () Variable camber wing driven by ultrasonic motors. Trans Nanjing Univ Aeronaut Astronaut
6. Shi Y,Li Y,Zhao C (2008) Optimum design of a linear ultrasonic motor based on in-plane modes.In: Proceedings of CSEE 2008, vol 30(28), pp 56–60(in Chinese)
7. Wen J (2009) Study on planar inertia piezoelectric moving mechanism.Jilin University, Jinlin, pp 6–13 (in Chinese)
8. Duan Z, Wang Q (2005) Development of a novel high precision piezoelectric linear stepper actuator. Sens Actuators A 118(2):285–291
9. Hii KF, Vallance RR, Pinar Mengüç M (2010) Design, operation, and motion characteristics of a precise piezoelectric linear motor. Prec Eng 34(2):231–241
10. Li J, Sedaghati R, Dargahi J et al (2005) Design and development of a new piezoelectric linear Inchworm actuator. Mechatronics 15(6):651–681
11. Weiqing H, Yimi M (2009) Design of a new type of piezoelectric linear motor based on non-resonant vibration. Chin Mech Eng 20(14):1717–1721 (in Chinese)

Lightning Source UK Ltd.
Milton Keynes UK
UKOW04n1612200814

237255UK00002B/6/P